Natural Polyphenols in Human Health-Volume I

Natural Polyphenols in Human Health-Volume I

Editor

Nour Eddine Es-Safi

MDPI

Basel • Beijing • Wuhan • Barcelona • Belgrade • Novi Sad • Cluj • Manchester

Editor
Nour Eddine Es-Safi
Mohammed V University in Rabat
Rabat, Morocco

Editorial Office
MDPI
St. Alban-Anlage 66
4052 Basel, Switzerland

This is a reprint of articles from the Special Issue published online in the open access journal *Molecules* (ISSN 1420-3049) (available at: https://www.mdpi.com/journal/molecules/special_issues/NPHH).

For citation purposes, cite each article independently as indicated on the article page online and as indicated below:

Lastname, A.A.; Lastname, B.B. Article Title. *Journal Name* **Year**, *Volume Number*, Page Range.

Volume I
ISBN 978-3-0365-9074-5 (Hbk)
ISBN 978-3-0365-9075-2 (PDF)
doi.org/10.3390/books978-3-0365-9075-2

Set
ISBN 978-3-0365-9072-1 (Hbk)
ISBN 978-3-0365-9073-8 (PDF)

Contents

molecules

MDPI

Article

Nanoformulation of Polyphenol Curcumin Enhances Cisplatin-Induced Apoptosis in Drug-Resistant MDA-MB-231 Breast Cancer Cells

Parastoo Karami [1], Goran Othman [2,3], Zjwan Housein [2], Abbas Salihi [4,5], Mohammad Ali Hosseinpour Feizi [1], Hewa Jalal Azeez [1] and Esmaeil Babaei [1,*]

[1] Department of Biology, School of Natural Sciences, University of Tabriz, Tabriz P.O. Box 5166616471, Iran; karamiparastoo431@yahoo.com (P.K.); pourfeizi@tabrizu.ac.ir (M.A.H.F.); hewa.azeez.ha@gmail.com (H.J.A.)
[2] Department of Medical Laboratory Technology, Erbil Health and Medical Technical College, Erbil Polytechnic University, Erbil 44001, Iraq; goran.othman@epu.edu.iq (G.O.); zjwan.h@epu.edu.iq (Z.H.)
[3] Department of Medical Laboratory Technology, Al-Qalam University College, Kirkuk 36001, Iraq
[4] Department of Biology, College of Science, Salahaddin University-Erbil, Erbil 44001, Iraq; abbas.salihi@su.edu.krd
[5] Center of Research and Strategic Studies, Lebanese French University, Erbil 44001, Iraq
* Correspondence: babaei@tabrizu.ac.ir; Tel.: +98-912-217-9167

Abstract: Triple Negative Breast Cancer (TNBC) is the aggressive and lethal type of breast malignancy that develops resistance to current therapies. Combination therapy has proven to be an effective strategy on TNBC. We aimed to study whether the nano-formulation of polyphenolic curcumin (Gemini-Cur) would affect the cisplatin-induced toxicity in MDA-MB-231 breast cancer cells. MDA-MB-231 cells were treated with Gemini-Cur, cisplatin and combination of Gemini-Cur/Cisplatin in a time- and dose-dependent manner. Cell viability was studied by using MTT, fluorescence microscopy and cell cycle assays. The mode of death was also determined by Hoechst staining and annexin V-FITC. Real-time PCR and western blotting were employed to detect the expression of BAX and BCL-2 genes. Our data demonstrated that Gemini-Cur significantly sensitizes cancer cells to cisplatin (combination index ≤ 1) and decreases IC50 values in comparison with Gemini-cur or cisplatin. Further studies confirmed that Gemini-Cur/Cisplatin suppresses cancer cell growth through induction of apoptosis ($p < 0.001$). In conclusion, the data confirm the synergistic effect of polyphenolic curcumin on cisplatin toxicity and provide attractive strategy to attain its apoptotic effect on TNBC.

Keywords: polyphenolic curcumin; cisplatin; combination therapy; breast cancer; apoptosis; gemini surfactant nanoparticles

Citation: Karami, P.; Othman, G.; Housein, Z.; Salihi, A.; Hosseinpour Feizi, M.A.; Azeez, H.J.; Babaei, E. Nanoformulation of Polyphenol Curcumin Enhances Cisplatin-Induced Apoptosis in Drug-Resistant MDA-MB-231 Breast Cancer Cells. *Molecules* **2022**, *27*, 2917. https://doi.org/10.3390/molecules27092917

Academic Editor: Nour Eddine Es-Safi

Received: 14 April 2022
Accepted: 26 April 2022
Published: 3 May 2022

1. Introduction

The International Agency for Research on Cancer (IARC) reported that in 2020, an estimated 19.3 million new cancer cases and 10.0 million cancer deaths has occurred. Breast cancer is the most common malignancy in women with nearly 2.3 million new cases (24.2% of all tumor malignancies in women) diagnosed in 2020 [1]. Despite recent advances in treatment, side effects and the development of drug resistance limit the usefulness of current therapies for triple negative breast cancer [2,3].

One of the classical drugs used as chemotherapeutic agent against TNBCs and related cell lines is cisplatin (Figure 1A). Cisplatin has been widely used as antitumor agent in the clinics ever since its discovery in the 1960s. It is estimated that at least half of current therapy protocols employ platinum-based anticancer drugs [4]. Binding of platinum to DNA results in structural deformation of the double-stranded structure, which leads to inhibition of DNA replication and transcription. Subsequently, the DNA damage response leads to apoptosis. The toxicity of cisplatin to normal tissues, like neurotoxicity and hepatotoxicity, along with the acquired therapeutic resistance of cancer cells, reduce the clinical efficacy of this drug [5].

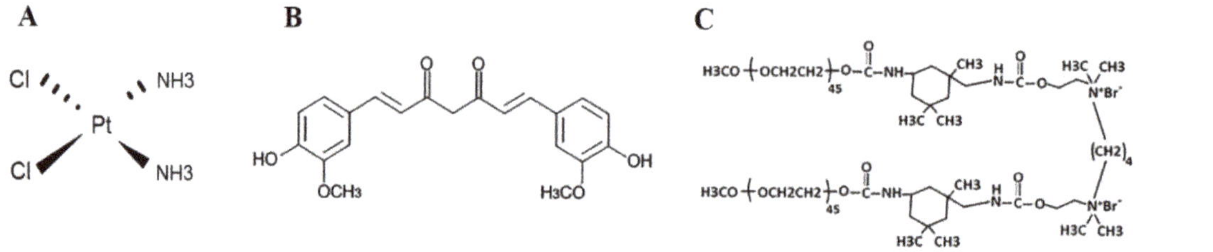

Figure 1. The molecular structure of (**A**) Cisplatin (**B**) Curcumin and (**C**) mPEG urethane gemini surfactant nanoparticle.

Curcumin, an extractive of turmeric root, is a natural polyphenol which has been widely reported to possess the anticancer properties (Figure 1B) [6]. The chemical groups of –OH and –OCH3 in curcumin structure, are reported to be responsible for the antioxidant and anti-proliferative properties, respectively [7]. However, the poor bioavailability of curcumin and its instability in physiological media have limited its therapeutic application in clinic [8]. The use of nano carriers in drug delivery systems may increase the solubility, reduce the required dose, attribute to the targeted delivery and, can prolong drug's maintenance in the systemic circulation [9]. Recent works by our lab have shown that the nano-based compounds of curcumin could be employed as an anticancer agents in vitro and in vivo [10,11].

Combination therapies of natural compounds can be employed as a novel strategy in promoting routine drug efficiency and reducing side effects [12]. Curcumin increases the sensitivity of breast cancer cells to cisplatin through down-regulation of FEN1 [13]. It has been demonstrated that curcumin in combination with carboplatin induces apoptosis and suppresses metastasis in human lung and hepatic cancer cells [14,15]. Montopoli et al. suggested that curcumin is an interesting natural polyphenol capable of limiting cell proliferation and possibly, increasing clinical impact of platinum drugs in ovarian cancer patients [16]. More recently, a study confirmed the suppressive effect of curcumin on cisplatin resistance in colorectal cancer cells [17]. These findings support the auxiliary role of curcumin as an adjunct to current chemotherapies and indicate that curcumin could be more effective in combination with chemotherapeutic drugs. This phytochemical in the form of gemini curcumin is more advantageous because of its improved stability, cellular uptake and cytotoxicity (Figure 1C). Gemini surfactants are a class of nano-sized materials consisting of two identical structures liked by a spacer that are highly effective in delivering gene and drugs. Gemini-Cur can trigger apoptosis in cancer cells through modulation of cell cycle and up regulation of apoptotic genes [18].

Here, we investigated the therapeutic effect of the combination of Gemini-Cur and cisplatin as a novel potential therapy on resistance TNBC cells. Gemini-Cur reduces the resistance of MDA-MB-231 to cisplatin through induction of apoptosis.

2. Results

2.1. Gemini-Cur and Cisplatin Have a Synergistic Toxic Effect on MDA-MB-231 Cells

Breast cancer cells were treated with different concentrations of Gemini-Cur and cisplatin and the combination index (CI) was analyzed by using Chou–Talalay equation method [17]. Our results showed that Gemini-Cur has an inhibitory effect on the viability of MDA-MB-231 cells in a time- and dose-dependent manner with IC50 values of 35.06 and 23.48 μM in 48 and 72 h, respectively (Figure 2A). However, cisplatin lonely suppressed the proliferation of MDA-MB-231 cells with IC50 of 58.32 μM for 48 h. Then, serial doses of cisplatin and Gemini-Cur were employed and concentrations with CI < 1 were selected as proper ratio for further studies. The IC50 of cisplatin and Gemini-Cur was adjusted to 13 μM and 20 μM, respectively (Figure 2C). As shown in Figure 2D, the combination index is <1, pretending synergism between Gemini-Cur and cisplatin in 20 and 13 μM concentrations, respectively.

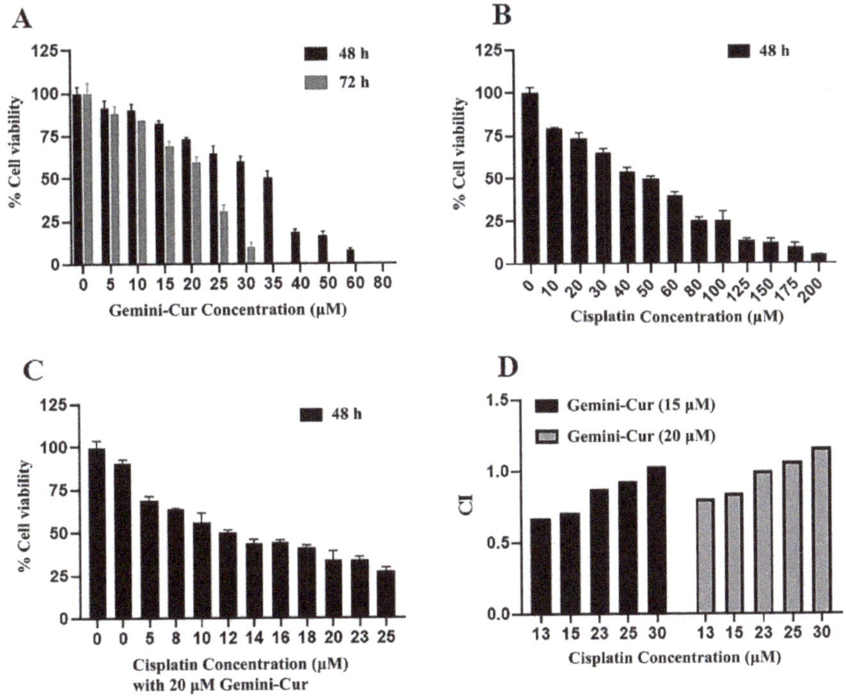

Figure 2. Effect of Gemini curumin and cisplatin on the viability of MDA-MB-231 cells in vitro. MDA-MB-231 cells were grown and treated with Gemini-Cur (**A**), Cisplatin (**B**) and Gemini-Cur/Cis (**C**,**D**). 15 and 20 μM of Gemini-Cur combined with serial concentrations of cisplatin. Data represent mean ± standard deviation of three independent experiments. CI: Combination index.

2.2. Morphological Visualization of Apoptosis

To visualize morphological changes in MDA-MB-231 cells, Hoechst staining was performed. As Figure 3 illustrates, TNBC cells undergo apoptosis after treatment with 20 μM Gemini-Cur and 13 μM cisplatin in both singular and combination forms. The microscopic visualization shows a uniformly light stain on the cells. However, a remarkable change in color intensity is clearly seen in treated cells with a significant difference in the number of apoptotic cells in Gemini-Cur/Cis treated group. The Hoechst staining clearly differentiate cells with nuclear fragmentation and DNA condensation as specified by arrows.

2.3. Gemini-Cur/Cis Modulates Cell Cycle Distribution in MDA-MB-231 Cells

Cell cycle distribution was studied by propidium iodide staining in flow cytometry. Our data showed that cell cycle distribution is modulated in all treated cells either in singular or combined forms. As Figure 4 shows, the percentage of SubG1 cells as a hallmark of apoptosis is significantly increased in Gemini-Cur/Cis group compared to curcumin or cisplatin groups ($p < 0.001$). Accordingly, the number of live cells (G1 phase) was decrease in combination treatment when compared to the Gemini-Cue and Cis groups ($p < 0.001$). There was a statistically significant interaction between the effects of Cis and Gemini-Cur on interest in sub G1 ($p = 0.0001$).

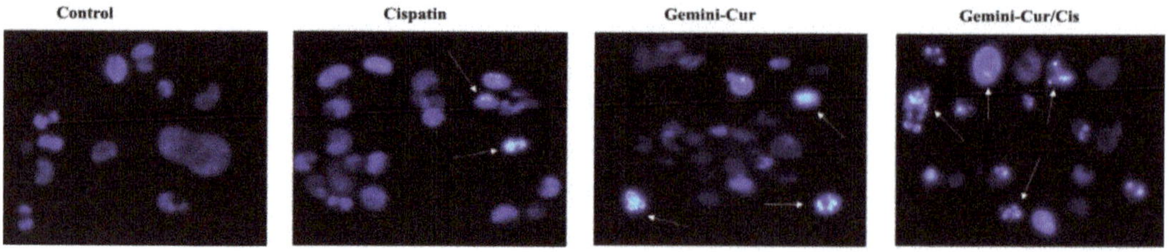

Figure 3. Morphological observation of apoptosis in Gemini-Cur and cisplatin treated cells by Hoechst staining in 48 h. Control shows a uniform exposure of live cells to stain. The color intensity and the number of cells indicating apoptotic features are significantly increased in treated groups. Arrows indicate apoptotic cells with cell shrinkage, nuclear fragmentation and DNA condensation. Magnification: 200×.

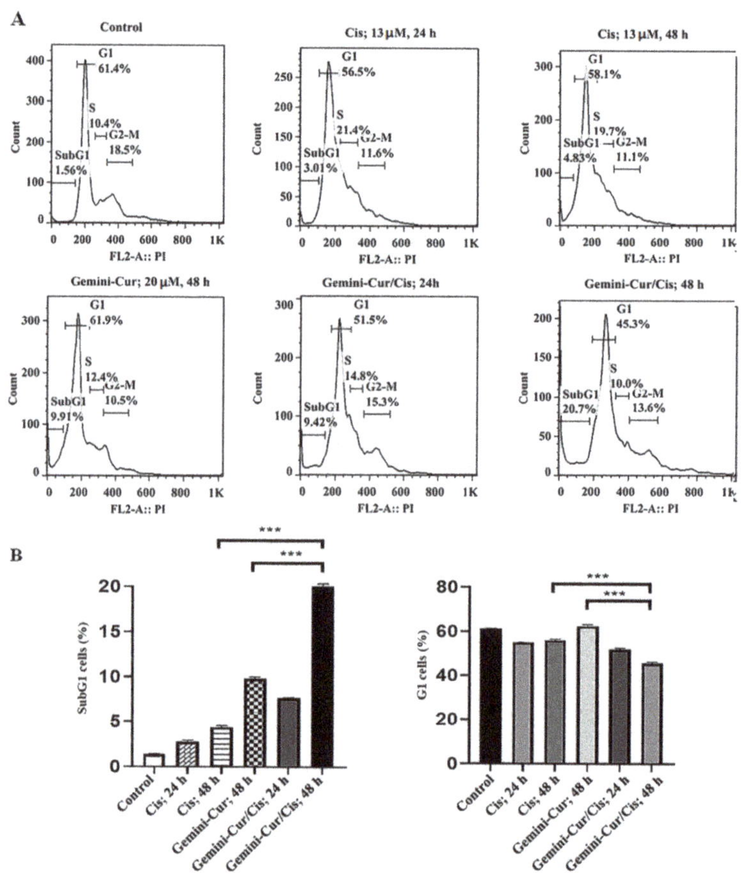

Figure 4. Analysis of cell cycle by flow cytometry in cancer cells treated with 13 µM cisplatin and 20 µM Gemini-Cur for 24 & 48 h. (**A**) Histograms for MDA-MB-231 cells. (**B**) The percentage of cells in sub-G1 and G1 phases. The data clearly illustrate an increase in the number of SubG1 cells in combination treatment compared to either void cisplatin or Gemini-Cur. Accordingly, the number of live cells is decrease in Gemini-Cur/Cis group versus Gemini-Cur and Cis groups. Data represent mean ± standard deviation of three independent experiments. (*** $p < 0.001$).

2.4. Annexin V-FITC/PI Assay Confirmed Apoptosis in MDA-MB-231 Treated Cells

Annexin V-FITC/PI was employed to further confirm the mode of death in treated cells. As Figure 5 shows, apoptosis is induced in all treatments including 20 μM Gemini-Cur, 13 μM cisplatin and Gemini-Cur/Cis. However, the proportion of cells in late apoptosis is meaningfully increased to 59.3% in combination form compared to 15 and 42.6% in cisplatin and Gemini-Cur, respectively ($p < 0.01$).

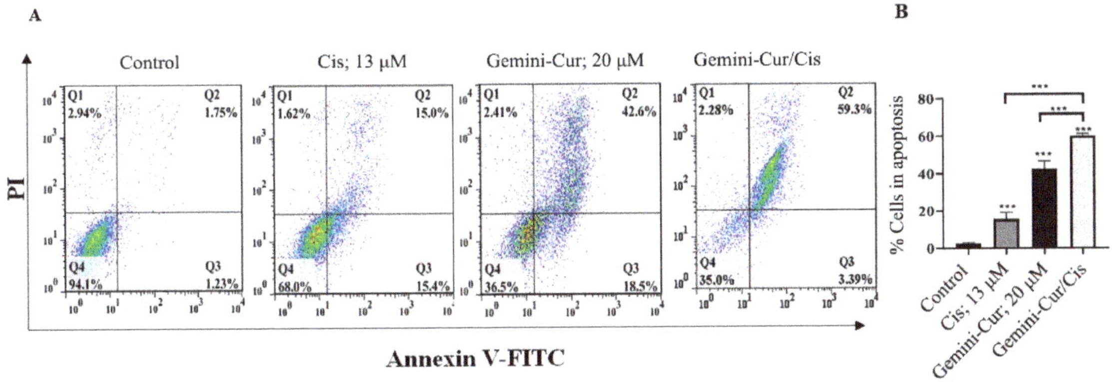

Figure 5. The combination of cisplatin and Gemini-Cur induces breast cancer cell apoptosis. MDA-MB-231 cells were treated with 13 μM cisplatin, 20 μM Gemini-Cur, and Gemini-Cur/Cis for 48 h. (**A**): plots show that live cells undergo apoptosis and the number of death cells is increased in treated cells, especially in Gemini-Cur/Cis group. (**B**) The proportion of apoptotic cells in late stage is significantly increased in combination treatment. ***: $p < 0.001$.

2.5. Expression of BAX, BCL-2 Genes Are Modulated in Treated Cells

The expression ratio of BAX/BCL-2 is usually considered as a hallmark of apoptosis. Real-time PCR demonstrated that BAX/BCL-2 expression is modulated in treated MDA-MB-231 cells. As Figure 6 shows, apoptotic BAX is upregulated while anti apoptotic BCL-2 is downregulated in all treatments. Furthermore, this modulation was significant in Gemini-Cur/Cis treatments rather than cisplatin or Gemini-Cur groups. This differential effect of combination treatment was clearly detected in protein level (Figure 7A). Further analysis demonstrated that the protein ratio of BAX/BCL-2 is significantly increased in combination treatments (p value < 0.01) and Gemini-Cur (p value < 0.001) groups.

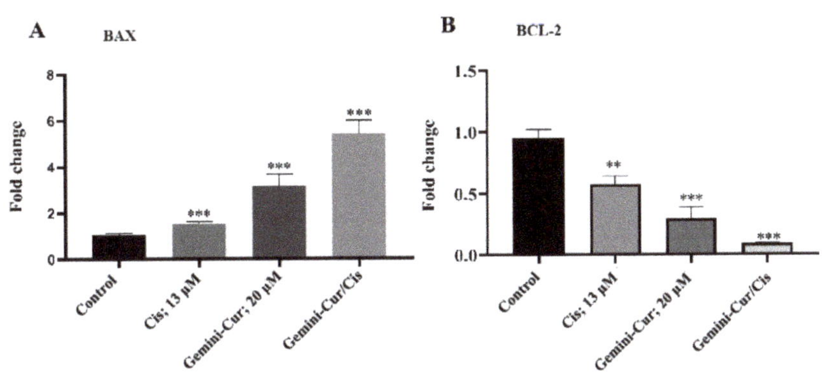

Figure 6. Quantitative analysis of the expression of apoptotic genes by real-time PCR. Relative expression for BAX (**A**) and BCL-2 (**B**) in MDA-MB-231 cells. Values represent mean ± standard deviation of three independent experiments; ** $p < 0.01$ and *** $p < 0.001$.

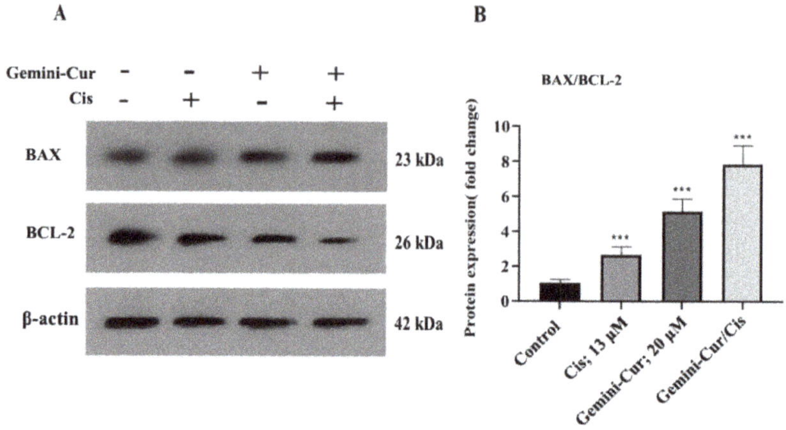

Figure 7. (**A**) Western blotting for BCL-2 and BAX (**A**) proteins. β-actin was used as internal control. (**B**) Data analysis showed that the protein ratio of BAX/BCL-2 is significantly increased in the cells treated with Gemini-Cur/Cis compared to the groups of Cis and Gemini-Cur. Values represent mean ± standard deviation of three independent experiments. *** $p < 0.001$.

3. Discussion

"Based on the St. Gallen/Vienna 2019 consensus discussion, 53% of the panelists didn't accept the use of a platinum-based regimen in the neoadjuvant treatment of TNBC patients" [19]. Presently, there are limited criteria for TNBC treatment, and what could help medical oncologists to cure TNBC or prolong the survival rate in TNBC patients are novel therapeutic strategies [20]. The present study aimed to sensitize TNBC MDA-MB-231 cells to cisplatin through employment of novel curcumin nano formulation (Gemini-Cur).

Our data showed that cisplatin modulates the proliferation of MDA-MB-231 cells in high concentrations. However, its combination with appropriate dose of Gemini-Cur not only reduced the IC50 value of cisplatin to 13 µM but also, boosted its suppressive effect on the growth of MDA-MB-23 cells. Due to the nephrotoxicity and hepatoxicity of cisplatin, decreasing its effective dose in cancer treatments is of interest in clinic [4,5]. Different therapeutic compounds have already been employed to improve the toxicity of cisplatin. More recently, it has been reported that curcumin reduces the resistance of colorectal cancer cells to cisplatin [21]. Here, the employment of Gemini-Cur, has significantly reduced the effective dose of cisplatin to 13 µM, a concentration that may not indicate side effects. Cell cycle analysis and microscopic visualization also demonstrated that the mode of death induced in treated-TNBC cells is of apoptosis.

Since the growth inhibitory effect of Gemini-Cur/Cis was more effective than singular treatments, it seems that Gemini-Cur increases the sensitivity of MDA-MB-231 cells to cisplatin. The anticancer activities of natural compounds such as curcumin has been examined in combination with numerous chemotherapy drugs such as cisplatin. It was observed that curcumin affects the toxic properties of cisplatin in different cancer cells [12,13,15]. It has also been shown that curcumin inhibits the function of P-gp efflux pump in cancer cells that might support the reversal effect of this phytochemical on therapeutic resistance of cancer cells [22].

Here, we indicated that the expression ratio of BAX/BCL-2 is modulated in apoptotic cells. These proteins are both essential gateways to cell death. Numerous studies have revealed that curcumin triggers apoptosis through BAX/BCL-2 mediated pathway in cancer cells [8,14]. The data is also in concordance with results of Karimpour et al. for the modulatory effect of Gemini-Cur on the expression of apoptotic genes in breast cancer cell lines [18]. The up regulation of BAX protein and down regulation of BCL2 was clearly detected in Gemini-Cur/Cis group compared to controls.

Taken together, the current results illustrate the synergistic property of Gemini-Cur as a novel nano formulation of curcumin on the therapeutic potential of cisplatin against drug-resistant MDA-MB-231 cells. Enhancing the sensitivity of TNBC cells to cisplatin is still an interesting topic of cancer therapy for breast cancer. Therefore, it is worth investigating the exact cellular mechanisms and pathways involved in Gemini-Cur/Cis toxicity on TNCB cells. Furthermore, co-encapsulation of curcumin and cisplatin in gemini nanopartilces can be considered as an effective strategy to target cancer.

4. Materials and Methods

4.1. Reagents and Cell Culture

Curcumin and mPEG urethane gemini surfactant nanoparticles were a kind gift from Dr. Farhood Najafi at the Institute for Color Science and Technology, Tehran, Iran. Cisplatin was obtained from Mylan Corporation, Germany. Gemini curcumin was formulated by dissolving 6 mg of curcumin and 100 mg of gemini nanoparticles in 3 mL methanol by using sonication at room temperature. Then, the organic phase was evaporated by a rotary evaporator in room temperature. The encapsulated nano curcumin was dissolved in 10 mL distilled water and stored at 4 °C for further analyses [11].

Human breast adenocarcinoma cell line, MDA-MB-231, was obtained from the national cell bank of Iran (Pasteur Institute, Tehran, Iran). The cells were cultured in DMEM-High glucose medium (Sigma-Aldrich, St. Louis, MO, USA) supplemented with 100 U/mL penicillin, 100 mg/mL streptomycin, and 10% fetal bovine serum (Gibco, New York, NY, USA), and incubated at 37 °C in a humidified atmosphere of 5% CO_2.

4.2. Cell Viability Assay

Cell viability was measured using the 3-[4,5-dimethylthiazol-2-yl] 2,5-iphenyltetrazolium bromide (MTT; Sigma-Aldrich, USA) assay. In brief, MDA-MB-231 cells were seeded into 96-well plates at a density of 10^4 cells per well and cultured overnight. After 24 h, cells were treated with various concentrations of cisplatin (0–200 µM), Gemini-Cur (0–100 µM) and combinations of Gemini-Cur and cisplatin (Gemini-Cur/Cis) for 24 and 48 h at 37 °C. Then, 20 µL of 5 mg/mL MTT was added to each well and further incubated for 4 h. The medium was replaced with 150 µL of DMSO and thoroughly mixed. Optical absorbance value was recorded at 570 nm using a plate reader (Biotek, ELX808, Winooski, VT, USA). The IC50 values of cisplatin, Gemini-Cur, and Gemini-Cur/Cis were calculated using the Graphpad prism software version 8.0.2. (Graphpad Prism, La Jolla, CA, USA).

4.3. Analysis of Synergistic Cytotoxicity

The synergistic effects of cisplatin and Gemini-Cur were quantitatively analyzed by the calculation of the combination index (CI) using Compusyn software program, which utilizes the Chou–Talalay equation method [23]. CI analysis calculates the interaction between Gemini-Cur and cisplatin agents. The CI value was calculated using the following equation:

$$CI = \frac{CA.x}{ICx.A} + \frac{CB.x}{ICx.B}$$

where CA.x and CB.x are the concentrations of agent A and agent B, respectively used in combination that inhibits cell growth by x%. The ICx.A and ICx.B are concentrations required for x% inhibition by agent A and agent B in singular form, respectively. Values of $CI < 1$, $CI = 1$, and $CI > 1$ indicate synergism, additive effect, and antagonism, respectively.

4.4. Visualization of Apoptosis by Hoechst Staining

MDA-MB-231 cells (4×10^5) were seeded into 6-well plates for 24 h. Then, the cells were treated with 13 µM cisplatin and 20 µM Gemini-Cur, both in singular and combination forms. After 48 h, treated cells were collected and fixed with 3.7% paraformaldehyde for 30 min at room temperature, washed and stained with 167 µmol/L Hoechst 33258 at 37 °C for 30 min. Finally, cells were observed under a fluorescence microscope (RX50, LABEX, London, England) equipped with a UV filter.

4.5. Cell Cycle Analysis by Flow Cytometry

MDA-MB-231 cells (4×10^5) were seeded into 6-well plates and treated with 13 μM cisplatin and 20 μM Gemini-Cur in singular and combination forms for 24 and 48 h. Treated cells were fixed using ice cold 70% ethanol, washed 2X with PBS and then resuspended with propidium iodide (10 mg/mL) and ribonuclease A (0.1%) for 30 min. Finally, the cells were incubated for 30 min in the dark place at room temperature. Fluorescent events from propidium iodide–DNA complexes were quantified by fluorescence-activated cell sorter (FACS) (BD Biosciences, Franklin Lakes, NJ, USA) with a count of 10,000 cells per sample. Finally, DNA contents at different phases of the cell cycle were determined by using FlowJo 7.6.1 Software.

4.6. Annexin V FITC/PI Assay

To further confirm apoptosis, annexin V/FITC assay (ApoFlowEx FITC Kit, Exbio, Czech RepubliC), was employed to detect the mode of death in MDA-MB-231 cells. Briefly, the cells were seeded on 6-well plates for 24 h. After 24 h, cisplatin (13 μM) and Gemini-Cur (20 μM) was added to different wells in both singular and combination forms. After 24 and 48 h, cells were collected, washed twice with PBS and suspended in 1oo μL binding buffer. Annexin V/FITC solution was added to the cells followed by addition of 10 μL PI. Stained cells were then detected with a flow cytometer (BD Biosciences, Franklin Lakes, NJ, USA).

4.7. Expression Studies by Real-Time PCR and Western Blotting

4.7.1. Real-Time PCR

Total RNA was extracted with a selfmade TRIzol reagent, BRIzol, and followed by DNaseI treatment to eliminate any DNA content. The integrity and purity of RNA were evaluated using Pico200 (Picodrop, Hinxton, UK) and agarose gel electrophoresis. Total RNA was reverse transcribed into cDNA by employing PrimeScript RT Reagent Kit (Takara Bio, Kusatsu City, Japan). Real-time polymerase chain reaction (real-time PCR) was used to quantify the expression of BAX, BCL-2 and beta-2 microglobin (*β2M)*. Table 1 shows the characteristics of primers used in PCR. The comparative CT ($2^{-\Delta\Delta CT}$) method was used to determine the relative gene expressions [24].

Table 1. Sequences of primers used in real-time PCR.

Genes	Primers	Size (bp)
β-actin	F: 5′-TGCCCATCTACGAGGGGTATG-3′ R: 5′-CTCCTTAATGTCACGCACGATTTC-3′	155
BAX	F: 5′-GCAAACTGGTGCTCAAGG-3′ R: 5′-ACTCCCGCCACAAAGA-3′	236
BCL-2	F: 5′-TGGGAAGTTTCAAATCAGC-3′ R: 5′-GCATTCTTGGACGAGGG-3′	298

4.7.2. Western Blotting

To confirm the data of quantitative PCR, we evaluated gene expressions in protein level. Briefly, total protein was extracted from cells before and after treatment. Equal amounts of extracted proteins (40 mg/sample) were separated by SDS polyacrylamide gel electrophoresis and transferred to 0.45-mm polyvinylidene difluoride membrane. The membrane was blocked with 5% (*w/v*) non-fat dried milk (Difco/Becton Dickinson, Franklin Lakes, NJ, USA) and incubated for 2 h at room temperature with primary antibody [1:200, BCL-2 (sc-492), BAX (sc-7480), and β-actin (sc-47778)]. Mouse anti-rabbit IgG-HRP in 5% defatted dry milk-TBS-0.1% Tween (Santa Cruz Biotechnology, Dallas, TX, United States) were employed to visualize proteins. All steps were performed at room temperature. All signals were visualized using enhanced Western Blotting Luminol Reagent (Santa Cruz Inc., Dallas, TX, USA).

4.8. Statistical Analysis

All experiments were repeated at least three times and the data were expressed as mean ± standard deviation. SPSS 19.0 software was used in statistical analysis. Significance was determined using the one-way or two-way ANOVA test with a post hoc multiple comparison test. Meanwhile, the IC50 and combination index values were calculated using Graphpad prism version 8.0.2 and Compusyn software.

5. Conclusions

In conclusion, our data show that Gemini-Cur can improve the apoptotic effect of cisplatin on cancer resistant cells. Further studies are needed to figure out the molecular mechanism of synergistic effect of Gemini-Cur/Cis in TNCB cells.

Author Contributions: Conceptualization, P.K., H.J.A. and E.B.; methodology, E.B., A.S. and M.A.H.F.; software, P.K., E.B. and G.O.; investigation, P.K. and H.J.A.; writing—original draft preparation, P.K., A.S. and E.B.; writing—review and editing, A.S., G.O. and Z.H.; supervision, E.B. All authors have read and agreed to the published version of the manuscript.

Funding: This research received no external funding.

Institutional Review Board Statement: Not applicable.

Informed Consent Statement: Not applicable.

Data Availability Statement: Data is contained within the article.

Acknowledgments: Authors would like to thank staff of Genomics & Molecular biology Lab at the University of Tabriz for their technical assistance.

Conflicts of Interest: The authors declare no conflict of interest.

Sample Availability: Samples of the compounds are available from the authors.

References

1. Sung, H.; Ferlay, J.; Siegel, R.L.; Laversanne, M.; Soerjomataram, I.; Jemal, A.; Bray, F. Global Cancer Statistics 2020: GLOBOCAN Estimates of Incidence and Mortality Worldwide for 36 Cancers in 185 Countries. *CA Cancer J. Clin.* **2021**, *71*, 209–249. [CrossRef]
2. Baranova, A.; Krasnoselskyi, M.; Starikov, V.; Kartashov, S.; Zhulkevych, I.; Vlasenko, V.; Oleshko, K.; Bilodid, O.; Sadchikova, M.; Vinnyk, Y. Triple-negative breast cancer: Current treatment strategies and factors of negative prognosis. *J. Med. Life* **2022**, *15*, 153–161. [CrossRef]
3. Ke, L.; Li, Z.; Fan, X.; Loh, X.J.; Cheng, H.; Wu, Y.-l.; Li, Z. Cyclodextrin-Based Hybrid Polymeric Complex to Overcome Dual Drug Resistance Mechanisms for Cancer Therapy. *Polymers* **2021**, *13*, 1254. [CrossRef]
4. Dasari, S.; Tchounwou, P.B. Cisplatin in cancer therapy: Molecular mechanisms of action. *Eur. J. Pharmacol.* **2014**, *740*, 364–378. [CrossRef]
5. Makovec, T. Cisplatin and beyond: Molecular mechanisms of action and drug resistance development in cancer chemotherapy. *Radiol. Oncol.* **2019**, *53*, 148–158. [CrossRef]
6. Panda, A.K.; Chakraborty, D.; Sarkar, I.; Khan, T.; Sa, G. New insights into therapeutic activity and anticancer properties of curcumin. *J. Exp. Pharmacol.* **2017**, *9*, 31–45. [CrossRef]
7. Priyadarsini, K.I. The chemistry of curcumin: From extraction to therapeutic agent. *Molecules* **2014**, *19*, 20091–20112. [CrossRef]
8. Tabanelli, R.; Brogi, S.; Calderone, V. Improving Curcumin Bioavailability: Current Strategies and Future Perspectives. *Pharmaceutics* **2021**, *13*, 1715. [CrossRef]
9. Souto, E.B.; Silva, G.F.; Dias-Ferreira, J.; Zielinska, A.; Ventura, F.; Durazzo, A.; Lucarini, M.; Novellino, E.; Santini, A. Nanopharmaceutics: Part II-Production Scales and Clinically Compliant Production Methods. *Nanomaterials* **2020**, *10*, 455. [CrossRef]
10. Babaei, E.; Sadeghizadeh, M.; Hassan, Z.M.; Feizi, M.A.H.; Najafi, F.; Hashemi, S.M. Dendrosomal curcumin significantly suppresses cancer cell proliferation in vitro and in vivo. *Int. Immunopharmacol.* **2012**, *12*, 226–234. [CrossRef]
11. Zibaei, Z.; Babaei, E.; Rezaie Nezhad Zamani, A.; Rahbarghazi, R.; Azeez, H.J. Curcumin-enriched Gemini surfactant nanoparticles exhibited tumoricidal effects on human 3D spheroid HT-29 cells in vitro. *Cancer Nano* **2021**, *12*, 1–15. [CrossRef]
12. Hussain, Y.; Islam, L.; Khan, H.; Filosa, R.; Aschner, M.; Javed, S. Curcumin-cisplatin chemotherapy: A novel strategy in promoting chemotherapy efficacy and reducing side effects. *Phytother. Res.* **2021**, *35*, 6514–6529. [CrossRef]

13. Zou, J.; Zhu, L.; Jiang, X.; Wang, Y.; Wang, Y.; Wang, X.; Chen, B. Curcumin increases breast cancer cell sensitivity to cisplatin by decreasing FEN1 expression. *Oncotarget* **2018**, *9*, 11268–11278. [CrossRef]
14. Kang, J.H.; Kang, H.S.; Kim, I.K.; Lee, H.Y.; Ha, J.H.; Yeo, C.D.; Kang, H.H.; Moon, H.S.; Lee, S.H. Curcumin sensitizes human lung cancer cells to apoptosis and metastasis synergistically combined with carboplatin. *Exp. Biol. Med.* **2015**, *240*, 1416–1425. [CrossRef]
15. Notarbartolo, M.; Poma, P.; Perri, D.; Dusonchet, L.; Cervello, M.; D'Alessandro, N. Antitumor effects of curcumin, alone or in combination with cisplatin or doxorubicin, on human hepatic cancer cells. Analysis of their possible relationship to changes in NF-kB activation levels and in IAP gene expression. *Cancer Lett.* **2005**, *224*, 53–65. [CrossRef]
16. Montopoli, M.; Ragazzi, E.; Froldi, G.; Caparrotta, L. Cell-cycle inhibition and apoptosis induced by curcumin and cisplatin or oxaliplatin in human ovarian carcinoma cells. *Cell Prolif.* **2009**, *42*, 195–206. [CrossRef]
17. Zheng, Z.H.; You, H.Y.; Feng, Y.J.; Zhang, Z.T. LncRNA KCNQ1OT1 is a key factor in the reversal effect of curcumin on cisplatin resistance in the colorectal cancer cells. *Mol. Cell Biochem.* **2021**, *476*, 2575–2585. [CrossRef]
18. Karimpour, M.; Feizi, M.A.H.; Mahdavi, M.; Krammer, B.; Verwanger, T.; Najafi, F.; Babaei, E. Development of curcumin-loaded gemini surfactant nanoparticles: Synthesis, characterization and evaluation of anticancer activity against human breast cancer cell lines. *Phytomedicine* **2019**, *57*, 183–190. [CrossRef]
19. Balic, M.; Thomssen, C.; Würstlein, R.; Gnant, M.; Harbeck, N. St. Gallen/Vienna 2019: A Brief Summary of the Consensus Discussion on the Optimal Primary Breast Cancer Treatment. *Breast Care* **2019**, *14*, 103–110. [CrossRef]
20. Chalakur-Ramireddy, N.K.R.; Pakala, S.B. Combined drug therapeutic strategies for the effective treatment of Triple Negative Breast Cancer. *Biosci. Rep.* **2018**, *38*, BSR20171357. [CrossRef]
21. Shehzad, A.; Lee, J.; Huh, T.L.; Lee, Y.S. Curcumin induces apoptosis in human colorectal carcinoma (HCT-15) cells by regulating expression of Prp4 and p53. *Mol. Cells.* **2013**, *35*, 526–532. [CrossRef]
22. Lopes-Rodrigues, V.; Sousa, E.; Vasconcelos, M.H. Curcumin as a Modulator of P-Glycoprotein in Cancer: Challenges and Perspectives. *Pharmaceuticals* **2016**, *9*, 71. [CrossRef]
23. Chou, T.C. Drug combination studies and their synergy quantification using the Chou-Talalay method. *Cancer Res.* **2010**, *70*, 440–446. [CrossRef]
24. Livak, K.J.; Schmittgen, T.D. Analysis of relative gene expression data using real-time quantitative PCR and the 2(-Delta Delta C(T)) Method. *Methods* **2001**, *25*, 402–408. [CrossRef]

molecules

Article

Mucuna pruriens Seed Aqueous Extract Improved Neuroprotective and Acetylcholinesterase Inhibitory Effects Compared with Synthetic L-Dopa

Narisa Kamkaen [1,*], Chuda Chittasupho [2,3,*], Suwanna Vorarat [4], Sarin Tadtong [5], Watoo Phrompittayarat [6], Siriporn Okonogi [2,3] and Pakakrong Kwankhao [7]

1 Department of Industrial Pharmacy, School of Pharmacy, Eastern Asia University, Pathum Thani 12110, Thailand
2 Department of Pharmaceutical Sciences, Faculty of Pharmacy, Chiang Mai University, Chiang Mai 50200, Thailand; siriporn.okonogi@cmu.ac.th
3 Research Center of Pharmaceutical Nanotechnology, Faculty of Pharmacy, Chiang Mai University, Chiang Mai 50200, Thailand
4 Department of Pharmaceutical Chemistry, Faculty of Pharmacy, Srinakharinwirot University, Nakhon Nayok 26120, Thailand; suwannav@g.swu.ac.th
5 Department of Pharmacognosy, Faculty of Pharmacy, Srinakharinwirot University, Nakhon Nayok 26120, Thailand; sarin@g.swu.ac.th
6 Faculty of Public Health, Naresuan University, Muang, Phitsanulok 65000, Thailand; watoop@nu.ac.th
7 Chao Phya Abhaibhubejhr Hospital, Ministry of Public Health, Prachin Buri 25000, Thailand; pakakrong2@gmail.com
* Correspondence: narisa.k@eau.ac.th (N.K.); chuda.c@cmu.ac.th (C.C.)

Citation: Kamkaen, N.; Chittasupho, C.; Vorarat, S.; Tadtong, S.; Phrompittayarat, W.; Okonogi, S.; Kwankhao, P. *Mucuna pruriens* Seed Aqueous Extract Improved Neuroprotective and Acetylcholinesterase Inhibitory Effects Compared with Synthetic L-Dopa. *Molecules* **2022**, *27*, 3131. https://doi.org/10.3390/molecules27103131

Academic Editors: Nour Eddine Es-Safi and Celestino Santos-Buelga

Received: 15 April 2022
Accepted: 11 May 2022
Published: 13 May 2022

Abstract: L-dopa, a dopaminergic agonist, is the gold standard for the treatment of Parkinson's disease. However, due to the long-term toxicity and adverse effects of using L-dopa as the first-line therapy for Parkinson's disease, a search for alternative medications is an important current challenge. Traditional Ayurvedic medicine has suggested the use of *Mucuna pruriens* Linn. (Fabaceae) as an anti-Parkinson's agent. The present study aimed to quantify the amount of L-dopa in *M. pruriens* seed extract by HPLC analysis. The cytotoxicity and neuroprotective properties of *M. pruriens* aqueous extract were investigated by two in vitro models including the serum deprivation method and co-administration of hydrogen peroxide assay. The results showed the significant neuroprotective activities of *M. pruriens* seed extracts at a concentration of 10 ng/mL. In addition, the effects of L-dopa and *M. pruriens* seed extract on in vitro acetylcholinesterase activities were studied. *M. pruriens* seed extract demonstrated acetylcholinesterase inhibitory activity, while synthetic L-dopa enhanced the activity of the enzyme. It can be concluded that the administration of *M. pruriens* seed might be effective in protecting the brain against neurodegenerative disorders such as Parkinson's and Alzheimer's diseases. *M. prurience* seed extract containing L-dopa has shown less acetylcholinesterase activity stimulation compared with L-dopa, suggesting that the extract might have a superior benefit for use in the treatment of Parkinson's disease.

Keywords: *Mucuna pruriens* seed; neuroprotective activity; Parkinson's disease; anti-acetylcholinesterase activity; L-dopa

1. Introduction

Due to the low efficiency and toxicity of current medications for Parkinson's disease treatment, interest is growing in phytochemicals as a potential treatment option. Phytochemicals with neuroprotective activity target various pathways due to their antioxidant, anti-inflammatory, and antiapoptotic properties [1]. *Mucuna pruriens* Linn. is a leguminous plant growing spontaneously in tropical and subtropical areas worldwide. The seeds have traditionally been used in India as a nerve tonic, and a male virility enhancement [2]. In addition, the pods have anthelmintic activity, and the seeds have anti-inflammatory activity. Powdered seeds possess anti-parkinsonism properties, possibly due to the existence of L-dopa, which is a precursor of the neurotransmitter dopamine. The dopamine content in brain tissue is reduced when the conversion of tyrosine to L-dopa is blocked. L-Dopa can cross the blood–brain barrier and undergoes conversion to dopamine, restoring neurotransmission [3]. Particularly, the hydro-alcoholic extract of *M. pruriens* seeds gave high yields of L-dopa, using ascorbic acid as a protector [4]. Surprisingly, n-propanol extract of *M. pruriens* seeds that contained a small amount of L-dopa, yielded the highest reaction for the growth and survival of dopaminergic culture neurons [2]. The significant neuroprotective activity of n-propanol extracts suggested that a whole extract could be potential for the treatment of Parkinson's disease [5].

Parkinson's disease is characterized by signs of major oxidative stress and mitochondrial damage in the pars compacta of the substantia nigra [6]. The previous study suggested that reactive oxygen species (ROS) played an important role in age-related neurodegenerative changes including Parkinson's disease [7]. Interestingly, the ethyl acetate and methanolic extract of the whole *M. pruriens* plant exhibited high antioxidant and free radical scavenging activities [2]. These in vitro assays indicated that the whole plant extract contained large amounts of phenolic compounds, which may be useful in preventing various oxidative stresses. Furthermore, it has been reported that methanolic extracts of *M. pruriens var.utilis* leaves have numerous biochemical and physiological activities, and contain pharmaceutically valuable compounds [8]. However, the in vitro neuroprotective activity of the *M. pruriens* seed aqueous extract obtained from the traditional extraction method has never been reported.

Although the first-line therapy for Parkinson's disease is L-dopa, long-term L-dopa use results in the development of significant clinical complications. L-dopa-induced dyskinesia (LID) and abnormal involuntary movements (AIMs) normally occur in a vast majority of Parkinson's disease patients with repeated administration of L-dopa [9]. AIMs occurring at the head, trunk, and extremities affect the function of daily living, and patients can become debilitated. These side effects are common with 30% incidence after 2 years of L-dopa administration. The incidence increased up to 40% and 90% for 5- and 10-year-treatment, respectively [10]. Parkinson's patients who early started high doses of levodopa are known to have risk factors for LID [11]. At present, few therapeutic options are available for the treatment of LIDs. Several strategies have been proposed to reduce these side effects including postponing the initiation of L-dopa dosing, adjusting the dose of L-dopa, and combining it with other drugs [12,13]. However, these strategies resulted in a decrease in the efficacy of L-dopa in the treatment of Parkinson's disease.

The seed of *M. pruriens* has been widely investigated for its pharmacological properties including Parkinson's disease because it contains significant amounts of L-dopa. The advantages of natural L-dopa in *M. pruriens* seed extract over the synthetic forms have been reported. The natural L-dopa is less toxic [14]. It has a shorter onset of action, but it has a longer therapeutic effect, which could delay the need for combination therapy [14–16]. In this study, we quantified the amount of L-dopa, total phenolic content, and total flavonoid contents in freeze-dried *M. pruriens* seed extract obtained from an aqueous extraction method. The cytotoxicity and neuroprotective activities of the freeze-dried extract were investigated in neuronal cells using two models. The effects of L-dopa and *M. pruriens* on acetylcholinesterase enzyme activity were studied.

2. Materials and Method

2.1. Materials

L-dopa and acetylcholinesterase activity assay kit (Cat# CS0003) were purchased from Sigma-Aldrich (St. Louis, MO, USA). Folin–Ciocalteu phenol reagent and aluminum chloride were obtained from Loba Chemie (Mumbai, India). Quercetin (98% purity) was purchased from Chanjao Longevity Co., Ltd. (Bangkok, Thailand). P19 cell line ATCC CRL-1857 was obtained from American Type Culture Collection, USA. Alpha minimal essential medium (α-MEM), fetal bovine serum (FBS), newborn calf serum (NCS), and antibiotics-antimycotic solution were purchased from Gibco, USA. All *trans*-retinoic acid, cytosine-1-β-D-arabinoside, 1:250 porcine trypsin, poly-L-lysine (MW > 300,000), XTT (2,3-bis(2-methoxy-4-nitro-5-sulphonyl)-2H-tetrazolium-5-carboxanilide sodium), and phenazine methosulfate (PMS) were obtained from Sigma, USA. Dimethylsulfoxide (DMSO) and methanol analytical grade were purchased from Merck, Germany. A total of 96-well plates were purchased from Corning, USA. A 100-mm Bacteriological culture dish was obtained from Hycon, USA.

2.2. Plant Sample Collection and Identification

The seeds of *M. pruriens* were purchased from a local market in Kanchanaburi province and kindly provided by *Chao-Phraya Abhaibhubejhr* Hospital *Foundation*. The plant materials were compared with the authentic specimens at the Bangkok Herbarium (BK), Botany Section, Botany and Weed Science Division, Department of Agriculture. The voucher specimen (RSU-MP-KP01) was kept in the Department of Pharmacognosy, College of Pharmacy, Rangsit University for future reference.

2.3. Plant Extraction

The seeds of *M. pruriens* (10 kg) were roasted by a roasting drum for 30 min at a temperature of 180 °C. The roasting temperature was measured by a sensor located in the drum. The roasted seeds were crushed to fine powders by passing through a stainless-steel sieve with a nominal mesh aperture of 180 µm. The fine powders were extracted with boiling water at 100 °C. The ratio of fine powder to hot water was 1:7. The water extract was mixed using the heating and stirring method with the heater and homogenizer for 15 min. After filtration with the white cloth, the filtrate was then freeze-dried to remove the solvent for 24 h. Finally, the concentrate crude extract was obtained.

2.4. HPLC Analysis of L-Dopa in M. pruriens Extract

Isocratic HPLC analysis was performed to analyze the amount of L-dopa remaining in the extract using Perkin Elmer series 200, USA, equipped with auto-sampler and a UV–Vis detector. The injection volume was 20 µL for all samples. The mobile phase for elution was 0.1 M KH_2PO_4 (pH 2.5). Samples were eluted on ACE-129-2546 (250 mm × 4.6 mm) C_{18} column with a flow rate of 1.0 mL/min. L-dopa peak was integrated at the wavelength of 283 nm. A stock solution of L-dopa was prepared to obtain L-dopa concentration of 1 mg/mL. The stock solution was diluted to 10, 20, 30, 40, and 50 µg/mL in 0.1 N HCl. *M. pruriens* aqueous extract was weighed and dissolved in 0.1 N HCl. Samples were then filtered through Whatman filter paper and 0.45 µm membrane filter.

2.5. Quantitative Analysis of Total Phenolic Content

Total phenolic content in *M. pruriens* seed extract was determined by the Folin–Ciocalteu reaction. Gallic acid solution (3.9–125 µg/mL) and *M. pruriens* seed extract solution (2 mg/mL) were mixed with 10% *v/v* of Folin–Ciocalteu phenol reagent (100 µL) for 1 min. After 4 min of incubation, 7.5% *w/v* Na_2CO_3 solution (50 µL) was added, and the mixture was further incubated in the dark at room temperature for 2 h. The absorbance was read using a UV–visible spectrophotometer (Spectramax M3, Thermo Scientific, Waltham, MA, USA) at a wavelength of 765 nm. Total phenolic contents were calculated from the gallic acid standard curve. Data were expressed as mg/g gallic acid equivalents (GAE) of dry crude extract.

2.6. Quantitative Analysis of Total Flavonoid Content

The total flavonoid content in *M. pruriens* seed extract was determined by the aluminum chloride colorimetric method. Quercetin solution (3.9–500 µg/mL) and *M. pruriens* seed extract solution (2 mg/mL, 100 µL/well) were mixed with 5% $NaNO_2$ (30 µL) and incubated for 5 min. Aluminum chloride (2% *w/v*, 50 µL) was added and incubated for

3.3. Cell Viability Assay

P19-derived neurons culture was treated with an increasing concentration of *M. pruriens* seed aqueous extract and L-Dopa (0–10 μg/mL). The neuronal viability was determined by XTT assay. The results were expressed as % cell viability. The effective concentration of *M. pruriens* seed aqueous extract at 10 ng/mL (% neuron viability = 114.48 ± 19.25%) that enhanced the survival of cultured neurons more than control (% neuron viability of the control = 100.47 ± 0.67%) were further investigated for the neuroprotective ability (Figure 1). Interestingly, L-dopa showed no enhancing cell viability effect on P19-derived neurons, suggesting that L-dopa was not responsible for the neuroprotective ability of *M. pruriens*. Therefore, only the extract was further investigated for its neuroprotective ability.

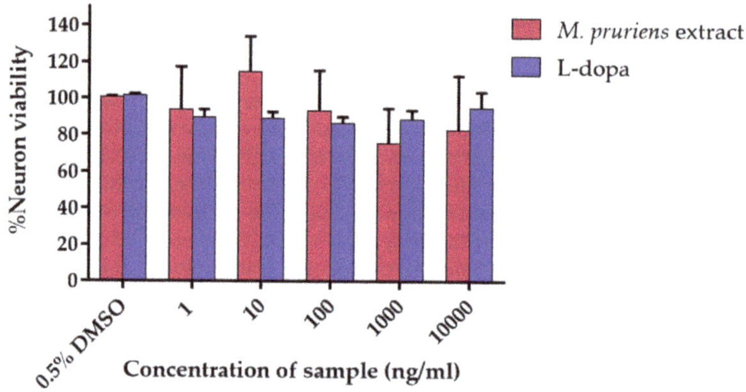

Figure 1. P19-derived neurons cell viability in various concentration (1–10,000 ng/mL) of *Mucuna pruriens* seed aqueous extract and L-dopa.

3.4. Serum Deprivation Method

P19-derived neurons culture was treated with 10 ng/mL of *M. pruriens* seed aqueous extract compared with toxic condition (0.5% *v/v* DMSO in α-MEM without serum) and positive control (1 nM quercetin). Cell viability was assayed by XTT reduction method. The results show significant cell viability in 10 ng/mL Mucuna (28.73 ± 1.90%) compared with the toxic condition (8.37 ± 0.73%) (Figure 2). The % neuron viability, when treated with 1 nM quercetin (positive control), was 34.73 ± 3.93%. No significant difference was found between the % neuron viability when treated with 10 ng/mL Mucuna and 1 nM quercetin.

3.5. Co-Administration of H_2O_2 Assay

P19-derived neurons culture was co-treated with 10 ng/mL of *M. pruriens* seed aqueous extract and 5 mM H_2O_2. The cell viability was compared with the toxic condition (5 mM H_2O_2). The cell viability was assayed by XTT reduction method. The results showed significant cell viability in 10 ng/mL *M. pruriens* seed extract co-treated with 5 mM H_2O_2 (25.01 ± 2.66%), compared with the toxic condition (11.61 ± 0.50%) (Figure 3). The % neuron viability when co-treated with 1 nM quercetin (positive control) and 5 mM H_2O_2 was 20.96 ± 3.67%. No significant difference was found between the % neuron viability when treated with 10 ng/mL *M. pruriens* seed extract and 1 nM quercetin.

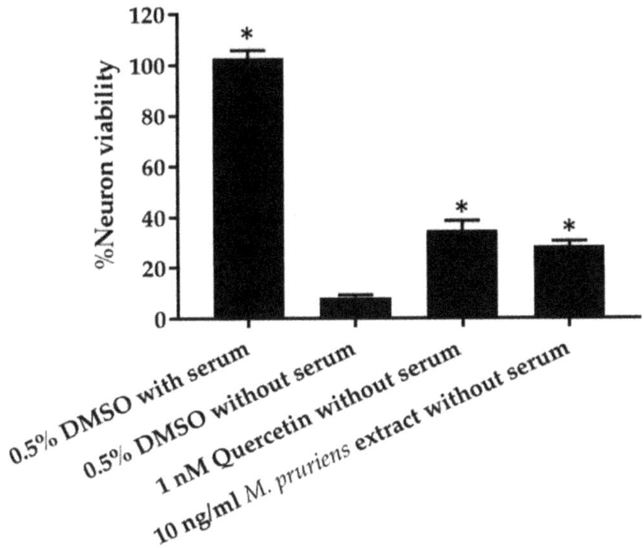

Figure 2. P19-derived neurons cell culture viability in 10 ng/mL *Mucuna pruriens* seed aqueous extract compared with the positive control (1 nM quercetin) and toxic condition (0.5% *v/v* DMSO in α-MEM without serum). * $p < 0.05$ compare with toxic condition (0.5% *v/v* DMSO in α-MEM without serum).

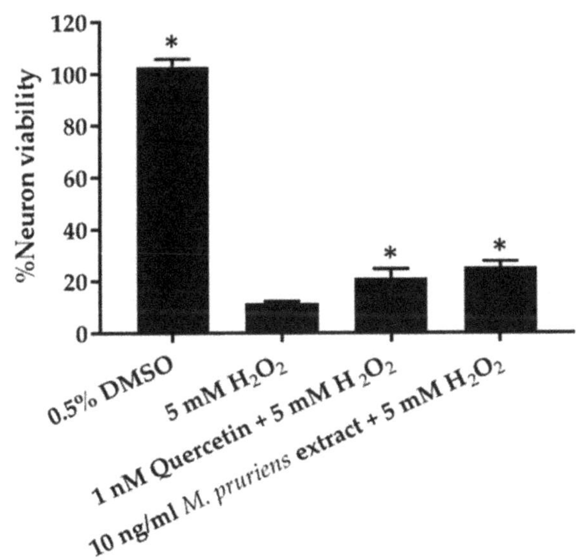

Figure 3. P19-derived neurons cell culture viability in 10 ng/mL *M. pruriens* seed aqueous extract co-treated with 5 mM hydrogen peroxide compared with the positive control (1 nM quercetin co-treated with 5 mM hydrogen peroxide) and toxic condition (5 mM hydrogen peroxide). * $p < 0.05$ compared with toxic condition (5 mM hydrogen peroxide).

3.6. Effects of L-Dopa and M. pruriens Seed Aqueous Extract on the Acetylcholinesterase Enzyme Activity

The effect of L-dopa and *M. pruriens* seed extract on acetylcholinesterase activity was investigated by the Ellman method, in which thiocholine produced by AChE reacted with 5,

5′-dithiobis(2-nitrobenzoic acid) to form a colorimetric product. The enzyme activity of the control without reaction with samples was 2.29 Units/L. One unit of AchE is the amount of enzyme that catalyzes the production of 1.0 µmole of tricholine per minute at pH 8.0 at room temperature. L-dopa was found to increase the activity of the acetylcholinesterase enzyme in a dose-dependent manner, whereas *M. pruriens* seed extract showed inhibitory activity against the acetylcholinesterase enzyme. Figure 4 demonstrated a kinetic curve plotted between the absorbance of the Ellman reaction product and time. The slopes of each plot were further used to calculate the AChE activity. The AChE activity increased with increasing concentrations of L-dopa (Figure 5). At the highest concentration of L-dopa (5 mg/mL), the AChE activity was activated up to 213%. In contrast, M. pruriens seed extract showed enzymatic inhibitory activity. The AChE inhibitory activity of *M. pruriens* seed extract was found to be 52.61% at an L-dopa equivalent concentration of 0.05 mg/mL. However, at the higher dose, the enzyme inhibitory effect was considerably lower to 11.02% at an L-dopa equivalent concentration of 3.5 mg/mL. This result was probably due to a higher concentration of L-dopa present in the extract.

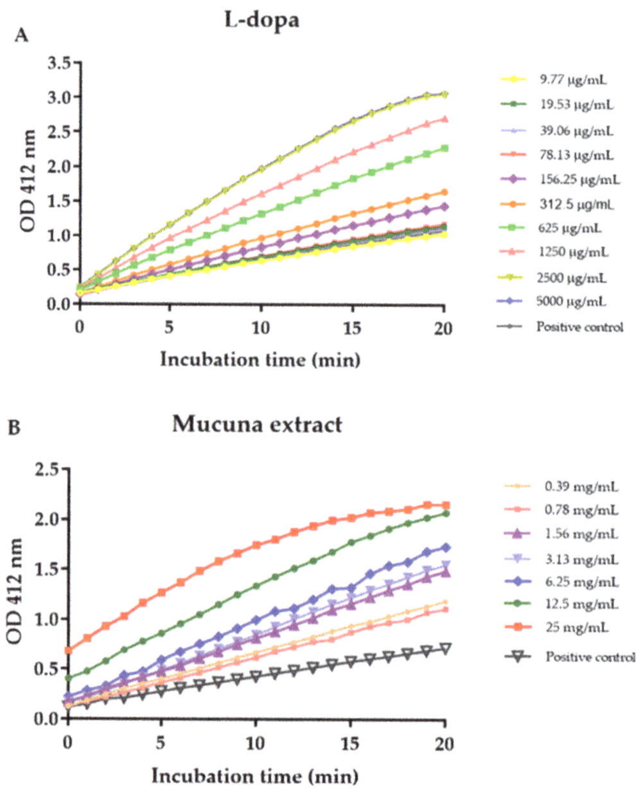

Figure 4. Plots of average O.D. readings subtracted from the mean blank (no enzyme) of each time point from its respective mean sample time points vs. time in the presence of (**A**) L-dopa and (**B**) *M. pruriens* seed extract.

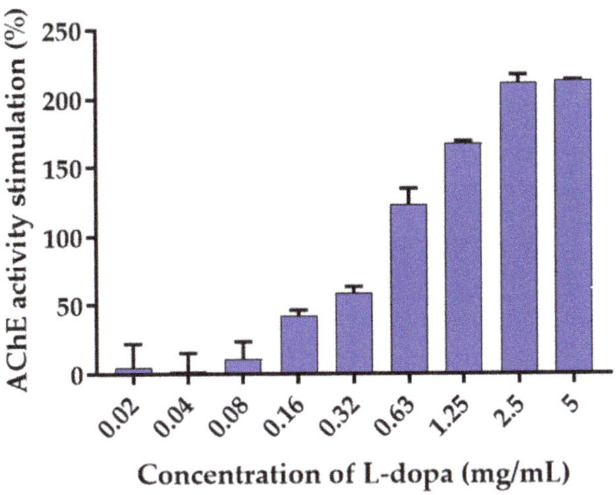

(A)

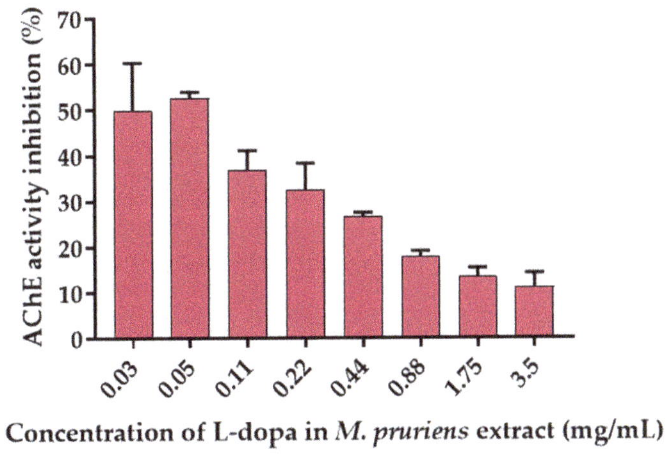

(B)

Figure 5. Acetylcholinesterase activity in the presence of (**A**) synthetic L-dopa and (**B**) *M. pruriens* seed aqueous extract.

4. Discussion

Parkinson's disease is a progressive neurodegenerative disorder resulting from the damaged dopaminergic neurons in the substantia nigra. The disease recently has no proven neuroprotective or neurorestorative therapy [27]. However, protecting neurons from premature cell death might be a promising alternative to managing this disease. Research directions include the investigation into animal models of the disease and the potential usefulness of gene therapy, stem cell transplant, neuroprotective agents, and herbals drugs.

The present study investigated the neuroprotective activities in P19-derived neurons using two models including the serum deprivation method and co-administration of hydrogen peroxide assay. Remarkably, the significant neuroprotective activities of standardized aqueous extracts of *M. pruriens* seed (L-dopa 7.05 ± 0.02 mg/100 g) were observed. The total phenolic compounds and total flavonoid content in *M. pruriens* seed aqueous extract were 106 GAE mg/g crude extract and 84 QE mg/g crude extract, respectively. Shin et al.

have reported that L-dopa and pramipexol (a dopamine agonist) had comparable neuroprotective properties in 1-methyl-4-phenyl-1,2,3,6-tetrahydropyridine (MPTP)-treated Parkinson's disease animal models through modulation of cell survival and apoptotic pathway [28]. The total phenolic compounds have been shown to exert neuroprotective effects against hydrogen peroxide-induced oxidative damage by blocking reactive oxygen species production and improving mitochondrial function [29]. Flavonoids have demonstrated neuroprotective effect via the inhibition of cholinesterase enzymes including acetylcholinesterase (AChE), butyrylcholinesterase (BChE), and β-secretase (BACE1) [30]. Therefore, our findings suggest that phenolic compounds and flavonoids in *M. pruriens* seed extract might play an important role in the neuroprotective effect against oxidative damage.

Regarding the serum deprivation method, the results demonstrated significant cell viability in 10 ng/mL *M. pruriens* extract (28.73 ± 1.90%) compared with the toxic condition (8.37 ± 0.73%). In addition, the results of the co-administration of hydrogen peroxide assay showed significant cell viability in 10 ng/mL *M. pruriens* extract (25.01 ± 2.66%) compared with the toxic condition (11.61 ± 0.50%). Our results agree with those of some previous studies. The evaluation of the anti-neuroinflammatory effects of *M. pruriens* extract was performed in murine microglia BV-2 cells. [16] *M. pruriens* extract was evaluated for neuroprotective effects in human neuroblastoma SH-SY5Y cells. The study showed that *M. pruriens* extract (at 1 µg/mL and 10 ng/mL) significantly reduced 6-hydroxydopamine-induced cytotoxicity and increased cell viability by 73.1 and 75.1%, respectively. The examination of ethanolic extract of *M. pruriens* on the level of nitric oxide in paraquat-induced Parkinson's disease mouse model and its subsequent contribution to lipid peroxidation results demonstrated that *M. pruriens* protects the dopaminergic neurons from the NO injury in substantia nigra [31]. The investigation of *M. pruriens* ethanolic seed extract in the Parkinsonian mouse model was performed [32]. A significant reduction in the activity of tyrosine hydroxylase positive neurons was observed in the substantia nigra region of the brain, after treatment with 1-methyl-4-phenyl-1,2,3,6-tetrahydropyridine.

Although the gold standard treatment of Parkinson's disease is L-dopa dopamine precursor, long-term L-dopa administration resulted in the development of serious side effects, including dyskinesia and the development of fluctuations in motor response. L-dopa-induced dyskinesia in Parkinson's disease patients significantly contributes to altered cholinergic signaling. Importantly, many studies showed that cholinergic receptor drugs including nicotine resulted in significant declines in L-dopa-induced dyskinesia [33–35]. Keeping the dose of L-dopa below 400 mg per day in early Parkinson's disease has been shown to reduce the risk of dyskinesia induction [36]. In this study, we showed that *M. pruriens* extract containing 7% natural L-dopa decreased the acetylcholine esterase stimulation compared with control. In contrast, the synthetic L-dopa significantly stimulated acetylcholinesterase activity. The results suggested that *M. pruriens* seed aqueous extract was probably able to modulate the cholinergic system and reduce L-dopa-induced dyskinesia and movement disorders resulting from L-dopa treatment. The present study demonstrated the neuroprotective activity of *M. pruriens* seed extract which may imply the therapeutic effects in various nervous system disorders including Parkinson's disease. Different mechanisms could underline this protection. Since *M. pruriens* seed extract showed neuroprotective activity against hydrogen peroxide-induced neurotoxicity, its proposed antioxidative therapeutics can be used for the protection of neuron injury induced by oxidative stress [37]. Serum deprivation-induced cell death is recognized as one of the standard models for the study of neurotoxicity. cGMP/PKG, PI13/Akt, and Bcl-2/Bax pathways have been shown to involve in the serum-deprivation-induced toxicity [38]. *M. pruriens* seed extract presented the potential of neuroprotective activity and thus may have a positive impact on aging and neurodegenerative diseases to retard the accelerated rate of neuronal degeneration.

M. pruriens seed extract contained substances including 5-hydroxytryptamine, alanine, arachidic acid, arginine, aspartic acid, behenic acid, beta-carboline, beta-sitosterol, bufotenine, choline, cis-12,13-epoxyoctadec-trans-9-cis-acid, cis-12,13-epoxyoctadec-trans-9-enoic-acid, cystine, gallic-acid, glutamic acid, glutathione, glycine, histidine, L-dopa, lecithin, leucine, linoleic acid, n, n-dimethyltryptamine, n, n-dimethyltryptamine-n-oxide, nicotine, phenylalanine, phosphorus, proline, protein, prurienidine, prurienine, saponins, serine, serotonin, threonine, tryptamine, tyrosine, valine, and vernolic acid, according to Dr. Duke's Phytochemical and Ethnobotanical Databases at Phytochemical Database, USDA-ARS-NGRL, Beltsville Agricultural Research Center [39]. Some of these compounds were AChE inhibitors, for example, physostigmine, which can enhance central levels of synaptic choline [40]. The aromatic amino acids such as tyrosine, phenylalanine, and tryptophan were identified as AChE inhibitors [41]. Pan et al. reported that flavonoid glycoside can inhibit AChE leading to a significant improvement in dyskinesia recovery rate in zebrafish [11]. Cognitive impairment and dementia are common complications of Parkinson's disease. Patients with Parkinson's disease with dementia often have significant cholinergic defects, which may be treated with cholinesterase inhibitors [42]. Several reports revealed that the use of an acetylcholinesterase inhibitor might improve cognitive function and reduce the risk of falls in patients with Parkinson's disease [43]. Therefore, *M. pruriens* seed extract with acetylcholinesterase inhibitory activity may be considered a prospective natural product for the treatment of Parkinson's disease, which may improve cognitive and motor functions of patients. This study reported the neuroprotective effect and anticholinesterase inhibition of *M. pruriens* seed aqueous extract, which contained large amounts of L-dopa (7%) along with other bioactive compounds including phenolic compounds and flavonoids. *M. pruriens* seed extract obtained in this study suggested more effectiveness and less toxicity than synthetic L-dopa for the treatment of Parkinson's disease. Our results were supported by Katzenschlager et al. showing the rapid onset of action and a long time without a concomitant increase in dyskinesias on *M. pruriens* seed powder formulation compared with L-dopa preparations in the long-term management of Parkinson's disease [14]. However, additional research is necessary to determine the agents responsible for other in vivo activities as well as the molecular mechanisms involved in their effects.

5. Conclusions

Long-term use of synthetic L-dopa can cause serious side effects such as dyskinesia and abnormal involuntary movement. A search for more effective treatment with lower side effects is significant for improving the quality of life of Parkinson's patients taking this drug. In this study, we reported the dual effects of *M. pruriens* seed extract on neuroprotective effect and acetylcholinesterase inhibitory activity at specific doses. Our findings suggested that *M. pruriens* seed extract is an attractive candidate for neuroprotection and preventing dyskinesia induced by cholinergic neuronal excitability in dopaminergic-depleted striatum.

Author Contributions: Conceptualization, N.K. and C.C.; methodology, N.K., C.C., S.T. and S.V.; formal analysis, C.C., S.T. and W.P.; investigation, N.K., C.C., S.T. and S.V.; resources, N.K., S.O. and P.K.; writing—original draft preparation, N.K., C.C. and S.T.; writing—review and editing, N.K., C.C., S.T., S.V., W.P., S.O. and P.K.; funding acquisition, N.K., C.C. and S.O. All authors have read and agreed to the published version of the manuscript.

Funding: This research was funded by Chaophraya Abhaibhubejhr Hospital and Agricultural Research Development Agency (Public Organization) grant number CRP6405030460. The APC was funded by Research Center of Pharmaceutical Nanotechnology, Faculty of Pharmacy, Chiang Mai University.

Institutional Review Board Statement: Not applicable.

Informed Consent Statement: Not applicable.

Acknowledgments: The authors are grateful to Chaophraya Abhaibhubejhr Hospital for funding the research. The authors acknowledge financial support from Agricultural Research Development Agency (Public Organization) (Grant No. CRP6405030460).

Conflicts of Interest: The authors declare no conflict of interest.

References

1. Surguchov, A.; Bernal, L.; Surguchev, A.A. Phytochemicals as Regulators of Genes Involved in Synucleinopathies. *Biomolecules* **2021**, *11*, 624. [CrossRef] [PubMed]
2. Lampariello, L.R.; Cortelazzo, A.; Guerranti, R.; Sticozzi, C.; Valacchi, G. The Magic Velvet Bean of *Mucuna pruriens*. *J. Tradit. Complement. Med.* **2012**, *2*, 331–339. [CrossRef]
3. Muthuraman, M.; Koirala, N.; Ciolac, D.; Pintea, B.; Glaser, M.; Groppa, S.; Tamás, G.; Groppa, S. Deep Brain Stimulation and L-DOPA Therapy: Concepts of Action and Clinical Applications in Parkinson's Disease. *Front. Neurol.* **2018**, *9*, 711. [CrossRef] [PubMed]
4. Misra, L.; Wagner, H. Extraction of bioactive principles from *Mucuna pruriens* seeds. *Indian J. Biochem. Biophys.* **2007**, *44*, 56–60. [PubMed]
5. Kumar, D.S.; Muthu, A.; Smith, A.A.; Manavalan, R. In vitro antioxidant activity of various extracts of whole plant of *Mucuna pruriens* (Linn). *Int. J. PharmTech Res.* **2010**, *2*, 2063–2070.
6. Emamzadeh, F.N.; Surguchov, A. Parkinson's Disease: Biomarkers, Treatment, and Risk Factors. *Front. Neurosci.* **2018**, *12*, 1–14. [CrossRef]
7. Manoharan, S.; Guillemin, G.J.; Abiramasundari, R.S.; Essa, M.M.; Akbar, M.; Akbar, M.D. The Role of Reactive Oxygen Species in the Pathogenesis of Alzheimer's Disease, Parkinson's Disease, and Huntington's Disease: A Mini Review. *Oxid. Med. Cell Longev.* **2016**, *2016*, 8590578. [CrossRef]
8. Ujowundu, C.O.; Kalu, F.N.; Emejulu, A.A.; Okafor, O.E.; Nkwonta, C.G.; Nwosunjoku, E.C. Evaluation of the chemical composition of Mucuna utilis leaves used in herbal medicine in Southeastern Nigeria. *Afr. J. Pharm. Pharmacol.* **2010**, *4*, 811–816.
9. Bastide, M.F.; Meissner, W.G.; Picconi, B.; Fasano, S.; Fernagut, P.O.; Feyder, M.; Francardo, V.; Alcacer, C.; Ding, Y.; Brambilla, R.; et al. Pathophysiology of L-dopa-induced motor and non-motor complications in Parkinson's disease. *Prog. Neurobiol.* **2015**, *132*, 96–168. [CrossRef]
10. Ahlskog, J.E.; Muenter, M.D. Frequency of levodopa-related dyskinesias and motor fluctuations as estimated from the cumulative literature. *Mov. Disord.* **2001**, *16*, 448–458. [CrossRef]
11. Pan, H.; Zhang, J.; Wang, Y.; Cui, K.; Cao, Y.; Wang, L.; Wu, Y. Linarin improves the dyskinesia recovery in Alzheimer's disease zebrafish by inhibiting the acetylcholinesterase activity. *Life Sci.* **2019**, *222*, 112–116. [CrossRef] [PubMed]
12. Fahn, S. How do you treat motor complications in Parkinson's disease: Medicine, surgery, or both? *Ann. Neurol.* **2008**, *64* (Suppl. S2), S56–S64. [CrossRef] [PubMed]
13. Salat, D.; Tolosa, E. Levodopa in the treatment of Parkinson's disease: Current status and new developments. *J. Parkinsons Dis.* **2013**, *3*, 255–269. [CrossRef] [PubMed]
14. Katzenschlager, R.; Evans, A.; Manson, A.; Patsalos, P.N.; Ratnaraj, N.; Watt, H.; Timmermann, L.; Van der Giessen, R.; Lees, A.J. *Mucuna pruriens* in Parkinson's disease: A double blind clinical and pharmacological study. *J. Neurol. Neurosurg. Psychiatry* **2004**, *75*, 1672–1677. [CrossRef] [PubMed]
15. Patil, S.A.; Apine, O.A.; Surwase, S.N.; Jadhav, J.P. Biological sources of L-DOPA: An alternative approach. *Adv. Parkinson's Dis.* **2013**, *2*, 81–87. [CrossRef]
16. Hussian, G.; Manyam, B.V. *Mucuna pruriens* proves more effective than L-DOPA in Parkinson's disease animal model. *Phytother. Res.* **1997**, *11*, 419–423. [CrossRef]
17. Jones-Villeneuve, E.M.; McBurney, M.W.; Rogers, K.A.; Kalnins, V.I. Retinoic acid induces embryonal carcinoma cells to differentiate into neurons and glial cells. *J. Cell Biol.* **1982**, *94*, 253–262. [CrossRef]
18. MacPherson, P.A.; McBurney, M.W. P19 Embryonal Carcinoma Cells: A Source of Cultured Neurons Amenable to Genetic Manipulation. *Methods* **1995**, *7*, 238–252. [CrossRef]
19. Jones-Villeneuve, E.M.; Rudnicki, M.A.; Harris, J.F.; McBurney, M.W. Retinoic acid-induced neural differentiation of embryonal carcinoma cells. *Mol. Cell Biol.* **1983**, *3*, 2271–2279.
20. Tadtong, S.; Kanlayavattanakul, M.; Lourith, N. Neuritogenic and neuroprotective activities of fruit residues. *Nat. Prod. Commun.* **2013**, *8*, 1583–1586. [CrossRef]
21. Supasuteekul, C.; Nonthitipong, W.; Tadtong, S.; Likhitwitayawuid, K.; Tengamnuay, P.; Sritularak, B. Antioxidant, DNA damage protective, neuroprotective, and α-glucosidase inhibitory activities of a flavonoid glycoside from leaves of *Garcinia gracilis*. *Rev. Bras. Farmacogn.* **2016**, *26*, 312–320. [CrossRef]
22. Tadtong, S.; Chatsumpun, N.; Sritularak, B.; Jongbunprasert, V.; Ploypradith, P.; Likhitwitayawuid, K. Effects of oxyresveratrol and its derivatives on cultured P19-derived neurons. *Trop. J. Pharm. Res.* **2016**, *15*, 2619–2628. [CrossRef]
23. Supasuteekul, C.; Tadtong, S.; Putalun, W.; Tanaka, H.; Likhitwitayawuid, K.; Tengamnuay, P.; Sritularak, B. Neuritogenic and neuroprotective constituents from *Aquilaria crassna* leaves. *J. Food Biochem.* **2017**, *41*, e12365. [CrossRef]
24. Puksasook, T.; Kimura, S.; Tadtong, S.; Jiaranaikulwanitch, J.; Pratuangdejkul, J.; Kitphati, W.; Suwanborirux, K.; Saito, N.; Nukoolkarn, V. Semisynthesis and biological evaluation of prenylated resveratrol derivatives as multi-targeted agents for Alzheimer's disease. *J. Nat. Med.* **2017**, *71*, 665–682. [CrossRef]
25. Tangsaengvit, N.; Kitphati, W.; Tadtong, S.; Bunyapraphatsara, N.; Nukoolkarn, V. Neurite Outgrowth and Neuroprotective Effects of Quercetin from Caesalpinia mimosoides Lamk. on Cultured P19-Derived Neurons. *Evid. Based Complement. Altern. Med.* **2013**, *2013*, 838051. [CrossRef]

26. Johnson, S.L.; Park, H.Y.; DaSilva, N.A.; Vattem, D.A.; Ma, H.; Seeram, N.P. Levodopa-Reduced *Mucuna pruriens* Seed Extract Shows Neuroprotective Effects against Parkinson's Disease in Murine Microglia and Human Neuroblastoma Cells, Caenorhabditis elegans, and Drosophila melanogaster. *Nutrients* **2018**, *10*, 1139. [CrossRef] [PubMed]
27. Gupta, A.; Dawson, V.L.; Dawson, T.M. What causes cell death in Parkinson's disease? *Ann. Neurol.* **2008**, *64* (Suppl. S2), S3–S15. [CrossRef]
28. Shin, J.Y.; Park, H.J.; Ahn, Y.H.; Lee, P.H. Neuroprotective effect of L-dopa on dopaminergic neurons is comparable to pramipexol in MPTP-treated animal model of Parkinson's disease: A direct comparison study. *J. Neurochem.* **2009**, *111*, 1042–1050. [CrossRef] [PubMed]
29. Wang, J.; Zhao, Y.M.; Zhang, B.; Guo, C.Y. Protective Effect of Total Phenolic Compounds from Inula helenium on Hydrogen Peroxide-induced Oxidative Stress in SH-SY5Y Cells. *Indian J. Pharm. Sci.* **2015**, *77*, 163–169.
30. Ayaz, M.; Sadiq, A.; Junaid, M.; Ullah, F.; Ovais, M.; Ullah, I.; Shahid, M. Flavonoids as Prospective Neuroprotectants and Their Therapeutic Propensity in Aging Associated Neurological Disorders. *Front. Aging Neurosci.* **2019**, *11*, 155. [CrossRef]
31. Yadav, S.K.; Rai, S.N.; Singh, S.P. *Mucuna pruriens* reduces inducible nitric oxide synthase expression in Parkinsonian mice model. *J. Chem. Neuroanat.* **2017**, *80*, 1–10. [CrossRef] [PubMed]
32. Manyam, B.V.; Dhanasekaran, M.; Hare, T.A. Neuroprotective effects of the antiparkinson drug *Mucuna pruriens*. *Phytother. Res.* **2004**, *18*, 706–712. [CrossRef] [PubMed]
33. Quik, M.; Cox, H.; Parameswaran, N.; O'Leary, K.; Langston, J.W.; Di Monte, D. Nicotine reduces levodopa-induced dyskinesias in lesioned monkeys. *Ann. Neurol.* **2007**, *62*, 588–596. [CrossRef] [PubMed]
34. Bordia, T.; Campos, C.; Huang, L.; Quik, M. Continuous and Intermittent Nicotine Treatment Reduces L-3,4-Dihydroxyphenylalanine (L-DOPA)-Induced Dyskinesias in a Rat Model of Parkinson's Disease. *J. Pharmacol. Exp. Ther.* **2008**, *327*, 239–247. [CrossRef] [PubMed]
35. Huang, L.Z.; Campos, C.; Ly, J.; Ivy Carroll, F.; Quik, M. Nicotinic receptor agonists decrease L-dopa-induced dyskinesias most effectively in partially lesioned parkinsonian rats. *Neuropharmacology* **2011**, *60*, 861–868. [CrossRef] [PubMed]
36. Rajput, A.H. Factors predictive of the development of levodopa-induced dyskinesia and Wearing-Off in Parkinson's disease. *Mov. Disord.* **2014**, *29*, 429. [CrossRef] [PubMed]
37. Dias, V.; Junn, E.; Mouradian, M.M. The role of oxidative stress in Parkinson's disease. *J. Parkinson's Dis.* **2013**, *3*, 461–491. [CrossRef]
38. Hsu, Y.Y.; Liu, C.M.; Tsai, H.H.; Jong, Y.J.; Chen, I.J.; Lo, Y.C. KMUP-1 attenuates serum deprivation-induced neurotoxicity in SH-SY5Y cells: Roles of PKG, PI3K/Akt and Bcl-2/Bax pathways. *Toxicology* **2010**, *268*, 46–54. [CrossRef]
39. Longhi, J.G.; Pérez, E.S.; Lima, J.J.D.; Cândido, L.M.B. In vitro evaluation of *Mucuna pruriens* (L.) DC. antioxidant activity. *Braz. J. Pharm. Sci.* **2011**, *47*, 535–544. [CrossRef]
40. Giarola, A.; Auber, A.; Chiamulera, C. Acetylcholinesterase inhibitors partially generalize to nicotine discriminative stimulus effect in rats. *Behav. Pharmacol.* **2011**, *22*, 1–6. [CrossRef]
41. Sousa, B.L.D.; Leite, J.P.; Mendes, T.A.; Varejão, E.V.; Chaves, A.; Silva, J.G.D.; Agrizzi, A.T.; Ferreira, P.G.; Pilau, E.J.; Silva, E.; et al. Inhibition of acetylcholinesterase by coumarin-linked amino acids synthetized via triazole associated with molecule partition coefficient. *J. Braz. Chem. Soc.* **2021**, *32*, 652–664. [CrossRef]
42. van Laar, T.; De Deyn, P.P.; Aarsland, D.; Barone, P.; Galvin, J.E. Effects of cholinesterase inhibitors in Parkinson's disease dementia: A review of clinical data. *CNS Neurosci. Ther.* **2011**, *17*, 428–441. [CrossRef] [PubMed]
43. Chung, K.A.; Lobb, B.M.; Nutt, J.G.; Horak, F.B. Effects of a central cholinesterase inhibitor on reducing falls in Parkinson disease. *Neurology* **2010**, *75*, 1263–1269. [CrossRef] [PubMed]

molecules

MDPI

Article

Anti-Inflammatory Effects of *Mitrephora sirikitiae* Leaf Extract and Isolated Lignans in RAW 264.7 Cells

Supachoke Mangmool [1], Chayaporn Limpichai [2], Khine Kyi Han [3], Vichai Reutrakul [4,5] and Natthinee Anantachoke [2,5,*]

1 Department of Pharmacology, Faculty of Science, Mahidol University, Bangkok 10400, Thailand; supachoke.man@mahidol.ac.th
2 Department of Pharmacognosy, Faculty of Pharmacy, Mahidol University, Bangkok 10400, Thailand; chayapornlim@gmail.com
3 Department of Pharmacology, University of Pharmacy, Yangon 11031, Myanmar; khinekyihanuop@gmail.com
4 Department of Chemistry, Faculty of Science, Mahidol University, Bangkok 10400, Thailand; vichai.reu@mahidol.ac.th
5 Center of Excellence for Innovation in Chemistry, Faculty of Science, Mahidol University, Bangkok 10400, Thailand
* Correspondence: natthinee.ana@mahidol.ac.th

Abstract: *Mitrephora sirikitiae* Weeras., Chalermglin & R.M.K. Saunders has been reported as a rich source of lignans that contribute to biological activities and health benefits. However, cellular anti-inflammatory effects of *M. sirikitiae* leaves and their lignan compounds have not been fully elucidated. Therefore, this study aimed to investigate the anti-inflammatory activities of methanol extract of *M. sirikitiae* leaves and their lignan constituents on lipopolysaccharide (LPS)-induced inflammation in RAW 264.7 mouse macrophage cells. Treatment of RAW 264.7 cells with the methanol extract of *M. sirikitiae* leaves and its isolated lignans, including (−)-phylligenin (**2**) and 3′,4-*O*-dimethylcedrusin (**6**) significantly decreased LPS-induced prostaglandin E_2 (PGE_2) and nitric oxide (NO) productions. These inhibitory effects of the extract and isolated lignans on LPS-induced upregulation of PGE_2 and NO productions were derived from the suppression of cyclooxygenase 2 (COX-2) and inducible nitric oxide synthase (iNOS) production, respectively. In addition, treatment with 2-(3,4-dimethoxyphenyl)-6-(3,5-dimethoxyphenyl)-3,7-dioxabicyclo[3.3.0]octane (**3**) and mitrephoran (**5**) was able to suppress LPS-induced tumor necrosis factor alpha (TNF-α) secretion and synthesis in RAW 264.7 cells. These results demonstrated that *M. sirikitiae* leaves and some isolated lignans exhibited potent anti-inflammatory activity through the inhibition of secretion and synthesis of PGE_2, NO, and TNF-α.

Keywords: *Mitrephora sirikitiae*; anti-inflammation; lignans; polyphenols; mRNA expression; cytokines

Citation: Mangmool, S.; Limpichai, C.; Han, K.K.; Reutrakul, V.; Anantachoke, N. Anti-Inflammatory Effects of *Mitrephora sirikitiae* Leaf Extract and Isolated Lignans in RAW 264.7 Cells. *Molecules* **2022**, *27*, 3313. https://doi.org/10.3390/molecules27103313

Academic Editor: Nour Eddine Es-Safi

Received: 23 April 2022
Accepted: 20 May 2022
Published: 21 May 2022

1. Introduction

Inflammation is a defensive mechanism that responds to foreign substances or pathogens. During inflammatory reactions, prostaglandin E_2 (PGE_2) and nitric oxide (NO) are the crucial proinflammatory mediators. Cyclooxygenase 2 (COX-2) and inducible nitric oxide synthase (iNOS), the key enzymes responsible for the production of inflammatory PGE_2 and NO, have been identified in activated macrophages [1–3]. Lipopolysaccharide (LPS; a pathogen- and host-derived molecule) stimulates macrophages to, in turn, upregulate inflammatory mediators such as PGE_2, tumor necrosis factor alpha (TNF-α), proinflammatory cytokines (e.g., interleukin (IL)-1, IL-6, IL-8, and IL-10), reactive oxygen species (ROS), and NO [1,4–6]. Upregulation of various types of inflammatory mediators leads to acute and chronic inflammation, which is associated with many chronic diseases such as gout, arthritis, diabetes, cancer, atherosclerosis, and neurodegenerative disease [1,2,7,8]. Therefore, inhibition of secretion and synthesis of these inflammatory mediators is a potential therapeutic treatment of inflammation-associated diseases.

2.2. Effects of the methanol Extract and Isolated Lignans on LPS-Induced PGE_2 and TNF-α Secretion

Lipopolysaccharide (LPS), known as a major factor involved in inflammation, is a component of the outer membrane of Gram-negative bacteria. Moreover, it is called endotoxin because of its endotoxic properties. LPS contributes to the pathogenicity of bacteria and can activate several immune cells, including macrophages [20]. In this study, LPS was used for inducing inflammation in RAW 264.7 macrophages. Incubation of the cells with LPS (5 μg/mL) for 24 h markedly increased PGE_2 and TNF-α secretion (Figure 3). The methanol extract of *M. sirikitiae* leaves was first assessed for the inhibitory effects on LPS-induced PGE_2 and TNF-α secretions. Treatment of the cells with the methanol extract of *M. sirikitiae* leaves resulted in a decrease in LPS-induced PGE_2 secretion in a dose-dependent manner, and the maximal inhibitory effect of the methanol extract was observed at a concentration of 10 μg/mL (Figure 3A). However, the methanol extract exhibited little effect on inhibition of LPS-induced TNF-α secretion at the same concentration and did not inhibit LPS-induced TNF-α secretion at a concentration of 1 μg/mL (Figure 3B). These results indicated that the compounds found in *M. sirikitiae* leaves might have anti-inflammatory effects by reducing the secretion of PGE_2 in RAW 264.7 macrophages.

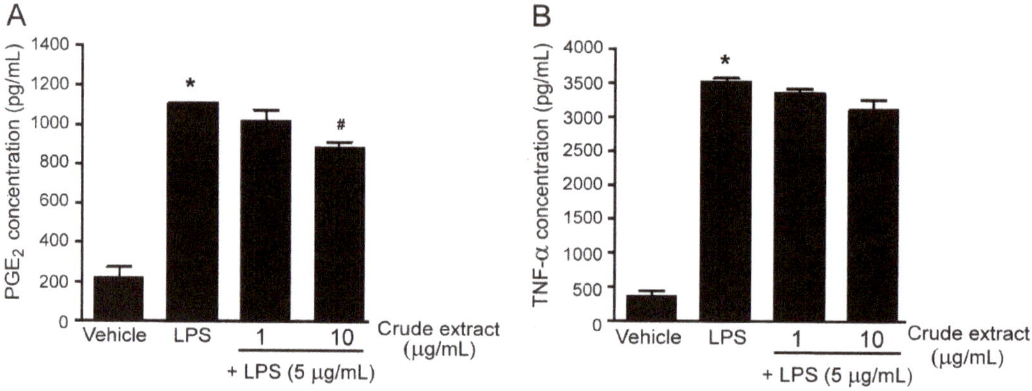

Figure 3. Effects of the methanol extract of *M. sirikitiae* leaves on LPS-induced PGE_2 and TNF-α secretion in RAW 264.7 cells; Serum-starved cells were pretreated with various concentrations of 0, 1, and 10 μg/mL of crude methanol extract for 3 h, and then stimulated with LPS (5 μg/mL) for 24 h at 37 °C. The levels of PGE_2 and TNF-α secreted into the medium were assessed by ELISA assay. The PGE_2 (**A**) and TNF-α (**B**) levels were quantified using a standard curve and expressed as the mean ± SEM (n = 3). *, $p < 0.05$ vs. vehicle; #, $p < 0.05$ vs. LPS.

Lignans are polyphenols, and the main components of *M. sirikitiae* leaves. Our previous study has revealed that lignans (−)-epieudesmin (**1**), (−)-phylligenin (**2**), 2-(3,4-dimethoxyphenyl)-6-(3,5-dimethoxyphenyl)-3,7-dioxabicyclo[3.3.0]octane (**3**), magnone A (**4**), mitrephoran (**5**), and 3′,4-*O*-dimethylcedrusin (**6**) are the main secondary metabolites in *M. sirikitiae* leaves [10]. As shown in Figure 4A, treatment of the cells with (−)-phylligenin (**2**) (10 μg/mL) significantly inhibited LPS-induced PGE_2 secretion. However, (−)-epieudesmin (**1**), 2-(3,4-dimethoxyphenyl)-6-(3,5-dimethoxyphenyl)-3,7-dioxabicyclo[3.3.0]octane (**3**), magnone A (**4**), mitrephoran (**5**), and 3′,4-*O*-dimethylcedrusin (**6**) had no effect on the inhibition of LPS-induced PGE_2 secretion. These results are consistent with previous studies, which reported that phylligenin and koreanaside A, lignans isolated from *Forsythia koreana* fruits and flowers, could inhibit PGE_2 secretion in LPS-stimulated RAW 264.7 cells in a dose-dependent manner [15,16]. Moreover, the anti-inflammation of koreanaside A was also represented by the inhibitory effects on NO, TNF-α, and IL-6 productions [16].

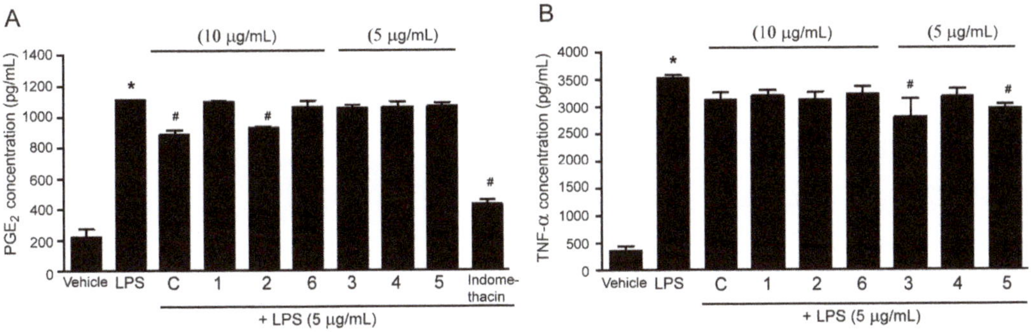

Figure 4. Effects of the methanol extract and isolated lignans **1–6** from *M. sirikitiae* leaves on LPS-induced PGE_2 and TNF-α secretion in RAW 264.7 cells; Serum-starved cells were pretreated with crude methanol extract (C) (10 μg/mL), lignans (5 or 10 μg/mL), or indomethacin (10 μM) for 3 h, and then stimulated with LPS (5 μg/mL) for 24 h at 37 °C. The levels of PGE_2 and TNF-α secreted into the medium were assessed by ELISA assay. The relative PGE_2 (**A**) and TNF-α (**B**) levels were quantified using a standard curve and expressed as the mean ± SEM (n = 3). *, $p < 0.05$ vs. vehicle; #, $p < 0.05$ vs. LPS.

Interestingly, all lignans isolated from the caulis of *Urceola rosea* exhibited anti-inflammatory activities by suppressing LPS-induced synthesis of NO, TNF-α, and/or IL-6 in RAW 264.7 cells. In addition, among those lignans, ecdysanol A and ecdysanol F exhibited potent anti-inflammatory activity against TNF-α secretion with IC_{50} values of 22.9 and 41.9 μM, respectively [17]. Consistent with this previous study, our results showed that among those six lignans isolated from *M. sirikitiae* leaves, 2-(3,4-dimethoxyphenyl)-6-(3,5-dimethoxyphenyl)-3,7-dioxabicyclo[3.3.0]octane (**3**) and mitrephoran (**5**) (5 μg/mL) significantly inhibited LPS-induced TNF-α secretion in RAW 264.7 cells (Figure 4B).

2.3. *Effects of the Methanol Extract and Isolated Lignans on LPS-Induced Nitric Oxide Production*

Nitric oxide (NO) is a free-radical signaling molecule playing a role in many biological processes, including inflammation. NO is synthesized from L-arginine and oxygen by using NOS as catalysts. There are three main isoforms of NOS, endothelial NOS (eNOS), neuronal NOS (nNOS), and inducible NOS (iNOS). The iNOS enzyme is a major isoform in the inflammatory process [3,21,22]. Thus, upregulation of iNOS might be reflected in NO production in the cells during inflammation. NO is immediately oxidized to generate nitrate and nitrite, both of which are used for measuring NO levels in the cells. Therefore, this study investigated the inhibitory effects of methanol extract and isolated lignans from *M. sirikitiae* leaves on LPS-mediated NO production by determining the levels of nitrate and nitrite in RAW 264.7 cells. Incubation of the cells with LPS (10 μg/mL) robustly increased nitrate and nitrite productions (Figure 5). Treatment with the methanol extract, (−)-epieudesmin (**1**), (−)-phylligenin (**2**), or 3′,4-*O*-dimethylcedrusin (**6**) at a concentration of 10 μg/mL resulted in a significant reduction of LPS-induced nitrate and nitrite levels in RAW 264.7 cells (Figure 5A and Figure 5B, respectively). In contrast, treatment with 2-(3,4-dimethoxyphenyl)-6-(3,5-dimethoxyphenyl)-3,7-dioxabicyclo[3.3.0]octane (**3**), magnone A (**4**), and mitrephoran (**5**) (5 μg/mL) had no effect (Figure 5). These results indicated that the lignans found in *M. sirikitiae* leaves, including (−)-epieudesmin (**1**), (−)-phylligenin (**2**), and 3′,4-*O*-dimethylcedrusin (**6**) exhibited anti-inflammatory activities by attenuating NO production in RAW 264.7 cells. Consistent with our present data, many previous studies have demonstrated that some lignans can suppress NO production during inflammation. For example, the methanol extract of *F. koreana* fruits and its isolated lignan, phylligenin, could inhibit NO synthesis in LPS-treated RAW 264.7 cells [15]. Furthermore, nine lignans isolated *Acanthopanax sessiliflorus* fruits [18] and two dimeric lignans isolated from *Zanthoxylum podocarpum* barks [19] have been reported to have inhibitory effects on

NO production in LPS-treated RAW 264.7 macrophages. Thus, natural phenols found in plants as lignans may serve as potential anti-inflammatory agents via inhibition of the NO signaling pathway.

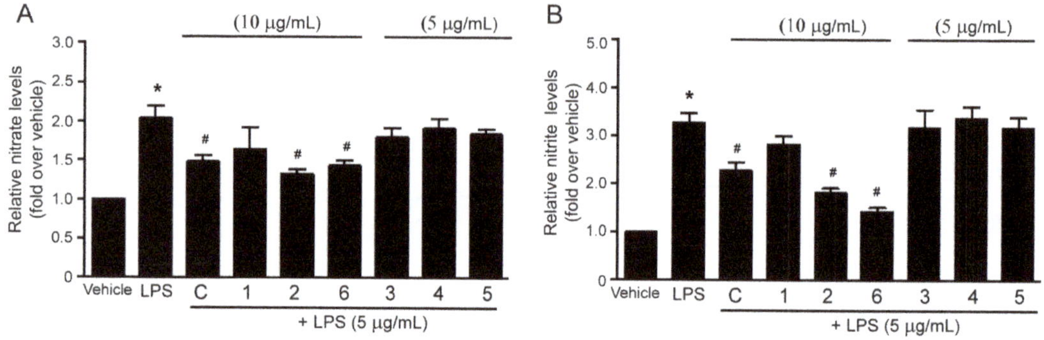

Figure 5. Effects of the methanol extract and isolated lignans **1–6** from *M. sirikitiae* leaves on LPS-induced nitrate and nitrite productions in RAW 264.7 cells; Serum-starved cells were pretreated with crude methanol extract (C) (10 µg/mL) or lignans (5 or 10 µg/mL) for 3 h, and then stimulated with LPS (5 µg/mL) for 24 h at 37 °C. The levels of nitrate and nitrite in the medium were assessed by a total nitric oxide assay. The relative nitrate (**A**) and nitrite (**B**) productions were quantified using a standard curve and expressed as the mean ± SEM (n = 3). *, $p < 0.05$ vs. vehicle; #, $p < 0.05$ vs. LPS.

2.4. Effects of the Methanol Extract and Isolated Lignans on LPS-Induced mRNA Expressions of Inflammatory Biomarkers

According to the methanol extract of *M. sirikitiae* leaves and some isolated lignans representing the inhibitory effect on LPS-mediated PGE_2, TNF-α, and nitric oxide secretions in RAW 264.7 cells, the mRNA expression of inflammatory biomarkers, including COX-2, iNOS, TNF-α, IL-6, IL-10, and NF-κB affected by the methanol extract and isolated lignans **1–6** was subsequently studied. In this study, incubation of the RAW 264.7 cells with LPS (5 µg/mL) markedly increased TNF-α, IL-6, IL-10, NF-κB, COX-2, and iNOS mRNA expression levels as compared with a control group (Figure 6A, Figure 6B, Figure 6C, Figure 6D, Figure 6E, and Figure 6F, respectively), indicating the induction of inflammation. The methanol extract of *M. sirikitiae* leaves was able to suppress LPS-induced mRNA expression of NF-κB, COX-2, and iNOS (Figure 6D, Figure 6E, and Figure 6F, respectively) while treatment of the cells with this methanol extract tended to inhibit LPS-induced TNF-α and IL-10 syntheses (Figure 6A and Figure 6C, respectively). However, the extract did not affect IL-6 mRNA expression (Figure 6B). These results demonstrated that the methanol extract of *M. sirikitiae* leaves possesses anti-inflammatory effects by suppressing the synthesis of several inflammatory mediators, indicating that the anti-inflammatory effects are derived from some active compounds contained in *M. sirikitiae* leaves.

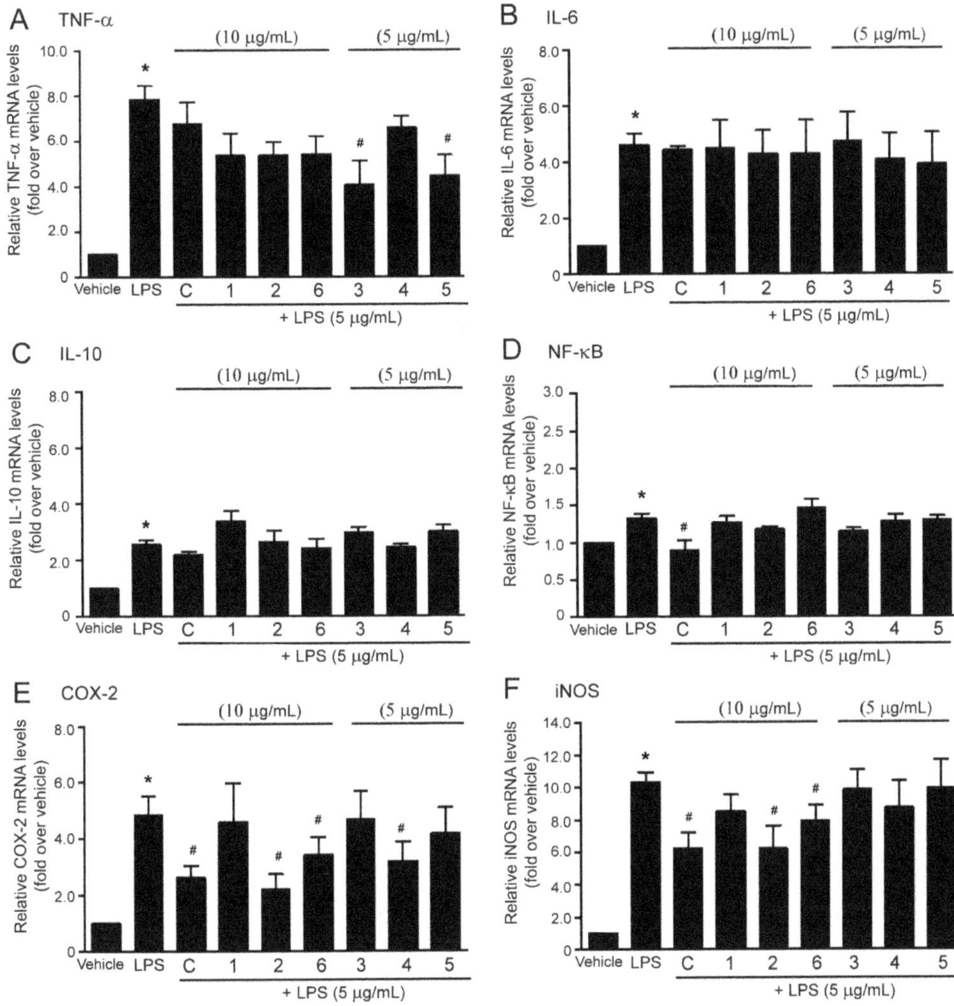

Figure 6. Effects of the methanol extract and isolated lignans **1–6** from *M. sirikitiae* leaves on LPS-induced mRNA expressions of inflammatory biomarkers in RAW 264.7 cells; Serum-starved cells were pretreated with crude methanol extract (C) (10 μg/mL) or lignans (5 or 10 μg/mL) for 3 h, and then stimulated with LPS for 6 h at 37 °C. After treatment, the total RNA was extracted from the cells and the mRNA expressions of TNF-α (**A**), IL-6 (**B**), IL-10 (**C**), NF-κB (**D**), COX-2 (**E**), and iNOS (**F**) were analyzed by RT-qPCR with gene-specific primers. The relative mRNA levels were quantified and shown as the mean ± SEM (n = 3). *, $p < 0.05$ vs. vehicle; #, $p < 0.05$ vs. LPS.

Moreover, all isolated lignans **1–6** were investigated for their effects on suppressing mRNA expressions of inflammatory mediators induced by LPS. As shown in Figure 6A, treatment of the cells with (−)-epieudesmin (**1**), (−)-phylligenin (**2**), magnone A (**4**), and 3′,4-*O*-dimethylcedrusin (**6**) tended to inhibit LPS-induced mRNA expression of TNF-α. In addition, treatment of the cells with 2-(3,4-dimethoxyphenyl)-6-(3,5-dimethoxyphenyl)-3,7-dioxabicyclo[3.3.0]octane (**3**) and mitrephoran (**5**) significantly inhibited LPS-induced TNF-α mRNA expression. Macrophage cells are the main source of proinflammatory cytokine TNF-α, which is used as one of the major markers of inflammation. This study reported for the first time that isolated lignans **3** and **5** from *M. sirikitiae* leaves exhibited the anti-inflammatory effects by suppressing LPS-induced TNF-α secretion (Figure 4B) and synthesis (Figure 6A) in RAW 264.7 cells.

However, lignans **1–6** did not show the inhibitory effects on LPS-induced mRNA expression of inflammatory cytokines, IL-6, IL-10, and NF-κB in RAW 264.7 cells (Figure 6B–D). After exposure to a pathogen (e.g., LPS), macrophages are activated to produce pro-inflammatory cytokines such as IL-1 and IL-6, which play an important role in the inflammatory response [23]. Although the lignans isolated from *M. sirikitiae* leaves did not show the inhibitory effects on LPS-induced IL synthesis, the previous study demonstrated that other types of lignans isolated from the caulis of *U. rosea* exhibited anti-inflammatory effects by reducing the LPS-induced IL-6 production in RAW 264.7 cells with the IC_{50} values ranging from 5 to 50 μM [17]. Thus, each lignan has its own anti-inflammatory properties and mechanisms.

NF-κB is one of the major regulators of inflammatory gene expression. The activation of NF-κB resulted in the upregulation of mRNA expression of cytokines such as TNF-α, IL-1β, IL-6, and IL-8, including COX-2 [24]. Even though the methanol extract of *M. sirikitiae* leaves was able to suppress LPS-induced NF-κB synthesis, all lignans isolated from the *M. sirikitiae* leaves had no effect on the inhibition of NF-κB synthesis induced by LPS (Figure 6D). Further studies are required to identify other constituents that have inhibitory effects on the synthesis of NF-κB.

Interestingly, treatment with (−)-phylligenin (**2**), magnone A (**4**), and 3′,4-*O*-dimethylcedrusin (**6**) significantly inhibited the mRNA expression of COX-2 in LPS-stimulated RAW 264.7 cells (Figure 6E). The effects of the methanol extract and (−)-phylligenin (**2**) on COX-2 mRNA expression were correlated with their inhibitory effects on LPS-induced PGE_2 secretion in RAW 264.7 cells. Our data are in concordance with other studies that some lignans, such as koreanaside A isolated from *F. koreana* flowers, could downregulate COX-2 mRNA expression and attenuate PGE_2 secretion in LPS-induced RAW 264.7 cells [16]. The methanol extracts of *Schisandra rubriflora* and *Schisandra chinensis* and their isolated lignans, including 6-obenzoylgomisin O, schisandrin, gomisin D, gomisin N, and schisantherin A, have been revealed to exhibit significant COX-2 inhibitory activity [14]. Our study demonstrated the inhibitory effects of magnone A (**4**), and 3′,4-*O*-dimethylcedrusin (**6**) on LPS-induced COX-2 synthesis for the first time. During inflammation, arachidonic acid converses into PGE_2 mediated through COX-2 catalytic reaction. Inhibition of COX-2 activity and synthesis has been proposed as a useful treatment for various inflammatory diseases [25]. Therefore, COX-2 is a well-known target of various anti-inflammatory drugs such as aspirin and other nonsteroidal anti-inflammatory drugs (NSAIDs). In our present study, the methanol extract of *M. sirikitiae* leaves could inhibit the synthesis of COX-2 and reduce PGE_2 secretion in LPS-treated macrophages. Thus, *M. sirikitiae* might be used as a medicinal plant for the prevention and restoration of inflammatory diseases.

The iNOS is an inducible isoform of NOS and upregulation of iNOS activity and synthesis occurred in response to inflammation [22]. Therefore, the inhibitory effect on iNOS enzyme, which subsequently suppresses NO generation was investigated. We found that treatments of the cells with the methanol extract of *M. sirikitiae* leaves, (−)-phylligenin (**2**), and 3′,4-*O*-dimethylcedrusin (**6**) significantly inhibited LPS-induced iNOS mRNA expression in RAW 264.7 macrophages, while (−)-epieudesmin (**1**), 2-(3,4-dimethoxyphenyl)-6-(3,5-dimethoxyphenyl)-3,7-dioxabicyclo[3.3.0]octane (**3**), magnone A (**4**), and mitrephoran (**5**) had no effects (Figure 6F). These inhibitory effects of the methanol extract, together with (−)-phylligenin (**2**), and 3′,4-*O*-dimethylcedrusin (**6**) on iNOS mRNA expression (Figure 5F) were correlated with their inhibitory effects on LPS-induced NO production (Figure 4) in RAW 264.7 cells. Interestingly, phylligenin (10–100 μM) and koreanaside A (20–80 μM) isolated from *F. koreana* have previously reported the inhibitory effects on LPS-induced iNOS synthesis in RAW 264.7 cells [15,16]. These findings from ours and previous studies confirmed the anti-inflammatory effects of (−)-phylligenin (**2**) by inhibiting iNOS synthesis and subsequent reducing NO production. Moreover, our study reported the inhibitory effects on iNOS synthesis of 3′,4-*O*-dimethylcedrusin (**6**), the lignan isolated from *M. sirikitiae* leaves, for the first time.

In this study, the structure–activity relationships (SARs) of the natural lignans for anti-inflammatory properties were considered in order to identify the specific structures and functional groups that play an important role in the activities. Based on a variety of chemical core skeletons and functional groups of lignans, isolated lignans **1–6** from *M. sirikitiae* leaves could be classified into three subgroups, including furofurans (lignans **1–3**), furans (lignans **4–5**), and benzofurans (lignan **6**) [26]. The obtained results from our present study indicated that the possible SARs of the anti-inflammatory lignans are different in each specific mechanism of action. For instance, substitution of a hydroxyl group on the phenyl ring and a furofuran moiety as found in lignan **2** might be important for the inhibitory effects against LPS-induced PGE_2 secretion as well as COX-2 mRNA expression. Moreover, the furofuran with hydroxy-substituted phenyl (lignan **2**) and benzofuran (lignan **6**) skeletons played a determinant role against NO production and iNOS mRNA expression induced by LPS in cells. Meanwhile, furofuran with non-hydroxy-substituted phenyl (lignan **3**) and furan with hydroxy-substituted phenyl (lignan **5**) moieties were found to be essential for the inhibition of LPS-induced TNF-α secretion and TNF-α mRNA expression. Although the SARs of lignans for the downregulation of LPS-induced mRNA expression of IL-6, IL-10, and NF-κB could not be discussed in this study, some previous studies have disclosed that furofuran skeleton is possibly related to the suppression of NF-κB signaling pathway [27]. Moreover, the SARs of coumarinolignans have reported that the substitution of hydroxyl groups on the phenyl ring is important to anti-inflammatory activity [28]. In order to fully understand SARs of lignans for anti-inflammatory properties, a number of lignans with different skeletal types and functional groups are required for the biological testing.

It should be noted that our data were obtained from LPS-induced inflammation in RAW 264.7 cells, and therefore may not reflect the actual systemic anti-inflammatory effects in the human body. Thus, further studies will be necessary to evaluate the anti-inflammatory activities in vivo studies.

3. Materials and Methods

3.1. Chemicals and Reagents

Lipopolysaccharide (LPS from *Salmonella enterica* serotype Typhimurium), indomethacin, and 3-[4,5-dimethylthiazol-2-yl]-2,5-diphenyltetrazolium bromide (MTT) were purchased from Sigma-Aldrich (St. Louis, MO, USA). Dulbecco's Modified Eagle Medium (DMEM), 0.25% trypsin-EDTA solution, penicillin/streptomycin (P/S) solution, and fetal bovine serum (FBS) were purchased from Gibco (Grand Island, NY, USA). Dimethyl sulfoxide (DMSO) was purchased from Merck (Darmstadt, Germany). PGE_2 ELISA assay kit was obtained from R&D Systems (Minneapolis, MN, USA). TNF-α ELISA assay kit (ab100747) and nitric oxide assay kit (ab65328) were purchased from Abcam (Waltham, MA, USA). The specific primers were obtained from Integrated DNA Technologies (Coralville, IA, USA). KAPA SYBR FAST One-step RT-qPCR kit was purchased from KAPA biosystems (Wilmington, MA, USA). RNeasy kit was obtained from Qiagen (Hilden, Germany).

3.2. Plant Materials and Isolated Lignans

The *M. sirikitiae* leaves were collected from Mae Surin Waterfall National Park, Mae Hong Sorn Province, Thailand (Latitude 18.94271; Longtitude 98.07122). A voucher specimen (BKF144972) of the plant has been deposited at the Forest Herbarium, Royal Forestry Department, Bangkok, Thailand. The methanol extract of *M. sirikitiae* leaves together with its lignans **1–6**, including (−)-epieudesmin (**1**), (−)-phylligenin (**2**), 2-(3,4-dimethoxyphenyl)-6-(3,5-dimethoxyphenyl)-3,7-dioxabicyclo[3.3.0]octane (**3**), magnone A (**4**), mitrephoran (**5**), 3′,4-*O*-dimethylcedrusin (**6**) were prepared as described in our previous study [10]. Briefly, 1 kg of ground dried *M. sirikitiae* leaves was extracted by maceration with methanol. After the solvent was evaporated, 102.0 g of methanol leaf extract was obtained. Solvent–

solvent partitioning of the methanol extract (90.0 g) resulted in four subfractions, including hexane (20.7 g), ethyl acetate (21.1 g), butanol (47.6 g), and aqueous (24.2 g) subfractions. Separation of the mixture of hexane and ethyl acetate subfractions by column chromatography using silica gel P60 40–63 μm (SiliaFlash®; Silicycle) and Sephadex™ LH-20 (GE Healthcare Life Sciences) as stationary phases afforded six isolated lignans, **1**–**6**. The colorless crystals of (−)-epieudesmin (**1**) (197.5 mg), (−)-phylligenin (**2**) (717.4 mg), magnone A (**4**) (24.0 mg), and mitrephoran (**5**) (12.4 mg) were obtained after recrystallization from mixed solvent of methylene chloride and methanol. Moreover, 2-(3,4-dimethoxyphenyl)-6-(3,5-dimethoxyphenyl)-3,7-dioxabicyclo[3.3.0]octane (**3**) (9.6 mg) and 3′,4-*O*-dimethylcedrusin (**6**) (141.2 mg) were isolated as pale-yellow semisolids. The chemical structures of the isolated lignans were elucidated by their spectroscopic and physical characteristic data (see Supplementary Materials).

3.3. *Cell Culture and Treatment*

Murine macrophage RAW 264.7 cells obtained from American Type Culture Collection (ATCC, #TIB-71; Rockville, MD, USA) were cultured in DMEM plus 10% FBS and 1% P/S and incubated under a temperature 37 °C in a humidified atmosphere of 5% CO_2 incubator as previously described [29]. The cells were grown in culture dishes. After the confluence of cells reached 80%, the cell density was calculated and used for the subsequent experiment. The cells were passaged by trypsinization every 3–5 days.

3.4. *Determination of Cytotoxicity by MTT Assay*

Cell viability was determined by MTT assay as previously described [30]. RAW 264.7 cells were seeded in 96-well culture plates at a 1×10^4 cells/well density in DMEM supplemented with 1% FBS and 1% P/S and incubated in a humidified 37 °C, 5% CO_2 incubator for 24 h. After incubation, the cultured cells were treated with various concentrations of the methanol extract of *M. sirikitiae* leaves and isolated lignans (0.05–250 μg/mL, diluted with DMEM), or dimethyl sulfoxide (DMSO; vehicle) and then incubated for 36 h. The culture medium was removed and added with MTT solution (1 mg/mL in DMEM). After incubated for 4 h, the MTT solution was removed before adding DMSO to dissolve the formed purple formazan crystals. Each experiment was performed in triplicate. The absorbance was detected using an Infinite M200 microplate reader (Tecan) at a wavelength of 570 nm. The results were shown in the graph of percentage of cell viability (% cell viability = $[A_{\text{treated cells}}/A_{\text{untreated cells}}] \times 100$) against concentrations.

3.5. *Enzyme-Linked Immunosorbent Assay (ELISA) for PGE_2 and TNF-α Measurement*

RAW 264.7 cells (2×10^5 cells/well) were plated in 6-well culture plates overnight. In a serum-free medium, cells were pretreated with the methanol extract (1, 10 μg/mL), isolated lignans (5, 10 μg/mL), or indomethacin (10 μM; NSAID) for 3 h and subsequently treated with LPS (5 μg/mL) for 24 h. The supernatants were collected and stored at −80 °C until analysis. The assessment of PGE_2 and TNF-α levels was quantified with the mouse PGE_2 and TNF-α ELISA kits following the manufacturer's instructions.

3.6. *Measurement of Nitric Oxide Production*

Nitrate and nitrite concentrations were assayed by Griess reagent using a nitric oxide assay kit as described previously [31]. RAW 264.7 cells (2×10^5 cells/well) were plated in 6-well culture plates overnight. Under the serum-free condition, cells were pretreated with the methanol extract (1, 10 μg/mL), or isolated lignans (5, 10 μg/mL) for 3 h and subsequently treated with LPS (5 μg/mL) for 24 h. The absorbance was measured at a wavelength of 540 nm using the Infinite M200 microplate reader (TECAN). Nitrate and nitrite concentrations in the samples were calculated from the standard curve following the manufacturer's instructions. The amounts of nitrate and nitrite accurately reflect the NO production in RAW 264.7 cells.

3.7. Measurement of mRNA Expression of Inflammatory Biomarkers

The effects of the methanol extract and the isolated lignans on the synthesis of inflammatory biomarkers were determined by measurement of mRNA expression. Under serum-free condition, RAW 264.7 cells (2×10^5 cells/well) were treated with the methanol extract (1, 10 µg/mL), or isolated lignans (5, 10 µg/mL) for 6 h. The total mRNA of RAW 264.7 cells was isolated with RNeasy kit (Qiagen). The mRNA expression of inflammatory biomarkers was determined by using the KAPA SYBR FAST One-step RT-qPCR kit with gene-specific primer sequences (Table 1) and Mx 3005p Real-Time PCR system (Stratagene) as previously described [32]. The mRNA expression level of inflammatory biomarkers was calculated by the comparative cycle threshold (CT) method. Glyceraldehyde-3-phosphate dehydrogenase (GAPDH) was used as the housekeeping gene.

Table 1. The gene-specific primers for RT-qPCR (mouse).

Gene-Specific Primers		Sequences
COX-2	Sense	5′-tgcatgtggctgtggatgtcatcaa-3′
	Antisense	5′-cactaagacagacccgtc atctcca-3′
IL-6	Sense	5′-gacaaagccagagtccttcagagag-3′
	Antisense	5′-ctaggtttgccgagtagatctc-3′
IL-10	Sense	5′-gctggacaacatactgctaacc-3′
	Antisense	5′-atttccgataaggcttggcaa-3′
iNOS	Sense	5′-gtgttccaccaggagatgttg-3′
	Antisense	5′-ctcctgcccactgagttcgtc-3′
NF-κB	Sense	5′-gaaattcctgatccagacaaaaac-3′
	Antisense	5′-atcacttcaatggcctctgtgtag-3′
TNF-α	Sense	5′-atgagcacagaaagcatgatc-3′
	Antisense	5′-tacaggcttgtcactcgaatt-3′
GAPDH	Sense	5′-gcctgcttcaccaccttc-3′
	Antisense	5′-ggctctccagaacatcatcc-3′

3.8. Statistical Analysis

Data are presented as the mean ± SEM from 3–5 independent experiments. Statistical analysis was obtained by using SPSS software (version 25). The difference between groups was evaluated using one-way ANOVA with a post hoc test. The value of *p*-value less than 0.05 ($p < 0.05$) was accepted significant.

4. Conclusions

This study revealed that the methanol extract of *M. sirikitiae* leaves and isolated lignans, including (−)-phylligenin (**2**) and 3′,4-*O*-dimethylcedrusin (**6**) possess the anti-inflammatory activity by suppressing LPS-induced iNOS and COX-2 synthesis which contribute to a reduction in PGE_2 and nitric oxide secretions in RAW 264.7 macrophages. Moreover, 2-(3,4-dimethoxyphenyl)-6-(3,5-dimethoxyphenyl)-3,7-dioxabicyclo[3.3.0] octane (**3**) and mitrephoran (**5**) elicit the anti-inflammatory effects through inhibition of LPS-induced the secretion and synthesis of TNF-α in RAW 264.7 cells. Hence, these findings suggested that several lignans isolated from *M. sirikitiae* leaves might be potential therapeutic candidates for the inhibition and prevention of various inflammatory diseases.

Supplementary Materials: The following supporting information can be downloaded at: https://www.mdpi.com/article/10.3390/molecules27103313/s1. Spectroscopic and physical data of isolated lignans 1–6 from *Mitrephora sirikitiae* Leaf extract.

Author Contributions: Conceptualization, N.A. and S.M.; methodology, N.A., S.M., C.L. and K.K.H.; formal analysis, N.A. and S.M.; writing, N.A. and S.M.; supervision, N.A., S.M. and V.R.; project administration, N.A. All authors have read and agreed to the published version of the manuscript.

Funding: This research was funded by Mahidol University, the Thailand Research Fund (MRG5580047 (to N.A.), the CIF and CNI Grant, Faculty of Science, Mahidol University (to S.M.), and the Center of Excellence for Innovation in Chemistry (PERCH-CIC).

Institutional Review Board Statement: Not applicable.

Informed Consent Statement: Not applicable.

Data Availability Statement: The data presented in this study are available in supplementary material.

Acknowledgments: We would like to acknowledge Narong Nuntasaen for collecting and identification of the plant.

Conflicts of Interest: The authors declare no conflict of interest.

References

1. Abdulkhaleq, L.A.; Assi, M.A.; Abdullah, R.; Zamri-Saad, M.; Taufiq-Yap, Y.H.; Hezmee, M. The crucial roles of inflammatory mediators in inflammation: A review. *Vet. World* **2018**, *11*, 627–635. [CrossRef] [PubMed]
2. Ricciotti, E.; FitzGerald, G.A. Prostaglandins and inflammation. *Arterioscler. Thromb. Vasc. Biol.* **2011**, *31*, 986–1000. [CrossRef] [PubMed]
3. Tripathi, P.; Tripathi, P.; Kashyap, L.; Singh, V. The role of nitric oxide in infamematory reactions. *FEMS Immunol. Med. Microbiol.* **2007**, *51*, 443–452. [CrossRef] [PubMed]
4. Kany, S.; Vollrath, J.T.; Relja, B. Cytokines in inflammatory disease. *Int. J. Mol. Sci.* **2019**, *20*, 6008. [CrossRef] [PubMed]
5. Liu, X.; Yin, S.; Chen, Y.; Wu, Y.; Zheng, W.; Dong, H.; Bai, Y.; Qin, Y.; Li, J.; Feng, S.; et al. LPS-induced proinflammatory cytokine expression in human airway epithelial cells and macrophages via NF-κB, STAT3 or AP-1 activation. *Mol. Med. Rep.* **2018**, *17*, 5484–5491. [CrossRef] [PubMed]
6. Hsu, H.Y.; Wen, M.H. Lipopolysaccharide-mediated reactive oxygen species and signal transduction in the regulation of interleukin-1 gene expression. *J. Biol. Chem.* **2002**, *277*, 22131–22139. [CrossRef]
7. Libby, P. Inflammatory mechanisms: The molecular basis of inflammation and disease. *Nutr. Rev.* **2007**, *65*, 140–146. [CrossRef]
8. Hansson, G.K. Inflammation, atherosclerosis, and coronary artery disease. *N. Engl. J. Med.* **2005**, *352*, 1685–1695. [CrossRef]
9. Weerasooriya, A.D.; Chalermglin, P.; Saunders, R.M.K. *Mitrephora sirikitiae* (Annonaceae): A remarkable new species endemic to northern Thailand. *Nord. J. Bot.* **2004**, *24*, 201–206. [CrossRef]
10. Anantachoke, N.; Lovacharaporn, D.; Reutrakul, V.; Michel, S.; Gaslonde, T.; Piyachaturawat, P.; Suksen, K.; Prabpai, S.; Nuntasaen, N. Cytotoxic compounds from the leaves and stems of the endemic Thai plant *Mitrephora sirikitiae*. *Pharm. Biol.* **2020**, *58*, 490–497. [CrossRef]
11. Barker, D. Lignans. *Molecules* **2019**, *24*, 1424. [CrossRef] [PubMed]
12. Rodríguez-García, C.; Sánchez-Quesada, C.; Toledo, E.; Delgado-Rodríguez, M.; Gaforio, J.J. Naturally lignan-rich foods: A dietary tool for health promotion? *Molecules* **2019**, *24*, 917. [CrossRef] [PubMed]
13. Xu, W.H.; Zhao, P.; Wang, M.; Liang, Q. Naturally occurring furofuran lignans: Structural diversity and biological activities. *Nat. Prod. Res.* **2019**, *33*, 1357–1373. [CrossRef] [PubMed]
14. Szopa, A.; Dziurka, M.; Warzecha, A.; Kubica, P.; Klimek-Szczykutowicz, M.; Ekiert, H. Targeted lignan profiling and anti-inflammatory properties of *Schisandra rubriflora* and *Schisandra chinensis* extracts. *Molecules* **2018**, *23*, 3103. [CrossRef] [PubMed]
15. Lim, H.; Lee, J.G.; Lee, S.H.; Kim, Y.S.; Kim, H.P. Anti-inflammatory activity of phylligenin, a lignan from the fruits of *Forsythia koreana*, and its cellular mechanism of action. *J. Ethnopharmacol.* **2008**, *118*, 113–117. [CrossRef] [PubMed]
16. Kim, T.W.; Shin, J.S.; Chung, K.S.; Lee, Y.G.; Baek, N.I.; Lee, K.T. Anti-inflammatory mechanisms of koreanaside A, a lignan isolated from the flower of *Forsythia koreana*, against LPS-induced macrophage activation and DSS-induced colitis mice: The crucial role of AP-1, NF-κB, and JAK/STAT signaling. *Cells* **2019**, *8*, 1163. [CrossRef]
17. Dong, D.D.; Li, H.; Jiang, K.; Qu, S.J.; Tang, W.; Tan, C.H.; Li, Y.M. Diverse lignans with anti-inflammatory activity from *Urceola rosea*. *Fitoterapia* **2019**, *134*, 96–100. [CrossRef]
18. Lee, D.Y.; Seo, K.H.; Jeong, R.H.; Lee, S.M.; Kim, G.S.; Noh, H.J.; Kim, S.Y.; Kim, G.W.; Kim, J.Y.; Baek, N.I. Anti-inflammatory lignans from the fruits of *Acanthopanax sessiliflorus*. *Molecules* **2013**, *18*, 41–49. [CrossRef]
19. Zhou, X.J.; Chen, X.L.; Li, X.S.; Su, J.; He, J.B.; Wang, Y.H.; Li, Y.; Cheng, Y.X. Two dimeric lignans with an unusual α,β-unsaturated ketone motif from *Zanthoxylum podocarpum* and their inhibitory effects on nitric oxide production. *Bioorg. Med. Chem. Lett.* **2011**, *21*, 373–376. [CrossRef]
20. Rhee, S.H. Lipopolysaccharide: Basic biochemistry, intracellular signaling, and physiological impacts in the gut. *Intest. Res.* **2014**, *12*, 90–95. [CrossRef]
21. Jang, D.; Murrell, G. Nitric oxide in arthritis. *Free Radic. Biol Med.* **1998**, *24*, 1511–1519. [CrossRef]
22. Moncada, S.; Palmer, R.M.; Higgs, E.A. Nitric oxide: Physiology, pathophysiology, and pharmacology. *Pharmacol. Rev.* **1991**, *43*, 109–142. [PubMed]
23. Ben-Baruch, A. Inflammation-associated immune suppression in cancer: The roles played by cytokines, chemokines and additional mediators. *Semin. Cancer Biol.* **2006**, *16*, 38–52. [CrossRef] [PubMed]

24. Tak, P.P.; Firestein, G.S. NF-kappaB: A key role in inflammatory diseases. *J. Clin. Investig.* **2001**, *107*, 7–11. [CrossRef]
25. Clària, J.; Romano, M. Pharmacological intervention of cyclooxygenase-2 and 5-lipoxygenase pathways. Impact on inflammation and cancer. *Curr. Pharm. Des.* **2005**, *11*, 3431–3447. [CrossRef]
26. Teponno, R.B.; Kusari, S.; Spiteller, M. Recent advances in research on lignans and neolignans. *Nat. Prod. Rep.* **2016**, *33*, 1044–1092. [CrossRef]
27. During, A.; Debouche, C.; Raas, T.; Larondelle, Y. Among plant lignans, pinoresinol has the strongest antiinflammatory properties in human intestinal Caco-2 cells. *J. Nutr.* **2012**, *142*, 1798–1805. [CrossRef]
28. Kumar, S.S.; Hira, K.; Ahil, S.B.; Kulkarni, O.P.; Araya, H.; Fujimoto, Y. New synthetic coumarinolignans as attenuators of pro-inflammatory cytokines in LPS-induced sepsis and carrageenan-induced paw oedema models. *Inflammopharmacology* **2020**, *28*, 1365–1373. [CrossRef]
29. Sato, V.H.; Sungthong, B.; Nuamnaichati, N.; Rinthong, P.; Mangmool, S.; Sato, H. In vivo and in vitro evidence for the antihyperuricemic, anti-inflammatory and antioxidant effects of a Traditional Ayurvedic Medicine, Triphala. *Nat. Prod. Commun.* **2017**, *12*, 1635–1638. [CrossRef]
30. Wongthai, N.; Tanticharakunsiri, W.; Mangmool, S.; Ochaikul, D. Characteristics and antioxidant activity of royal lotus pollen, butterfly pea flower, and oolong tea kombucha beverages. *Asia Pac. J. Sci. Technol.* **2021**, *26*, APST-26-04-17. [CrossRef]
31. Nuamnaichati, N.; Parichatikanond, W.; Mangmool, S. Cardioprotective effects of glucagon-like peptide-1 (9-36) against oxidative injury in H9c2 cardiomyoblasts: Potential role of the PI3K/Akt/NOS pathway. *J. Cardiovasc. Pharmacol.* **2022**, *79*, e50–e63. [CrossRef] [PubMed]
32. Mangmool, S.; Kunpukpong, I.; Kitphati, W.; Anantachoke, N. Antioxidant and anticholinesterase activities of extracts and phytochemicals of *Syzygium antisepticum* leaves. *Molecules* **2021**, *26*, 3295. [CrossRef] [PubMed]

effects through a multi-targeted approach, which is a characteristic that is highly desirable in cancer malignancies [6]. Phytochemicals may illustrate their antitumor properties through promoting apoptosis, suppressing the cell cycle, inhibiting angiogenesis and regulating antioxidant activities. Moreover, numerous naturally bioactive compounds have been shown to modulate immune checkpoints and affect the activities of immune cells including T and B cells, Treg cells and NK cells [8]. More interestingly, these natural products have gained competing interests due to the absence of toxicity and harmful side effects commonly associated with current therapies [9].

Curcumin is a polyphenolic derivative of turmeric known to have dramatic anticancer effects on cancer cells rather than normal ones. It has been demonstrated that curcumin exerts its toxic effects through modulation of the function of multiple genes including apoptotic, metastatic, cell proliferation and transcription factors [10]. Studies reported that curcumin modulates cellular pathways involved in cancer pathogenesis including NF-kB, MAPK, PTEN, P53 and wnt [11]. Despite these tempting advantages of curcumin, the poor bioavailability limits its exploitation as a therapeutic compound [12]. Our team has recently formulated and characterized a nano-based encapsulated curcumin, gemini curcumin (Gemini-Cur), with significant anticancer effects on ovarian, gastric, breast and colorectal cancer [13–16]. Briefly, gemini surfactant nanoparticles belong to a surfactant family with two identical structures that are linked by a rigid or flexible spacer that could harbor and deliver drugs and genes into the cells and tissues. Gemini curcumin nanoparticles are spherical and well-dispersed vesicles that easily enter the cancer cells [15].

Exploration of the genes with abnormal expression during the treatment of colon cancer with Gemini-Cur is essential to provide a deeper understanding of the mechanisms involved. Because regulatory genes are affected by dietary compounds, the ability of curcumin to modulate the transcriptome profile has attracted much attention [8,17]. Based on our recent findings on the significant toxic properties of Gemini-Cur on cancer cells, here, we employed RNA sequencing and bioinformatics analysis to identify the key genes and related pathways modulated in colorectal HCT-116 cells treated with Gemini-Cur. The data of the current study help us to determine top Differentially Expressed Genes (DEGs) as possible cellular targets and figure out potential biological pathways in colon cancer that are modulated by curcumin.

2. Materials and Methods

2.1. *In Vitro Studies*

2.1.1. Cell Culture and Reagents

The colorectal cancer HCT-116 cell line was purchased from the Iranian national cell bank (Pasteur institute, Tehran, Iran). The cells were cultured in Dulbecco's Modified Eagle's Medium (DMEM, Sigma-Aldrich, St. Louis, MO, USA) supplemented with 10% (*v*/*v*) fetal bovine serum and 1% (*v*/*v*) penicillin–streptomycin solution (both from GIBCO, USA) at 37 °C in a humidified environment with 5% CO_2. Curcumin (CAS Number 458-37-7; Sigma-Aldrich, USA) and mPEG urethane gemini surfactant nanoparticles were a kind gift from the Institute for Color Science and Technology, Tehran, Iran.

2.1.2. Synthesis of Gemini-Cur nanoparticles

Gemini-Cur nanoparticles were prepared by a nanoprecipitation method previously reported by our lab [16]. Briefly, we added 6 mg of Cur and 100 mg of gemini surfactants to 3 mL of methanol. Then, the solution was diluted twice in PBS under gently stirring condition, and the methanol was evaporated by using a rotary evaporator. The remaining solution was passed through a 0.22 μM syringe filter to remove possible contaminations and stored at 4 °C until use.

2.1.3. Gemini-Cur Treatments

We have previously reported the IC50 values for Gemini-Cur on HCT-116 cells. Furthermore, we demonstrated that Gemini-Cur modulates the cell cycle and induces apoptosis in HCT-116 cells compared to controls [14]. To further confirm cellular toxicity on the cells treated with Gemini-Cur, we used ethidium bromide/acridine orange (EB/AO) staining. Briefly, the cells were left untreated or treated with Gemini-Cur at an IC50 dose onto 6-well plates. After 24, 48 and 72 h, the cells were detached by trypsin (0.25%; Sigma-Aldrich, USA) and transferred to glass slides. Staining solution (1 μL) containing 100 μg/mL acridine orange and 100 μg/mL ethidium bromide (Sigma-Aldrich, USA) was added to a suspension of HCT-116 cells. The cells were visualized under fluorescence microscopy (RX50, LABEX, England), and representative photographs were taken for further qualitative analysis. Fluorouracil (5-FU) apoptotic images were also provided as positive control.

For RNA sequencing, the cells were seeded on 6-well plates for 24 h and subsequently treated with Gemini-Cur. After 24 h, the cells were processed for RNA sequencing.

2.1.4. RNA Extraction and Preparation

According to the protocol of TRIzol reagent (Thermo Fisher Scientific, USA), total RNA was extracted from treated and non-treated HCT-116 cells. A total of six samples including three controls and three treated cells were incorporated in the study. After validating the integrity and purity (NanoDropTM, ThermoFisher, USA), all RNAs were treated with RNase-free DNase I to remove any DNA contamination. Then, RNAs were transferred to GeneTegra-RNA tubes (GenTegra Co., Seoul, Korea), dried in a freezer dryer (Sartorius Co., Germany) and sent to Macrogen Co., for sequencing (Macrogen Co., Seoul, Korea).

2.1.5. Library Construction and RNA Sequencing

Approximately 1 μg of RNA from each sample was used to generate RNA-Seq cDNA libraries for sequencing using the TruSeq RNA Sample Prep Kit v2 (Illumina, Inc., San Diego, CA, USA). Sample preparation followed the manufacturer's protocol with a workflow that included isolation of polyadenylated RNA molecules using poly-T oligo-attached magnetic beads, enzymatic RNA fragmentation, cDNA synthesis, ligation of bar-coded adapters, and PCR amplification. Ambion External RNA Controls Consortium (ERCC) RNA Spike-In Control Mix 1 (Life Technologies Corporation, Carlsbad, CA, USA) was added to the samples. The amplified cDNA fragments were sequenced using a HiSeq 2000 sequencing system (Illumina, San Diego, CA, USA). Finally, sequencing data were converted to raw data in FASTQ format utilizing illumina package bcl2fastq.

2.2. Bioinformatics Studies

2.2.1. Quality Assessment of RNA-seq Data, Mapping and Read Annotation

All processing and analysis on raw data were performed using Ubuntu 20.00 (64-bit) and open-source software available through the R/Bioconductor. After check and quality control of paired-end reads with the final version of MultiQC (https://github.com/ewels/MultiQC, 25 January 2022) and Trimmomatic (http://www.usadellab.org/cms/?page=trimmomatic, 25 January 2022), the remaining reads as clean reads were mapped to the genome reference GRCh37 (hg19) using the star (https://github.com/alexdobin/STAR/releases, 10 January 2022) package, and mapping efficiencies accounted for 98.50%. The counting of transcripts was also performed with Htseq-count (https://htseq.readthedocs.io/en/release_0.11.1/count.html, 10 January 2022).

2.2.2. Normalization of Read Counts, Differentially Expression Analysis (DEA) and Network Construction

In order to normalize and perform differential expression analysis on counts, the standard Bioconductor RNA-seq workflow (DESeq2) was used to detect differentially expressed genes (DEGs). The distribution of expression values across all samples (normal and treatment) before and after normalization was applied to ensure that expression values

3.2. Raw Data Statistics and Quality Assessment

Three replicates of treated samples and corresponding controls were subjected to RNA sequencing. On average, more than 40 million reads per sample were recorded (Figure 2A). The proportion of bases with high quality (Q30) was more than 90, indicating that the quality of RNA sequencing was proper for further analysis. After quality control and the removal of adaptors, more than 8 million reads were produced for each sample (Figure 2B).

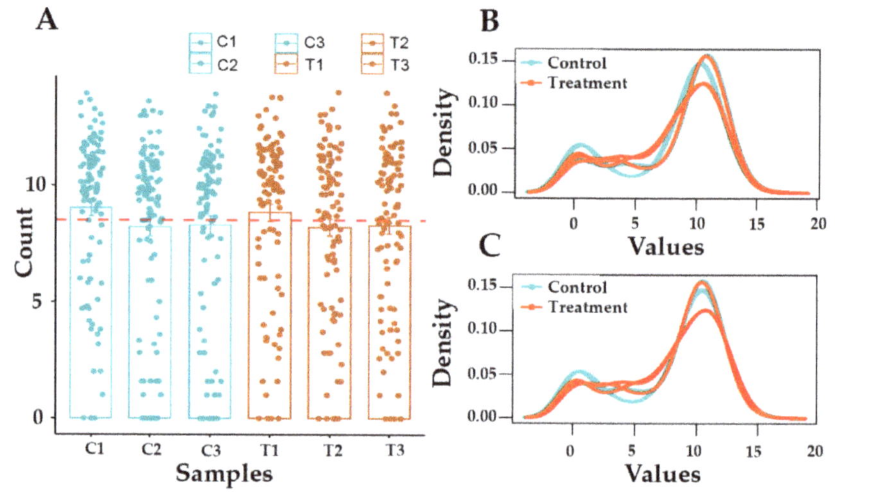

Figure 2. Distribution of gene counts. As shown in bar plot (**A**), more than 8 million reads are produced for each sample. According to density plot, raw read Counts (log2 (counts + 1)) are not non-normalized distributed (**B**). The density plot of normalized count (log2 (normalized counts)) per sample is shown in plot (**C**). C1-3: controls; T1-3: Treated cells with Gemini-Cur.

3.3. Identification of Differentially Expressed Genes (DEGs)

In order to determine DEGs in treated samples versus controls, the standard Bioconductor RNA-seq workflow (DESeq2) was employed. In total, 3892 DEGs (padj < 0.01) including 244 upregulated (padj < 0.01, log2 FC > 2) and 198 downregulated (padj < 0.01, log2 FC < −2) genes were obtained in this study (Supplementary Materials). As Figure 3A shows, all genes were categorized in non-significant (dark gray) and significant (light blue) DEGs as well as top down/upregulated genes (red). To further illustrate top-modulated genes, a heatmap with a color range of blue (down) to red (up) was employed for non-normalized (Figure 3B) and normalized genes (Figure 3C). The heatmap demonstrates the differential expression of genes in Gemini-Cur-treated cells in comparison with non-treated samples. Subsequently, a list of top ten up and downregulated DEGs (padj < 0.01) was obtained with padj < 0.01, log2 FC > 2 and padj < 0.01, log2 FC < −2, respectively (Table 2).

3.4. Exploration of DEGs in Protein–Protein Interaction (PPI) Network and Subnetworks (Modules)

To find out the potential interactions at the protein level, DEGs were mapped in STRING, and the PPI network (Figure 4A) was constructed for top 300 DEGs (up/downregulated genes). The significant pairs of the network (p-value: 1.0×10^{-164}) in 2736 interactions between 180 nodes based on confidence score (0.007) were visualized by Cytoscape software and the network analyzer plug-in with a cut-off value for BC > 0 and K > 8. After analyzing PPI network, subnetworks (modules) with a MCODE plug-in of Cytoscape were extracted, and generally, 10 modules obtained. Three significant modules were identified with MCODE score $\geq$ 3 and nodes $\geq$ 3 (Figure 4B–D). CDK1, CCNA2, and BUB1B with the highest BC and K in the PPI network and with the highest score in the MCODE plug-in were significantly identified in Module 1 (Figure 4D and Table 3). Accordingly, it is clearly obvious that three of the DEGs (CDK1, CCNA2, and BUB1B) in the PPI network and module 1 are the genes that interact the most with others in the PPI network.

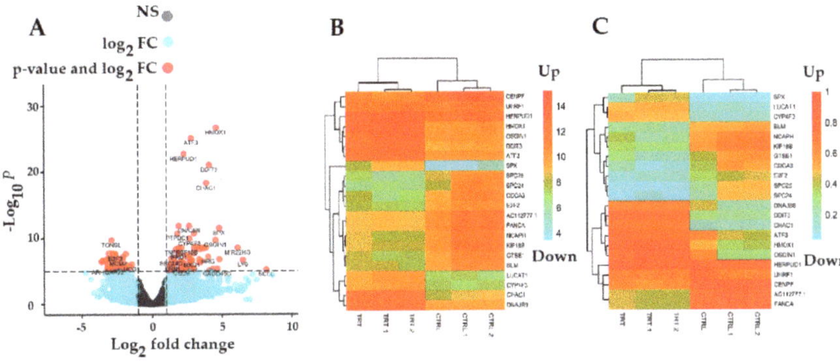

Figure 3. Normalization of raw read counts and Differential Expression Analysis (DEA). The volcano plot (**A**) of non-significant DEGs (colored in dark gray), significant DEGs (padj < 0.01) (colored in light blue), and up/downregulated genes (Colored in red). The x-axis represents the log2 FC, while the y-axis represents the statistical significance for each gene based on −log10 (*p* value). Heatmap of top 22 up/downregulated genes associated with non-normalized counts (**B**). Heatmap of top 22 up/downregulated DEGs associated with normalized counts (**C**). From the heatmap (downregulated genes are colored in light blue and upregulated genes are colored in orange in both plots). NS: non-significant; FC: fold change.

Table 2. The list of top 10 up/downregulated Differentially Expressed Genes (DEGs) based on RNA-seq data analysis. The DEGs (padj < 0.01) between control and treatment groups are shown with top 10 upregulated genes (padj < 0.01, log2 FC > 2), and top 10 downregulated genes (padj < 0.01, log2 FC < −2). Padj: adjusted *p* value.

Gene Symbol	STATUS	Base Mean	log2 Fold Change	lfcSE	Stat	*p* Value	padj
HMOX1	Up	14082.63	4.581563	0.393634	11.63913	2.61×10^{-31}	9.29×10^{-27}
DDIT3	Up	7029.257	4.083832	0.394679	10.34723	4.31×10^{-25}	7.68×10^{-21}
ATF3	Up	5160.583	2.807268	0.27917	10.05577	8.66×10^{-24}	1.01×10^{-19}
CHAC1	Up	2994.215	3.896157	0.400672	9.724061	2.38×10^{-22}	2.12×10^{-18}
HERPUD1	Up	12540.28	2.267609	0.254122	8.92332	4.53×10^{-19}	3.23×10^{-19}
SPX	Up	260.3101	4.781735	0.578882	8.260295	1.45×10^{-16}	8.64×10^{-13}
DNAJB9	Up	2736.219	2.665123	0.342986	7.770348	7.83×10^{-15}	3.99×10^{-11}
OSGIN1	Up	13215.95	4.534403	0.586017	7.737659	1.01×10^{-14}	4.51×10^{-11}
LUCAT1	Up	608.1075	3.02517	0.396356	7.632456	2.30×10^{-14}	9.13×10^{-11}
CYP4F3	Up	547.2113	2.671816	0.369558	7.22977	4.84×10^{-13}	1.57×10^{-09}
TONSL	Down	2059.716	−2.89597	0.395525	−7.32183	2.45×10^{-13}	8.72×10^{-10}
MCM4	Down	5519.294	−2.8152	0.421581	−6.67772	2.43×10^{-11}	$4.5\ 5\times 10^{-08}$
AC112777.1	Down	221.6533	−3.06376	0.459517	−6.66735	2.60×10^{-11}	4.64×10^{-08}
FANCA	Down	1932.18	−2.67477	0.406894	−6.57362	4.91×10^{-11}	7.47×10^{-08}
E2F2	Down	629.9052	−2.62378	0.402383	−6.52061	7.00×10^{-11}	9.25×10^{-08}
BLM	Down	995.8815	−2.69881	0.413601	−6.52516	6.79×10^{-11}	9.25×10^{-08}
NCAPH	Down	1459.514	−3.16292	0.490122	−6.45332	1.09×10^{-10}	1.32×10^{-07}
KIF18B	Down	1583.582	−3.07757	0.477409	−6.44639	1.15×10^{-10}	1.32×10^{-07}
UHRF1	Down	2198.428	−2.78382	0.432416	−6.43782	1.21×10^{-10}	1.32×10^{-07}
CDCA3	Down	933.7762	−3.48616	0.542823	−6.42228	1.34×10^{-10}	1.41×10^{-07}

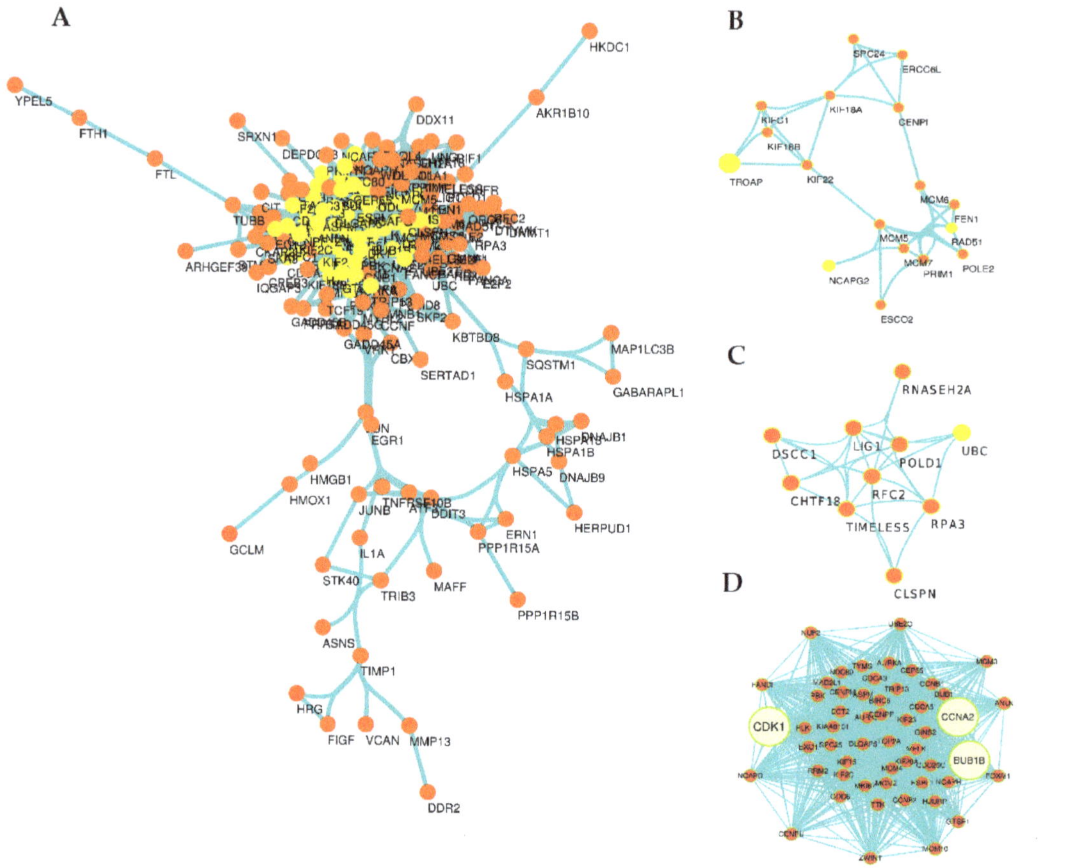

Figure 4. Overview of PPI network and subnetworks (modules). The PPI network (**A**); the yellow colors represent key nodes (highest BC and K) in PPI networks. Generally, 10 modules were obtained from the main network and module 1 (**D**), module 2 (**B**) and module 3 (**C**) are significant modules. Module 1 with the score of 47.057 and 54 nodes is the more significant module, covering three genes CDK1, CCNA2, and BUB1B with the highest BC and K in the PPI network and with the highest score in the MCODE plug-in.

Table 3. The position of 5 CRC-related candidate genes in the PPI network and subnetworks (modules). As shown, CDK1, CCNA2 and BUB1B genes with the highest degree (K) and betweenness (BC) are clustered in subnetwork 1, demonstrating their significant influence in the network. In contrast, GADD45G and ATF3 genes are classified in clusters 5 and 8, perhaps indicating their independence to the networks in CRC.

Gene Symbol	Expression Based on RNA-seq	Degree (K)	Betweenness (BC)	Closeness Centrality (CC)	MCODE: Score	No; Cluster
CDK1	Down	106	0.94541487	0.885964912	34.77520814	1
CCNA2	Down	96	0.299119893	0.789115646	34.77520814	1
BUB1B	Down	90	0.296979671	0.724550898	34.77520814	1
GADD45G	Up	5	0	0	5	5
ATF3	Up	6	0	0.411764706	2.7	8

3.5. KEGG Pathway Analysis of DEGs

For the characterization of DEGs, we performed pathway enrichment analysis in KEGG. In total, 300 upregulated (padj < 0.01, log2 FC > 2) and downregulated (padj < 0.01, log2 FC < −2) genes were mapped to 117 KEGG pathways. In order to identify which gene is associated with biological pathways, a gene-pathway annotated network was created using 300 DEGs (up/downregulated genes) alongside the most significant KEGG pathways using Cytoscape software.

Generally, the gene-pathway annotated network showed that DEGs were significantly enriched in the cell cycle, E2F-mediated regulation of DNA replication, G1/S-specific transcription, cell cycle checkpoints, G2/M checkpoints, FOXM1 transcription factor network, and G1 to S cell cycle control (Table 4). Surprisingly, a vital gene of our study, CDK1, was significantly enriched in all significant pathways that may act as a bridge between both the CRC-related genes (CCNA2, BUB1B, GADD45G and ATF3) and other gene-pathway annotated network genes (Figure 5).

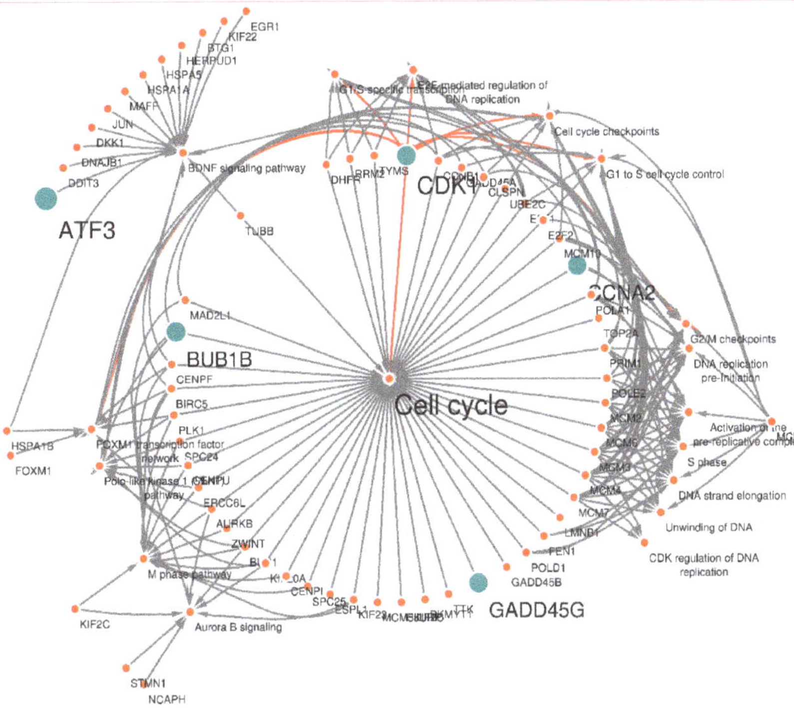

Figure 5. Overview of the gene-pathway network (annotated network) created by cystoscope. The blue nodes represent the CRC candidate genes connected with the pathways and other nodes in the network. The orange nodes represent other pathways in connection with CRC genes. According to this network, it can be demonstrated that CDK1 is involved in all significant first-level pathways; it is associated with related genes in each of the pathways and with other most significant CRC-related genes (CCNA2, BUB1B, GADD45G, ATF3). The dominant pathway is cell cycle with padj 1.82×10^{-34}.

Table 4. Status of CRC-related genes—CDK1, CCNA2, BUB1B, GADD45G, and ATF3—in gene-pathway annotated networks. The table shows that CDK1 is prominently involved in different pathways. The gene pathways are significantly enriched in cell cycle-related networks. Padj: adjusted *p*-value.

Pathway	Gene Symbol	Padj
Cell cycle	TOP2A; ERCC6L; MCM7; PRIM1; HJURP; **BUB1B**; MCM10; TTK; PKMYT1; TYMS; AURKB; LMNB1; CCNB1; POLD1; E2F1; E2F2; CLSPN; BUB1; CENPURRM2; GADD45B; GADD45A; UBE2C; TUBB; PLK1; KIF23; ZWINT; **GADD45G**; DHFR, **CCNA2**; POLA1; CENPF; ESPL1; CENPI; POLE2; **CDK1**; MCM3; MCM4; BIRC5; MCM5; KIF2C; KIF20A; MCM6; SPC24; MCM2; SPC25; MAD2L1	1.82×10^{-34}
E2F-mediated regulation of DNA replication	DHFR; POLA1; CCNB1; RRM2; PRIM1; E2F1; **CDK1**; TYMS	1.67×10^{-08}
G1/S-specific transcription	DHFR; POLA1; RRM2; **CDK1**; E2F1; TYMS	1.79×10^{-07}
Cell cycle checkpoints	MCM7; UBE2C; **BUB1B**; MCM10; CCNB1; **CDK1**; MCM3; MCM4; MCM5; CLSPN; MCM6; MCM2; MAD2L1	2.84×10^{-09}
G2/M checkpoints	CCNB1; MCM7; **CDK1**; MCM3; MCM4; MCM10; MCM5; CLSPN; MCM6; MCM2	2.89×10^{-10}
FOXM1 transcription factor network	**CCNA2**; CCNB1; CENPF; PLK1; **CDK1**; BIRC5; FOXM1; AURKB; HSPA1B	4.54×10^{-09}
G1 to S cell cycle control	DHFR; POLA1; RRM2; **CDK1**; E2F1; TYMS	3.83×10^{-12}

3.6. Functional Enrichment and Gene Ontology (GO) Analysis

To better understand the molecular role of selected genes involved in the suppressive effect of Gemini-Cur on HCT-116 cells, DEGs were mapped in the GO database using an online web tool (https://maayanlab.cloud/Enrichr, 25 January 2021), and a threshold of Padj < 0.05 and gene counts > 5 was considered. Here, gene ontology was performed for 198 downregulated and 244 upregulated genes separately, and the status of all five selected genes in all three categories (BP; biological process, MF; molecular function, and CC; cellular component) was assessed.

Interestingly, the GO results of the integrated group showed that upregulated ATF3 is significantly enriched in BP category including response to endoplasmic reticulum stress, regulation of transcription from RNA polymerase II promoter in response to stress, and MF category with protein hetero-dimerization activity. In addition, GADD45G was enriched in the regulation of p38MAPK cascade as well as control of the p38MAPK pathway, both in BP process (Table 5). In contrast, downregulated genes (CCNA2, BUB1B, and CDK1) were involved in almost all three categories and different cellular cascades.

In Vitro Validation Study by Real-Time PCR and Western Blotting

To further validation of the RNA sequencing data generated by the above-mentioned parameters, the differential expression of top five genes including two upregulated (GADD45G and ATF3) and three downregulated (BUB1B, CCNA2 and CDK1) were evaluated on all treated and non-treated samples in mRNA and protein levels. As Figure 6A shows, the expression of BUB1B, CCNA2 and CDK1 is significantly down-expressed in treated cells. Accordingly, the data illustrated that ATF3 (p value < 0.05) and GADD45G are upregulated in Gemini-Cur treated cells, although this was not significant for GADD45G. These modulations were confirmed in protein level as shown in Figure 6B.

Table 5. The top significantly gene ontology (GO) categories (BP: biological process; CC: cellular component; MF; molecular function) based on upregulated genes with the threshold of | log fold change (FC) | ≥ 2 and a Bonferroni $p < 0.05$. Padj: adjusted p value.

GO Term	Source	Padj	Gene Symbol
microtubule cytoskeleton organization involved in mitosis	BP	3.94×10^{-23}	ERCC6L; BUB1B; CDCA8; TTK; CENPA; TACC3; BIRC5; CENM; KIF2C; SPC24; MAD2L1;
mitotic spindle organization	BP	1.75×10^{-22}	ERCC6L; BUB1B; CDCA8; TTK; CENPA; KIF2C; SPC24; MAD2L1; SPC25
mitotic sister chromatid segregation	BP	1.37×10^{-21}	SPAG5; CDCA5; P LK1; NCAPG2; CDCA8; NCAPG; PSRC1; ESPL1; KIFC1; PRC1; CDK1; KIF2C
DNA metabolic process	BP	8.96×10^{-20}	TOP2A; BLM; FEN1; RNASEH2A; MCM7; UHRF1; HMGB2; MCM10; TYMS; CDK1; MCM4; MCM5; MCM6; MCM2
positive regulation of cell cycle process	BP	9.66×10^{-10}	UBE2C; TUBB; CCNF; PLK1; CDC7; CDC25C; PKMYT1; FOXM1; AURKA; CCNA2; CCNB2; CCNB1; CDK1; E2F1; TACC3; NEK2; CDKN3
spindle	CC	2.01×10^{-18}	SPAG5; CKAP2L; PLK1; BUB1B; CDC7; KIF23; TTK; KIF22; SKA3; AURKB; AURKA; CDC20; CENPF; CDK1; TACC3; BIRC5; KIF2C; KIF20A
nucleus	CC	2.20×10^{-13}	TOP2A; ARHGAP11A; FEN1; MCM7; DSCC1; CCNF; NCAPG2; HMGB2; MCM10; CCNA2; ASPM; MCM5; KIF20A; MCM6; PRR11; MCM2; BLM; RAD51; PRC1; UBE2T; CDK1; TRIP13; MAD2L1
intracellular non-membrane-bounded organelle	CC	4.72×10^{-10}	TOP2A; FEN1; MCM7; CDCA5; HJURP; HMGB2; BUB1B; CDCA8; MCM10; TTK; MKI67; PKMYT1; AURKB; PLK1; VRK1; KIF23; ESCO2; PIMREG; CIT; CENPF; PSRC1; PRC1; UBE2T; KIF2C; KIF20A; TRIP13; SPC24; MCM2
cyclin-dependent protein kinase holoenzyme complex	CC	4.89×10^{-05}	CCNA2; CCNB2; CCNB1; CCNF; CDK1
serine/threonine protein kinase complex	CC	1.32×10^{-04}	CCNA2; CCNB2; CCNB1; CCNF; CDK1
histone kinase activity	MF	7.17×10^{-04}	CDK1; AURKB; AURKA
protein serine/threonine kinase activity	MF	0.001470884	PLK1; CDK1; PBK; NEK2; VRK1; CDC7; TTK; PASK; PKMYT1; AURKB; CIT; AURKA
kinase binding	MF	0.004957845	CAV1; PLK1; VRK1; CDC25C; FOXM1; AURKB; CIT; AURKA; ARHGAP33; CCNA2
cyclin-dependent protein serine/threonine kinase regulator activity	MF	0.00820989	CCNA2; CCNB2; CCNB1; CCNF

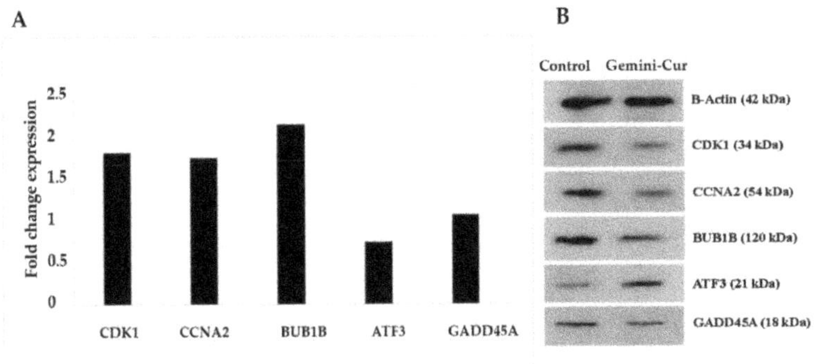

Figure 6. Validation of the expression of selected genes by real-time PCR and Western blotting. Nearly all genes except GADD45A validate findings from RNA seq. data. (**A**) The data show relative expression of genes compared to non-treated controls. (**B**) Blots illustrate protein expression in control and treated cells.

4. Discussion

Despite recent advances in early diagnosis and proper treatments, colorectal cancer (CRC) is still considered as the third leading cause of cancer-related deaths worldwide [1,2]. Gemini-Cur is one of the latest nano-formulations of curcumin with a significant anticancer effect that has recently been developed by our group [15]. We have already reported that Gemini-Cur inhibits the proliferation of different cancer cells through the induction of apoptosis [13–16]. Due to the limited information on the global effect of curcumin on transcriptome profiling, we employed RNA sequencing to uncover the differentially expressed genes and cellular pathways that are affected by Gemini-Cur in colorectal cancer cells. Here, our data not only confirm the suppressive effect but also demonstrate that numerous genes in different cellular pathways are modulated by Gemini-Cur in HCT-116 cells.

Along with all the steps of quality control and RNA-seq raw data normalization, the DEA (Differentially Expressed Analysis) process introduced about 3892 genes as DEGs (padj < 0.01) with 244 upregulated (padj < 0.01, log2 FC ≥ 2), and 198 downregulated genes (padj < 0.01, log2 FC ≤ -2). The PPI network was also created by Cytoscape to better figure out the possible correlations between DEGs. Consequently, three of the top significant downregulated genes including cyclin-dependent kinase 1 (CDK; padj = 1.08×10^{-06}, log2 FC = -3.15), cyclin A2 (CCNA2; padj = 2.87×10^{-05}, log2 FC = -3.056) and BUB1 mitotic checkpoint serine/threonine kinase B (BUB1B; padj = 2.72×10^{-06}, log2 FC = -2.99) as well as two upregulated genes including growth arrest and DNA damage inducible gamma (GADD45G; padj = 1.47×10^{-06}, log2 FC = 4) and activating transcription factor 3 (ATF3; padj = 1.03×10^{-19}, log2 FC = 4) were selected from top ten DEG lists for further validation of RNA sequencing results.

Numerous reports have shown that ATF3 and GADD45G play critical roles in cancer. Inoue et al. illustrated that the ATF3 transcription factor inhibits the migration and invasion of HCT-116 cells. ATF3 is also involved in the process of cellular stress response [18]. Here, RNA sequencing and experimental studies show that transcription factor ATF3 as a key regulator of cellular stress response is upregulated in cancer cells after treatment with Gemini-Cur. These findings demonstrate increasingly the potential of ATF3 as a therapeutic candidate in colorectal cancer. In another work, Guo et al. demonstrated that the overexpression of GADD45G acts as tumor suppressor in AML [19]. Although RNA sequencing showed that GADD45G is upregulated in treated cells, this overexpression was not detected in PCR and Western blotting. This contradictory finding may be due to the presence of different RNA reads including RNAs from paralogous genes and pseudogenes. However, PCR and Western blotting only detect main predominant variant of GADD45G.

Accordingly, CCNA2, BUB1B, and CDK1 genes have been downregulated in Gemini-Cur treated cells versus the non-treated group. According to the gene-pathway annotated network, it is revealed that CDK1 is involved in the cell cycle and associated with other CRC-related genes (CCNA2, BUB1B, GADD45G, ATF3). CDK1 contributes to the cell proliferation, apoptosis, and cell migration [20,21]. It was shown that the upregulation of CDK1 leads to poor prognosis in patients with CRC [22]. Zhu et al. and Thoma et al. indicated that the downregulation of CDK1 inhibited fluorouracil-resistant CRC cell proliferation [23,24].

Regarding the BUB1B and CCNA2, both genes have been reported as upregulated genes in CRC [25–27]. Gan et al. indicated that the CCNA2 knockdown could significantly suppress CRC cell growth by impairing cell cycle progression and inducing cell apoptosis [20]. Ding et al. also reported that the upregulation of BUB1B, CDK1, and CCNA2 genes contribute to the progression of tumor growth and metastasis of CRC [28]. These results further validate our analysis methods and the accuracy of RNA sequencing for exploring the novel top up/downregulated genes in different conditions.

The joint cooperation of GADD45G and CDK1 in the p53 activity regulation pathway is another attractive result of this study. GADD45G plays a role in the activation of S and G2/M checkpoints during p53-related DNA damage responses [29,30]. The modulatory effect of Gemini-Cur on these genes may highlight their collaborative role in cell stress response.

5. Conclusions

Taken together, our data show that Gemini-Cur comprehensively modulates gene expression in colorectal cancer HCT-116 cells. Gene ontology annotations related to DEGs found here include DNA-dependent ATPase (MF) during metabolic processes (BP) in nucleus (CC). Further analysis also demonstrated that Gemini-Cur dominantly affects cell cycle-related pathways. These RNA sequencing data greatly expand our understanding of the molecular and cellular targets of curcumin in cancer. We also reported novel potential targets for curcumin as listed in the top ten DEGs. Further investigations on the top up/downregulated genes especially in different cancer cell lines and non-cancerous controls will facilitate the findings of curcumin targets in colon cancer.

Supplementary Materials: The following supporting information can be downloaded at: https://www.mdpi.com/article/10.3390/molecules27113470/s1, supplementary materials: Transcriptome Profiling of HCT-116 Colorectal Cancer Cells with RNA Sequencing Reveals Novel Targets for Polyphenol Nano Curcumin. The standard Bioconductor RNA-seq workflow (DESeq2), was employed to detect differentially expressed genes (DEGs) in Gemini-Cur treated cell compared to controls. Regarding the parameters *p* values and Log2 fold change, 3892 DEGs were selected as listed.

Author Contributions: Conceptualization, H.J.A. and E.B.; methodology, E.B. and F.N.; software, E.B., M.A.H.F. and E.B.; investigation, H.J.A.; writing—original draft preparation, E.B. and H.J.A.; writing—review and editing, F.N., E.B. and M.A.H.F.; supervision, E.B. All authors have read and agreed to the published version of the manuscript.

Funding: This research received no external funding.

Institutional Review Board Statement: Not applicable.

Informed Consent Statement: Not applicable.

Data Availability Statement: Data are contained within the article.

Acknowledgments: The authors would like to thank the staff of Genomics & Molecular biology Lab at the University of Tabriz for their technical assistance.

Conflicts of Interest: The authors declare no conflict of interest.

Sample Availability: Samples of the compounds are available from the authors.

References

1. Singh, M.P.; Rai, S.; Singh, N.K.; Srivastava, S. Transcriptomic landscape of early age onset of colorectal cancer identifies novel genes and pathways in Indian CRC patients. *Sci. Rep.* **2021**, *11*, 11765. [CrossRef] [PubMed]
2. Ferlay, J.; Colombet, M.; Soerjomataram, I.; Mathers, C.; Parkin, D.M.; Piñeros, M.; Znaor, A.; Bray, F. Estimating the global cancer incidence and mortality in 2018: GLOBOCAN sources and methods. *Int. J. Cancer* **2019**, *144*, 1941–1953. [CrossRef] [PubMed]
3. Guo, Y.; Wu, R.; Gaspar, J.M.; Sargsyan, D.; Su, Z.-Y.; Zhang, C.; Gao, L.; Cheng, D.; Li, W.; Wang, C.; et al. DNA methylome and transcriptome alterations and cancer prevention by curcumin in colitis-accelerated colon cancer in mice. *Carcinogenesis* **2018**, *39*, 669–680. [CrossRef] [PubMed]
4. Passirani, C.; Vessières, A.; La Regina, G.; Link, W.; Silvestri, R. Modulating undruggable targets to overcome cancer therapy resistance. *Drug Resist. Updat.* **2022**, *60*, 100788. [CrossRef]
5. Obrand, D.I.; Gordon, P.H. Incidence and patterns of recurrence following curative resection for colorectal carcinoma. *Dis. Colon Rectum* **1997**, *40*, 15–24. [CrossRef]
6. André, T.; Boni, C.; Navarro, M.; Tabernero, J.; Hickish, T.; Topham, C.; Bonetti, A.; Clingan, P.; Bridgewater, J.; Rivera, F.; et al. Improved Overall Survival with Oxaliplatin, Fluorouracil, and Leucovorin as Adjuvant Treatment in Stage II or III Colon Cancer in the MOSAIC Trial. *J. Clin. Oncol.* **2009**, *27*, 3109–3116. [CrossRef]
7. Pallag, A.; Roşca, E.; Ţiţ, D.M.; Muţiu, G.; Bungau, S.G.; Pop, O.L. Monitoring the effects of treatment in colon cancer cells using immunohistochemical and histoenzymatic techniques. *Rom. J. Morphol. Embryol.* **2015**, *56*, 1103–1109.
8. Zhong, Z.; Vong, C.T.; Chen, F.; Tan, H.; Zhang, C.; Wang, N.; Cui, L.; Wang, Y.; Feng, Y. Immunomodulatory potential of natural products from herbal medicines as immune checkpoints inhibitors: Helping to fight against cancer via multiple targets. *Med. Res. Rev.* **2022**, *42*, 1246–1279. [CrossRef]

9. Sivasankarapillai, V.S.; Nair, R.M.K.; Rahdar, A.; Bungau, S.; Zaha, D.C.; Aleya, L.; Tit, D.M. Overview of the anticancer activity of withaferin A, an active constituent of the Indian ginseng *Withania somnifera*. *Environ. Sci. Pollut. Res.* **2020**, *27*, 26025–26035. [CrossRef]
10. Kumar, A.; Harsha, C.; Parama, D.; Girisa, S.; Daimary, U.D.; Mao, X.; Kunnumakkara, A.B. Current clinical developments in curcumin-based therapeutics for cancer and chronic diseases. *Phytother. Res.* **2021**, *35*, 6768–6801. [CrossRef]
11. Nocito, M.C.; De Luca, A.; Prestia, F.; Avena, P.; La Padula, D.; Zavaglia, L.; Sirianni, R.; Casaburi, I.; Puoci, F.; Chimento, A.; et al. Antitumoral Activities of Curcumin and Recent Advances to ImProve Its Oral Bioavailability. *Biomedicines* **2021**, *9*, 1476. [CrossRef]
12. D'Angelo, N.A.; Noronha, M.A.; Kurnik, I.S.; Câmara, M.C.; Vieira, J.M.; Abrunhosa, L.; Martins, J.T.; Alves, T.F.; Tundisi, L.L.; Ataide, J.A.; et al. Curcumin encapsulation in nanostructures for cancer therapy: A 10-year overview. *Int. J. Pharm.* **2021**, *604*, 120534. [CrossRef]
13. Ghaderi, S.; Babaei, E.; Hussen, B.M.; Mahdavi, M.; Azeez, H.J. Gemini Curcumin Suppresses Proliferation of Ovarian Cancer OVCAR-3 Cells via Induction of Apoptosis. *Anti-Cancer Agents Med. Chem.* **2021**, *21*, 775–781. [CrossRef]
14. Emami, A.; Babaei, E.; Nagishbandi, A.; Azeez, H.J.; Feizi, M.A.H.; Golizadeh, A. Cellular uptake and apoptotic properties of gemini curcumin in gastric cancer cells. *Mol. Biol. Rep.* **2021**, *48*, 7215–7222. [CrossRef]
15. Karimpour, M.; Feizi, M.A.H.; Mahdavi, M.; Krammer, B.; Verwanger, T.; Najafi, F.; Babaei, E. Development of curcumin-loaded gemini surfactant nanoparticles: Synthesis, characterization and evaluation of anticancer activity against human breast cancer cell lines. *Phytomedicine* **2019**, *57*, 183–190. [CrossRef]
16. Ebrahimi, M.; Babaei, E.; Neri, F.; Feizi, M.A.H. Anti-proliferative and apoptotic effect of gemini curcumin in p53-wild type and p53-mutant colorectal cancer cell lines. *Int. J. Pharm.* **2021**, *601*, 120592. [CrossRef]
17. Yu, L.-L.; Wu, J.-G.; Dai, N.; Yu, H.G.; Si, J.M. Curcumin reverses chemoresistance of human gastric cancer cells by downregulating the NF-?B transcription factor. *Oncol. Rep.* **2011**, *26*, 1197–1203. [CrossRef]
18. Inoue, M.; Uchida, Y.; Edagawa, M.; Hirata, M.; Mitamura, J.; Miyamoto, D.; Taketani, K.; Sekine, S.; Kawauchi, J.; Kitajima, S. The stress response gene ATF3 is a direct target of the Wnt/β-catenin pathway and inhibits the invasion and migration of HCT116 human colorectal cancer cells. *PLoS ONE* **2018**, *13*, e0194160. [CrossRef]
19. Guo, W.; Dong, Z.; Guo, Y.; Chen, Z.; Kuang, G.; Yang, Z. Methylation-mediated repression of GADD45A and GADD45G expression in gastric cardia adenocarcinoma. *Int. J. Cancer* **2013**, *133*, 2043–2053. [CrossRef]
20. Gan, W.; Zhao, H.; Li, T.; Liu, K.; Huang, J. CDK1 interacts with iASPP to regulate colorectal cancer cell proliferation through p53 pathway. *Oncotarget* **2017**, *8*, 71618–71629. [CrossRef]
21. Tong, Y.; Huang, Y.; Zhang, Y.; Zeng, X.; Yan, M.; Xia, Z.; Lai, D. Correction: DPP3/CDK1 contributes to the progression of colorectal cancer through regulating cell proliferation, cell apoptosis, and cell migration. *Cell Death Dis.* **2021**, *12*, 529, Erratum in *Cell Death Dis.* **2021**, *12*, 623. [CrossRef]
22. Sung, W.-W.; Lin, Y.-M.; Wu, P.-R.; Yen, H.-H.; Lai, H.-W.; Su, T.-C.; Huang, R.-H.; Wen, C.-K.; Chen, C.-Y.; Chen, C.-J.; et al. High nuclear/cytoplasmic ratio of Cdk1 expression predicts poor prognosis in colorectal cancer patients. *BMC Cancer* **2014**, *14*, 951. [CrossRef]
23. Zhu, Y.; Li, K.; Zhang, J.; Wang, L.; Sheng, L.; Yan, L. Inhibition of CDK1 Reverses the Resistance of 5-Fu in Colorectal Cancer. *Cancer Manag. Res.* **2020**, *12*, 11271–11283. [CrossRef]
24. Thoma, O.-M.; Neurath, M.F.; Waldner, M.J. Cyclin-Dependent Kinase Inhibitors and Their Therapeutic Potential in Colorectal Cancer Treatment. *Front. Pharmacol.* **2021**, *12*, 757120. [CrossRef]
25. Dai, S.; Mo, Y.; Wang, Y.; Xiang, B.; Liao, Q.; Zhou, M.; Li, X.; Li, Y.; Xiong, W.; Li, G.; et al. Chronic Stress Promotes Cancer Development. *Front. Oncol.* **2020**, *10*, 1492. [CrossRef]
26. Gan, Y.; Li, Y.; Li, T.; Shu, G.; Yin, G. CCNA2 acts as a novel biomarker in regulating the growth and apoptosis of colorectal cancer. *Cancer Manag. Res.* **2018**, *10*, 5113–5124. [CrossRef]
27. Li, J.; Zhou, L.; Liu, Y.; Yang, L.; Jiang, D.; Li, K.; Xie, S.; Wang, X.; Wang, S. Comprehensive Analysis of Cyclin Family Gene Expression in Colon Cancer. *Front. Oncol.* **2021**, *11*, 674394. [CrossRef]
28. Ding, X.; Duan, H.; Luo, H. Identification of Core Gene Expression Signature and Key Pathways in Colorectal Cancer. *Front. Genet.* **2020**, *11*, 45. [CrossRef]
29. Vairapandi, M.; Balliet, A.G.; Hoffman, B.; Liebermann, D.A. GADD45b and GADD45g are cdc2/cyclinB1 kinase inhibitors with a role in S and G2/M cell cycle checkpoints induced by genotoxic stress. *J. Cell. Physiol.* **2002**, *192*, 327–338. [CrossRef]
30. Taylor, W.R.; Stark, G.R. Regulation of the G2/M transition by p53. *Oncogene* **2001**, *20*, 1803–1815. [CrossRef]

molecules MDPI

Article

Determination of the Phenolic Profile by Liquid Chromatography, Evaluation of Antioxidant Activity and Toxicity of Moroccan *Erica multiflora*, *Erica scoparia*, and *Calluna vulgaris* (*Ericaceae*)

Douaa Bekkai [1,2], Yassine Oulad El Majdoub [2], Hamid Bekkai [3], Francesco Cacciola [4,*], Natalizia Miceli [2], Maria Fernanda Taviano [2], Emilia Cavò [2], Tomader Errabii [1], Roberto Laganà Vinci [2], Luigi Mondello [2,5,6] and Mohammed L'Bachir El Kbiach [1]

1 Team of Plant Biotechnology, Biology Department, Abdelmalek Essaadi University, Tetouan 93000, Morocco; bekkai.douaa@gmail.com (D.B.); t.errabii@uae.ac.ma (T.E.); melkbiach@uae.ac.ma (M.L.E.K.)
2 Department of Chemical, Biological, Pharmaceutical and Environmental Sciences, University of Messina, 98166 Messina, Italy; youladelmajdoub@unime.it (Y.O.E.M.); nmiceli@unime.it (N.M.); mtaviano@unime.it (M.F.T.); ecavo@unime.it (E.C.); robertolaganavinci@gmail.com (R.L.V.); lmondello@unime.it (L.M.)
3 Chemistry Department, Abdelmalek Essaadi University, Tetouan 93000, Morocco; bekhamid1@hotmail.com
4 Department of Biomedical, Dental, Morphological and Functional Imaging Sciences, University of Messina, 98125 Messina, Italy
5 Chromaleont s.r.l., c/o Department of Chemical, Biological, Pharmaceutical and Environmental Sciences, University of Messina, 98166 Messina, Italy
6 Department of Sciences and Technologies for Human and Environment, University Campus Bio-Medico of Rome, 00128 Rome, Italy
* Correspondence: cacciolaf@unime.it; Tel.: +39-0906766570

Citation: Bekkai, D.; Oulad El Majdoub, Y.; Bekkai, H.; Cacciola, F.; Miceli, N.; Taviano, M.F.; Cavò, E.; Errabii, T.; Laganà Vinci, R.; Mondello, L.; et al. Determination of the Phenolic Profile by Liquid Chromatography, Evaluation of Antioxidant Activity and Toxicity of Moroccan *Erica multiflora*, *Erica scoparia*, and *Calluna vulgaris* (*Ericaceae*). *Molecules* **2022**, *27*, 3979. https://doi.org/10.3390/molecules27133979

Academic Editor: Nour Eddine Es-Safi

Received: 1 June 2022
Accepted: 18 June 2022
Published: 21 June 2022

Abstract: This study aimed to investigate the phenolic profile and selected biological activities of the leaf and aerial extracts of three *Ericaceae* species, namely *Erica multiflora*, *Erica scoparia*, and *Calluna vulgaris*, collected from three different places in the north of Morocco. The phenolic composition of all extracts was determined by LC coupled with photodiode array and mass spectrometry detection. Among the investigated extracts, that of *E. scoparia* aerial parts was the richest one, with a total amount of polyphenols of 9528.93 mg/kg. Up to 59 phenolic compounds were detected: 52 were positively identified and 49 quantified—11 in *C. vulgaris*, 14 in *E. multiflora*, and 24 in *E. scoparia*. In terms of chemical classes, nine were phenolic acids and 43 were flavonoids, and among them, the majority belonged to the class of flavonols. The antioxidant activity of all extracts was investigated by three different in vitro methods, namely DPPH, reducing power, and Fe^{2+} chelating assays; *E. scoparia* aerial part extract was the most active, with an IC_{50} of 0.142 ± 0.014 mg/mL (DPPH test) and 1.898 ± 0.056 ASE/mL (reducing power assay). Further, all extracts were non-toxic against *Artemia salina*, thus indicating their potential safety. The findings attained in this work for such Moroccan *Ericaceae* species, never investigated so far, bring novelty to the field and show them to be valuable sources of phenolic compounds with interesting primary antioxidant properties.

Keywords: *Ericaceae*; LC–DAD/ESI–MS; phenolic compounds; flavonoids; antioxidant activity; *Artemia salina* Leach

1. Introduction

Ericaceae is a cosmopolitan family, represented by 124 genera and 4100–4250 species that are widely distributed around the world, particularly in the Mediterranean area, in deficient and non-calcic soils, as well as in high mountains [1–4]. Within this family, *Erica* and *Calluna* are the most abundant and widely spread genera. In Northern Morocco, *E. multiflora*, *E. scoparia*, and *C. vulgaris* are traditionally consumed by local people in the form of infusions, and are well known for their therapeutic properties [5–7]. In Morocco, *Erica multiflora* L. and *Erica scoparia* L. are considered among the most well-known species of the *Erica* genus [1,8]. According to popular knowledge, both species might have anti-inflammatory and analgesic properties when it comes to urinary diseases [5,6]. Moreover, *E. multiflora* has shown antihyperlipidemic and liver function repair effects [8,9], and effective antilithiatic activity [10].

Calluna vulgaris (L.) Hull belongs to the monotypic genus of *Calluna*, also known for its powerful bioactive compounds. It is widely used to treat kidney and urinary system disorders, particularly inflammatory diseases of the bladder, prostate, and urinary tract [7,11–15]. It is also important to note that heather honey obtained from *C. vulgaris* nectar is a special type of honey that is highly appreciated by consumers, not only for its distinctive flavor and dietary value but also for its therapeutic purposes [12,15].

These biological effects are closely related to their composition in bioactive compounds such as flavonoids, tannins, anthocyanins, vitamins C and E, triterpenoids, saponins, proteins, steroids, coumarins, ascorbic acid, hydroquinone, etc. [4,16–19]. In the human body, the accumulation of free radicals induces numerous illnesses and health issues. Therefore, research within plants for natural antioxidant sources might be a promising alternative to lower the incidence of multiple diseases that are due to oxidative stress [20,21]. Polyphenols are an important class of secondary metabolites in plants, characterized by one or more hydroxyl groups binding to one or more aromatic rings, and are divided into two groups: flavonoids and non-flavonoids [22]. The biological and medicinal proprieties of antioxidant compounds such as plant polyphenols have been widely reported in the scientific literature [23]. Indeed, the protective role of polyphenols, especially as free radical scavengers, has been well established, and these molecules may play a prominent role in the prevention and/or the treatment of oxidative stress-induced diseases [24].

In the current study, *E. multiflora*, *E. scoparia*, and *C. vulgaris*, collected from Northern Morocco, were investigated for their phenolic composition and were further tested for their antioxidant properties as well as for their potential toxicity. In particular, the qualitative–quantitative profile of the phenolic constituents contained in the hydroalcoholic extracts obtained from the leaves and aerial parts of both *Erica* species and from the leaves of *C. vulgaris* was determined by LC–DAD/ESI–MS analyses. In order to provide a comprehensive view of the antioxidant profiles, the in vitro antioxidant effectiveness of the extracts was assessed by using three different methods: the DPPH (1,1-diphenyl-1-picrylhydrazyl) test and the reducing power and ferrous ion chelating assays. Moreover, the brine shrimp (*Artemia salina* Leach) lethality bioassay was utilized to evaluate the toxicity.

The phenolic content of *E. multiflora* has been already evaluated in other works [2,8,25,26]; however, either the leaves [2,8] or flowers [25]/entire plant [26] have been investigated. Notably, ref. [2] refers to an Algerian species, whereas ref. [25] refers to a Tunisian species. No data are available in the literature on the chemical composition and biological properties of *E. scoparia*; on the other hand, for *C. vulgaris*, only the inflorescences of a Portuguese species [27] have been reported so far.

2. Results and Discussion

2.1. Polyphenol Composition

The phenolic compounds present in the aerial parts and leaves of *C. vulgaris*, *E. multiflora*, and *E. scoparia* were identified by using an HPLC chromatogram at 330 nm (Figure 1). The main phenolic compounds were recognized by combining the retention times, UV spectra, and mass spectra of each peak with its standard, when available, and with literature data. The results revealed different quali-quantitative profiles among the studied parts, as shown in Figure 1. A total of 59 phenolic compounds were detected: 14 in *C. vulgaris*, 18 in *E. multiflora*, and 27 in *E. scoparia* (Table 1). Among them, 52 were positively identified (11 in *C. vulgaris*, 14 in *E. multiflora*, and 24 in *E. scoparia*). In terms of chemical classes, nine were phenolic acids and 43 were flavonoids, and among them, the majority belonged to the class of flavonols, mainly derivates of quercetin, myricetin, isorhamnetin, and kaempferol, while the rest of the compounds belonged to the class of flavanones, specifically eriodictyol and taxifolin. It is worth mentioning that, to the best of our knowledge, no previous studies have investigated the chemical composition of *E. scoparia*.

Calluna vulgaris leaves contained a total amount of phenolic compounds of 1567.78 mg/kg, comprising caffeoylquinic acid, which was the most abundant phenolic compound (1180 ± 8.18 mg/kg), followed by myricetin-*O*-rhamnoside (232.98 ± 0.30 mg/kg), myricetin-*O*-pentoside (48.81 ± 2.22 mg/Kg), and myricetin-*O*-hexoside (41.66 ± 1.88 mg/kg), whereas quercetin-*O*-hexoside (2.82 ± 3.24 mg/kg) was the lowest one. The results are in accordance with those presented by Mandim et al. [27] at the qualitative level, except for catechin, isorhamnetin-3-*O*-glucoside, and isorhmnetin-*O*-rhamnoside, which were absent in this studied species. However, a notable difference has been shown at the quantitative level, which could be, at least in part, attributed to the different organ of the plant used in this study, viz. leaves instead of inflorescences.

The leaves of *E. multiflora* contained 399.01 mg/kg of phenolic compounds, and were characterized by the presence of a quercetin derivative, myricetin-*O*-hexoside, and quercetin-*O*-(6″-cinnamoyl)-hexoside, while the aerial parts contained 227.6 mg/kg of phenolic compounds, and were distinguished by the presence of 4-caffeoylquinic acid, methyl-ellagic acid hexoside, and eriodictyol-*O*-hexoside, wherein 4-caffeoylquinic acid was the main compound in the aerial parts, with 83.75 ± 0.74 mg/kg, and where kaempferol was the least prevalent compound, with 0.95 ± 1.84 mg/kg. According to these results, it can be concluded that *E. multiflora* leaves presented higher phenolic compound content when compared to the aerial parts. The output of heat map analysis showed that the leaves and aerial parts of *E. multiflora* were clustered together into the same group and displayed the following main compounds in common: quercetin-*O*-hexoside, kaempferol-rhamnosyl-hexoside, rutin, caffeoylquinic acid, and kaempferol-hexoside. Moreover, in both parts, the presence of small amounts of three other compounds, quercetin, dimethylquercetin, and kaempferol, was noted. These results contradict those obtained by Mandim et al. [27], where quercetin was the most abundant compound, followed by kaempferol. This discordance could be partially related to the time and the location of the harvest, and/or the extraction method. *Erica scoparia* aerial parts presented a total amount of polyphenols of 9528.93 mg/kg. The most abundant compounds identified were myricetin-*O*-hexoside (2130.25 ± 0.78 mg/kg), myricetin-*O*-rhamnoside (1625.89 ± 0.39 mg/kg), and myricetin-*O*-pentoside (852.85 ± 1.97 mg/kg), whereas quercetin-*O*-(6″-p-hydroxybenzoyl)-hexoside (91.34 ± 1.22 mg/kg) was the least abundant one. Notably, myricetin-*O*-hexoside was shown to be the greatest phenolic compound in the leaves of *E. scoparia* (184.38 ± 0.26 mg/kg), while the smallest content was recorded for quercetin-*O*-(malonyl)-hexoside (18.52 ± 0.27 mg/kg). Thus, a remarkable discrepancy in the phenolic composition between the leaves and aerial parts of *E. scoparia* was observed. In addition, some phenolic compounds contained in the aerial parts seemed to be entirely absent in the leaves, such as taxifolin, digalloyl-quinic acid, and kaempferol.

A principal component analysis (PCA) alongside a heat map analysis were carried out on the phenolic compounds as variables to identify the connection between all the plant parts under observation (Figures 2 and 3). The PCA results presented two main components (F1 × F2) that determine 68.94%, whereas (F1 × F3) showed a contribution of 62.60%.

Table 1. Phenolic compounds detected in *C. vulgaris*, *E. multiflora*, and *E. scoparia*.

Peak No	Compound	t_R (min)	UV max (nm)	[M-H]-	*E. multiflora* (mg/Kg ± RSD%)		*E. scoparia* (mg/Kg ± RSD%)		*C. vulgaris* (mg/Kg ± RSD%)
					Leaves	Aerial Parts	Leaves	Aerial Parts	Leaves
1	Taxifolin-*O*-hexoside	4.11	288	465, 303, 313				332.96 ± 0.68	
2	Taxifolin-*O*-hexoside isomer	4.24	284	465, 303, 313				214.93 ± 1.49	
3	Digalloyl-quinic acid	4.61	274	495				Nq	
4	Caffeoylquinic acid	4.81	297sh, 326	353, 191, 179	53.93 ± 0.11	61.11 ± 0.18			
5	4-*O*-Caffeoylquinic acid	4.91	297sh, 326	353, 191, 179		83.75 ± 0.74			
6	Caffeoylquinic acid	4.99	290, 325	353, 191,137					626.40 ± 0.77
7	Myricetin-*O*-hexoside	5.38	258, 358	479, 317				2130.25 ± 0.78	
8	Eriodictyol-*O*-hexoside	5.42	297, 321	449, 287		Nq			
9	Caffeoylquinic acid	5.42	290, 325	353, 191,137					138.37 ± 0.23
10	Caffeoylquinic acid	5.47	290, 325	353, 191,137					231.54 ± 1.68
11	Quercetin derivative	5.63	260, 356	615, 463, 301	2.89 ± 0.83				
12	Myricetin-*O*-hexoside isomer	5.67	356	479, 317	43.46 ± 0.35				
13	Myricetin-*O*-pentoside	5.70	259, 357	449, 317				852.85 ± 1.97	
14	Myricetin-*O*-rhamnoside	5.74	260, 357	463, 317				1625.89 ± 0.39	
15	Quercetin-*O*-hexoside	5.83	255, 353	463, 301				213.14 ± 0.43	
16	Rutin	5.87	257, 354	609, 301	55.44 ± 2.59	14.16 ± 0.18			
17	Caffeoylquinic acid	5.87	290, 325	353, 191,137					184.69 ± 0.95
18	Methoxy-myricetin-*O*-rhamnoside	5.88	254, 358	493				810.78 ± 0.43	
19	p-Coumaroylquinic acid	6.07	312	337					Nq
20	Quercetin-*O*-hexoside	6.08	255, 355	463, 301	117.43 ± 0.48	29.48 ± 1.76			
21	Quercetin-*O*-hexoside	6.13	354	463, 301	4.78 ± 0.67	0.10 ± 2.51			

Table 1. *Cont.*

Peak No	Compound	t_R (min)	UV max (nm)	[M-H]-	*E. multiflora* (mg/Kg ± RSD%)		*E. scoparia* (mg/Kg ± RSD%)		*C. vulgaris* (mg/Kg ± RSD%)
					Leaves	Aerial Parts	Leaves	Aerial Parts	Leaves
22	Kaempferol-*O*-(6″-galloyl)hexoside	6.17	253, 358	599, 285				564.64 ± 0.19	
23	Myricetin-*O*-rhamnoside	6.20	358	463				268.52 ± 0.08	
24	Myricetin-*O*-hexoside	6.21	356	479, 317			184.38 ± 0.26		
25	Kaempferol-rhamnosyl-hexoside	6.24	264, 347	593, 447, 285	90.76 ± 1.19	15.24 ± 0.21			
26	Myricetin-*O*-hexoside	6.25	356	479, 317					41.66 ± 1.88
27	Isorhamnetin-*O*-hexoside	6.32	252, 357	477				683.43 ± 0.93	
28	Kaempferol-hexoside	6.51	264, 348	447, 285	4.83 ± 1.27	5.55 ± 2.06			
29	Myricetin-*O*-pentoside	6.55	260, 357	449, 317			72.79 ± 0.05		
30	Quercetin galloyl hexoside derivative	6.56	357	615				160.67 ± 1.25	
31	Myricetin-*O*-pentoside	6.59	281, 349	449, 317					48.81 ± 2.22
32	Kaempferol-hexoside isomer	6.61	264, 348	447, 285	17.08 ± 0.35	14.53 ± 0.44			
33	Myricetin-*O*-rhamnoside	6.65	260, 357	463, 317			153.65 ± 1.13		
34	Quercetin-*O*-hexoside	6.68	255, 353	463, 301			64.25 ± 1.47		2.82 ± 3.24
35	Myricetin-*O*-(6″-benzoyl)hexoside	6.70	265,316, 358	583, 316				200.83 ± 0.20	
36	Methyl-ellagic acid hexoside	6.72	283	477		Nq			
37	Myricetin-*O*-rhamnoside	6.72	260, 357	463, 317					232.98 ± 0.35

Table 1. *Cont.*

Peak No	Compound	t_R (min)	UV max (nm)	[M-H]-	*E. multiflora* (mg/Kg ± RSD%)		*E. scoparia* (mg/Kg ± RSD%)		*C. vulgaris* (mg/Kg ± RSD%)
					Leaves	Aerial Parts	Leaves	Aerial Parts	Leaves
38	Unknown	6.97	344	649		Nq			
39	Quercetin-*O*-(malonyl)hexoside	7.03	356	549			18.52 ± 0.27		
40	Quercetin-*O*-pentoside	7.06	255, 354	433, 301					9.44 ± 0.28
41	Unknown	7.11	358	599, 507, 463			Nq		
42	Quercetin-*O*-(6″-p-hydroxybenzoyl) hexoside	7.17	269, 356	583, 316				91.34 ± 1.22	
43	Unknown	7.22	350	723, 677, 477			Nq		
44	Quercetin-*O*-rhamnoside	7.22	255, 342	447, 301					32.30 ± 0.02
45	Kaempferol-*O*-rhamnoside	7.77	263, 341	431, 285					18.77 ± 0.55
46	Unknown	7.89	312	731		Nq			
47	Myricetin-*O*-(6″-cinnamoyl)hexoside	8.14	265, 359	609, 317, 301				757.33 ± 1.96	
48	Unknown	8.22	288, 308	289					Nq
49	Unknown	8.30	309	483, 289					Nq
50	Quercetin-*O*-(6″-cinnamoyl)hexoside	8.41	281	593, 447, 301	3.72 ± 1.10				
51	Quercetin	8.66	268, 370	301	3.34 ± 2.11	1.39 ± 5.97			
52	Myricetin-*O*-(6″-p-coumaroyl)hexoside	8.77	265, 360	624				509.39 ± 0.94	
53	Isorhamnetin-*O*-(6″-caffeoyl)hexoside	9.45	264, 359	639				111.98 ± 0.50	

Table 1. *Cont.*

Peak No	Compound	t_R (min)	UV max (nm)	[M-H]-	*E. multiflora* (mg/Kg ± RSD%)		*E. scoparia* (mg/Kg ± RSD%)		*C. vulgaris* (mg/Kg ± RSD%)
					Leaves	**Aerial Parts**	**Leaves**	**Aerial Parts**	**Leaves**
54	Dimethylquercetin	9.46	227, 344	329, 301	0.86 ± 8.95	1.38 ± 0.62			
55	Myricetin-*O*-(6″-cinnamoyl)hexoside	9.53	264, 359	609, 317, 301			23.46 ± 1.49		
56	Kaempferol	10.22	366	285	0.49 ± 1.89	0.91 ± 1.84			
57	Isorhamnetin-*O*-hexoside-*O*-rhamnoside	10.22	264, 359	623			9.76 ± 0.34		
58	Quercetin-*O*-(6″-cinnamoyl)hexoside	10.40	356	593, 447, 301			0.79 ± 1.12		
59	Unknown	10.97	356	637, 347			Nq		
	Total				399.01 ± 1.46	227.6 ± 0.15	527.6 ± 1.55	9528.93 ± 54.32	1567.78 ± 13.01

Nq: Not quantified.

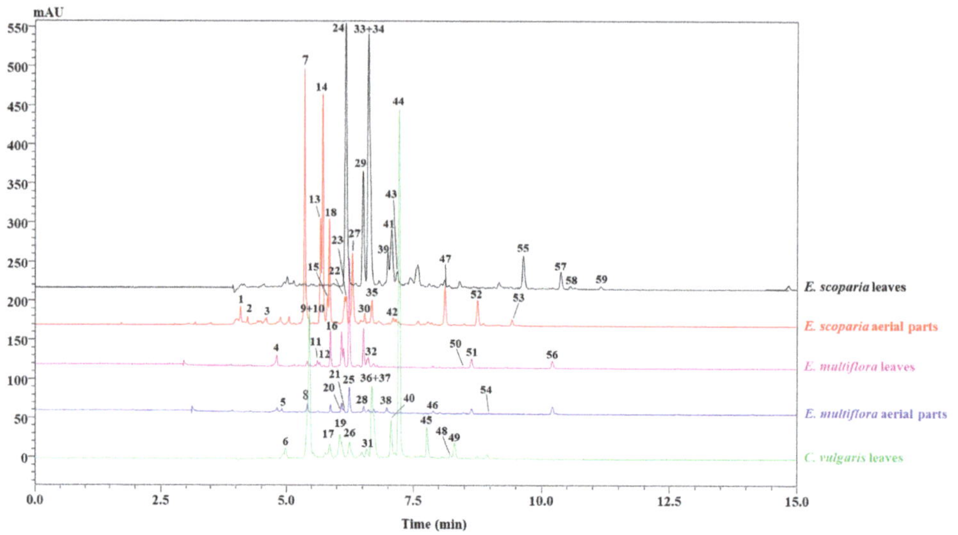

Figure 1. Chromatographic profile of hydroalcoholic extracts from leaves and aerial parts of 3 different *Ericaceae* taxa at λ = 330 nm.

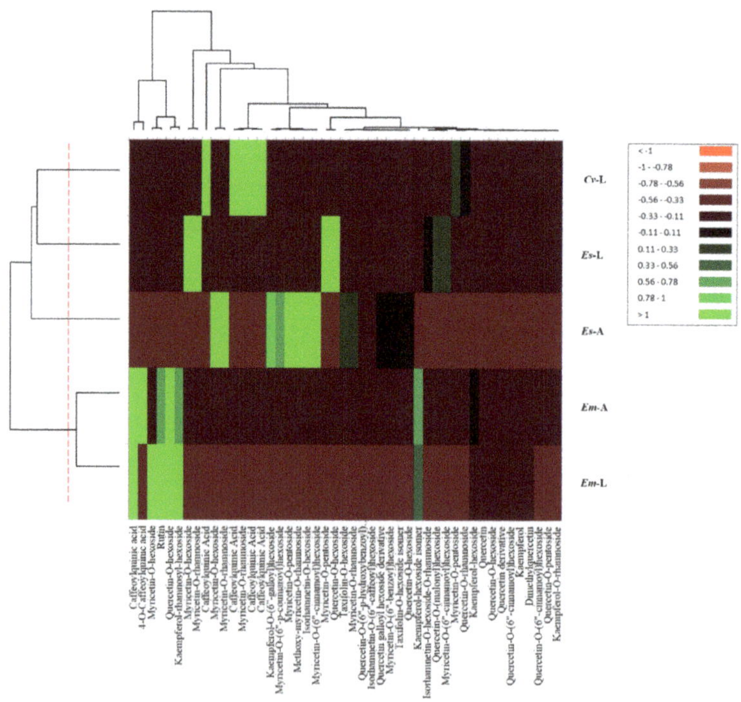

Figure 2. Heat map analysis of phenolic compounds (mean, N = 3) in leaves and aerial parts of 3 different *Ericaceae* taxa: *C. vulgaris* leaves (*Cv*-L), *E. scoparia* leaves (*Es*-L), *E. scoparia* aerial parts (*Es*-A), *E. multiflora* aerial parts (*Em*-A), *E. multiflora* leaves (*Em*-L).

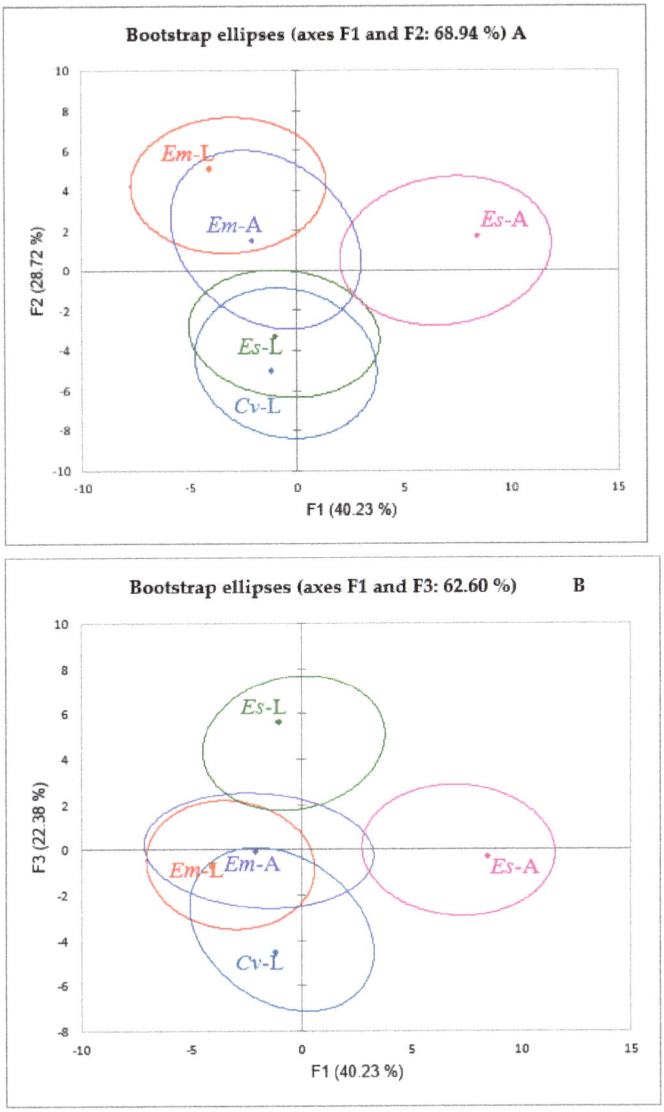

Figure 3. The correlation between phenolic compounds (variables) and plant parts of *Ericaceae* taxa (observations) through PCA. (**A**) represents the first two factorials F1xF2. (**B**) represents the second two factorials F1xF3.

Both statistical analyses confirmed the presence of four different clusters: the first cluster regrouped both parts of *E. multiflora*, and the second and the third clusters were attributed to *E. scoparia* parts, while a completely distinguished fourth cluster was ascribed to *C. vulgaris* leaves. According to the principal components F1 and F2, the leaves of *E. scoparia* and *C. vulgaris* showed a false positive correlation, resulting in a unique cluster, whereas F1 and F3 led to the rejection of the previous correlation and the presence of two different clusters.

2.2. *Antioxidant and Cytotoxic Activities*

2.2.1. Antioxidant Activity

The human body is constantly dealing with the formation of free radicals. When produced in excess, the latter trigger oxidative stress, causing serious tissue injuries. It is well known that many diseases are closely related to oxidative stress, mainly cancer and neurodegenerative disorders (Alzheimer's, Parkinson's, etc.). To cope with these health issues, plants provide a cheap and affordable source of natural antioxidants to prevent free radical-induced diseases, especially in countries with low incomes and limited healthcare resources [28]. Many primary antioxidant chemistry reactions can be grouped into the categories of hydrogen-atom transfer (HAT) and single-electron transfer (SET). The HAT mechanism occurs when an antioxidant compound scavenges free radicals by donating hydrogen atoms; the SET mechanism is based on the transfer of a single electron to reduce any compound, including metals, carbonyls, and free radicals [29,30]. It has been reported that, even if many antioxidant reactions are characterized as following either HAT or SET chemical processes, these reaction mechanisms can simultaneously occur [29,31,32].

Due to the complex nature of phytochemicals and their interactions, the importance of using various methods based on different mechanisms for a comprehensive study of the antioxidant properties of plant extracts has been argued. Therefore, the antioxidant activity of *Em*-L, *Em*-A, *Es*-L, *Es*-A, and *Cv*-L extracts was investigated by three different in vitro methods: in order to establish the primary antioxidant properties, the 1,1-diphenyl-1-picrylhydrazyl (DPPH) test, involving HAT and SET mechanisms, and the reducing power, a SET-based assay, were used. The secondary antioxidant properties were determined through the estimation of the ferrous ion (Fe^{2+}) chelating activity.

The DPPH test is a rapid, simple, inexpensive, and widely used method to measure the free radical scavenging ability of pure compounds or phytocomplexes. Based on the results shown in Figure 4, all extracts, except for *Em*-A, demonstrated valuable radical scavenging activity, reaching approximately 90% of inhibition at the concentration of 0.5 mg/mL. Among the tested extracts, *Es*-A was the most active, as confirmed also by the lowest IC_{50} value ($p < 0.001$); at the concentration of 0.25 mg/mL, it showed activity higher than that of BHT, used as a standard drug, displaying radical scavenging activity superimposable to that of the standard (around 100%) at the concentrations of 1 and 2 mg/mL (Figure 4).

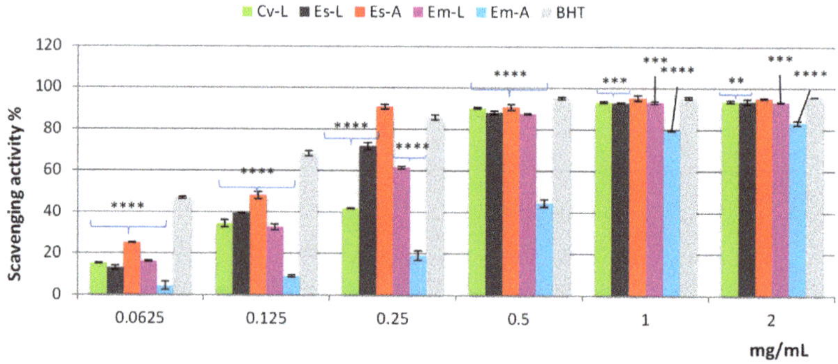

Figure 4. Free radical scavenging activity (DPPH test) of hydroalcoholic extracts from leaves and aerial parts of 3 different *Ericaceae* taxa: *C. vulgaris* leaves (*Cv*-L), *E. scoparia* leaves (*Es*-L), *E. scoparia* aerial parts (*Es*-A), *E. multiflora* leaves (*Em*-L), *E. multiflora* aerial parts (*Em*-A). Data are expressed as the mean ± SD of three independent experiments ($n = 3$) and were analyzed by one-way ANOVA followed by Dunnett's post-hoc test. **** $p < 0.0001$, *** $p < 0.001$, ** $p < 0.05$ vs. BHT.

Based on the IC_{50} values, the efficacy of the extracts and the standard decreases in the order *Es*-A > BHT > *Es*-L > *Em*-L > *Cv*-L > *Em*-A (Table 2); however, at 1 mg and 2 mg/mL, *Es*-L, *Em*-L and *Cv*-L exhibited radical scavenging activity close to that of BHT, while only *Em*-A reached about 80% of inhibition (Figure 4).

Table 2. Free radical scavenging activity (DPPH test), reducing power, and ferrous ion (Fe^{2+}) chelating activity of hydroalcoholic extracts from leaves and aerial parts of 3 different *Ericaceae* taxa.

Ericaceae Taxa	DPPH Test IC_{50} (mg/mL)	Reducing Power ASE/mL	Fe^{2+} Chelating Activity IC_{50} (mg/mL)
Cv-L	0.212 ± 0.061 [a]	2.790 ± 0.100 [a]	NA
Es-L	0.189 ± 0.051 [a]	2.721 ± 0.062 [a]	NA
Es-A	0.142 ± 0.014 [b]	1.898 ± 0.056 [b]	>2
Em-L	0.200 ± 0.001 [a]	3.814 ± 0.091 [c]	NA
Em-A	0.611 ± 0.017 [c]	5.538 ± 0.148 [d]	>2
Standard	BHT 0.154 ± 0.001 [b]	BHT 1.131 ± 0.037 [e]	EDTA 0.0067 ± 0.0003

C. vulgaris leaves (*Cv*-L), *E. scoparia* leaves (*Es*-L), *E. scoparia* aerial parts (*Es*-A), *E. multiflora* leaves (*Em*-L), *E. multiflora* aerial parts (*Em*-A). NA: no activity. Data are expressed as the mean ± SD of three independent experiments ($n = 3$) and were analyzed by one-way ANOVA followed by Tukey–Kramer multiple comparisons test. [a–e] Different letters within the same column indicate significant differences between mean values ($p < 0.001$).

The reducing power reflects the ability to stop the radical chain reaction. In this assay, the presence of antioxidant compounds in the sample determines the reduction of Fe^{3+} to the ferrous form (Fe^{2+}). As shown in Figure 5, all the extracts, except *Em*-A, displayed good reducing power, which was dose-dependent. Among the tested extracts, those of *E. scoparia* were the most active. In fact, at the concentration of 1 mg/mL, *Es*-A showed activity close to that of BHT; at 2 mg/mL, the reducing power of both *Es*-A and *Es*-L was higher than that of the standard. Based on the ASE/mL values, the efficacy of the extracts and the standard decreases in the order BHT > *Es*-A > *Es*-L > *Cv*-L > *Em*-L > *Em*-A (Table 2).

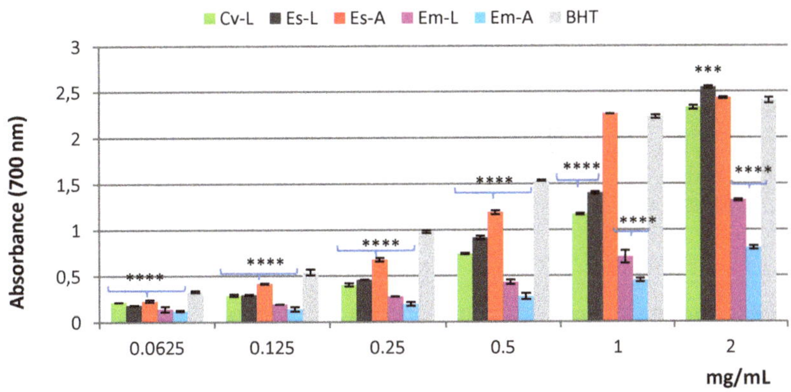

Figure 5. Reducing power of hydroalcoholic extracts from leaves and aerial parts of 3 different *Ericaceae* taxa evaluated by spectrophotometric detection of Fe^{3+}-Fe^{2+} transformation method. *C. vulgaris* leaves (*Cv*-L), *E. scoparia* leaves (*Es*-L), *E. scoparia* aerial parts (*Es*-A), *E. multiflora* leaves (*Em*-L), *E. multiflora* aerial parts (*Em*-A). Data are expressed as the mean ± SD of three independent experiments ($n = 3$) and were analyzed by one-way ANOVA followed by Dunnett's post-hoc test. **** $p < 0.0001$, *** $p < 0.001$, vs. BHT.

The Fe^{2+} chelating activity of *Em*-L, *Em*-A, *Es*-L, *Es*-A, and *Cv*-L extracts was estimated by monitoring the formation of the Fe^{2+}-ferrozine complex. In this assay, only *Es*-A and *Em*-A displayed weak chelating properties, whereas all the other extracts were not active (Table 2).

From our findings, it is evident that all the extracts possess strong primary antioxidant properties; interestingly, that obtained from the aerial parts of *E. scoparia* is the most powerful. HPLC analysis revealed, for this extract, the highest content of flavonoid compounds, represented mainly by flavonols such as several myricetin glycosides, but also kaempferol, quercetin, and isorhamnetin glycosides. The flavonols, containing more hydroxyl groups (one to six OH groups), have a very strong ability to scavenge DPPH radicals and they are well-known, potent antioxidants. These compounds have a 3-hydroxyl group in the C-ring and *3′,4′*-dihydroxy groups (catechol structure) in the B-ring, but also possess the 2,3-double bond in conjugation with the 4-oxo function in the C-ring, which are the essential structural elements for potent radical scavenging activity [33].

Erica scoparia aerial part extract is rich in myricetin glycosides, which have been shown to possess strong primary antioxidant activity [34,35]. Thus, the best activity observed for *Es*-A could be correlated primarily to these compounds, but also to kaempferol, isorhamnetin, and quercetin glycosides.

2.2.2. *Artemia salina* Lethality Bioassay

The toxicity of *Em*-L, *Em*-A, *Es*-L, *Es*-A, and *Cv*-L extracts was assessed by the *Artemia salina* lethality bioassay, extensively utilized as an alternative model for toxicity evaluation. This simple method offers numerous advantages, such as rapidity, low cost, continuous availability of cysts (eggs), and ease of maintenance under laboratory conditions [36]. It is a useful system for predicting the toxicity of plant extracts in order to consider their safety. The results of the bioassay showed that the median lethal concentration values were higher than 1000 μg/mL for all the tested extracts, thus indicating the lack of toxicity against brine shrimp larvae based on Clarkson's toxicity criterion [37].

3. Materials and Methods

3.1. Chemicals and Reagents

LC–MS-grade water (H_2O), acetonitrile (ACN), formic acid, methanol, and DMSO were purchased from Merck Life Science (Merck KGaA, Darmstadt, Germany). Taxifolin, rutin, 4-caffeoylquinic acid, ishorhamnetin, quercetin, and kaempferol-3-glucoside were also obtained from Merck Life Science (Merck KGaA, Darmstadt, Germany). Unless indicated otherwise, all chemicals were purchased from Sigma-Aldrich (Milan, Italy).

3.2. Plant Materials

Three *Ericaceae* taxa, *Erica multiflora*, *Erica scoparia*, *Calluna vulgaris*, were collected in December 2019 from three different places in Northern Morocco; Khemiss anjra (Tetouan province) with longitude −5.5125257, latitude 35.6632287; Ben karrich (Tetouan province), longitude −5.4279948, latitude 35.5068513; Souq l'qolla (Chefchouaen), longitude −5.59873, latitude 35.12112 35, respectively. The taxonomic identification was confirmed by Prof. Kadiri Mohamed, Abdelmalek Essaadi University, Tetouan, Morocco.

The plant material was collected in December according to their flourishing stage. The selected samples for the preparation of the extracts consisted of leaves and aerial parts, for both *Erica* species that bloomed in this month, while, for *C. vulgaris*, only the leaves were used because, in the same period, this species had not yet flowered.

The selected parts were dried in darkness at room temperature for 2 weeks, and then crushed in an electrical grinder to a particle size less than 4 mm; the grounded parts were stored in a refrigerator under 4 °C in amber glass vials to avoid oxidation effects.

3.3. Extraction Procedure

One hundred milligrams of different powdered plant material of the three studied species was extracted, in a 50 mL volumetric flask, with 10 mL of ethanol:water, 96:4 (*v:v*), followed by sonication (60 W, 25 °C, 37 Hz) for 20 min. The obtained extracts were centrifugated for 10 min under 3000 rpm and filtered using Whatman filter paper (Merck Life Science, Merck KGaA, Darmstadt, Germany). The extraction procedure was repeated three times, and then the filtrates were combined, evaporated to dryness by a rotavapor and stored at 4 °C. The yields of the extracts, referring to 100 g of dried plant material, were 31.37% for *E. multiflora* leaves (*Em*-L), 33.26% for *E. multiflora* aerial parts (*Em*-A), 37.97% for

E. scoparia leaves (*Es*-L), 46.76% for *E. scoparia* aerial parts (*Es*-A), and 33.72% for *C. vulgaris* leaves (*Cv*-L).

3.4. LC–DAD/ESI–MS Analyses

The hydroalcoholic extracts (*Em*-L, *Em*-A, *Es*-L, *Es*-A, and *Cv*-L) were analyzed through the LC–MS technique using a Shimadzu liquid chromatography system (Kyoto, Japan), composed of a CBM-20A controller, two LC-30AD dual-plunger parallel-flow pumps, a DGU-20A5R degasser, a CTO-40C column oven, a SIL-40C autosampler, an SPD-M40 photo diode array detector, and an LCMS-8050 mass spectrometer, through an ESI source (Shimadzu, Kyoto, Japan).

Separation analyses were performed on a 150 × 4.6 mm; 2.7 µm Ascentis Express RP C18 column (Merck Life Science, Merck KGaA, Darmstadt, Germany). The mobile phase was composed of two solvents, water (solvent A) and acetonitrile (solvent B), both acidified with formic acid at 0.1% *v*/*v*. The flow rate was set at 1 mL/min and a simplified linear gradient of elution program was followed: 0–5 min, 0–30% B, 5–30 min, 30–100% B, 35 min, 100% B. PDA range: 200–400; λ = 280 nm (sampling frequency: 40.0 Hz, time constant: 0.08 s).

The applied mass spectrometry conditions were as follows: scan range, *m/z* 100–1200; scan speed, 2500 amu/s; event time, 0.3 s; nebulizing gas (N_2) flow rate, 1.5 L/min; drying gas (N_2) flow rate, 15 L/min; interface temperature, 350 °C; heat block temperature, 300 °C; DL (desolvation line) temperature, 300 °C; DL voltage, 1 V; interface voltage, −4.5 kV.

3.5. Preparation of Calibration Curves

Calibration curves of six polyphenolic standards (R^2 > 0.9989) were used for the quantification of the polyphenolic content in sample extracts by using different concentration levels: 4-caffeoylquinic acid (y = 3450.1x − 26,363; LoD = 0.034, LoQ = 0.104), taxifolin (y = 18,001x − 35,329; LoD = 0.071, LoQ = 0.215), rutin (y = 10,066x + 2176.5; LoD = 0.014, LoQ = 0.042), isorhamnetin (y = 25,334x + 1890.3; LoD = 0.116, LoQ = 0.353), quercetin (y = 20,376x + 7053.8, LoD = 0.007, LoQ = 0.022), kaempferol-3-glucoside (y = 13,848x + 2354.1, LoD = 0.090, LoQ = 0.274). Each analysis was performed in triplicate.

3.6. Antioxidant and Cytotoxic Activities

3.6.1. Free Radical Scavenging Activity

The free radical scavenging activity of *Em*-L, *Em*-A, *Es*-L, *Es*-A, and *Cv*-L extracts was determined using the DPPH (1,1-diphenyl-1-picrylhydrazyl) method [38]. The samples were tested at different concentrations (0.0625–2 mg/mL). An aliquot (0.5 mL) of solution containing different amounts of sample was added to 3 mL of daily prepared methanol DPPH solution (0.1 mM). The optical density change at 517 nm was measured, 20 min after the initial mixing, with a model UV-1601 spectrophotometer (Shimadzu). Butylated hydroxytoluene (BHT) was used as reference.

The scavenging activity was measured as the decrease in the absorbance of the samples versus DPPH standard solution. Results were expressed as the radical scavenging activity percentage (%) of the DPPH, defined by the formula [(Ao − Ac)/Ao] × 100, where Ao is the absorbance of the control and Ac is the absorbance in the presence of the sample or standard.

The results, obtained from the average of three independent experiments, are reported as mean radical scavenging activity percentage (%) ± standard deviation (SD) and mean 50% inhibitory concentration (IC_{50}) ± SD. The IC_{50} value is a parameter calculated as the concentration of extract needed to decrease the initial DPPH concentration by 50%. Thus, the lower IC_{50} value, the higher the antioxidant activity of the sample.

3.6.2. Reducing Power Assay

The reducing power of *Em*-L, *Em*-A, *Es*-L, *Es*-A, and *Cv*-L extracts was evaluated by the spectrophotometric detection of Fe^{3+}-Fe^{2+} transformation method [39]. The extracts were tested at different concentrations ranging from 0.0625 to 2 mg/mL. Solutions of different concentrations of extracts in 1 mL solvent were mixed with 2.5 mL of phosphate buffer (0.2 M, pH 6.6) and 2.5 mL of 1% potassium ferricyanide [$K_3Fe(CN)_6$], and the resulting mixture was incubated at 50 °C for 20 min. The solution was cooled rapidly, mixed with 2.5 mL of 10% trichloroacetic acid, and centrifuged at 3000 rpm for 10 min. After centrifugation, the supernatant (2.5 mL) was mixed with 2.5 mL of distilled water and 0.5 mL of 0.1% fresh ferric chloride ($FeCl_3$). The absorbance of the solution was measured at a wavelength of 700 nm after 10 min. An increase in the absorbance of the reaction mixture indicates an increase in its reducing power. An equal volume (1 mL) of water mixed with a solution prepared as described above was used as a blank. Ascorbic acid and BHT were used as references. The results averaged from three independent experiments were expressed as mean absorbance values ± SD. The reducing power was also expressed as ascorbic acid equivalent (ASE/mL); when the reducing power is 1 ASE/mL, the reducing power of 1 mL extract is equivalent to 1 µmol ascorbic acid.

3.6.3. Ferrous Ion (Fe^{2+}) Chelating Activity

The Fe^{2+} chelating activity of *Em*-L, *Em*-A, *Es*-L, *Es*-A, and *Cv*-L extracts was estimated according to the method reported by Decker and Welch [40]. The samples were tested at different concentrations (0.0625–2 mg/mL). Briefly, different concentrations of each sample in 1 mL solvent were mixed with 0.5 mL of methanol and 0.05 mL of 2 mM $FeCl_2$. The reaction was initiated by the addition of 0.1 mL of 5 mM ferrozine. Then, the mixture was shaken vigorously and left standing at room temperature for 10 min. The absorbance of the solution was measured spectrophotometrically at 562 nm. The control contained $FeCl_2$ and ferrozine, complex formation molecules. Ethylenediaminetetraacetic acid (EDTA) was used as a reference. The percentage of inhibition of the ferrozine—(Fe^{2+}) complex formation was calculated by the formula $[(Ao - Ac)/Ao] \times 100$, where Ao is the absorbance of the control and Ac is the absorbance in the presence of the sample or standard. The results, obtained from the average of three independent experiments, are reported as mean inhibition of the ferrozine—(Fe^{2+}) complex formation (%) ± SD and IC_{50} ± SD.

3.6.4. *Artemia salina* Lethality Bioassay

The potential toxicity of *Em*-L, *Em*-A, *Es*-L, *Es*-A, and *Cv*-L L extracts was investigated in brine shrimp (*Artemia salina* Leach) [41]. Ten brine shrimp larvae, taken 48 h after initiation of hatching in artificial seawater, were transferred to each sample vial, and then artificial seawater was added to obtain a final volume of 5 mL. Different concentrations of each extract were added (10–1000 µg/mL) and the brine shrimp larvae were incubated for 24 h at 25–28 °C. Then, the surviving larvae were counted using a magnifying glass. The assay was carried out in triplicate, and median lethal concentration (LC_{50}) values were determined by Litchfield and Wilcoxon's method. Extracts giving LC_{50} values greater than 1000 µg/mL were considered non-toxic.

3.7. *Statistical Analysis*

The heat map and PCA were established to provide an easier comparison of the phenolic compounds between the plant parts; the results were expressed as mean values ± relative standard deviation (RSD). All data were processed with principal component analysis (PCA) and collected in a heat map; the phenolic compounds were considered as variables in these plots to identify the connections between all the plant parts as observations. Principal component analysis (PCA) and heat map were generated using XLSTAT software ver. 2019.2.2.

Statistical comparison of the antioxidant activity data was carried out by using one-way analysis of variance (ANOVA) (GraphPAD Prism Version 9.4.0. Software for Science). p-values lower than 0.05 were considered statistically significant.

4. Conclusions

In this contribution, three Moroccan *Ericaceae* species, namely *Erica multiflora, Erica scoparia*, and *Calluna vulgaris*, were investigated. The phenolic profiles of the leaf and aerial extracts revealed a quite complex pattern, with up to 52 phenolic compounds positively identified, including phenolic acids and flavonoids. The antioxidant properties of the extracts were evaluated by means of three different methods, namely DPPH, reducing power, and Fe^{2+} chelating assays, demonstrating their high potential. On the basis of the phenolic profile and remarkable results achieved for the antioxidant activity, such species could be considered as a potential safe source of bioactive compounds to be advantageously employed in traditional Moroccan medicine. Interestingly, myricetin derivates might have important therapeutic potential, e.g., antioxidant, anti-inflammatory, anti-diabetes, anticancer, and protective effects against Alzheimer's disease [42]; furthermore, the efficacy kaempferol and rutin can be exploited against doxorubicin-induced cardiotoxicity [43], while quercetin could be employed for its interesting anticancer effects against prostate and breast cancers [44].

Author Contributions: Conceptualization, D.B. and F.C.; Methodology, D.B., F.C., H.B., T.E. and M.L.E.K.; Investigation, D.B., Y.O.E.M., N.M., M.F.T. and E.C.; Writing—Original Draft Preparation, D.B., Y.O.E.M., N.M. and M.F.T.; Writing—Review and Editing, F.C., R.L.V. and H.B.; Supervision, F.C. and M.L.E.K.; Project Administration, L.M. All authors have read and agreed to the published version of the manuscript.

Funding: This research received no external funding.

Institutional Review Board Statement: Not applicable.

Informed Consent Statement: Not applicable.

Data Availability Statement: Not applicable.

Acknowledgments: The authors thank Merck Life Science and Shimadzu Corporations for their continuous support.

Conflicts of Interest: The authors declare no conflict of interest.

References

1. Amezouar, F.; Badri, W.; Hsaine, M.; Bourhim, N.; Fougrach, H. Évaluation des activités antioxydante et anti-inflammatoire de *Erica arborea* L. du Maroc. *Pathol. Biol.* **2013**, *61*, 254–258. [CrossRef] [PubMed]
2. Guendouze-Bouchefa, N.; Madani, K.; Chibane, M.; Boulekbache-Makhlouf, L.; Hauchard, D.; Kiendrebeogo, M.; Stévigny, C.; Okusa, P.N.; Duez, P. Phenolic compounds, antioxidant and antibacterial activities of three Ericaceae from Algeria. *Ind. Crops Prod.* **2015**, *70*, 459–466. [CrossRef]
3. Ştefănescu, B.E.; Szabo, K.; Mocan, A.; Crişan, G. Phenolic compounds from five Ericaceae species leaves and their related bioavailability and health benefits. *Molecules* **2019**, *24*, 2046. [CrossRef] [PubMed]
4. Hamim, A.; Miché, L.; Douaik, A.; Mrabet, R.; Ouhammou, A.; Duponnois, R.; Hafidi, M. Diversity of fungal assemblages in roots of Ericaceae in two Mediterranean contrasting ecosystems. *Comptes Rendus Biol.* **2017**, *340*, 226–237. [CrossRef]
5. Sadki, C.; Hacht, B.; Souliman, A.; Atmani, F. Acute diuretic activity of aqueous *Erica multiflora* flowers and *Cynodon dactylon* rhizomes extracts in rats. *J. Ethnopharmacol.* **2010**, *128*, 352–356. [CrossRef]
6. Chaachouay, N.; Benkhnigue, O.; Zidane, L. Ethnobotanical Study Aimed at Investigating the Use of Medicinal Plants to Treat Nervous System Diseases in the Rif of Morocco. *J. Chiropract. Med.* **2020**, *19*, 70–81. [CrossRef]
7. Djahafi, A.; Taïbi, K.; Abderrahim, L.A. Aromatic and medicinal plants used in traditional medicine in the region of Tiaret, North West of Algeria. *Medit. Bot.* **2021**, *42*, e71465. [CrossRef]
8. Harnafi, H.; Bouanani, N.E.H.; Aziz, M.; Serghini-Caid, H.; Ghalim, N.; Amrani, S. The hypolipidaemic activity of aqueous *Erica multiflora* flowers extract in triton WR-1339 induced hyperlipidaemic rats: A comparison with fenofibrate. *J. Ethnopharmacol.* **2007**, *109*, 156–160. [CrossRef]
9. Khlifi, R.; Lahmar, A.; Dhaouefi, Z.; Kalboussi, Z.; Maatouk, M.; Kilani-Jaziri, S.; Ghedira, K.; Chekir-Ghedira, L. Assessment of hypolipidemic, anti-inflammatory and antioxidant properties of medicinal plant *Erica multiflora* in triton WR-1339-induced hyperlipidemia and liver function repair in rats: A comparison with fenofibrate. *Regul. Toxicol. Pharmacol.* **2019**, *107*, 104404. [CrossRef]
10. Sadki, C.; Atmani, F. Évaluation de l'effet antilithiasique, oxalo-calcique et phospho-ammoniaco-magnésien d'extrait aqueux d'*Erica multiflora* L. *Prog. Urol.* **2017**, *27*, 1058–1067. [CrossRef]

11. Chepel, V.; Lisun, V.; Skrypnik, L. Changes in the content of some groups of phenolic compounds and biological activity of extracts of various parts of heather (*Calluna vulgaris* (L.) hull) at different growth stages. *Plants* **2020**, *9*, 926. [CrossRef] [PubMed]
12. Rodrigues, F.; Moreira, T.; Pinto, D.; Pimentel, F.B.; Costa, A.S.G.; Nunes, M.A.; Gonçalves Albuquerque, T.; Costa, H.S.; Palmeira-de-Oliveira, A.; Oliveira, A.I.; et al. The phytochemical and bioactivity profiles of wild *Calluna vulgaris* L. *flowers. Food Res. Int.* **2018**, *111*, 724–731. [CrossRef] [PubMed]
13. Monschein, M.; Iglesias Neira, J.; Kunert, O.; Bucar, F. Phytochemistry of heather (*Calluna Vulgaris* (L.) Hull) and its altitudinal alteration. *Phytochem. Rev.* **2010**, *9*, 205–215. [CrossRef]
14. Tunón, H.; Olavsdotter, C.; Bohlin, L. Evaluation of anti-inflammatory activity of some Swedish medicinal plants. Inhibition of prostaglandin biosynthesis and PAF-induced exocytosis. *J. Ethnopharmacol.* **1995**, *48*, 61–76. [CrossRef]
15. Dezminieran, D.; Al Marghitas, L.; Fit, N.; Chirilaz, F.; Gherman, B.; Margaouan, R.; Aurori, A.; Bobis, O. Antibacterial Effect of Heather Honey (*Calluna vulgaris*) against Different Microorganisms of Clinical Importance. *Bull. UASVM Anim. Sci. Biotechnol.* **2015**, *72*, 72–77.
16. Kumarasamy, Y.; Cox, P.J.; Jaspars, M.; Nahar, L.; Sarker, S.D. Screening seeds of Scottish plants for antibacterial activity. *J. Ethnopharmacol.* **2002**, *1–2*, 73–77. [CrossRef]
17. Kiruba, S.; Mahesh, M.; Nisha, S.R.; Miller Paul, Z.; Jeeva, S. Phytochemical analysis of the flower extracts of *Rhododendron arboreum* Sm. ssp. *nilagiricum (Zenker) Tagg. Asian Pac. J. Trop. Biomed.* **2011**, *1*, S284–S286. [CrossRef]
18. Shamilov, A.A.; Bubenchikova, V.N.; Chernikov, M.V.; Pozdnyakov, D.I.; Garsiya, E.R. *Vaccinium Vitis-Idaea* L.: Chemical contents, pharmacological activities. *Pharm. Sci.* **2020**, *26*, 344–362. [CrossRef]
19. Madhvi, S.K.; Sharma, M.; Iqbal, J.; Younis, M. Phytochemistry, traditional uses and pharmacology of *Rhododendron Arboreum*: A review. *Res. J. Pharm. Technol.* **2019**, *12*, 4565. [CrossRef]
20. Lobo, V.; Patil, A.; Phatak, A.; Chandra, N. Free radicals, antioxidants and functional foods: Impact on human health. *Pharmacogn. Rev.* **2010**, *4*, 118. [CrossRef]
21. Kasote, D.M.; Katyare, S.S.; Hegde, M.V.; Bae, H. Significance of antioxidant potential of plants and its relevance to therapeutic applications. *Int. J. Biol. Sci.* **2015**, *11*, 982–991. [CrossRef] [PubMed]
22. Arigò, A.; Česla, P.; Šilarová, P.; Calabrò, M.L.; Česlová, L. Development of extraction method for characterization of free and bonded polyphenols in barley (*Hordeum vulgare* L.) grown in Czech Republic using liquid chromatography-tandem mass spectrometry. *Food Chem.* **2018**, *245*, 829–837. [CrossRef] [PubMed]
23. Česla, P.; Fischer, J.; Jandera, P. Improvement of the sensitivity of 2D LC-MEKC separation of phenolic acids and flavonoids natural antioxidants using the on-line preconcentration step. *Electrophoresis* **2012**, *33*, 2464–2473. [CrossRef] [PubMed]
24. Engwa, G.A. Free radicals and the role of plant phytochemicals as antioxidants against oxidative stress-related diseases. In *Phytochemicals—Source of Antioxidants and Role in Disease Prevention*; Asao, T., Asaduzzaman, M., Eds.; IntechOpen: London, UK, 2018.
25. Khlifi, R.; Dhaouefi, Z.; Toumia, I.B.; Lahmara, A.; Siouda, F.; Bouhajeba, R.; Bellalah, A.; Chekir-Ghedira, L. *Erica multiflora* extract rich in quercetin-3-O-glucoside and kaempferol-3-O-glucoside alleviates high fat and fructose diet-induced fatty liver disease by modulating metabolic and inflammatory pathways in Wistar rats. *J. Nutr. Biochem.* **2020**, *86*, 108490. [CrossRef] [PubMed]
26. Guesmi, F.; Ben Hadj, A.S.; Landoulsi, A. Investigation of Extracts from Tunisian Ethnomedicinal Plants as Antioxidants, Cytotoxins, and Antimicrobials. *Biomed. Environ. Sci.* **2017**, *30*, 811–824.
27. Mandim, F.; Barros, L.; Calhelha, R.C.; Abreu, R.M.V.; Pinela, J.; Alves, M.J.; Heleno, S.; Santos, P.F.; Ferreira, I.C.F.R. *Calluna vulgaris* (L.) Hull: Chemical characterization, evaluation of its bioactive properties and effect on the vaginal microbiota. *Food Funct.* **2019**, *10*, 78–89. [CrossRef]
28. Ahmed, M.; Khan, M.I.; Khan, M.R.; Muhammad, N.; Khan, A.U.; Khan, R.A. Role of medicinal plants in oxidative stress and cancer. *Open Access Sci. Rep.* **2013**, *2*, 641.
29. Craft, B.D.; Kerrihard, A.L.; Amarowicz, R.; Pegg, R.B. Phenol-based antioxidants and the in vitro methods used for their assessment. *Compr. Rev. Food Sci. Food Saf.* **2012**, *11*, 148–173. [CrossRef]
30. Prior, R.L.; Wu, X.; Schaich, K. Standardized methods for the determination of antioxidant capacity and phenolics in foods and dietary supplements. *J. Agric. Food Chem.* **2005**, *53*, 4290–4302. [CrossRef]
31. Wright, J.S.; Johnson, E.R.; DiLabio, G.A. Predicting the Activity of Phenolic Antioxidants: Theoretical Method, Analysis of Substituent Effects, and Application to Major Families of Antioxidants. *J. Am. Chem. Soc.* **2001**, *123*, 1173–1183. [CrossRef]
32. Leopoldini, M.; Marino, T.; Russo, N.; Toscano, M. Antioxidant properties of phenolic compounds: H-atom versus electron transfer mechanism. *J. Phys. Chem. A* **2004**, *108*, 4916–4922. [CrossRef]
33. Cai, Y.-Z.; Sun, M.; Xing, J.; Luo, Q.; Corke, H. Structure–radical scavenging activity relationships of phenolic compounds from traditional Chinese medicinal plants. *Life Sci.* **2006**, *78*, 2872–2888. [CrossRef]
34. Hayder, N.; Bouhlel, I.; Skandrani, I.; Kadri, M.; Steiman, R.; Guiraud, P.; Mariotte, A.M.; Ghedira, K.; Dijoux-Franca, M.G.; Chekir-Ghedira, L. In vitro antioxidant and antigenotoxic potentials of myricetin-3-*O*-galactoside and myricetin-3-*O*-rhamnoside from *Myrtus communis*: Modulation of expression of genes involved in cell defence system using cDNA microarray. *Toxicol. In Vitro* **2008**, *22*, 567–581. [CrossRef] [PubMed]
35. Arumugam, B.; Palanisamy, U.D.; Chua, K.H.; Kuppusamy, U.R. Protective effect of myricetin derivatives from *Syzygium malaccense* against hydrogen peroxide-induced stress in ARPE-19 cells. *Mol. Vis.* **2019**, *25*, 47–59.
36. Libralato, G.; Prato, E.; Migliore, L.; Cicero, A.M.; Manfra, L. A review of toxicity testing protocols and endpoints with *Artemia* spp. *Ecol. Indic.* **2016**, *69*, 35–49. [CrossRef]

37. Clarkson, C.; Maharaj, V.J.; Crouch, N.R.; Grace, O.M.; Pillay, P.; Matsabisa, M.G.; Bhagwandin, N.; Smith, P.J.; Folb, P.I. *In vitro* antiplasmodial activity of medicinal plants native to or naturalised in South Africa. *J. Ethnopharmacol.* **2004**, *92*, 177–191. [CrossRef] [PubMed]
38. Ohnishi, M.; Morishita, H.; Iwahashi, H.; Shitzuo, T.; Yoshiaki, S.; Kimura, M.; Kido, R. Inhibitory effects of chlorogenic acid on linoleic acid peroxidation and haemolysis. *Phytochemistry.* **1994**, *36*, 579–583. [CrossRef]
39. Oyaizu, M. Studies on products of browning reaction: Antioxidative activities of products of browning reaction prepared from glucosamine. *Jpn. J. Nutr. Diet.* **1986**, *44*, 307–315. [CrossRef]
40. Decker, E.A.; Welch, B. Role of ferritin as a lipid oxidation catalyst in muscle food. *J. Agric. Food Chem.* **1990**, *38*, 674–677. [CrossRef]
41. Meyer, B.N.; Ferrigni, N.R.; Putnam, J.E.; Jacobsen, L.B.; Nichols, D.E.; McLaughlin, J.L. Brine shrimp: A convenient general bioassay for active plant constituents. *Planta Med.* **1982**, *45*, 31–34. [CrossRef]
42. Park, K.-S.; Chong, Y.; Kim, M.K. Myricetin: Biological activity related to human health. *Appl. Biol. Chem.* **2016**, *59*, 259–269. [CrossRef]
43. Repo-Carrasco-Valencia, R.; Hellström, J.K.; Pihlava, J.M.; Mattila, P.H. Flavonoids and other phenolic compounds in Andean indigenous grains: Quinoa (*Chenopodium quinoa*), kañiwa (*Chenopodium pallidicaule*) and kiwicha (*Amaranthus caudatus*). *Food Chem.* **2010**, *120*, 128–133. [CrossRef]
44. Kumar, S.; Pandey, A.K. Chemistry and biological activities of flavonoids: An overview. *Sci. World J.* **2013**, *2013*, 162750. [CrossRef] [PubMed]

molecules

MDPI

Article

Protective Effect of Resveratrol against Hexavalent Chromium-Induced Genotoxic Damage in Hsd:ICR Male Mice

Tonancy Nicolás-Méndez [1,2], Sam Kacew [3], Alda Rocío Ortiz-Muñiz [4], Víctor Manuel Mendoza-Núñez [5] and María del Carmen García-Rodríguez [1,*]

1 Laboratorio de Antimutagénesis, Anticarcinogénesis y Antiteratogénesis Ambiental, Facultad de Estudios Superiores—Zaragoza, Universidad Nacional Autónoma de México (UNAM), Mexico City 09230, Mexico; tonic_1986@comunidad.unam.mx
2 Posgrado en Ciencias Biológicas, Universidad Nacional Autónoma de México (UNAM), Mexico City 04510, Mexico
3 McLaughlin Centre for Population Health Risk Assessment, University of Ottawa, Ottawa, ON K2G 3G8, Canada; skacew@uottawa.ca
4 Departamento de Ciencias de la Salud, Universidad Autónoma Metropolitana (UAM), Mexico City 09310, Mexico; arom@xanum.uam.mx
5 Unidad de Investigación en Gerontología, Facultad de Estudios Superiores—Zaragoza, Universidad Nacional Autónoma de México (UNAM), Mexico City 09230, Mexico; mendovic@unam.mx
* Correspondence: carmen.garcia@unam.mx; Tel.: +52-55-5623-0772

Abstract: The aim of this study is to examine the ability of resveratrol to counteract hexavalent chromium [Cr(VI)]-induced genetic damage, as well as the possible pathways associated with this protection. Hsd:ICR male mice are divided into groups of the following five individuals each: (a) control 1, distilled water; (b) control 2, ethanol 30%; (c) resveratrol, 50 mg/kg by gavage; (d) CrO_3, 20 mg/kg intraperitoneally; (e) resveratrol + CrO_3, resveratrol administered 4 h prior to CrO_3. The assessment is performed on peripheral blood. Micronuclei (MN) kinetics are measured from 0 to 72 h, while 8-hydroxydeoxyguanosine (8-OHdG) adduct repair levels, endogenous antioxidant system biomarkers, and apoptosis frequency were quantified after 48 h. Resveratrol reduces the frequency of Cr(VI)-induced MN and shows significant effects on the 8-OHdG adduct levels, suggesting that cell repair could be enhanced by this polyphenol. Concomitant administration of resveratrol and Cr(VI) results in a return of the activities of glutathione peroxidase and catalase to control levels, accompanied by modifications of superoxide dismutase activity and glutathione levels. Thus, antioxidant properties might play an important role in resveratrol-mediated inhibition of Cr(VI)-induced oxidant genotoxicity. The increase in apoptotic cells and the decrease in necrosis further confirmed that resveratrol effectively blocks the actions of Cr(VI).

Keywords: resveratrol; hexavalent chromium; 8-hydroxydeoxyguanosine adduct repair; apoptosis; endogenous antioxidant system; antigenotoxic

Citation: Nicolás-Méndez, T.; Kacew, S.; Ortiz-Muñiz, A.R.; Mendoza-Núñez, V.M.; García-Rodríguez, M.d.C. Protective Effect of Resveratrol against Hexavalent Chromium-Induced Genotoxic Damage in Hsd:ICR Male Mice. *Molecules* **2022**, *27*, 4028. https://doi.org/10.3390/molecules27134028

Academic Editor: Nour Eddine Es-Safi

Received: 20 May 2022
Accepted: 21 June 2022
Published: 23 June 2022

1. Introduction

Resveratrol (3,4′,5-trihydroxy-trans-stilbene) is a polyphenol found in a group of stilbenes. It has high antioxidant potential associated with beneficial health effects in the context of neurodegenerative and cardiovascular diseases, as well as some types of cancer, diabetes, and obesity-related disorders [1,2]. The antioxidant effects of resveratrol have been attributed to its ability to scavenge reactive oxygen species (ROS), activate repair mechanisms, and induce apoptosis [3,4]. Notably, resveratrol has been found to prevent DNA damage [5,6]. Although the effects of resveratrol on toxicity induced by metals (e.g., arsenic trioxide, sodium arsenite, copper oxide, chromic chloride, and potassium dichromate) have been examined in rodent hearts, livers, kidneys, thymus, and ovaries [7–11], there are no studies evaluating the effects of this polyphenol on hexavalent chromium [Cr(VI)] compound-induced genotoxicity.

Cr(VI) is largely released into the environment due to industrial activities, mainly including electroplating, welding, leather tanning, and pigment manufacturing, or found in automobile exhaust and tobacco products [12,13]. Both acute and chronic exposure to Cr(VI) compounds have been associated with cancer induction in different organs and tissues [14,15]. The genotoxic damage produced during its intracellular reduction may initiate and promote Cr(VI)-induced carcinogenesis by the formation of DNA adducts, cross-linking (DNA-protein and DNA-DNA), abasic sites, and oxidized DNA bases [16]. It is important to highlight that the induction of apoptosis, the inhibition of repair mechanisms, and gene expression play a crucial role in the genotoxic damage generated by exposure to Cr(VI) compounds [15,17]. Several studies have shown that antioxidants can counteract the effects of ROS and free radicals [18], such that antioxidant-rich substances have emerged as potential agents for preventing and adjuvating oxidative stress and DNA damage [19–21]. More precisely, compounds such as polyphenols have been shown to play a direct role as radical scavengers and metal chelators and to exert indirect effects by modulating levels of transcription factors and enzymes [22–24]. Therefore, the aim of the present study was to examine the protective effects of resveratrol against Cr(VI)-induced genotoxicity in vivo and the underlying metabolic processes including 8-hydroxydeoxyguanosine (8-OHdG, 7,8-dihydro-8-oxodeoxyguanosine) adduct repair, the endogenous antioxidant component system, and apoptosis, which may be involved in preventing Cr(VI)-induced DNA damage.

2. Results

2.1. *Effect of Resveratrol on MN Induced by* CrO_3

The genotoxic damage caused by Cr(VI) was evaluated using the micronuclei (MN) assay in erythrocytes of peripheral blood using acridine orange (AO)-coated slides. Differential AO staining distinguished polychromatic erythrocytes (PCE) from normochromatic erythrocytes (NCE) because PCE were stained, showing orange fluorescence due to the presence of ribosomal RNA (Figure 1A(i)), while NCE did not stain at all (shadow) (Figure 1A(ii)). The AO also enabled the identification of MN, which exhibited yellow fluorescence due to their DNA content (Figure 1A(iii)). To compare the kinetics of MN induction in treatment groups, data were analyzed by calculating the net induction frequency (NIF) using Equation (1) as follows:

$$\text{NIF} = |\text{MN frequencies measured at } \textit{time } x_i - \text{MN frequencies measured at } \textit{time } 0| / n \quad (1)$$

where x_i is the evaluation at 24, 48, or 72 h per group, *time 0* is the evaluation at 0 h (before treatment) per group, and n is the number of mice per group.

Calculating the NIF enhanced the ability to determine net MN induction by eliminating baseline MN variability among treated groups at *time 0*. Figure 1B illustrates the NIF of MN values for all treatments at 24, 48, and 72 h after administration. Treatments had a significant effect ($p < 0.0001$) on the MN frequencies according to the two-way repeated-measures analysis of variance (RM-ANOVA). In the chromium trioxide (CrO_3) group, an increase of about 7, 10, and 5 MN was observed at 24, 48, and 72 h, respectively, which was significantly higher than the control C1 ($p < 0.001$, $p < 0.0001$, and $p < 0.015$, respectively). The group treated with resveratrol and CrO_3 (resveratrol + CrO_3) had lower MN frequency than the CrO_3 only treatment at all times examined and was highest at 48 h ($p < 0.001$), though this reduction was no longer significant at 72 h ($p > 0.05$). However, in this group, the MN frequencies observed at 48 h were significantly different from the control groups (C1, $p < 0.001$; C2, $p < 0.040$) and the group treated with resveratrol alone ($p < 0.006$). There was a significant effect of time on the frequency of MN in the two-way RM-ANOVA ($p < 0.0001$). In the CrO_3 group, the frequency of MN increased at 48 h with respect to the initial time ($p < 0.038$) and decreased at 72 h ($p < 0.004$). In the resveratrol + CrO_3 group, the MN frequency decreased at 72 h ($p < 0.009$), and values at the initial and final times were similar. Treatment with resveratrol alone did not significantly affect the frequency of MN compared to control group C2 ($p > 0.05$).

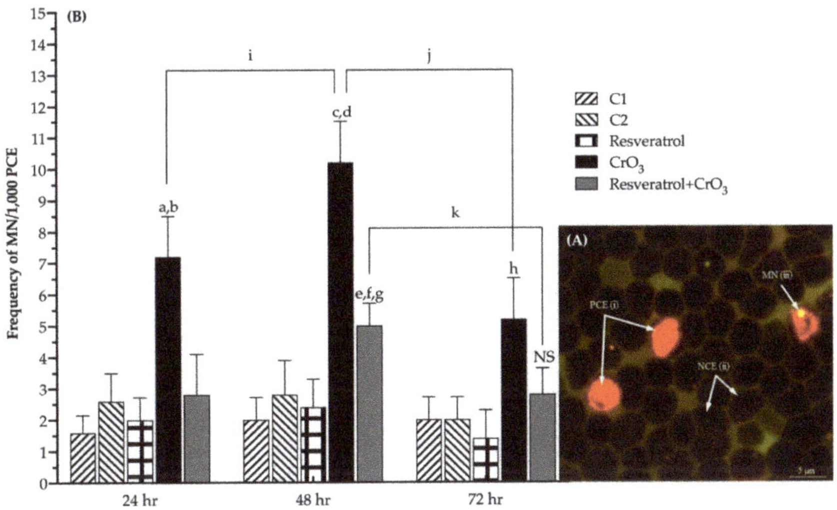

Figure 1. Effect of resveratrol and CrO_3 on the frequency of micronuclei (MN) were evaluated in the peripheral blood of mice. (**A**) Fluorescent microphotograph (1000×) of peripheral blood cells using the AO coating method. Polychromatic erythrocytes (PCE) stain fluorescent orange (**i**), normochromatic erythrocytes (NCE) do not stain at all (shadow) (**ii**), and MN fluoresces yellow (**iii**). (**B**) Data show the MN frequency at 24, 48, and 72 h minus the MN frequency at 0 h (net induction frequency: NIF, see results text). The frequencies of MN in the resveratrol + CrO_3 group decreased by 63, 50, and 47% at 24, 48, and 72 h, respectively, compared with those in the CrO_3 group. A total of 4000 PCE were evaluated in each mouse (n = 5 mice/group). Statistical significance was determined using two-way repeated measures-ANOVA followed by Tukey's post-hoc test. Analysis by treatments: ^{a} $p < 0.001$ vs. C1, 24 h; ^{b} $p < 0.004$ vs. resveratrol + CrO_3, 24 h; ^{c} $p < 0.0001$ vs. C1, 48 h; ^{d} $p < 0.001$ vs. resveratrol + CrO_3, 48 h; ^{e} $p < 0.001$ vs. C1, 48 h; ^{f} $p < 0.040$ vs. C2, 48 h; ^{g} $p < 0.006$ vs. resveratrol, 48 h; ^{h} $p < 0.015$ vs. C1, 72 h; ^{NS} $p > 0.05$ vs. CrO_3, 72 h. Analysis by time of evaluations: ^{i} $p < 0.038$, CrO_3; ^{j} $p < 0.004$, CrO_3; ^{k} $p < 0.009$, resveratrol + CrO_3. C1, Control 1, vehicle only (distilled water). C2, Control 2, vehicle only (ethanol 30%). CrO_3, chromium trioxide. AO, acridine orange.

2.2. *Effect of Resveratrol and CrO_3 on 8-OHdG Adduct Levels*

The 8-OHdG adduct was measured at 48 h in blood plasma since this method does not require sacrificing the animals. It is generally accepted that the excretion of the oxidized nucleosides 8-oxodG and 8-oxoGuo can be measured in fluids such as plasma, under the assumption that an organism maintains a steady-state with no changes in the rate of oxidation [25]. In that situation, the number of oxidized guanine moieties in the nucleic acid and its precursor pool must be equal to the number removed/excreted from the cell. Thus, the levels of 8-OHdG evaluated in fluids represent the balance between formation and repair rates. When evaluating oxidative damage to DNA using the 8-OHdG adduct repair levels in peripheral blood plasma analyzed by one-way ANOVA, there was a significant effect of treatment ($p < 0.0015$). The group treated with resveratrol prior to CrO_3 (resveratrol + CrO_3) had higher 8-OHdG levels than the control groups (C1, $p < 0.012$; C2, $p < 0.007$) and the CrO_3 group ($p < 0.002$) (Figure 2). Although the resveratrol group displayed numerically higher 8-OHdG levels than the control C2 and CrO_3 had lower levels than control C1, neither comparison was statistically significant.

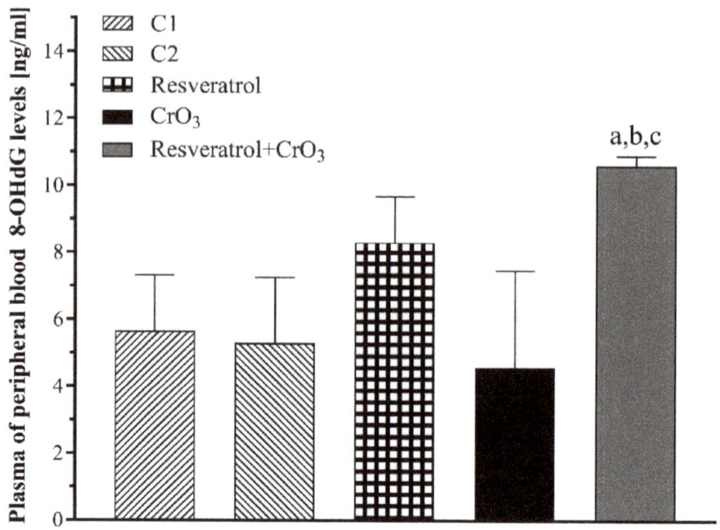

Figure 2. Effect of resveratrol and CrO_3 on 8-hydroxydeoxyguanosine (8-OHdG, 7,8-dihydro-8-oxodeoxyguanosine) levels evaluated in peripheral blood plasma 48 h after treatments ($n = 4$ mice/group). Statistical significance was determined using one-way ANOVA followed by Tukey's post-hoc test: [a] $p < 0.012$ vs. C1; [b] $p < 0.007$ vs. C2; [c] $p < 0.002$ vs. CrO_3. C1, Control 1, vehicle only (distilled water); C2, Control 2, vehicle only (ethanol 30%); CrO_3, chromium trioxide.

2.3. *Effect of Resveratrol and CrO_3 on the Antioxidant System*

The effect of the treatments on the antioxidant system was determined by evaluating glutathione (GSH) levels and the enzymatic activity of superoxide dismutase (SOD), glutathione peroxidase (GPx), and catalase (CAT). During the reduction of Cr(VI) to trivalent chromium [Cr(III)] superoxide radical ($O_2^{\bullet-}$) is generated, which can be dismutated by SOD. While GPx and CAT, when interacting with hydrogen peroxide (H_2O_2), can inhibit the production of the hydroxyl radical (•OH), and GSH participates in one of the Cr(VI) reduction pathways [15]. The enzymatic activities of SOD, GPx, and CAT are shown in Figure 3. Data were analyzed with a one-way ANOVA. Treatment had a significant effect on SOD activity ($p < 0.0001$). The resveratrol group had higher SOD activity than the control group (C2) ($p < 0.0001$), and the resveratrol + CrO_3 group showed an increase compared to the control groups (C1, $p < 0.016$; C2, $p < 0.001$), the resveratrol group ($p < 0.019$) and the CrO_3 group ($p < 0.0001$). GPx activity was also significantly affected by treatment ($p < 0.0001$). The CrO_3 treatment increased GPx activity compared to the control group C1 ($p < 0.019$) and the resveratrol + CrO_3 group ($p < 0.004$). Resveratrol treatment alone also increased GPx activity compared to control group C2 ($p < 0.0001$). The resveratrol + CrO_3 group had lower GPx activity than the resveratrol only group ($p < 0.0001$). There was also a significant effect of treatment on CAT activity ($p < 0.0001$). Resveratrol treatment increased CAT activity relative to the C2 control ($p < 0.005$), while treatment with CrO_3 increased it compared to control group C1 ($p < 0.034$). CAT activity was lower in the resveratrol + CrO_3 group than in the resveratrol group ($p < 0.001$) and the CrO_3 group ($p < 0.0002$). Together, these results demonstrate that in the resveratrol + CrO_3 mice, resveratrol restored GPx and CAT activity to levels similar to the controls. In the resveratrol + CrO_3 group, SOD levels were higher than those of the control and CrO_3 groups.

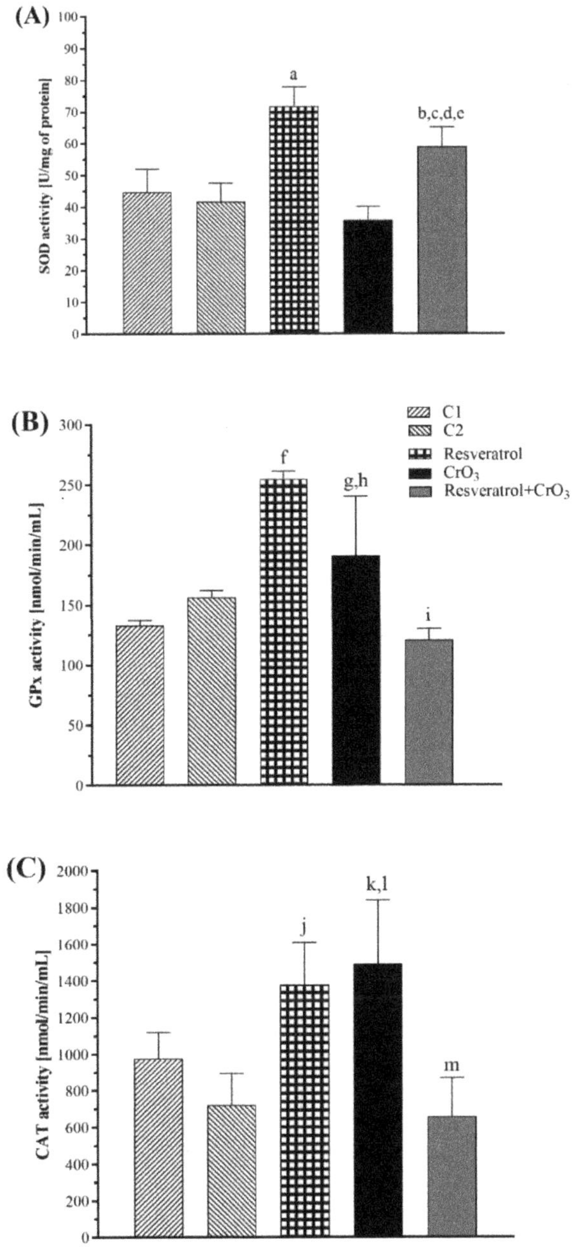

Figure 3. (**A**) Superoxide dismutase (SOD), (**B**) glutathione peroxidase (GPx), and (**C**) catalase (CAT) activities evaluated in peripheral blood at 48 h after treatments with resveratrol and CrO_3 (n = 5 mice/group). Statistical significance was determined using one-way ANOVA followed by Tukey's post-hoc test: ^{a} $p < 0.0001$ vs. C2; ^{b} $p < 0.016$ vs. C1; ^{c} $p < 0.001$ vs. C2; ^{d} $p < 0.019$ vs. resveratrol; ^{e} $p < 0.0001$ vs. CrO_3; ^{f} $p < 0.0001$ vs. C2; ^{g} $p < 0.019$ vs. C1; ^{h} $p < 0.004$ vs. resveratrol + CrO_3; ^{i} $p < 0.0001$ vs. resveratrol; ^{j} $p < 0.005$ vs. C2; ^{k} $p < 0.034$ vs. C1; ^{l} $p < 0.0002$ vs. resveratrol + CrO_3; ^{m} $p < 0.001$ vs. resveratrol. C1, Control 1, vehicle only (distilled water); C2, Control 2, vehicle only (ethanol 30%); CrO_3, chromium trioxide.

The GSH levels are shown in Figure 4. There was a significant effect of treatment on GSH levels ($p < 0.0001$). There were no significant differences in GSH levels between CrO_3-treated animals and any other group ($p > 0.05$), while resveratrol + CrO_3 mice had significantly lower GSH levels than the controls (C1, $p < 0.0001$; C2, $p < 0.0001$), the resveratrol group ($p < 0.0001$), and the CrO_3 only group ($p < 0.0001$).

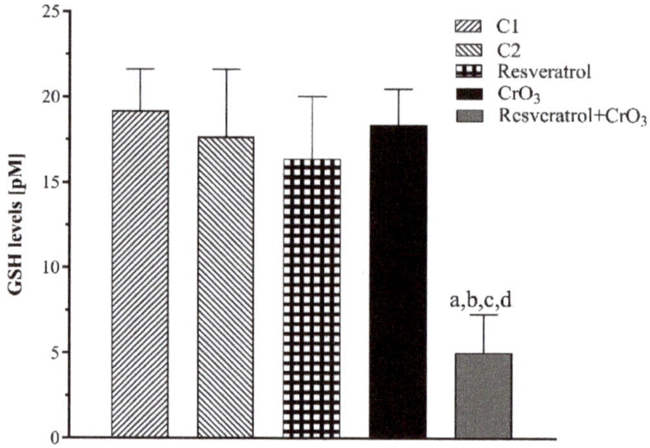

Figure 4. Average glutathione (GSH) levels evaluated in peripheral blood erythrocytes 48 h after treatment with resveratrol and CrO_3 (n = 5 mice/group). Statistical significance was determined using one-way ANOVA followed by Tukey's post-hoc test: ^{a} $p < 0.0001$ vs. C1; ^{b} $p < 0.0001$ vs. C2; ^{c} $p < 0.0001$ vs. resveratrol; ^{d} $p < 0.0001$ vs. CrO_3. C1, Control 1, vehicle only (distilled water); C2, Control 2, vehicle only (ethanol 30%); CrO_3, chromium trioxide.

2.4. *Effect of Resveratrol and CrO_3 on Apoptotic and Necrotic Cells*

Apoptotic and necrotic cells as well as cell viability were evaluated using differential acridine orange/ethidium bromide (AO/EB) staining (Figure 5A). The dual fluorochrome assay is capable of distinguishing between viable and nonviable cells based on membrane integrity. When cells are still viable, they keep the plasma membrane intact, allowing only AO to intercalate into DNA, which causes the nucleus to fluoresce green (Figure 5A(i,iii)). However, in nonviable cells, membrane integrity is lost, causing, ethidium bromide (EB) to also intercalate into DNA, making the nucleus fluoresce red since EB overwhelms AO staining (Figure 5A(ii,iv)). The color of the nucleus depends on the viability of the cell, not the state of the nucleus. Early apoptotic cells that have intact membranes but in which the DNA has begun to fragment still exhibit green nuclei because the EB cannot enter the cell, but chromatin condensation is visible as bright green patches in the nuclei (Figure 5A(iii)). As the cell progresses through the apoptotic pathway and membrane blebbing begins to occur, EB permeates the cell, producing a red-stained cell. Late apoptotic cells show bright red patches of condensed chromatin in the nuclei (Figure 5A(iv)); this distinguishes them from necrotic cells, which stain uniformly red (Figure 5A(ii)). When comparing the effect of treatments on apoptosis by one-way ANOVA, there was a significant effect of treatment on the frequency of healthy, total, early and late apoptotic cells as well as necrotic cells when compared to their control groups ($p < 0.0001$). Resveratrol reduced the frequency of total and early apoptotic cells compared to control groups ($p < 0.022$ and $p < 0.015$, respectively), while CrO_3 induced an increased number of total, early, late apoptotic, and necrotic cells compared to control groups ($p < 0.0001$). In the resveratrol + CrO_3 mice, there were fewer late apoptotic and necrotic cells compared to the CrO_3 group ($p < 0.001$ and $p < 0.0001$, respectively) and an increase in total and early apoptotic cells compared to the control group ($p < 0.0001$) and the resveratrol group ($p < 0.0001$) (Figure 5B).

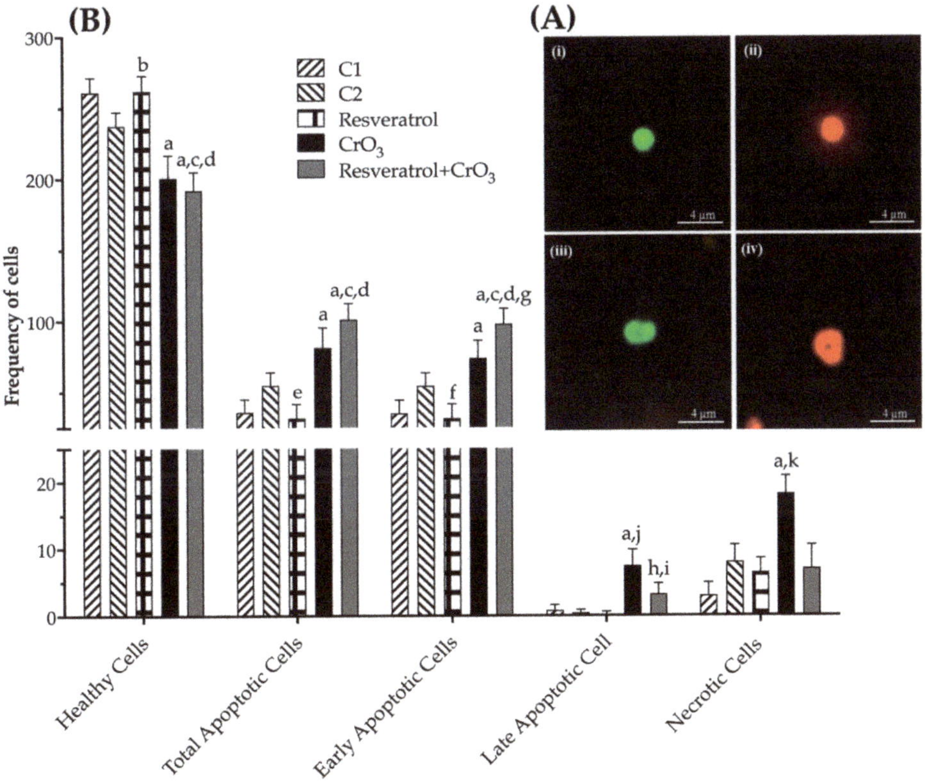

Figure 5. (**A**) Fluorescent microphotograph (400×) of peripheral blood cells using differential acridine orange/ethidium bromide (AO/EB) staining. (**i**) Healthy cell. (**ii**) Necrotic cell. (**iii**) Early apoptotic cell. (**iv**) Late apoptotic cell. (**B**) Effect of resveratrol and CrO_3 on the frequencies of healthy, apoptotic (total, early, and late), and necrotic cells in peripheral blood, evaluated 48 h after treatments. A total of 300 nucleated cells were evaluated in each mouse (n = 5 mice/group). Statistical significance was determined using one-way ANOVA followed by Tukey's post-hoc test: ^{a} $p < 0.0001$ vs. C1; ^{b} $p < 0.03$ vs. C2; ^{c} $p < 0.0001$ vs. C2; ^{d} $p < 0.0001$ vs. resveratrol; ^{e} $p < 0.022$ vs. C2; ^{f} $p < 0.015$ vs. C2; ^{g} $p < 0.011$ vs. CrO_3; ^{h} $p < 0.034$ vs. C2; ^{i} $p < 0.02$ vs. resveratrol; ^{j} $p < 0.001$ vs. resveratrol + CrO_3; ^{k} $p < 0.0001$ vs. resveratrol + CrO_3. C1, Control 1, vehicle only (distilled water); C2, Control 2, vehicle only (ethanol 30%); CrO_3, chromium trioxide.

Table 1 shows the PCE/NCE ratio. These evaluations were performed on the same samples and times used for MN. There were no significant effects in any of the treatments compared to their control groups (C1, C2) or time 0. However, when cell viability was compared in nucleated peripheral blood cells (48 h) using the dual fluorochrome assay, a significant effect of treatment on viable and nonviable cells ($p < 0.0001$) was observed (one-way ANOVA). The dual fluorochrome assay is an indicator of cell metabolism and death caused by cell membrane injury. Viable cells included those with an intact membrane, and thus they exhibited a nucleus fluoresced green by AO intercalation (healthy and early apoptotic cells; Figure 5A(i,iii), respectively). Moreover, nonviable cells included those in which the integrity of the membrane had been lost and that, therefore, presented a nucleus fluoresced red due to the intercalation of the EB (late apoptotic and necrotic cells; Figure 5A(ii,iv), respectively). Treatment with CrO_3 increased nonviable cells compared to the control group ($p < 0.0001$), while treatment with resveratrol prior to CrO_3 exposure decreased the nonviable cells observed in the group treated with CrO_3 alone ($p < 0.0001$). Resveratrol treatment alone had no significant effect on cell viability (Figure 6).

Table 1. PCE/NCE ratio in peripheral blood of mice treated with resveratrol and CrO_3.

Treatment	Dose (mg/kg)	Time Analysis (h)	n	PCE/NCE 1000 Cells (mean ± SD)
C1	0	0	5	48.5 ± 8.6
		24		46.7 ± 4.6
		48		48.2 ±11.9
		72		52.0 ± 8.2
C2	60	0	5	46.7 ± 7.0
		24		47.2 ± 9.6
		48		53.0 ± 7.5
		72		43.4 ±13.2
Resveratrol	50	0	5	54.8 ±12.2
		24		47.5 ± 9.5
		48		48.7 ± 5.9
		72		50.5 ± 6.4
CrO_3	20	0	5	46.5 ±10.0
		24		45.8 ± 7.0
		48		51.1 ±12.8
		72		39.8 ±12.9
Resveratrol + CrO_3	50 + 20	0	5	45.6 ± 7.9
		24		46.4 ± 2.8
		48		53.8 ±12.7
		72		44.6 ± 5.8

A total of 2000 erythrocytes were evaluated in each mouse (n = 5 mice/group). C1, Control 1, vehicle only (distilled water); C2, Control 2, vehicle only (ethanol 30%). CrO_3, chromium trioxide; PCE, polychromatic erythrocytes; NCE, normochromatic erythrocytes.

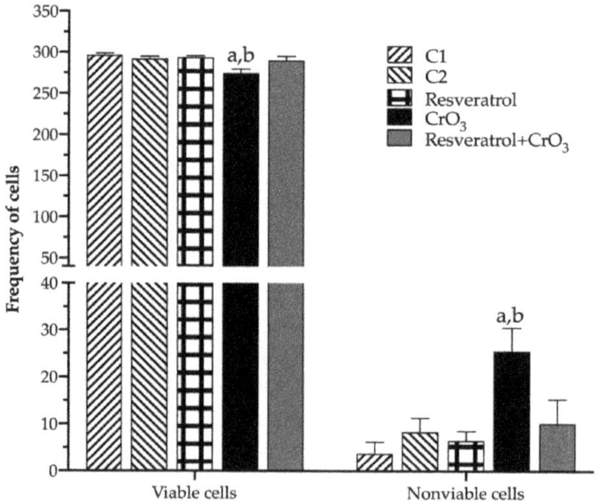

Figure 6. Effect of resveratrol and CrO_3 on the frequencies of viable and nonviable cells in the peripheral blood of mice evaluated 48 h after treatment using differential acridine orange/ethidium bromide (AO/EB) staining. Viable cells include healthy and early apoptotic cells. Nonviable cells include late apoptotic and necrotic cells. A total of 300 nucleated cells were evaluated for each mouse (n = 5 mice/group). Statistical significance was determined using one-way ANOVA followed by Tukey's multiple comparisons post hoc test: ^{a} $p < 0.0001$ vs. C1; ^{b} $p < 0.0001$ vs. resveratrol + CrO_3. C1, Control 1, vehicle only (distilled water); C2, Control 2, vehicle only (ethanol 30%); CrO_3, chromium trioxide.

The mice in the CrO_3 group showed clinical signs of toxicity, including bristling hair, decreased mobility, and loss of appetite. The dose of 50 mg/kg of resveratrol did not exhibit any apparent clinical signs of toxicity. None of the mice exposed to resveratrol, CrO_3, or both treatments died.

3. Discussion

The aim of this study was to (1) examine the protective effect of resveratrol against Cr(VI)-induced genotoxic damage in vivo and, (2) explore the possible protective pathways of resveratrol at the time of greatest induction of genotoxic damage by CrO_3 (48 h). The evaluations were carried out on the same peripheral blood samples, in which the protection from genotoxic damage (MN) was studied. The pathways explored were adduct repair 8-OHdG levels, antioxidant system GSH levels, and enzymatic activities of SOD, GPx, and CAT. Our findings showed that the administration of resveratrol 4 h prior to exposure to CrO_3 reduced the frequency of MN induced by this metal in Hsd:ICR male mice. Similarly, an approximation of the possible pathways involved in the protection of genotoxic damage induced by CrO_3 was achieved.

To evaluate the genotoxic damage attributed to Cr(VI), a dose of 20 mg/kg CrO_3 was administered intraperitoneally (ip). The 20 mg/kg dosage was based upon a previous study, in which this dose induced the formation of MN in the peripheral blood of mice [19,26]. Cr(VI) detoxification is relatively fast (no more than 48 h) when administered ip, in contrast to the effects observed with long-term oral and inhalation exposure to Cr(VI) [27]. Although the ip route is an artificial exposure route, it is useful for detecting genotoxic damage in short-term protocols, such as the MN assay, when testing compounds with potential clastogenic properties. Similarly, it is a more sensitive and direct route than inhalation or oral exposure [28,29]. Thus, a short-term protocol using the ip route of administration was selected to examine direct genotoxic damage induced by CrO_3. Resveratrol bioavailability studies, which have described peak plasma concentrations from approximately 1 to 6 h after treatment [30], also support the use of a short-term protocol to assess genotoxic damage.

The increase in MN is an indication of the genotoxic effects exerted by Cr(VI). According to the guidelines of the Organization for Economic Cooperation and Development (OECD) and the Environmental Protection Agency (EPA), a substance or compound that induces more than 4 MN/1000 PCE should be considered a genotoxic agent [29,31]. Our results are consistent with the genotoxic damage reported for Cr(VI) compounds and particularly CrO_3 [26,32], as increases greater than 5 MN were observed at all times evaluated. Administration of resveratrol 4 h before exposure to CrO_3 reduced these frequencies of MN in vivo. When evaluating the levels of 8-OHdG in blood plasma at 48 h after CrO_3 treatment, no significant alterations in adduct repair were detected. Notably, Maeng et al. [33] reported that inhalation of 18 mg/m^3 of sodium chromate resulted in significantly elevated 8-OHdG levels in the lungs after 1 week. However, after 2 weeks of exposure, this dose produced no significant differences in pulmonary 8-OHdG levels, with full recovery after 3 weeks. Maeng et al. [33] also demonstrated that inhalation of higher sodium chromate levels did not significantly alter pulmonary 8-OHdG levels. Similarly, Thompson et al. [34] noted that the in vitro genotoxicity of Cr(VI) is primarily oxidative in nature at low concentrations. They observed that 8-OHdG reaches non-cytotoxic concentrations at 24 h in cell cultures treated with different doses of sodium dichromate. It is conceivable that ip administration of 20 mg/kg CrO_3 might be too high to significantly affect 8-OHdG levels in the blood, in agreement with in vivo findings of Maeng et al. [33] and in vitro observations of Thompson et al. [34]. However, it should not be ruled out that Cr(VI) might reduce the levels of protein expression initiating DNA mismatch repair by inhibiting the hMLH1 and hMLH2 genes and the 8-oxoguanine DNA glycosylase1 (OGG1) repair enzyme involved in base excision repair (BER) [35–37]. Mice treated with resveratrol prior to CrO_3 showed an elevation in 8-OHdG levels. There are the following two possible explanations for these results: (1) resveratrol activated repair mechanisms that counteract oxidative damage in DNA, and/or (2) resveratrol contributed to the elimination of 8-OHdG adducts formed by the oxidative damage. The 8-OHdG levels are known to be related to the balance between oxidative DNA damage and the rate at which it is repaired [25]. Yan et al. [38] observed in vitro that resveratrol activates the BER pathway, increasing the expression of OGG1. Further, Mikuła-Pietrasik et al. [39] noted that resveratrol enhanced the activity of the repair enzyme OGG1 in senescent human cells. In our study, resveratrol provided approximately 50% protection against genotoxic damage from CrO_3 at all evaluation times. This effect might be attributed to the antioxidant properties of resveratrol, which enable this substance to interact with H_2O_2 and $O_2^{\bullet-}$ and •OH radicals [40]. Previously, Leonard et al. [41] demonstrated in vitro that resveratrol scavenged •OH in JB6 cells exposed to Cr(VI). In a previous study in vivo with (-)-epigallocatechin-3-gallate (EGCG), García-Rodríguez et al. [42] reported that 8-OHdG levels returned to control levels when EGCG and Cr(VI) were co-administered, contrasting with the findings of this study. They also found that co-administration of EGCG and Cr(VI) decreased the magnitude of MN increase compared to Cr(VI) alone [42], similar to the effect observed in this study with resveratrol. Hence, it is possible that when resveratrol was administered in combination with CrO_3, the repair mechanisms were enhanced by this polyphenol, contributing to a reduction in MN levels.

In the group treated with resveratrol and CrO_3, MN frequencies were reduced by 60, 51, and 46% at 24, 48, and 72 h, respectively. However, the reduction at 72 h was no longer significant. This may be due to the pharmacokinetics of CrO_3. It has been reported that Cr(VI) compounds might be excreted within 48 h of exposure [27,43]. Hence, the greatest damage to DNA occurs during that period. Another possibility is that the micronucleated PCE induced over 24 h matured into NCE by 72 h, such that these were not quantified at that time. When leaving the bone marrow, PCE degrades ribosomal RNA in 24 h [44]. In other studies, this same trend was also observed in the reduction of MN at 72 h after administration of Cr(VI) [19,45].

The evaluations of SOD, GPx, CAT, and GSH activities were performed in peripheral blood samples obtained at 48 h because this did not require sacrificing the animals, which was necessary to continue with the evaluation of MN kinetics in the same individuals. Further, in vitro studies showed that resveratrol's antioxidant properties may neutralize oxidative capacity in human erythrocytes [46], and SOD and GSH play important roles in the antioxidant system of erythrocytes [47]. In addition, altered functions of extracellular antioxidants may be assessed by the evaluation of antioxidant molecules in plasma [48]. SOD and GSH measurements were carried out in erythrocytes while CAT and GPx were carried out in plasma.

In the group treated with CrO_3, there was a decrease in SOD activity accompanied by an elevation in GPx and CAT activities, which is consistent with previous findings. Both in vitro [49] and in vivo studies demonstrated the effects of oral [50] and ip [42] administration of Cr(VI) compounds on endogenous antioxidants, such as activities of SOD, CAT, GPx, heme oxygenase-1 (HO-1) and levels of GSH. Matés [51] proposed that SOD plays an important role as a first-line antioxidant defense enzyme by catalyzing the dismutation of $O_2^{\bullet-}$ to form H_2O_2, which is subsequently reduced to H_2O by GPx and CAT. The decrease in SOD activity observed in the group treated with CrO_3 might be related to its depletion by reacting with the $O_2^{\bullet-}$ radicals that are generated in excess during reduction to Cr(III). Meanwhile, the H_2O_2 generated by the activity of SOD may be increased due to the activity of GPx and CAT that was observed in this group. Regarding GSH levels, no significant changes were observed with CrO_3 treatment, suggesting that reduction of Cr(VI) was not primarily mediated by this pathway. In this sense, in vivo studies noted that the reduction of Cr(VI) compounds is predominantly via nicotinamide adenine dinucleotide phosphate (NADPH) and other reducing agents such as ascorbate, cysteine, lipoic acid, fructose, and ribose [15]. In the group treated with resveratrol and CrO_3, GSH levels were significantly decreased. GSH is a primary antioxidant molecule that plays a fundamental role in reducing cellular oxidative stress. GSH might act in the following different ways: (1) directly as an electron donor by eliminating $O_2^{\bullet-}$, (2) through

GPx catalysis by reducing H_2O_2 levels, or (3) by forming complexes with detoxifying enzymes such as glutathione S-transferase (GST) [52]. GSH, GPx, and GST are some of the major antioxidant defense systems that scavenge ROS [53]. Resveratrol was found to induce phase II detoxification enzymes in in vitro and in vivo systems [54], and to increase GST activity in cultured aortic smooth muscle cells [55]. Further, several investigators have shown that polyphenols elevate GSH levels and stimulate the transcription of genes that are relevant for the synthesis of endogenous antioxidants, thus counteracting oxidative stress [56]. On the other hand, in this same group (resveratrol + CrO_3), the activities of GPx and CAT were restored, and SOD activity increased. Data suggest that resveratrol might counteract CrO_3-induced oxidative stress by an indirect antioxidant effect related to the regulation of the endogenous antioxidant system. In in vitro studies, Yao et al. [57] observed that resveratrol protected against oxidative damage induced by sodium sulfate dextran, while Chen et al. [8] found that polyphenols diminished damage mediated by sodium arsenite. In both cases, evidence indicated that the observed protection was related to increased SOD activity. SOD is a phase 2 enzyme that can be activated through the Nrf2/Keap1 signaling pathway. Nrf2 is a fundamental sensor of oxidative stress that plays a central role in the regulation of phase 2 antioxidant and detoxifying enzymes and related proteins [58]. Zhuang et al. [59] found that resveratrol regulates p-Nrf2 levels in a dose-dependent manner. This suggests that resveratrol attenuates the oxidative state, probably by activating the Nrf2 signaling pathway, which in turn elevates SOD activity. Resveratrol was also found to maintain the cellular redox balance by enhancing the activity of antioxidant enzymes, including HO-1, CAT, GPx, and SOD, in rat arterial endothelial cells [60]. Banu et al. [10] showed that 10 mg/kg resveratrol protected against potassium dichromate-induced oxidative stress in rat ovarian tissue by enhancing the activities of GPx, CAT, SOD, peroxiredoxin, and thioredoxin, and by lowering the concentration of H_2O_2. Therefore, it is possible that resveratrol also removed H_2O_2 generated by SOD activity, which reduced the need for GPx and CAT activation in the group treated with resveratrol and CrO_3. Nevertheless, it is important to keep in mind that the endogenous antioxidant system is dynamic, and thus, it is possible that our results may depend upon the evaluation time (48 h), the dose of resveratrol used, and the experimental model.

Although the PCE/NCE ratio is included in the MN assay as an indicator of cytotoxicity [29], no marked changes in the PCE/NCE ratio were observed in any of the treated groups in this study. These results need to be interpreted with caution, since when toxicity occurs during erythropoiesis, activation of cell division mechanisms may mask the effects on the PCE/NCE ratio [61]. For this reason, cytotoxicity was assessed by analyzing cell viability using the AO/EB differential staining technique, which allows us to distinguish between viable and non-viable cells according to the integrity of the membrane [45]. When cell viability, apoptosis, and necrosis were determined, treatment with CrO_3 significantly increased the numbers in total, early and late apoptotic cells, as well as necrotic and nonviable cells. These results corroborate the cytotoxicity reported for Cr(VI) compounds [26,27,45]. ROS generation and DNA damage induced by Cr(VI) exposure might play an essential role in cytotoxicity and the apoptotic signaling pathway [62]. It has been proposed that apoptosis induction is mediated by DNA damage sensors that directly activate p53 through proteins such as DNA-dependent kinase (DNA-PK) or indirectly through mutated ataxia-telangiectasia (ATM) and ATM-Rad3 (ATR) with Chk1 or Chk2 [63]. It has also been reported that apoptosis induction may be mediated through p53-independent pathways such as cleaved caspase 3 and cytochrome C [10,64]. However, when polyphenol was administered 4 h prior to CrO_3 in mice, a significant increase in early apoptotic cell number was observed compared to the group treated with CrO_3 alone. This effect was masked when early and late apoptotic cells were summed, as late apoptotic cells also showed a significant decrease in this group. Hence, it is possible that the enhanced induction of early apoptotic cells following combined CrO_3 resveratrol treatment may contribute to the elimination of cells containing Cr(VI)-induced DNA (MN) damage. Although there are no apparent studies of the effects of resveratrol on Cr(VI)-induced apoptosis pathways, resveratrol was reported to induce apoptosis as a mechanism of elimination of damaged cells in cancer cell lines [4]. Further, alterations in the expression of the Bcl2 protein, the loss of mitochondrial function, the release of cytochrome c, and the activation of caspases trigger the response for the activation of apoptosis [65]. Mirzapur et al. [66] reported that in breast cancer cells, resveratrol elevated the levels of the Bcl2/Bax protein, as well as the expression of p53 genes and caspases 3 and 8. Therefore, based upon these observations and the results reported in the present study, there is clearly a need to conduct studies that aim to reach a more detailed understanding of how resveratrol interacts with proteins such as p53, DNA-PK, ATM, ATR, Bax, Bcl2, caspases (3 and 8), among others. Indeed, these studies may greatly contribute toward understanding the mechanisms by which polyphenols such as resveratrol might contribute to the elimination of cells with genotoxic damage induced by compounds with carcinogenic potential such as Cr(VI).

The administration of resveratrol reduced the frequency of MN induced by CrO_3 and resveratrol treatment itself did not produce DNA damage (MN induction). The reduction in GSH and elevation in apoptotic cell number with both treatments (resveratrol + CrO_3), as well as increases in SOD, GPx, CAT, and 8-OHdG (the latter non-significant) with resveratrol alone, suggest a toxic effect. In in vitro and in vivo studies, it has been observed that resveratrol exhibits biphasic effects (antioxidant and prooxidant). Meira-Martin et al. [67] considered that the increases in SOD and SOD/CAT activity observed in vitro with different doses of resveratrol are generated to maintain the cellular redox balance. Hence, it has been proposed that its prooxidant activity contributes to the activation of the endogenous antioxidant system [68]. Sinha et al. [69] found that the prooxidative effects of resveratrol are associated with the generation of the $O_2^{\bullet-}$ radical, H_2O_2, and a complex mixture of semiquinones and quinones. However, in this study, no marked effects on viable cell numbers were observed in the group treated with resveratrol and CrO_3, and this group exhibited a significant decrease in necrotic cell frequency, suggesting that polyphenols diminished the cytotoxicity produced by CrO_3. Other in vivo studies also noted that resveratrol diminished the toxicity induced by metals, such as arsenic [70], copper, and zinc [71], contributing to the balance of the cellular redox system and reducing the expression of proinflammatory cytokines. On the other hand, the administration of resveratrol alone significantly decreased total and early apoptotic cells when compared to its control, suggesting that resveratrol alone does not induce toxicity and that it reduced the potential toxic effect of the vehicle (30% ethanol). Although ethanol is a less toxic polar vehicle than other vehicles such as dimethyl sulfoxide (DMSO) [72], ad libitum administration (11%) was shown to increase serum ROS in treated mice for 60 days [73]. Based on our results, it is suggested to extend these studies by using more diluted doses of resveratrol and even administering it in repeated doses, to reduce the possible toxic effects and improve the protection against the genotoxic damage observed in the present study.

4. Materials and Methods

4.1. Chemicals and Reagents

Cr(VI) [CrO_3, purity grade 99.9%; CAS 1333-82-0], 3,4′,5-trihydroxy-trans-stilbene [resveratrol, purity grade $\geq$ 98%; CAS 501-36-0], AO [CAS 10127-02-3], and EB [CAS 1239-45-8] were obtained from Sigma Chemical Co. (St. Louis, MO, USA).

4.2. Animals

A group of 25 adult male Hsd:ICR mice (8–12 weeks old, 28–35 g) were used in the experiment. The animals were obtained from Harlan® (Mexico City, CDMX, Mexico) at the "Facultad de Química, Universidad Nacional Autónoma de México-UNAM" and acclimated for two weeks prior to initiating the experiments. During the acclimation period, the groups of five mice were kept in a plastic cage at a controlled room temperature (22 $\pm$ 2 °C) with a 12-h light-dark cycle (the lights came on at 7:00 a.m. and went off at 7:00 p.m.). Mice had free access to food (Purina-Mexico®, Mexico City, CDMX, Mexico; small rodent chow) and water. Considering that in previous studies there were no differences between males and females in the genotoxic effects of CrO_3 administered by ip injection [16,26,45], this study was carried out using only male mice, in accordance with guidelines for the testing of chemicals (mammalian erythrocyte micronucleus test) of the OECD and the EPA [29,31].

The mice were randomly divided into five groups of five individuals each. Two control groups were used (C1: mice treated ip with sterile distilled water and C2: mice treated with 30% ethanol by gavage) because the CrO_3 solution was prepared by dissolving the compound in water, whereas resveratrol was dissolved in 30% ethanol. The resveratrol group was treated with a single dose of 50 mg/kg by gavage, and the CrO_3 group was treated with a single dose of 20 mg/kg ip. The last group received combined resveratrol and Cr(VI) treatments (resveratrol + CrO_3); the mice were treated with resveratrol at 50 mg/kg by gavage 4 h prior to CrO_3 ip injection (20 mg/kg).

The assessment was carried out on peripheral blood obtained from the tail vein since this does not require animals to be sacrificed.

4.3. *Micronuclei Assay*

For the MN evaluations, sequential peripheral blood samples (5 μL) were obtained from the same individuals (0 to 72 h), and 0–h samples were designated as a negative control. The samples were placed directly onto slides previously treated with AO, as described by Hayashi et al. [74]. Two slides were prepared for each mouse and were stored in the dark at 4 °C for 24 h. The assessments were performed by identifying PCE, NCE, and MN in PCE using a fluorescence microscope (Nikon™ OPTIPHOT-2; Tokyo, Japan) with blue excitation (480 nm) and a barrier filter emission (515–530 nm) at 100× magnification. MN analysis was based upon 4000 PCE per mouse, and the presence of MN was considered to indicate genotoxic damage. The relative proportion of PCE to NCE was also analyzed for 2000 erythrocytes.

In this study, underlying metabolic processes such as 8-OHdG adduct repair, endogenous antioxidant component system, apoptosis, and cell viability analyses were also evaluated because these processes may be involved in preventing Cr(VI)-induced DNA damage. These parameters were measured using the same peripheral blood samples obtained at 48 h after treatments, in which MN were measured since this is the time when the greatest genotoxic damage induced by Cr(VI) has been observed [26].

4.4. *Plasma 8-Hydroxydeoxyguanosine Levels*

Plasma 8-OHdG levels were determined using an enzyme-linked immunosorbent assay. Peripheral blood samples (50 μL) were centrifuged (15 min at 2500× *g*) at room temperature. The plasma was collected and immediately analyzed according to the manufacturer's instructions using Trevigen's HT 8-oxo-dG ELISA Kit II (No. 4380-192-K; Gaithersburg, MD, USA). The absorbance at 450 nm of each well was determined using a Multiskan™ FC microplate reader (Thermo Scientific™, Vantaa, Finland). Product formation is inversely proportional to the amount of 8-OHdG present in the sample. The 8-OHdG levels were determined in duplicate (per sample) and according to the 8-OHdG standard curve.

4.5. *Antioxidant System*

The antioxidant system was evaluated by determining the activities of SOD, GPx, and CAT, as well as the GSH levels.

4.5.1. Superoxide Dismutase Activity

SOD activity was evaluated in peripheral blood erythrocytes. Fifty-μL samples were diluted in phosphate-buffered saline. The erythrocytes were separated using FicollPaque™ (Sigma Chemical Co., St. Louis, MO, USA). (800× *g* for 25 min at 12 °C). The precipitate was separated, and cold distilled water (4 °C) was added (10:1). Then, it was incubated (0 °C for 15 min) to lyse. Hemoglobin was then precipitated by adding ethanol and chloroform (10,000× *g* for 10 min at 4 °C). The SOD activity was determined according to the manufacturer's instructions using Trevigen's HT Superoxide Dismutase Assay Kit (No. 7501-500-K; Gaithersburg, MD, USA). The absorbance was read at 450 nm at 1-min intervals for 10 min in a Multiskan™ FC microplate reader (Thermo Scientific™, Vantaa, Finland). One unit of SOD activity was defined as the amount of protein that inhibited tetrazolium salt (WST-1)-formazan, up to a maximum of 50%. The protein concentration was determined according to the instructions for the Cayman Chemicals Protein Determination Kit (No. 704002; Ann Arbor, MI, USA). SOD activity was measured in duplicate (per sample) and according to the SOD standard curve.

4.5.2. Glutathione Peroxidase Activity

GPx activity was detected in the plasma of peripheral blood. Twenty-five µL samples were centrifuged at 1000× *g* for 10 min at 4 °C. The plasma was collected and diluted with the GPx sample buffer (1:2) included in the kit. GPx activity was determined according to the instructions for the Cayman Chemicals Glutathione Peroxidase Assay Kit (No. 703102; Ann Arbor, MI, USA). One unit of GPx was defined as the amount of enzyme that oxidized 1 nmol of NADPH/min. The absorbance was read at 340 nm in a Multiskan™ FC microplate reader (Thermo Scientific™, Vantaa, Finland). GPx activity was determined in duplicate (per sample) and according to the GPx standard curve.

4.5.3. Catalase Activity

CAT activity was evaluated in the plasma of peripheral blood. Twenty-five µL samples were centrifuged at 1000× *g* for 10 min at 4 °C to obtain plasma. The activity of CAT was evaluated according to the instructions for the Cayman Chemicals Catalase Assay (No. 707002; Ann Arbor, MI, USA). This kit uses the peroxidation function of CAT to determine enzyme activity. The absorbance was read at 540 nm in a Multiskan™ FC microplate reader (Thermo Scientific™, Vantaa, Finland). One unit of CAT was defined as the amount of enzyme that induced the formation of 1 nmol of formaldehyde/min. CAT activity was measured in duplicate (per sample) and according to the CAT standard curve.

4.5.4. Glutathione Levels

GSH levels were evaluated in the erythrocytes of peripheral blood. Fifty µL samples were centrifuged at 3000× *g* for 15 min at 0 °C. The erythrocytes were suspended in 5% cold (*w*/*v*) metaphosphoric acid, mixed and stored at 0 °C for 15 min. Subsequently, the suspension was centrifuged at 14,000× *g* for 10 min at 4 °C. The clarified supernatant was collected, and GSH levels were analyzed according to the instructions using Trevigen's HT Glutathione Assay Kit (Item No. 7511-100-K; Gaithersburg, MD, USA). The absorbance was read at 405 nm in a Multiskan™ FC microplate reader (Thermo Scientific™, Vantaa, Finland). The levels of GSH were determined in duplicate (per sample) according to the GSH standard curve.

4.6. *Apoptosis and Cell Viability*

To evaluate apoptosis, necrosis, and cell viability, differential acridine AO/EB staining was performed using a technique previously adapted for peripheral blood [45]. Ten µL samples were centrifuged at 4500× *g* for 5 min. The cell pellet was resuspended in 20 µL of AO/EB dye mix and plated on a clean slide. Two slides were prepared per mouse, and the analysis was performed immediately. The assessments were based upon 300 cells per mouse. Apoptotic, necrotic, viable, and nonviable cells were identified using a fluorescence microscope OPTIPHOT-2 (Nikon™; Tokyo, Japan) with blue excitation (480 nm) and a barrier filter emission (515–530 nm) at 40× magnification.

4.7. *Statistical Analysis*

Each mouse was considered an independent replicate according to the OECD and EPA guidelines [29,31]. Individual samples were averaged for each experimental group. The MN frequencies, PCE/NCE ratio, viability (viable/nonviable cells), number of apoptotic and necrotic cells, levels of 8-OHdG and GSH, and activities of SOD, GPx, and CAT are expressed as the mean ± standard deviation (SD). The data were checked for normality using the Shapiro–Wilk test. Statistical significance between the groups for MN was determined by using two-way RM-ANOVA because MN depends on two factors (i.e., treatment and time). The treatment is independent, while the evaluations at each time are considered dependent since the samples were obtained from the same mouse. For the other parameters, one-way ANOVA was used because the evaluations depended only on one factor (i.e., treatment). In the analysis of ANOVA, post hoc Tukey multiple comparisons were carried out. GraphPad Prism 8.0 (GraphPad Software, San Diego, CA, USA) was used for all analyses. Differences were considered significant at $p < 0.05$.

5. Conclusions

Our findings demonstrate a protective effect of resveratrol against Cr(VI)-induced genotoxic damage, by reducing the frequency of MN induced by CrO_3 in vivo. Likewise, an approximation of the possible pathways involved in the protection of genotoxic damage induced by these compounds with carcinogenic potential, such as Cr(VI), was achieved. Resveratrol showed effects on the modulation of the endogenous antioxidant system, 8-OHdG adduct repair, and apoptosis when administered 4 h prior to Cr(VI) exposure. These effects suggest that these pathways might be involved in the protection provided by this polyphenol against genotoxic damage induced by Cr(VI). Although resveratrol treatment modified endogenous antioxidant system constituents, the dose of 50 mg/kg alone did not alter MN frequencies, suggesting that it is not related to the induction of DNA damage. In vivo studies using more diluted doses of resveratrol and even administering it in repeated doses, as well as direct evaluations in target organs, could help determine the specific mechanisms, by which resveratrol counteracts Cr(VI)-induced genotoxicity. These studies contribute to the understanding of the potential antigenotoxic value of polyphenols such as resveratrol, and to the exploration of their possible use as chemotherapeutic agents in the prevention and treatment of diseases related to genotoxic damage.

Author Contributions: Conceptualization, M.d.C.G.-R.; Methodology, T.N.-M.; Validation, M.d.C.G.-R., V.M.M.-N., A.R.O.-M.; Data Curation, T.N.-M., M.d.C.G.-R., V.M.M.-N. and A.R.O.-M.; Investigation, M.d.C.G.-R., T.N.-M.; Writing—Original Draft Preparation, TN-M.; Writing—Review and Editing, M.d.C.G.-R., S.K. and T.N.-M.; Supervision, M.d.C.G.-R., S.K., A.R.O.-M. and V.M.M.-N.; Project Administration, M.d.C.G.-R.; Funding Acquisition, M.d.C.G.-R. All authors have read and agreed to the published version of the manuscript.

Funding: This research was funded by DGAPA-UNAM, Support Program for Research and Technological Innovation Projects, PAPIIT-IN224719; IN216122. The National Council of Science and Technology granted a scholarship for postgraduate studies (CONACyT, No. 703847) to Tonancy Nicolás-Méndez.

Institutional Review Board Statement: The Bioethics Committee of the "Facultad de Estudios Superiores-Zaragoza, UNAM" approved the experimental conditions and protocols (Code: FESZ/DEPI/363/14; FESZ/DEPI/CE/016/21).

Informed Consent Statement: Not applicable.

Data Availability Statement: The data presented in this study are available on request from the corresponding author.

Conflicts of Interest: The authors declare no conflict of interest.

Sample Availability: Samples of the compounds are not available from the authors.

References

1. Rauf, A.; Imran, M.; Suleria, H.A.R.; Ahmad, B.; Peters, D.G.; Mubarak, M.S. A comprehensive review of the health perspectives of resveratrol. *Food Funct.* **2017**, *8*, 4284–4305. [CrossRef] [PubMed]
2. Nicolás-Méndez, T.; Ortiz-Muñiz, A.R.; Mendoza-Núñez, V.M.; García-Rodríguez, M.C. The role of resveratrol on heavy metal-induced oxidative stress. *Nutr. Hosp.* **2020**, *37*, 374–383. [CrossRef] [PubMed]

3. Gülçin, İ. Antioxidant properties of resveratrol: A structure–activity insight. *Innov. Food Sci. Emerg. Technol.* **2010**, *11*, 210–218. [CrossRef]
4. Repossi, G.; Das, U.N.; Eynard, A.R. Molecular Basis of the Beneficial Actions of Resveratrol. *Arch. Med. Res.* **2020**, *51*, 105–114. [CrossRef] [PubMed]
5. Quincozes-Santos, A.; Andreazza, A.C.; Nardin, P.; Funchal, C.; Gonçalves, C.A.; Gottfried, C. Resveratrol attenuates oxidative-induced DNA damage in C6 Glioma cells. *Neurotoxicology* **2007**, *28*, 886–891. [CrossRef]
6. Zhang, Y.; Guo, L.; Law, B.Y.; Liang, X.; Ma, N.; Xu, G.; Wang, X.; Yuan, X.; Tang, H.; Chen, Q.; et al. Resveratrol decreases cell apoptosis through inhibiting DNA damage in bronchial epithelial cells. *Int. J. Mol. Med.* **2020**, *45*, 1673–1684. [CrossRef] [PubMed]
7. Burkhardt, S.; Reiter, R.J.; Tan, D.X.; Hardeland, R.; Cabrera, J.; Karbownik, M. DNA oxidatively damaged by chromium(III) and H_2O_2 is protected by the antioxidants melatonin, N^1-acetyl-N^2-formyl-5-methoxykynuramine, resveratrol and uric acid. *Int. J. Biochem. Cell Biol.* **2001**, *33*, 775–783. [CrossRef]
8. Chen, C.; Jiang, X.; Zhao, W.; Zhang, Z. Dual role of resveratrol in modulation of genotoxicity induced by sodium arsenite via oxidative stress and apoptosis. *Food Chem. Toxicol.* **2013**, *59*, 8–17. [CrossRef]
9. Zhao, X.; Zhang, K.P.; Huang, T.; Yan, C.C.; Liu, L.R.; Zhu, Q.L.; Guo, F.F.; Liu, C.; Li, B.X. The rescuable function and mechanism of resveratrol on As_2O_3-induced hERG K^+ channel deficiency. *Naunyn Schmiedebergs Arch. Pharmacol.* **2014**, *387*, 1079–1089. [CrossRef]
10. Banu, S.K.; Stanley, J.A.; Sivakumar, K.K.; Arosh, J.A.; Burghardt, R.C. Resveratrol protects the ovary against chromium-toxicity by enhancing endogenous antioxidant enzymes and inhibiting metabolic clearance of estradiol. *Toxicol. Appl. Pharmacol.* **2016**, *303*, 65–78. [CrossRef]
11. Khalid, S.; Afzal, N.; Khan, J.A.; Hussain, Z.; Qureshi, A.S.; Anwar, H.; Jamil, Y. Antioxidant resveratrol protects against copper oxide nanoparticle toxicity in vivo. *Naunyn Schmiedebergs Arch. Pharmacol.* **2018**, *391*, 1053–1062. [CrossRef] [PubMed]
12. Rowbotham, A.L.; Levy, L.S.; Shuker, L.K. Chromium in the environment: An evaluation of exposure of the UK general population and possible adverse health effects. *J. Toxicol. Environ. Health B Crit. Rev.* **2000**, *3*, 145–178. [CrossRef] [PubMed]
13. EPA. Environmental Protection Agency. Toxicological review of hexavalent chromium. CAS No. 18540-29-9. In *Support of Summary Information on the Integrated Risk Information System (IRIS). External Review Draft*; EPA/635/R-10/004ª; Office of Research and Development: Washington, DC, USA, 2010; Volume 635. Available online: https://cfpub.epa.gov/ncea/iris_drafts/recordisplay.cfm?deid=221433# (accessed on 1 January 2022).
14. Shi, X.; Chiu, A.; Chen, C.T.; Halliwell, B.; Castranova, V.; Vallyathan, V. Reduction of chromium(VI) and its relationship to carcinogenesis. *J. Toxicol. Environ. Health B Crit. Rev.* **1999**, *2*, 87–104. [CrossRef] [PubMed]
15. Valko, M.; Rhodes, C.J.; Moncol, J.; Izakovic, M.; Mazur, M. Free radicals, metals and antioxidants in oxidative stress-induced cancer. *Chem. Biol. Interact.* **2006**, *160*, 1–40. [CrossRef]
16. O'Brien, T.J.; Ceryak, S.; Patierno, S.R. Complexities of chromium carcinogenesis: Role of cellular response, repair and recovery mechanisms. *Mutat. Res.* **2003**, *533*, 3–36. [CrossRef]
17. Mishra, S.; Bharagava, R.N. Toxic and genotoxic effects of hexavalent chromium in environment and its bioremediation strategies. *J. Environ. Sci. Health C Environ. Carcinog. Ecotoxicol. Rev.* **2016**, *34*, 1–32. [CrossRef]
18. Patlolla, A.K.; Barnes, C.; Yedjou, C.; Velma, V.R.; Tchounwou, P.B. Oxidative stress, DNA damage, and antioxidant enzyme activity induced by hexavalent chromium in Sprague-Dawley rats. *Environ. Toxicol.* **2009**, *24*, 66–73. [CrossRef]
19. García-Rodríguez, M.C.; Nicolás-Méndez, T.; Montaño-Rodríguez, A.R.; Altamirano-Lozano, M.A. Antigenotoxic effects of (-)-epigallocatechin-3-gallate (EGCG), quercetin, and rutin on chromium trioxide-induced micronuclei in the polychromatic erythrocytes of mouse peripheral blood. *J. Toxicol. Environ. Health A* **2014**, *77*, 324–336. [CrossRef]
20. de Freitas, K.S.; Squarisi, I.S.; Acésio, N.O.; Nicolella, H.D.; Ozelin, S.D.; Reis Santos de Melo, M.; Guissone, A.; Fernandes, G.; Silva, L.M.; da Silva Filho, A.A.; et al. Licochalcone A, a licorice flavonoid: Antioxidant, cytotoxic, genotoxic, and chemopreventive potential. *J. Toxicol. Environ. Health A* **2020**, *83*, 673–686. [CrossRef]
21. Sousa, H.G.; Uchôa, V.T.; Cavalcanti, S.; de Almeida, P.M.; Chaves, M.H.; Lima Neto, J.S.; Nunes, P.; da Costa Júnior, J.S.; Rai, M.; Do Carmo, I.S.; et al. Phytochemical screening, phenolic and flavonoid contents, antioxidant and cytogenotoxicity activities of *Combretum leprosum* Mart. (Combretaceae). *J. Toxicol. Environ. Health A* **2021**, *84*, 399–417. [CrossRef]
22. Gu, H.F.; Mao, X.Y.; Du, M. Prevention of breast cancer by dietary polyphenols-role of cancer stem cells. *Crit. Rev. Food Sci. Nutr.* **2020**, *60*, 810–825. [CrossRef]
23. Majolo, F.; Bitencourt, S.; Wissmann Monteiro, B.; Viegas Haute, G.; Alves, C.; Silva, J.; Pinteus, S.; Santos, R.; Torquato, H.; Paredes-Gamero, E.J.; et al. Antimicrobial and antileukemic effects: In vitro activity of *Calyptranthes grandifolia* aqueous leaf extract. *J. Toxicol. Environ. Health A* **2020**, *83*, 289–301. [CrossRef] [PubMed]
24. Wang, S.; Wang, H.; Wang, Y.; Chen, J.; Liu, J.; He, X.; Huang, D.; Wu, Y.; Chen, Y.; Weng, Z. Protective effects of (-)-epigallocatechin gallate and curcumin against acrylamide toxicity. *Toxicol. Environ. Chem.* **2021**, *103*, 199–218. [CrossRef]
25. Poulsen, H.E.; Nadal, L.L.; Broedbaek, K.; Nielsen, P.E.; Weimann, A. Detection and interpretation of 8-oxodG and 8-oxoGua in urine, plasma and cerebrospinal fluid. *Biochim. Biophys. Acta* **2014**, *1840*, 801–808. [CrossRef] [PubMed]
26. García-Rodríguez, M.C.; López-Santiago, V.; Altamirano-Lozano, M.A. Effect of chlorophyllin on chromium trioxide-induced micronuclei in polychromatic erythrocytes in mouse peripheral blood. *Mutat. Res.* **2001**, *496*, 145–151. [CrossRef]

27. Jomova, K.; Valko, M. Advances in metal-induced oxidative stress and human disease. *Toxicology* **2011**, *283*, 65–87. [CrossRef] [PubMed]
28. Hayashi, M.; Sutou, S.; Shimada, H.; Sato, S.; Sasaki, Y.F.; Wakata, A. Difference between intraperitoneal and oral gavage application in the micronucleus test. The 3rd collaborative study by CSGMT/JEMS.MMS. Collaborative Study Group for the Micronucleus Test/Mammalian Mutagenesis Study Group of the Environmental Mutagen Society of Japan. *Mutat. Res.* **1989**, *223*, 329–344. [CrossRef] [PubMed]
29. OECD. *Test No. 474: Mammalian Erythrocyte Micronucleus Test, OECD Guidelines for the Testing of Chemicals, Section 4*; OECD: Paris, France, 2016. [CrossRef]
30. Pannu, N.; Bhatnagar, A. Resveratrol: From enhanced biosynthesis and bioavailability to multitargeting chronic diseases. *Biomed. Pharmacother.* **2019**, *109*, 2237–2251. [CrossRef]
31. EPA. Environmental Protection Agency. Health Effects Test Guidelines OPPTS 870.5395. Mammalian Erythrocyte Micronucleus Test, Office of Prevention, Pesticides and Toxic Substances (7101). US. 1998. EPA 712-C-98-226. Available online: https://www.regulations.gov/document/EPA-HQ-OPPT-2009-0156-0032 (accessed on 1 January 2022).
32. Hu, G.; Long, C.; Hu, L.; Xu, B.P.; Chen, T.; Gao, X.; Zhang, Y.; Zheng, P.; Wang, L.; Wang, T.; et al. Circulating lead modifies hexavalent chromium-induced genetic damage in a chromate-exposed population: An epidemiological study. *Sci. Total Environ.* **2021**, *752*, 141824. [CrossRef]
33. Maeng, S.H.; Chung, H.W.; Yu, I.J.; Kim, H.Y.; Lim, C.H.; Kim, K.J.; Kim, S.J.; Ootsuyama, Y.; Kasai, H. Changes of 8-OH-dG levels in DNA and its base excision repair activity in rat lungs after inhalation exposure to hexavalent chromium. *Mutat. Res.* **2003**, *539*, 109–116. [CrossRef]
34. Thompson, C.M.; Fedorov, Y.; Brown, D.D.; Suh, M.; Proctor, D.M.; Kuriakose, L.; Haws, L.C.; Harris, M.A. Assessment of Cr(VI)-induced cytotoxicity and genotoxicity using high content analysis. *PLoS ONE* **2012**, *7*, e42720. [CrossRef] [PubMed]
35. Urbano, A.M.; Ferreira, L.M.; Alpoim, M.C. Molecular and cellular mechanisms of hexavalent chromium-induced lung cancer: An updated perspective. *Curr. Drug Metab.* **2012**, *13*, 284–305. [CrossRef] [PubMed]
36. Wise, S.S.; Wise, J.P.S. Chromium and genomic stability. *Mutat. Res.* **2012**, *733*, 78–82. [CrossRef]
37. Xia, H.; Ying, S.; Feng, L.; Wang, H.; Yao, C.; Li, T.; Zhang, Y.; Fu, S.; Ding, D.; Guo, X.; et al. Decreased 8-oxoguanine DNA glycosylase 1 (hOGG1) expression and DNA oxidation damage induced by Cr (VI). *Chem. Biol. Interact.* **2019**, *299*, 44–51. [CrossRef]
38. Yan, Y.; Yang, J.Y.; Mou, Y.H.; Wang, L.H.; Zhou, Y.N.; Wu, C.F. Differences in the activities of resveratrol and ascorbic acid in protection of ethanol-induced oxidative DNA damage in human peripheral lymphocytes. *Food Chem. Toxicol.* **2012**, *50*, 168–174. [CrossRef] [PubMed]
39. Mikuła-Pietrasik, J.; Kuczmarska, A.; Rubiś, B.; Filas, V.; Murias, M.; Zieliński, P.; Piwocka, K.; Książek, K. Resveratrol delays replicative senescence of human mesothelial cells via mobilization of antioxidative and DNA repair mechanisms. *Free Radic. Biol. Med.* **2012**, *52*, 2234–2245. [CrossRef] [PubMed]
40. Truong, V.L.; Jun, M.; Jeong, W.S. Role of resveratrol in regulation of cellular defense systems against oxidative stress. *Biofactors* **2018**, *44*, 36–49. [CrossRef]
41. Leonard, S.S.; Xia, C.; Jiang, B.H.; Stinefelt, B.; Klandorf, H.; Harris, G.K.; Shi, X. Resveratrol scavenges reactive oxygen species and effects radical-induced cellular responses. *Biochem. Biophys. Res. Commun.* **2003**, *309*, 1017–1026. [CrossRef]
42. García-Rodríguez, M.C.; Serrano-Reyes, G.; Hernández-Cortés, L.M.; Altamirano-Lozano, M.A. Antigenotoxic effects of (-)-epigallocatechin-3-gallate (EGCG) and its relationship with the endogenous antioxidant system, 8-hydroxydeoxyguanosine adduct repair (8-OHdG), and apoptosis in mice exposed to chromium(VI). *J. Toxicol. Environ. Health A* **2021**, *84*, 331–344. [CrossRef]
43. O'Flaherty, E.J. A pharmacokinetic model for chromium. *Toxicol. Lett.* **1993**, *68*, 145–158. [CrossRef]
44. Moras, M.; Lefevre, S.D.; Ostuni, M.A. From Erythroblasts to Mature Red Blood Cells: Organelle Clearance in Mammals. *Front. Physiol.* **2017**, *8*, 1076. [CrossRef] [PubMed]
45. García-Rodríguez, M.C.; Carvente-Juárez, M.M.; Altamirano-Lozano, M.A. Antigenotoxic and apoptotic activity of green tea polyphenol extracts on hexavalent chromium-induced DNA damage in peripheral blood of CD-1 mice: Analysis with differential acridine orange/ethidium bromide staining. *Oxidative Med. Cell Longev.* **2013**, *2013*, 486419. [CrossRef] [PubMed]
46. Gallardo, M.J.; Suwalsky, M.; Ramírez, D.; Tapia, J.; Sepulveda, B. Antioxidant effect of resveratrol in single red blood cells measured by thermal fluctuation spectroscopy. *Arch. Biochem. Biophys.* **2019**, *665*, 30–35. [CrossRef]
47. Nwose, E.U.; Jelinek, H.F.; Richards, R.S.; Kerr, P.G. Erythrocyte oxidative stress in clinical management of diabetes and its cardiovascular complications. *Br. J. Biomed. Sci.* **2007**, *64*, 35–43. [CrossRef]
48. Zeitz, J.O.; Mohrmann, S.; Fehse, L.; Most, E.; Helmbrecht, A.; Saremi, B.; Eder, K. Tissue and plasma antioxidant status in response to dietary methionine concentration and source in broilers. *J. Anim. Physiol. Anim. Nutr.* **2018**, *102*, 999–1011. [CrossRef] [PubMed]
49. Wani, P.A.; Hussaini, N.A.; Garba, S.H.; Wahid, S.; Damilola, F.K.; Adeola, A.A.; Wasiu, I.A. Prospective of chromium (VI) reduction under in vitro and in vivo conditions and stimulation of antioxidant defense of cowpea under the exposure of Cr (VI). *Appl. Soil Ecol.* **2018**, *132*, 187–193. [CrossRef]
50. Wang, X.F.; Xing, M.L.; Shen, Y.; Zhu, X.; Xu, L.H. Oral administration of Cr(VI) induced oxidative stress, DNA damage and apoptotic cell death in mice. *Toxicology* **2006**, *228*, 16–23. [CrossRef]

51. Matés, J.M. Effects of antioxidant enzymes in the molecular control of reactive oxygen species toxicology. *Toxicology* **2000**, *153*, 83–104. [CrossRef]
52. Franco, R.; Cidlowski, J.A. Apoptosis and glutathione: Beyond an antioxidant. *Cell Death Differ.* **2009**, *16*, 1303–1314. [CrossRef]
53. Upadhyay, G.; Singh, A.K.; Kumar, A.; Prakash, O.; Singh, M.P. Resveratrol modulates pyrogallol-induced changes in hepatic toxicity markers, xenobiotic metabolizing enzymes and oxidative stress. *Eur. J. Pharmacol.* **2008**, *596*, 146–152. [CrossRef]
54. Chow, H.H.; Garland, L.L.; Hsu, C.H.; Vining, D.R.; Chew, W.M.; Miller, J.A.; Perloff, M.; Crowell, J.A.; Alberts, D.S. Resveratrol modulates drug- and carcinogen-metabolizing enzymes in a healthy volunteer study. *Cancer Prev. Res.* **2010**, *3*, 1168–1175. [CrossRef] [PubMed]
55. Li, Y.; Cao, Z.; Zhu, H. Upregulation of endogenous antioxidants and phase 2 enzymes by the red wine polyphenol, resveratrol in cultured aortic smooth muscle cells leads to cytoprotection against oxidative and electrophilic stress. *Pharmacol. Res.* **2006**, *53*, 6–15. [CrossRef] [PubMed]
56. Rodrigo, R.; Gil-Becerra, D. Implications of polyphenols on endogenous antioxidant defense systems in human diseases. In *Polyphenols in Human Health and Disease*; Watson, R.R., Preedy, V.R., Zibadi, S., Eds.; Academic Press: San Diego, CA, USA, 2014; pp. 201–217. [CrossRef]
57. Yao, J.; Wang, J.Y.; Liu, L.; Li, Y.X.; Xun, A.Y.; Zeng, W.S.; Jia, C.H.; Wei, X.X.; Feng, J.L.; Zhao, L.; et al. Anti-oxidant effects of resveratrol on mice with DSS-induced ulcerative colitis. *Arch. Med. Res.* **2010**, *41*, 288–294. [CrossRef] [PubMed]
58. Lau, W.L.; Liu, S.M.; Pahlevan, S.; Yuan, J.; Khazaeli, M.; Ni, Z.; Chan, J.Y.; Vaziri, N.D. Role of Nrf2 dysfunction in uremia-associated intestinal inflammation and epithelial barrier disruption. *Dig. Dis. Sci.* **2015**, *60*, 1215–1222. [CrossRef]
59. Zhuang, Y.; Wu, H.; Wang, X.; He, J.; He, S.; Yin, Y. Resveratrol Attenuates Oxidative Stress-Induced Intestinal Barrier Injury through PI3K/Akt-Mediated Nrf2 Signaling Pathway. *Oxidative Med. Cell Longev.* **2019**, *2019*, 7591840. [CrossRef]
60. Ungvari, Z.; Orosz, Z.; Rivera, A.; Labinskyy, N.; Xiangmin, Z.; Olson, S.; Podlutsky, A.; Csiszar, A. Resveratrol increases vascular oxidative stress resistance. *Am. J. Physiol. Heart Circ. Physiol.* **2007**, *292*, H2417–H2424. [CrossRef]
61. Krishna, G.; Hayashi, M. In vivo rodent micronucleus assay: Protocol, conduct and data interpretation. *Mutat. Res.* **2000**, *455*, 155–166. [CrossRef]
62. Hu, G.; Zheng, P.; Feng, H.; Jia, G. Imbalance of oxidative and reductive species involved in chromium(VI)-induced toxic effects. *React. Oxyg. Species* **2017**, *3*, 1–11. [CrossRef]
63. Chiu, A.; Shi, X.L.; Lee, W.K.; Hill, R.; Wakeman, T.P.; Katz, A.; Xu, B.; Dalal, N.S.; Robertson, J.D.; Chen, C.; et al. Review of chromium (VI) apoptosis, cell-cycle-arrest, and carcinogenesis. *J. Environ. Sci. Health C Environ. Carcinog. Ecotoxicol. Rev.* **2010**, *28*, 188–230. [CrossRef]
64. Wu, Y.H.; Lin, J.C.; Wang, T.Y.; Lin, T.J.; Yen, M.C.; Liu, Y.H.; Wu, P.L.; Chen, F.W.; Shih, Y.L.; Yeh, I.J. Hexavalent chromium intoxication induces intrinsic and extrinsic apoptosis in human renal cells. *Mol. Med. Rep.* **2020**, *21*, 851–857. [CrossRef]
65. Takashina, M.; Inoue, S.; Tomihara, K.; Tomita, K.; Hattori, K.; Zhao, Q.L.; Suzuki, T.; Noguchi, M.; Ohashi, W.; Hattori, Y. Different effect of resveratrol to induction of apoptosis depending on the type of human cancer cells. *Int. J. Oncol.* **2017**, *50*, 787–797. [CrossRef]
66. Mirzapur, P.; Khazaei, M.R.; Moradi, M.T.; Khazaei, M. Apoptosis induction in human breast cancer cell lines by synergic effect of raloxifene and resveratrol through increasing proapoptotic genes. *Life Sci.* **2018**, *205*, 45–53. [CrossRef]
67. Martins, L.A.; Coelho, B.P.; Behr, G.; Pettenuzzo, L.F.; Souza, I.C.; Moreira, J.C.; Borojevic, R.; Gottfried, C.; Guma, F.C. Resveratrol induces pro-oxidant effects and time-dependent resistance to cytotoxicity in activated hepatic stellate cells. *Cell Biochem. Biophys.* **2014**, *68*, 247–257. [CrossRef]
68. Shaito, A.; Posadino, A.M.; Younes, N.; Hasan, H.; Halabi, S.; Alhababi, D.; Al-Mohannadi, A.; Abdel-Rahman, W.M.; Eid, A.H.; Nasrallah, G.K.; et al. Potential Adverse Effects of Resveratrol: A Literature Review. *Int. J. Mol. Sci.* **2020**, *21*, 2084. [CrossRef]
69. Sinha, D.; Sarkar, N.; Biswas, J.; Bishayee, A. Resveratrol for breast cancer prevention and therapy: Preclinical evidence and molecular mechanisms. *Semin. Cancer Biol.* **2016**, *40–41*, 209–232. [CrossRef]
70. Yu, M.; Xue, J.; Li, Y.; Zhang, W.; Ma, D.; Liu, L.; Zhang, Z. Resveratrol protects against arsenic trioxide-induced nephrotoxicity by facilitating arsenic metabolism and decreasing oxidative stress. *Arch. Toxicol.* **2013**, *87*, 1025–1035. [CrossRef]
71. Asadi, S.; Moradi, M.N.; Khyripour, N.; Goodarzi, M.T.; Mahmoodi, M. Resveratrol Attenuates Copper and Zinc Homeostasis and Ameliorates Oxidative Stress in Type 2 Diabetic Rats. *Biol. Trace Elem. Res.* **2017**, *177*, 132–138. [CrossRef]
72. Marques, F.Z.; Morris, J.B. Commentary on resveratrol and hormesis: Resveratrol—A hormetic marvel in waiting? *Hum. Exp. Toxicol.* **2010**, *29*, 1026–1028. [CrossRef]
73. Petrella, C.; Carito, V.; Carere, C.; Ferraguti, G.; Ciafrè, S.; Natella, F.; Bello, C.; Greco, A.; Ralli, M.; Mancinelli, R.; et al. Oxidative stress inhibition by resveratrol in alcohol-dependent mice. *Nutrition* **2020**, *79–80*, 110783. [CrossRef]
74. Hayashi, M.; Morita, T.; Kodama, Y.; Sofuni, T.; Ishidate, M., Jr. The micronucleus assay with mouse peripheral blood reticulocytes using acridine orange-coated slides. *Mutat. Res.* **1990**, *245*, 245–249. [CrossRef]

molecules

MDPI

Review

Phenolic Compounds' Occurrence in *Opuntia* Species and Their Role in the Inflammatory Process: A Review

Walid Zeghbib [1,2], Fares Boudjouan [3,4], Vitor Vasconcelos [2,5] and Graciliana Lopes [2,*]

1 Laboratoire de Biochimie Appliquée, Faculté des Sciences de la Nature et de la Vie, Université de Bejaia, Bejaia 06000, Algeria; walidzeghbib1993@gmail.com
2 CIIMAR—Interdisciplinary Centre of Marine and Environmental Research, Terminal de Cruzeiros do Porto de Leixões, Avenida General Norton de Matos s/n, 4450-208 Matosinhos, Portugal; vmvascon@fc.up.pt
3 Laboratoire de Génie de l'Environnement, Faculté de Technologie, Université de Bejaia, Bejaia 06000, Algeria; fares0501@gmail.com
4 Faculté des Sciences de la Nature et de la Vie, Université de Bejaia, Bejaia 06000, Algeria
5 FCUP—Faculty of Sciences, University of Porto, Rua do Campo Alegre s/n, 4169-007 Porto, Portugal
* Correspondence: glopes@ciimar.up.pt; Tel.: +351-223401830

Abstract: Within the *Cactaceae* family, *Opuntia* comprises the most widespread species, with a recognized importance in human life, including feeding, domestic use as home natural barriers, and as a traditional remedy for diverse diseases and conditions such asthma, edema, and burns. Indeed, scientific reports have stated that these health benefits may be due to various active compounds, particularly polyphenols, which are ubiquitously found in plants and have proven their pharmacological efficiency by displaying antimicrobial, anti-cancer, and anti-inflammatory activities, among others. *Opuntia* species contain different classes of phenolic compounds that are recognized for their anti-inflammatory potential. Among them, quercetin, isorhamnetin, and kaempferol derivates were reported to greatly contribute to modulate cells' infiltration and secretion of soluble inflammatory mediators, with key implications in the inflammatory process. In this review, we make a summary of the different classes of phenolic compounds reported in *Opuntia* species so far and explore their implications in the inflammatory process, reported by in vitro and in vivo bioassays, supporting the use of cactus in folk medicine and valorizing them from the socio-economic point of view.

Keywords: *Opuntia* sp.; prickly pear; phenolic compounds; flavonoids; inflammation

Citation: Zeghbib, W.; Boudjouan, F.; Vasconcelos, V.; Lopes, G. Phenolic Compounds' Occurrence in *Opuntia* Species and Their Role in the Inflammatory Process: A Review. *Molecules* **2022**, *27*, 4763. https://doi.org/10.3390/molecules27154763

Academic Editor: Nour Eddine Es-Safi

Received: 5 July 2022
Accepted: 18 July 2022
Published: 25 July 2022

1. Introduction

Cacti belong to the plant family *Cactaceae*, which includes around 1500 species with recognized and very high phenotypic variations. Although the taxonomical characterization of the species is not always easy [1,2], the main subfamilies are currently well classified (Figure 1). Among them, the genus *Opuntia*, from the *Opuntioideae* subfamily, is the most recognized, being widely distributed across the globe [3,4]. *Opuntia* cacti (prickly pear) are among the plants with greater recognition in quotidian life, with archeological evidence encompassing the analysis of human coprolites dating back between 6500 to 10,000 years ago and suggesting its use as foodstuff [5,6]. The first discovery of cacti dates back to the times of the conquest of the new world by Spanish conquistadors who, besides being amazed by its attractive and delicious fruits, also noticed its granted economic and cultural importance in the daily life of the ancient Mesoamerican population [7,8] (Figure 2). After this, the plant became widely spread worldwide through cultivation and trade, nowadays being abundant in many arid and semi-arid regions of America, Africa, Asia, Europe, and Oceania [9,10].

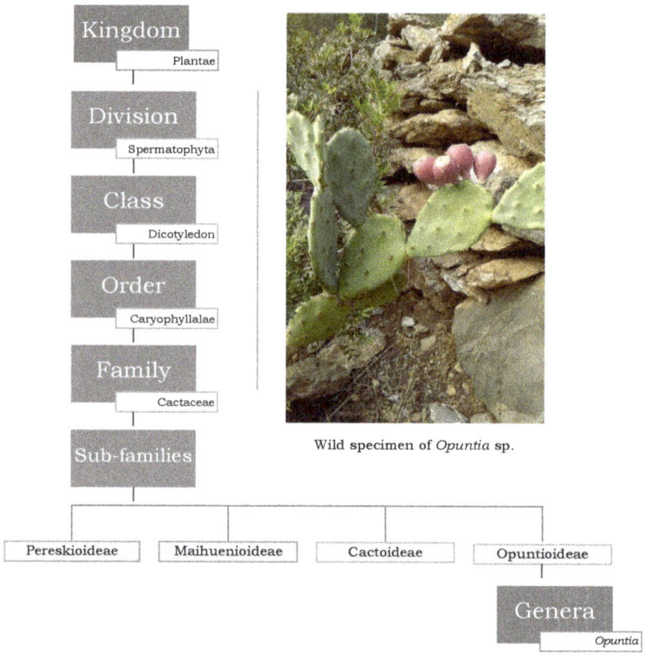

Figure 1. Taxonomic classification of *Opuntia* species (photograph from the author (W.Z.): wild *Opuntia* sp., Bejaia, Algeria).

Figure 2. Representation of the emblem of Tenochtitlan from the Codex Mendoza, with the prickly pear as the center of the universe. Photo: © Bodleian Libraries, University of Oxford; Shelfmark: Bodleian Library MS. Arch. Selden. A. 1; Holding Institution: Bodleian Libraries, University of Oxford; Terms of use: CC-BY-NC 4.0.

The fast growth of cactus species and their good adaptation to poor soil make them very important plants for these populations, being part of their nourishment, livestock feed, and as a home natural barrier [11,12]. In addition, cacti became important in folk medicine for their capacity to alleviate diverse health conditions, such as diarrhea, asthma, hemorrhoids, ulcers, burns, and edema [2,13,14]. These beneficial properties to human health led to an increase in scientific research focused in cacti plants, particularly on species of the genus *Opuntia*, due to their recognized pharmacological properties [6]. Indeed, it was revealed that all parts of cacti (flowers, fruits, cladodes, and peels) constitute undeniable sources of valuable nutritional elements and biologically active primary and secondary metabolites, such as vitamins, carotenoids, betalains, polyunsaturated fatty acids, and polyphenols, leading people to consider them as important functional foods, with interesting nutraceutical and pharmacological properties [15–17].

Phenolic compounds, including flavonoids and phenolic acids, are ubiquitous molecules found in nature, particularly in plants, with more than 8000 compounds described so far, and divided into different classes [18]. Among other factors, the phenolics' qualitative and quantitative profiles vary with plants' genus, species, ripeness, cultivar, growth region, and kind of plant tissue [19–21]. The literature has reported a multitude of phenolic compounds in all *Opuntia* species [6,22], with a particular prevalence of phenolic acids and flavonoids, such as dihydroquercetin, quercetin, isorhamnetin, and kaempferol, known for their efficient antioxidant activity and ability to protect human organisms from the deleterious effects of free radicals through diverse mechanisms of action. It is widely known that oxidative stress appears as a consequence of tilting the balance in favor of free radicals compared to the antioxidant system. This imbalance stands at the base of many diseases, including different types of cancer, arteriosclerosis, myocardial infarction, diabetes, inflammatory diseases, central nervous system disorders, and cells' aging [23,24]. In addition to the antioxidant power of polyphenols, recent investigations also recognized their antimicrobial, hepato-protective, anti-carcinogenic, and anti-inflammatory properties [22,25]. As a matter of fact, research devoted to phenolic compounds' valorization is ever present in the majority of natural matrices, from the optimization of the extraction processes to their biotechnological exploitation in food and cosmetic industries and elucidation of mechanisms of action in a wide array of pharmacological targets. Following the traditional use of cacti in acute health conditions where inflammation plays a central role, inflammatory mediators and enzymes appear among the main targets. Inflammation is a body's natural response to a pathogen invasion, toxin, or physical damage (chemical or traumatic), which involves the generation of a wide array of inflammatory mediators, such as reactive oxygen species (ROS), by inflammatory and immune cells. When the inflammatory process is uncontrolled or when the endogenous defense systems fail to establish homeostasis, inflammation can become chronic, leading to tissue damage and often preceding the establishment of chronic diseases [26,27]. There are many pharmacological treatments for inflammation based on steroidal and non-steroidal compounds; however, they present significant undesirable side effects and resistances, leading to an increasing interest in the search for bioactive compounds from natural sources as a potential effective and alternative non-pharmacological approach [28]. In this regard, the present work provides a general review on the different classes of phenolic compounds found in *Opuntia* sp. and on the anti-inflammatory activity reported for the genus so far.

2. Phenolic Compounds

2.1. General Overview

After cellulose, phenolic compounds represent the most abundant group of secondary metabolites of the plant kingdom. This large family ranges from simple compounds with low molecular weight to large and complex polyphenols mainly found conjugated with sugars and organic acids [29]. In plants, phenolic compounds are biosynthesized by the shikimate pathway, which is localized in the chloroplasts. These aromatic molecules have important roles in plants, being implicated in the regulation of their growth, signaling, defense, and in conferring color to their fruits, leaves, and flowers [24,30,31]. Their chemical structure is characterized by the presence of at least one aromatic ring containing one or more hydroxyl groups [32]. From the chemical point of view, phenolic compounds are characterized by an acidic behavior, since the oxygen of the hydroxyl function is strongly linked to the ring, while the connection to the hydrogen atom is weak, allowing the proton dissociation into the medium and giving origin to a negatively charged phenolate ion [33].

The medicinal properties reported for polyphenols over the years aroused scientists' interest in improving their extraction methodologies by using different solvents and extraction methods, the most widely explored being infusion, decoction, and maceration as well as Soxhlet, ultrasound, and microwave-assisted extraction [34]. There are more than 8000 phenolic compounds described in plants, with a high structural variability [18,35]. According to the classification system followed by De la Rosa et al. [30], phenolic compounds can be divided according to their chemical structure into two main classes: flavonoids and non-flavonoids. The first category is characterized by its structure complexity and known for its efficient bioactivity, accounting for nearly two-thirds of dietary polyphenols [30]. Its basic structure consists of a 15-carbon structure with two phenyl rings (A and B) connected by a three-carbon bridge, forming a heterocyclic pyran ring (ring C) skeleton (Figure 3a). The differences in the pyran ring substituents and the extent of hydrogenation allows defining six subcategories: flavones, flavonols, flavanols, isoflavones, flavanones, and anthocyanidins [29,31].

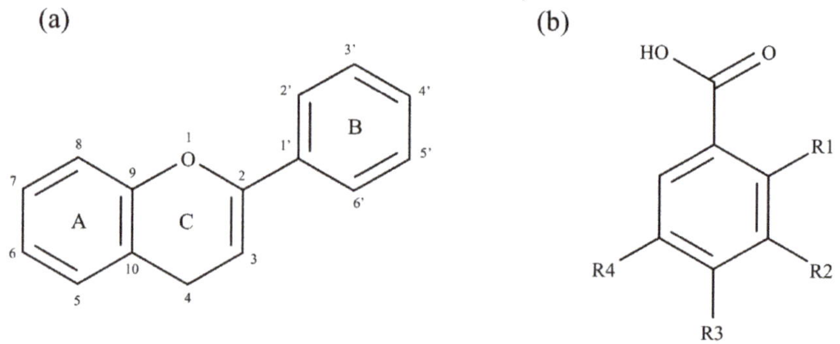

Figure 3. Basic structures of flavonoids (**a**) and phenolic acids (**b**).

The non-flavonoids' category includes smaller and simpler compounds. The principal molecules in this category are phenolic acids, particularly present in fruits and vegetables, accounting for one-third of dietary phenolic compounds. Phenolic acids are structurally composed by a single benzene unit, substituted by one carboxylic group and at least one hydroxyl group. Thus, many compounds can be considered simple phenols, and they are generally classified according to the number of carbons they have, the most common being hydroxybenzoic acids with a basic skeleton C_6–C_1 (Figure 3b) [30,33]. However, the non-flavonoids' category also includes other compounds with a complex structure and with a high molecular weight, which are characteristic and major components of some plants, such as lignans, chalcones, and stilbenes [30].

2.2. Phenolic Compounds' Occurrence in Opuntia sp.

Different classes of phenolic compounds can be found in cactus plants, with different qualitative and quantitative profiles, which mainly depends on environmental factors, plant origin, species, developmental stage, and age [2]. From the literature, there are no records about a specific extraction method or the most suitable solvent for polyphenols' recovery from *Opuntia* sp. samples; hence, researchers tend to choose the procedure that suits them according to their objectives and equipment availability. Accordingly, due to the polarity of polyphenols, some high polar solvents have been generally used, such as methanol, acetone, ethanol, water, or even mixtures between them [36,37]; several extraction methods have been followed, including conventional ones, such as maceration, infusion, and decoction [10,38,39], and non-conventional ones, such as microwave-assisted extraction and ultrasonic-assisted extraction [40,41].

The occurrence of phenolic compounds in cactus has been described in seeds, flowers, cladodes, pulp, and peel, as detailed in Table 1. According to Santos-Díaz et al. [6], the phenolic profile in *Opuntia* genera is complex, with more than 40 compounds described in both pulp and cladodes and more than 20 in seeds of different species. It seems important to highlight that the majority of the available reports do not specify the phenolic composition of each vegetative tissue of *Opuntia* sp., which makes it difficult to establish a defined qualitative profile. However, some compounds such as delphinidin, petunidin, and malvidin seem to be exclusively found in *Opuntia* sp. cladodes, while phloretin, psoralen, pinoresinol, and epigallocatechin are found in seeds.

Furthermore, the significant differences between species reported in the literature make it difficult to establish the qualitative and quantitative phenolic profiles of *Opuntia* sp. These differences are mostly devoted to the different extraction methods followed by the authors, which include both the use of solvents of different polarities, different extraction times, temperatures, and equipment. Differences in the *Opuntia* raw material used for extraction are also worth considering: phenolics' extractions can be performed using dry or fresh material, of different maturation states, and from different geographic locations, which modifies the abiotic factors to which species are exposed and, consequently, their phenolic profile. For instance, Chahdoura et al. [60] found that ferulic acid derivatives were the most abundant compounds in *Opuntia* sp. seeds, reaching about 0.36 and 0.95 mg/g for *Opuntia microdasys* (Lehm.) N.E. Pfeiffer and *Opuntia macrorhiza* Engelm., respectively, while Amrane-Abider et al. [46] found a higher chlorogenic acid content, with 0.89 mg/g in *Opuntia ficus-indica* L. (Mill) seeds. The most abundant phenolic acid from the pulp of different varieties of *O. ficus-indica* found by the authors was piscidic acid, with 8.70–22.31 mg/g, and quercetin was the most abundant flavonoid with 0.08–0.26 mg/g (dry weight, DW). Contrarily, Zenteno-Ramírez et al. [17] reported gallic acid and epicatechin as the most abundant compounds in the pulp of different *Opuntia* species. In the peel of different *O. ficus-indica* varieties, García-Cayuela et al. [57] found that piscidic acid was the most abundant phenolic acid with 27.53–44.62 mg/g DW and isorhamnetin derivates were the most representative flavonoids (1.48–2.54 mg/g DW). For the cladodes, some reports indicated that quinic acid and myricetin were the most common in *Opuntia dillenii* (Ker Gawl.) Haw. cladodes [39,56], while Missaoui et al. [41] reported for *O. ficus-indica* a higher abundance for piscidic acid and isorhamnetin derivates, with 9.67 mg/g and 3.93 mg/g, respectively. Regarding the flowers, a study by Chahdoura et al. [59] on the flowering stage of *O. microdasys* demonstrated that ferulic acid and isorhamnetin derivates were the most frequent, ranging from 1.24–2.95 mg/g and 4.68–23.04 mg/g, respectively, while Ammar et al. [65] found a higher content in quinic acid and quercetin derivates, with 1.32 and 8.50 mg/g, respectively, for *O. ficus-indica* flowers. Moreover, Ouerghemmi et al. [54] reported that ferulic acid and quercetin were in higher amounts when compared to other phenolic compounds in *O. ficus-indica* flowers. It seems evident, based on the available studies, that besides the abiotic factors, the species-specific ones have a significant role in phenolics' concentration and distribution throughout the different plant tissues. It seems difficult to establish a tissue fingerprint for *Opuntia* sp. since both flavonoids and phenolic acids present a wide distribution throughout all the studied plant parts. Based on the available studies, it seems that seeds present the widest variety of phenolic compounds, while the lowest variability has been observed for flowers. It was also observed that anthocyanidins and hydroxycinnamic acid were almost exclusive of cladodes; however, the fact that this was reported in only one study is not enough to state it as a tissue fingerprint or a species-specific characteristic. The same line of thought can be followed for phloretin, psoralen, and pinoresinol, which were only reported in the seeds of *Opuntia stricta* (Haw.) Haw.

Table 1. Phenolic compounds reported in the different vegetative parts of the genus *Opuntia* [1].

Phenolic Compounds	Plant Tissue	Concentration (μg/g)	*Opuntia* Species	References
Flavonoids				
Flavones				
Apigenin	Seeds Cladodes Flowers	NS 0.19–0.65 NS	*O. stricta* *O. ficus-indica* *O. hyptiacantha* *O. streptacantha* *O. megacantha* *O. albicarpa*	[38,42–44]
Luteolin	Pulp Peel	NS NS	*O. ficus-barbarica* *O. robusta*	[45]
Flavonols				
Myricetin	Seeds Pulp Peel Cladodes	198.19–428.14 NS NS 8.52	*O. ficus-indica* *O. ficus-barbarica* *O. robusta*	[45–47]
Rutin	Seeds Pulp Peel Cladodes	8.00–100.00 9.70–12.50 65.70–103.40 2.11–4.95	*O. ficus-indica* *O. ficus-barbarica* *O. hyptiacantha* *O. streptacantha* *O. megacantha* *O. albicarpa*	[42,44,45,48,49]
Quercetin and derivates	Seeds Pulp Peel Cladodes Flowers	4.37–18.77 84.20–599.20 715.70–1316.20 8.97–75.13 NS	*O. ficus-indica* *O. ficus-barbarica* *O. robusta* *O. engelmannii* *O. streptacantha* *O. hyptiacantha* *O. megacantha* *O. albicarpa*	[12,38,42,44,45,47,49–58]
Kaempferol and derivates	Pulp Peel Cladodes Flowers	207.10–529.10 52.90–675.50 72.97–241.68 321.00–708.00	*O. ficus-indica* *O. engelmannii* *O. streptacantha* *O. hyptiacantha* *O. megacantha* *O. albicarpa* *O. microdasys*	[38,47–50,52–55,57,59]
Isorhamnetin and derivates	Seeds Pulp Peel Cladodes Flowers	67.14–288.58 29.30–58.40 1484.70–2213.70 1250.00–4140.00 NS	*O. ficus-indica* *O. microdasys* *O. stricta* *O. streptacantha* *O. hyptiacantha* *O. megacantha* *O. albicarpa*	[12,19,38,44,46–48,50–57,59,60]
Flavanones				
Naringenin	Pulp Peel	210.00 20.00–180.00	*O. ficus-indica* *O. ficus-barbarica* *O. robusta*	[45,56]
Flavanols				
Catechin	Seeds Pulp Peel Cladodes Flowers	NS 14.44–27.89 NS 180.00 NS	*O. stricta* *O. ficus-indica* *O. megacantha* *O. streptacantha* *O. robusta*	[17,38,43,49,52,54,61]
Epicatechin	Seeds Pulp Peel	NS 19.16–90.81 NS	*O. ficus-indica* *O. albicarpa* *O. megacantha* *O. streptacantha* *O. robusta*	[17,42,61]
Gallocatechin	Seeds Pulp Peel	NS 116.60–178.20 120.40–334.70	*O. stricta* *O. ficus-indica*	[43,49]
Epigallocatechin	Seeds	NS	*O. stricta* *O. ficus-indica*	[42,43]
Anthocyanidins				
Pelargonidin	Seeds Cladodes	NS 187.97	*O. stricta* *O. ficus-indica*	[43,47]

Table 1. *Cont.*

Phenolic Compounds	Plant Tissue	Concentration (μg/g)	*Opuntia* Species	References
Flavonoids				
Flavones				
Cyanidin	Seeds Cladodes	NS 1058.57	*O. stricta* *O. ficus-indica*	[43,47]
Delphinidin	Cladodes	2.81	*O. ficus-indica*	[47]
Petunidin	Cladodes	186.55	*O. ficus-indica*	[47]
Malvidin	Cladodes	4.31	*O. ficus-indica*	[47]
Phenolic Acids				
Gallic acid and derivates	Seeds Pulp Peel Cladodes Flowers	NS 32.60–81.20 NS 20.53–38.96 NS	*O. ficus-indica* *O. stricta* *O. ficus-barbarica* *O. robusta* *O. albicarpa* *O. megacantha* *O. streptacantha* *O. hyptiacantha*	[17,42–45,49,54,61–63]
Ferulic acid and derivates	Seeds Pulp Peel Cladodes Flowers	96.33–1366.24 80.00 150.00–390.00 130.00–370.00 291.00–786.00	*O. ficus-indica* *O. stricta* *O. ficus-barbarica* *O. microdasys* *O. hyptiacantha* *O. streptacantha* *O. megacantha* *O. albicarpa*	[12,43–46,48,49,51,52,55,56,59,61,64]
Caffeic acid and derivates	Seeds Pulp Peels Cladodes Flowers	NS NS NS NS 255.00–469.00	*O. ficus-indica* *O. ficus-barbarica* *O. robusta* *O. microdasys* *O. hyptiacantha* *O. streptacantha* *O. megacantha* *O. albicarpa*	[12,38,42,44,45,48,49,51,54,59,61]
Sinapic acid	Seeds Pulp Peel Cladodes	NS 100.00–4100.00 820.00–2350.00 40.00–750.00	*O. stricta* *O. ficus-indica*	[43,49,56]
p-Coumaric acid	Seeds Pulp Peel Cladodes Flowers	NS NS NS 20.91 65.00–178.00	*O. ficus-indica* *O. ficus-barbarica* *O. robusta* *O. microdasys* *O. hyptiacantha* *O. streptacantha* *O. megacantha* *O. albicarpa*	[42,44,45,48,49,52,59]
Hydroxycinnamic acid	Cladodes	8.45–1248.24	*O. ficus-indica* *O. hyptiacantha* *O. streptacantha* *O. megacantha* *O. albicarpa*	[44,47]
Chlorogenic acid	Seeds Cladodes	885.31–1148.41 5.00–26.49	*O. ficus-indica* *O. streptacantha* *O. hyptiacantha* *O. megacantha* *O. albicarpa*	[42,44,46,50,52]
Ellagic acid	Seeds Pulp Peel	73.74–74.38 25.00–73.20 NS	*O. ficus-indica* *O. megacantha* *O. streptacantha* *O. robusta* *O. ficus-indica*	[17,46,61]
Vanillic acid	Seeds Pulp Peel Cladodes Flowers	NS NS NS 0.11–24.30 NS	*O. stricta* *O. ficus-barbarica* *O. robusta* *O. ficus-indica* *O. hyptiacantha* *O. streptacantha* *O. megacantha* *O. albicarpa*	[43–45,49,54,61]

Table 1. *Cont.*

Phenolic Compounds	Plant Tissue	Concentration (μg/g)	*Opuntia* Species	References
Syringic acid	Seeds Pulp Peel Cladodes Flowers	NS 13.60–66.50 NS 2.34–13.99 NS	*O. ficus-indica* *O. robusta* *O. albicarpa* *O. megacantha* *O. streptacantha* *O. hyptiacantha* *O. stricta*	[12,17,42–45,54,61]
Protocatechuic acid	Seeds Pulp Peel	4.57–22.36 NS NS	*O. ficus-indica* *O. ficus-barbarica* *O. robusta* *O. stricta*	[45,46,49,58,61–63]
Hydroxybenzoic acid	Pulp Peel Cladodes	200.90–816.80 964.00–1718.20 114.01	*O. ficus-indica* *O. hyptiacantha* *O. streptacantha* *O. megacantha* *O. albicarpa*	[44,47,49,57,62]
Piscidic acid	Seeds Pulp Peel Cladodes	NS NS NS NS	*O. ficus-indica* *O. stricta*	[48,51,53,55,57,58,63]
Eucomic acid	Seeds Pulp Peel Cladodes	NS NS NS NS	*O. ficus-indica* *O. streptacantha* *O. hyptiacantha* *O. megacantha* *O. albicarpa* *O. stricta*	[48,50,51,53,55,58,63]
Gentisic acid	Pulp Peel	NS NS	*O. ficus-barbarica* *O. robusta*	[45]
Rosmarinic acid	Peel Flowers	NS NS	*O. ficus-indica*	[49,54]
Catechol	Seeds Pulp Peel	NS NS NS	*O. stricta* *O. ficus-barbarica* *O. robusta* *O. ficus-indica*	[43,45,61]
Other Phenolics				
Phloretin Psoralen Pinoresinol	Seeds	NS	*O. stricta*	[43]

[1] NS, not specified.

3. *Opuntia* sp. in Inflammation

Inflammation is a physiological, self-limiting process occurring in mammalian tissues as a response to harmful situations such as microorganism invasion, physical damage, or exposition to toxic chemicals. The inflammatory process tends to eliminate primary triggers and contributes to initiating the regeneration of injured tissues by mediating an organized immune response, involving particularly macrophages and mast cells [66,67]. However, in some situations, the mechanisms involved in restoring tissues' homeostasis fail, generating a deregulated response that often results in a chronic inflammatory response, which is ever present in a wide variety of diseases and metabolic disorders such as diabetes, obesity, cancer, arthritis, and neurodegenerative and cardiovascular diseases [66,68].

The inflammatory framework involves a complex cascade of events with a coordinated action between pro- and anti-inflammatory mediators and biological systems including different cell lines (macrophages, neutrophils) and signaling molecules [69,70]. The NF-κB transcription factors have been long recognized for constituting a prototypical pro-inflammatory signaling pathway. In fact, these proteins are normally retained in the cytoplasm, being bound to a class of inhibitory proteins known as the IκB family. However, after stimulation, the activation of specific enzymes, known as IκB kinases (IKK), may phosphorylate the inhibitory protein, leading to the dissociation of the IκB/NF-κB complex. This results in a proteasomal degradation of IκB protein, while NF-κB can then translocate to the nucleus, binding DNA and activating the transcription of some targeted genes for cytokines, such as tumor necrosis factor-α (TNF-α) and interleukins (IL) (IL-1β and IL-6), as well as the production of several enzymes including cyclooxygenase (COX) and lipoxygenase (LOX) (Figure 4) [71–73]. These latter enzymes have a key role in the inflammatory process through the transformation of the arachidonic acid released from the phospholipid membrane into a spectrum of pro-inflammatory bioactive mediators including prostanoids and leukotrienes, which act by enhancing edema formation, increasing vascular permeability and leukocytes' infiltration into the injured tissue [74–76].

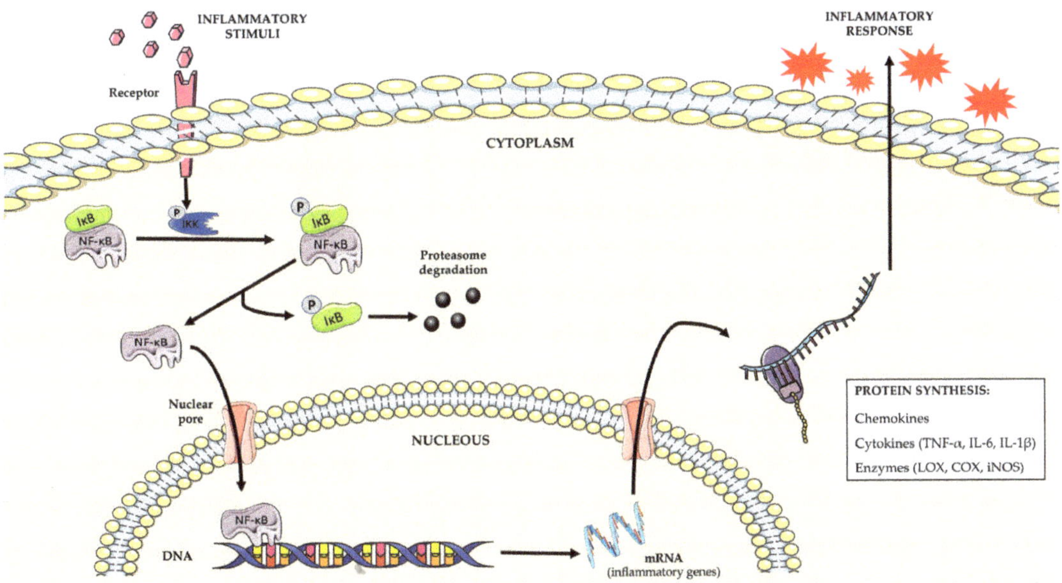

Figure 4. The implication of NF-κB signaling pathway in inflammation. The inflammatory stimulation of a cell may lead to IKK phosphorylation and activation which, in turn, may lead to the phosphorylation of the IκB/NF-κB complex retained in the cytoplasm. The IκB is degraded by the proteasome, while the NF-κB transcription factor can enter into the nucleus and bind DNA to initiate the transcription of some targeted genes implicated in the inflammatory response.

Over the years, many studies have been conducted in order to explore the potential targets of *Opuntia* sp. extracts and isolated phenols regarding the inflammatory process. From the anti-inflammatory investigations of *Opuntia* sp. reported in the literature, some species were the subject of greater scrutiny: *O. stricta*, *Opuntia humifusa* (Raf.) Raf., *Opuntia elatior* Mill., *O. dillenii*, and, mostly, *O. ficus-indica* [65,77–81]. Different parts of the plant (flowers, cladodes, seeds, and fruits) were explored, and different extraction methods and solvents were used according to the desired purpose. However, most of the works exploring the anti-inflammatory potential of *Opuntia* sp. were conducted using crude extracts, and only a few of them provided information on the anti-inflammatory potential of isolated phenolic molecules, especially flavonoids from *Opuntia* sp., such as isorhamnetin and kaempferol derivates [77,82,83].

3.1. Modulation of Inflammatory Mediators and Enzymes

Most of the available studies exploring the anti-inflammatory activity of *Opuntia* sp. extracts and isolated compounds were undertaken in vitro and explored specific mediators and enzymes involved in the inflammatory process (Table 2). Among them, nitric oxide (NO), produced by the inducible nitric oxide synthase (iNOS) upon inflammatory stimuli, is among the most explored. Gómez-Maqueo and co-workers [84] reported that the pulps and peels of two varieties of prickly pears presented anti-inflammatory potential by scavenging NO radicals. Indeed, besides being implicated in many physiological processes, the high levels of NO also play a key role in the pathogenesis of inflammation, by upregulating iNOS and pro-inflammatory cytokines' production (TNF-α and IL-8), leading at the end to serious tissue damage [78,85]. Following this train of thought, the reduction in the inflammatory response can benefit from the well-known free radical scavenging ability of phenolic compounds. The same authors followed their study and found that the different prickly pears' parts were also able to inhibit hyaluronidase activity. This enzyme is implicated in both physiological and pathological processes by hydrolyzing hyaluronic acid (HA), one of the most important compounds in the extracellular matrix, with more than 50% found in human skin [86]. The degradation of HA leads to the breakdown of tissues' structural integrity and, consequently, to an increase in their permeability, favoring the progression of inflammatory mediators. Thus, a balanced regulation of hyaluronic acid metabolism is important to maintain normal tissue organization and structure of the extracellular matrix. On the other hand, an over-activation of hyaluronidase may contribute to degenerative changes in connective tissues; therefore, enzyme inhibitors may play a beneficial role during the inflammatory process, presenting potential beneficial health effects such as hindering the exacerbation of the inflammatory response [86–88]. Similarly, another study of Gómez-Maqueo et al. [89], using other varieties of prickly pears, explored the same bioactivities in vitro, with a notably higher efficiency for peels when compared to pulps. The authors attributed this behavior to some compounds found in the species under study, particularly to indicaxanthin and to the phenolic compounds isorhamnetin glycosides, kaempferol glycoside, and quercetin, which mainly contributed to the observed anti-inflammatory potential by presenting a higher hyaluronidase inhibitory activity when compared to other purified standards.

A study conducted by Chaalal et al. [90] demonstrated that polyphenols extracted from different parts of *O. ficus-indica* fruits (seeds, pulp, and entire fruit) presented an anti-inflammatory potential and could exert a neuroprotective effect by decreasing the transcriptional expression of pro-inflammatory mediators such as TNF-α, IL-1β, and iNOS in N13-microglial cells after lipopolysaccharide (LPS) stimulation, pointing out the potential health benefit of these compounds in case of neuronal damage. In another study, Cho et al. [78] reported that chloroform and ethyl acetate fractions of *O. humifusa* cladodes were able to decrease NO production in LPS-stimulated RAW 264.7 macrophages, pointing to quercetin as one of the main compounds responsible for this behavior. Moreover, they also noticed that both fractions differently modulated cytokines' gene expression, especially for iNOS, IL-6, and IL-1β, suggesting the implication of other bioactive components from the species in the anti-inflammatory potential, while Yeo et al. [91] showed that methanolic extracts obtained from the seeds of the same species reduced NO production from LPS-stimulated macrophages' RAW 264.7. The isolated compound americanin A was found to be responsible for reducing iNOS and pro-inflammatory cytokines' (TNF-α and IL-6) expression levels, which resulted principally by preventing NF-κB translocation into the nucleus.

Table 2. Anti-inflammatory potential and mechanism of action of phenolic compounds extracted from *Opuntia* species [1].

	Species (Tissue)	Compounds	Dose	Model	Mechanism of Action	Ref.
In Vitro Studies	*Opuntia ficus-indica* (seeds, pulp, fruits)	Phenolic compounds from crude extracts	10 mg/mL	LPS-stimulated murine N13 microglial cells	- Downregulation of TNF-α, IL-1β, and iNOS expression	[90]
	Opuntia humifusa (cladodes)	Phenolic compound from crude extracts	0.05/0.1 mg/mL	LPS-stimulated Macrophages' RAW 264.7	- ↓ NO production - Downregulation of iNOS, IL-1β, and IL-6 genes' expression	[78]
	Opuntia humifusa (seeds)	Isoamericanin A	1.0–4.0 μg/mL	LPS-stimulated Macrophages' RAW 264.7	- ↓ TNF-α, IL-6, and iNOS expression levels - ↓ NF-Kb levels in the nucleus by inhibition of IκB phosphorylation	[91]
	Opuntia ficus-indica (cladodes)	Crude extract	10 mg/mL	Human intestinal Caco-2/TC7 cells	- ↓ NO, TNF-α, and IL-8 production - Intracellular reduction in reactive species - ↓ Prostaglandin synthesis	[92]
	Opuntia ficus-indica (fruits)	Polyphenols from crude extract	0.05 mg/mL	Human colon carcinoma Caco-2 cells	- ↓ H_2O_2-induced reactive species - Prevention of H_2O_2–induced protein oxidation - Cell protection from barrier dysfunction - ↓ NO, TNF-α, and IL-8 secretion - ↓ IκBα depletion	[82]
In Vivo Studies	*Opuntia dillenii* (stems, flowers, fruits)	Kaempferol 3-*O*-α-arabinoside Isorhamnetin-3-*O*-β-D-glucopyranoside Isorhamnetin-3-*O*-β-D-rutinoside	50 mg/kg BW	Carrageenan-induced paw edema in Albino rats	- ↓ Edema formation	[77]
	Opuntia ficus-indica (flowers)	Phenolic compounds from crude extract	400 mg/kg BW	Carrageenan-induced paw edema in Wistar rats	- ↓ Edema formation - ↓ Amount of immune cells - Neutralization of lipid peroxidation induced by reactive species - ↑ CAT, SOD, and GSH activities	[65]
	Opuntia ficus-indica (cladodes)	Isorhamnetin-3-*O*-glucosyl-rhamnoside Isorhamnetin-3-*O*-glucosyl-rhamnosyl-rhamnoside	5 mg/kg BW	Carrageenan-induced air-pouch inflammation in Wistar rats	- ↓ Edema formation - ↓ Total leukocytes' amount - Inhibition of COX-2 activity - ↓ NO, TNF-α, and IL-6 production	[83]

[1] BW, body weight; CAT, catalase; COX, cyclooxygenase; GSH, glutathione; H_2O_2, hydrogen peroxide; IL, interleukin; iNOS, inducible nitric oxide synthase; LPS, lipopolysaccharide; NO, nitric oxide; SOD, superoxide dismutase; TNF, tumor necrosis factor.

In a study conducted with the intestinal Caco-2/TC7 cell line, Filannino and co-workers [92] reported that raw material and fermented extracts from *O. ficus-indica* cladodes presented an anti-inflammatory effect by significantly reducing NO and chemokines' (IL-8 and TNF-α) production, which are important effectors in the inflammatory process contributing to the recruitment and activation of different inflammatory cells. In addition to decreasing the intracellular ROS generated during cells' stimulation, flavonoids (especially kaempferol and isorhamnetin), were considered to be the main responsible for this anti-inflammatory modulation. They displayed a significant decrease in prostaglandin E2 accumulation, which is a pro-inflammatory product from COX-2 and prostaglandin synthase metabolism, generally implicated in promoting local vasodilatation, and attraction and regulation of different immune cells' functions. Similarly, a study conducted by Matias et al. [82] found that flavonoid-rich concentrate from *O. ficus-indica* fruits prevented oxidative stress through the neutralization of H_2O_2-induced free radicals and also prevented protein oxidation in inflamed Caco-2 cells. Otherwise, the extract allowed protecting the intestinal barrier dysfunction, which was correlated with the ability of some flavonoids to decrease TNF-α secretion. They also found that the incubation of inflamed Caco-2 cells with the extract significantly modulated cytokines' secretion, leading particularly to a decrease in IL-8 and NO production, which are linked to the activation of the NF-κB pathway. Indeed, it was proven that the extract reduced the degradation of IκBα, an important inhibitor of NF-κB, preventing its migration from cytosol to the nucleus, where it could promote the transcription of pro-inflammatory genes.

3.2. *Anti-Inflammatory Activity In Vivo*

Most of the in vivo works exploring the anti-inflammatory activity of *Opuntia* species were conducted using rats as animal models, following the carrageenan-induced inflammation method, which is frequently used to evaluate the anti-edematous effect of natural products (Table 2) [79].

A study by Ahmed et al. [77] reported the in vivo anti-inflammatory potential of different parts of *O. dillenii* with a high efficiency verified for the flowers, from which three isolated compounds, namely, kaempferol 3-*O*-α-arabinoside, isorhamnetin-3-*O*-β-D-glucopyranoside, and isorhamnetin-3-*O*-β-D-rutinoside, were characterized as the active principles contributing significantly to reducing paw edema in albino rats. Ammar et al. [65] also reported the anti-inflammatory potential for *Opuntia* sp. flowers; the authors found that the methanolic extracts of *O. ficus-indica* flowers exhibited an anti-inflammatory potential by reducing the paw edema size in Wistar rats, with the same efficiency as the non-steroidal anti-inflammatory drug indomethacin. This effect was confirmed by a significant decrease in the number of inflammatory cells, including leukocytes and lymphocytes, a decrease in malondialdehyde (MDA) levels, which is correlated with the decrease in the lipid peroxidation process occurring at inflammatory sites, and a restoration of some antioxidant enzymes' activities, including superoxide dismutase (SOD), catalase (CAT), and glutathione (GSH), which contribute to neutralize free radicals' overproduction. According to the phytochemical analysis, the authors suggested the implication of phenolic compounds, particularly quercetin, isorhamnetin, and kaempferol, which could scavenge free radicals and decrease inflammation. Moreover, Antunes-Ricardo et al. [83] found that *O. ficus-indica* cladodes' extract and its isolated isorhamnetin derivatives (isorhamnetin-3-*O*-glucosyl-rhamnosyl-rhamnoside and isorhamnetin-3-*O*-glucosyl-rhamnoside) decreased the amount of neutrophils' infiltration into the inflammatory site (carrageenan-induced air-pouch inflammation in rats) with a decrease in NO production more efficient than that obtained with the standard drug used, indomethacin. The authors also found that cladodes' extract and isolated compounds were able to inhibit COX-2 activity and cytokines' production, particularly TNF-α and IL-6, with a better efficiency for the crude sample, probably due to a synergistic effect of its different phytochemicals.

4. Conclusions

The present review shows the richness of *Opuntia* species as producers of a wide variety of phenolic compounds, with an important role in the inflammatory process. The anti-inflammatory studies conducted until now demonstrated the benefit of different species to reduce the oxidative stress occurring at the site of an injury, decreasing the amount of neutrophils' infiltration, as well as pro-inflammatory mediators' production, such as NO, TNF-α, and interleukins. These effects were correlated with the presence of flavonoids in the different tissues of cactus, namely, quercetin, isorhamnetin, and kaempferol derivates, as the important bioactive components. Even though the *Opuntia* genus regroups a lot of species throughout the world, the available studies are limited to a few of them, with *O. ficus-indica* being by far the most explored. Nevertheless, reports on cactus plants reveal their potential anti-inflammatory application in the pharmaceutical industry, supporting the traditional use of these species in folk medicine and enhancing their economic value worldwide and for local communities.

Author Contributions: Conceptualization, G.L.; methodology, G.L. and W.Z.; investigation, W.Z.; writing—original draft preparation, W.Z.; writing—review and editing, F.B., V.V., and G.L.; supervision, F.B. and G.L. All authors have read and agreed to the published version of the manuscript.

Funding: This research was partially funded by the strategic funding from UIDB/04423/2020 and UIDP/04423/2020 of CIIMAR.

Institutional Review Board Statement: Not applicable.

Informed Consent Statement: Not applicable.

Data Availability Statement: Not applicable.

Acknowledgments: Graciliana Lopes thanks the Portuguese Foundation for Science and Technology (FCT) for the financial support for her work contract through the Scientific Employment Stimulus-Individual Call (CEECIND/01768/2021).

Conflicts of Interest: The authors declare no conflict of interest.

References

1. Sáenz, C. Opuntias as a natural resource. In *Agro-Industrial Utilization of Cactus Pear*; Sáenz, C., Berger, H., Rodríguez-Félix, A., Galletti, L., García, J.C., Sepúlveda, E., Varnero, M.T., García de Cortázar, V., García, R.C., Arias, E., et al., Eds.; Rural Infrastructure and Agro-Industries Division: Rome, Italy, 2013; pp. 17–21.
2. El-Mostafa, K.; El-Kharrassi, Y.; Badreddine, A.; Andreoletti, P.; Vamecq, J.; El-Kebbaj, M.S.; Latruffe, N.; Lizard, G.; Nasser, B.; Cherkaoui-Malki, M. Nopal Cactus (*Opuntia ficus-indica*) as a Source of Bioactive Compounds for Nutrition, Health and Disease. *Molecules* **2014**, *19*, 14879–14901. [CrossRef] [PubMed]
3. Labra, M.; Grassi, F.; Bardini, M.; Imazio, S.; Guiggi, A.; Citterio, S.; Banfi, E.; Sgorbati, S. Genetic Relationships in *Opuntia* Mill. Genus (Cactaceae) Detected by Molecular Marker. *Plant Sci.* **2003**, *165*, 1129–1136. [CrossRef]
4. Hahm, S.-W.; Park, J.; Oh, S.-Y.; Lee, C.-W.; Park, K.-Y.; Kim, H.; Son, Y.-S. Anticancer Properties of Extracts from *Opuntia humifusa* against Human Cervical Carcinoma Cells. *J. Med. Food* **2015**, *18*, 31–44. [CrossRef] [PubMed]
5. Callen, E.O. Analysis of the Tehuacan coprolites. In *The Prehistory of the Tehuacan Valley*; Byers, D.S., Ed.; University of Texas Press: London, UK, 1967; Volume 1, pp. 261–289.
6. Santos-Díaz, M.S.; Balch, E.P.M.; Ramírez-Malagón, R.; Nuñez-Palenius, H.G.; Ochoa-Alejo, N. Mexican threatened cacti: Current status and strategies for their conservation. In *Species Diversity and Extinction*; Nova Science Publishers: Hauppauge, NY, USA, 2011; pp. 1–59.
7. Kamble, S.M.; Debaje, P.P.; Ranveer, R.C.; Sahoo, A. Nutritional Importance of Cactus: A Review. *Trends Biosci.* **2017**, *10*, 7668–7677.
8. Ochoa, M.J.; Barbera, G. History and economic and agro-ecological importance. In *Crop Ecology, Cultivation and Uses of Cactus Pear*; Inglese, P., Mondragon, C., Nefzaoui, A., Sáenz, C., Eds.; Food and Agriculture Organization of the United Nations (FAO): Rome, Italy, 2017; pp. 1–11.
9. Ganopoulos, I.; Kalivas, A.; Kavroulakis, N.; Xanthopoulou, A.; Mastrogianni, A.; Koubouris, G.; Madesis, P. Genetic Diversity of Barbary Fig (*Opuntia ficus-indica*) Collection in Greece with ISSR Molecular Markers. *Plant Gene* **2015**, *2*, 29–33. [CrossRef]
10. Zeghbib, W.; Boudjouan, F.; Bachir-bey, M. Optimization of Phenolic Compounds Recovery and Antioxidant Activity Evaluation from *Opuntia ficus indica* Using Response Surface Methodology. *J. Food Meas. Charact.* **2022**, *16*, 1354–1366. [CrossRef]
11. Gurrieri, S.; Miceli, L.; Lanza, C.M.; Tomaselli, F.; Bonomo, R.P.; Rizzarelli, E. Chemical Characterization of Sicilian Prickly Pear (*Opuntia ficus indica*) and Perspectives for the Storage of Its Juice. *J. Agric. Food Chem.* **2000**, *48*, 5424–5431. [CrossRef]

12. Benayad, Z.; Martinez-Villaluenga, C.; Frias, J.; Gomez-Cordoves, C.; Es-Safi, N.E. Phenolic Composition, Antioxidant and Anti-Inflammatory Activities of Extracts from Moroccan *Opuntia ficus-indica* Flowers Obtained by Different Extraction Methods. *Ind. Crops Prod.* **2014**, *62*, 412–420. [CrossRef]
13. Kaur, M.; Kaur, A.; Sharma, R. Pharmacological Actions of *Opuntia ficus indica*: A Review. *J. App. Pharm. Sci.* **2012**, *2*, 15–18. [CrossRef]
14. Leem, K.-H.; Kim, M.-G.; Hahm, Y.-T.; Kim, H.K. Hypoglycemic Effect of Opuntia Ficus-Indica Var. Saboten Is Due to Enhanced Peripheral Glucose Uptake through Activation of AMPK/P38 MAPK Pathway. *Nutrients* **2016**, *8*, 800. [CrossRef]
15. Hernández-Urbiola, M.I.; Pérez-Torrero, E.; Rodríguez-García, M.E. Chemical Analysis of Nutritional Content of Prickly Pads (*Opuntia ficus indica*) at Varied Ages in an Organic Harvest. *Int. J. Environ. Res. Public Health* **2011**, *8*, 1287–1295. [CrossRef] [PubMed]
16. Slimen, I.B.; Mabrouk, M.; Hanène, C.; Najar, T.; Abderrabba, M. LC-MS Analysis of Phenolic Acids, Flavonoids and Betanin from Spineless *Opuntia ficus-indica* Fruits. *Cell Biol.* **2017**, *5*, 17–28. [CrossRef]
17. Zenteno-Ramírez, G.; Juárez-Flores, B.I.; Aguirre-Rivera, J.R.; Monreal-Montes, M.; García, J.M.; Serratosa, M.P.; Santos, M.Á.V.; Pérez, M.D.O.; Rendón-Huerta, J.A. Juices of Prickly Pear Fruits (*Opuntia* spp.) As Functional Foods. *Ital. J. Food Sci.* **2018**, *30*, 614–627. [CrossRef]
18. Ćavar Zeljković, S.; Šišková, J.; Komzáková, K.; de Diego, N.; Kaffková, K.; Tarkowski, P. Phenolic Compounds and Biological Activity of Selected *Mentha* Species. *Plants* **2021**, *10*, 550. [CrossRef]
19. Yeddes, N.; Chérif, J.K.; Guyot, S.; Sotin, H.; Ayadi, M.T. Comparative Study of Antioxidant Power, Polyphenols, Flavonoids and Betacyanins of the Peel and Pulp of Three Tunisian *Opuntia* Forms. *Antioxidants* **2013**, *2*, 37–51. [CrossRef] [PubMed]
20. Pérez-Loredo, M.G.; García-Ochoa, F.; Barragán-Huerta, B.E. Comparative Analysis of Betalain Content in *Stenocereus Stellatus* Fruits and Other Cactus Fruits Using Principal Component Analysis. *Int. J. Food Prop.* **2016**, *19*, 326–338. [CrossRef]
21. Murevanhema, Y.Y.; Jideani, V.A.; Oguntibeju, O.O. Review on potential of seeds and value-added products of bambara groundnut (*Vigna Subterranea*): Antioxidant, anti-inflammatory, and anti-oxidative stress. In *Bioactive Compounds of Medicinal Plants*; Goyal, M.R., Ayeleso, A.O., Eds.; Apple Academic Press: New York, NY, USA, 2018; pp. 102–141. ISBN 978-1-315-14747-5.
22. De Santiago, E.; Gill, C.I.R.; Carafa, I.; Tuohy, K.M.; de Peña, M.-P.; Cid, C. Digestion and Colonic Fermentation of Raw and Cooked *Opuntia ficus-indica* Cladodes Impacts Bioaccessibility and Bioactivity. *J. Agric. Food Chem.* **2019**, *67*, 2490–2499. [CrossRef]
23. Osuna-Martínez, L.; Reyes Esparza, J.; Rodríguez-Fragoso, L. Cactus (*Opuntia ficus-indica*): A Review on Its Antioxidants Properties and Potential Pharmacological Use in Chronic Diseases. *Stud. Nat. Prod. Chem.* **2014**, *2*, 153–160. [CrossRef]
24. Santos-Sánchez, N.F.; Salas-Coronado, R.; Villanueva-Cañongo, C.; Hernández-Carlos, B. Antioxidant compounds and their antioxidant mechanism. In *Antioxidants*; Shalaby, E., Ed.; IntechOpen: London, UK, 2019; ISBN 978-1-78923-920-1.
25. Shan, S.; Huang, X.; Shah, M.H.; Abbasi, A.M. Evaluation of Polyphenolics Content and Antioxidant Activity in Edible Wild Fruits. *Biomed. Res. Int.* **2019**, *2019*, e1381989. [CrossRef]
26. Biswas, S.K. Does the Interdependence between Oxidative Stress and Inflammation Explain the Antioxidant Paradox? *Oxid. Med. Cell Longev.* **2016**, *2016*, 5698931. [CrossRef]
27. Chatterjee, S. Chapter two—Oxidative stress, inflammation, and disease. In *Oxidative Stress and Biomaterials*; Dziubla, T., Butterfield, D.A., Eds.; Academic Press: Cambridge, MA, USA, 2016; pp. 35–58. ISBN 978-0-12-803269-5.
28. Maroon, J.C.; Bost, J.W.; Maroon, A. Natural Anti-Inflammatory Agents for Pain Relief. *Surg. Neurol. Int.* **2010**, *1*, 80. [CrossRef] [PubMed]
29. Crozier, A.; Jaganath, I.B.; Clifford, M.N. Phenols, polyphenols and tannins: An overview. In *Plant Secondary Metabolites*; Crozier, A., Clifford, M.N., Ashihara, H., Eds.; John Wiley & Sons, Ltd.: Hoboken, NJ, USA, 2006; pp. 1–24. ISBN 978-0-470-98855-8.
30. De la Rosa, L.A.; Alvarez-Parrilla, E.; Shahidi, F. Phenolic Compounds and Antioxidant Activity of Kernels and Shells of Mexican Pecan (*Carya Illinoinensis*). *J. Agric. Food Chem.* **2011**, *59*, 152–162. [CrossRef] [PubMed]
31. Chu, K.O.; Chan, S.-O.; Pang, C.P.; Wang, C.C. Pro-Oxidative and Antioxidative Controls and Signaling Modification of Polyphenolic Phytochemicals: Contribution to Health Promotion and Disease Prevention? *J. Agric. Food Chem.* **2014**, *62*, 4026–4038. [CrossRef] [PubMed]
32. Soto-Vaca, A.; Gutierrez, A.; Losso, J.N.; Xu, Z.; Finley, J.W. Evolution of Phenolic Compounds from Color and Flavor Problems to Health Benefits. *J. Agric. Food Chem.* **2012**, *60*, 6658–6677. [CrossRef]
33. Pabón-Baquero, L.C.; Otálvaro-Álvarez, Á.M.; Fernández, M.R.R.; Chaparro-González, M.P. Plant extracts as antioxidant additives for food industry. In *Antioxidants in Foods and Its Applications*; Shalaby, E., Azzam, G.M., Eds.; IntechOpen: London, UK, 2018; pp. 87–116. ISBN 978-1-78923-379-7.
34. Ali Redha, A. Review on Extraction of Phenolic Compounds from Natural Sources Using Green Deep Eutectic Solvents. *J. Agric. Food Chem.* **2021**, *69*, 878–912. [CrossRef]
35. Wang, Z.; Li, S.; Ge, S.; Lin, S. Review of Distribution, Extraction Methods, and Health Benefits of Bound Phenolics in Food Plants. *J. Agric. Food Chem.* **2020**, *68*, 3330–3343. [CrossRef]
36. Haminiuk, C.W.I.; Plata-Oviedo, M.S.V.; de Mattos, G.; Carpes, S.T.; Branco, I.G. Extraction and Quantification of Phenolic Acids and Flavonols from Eugenia Pyriformis Using Different Solvents. *J. Food Sci. Technol.* **2014**, *51*, 2862–2866. [CrossRef]
37. Alara, O.R.; Abdurahman, N.H.; Ukaegbu, C.I. Extraction of Phenolic Compounds: A Review. *Curr. Res. Nutr. Food Sci.* **2021**, *4*, 200–214. [CrossRef]

38. Ammar, I.; Ennouri, M.; Bouaziz, M.; Ben Amira, A.; Attia, H. Phenolic Profiles, Phytchemicals and Mineral Content of Decoction and Infusion of *Opuntia ficus-indica* Flowers. *Plant Foods Hum. Nutr.* **2015**, *70*, 388–394. [CrossRef]
39. Ben Lataief, S.; Zourgui, M.-N.; Rahmani, R.; Najjaa, H.; Gharsallah, N.; Zourgui, L. Chemical Composition, Antioxidant, Antimicrobial and Cytotoxic Activities of Bioactive Compounds Extracted from *Opuntia dillenii* Cladodes. *J. Food Meas. Charact.* **2021**, *15*, 782–794. [CrossRef]
40. Melgar, B.; Dias, M.I.; Barros, L.; Ferreira, I.C.F.R.; Rodriguez-Lopez, A.D.; Garcia-Castello, E.M. Ultrasound and Microwave Assisted Extraction of *Opuntia* Fruit Peels Biocompounds: Optimization and Comparison Using RSM-CCD. *Molecules* **2019**, *24*, 3618. [CrossRef] [PubMed]
41. Missaoui, M.; D'Antuono, I.; D'Imperio, M.; Linsalata, V.; Boukhchina, S.; Logrieco, A.F.; Cardinali, A. Characterization of Micronutrients, Bioaccessibility and Antioxidant Activity of Prickly Pear Cladodes as Functional Ingredient. *Molecules* **2020**, *25*, 2176. [CrossRef] [PubMed]
42. Tounsi, M.S.; Ouerghemmi, I.; Ksouri, R.; Wannes, W.A.; Hammrouni, I.; Marzouk, B. HPLC-Determination of Phenolic Composition and Antioxidant Capacity of Cactus Prickly Pears Seeds. *Asian J. Chem.* **2011**, *23*, 1006–1010.
43. Koubaa, M.; Mhemdi, H.; Barba, F.J.; Angelotti, A.; Bouaziz, F.; Chaabouni, S.E.; Vorobiev, E. Seed Oil Extraction from Red Prickly Pear Using Hexane and Supercritical CO_2: Assessment of Phenolic Compound Composition, Antioxidant and Antibacterial Activities. *J. Sci. Food Agric.* **2017**, *97*, 613–620. [CrossRef]
44. López-Palacios, C.; Peña-Valdivia, C.B. Screening of Secondary Metabolites in Cladodes to Further Decode the Domestication Process in the Genus *Opuntia* (Cactaceae). *Planta* **2020**, *251*, 74. [CrossRef]
45. Kıvrak, Ş.; Kıvrak, İ.; Karababa, E. Analytical Evaluation of Phenolic Compounds and Minerals of *Opuntia robusta* J.C. Wendl. and *Opuntia ficus-barbarica* A. Berger. *Int. J. Food Prop.* **2018**, *21*, 229–241. [CrossRef]
46. Amrane-Abider, M.; Nerin, C.; Canellas, E.; Benkerrou, F.; Louaileche, H. Modeling and Optimization of Phenolic Compounds Extraction from Prickly Pear (*Opuntia ficus-indica*) Seeds via Ultrasound-Assisted Technique. *Ann. Univ. Dunarea Jos Galati. Fascicle VI Food Technol.* **2018**, *42*, 109–121.
47. Rocchetti, G.; Pellizzoni, M.; Montesano, D.; Lucini, L. Italian *Opuntia ficus-indica* Cladodes as Rich Source of Bioactive Compounds with Health-Promoting Properties. *Foods* **2018**, *7*, 24. [CrossRef]
48. Aruwa, C.E.; Amoo, S.; Kudanga, T. Phenolic Compound Profile and Biological Activities of Southern African *Opuntia ficus-indica* Fruit Pulp and Peels. *LWT* **2019**, *111*, 337–344. [CrossRef]
49. Ortega-Hernández, E.; Nair, V.; Welti-Chanes, J.; Cisneros-Zevallos, L.; Jacobo-Velázquez, D.A. Wounding and UVB Light Synergistically Induce the Biosynthesis of Phenolic Compounds and Ascorbic Acid in Red Prickly Pears (*Opuntia ficus-indica* Cv. Rojo Vigor). *Int. J. Mol. Sci.* **2019**, *20*, 5327. [CrossRef]
50. Astello-García, M.G.; Cervantes, I.; Nair, V.; del Santos-Díaz, M.S.; Reyes-Agüero, A.; Guéraud, F.; Negre-Salvayre, A.; Rossignol, M.; Cisneros-Zevallos, L.; Barba de la Rosa, A.P. Chemical Composition and Phenolic Compounds Profile of Cladodes from *Opuntia* spp. Cultivars with Different Domestication Gradient. *J. Food Compos. Anal.* **2015**, *43*, 119–130. [CrossRef]
51. Mata, A.; Ferreira, J.P.; Semedo, C.; Serra, T.; Duarte, C.M.M.; Bronze, M.R. Contribution to the Characterization of *Opuntia* spp. Juices by LC-DAD-ESI-MS/MS. *Food Chem.* **2016**, *210*, 558–565. [CrossRef] [PubMed]
52. Lanuzza, F.; Occhiuto, F.; Monforte, M.T.; Tripodo, M.M.; D'Angelo, V.; Galati, E.M. Antioxidant Phytochemicals of *Opuntia ficus-indica* (L.) Mill. Cladodes with Potential Anti-Spasmodic Activity. *Pharmacogn. Mag.* **2017**, *13*, S424–S429. [CrossRef] [PubMed]
53. Melgar, B.; Dias, M.I.; Ciric, A.; Sokovic, M.; Garcia-Castello, E.M.; Rodriguez-Lopez, A.D.; Barros, L.; Ferreira, I. By-Product Recovery of *Opuntia* spp. Peels: Betalainic and Phenolic Profiles and Bioactive Properties. *Ind. Crops Prod.* **2017**, *107*, 353–359. [CrossRef]
54. Ouerghemmi, I.; Harbeoui, H.; Aidi Wannes, W.; Bettaieb Rebey, I.; Hammami, M.; Marzouk, B.; Saidani Tounsi, M. Phytochemical Composition and Antioxidant Activity of Tunisian Cactus Pear (*Opuntia ficus indica* L.) Flower. *J. Food Biochem.* **2017**, *41*, 1–10. [CrossRef]
55. De Santiago, E.; Pereira-Caro, G.; Moreno-Rojas, J.M.; Cid, C.; de Peña, M.-P. Digestibility of (Poly)phenols and Antioxidant Activity in Raw and Cooked Cactus Cladodes (*Opuntia ficus-indica*). *J. Agric. Food Chem.* **2018**, *66*, 5832–5844. [CrossRef]
56. Mena, P.; Tassotti, M.; Andreu, L.; Nuncio-Jáuregui, N.; Legua, P.; del Rio, D.; Hernández, F. Phytochemical Characterization of Different Prickly Pear (*Opuntia ficus-indica* (L.) Mill.) Cultivars and Botanical Parts: UHPLC-ESI-MS^n Metabolomics Profiles and Their Chemometric Analysis. *Food Res. Int.* **2018**, *108*, 301–308. [CrossRef]
57. García-Cayuela, T.; Gómez-Maqueo, A.; Guajardo-Flores, D.; Welti-Chanes, J.; Cano, M.P. Characterization and Quantification of Individual Betalain and Phenolic Compounds in Mexican and Spanish Prickly Pear (*Opuntia ficus-indica* L. Mill) Tissues: A Comparative Study. *J. Food Compos. Anal.* **2019**, *76*, 1–13. [CrossRef]
58. Kolniak-Ostek, J.; Kita, A.; Miedzianka, J.; Andreu-Coll, L.; Legua, P.; Hernandez, F. Characterization of Bioactive Compounds of *Opuntia ficus-indica* (L.) Mill. Seeds from Spanish Cultivars. *Molecules* **2020**, *25*, 5734. [CrossRef]
59. Chahdoura, H.; Barreira, J.C.M.; Barros, L.; Santos-Buelga, C.; Ferreira, I.C.F.R.; Achour, L. Phytochemical Characterization and Antioxidant Activity of *Opuntia microdasys* (Lehm.) Pfeiff Flowers in Different Stages of Maturity. *J. Funct. Foods.* **2014**, *9*, 27–37. [CrossRef]
60. Chahdoura, H.; Barreira, J.C.M.; Barros, L.; Santos-Buelga, C.; Ferreira, I.C.F.R.; Achour, L. Seeds of *Opuntia* spp. as a Novel High Potential by-Product: Phytochemical Characterization and Antioxidant Activity. *Ind Crops Prod.* **2015**, *65*, 383–389. [CrossRef]

61. Anwar, M.M.; Sallam, E.M. Utilization of Prickly Pear Peels to Improve Quality of Pan Bread. *Arab. J. Nucl. Sci. Appl.* **2016**, *49*, 151–163.
62. Cruz-Bravo, R.K.; Guzmán-Maldonado, S.H.; Araiza-Herrera, H.A.; Zegbe, J.A. Storage Alters Physicochemical Characteristics, Bioactive Compounds and Antioxidant Capacity of Cactus Pear Fruit. *Postharvest Biol. Technol.* **2019**, *150*, 105–111. [CrossRef]
63. Gómez-López, I.; Lobo-Rodrigo, G.; Portillo, M.P.; Cano, M.P. Characterization, Stability, and Bioaccessibility of Betalain and Phenolic Compounds from *Opuntia stricta* var *dillenii* Fruits and Products of Their Industrialization. *Foods* **2021**, *10*, 1593. [CrossRef]
64. Chougui, N.; Tamendjari, A.; Hamidj, W.; Hallal, S.; Barras, A.; Richard, T.; Larbat, R. Oil Composition and Characterisation of Phenolic Compounds of *Opuntia ficus-indica* Seeds. *Food Chem.* **2013**, *139*, 796–803. [CrossRef]
65. Ammar, I.; Ben Salem, M.; Harrabi, B.; Mzid, M.; Bardaa, S.; Sahnoun, Z.; Attia, H.; Ennouri, M. Anti-Inflammatory Activity and Phenolic Composition of Prickly Pear (*Opuntia ficus-indica*) Flowers. *Ind. Crops Prod.* **2018**, *112*, 313–319. [CrossRef]
66. Geronikaki, A.A.; Gavalas, A.M. Antioxidants and Inflammatory Disease: Synthetic and Natural Antioxidants with Anti-Inflammatory Activity. *Comb. Chem. High Throughput Screen* **2006**, *9*, 425–442. [CrossRef]
67. Linus, L.O.; Hanson, C.; Alolga, R.N.; Zhou, W.; Qi, L. Targeting the Key Factors of Inflammation in Cancer: Plant Intervention. *Int. J. Clin. Exp. Med.* **2017**, *10*, 15834–15865.
68. Yahfoufi, N.; Alsadi, N.; Jambi, M.; Matar, C. The Immunomodulatory and Anti-Inflammatory Role of Polyphenols. *Nutrients* **2018**, *10*, 1618. [CrossRef]
69. Jones, S.L.; Blikslager, A. The Future of Anti-inflammatory Therapy. *Vet. Clin. N. Am.* **2001**, *17*, 245–262.
70. Joseph, S.V.; Edirisinghe, I.; Burton-Freeman, B.M. Berries: Anti-Inflammatory Effects in Humans. *J. Agric. Food Chem.* **2014**, *62*, 3886–3903. [CrossRef] [PubMed]
71. Lawrence, T. The Nuclear Factor NF-κB Pathway in Inflammation. *Cold Spring Harb. Perspect. Biol.* **2009**, *1*, a001651. [CrossRef] [PubMed]
72. Hoesel, B.; Schmid, J.A. The Complexity of NF-κB Signaling in Inflammation and Cancer. *Mol. Cancer* **2013**, *12*, 86. [CrossRef] [PubMed]
73. Liu, T.; Zhang, L.; Joo, D.; Sun, S.-C. NF-κB Signaling in Inflammation. *Signal Transduct. Target Ther.* **2017**, *2*, 17023. [CrossRef] [PubMed]
74. Li, C.; Chen, R.; Jiang, C.; Chen, L.; Cheng, Z. Correlation of LOX-5 and COX-2 Expression with Inflammatory Pathology and Clinical Features of Adenomyosis. *Mol. Med. Rep.* **2019**, *19*, 727–733. [CrossRef]
75. Wang, B.; Wu, L.; Chen, J.; Dong, L.; Chen, C.; Wen, Z.; Hu, J.; Fleming, I.; Wang, D.W. Metabolism Pathways of Arachidonic Acids: Mechanisms and Potential Therapeutic Targets. *Sig. Transduct. Target Ther.* **2021**, *6*, 1–30. [CrossRef]
76. Cuzzo, B.; Lappin, S.L. Physiology, leukotrienes. In *StatPearls*; StatPearls Publishing: Treasure Island, FL, USA, 2022.
77. Ahmed, M.S.; El Tanbouly, N.D.; Islam, W.T.; Sleem, A.A.; El Senousy, A.S. Antiinflammatory Flavonoids from *Opuntia dillenii* (Ker-Gawl) Haw. Flowers Growing in Egypt. *Phytother. Res.* **2005**, *19*, 807–809. [CrossRef]
78. Cho, J.Y.; Park, S.C.; Kim, T.W.; Kim, K.S.; Song, J.-C.; Kim, S.K.; Lee, H.M.; Sung, H.J.; Park, H.J.; Song, Y.B.; et al. Radical Scavenging and Anti-Inflammatory Activity of Extracts from *Opuntia humifusa* Raf. *J. Pharm. Pharmacol.* **2006**, *58*, 113–119. [CrossRef]
79. Chauhan, S.P.; Sheth, N.R.; Suhagia, B.N. Analgesic and Anti-Inflammatory Action of *Opuntia elatior* Mill Fruits. *J. Ayurveda Integr. Med.* **2015**, *6*, 75–81. [CrossRef]
80. Benattia, F.K.; Arrar, Z.; Khabbal, Y. Evaluation of the Anti-Inflammatory Activity of the Seeds Extracts of Prickly Pear (*Opuntia ficus-indica* L.). *Der Pharma Chem.* **2017**, *9*, 14–17.
81. Izuegbuna, O.; Otunola, G.; Bradley, G. Chemical Composition, Antioxidant, Anti-Inflammatory, and Cytotoxic Activities of *Opuntia stricta* Cladodes. *PLoS ONE* **2019**, *14*, e0209682. [CrossRef] [PubMed]
82. Matias, A.; Nunes, S.L.; Poejo, J.; Mecha, E.; Serra, A.T.; Madeira, P.J.A.; Bronze, M.R.; Duarte, C.M.M. Antioxidant and Anti-Inflammatory Activity of a Flavonoid-Rich Concentrate Recovered from *Opuntia ficus-indica* Juice. *Food Funct.* **2014**, *5*, 3269–3280. [CrossRef] [PubMed]
83. Antunes-Ricardo, M.; Gutiérrez-Uribe, J.A.; López-Pacheco, F.; Alvarez, M.M.; Serna-Saldívar, S.O. In Vivo Anti-Inflammatory Effects of Isorhamnetin Glycosides Isolated from *Opuntia ficus-indica* (L.) Mill Cladodes. *Ind. Crops Prod.* **2015**, *76*, 803–808. [CrossRef]
84. Gómez-Maqueo, A.; García-Cayuela, T.; Fernández-López, R.; Welti-Chanes, J.; Cano, M.P. Inhibitory Potential of Prickly Pears and Their Isolated Bioactives against Digestive Enzymes Linked to Type 2 Diabetes and Inflammatory Response. *J. Sci. Food Agric.* **2019**, *99*, 6380–6391. [CrossRef]
85. Sharma, J.N.; Al-Omran, A.; Parvathy, S.S. Role of Nitric Oxide in Inflammatory Diseases. *Inflammopharmacology* **2007**, *15*, 252–259. [CrossRef] [PubMed]
86. McCook, J.P.; Dorogi, P.L.; Vasily, D.B.; Cefalo, D.R. In Vitro Inhibition of Hyaluronidase by Sodium Copper Chlorophyllin Complex and Chlorophyllin Analogs. *Clin. Cosmet. Investig. Dermatol.* **2015**, *8*, 443. [CrossRef] [PubMed]
87. Bralley, E.; Greenspan, P.; Hargrove, J.L.; Hartle, D.K. Inhibition of Hyaluronidase Activity by *Vitis rotundifolia*. (Muscadine) Berry Seeds and Skins. *Pharm. Biol.* **2007**, *45*, 667–673. [CrossRef]
88. González-Peña, D.; Colina-Coca, C.; Char, C.D.; Cano, M.P.; de Ancos, B.; Sánchez-Moreno, C. Hyaluronidase Inhibiting Activity and Radical Scavenging Potential of Flavonols in Processed Onion. *J. Agric. Food Chem.* **2013**, *61*, 4862–4872. [CrossRef]

89. Gómez-Maqueo, A.; García-Cayuela, T.; Welti-Chanes, J.; Cano, M.P. Enhancement of Anti-Inflammatory and Antioxidant Activities of Prickly Pear Fruits by High Hydrostatic Pressure: A Chemical and Microstructural Approach. *Innov. Food Sci. Emerg. Technol.* **2019**, *54*, 132–142. [CrossRef]
90. Chaalal, M.; Gavilán, E.; Louaileche, H.; Ruano, D.; Parrado, J.; Castaño, A. Anti-Inflammatory Activity of Phenolic Extracts from Different Parts of Prickly Pear on Lipopolysaccharide-Stimulated N13 Microglial Cells. *Int. J. Phytomed.* **2016**, *7*, 411–419.
91. Yeo, J.Y.; Hwang, K.W.; Park, S.-Y. Anti-Inflammatory Effect of Neo-Lignan Isoamericanin A via Suppression of NF-κB in Liposaccharide-Stimulated RAW 264.7 Cells. *Trop. J. Pharm. Res.* **2020**, *19*, 1857–1862. [CrossRef]
92. Filannino, P.; Cavoski, I.; Thlien, N.; Vincentini, O.; Angelis, M.D.; Silano, M.; Gobbetti, M.; Cagno, R.D. Lactic Acid Fermentation of Cactus Cladodes (*Opuntia ficus-indica* L.) Generates Flavonoid Derivatives with Antioxidant and Anti-Inflammatory Properties. *PLoS ONE* **2016**, *11*, e0152575. [CrossRef]

molecules

MDPI

Article

Thermal Control Using Far-Infrared Irradiation for Producing Deglycosylated Bioactive Compounds from Korean Ginseng Leaves

Shucheng Duan [1,2,†], Jia Rui Liu [1,†], Xin Wang [1], Xue Mei Sun [1], Han Sheng Gong [1], Cheng Wu Jin [1,*,‡] and Seok Hyun Eom [2,*,‡]

1 College of Food Engineering, Ludong University, Yantai 264025, China; dsc97@khu.ac.kr (S.D.); ljr19980227@163.com (J.R.L.); 17616196867@163.com (X.W.); xuemei0110@163.com (X.M.S.); hsgong_221@163.com (H.S.G.)

2 Department of Smart Farm Science, College of Life Sciences, Kyung Hee University, Yongin 17104, Korea

* Correspondence: jinchwu@ldu.edu.cn (C.W.J.); se43@khu.ac.kr (S.H.E.)

† These authors contributed equally to this work.

‡ These authors are equally contributed as the corresponding authors.

Abstract: Although ginseng leaf is a good source of health-beneficial phytochemicals, such as polyphenols and ginsenosides, few studies have focused on the variation in compounds and bioactivities during leaf thermal processing. The efficiency of far-infrared irradiation (FIR) between 160 °C and 200 °C on the deglycosylation of bioactive compounds in ginseng leaves was analyzed. FIR treatment significantly increased the total polyphenol content (TPC) and kaempferol production from panasenoside conversion. The highest content or conversion ratio was observed at 180 °C (FIR-180). Major ginsenoside contents gradually decreased as the FIR temperature increased, while minor ginsenoside contents significantly increased. FIR exhibited high efficiency to produce dehydrated minor ginsenosides, of which F4, Rg6, Rh4, Rk3, Rk1, and Rg5 increased to their highest levels at FIR-190, by 278-, 149-, 176-, 275-, 64-, and 81-fold, respectively. Moreover, significantly increased antioxidant activities were also observed in FIR-treated leaves, particularly FIR-180, mainly due to the breakage of phenolic polymers to release antioxidants. These results suggest that FIR treatment is a rapid and efficient processing method for producing various health-beneficial bioactive compounds from ginseng leaves. After 30 min of treatment without leaf burning, FIR-190 was the optimum temperature for producing minor ginsenosides, whereas FIR-180 was the optimum temperature for producing polyphenols and kaempferol. In addition, the results suggested that the antioxidant benefits of ginseng leaves are mainly due to polyphenols rather than ginsenosides.

Keywords: ginseng leaf; far-infrared irradiation; polyphenols; ginsenosides; antioxidants; health benefits

Citation: Duan, S.; Liu, J.R.; Wang, X.; Sun, X.M.; Gong, H.S.; Jin, C.W.; Eom, S.H. Thermal Control Using Far-Infrared Irradiation for Producing Deglycosylated Bioactive Compounds from Korean Ginseng Leaves. *Molecules* **2022**, *27*, 4782. https://doi.org/10.3390/molecules27154782

Academic Editor: Nour Eddine Es-Safi

Received: 28 June 2022
Accepted: 23 July 2022
Published: 26 July 2022

1. Introduction

Ginseng (*Panax ginseng* Meyer) roots and their processed products are widely consumed because of their excellent health benefits, which are attributed to their bioactive ginsenosides [1]. Qualified ginseng roots require a relatively long cultivation period and are typically harvested between 4 and 6 years. Consequently, ginseng leaves are produced in large quantities annually and disposed of as waste at the end of the growing season. However, ginseng leaves possess similar pharmacological activities to the roots [2]. The distribution of ginsenosides in ginseng leaves has been investigated, which suggests that the leaves have similar ginsenoside compositions to the roots [3–5]. In addition, Chung et al. [6] reported an extremely high total phenolic content of ginseng leaves compared to the roots. Yin et al. [7] suggested that ginseng leaves contain higher total flavonoids compared to the roots, especially flavonol glycosides of panasenoside and kaempferol-3-O-glucoside. However, although ginseng leaves are potentially good sources for producing phytochemicals, especially polyphenols and ginsenosides, and have advantages in annual yield production, ginseng leaves have not been fully studied and used.

Thermal processing methods play a vital role in the global ginseng market. Numerous studies have shown significant processing results related to the deglycosylation pattern of ginsenosides and various health-beneficial biological activities, exhibiting bioconversion from major ginsenosides to minor ginsenosides, as well as antioxidant, antitumor, and anti-inflammatory effects [8–11]. Nevertheless, considerable efforts are required to reasonably use ginseng leaves, because few studies have focused on phytochemical variations, especially polyphenols, during dry-heat thermal processing.

Far-infrared irradiation (FIR) treatment is widely used in the food-processing industry because it is easily applicable in terms of fast and efficient heating of plant materials for inducing the decomposition of multi-chain molecular clusters [12]. Our previous studies have suggested that FIR is an efficient method for accelerating ginsenoside conversions in ginseng roots and leaves [10,13], which is significantly different from traditional steaming treatment at the same temperature [13]. An increase in the content of bioactive compounds, such as phenolic acids and flavonoids, by FIR has been reported in rice, angelica, gamguk flowers, grapes, and buckwheat sprouts [14–18]. However, there is still no detailed study related to defining an adequate FIR temperature for producing the maximum range of bioactive compounds from ginseng leaves.

When considering cost, source availability, and sustainability, ginseng leaves are a more appropriate choice for obtaining phytochemicals, especially polyphenols and ginsenosides, compared to their roots. The aim of this study was to investigate the effects of FIR treatment on the bioactive compounds (total phenolics, flavonoids, and ginsenosides) and human health benefits (antioxidant activities) of ginseng leaves. Our results will provide useful information for the further use and application of ginseng leaves in the health product industry and food science.

2. Results and Discussion

2.1. *Effects of FIR Treatment on TPC in Ginseng Leaves*

The effects of FIR treatment on TPC in ginseng leaves are shown in Figure 1. As the FIR temperature increased, the TPC gradually increased up to 180 °C and then decreased. The highest TPC in ginseng leaves was 25.27 mg/g d.w. at FIR-180, which is about 1.56 times that of the untreated control. The lowest TPC (15.16 mg/g d.w.) was observed at FIR-200, with no significant difference being observed between FIR-200 and the untreated control. It can be concluded that suitable FIR treatment conditions can increase the TPC in ginseng leaves. The effect of FIR treatment on TPC improvement has also been reported in *Angelica gigas* Nakai, *Arachis hypogaea* L., *Camellia sinensis* var. sinensis, and *Hibiscus cannabinus* L. [15,19–21]. The increase in the TPC can be explained by the release of small polyphenols due to the breaking of molecular bonds in large polyphenols at high FIR temperatures [12,22]. However, relatively higher temperatures (FIR-190 and FIR-200) caused a decrease in the TPC in ginseng leaves compared to low temperatures (FIR-160, FIR-170, and FIR-180), which might be explained by the destruction of polyphenol structures by high FIR energy [18]. Based on our results, it can be concluded that 180 °C is a suitable temperature for FIR treatment for the processing of ginseng leaves to obtain more polyphenols. Polyphenols are an important phytochemical group in edible plants due to their various biological activities and health benefits, gaining increasingly more attention of researchers to investigate their content variation in plant tissues during different processing methods [23–25]. For ginseng, previous studies have mainly focused on their root polyphenol content changes during processing [26,27]. To the best of our knowledge, this is the first study to show the excellent effect of suitable FIR treatment on improving the TPC in ginseng leaves.

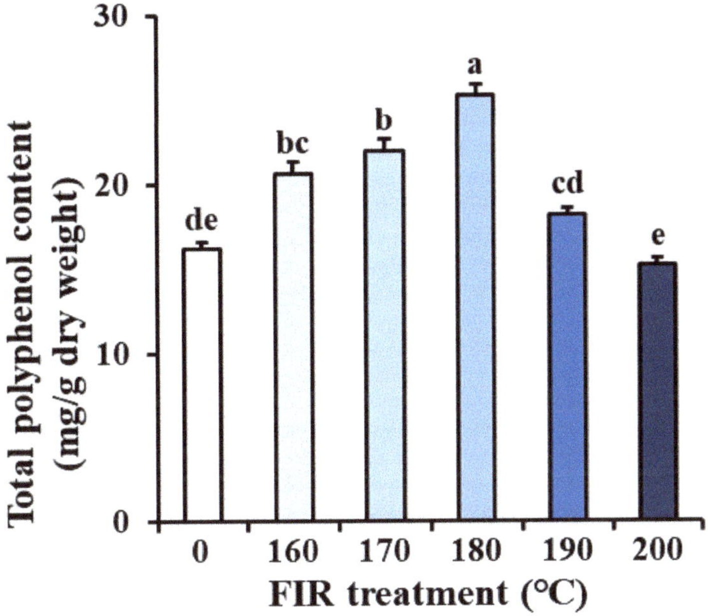

Figure 1. Changes in the total polyphenol content of ginseng leaves treated by FIR at different temperatures. The values are expressed as the mean with standard error ($n = 3$). Different letters (a–e) above the bar graphs indicate significant differences at $p < 0.05$ on Tukey's HSD test.

2.2. Effects of FIR Treatment on Panasenoside and Kaempferol Contents of Ginseng Leaves

Panasenoside (kaempferol 3-*O*-glucosyl-(1-2)-galactoside) is a kaempferol glycoside and has been found to be the main flavonoid in ginseng leaves [7]. Figure 2A,B shows the variation in the contents of panasenoside and its aglycone (kaempferol) before and after FIR treatment. Before FIR treatment (0), ginseng leaves had 10.53 mg/g d.w. of panasenoside and only 0.03 mg/g d.w. of kaempferol. Similarly, Yin et al. [7] reported contents of 15.84 mg/g d.w. and 0.43 mg/g d.w of panasenoside and kaempferol, respectively, in 4-year-old ginseng leaves. After FIR treatment, a significant decrease in the panasenoside content of ginseng leaves was observed with increasing FIR temperature, with only 2.29 mg/g d.w. remaining in the leaves after FIR-200 treatment. In contrast, our results showed a significant increase in kaempferol. Its content increased sharply to 0.42 mg/g d.w. until FIR-180, which was 14.3 times that of the untreated control (0). The kaempferol content did not significantly increase at temperatures greater than 180 °C, such as FIR-190 and FIR-200.

According to the structures of panasenoside and kaempferol (Figure 2C), it is assumed that FIR energy causes a significant increase in the kaempferol content due to the deglycosylation of panasenoside. It has been suggested that FIR has the capacity to transfer heat energy, cleave covalent bonds, and liberate low-molecular-weight compounds [12,22]. However, the content of kaempferol did not significantly vary when the temperature exceeded 180 °C, despite the continuous decrease in the panasenoside content with increasing FIR temperature (Figure 2A,B). Furthermore, the increased aglycone content was not as high as that of panasenoside. The effects of FIR heat energy on ginseng leaf flavonoids were assumed in two situations according to temperature variation: (1) The relatively fast deglycosylation of panasenoside accompanied by the slow degradation of kaempferol occurred when FIR temperatures were below 180 °C, which explains the relatively slow decrease in the panasenoside content and the sharp increase in kaempferol, and (2) the rapid deglycosylation of panasenoside accompanied with the faster degradation of kaempferol when the FIR temperature exceeded 180 °C, which explains the rapid decrease in panasenoside and the lack of significant change in the kaempferol content. These assumptions were supported by Oliveira et al. [28]. They reported that the kaempferol content of *Poincianella pyramidalis* decreased significantly when the air inlet temperature was increased from 160 to 180 °C during spray-drying. Deng et al. [29] suggested that flavonol glycosides (kaempferol and quercetin derivatives) are degraded during the roasting process to correspond to an increase in aglycones in noni leaves (roasted at 175 °C from 10 to 60 min or roasted for 20 min from 100 to 250 °C, respectively). Moreover, FIR treatment usually provides higher energy during the material dry process compared to common dry processes under the same temperature control. In addition, it is expected that FIR treatment may lead to an intermediate compound by cleaving one glucoside of panasenoside rather than directly deglycosylating two glucosides to kaempferol. However, this intermediate compound was not detected in this study (data not shown). This result also supports our assumption. Previous studies on flavonoids in ginseng leaves have mainly focused on the types and contents of flavonoids in raw materials [7]. To the best of our knowledge, this is the first study to investigate flavonoid variation patterns in ginseng leaves during processing. According to our results, FIR treatment significantly increases the content of kaempferol via the deglycosylation of panasenoside (Figure 2), which has been found to possess higher bioactivities, such as antitumor, antioxidant, and anti-inflammatory activities, compared to its glycosides [30]. Thus, it can be concluded that FIR treatment (FIR-180) is a promising method for processing ginseng leaves to obtain more kaempferol. Here, it is also important to note that leaf burning was observed when the FIR temperature exceeded 200 °C during the 30 min treatment, suggesting that higher FIR temperatures are risky for ginseng leaf processing.

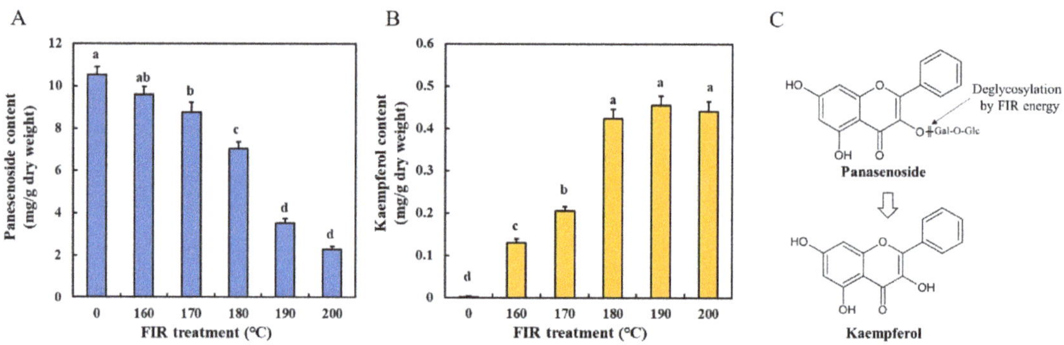

Figure 2. Changes in panasenoside (**A**) and kaempferol (**B**) contents and structural conversion pattern (**C**) from panasenoside to kaempferol in ginseng leaves treated by FIR at different temperatures. The values are expressed as the mean with standard error ($n = 3$). Different letters (a–d) above the bar graphs indicate significant differences at $p < 0.05$ on Tukey's HSD test.

2.3. *Effects of FIR Treatment on Ginsenoside Contents of Ginseng Leaves*

Ginsenosides can be roughly divided into two types according to the position and amount of glycol groups in the glycosides: (1) protopanaxadiol (PPD) type and (2) protopanaxatriol (PPT) type [8]. Here, the variations in nine PPD-type and eight PPT-type ginsenosides in ginseng leaves during FIR treatment were qualified and quantified using HPLC, respectively (Figure 3 and Figure S1, and Table S1).

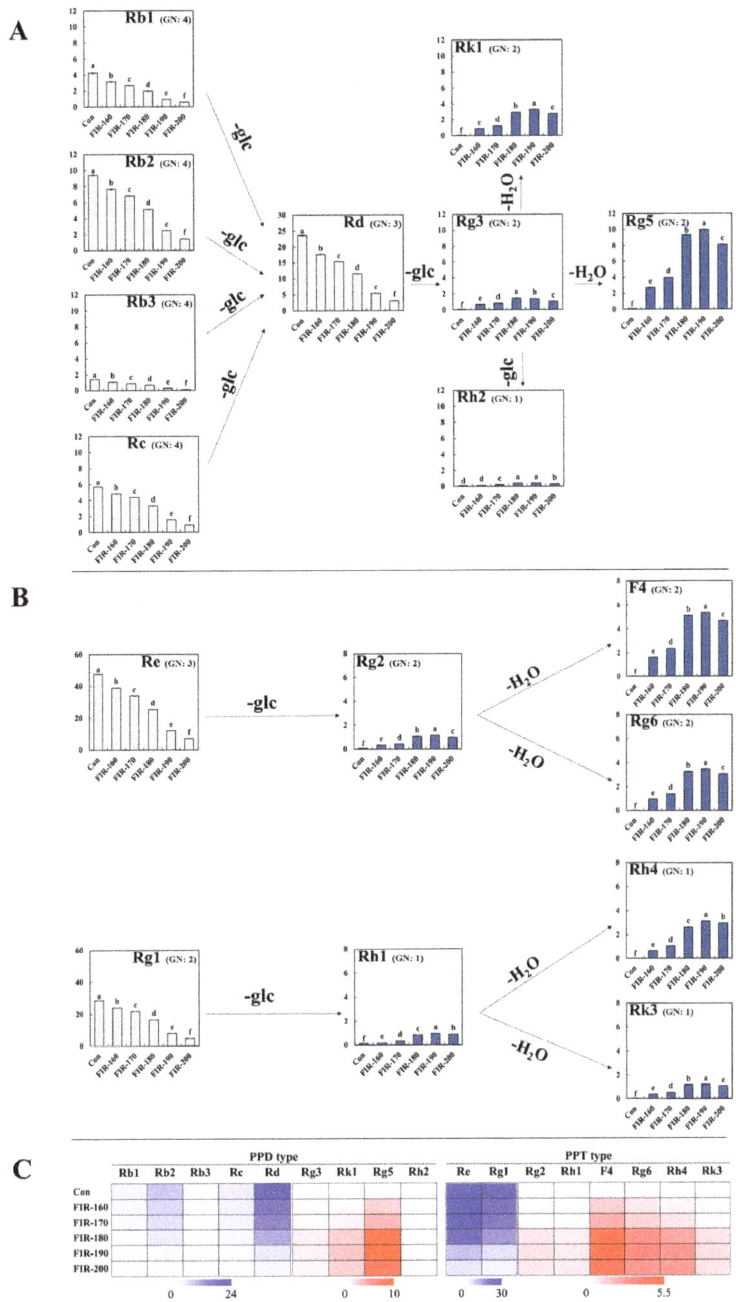

Figure 3. Flow schemes of the conversion patterns of PPD type (**A**) and PPT type (**B**) of ginsenosides in ginseng leaves treated by FIR at different temperatures. A heat map (**C**) of 17 kinds of ginsenoside changes in ginseng leaves treated by FIR at different temperatures. The unit of ginsenoside content is mg/g dry weight. GN indicates the glycosyl number attached on ginsenoisde. The values in bar graphs are expressed as the mean with standard error ($n = 3$). Different letters (a–f) above the bar graphs indicate significant differences at $p < 0.05$ on Tukey's HSD test.

2.3.1. PPD-Type Ginsenosides

The effects of FIR treatment on the transformation of PPD-type ginsenosides in ginseng leaves are shown in Figure 3A,C, and Table S1. Overall, FIR treatment reduced the major ginsenoside content and increased the minor ginsenoside content. In detail, the ginsenosides Rb1 (4.25 mg/g d.w.), Rb2 (9.41 mg/g d.w.), Rb3 (1.42 mg/g d.w.), Rc (5.72 mg/g d.w.), and Rd (23.65 mg/g d.w.) were the predominant compounds in untreated ginseng leaves, whereas their contents gradually decreased after FIR treatment (Figure 3, light yellow). Each of the aforementioned ginsenosides showed a relatively low amount at FIR-200, with 0.58, 1.45, 0.13, 0.90, and 3.11 mg/g d.w. of Rb1, Rb2, Rb3, Rc, and Rd being observed, respectively. In contrast, non-FIR-treated ginseng leaves only had small amounts of minor ginsenosides (Figure 3, blue column), such as Rg3 (0.07 mg/g d.w.), Rk1 (0.04 mg/g d.w.), Rg5 (0.10 mg/g d.w.), and Rh2 (0.12 mg/g d.w.), whereas they significantly increased after FIR treatment. The ginsenoside Rg3 increased to its highest amount at FIR-180, presenting a 22-fold (1.49 mg/g d.w.) higher content than the untreated control. The highest amounts of Rk1, Rg5, and Rh2 were observed at FIR-190, which increased 77-fold (3.30 mg/g d.w. Rk1), 100-fold (10.01 mg/g d.w. Rg5), and 3-fold (0.41 mg/g d.w. Rh2), respectively, compared to the non-FIR-treated control.

FIR treatment significantly increased the content of minor PPD-type ginsenosides (Rg3, Rk1, Rg5, and Rh2) in ginseng leaves by achieving the deglycosylation and dehydration of the major ginsenosides. The major ginsenosides Rb1, Rb2, Rb3, Rc, and Rd had their glycosyl residue(s) at C-20 removed, causing a significant increase in the content of the minor ginsenoside Rg3. Furthermore, Rg3 was then converted to Rk1 and Rg5 via dehydration at C-20 and C-21/C-22 or to Rh2 by deglycosylation at C-3. Similar results have been reported for ginseng roots and observed in the ginsenoside variation patterns after different thermal processing methods, such as microwave heating, puffing, and steaming [9,31,32]. However, in our results, it is important to point out that most of the Rg3 was further converted to the dehydrated type (Rk1 and Rg5) instead of the deglycosylated type (Rh2). According to the carbenium ion mechanism [33], the deglycosylation of ginsenosides at C-20 is easy than at other positions because it can generate tertiary carbenium ions with more stability. Then, the active tertiary alcohol is further eliminated during thermal processing. In addition, Zaitsev's rule can potentially give the reason why more Rg5 is enhanced than Rk1 during FIR treatment [34,35]. Our previous study suggested that 10 min of FIR treatment (between 60 and 120 °C) significantly increases the major ginsenoside (Rb1, Rb2, Rc, Rd) content of ginseng leaves, whereas it has no effect on minor ginsenosides (Rg3, Rh2) [13]. These results indicate that both the temperature and time of FIR treatment are important factors affecting ginsenoside variation in ginseng leaves. In addition, Chen et al. [36] reported that polar PPD-type ginsenosides (Rb1, Rb2, Rb3, Rc, and Rd) can be converted to Rg3, F2, Rh2, C-K, PPD, and Rk2 during the creation of black ginseng leaves. Thus, it is important to select suitable processing methods in consideration of the final products before processing ginseng leaves. Compared to FIR-190, a decreased pattern of each PPD-type minor ginsenoside was observed in FIR-200-treated ginseng leaves, which can be explained by the destruction of ginsenoside structures by high FIR energy.

2.3.2. PPT-Type Ginsenosides

Re (47.71 mg/g d.w.) and Rg1 (28.47 mg/g d.w.) were the major PPT-type ginsenosides (Figure 3B,C, and Table S1) found in non-FIR-treated ginseng leaves. During FIR treatment, the Re and Rg1 contents gradually decreased as the FIR temperature increased, exhibiting 85% and 83% reductions at FIR-200, respectively. However, the content of minor ginsenosides, Rg2, Rh1, F4, Rg6, Rh4, and Rk3, significantly increased as the FIR temperature increased, presenting their highest amounts at FIR-190. Hundred-fold increases in most minor ginsenosides described above were found at FIR-190, namely 319-fold (5.34 mg/g d.w.) in F4, 169-fold (3.48 mg/g d.w.) in Rg6, 187-fold (3.14 mg/g d.w.) in Rh4, and 314-fold (1.21 mg/g d.w.) in Rk3 compared to the untreated control (less than 0.03 mg/g d.w. of each ginsenoside). The increases in Rg2 and Rh1 at FIR-190 were relatively small, exhibiting 8.07- and 3.11-fold increases compared to the untreated control, respectively.

The conversion patterns of the PPT-type ginsenosides are shown in Figure 3B. The increase in the minor ginsenoside Rg2 was as a result of the glycosyl residue at C-20 of the major ginsenoside Re being detached due to high FIR energy [8,22]. As we mentioned in Section 2.3.1, the fact that the glycosyl residue at C-20 was easily deglycosylated than at C-6 can be explained by the carbenium ion mechanism [33]. Rg2 was then converted to F4 via dehydration between C-20 and C-22 positions or to Rg6 via dehydration between C-20 and C-21 positions. Similar results were reported by Kim et al. [11] using standard ginsenosides. The standard ginsenoside Re was deglycosylated and transformed to the less polar ginsenosides Rg2, F4, and Rg6 after heat processing at 120 °C. Likewise, the deglycosylation of Rg1 at C-20 caused an increase in the Rh1 content, followed by an increase in Rk3 and Rh4, resulting from the dehydration of Rh1 between C-20 and C-21/C-22. A similar result was reported by Hwang et al. [31]. They showed the variation in Rg1, Rh1, Rh4, and Rk3 in white ginseng heated at different temperatures at 20 MPa for 2 h. In our results, relatively higher amounts of F4 and Rg6 were transformed compared to Rh4 and Rk3. Differently, Chen et al. [36] investigated ginsenoside changes in black ginseng leaf products and suggested that the degradation of Re and Rg1 to minor ginsenosides, such as Rk3, Rh4, and PPT, is the main transforming PPT-type pattern. This can be explained by the different progressions of thermal processing (methods, time, temperature, etc.) affecting different substances [13,37–41], as Rk3 and Rh4 are the deglycosylated products of Rg6 and F4 at C-6, respectively. Ginsenoside mutations in ginseng leaves during heat treatment have not been widely studied in the literature, especially in terms of dry-heat treatment. In our results, it is important to point out that FIR treatment (especially FIR-190) significantly increased the content of each dehydrated minor ginsenoside compared to unprocessed ginseng leaves (Figure 3C). The moderate FIR energy that showed high dehydration efficiency at C-20 instead of deglycosylation at C-6 was considered a response to the significant increased minor ginsenosides F4, Rg6, Rh4, and Rk3. A decreased pattern of each minor ginsenoside at FIR-200 compared to those at FIR-190 was also observed in our results, which may be due to ginsenoside degradation caused by excessive FIR energy.

These results suggest that a higher FIR temperature is more advantageous for ginseng leaf processing and indicate that 190 °C is a suitable temperature. It is also recommended for obtaining more dehydrated minor ginsenosides based on the processing conditions. Compared to major ginsenosides, minor ginsenosides show stronger or even new biological activities [42], so numerous studies have been devoted to promote the transformation of major ginsenosides into profitable ginsenosides through thermal processing [31,43]. Although our previous results have shown that FIR treatment can promote the conversion of major ginsenosides to minor ginsenosides in ginseng roots and leaves [10,17], this study optimized the processing technology and significantly improved the conversion efficiency in ginseng leaves.

2.4. *Effects of FIR Treatment on Antioxidant Activities of Ginseng Leaves*

The effects of FIR treatment on the antioxidant activities of ginseng leaves are shown in Table 1. Non-FIR-processed ginseng leaves (0; IC50 = 2.33) showed relative lower DPPH free-radical-scavenging ability compared to FIR-treated leaves, especially FIR-180 (IC50 = 1.52; Table 1.) Similar to the variations in DPPH radical-scavenging ability, ABTS radical-scavenging ability initially increased from FIR-160 to FIR-180 and then decreased (Table 1). These patterns show a distinct indication that appropriate FIR treatment can significantly improve antioxidant activities of ginseng leaves. Similar results were also observed in *Oriza sativa* L. [14], *Camellia sinensis* var. sinensis [20], and *Hibiscus cannabinus* L. [21], where FIR treatment increased antioxidant activities compared to non-treated controls.

Table 1. Changes in antioxidant activities (DPPH/ABTS radical-scavenging ability) in ginseng leaves treated by FIR at different temperatures.

FIR Treatment (°C)	IC50 (mg/mL)	
	DPPH Radical-Scavenging Ability	**ABTS Radical-Scavenging Ability**
0	2.33 ± 0.03 e	2.61 ± 0.04 e
160	1.72 ± 0.02 c	1.88 ± 0.04 c
170	1.61 ± 0.02 d	1.75 ± 0.02 b
180	1.52 ± 0.02 a	1.66 ± 0.02 a
190	1.98 ± 0.02 b	2.15 ± 0.03 d
200	2.47 ± 0.03 f	2.47 ± 0.03 e

The values are expressed as the mean ± SE (n = 3). Different letters within the same column indicate significant differences at $p < 0.05$ on Tukey's HSD test.

In our results, extremely consistent variation patterns were exhibited between antioxidant activities and the TPC (Figure 1 and Table 1), which suggested that the antioxidant activities of ginseng leaves are most likely contributed by polyphenols rather than other compounds. The reason why stronger antioxidant activities in ginseng leaves were exhibited at FIR-160 to FIR-180 may be because FIR accelerates the release of antioxidants or leads to the transformation of certain substances to stronger-antioxidant-activity compounds, such as breakage of phenolic polymers [12,17,21]. The broken polyphenols exhibit better antioxidant activities. However, the reason why patterns of antioxidant activities decreased at higher FIR temperatures, FIR-190 to FIR-200, may be explained by the destruction of phenolic polymers and the degradation of simple formed phenolics. The degradation of phenolics may have been increasingly accelerated at temperatures higher than 180 °C in this study. It has been reported that ginseng leaves show stronger antioxidant activity compared to the roots [6,40]. In addition, several studies have evaluated the effects of different types of thermal processing on ginseng leaf antioxidant activities [43,44]. However, to the best of our knowledge, this is the first study to investigate the effect of FIR treatment on the antioxidant activity of ginseng leaves. Proper application of FIR can be considered a good method of improving the antioxidant activities and health benefits of herbal plants by increasing active polyphenols.

3. Materials and Methods

3.1. Chemicals

Organic solvents (HPLC grade) were purchased from Merck KGaA (Darmstadt, Germany). Folin–Ciocalteu reagent was purchased from Wako Pure Chemicals (Osaka, Japan). The standard compounds of panasenoside and kaempferol were purchased from Sigma Chemical Co. (St. Louis, MO, USA). Pure ginsenoside standards (98%) were purchased from ChromaDex (Santa Ana, CA, USA) and Ambo Institute (Daejeon, Korea).

3.2. Sample Collection and FIR Treatment

Six-year-old Korean ginseng leaves were used, which were collected from a ginseng farm in Wonju, Korea. Intact and undamaged leaves were washed with distilled water and then wiped with gauze. Fresh ginseng leaves were fully dried in an oven at 50 °C for 24 h and then ground using a grinder. The powder was meshed using a 200 μm sieve to obtain a uniform particle size of the powder. The powder was divided into two portions: (1) non-FIR-treated control (Con; 0) and (2) FIR treatments in an FIR dryer (HKD-10; Korea Energy Technology, Seoul, Korea) for 30 min at 160 °C (FIR-160), 170 °C (FIR-170), 180 °C (FIR-180), 190 °C (FIR-190), and 200 °C (FIR-200).

3.3. Bioactive Compound Analysis

3.3.1. Sample Extraction

First, 2 g of each sample was weighed and added to 100 mL of 80% methanol solution (*v/v*). Sample extraction was performed in a shaking incubator for 24 h at 30 °C. After extraction, the solution was centrifuged (05PR-22 centrifuge, Hitachi, Tokyo, Japan) at 500× *g* for 10 min at room temperature. The supernatants were gathered after centrifugation and filtered through Whatman No. 42 filter paper (Whatman Inc., Clifton, NJ, USA). The filtrate was concentrated using a vacuum rotary evaporator (Eyela Co., Tokyo, Japan) at 40 °C. The samples were freeze-dried in a vacuum freeze-dryer (Christ Alpaha 1–4, Germany). Finally, the samples were stored in a refrigerator at −20 °C for subsequent experiments.

3.3.2. Determination of TPC

The TPC was determined by the Folin–Ciocalteu method as follows: (1) Mix 1.9 mL of distilled water and 1.0 mL of Folin–Ciocalteu reagent in a tube, (2) prepare 0.1 mL of the sample solution (2 mg/mL, dissolved by 80% methanol) and add it to the tube, and (3) add 1.0 mL of 20% Na_2CO_3. The reaction mixture was incubated for 2 h at 25 °C. After incubation, the absorbance of the sample was recorded at 765 nm. The results were expressed as milligrams of gallic acid equivalents (GAE) per gram of dry weight (d.w.).

3.3.3. High-Performance Liquid Chromatography (HPLC) Analysis of Panasenoside and Kaempferol

Panasenoside and kaempferol contents of ginseng leaves were determined using HPLC. The equipment was an HPLC system (CBM-20A; Shimadzu Co, Ltd., Kyoto, Japan) with 2 gradient pump systems (LC-20AT; Shimadzu, Japan), an auto sample injector (SIL-20A; Shimadzu), a UV detector (SPD-20A; Shimadzu), and a column oven (CTO-20A; Shimadzu). Each sample extract (0.1 mg) was dissolved in 1 mL of 80% methanol (*v*/*v*) and filtered through a 0.22 μm membrane filter before sample injection into the HPLC system. HPLC separation was performed on an Inertsil ODS-SP C18 column (250 mm × 4.6 mm, 5 μm; GL Sciences, Tokyo, Japan). The injection volume of samples was 10 μL. The gradient running phase was programmed with the combination of solvent A (water with 0.1% trifluoroacetic acid) and solvent B (acetonitrile), in which solvent B was sequentially increased from 14% to 18% (0 to 10 min), 18% to 30% (10 to 20 min), 30% to 60% (20 to 30 min), 60% to 65% (30 to 33 min), 65% to 100% (33 to 40 min), and 100% to 100% (40 to 50 min) and then finally adjusted from 100% to 14% (50 to 65 min). The operating temperature was set at 35 °C. The flow rate of the mobile phase was kept at 1.0 mL per min. The detector was set at 355 nm for monitoring panasenoside and kaempferol.

3.3.4. HPLC analysis of Ginsenosides

The ginsenoside contents were determined by HPLC, the same equipment as mentioned before in Section 3.3.3. The prepared sample solution was filtered through a 0.22 μm membrane filter. The injection volume was 10 μL. A Kinetex C18 column (100 mm × 4.6 mm, 2.6 μm; Torrance, CA, USA) was used. The gradient running phase was programmed with the combination of solvent A (water) and solvent B (acetonitrile), in which solvent B was sequentially increased from 17% to 23% (0 to 30 min), 23% to 24% (30 to 35 min), 24% to 32% (35 to 45 min), 32% to 44% (45 to 48 min), 44% to 44% (48 to 52 min), 44% to 55% (52 to 65 min), 55% to 100% (65 to 85 min), and 100% to 100% (85 to 95 min) and finally adjusted from 100% to 17% (95 to 105 min). The operating temperature and flow rate were the same as those in the procedure mentioned in Section 3.3.3. The detector was set at 203 nm for monitoring ginsenosides.

3.4. Determination of Antioxidant Activities

The ginseng leaf powder (90 mg) was weighed, added to 30 mL of 80% methanol solution in a 50 mL tube, and extracted at 30 °C for 30 min with sonication. Then, the sample was centrifuged at 3500 r/min for 15 min. The supernatant was collected and filtered through a 0.22 μm membrane filter. The DPPH radical-scavenging activity was measured, as described by Eom et al. [17] with some modifications. Roughly, 1 mL of the sample solutions at different concentrations were mixed with 3 mL of DPPH solution. After the reaction for 30 min in the dark, the absorbance was measured at 517 nm. Next, the

ABTS radical-scavenging ability was measured, as described by Lim et al. [45] with some modifications. Briefly, 0.5 mL of sample solutions at different concentrations were taken in a test tube and 5 mL of the prepared ABTS solution was added. After the reaction at room temperature for 10 min in the dark, the absorbance was measured at 734 nm. IC50 values denote the concentration of the sample, which is required to scavenge 50% of DPPH and ABTS free radicals.

3.5. Statistical Analysis

The data were analyzed statistically using SAS software (Enterprise Guide version 7.1; SAS Institute Inc., Cary, NC, USA). The data between non-FIR-treated control and FIR treatment groups were analyzed by one-way analysis of variance. The significance between experimental groups was evaluated using Tukey's honestly significant difference (HSD) test at a $p < 0.05$ significance level.

4. Conclusions

This study studied in detail the effect of FIR on the representative health-beneficial compounds and antioxidant activities of ginseng leaves. Our results demonstrate that FIR treatment is a rapid and efficient method for producing deglycosylated bioactive compounds from ginseng leaves. FIR treatment at 180 °C is recommended to obtain more polyphoneols and kaempferol. However, the ideal FIR treatment temperature for producing deglycosylated minor ginsenosides, such as F4, Rg6, Rh4, Rk3, Rk1, and Rg5, is 190 °C. Both polyphenols and ginsenosides are compounds that are beneficial to human health, with different bioactivities that vary between individual compounds. However, with regard to the antioxidant activity of ginseng leaves, it seems to be mainly contributed by the highly accumulated polyphenols rather than ginsenosides. These findings will help the further use of ginseng leaves in health care products.

Supplementary Materials: The following are available online at https://www.mdpi.com/article/10.3390/molecules27154782/s1, Figure S1. The chemical structures of analyzed ginsensides in this study. PPD, protopanaxadiol; PPT, protopanaxadiol; glc, β-D-glucose; arap, α-L-arabinopyranosyl; xyl, β-D-xylose; araf, α-L-arabinofuranosyl; Table S1. Changes of PPT and PPD ginsenoside contents (mg/g dry weight) in ginseng leaves treated to different FIR temperatures.

Author Contributions: Conceptualization, C.W.J. and S.H.E.; methodology, S.D. and J.R.L.; software, S.D.; validation, S.D. and S.H.E.; formal analysis, S.D. and J.R.L.; investigation, X.W. and C.W.J.; resources, C.W.J. and S.H.E.; data curation, S.D., X.W., X.M.S., and H.S.G.; writing—original draft preparation, S.D.; writing—review and editing, S.H.E.; visualization, S.D.; supervision, C.W.J. and S.H.E. All authors have read and agreed to the published version of the manuscript.

Funding: This study was supported by the College Student Challenge Cup of Ludong University, Yantai, China; the New Talent Introduction Project (LY2013021) of Ludong University; and the Scientific Research Foundation for the Returned Overseas Chinese Scholars, State Education Ministry. This research was also funded by the National Research Foundation of Korea (NRF; Grant NRF-2022R1A2C100769511).

Institutional Review Board Statement: Not applicable.

Informed Consent Statement: Not applicable.

Data Availability Statement: The data presented in this study are available in this article.

Conflicts of Interest: The authors declare no conflict of interest.

References

1. Kim, J.H. Pharmacological and medical applications of *Panax ginseng* and *ginsenosides*: A review for use in cardiovascular diseases. *J. Ginseng Res.* **2018**, *42*, 264–269. [CrossRef]
2. Wang, H.; Peng, D.; Xie, J. Ginseng leaf-stem: Bioactive constituents and pharmacological functions. *Chin. Med.* **2009**, *4*, 1–8. [CrossRef]
3. Zhang, F.; Tang, S.; Zhao, L.; Yang, X.; Yao, Y.; Hou, Z.; Xue, P. Stem-leaves of *Panax* as a rich and sustainable source of less-polar ginsenosides: Comparison of ginsenosides from *Panax ginseng*, American ginseng and *Panax notoginseng* prepared by heating and acid treatment. *J. Ginseng Res.* **2021**, *45*, 163–175. [CrossRef]
4. Li, X.G.; Yan, Y.Z.; Jin, X.J.; Kim, Y.K.; Uddin, M.R.; Kim, Y.B.; Bae, H.H.; Kim, Y.C.; Lee, S.W.; Park, S.U. Ginsenoside content in the leaves and roots of *Panax ginseng* at different ages. *Life Sci. J.* **2012**, *9*, 679–683.
5. Kang, O.J.; Kim, J.S. Comparison of ginsenoside contents in different parts of Korean ginseng (*Panax ginseng* C.A. Meyer). *Prev. Nutr. Food Sci.* **2016**, *21*, 389–392. [CrossRef]
6. Chung, I.M.; Lim, J.J.; Ahn, M.S.; Jeong, H.N.; An, T.J.; Kim, S.H. Comparative phenolic compound profiles and antioxidative activity of the fruit, leaves, and roots of Korean ginseng (*Panax ginseng* Meyer) according to cultivation years. *J. Ginseng Res.* **2016**, *40*, 68–75. [CrossRef] [PubMed]
7. Yin, Q.; Han, X.; Chen, J.; Han, Z.; Shen, L.; Sun, W.; Chen, S. Identification of specific glycosyltransferases involved in flavonol glucoside biosynthesis in ginseng using integrative metabolite profiles, DIA proteomics, and phylogenetic analysis. *J. Agric. Food Chem.* **2021**, *69*, 1714–1726. [CrossRef]
8. Piao, X.M.; Huo, Y.; Kang, J.P.; Mathiyalagan, R.; Zhang, H.; Yang, D.U.; Kim, M.; Yang, D.C.; Kang, S.C.; Wang, Y.P. Diversity of ginsenoside profiles produced by various processing technologies. *Molecules* **2020**, *25*, 4390. [CrossRef] [PubMed]
9. An, Y.E.; Ahn, S.C.; Yang, D.C.; Park, S.J.; Kim, B.Y.; Baik, M.Y. Chemical conversion of ginsenosides in puffed red ginseng. *LWT* **2011**, *44*, 370–374. [CrossRef]
10. Jin, C.W.; Ghimeray, A.K.; Kim, W.W.; Kang, W.S.; Lim, H.T.; Lee, B.G.; Cho, D.H. Enhancement of ginsenoside compounds by far infrared irradiation in ultra fine powdered red ginseng (*Panax ginseng* C.A. Meyer). In Proceedings of the 2012 12th IEEE International Conference on Nanotechnology (IEEE-NANO), Birmingham, UK, 20–23 August 2012; pp. 1–4.
11. Kim, Y.J.; Yamabe, N.; Choi, P.; Lee, J.W.; Ham, J.; Kang, K.S. Efficient thermal deglycosylation of ginsenoside Rd and its contribution to the improved anticancer activity of ginseng. *J. Agric. Food Chem.* **2013**, *61*, 9185–9191. [CrossRef] [PubMed]
12. Aboud, S.A.; Altemimi, A.B.; Al-Hiiphy, A.; Lee, Y.C.; Cacciola, F. A comprehensive review on infrared heating applications in food processing. *Molecules* **2019**, *24*, 4125. [CrossRef]
13. Eom, S.H.; Seo, S.H.; Gimery, A.K.; Jin, C.W.; Kango, E.Y.; Kang, W.S.; Chung, I.M.; Cho, D.H. Changes of protopanaxadiol ginsenosides in ginseng leaves by far infrared and steaming heat treatments. *Korean J. Med. Crop Sci.* **2008**, *16*, 332–336.
14. Ratseewo, J.; Meeso, N.; Siriamornpun, S. Changes in amino acids and bioactive compounds of pigmented rice as affected by far-infrared radiation and hot air drying. *Food Chem.* **2020**, *306*, 125644. [CrossRef]
15. Azad, M.; Piao, J.P.; Park, C.H.; Cho, D.H. Far infrared irradiation enhances nutraceutical compounds and antioxidant properties in *Angelica gigas* Nakai powder. *Antioxidants* **2018**, *7*, 189. [CrossRef]
16. Kim, W.W.; Ghimeray, A.K.; Jin, C.W.; Eom, S.H.; Lee, B.G.; Kang, W.S.; Cho, D.H. Effect of far infrared drying on antioxidant property, anti-inflammatory activity, and inhibitory activity in A549 cells of Gamguk (*Chrysanthemum indicum* L.) flower. *Food Sci. Biotechnol.* **2012**, *21*, 261–265. [CrossRef]
17. Eom, S.H.; Park, H.J.; Seo, D.W.; Kim, W.W.; Cho, D.H. Stimulating effects of far-infrared ray radiation on the release of antioxidative phenolics in grape berries. *Food Sci. Biotechnol.* **2009**, *18*, 362–366.
18. Ghimeray, A.K.; Sharma, P.; Phoutaxay, P.; Salitxay, T.; Woo, S.H.; Park, S.U.; Park, C.H. Far infrared irradiation alters total polyphenol, total flavonoid, antioxidant property and quercetin production in tartary buckwheat sprout powder. *J. Cereal Sci.* **2014**, *59*, 167–172. [CrossRef]
19. Rim, A.R.; Jung, E.S.; Jo, S.C.; Lee, S.C. Effect of far-infrared irradiation and heat treatment on the antioxidant activity of extracts from peanut (*Arachis hypogaea*) shell. *J. Korean Soc. Food Sci. Nutr.* **2005**, *34*, 1114–1117.
20. Park, J.H.; Lee, J.M.; Cho, Y.J.; Kim, C.T.; Kim, C.J.; Nam, K.C.; Lee, S.C. Effect of far-infrared heater on the physicochemical characteristics of green tea during processing. *J. Food Biochem.* **2010**, *33*, 149–162. [CrossRef]
21. Jin, C.W.; Ghimeray, A.K.; Wang, L.; Xu, M.L.; Piao, J.P.; Cho, D.H. Far infrared assisted kenaf leaf tea preparation and its effect on phenolic compounds, antioxidant and ACE inhibitory activity. *J. Med. Plants Res.* **2013**, *7*, 1121–1128.
22. Escobedo, R.; Miranda, R.; Martínez, J. Infrared irradiation: Toward green chemistry, A review. *Int. J. Mol. Sci.* **2016**, *17*, 453. [CrossRef]
23. Tsikrika, K.; O'Brien, N.; Rai, D.K. The effect of high pressure processing on polyphenol oxidase activity, phytochemicals and proximate composition of irish potato cultivars. *Foods* **2019**, *8*, 517. [CrossRef]
24. Xu, L.; Cheng, J.R.; Liu, X.M.; Zhu, M.J. Effect of microencapsulated process on stability of mulberry polyphenol and oxidation property of dried minced pork slices during heat processing and storage. *LWT* **2019**, *100*, 62–68. [CrossRef]
25. Schmid, V.; Mayer-Miebach, E.; Behsnilian, D.; Briviba, K.; Karbstein, H.P.; Emin, M. Enrichment of starch-based extruded cereals with chokeberry (*Aronia melanocarpa*) pomace: Influence of processing conditions on techno-functional and sensory related properties, dietary fibre and polyphenol content as well as in vitro digestibility. *LWT* **2022**, *154*, 112610. [CrossRef]

26. Jo, J.E.; Kim, K.H.; Kim, M.S.; Choi, J.E.; Byun, M.W.; Yook, H.S. Antioxidant activity from different root parts of 6-year-old *Panax ginseng* C.A. Meyer (Yun-poong). *J. Korean Soc. Food Sci. Nutr.* **2011**, *40*, 493–499. [CrossRef]
27. Zhang, Y.J.; Zhang, Y.; Taha, A.A.; Ying, Y.; Li, X.P.; Chen, X.Y.; Ma, C. Subcritical water extraction of bioactive components from ginseng roots (*Panax ginseng* C.A. Mey). *Ind. Crops Prod.* **2018**, *117*, 118–127. [CrossRef]
28. Oliveira, A.; Leite, R.; Dantas, F.; Souza, V.; Júnior, J.; Souza, F.; Macedo, R. Thermal degradation kinetics of kaempferol and quercetin in the pre-formulated of the standardized extracts of *poincianella pyramidalis* (Tul.) LP Queiroz obtained by spray dryer. *Int. J. Pharm. Pharm. Sci.* **2017**, *9*, 123–128. [CrossRef]
29. Deng, S.; West, B.J.; Jensen, C.J. Thermal degradation of flavonol glycosides in noni leaves during roasting. *Adv. J. Food. Sci. Technol.* **2011**, *3*, 155–159.
30. Wang, J.; Fang, X.; Ge, L.; Cao, F.; Zhao, L.; Wang, Z.; Xiao, W. Antitumor, antioxidant and anti-inflammatory activities of kaempferol and its corresponding glycosides and the enzymatic preparation of kaempferol. *PLoS ONE* **2018**, *13*, e0197563. [CrossRef]
31. Hwang, I.G.; Kim, H.Y.; Joung, E.M.; Woo, K.S.; Jeong, J.H.; Yu, K.W.; Lee, J.; Jeong, H.S. Changes in ginsenosides and antioxidant activity of Korean ginseng (*Panax ginseng* C.A. Meyer) with heating temperature and pressure. *Food Sci. Biotechnol.* **2010**, *19*, 941–949. [CrossRef]
32. Jin, Y.; Kim, Y.J.; Jeon, J.N.; Wang, C.; Min, J.W.; Noh, H.Y.; Yang, D.C. Effect of white, red and black ginseng on physicochemical properties and ginsenosides. *Plant Foods Hum. Nutr.* **2015**, *70*, 141–145. [CrossRef] [PubMed]
33. Xiu, Y.; Zhao, H.X.; Gao, Y.; Liu, W.L.; Liu, S.Y. Chemical transformation of ginsenoside Re by a heteropoly acid investigated using HPLC-MS^n/HRMS. *New J. Chem.* **2016**, *40*, 9073. [CrossRef]
34. Rosado-Reyes, C.M.; Tsang, W.; Alecu, L.M.; Merchant, S.S.; Green, W.H. Dehydration of isobutanol and the elimination of water from fuel alcohols. *J. Phys. Chem. A* **2013**, *117*, 6724–6736. [CrossRef] [PubMed]
35. Toteva, M.M.; Richard, J.P. Mechanism for nucleophilic substitution and elimination reactions at tertiary carbon in largely aqueous solutions: Lifetime of a simple tertiary carbocation. *J. Am. Chem. Soc.* **1996**, *118*, 11434–11445. [CrossRef]
36. Chen, W.; Balan, P.; Popovich, D.G. Changes of ginsenoside composition in the creation of black ginseng leaf. *Molecules* **2020**, *25*, 2809. [CrossRef]
37. Shin, J.H.; Park, Y.J.; Kim, W.; Kim, D.O.; Kim, B.Y.; Lee, H.; Baik, M.Y. Change of ginsenoside profiles in processed ginseng by drying, steaming, and puffing. *J. Microbiol. Biotechnol.* **2019**, *29*, 222–229. [CrossRef]
38. Duan, S.C.; Kwon, S.J.; Eom, S.H. Effect of thermal processing on color, phenolic compounds, and antioxidant activity of faba bean (*Vicia faba* L.) leaves and seeds. *Antioxidants* **2021**, *10*, 1207. [CrossRef]
39. Qu, S.S.; Kwon, S.J.; Duan, S.C.; Lim, Y.J.; Eom, S.H. Isoflavone changes in immature and mature soybeans by thermal processing. *Molecules* **2021**, *26*, 7471. [CrossRef]
40. Lee, S.E.; Lee, S.W.; Bang, J.K.; Yu, Y.J.; Seong, N.S. Antioxidant activities of leaf, stem and root of *Panax ginseng* CA Meyer. *Korean J. Med. Crop Sci.* **2004**, *12*, 237–242.
41. Duan, S.; Kwon, S.-J.; Gil, C.S.; Eom, S.H. Improving the antioxidant activity and flavor of faba (*Vicia faba* L.) leaves by domestic cooking methods. *Antioxidants* **2022**, *11*, 931. [CrossRef]
42. He, Y.; Hu, Z.Y.; Li, A.; Zhu, Z.Z.; Yang, N.; Ying, Z.X.; He, J.R.; Wang, C.T.; Yin, S.; Cheng, S.Y. Recent advances in biotransformation of saponins. *Molecules* **2019**, *24*, 2365. [CrossRef]
43. Hwang, C.R.; Lee, S.H.; Jang, G.Y.; Hwang, I.G.; Kim, H.Y.; Woo, K.S.; Lee, J.; Jeong, H.S. Changes in ginsenoside compositions and antioxidant activities of hydroponic-cultured ginseng roots and leaves with heating temperature. *J. Ginseng Res.* **2014**, *38*, 180–186. [CrossRef]
44. Bae, M.J.; Kim, S.J.; Ye, E.J.; Nam, H.S.; Park, E.M. Antioxidant activity of tea made from Korean mountain-cultivated ginseng leaves and its influence on lipid metabolism. *J. Korean Soc. Food Cult.* **2009**, *24*, 77–83.
45. Lim, Y.J.; Kwon, S.J.; Qu, S.; Kim, D.G.; Eom, S.H. Antioxidant contributors in seed, seed coat, and cotyledon of γ-ray-induced soybean mutant lines with different seed coat colors. *Antioxidants* **2021**, *10*, 353. [CrossRef]

molecules

MDPI

Review

Polysaccharides as Carriers of Polyphenols: Comparison of Freeze-Drying and Spray-Drying as Encapsulation Techniques

Ivana Buljeta [1], Anita Pichler [1], Josip Šimunović [2] and Mirela Kopjar [1,*]

1 Faculty of Food Technology Osijek, Josip Juraj Strossmayer University of Osijek, F. Kuhača 18, 31000 Osijek, Croatia
2 Department of Food, Bioprocessing and Nutrition Sciences, North Carolina State University, Raleigh, NC 27695, USA
* Correspondence: mirela.kopjar@ptfos.hr; Tel.: +385-3122-4309

Abstract: Polyphenols have received great attention as important phytochemicals beneficial for human health. They have a protective effect against cardiovascular disease, obesity, cancer and diabetes. The utilization of polyphenols as natural antioxidants, functional ingredients and supplements is limited due to their low stability caused by environmental and processing conditions, such as heat, light, oxygen, pH, enzymes and so forth. These disadvantages are overcome by the encapsulation of polyphenols by different methods in the presence of polyphenolic carriers. Different encapsulation technologies have been established with the purpose of decreasing polyphenol sensitivity and the creation of more efficient delivery systems. Among them, spray-drying and freeze-drying are the most common methods for polyphenol encapsulation. This review will provide an overview of scientific studies in which polyphenols from different sources were encapsulated using these two drying methods, as well as the impact of different polysaccharides used as carriers for encapsulation.

Keywords: polysaccharides; polyphenols; freeze-drying; spray-drying

Citation: Buljeta, I.; Pichler, A.; Šimunović, J.; Kopjar, M. Polysaccharides as Carriers of Polyphenols: Comparison of Freeze-Drying and Spray-Drying as Encapsulation Techniques. *Molecules* **2022**, *27*, 5069. https://doi.org/10.3390/molecules27165069

Academic Editor: Nour Eddine Es-Safi

Received: 21 July 2022
Accepted: 7 August 2022
Published: 9 August 2022

1. Introduction

Polyphenols are secondary plant metabolites consisting of an aromatic ring to which one or more hydroxyl groups are attached [1]. These compounds are synthesized through plant development and/or as a plant's response to environmental stress conditions [2]. Even when they are at low concentrations in plants, polyphenols protect them from predators or ultraviolet damage [3]. They are known as natural antioxidants and therefore have many beneficial effects on health (e.g., antimicrobial, anti-inflammatory, antioxidant effect, etc.) [4–6]. The health benefits of polyphenols are influenced by the matrix in which they are processed and ultimately consumed [7]. One well-known property of polyphenols is the positive influence on diabetes and obesity due to the possibility of inhibition of digestive enzymes such as α-glucosidase and α-amylase [8]. Anthocyanins are a group of polyphenols responsible for the red, blue and purple color of fruit and vegetables. The major anthocyanin found in most plants is cyanidin-3-glucoside, correlated with reduced reactive oxygen species (ROS) levels and antioxidant potential in in vitro conditions [9,10]. Flavan-3-ols are a subgroup of flavonoids and their main representatives are catechin, epicatechin, epigallocatechin and epigallocatechin-3-gallate. These compounds are abundantly found in green tea, strawberries and black grapes. Studies showed a positive effect of catechin in Alzheimer's and Parkinson's diseases, diabetes and in cancer treatment [11]. Gallic acid, a polyphenol from a group of phenolic acids, a subgroup of hydroxybenzoic acids was the subject of many studies that have proven its significant antioxidant, anticarcinogenic, antimicrobial and antimutagenic effects [12].

Due to the presence of unsaturated bonds in their structures, polyphenols are sensitive to various environmental conditions such as the presence of oxygen, light and water [2]. The presence of water is the most important factor, due to its essentiality in most chemical reactions [13,14]. In the food industry, thermal processes are mostly used to obtain edible, microbiologically safe foods, to improve digestibility, and to modulate their textures, flavors and colors [14]. During these processes, structural changes occur leading to the degradation of polyphenols which are often ignored [14]. In order to maintain their stability, they need to be protected, and one of the possible ways is encapsulation in which polysaccharides, proteins, lipids or combinations thereof can be utilized as carriers. In that way, the preservation of polyphenolic properties is achieved over longer periods because the carrier materials represent a barrier to oxygen and water [2]. By encapsulation of polyphenols, besides increased stability, mitigation of unpleasant tastes or flavors, controlled release, improved aqueous solubility and bioavailability can be achieved [3]. Drying has effect on the material's appearance and chemical composition. It also prolongs shelf life and inhibits enzymatic degradation and microbial growth of materials or foods [15]. Adequate selection of a drying method and operating conditions yields foods with slight changes in appearance and maximum retention of bioactive compounds [15].

Spray-drying is a commonly used method for encapsulating due to its simple regulation and control, limited cost, and continuous operation [16]. Freeze-drying is also often used for encapsulation of thermosensitive compounds and materials with some disadvantages such as higher unit cost and long processing time [17]. Suitable selection of carrier and encapsulation technique leads to successful incorporation and retention of bioactive compounds [2]. Food enriched with encapsulated polyphenols can be a versatile and cost-effective approach [14]. In addition, this approach enables other features such as controlled release, improved bioaccessibility and bioavailability for absorption [14].

This paper will provide the literature review of spray-drying and freeze-drying for the encapsulation of polyphenols from different sources. Moreover, with emphasis on polysaccharides, the influence of carrier materials on polyphenol encapsulation will be reviewed.

2. Polyphenols

Polyphenols include various different compounds divided into several classes: phenolic acids, flavonoids, stilbenoids, tannins, coumarins, and polymeric lignans (Figure 1) [18]. Their chemical structure may vary from simple to complex. The largest group of polyphenols is that of flavonoids, divided into several subgroups (flavonols, flavones, flavanols, flavanones, anthocyanidins, isoflavonoids, and chalcones) [19]. Their structure consists of two phenyl groups linked with a three-carbon bridge. According to the degree of oxidation and unsaturation of the three-carbon segment, they differ from each other. Different sugar molecules can be attached to the hydroxyl groups of flavonoids. They are usually in a glycosidic form which improves their solubility in water. Acylation of the glycosides where sugar hydroxyls are derivatized with acid (such as ferulic and acetic acids) is also common. The interconnection between several basic units of polyphenols makes larger and more complex structures, such as hydrolysable tannins and condensed tannins [20]. Phenolic acids represent a large group of hydrophilic polyphenols and they are constituted of a single phenyl ring [14]. Due to the diversity in the structures of polyphenols, they possess different properties (such as solubility and polarity) [20].

Polyphenols exist ubiquitously in vegetables and fruits and their consummation is very desirable [14]. Various positive bioactivities of polyphenols toward pathologic conditions are known, through their antioxidant properties. These molecules are capable of donating hydrogen atoms and electrons [14]. One such activity is the anticancer property of polyphenols [21]. Hollman et al. [22] reviewed the antioxidant activities of polyphenols within the organism and their positive impact on cardiovascular health. In addition, polyphenols contribute to the sensory quality of the products (wine, jellies, juices, chocolate, etc.). They affect color, bitterness, turbidity, etc. [6,23]. Due to their many functional properties, polyphenols are of great interest to the food, chemical and pharmaceutical industries [2].

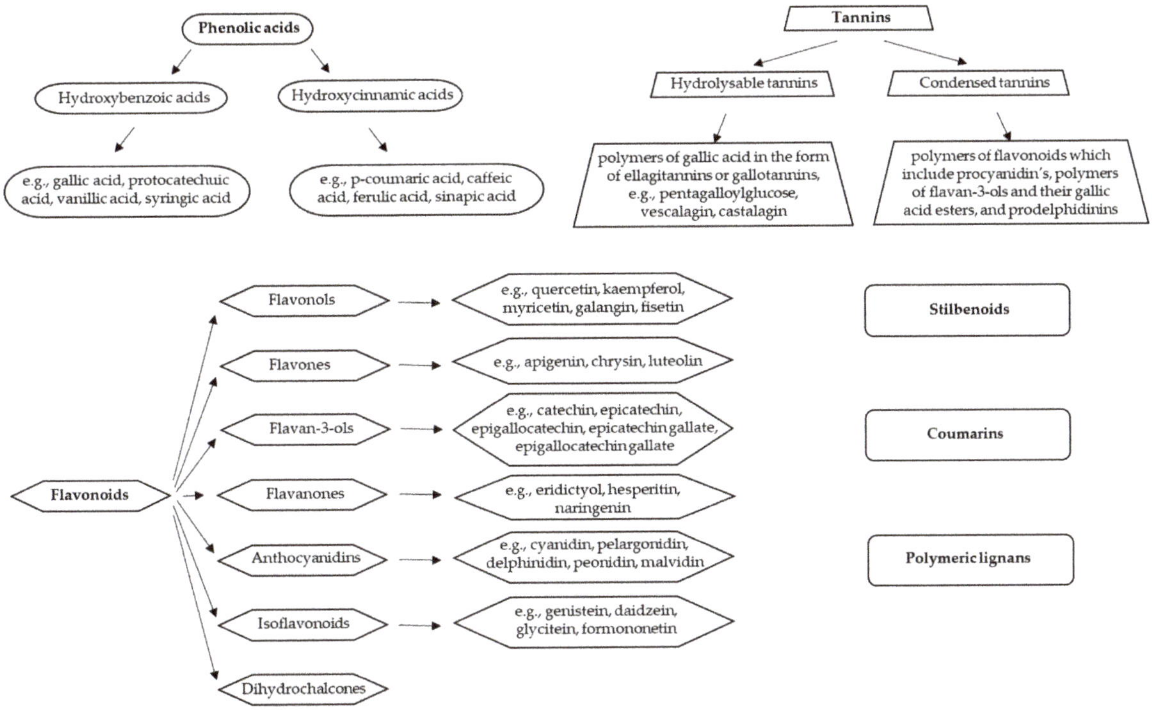

Figure 1. Classification of polyphenols and the most important representatives of individual groups (adapted from Dobson et al. [18]).

3. Encapsulation

Encapsulation is the process of entrapment of active compounds into particles to enable isolation or controlled release of given compounds. The main role of encapsulation is the protection of sensitive active compounds from degradation [17]. Various techniques can be used for encapsulation depending on core material, size of required particles, physical state, or sensitivity to high temperature [24]. These techniques include freeze-drying, spray-draying, extrusion, fluidized bed coating, spray-cooling/chilling, coacervation, liposome entrapment, cocrystallization, vacuum-drying, centrifugal suspension separation, nanoencapsulation, molecular inclusion and emulsification [13]. Encapsulation with coacervation is based on phase separation of a hydrocolloid from an initial solution and subsequent deposition of the formed coacervate phase around an active ingredient suspended in media. It is considered an expensive technique but very beneficial for high-value compounds [25]. Extrusion is a process based on passing a polymer solution with active ingredients through a nozzle (or syringes) into a gelling solution. Usually the used wall material is sodium alginate while calcium chloride solution serves for capsule forming. It is easy to execute on a laboratory scale with long shelf-life capsules, while scale-up of this technique is expensive and demanding with a limited choice of wall materials [26]. Emulsification includes the dispersion of one liquid into the other (two immiscible liquids) in the form of droplets. An emulsifier is required for stabilization and by application of drying, a powder form of encapsulates can be achieved [26]. The molecular inclusion method is also known as host-guest complexation. Apolar guest molecules are trapped inside the apolar cavity of host molecules (such as cyclodextrins) through non-covalent bonds [26]. Cocrystallization techniques include modification of the crystalline structure of sucrose to an irregular agglomerated crystal with porous structure in which active ingredient can be incorporated [25]. Fluidized bed coating is also referred to as fluidized bed processing, air suspension coating, or spray coating. The principle is that coating is applied to the particles suspended in the air. For this technique a wide range of coating materials such as aqueous solutions of cellulose, starch derivates, gums, or proteins is suitable [27]. Spray chilling is also known as spray cooling, prilling, or spray congealing. The basic principle is similar to spray drying with the key difference of using a cooling chamber instead of a drying chamber. Regarding coating materials, only lipid-based materials (fats, waxes, fatty acids, fatty alcohols and polyethylene glycols) are used [27]. Nanoencapsulation is an innovative trend in the field of food technologies. The final results are particles of diameters ranging from 1 to 1000 nm. The term nanoparticles includes nanospheres and nanocapsules. The first one has a matrix-type structure where the active ingredients can be adsorbed at the sphere surface or encapsulated in particles. In nanocapsules, the active ingredient is limited to a cavity with an inner liquid core surrounded by a polymeric membrane. Nanoparticles have a larger surface area, increased solubility, enhanced bioavailability and improved controlled release [25]. Freeze-drying and spray-drying are the most frequently employed methods for removing water from foods with encapsulation effects [28]. A schematic view of these methods is presented in Figure 2.

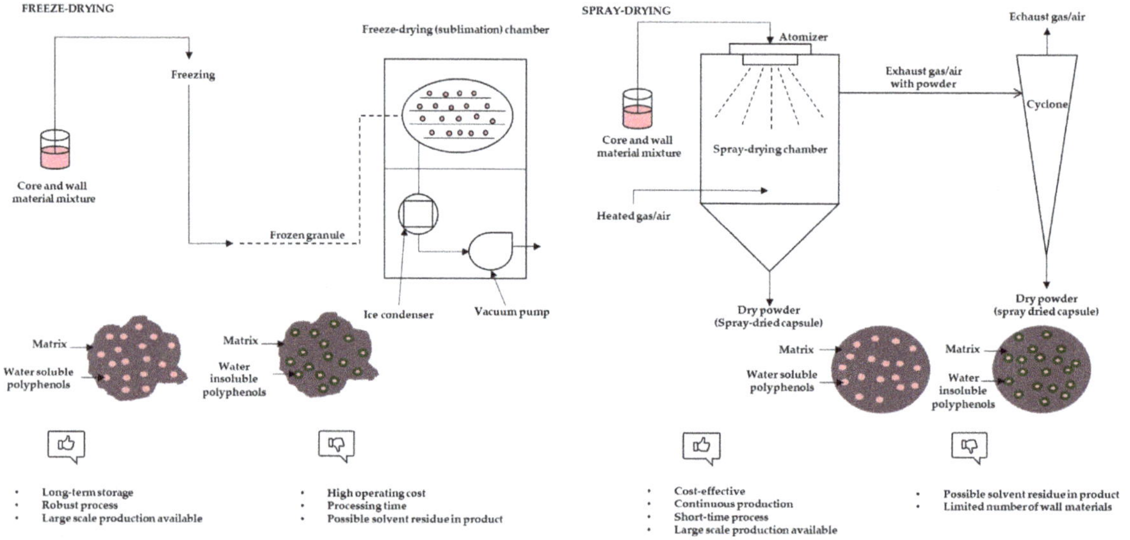

Figure 2. Schematic view of freeze-drying and spray-drying process, characteristics of particles and advantages and disadvantages of methods (adapted from Fang and Bhandari, [25]; Grgić et al., [26]).

The wide range of polyphenol biological activities can be restricted due to their low stability, low bioavailability, and unpleasant flavor [14]. Encapsulation of polyphenols improves their stability during storage and can also be used to achieve the masking of unpleasant flavors in foods (such as bitter taste and astringency) [2]. Food with high intensity of bitterness and astringency elicit negative consumer reactions. Compounds responsible for that are flavanols and flavonols. Flavan-3-ol monomers such as catechin, epicatechin, epigallocatechin, epicatechin gallate and epigallocatechin gallate as well as their oligomers proanthocyanidins (condensed tannins) are abundant in wine and tea. It can be said that flavanols are the main compounds that cause bitterness and astringency in tea and red wine (with the exception of caffeine in tea) [29]. The method selected for encapsulation must be based on the polyphenol's characteristics such as chemical structure, thermophysical stability, solubility, affinity to coating material, target properties such as particle size and morphology, among others [14].

3.1. Freeze-Drying

Freeze-drying, also known as lyophilization, is a drying technique based on the phenomenon of sublimation. This technique allows the long-term preservation of heat-sensitive and oxidation-prone compounds, as well as foods and other biological materials, since it is conducted at low temperatures and under vacuum [30]. This method does have disadvantages such as higher unit cost and processing time [17]. Despite those issues, freeze-drying is widely used to obtain high-value food products and is considered a standard method for encapsulation in most research studies [30].

The freeze-drying process includes the complete freezing of samples, ice sublimation (primary drying) and desorption of remaining unfrozen/bond water (secondary drying). In the first step, the freezing rate determines the formation and size of ice crystals. Large ice crystals formed by the slow rate of freezing can sublimate easily and increase the primary drying rate. In primary drying, the shelf temperature will increase using a vacuum to start the sublimation. It is necessary that product temperature is 2–3 °C below the temperature level of collapse, at which the product can lose its macroscopic structure. The endpoint of the primary drying phase is a key parameter to determine because the increased temperature (in the secondary drying) before the sublimation of all ice could collapse the final product quality [30].

3.2. Spray-Drying

In the food industry spray-drying is the most frequently used technique. It is economical, flexible and can be used continuously with easy scale-up [13]. Advantages over the other methods are higher effectiveness and shorter drying time [16]. This technique is based on transforming a material from liquid form into powder form [24]. However, when spray-drying is used for polyphenol encapsulation, conditions must be optimized in order to avoid an accelerated degradation [17]. Depending on process conditions and formulations, spray-drying micrometric capsules can be core-and-shell, multiple core or matrix type [14]. Before spray-drying, a solution or suspension of polyphenols with carriers must be obtained, followed by atomizing into a hot air stream.

3.3. Particle Morphology

Spray-dried particles are usually spherical with varying diameters and concavities, regardless of the choice of coating material [17]. Mean size range of such particles range from 10 μm to 100 μm [25]. The formation of concavities is associated with shrinkage of the particles due to dramatic loss of moisture after cooling. On the other hand, powders obtained by freeze-drying have a flake-like structure or one that resemble broken glass. The reason could be the low temperature of the process which results in the absence of forces to break the frozen liquid into droplets. Differences in surface morphology of freeze-dried powders could be due to different coating agents. For example, powder with soybean protein and maltodextrin as coating materials had a spherical porous structure, while powders with only maltodextrin lost their porous structure [17]. The final particle size of freeze-dried powders depends on the grinding procedure, not on the drying process [31].

3.4. Polyphenol Carriers

Materials used for encapsulation, which makes a protective shell, should be biodegradable and food-grade. They also must be able to establish a barrier between an internal phase and its surroundings [13]. Selection of coating material influences encapsulation efficiency and encapsulates stability [24]. Commonly used materials are carbohydrates such as maltodextrin, cyclodextrins, gum Arabic and modified starch. These materials lead to an increase in the glass transition temperature of the dried product. By trapping a bioactive compound, they enable its preservation against stickiness, temperature, enzymatic and chemical changes [13].

Maltodextrin is one of the most often used carriers, having a low bulk density and viscosity and high solubility at high solids contents [16,24]. It is obtained by partial hydrolysis of starch using an enzyme or acid. Maltodextrin has the ability to retain volatile compounds. Its disadvantages, such as low emulsibility, are often overcome by combining maltodextrin with other materials [16].

Cyclodextrins are safe for food applications and broadly studied as hosts for encapsulation [3]. The commonly used cyclodextrins are α-, β- and γ-cyclodextrin. They possess a hydrophobic central cavity and a hydrophilic external part [32]. Due to this structure, guest molecules (different organic and inorganic molecules) are able to be accommodated into the cavity. The hydrophilic external surface provides aqueous solubility [3,24]. The application of cyclodextrins in spray-drying is limited due to their low water solubility (1.8%). This limitation can be overcome by cyclodextrin modification. An example of such modification is hydroxypropyl-β-cyclodextrin which has a higher water solubility (60%) and thus can be subjected to drying techniques [24].

Gum Arabic consists of galactose, rhamnose, arabinose, 4-O-methyglucuronic acid and glucuronic acid and its natural source is the acacia plant (stems and branches). Furthermore, its low viscosity and high solubility, stable emulsions and high retention rates of volatile compounds have enabled a broad range of applications of this polymer [16]. On the other hand, some disadvantages such as low production yield and a consequently higher price make it less accessible [33].

4. Application of Spray-Drying and Freeze-Drying for Encapsulation of Polyphenols

During freeze-drying of foods rich in polyphenols, cells are disrupted, and therefore exposed to an increased enzyme activity (polyphenol oxidase and peroxidase enzyme) upon thawing, and the degradation of the polyphenols can occur [34,35]. However, some studies have shown that the amount of polyphenols may increase after this process. The flavonol content in freeze-dried onions increased, which can be attributed to the release of polyphenols from the matrix [36]. Wilkowska et al. [24] observed that freeze-dried powders had 1.5 times higher retention of anthocyanins than spray-dried ones. From studying the values of total polyphenols content, it has been noticed that when applying spray-drying, 73% of compounds were lost. Encapsulates of polyphenols in coffee grounds extract achieved by freeze-drying and spray-drying with maltodextrin, gum Arabic and maltodextrin:gum Arabic (1:1) as carrier materials were evaluated for total polyphenols content and flavonoid content. The results showed that freeze-drying was a more effective technique for retention of polyphenols and flavonoids and maltodextrin a more efficient carrier. On the other hand, spray-dried particles possessed a higher antioxidant activity than freeze-dried ones [2]. One interesting investigation was conducted on developing novel protein ingredients fortified with blackcurrant concentrate. As a source of protein, they used whey protein isolate and freeze-drying and spray-drying techniques. Encapsulates obtained by spray-drying possessed a higher total polyphenols content, anthocyanins content and encapsulation efficiency compared to the freeze-dried ones [28]. Robert et al. [37] encapsulated polyphenols from pomegranate juice and ethanolic extract by spray-drying, and observed a higher encapsulation efficiency when soy protein isolates were used compared to the maltodextrin. On the other hand, capsules with maltodextrin, stored at 60 °C in an oven for 56 days, resulted in a lower degradation of polyphenols and anthocyanins. Wu et al. [38] investigated the physicochemical properties and nutritional characteristics of functional cookies with incorporated encapsulated blackcurrant polyphenols. For the preparation of encapsulates, whey protein isolate and blackcurrant concentrate were used and the applied encapsulating techniques used were freeze-drying and spray-drying. The results of total polyphenols content were higher for enriched cookies with freeze-dried encapsulates than for those with spray-dried encapsulates.

Ersus and Yurdagel [39] observed that during spray-drying, maltodextrins with higher DE (equivalents of dextrose) are more sensitive to higher outlet air temperatures. Heating could lead to structural deformations due to shorter chains and oxidation of free glucose functional groups at the open ends. Their results confirmed the encapsulation of black carrot anthocyanins using maltodextrin DE 20-21 for 20% feed solid content and

160–180 °C drying temperatures. Gomes et al. [40] obtained a higher retention of papaya pulp polyphenol and flavonoid compounds in spray-dried products than in freeze-dried products. Vanillic acid had an enormous decrease of 76% after freeze-drying. Enzymatic reactions with the action of peroxidase and polyphenol oxidase are likely to occur in the freeze-drying process. Moreover, the disrupted material structure caused by the formation of ice crystals and lower exposure to oxygen can cause the liberation of these enzymes. Furthermore, artepillin C concentration increased three times after drying, which can lead to the false-positive results of the spray-drying technique. It is known that processes at high temperatures may release more bound polyphenols which cannot be detected in fresh samples [40]. Saikia et al. [41] encapsulated polyphenols from *Averrhoa carambola* pomace using maltodextrin and freeze-drying and spray-drying methods. The authors obtained a higher encapsulation efficiency in freeze-dried encapsulates. It might be that some polyphenols are destroyed during spray-drying, due to their sensitivity to heat. During the spray-drying process, fine misty droplets with increased surface are obtained and due to that higher surface, they are more exposed to heat. Moreover, during atomization, some amount of carrier can be eliminated from the core material and the partially covered capsules thus obtained can be destroyed by heat. During the freeze-drying process, no atomization or heat exposure are present. These authors also observed a decrease in surface polyphenols with higher maltodextrin content [41]. On the other hand, high inlet temperatures applied in spray-drying are short-lived, so this technique is less destructive for bioactive compounds compared to the other conventional thermal processes [42]. It has to be taken into consideration that in the freeze-drying process, obtained powders are ground and this process increases the possibility of contact with air, resulting in oxidation reactions [42,43]. In the freeze-drying process, the formation of a sawdust-like form is usual, leading to a lower surface area/volume ratio. Additionally, by the spray-drying process, smaller sized microspheres with a larger surface area were obtained using the spray-drying process (for the same amount of material as for freeze-drying), which led to the deterioration of the surface polyphenols [2]. Considering the short time of exposure to high temperatures during spray-drying, this technique seems good for the encapsulation of polyphenols. Dealing with thermosensitive and highly valuable materials, freeze-drying is a suitable method [33]. Tables 1–3 present the studies on the encapsulation of polyphenols using freeze-drying and spray-drying techniques.

Table 1. Studies of spray-drying application for encapsulation of polyphenols from different sources.

Bioactive Material		Reference
Roselle (*Hibiscus sabdariffa* L.) extract		[44]
Carrier	Maltodextrin, gelatin, pectin, carboxymethyl cellulose, carrageenan, gum Arabic and whey protein	
Conditions	0.5 mm—nozzle diameter; 400 kPa—compressor air pressure; 7.5 mL/min—feed flow rate; 155 °C—inlet temperature; 55 °C—outlet temperature; 56 m^3/h—air flow rate	
Observations	Pectin can be a suitable carrier of roselle polyphenols because it retained the highest amount of polyphenols (98.20 mg/100 mg)	
Plum (*Prunus salicina* Lindl.) polyphenols		[45]
Carrier	Maltodextrin, gum Arabic, gelatin, chitosan and β-cyclodextrin	
Conditions	150 °C—inlet temperature; 85 °C—outlet temperature; 850 mL/h—feed flow rate; 90 MPa—atomization pressure; 1.15 m^3/min—blower rate; 0.7 mm—nozzle diameter	
Observations	Powders obtained with maltodextrin/chitosan possessed the highest stability as well as retention of total polyphenols (94%) after storage (60 days at 25 °C)	

Table 1. *Cont.*

Bioactive Material		Reference
Pomegranate juice anthocyanins		[46]
Carrier	Gum Arabic, modified starch from waxy maize and maltodextrin	
Conditions	1 mm—nozzle diameter; 1 kg/h—mass flow rate; 162–170 °C—inlet temperatures; 89–93 °C—outlet temperatures; 500 m^3/h—air flow rate	
Observations	By using gum Arabic:modified starch from waxy maize (1:1) mixture, up to 70% of total monomeric anthocyanins were retained	
Cinnamon (*Cinnamomum zeylanicum*) proanthocyanidins		[47]
Carrier	Maltodextrin	
Conditions	1.2 mm—nozzle diameter; 130 °C and 160 °C—drying temperatures; 40 mL/min—feed flow rate	
Observations	The highest retention of proanthocyanidins (100%) was obtained in the sample with 20% of maltodextrin and dried at 160 °C; during 90 days of storage, maltodextrin contributed to stability of proanthocyanidins	
Polyphenols of *Orthosiphon stamineus* leaves		[48]
Carrier	Maltodextrin and whey protein isolate	
Conditions	0.5 mm—atomizer; 180 °C—inlet temperature; 407 mL/h—feed flow rate	
Observations	By using 5.33 wt.% of maltodextrin, the highest retention of sinensetin (82.24%), rosmarinic acid (82.67%) and eupatorine (80.19%) was obtained	
Fadogia ancylantha, Tussilago farfara and *Melissa officinalis* extracts		[49]
Carrier	Maltodextrin and apple pectin	
Conditions	120 °C—inlet temperature; 69–71 °C—outlet temperature; 5 mL/min—feed flow rate; 0.5 mm—nozzle diameter; 500 L/h—drying air flow; 6 bar—air pressure; 100%—aspiration	
Observations	By using 10:1 maltodextrin:pectin ratio, 3% w/v of raw dried extract was encapsulated; loading efficiency values were very high: 80.1%, 91.5% and 97.2% for Tussilago, Melissa and Fadogia polyphenols, respectively	
Black carrot (*Daucus carota* L.)		[39]
Carrier	Maltodextrins (DE 10, DE 28–31 and DE 20–23)	
Conditions	160 °C, 180 °C and 200 °C—inlet temperatures; 107 °C, 118 °C and 131 °C—outlet temperatures; 5 mL/min—feed flow rate	
Observations	Powders with maltodextrin DE 20–23 contained the highest anthocyanin concentration (630 mg anthocyanin/100 g dry matter of powder)	
Mountain tea (*Sideritis stricta*) extract		[50]
Carrier	β-cyclodextrin, maltodextrin and gum Arabic	
Conditions	145 °C, 155 °C and 165 °C—inlet temperatures; 75 °C—outlet temperature; 500 L/h—flow rate; 70%—aspiration rate; 240–640 mL/h—feed rate	
Observations	By increasing inlet air temperature from 145 °C to 155 °C, a 4% increase in total polyphenols content occurred, but a further increase in inlet temperature had a negative effect; an increase in the concentration of carriers resulted in a decrease in total polyphenols content (18.96 g/100 g—without carrier, 7.61 g/100 g—for 3 g/100 g of carrier, and 5.43 g/100 g—for 5 g/100 g of carrier)	

Table 1. *Cont.*

Bioactive Material		Reference
Cactus pear (*Opuntia ficus-indica*)		[51]
Carrier	Maltodextrin and inulin	
Conditions	140–160 °C and 120–160 °C—inlet temperatures for maltodextrin and inulin, respectively; 600 L/h—air flow; 10 mL/min—feed rate; 20 psi—atomization pressure	
Observations	Optimal conditions for encapsulation with maltodextrin were 3:1 ratio of core/coating material and 140 °C inlet temperature while for inulin the optimal ratio was 3:1 for encapsulation of pulp extract and 5:1 for encapsulation of ethanolic extract and 120 °C inlet temperature; concentration of polyphenols after encapsulation ranged as follows: cactus ethanolic extract + inulin (2410 mg GAE/g powder) > cactus pulp + maltodextrin (2135 mg GAE/g powder) > cactus pulp + inulin (2028 mg GAE/g powder) > cactus ethanolic extract + maltodextrin (1812 mg GAE/g powder)	
Açai (*Euterpe oleracea* Mart.) juice		[52]
Carrier	Maltodextrin (DE 10 and DE 20), gum Arabic, and tapioca starch	
Conditions	1.5 mm—nozzle diameter; 0.06 MPa—compressor air pressure; 15 g/min—feed flow rate; 140 °C—inlet temperature; 78 °C—outlet temperature; 73 m^3/h—air flow rate	
Observations	Powders obtained with tapioca starch possessed the lowest anthocyanin content (3247.15 mg/100 g) while between the two types of maltodextrin (for DE 10 3436.85 mg/100 g and for DE 20 3402.30 mg/100 g) and gum Arabic (3415.96 mg/100 g) no statistical difference was determined in anthocyanin content	
Jaboticaba (*Myrciaria jaboticaba*) peel extracts		[53]
Carrier	Maltodextrin, gum Arabic, and modified starch from waxy maize	
Conditions	140 °C, 160 °C and 180 °C—inlet temperatures; 360 mL/h—feed flow rate; 20 mbar—vacuum; 28 m^3/h—aspiration	
Observations	The highest anthocyanins retention was obtained at 160 °C air drying temperature using maltodextrin (99.02%) and gum Arabic/maltodextrin (100%) as carriers; a combination of maltodextrin with starch resulted in the lowest anthocyanins retention (around 80%) regardless of air drying temperatures	
Pomegranate (*Punica granatum*) polyphenols and anthocyanins of juice and ethanolic extracts		[37]
Carrier	Maltodextrin and soybean protein isolates	
Conditions	140–160 °C and 100–140 °C—inlet temperatures for maltodextrin and soybean protein isolates, respectively; 600 L/h—air flow; 10 mL/min—feed rate; 20 psi—atomization pressure	
Observations	Encapsulation efficiency of polyphenols was better in the soybean protein isolates matrix (76.2% for pomegranate juice and 82.9% for pomegranate ethanolic extract) than for maltodextrin (53.3% for pomegranate juice and 71.0% for pomegranate ethanolic extract); the highest encapsulation efficiency of anthocyanins from pomegranate ethanolic extract was in the soybean protein isolates matrix (100%) while the highest efficiency of pomegranate juice anthocyanins was in the maltodextrin matrix (86.6%)	
Rosemary (*Rosmarinus officinalis* L.) leaves polyphenols		[54]
Carrier	Maltodextrins (MDE 10 and 21), whey protein isolates and polyglycerol polyricinoleate	
Conditions	175 °C—inlet temperature; 90 °C—outlet temperature; 600 L/h—drying air flow; 18%—feed flow rate; 700 kPa compressor air pressure	
Observations	Higher encapsulation efficiency (around 42%) of total polyphenols was obtained when higher amounts of protein (4%) were used, pointing to the important impact of protein in that target delivery system; encapsulation efficiency with 2% of whey protein isolates was around 28%; type of maltodextrin showed no significant impact on encapsulation efficiency	

Table 1. *Cont.*

Bioactive Material		Reference
Epigallocatechin-3-gallate (EGCG)		[55]
Carrier	Gum Arabic and maltodextrin	
Conditions	160 °C—inlet temperature; 60 °C—outlet temperature	
Observations	Concentration of EGCG at surface was 8% and loading efficiency (EGCG in the inner space) was 85%	
Turmeric oleoresin (curcumin)		[56]
Carrier	Maltodextrin and gum Arabic	
Conditions	5 mL/min—feed flow rate; 150 °C, 175 °C and 200 °C—inlet air temperatures; 90 °C—outlet air temperature	
Observations	Gum Arabic used as a carrier at the inlet air temperature of 175 °C showed the highest encapsulation efficiency (71.74%) and total curcumin content (3.41 g/100 g)	
Barberry (*Berberis vulgaris*) extract		[57]
Carrier	Gum Arabic, gelatin and maltodextrin	
Conditions	800 mL/h—flow rate; 150 °C—inlet temperature; 100—outlet temperature	
Observations	Samples produced with gum Arabic and maltodextrin (core/wall material ratio 25%) in combination possessed the highest microencapsulation efficiency (96.22%)	
Roselle (*Hibiscus sabdariffa* L.) extract		[58]
Carrier	Maltodextrin and gum Arabic	
Conditions	180 °C—inlet air temperature; 80 °C—outlet air temperature; 12 mL/min—feed flow rate; 0.8 bar—atomizing-air pressure	
Observations	In sample where maltodextrin:gum Arabic ratio was 70:30, improved retention of polyphenols (465.80 mg GAE/100 g), anthocyanins (171.21 mg cyanidn-3-glucoside/100 g) and antioxidant activity (3.81 mmol Trolox/kg) was achieved	
Blackberry pulp		[59]
Carrier	Maltodextrin and gum Arabic	
Conditions	0.49 kg/h—flow rate; 145 °C—inlet temperature; 75–80 °C outlet temperature; 0.36 m^3/h—drying air flow rate; 35 m^3/h—aspirator flow rate	
Observations	Samples with maltodextrin and a combination of both carriers possessed higher anthocyanin retention (around 85%) than gum Arabic (around 78%)	
Chokeberry anthocyanins		[60]
Carrier	Maltodextrin, guar gum, gum Arabic, inulin, pectin and β-glucan	
Conditions	140 °C—drying temperature; 25%—pump flow; 600 L/h—air flow	
Observations	Capsules with β-glucan had the highest content of anthocyanins (first day and after 7 days of storage) while the ones with gum Arabic had the lowest content; the following encapsulation efficiency was observed: maltodextrin + gum Arabic—78.61%, maltodextrin + inulin—88.37%, maltodextrin + β-glucan—92.78%, maltodextrin + pectin—91.85% and maltodextrin + guar gum—92.98%	
Berries and roselle anthocyanins		[61]
Carrier	Gum Arabic, maltodextrin, whey protein isolate and agave fructans	
Conditions	180 °C—inlet temperature; 80 °C—outlet temperature; 13%—feed rate; 94%—air flow	
Observations	Retention of anthocyanins was 10.71–86.09%; retention of total polyphenols ranged from 34.71–100%; whey protein isolate was the best carrier agent for anthocyanins and phenolic compounds	

Table 2. Studies of freeze-drying application for encapsulation of polyphenols from different sources.

Bioactive Material		Reference
Red wine polyphenols		[62]
Carrier	Maltodextrin 20% DE 10	
Conditions	Freezing plate and condenser at −40 °C, vacuum below 200 µm Hg; process duration—40 h	
Observations	The freeze-drying process resulted with 97% polyphenol retention; there was no significant changes in polyphenols during 15 days of storage at 38 °C	
Fermented Miang wastewater bioactive compounds		[63]
Carrier	Maltodextrin, gum Arabic and modified starch	
Conditions	Samples were frozen at −18 °C for 24 h and then freeze-dried at −45 °C under a pressure of 0.133 mbar for 72 h	
Observations	Between used carriers, there were no statistical difference in total polyphenols content but in surface polyphenols content, differences were observed: by using gum Arabic in 10:1 core:coating material ratio (% *w*/*w*), the lowest concentration of surface polyphenols was observed with the highest encapsulation efficiency (98.05%); encapsulation efficiency for maltodextrin and modified starch was 89.07% and 81.58%, respectively	
Polyphenols of wastewater from Miang (fermented tea leaf) production		[64]
Carrier	Maltodextrin and gum Arabic	
Conditions	Samples were frozen at −18 °C for 24 h and then freeze-dried under a pressure of 0.133 mbar for 72 h	
Observations	The weight ratio of maltodextrin:gum Arabic mixture to concentrated fermented Miang water of 1:10 was the best for polyphenols encapsulation with 99.4% efficiency	
Cloudberry (*Rubus chamaemorus*) polyphenols		[65]
Carrier	Maltodextrins DE5–8 and DE18.5	
Conditions	Pressure < 0.1 mbar and duration of 48 h	
Observations	Maltodextrin DE 5–8 resulted in higher encapsulation efficiency for all polyphenols, especially for ellagitannins (99%), proanthocyanidins (94%) and flavonols (90%); the highest encapsulation efficiency values of maltodextrin DE 18.8 were for hydroxycinnamic acids and hydroxybenzoic acids (69% and 68%, respectively)	
Red onion peel polyphenols		[66]
Carrier	Maltodextrin and soybean protein isolate	
Conditions	Not defined	
Observations	The combination of maltodextrin with soybean protein isolate showed higher encapsulation efficiency (94.30%) than for the carriers individually (maltodextrin 91.5% and soybean protein isolate 89.83%)	
Yellow onions skins flavonoids extract		[67]
Carrier	Maltodextrin, pectin and whey proteins hydrolysates	
Conditions	Samples were frozen at −70 °C and then freeze-dried at −42 °C under a pressure of 0.1 mbar for 48 h	
Observations	Maltodextrin:pectin:whey proteins hydrolysates in a ratio of 2:1:0.4 resulted in the highest flavonoid encapsulation (66.46%)	
Phoenix dactylifera L pit polyphenols		[68]
Carrier	Gum Arabic and egg yolk protein	
Conditions	Samples were frozen at −80 °C for 12 h and then freeze-dried for 48 h	
Observations	Microparticles with a higher amount of egg yolk protein showed the highest encapsulation efficiency (99.75%); the lowest encapsulation efficiency was observed when gum Arabic and gum Arabic:egg yolk protein (3:1) were used (44.06% and 43.04%, respectively)	

Table 2. *Cont.*

Bioactive Material		Reference
Elsholtzia ciliate ethanolic extract		[69]
Carrier	Gum Arabic, maltodextrin, beta-maltodextrin, resistant-maltodextrin, skim milk and sodium caseinate	
Conditions	Samples were frozen at −80 °C for 24 h and then freeze-dried at −50 °C at 0.05 mbar for 24 h	
Observations	The highest encapsulation efficiency of total polyphenols content were observed with sodium caseinate (83.02%) while the lowest were with maltodextrin (21.17%)	
Blackberry juice polyphenols		[70]
Carrier	Apple fibers	
Conditions	Samples were frozen at −18 °C for 24 h and then freeze-dried under the following conditions: −55 °C—freezing temperature; −35–0 °C—temperature of sublimation; 0.220 mbar—vacuum level; 0–21 °C—isothermal desorption temperatures; 12 h—process duration	
Observations	Different amounts of apple fibers (1%, 2%, 4%, 6%, 8%, and 10%) were used for polyphenol encapsulation, and results showed the best adsorption of total polyphenols when 1% (1.82 g GAE/100 g) and 2% (1.79 g GAE/100 g) of fiber was used	
Blackberry juice polyphenols		[71]
Carrier	Citrus fibers	
Conditions	Samples were frozen at −18 °C for 24 h and then freeze-dried under the following conditions: −55 °C—freezing temperature; −35–0 °C—temperature of sublimation; 0.220 mbar—vacuum level; 0–21 °C—isothermal desorption temperatures; 12 h—process duration	
Observations	By increasing the amount of fiber above 1%, a decrease in the concentration of adsorbed polyphenols occurred; complexes with higher amounts of fiber (2% and 4%) had higher retention levels of polyphenols after eight months' storage (70% and 79%, respectively)	
Raspberry juice polyphenols		[72]
Carrier	Cellulose	
Conditions	Samples were frozen at −18 °C for 24 h and then freeze-dried under the following conditions: −55 °C—freezing temperature; −35–0 °C—temperature of sublimation; 0.220 mbar—vacuum level; 0–21 °C—isothermal desorption temperatures; 12 h—process duration	
Observations	The complex with 2.5% of cellulose resulted in the highest concentration of polyphenols (2.43 g/kg for 15 min of complexation and 1.96 g/kg for 60 min of complexation); higher amounts of the carrier (5%, 7.5% and 10%) negatively affected polyphenols adsorption; the highest retention of polyphenols during storage was observed in powders with 5% and 7.5% of cellulose (from 90–100%)	

Table 3. Studies of spray-drying (SD) and freeze-drying (FD) applications for encapsulation of polyphenols from different sources.

Bioactive Material		Reference
Roselle (*Hibiscus sabdariffa* L.) anthocyanins		[16]
Carrier	Maltodextrin, gum Arabic, inulin and konjac	
Conditions	SD: 500 mL/h—flow rate; 150 °C—inlet temperature; 91 °C—outlet temperature FD: Not defined	
Observations	The freeze-dried sample with 100% konjac had the highest antioxidant content but with low encapsulation efficiency of 43.6% (anthocyanins located on the surface); a mixture of maltodextrin and gum Arabic provided powders with high antioxidant content and efficiency for both methods of drying (around 95%)	

Table 3. *Cont.*

Bioactive Material		Reference
Grape (*Vitis labrusca* var. Bordo) skin phenolic extract		[73]
Carrier	Gum Arabic, polydextrose, and partially hydrolyzed guar gum	
Conditions	SD: 0.60 L/h—flow rate; 140 °C—drying air temperature; 3.5 kg/cm^2—air pressure; 40.5 L/h—air flow rate FD: Dispersions were frozen at −68 °C for 24 h and then freeze-dried at −57 °C for 48 h at vacuum pressure of less than 20 μm Hg	
Observations	Between these drying methods, there was no difference in polyphenol retention; the spray-drying method using 10% gum Arabic had the highest retention of phenolic compounds (25.03 mg GAE/g) as well as freeze-drying with 5% gum Arabic and 5% of polydextrose (24.57 mg GAE/g)	
Coffee grounds polyphenols extract		[2]
Carrier	Maltodextrin and gum Arabic	
Conditions	SD: 108 mL/h—flow rate; 100 °C—air inlet temperature; 75% (28 m^3/h)—aspiration FD: The samples were previously frozen and then freeze-dried at −60 °C at 0.05 bar for 48 h	
Observations	In spray-drying, using maltodextrin as wall material achieved the best encapsulation of flavonoids (around 52%) while a combination of maltodextrin and gum Arabic was the best for encapsulation of total phenolic compounds (around 65%); In freeze-drying, 100% maltodextrin as wall material was the best for encapsulation of total polyphenols and flavonoids (62% and 73%, respectively)	
Model fruit juice (0.1% citrus pectin, 10% sucrose and 0.5% gallic acid)		[13]
Carrier	Maltodextrin and gum Arabic	
Conditions	SD: 72–144 mL/h—flow rate; 80–120 °C—inlet temperature; 600 L/h—nozzle air flow rate; 75% (28 m^3/h)—aspiration FD: 300-500mTorr—chamber pressure; 0.3–0.7 °C—freezing rate; 16 ± 0.5 h—process duration	
Observations	The higher concentrations of gallic acid in freeze-dried samples were achieved with close to 100% gum Arabic and encapsulant concentration of 10–20% and with maltodextrin concentration of 80–100%; for spray-dried samples the best conditions also included 10–20% encapsulant concentration and maltodextrin:gum Arabic ratio of 50–80%	
Lemon by-product aqueous extract		[17]
Carrier	Maltodextrin, soybean protein and ȷ-carrageenan	
Conditions	SD: 125 °C—inlet temperature; 55 °C—maximum outlet temperature; 601 L/h—atomization air flow rate; 4 mL/min—liquid feed pump rate; 38 m^3/h—main drying air flow rate; 70 °C—feed solution temperature; 70 mL—feed solution FD: Using liquid nitrogen, samples were initially frozen and then freeze-dried (48 h)	
Observations	Freeze-dried samples obtained with a combination of maltodextrin and soybean protein achieved the highest encapsulation productivity of total polyphenols content and total flavonoids content (74%); in spray-drying, the best encapsulation productivity for total polyphenols content (67%) was achieved with the same combination of wall materials as for freeze-drying, while for encapsulation productivity of total flavonoids content (58%), no statistical difference was observed between wall materials	
Acerola (*Malpighia emarginata* DC) pulp and residue		[74]
Carrier	Gum Arabic and maltodextrin mixture	
Conditions	SD: 1 mm—feed nozzle diameter; 4 m^3/min—drying air flow rate; 0.36 L/h—feed rate; 30 L/min—compressed air flow; 3.5 kgf/cm^2—air pressure; 170 °C—inlet temperature; 82 °C—outlet temperature FD: Samples were previously frozen at −18 °C for 48 h and then freeze-dried at −58.8 °C for 48 h at 0.42 a mbar vacuum	
Observations	Microencapsulation efficiency for total polyphenols content of freeze-dried samples of both pulp and residue was higher (around 68%) than for spray-dried samples; for microencapsulation efficiency of total flavonoids content, the best results (around 59%) were obtained for freeze-dried acerola residue	

Table 3. *Cont.*

Bioactive Material		Reference
Star fruit (*Averrhoa carambola*) pomace polyphenols		[41]
Carrier	Maltodextrin	
Conditions	SD: 185 °C—inlet temperature; 88 °C—outlet temperature; 6 mL/min—feed rate; 0.1 mm—nozzle size FD: Frozen samples (−40 °C) were freeze-dried at −55 °C for 24 h	
Observations	Freeze-dried encapsulates had a higher encapsulation efficiency (78–97%) than spray-dried ones (63–79%); an increase in maltodextrin concentration led to an increase in core polyphenols content	
Gallic acid		[33]
Carrier	Acid-hydrolyzed low dextrose equivalent potato starch	
Conditions	SD: 160 °C—inlet temperature; 75 °C—outlet temperature FD: Not defined	
Observations	Encapsulation efficiency for freeze-dried samples ranged from 70–84% and for spray-dried from 65–79% without statistically significant differences between methods	
Hydroxytyrosol		[3]
Carrier	β-cyclodextrin	
Conditions	SD: 100 °C—gas inlet temperature; 100 L/min—drying gas (air) flow rate; 0.5 mL/min—feed rate; 35 mbar—inside pressure; 100%—spray rate FD: Frozen samples were slowly dried at −50 °C under 0.06 mbar	
Observations	Spray-dried particles had a spherical and smooth surface with encapsulation efficiency of 84.4%; freeze-dried particles had an irregular shape and encapsulation efficiency was 89.6%	
Papaya pulp		[40]
Carrier	Maltodextrin	
Conditions	SD: 150 °C—inlet temperature; 4 m^3/min—air flow; 3 kgf/cm^2—air pressure; 0.4 L/h—feed flow FD: −62 °C—processing temperature; 6.11 mbar—vacuum degree; 48 h—process duration	
Observations	Freeze-dried products possessed lower retention of vanillic, ferulic, and caffeic acids (15.85 ng/g, under detection limit, and under detection limit, respectively) than spray-dried (30.73 ng/g, 11.26 ng/g, and 9.45 ng/g, respectively)	
Moringa stenopetala leaves extract		[75]
Carrier	Maltodextrin and high methoxyl pectin	
Conditions	SD: 0.5 mm—nozzle diameter; 485 mL/h—flow rate; 140 °C—inlet temperature; 78–81 °C—outlet temperature FD: Samples were frozen at −70 °C for 2 h and then freeze-dried for 72 h	
Observations	Total polyphenols content and total flavonoid content of freeze-dried samples were higher than in spray-dried ones, but encapsulation efficiency and storage stability were better in spray-dried samples; encapsulation efficiency for spray-dried powders with maltodextrin and maltodextrin/high methoxyl pectin was 83.52% and 87.93%, while for freeze-dried ones, it was 71.44% and 82.12%, respectively	
Cranberry juice		[76]
Carrier	Maltodextrin	
Conditions	SD: 50%—pump capacity; 35 m^3/h—air flow FD: 0.03 mbar	
Observations	Powders obtained with sugar free cranberry juice and without maltodextrin had almost five times higher contents of polyphenols (6423 mg/kg—FD and 6433 mg/kg—SD) than powders with 15% maltodextrin and cranberry juice (961 mg/kg—FD and 807 mg/kg—SD); in samples with maltodextrin, *p*-coumaroyl-hexose concentration was higher when spray-drying was applied (180 mg/kg); considering anthocyanins, spray- and freeze-drying equally affected their retention	

Table 3. *Cont.*

Bioactive Material		Reference
Rose (*Rosa rugosa*) anthocyanins		[42]
Carrier	Gum Arabic and maltodextrin	
Conditions	SD: 170 °C—inlet temperature; 5 m^3/min—drying airflow rate; 0.36 L/h—feed rate FD: −52 °C—processing temperature; 0.45 mbar—vacuum degree; 48 h—process duration	
Observations	The retention rate of total polyphenols content and anthocyanin content was 86% and 75.85% for spray-dried powder, and 91.44% and 95.12% for freeze-dried powder, respectively	
Lowbush *Vaccinium myrtillus* blueberry fruit juice		[24]
Carrier	Hydroxypropyl-β-cyclodextrin and maltodextrin	
Conditions	SD: 140 °C—inlet temperature; 70 °C—outlet temperature; 75%—air flow rate; 0.7 mm—nozzle diameter FD: Slowly freezing at −50 °C; pre-drying at 0.42 mbar and 30 °C; secondary drying—reducing pressure to 0.05 mbar and increasing temperature to 40 °C	
Observations	Spray-dried microparticles with maltodextrin and hydroxypropyl-β-cyclodextrin microcapsules had higher total polyphenols content and total anthocyanins content (around 1.65 g/100 g and 1.3 g/100 g, respectively) than freeze-dried microparticles with β-cyclodextrin (around 1.45 g/100 g and 1.1 g/100 g, respectively); total losses of anthocyanins and total polyphenols during drying were lower in the freeze-drying process	
Apple peel polyphenols		[77]
Carrier	Maltodextrin, gum Arabic, and whey protein concentrate	
Conditions	SD: 150 °C—inlet temperature; 50 °C—outlet temperature FD: Samples were frozen at −20 °C for 24 h and then freeze-dried at −45 °C under a pressure of less than 0.12 mbar for more than 48 h	
Observations	Freeze-dried samples homogenized by ultrasonication possessed higher encapsulation efficiency values for phenolic content (83.69%), flavonoid content (85.47%), and antioxidant activity (86.85%) compared to the spray-dried samples previously homogenized by ultra turrax (83.58%, 48.31% and 80.21%, respectively)	

5. Application of Encapsulated Polyphenols in Food Products

Due to the increasing awareness of consumers toward health and the consumption of food that promotes health, the enrichment of food products with encapsulated polyphenols and replacement of artificial food additives with natural ones is strongly supported [78]. Over the last few years, applying encapsulated polyphenols in food products has been on the rise. In reviewing scientific papers, those dealing with encapsulated polyphenols with polysaccharides were singled out.

Yogurt has high water content and a low pH value which makes it challenging to incorporate polyphenols with poor solubility. Encapsulated polyphenols into hydrophilic wall materials can overcome these shortcomings [78]. Robert et al. [37] encapsulated polyphenols from pomegranate with maltodextrin and soybean protein isolates and incorporated them in yogurt. Encapsulates with maltodextrin had the lower degradation rate during storage. Moreover, mushroom extract rich in polyphenols, encapsulated with maltodextrin crosslinked with citric acid, was incorporated in yogurt [79]. In a study of El-Messery et al. [77], polyphenols extracted from apple peel were encapsulated with maltodextrin, whey protein and gum Arabic using spray-drying and freeze-drying. The obtained powders were used in supplementing yogurt. Results showed no significant influence of powders on the physiochemical and texture properties of samples. The authors suggested that those encapsulated polyphenols can be used as a functional food ingredient for yogurt. One interesting study dealt with encapsulation of eugenol-rich clove extract in maltodextrin and gum Arabic by spray-drying. These encapsulates were incorporated into soybean oil for increasing antioxidant activity. Potatoes fried in that oil had better sensorial properties than ones fried in oil with butylated hydroxytoluene (synthetic antioxidant) [80].

Encapsulated polyphenols can also be incorporated into bread. Ezhilarasi et al. [81] enriched bread with encapsulated *Garcinia* fruit polyphenols with maltodextrin and whey protein isolates. Furthermore, green tea polyphenols, encapsulated using β-cyclodextrin and maltodextrin by freeze-drying and spray-drying, were added to bread. Bread quality (volume and crumb firmness) didn't change compared to the control sample [82]. Furthermore, anthocyanins from red onion skins were encapsulated using gum Arabic, soy protein isolate and carboxymethyl cellulose as wall materials, and applying the gelation and freeze-drying techniques. The powder, which has the highest encapsulation efficiency, has been added to crackers and results showed improved antioxidant activity of these enriched crackers. The authors suggest the suitability of applying such additives in bakery products [83]. Table 4 presents some other studies that dealt with incorporating encapsulated polyphenols with polysaccharides using freeze-drying and spray-drying techniques into food products.

Table 4. Selected studies on the incorporation of encapsulated polyphenols into food products.

Food Product	Source of Polyphenols	Wall Material	Encapsulation Technique	Major Findings	Reference
Biscuit	Italian black rice polyphenols	Maltodextrin Gum Arabic	Spray-drying Freeze-drying	Spray-dried encapsulates, added to biscuits, were the most stable during storage and were partially protected during the baking; enriched biscuits showed a higher content of polyphenols, anthocyanins and antioxidant activity than control biscuits	[84]
Biscuit	Cocoa hulls polyphenols	Maltodextrin Gum Arabic	Spray-drying	By using powder with maltodextrin, the most stable sample with unaffected total polyphenols content after baking was obtained	[85]
Cake	Sour cherry polyphenols	Maltodextrin Gum Arabic	Freeze-drying	The incorporation of encapsulated polyphenols didn't impair the sensory or quality properties of cakes; a positive effect on hygroscopicity, baking stability, storage and digestibility was observed	[86]
Chocolate	Peanut skins polyphenols	Maltodextrin	Spray-drying	Antioxidant activity increased after the addition of encapsulates; with 9% of additives antioxidant activity was similar to dark chocolate while flavor was similar to milk chocolate	[87]
Jelly	Barberry polyphenols	Maltodextrin Gum Arabic Gelatin	Spray-drying	Gum Arabic/maltodextrin was the best wall material; jelly with 7% of powder showed better consumer acceptability than commercial jelly; antioxidant activity was increased	[88]

6. Conclusions

Bioactive compounds, such as polyphenols attract a lot of attention from scientists, functional food product developers and consumers due to their health-promoting effects. Most of these compounds are chemically unstable and encapsulation techniques have been widely applied in order to enhance their stability. Nevertheless, freeze-drying is still assumed to be the most suitable for heat-sensitive compounds. The adequate method of encapsulation depends on the type of polyphenol and material for encapsulation, however. Therefore, it cannot be generally said that freeze-drying is better than spray-drying or vice versa. Choosing a polyphenol carrier is important in order to achieve effective encapsulation. While freeze-drying will definitely result in a higher quality and better bioactivity retention compared with conventional spray drying methods, its application is excessively expensive, time consuming and limited in throughput capacity for most commercial applications. Therefore, in order to increase the number and variety of products on the consumer markets, as well as the application range of encapsulated polyphenols, new technologies will need to be researched and if necessary developed.

This paper contributes to the insights into previously researched and optimized encapsulating conditions for a large number of polyphenol-rich materials.

Author Contributions: Conceptualization, M.K., A.P. and J.Š.; methodology, I.B. and A.P.; investigation, I.B. and A.P.; writing—I.B.; writing—review and editing, M.K., A.P. and J.Š.; project administration, M.K.; funding acquisition, M.K. All authors have read and agreed to the published version of the manuscript.

Funding: This work was part of the project PZS-2019-02-1595 which has been fully supported by the "Research Cooperability" Program of the Croatian Science Foundation, funded by the European Union from the European Social Fund under the Operational Program for Efficient Human Resources 2014–2020.

Institutional Review Board Statement: Not applicable.

Informed Consent Statement: Not applicable.

Data Availability Statement: Not applicable.

Conflicts of Interest: The authors declare no conflict of interest.

References

1. Quirós-Sauceda, A.E.; Palafox-Carlos, H.; Sáyago-Ayerdi, S.G.; Ayala-Zavala, J.F.; Bello-Perez, L.A.; Álvarez-Parrilla, E.; de la Rosa, L.A.; Gonzáles-Córdova, A.F.; González-Aguilar, G.A. Dietary fiber and phenolic compunds as functional ingredients: Interaction and possible effect after ingestion. *Food Funct.* **2014**, *5*, 1063–1072. [CrossRef] [PubMed]
2. Ballesteros, L.F.; Ramirez, M.J.; Orrego, C.E.; Teixeira, J.A.; Mussatto, S.I. Encapsulation of antioxidant phenolic compounds extracted from spent coffee grounds by freeze-drying and spray-drying using different coating materials. *Food Chem.* **2017**, *237*, 623–631. [CrossRef] [PubMed]
3. Malapert, A.; Reboul, E.; Tourbin, M.; Dangles, O.; Thiéry, A.; Ziarelli, F.; Tomao, V. Characterization of hydroxytyrosol-β-cyclodextrin complexes in solution and in the solid state, a potential bioactive ingredient. *LWT-Food Sci. Technol.* **2019**, *102*, 317–323. [CrossRef]
4. Velderrain-Rodríguez, G.R.; Palafox-Carlos, H.; Wall-Medrano, A.; Ayala-Zavala, J.F.; Chen, C.Y.O.; Robles-Sánchez, M.; Astiazaran-Garcia, H.; Alvarez-Parrilla, E.; González-Aguilar, G.A. Phenolic compounds: Their journey after intake. *Food Funct.* **2014**, *5*, 189–197. [CrossRef] [PubMed]
5. Selma, M.V.; Espin, J.C.; Tomás-Barberán, F.A. Interaction between phenolics andgut microbiota: Role in human health. *J. Agric. Food Chem.* **2009**, *57*, 6485–6501. [CrossRef]
6. Zhu, F. Interactions between cell wall polysaccharides and polyphenols. *Crit. Rev. Food Sci. Nutr.* **2017**, *58*, 1808–1831. [CrossRef]
7. Kardum, N.; Glibetic, M. Polyphenols and their interactionswith other dietary compounds: Implications for human health. *Adv. Food Nutr. Res.* **2018**, *84*, 103–144. [CrossRef]
8. Sahiner, M.; Blake, D.A.; Fullerton, M.L.; Suner, S.S.; Sunol, A.K.; Sahiner, N. Enhancement of biocompatibility and carbohydrate absorption control potential of rosmarinic acid through crosslinking into microparticles. *Int. J. Biol. Macromol.* **2019**, *137*, 836–843. [CrossRef]
9. Khoo, H.E.; Azlan, A.; Tang, S.T.; Lim, S.M. Anthocyanidins and anthocyanins: Colored pigments as food, pharmaceutical ingredients, and the potential health benefits. *Food Nutr. Res.* **2017**, *61*, 1361779. [CrossRef]
10. Rahman, S.; Mathew, S.; Nair, P.; Ramadan, W.S.; Vazhappilly, C.G. Health benefits of cyanidin-3-glucoside as a potent modulator of Nrf2-mediated oxidative stress. *Inflammopharmacology* **2021**, *29*, 907–923. [CrossRef]
11. Suner, S.S.; Sahiner, M.; Mohapatra, S.; Ayyala, R.S.; Bhethanabotla, V.R.; Sahiner, N. Degradable poly(catechin) nanoparticles as a versatile therapeutic agent. *Int. J. Polym. Mater. Polym. Biomater.* **2021**, *71*, 1–12. [CrossRef]
12. Dludla, P.V.; Nkambule, B.B.; Jack, B.; Mkandla, Z.; Mutize, T.; Silvestri, S.; Orlando, P.; Tiano, L.; Louw, J.; Mazibuko-Mbeje, S.E. Inflammation and oxidative stress in an obese state and the protective effects of gallic acid. *Nutrients* **2019**, *11*, 23. [CrossRef]
13. Ramírez, M.J.; Giraldo, G.I.; Orrego, C.E. Modeling and stability of polyphenol in spray-dried and freeze-dried fruit encapsulates. *Powder Technol.* **2015**, *277*, 89–96. [CrossRef]
14. Cao, H.; Saroglu, O.; Karadag, A.; Diaconeasa, Z.; Zoccatelli, G.; Conte-Junior, C.A.; Gonzales-Aguilar, G.A.; Ou, J.; Bai, W.; Zamarioli, C.M.; et al. Available technologies on improving the stability of polyphenols in food processing. *Food Front.* **2021**, *2*, 109–139. [CrossRef]
15. Wojdyło, A.; Lech, K.; Nowicka, P.; Hernandez, F.; Figiel, A.; Carbonell-Barrachina, A.A. Influence of Different Drying Techniques on Phenolic Compounds, Antioxidant Capacity and Colour of *Ziziphus jujube* Mill. Fruits. *Molecules* **2019**, *24*, 2361. [CrossRef]
16. Nguyen, Q.D.; Dang, T.T.; Nguyen, T.V.L.; Dung Nguyen, T.T.; Mguyen, N.N. Microencapsulation of roselle (*Hibiscus sabdariffa* L.) anthocyanins: Effects of different carriers on selected physicochemical properties and antioxidant activities of spray-dried and freeze-dried powder. *Int. J. Food Prop.* **2022**, *25*, 359–374. [CrossRef]

17. Papoutsis, K.; Golding, J.; Vuong, Q.; Pristijono, P.; Stathopoulos, C.; Scarlett, C.; Bowyer, M. Encapsulation of Citrus By-Product Extracts by Spray-Drying and Freeze-Drying Using Combinations of Maltodextrin with Soybean Protein and ι-Carrageenan. *Foods* **2018**, *7*, 115. [CrossRef]
18. Dobson, C.C.; Mottawea, W.; Rodrigue, A.; Buzati Pereira, L.B.; Hammami, R.; Power, A.K.; Bordenave, N. Impact of molecular interactions with phenolic compounds on food polysaccharides functionality. *Adv. Food Nutr. Res.* **2019**, *90*, 135–181. [CrossRef]
19. Crozier, A.; Jaganath, I.B.; Clifford, M.N. Dietary phenolics: Chemistry, bioavailability and effects on health. *Nat. Prod. Rep.* **2009**, *26*, 1001–10043. [CrossRef]
20. Jakobek, L. Interactions of polyphenols with carbohydrates, lipids and proteins. *Food Chem.* **2015**, *175*, 556–567. [CrossRef]
21. Bellion, P.; Digles, J.; Will, F.; Dietrich, H.; Baum, M.; Eisenbrand, G.; Janzowski, C. Polyphenolic apple extracts: Effects of raw material and production method on antioxidant effectiveness and reduction of DNA damage in Caco-2 cells. *J. Agric. Food Chem.* **2010**, *58*, 6636–6642. [CrossRef]
22. Hollman, P.C.H.; Cassidy, A.; Comte, B.; Heinonen, M.; Richelle, M.; Richling, E.; Richling, E.; Serafini, M.; Scalbert, A.; Sies, H.; et al. The biological relevance of direct antioxidant effects of polyphenols for cardiovascular health in humans is not established. *J. Nutr.* **2011**, *141*, 989S–1009S. [CrossRef] [PubMed]
23. Le Bourvellec, C.; Renard, C.M.G.C. Interactions between Polyphenols and Macromolecules: Quantification Methods and Mechanisms. *Crit. Rev. Food Sci. Nutr.* **2012**, *52*, 213–248. [CrossRef] [PubMed]
24. Wilkowska, A.; Ambroziak, W.; Czyżowska, A.; Adamiec, J. Effect of Microencapsulation by Spray-drying and Freeze Drying Technique on the Antioxidant Properties of Blueberry (*Vaccinium myrtillus*) Juice Polyphenolic Compounds. *Pol. J. Food Nutr. Sci.* **2016**, *66*, 11–16. [CrossRef]
25. Fang, Z.; Bhandari, B. Encapsulation of polyphenols—A review. *Trends Food Sci. Technol.* **2010**, *21*, 510–523. [CrossRef]
26. Grgić, J.; Šelo, G.; Planinić, M.; Tišma, M.; Bucić-Kojić, A. Role of the Encapsulation in Bioavailability of Phenolic Compounds. *Antioxidants* **2020**, *9*, 923. [CrossRef]
27. Saifullah, M.; Islam Shishir, M.R.; Ferdowsi, R.; Tanver Rahman, M.R.; Van Vuong, Q. Micro and nano encapsulation, retention and controlled release of flavor and aroma compounds: A critical review. *Trends Food Sci. Technol.* **2019**, *86*, 230–251. [CrossRef]
28. Wu, G.; Hui, X.; Mu, J.; Brennan, M.A.; Brennan, C.S. Functionalization of whey protein isolate fortified with blackcurrant concentrate by spray-drying and freeze-drying strategies. *Food Res. Int.* **2021**, *141*, 110025. [CrossRef]
29. Lesschaeve, I.; Noble, A.C. Polyphenols: Factors influencing their sensory properties and their effects on food and beverage preferences. *Am. J. Clin. Nutr.* **2005**, *81*, 330S–335S. [CrossRef]
30. Bhatta, S.; StevanovicJanezic, T.; Ratti, C. Freeze-Drying of Plant-Based Foods. *Foods* **2020**, *9*, 87. [CrossRef]
31. Guo, J.; Li, P.; Kong, L.; Xu, B. Microencapsulation of curcumin by spray drying and freeze drying. *LWT* **2020**, *132*, 109892. [CrossRef]
32. Demirci, S.; Khiev, D.; Can, M.; Sahiner, M.; Biswal, M.R.; Ayyala, R.S.; Sahiner, N. Chemically cross-linked poly(β-cyclodextrin) particles as promising drug delivery materials. *Appl. Polym. Mater.* **2021**, *3*, 6238–6251. [CrossRef]
33. Sepelevs, I.; Stepanova, V.; Galoburda, R. Encapsulation of Gallic Acid with Acid-Modified Low Dextrose Equivalent Potato Starch Using Spray-and Freeze-Drying Techniques. *Polish J. Food Nutr. Sci.* **2018**, *68*, 273–280. [CrossRef]
34. Shofian, N.M.; Hamid, A.A.; Osman, A.; Saari, N.; Anwar, F.; Dek, M.S.; Hairuddin, M.R. Effect of Freeze-Drying on the Antioxidant Compounds and Antioxidant Activity of Selected Tropical Fruits. *Int. J. Mol. Sci.* **2011**, *12*, 4678–4692. [CrossRef] [PubMed]
35. Tan, J.J.Y.; Lim, Y.Y.; Siow, L.F.; Tan, J.B.L. Effects of drying on polyphenol oxidase and antioxidant activity of *Morus alba* leaves. *J. Food Process. Preserv.* **2015**, *39*, 2811–2819. [CrossRef]
36. Pérez-Gregorio, M.R.; Regueiro, J.; González-Barreiro, C.; Rial-Otero, R.; Simal-Gándara, J. Changes in antioxidant flavonoids during freeze-drying of red onions and subsequent storage. *Food Control* **2011**, *22*, 1108–1113. [CrossRef]
37. Robert, P.; Gorena, T.; Romero, N.; Sepulveda, E.; Chavez, J.; Saenz, C. Encapsulation of polyphenols and anthocyanins from pomegranate (*Punica granatum*) by spray-drying. *Int. J. Food Sci.* **2010**, *45*, 1386–1394. [CrossRef]
38. Wu, G.; Hui, X.; Stipkovits, L.; Rachman, A.; Tu, J.; Brennan, M.A.; Brennan, C.S. Whey protein-blackcurrant concentrate particles obtained by spray-drying and freeze-drying for delivering structural and health benefits of cookies. *Innov. Food Sci. Emerg. Technol.* **2021**, *68*, 102606. [CrossRef]
39. Ersus, S.; Yurdagel, U. Microencapsulation of anthocyanin pigments of black carrot (*Daucus carota* L.) by spray drier. *J. Food Eng.* **2007**, *80*, 805–812. [CrossRef]
40. Gomes, W.F.; França, F.R.M.; Denadai, M.; Andrade, J.K.S.; da Silva Oliveira, E.M.; de Brito, E.S.; Rodrigues, S.; Narain, N. Effect of freeze- and spray-drying on physico-chemical characteristics, phenolic compounds and antioxidant activity of papaya pulp. *J. Food Sci. Technol.* **2018**, *55*, 2095–2102. [CrossRef]
41. Saikia, S.; Mahnot, N.K.; Mahanta, C.L. Optimisation of phenolic extraction from *Averrhoa carambola* pomace by response surface methodology and its microencapsulation by spray and freeze drying. *Food Chem.* **2015**, *171*, 144–152. [CrossRef] [PubMed]
42. Yu, Y.; Lv, Y. Degradation kinetic of anthocyanins from rose (*Rosa rugosa*) as prepared by microencapsulation in freeze-drying and spray-drying. *Int. J. Food Prop.* **2019**, *22*, 2009–2021. [CrossRef]
43. Hussain, S.A.; Hameed, A.; Nazir, Y.; Naz, T.; Wu, Y.; Suleria, H.; Song, Y. Microencapsulation and the Characterization of Polyherbal Formulation (PHF) Rich in Natural Polyphenolic Compounds. *Nutrients* **2018**, *10*, 843. [CrossRef] [PubMed]

44. Díaz-Bandera, D.; Villanueva-Carvajal, A.; Dublán-García, O.; Quintero-Salazar, B.; Dominguez-Lopez, A. Assessing release kinetics and dissolution of spray-dried Roselle (*Hibiscus sabdariffa* L.) extract encapsulated with different carrier agents. *LWT-Food Sci. Technol.* **2015**, *64*, 693–698. [CrossRef]
45. Yibin, L.I.; Wu, L.; Weng, M.; Tang, B.; Lai, P.; Chen, J. Effect of different encapsulating agent combinations on physicochemical properties and stability of microcapsules loaded with phenolics of plum (*Prunus salicina lindl.*). *Powder Technol.* **2018**, *340*, 459–464. [CrossRef]
46. Santiago, M.C.P.A.; Nogueira, R.I.; Paim, D.R.S.F.; Gouvêa, A.C.M.S.; Godoy, R.L.O.; Peixoto, F.M.; Pacheco, S.; Freitas, S.P. Effects of encapsulating agents on anthocyanin retention in pomegranate powder obtained by the spray-drying process. *LWT* **2016**, *73*, 551–556. [CrossRef]
47. Ostroschi, L.C.; Brito de Souza, V.; Echalar-Barrientos, M.A.; Tulini, F.L.; Comunian, T.A.; Thomazini, M.; Baliero, C.C.; Roudaut, G.; Genovese, M.I.; Favaro-Trindade, C.S. Production of spray-dried proanthocyanidin-rich cinnamon (*Cinnamomum zeylanicum*) extract as a potential functional ingredient: Improvement of stability, sensory aspects and technological properties. *Food Hydrocoll.* **2018**, *79*, 343–351. [CrossRef]
48. Pang, S.F.; Yusoff, M.M.; Gimbun, J. Assessment of phenolic compounds stability and retention during spray-drying of *Orthosiphon stamineus* extracts. *Food Hydrocoll.* **2014**, *37*, 159–165. [CrossRef]
49. Sansone, F.; Mencherini, T.; Picerno, P.; d'Amore, M.; Aquino, R.P.; Lauro, M.R. Maltodextrin/pectin microparticles by spray-drying as carrier for nutraceutical extracts. *J. Food Eng.* **2011**, *105*, 468–476. [CrossRef]
50. Nadeem, H.S.; Torun, M.; Özdemir, F. Spray-drying of the mountain tea (*Sideritis stricta*) water extract by using different hydrocolloid carriers. *LWT-Food Sci. Technol.* **2011**, *44*, 1626–1635. [CrossRef]
51. Saénz, C.; Tapia, S.; Chávez, J.; Robert, P. Microencapsulation by spray-drying of bioactive compounds from cactus pear (*Opuntia ficus-indica*). *Food Chem.* **2009**, *114*, 616–622. [CrossRef]
52. Tonon, R.V.; Brabet, C.; Hubinger, M.D. Anthocyanin stability and antioxidant activity of spray-dried açai (*Euterpe oleracea* Mart.) juice produced with different carrier agents. *Int. Food Res. J.* **2010**, *43*, 907–914. [CrossRef]
53. Silva, P.I.; Stringheta, P.C.; Teófilo, R.F.; de Oliveira, I.R.N. Parameter optimization for spray-drying microencapsulation of jaboticaba (*Myrciaria jaboticaba*) peel extracts using simultaneous analysis of responses. *J. Food Eng.* **2013**, *117*, 538–544. [CrossRef]
54. Bušić, A.; Komes, D.; Belšćak-Cvitanović, A.; VojvodićCebin, A.; Špoljarić, I.; Mršić, G.; Miao, S. The Potential of Combined Emulsification and Spray-drying Techniques for Encapsulation of Polyphenols from Rosemary (*Rosmarinus officinalis* L.) Leaves. *Food Technol. Biotechnol.* **2018**, *56*, 494–505. [CrossRef] [PubMed]
55. Rocha, S.; Generalov, R.; Pereira, M.C.; Peres, I.; Juzenas, P.; Coelho, M.A. Epigallocatechin gallate-loaded polysaccharide nanoparticles for prostate cancer chemoprevention. *Nanomed. J.* **2011**, *6*, 79–87. [CrossRef]
56. AniesraniDelfiya, D.S.; Thangavel, K.; Natarajan, N.; Kasthuri, R.; Kailappan, R. Microencapsulation of Turmeric Oleoresin by Spray-drying and In Vitro Release Studies of Microcapsules. *J. Food Process Eng.* **2015**, *38*, 37–48. [CrossRef]
57. Akhavan Mahdavi, S.; Jafari, S.M.; Assadpoor, E.; Dehnad, D. Microencapsulation optimization of natural anthocyanins with maltodextrin, gum Arabic and gelatin. *Int. J. Biol. Macromol.* **2016**, *85*, 379–385. [CrossRef]
58. Archaina, D.; Vasile, F.; Jiménez-Guzmán, J.; Alamilla-Beltrán, L.; Schebor, C. Physical and functional properties of roselle (*Hibiscus sabdariffa* L.) extract spray-dried with maltodextrin-gum arabic mixtures. *J. Food Process. Preserv.* **2019**, *43*, e14065. [CrossRef]
59. Ferrari, C.C.; Germer, S.P.M.; Alvim, I.D.; Vissotto, F.Z.; de Aguirre, J.M. Influence of carrier agents on the physicochemical properties of blackberry powder produced by spray-drying. *Int. J. Food Sci.* **2012**, *47*, 1237–1245. [CrossRef]
60. Pieczykolan, E.; Kurek, M.A. Use of guar gum, gum arabic, pectin, beta-glucan and inulin for microencapsulation of anthocyanins from chokeberry. *Int. J. Biol. Macromol.* **2019**, *129*, 665–671. [CrossRef]
61. Farias-Cervantes, V.S.; Chávez-Rodríguez, A.; García-Salcedo, P.A.; García-López, P.M.; Casas-Solís, J.; Andrade-González, I. Antimicrobial effect and in vitro release of anthocyanins from berries and Roselle obtained via microencapsulation by spray-drying. *J. Food Process. Preserv.* **2018**, *42*, e13713. [CrossRef]
62. Sanchez, V.; Baeza, R.; Galmarini, M.V.; Zamora, M.C.; Chirife, J. Freeze-Drying Encapsulation of Red Wine Polyphenols in an Amorphous Matrix of Maltodextrin. *Food. Bioproc. Tech.* **2011**, *6*, 1350–1354. [CrossRef]
63. Ravichai, K.; Muangrat, R. Effect of different coating materials on freeze-drying encapsulation of bioactive compounds from fermented tea leaf wastewater. *J. Food Process. Preserv.* **2019**, *43*, e14145. [CrossRef]
64. Muangrat, R.; Ravichai, K.; Jirarattanarangsri, W. Encapsulation of polyphenols from fermented wastewater of Miang processing by freeze drying using a maltodextrin/gum Arabic mixture as coating material. *J. Food Process. Preserv.* **2019**, *43*, e13908. [CrossRef]
65. Laine, P.; Kylli, P.; Heinonen, M.; Jouppila, K. Storage Stability of Microencapsulated Cloudberry (*Rubus chamaemorus*) Phenolics. *J. Agric. Food Chem.* **2008**, *56*, 11251–11261. [CrossRef] [PubMed]
66. Elsebaie, E.M.; Essa, R.Y. Microencapsulation of red onion peel polyphenols fractions by freeze drying technicality and its application in cake. *J. Food Process. Preserv.* **2018**, *42*, e13654. [CrossRef]
67. Milea, Ş.A.; Aprodu, I.; Vasile, A.M.; Barbu, V.; Râpeanu, G.; Bahrim, G.E.; Stănciuc, N. Widen the functionality of flavonoids from yellow onion skins through extraction and microencapsulation in whey proteins hydrolysates and different polymers. *J. Food Eng.* **2019**, *251*, 29–35. [CrossRef]

68. Ben Sassi, C.; Marcet, I.; Rendueles, M.; Díaz, M.; Fattouch, S. Egg yolk protein as a novel wall material used together with gum Arabic to encapsulate polyphenols extracted from *Phoenix dactylifera* L pits. *LWT* **2020**, *131*, 109778. [CrossRef]
69. Pudziuvelyte, L.; Marksa, M.; Sosnowska, K.; Winnicka, K.; Morkuniene, R.; Bernatoniene, J. Freeze-Drying Technique for Microencapsulation of *Elsholtziaciliata* Ethanolic Extract Using Different Coating Materials. *Molecules* **2020**, *25*, 2237. [CrossRef]
70. Buljeta, I.; Nosić, M.; Pichler, A.; Ivić, I.; Šimunović, J.; Kopjar, M. Apple Fibers as Carriers of Blackberry Juice Polyphenols: Development of Natural Functional Food Additives. *Molecules* **2022**, *27*, 3029. [CrossRef]
71. Buljeta, I.; Pichler, A.; Šimunović, J.; Kopjar, M. Polyphenols and Antioxidant Activity of Citrus Fiber/Blackberry Juice Complexes. *Molecules* **2021**, *26*, 4400. [CrossRef] [PubMed]
72. Vukoja, J.; Buljeta, I.; Pichler, A.; Šimunović, J.; Kopjar, M. Formulation and Stability of Cellulose-Based Delivery Systems of Raspberry Phenolics. *Processes* **2021**, *9*, 90. [CrossRef]
73. Kuck, L.S.; Noreña, C.P.Z. Microencapsulation of grape (*Vitis labrusca* var. Bordo) skin phenolic extract using gum Arabic, polydextrose, and partially hydrolyzed guar gum as encapsulating agents. *Food Chem.* **2016**, *194*, 569–576. [CrossRef] [PubMed]
74. Rezende, Y.R.R.S.; Nogueira, J.P.; Narain, N. Microencapsulation of extracts of bioactive compounds obtained from acerola (*Malpighia emarginata* DC) pulp and residue by spray and freeze drying: Chemical, morphological and chemometric characterization. *Food Chem.* **2018**, *254*, 281–291. [CrossRef]
75. Dadi, D.W.; Emire, S.A.; Hagos, A.D.; Eun, J.B. Physical and Functional Properties, Digestibility, and Storage Stability of Spray- and Freeze-Dried Microencapsulated Bioactive Products from *Moringa stenopetala* Leaves Extract. *Ind. Crops. Prod.* **2020**, *156*, 112891. [CrossRef]
76. Michalska, A.; Wojdyło, A.; Honke, J.; Ciska, E.; Andlauer, W. Drying-induced physico-chemical changes in cranberry products. *Food Chem.* **2018**, *240*, 448–455. [CrossRef]
77. El-Messery, T.M.; El-Said, M.M.; Demircan, E.; Ozçelik, B. Microencapsulation of natural polyphenolic compounds extracted from apple peel and its application in yoghurt. *Acta Sci. Pol. Technol. Aliment.* **2019**, *18*, 25–34. [CrossRef]
78. Delfanian, M.; Ali Sahari, M. Improving functionality, bioavailability, nutraceutical and sensory attributes of fortified foods using phenolics-loaded nanocarriers as natural ingredients. *Food Res. Int.* **2020**, *137*, 109555. [CrossRef]
79. Francisco, C.R.L.; Heleno, S.A.; Fernandes, I.P.M.; Barreira, J.C.M.; Calhelha, R.C.; Barros, L.; Barreiro, M.F. Functionalization of yogurts with *Agaricusbisporus* extracts encapsulated in spray-dried maltodextrin crosslinked with citric acid. *Food Chem.* **2018**, *245*, 845–853. [CrossRef]
80. Chatterjee, D.; Bhattacharjee, P. Comparative evaluation of the antioxidant efficacy of encapsulated and un-encapsulated eugenol-rich clove extracts in soybean oil: Shelf-life and frying stability of soybean oil. *J. Food Eng.* **2013**, *117*, 545–550. [CrossRef]
81. Ezhilarasi, P.N.; Indrani, D.; Jena, B.S.; Anandharamakrishnan, C. Microencapsulation of Garciniafruit extract by spray drying and its effect on bread quality. *J. Sci. Food Agric.* **2013**, *94*, 1116–1123. [CrossRef] [PubMed]
82. Pasrija, D.; Ezhilarasi, P.N.; Indrani, D.; Anandharamakrishnan, C. Microencapsulation of green tea polyphenols and its effect on incorporated bread quality. *LWT* **2015**, *64*, 289–296. [CrossRef]
83. Stoica, F.; Condurache, N.N.; Horincar, G.; Constantin, O.E.; Turturică, M.; Stănciuc, N.; Aprodu, I.; Croitoru, C.; Râpeanu, G. Value-Added Crackers Enriched with Red Onion Skin Anthocyanins Entrapped in Different Combinations of Wall Materials. *Antioxidants* **2022**, *11*, 1048. [CrossRef] [PubMed]
84. Papillo, V.A.; Locatelli, M.; Travaglia, F.; Bordiga, M.; Garino, C.; Arlorio, M.; Coïsson, J.D. Spray-dried polyphenolic extract from Italian black rice (*Oryza sativa* L., var. Artemide) as new ingredient for bakery products. *Food Chem.* **2018**, *269*, 603–609. [CrossRef] [PubMed]
85. Papillo, V.A.; Locatelli, M.; Travaglia, F.; Bordiga, M.; Garino, C.; Coïsson, J.D.; Arlorio, M. Cocoa hulls polyphenols stabilized by microencapsulation as functional ingredient for bakery applications. *Food Res. Int.* **2018**, *115*, 511–518. [CrossRef] [PubMed]
86. Luca, A.; Cilek, B.; Hasirci, V.; Sahin, S.; Sumnu, G. Storage and Baking Stability of Encapsulated Sour Cherry Phenolic Compounds Prepared from Micro- and Nano-Suspensions. *Food Bioprocess Technol.* **2014**, *7*, 204–211. [CrossRef]
87. Dean, L.L.; Klevorn, C.M.; Hess, B.J. Minimizing the Negative Flavor Attributes and Evaluating Consumer Acceptance of Chocolate Fortified with Peanut Skin Extracts. *J. Food Sci.* **2016**, *81*, S2824–S2830. [CrossRef]
88. Mahdavi, S.A.; Jafari, S.M.; Assadpour, E.; Ghorbani, M. Storage stability of encapsulated barberry's anthocyanin and its application in jelly formulation. *J. Food Eng.* **2016**, *181*, 59–66. [CrossRef]

molecules MDPI

Article

Anti-Obesity Evaluation of *Averrhoa carambola* L. Leaves and Assessment of Its Polyphenols as Potential α-Glucosidase Inhibitors

Nehal S. Ramadan [1], Nabil H. El-Sayed [1], Sayed A. El-Toumy [1], Doha Abdou Mohamed [2], Zeinab Abdel Aziz [3], Mohamed Sobhy Marzouk [1], Tuba Esatbeyoglu [4,*], Mohamed A. Farag [3,*] and Kuniyoshi Shimizu [5]

1 Chemistry of Tanning Materials and Leather Technology Department, National Research Centre, Dokki, Cairo 12622, Egypt
2 Nutrition and Food Sciences Department, National Research Centre, Dokki, Cairo 12622, Egypt
3 Pharmacognosy Department, College of Pharmacy, Cairo University, Kasr El Aini St., Cairo 11562, Egypt
4 Department of Food Development and Food Quality, Institute of Food Science and Human Nutrition, Gottfried Wilhelm Leibniz University Hannover, Am Kleinen Felde 30, 30167 Hannover, Germany
5 Department of Agro-Environmental Sciences, Graduate School of Bioresource and Bioenvironmental Sciences, Kyushu University, Fukuoka 819-0395, Japan
* Correspondence: esatbeyoglu@lw.uni-hannover.de (T.E.); mohamed.farag@pharma.cu.edu.eg (M.A.F.); Tel.: +49-511-762-5589 (T.E.); +011-202-2362245 (M.A.F.)

Abstract: *Averrhoa carambola* L. is reported for its anti-obese and anti-diabetic activities. The present study aimed to investigate its aqueous methanol leaf extract (CLL) in vivo anti-obese activity along with the isolation and identification of bioactive compounds and their in vitro α-glucosidase inhibition assessment. CLL improved all obesity complications and exhibited significant activity in an obese rat model. Fourteen compounds, including four flavone glycosides (**1–4**) and ten dihydrochalcone glycosides (**5–12**), were isolated and identified using spectroscopic techniques. New compounds identified in planta included (**1**) apigenin 6-*C*-(2-deoxy-*β*-D-galactopyranoside)-7-*O*-*β*-D-quinovopyranoside, (**8**) phloretin 3′-*C*-(2-*O*-(*E*)-cinnamoyl-3-*O*-*β*-D-fucopyranosyl-4-*O*-acetyl)-*β*-D-fucopyranosyl-6′-*O*-*β*-D fucopyranosyl-(1/2)-α-L arabinofuranoside, (**11a**) phloretin3′-*C*-(2-*O*-(*E*)-p-coumaroyl-3-*O*-*β*-D-fucosyl-4-*O*-acetyl)-*β*-D-fucosyl-6′-*O*-(2-*O*-*β*-D-fucosyl)-α-L-arabinofuranoside, (**11b**) phloretin3′-*C*-(2-O-(*Z*)-*p*-coumaroyl-3-*O*-*β*-D-fucosyl-4-*O*-acetyl)-*β*-D-fucosyl-6′-*O*-(2-*O*-*β*-D-fucosyl)-α-L-arabinofuranoside. Carambolaside M (**5**), carambolaside Ia (**6**), carambolaside J (**7**), carambolaside I (**9**), carambolaside P (**10a**), carambolaside O (**10b**), and carambolaside Q (**12**), which are reported for the first time from *A. carambola* L. leaves, whereas luteolin 6-*C*-α-L-rhamnopyranosyl-(1-2)-*β*-D-fucopyranoside (**2**), apigenin 6-*C*-*β*-D-galactopyranoside (**3**), and apigenin 6-*C*-α-L-rhamnopyranosyl-(1-2)-*β*-L-fucopyranoside (**4**) are isolated for the first time from Family. Oxalidaceae. In vitro α-glucosidase inhibitory activity revealed the potential efficacy of flavone glycosides, viz., **1**, **2**, **3**, and **4** as antidiabetic agents. In contrast, dihydrochalcone glycosides (**5–11**) showed weak activity, except for compound **12**, which showed relatively strong activity.

Keywords: antidiabetic; *Averrhoa carambola* L.; dihydrochalcone; flavone glycosides; obesity; Oxalidaceae; type 2 diabetes

Citation: Ramadan, N.S.; El-Sayed, N.H.; El-Toumy, S.A.; Mohamed, D.A.; Aziz, Z.A.; Marzouk, M.S.; Esatbeyoglu, T.; Farag, M.A.; Shimizu, K. Anti-Obesity Evaluation of *Averrhoa carambola* L. Leaves and Assessment of Its Polyphenols as Potential α-Glucosidase Inhibitors. *Molecules* **2022**, *27*, 5159. https://doi.org/10.3390/molecules27165159

Academic Editor: Nour Eddine Es-Safi

Received: 25 July 2022
Accepted: 7 August 2022
Published: 12 August 2022

1. Introduction

The radical shift from malnutrition to overnutrition, as well as the increase in sedentary behaviour, has led to the increasing incidence of obesity, a complex chronic nutritional disorder characterized by an energy expenditure and intake imbalance. With estimates of 2.3 billion overweight individuals and 700 million obese adults, obesity with its comorbidities is considered the fifth-largest cause of death worldwide [1,2]. Insulin resistance is one of the most prevalent obesity-related changes [3] and hence, obesity is a key predisposal to type 2 diabetes [1]. Furthermore, some white fat storage areas in the body are more directly associated to metabolic consequences of obesity, such as diabetes, than others [2]. Obesity-related problems, viz., diabetes has been associated with decreased life expectancy [4] as well as imparting various clinical disorders such as renal disease, blindness, and amputation of lower limbs, among others [5].

Moreover, obesity and associated symptoms are predicted to cost the global economy USD 2 trillion per year nearly as much as smoking, armed conflict, and terrorism [4]. The common synthetic anti-obesity medicine orlistat is successful in treating obesity, however, it exerts serious gastrointestinal side effects [6].

Likewise, α-glucosidase inhibitors as typical anti-diabetic medicines reported to possess side effects despite their important function in lowering blood glucose levels [7]. Only three α-glucosidase inhibitors are currently used in clinical practice including acarbose, miglitol, and voglibose [5,7], warranting for the development of natural medicines that comprise medicinal herbs, either as pure components or as extracts, as an alternative therapy for obesity [3,7].

For decades, *Averrhoa carambola* L., commonly known as starfruit, a member of the Oxalidaceae family, indigenous to the tropical southeast, is planted across the tropics for its edible fruit as well as its ornamental traits, and it was recently domesticated in other countries, including Ecuador and Egypt [8]. *A. carambola* L. flesh is reported for its potential in the treatment of diabetes [9] as well as its confirmed hypoglycemic and porcine pancreatic lipase inhibitory effects [10,11].

In spite of being edible with several health benefits, starfruit is contraindicated in uremic patients owing to its high oxalate content in addition to its negative inotropic and chronotropic effects [8].

Besides, *A. carambola* leaves were reported for their traditional uses in treatment of hyperglycemia, diabetes, and its related diseases [12,13]. Biological studies further confirmed the hypoglycemic activities of leaves and some of its isolated compounds [13,14] Moreover, leaves were reported to possess potential antioxidant activity [15]. Further, leaf decoction are reported to be used for treatment of aphthous stomatitis and angina [16].

With regards to chemical composition and compared to fruits, *A. carambola* leaves are less investigated. Flavone C-glycosides have been previously isolated from leaves, viz., isovitexin, carambolaflavones A and B, and apigenin 6-C-(2″-O-α-L-rhamnopyranosyl)-β-D-glucopyranoside [15]. Both carambolaflavones were reported for their hypoglycemic effect in rats [12]. Recently, 12 dihydrochalcone C-glycosides were reported from leaves [17]. Dihydrochalcones are natural phenolics with a C6–C3–C6 skeleton structure, where two aromatic rings are connected via a C3 chain [18] and abundant in *A. carambola*.

In the context of the overall strategy to control obesity and its complications using functional foods, an *A. carambola* crude methanol-leaf extract (CLL) anti-obese effect was assessed using an in vivo high fat diet (HFD)-induced obesity rat model. CLL extract as well as Orly (as reference drug) were orally administered as interventions for the management of obesity showing significant improvement in obesity and its associated complications.

Where, rats were fed on high fat diet for eight weeks resulting in dyslipidemia, hyperglycemia, hyperleptinemia, insulin resistance, oxidative stress, and abnormalities in liver and kidney functions. To confirm development of the obesity model and to assess different treatment actions, both physical and biochemical parameters were monitored. Further, leaf extract was subjected to detailed phytochemical isolation to identify active agent(s) using NMR and MS spectroscopy, with pure compounds assessed for their α-glucosidase inhibitory activity.

2. Results and Discussion

2.1. *In Vivo Assay of A. carambola Leaf Extract against HFD-Induced Obesity Model in Rats*

A. carambola L. flesh has been reported for the treatment of diabetes in folk medicine [9]. Besides, pharmacological assays confirmed its hypoglycemic effect [10] as well as its porcine pancreatic lipase inhibitory effect [11]. Starfruit is known for its richness in phenolics, especially flavonoids [19], its potential for preventing and curing metabolic disorders, viz., obesity and obesity-related metabolic syndrome [20]. The existence of bioactive phytochemicals, i.e., flavan-3-ols and 2-diglycosyloxybenzoates in carambola leaf, with reported lipase and α-glucosidase inhibitory activities [17], might participate in the anti-obese activity of leaves, warranting their assessment. The effect of CLL extract was assessed against different parameters in HFD-induced obese rats including body weight, dyslipidemia, effect on leptin, α-amylase, plasma glucose, insulin levels and insulin resistance, oxidative stress, and lipid peroxidation as well as the effect on liver and kidney functions, as detailed in the next subsections.

2.1.1. Body Weight and Biochemical Markers Determination

CLL extracts showed a significant reduction in rat body weight (258 g) compared to obese rats (291 g), however, it was still higher than the normal rats group (246 g) (Figure 1A and Table 1). Although Orly caused a significant decrease in BW gain ($p < 0.05$), as compared with the CLL and obese groups, several biochemical markers were measured as the index for obesity status, revealing the excelling of CLL over Orly, as detailed in the next subsections.

Table 1. List of nutritional and biochemical parameters in obese, normal, and treated animal groups ($n = 3$).

No.	Measured Parameters	Tested Groups			
		Normal	Obese	Orly	CLL
1	Initial BW (g)	115.7 ^{a} ± 1.7	115.8 ^{a} ± 2.6	115.8 ^{a} ± 2.7	115.7 ^{a} ± 5.1
2	BW after induction of obesity (g)	217.7 ^{a} ± 7.8	248.7 ^{b} ± 8.3	248.8 ^{b} ± 7.5	249 ^{b} ± 3.8
3	Final BW (g)	246.5 ^{b} ± 10.1	291 ^{d} ± 4.9	233.2 ^{a} ± 11.2	258 ^{c} ± 8.7
4	Leptin (ng/mL)	12.8 ^{a} ± 0.3	24.1 ^{d} ± 0.4	21.09 ^{c} ± 0.1	18.0 ^{b} ± 0.2
5	Insulin (μg/L)	6.4 ^{a} ± 0.1	12.2 ^{e} ± 0.3	10.1 ^{d} ± 0.1	9.4 ^{b} ± 0.2
6	Glucose (mg/dL)	69.8 ^{a} ± 2.01	106.6 ^{e} ± 3.3	89.9 ^{c} ± 2.1	80.0 ^{d} ± 1.7
7	IR	1.1 ^{a}± 0.1	3.2 ^{e}± 0.1	2.2 ^{d} ± 0.1	1.9 ^{c} ± 0.1
8	BChE (U/L)	250.7 ^{a} ± 5.2	415.7 ^{e} ± 8.5	285.2 ^{d} ± 7.3	274.8 ^{c} ± 5.8
9	α-amylase (U/L)	9.2 ^{a} ± 0.4	15.33 ^{e} ± 0.3	13.7 ^{d} ± 0.3	12.4 ^{c} ± 0.2
10	MDA (nmol/mL)	5.6 ^{a} ± 0.2	15.1 ^{e} ± 0.6	12.9 ^{d} ± 0.6	10.4 ^{c} ± 0.4
11	CAT (U/L)	598.1 ^{a} ± 14.8	319.5 ^{e} ± 0.4	329.3 ^{d} ± 12.7	473.2 ^{c} ± 10.7
12	T-Ch (mg/dL)	70.4 ^{a} ± 2.1	136.6 ^{e} ± 5.4	97.9 ^{d} ± 3.9	89.7 ^{c} ± 3.8
13	TG (mg/dL)	67.7 ^{a} ± 1.9	113.3 ^{e} ± 2.4	99.5 ^{d} ± 4.2	87.6 ^{c} ± 2.4
14	HDL-Ch (mg/dL)	42.8 ^{a} ± 0.7	27.8 ^{e} ± 1.2	31.5 ^{d} ± 0.8	37 ^{c} ± 0.6
15	LDL-Ch (mg/dL)	19.5 ^{a}± 0.7	85.5 ^{d} ± 3.5	69.2 ^{c} ± 1.8	49.8 ^{c} ± 1.4
16	T-Ch/HDL-Ch ratio	1.64 ^{a} ± 0.03	4.9 ^{e} ± 0.2	3.1 ^{d} ± 0.2	2.4 ^{c} ± 0.1
17	ALT (IU/L)	18.6 ^{a} ± 0.6	25.3 ^{d} ± 1.1	21.0 ^{c} ± 0.6	19.7 ^{d} ± 0.4
18	AST (IU/L)	41.9 ^{a} ± 0.7	49.7 ^{b} ± 1.1	43.8 ^{c} ± 0.6	43.2 ^{d} ± 0.7
19	Creatinine (mg/dL)	0.624 ^{a} ± 0.01	0.76 ^{d} ± 0.02	0.68 ^{c} ± 0.02	0.617 ^{b} ± 0.02
20	Urea (mg/dL)	24.6 ^{a} ± 1.1	33.2 ^{d} ± 0.9	26.8 ^{c} ± 0.6	25.0 ^{b} ± 0.6
21	Uric acid (mg/dL)	0.8 ^{a} ± 0.04	1.3 ^{b} ± 0.09	1.4 ^{c} ± 0.1	1.9 ^{c} ± 0.1

Results are expressed as mean ± S.E.M. Values with different superscript letters in the same raw are significantly different at $p < 0.05$ levels. ^{b,c,d} and ^{e} are significantly higher than ^{a}.

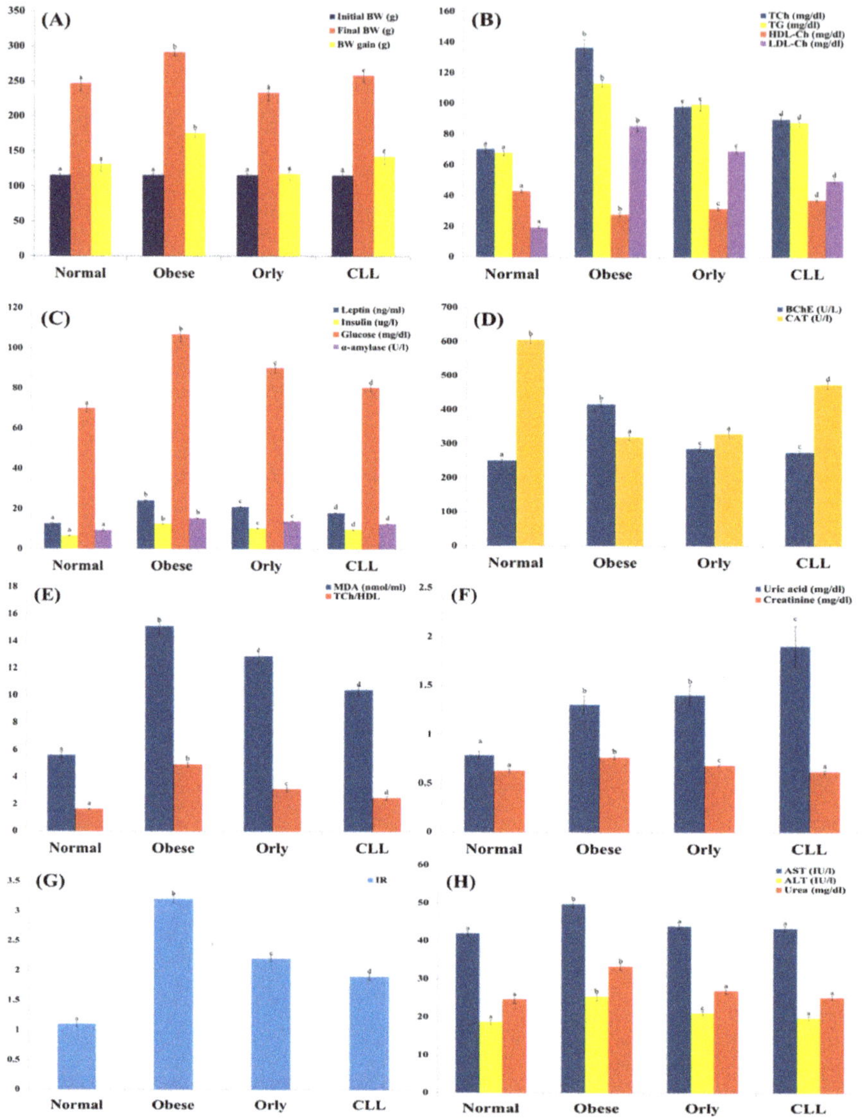

Figure 1. Body weight and biochemical parameters in the 4th week of treatment of experimental animals with Orly and CLL compared to normal and obese groups. (**A**) The initial, final body weights (BW) and BW gain (g). (**B**) The level of plasma total cholesterol (TCh mg/dL), triglycerides (TG mg/dL), high-density lipoprotein cholesterol (HDL-Ch mg/dL), and low-density lipoprotein cholesterol (LDL-Ch mg/dL). (**C**) The level of leptin (ng/mL), insulin (µg/L), glucose (mg/dL), and α-amylase (U/L). (**D**) The level of plasma butyrylcholinesterase (BChE (U/L)) and plasma catalase activity (CAT (U/L)). (**E**) The level of malondialdehyde (MDA (nmol/mL) and TCh/HDL. (**F**) The level of plasma uric acid (mg/dL) and plasma creatinine (mg/dL). (**G**) Calculated insulin resistance (IR). (**H**) The activity of aspartate transaminase (AST (IU/L)), alanine transaminase (ALT (IU/L)), and urea (mg/dL). Each bar graph represents the mean replicate measurement ($n = 6$) expressed as mean ± S.E. The bar graphs with a similar lower-case letter (such as 'a') among experimental groups are not significantly different from each other ($p > 0.05$). The bar graphs with different lower-case letters (such as a, b, c, d, and e) are statistically different from each other ($p < 0.05$).

2.1.2. Effect of CLL on Dyslipidemia

Dyslipidemia, a metabolic complication of obesity manifested by hypertriglyceridemia [21], was observed in obese rats compared to normal rats, as shown by the elevation of plasma total cholesterol (two-fold increase), triglycerides (1.7-fold increase), LDL cholesterol (four-fold increase), and the ratio of T-Ch/HDL-Ch (three-fold increase) (Figure 1A, Table 1) concurrent with the reduction in plasma level of HDL-Ch (1.5-fold decrease) (Figure 1E, Table 1). CLL significantly improved dyslipidemia compared to obese control rats as well as rats treated with Orly ($p < 0.05$), however, it was still higher than normal rats (Figure 1A,E, Table 1).

2.1.3. Effect of CLL on Leptin, α-Amylase, Plasma Glucose, Insulin Levels, and Insulin Resistance

Obese rats are also reported to exhibit increment in the plasma levels of plasma glucose, insulin, insulin resistance, leptin, and α-amylase [22]. A significant elevation in plasma levels of glucose (1.5-fold increase), insulin (two-fold increase), and insulin resistance (three-fold increase) was noted in obese rats compared to normal rats.

Oral administration of Orly and CLL improved plasma levels of glucose (89.9 and 80 mg/dL, respectively), insulin (10.1 and 9.4 μg/L, respectively), and insulin resistance (2.2 and 1.9, respectively) with different degrees (Figure 1C,G, Table 1). Obese rats exhibited significantly elevated levels of plasma leptin, a key hormone in the control of food intake and body weight and a target in obesity management [23,24] (two-fold increase) comparable to those in normal rats. Orly and CLL significantly reduced plasma levels of leptin (21.1 and 18.0 ng/mL, respectively) compared to obese control (24.1 ng/mL). Another drug target in obesity is the inhibition of the digestive enzyme α-amylase [25]. In this study, significant increase in α-amylase activity in obese rats (15.3 U/L) was dramatically reduced upon administration of both Orly and CLL (13.7 and 12.4 U/L, respectively) (Figure 1C, Table 1). Hence, CLL is significantly excelling over Orly in decreasing leptin, insulin, glucose, and α-amylase levels ($p < 0.05$).

2.1.4. Effect of CLL on Oxidative Stress and Lipid Peroxidation

In the current study, elevated plasma levels of butyrylcholinesterase (BChE) were observed in the obese control (415.7 U/L) in agreement with [22], compared to different experimental groups (250.7, 274.8, and 285.2 U/L in normal, CLL and Orly treated groups, respectively). High plasma BChE activity is associated with aberrant lipid profiles, insulin resistance, and hypertension [23], suggestive for BChE role in many metabolic functions [24]. Oral administration of Orly as well as CLL reduced BchE plasma elevation significantly at different levels (285.2 and 274.8 U/L, respectively). Malondialdehyde (MDA), a biomarker used for assessing oxidative stress, was significantly enhanced in obese rats (15.1 nmol/mL) compared to those of normal ones (5.6 nmol/mL), as an indicator of lipid peroxidation, while catalase enzyme activity, an indicator of antioxidant status, showed reduction by 1.9 fold. Rats treated with Orly and CLL exhibited improved oxidative stress markers at different levels (Figure 1D,E, Table 1). CLL revealed significant improvement in oxidative stress markers and lipid peroxidation profiles, better than Orly ($p < 0.05$).

2.1.5. Effect of CLL on Kidney and Liver Functions

Kidney function indicators (creatinine, urea, and uric acid) as well as plasma transaminases (AST and ALT) revealed significant elevation in obese rats compared to normal rats, in agreement with [25]. Treatment with Orly and CLL significantly improved kidney and liver functions, except for the significant elevation of uric acid content in the case of CLL, which is most probably attributed to the high oxalate level in the leaves [26] (Figure 1F,H, Table 1). Hence, CLL was significantly better than Orly in terms of kidney- and liver-function improvement, except for an elevated uric acid level ($p < 0.05$).

Overall, despite the better effect of Orly in reducing body weight gain compared to CLL, the latter revealed better improvement in mostly all tested biochemical parameters, except for an elevated uric acid level.

2.2. *Isolation and Structure Elucidation*

To identify anti-obese agents in the CLL extract, the extract was subjected to fractionation using column chromatography (CC) and liquid chromatography (LC), to afford 14 compounds (C1–C12) including 4 flavone glycosides, i.e., **1** (Figures S1–S6), **2** (Figures S7–S11), **3** (Figures S12–S14), and **4** (Figures S15–S19) as well as 10 dihydrochalcone glycosides, i.e., **5** (Figures S20–S24), **6** (Figures S25–S29), **7** (Figures S30–S36), **8** (Figures S37–S43), **9** (Figures S44–S48), **10a** and **10b** (Figures S49–S54), **11a** and **11b** (Figures S55–S60), and **12** (Figures S61–S67). All compounds were checked for their purity using HPLC (Figure S68).

Isolated Compounds Structure Determination Using NMR and MS

Fourteen compounds were isolated and identified using different spectroscopic techniques including (**1**) apigenin 6-*C*-(2-deoxy-*β*-D-galactopyranoside)-7-*O*-*β*-D-quinovopyranoside, (**2**) luteolin 6-*C*-α-L-rhamnopyranosyl-(1-2)-*β*-D-fucopyranoside, (**3**) apigenin 6-*C*-*β*-D-galactopyranoside, (**4**) apigenin 6-*C*-*α*-L-rhamnopyranosyl-(1-2)-*β*-L-fucopyranoside, (**5**) carambolaside M, (**6**) carambolaside Ia, (**7**) carambolaside J, (**8**) phloretin 3′-*C*-(2-*O*-(*E*)-cinnamoyl-3-*O*-*β*-D-fucopyranosyl-4-*O*-acetyl)-*β*-D-fucopyranosyl-6′-*O*-*β*-D fucopyranosyl-(1/2)-α-L arabinofuranoside, (**9**) carambolaside I, (**10a**) carambolaside P, (**10b**) carambolaside O, (**11a**) phloretin3′-*C*-(2-*O*-(*E*)-p-coumaroyl-3-*O*-*β*-D-fucosyl-4-*O*-acetyl)-*β*-D-fucosyl-6′-*O*-(2-*O*-*β*-D-fucosyl)-α-L-arabinofuranoside, (**11b**) phloretin3′-*C*-(2-*O*-(*E*)-p-coumaroyl-3-*O*-*β*-D-fucosyl-4-*O*-acetyl)-*β*-D-fucosyl-6′-*O*-(2-*O*-*β*-D-fucosyl)-α-L-arabinofuranoside, and (**12**) carambolaside Q.

New compounds for the first time to be identified *in nature*, including compounds **1**, **8**, **11a**, and **11b**, are discussed in detail in this section. All spectral data are provided in supplementary file. Compound **1** (Figure 2) was isolated as a yellowish amorphous powder soluble in 100% MeOH. The molecular formula of compound **1** was calculated as $C_{27}H_{30}O_{13}$, based on a deprotonated ion peak calculated at *m/z* 561.16137, detected at *m/z* 561.1614 $[M\text{-}H]^-$ (calculated $C_{27}H_{29}O_{13}{}^-$, error −0.1 ppm) in the HR-ESI-MS spectrum (Figure S1). Compound 1 showed two UV maximums (λ_{max}) (MeOH) at 270 nm (Band II) and 334 nm (Band I), characteristic for a flavone skeleton [27]. The IR spectrum of compound 1 illustrated a broad band at 3431.4 cm^{-1} and 1623 cm^{-1}, consistent with the presence of hydroxy group and carbonyl functions [28].

The full assignment of ^1H and ^{13}C NMR data (Figures S2 and S3, Table 2) was adopted based on the analysis of the ^1H-^1H COSY, HSQC, and HMBC spectra (Figures S4–S6). The existence of a flavone unit could be easily assigned from the ^1H NMR and ^{13}C NMR spectra (Figures S2 and S3, Table 2) from key signals of 4 A_2B_2-type aromatic protons at δ 7.80 (2H, d, *J* = 8.8 Hz, H-2′/6′) and at δ 6.77 (2H, d, *J* = 8.8 Hz, H-3′/5′) for a *p*-disubstituted benzene ring, together with two aromatic singlets at δ 7.02 (H-8) and at δ 6.58 (H-3), referring to 6,7-disubstituted apigenin [29]. Moreover, ^{13}C NMR (Figure S3) revealed 27 carbon resonances, which may be typical for di-glycosylated apigenin as follows: a carbon signal at δ 184.0 (C-4) assignable for a ketonic carbonyl, carbon resonances at δ 102.4 (C-3), δ 159.9 (C-5), δ 113.6 (C-6), δ 164.5 (C-7), and δ 96.4 (C-8) were similar to those reported for 6,7-disubstituted apigenin [29]. Excluding carbons of flavone unit, another 12 carbons were left assigned to two sugar moieties for deoxy-hexopyranosyl units. δ (Chemical shift) and *J* (coupling constant) values as well as the ^1H-^1H COSY spectrum (Figure S4) identified the first hexose moiety as 2-deoxy-β-D-galactose. The signals for an anomeric proton at δ 5.10 (dd, *J* = 12.1, 2.4 Hz, H-1″), two protons at δ 2.83 (q, *J* = 12.1 Hz, H_1-2″) and at δ 1.59 (m, H_2-2″), two protons at δ 4.02 (dd, *J* = 12.1, 2.1 Hz, H_1-6″) and at δ 3.74 (dd, *J* = 12.1, 6.4 Hz, H_2-6″), six carbons at δ 70.5 (C-1″), 32.3 (C-2″), 71.6 (C-3″), 78.7 (C-4″), 76.1 (C-5″), and δ 62.8 (C-6″) are consistent with those of a 2-deoxy-β-D-galactose [30] attached at the C-6 position in apigenin via a C-glycosidic linkage, confirmed via HMBC correlations (Figure 3). The second sugar was assigned as β-quinovopyranose attached to carbon 7 via an oxygen bridge, based on its anomeric proton and carbon at δ 4.92 (1H, d, *J* = 7.7 Hz, H-1‴) and δ 103.8 (C-1‴). Further, methyl protons at δ 1.26 (3H, d, *J* = 6.5 Hz, H-6‴), four oxymethine carbons at δ 75.0 (C-2‴), 77.1 (C-3‴), 71.8 (C-4‴), and 72.1 (C-5‴), and a methyl carbon at δ 17.9 (C-6‴) confirmed sugar constitution. This sugar unit was identified from large axial–axial coupling constants revealing the axial orientation of all the ring protons of this unit, in agreement with the literature [31]. Hence, compound **1** was identified as apigenin 6-*C*-(2-deoxy-β-D-galactopyranoside)-7-*O*-β-D-quinovopyranoside, a new compound first time to be isolated in planta.

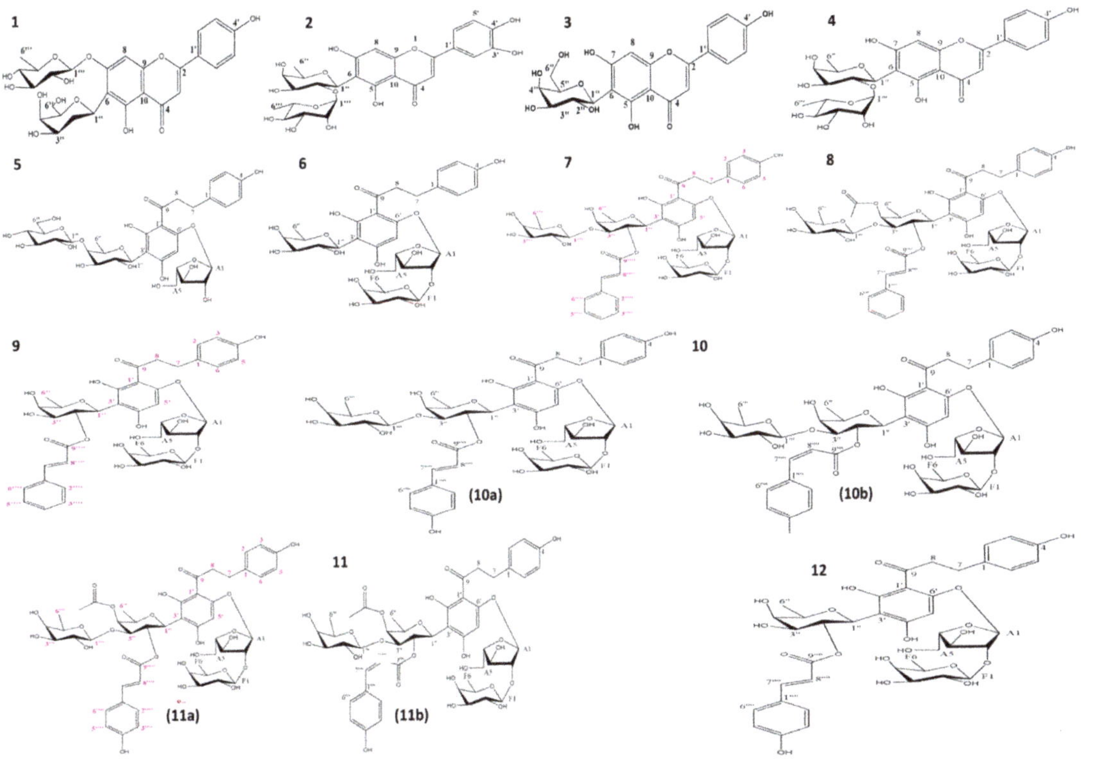

Figure 2. Chemical structures of compounds **1–12** isolated from CLL extract.

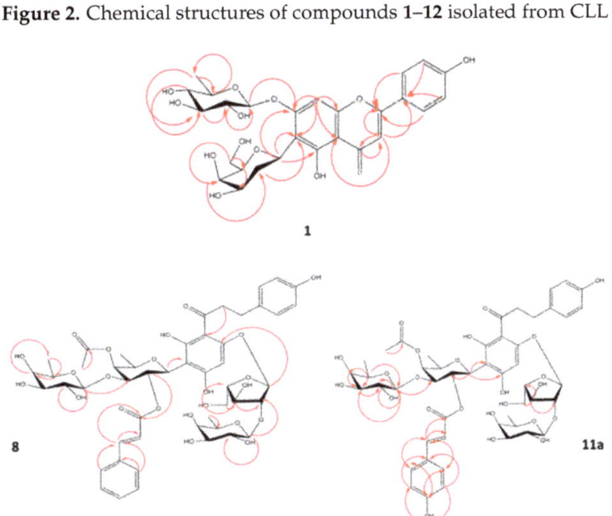

Figure 3. Key HMBC correlations of compound **1**, **8**, and **11a**.

Table 2. ^{1}H (600 MHz) and ^{13}C NMR (150 MHz) data of compounds **1–4** in CD_3OD.

H/C	**1**		**2**		**3**		**4**	
	δ_H (*J* in Hz)	**δ_C**	**δ_H (*J* in Hz)**	**δ_C**	**δ_H (*J* in Hz)**	**δ_C**	**δ_H (*J* in Hz)**	**δ_C**
2		168.0		166.4		166.2		166.4
3	6.58, s	102.4	6.55, s	104.1	6.58, s	103.9	6.57, s	103.2
4		184.0		184.2		184.1		184.0
5		159.9		160.8		162.1		160.8
6		113.6		110.0		109.3		110.3
7		164.5		164.7		165.2		163.5
8	7.02, s	96.4	6.52, s	96.2	6.49, s	95.4	6.49, s	96.6
9		158.5		158.9		158.8		159.1
10		106.9		105.5		105.2		104.9
1″		118.8		123.7		123.2		122.9
2′	7.8, d (8.8)	129.8	7.38, br.s	114.2	7.82, d (8.4)	129.5	7.83, d (8.4)	129.4
3′	6.77, d (8.8)	119.3		147.1	6.92, d (8.5)	117.1	6.91, d (8.5)	117.3
4′		170.2		151.1		162.9		162.3
5′	6.77, d (8.8)	119.3	6.90, d (8.3)	116.9	6.92, d (8.5)	117.1	6.91, d (8.5)	117.3
6′	7.8, d (8.8)	129.8	7.39, br.s	120.4	7.82, d (8.4)	129.5	7.83, d (8.4)	129.4
1″	5.10, dd (12.1, 2.4)	70.5	4.92, d (9.8)	73.7	4.9, d (9.9)	75.4	4.91, d (9.5)	73.7
2″	2.83, q (12.1), 1.59, m	32.3	4.27, t-like (9.3)	75.8	4.16, t (8.9)	72.7	4.31, t (9.3)	76.1
3″	3.80, ddd (11.7, 4.9. 2.8)	71.6	3.75, dd (9.1, 6.2)	77.8	3.8, m	80.2	3.73, d (8.5)	77.9
4″	3.59, d (2.1)	78.7	3.70, br.s	74.1	3.7, dd (12.2, 5.4)	71.8	3.68, br.s	74.2
5″	3.60, d (2.1)	76.1	3.78, m	76.3	3.42, br.s	82.7	3.77, d (6.3)	76.1
6″	4.02, dd (12.1, 2.1), 3.74, dd (12.1, 6.4)	62.8	1.29, d (6.2)	17.2	3.48, m	62.9	1.28, d (5.9)	18.0
1‴	4.92, d (7.7)	103.8	5.19, br.s	102.4			5.17, s	102.4
2‴	3.64, dd (9.5, 7.8)	75	3.85, br.d	72.4			3.36, s	72.4
3‴	3.53, t (9.1)	77.1	3.47, d (8.2)	72.1			3.48, dd (3.2, 9.5)	72.2
4‴	3.40, t (9.4)	71.8	3.10, t-like (9.4)	73.4			3.10, t (9.5)	73.5
5‴	3.62, d (6.7)	72.1	2.56, m	69.9			2.63, m	69.9
6‴	1.26, d (3H, 6.5)	17.9	0.72, d (3H, 6)	18.0			0.73, d (5.9)	17.3

Another novel dihydrochalcone reported for the first time is compound **8** (Figure 2), isolated as yellowish amorphous powder soluble in 100% MeOH, with an estimated molecular formula of $C_{49}H_{60}O_{23}$, based on a deprotonated ion peak calculated at m/z 1015.34526 and detected at m/z 1015.3460 $[M\text{-}H]^-$ (calculated $C_{49}H_{59}O_{23}^-$, error −0.7 ppm) in the HR-ESI-MS spectrum. The IR spectrum of compound 8 showed absorption bands ascribable to hydroxyl (3432 cm^{-1}) and carbonyl moeity (1616 cm^{-1}) [28]. The ^{13}C/DEPT spectrum (Figure S40) revealed the presence of 49 carbon atoms with 48 directly attached protons (3 × CH_2, 11 × C, 31 × CH, 4 × CH_3). Analysis of the ^{1}H-^{1}H COSY, HSQC, and HMBC spectra led to their assignment as the dihydrochalcone skeleton [28] (Figures S41–S43, Table 3). Typical signals for the dihydrochalcone of four A_2B_2-type aromatic protons were detected at δ 6.91 (1H, br.s, H-2), δ 7.09 (1H, d, J = 8.4 Hz, H-6), 6.62 (1H, br.s, H-3), and δ 6.70 (1H, d, J = 8.4 Hz, H-5), with an aromatic proton singlet at δ 6.00 (H-5′) and four aliphatic methylene protons at δ 2.73 and δ 2.68 (H_2-7) and δ 3.19 and δ 3.35 (H_2-8) (Figure S38, Table 3). Further, ^{13}C NMR (Figure S39, Table 3) revealed the signal at δ 205.1 (C-9) of a ketonic carbonyl and two aliphatic methylene carbons at δ 30.8

(C-7) and 49.1 (C-8), confirming the 3′-substituted phloretin structure. The presence of two trans-olefinic protons at δ 7.51 (1H, d, J = 16.0 Hz, H-7′′′′) and 6.26 (1H, d, J = 16.0 Hz, H-8′′′′), five aromatic protons at δ 7.51 (2H, d, J = 3.8 Hz, H-2′′′′/6′′′′) and 7.38 (3H, m, H-3′′′′/4′′′′/5′′′′), a carboxyl carbon at δ 168 (C-9′′′′′′′), two olefinic carbons at δ 146 (C-7′′′′) and 119.3 (C-8′′′′), six aromatic carbons, typical for a cinnamoyl unit [32] and a resonance for a single acetate methyl singlet at δ 2.02, and two carbons of an acetyl moiety at δ 173.6 and 20.5 [33] that suggested a phloretin-acetylated cinnamate. However, the HMBC experiment (Figure S43) could not clarify the connection of the acetyl group, most probably due to the need for using the low-temperature NMR technique [34]. Excluding carbons of dihydrochalcone, acetyl moiety, and trans cinnamoyl units, 23 carbons remained in the ^{13}C NMR spectrum, assigned for four sugar moieties including three hexoses and a pentose. Sugars δ (chemical shift) and J (coupling constant) values as well as ^{1}H-^{1}H-COSY cross peaks and three hexose moieties were determined to be β-fucopyranosyls. The signals for an anomeric proton at δ 5.09 (1H, d, J = 9.9 Hz, H-1′′), a methyl proton doublet at δ 1.31 (3H, d, J = 4.2 Hz, H_3-6′′), five oxymethine carbons at δ 74.1 (C-1′′), 71.9 (C-2′′), 84.5 (C-3′′), 74.1 (C-4′′), and 76.3 (C-5′′), and a methyl carbon at δ 17.2 (C-6′′) are consistent with those of a β-fucosyl moiety attached to C-3′ of the phloretin moiety via a C-glycosidic linkage [35]. The HMBC spectrum could not though confirm this linkage, as no correlation appeared between (H-1′′) and C-2′, C-3′, or C-4′, which is likely attributed to the phenomenon of coexistence of two conformationally variant rotamers, due to restricted rotation around the single bond between C-9 and C-1 resulting from the steric hindrance of the cinnamoyl moiety [34]. Further, the change pattern of δ values at C-1′′ (−1.7 ppm), C-2′′ (+1.9), and C-3′′ (−1.7), due to the esterification in comparison to the unesterified analog (carambolaside Ja), confirmed the connection of a cinnamoyl unit at C-2′′ [34]. The downfield shift in C-3′′ (Δδ + 1.6) and C-4′′ (Δδ +0.5), relative to those in compound 7 (Figure S32), located the acetyl moiety at C-4′′. A second hexose was assigned based on its anomeric signals at δ 4.36 (1H, d, J = 7.6 Hz, H-1′′′) and δ 106 (C-1′′′), a methyl proton doublet at δ 1.26 (3H, d, J = 6.4 Hz, H_3-6′′′), four oxymethine carbons at δ 72.3 (C-2′′′), 74.7 (C-3′′′), 73 (C-4′′′), and 72 (C-5′′′), and a methyl carbon at δ 16.8 (C-6′′′), as β-fucopyranose connected via an oxygen linkage [35] between C-1′′′ and C-3′′, confirmed by the HMBC correlations from H-1′′′ to C-3′′ (Figure 3 and Figure S43). The third sugar signals were typical for a pentose from its anomeric proton at δ 5.71 (1H, s, H-A_1), four oxymethine carbons at δ 106.6 (C-A_1), 92.9 (C-A_2), 76.3 (C-A_3), and 84.1 (C-A_4), and an oxymethylene carbon at δ 62 (C-A_5) annotated as a α-arabinofuranosyl moiety [36]. Lastly, signals of a third β-fucopyranosyl moiety were assigned from its anomeric signal at δ 3.97 (1H, br.s, H-F_1) and δ 105.6 (C-F_1), methyl protons at δ 1.29 (1H, s, H_1-F_6) and δ 0.75 (2H, s, H_2-F_6), four oxymethine carbons at δ 72.9 (C-F_2), 72.2 (C-F_3), 75 (F_4), and 71.9 (F_5), and a methyl carbon at δ 16.9 (F_6).

The HMBC experiment could not clarify the connection of the acetyl group, α-arabinofuranosyl moiety, or the last *β*-fucopyranosyl moiety, which warranted using the low-temperature NMR technique [34]. Altogether, compound **8** was identified as phloretin 3′-C-(2-*O*-*(E)*-cinnamoyl-3-*O*-β-D-fucopyranosyl-4-*O*-acetyl)-β-D-fucopyranosyl-6′-*O*-β-D fucopyranosyl-(1/2)-α-L arabinofuranoside.

Compound **11** (Figure 2), another novel dihydrochalcone, was isolated as a yellowish amorphous powder soluble in 100% MeOH. The molecular formula of compound **11** was established as $C_{49}H_{60}O_{24}$, based on a deprotonated mol. ion peak calculated at m/z 1031.34018 and detected at m/z 1031.3389 $[M-H]^-$ (calculated $C_{49}H_{59}O_{24}{}^-$, error +1.2 ppm) in its HR-ESI-MS spectrum (Figure S55). The IR spectrum of compound 11 revealed two major absorption bands at 3432 cm^{-h} and 1616 $cm^{-\&}$, consistent to hydroxyl and carbonyl moieties, respectively [28].

Table 3. ^{1}H (600 MHz) and ^{13}C NMR (150 MHz) data of compounds **5–9** in CD_3OD isolated from CLL extract.

H/C	5		6		7		8		9	
	δ_H (*J* in Hz)	δ_C	δ_H (*J* in Hz)	δ_C	δ_H (*J* in Hz)	δ_C	δ_H (*J* in Hz)	δ_C	δ_H (*J* in Hz)	δ_C
1		134.2		134.2		134.1		134.2		134.1
2	7.07, d (8.4)	130.5	7.11, d (8.4)	130.5	6.89, br.s	130.5	6.91, br.s	130.5	7.09, d (8.3)	130.5
3	6.67, d (8.4)	116.1	6.72, d (8.4)	116.3	6.61, br.s	116.4	6.62, br.s	116.3	6.70, d (8.4)	116.3
4		156.5		156.7		156.6		156.6		156.6
5	6.67, d (8.4)	116.1	6.72, d (8.4)	116.3	6.61, br.s	116.4	6.70, d (8.4)	116.3	6.70, d (8.4)	116.3
6	7.07, d (8.4)	130.5	7.11, d (8.4)	130.5	6.89, br.s	130.5	7.09, d (8.4)	130.5	7.09, d (8.3)	130.5
7	2.87, t (7.4)	31.4	2.91, d (5.7) 1.31, s	30.8	2.73/2.65, br.s	30.9	2.73/2.68, br.s	30.8	2.91, m/1.29, s	30.8
8	3.36, t (7.4)	46.6	3.41 unresolved	45.8	3.35/3.06	47.6	3.35/3.19, br.s	49.1	3.35/3.41	46.2
9		204.8		203.6		205.4		205.1		205.2
1′		106.4		106.5		106.0		106.0		106.8
2′		167.5		168		165.5		168		165.8
3′		107.1		106.5		106.0		106.0		106.8
4′		167.5		168		164.6		166.3		165.1
5′	6.05, s	98.3	6.03, s	99.5	6.12, s	97.4	6.00, S	98.7	6.13, S	97.6
6′		161.5		161.6		161.8		161.8		161.5
1″	4.77, d (9.9)	76.8	4.78, d (9.8)	76.1	5.11, d (9.5)	74.1	5.09, d (9.9)	74.1	5.07, d (9.7)	73.9
2″	4.36, br.s	71.3	4.43, t (9.4)	70.3	5.78, br.s	71.8	5.88, br.s	71.9	4.78, br.s	75.0
3″	3.63, dd (9.7, 3.3)	77.5	3.52, dd (9.4, 3.2)	77.7	3.96, br.s	82.9	3.97, br.s	84.5	3.53, dd (9.3, 3)	77.1
4″	3.95, d (3)	84.2	3.48, d (1.8)	74.4	3.95, d (2.5)	73.6	3.97, br.s	74.1	3.69, d (2.9)	73.7
5″	3.77, q (6.4)	76.1	3.74, q (6.7)	76	3.88, br.s	76.5	3.85, d (5.8)	76.3	3.73, q (6.5)	76.0
6″	1.33, d (3H, 6.4)	17.6	1.27, d (3H, 6.5)	17.2	1.33, d (3H, 6)	17.2	1.31, d (3H, 4.2)	17.2	1.32, d (6), 1.26, d (2H, 6.5)	17.2
1‴	4.58, d (7.7)	106.4			4.37, d (7.6)	105.7	4.36, d (7.6)	106.0		
2‴	3.34, m	76.2			3.48, dd (9.7, 7.7)	72.3	3.49, dd (9.7, 7.7)	72.3		
3‴	3.28, m	78.2			3.38, dd (9.7, 3.4)	74.8	3.38, dd (9.8, 3.4)	74.7		
4‴	3.35, m	71.3			3.55, d (3.4)	73.0	3.55, d (3.2)	73.0		
5‴	3.40, m	78.3			3.62, q (6.9)	72.0	3.62, q (6.5)	72.0		
6‴	3.85, dd (11.9, 2.1) 3.71, dd (11.8, 5.2)	62.7			1.26, d (3H, (6.5)	16.9	1.26, d (3H, 6.4)	16.8		
1⁗						136.0		136.0		136.4
2⁗					7.50, d (6.1)	129.3	7.51, d (3.8)	129.3	7.52, dd (7.4, 3.5)	129.2
3⁗					7.39, br.s	130.1	7.38, br.s	130.0	7.40, m	130.0
4⁗					7.39, br.s	131.4	7.38, br.s	131.4	7.40, m	131.5
5⁗					7.39, br.s	130.1	7.38, br.s	130.0	7.40, m	130.0
6⁗					7.50, d (6.1)	129.3	7.51, d (3.8)	129.3	7.60, dd (7.4, 3.5)	129.2
7⁗					7.51, d (16.0)	146.1	7.51, d (16.0)	146.0	7.69, d (16.0)	146.3
8⁗					6.27, d (16.0)	119.2	6.26, d (13.1)	119.3	6.53, d (16.0)	118.6
9⁗						167.7		168.0		167.7
A1	5.59, d (1.1)	108.1	5.78, d (1.2)	107.5	5.73, s	106.9	5.71, s	106.6	5.78, s	106.8
A2	4.04, dd (9.5, 6)	87.1	4.31, dd (4.6, 1.6)	92.6	4.19, br.s	92.9	4.19, br.s	92.9	4.29, m	92.5
A3	4.00, dd (6, 3.6)	78.2	4.17, dd (7.7, 4.9)	76.3	4.11, dd (7.9, 4.9)	76.3	4.10, dd (8, 5)	76.3	4.11, dd (7.9, 4.9)	76.0

Table 3. *Cont.*

H/C	5		6		7		8		9	
	δ_H (*J* in Hz)	δ_C	δ_H (*J* in Hz)	δ_C	δ_H (*J* in Hz)	δ_C	δ_H (*J* in Hz)	δ_C	δ_H (*J* in Hz)	δ_C
A4	4.25, dd (3.6, 1.7)	83.5	4, m	84.5	3.99, br.s	84.4	3.97, br.s	84.1	4.00, br.s	84.2
A5	3.65, dd (12.1, 4.7) 3.73, dd (8.1, 3.1)	62.7	3.67, dd (12.5, 4.6) 3.8, dd (12.5, 2.8)	62.2	3.62, dd (13.3, 6.4), 3.77, dd (13.4, 6.3)	62.1	3.62, dd (13.4, 6.5), 3.78, br.s	62.0	3.66, dd (12.4, 4.8), 3.79, dd (12.4, 2.9)	62.2
F1			4.14, s	105.4	3.99, br.s	105.2	3.97, br.s	105.6	4.15, br.s	105.4
F2			3.48, m	72.9	3.41, br.s	72.9	3.41, br.s	72.9	3.47, dd (9.2, 3.4)	72.9
F3			3.37, dd (9.7, 3.4)	72	3.25, br.s	72.2	3.23, br.s	72.2	3.25, br.s	72.2
F4			3.37, s	75	3.43, br.s	75	3.42, br.s	75	3.36, d (3.4)	75
F5			3.26, q (6.5)	72.2	2.93, br.s	71.8	2.94, br.s	71.9	3.25, q (6.3)	72.0
F6			1.01, d (6.5)	16.7	1.29, 0.78, s (3H)	16.9	1.29, s (2H), 0.75, br.s	16.9	1.00, d (3H, 6.4)	16.9
CH_3 CO-4″							2.02, S	20.5		
CO acetyl								173.6		

NMR spectral analysis confirmed that compound **11** existed in the form of a mixture of two diastereoisomers (**11a** and **11b**). ^{1}H and ^{13}C NMR data, in addition to the HSQC spectrum (Figures S56–S58), indicated a structure closely related to that of compound **8**, with an extra hydroxyl group characteristic for a p-coumaroyl moiety, instead of the cinnamoyl moiety in compound **8** existing in two diastereomers, i.e., *(E)* and *(Z)* isomers. Signals for *(E)*-isomer were assigned for the two olefinics at δ 7.45 (d, *J* = 14.3 Hz, H-7⁗) and at 6.05 (d, *J* = 14.3 Hz, H-8⁗), characteristic for an *(E)-p*-coumaroyl moiety in addition to four *p*-coupled aromatic protons at δ 7.36 (2H, d, *J* = 8.6 Hz, H-2⁗/6⁗) and 6.67 (2H, d, *J* = 8.5 Hz, H-3⁗/5⁗). Moreover, ^{13}C NMR (Figure S57) exhibited a carboxyl carbon at δ 168.9 (C-9⁗), two olefinic carbons at δ 145.2 (C-7⁗) and 113.2 (C-8⁗), and six aromatic carbons, typical for a *(E)*-coumaroyl unit [37]. In contrast, (Z)-*p*-coumaroyl moiety exhibited signals of four para-coupled aromatic protons at δ 7.19 and 7.4 (2H, s, H-2⁗/6⁗), unresolved peaks corresponding to H-3⁗/5⁗, and two olefinic protons at δ 6.69 (d, *J* = 11.2 Hz, H-7⁗) and 5.60 (d, *J* = 11.6 Hz, H-8⁗), coupled with a characteristic constant of *J* = 11.2 Hz. Then, ^{13}C-NMR exhibited a carboxyl carbon at δ 168.9 (C-9⁗), two olefinic carbons at δ 142.4 (C-7⁗) and 114.8 (C-8⁗), and six aromatic carbons typical for an (Z)-coumaroyl unit [37]. A resonance for a single acetate methyl singlet at δ 2.04 and two carbons of an acetyl moiety at (δ 172.0 and 19.1) were detected [33], as in compound **8**. The downfield-shifted C-3″ (Δδ + 0.3) and C-4″ (Δδ − 2), relative to those in compound **10** (Figure S51, Table 4), located the acetyl moiety at C-4″. The HMBC experiment (Figure 3) could not confirm the connection of the acetyl group most probably due to the need for using low-temperature NMR technique, as in **8** [34]. Consequently, compound **11a** was established as phloretin 3′-C-(2-*O*-*(E)*-p-coumaroyl-3-*O*-β-D-fucosyl-4-*O*-acetyl)-β-D-fucosyl-6′-*O*-(2-*O*-β-D-fucosyl)-α-L-arabinofuranoside. Whereas **11b** was assigned as phloretin 3′-C-(2-*O*-*(Z)*-p-coumaroyl-3-*O*-β-D-fucosyl-4-*O*-acetyl)-β-D-fucosyl-6′-*O*-(2-*O*-β-D-fucosyl)-α-L-arabinofuranoside. These compounds are reported for the first time in nature. Other identified compounds reported in the literature included carambolaside M (**5**) [11] (Figures S20–S24), carambolaside Ia (**6**) [11] (Figures S25–S29), carambolaside J (**7**) [11] (Figures S30–S36), carambolaside I (**9**) [34] (Figures S44–S48), carambolaside P and O (**10**) [11] (Figures S49–S54), carambolaside Q (**12**) [11] (Figures S61–S67), luteolin 6-C-α-L-rhamnopyranosyl-(1-2)-β-D-fucopyranoside (**2**) [38] (Figures S7–S11), apigenin 6-C-β-D-galactopyranoside (**3**) [39] (Figures S12–S14) and apigenin 6-C-α-L-rhamnopyranosyl-(1-2)-β-L-fucopyranoside (**4**) [38] (Figures S15–S19), by comparison of their spectroscopic data to those in previous references, but isolated for the first time from starfruit leaves.

Table 4. ^{1}H (600 MHz) and ^{13}C NMR (150 MHz) data of compounds **10–12** in CD_3OD.

H/C	10				11				12	
	10a (*Z*-isomer)		**10b (*E*-isomer)**		**11a (*Z*-isomer)**		**11b (*E*-isomer)**			
	δ_H (*J* in Hz)	δ_C	δ_H (*J* in Hz)	δ_C	δ_H (*J* in Hz)	δ_C	δ_H (*J* in Hz)	δ_C	δ_H (*J* in Hz)	δ_C
1		134.2		133.8		134.7		134.4		134.2
2	6.93/7.00, br.s	130.5	7.09, d (8.2)/7.18, s	130.6	6.93/7.02, br.s	129.0	7.11, d (8.5)/7.23, d (8.2)	129.1	6.90, br.s	130.5
3	6.75, d (6.1)	117.6	6.70, d (8.5)	116.4	6.79, d (8.4)	115.9	6.72, d (8.5)	114.9	6.63, d (7.4)	116.4
4		156.6		156.5		155.1		155.1		156.6
5	6.75, d (6.1)	117.6	6.70, d (8.5)	116.4	6.79, d (8.4)	115.9	6.72, d (8.5)	114.9	6.63, d (7.4)	116.4
6	6.93/7.00, br.s	130.5	7.09, d (8.2)/7.18 s	130.6	6.93/7.02, br.s	129.0	7.11, d (8.5)/7.23, d (8.2)	129.1	6.90, br.s	130.5
7	2.77/2.70, br.s	30.8	2.77/2.89, br.s	30.8	2.78/2.68, br.s	29.3	2.78/2.91, br.s	29.3	2.65/2.75, br.s	30.4
8	3.35/3.07	46.6	3.35/3.16	46.6	3.37/unresolved	45.0	3.37/unresolved	45.0	3.09/3.36, br.s	47.5
9		204.9		204.9		204.1		204.1		206
1′		105.8		105.8		104.5		104.5		105.7
2′		165.8		165.8		167.1		167.1		165.5
3′		105.6		106.5		104.5		104.5		106
4′		165.8		165.8		167.1		167.1		164.9
5′	6.09, s	97.7	5.95, s	97.7	6.16, s	95.7	6.11, s	95.7	6.15, s	96.3
6′		161.8		161.8		160.3		160.3		161.7
1″	5.09, d (9.9)	74.1	5.02, d (9.8)	74.1	5.04, d (9.9)	72.6	5.12, d (9.5)	72.6	5.07, d (7.8)	74
2″	5.79, br.s	71.9	6.23, br.s	71.5	5.78, br.s	70	6.26, br.s	70	5.52, br.s	72.9
3″	3.91, br.d (9.7)	82.5	3.91, br.d (9.7)	82.5	3.94, br.s	82.8	3.94, br.s	82.8	3.85, br.s	74.9
4″	3.96, br.s	73.6	3.96, br.s	73.6	3.98, s	71.6	3.98, s	71.6	3.78, m	73.6
5″	3.82, m	76.3	3.86, br.s	76.3	3.85, br.d (6.7)	74.8	3.88, br.d (6.6)	74.8	3.82, m	76.9
6″	1.31, d (3H, 6.8)	17.2	1.29, br.s	17.2	1.31, d (3H, 6.8)	17.2	1.29, br.s	17.2	1.33, d (3H, 6.4)	17.2
1‴	4.36, d (7.6)	105.9	4.30, d (6.8)	105.9	4.38, d (7.6)	104.5	4.32, d (6.8)	104.5		
2‴	3.48, dd (9.7, 7.7)	72.2	3.46, br.d (7.2)	72.2	3.50, dd (9.7, 7.7)	70.8	3.46, br.d (7.2)	70.7		
3‴	3.39, dd (9.8, 3.4)	75.0	3.39, dd (9.8, 3.4)	74.9	3.41, dd (9.8, 3.4)	73.6	3.41, dd (9.8, 3.4)	73.5		
4‴	3.55, d (3.5)	73.0	3.56, d (4.0)	72.9	3.57, d (3.6)	71.4	3.58, d (4.1)	71.5		
5‴	3.62, q (6.9)	72.0	3.62, q (6.9)	72.0	3.63, q (6.6)	70.5	3.63, q (6.6)	70.4		
6‴	1.26, d (3H, 6.4)	16.9	1.26, d (3H, 6.4)	16.9	1.26, d (3H, 6.4)	15.4	1.26, d (3H, 6.4)	15.4		
1⁗		127.3		127.3		125.9		125.9		127.2
2⁗	7.33, d (8.4)	131.3	7.18/7.37	133.3	7.36, d (8.6)	129.8	7.19/7.4	132.7	7.37, d (8.5)	131.2
3⁗	6.65, d (8.4)	116.1	6.95, d (8.9)	116.3	6.67, d (8.5)	114.5			6.81, d (8.5)	117
4⁗		163.7		163.7		164.1		164.1		161.5
5⁗	6.65, d (8.4)	116.1	6.95, d (8.9)	116.3	6.67, d (8.5)	114.5			6.81, d (8.5)	117
6⁗	7.33, d (8.4)	131.3	7.18/7.37	133.3	7.36, d (8.6)	129.8	7.19/7.4	132.7	7.37, d (8.5)	131.2
7⁗	7.43, d (15.9)	146.8	6.66, d (11.2)	144.7	7.45, d (14.3)	145.2	6.69, d (11.2)	142.4	7.45, d (15.9)	146.4
8⁗	6.01, d (15.0)	114.4	5.59, d (12.4)	116.4	6.05, d (14.3)	113.2	5.60, d (11.6)	114.8	6.08, d (15.9)	115.3
9⁗		168.8		168.8		168.9		168.9		168.4
A1	5.71, s	105.9		105.9	5.74, s	104.5	5.92, s	104.5	5.73, br.s	107.1
A2	4.15, m	93.1	4.31, br.s	93.1	4.18, m	91.5	4.31, br.s	91.5	4.17, br.s	93
A3	4.10, dd (7.9, 4.9)	76.1	4.16, dd (7.7, 3.8)	76.3	4.13, dd (7.9, 4.9)	74.8	4.18, dd (6.7, 4.4)	74.8	4.11, dd (7.9, 4.8)	76.3

Table 4. *Cont.*

H/C	10				11				12	
	10a (*Z*-isomer)		**10b (*E*-isomer)**		**11a (*Z*-isomer)**		**11b (*E*-isomer)**			
	δ_H (*J* in Hz)	δ_C	δ_H (*J* in Hz)	δ_C	δ_H (*J* in Hz)	δ_C	δ_H (*J* in Hz)	δ_C	δ_H (*J* in Hz)	δ_C
A4	3.94, m	84.1	3.94, m	84.1	3.94, m	84.3	3.94, m	84.3	3.95, ddd	84.5
A5	3.62, m, 3.77, br.s	62.0	3.62, m, 3.68, br.s	62.0	3.65, d (11.8, 5.3), 3.79, m	60.6	3.65, d (11.8, 5.3), 3.79, m	60.6	3.63, dd (12.1, 4.7), 3.83, br.d (12.3)	62
F1	3.97, d (5.6)	105.1	4.06, br.s	105.1	4.00, d (5.2)	104.5	4.08, br.s	104.5	3.96, d (7.6)	105.4
F2	3.38, dd (9.8, 6.4)	72.9	3.53, br.d (6.6)	72.9	3.4, dd (9.8, 6.4)	73.2	3.54, m	71.4	3.41, dd (9.6, 7.9)	72.2
F3	3.23, br.s	74.7	3.33, m	74.8	3.25, m	73.2	3.35, m	73.4	3.24, br.d (6.7)	74.90
F4	3.45, d (3.4)	72.3	3.42, d (3.2)	72.4	3.48, br.s	70.9	3.44, d (3.1)	70.8	3.40, d (3.3)	72.8
F5	2.89, br.s	71.5	3.19, m	71.5	2.91, br.s	70.4	3.21, m	70.5	2.9, br.s	71.8
F6	0.89, 0.76, s (3H)	16.9	0.94, 0.76 (3H)	16.9	0.89, 0.76, s (3H)	15.7	0.94, 0.76, s (3H)	15.7	0.79, br.s (3H)	16.9
CH_3 CO-4″					2.04, s	19.1	2.04, s	19.1		
CO acetyl						172.0		172.0		

2.3. *Structure-Activity Relationship Assessment of Isolated Compounds as α-Glucosidase Inhibitors*

To further confirm potential efficacy of CLL compounds, isolated compounds were tested for their in vitro α-glucosidase inhibitory activity, to assess their efficacy. Considering the limitation of yield, in vivo assay was not possible to be performed. The efficacy of the isolated compounds was measured and discussed in relationship to the flavonoid structures, as discussed in the next subsections for each class separately, to identify the most crucial motifs within each for activity.

2.3.1. Structure-Activity Relationship Assessment of Flavones as α-Glucosidase Inhibitors

Tested flavone compounds **1**, **2**, **3**, and **4**, along with flavone standard aglycones, i.e., apigenin and luteolin, exhibited strong α-glucosidase inhibition, where IC_{50} values were determined at 613, 328, 439, and 390 μM, respectively, exceeding that of acarbose determined at 662 μM, a commercial α-glucosidase inhibitor anti-diabetic drug. The order of activity of the isolated flavone glycosides was as such, with IC_{50} values at 327.9, 390.4, 439.2, and 612.9 μM for compounds, viz., **2**, **4**, **3**, and **1**, which are much higher than those of acarbose. However, flavone glycosides were less potent than their corresponding aglycones, i.e., apigenin and luteolin, with IC_{50} values at 85.6 and 48.2 μM, respectively (Figure 4A,B, Table 5).

Among glycosides, compound **2**, identified as luteolin 6-C-α-L-rhamnopyranosyl-(1-2)-β-D-fucopyranoside, showed the highest inhibitory activity among all isolated flavone glycosides in line with its aglycone, suggestive for the improved efficacy of C-glycosyl flavone against α-glucosidase enzyme, which is in agreement with reports that sugar moiety attached at C-6 position improved efficacy against pancreatic lipase inhibitory activity [40], extended herein to include the α-glucosidase inhibition effect (Figure 4B).

In contrast, compound **1**, identified as apigenin 6-C-(2-deoxy-β-D-galactopyranoside)-7-O-β-D-quinovopyranoside, showed the weakest inhibitory activity among all isolated flavone glycosides, with IC_{50} 612.9 μM, likely attributed to the glycosylation of hydroxy group at the C-7 position [41] (Figure 4B) and absent in compounds **2**, **3**, and **4**. Compounds **3** and **4** exhibited strong inhibition with an IC_{50} value of 439.2 μM and 390.4 μM, respectively, in line with previously published data on the efficacy of apigenin 6-C-(2″-O-α-rhamnopyranosyl)-β-fucopyranoside in lowering the glucose level in hyperglycemic rats [40]. Standard apigenin and luteolin showed the strongest inhibitory activity, with IC_{50} values at 85.6 and 48.2 μM, respectively, compared to that of acarbose (661.6 μM), with luteolin showing the stronger inhibitory activity compared to that of acarbose, which is in accordance with the previously reported α-glucosidase inhibitory activity [42]. In line with our findings, hydroxylation at C-3′ of the B-ring of apigenin, in particular, was reported to enhance the α-glucosidase inhibition activity [41], whereas glycosylation affected it negatively compared to the aglycones [43].

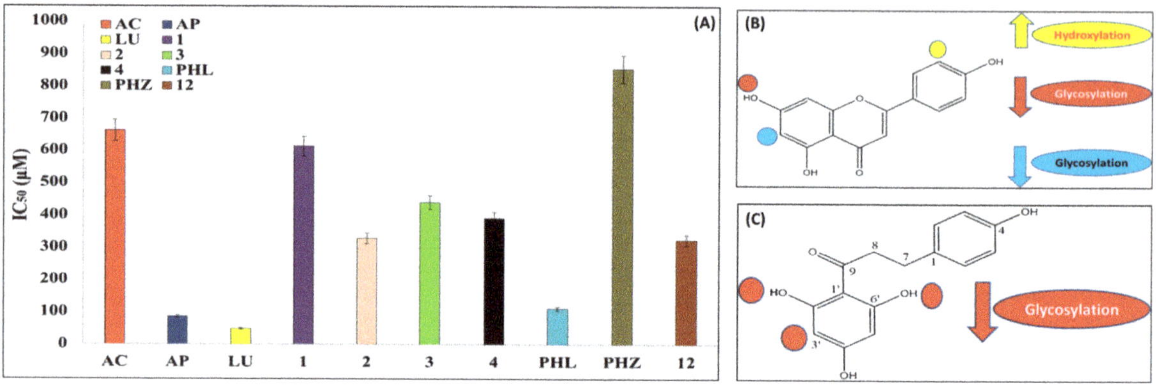

Figure 4. (**A**) IC_{50} (µM) of tested compounds against α-glucosidase enzymes in vitro. (**B**) The potential sites of flavone *C*-glycosides affecting α-glucosidase inhibitory potential. (**C**) The potential sites of dihydrochalcone *C*-glycosides affecting the α-glucosidase inhibitory potential. The up arrows represent increased inhibition, whereas the down arrows represent decreased inhibition activity. Results are expressed as mean ± SE (*n* = 3).

Table 5. IC_{50} of tested compounds using in vitro α-glucosidase inhibition assay. (-) indicates inactive compounds.

Class	Code	Compound Name	IC_{50} (µM)
Standard	AC	Acarbose	661.6 ± 0.01
Flavone	Ap	Apigenin	85.6 ± 0.01
	LU	Luteolin	48.2 ± 0.02
	1	Apigenin 6-C-(2-deoxy-β-D-galactopyranoside)-7-*O*-β-D-quinovopyranoside	612.9 ± 0.03
	2	Luteolin 6-*C*-α-L-rhamnopyranosyl-(1-2)-β-D-fucopyranoside	327.9 ± 0.05
	3	Apigenin 6-C-β-D-galactopyranoside	439.2 ± 0.01
	4	Apigenin 6-*C*-α-L-rhamnopyranosyl-(1-2)-β-L-fucopyranoside	390.4 ± 0.2
Dihydrochalcone	PHL	Phloretin	110.4 ± 0.06
	PHZ	Phloridzin	853.1 ± 0.02
	5	carambolaside M	-
	6	carambolaside Ia	-
	7	carambolaside J	-
	8	4″-*O*-acetyl-carambolaside J	-
	9	carambolaside I	-
	10	mix of carambolaside P and carambolaside O	-
	11	Mix of 4″-*O*-acetyl-carambolaside P and 4″-*O*-acetyl-carambolaside O	-
	12	carambolaside Q	323.6 ± 0.06

2.3.2. Structure-Activity Relationship of Dihydrochalcones and Their Glycosides as α-Glucosidase Inhibitors

The isolated dihydrochalcone glycosides, viz., compounds **5**, **6**, **7**, **8**, **9**, **10**, **11**, and **12**, were assessed for their α-glucosidase inhibitory activity in comparison to acarbose, together with two standard dihydrochalcones, phloretin and its glucoside phloretin-2′-glucose, commonly named phloridzin. All dihydrochalcone glycosides, viz., **5**, **6**, **7**, **8**, **9**, **10**, and **11**, were found inactive except for compound **12**, which showed relatively strong activity with IC_{50} at 323.6 μM (Figure 4A, Table 5). The reported weak α-glucosidase inhibitory activity of diglycosylated chalcones [41] clarified the inactivity of all isolated compounds and suggested that with regards to α-glucosidase inhibition, glycosylated forms of flavones are more active than dihydrochalcones. Further study is warranted, to carefully assess the α-glucosidase enzyme kinetic analysis of the dihydrochalcone carambolaside Q.

Regarding standard dihydrochalcones, phloretin was reported as a strong α-glucosidase inhibitor [44] and as a glucose transporter inhibitor [45], with a measured IC_{50} value at 110.4 μM (Figure 4A). Further, phloridzin revealed moderate inhibitory activity, with an IC_{50} of 853.1 μM (Figure 4A), in accordance with the reported dose-dependent α-glucosidase inhibition [46] and confirming that the inhibitory activity of monoglycosyl chalcones is lower than its aglycones [41] (Figure 4C). These results suggests that α-glucosidase inhibitory activity of *A. carambola* L. extract is mainly mediated by flavone glycosides composition, with a smaller contribution coming from dihydrochalcone glycosides being less active, except for compound **12**.

3. Materials and Methods

3.1. Plant Material

A. carambola, fresh leaf was collected from Groppy Arboretum, Giza, Egypt, in May 2021. The soil is of clay type with high humidity up to 90%. The tree grows in shade and is irrigated every 15 days. Plant material was authenticated by plant taxonomist Dr. Mohamed Gibali, Senior Botanist, Orman Botanic Garden (Giza, Egypt), and Mrs. Therese Labib, Consultant of Plant Taxonomy at the Ministry of Agriculture and Orman Botanic Garden, Giza, Egypt. A voucher specimen number (4754) was deposited in the (CAIM) Herbarium of Flora and Phytotaxonomy Researches Department, Horticultural Research Institute, Agricultural Research Center, Egypt.

Shade-dried powdered sample of *A. carambola* leaf (2 kg) was repeatedly extracted with 70% MeOH of analytical grade (Sigma Aldrich, St. Louis, MO, USA) in a water bath at 40 °C (3 × 5 L, each 48 h) until exhaustion and then filtered off. The filtrate was concentrated under reduced pressure to dryness at 55 °C to yield 500 g (25%) crude extract of *A. carambola* leaves. The obtained extract was kept at 4 °C for further phytochemical and biological assessments.

3.2. Chemicals

Biodiagnostic kits were purchased from Biodiagnostic Co. (Dokki, Giza, Egypt) for measurement of AST, ALT, urea, uric acid, creatinine, total cholesterol, HDL, LDL, MDA, leptin, insulin, glucose, α-amylase, BChe, and CAT levels. The enzyme α-glucosidase was purchased from Oriental Yeast Co. (Tokyo, Japan), while HEPES for making buffer solution was purchased from EMD Millipore Corp (Billerica, MA USA). Phenolic standards, i.e., luteolin, apigenin, phloretin, and phloridzin, and 5-fluorouracil as reference cytotoxic drug were purchased from Wako Pure Chemical Industries (Osaka, Japan). Orly as a reference anti-obese drug for in vivo experiment was obtained from Eva Pharma, Egypt. Acarbose as a reference antidiabetic for in vitro experiments was purchased from Wako Pure Chemical Industries (Tokyo, Japan).

3.3. Chromatographic and Spectroscopic Techniques

Polyamide 6S, Silica Gel 60 (60–120 mesh), and Sephadex LH-20 (Riedel-de Haën AG, Seelze, Germany) were used for column chromatography (CC). Medium-pressure liquid chromatography (MPLC) was performed using Reveleris Prep System set (Buchi, Flawel, Switzerland) with a UV-ELSD detector, a C-18 flash column (FP ID C18, 35–45 μM, 40 g). Celite No. 545 from Wako (Japan) was used for loading sample. Analytical pre-coated Silica Gel 60 F245 plates (NP-TLC), preparative reversed-phase silica gel 60 RP-18 F254S (RP_{18}-PTLC) thin layer chromatography plates (Merck, Germany), and preparative normal phase silica Gel 70 FM (NP-PTLC) thin layer chromatography plates (Wako, Japan) were used for the final purification of compounds. Thin layer chromatography (TLC) plates were visualized under UV light at (254 and 365 nm) and sprayed with 10% MeOH-H_2SO_4 reagent, followed by heating for 2–3 min. Methanol used for extraction in CC was of analytical grade. Methanol and formic acid for MS and HPLC analyses were of HPLC grade. HPLC analysis was employed using an Agilent 1220 Infinity LC system, equipped with ELSD detector, a binary solvent delivery system, and an autosampler and connected to YMC column (5 μM, 4.6 × 150 mm, Japan). Aqueous formic acid (0.1%) and acetonitrile were used as mobile phases A and B, respectively, with the total flow rate at 1.0 mL/min for 35 min.

Detection of UV absorption of isolated compounds was done using a Shimadzu ultraviolet–visible (UV–Vis) 1601 recording spectrophotometer (P/N 206-67001, Kyoto, Japan) over the range of 190–500 nm was used for all measurements. Path length of cuvettes used was 1 cm. Manipulation of spectra was performed using UVProbe 2.42 software.

Optical rotation was measured on a Jasco DIP-370 polarimeter.

The NMR 1D and 2D spectra were recorded in CD_3OD, using TMS as internal standard, and chemical shift values were recorded in δ ppm on a Bruker DRX 600 NMR spectrometer. Sample was completely dried to remove any residual solvent, resuspended in 600 μL deuterated methanol (CD_3OD), and centrifuged prior to NMR analysis.

The HR-ESI-MS was acquired on an Agilent 6545 Q-TOF LC–MS system with dual electrospray ionization (ESI) (Santa Clara, CA, USA) in negative ionization mode, as it is more sensitive for the detection of phenolics, due to their acidic nature making it easier for them to lose protons. IR was recorded on an FTIR-6700 (JASCO, Tokyo, Japan). The sample was ground with KBr in a ratio of (1:10); the mixture is then pressed in disc form and placed into the sample hold, and the IR spectrum was run.

3.4. *In Vivo Assessment of CLL Extract in an HFD Rat Anti-Obesity Activity*

3.4.1. Experimental Animals

Male albino rats of Sprague Dawley strain weighing 100–134 g (115.7 ± 7.6 g as mean ± SD) obtained from Animal House of National Research Centre, Cairo, Egypt, were kept on standard chow diet (8% fat) or high-fat diet (HFD) (30% saturated fat), provided with water ad libitum. Animals were kept individually in stainless steel metabolic cages at 25 °C; water and food were given ad libitum. All experiments were carried out according to the research protocols established by Research Ethics Committee in Faculty of Pharmacy, Cairo University, and by Medical Research Ethics Committee (MREC) in NRC, which follow the recommendations of the National Institutes of Health Guide for the Care and Use of Laboratory Animals Ethical Approval Certificate No. MP (1959).

3.4.2. In Vivo Assay Experimental Design

Twenty-four rats were randomized into two groups and received either standard chow diet (8% fat, $n = 6$), as a normal control, or an HFD (30% saturated fat, $n = 18$) to induce obesity for 8 weeks [47]. Body weight and food intake were recorded every week. After induction of obesity, rats were divided into three subgroups and still fed on HFD. For subgroup 1, rats were fed on HFD and given the vehicle as obese control. For subgroup 2, rats were fed on HFD and given oral dose of Orly (10 mg/kg RBW/day) for 4 weeks as anti-obesity drug group. Rats in subgroup 3 were fed on HFD and given oral administration of crude methanol extract of *A. carambola* leaf (CLL), prepared as described in Section 3.1 (300 mg/kg RBW/day) for four weeks. Normal control rats were continued to be fed on standard chow diet for four weeks. Body weight and food intake were recorded every week.

At the end of the experiment, blood samples were collected for determination of plasma total cholesterol (T-Ch) [48], high-density lipoprotein cholesterol (HDL-Ch) [49], low-density lipoprotein cholesterol (LDL-Ch) [50], and triglycerides (TG) [51]. T-Ch/HDL-Ch ratio was calculated as an indicator of cardiovascular risk. Plasma butyrylcholinesterase (BChE) [52], plasma α-amylase activity [53] was assessed. Plasma malondialdehyde (MDA) [54] and plasma catalase activity (CAT) [55] were estimated as indicators of lipid peroxidation and oxidative stress, respectively. For assessment of liver functions, the activity of plasma transaminases aspartate transaminase (AST) and alanine transaminase (ALT) were estimated, according to the method of [56]. Plasma level of creatinine [57], urea [58], and uric acid [59] were determined to assess changes in kidney functions. Plasma insulin [60] and blood glucose levels [61] were determined. Insulin resistance was calculated based on homeostasis model assessment of insulin resistance (HOMA-IR), according to [62]: (fasting plasma glucose (FPG) (mmoL/L) $\times$ fasting plasma insulin (FPI) (μU/mL))/22.5.

The animal experiment has been carried out according to Ethics Committee, National Research Centre, Cairo, Egypt, following the recommendations of the National Institutes of Health Guide for Care and Use of Laboratory Animals (Publication No. 85-23, revised 1985).

3.4.3. Statistical Analysis

Statistical analyses were done using SPSS version 22. The results were expressed as mean $\pm$ standard error (SE) and analyzed statistically using one-way analysis of variance (ANOVA) followed by Duncan test. The statistical significance of difference was taken as $p \leq 0.05$.

3.5. *Isolation and Structural Elucidation*

An amount of 150 g from aqueous methanol leaf extract (CLL; see Section 3.1) was fractionated using a polyamide column (Figure S69). Elution started with distilled H_2O followed by H_2O/MeOH, with gradual increase until reaching pure MeOH. The obtained fractions from the column (500 mL each) were examined using PC and TLC and observed under UV light. Similar fractions were pooled together, according to their TLC and PC profiles, to furnish 9 major fractions (A~I). Fraction B was the selected fraction for further purification based on TLC and PC detection. Fraction B (100% H_2O, 40 g) was subjected to column chromatography (CC) on Sephadex LH-20 with aqueous MeOH for elution (30–100%). Similar fractions were pooled together, according to their TLC profiles, to furnish 7 fractions, B-1~B-7.

Fraction B-4 (30% MeOH, 700 mg) was subjected to reversed-phase flash column chromatography using a Buchi MPLC eluted with a gradient solvent mixture of MeOH/H_2O (v/v, 4:6$\rightarrow$ 5:5$\rightarrow$ 6:4$\rightarrow$ 7:3$\rightarrow$8:2$\rightarrow$9:1, each 400 mL, and flushed with 600 mL 100% MeOH). Similar fractions were pooled together, according to their TLC profiles, to afford 15 fractions, B-4-1~B-4-15. Fraction B-4-4 (v/v, 4:6 MeOH/H_2O, 42 mg) was repeatedly purified with an n-PTLC to yield compound 5 (4 mg). Further, fraction B-4-11 (v/v, 5:5 MeOH/H_2O, 150 mg) was repeatedly purified with an n-PTLC to yield compound 6 (3 mg). In addition, fraction B-4-13 (v/v, 5:5 MeOH/H_2O, 25 mg) was repeatedly purified with an n-PTLC to yield compounds 7 and 8 (7 and 6 mg, respectively) (Figure S69).

Fraction B-6 (50% MeOH, 5.5 gm) was subjected to chromatographic separation using Sephadex LH-20 column eluted with *n*-butanol/H_2O (1:1), to afford fractions B-6-1~B-6-9. Fraction B-6-6 (700 mg) was subjected to reversed-phase flash column chromatography using a Buchi MPLC and eluted with a gradient solvent mixture of MeOH/H_2O (v/v, 20:80–100:0, each 400 mL) MPLC to afford 17 subfractions B-6-6-1~B-6-6-17. Subfraction B-6-6-3 (v/v, 4:6 MeOH/H_2O, 16 mg) was repeatedly purified with an n-PTLC to yield compound 1 (7 mg). Further, subfraction B-6-6-12 (v/v, 6:4 MeOH/H_2O, 40 mg) was repeatedly purified with an n-PTLC to afford compounds 9, 10 and 11 (6, 8.5 and 10 mg, respectively) (Figure S69).

Similarly, Fraction B-6-7 (2.2 g) was fractionated on a Sephadex LH-20 column, using *n*-butanol/H_2O (1:1) to afford fractions B-6-7-1~B-6-7-3. Thereafter, subfraction B-6-7-3

(76 mg) was separated by reversed-phase flash column chromatography using a Buchi MPLC and eluted with a gradient solvent mixture of MeOH/H_2O (*v*/*v*, 40:60–100:0, each 400 mL and flushed with 1000 mL 100% MeOH) to yield compound 2 (*v*/*v*, 5:5 MeOH/H_2O, 7.4 mg) (Figure S69).

Fraction B-7 (50% MeOH, 1.14 g) was subjected to Sephadex eluted with butanol/H_2O (1:1; *v*/*v*) to furnish fractions B-7-1~B-7-7. Moreover, the fractionation of fraction B-7-5 (146 mg) on reversed-phase flash column chromatography using a Buchi MPLC and eluted with a gradient solvent mixture of MeOH/H_2O (*v*/*v*, 40:60 to 100:0, each 400 mL), which resulted in pure compounds 3, 4, and 12 (8, 4, and 5 mg, respectively).

Compound **1**: Yellowish amorphous powder (MeOH); [α]D^{25} −2.72 (c 0. 0044, MeOH); UV (MeOH) λ_{max} nm (log ε) 225 (2.39), 271 (2.43) and 333 (2.49) (Figure S70); IR (FTIR): ν = 3431, 2360, 1623 cm^1 (Figure S71); R_t from HPLC 14.76 min; HR-ESI-MS detected at *m*/*z* 561.1614 $[M\text{-}H]^-$ (calculated at *m*/*z* 561.16137, $C_{27}H_{29}O_{13}^-$, error −0.1 ppm); ^1H (600 MHz) and ^{13}C (150 MHz) NMR data in CD_3OD, see Table 2.

Compound **2**: Yellowish amorphous powder (MeOH); [α]D^{25} −2.2 (c 0. 005, MeOH); UV (MeOH) λ_{max} nm (log ε) 226 (2.23), 272 (2.25) and 329 (2.37) (Figure S70); Rt from HPLC 14.59 min; HRESIMS *m/z* 577.1600 $[M\text{-}H]^-$ (calculated $C_{27}H_{29}O_{14}^-$, error 6.53 ppm); ^1H (600 MHz) and ^{13}C (150 MHz) NMR data in CD_3OD, see Table 2.

Compound **3**: Yellowish amorphous powder (MeOH); [α]D^{25} 15.0289 (c 0. 00519, MeOH); UV (MeOH) λ_{max} nm (log ε) 269 (1.33) and 335 (1.3) (Figure S70); R_t from HPLC 12.44 min; HRESIMS *m*/*z* 431.0986 $[M\text{-}H]^-$ (calculated $C_{21}H_{19}O_{10}^-$, error 0.62 ppm); ^1H (600 MHz) and ^{13}C (150 MHz) NMR data in CD_3OD, see Table 2.

Compound **4**: Yellowish amorphous powder (MeOH); [α]D^{25} 1.689 (c 0. 00296, MeOH); UV (MeOH) λ_{max} nm (log ε) 269 (2.73) and 336 (2.75) (Figure S70); R_t from HPLC 15.42 min; HRESIMS *m*/*z* 561.1661 $[M\text{-}H]^-$ (calculated $C_{27}H_{29}O_{13}^-$, error 8.03 ppm); ^1H (600 MHz) and ^{13}C (150 MHz) NMR data in CD_3OD, see Table 2.

Compound **5**: Yellowish amorphous powder (MeOH); [α]D^{25} +8.9 (c 0.003, MeOH); UV (MeOH) λ_{max} nm (log ε) 232 (1.92) and 286 (1.94) (Figure S70); R_t from HPLC 12.88 min; HRESIMS *m*/*z* 713.2272 $[M\text{-}H]^-$ (calculated $C_{32}H_{41}O_{18}^-$, error 3.69 ppm); ^1H (600 MHz) and ^{13}C (150 MHz) NMR data in CD_3OD, see Table 3.

Compound **6**: Yellowish amorphous powder (MeOH); [α]D^{25} −20.69 (c 0.0003, MeOH); UV (MeOH) λ_{max} nm (log ε) 228 (0.63) and 284 (0.54) (Figure S70); R_t from HPLC 14.37 min; HRESIMS *m*/*z* 697.2350 $[M\text{-}H]^-$ (calculated $C_{32}\ H_{41}O_{17}^-$, error 0.06 ppm); ^1H (600 MHz) and ^{13}C (150 MHz) NMR data in CD_3OD, see Table 3.

Compound **7**: Yellowish amorphous powder (MeOH); [α]D^{25} −66.41 (c 0.003, MeOH); UV (MeOH) λ_{max} nm (log ε) 231 (2.53) and 284 (2.58) (Figure S70); R_t from HPLC 18.15 min; HRESIMS *m*/*z* 973.3340 $[M\text{-}H]^-$ (calculated $C_{47}\ H_{57}O_{22}^-$, error 1.1 ppm); ^1H (600 MHz) and ^{13}C (150 MHz) NMR data in CD_3OD, see Table 3.

Compound **8**: Yellowish amorphous powder (MeOH); [α]D^{25} −40.39 (c 0.003, MeOH); UV (MeOH) λ_{max} nm (log ε) 230 (1.60) and 285 (1.65) (Figure S70); IR (FTIR): ν = 3433, 2921, 1616, 1516, 1449, 1382, 1222, 1173, 1073 cm^1 (Figure S71); R_t from HPLC 18.14 min; HRESIMS detected at *m*/*z* 1015.3460 $[M\text{-}H]^-$ (calculated at *m*/*z* 1015.34526, $C_{49}H_{59}O_{23}^-$, error −0.7 ppm); ^1H (600 MHz) and ^{13}C (150 MHz) NMR data in CD_3OD, see Table 3.

Compound **9**: Yellowish amorphous powder (MeOH); [α]D^{25} −31.379 (c 0. 0015, MeOH); UV (MeOH) λ_{max} nm (log ε) 232 (1.99) and 290 (2.03) (Figure S70); R_t from HPLC 18.10 min; HRESIMS *m*/*z* 827.2768 $[M\text{-}H]^-$ (calculated $C_{41}H_{47}O_{18}^-$, error 0.78 ppm); ^1H (600 MHz) and ^{13}C (150 MHz) NMR data in CD_3OD, see Table 3.

Compound **10**: Yellowish amorphous powder (MeOH); [α]D^{25} −102.778 (c 0. 0014, MeOH); UV (MeOH) λ_{max} nm (log ε) 230 (1.94), 290 (1.99) and 310 (2.05) (Figure S70); R_t from HPLC 17.50 min; HRESIMS *m*/*z* 989.3289 $[M\text{-}H]^-$ (calculated $C_{47}H_{57}O_{23}^-$, error 0.67 ppm); ^1H (600 MHz) and ^{13}C (150 MHz) NMR data in CD_3OD, see Table 4.

Compound **11**: Yellowish amorphous powder (MeOH); [α]D^{25} −95.42 (c 0. 0024, MeOH); UV (MeOH) λ_{max} nm (log ε) 226 (1.27) and 286 (1.26) (Figure S70); IR (FTIR): ν = 3433, 2921, 1616, 1516, 1449, 1382, 1222, 1173, 1073 cm^1 (Figure S71); R_t from HPLC

17.60 min; HRESIMS detected at m/z 1031.3389 [M-H]$^-$ (calculated at m/z 1031.34018, $C_{49}H_{59}O_{24}^-$, error + 1.2 ppm); ^1H (600 MHz) and ^{13}C (150 MHz) NMR data in CD_3OD, see Table 4.

Compound **12**: Yellowish amorphous powder (MeOH); $[\alpha]D^{25}$ −112.25 (c 0. 004, MeOH); λ_{max} nm (log ε) 226 (2.17), 290 (2.21) and 310 (2.19) (Figure S70); R_t from HPLC 16.94 min; HRESIMS m/z 843.2782 [M-H]$^-$ (calculated $C_{41}H_{47}O_{19}^-$, error 7.27 ppm); ^1H (600 MHz) and ^{13}C (150 MHz) NMR data in CD_3OD, see Table 4.

3.6. *In Vitro α-Glucosidase Inhibitory Assay*

The assay of α-glucosidase inhibitory activity of compounds was adopted from [63]. Briefly, 100 µL of DMSO and 100 µL of α-glucosidase enzyme (5 U/mL in 0.15 M HEPES buffer) were added to 100 µL substrate (0.1 M sucrose solution dissolved into 0.15 M HEPES buffer). The mixture was vortexed for 5 sec and then incubated at 37 °C for 30 min to allow for enzymatic reaction. After incubation, the reaction was stopped by heating at 100 °C for 10 min in a block incubator. The formation of glucose was determined by means of glucose oxidase method, using a BF-5S Biosensor (Oji Scientific Instruments, Hyogo, Japan). Mathematically, α-glucosidase inhibitory activity of each sample was calculated according to this equation: (Average value of control (Ac) − average value of the sample (As))/Ac × 100.

The IC_{50} values were calculated from plots of log concentration of inhibitor concentration against the percentage inhibition curves, using Microsoft Excel 2016. The data were expressed as mean ± standard deviation (SD) of at least three independent experiments (n = 3).

4. Conclusions

The global quest for anti-obesity as well as anti-diabetic drugs is currently ongoing, as obesity and its complications continue to afflict the world's population, warranting the discovery of new therapeutic regimens. A high-fat diet induced obesity model in rats was used for the assessment of anti-obese activity of *A. carambola* leaf extract, in relation to its phenolic composition. To the best of our knowledge, this study presents the first comprehensive attempt to reveal the in vivo anti-obese activity of *A. carambola* leaf extract, leading to the isolation of new bioactive components. Oral administration of *A. carambola* leaf extract enhanced all obesity complications, viz., dyslipidemia, hyperglycemia, insulin resistance, and oxidative stress, and exhibited significant anti-obesity activity in obese rats (Figure 5). Further, the effect of CLL was significantly better than Orly in almost all tested biochemical parameters, except for elevated uric acid level, although Orly revealed better reduction in body weight gain.

Multiple chromatographic approaches of the leaf extract led to the isolation of 14 compounds, including 4 flavone glycosides (**1–4**) and 10 dihydrochalcone glycosides (**5–12**) with two non-separable mixtures, including four newly described compounds, i.e., **1**, **8**, **11a**, and **11b** were reported for the first time in the literature. Further, in vitro α-glucosidase inhibitory activity assessment of isolated compounds revealed the strong potency of isolated flavone glycosides, viz., compounds **1**, **2**, **3**, and **4**, as α-glucosidase inhibitors, compared to dihydrochalcone glycosides, except for compound **12**. These results suggest for the role of flavone glycosides in alleviation of the major obesity comorbidity, i.e., diabetes via α-glucosidase inhibition, and has yet to be confirmed for other action mechanisms. An extended approach utilizing detailed studies on the molecular mechanisms of *A. carambola* leaf effect should now follow, together with subclinical and clinical trials on leaf crude extract, to be more conclusive, especially considering the known negative impact of its fruit on kidney functions. Moreover, assessment of the isolated phytoconstituents for their anti-obese activity using in vivo model or targeting other enzymes, i.e., lipases, etc., should now follow to correlate for the extract's potential anti-obesity effect. This study poses *A. carambola* leaf as a new anti-obesity functional food and adds to its effects aside from its fruit's more explored uses.

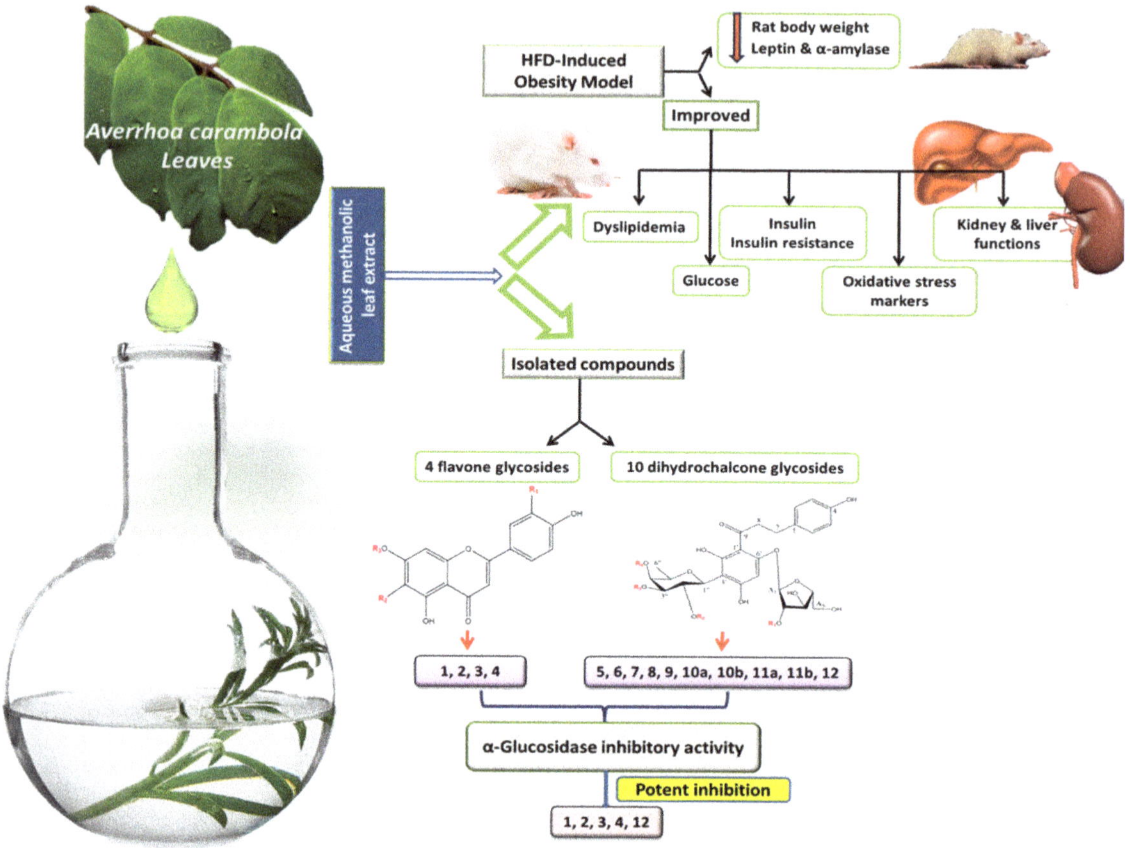

Figure 5. Collective scheme for extraction, isolation, and biological activities performed on *A. carambola* leaves.

Supplementary Materials: The following are available online at https://www.mdpi.com/article/10.3390/molecules27165159/s1. HRESIMS: ^{1}H NMR: ^{13}C NMR, ^{1}H-^{1}H COSY, HSQC, and HMBC spectra of compound **1** (Figures S1–S6); HRESIMS, ^{1}H NMR, ^{13}C NMR, HSQC, and HMBC spectra of compound **2** (Figures S7–S11); HRESIMS, ^{1}H NMR, and ^{13}C NMR spectra of compound **3** (Figures S12–S14); HRESIMS, ^{1}H NMR, ^{13}C NMR, HSQC, and HMBC spectra of compound **4** (Figures S15–S19); HRESIMS, ^{1}H NMR, ^{13}C NMR, HSQC, and HMBC spectra of compound **5** (Figures S20–S24); HRESIMS, ^{1}H NMR, ^{13}C NMR, HSQC, and HMBC spectra of compound **6** (Figures S25–S29); HRESIMS, ^{1}H NMR, ^{13}C NMR, DEPT-135, ^{1}H-^{1}H COSY, HSQC, and HMBC spectra of compound **7** (Figures S30–S36); HRESIMS, ^{1}H NMR, ^{13}C NMR, ^{1}H-^{1}H COSY, DEPT-135, HSQC, and HMBC spectra of compound **8** (Figures S37–S43); HRESIMS, ^{1}H NMR, ^{13}C NMR, HSQC, and HMBC spectra of compound **9** (Figures S44–S48); HRESIMS, ^{1}H NMR, ^{13}C NMR, HSQC, and HMBC spectra of compound **10** (Figures S49–S54); HRESIMS, ^{1}H NMR, ^{13}C NMR, HSQC, and HMBC spectra of compound **11** (Figures S55–S60); HRESIMS, ^{1}H NMR, ^{13}C NMR, HSQC, ^{1}H-^{1}H COSY, DEPT-135, and HMBC spectra of compound **12** (Figures S61–S67); Scheme of isolation (Figure S69); UV spectra of isolated compounds (Figure S70); IR spectra of new compounds **1**, **8**, **11** (Figure S71).

Author Contributions: Conceptualization, N.S.R., N.H.E.-S., S.A.E.-T. and M.A.F.; data curation, N.S.R. and M.A.F.; formal analysis, N.S.R.; funding acquisition, N.S.R., K.S., M.A.F. and T.E.; investigation, N.S.R., D.A.M., M.S.M., M.A.F. and T.E.; methodology, N.S.R., K.S., S.A.E.-T. and M.A.F.; project administration, N.S.R., K.S., N.H.E.-S., S.A.E.-T. and M.A.F.; resources, N.S.R. and K.S.; software, N.S.R. and K.S.; supervision, K.S., N.H.E.-S., S.A.E.-T., D.A.M., Z.A.A. and M.A.F.; validation,

N.S.R., D.A.M. and M.A.F.; visualization, N.S.R., M.S.M. and M.A.F.; writing—original draft, N.S.R.; writing—review and editing, N.S.R., K.S., N.H.E.-S., S.A.E.-T., M.S.M., Z.A.A., M.A.F. and T.E. All authors have read and agreed to the published version of the manuscript.

Funding: This research was funded by a fellowship supported by the Egyptian Cultural Affairs and Missions sector, Ministry of Higher Education, Egypt. The publication of this article was funded by the Open Access Fund of Leibniz Universität Hannover.

Institutional Review Board Statement: All animal experiments were carried out according to the research protocols established by Research Ethics Committee in Faculty of Pharmacy, Cairo University, and by Medical Research Ethics Committee (MREC) in NRC, which follow the recommendations of the National Institutes of Health Guide for the Care and Use of Laboratory Animals Ethical Approval Certificate No. MP (1959).

Acknowledgments: The Egyptian Cultural Affairs and Missions sector, Ministry of Higher Education, Egypt, is acknowledged for the fellowship and financial support offered to N.S.R.

Conflicts of Interest: The authors declare no conflict of interest.

References

1. Venkatakrishnan, K.; Chiu, H.-F.; Wang, C.K. Extensive Review of Popular Functional Foods and Nutraceuticals against Obesity and Its Related Complications with a Special Focus on Randomized Clinical Trials. *Food Funct.* **2019**, *10*, 2313–2329. [CrossRef] [PubMed]
2. Gesta, S.; Tseng, Y.-H.; Kahn, C.R. Developmental Origin of Fat: Tracking Obesity to Its Source. *Cell* **2007**, *131*, 242–256. [CrossRef] [PubMed]
3. Gamboa-Gómez, C.I.; Rocha-Guzmán, N.E.; Gallegos-Infante, J.A.; Moreno-Jiménez, M.R.; Vázquez-Cabral, B.D.; González-Laredo, R.F. Plants with Potential Use on Obesity and Its Complications. *EXCLI J.* **2015**, *14*, 809–831. [CrossRef] [PubMed]
4. Kinlen, D.; Cody, D.; O'Shea, D. Complications of Obesity. *QJM Int. J. Med.* **2018**, *111*, 437–443. [CrossRef] [PubMed]
5. Gani, R.S.; Kudva, A.K.; Timanagouda, K.; Raghuveer; Mujawar, S.B.H.; Joshi, S.D.; Raghu, S.V. Synthesis of Novel 5-(2,5-Bis(2,2,2-Trifluoroethoxy)Phenyl)-1,3,4-Oxadiazole-2-Thiol Derivatives as Potential Glucosidase Inhibitors. *Bioorg. Chem.* **2021**, *114*, 105046. [CrossRef] [PubMed]
6. Shang, A.; Gan, R.-Y.; Xu, X.-Y.; Mao, Q.-Q.; Zhang, P.-Z.; Li, H.-B. Effects and Mechanisms of Edible and Medicinal Plants on Obesity: An Updated Review. *Crit. Rev. Food Sci. Nutr.* **2021**, *61*, 2061–2077. [CrossRef]
7. Dirir, A.M.; Daou, M.; Yousef, A.F.; Yousef, L.F. A Review of Alpha-Glucosidase Inhibitors from Plants as Potential Candidates for the Treatment of Type-2 Diabetes. *Phytochem. Rev.* **2021**, *21*, 1049–1079. [CrossRef] [PubMed]
8. Ramadan, N.S.; Wessjohann, L.A.; Mocan, A.; Vodnar, D.C.; El-Sayed, N.H.; El-Toumy, S.A.; Abdou Mohamed, D.; Abdel Aziz, Z.; Ehrlich, A.; Farag, M.A. Nutrient and Sensory Metabolites Profiling of *Averrhoa carambola* L. (Starfruit) in the Context of Its Origin and Ripening Stage by GC/MS and Chemometric Analysis. *Molecules* **2020**, *25*, 2423. [CrossRef]
9. Rashid, A.M.; Lu, K.; Yip, Y.M.; Zhang, D. *Averrhoa carambola* L. Peel Extract Suppresses Adipocyte Differentiation in 3T3-L1 Cells. *Food Funct.* **2016**, *7*, 881–892. [CrossRef]
10. Saghir, S.A.M.; Sadikun, A.; Khaw, K.Y.; Murugaiyah, V. Star Fruit (*Averrhoa carambola* L.): From Traditional Uses to Pharmacological Activities Fruta de La Estrella (*Averrhoa carambola* L.): Desde Los Usos Tradicionales a Las Actividades Farmacologicas]. *Bol. Latinoam. y del Caribe Plantas Med. y Aromat.* **2013**, *12*, 209–219.
11. Jia, X.; Xie, H.; Jiang, Y.; Wei, X. Flavonoids Isolated from the Fresh Sweet Fruit of *Averrhoa carambola*, Commonly Known as Star Fruit. *Phytochemistry* **2018**, *153*, 156–162. [CrossRef] [PubMed]
12. Yang, Y.; Xie, H.; Jiang, Y.; Wei, X. Flavan-3-Ols and 2-Diglycosyloxybenzoates from the Leaves of *Averrhoa carambola*. *Fitoterapia* **2020**, *140*, 104442. [CrossRef] [PubMed]
13. Luan, F.; Peng, L.; Lei, Z.; Jia, X.; Zou, J.; Yang, Y.; He, X.; Zeng, N. Traditional Uses, Phytochemical Constituents and Pharmacological Properties of *Averrhoa carambola* L.: A Review. *Front. Pharmacol.* **2021**, *12*, 1814. [CrossRef] [PubMed]
14. Lakmal, K.; Yasawardene, P.; Jayarajah, U.; Seneviratne, S.L. Nutritional and Medicinal Properties of Star Fruit (*Averrhoa carambola*): A Review. *Food Sci. Nutr.* **2021**, *9*, 1810–1823. [CrossRef]
15. Moresco, H.H.; Queiroz, G.S.; Pizzolatti, M.G.; Brighente, I. Chemical Constituents and Evaluation of the Toxic and Antioxidant Activities of *Averrhoa carambola* Leaves. *Rev. Bras. Farmacogn.* **2012**, *22*, 319–324. [CrossRef]
16. Dasgupta, P.; Chakraborty, P.; Bala, N.N. *Averrhoa carambola*: An Updated Review. *Int. J. Pharma Res. Rev.* **2013**, *2*, 54–63.
17. Yang, Y.; Jia, X.; Xie, H.; Wei, X. Dihydrochalcone C-Glycosides from *Averrhoa carambola* Leaves. *Phytochemistry* **2020**, *174*, 112364. [CrossRef]
18. Ninomiya, M.; Koketsu, M. Minor Flavonoids (Chalcones, Flavanones, Dihydrochalcones, and Aurones). In *BT-Natural Products;* Phytochemistry, Botany and Metabolism of Alkaloids, Phenolics and Terpenes; Ramawat, K.G., Mérillon, J.-M., Eds.; Springer: Berlin/Heidelberg, Germany, 2013; pp. 1867–1900. ISBN 978-3-642-22144-6.
19. Gregoris, E.; Lima, G.P.P.; Fabris, S.; Bertelle, M.; Sicari, M.; Stevanato, R. Antioxidant Properties of Brazilian Tropical Fruits by Correlation between Different Assays. *Biomed Res. Int.* **2013**, *2013*, 132759. [CrossRef]

20. Jelodar, G.; Mohammadi, M.; Akbari, A.; Nazifi, S. Cyclohexane Extract of Walnut Leaves Improves Indices of Oxidative Stress, Total Homocysteine and Lipids Profiles in Streptozotocin-Induced Diabetic Rats. *Physiol. Rep.* **2020**, *8*, e14348. [CrossRef]
21. Klop, B.; Elte, J.W.F.; Cabezas, M.C. Dyslipidemia in Obesity: Mechanisms and Potential Targets. *Nutrients* **2013**, *5*, 1218–1240. [CrossRef]
22. Çelik, M.N.; Söğüt, M.Ü. Probiotics Improve Chemerin Levels and Metabolic Syndrome Parameters in Obese Rats. *Balkan Med. J.* **2019**, *36*, 270. [PubMed]
23. Sridhar, G.R.; Rao, A.A.; Srinivas, K.; Nirmala, G.; Lakshmi, G.; Suryanarayna, D.; Rao, P.V.N.; Kaladhar, D.G.; Kumar, S.V.; Devi, T.U.; et al. Butyrylcholinesterase in Metabolic Syndrome. *Med. Hypotheses* **2010**, *75*, 648–651. [CrossRef] [PubMed]
24. Chen, V.P.; Gao, Y.; Geng, L.; Brimijoin, S. Butyrylcholinesterase Regulates Central Ghrelin Signaling and Has an Impact on Food Intake and Glucose Homeostasis. *Int. J. Obes.* **2017**, *41*, 1413–1419. [CrossRef] [PubMed]
25. Tvarijonaviciute, A.; Barić-Rafaj, R.; Horvatic, A.; Muñoz-Prieto, A.; Guillemin, N.; Lamy, E.; Tumpa, A.; Ceron, J.J.; Martinez-Subiela, S.; Mrljak, V. Identification of Changes in Serum Analytes and Possible Metabolic Pathways Associated with Canine Obesity-Related Metabolic Dysfunction. *Vet. J.* **2019**, *244*, 51–59. [CrossRef] [PubMed]
26. Sá, R.D.; Vasconcelos, A.L.; Santos, A.V.; Padilha, R.J.R.; Alves, L.C.; Soares, L.A.L.; Randau, K.P. Anatomy, Histochemistry and Oxalic Acid Content of the Leaflets of *Averrhoa bilimbi* and *Averrhoa carambola*. *Rev. Bras. Farmacogn.* **2019**, *29*, 11–16. [CrossRef]
27. Malikov, V.M.; Yuldashev, M.P. Phenolic Compounds of Plants of the Scutellaria L. Genus. Distribution, Structure, and Properties. *Chem. Nat. Compd.* **2002**, *38*, 358–406. [CrossRef]
28. She, G.; Wang, S.; Liu, B. Dihydrochalcone Glycosides from Oxytropis Myriophylla. *Chem. Cent. J.* **2011**, *5*, 71. [CrossRef]
29. Nassar, M.I. Flavonoid Triglycosides from the Seeds of Syzygium Aromaticum. *Carbohydr. Res.* **2006**, *341*, 160–163. [CrossRef]
30. De Bruyn, A.; Anteunis, M. 1H NMR Study of 2-Deoxy-D-Arabino-Hexopyranose (2-Deoxy Glucopyranose), 2-Deoxy-D-Lyxo-Hexopyranose (2-Deoxy Galactopyranose) and 2′-Deoxy Lactose. Shift Increment Studies in 2-Deoxy Carbohydrates. *Bull. des Sociétés Chim. Belges* **1975**, *84*, 1201–1209. [CrossRef]
31. Rayyan, S.; Fossen, T.; Andersen, Ø.M. Flavone C-Glycosides from Seeds of Fenugreek, *Trigonella foenum-graecum* L. *J. Agric. Food Chem.* **2010**, *58*, 7211–7217. [CrossRef]
32. Latza, S.; Gansser, D.; Berger, R.G. Identification and Accumulation of 1-O-Trans-Cinnamoyl-β-d-Glucopyranose in Developing Strawberry Fruit (*Fragaria ananassa* Duch. Cv. Kent). *J. Agric. Food Chem.* **1996**, *44*, 1367–1370. [CrossRef]
33. Torres-Mendoza, D.; González, J.; Ortega-Barría, E.; Heller, M.V.; Capson, T.L.; McPhail, K.; Gerwick, W.H.; Cubilla-Rios, L. Weakly Antimalarial Flavonol Arabinofuranosides from Calycolpus w Arszewiczianus. *J. Nat. Prod.* **2006**, *69*, 826–828. [CrossRef] [PubMed]
34. Yang, D.; Jia, X.; Xie, H.; Wei, X. Further Dihydrochalcone C-Glycosides from the Fruit of *Averrhoa carambola*. *LWT Food Sci. Technol.* **2016**, *65*, 604–609. [CrossRef]
35. Yang, D.; Xie, H.; Jia, X.; Wei, X. Flavonoid C-Glycosides from Star Fruit and Their Antioxidant Activity. *J. Funct. Foods* **2015**, *16*, 204–210. [CrossRef]
36. Mizutani, K.; Kasai, R.; Nakamura, M.; Tanaka, O.; Matsuura, H. NMR Spectral Study of α- and β-l-Arabinofuranosides. *Carbohydr. Res.* **1989**, *185*, 27–38. [CrossRef]
37. Ichiyanagi, T.; Kashiwada, Y.; Shida, Y.; Ikeshiro, Y.; Kaneyuki, T.; Konishi, T. Nasunin from Eggplant Consists of Cis–Trans Isomers of Delphinidin 3-[4-(p-Coumaroyl)-l-Rhamnosyl (1→6)Glucopyranoside]-5-Glucopyranoside. *J. Agric. Food Chem.* **2005**, *53*, 9472–9477. [CrossRef]
38. Araho, D.; Miyakoshi, M.; Chou, W.H.; Kambara, T.; Mizutani, K.; Ikeda, T. A New Flavone C-Glycoside from the Leaves of *Averrhoa carambola*. *Nat. Med.* **2005**, *59*, 113–116.
39. Ghada, A.F.; Areej, M.A.T.; Nayira, A.A.B.; Mohamed, S.M. Cytotoxic and Renoprotective Flavonoid Glycosides from *Horwoodia dicksoniae*. *Afr. J. Pharm. Pharmacol.* **2012**, *6*, 1166–1175.
40. Zeng, P.; Zhang, Y.; Pan, C.; Jia, Q.; Guo, F.; Li, Y.; Zhu, W.; Chen, K. Advances in Studying of the Pharmacological Activities and Structure–Activity Relationships of Natural C-Glycosylflavonoids. *Acta Pharm. Sin. B* **2013**, *3*, 154–162. [CrossRef]
41. Xiao, J.; Kai, G.; Yamamoto, K.; Chen, X. Advance in Dietary Polyphenols as α-Glucosidases Inhibitors: A Review on Structure-Activity Relationship Aspect. *Crit. Rev. Food Sci. Nutr.* **2013**, *53*, 818–836. [CrossRef]
42. Yan, J.; Zhang, G.; Pan, J.; Wang, Y. α-Glucosidase Inhibition by Luteolin: Kinetics, Interaction and Molecular Docking. *Int. J. Biol. Macromol.* **2014**, *64*, 213–223. [CrossRef] [PubMed]
43. Nicolle, E.; Souard, F.; Faure, P.; Boumendjel, A. Flavonoids as Promising Lead Compounds in Type 2 Diabetes Mellitus: Molecules of Interest and Structure-Activity Relationship. *Curr. Med. Chem.* **2011**, *18*, 2661–2672. [CrossRef] [PubMed]
44. Han, L.; Fang, C.; Zhu, R.; Peng, Q.; Li, D.; Wang, M. Inhibitory Effect of Phloretin on α-Glucosidase: Kinetics, Interaction Mechanism and Molecular Docking. *Int. J. Biol. Macromol.* **2017**, *95*, 520–527. [CrossRef] [PubMed]
45. Castro-Acosta, M.L.; Stone, S.G.; Mok, J.E.; Mhajan, R.K.; Fu, C.-I.; Lenihan-Geels, G.N.; Corpe, C.P.; Hall, W.L. Apple and Blackcurrant Polyphenol-Rich Drinks Decrease Postprandial Glucose, Insulin and Incretin Response to a High-Carbohydrate Meal in Healthy Men and Women. *J. Nutr. Biochem.* **2017**, *49*, 53–62. [CrossRef] [PubMed]
46. Lv, Q.; Lin, Y.; Tan, Z.; Jiang, B.; Xu, L.; Ren, H.; Tai, W.C.-S.; Chan, C.-O.; Lee, C.-S.; Gu, Z.; et al. Dihydrochalcone-Derived Polyphenols from Tea Crab Apple (Malus Hupehensis) and Their Inhibitory Effects on α-Glucosidase In Vitro. *Food Funct.* **2019**, *10*, 2881–2887. [CrossRef] [PubMed]

47. Jiang, T.; Gao, X.; Wu, C.; Tian, F.; Lei, Q.; Bi, J.; Xie, B.; Wang, H.Y.; Chen, S.; Wang, X. Apple-Derived Pectin Modulates Gut Microbiota, Improves Gut Barrier Function, and Attenuates Metabolic Endotoxemia in Rats with Diet-Induced Obesity. *Nutrients* **2016**, *8*, 126. [CrossRef]
48. Watson, D. A Simple Method for the Determination of Serum Cholesterol. *Clin. Chim. Acta* **1960**, *5*, 637–643. [CrossRef]
49. Burstein, M.; Scholnick, H.R.; Morfin, R. Rapid Method for the Isolation of Lipoproteins from Human Serum by Precipitation with Polyanions. *J. Lipid Res.* **1970**, *11*, 583–595. [CrossRef]
50. Schriewer, H.; Kohnert, U.; Assmann, G. Determination of LDL Cholesterol and LDL Apolipoprotein B Following Precipitation of VLDL in Blood Serum with Phosphotungstic Acid/$MgCl_2$. *Clin. Chem. Lab. Med.* **1984**, *22*, 35–40. [CrossRef]
51. Mcgraw, R.E.; Dunn, D.E.; Biggs, H.G. Manual and Continuous-Flow Colorimetry of Triacylglycerols by a Fully Enzymatic Method. *Clin. Chem.* **1979**, *25*, 273–278. [CrossRef]
52. Vaisi-Raygani, A.; Rahimi, Z.; Kharazi, H.; Tavilani, H.; Aminiani, M.; Kiani, A.; Vaisi-Raygani, A.; Pourmotabbed, T. Determination of Butyrylcholinesterase (BChE) Phenotypes to Predict the Risk of Prolonged Apnea in Persons Receiving Succinylcholine in the Healthy Population of Western Iran. *Clin. Biochem.* **2007**, *40*, 629–633. [CrossRef] [PubMed]
53. De Melo, C.L.; Queiroz, M.G.R.; Fonseca, S.G.C.; Bizerra, A.M.C.; Lemos, T.L.G.; Melo, T.S.; Santos, F.A.; Rao, V.S. Oleanolic Acid, a Natural Triterpenoid Improves Blood Glucose Tolerance in Normal Mice and Ameliorates Visceral Obesity in Mice Fed a High-Fat Diet. *Chem. Biol. Interact.* **2010**, *185*, 59–65. [CrossRef] [PubMed]
54. Poudyal, H.; Campbell, F.; Brown, L. Olive Leaf Extract Attenuates Cardiac, Hepatic, and Metabolic Changes in High Carbohydrate—, High Fat—Fed Rats. *J. Nutr.* **2010**, *140*, 946–953. [CrossRef] [PubMed]
55. Aebi, H. Catalase In Vitro. In *Methods in Enzymology*; Elsevier: Amsterdam, The Netherlands, 1984; Volume 105, pp. 121–126, ISBN 0076-6879.
56. Reitman, S.; Frankel, S. A Colorimetric Method for the Determination of Serum Glutamic Oxalacetic and Glutamic Pyruvic Transaminases. *Am. J. Clin. Pathol.* **1957**, *28*, 56–63. [CrossRef] [PubMed]
57. Owen, J.A.; Iggo, B.; Scandrett, F.J.; Stewart, C.P. The Determination of Creatinine in Plasma or Serum, and in Urine; a Critical Examination. *Biochem. J.* **1954**, *58*, 426–437. [CrossRef] [PubMed]
58. Fawcett, J.K.; Scott, J.E. A Rapid and Precise Method for the Determination of Urea. *J. Clin. Pathol.* **1960**, *13*, 156–159. [CrossRef]
59. Watts, R.W.E. Determination of Uric Acid in Blood and in Urine. *Ann. Clin. Biochem.* **1974**, *11*, 103–111. [CrossRef]
60. Turkington, R.W.; Estkowski, A.; Link, M. Secretion of Insulin or Connecting Peptide: A Predictor of Insulin Dependence of Obese 'Diabetics'. *Arch. Intern. Med.* **1982**, *142*, 1102–1105. [CrossRef]
61. Trinder, P. Determination of Glucose in Blood Using Glucose Oxidase with an Alternative Oxygen Acceptor. *Ann. Clin. Biochem.* **1969**, *6*, 24–27. [CrossRef]
62. Cacho, J.; Sevillano, J.; de Castro, J.; Herrera, E.; Ramos, M.D.P. Validation of Simple Indexes to Assess Insulin Sensitivity during Pregnancy in Wistar and Sprague-Dawley Rats. *Am. J. Physiol. Metab.* **2008**, *295*, E1269–E1276. [CrossRef]
63. Fatmawati, S.; Shimizu, K.; Kondo, R. Ganoderol B: A Potent α-Glucosidase Inhibitor Isolated from the Fruiting Body of *Ganoderma lucidum*. *Phytomedicine* **2011**, *18*, 1053–1055. [CrossRef] [PubMed]

provided a detailed explanation of the SARS-CoV-2 virus's entrance into the host cell [1]. This is a complex process since it requires multiple enzymatic structures from the host cell to be activated in stages. Coronavirus' S glycoprotein is a structural component required for interaction with the host receptor. Previous research has shown that entry glycoproteins are typically split into two subunits before being internalized by the host cell. SARS-CoV-2 S glycoprotein is made up of two subunits: S1 is in charge of making contact with ACE2 and S2 attaches virus glycoprotein to the host cell's membrane [2]. Then, this initiates a multi-step process that involves, furin convertase and transmembrane protease serine 2 [3]. However, it became obvious that an alternative method of internalization exists once it was shown that viruses may infiltrate host cells without using the ACE2 receptor. This method for virus internalization has recently been discovered to include neuropilin-1 receptors [4].

The anti-SARS-CoV-2 activity was studied in a number of medicinal plants with a documented history of antiviral use in traditional medicine [5]. One of their shared properties is the abundant presence of compounds with polyphenol chemical moiety. Plant polyphenols are a diverse group of molecules, and their substantial presence in plant tissue is associated with many medicinal plants' health-beneficial properties (antioxidant, antidiabetic, antibacterial, etc.) [6,7]. It has been shown that phenolic compounds can block viral attachment to the human angiotensin-converting enzyme 2 (ACE2) receptor by interacting with the spike (S)-glycoprotein's receptor-binding region [8]. One of the most promising modes of action for natural compounds has been identified as the interaction between S-glycoprotein and the ACE2 receptor [9]. In fact, numerous naturally occurring substances with known antiviral properties, such as hesperidin, punicalin, and punicalagin demonstrated potent anti-SARS-CoV-2 activity in a variety of in vitro and in silico studies. This was attributed to contact interference between the virus and the ACE2 receptor on a host cell [10–13].

The traditional medical usage of many *Alchemilla* species for treating viral infections has been supported by recent studies that demonstrated virucidal activity against influenza and orthopoxviruses [14,15]. Recent investigations have revealed that *Alchemilla viridiflora* Rothm. (Figure 1) polar extract possessed a strong ACE inhibitory effect, with particular components, such as miquelianin, being emphasized for their individual contributions to this activity [16].

(a) (b)

Figure 1. *Alchemilla viridiflora* Rothm: (**a**) plant at natural habitat; (**b**) magnified flowering parts.

This could be yet another link between *A. viridiflora* constituents and SARS-CoV-2 given that the incidence of COVID-19 disease requiring hospital admission is significantly reduced while taking ACE inhibitors [17]. Even though immunization is the most effective strategy to avoid SARS-CoV-2 infection, there are some circumstances where people are unable to get vaccinated due to medical reasons. Therefore, there is a need to develop alternative strategies to prevent and treat SARS-CoV-2 infection in these individuals. To avoid infection or at least reduce viral load, one strategy is to use natural compounds in appropriate pharmaceutical dosage forms to block early contact between the virus and ACE2 and NRP1 receptors.

To the best of the authors' knowledge, *Alchemilla* isolates have not yet been investigated for their capacity to prevent SARS-CoV-2 infection despite being a rich source of bioactive polyphenols with demonstrated antiviral activity. The overall aim of this study is to clarify *A. viridiflora* methanol extract's real potential for SARS-CoV-2 internalization through two main mechanisms by applying in silico and in vitro studies.

2. Results and Discussion

2.1. LC-MS Chemical Analysis (Phytochemical Analysis)

Given that polyphenols are thought to be the primary bioactive chemicals in *Alchemilla* species, the polyphenolic profile has been thoroughly examined. According to our previous study conducted on the same extract, 23% of the polyphenolic components of *A. viridiflora* are present in the dry methanol extract [16]. Similarly, ellagitannins and flavonoids are two of the most abundant polyphenol classes in the sample used in this study. A total of 17 compounds were identified using LC-MS analysis (mass spectra and chromatograms are shown in Figures S1 and S2 of the Supplementary Materials, respectively) of *A. viridiflora* methanol extract with subsequent MS data processed and analyzed using MestReNova software (Table 1). Pedunculagin, tellimagrandin I, tellimagrandin II and galloyl-bis-hexahydroxydiphenoyl (HHDP) hexose constituted the major ellagitannin fraction of *A. viridiflora* extract while flavonoid fraction comprised quercetin, quercetin derivates and kaempferol glycosides. The recent chemical characterization of *A. viridiflora* extract provided by Radovic et al. (2022) is consistent with the present phytochemical analysis [16].

Table 1. Polyphenols identified in *A. viridiflora* methanol extract sample by LC-MS method.

Compound	Formula:	Molecular Weight:	Match Score:	RT:	Adduct/Loss:
Pedunculagin	$C_{34}H_{24}O_{22}$	784.076	0.998	9.35	−/H+
Galloyl-HHDP hexose	$C_{27}H_{22}O_{17}$	618.086	0.999	12.41	Na+/−
Isoquercitrin	$C_{21}H_{20}O_{12}$	464.095	0.999	12.44	H+/−
Quercetin 3-(6″-ferulylglucoside)	$C_{31}H_{28}O_{15}$	640.143	0.993	12.46	−/H+
Tellimagrandin I	$C_{34}H_{26}O_{22}$	786.092	0.993	16.4	−/H+
Brevifolin carboxylic acid	$C_{13}H_8O_8$	292.022	0.997	21.5	−/H_2OH+
Myricetin 3-*O*-glucuronide	$C_{21}H_{18}O_{14}$	494.07	0.973	22.9	CH_3OHH+/−
Tellimagrandin II	$C_{41}H_{30}O_{26}$	938.103	0.992	23.97	−/H+
Pentagalloylglucose	$C_{41}H_{32}O_{26}$	940.118	0.879	29.36	−/H+
Kaempferol 7-*O*-glucuronide	$C_{21}H_{18}O_{12}$	462.08	0.996	30.97	Na+/−
HHDP-hexoside	$C_{20}H_{18}O_{14}$	482.07	0.961	31.1	CH_3OHH+/−
Quercetin 3-methyl ether 7-glucuronide	$C_{22}H_{20}O_{13}$	492.09	0.985	31.13	−/H+
Kaempferol 7-*O*-glucoside	$C_{21}H_{20}O_{11}$	448.101	0.981	33.06	Na+/−
Di-*O*-methylquercetin	$C_{17}H_{14}O_7$	330.074	0.999	33.82	−/H+
Tiliroside	$C_{30}H_{26}O_{13}$	594.137	0.996	37.7	−/H+
Isorhamnetin-3-*O*-glucoside	$C_{22}H_{22}O_{12}$	478.111	0.963	39.37	NH_4+/−
Miquelianin	$C_{21}H_{18}O_{13}$	478.075	0.96	39.37	NH_4+/−

Pedunculagin is a significant monomeric ellagitannin commonly found in *Alchemilla* species that has been related to various biological activities such as antitumor, antioxidant, gastroprotective, hepatoprotective, and anti-inflammatory properties [18]. Additionally, tellimagrandin I, a compound that was just recently identified as a constituent of *Alchemilla* species, and brevifolin carboxylic acid, an *Alchemilla* ellagitannin product of hydrolysis that is also typically found in various pomegranate parts, are both recognized for their bioactivity and significant antiviral activity [16,19,20]. Compounds from the flavonoid class are equally important in terms of biological function. Numerous activities, including those related to the prevention of SARS-CoV-2, have been linked to quercetin, its derivatives and isorhamnetin [21].

2.2. Molecular Docking Studies

To investigate the individual effects of the positively identified components of the tested *A. viridiflora* sample, we used molecular docking simulations. Starting with the most active compound, quercetin 3-(6″-ferulylglucoside), all compounds are listed in Table 2 in order of their binding affinity for the S-glycoprotein receptor (PDB ID: 7BZ5). This target's binding pocket is depicted in Figure S3, and its constituent residues are given in Table S1 (Supplementary Materials).

Table 2. Molecular docking simulation results of *A. viridiflora* constituents and positive controls against wild type S-glycoprotein target (PDB ID: 7BZ5).

Compound	Bind Energy [kcal/mol]	Interacting Residues *
Quercetin 3-(6″-ferulylglucoside)	−8.035	Gln160, **Glu151** (1.63 Å), **Phe157** (2.63 Å), **Ser161** (1.84 Å), **Tyr 162** (2.83 Å)
Tellimagrandin I	−8.022	**Gln160** (2.82 Å), **Glu151** (1.57 Å, 1.71 Å), **Phe157** (2.82 Å, 2.84 Å)
Tellimagrandin II	−7.955	Gln160, **Glu151** (1.59 Å, 1.64 Å), **Gly163** (3.10 Å), Tyr116, **Tyr116** (1.73 Å, 1.94 Å), **Tyr162** (2.24 Å)
Pedunculagin	−7.848	**Gln160** (2.23 Å), **Glu151** (1.64 Å, 2.15 Å), **Gly163** (2.70 Å), Leu119, Phe157, **Tyr116** (1.69 Å), **Tyr162** (2.46 Å)
Isorhamnetin-3-*O*-glucoside	−7.761	**Glu151** (1.63 Å, 1.76 Å), Leu119, **Leu159** (1.60 Å), Phe157, **Ser161** (2.92 Å), **Tyr162** (2.79 Å)
Tiliroside	−7.633	Arg70, **Gln73** (1.50 Å), Gln160, **Glu151** (1.61 Å), **Lys84** (2.73 Å), **Phe157** (2.42 Å), Tyr120
Pentagalloylglucose	−7.601	**Gln160** (2.52 Å), Gln160, **Glu151** (1.54 Å, 1.62 Å), **Ser161** (2.64 Å), Tyr156, **Tyr162** (2.02 Å)
Kaempferol 7-*O*-glucuronide	−7.519	**Glu151** (1.92 Å), **Phe157** (2.43 Å), **Tyr120** (1.20 Å)
Di-*O*-methylquercetin	−7.515	**Gln160** (2.38 Å), **Glu151** (1.59 Å, 1.60 Å), Phe123, **Phe157** (2.33 Å), Tyr156
HHDP-hexoside	−7.506	**Glu151** (1.71 Å, 2.00 Å), Phe157, **Ser161** (2.39 Å), **Tyr116** (2.34 Å)
Miquelianin	−7.406	**Gln160** (3.09 Å), **Glu151** (1.72 Å, 1.92 Å), Leu119, Phe157, **Ser161** (1.49 Å)
Myricetin 3-*O*-glucuronide	−7.404	**Glu151** (1.67 Å, 1.94 Å), Leu119, **Leu159** (1.73 Å), Phe157, **Ser161** (1.46 Å)
Umifenovir **	−7.384	**Glu151** (1.65 Å), **Ser161** (1.96 Å), Tyr116
Quercetin **	−7.189	**Gln160** (2.07 Å), **Glu151** (1.63 Å, 1.78 Å), Phe123, **Phe157** (2.02 Å), Tyr156
Kaempferol 7-*O*-glucoside	−7.121	**Gln160** (2.48 Å, 3.02 Å), **Glu151** (1.66 Å, 1.70 Å), **Phe157** (1.97 Å), **Tyr162** (1.90 Å)
Galloyl-HHDP hexose	−6.964	**Gln160** (2.56 Å, 2.77 Å), **Glu151** (1.56 Å, 1.82 Å), **Leu159** (1.70 Å, 1.96 Å)
Isoquercitrin	−6.953	**Gln160** (3.03 Å), **Glu151** (1.63 Å, 1.64 Å), **Leu159** (1.89 Å), **Ser161** (1.80 Å, 2.32 Å)
Quercetin 3-methyl ether 7-glucuronide	−6.579	**Glu151** (1.58 Å, 1.62 Å), **Lys84** (3.06 Å), Lys84, **Tyr120** (1.62 Å)
Brevifolin carboxylic acid	−6.359	**Arg70** (1.53 Å), **Tyr162** (1.84 Å), **Tyr172** (1.63 Å)

* In the interacting residues column residues involved in hydrogen bonding are denoted in bold font with the interaction distances enclosed in brackets. ** Positive control compounds are bordered with frame.

These findings demonstrated that the observed inhibitory activity was a result of contributions from both polyphenolic groups. Quercetin 3-(6″-ferulylglucoside) demonstrated the highest binding affinity (−8.035 kcal/mol). The most favorable binding orientation of this compound is presented in Figure 2. Preliminary 12.50 ns molecular dynamic simulation results for the quercetin 3-(6″-ferulylglucoside)-S-glycoprotein complex presented in Figures S4–S7 confirm the stability of the observed system. Radius gyration trajectory (Figure S5) deviations between 18.40 Å and 18.85 Å indicate a stable secondary protein structure with a high complexing potential for the studied ligand. Observing the root mean square deviation (RMSD) trajectory (Figure S6) reveals that after 2 ns, the complex reaches a stable state. The complex exhibits simulation-based deviations after that point that do not compromise the system's stability because oscillations between the mean and maximum value do not exceed 2.5 Å.

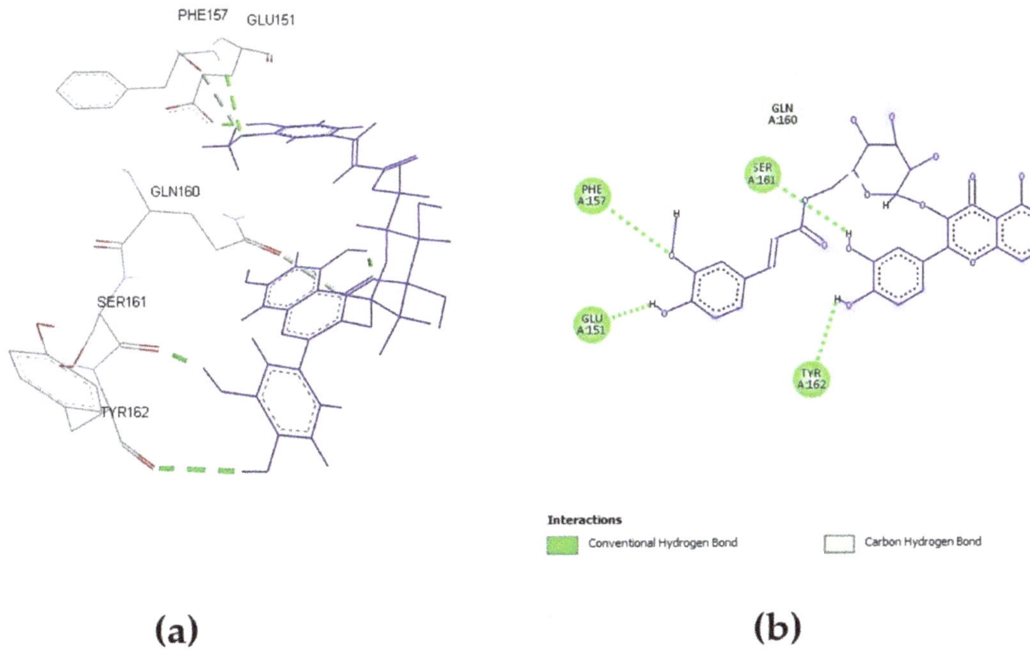

Figure 2. Quercetin 3-(6″-ferulylglucoside) interactions with S-glycoprotein (wild); (**a**) the most favorable binding pose of the compound; (**b**) 2D illustration of interaction types between the compound and target residues.

Two compounds from the ellagitannin class, tellimagrandin I and II, showed only a somewhat lower affinity for interacting with S-glycoprotein, with binding affinity energies of −8.022 and −7.955 kcal/mol, respectively. Additionally, every compound that was tested produced complexes with the target protein that were stabilized by regular hydrogen bonds on distances lower than 2 Å. One of the key interacting residues, Gln160, was previously identified as one of 23 virus residues that participate in stable hydrogen bonds, which let the virus bind to the ACE2 receptor. Most of the tested *A. viridiflora* polyphenols showed interaction with this residue, and the second-most potent compound, tellimagrandin I, interacted with it via an H bond at a distance of 2.82 Å (Figure 3) [22].

When complexing with the target S-glycoprotein, the two positive controls utilized in this investigation showed a very slight energy difference, with umifenovir forming a more stable complex (−7.384 kcal/mol) than quercetin (−7.189 kcal/mol).

According to a recent study, umifenovir inhibits the internalization of SARS-CoV-2 and its variants by directly binding to the S-glycoprotein {Shuster, 2021 #41}. However, 12 polyphenolic *Alchemilla* constituents exhibited more affinity for S-glycoprotein as a target than umifenovir, indicating more effective infection prevention (Table 2).

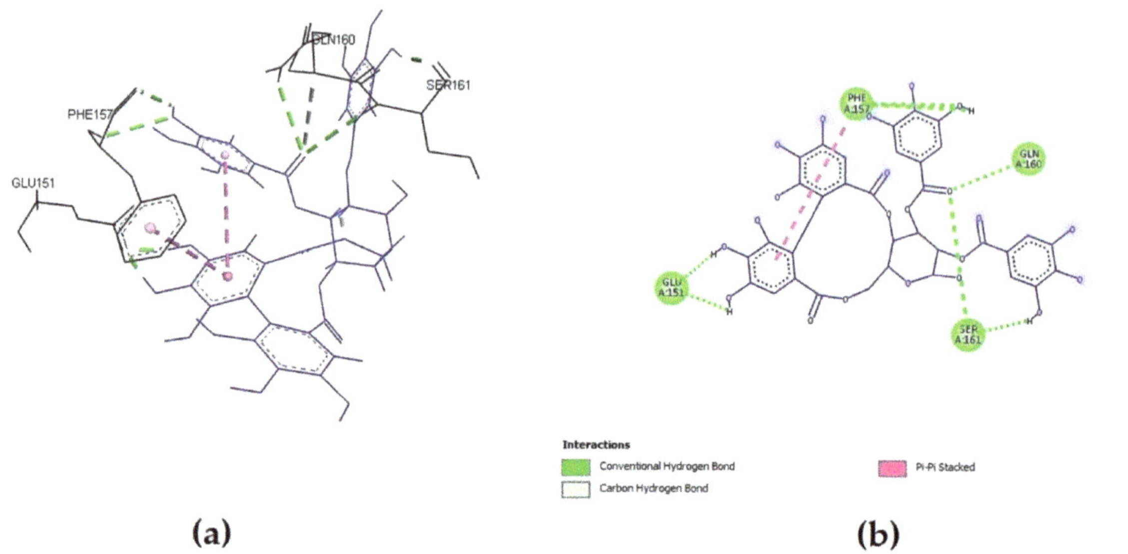

Figure 3. Tellimagrandin I interactions with S-glycoprotein (wild): (**a**) the most favorable binding pose of the compound; (**b**) 2D illustration of interaction types between the compound and target residues.

In addition to docking against wild strain-specific S-glycoprotein more docking simulations were performed on the V483A, N501Y417NE484K, N501Y, N439K, L452RT478K, K417N, G476S, F456L, and E484K strains to evaluate the stability of the observed inhibitory action on other viral strains resulting from mutations. The binding energy fluctuations curve for S-glycoprotein revealed that different drugs had varying binding affinities. In particular, the mutated strains identified in South Africa lineage B.1.351 (also known as 501Y.V2 variant) and P.1 lineage (a descendant of B.1.1.28) identified in December 2020 (in Manaus, Amazonas State, North Brazil) showed increased affinity for quercetin and tellimagrandin II, compounds with binding affinity on first and third place for wild type virus S-glycoprotein (Figure 4) [23,24]. The increased binding affinity seen in Figure 4 for the positive control umifenovir is also consistent with the findings of Shuster et al. (2021) about its maintained activity against new virus strains [25].

Although variations in binding affinity were observed for all identified constituents of *A. viridiflora* extract overall conclusion is that the range of the complex energies between −6.0 and −9.0 kcal/mol for all compounds proves they maintained significant inhibitory potential regardless of mutation changes in S-glycoprotein. According to these results, the tested extract should retain its efficacy against other virus strains, which is necessary given the significant mutational potential identified for the SARS-CoV-2 virus.

Molecular docking simulation results of *A. viridiflora* constituents and positive control against the NRP1 target are presented in Table 3. Seven compounds displayed a greater affinity for NRP1 than the positive control brevifolin carboxylic acid, showing that other polyphenols also significantly contribute to the inhibitory activity. Figure 5 presents the binding position and active site of pentagalloylglucose (compound with the highest binding activity) interaction with NRP1.

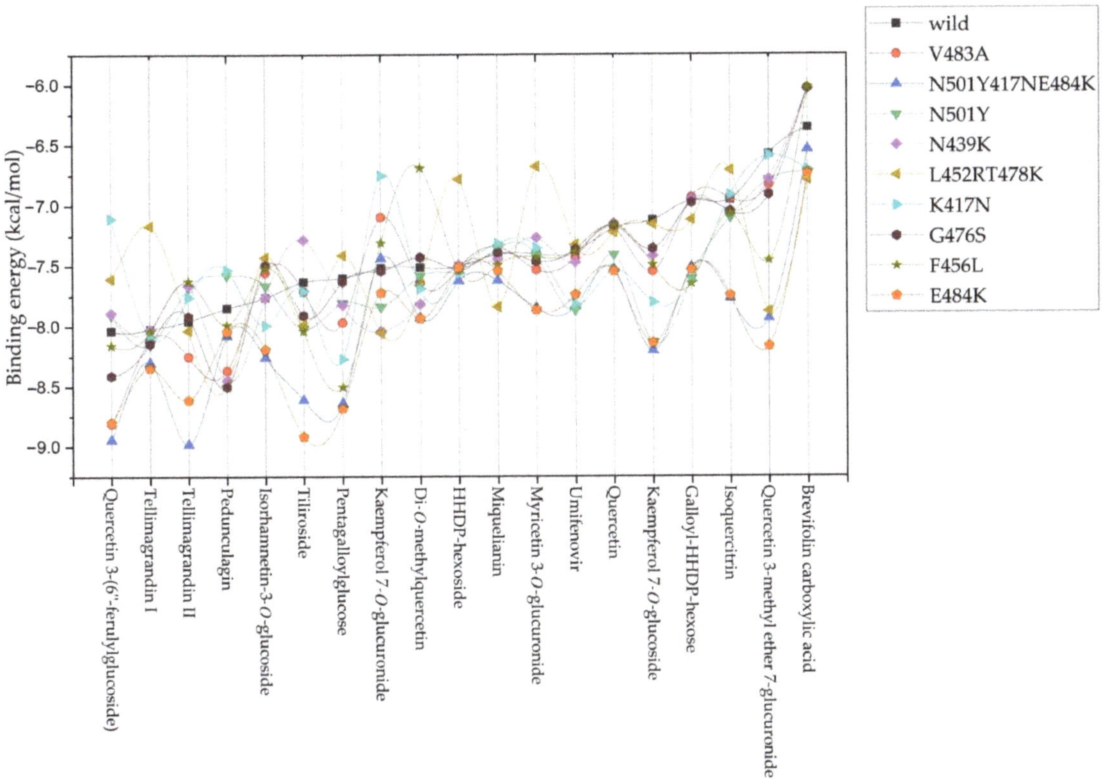

Figure 4. Binding energy (kcal/mol) curves for *A. viridiflora* constituents and positive controls against all tested S-glycoprotein structural variants.

Table 3. Molecular docking simulation results of *A. viridiflora* constituents and positive control against NRP1 target (PDB ID: 2QQI).

Compound	The Most Favorable Binding Pose **	Bind Energy [kcal/mol]	Interacting Residues *
Pentagalloylglucose	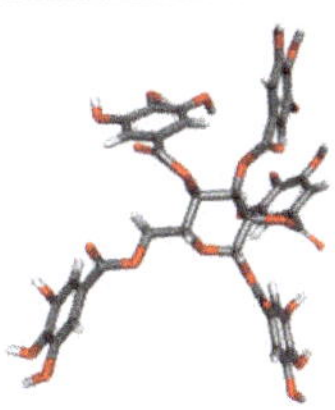	−7.685	**Asp320** (2.29 Å, 2.50 Å), **Lys351** (2.29, 2.32 Å), Lys351, **Lys352** (2.04 Å), **Pro317** (2.50 Å), **Thr413** (2.61 Å, 2.68 Å), Tyr297, **Tyr353** (2.41 Å, 2.71 Å)
Quercetin methyl ether glucuronide	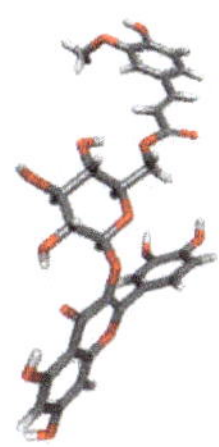	−7.667	**Asp320** (2.72 Å), **Glu348** (2.09 Å), **Lys351** (2.07 Å), Thr413, **Thr413** (3.01 Å), **Trp301** (2.00 Å), Trp411,Tyr297, **Tyr353** (2.51 Å)

Table 3. *Cont.*

Compound	The Most Favorable Binding Pose **	Bind Energy [kcal/mol]	Interacting Residues *
Tiliroside	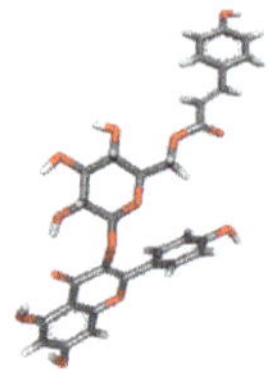	−7.594	Asp320, **Glu348** (2.15 Å), **Lys351** (2.05 Å), Thr413, Tyr297, **Tyr353** (2.82 Å)
Kaempferol 7-*O*-glucuronide	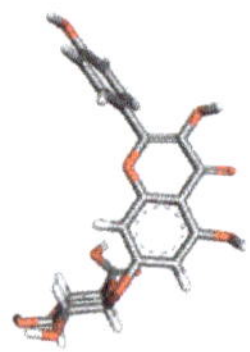	−7.452	**Asn300** (1.82 Å), Asp320, **Lys351** (2.05 Å), **Trp301** (2.36 Å), Tyr297
Kaempferol 7-*O*-glucoside	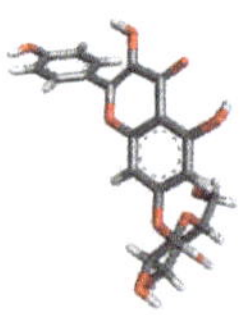	−7.264	**Asn300** (1.85 Å), Asp320, **Lys351** (2.02 Å), **Trp301** (2.30 Å), Tyr297
Quercetin	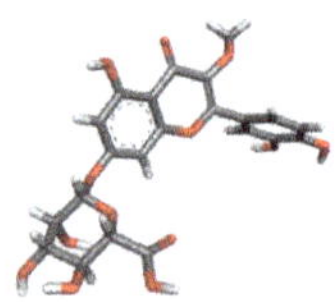	−7.205	**Asn300** (2.60 Å), Asp320, **Lys351** (1.94 Å, 2.85 Å), **Thr349** (2.72 Å), **Trp301** (2.28 Å), **Tyr297** (2.51 Å), Tyr353
Miquelianin	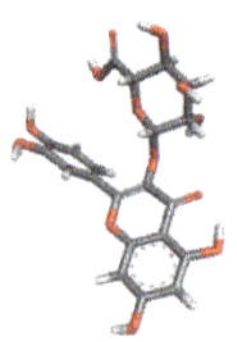	−6.986	**Asp320** (2.44 Å), **Glu348** (2.65 Å), **Lys351** (1.98 Å), **Thr349** (2.60 Å), Thr413
Brevifolin carboxylic acid ***	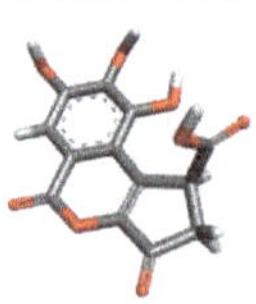	−6.976	**Lys351** (2.00 Å), Thr316, **Trp301** (2.32 Å), Trp301, Tyr297
Quercetin 3-(6″-ferulylglucoside)	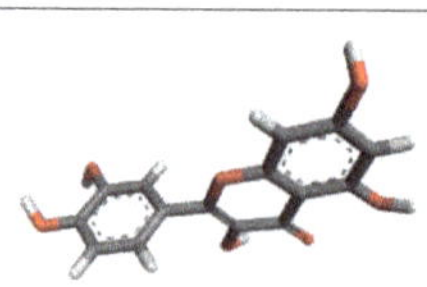	−6.875	**Arg418** (2.65 Å), Arg418, **Asn309** (2.90 Å), **Asn313** (2.37 Å),Glu312, Ile345, **Lys350** (2,84 Å), **Ser346** (2.83 Å), Thr388

Table 3. *Cont.*

Compound	The Most Favorable Binding Pose **	Bind Energy [kcal/mol]	Interacting Residues *
Pedunculagin	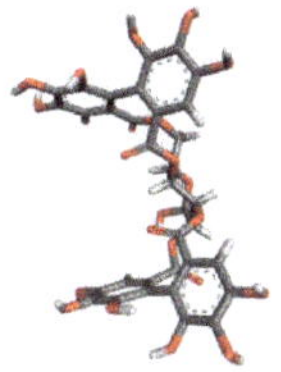	−6.855	**Asn300** (2.78 Å), Asp320, Ser298, **Tyr297** (2.33 Å, 2.49 Å)
Myricetin 3-*O*-glucuronide		−6.802	Asp320, **Gly318** (2.17 Å), **Lys351** (1.99 Å), **Thr349** (2.72 Å),Thr413, Tyr297
Isoquercitrin	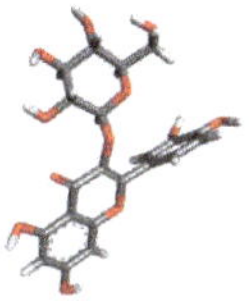	−6.799	Asp320, **Gly318** (2.52 Å), **Lys351** (2.03 Å, 2.52 Å), **Ser346** (3.02 Å), Thr413, Tyr297
Tellimagrandin I	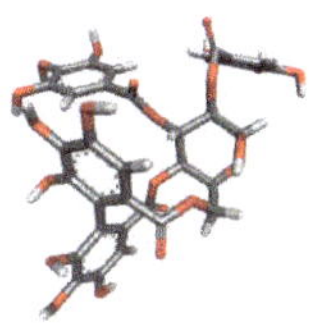	−6.76	Asp320, **Glu319** (2.77 Å), Gly318, Ser321, **Thr413** (1.93 Å), Tyr297
Di-*O*-methylquercetin	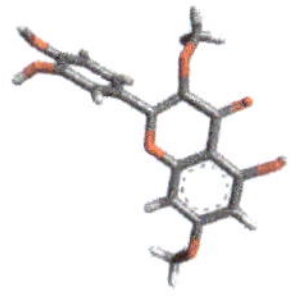	−6.713	**Asn300** (2.05 Å), Glu348, Tyr297, Tyr301, Tyr353
Isorhamnetin-3-*O*-glucoside	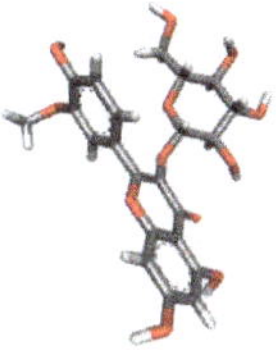	−6.467	Asp320, Asp320 (2.71 Å), Gly318, **Lys351** (2.25 Å, 2.40 Å), **Ser346** (2.65 Å), Thr413, **Thr413** (2.56 Å), Tyr297
Galloyl-HHDP-hexose	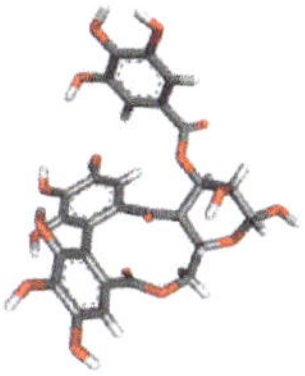	−6.365	Asp320, **Asp320** (2.07 Å, 2.92 Å), Thr413, Trp301, **Trp411** (2.51 Å), Tyr297, **Tyr353** (2.67 Å)

Table 3. *Cont.*

Compound	The Most Favorable Binding Pose **	Bind Energy [kcal/mol]	Interacting Residues *
Tellimagrandin II	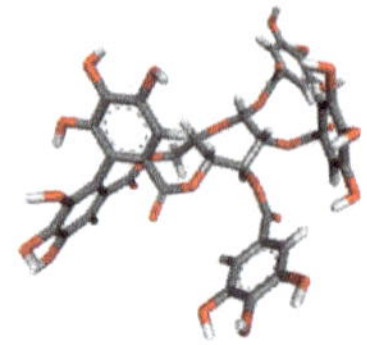	−6.051	Arg323, **Asp320** (2.76 Å), Tyr297
HHDP-hexoside	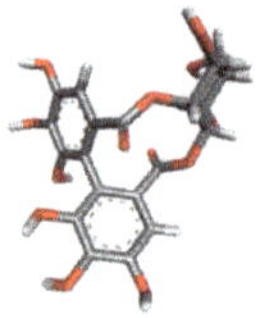	−5.897	**Lys351** (2.02 Å), **Thr413** (2.46 Å), Tyr353

* In the interacting residues column residues involved in hydrogen bonding are denoted in bold font with the interaction distances enclosed in brackets. ** 3D structure atoms color legend; red-oxygen, grey-carbon, light grey-hydrogen. *** positive control compound is bordered with frame.

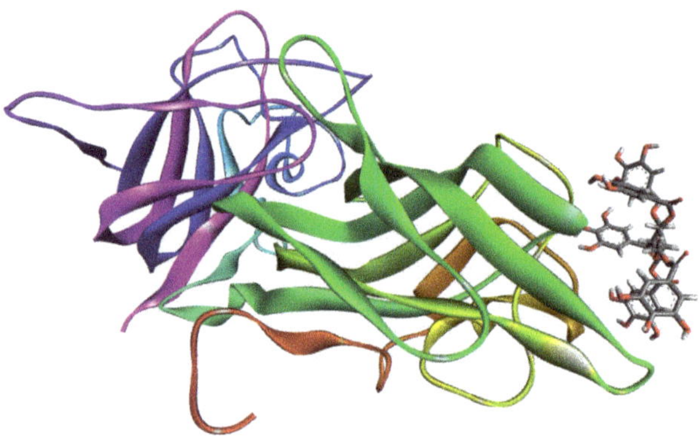

Figure 5. Pentagalloylglucose binding pose and site in complex with NRP1.

In Jin et al. study from 2022, pentagalloylglucose, the molecule listed first for its affinity to NRP1, was already recognized as a plant dietary polyphenol with substantial in vitro inhibitory effect against SARS-CoV-2 infection in Vero cells. This study confirmed that part of this inhibitory potential could be attributed to the inhibition of SARS-CoV-2 main- and RNA-dependent RNA-polymerase. Additionally, researchers found efficacy against the SARS-CoV and MERS-CoV viruses, indicating pentagalloylglucose has broad-spectrum anticoronaviral potential [26].

Other compounds with energy values below −6.5 kcal/mol, apart from pedunculagin, were from the flavonoid class, indicating that the flavonoid subclass of *A. viridiflora* polyphenols may contribute more significantly to inhibition activity when the virus primarily uses the NRP1 receptor for its internalization. Flavonoid potential for interaction with NRP1 has been already discussed in Yasmin et al. (2017) study where they proposed quercetin and diosmin as small ligands with promising potential for targeting NRP1 receptors with implications for therapeutic benefits in neurology and oncology [27]. Multiple interaction types contribute to the stabilization of ligand-target complexes, and it is noteworthy that all tested ligands were stabilized by at least one conventional hydrogen bond at a distance closer to 3 Å. The pharmacophore model highlighted ligand interactions with NRP1 residues: Tyr353, Thr349, Tyr297, Asn300, and Ser298 as critical for inhibitory activity in the Perez-Miller et al. (2020) investigation that found and confirmed inhibitors of the interaction between NRP1 and SARS-CoV-2 S-glycoprotein [28]. All these significant NRP1 residue interactions are also identified in the most favorable binding poses of *A. viridiflora* polyphenol constituents. In addition to H bonds with Tyr353, pedunculagin interacts with the same type of interaction with Asp320, a crucial residue for interaction with vascular endothelial growth factor C-terminal arginine [29]. In addition to conventional H-bond interactions, ligands, particularly those with a flavonoid structure, were stabilized by hydrophobic interactions. These interactions included ring B from the flavonoid structure and the amino acid residues Tyr297, Thr316, Tyr353 and Trp411 from the binding pocket presented in Figure S7.

2.3. In Vitro SARS-CoV-2 Internalization Inhibition Assays

The binding inhibitory effects of an extract of *A. viridiflora* on S glycoprotein-ACE2 and S glycoprotein-NRP were explored as a mechanism of their anti-SARS-CoV-2 potential in this work. Umifenovir and quercetin served as positive controls for the S glycoprotein-ACE2 inhibition assay, and brevifolin carboxylic acid served as a positive control for the S glycoprotein-NRP inhibition assay, both at the same concentrations as the samples. The extract was tested at concentrations ranging from 0.0625 to 1.00 mg/mL. The results indicated that the tested *A. viridiflora* extract was able to inhibit S-glycoprotein interactions with both receptor targets in a dose-dependent manner. The inhibition values for S-glycoprotein binding to NRP1 and ACE2 were 56.3% and 87.1%, respectively, at the highest tested concentration of *A. viridiflora* methanol extract (1.00 mg/mL). Positive controls umifenovir and quercetin inhibited S-glycoprotein and ACE2 contact at the highest tested concentrations (1 mg/mL), with inhibition values of 5.22% and 2.10%, respectively, whereas brevifolin carboxylic acid at the same concentration inhibited contact between S-glycoprotein and NRP1 by 63.07%. These in vitro results are consistent with the docking study simulation, which indicated that umifenovir had a higher inhibtion potential than quercetin, another positive control and that *A. viridiflora* methanol extract could have an even more potent antiviral effect considering 12 constituents with a higher affinity for the same target than umifenovir. Brevifolin carboxylic acid suppressed S-glycoprotein and NRP1 interactions more effectively than *A. viridiflora* extract, whose constituent it is. Possible cause could be its lower concentration in extract relative to less potent polyphenols. Using the OriginPro v. 9.8.0.200 program (OriginLab Corp.), the doses that resulted in a 50% inhibition of binding interactions between S-glycoprotein and receptors for internalization were determined to be 0.18 and 0.22 mg/mL for NRP1 and ACE2, respectively (Figure 6).

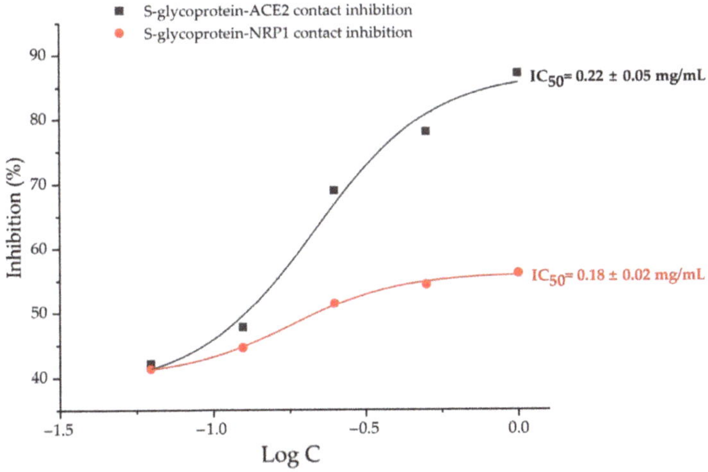

Figure 6. Concentration-inhibition curves of *A. viridiflora* methanol extract on S-glycoprotein-ACE2 (black line) and S-glycoprotein-NRP1(red line) contacts.

3.6. *In Vitro SARS-CoV-2 Internalization Inhibition Assays*

To investigate the in vitro effects of *A. viridiflora* polyphenols on SARS-CoV-2 binding activity to ACE2 and NRP1 the MBS669459 screening kit (https://www.mybiosource.com/covid-19-assay-kits/covid-19-coronavirus/669459 accessed on 29 June 2022) and RayBio COVID-19 Spike-NRP1 Binding assay kit (https://doc.raybiotech.com/pdf/Manual/CoV-NRP1S1_2021.10.06.pdf accessed on 29 June 2022) were employed. Both assays were based on a colorimetric ELISA kit that measures the binding of RBD of the S-glycoprotein from SARS-CoV-2 (wild strain) to its human receptors ACE2 and NRP1, respectively. All tested samples were dissolved in phosphate buffer solution or DMSO the final concentration of which did not exceed 0.1%. Reagents preparation and assay procedure steps were conducted strictly following the provided protocols for the default configuration.

4. Conclusions

The results of in vitro research, as well as in silico, showed that methanol extract of *A. viridiflora* and its components were capable of considerably inhibiting the internalization of SARS-CoV-2 through two of its currently most significant receptors. Ellagitannins more clearly blocked S-glycoprotein's interactions with ACE2, whilst flavonoids showed more affinity for interactions with the NRP1 receptor. Additionally, the structural changes to the S-glycoprotein brought on by mutations had a minor impact on the *A. viridiflora* constituens' activity. Lastly, the polyphenols found in the methanol extract of *A. viridiflora* offer intriguing starting points for future in vitro and in vivo anti-SARS-CoV-2 research, particularly considering their potential synergistic activity.

Supplementary Materials: The following supporting information can be downloaded at: https://www.mdpi.com/article/10.3390/molecules27165174/s1. Figure S1: *A. viridiflora* methanol extract mass spectrum; Figure S2: *A. viridiflora* methanol extract base peak chromatogram; Figure S3: Graphical presentation of binding pockets with constituent amino acid residues (ball and stick display style) for protein targets used in docking simulations (a) S-glycoprotein (PDB:7BZ5) with binding pocket residues marked red; (b) Neuropilin-1 (PDB:2QQI) with binding pocket residues marked blue; Figure S4: A ray-traced picture of the simulated system. The simulation cell boundary is set to periodic; Figure S5: Total potential energy of the system [vertical axis] as a function of simulation time [horizontal axis]; Figure S6: Radius of gyration of the solute [vertical axis] as a function of simulation time [horizontal axis]; Figure S7: Ligand movement root mean square deviation (RMSD) after superposing on the receptor [vertical axis] as a function of simulation time [horizontal axis]; Table S1: Binding pocket residues list for protein targets used in docking simulations; Table S2: Composition of the simulated system.

Author Contributions: Conceptualization, design, acquisition of data, analysis and interpretation of data, writing—original draft preparation R.S. and R.Š.; acquisition of data, analysis and interpretation of data J.R.S., T.K.-V., L.S., B.L. and M.T. All authors have participated in reviewing and editing the article critically for important intellectual content. All authors have read and agreed to the published version of the manuscript.

Funding: This research was funded by the Ministry of Education, Science and Technological Development, Republic of Serbia through Grant Agreement with University of Belgrade-Faculty of Pharmacy No: 451-03-9/2021-14/200161.

Institutional Review Board Statement: Not applicable.

Informed Consent Statement: Not applicable.

Data Availability Statement: All data are available upon reasonable request.

Conflicts of Interest: The authors declare no conflict of interest.

References

1. Jackson, C.B.; Farzan, M.; Chen, B.; Choe, H. Mechanisms of SARS-CoV-2 entry into cells. *Nat. Rev. Mol. Cell Biol.* **2022**, *23*, 3–20. [CrossRef]
2. Hoffmann, M.; Kleine-Weber, H.; Pohlmann, S. A Multibasic Cleavage Site in the Spike Protein of SARS-CoV-2 Is Essential for Infection of Human Lung Cells. *Mol. Cell* **2020**, *78*, 779–784.e5. [CrossRef]
3. Glowacka, I.; Bertram, S.; Muller, M.A.; Allen, P.; Soilleux, E.; Pfefferle, S.; Steffen, I.; Tsegaye, T.S.; He, Y.; Gnirss, K.; et al. Evidence that TMPRSS2 activates the severe acute respiratory syndrome coronavirus spike protein for membrane fusion and reduces viral control by the humoral immune response. *J. Virol.* **2011**, *85*, 4122–4134. [CrossRef] [PubMed]
4. Cantuti-Castelvetri, L.; Ojha, R.; Pedro, L.D.; Djannatian, M.; Franz, J.; Kuivanen, S.; van der Meer, F.; Kallio, K.; Kaya, T.; Anastasina, M.; et al. Neuropilin-1 facilitates SARS-CoV-2 cell entry and infectivity. *Science* **2020**, *370*, 856–860. [CrossRef] [PubMed]
5. Harwansh, R.K.; Bahadur, S. Herbal Medicines to Fight Against COVID-19: New Battle with an Old Weapon. *Curr. Pharm. Biotechnol.* **2022**, *23*, 235–260. [CrossRef]
6. Živković, J.; Suručić, R.; Arsenijević, J. Beneficial Effects of Polyphenolics from Fruit Species in Prevention and Management of Type 2 Diabetes. In *A Closer Look at Polyphenolics*; Bertollini, P., Ed.; Nova Science Publishers: Hauppauge, NY, USA, 2022.
7. Grabez, M.; Skrbic, R.; Stojiljkovic, M.P.; Vucic, V.; Rudic Grujic, V.; Jakovljevic, V.; Djuric, D.M.; Surucic, R.; Savikin, K.; Bigovic, D.; et al. A prospective, randomized, double-blind, placebo-controlled trial of polyphenols on the outcomes of inflammatory factors and oxidative stress in patients with type 2 diabetes mellitus. *Rev. Cardiovasc. Med.* **2022**, *23*, 57. [CrossRef] [PubMed]
8. Goc, A.; Sumera, W.; Rath, M.; Niedzwiecki, A. Phenolic compounds disrupt spike-mediated receptor-binding and entry of SARS-CoV-2 pseudo-virions. *PLoS ONE* **2021**, *16*, e0253489. [CrossRef] [PubMed]
9. Xiu, S.; Dick, A.; Ju, H.; Mirzaie, S.; Abdi, F.; Cocklin, S.; Zhan, P.; Liu, X. Inhibitors of SARS-CoV-2 Entry: Current and Future Opportunities. *J. Med. Chem.* **2020**, *63*, 12256–12274. [CrossRef] [PubMed]
10. Cheng, F.J.; Huynh, T.K.; Yang, C.S.; Hu, D.W.; Shen, Y.C.; Tu, C.Y.; Wu, Y.C.; Tang, C.H.; Huang, W.C.; Chen, Y.; et al. Hesperidin Is a Potential Inhibitor against SARS-CoV-2 Infection. *Nutrients* **2021**, *13*, 2800. [CrossRef]
11. Tito, A.; Colantuono, A.; Pirone, L.; Pedone, E.; Intartaglia, D.; Giamundo, G.; Conte, I.; Vitaglione, P.; Apone, F. Pomegranate Peel Extract as an Inhibitor of SARS-CoV-2 Spike Binding to Human ACE2 Receptor (in vitro): A Promising Source of Novel Antiviral Drugs. *Front. Chem.* **2021**, *9*, 638187. [CrossRef] [PubMed]
12. Surucic, R.; Tubic, B.; Stojiljkovic, M.P.; Djuric, D.M.; Travar, M.; Grabez, M.; Savikin, K.; Skrbic, R. Computational study of pomegranate peel extract polyphenols as potential inhibitors of SARS-CoV-2 virus internalization. *Mol. Cell Biochem.* **2021**, *476*, 1179–1193. [CrossRef] [PubMed]
13. Surucic, R.; Travar, M.; Petkovic, M.; Tubic, B.; Stojiljkovic, M.P.; Grabez, M.; Savikin, K.; Zdunic, G.; Skrbic, R. Pomegranate peel extract polyphenols attenuate the SARS-CoV-2 S-glycoprotein binding ability to ACE2 Receptor: In silico and in vitro studies. *Bioorg. Chem.* **2021**, *114*, 105145. [CrossRef]
14. Makau, J. Anti-influenza activity of Alchemilla mollis extract: Possible virucidal activity against influenza virus particles. *Drug Discov. Ther.* **2013**, *7*, 189–195.
15. Filippova, E.I. Antiviral Activity of Lady's Mantle (*Alchemilla vulgaris* L.) Extracts against Orthopoxviruses. *Bull. Exp. Biol Med.* **2017**, *163*, 374–377. [CrossRef] [PubMed]
16. Radovic, J.; Surucic, R.; Niketic, M.; Kundakovic-Vasovic, T. *Alchemilla viridiflora* Rothm.: The potent natural inhibitor of angiotensin I-converting enzyme. *Mol. Cell Biochem.* **2022**, *477*, 1893–1903. [CrossRef]
17. Hippisley-Cox, J.; Young, D.; Coupland, C.; Channon, K.M.; Tan, P.S.; Harrison, D.A.; Rowan, K.; Aveyard, P.; Pavord, I.D.; Watkinson, P.J. Risk of severe COVID-19 disease with ACE inhibitors and angiotensin receptor blockers: Cohort study including 8.3 million people. *Heart* **2020**, *106*, 1503–1511. [CrossRef] [PubMed]
18. Silva Fernandes, A.; Hollanda Veras, J.; Silva, L.S.; Puga, S.C.; Luiz Cardoso Bailao, E.F.; de Oliveira, M.G.; Cardoso, C.G.; Carneiro, C.C.; Costa Santos, S.D.; Chen-Chen, L. Pedunculagin isolated from Plinia cauliflora seeds exhibits genotoxic, antigenotoxic and cytotoxic effects in bacteria and human lymphocytes. *J. Toxicol. Environ. Health A* **2022**, *85*, 353–363. [CrossRef]
19. Tamura, S.; Yang, G.M.; Yasueda, N.; Matsuura, Y.; Komoda, Y.; Murakami, N. Tellimagrandin I, HCV invasion inhibitor from Rosae Rugosae Flos. *Bioorg. Med. Chem. Lett.* **2010**, *20*, 1598–1600. [CrossRef]
20. Tian, J.; Xie, Y.; Zhao, Y.; Li, C.; Zhao, S. Spectroscopy characterization of the interaction between brevifolin carboxylic acid and bovine serum albumin. *Luminescence* **2011**, *26*, 296–304. [CrossRef]
21. Zhan, Y.; Ta, W.; Tang, W.; Hua, R.; Wang, J.; Wang, C.; Lu, W. Potential antiviral activity of isorhamnetin against SARS-CoV-2 spike pseudotyped virus in vitro. *Drug Dev. Res.* **2021**, *82*, 1124–1130. [CrossRef]
22. Fakih, T.M.; Dewi, M.L. In silico Identification of Characteristics Spike Glycoprotein of SARS-CoV-2 in the Development Novel Candidates for COVID-19 Infectious Diseases. *J. Biomed. Transl. Res.* **2020**, *6*, 48–52. [CrossRef]
23. Tegally, H.; Wilkinson, E.; Giovanetti, M.; Iranzadeh, A.; Fonseca, V.; Giandhari, J.; Doolabh, D.; Pillay, S.; San, E.J.; Msomi, N.; et al. Emergence and rapid spread of a new severe acute respiratory syndrome-related coronavirus 2 (SARS-CoV-2) lineage with multiple spike mutations in South Africa. *MedRxiv* **2020**. [CrossRef]
24. Available online: https://virological.org/t/genomic-characterisation-of-an-emergent-sars-cov-2-lineage-in-manaus-preliminary-findings/586 (accessed on 27 January 2022).

25. Shuster, A.; Pechalrieu, D.; Jackson, C.B.; Abegg, D.; Choe, H.; Adibekian, A. Clinical Antiviral Drug Arbidol Inhibits Infection by SARS-CoV-2 and Variants through Direct Binding to the Spike Protein. *ACS Chem. Biol.* **2021**, *16*, 2845–2851. [CrossRef] [PubMed]
26. Jin, Y.H.; Lee, J.; Jeon, S.; Kim, S.; Min, J.S.; Kwon, S. Natural Polyphenols, 1,2,3,4,6-O-Pentagalloyglucose and Proanthocyanidins, as Broad-Spectrum Anticoronaviral Inhibitors Targeting Mpro and RdRp of SARS-CoV-2. *Biomedicines* **2022**, *10*, 1170. [CrossRef] [PubMed]
27. Yasmin, T.; Ali, M.T.; Haque, S.; Hossain, M. Interaction of Quercetin of Onion with Axon Guidance Protein Receptor, NRP-1 Plays Important Role in Cancer Treatment: An In Silico Approach. *Interdiscip. Sci.* **2017**, *9*, 184–191. [CrossRef] [PubMed]
28. Perez-Miller, S.; Patek, M.; Moutal, A.; Cabel, C.R.; Thorne, C.A.; Campos, S.K.; Khanna, R. In silico identification and validation of inhibitors of the interaction between neuropilin receptor 1 and SARS-CoV-2 Spike protein. *bioRxiv* **2020**. [CrossRef]
29. Parker, M.W.; Xu, P.; Li, X.; Vander Kooi, C.W. Structural basis for selective vascular endothelial growth factor-A (VEGF-A) binding to neuropilin-1. *J. Biol. Chem.* **2012**, *287*, 11082–11089. [CrossRef] [PubMed]
30. Fischer, U.A.; Carle, R.; Kammerer, D.R. Identification and quantification of phenolic compounds from pomegranate (*Punica granatum* L.) peel, mesocarp, aril and differently produced juices by HPLC-DAD-ESI/MS(n). *Food Chem.* **2011**, *127*, 807–821. [CrossRef]
31. Liu, X.; Raghuvanshi, R.; Ceylan, F.D.; Bolling, B.W. Quercetin and Its Metabolites Inhibit Recombinant Human Angiotensin-Converting Enzyme 2 (ACE2) Activity. *J. Agric. Food Chem.* **2020**, *68*, 13982–13989. [CrossRef]
32. Daly, J.L.; Simonetti, B.; Klein, K.; Chen, K.E.; Williamson, M.K.; Anton-Plagaro, C.; Shoemark, D.K.; Simon-Gracia, L.; Bauer, M.; Hollandi, R.; et al. Neuropilin-1 is a host factor for SARS-CoV-2 infection. *Science* **2020**, *370*, 861–865. [CrossRef] [PubMed]
33. Jarvis, A.; Allerston, C.K.; Jia, H.; Herzog, B.; Garza-Garcia, A.; Winfield, N.; Ellard, K.; Aqil, R.; Lynch, R.; Chapman, C.; et al. Small molecule inhibitors of the neuropilin-1 vascular endothelial growth factor A (VEGF-A) interaction. *J. Med. Chem.* **2010**, *53*, 2215–2226. [CrossRef] [PubMed]
34. Guedes, I.A.; Costa, L.S.C.; Dos Santos, K.B.; Karl, A.L.M.; Rocha, G.K.; Teixeira, I.M.; Galheigo, M.M.; Medeiros, V.; Krempser, E.; Custodio, F.L.; et al. Drug design and repurposing with DockThor-VS web server focusing on SARS-CoV-2 therapeutic targets and their non-synonym variants. *Sci. Rep.* **2021**, *11*, 5543. [CrossRef]
35. Trott, O.; Olson, A.J. AutoDock Vina: Improving the speed and accuracy of docking with a new scoring function, efficient optimization, and multithreading. *J. Comput. Chem.* **2010**, *31*, 455–461. [CrossRef] [PubMed]
36. Krieger, E.; Dunbrack, R.L., Jr.; Hooft, R.W.; Krieger, B. Assignment of protonation states in proteins and ligands: Combining pKa prediction with hydrogen bonding network optimization. *Methods Mol. Biol.* **2012**, *819*, 405–421. [PubMed]

molecules

MDPI

Article

Determination of the Phenolic Profile, and Evaluation of Biological Activities of Hydroethanolic Extract from Aerial Parts of *Origanum compactum* from Morocco

Mounia Chroho [1,*], Aziz Bouymajane [2], Mustapha Aazza [3], Yassine Oulad El Majdoub [4], Francesco Cacciola [5,*], Luigi Mondello [4,6,7], Touriya Zair [8] and Latifa Bouissane [1]

1 Molecular Chemistry, Materials and Catalysis Laboratory, Faculty of Sciences and Technologies, Sultan Moulay Slimane University, Beni-Mellal 23000, Morocco
2 Team of Microbiology and Health, Laboratory of Chemistry-Biology Applied to the Environment, Faculty of Sciences, Moulay Ismail University, Zitoune, Meknes 11201, Morocco
3 Laboratory of Chemistry-Biology Applied to the Environment, Faculty of Sciences, Moulay Ismail University, Zitoune, Meknes 11201, Morocco
4 Department of Chemical, Biological, Pharmaceutical and Environmental Sciences, University of Messina, 98168 Messina, Italy
5 Department of Biomedical, Dental, Morphological and Functional Imaging Sciences, University of Messina, 98125 Messina, Italy
6 Chromaleont s.r.l., c/o Department of Chemical, Biological, Pharmaceutical and Environmental Sciences, University of Messina, 98168 Messina, Italy
7 Department of Sciences and Technologies for Human and Environment, University Campus Bio-Medico of Rome, 00128 Rome, Italy
8 Research Team Chemistry of Bioactive Molecules and Environment, Laboratory of Innovative Materials and Biotechnologies of Natural Resources, Faculty of Sciences, Moulay Ismail University, Zitoune, Meknes 11201, Morocco
* Correspondence: chroho_mounia@yahoo.fr (M.C.); cacciolaf@unime.it (F.C.)

Citation: Chroho, M.; Bouymajane, A.; Aazza, M.; Oulad El Majdoub, Y.; Cacciola, F.; Mondello, L.; Zair, T.; Bouissane, L. Determination of the Phenolic Profile, and Evaluation of Biological Activities of Hydroethanolic Extract from Aerial Parts of *Origanum compactum* from Morocco. *Molecules* **2022**, *27*, 5189. https://doi.org/10.3390/molecules27165189

Academic Editors: Nour Eddine Es-Safi and H. P. Vasantha Rupasinghe

Received: 26 July 2022
Accepted: 13 August 2022
Published: 15 August 2022

Abstract: *Origanum compactum* belonging to the family Lamiaceae is widely used in food and pharmaceutical fields due to its biologically active substances. We aimed to investigate the total phenol and flavonoid contents and the phenolic composition, and to evaluate the antioxidant and antibacterial properties of hydroethanolic extract from of *Origanum compactum*. Total phenol and flavonoid contents were evaluated using gallic acid and quercetin as standards, respectively, and the phenolic profile was characterized using high-performance liquid chromatography coupled to a photodiode array and electrospray ionization mass spectrometry (HPLC-PDA-ESI/MS). The antioxidant activity was determined by two methods: ferric reducing power (FRAP) assay and the phosphomolybdate method. The antibacterial effect was evaluated against four bacteria (*Escherichia coli*, *Salmonella typhimurium*, *Staphylococcus aureus* and *Listeria monocytogenes*) using the broth microdilution method. The findings show that the total phenolic and flavonoid contents were 107.789 $\pm$ 5.39 mg GAE/g dm and 14,977 $\pm$ 0.79 mg QE/g dm, respectively. A total of sixteen phenolic compounds belonging to phenolic acids and flavonoids were detected. Furthermore, the extract showed strong antioxidant activity, and displayed a bacteriostatic effect against *Escherichia coli* and *Salmonella typhimuriumn*, and a bactericidal effect against *Staphylococcus aureus* and *Listeria monocytogenes*. Therefore, this study reveals that *Origanum compactum* extracts display potential as antibacterial and natural antioxidant agents for fighting against pathogenic bacteria and preventing oxidative stress.

Keywords: *Origanum compactum*; phenolic compounds; HPLC-PDA-ESI/MS; antioxidant activity; antibacterial activity

1. Introduction

The *genus Origanum* belonging to the family Lamiaceae (tribe Mentheae) includes 42 species and 18 hybrids found throughout North Africa and Eurasia [1]. In Morocco, the *genus Origanum* is represented by five taxa, namely, *O. elongatum*, *O. grosii*, *O. frontqueri* and *O. vulgare* and *O. compactum*, the most commonly used taxa of *Origanum* [2].

Origanum compactum is an endemic plant of Morocco, widely used for its therapeutic and culinary properties. In Moroccan traditional medicine, it is frequently employed in the form of infusions and decoctions to treat a variety of infections (gastrointestinal disorders, gastric acidity, and bronchopulmonary ailments). In traditional Moroccan cuisine, it is the first aromatic ingredient chosen for flavoring some traditional dishes due to its pleasant flavor and spicy fragrance [2,3]. *Origanum compactum's* vernacular name is "Zaâtar" and is widespread in Morocco. It is reported in the Middle Atlas, Rif and Northern regions [2].

Origanum compactum is a member of the compactum section where successive verticillasters are reconciled fake ears contracted terminal, short and globular. The principal morphological characteristic is that it secretes essential oils with a unique flavor due to the presence of secretary organs (glandular and non-glandular trichomes) [4].

Origanum species have been investigated for their secondary metabolites, such as phenolic compounds. *Origanum* extracts have attracted more attention recently, and investigations of the total phenolic contents, phenolic profiles and their relation to biological activities have been conducted. The main types of phenolic compounds found in *Origanum* extracts are phenolic acids and flavonoids. Such molecules are compounds with a variety of structures, characterized by having at least one aromatic ring linked with one or more hydroxyl groups. Phenolic compounds are important due to their various physiological functions that help plants adjust to environmental changes and survive (UV protection, disease resistance, pigmentation and growth regulation) [5].

The phytochemical content of *Origanum* species is responsible for its benefits for human health; in vitro and in vivo assays have proved a wide range of pharmacological properties (antioxidant, antibacterial, anti-inflammatory, anti-cancer, antifungal, antiviral, antileishmanial, anti-asthmatic, anti-ulcer, anti-diabetic, and decreased risk of cardiovascular diseases) [3,5].

Despite the numerous studies that have highlighted the phenolic profile and antioxidant and antibacterial powers of *Origanum* extracts, as far as we know, studies of the ethanolic extract of *Origanum compactum* are few or nonexistent. Thus, in the present study, we aimed to investigate the phenolic compounds and the antioxidant and antibacterial activities of hydroethanolic extract from aerial parts of *Origanum compactum* from the Middle Atlas of Morocco (Khenifra).

2. Results and Discussion

2.1. Phytochemical Screening

Phytochemical screening indicated the presence of secondary metabolites families in *Origanum compactum* aerial parts: alkaloids, catechic tannins, sterols and triterpenes, flavonoids, saponosides, leucoanthocyanins, oses and holosides, and mucilages. On the other hand, gallic tannins and reducing compounds were almost absent.

Gallic acids, tannins, anthocyanes and flavonoids were previously reported as constituents of *Origanum compactum* [6]. Other species belonging to the same genus, e.g., *Origanum vulgare*, have also been proven to contain tannins, mucilages, proteins, alkaloids, steroids, flavanoids, starch and anthraquinones [7].

2.2. Polyphenols Extractions Yield

The extraction yield obtained using hydroethanolic solvent at 70% for *Origanum compactum* was 30.60%. A similar yield of polyphenol extraction by methanol from *Origanum compactum* was obtained by Zeroual et al. (31.70%), while the lowest yield (10.30%) was obtained for n-hexane extraction [3]. It has been reported that ethanol and methanol present similar yields in most cases [8].

The yield of extraction depends on the solvent used. A solvent's effectiveness is mostly determined by its capacity to dissolve particular phenolic groups [8]. Ethanol is one of the solvents that have been used for the extraction of polyphenols from different plants, whether as an aqueous mixture or as absolute ethanol [9], as it provides better results, being also safer for human health [10].

2.3. *Total Polyphenols and Flavonoids Contents in Origanum compactum Extract*

Total phenolic content for the ethanolic extract of *Origanum compactum* was 107.79 ± 5.39 mg GAE/g dm and flavonoid content was 14.98 ± 0.79 mg QE/g dm (Table 1). The total phenolic content recorded was good compared to the total phenolic content in other *Origanum* extracts. The hydro-methanolic extract of *Origanum vulgare* presented total phenolic content of 79–147 mg GAE/g DW [11]. Bower et al. cited total phenolic content of 430 µg of GAE/mg dm for methanolic extract of *Origanum vulgare* leaves [12]. A relatively low total phenolic content (38 mg GAE/200 mL) was reported for the infusion of the leaves and flowers from *Origanum microphyllum* [13].

Table 1. Total phenols content, flavonoids content and antioxidant activity of hydroethanolic extract from *Origanum compactum*.

Extraction Yield	Total Phenols Content	Flavonoids Content	EC_{50} (FRAP)	TAC
30.60%	107.79 ± 5.39 mg GAE/g dm	14.98 ± 0.79 mg QE/g dm	0.017 ± 0.00085 mg/mL	470.90 mg EAA/g E

2.4. *HPLC-PDA/ESI-MS Analysis*

The phenolic profile analysis was carried out by using high-performance liquid chromatography coupled to a photodiode array and electrospray ionization mass spectrometry (HPLC-PDA/ESI-MS) (Figure 1). As listed in Table 2, a total of sixteen phenolic compounds were detected in *Origanum* extract, according to standards, retention times, mass spectrometry and literature data. The compounds were assigned to phenolic acids (syringic acid, caffeic acid, lithospermic acid A isomer, salvianolic acid, rosmarinic acid and melitric acid) and to flavonoids (apigenin-6,8-di-C-glucoside, luteolin glucoside, luteolin glucuronide, diosmetin jaceosidin, apigenin and cirsilineol). In terms of quantification, peak No. 9, rosmarinic acid, turned out to be the most abundant one in the studied plant extract (48,128.62 mg/Kg extract).

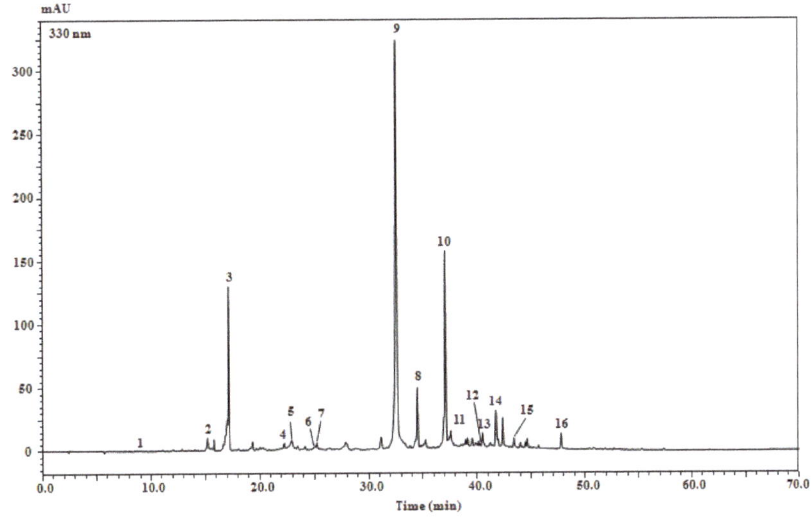

Figure 1. Chromatographic profile of phenolic compounds in *Origanum compactum* extract (EtOH:H_2O 7:3 *v*/*v*) acquired at 330 nm.

Table 2. Tentative characterization of phenolic compounds in (EtOH:H_2O 7:3 *v*/*v*) extract of *Origanum compactum* by HPLC-PDA/MS.

Peak N°	Compound	t_R (min)	UV (nm)	$[M-H]^-$	$[M+H]^+$	Fragments	Standards	Quantity (mg/Kg) Extract ± SD	References
1	Syringic acid	9.23	280	197	-	-	-	Nq	[14]
2	Caffeic acid	15.31	322	179	-	-	-	Nq	[14]
3	Apigenin-6,8-di-C-glucoside	17.20	270, 336	593	595	-	Apigenin	1623.53 ± 288.42	[15,16]
4	Lithospermic acid A isomer	22.25	284, 344	537	-	339	-	Nq	[17]
5	Unknown	22.93	287, 331	555	-	359(+)	-	-	[18]
6	Luteolin glucoside	25.06	286, 336	447	449	287(+)	Kaempferol-glucoside	185.92 ± 37.27	[5,18]
7	Luteolin glucuronide	25.27	253, 343	461	463	287(+)	Kaempferol-glucoside	258.28 ± 50.75	[15,18]
8	Salvianolic acid I	34.50	309	537	493, 341	297(+)	-	Nq	[19]
9	Rosmarinic acid	32.55	289, 328	359	-	-	Rosmarinic acid	48128.62 ± 8077.44	[15,18,20,21]
10	Melitric acid B	37.03	286, 310	519	521	-	-	Nq	[18]
11	Melitric acid A	37.52	287, 312	537	539	-	-	Nq	[18]
12	Unknown	40.27	287, 327	605	607	271(+)	-	-	[20]
13	Diosmetin	40.50	286, 332	299	301	-	Apigenin	263.56 ± 26.02	[18]
14	Jaceosidin	41.76	283, 341	329	331		-	Nq	[21]
15	Apigenin	44.01	288, 332	269	271	-	Apigenin	89.64 ± 14.49	[15,18,20]
16	Cirsilineol	47.71	284, 339	-	345	-	-	Nq	[5,18,22]

As it has been already reported in previous studies, the primary classes of phenolic chemicals in oregano are phenolic acids and flavonoids [5]. The results achieved in this study are in agreement with such studies [5]. For phenolic acids, the majority of them were previously cited in *Origanum* composition. Syringic acid and caffeic acid were previously reported in plants of the Lamiaceae family, including *Oregano* [14], and caffeic acid plays an important role in the biochemistry of this family [23]. Lithospermic acid and caffeic acid were identified in 80% methanol extract of *Origanum vulgare* ssp. *Hirtum* [24]. Lithospermic acid A and B were isolated from the aerial parts of *Origanum. Vulgare* ssp. *Hirtum* by Koukoulitsa et al. [23,25,26]. Salvianolic acid was reported in *Origanum majorana* methanol extract [19], and several studies have reported rosmarinic acid in the *Origanum* phenolic composition [15,20,23,27–29]. Additionally, the obtained results showing the high presence of rosmarinic acid in *Origanum compactum* extract are similar to those reported by Boutahiri et al. for the same plant originating from another site in Morocco [29].

It has been reported that rosmarinic acid and derivatives appear to constitute the main phenolic acids in oregano [24]. This is applicable in our study. Rosmarinic acid derivatives combine one or more rosmarinic acids with additional aromatic groups, which include lithospermic acid, salvianolic acid and melitric acid [30].

Regarding melitric acid, it was cited in the phenolic profile of some plants such as *Satureja biflora* [31] and *Melissa officinalis* [30,32], and to our knowledge, no previous studies have reported it as an *Origanum* phenolic component.

Concerning flavonoids, apigenin and lutelin are among the most abundant individual flavonoids found in different extracts of oregano species [5,33]. Cirsilineol was identified in *sicilian oregano* from Italy by Tuttolomondo et al. [34], and diosmetin was cited in flavonoid components of *Origanum vulgare* [35,36]. Jaceosidin was cited in relation to other plants' phenolic composition; it was isolated from the ethanolic extract of *Centaurea nicaeensis* [37] and was identified as a major phenolic compound in *Artemisia argyi* [38].

2.5. Antioxidant Activity

2.5.1. Antioxidant Activity of Hydro-Ethanolic Fractions by Frap (Ferric Reducing Power Assay)

Ethanolic extract of *Origanum compactum* aerial parts (Figure 2) showed powerful antioxidant potential. The parameter EC_{50} (effective concentration), which corresponds to an absorbance equal to 0.5, was equal to 0.017 ± 0.00085 mg/mL (Table 1). For ascorbic acid tested under the same conditions, the EC_{50} was equal to 0.031 mg/mL. The antioxidant power of *Origanum compactum* extract was more powerful than ascorbic acid.

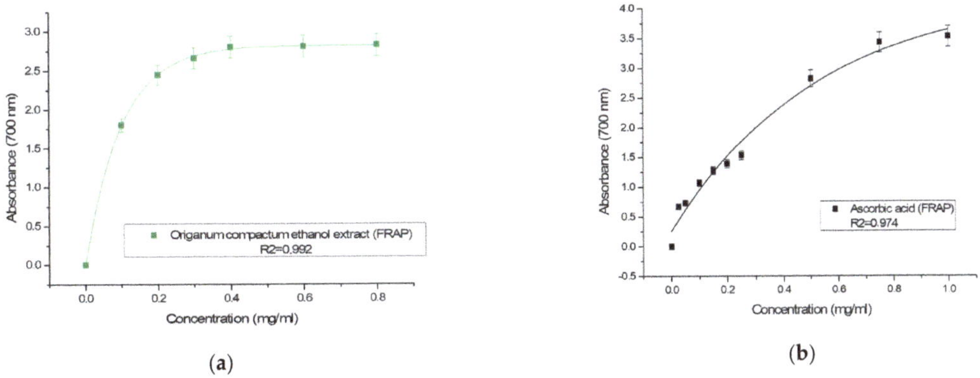

Figure 2. (**a**) Antioxidant activity of hydroethanolic extract of *Origanum compactum* by FRAP; (**b**) antioxidant activity of ascorbic acid by FRAP.

Lagouri et al. compared the ferric reducing antioxidant properties of the aqueous and methanol extracts of *Origanum dictamnus*; the obtained EC_{50} values were 0.028 and 0.038 mg/mL, respectively [39]. In another study, methanolic extract of *Origanum vulgare* showed the largest ferric reducing ability expressed by 1746.76 ± 45.11 μmol AAE/g extract [40].

The powerful ferric reducing power of *Origanum compactum* ethanol extract highlighted in this work is attributed to the phenolic contents of the extract, which are mainly represented by rosmarinic acid and its derivatives. It has been found that rosmarinic acid is the phenolic acid that provides the strongest antioxidant activity by ferric reducing power (FRAP) assay on hydromethanolic extract from *Origanum majorana* [41]. Similarly, Gonçalves et al. [10] partially attributed the high antioxidant capacity (FRAP) of the methanolic extract of *Origanum vulgare* to the large quantity of rosmarinic acid (23.53 mg/g of dry extract) [40].

2.5.2. Total Antioxidant Capacity by Phosphomolybdate Method

The phosphomolybdate method is based on the ability of the extract to convert molybdenum molybdate ions MoO_4^{2-} into molybdenum molybdate ions MoO_4^{2+}, and the consequent creation of a green phosphate/molybdenum (V) complex at acidic pH. The amount of ascorbic acid equivalents in one gram of dry extract (mg EAA/1 gE) was used to express the total antioxidant capability of the examined extract.

Total antioxidant capacity of *Origanum compactum* ethanol extract was found to be 470.905 mg EAA/1 gE (Table 1). According to a different study, the stem's methanolic extract and leaves' methanolic extract of Cuban oregano (*Plectranthus amboinicus*) showed important antioxidant potential, with TAC values of 634 μM AAE/g of extract and 782.56 μM AAE/g of extract, respectively [42]. Koldaş et al. noted that Kolda the antioxidant activities of water and methanol extracts of *Origanum vulgare* L. ssp. *Viride* were higher than those of ethyl acetate and hexane extracts in terms of phosphomolybdenum reduction potential [43]. Indeed, it has been reported that the antioxidant power of different *Origanum* extracts depends on the solvents used during their extraction, which has been correlated with the phenolic yield during the process [5].

The biological power of plant extracts is related to their content of polyphenols and flavonoids. Rosmarinic acid and its derivatives salvianolic acid and melitric acid are the main phenolic acids in *Origanum compactum* ethanol extract, while the main flavonoid is apigenin glucoside. The biological effects of rosmarinic acid and its derivatives, which include antioxidant, antibacterial, anti-inflammatory and anti-tumor actions, have been recently highlighted [30,44]. The antioxidant effect of rosmarinic acid is linked to its ability to scavenge free radicals, which increases membrane stability and protection against oxidative damage [45].

2.6. Antibacterial Activity

The antibacterial effectiveness of *Origanum compactum* hydroethanolic extract was tested against two Gram-negative bacteria (*Escherichia coli* and *Salmonella typhimirium*) and two Gram-positive bacteria (*Staphyloccocus aureus* and *Listeria monocytogenes*) through a broth microdilution assay. Tested concentrations for MIC and MBC were 166.66, 83.33, 41.66, 20.83, 10.41, 5.20, 2.60, 1.30, 0.65 and 0.32 mg/mL. Table 3 summarizes the results. The MIC values for *Origanum compactum* hydro-ethanolic extract against all tested bacteria ranged from 1.30 $\pm$ 0.11 to 41.66 $\pm$ 0.19 mg/mL. The extract displayed a bacteriostatic activity against *Escherichia coli* and *Salmonella typhimirium* with MBC/MIC values of 32 and 8, respectively, and a bactericidal activity against and *Staphyloccocus aureus* and *Listeria monocytogenes* with MBC/MIC values of 4 and 2, respectively.

Table 3. Determination of minimum inhibitory concentration and minimum bactericidal concentration exhibited by *Origanum compactum* ethanolic extract against bacterial strains (mg/mL).

Bacteria	MIC	MBC	MBC/MIC
Escherichia coli	1.30 $\pm$ 0.11	41.66 $\pm$ 0.20	32
Salmonella typhimirium	20.83 $\pm$ 0.20	166.66 $\pm$ 0.18	8
Staphyloccocus aureus	41.66 $\pm$ 0.15	166.66 $\pm$ 0.11	4
Listeria monocytogenes	41.66 $\pm$ 0.19	83.33 $\pm$ 0.15	2

The ethanolic extract of *Origanum compactum* was previously tested against *Escherichia coli* and *Staphyloccocus aureus*, and both strains were sensible to the extract [3].

The resistance of *Escherichia coli* and *Salmonella typhimirium* is related to the fact that Gram-negative bacteria are more resistant than Gram-positive bacteria to the majority of antibacterial agents. The resistance of Gram-negative bacteria is attributed to the hydrophilic nature of the membrane, which prevents the passage of hydrophobic molecules such as polyphenols. *Escherichia coli* have a bacterial wall where lipopolysaccharides are abundant and inhibit hydrophobic molecules from passing through the membrane [3].

The antibacterial effect of the ethanolic extract of *Origanum compactum* could be attributed to its phenolic composition. Indeed, the antibacterial activity of rosmarinic acid has been studied and proved against several bacterial strains. Even so, the molecular mechanisms and pathways explaining its biological activities have not been thoroughly investigated [30].

3. Materials and Methods

3.1. Plant Material

The plant was harvested in the region of Khenifra (El Hammam) in late June 2019. It was cultivated in 2009 by the cooperative El Hammam for the valuation of Medicinal and Aromatic Plants, and it was not chemically treated. The site is located in the Middle Atlas of Morocco at a height of 1182 m. Identification of the species was confirmed at the Scientific Institute of Rabat. The collected samples were dried for ten days at room temperature in the shade.

3.2. Phytochemical Screening

Phytochemical screening is a qualitative study based on coloring and/or precipitation reactions from different plant extracts. It aims to highlight the important families of secondary metabolites contained in the plant. These extracts were obtained by decoction, infusion or maceration by solvents. Phytochemical screening is also based on the use of several reagents. Table 4 contains the reagents and names of reactions used to recognize different groups.

Table 4. Reagents used in the search for secondary metabolites families.

Family Sought	Reagents and/or Reactions Used
Alcaloids	Dragendorff reagent and Mayer reagent
Tannins	Ferric chloride reaction and Stiasny reagent
Sterols and Triterpenes	Liebermann-Buchard reaction
Flavonoids	cyanidin reaction
Reducing compounds	Fehling's solution
Saponosides	stirring the aqueous solution

3.3. *Extraction of Polyphenols*

Extraction of polyphenols from aerial parts of *Origanum compactum* was performed by Soxhlet apparatus using an aqueous ethanolic mixture (70:30, *v*/*v*). First, 30 g of the plant powder was placed in a filter paper cartridge and introduced into the Soxhlet. The flask was filled with 350 mL of the hydroethanolic solvent and extracted several times until the plant was exhausted and the coloring disappeared. Afterwards, the solvent was evaporated and the extract was recovered with warm distilled water.

The yield of the crude extract was calculated in relation to m_0, the mass of the dry matter at which the solid-liquid extraction was carried out.

$$R\% = \frac{\text{Mass of crude extract}}{\text{Mass of dry matter powder}} \times 100 = \frac{m_0}{30} \times 100$$

3.3.1. Determination of Total Polyphenolic Contents in *Origanum compactum*

The total phenol content of the extract was determined by the Folin-Ciocalteu method [46]. In a 100 mL volumetric flask, 5 µL of the extract was mixed with 1.5 mL of Folin-Ciocalteu reagent (10%) and 1.5 mL of sodium carbonate (Na_2CO_3) at 7.5% (*m*/*v*). Then, the flask was filled with distilled water. The solution was left for 30 min at room temperature. The absorbance was measured at 760 nm

Gallic acid was used as a positive control. The polyphenol content of the studied extract is calculated from the regression equation of gallic acid calibration ($y = 0.095x + 0.003$). The results are expressed in milligrams of gallic acid equivalent per gram of dry matter (mg GAE/g dm). The polyphenol content is calculated using the following formula:

$$T = \frac{C \times V}{m\ (\text{dry matter})} \times D$$

where C is the concentration measured by the regression equation of gallic acid calibration, V is the sample volume and D is the dilution factor.

3.3.2. Determination of Flavonoids Contents in *Origanum compactum*

Flavonoid content was estimated by the aluminum trichloride ($AlCl_3$) method [47]. First, 10 µL of each fraction was mixed with 0.1 ml of aluminum trichloride 10%, followed by 20 mL of distilled water and supplemented at 50 mL with absolute methanol. The solution was incubated in darkness at room temperature for two hours, and the absorbance was measured at 430 nm. Flavonoids were quantified using a calibration curve performed with the quercetin standard ($y = 0.073x - 0.081$). The flavonoid content is expressed in milligrams of quercetin equivalent per gram of dry matter (mg QE/g dm).

3.4. *HPLC-PDA-ESI/MS Analysis*

The ethanolic extract of *Origanum compactum* was analyzed by high-performance liquid chromatography coupled to a photodiode array and electrospray ionization mass spectrometry (HPLC-PDA-ESI/MS). Identification of compounds was performed by using standard compounds (target identification) or by comparing mass spectrum obtained with literature data (tentative identification).

3.4.1. Sample Preparation

The crude ethanolic extract of *Origanum compactum* was redissolved in the same organic solvents and diluted 1:40 (*v*/*v*). For the chromatographic separation, an injection volume of 5 μL was employed, and the analysis was performed in triplicate.

3.4.2. HPLC-MS Analysis Condition

Chromatographic analysis was accomplished by means of a Shimadzu HPLC system (Kyoto, Japan) equipped with a CBM-20A controller, two LC-20AD dual-plunger parallel-flow pumps, a DGU20A5R degasser, a CTO-20AC column oven, a SIL-30AC autosampler, an SPD-M20A photodiode array detector and an LCMS-2020 single quadrupole mass spectrometer, with the employment of ESI source operated in negative and positive ionization modes.

Chromatographic separations were carried out on Ascentis Express RP C18 columns (150 × 4.6 mm; 2.7 μm) (Merck Life Science, Merck KGaA, Darmstadt, Germany). The employed mobile phase was composed of two solvents, water (solvent A) and ACN (solvent B), both acidified with 0.10% of formic acid *v*/*v*. The flow rate was set to 0.8 mL/min, and was split into 0.2 mL/min prior to ESI-MS detection, under gradient elution 0–15 min, 0–15% B; 30 min, 20% B; 60 min, 50% B; 70 min, 100% B; 79 min, 100% B. The injection volume was 5 μL. Diode array detection (DAD) was applied in the range of 200–400 nm and monitored at a wavelength of 330 nm (sampling frequency: 40.0 Hz, time constant: 0.08 s). MS conditions were as follows: scan range and scan speed were set to a mass-to-charge ratio (*m*/*z*) of 100–1000 and 2500 amu/s, respectively; event time was 0.3 s; nebulizing gas (N2) flow rate was 1.5 L/min; drying gas (N_2) flow rate was 15 L/min; interface temperature was 350 °C; heat block temperature was 300 °C; desolvation line temperature was 300 °C; desolvation line voltage was 1 V; interface voltage was −4.5 kV.

3.4.3. Standards Employed

Calibration curves of three polyphenolic standards (kaempferol-3-glucoside, apigenin, rosmarinic acid) were employed for the quantification of the polyphenolic content in sample extracts. Each analysis was performed in 6 repetitions. Data acquisition was performed by Shimadzu LabSolution software ver. 5.99. Kaempferol-3-glucoside (1, 10, 20, 50, 150) ppm; y = 13848x + 2354.1; R^2 = 0.9995; LoD = 0.090; LoQ = 0.274. Apigenin (1, 10, 25, 50, 100) ppm; y = 25625x + 608.11; R^2 = 0.9997; LoD = 0.0263; LoQ = 0.0798. Rosmarinic acid (5, 10, 25, 50, 100) ppm; y = 7150.6x + 9821.3; R^2 = 0.9993; LoD = 0.14; LoQ = 0.43.

3.5. *Antioxidant Activity*

The antioxidant activity was estimated by two methods: ferric reducing power (FRAP) assay and the phosphomolybdate method (total antioxidant capacity, TAC).

3.5.1. Ferric Reducing Power (FRAP) Assay

Ferric reducing power assay is a simple and inexpensive procedure that estimates the antioxidant level of a sample. It is based on the reducing potential of antioxidants in the extract that react with ferric ions (Fe^{3+}) provided by potassium ferricyanide $K_3Fe(CN)_6$ and reduce them to ferrous ions (Fe^{2+}). The method used is that described by Zovko Koncic [48].

Extract dilutions with concentrations ranging from 0 to 5 mg/mL were prepared. First, 0.5 mL of each solution was mixed with 2.5 mL of a phosphate buffer solution (0.2 M, pH 6.6) and 2.5 mL of a potassium ferricyanide solution $K_3Fe(CN)_6$ (1%). The mixtures were incubated in a water bath at 50 °C for 20 min. Afterwards, 2.5 mL of trichloroacetic acid (10%) was added to stop the reaction. The mixtures were then centrifuged at 3000 turns for 10 min. At the end, 2.5 mL of the supernatant of each concentration was mixed with

2.5 mL of distilled water and 0.5 mL of an aqueous solution of $FeCl_3$ at 0.1%. The absorbance was measured at 700 nm. The increase in absorbance in the reaction medium indicates the increase in the reducing power of the sample. Ascorbic acid was used as a positive control, and its absorbance was measured under the same conditions as the sample.

The antioxidant capacity was expressed by the determination of the effective concentration (EC_{50}), which corresponds to an absorbance equal to 0.5. This parameter was used to compare the reducing activity of the sample and the control.

3.5.2. Phosphomolybdate Method (Total Antioxidant Capacity, TAC)

The total antioxidant capacity (TAC) of the ethanolic extract from *Origanum compactum* aerial parts was assessed using the phosphomolybdate method according to Prieto et al., 1999 [49]. This method is based on the use of the plant extract to reduce molybdenum (VI) into molybdenum (V) in an acidic medium. To tubes containing 10 μL of the plant extract with different concentrations, 1 mL of the phosphomolybdate reagent (0.6 M sulfuric acid, 28 mM sodium phosphate and 4 mM ammonium molybdate) was added. After resting at room temperature for 20 min, the tubes were incubated for 90 min at 95 °C. At 695 nm, the absorbance was measured. The data are given as milligrams of ascorbic acid equivalent per gram of extract (mg EAA/g E).

The total antioxidant capacity (TAC) concentration of the analyzed extract was determined using the ascorbic acid calibration curve ($y = 0.0411x + 0.0159$. $R^2 = 0.9966$) and the results are expressed in milligrams of ascorbic acid equivalent per gram of dry extract (mg EAA/1 g E).

3.6. *Antibacterial Activity*

3.6.1. Bacterial Strains and Growth Conditions

The bacterial strains (*Escherichia coli*, *Salmonella typhimirium*, *Staphyloccocus aureus* and *Listeria monocytogenes*) used in this study were obtained from the Laboratory of Microbiology and Health, Faculty of Sciences at Moulay Ismail University of Morocco. Bacterial strains from the frozen stock (−80 °C) were spread on Mueller Hinton agar (Merck Life Science, Merck KGaA, Darmstadt, Germany) and incubated at 37 °C for 24 h. Then, bacterial suspensions were prepared in sterile distilled water and adjusted to the equivalent of 0.5 McFarland standard (108 cfu/mL).

3.6.2. Broth Microdilution Method

Minimum inhibitory concentration (MIC) and minimum bactericidal concentration (MBC) of extract against four bacterial strains were determined by the broth microdilution method as described by Bouymajane et al. [21]. To sterile, flat-bottom 96-well microplates, 50 μL of Mueller Hinton broth and dimethyl sulfoxide (MHB-DMSO) was added. Then, 50 μL of dried extract mixed with DMSO (500 mg/mL) of *Origanum compactum* was added to the first microplate and mixed in order to determine cascade dilutions. Then, 50 μL of bacterial suspensions and 50 μL of MHB-DMSO were added to each well. The well containing the mixture of bacterial suspensions and MHB-DMSO served as a control, and the well containing the mixture of extract and MHB-DMSO was used as a blank. All microplates were incubated at 37 °C for 24 h. Afterward, 50 μL of TTC (2, 3, 5-triphenyl tetrazolium chloride) was added to each well of the microplates and re-incubated at 37 °C for 30 min. The MIC was determined as the lowest concentration of the extract that showed no visible bacterial growth. The MBC was determined as the lowest concentration of extract that did not produce any bacterial colony. The wells that showed no visible bacterial growth were streaked on Petri dishes containing MHA and incubated 37 °C for 30 min. The MBC/MIC ratio was used to determine the bacteriostatic and bactericidal effects of the extract. If $MBC/MIC \leq 4$, the extract effect is bactericidal, and if $MBC/MIC > 4$, the extract effect is bacteriostatic. All the experiments were carried out in triplicate.

3.7. Statistical Analysis

The results are expressed as means ± SD. Statistical analysis was performed by one-way analysis of variance (*ANOVA*) using the SPSS package. All experiments were performed in triplicate and the differences were considered significant at $p < 0.05$.

4. Conclusions

We aimed to characterize the phenolic composition and evaluate the antioxidant and antibacterial activities of the hydroethanolic extract from aerial parts of *Origanum compactum*, collected in Morocco. The obtained results reveal the richness of the hydroethanolic extract of *Origanum compactum* in flavonoids and phenolic acids. Furthermore, this extract showed a strong antioxidant capacity and an antibacterial effect, probably due to the presence of luteolin, apigenin and their derivatives, rosmarinic acid and diosmetin. Therefore, based on the obtained results, the aerial parts of *Origanum compactum* are promising as a source of natural antibacterial and antioxidant agents that can be used in food and pharmaceutical fields.

Author Contributions: Conceptualization, M.C., T.Z. and L.B.; methodology, M.C., T.Z. and L.B.; investigation, M.C., T.Z., L.B., Y.O.E.M., M.A. and A.B.; resources, F.C. and L.M.; data curation, Y.O.E.M. and A.B.; writing—original draft preparation, M.C., T.Z., L.B. and A.B.; writing—review and editing, F.C.; supervision, F.C. and A.B.; project administration, L.M. and F.C. All authors have read and agreed to the published version of the manuscript.

Funding: This research received no external funding.

Institutional Review Board Statement: Not applicable.

Informed Consent Statement: Not applicable.

Data Availability Statement: Not applicable.

Acknowledgments: The authors are thankful to Shimadzu and Merck Life Science Corporations for the continuous support.

Conflicts of Interest: The authors declare no conflict of interest.

Sample Availability: Samples of the extract and the plant are available from the authors.

References

1. Chishti, S.; Kaloo, Z.A.; Sultan, P. Medicinal Importance of Genus Origanum: A Review. *J. Pharmacogn. Phytother.* **2013**, *5*, 170–177.
2. Aboukhalid, K.; Lamiri, A.; Agacka-Mołdoch, M.; Doroszewska, T.; Douaik, A.; Bakha, M.; Casanova, J.; Tomi, F.; Machon, N.; Faiz, C.A. Chemical Polymorphism of *Origanum compactum* Grown in All Natural Habitats in Morocco. *Chem. Biodivers.* **2016**, *13*, 1126–1139. [CrossRef]
3. Zeroual, A.; Eloutassi, N.; Chaouch, M.; Chaqroune, A. Antimicrobial, Antioxidant Activity, and Chemical Composition of *Origanum compactum* Benth from Taounate Province, North Morocco. *Asian J. Pharm. Clin. Res.* **2020**, *3*, 126–131. [CrossRef]
4. Bouyahya, A.; Et-Touys, A.; Abrini, J.; Talbaoui, A.; Fellah, H.; Bakri, Y.; Dakka, N. Lavandula Stoechas Essential Oil from Morocco as Novel Source of Antileishmanial, Antibacterial and Antioxidant Activities. *Biocatal. Agric. Biotechnol.* **2017**, *12*, 179–184. [CrossRef]
5. Gutiérrez-Grijalva, E.P.; Picos-Salas, M.A.; Leyva-López, N.; Criollo-Mendoza, M.S.; Vazquez-Olivo, G.; Heredia, J.B. Flavonoids and Phenolic Acids from Oregano: Occurrence, Biological Activity and Health Benefits. *Plants* **2018**, *7*, 2. [CrossRef] [PubMed]
6. El Babili, F.; Bouajila, J.; Souchard, J.P.; Bertrand, C.; Bellvert, F.; Fouraste, I.; Moulis, C.; Valentin, A. Oregano: Chemical Analysis and Evaluation of Its Antimalarial, Antioxidant, and Cytotoxic Activities. *J. Food Sci.* **2011**, *76*, C512–C518. [CrossRef] [PubMed]
7. Prathyusha, P.; Subramanian, M.S.; Nisha, M.C.; Santhanakrishnan, R.; Seena, M.S. Pharmacognostical and Phytochemical Studies on *Origanum vulgare* L. (Laminaceae). *Anc. Sci. Life* **2009**, *29*, 17–23. [PubMed]
8. Oreopoulou, A.; Tsimogiannis, D.; Oreopoulou, V. Extraction of Polyphenols From Aromatic and Medicinal Plants: An Overview of the Methods and the Effect of Extraction Parameters. In *Polyphenols in Plants*, 2nd ed.; Elsevier: Amsterdam, The Netherlands, 2019; pp. 243–259.
9. Turkmen, N.; Sari, F.; Velioglu, Y.S. Effects of Extraction Solvents on Concentration and Antioxidant Activity of Black and Black Mate Tea Polyphenols Determined by Ferrous Tartrate and Folin–Ciocalteu Methods. *Food Chem.* **2006**, *99*, 835–841. [CrossRef]
10. Do, Q.D.; Angkawijaya, A.E.; Tran-Nguyen, P.L.; Huynh, L.H.; Soetaredjo, F.E.; Ismadji, S.; Ju, Y.-H. Effect of Extraction Solvent on Total Phenol Content, Total Flavonoid Content, and Antioxidant Activity of Limnophila Aromatica. *J. Food Drug Anal.* **2014**, *22*, 296–302. [CrossRef] [PubMed]

11. Yan, F.; Azizi, A.; Janke, S.; Schwarz, M.; Zeller, S.; Honermeier, B. Antioxidant Capacity Variation in the Oregano (*Origanum vulgare* L.) Collection of the German National Genebank. *Ind. Crops Prod.* **2016**, *92*, 19–25. [CrossRef]
12. Bower, A.M.; Real Hernandez, L.M.; Berhow, M.A.; de Mejia, E.G. Bioactive Compounds from Culinary Herbs Inhibit a Molecular Target for Type 2 Diabetes Management, Dipeptidyl Peptidase IV. *J. Agric. Food Chem.* **2014**, *62*, 6147–6158. [CrossRef] [PubMed]
13. Kogiannou, D.A.A.; Kalogeropoulos, N.; Kefalas, P.; Polissiou, M.G.; Kaliora, A.C. Herbal Infusions; Their Phenolic Profile, Antioxidant and Anti-Inflammatory Effects in HT29 and PC3 Cells. *Food Chem. Toxicol.* **2013**, *61*, 152–159. [CrossRef] [PubMed]
14. Hossain, M.B.; Rai, D.K.; Brunton, N.P.; Martin-Diana, A.B.; Barry-Ryan, C. Characterization of Phenolic Composition in Lamiaceae Spices by LC-ESI-MS/MS. *J. Agric. Food Chem.* **2010**, *58*, 10576–10581. [CrossRef] [PubMed]
15. Kaiser, A.; Carle, R.; Kammerer, D.R. Effects of Blanching on Polyphenol Stability of Innovative Paste-like Parsley (*Petroselinum crispum* (Mill.) Nym Ex A. W. Hill) and Marjoram (*Origanum majorana* L.) Products. *Food Chem.* **2013**, *138*, 1648–1656. [CrossRef] [PubMed]
16. Engel, R.; Szabó, K.; Abrankó, L.; Rendes, K.; Füzy, A.; Takács, T. Effect of Arbuscular Mycorrhizal Fungi on the Growth and Polyphenol Profile of Marjoram, Lemon Balm, and Marigold. *J. Agric. Food Chem.* **2016**, *64*, 3733–3742. [CrossRef] [PubMed]
17. Barros, L.; Dueñas, M.; Dias, M.I.; Sousa, M.J.; Santos-Buelga, C.; Ferreira, I.C.F.R. Phenolic Profiles of Cultivated, in Vitro Cultured and Commercial Samples of *Melissa officinalis* L. Infusions. *Food Chem.* **2013**, *136*, 1–8. [CrossRef] [PubMed]
18. FoodDB. FooDB Version 1.0. 2020. Available online: http://www.foodb.ca (accessed on 17 July 2022).
19. Taamalli, A.; Arráez-Román, D.; Abaza, L.; Iswaldi, I.; Fernández-Gutiérrez, A.; Zarrouk, M.; Segura-Carretero, A. LC-MS-Based Metabolite Profiling of Methanolic Extracts from the Medicinal and Aromatic Species Mentha Pulegium and *Origanum majorana*. *Phytochem. Anal.* **2015**, *26*, 320–330. [CrossRef] [PubMed]
20. Zengin, G.; Cvetanović, A.; Gašić, U.; Dragićević, M.; Stupar, A.; Uysal, A.; Şenkardes, I.; Sinan, K.I.; Picot-Allain, M.C.N.; Ak, G.; et al. UHPLC-LTQ OrbiTrap MS Analysis and Biological Properties of *Origanum vulgare* subsp. Viridulum Obtained by Different Extraction Methods. *Ind. Crops Prod.* **2020**, *154*, 112747. [CrossRef]
21. Bouymajane, A.; Filali, F.R.; El Majdoub, Y.O.; Ouadik, M.; Abdelilah, R.; Cavò, E.; Miceli, N.; Taviano, M.F.; Mondello, L.; Cacciola, F. Phenolic Compounds, Antioxidant and Antibacterial Activities of Extracts from Aerial Parts of Thymus Zygis Subsp. Gracilis, Mentha Suaveolens and Sideritis Incana from Morocco. *Chem. Biodivers.* **2022**, *19*, e202101018. [CrossRef] [PubMed]
22. Skoula, M.; Grayer, R.J.; Kite, G.C.; Veitch, N.C. Exudate Flavones and Flavanones in Origanum Species and Their Interspecific Variation. *Biochem. Syst. Ecol.* **2008**, *36*, 646–654. [CrossRef]
23. Koukoulitsa, C.; Karioti, A.; Bergonzi, M.C.; Pescitelli, G.; Di Bari, L.; Skaltsa, H. Polar Constituents from the Aerial Parts of *Origanum vulgare* L. Ssp. Hirtum Growing Wild in Greece. *J. Agric. Food Chem.* **2006**, *54*, 5388–5392. [CrossRef] [PubMed]
24. Grevsen, K.; Fretté, X.C.; Christensen, L.P. Content and Composition of Volatile Terpenes, Flavonoids and Phenolic Acids in Greek Oregano (*Origanum vulgare* L. Ssp. Hirtum) at Different Development Stages during Cultivation in Cool Tem-Perate Climate. *Eur. J. Hortic. Sci.* **2009**, *74*, 193–203.
25. Koukoulitsa, C.; Zika, C.; Geromichalos, G.D.; Demopoulos, V.J.; Skaltsa, H. Evaluation of Aldose Reductase Inhibition and Docking Studies of Some Secondary Metabolites, Isolated from *Origanum vulgare* L. ssp. Hirtum. *Bioorg. Med. Chem.* **2006**, *14*, 1653–1659. [CrossRef] [PubMed]
26. Koukoulitsa, C.; Hadjipavlou-Litina, D.; Geromichalos, G.D.; Skaltsa, H. Inhibitory Effect on Soybean Lipoxygenase and Docking Studies of Some Secondary Metabolites, Isolated from *Origanum vulgare* L. ssp. *Hirtum*. *J. Enzym. Inhib. Med. Chem.* **2007**, *22*, 99–104. [CrossRef] [PubMed]
27. Baranauskaite, J.; Kopustinskiene, D.M.; Masteikova, R.; Gajdziok, J.; Baranauskas, A.; Bernatoniene, J. Effect of Liquid Vehicles on the Enhancement of Rosmarinic Acid and Carvacrol Release from Oregano Extract Liquisolid Compacts. *Colloids Surf. A Physicochem. Eng. Asp.* **2018**, *539*, 280–290. [CrossRef]
28. Desam, N.R.; Al-Rajab, A.J.; Sharma, M.; Mylabathula, M.M.; Gowkanapalli, R.R.; Albratty, M. Chemical Constituents, in Vitro Antibacterial and Antifungal Activity of *Mentha×Piperita* L. (Peppermint) Essential Oils. *J. King Saud Univ. Sci.* **2019**, *31*, 528–533. [CrossRef]
29. Boutahiri, S.; Eto, B.; Bouhrim, M.; Mechchate, H.; Saleh, A.; Al kamaly, O.; Drioiche, A.; Remok, F.; Samaillie, J.; Neut, C.; et al. Lavandula Pedunculata (Mill.) Cav. Aqueous Extract Antibacterial Activity Improved by the Addition of Salvia Rosmarinus Spenn., Salvia Lavandulifolia Vahl and Origanum compactum Benth. *Life* **2022**, *12*, 328. [CrossRef] [PubMed]
30. Kim, G.-D.; Park, Y.S.; Jin, Y.-H.; Park, C.-S. Production and Applications of Rosmarinic Acid and Structurally Related Compounds. *Appl. Microbiol. Biotechnol.* **2015**, *99*, 2083–2092. [CrossRef]
31. Moghadam, S.E.; Ebrahimi, S.N.; Gafner, F.; Ochola, J.B.; Marubu, R.M.; Lwande, W.; Frei Haller, B.; Salehi, P.; Hamburger, M. Metabolite Profiling for Caffeic Acid Oligomers in Satureja Biflora. *Ind. Crops Prod.* **2015**, *76*, 892–899. [CrossRef]
32. Isao, A.; Hijiri, K.; Tsutomu, H.; Sansei, N.; Takuo, O. Melitric Acids A and B, New Trimeric Caffeic Acid Derivatives from *Melissa officinalis*. *Chem. Pharm. Bull.* **1993**, *41*, 1608–1611.
33. Leyva-López, N.; Nair, V.; Bang, W.Y.; Cisneros-Zevallos, L.; Heredia, J.B. Protective Role of Terpenes and Polyphenols from Three Species of Oregano (Lippia Graveolens, Lippia Palmeri and Hedeoma Patens) on the Suppression of Lipopolysaccharide-Induced Inflammation in RAW 264.7 Macrophage Cells. *J. Ethnopharmacol.* **2016**, *187*, 302–312. [CrossRef]
34. Tuttolomondo, T.; La Bella, S.; Licata, M.; Virga, G.; Leto, C.; Saija, A.; Trombetta, D.; Tomaino, A.; Speciale, A.; Napoli, E.M.; et al. Biomolecular Characterization of Wild Sicilian Oregano: Phytochemical Screening of Essential Oils and Extracts, and Evaluation of Their Antioxidant Activities. *Chem. Biodivers.* **2013**, *10*, 411–433. [CrossRef]

Mediterranean pomegranates are mainly based on local cultivars, and their composition can differ from those of Eastern varieties, displaying a large variety of physical–chemical traits and distinct flavor profiles. Wonderful is the most widespread commercial pomegranate cultivar planted in Mediterranean countries and represents the industry standard variety. In the last few decades, around the world, there has been an increasing interest in the use of pomegranate and its parts, justified by an increasing demand from health care consumers and the pharmaceutical and cosmetic industries [13]. Generally, the edible part of pomegranate is directly consumed as food, or used for the preparation of juices, canned beverages, jams, and for the flavoring and coloring of drinks; conversely, pomegranate peel (approximatively 26–30% of the total fruit weight), which currently still represents a waste to be disposed of, is attracting the attention of the scientific community for its high content of phytochemicals that allow it to individuate as a new source of bioactive compounds, such as flavonoids, phenolic acids, and tannins with well-ascertained antioxidant capacity [16–18]. It has been reported that pomegranate by-products, and punicalagins in particular, decrease the level of fats in the blood and have anticancer, antiviral, and anti-inflammatory properties [15,19–21]. The scientific community was previously focused on the chemical characterization and health effects of pomegranate as a fruit or juice and only few studies were recently focused on the amount and composition of the bioactive compounds present in the pomegranate peel, which usually are a mixture, the synergistic effect of which can often cause different physiological responses acting on different organ targets contemporarily.

Based on the above considerations, the aim of the present study was to analyze the pomegranate peel composition of the variety Wonderful cultivated in Southern Italy and treated with an innovative system of dry concentration by the Gioia Succhi food industry. The specific aim was to verify if the growth area, cultivar type, and dry procedure influencing the chemical composition of the peels in terms of valuable bioactive compounds with beneficial effects in the prevention of numerous diseases or metabolic disorders. A comparison between the peel chemical composition of Wonderful cultivated in Southern Italy and treated with a spray-dry system and the peel chemical composition of freeze-dried Wonderful cultivated in Southern Italy, freeze-dried Wonderful cultivated in South Africa [15,22], and freeze-dried Kullu and Himachal [23,24] cultivated in India and already used in a pharmaceutical scope was carried out.

2. Materials and Methods

2.1. Chemicals

Metaphosphoric acid, 2,2-diphenyl-1-picrylhydrazyl (DPPH), NaOH, nitro-blue tetrazolium, dichlorophenol-indophenol (DCPID), 2,2′-azino-bis (3-ethylbenzothiazoline-6-sulfonic acid) di-ammonium salt ($ABTS^{\bullet+}$), 6-hydroxy-2,5,7,8-tetramethylchromane-2-carboxyl acid (Trolox), phenazine methosulphate, ethanol, gallic acid, ethylene-diamine-tetra acetic acid (EDTA), ferrozine, 2,4,6-tris (2-pyridyl)-s-triazine (TPTZ), and iron sulphate heptahydrate were purchased from Sigma Chemical Co. (St. Louis, MO, USA). HPLC-grade methanol and acetonitrile (Sigma Aldrich, St. Louis, MO, USA, 99.99%), acetone (Sigma Aldrich, 99.5%), deionized water, formic acid (Carlo Erba, 95%), and hydrochloric acid (Carlo Erba, 37%) were used for sample extraction and HPLC analysis. All chemical standards—gallic acid; protocatechuic acid; procyanidin 1 and 2; syringic acid; *p*, *m*, *o*-coumaric acids; pelargonidin; *trans*-cinnamic acid; bergamottin; cyanidine 3 *O*-glucoside; catechin; vanillic acid; epicatechin; delphinidin; *trans*-4-hydroxycinamic acid; sinapinic acid; 3-hydroxycinnamic acid; myricetin; luteolin; punicalagin; 2,5 dihydroxy benzoic acid; caffeic acid; ellagic acid; naringin; apigenin-7-neohesperosside; spiraeoside; quercetin; kaempferol; tocopherol; chlorogenic acid; vicenin 2; eriocitrine; rutin; vitexin; quercitin-3 beta-D-glucoside; ferulic acid; and apigenin were purchased from Sigma Aldrich, St. Louis, MO, USA. Other chemicals were of analytical grade and purchased from Carlo Erba Reagents s.r.l. (Milan, Italy).

2.2. Pomegranate Peel Preparation and Extraction

Fruits of the cultivar Wonderful grown in Calabria were washed and hand peeled. The peel of mature Wonderful fruits was dried in different ways: (1) with an innovative process by Gioia Succhi, a Calabrian food transformation industry, that used a physical spray-dried innovative process (PDS) that is held as an industry secret, and (2) stored a −80 °C for two days and then lyophilized with a freeze-dry system (Cheimika, SH Top, Pellezzano SA, Italy) at −56 °C for 96 h (FD). The dried peels were analyzed for chemical characteristics.

2.3. Sample Extract Preparation

The extracts were obtained using the method described in Muscolo et al. (2020) [25]. Briefly, lyophilized pomegranate peels were extracted at room temperature (22–25 °C) with continuous stirring for 90 min with 15 mL 95% ethanol. The samples were centrifuged (Unicen 21 RT167, Ortoalresa Inc., Madrid, Spain) at 2370× *g* (4000 rpm) for 15 min and the supernatants were filtered with 1 mm Whatman 185 filter paper (Merck, Darmstadt, Germany), evaporated to dryness in a rota-vapor (Diagonal condenser RE 400, Stuart Equipment, ST15, Stone, UK), and re-suspended in a final volume of 3.0 mL 95% ethanol.

Lyophilized pomegranate peels were extracted at room temperature with continuous stirring for 60 min with 2.0 mL dH20 (Intercontinental Mod still 3/ES, Bioltecnical Service, s.n.c., Rome, Italy). The samples were then centrifuged at 590× *g* (2000 rpm) for 10 min and the supernatants were filtered with Whatman 1 filter paper and used for the determination of protein, carbohydrates, and ferrous chelating activity.

2.4. Determination of Total Phenolic Compounds, Total Flavonoids, and Vitamins A, C, and E in Pomegranate Peel

Total phenol content was determined with Folin–Ciocalteu reagent according to Muscolo et al. [25]. Briefly, 500 µL of the aqueous extract was mixed with 250 µL of Folin–Ciocalteu reagent and 2 mL of a 20% Na_2CO_3 aqueous solution, and the mixture was filled up to 50 mL with deionized water and placed in the dark for 1 h. The absorbance was measured at 765 nm using a UV-Vis Agilent 8453 spectrophotometer (Agilent Technologies, Agilent Technologies, Santa Clara, CA, USA). The results were expressed as mg/L of gallic acid equivalents.

Total flavonoid content was determined according to the colorimetric method as reported in Muscolo et al. [25]. The absorbance was measured at 510 nm using a UV-Vis Agilent 8453 spectrophotometer (Agilent Technologies, Agilent Technologies, Santa Clara, CA, USA). The results were expressed as rutin equivalents (mg/L) using a calibration curve.

Vitamin A was detected as reported in Aremu and Nweze [26]. Absorbance was read at 436 nm and vitamin A was expressed as retinol equivalent (RE).

For vitamin C (ascorbic acid) determination, the method of Davies and Masten [27] was used. Pomegranate powders (0.10 g) were extracted with 10 mL of 3% meta-phosphoric acid—98% acetic acid centrifuged at 2370× *g* (4000 rpm) for 10 min, and the supernatant was used for the determination of ascorbic acid.

For vitamin E (α-tocopherol) analysis, pomegranate powder (0.10 g) was extracted with 10 mL of hexane:isopropanol solution (3:2 *v*/*v*) with agitation for 5 h, and centrifuged at 1330× *g* (3000 rpm) for 10 min. The supernatant was used for the determination of vitamin E [28].

2.5. Protein and Carbohydrate Detection in Pomegranate Peel

Soluble protein was determined using the Bradford method as reported in Muscolo et al. [25] by using Coomassie Brilliant Blue G-250. The absorbance of each sample was measured at 595 nm using an 1800 UV-Vis spectrophotometer (Shimadzu, Kyoto, Japan). Bovine serum albumin > 99% purity (Sigma) was used as standard, and soluble proteins were estimated as mg BSA/g DW.

The total available carbohydrates were measured using the anthrone method with minor modifications as reported in Muscolo et al. [25]. The amount of available carbohydrates was calculated using a glucose calibration curve (range of 10–100 mg/mL). The results were reported as mg/g DW.

2.6. Determination of Antioxidant Activities in Pomegranate Peel

The antioxidant activity against DPPH radical (2,2-diphenyl-1-picryl-hydrazyl-hydrate) was determined with the method reported in Muscolo et al. [25]. The DPPH concentration in the cuvette was chosen to give absorbance values of ∼1.0. Absorbance changes in the violet solution were recorded at 517 nm after 30 min of incubation at 37 °C. The inhibition I (%) of radical-scavenging activity was calculated as:

$$I\,(\%) = [(A0 - AS)/A0] \times 100 \quad (1)$$

where A0 is the absorbance of the control and AS is the absorbance of the sample after 30 min of incubation. Results were expressed as µmol Trolox/g DW.

The 2,2′-azino-bis-3-ethylbenzothiazoline-6-sulfonic acid assay (ABTS) was carried out according to Muscolo et al. [25] using a solution of 7 mM of ABTS in phosphate buffered saline (PBS). Aliquots of ethanol extracts (25, 50, and 100 µL) were added to 0.5 mL of $ABTS^{+\bullet}$ solution and brought to a final volume of 600 µL with PBS. After 6 min of incubation in the dark at room temperature the absorbance of the samples was measured at 734 nm. Results were expressed as µmol Trolox/g DW.

The total antioxidant capacity (TAC) was performed according to Muscolo et al. [25]. Sample absorbance was measured at 695 nm using UV-Vis spectrophotometer. Methanol (0.3 mL) in place of the extract was used as blank. The antioxidant activity was expressed as µg of α-tocopherol g^{-1} DW on a calibration curve.

2.7. RP-DAD-HPLC Identification of Phenolic and Flavonoid Components

Pomegranate peel was finely ground for analysis. By-product samples were subjected to solvent extraction before HPLC analysis for determination of the single phenolic and flavonoid compounds. Each sample was extracted in two different ways—0.1 g of previously lyophilised pomegranate peels was dissolved in 10 mL of 1% of HCl in methanol and 0.1 g of sample was dissolved in 10 mL of acetone solution: 1% of HCl in methanol (1:1). Each sample was analyzed in six independent replicates [25]. Reverse-phase–diode array detector–high-performance liquid chromatography (RP-DAD-HPLC) analyses of samples was carried out with a Shimadzu system (Kyoto, Japan), consisting of an LC-10AD pump system, a vacuum degasser, a quaternary solvent mixer, an SPD-M10A diode array detector, and a Rheodyne 7725i injector (Merck KGaA, Darmstadt, Germany). Separation of each compound was done on a 250 × 4.6 mm i.d. 5 µm Discovery C18 column supplied by Supelco Park (Bellefonte, PA, USA) and equipped with a 4.0 × 20 mm guard column. The column was placed in a column oven set at 25 °C. The injection loop was 20 µL and the flow rate was 1.0 mL/min. The mobile phase consisted of a linear gradient of solvent A (acetonitrile) in 2% acidified water (acetic acid:H_2O, 2:98) as follows: 0–80% (0–55 min), 90% (55–70 min), 95% (70–80 min), 100% (80–90 min), and 0% (90–110 min). UV-Vis spectra were measured between 200 and 600 nm and simultaneous detection using a diode array at 278 and 325 nm. Compounds were measured using their retention time and UV spectra (Dueñas and Estrella, 2002), through comparison with purified standards (Sigma Chemical Co., Saint Louis, MO, USA).

2.8. Statistical Analysis

Analysis of variance was carried out for all the data sets. One-way ANOVA with Tukey's Honestly Significant Difference tests were carried out to analyze the effects of treatment/cultivar on each of the various parameters measured. ANOVA and a *T*-test were carried out using SPSS software (IBM Corp. 2012, New York, NY, USA). Effects were significant at $p \leq 0.05$. To explore relationships among different treatments/cultivars and chemical parameters, datasets were analyzed using principal component analysis (PCA).

3. Results and Discussion

Results evidenced that total carbohydrates were contained in a higher quantity in peels of other cultivars than in peels of Wonderful. The lowest carbohydrate content was found in spray-dried Wonderful peel (PSD) (Figure 1). Total proteins had an opposite trend, with the lowest in the peels of the other cultivars and the highest in PSD (Figure 1A). Total phenols and total flavonoids were present in the highest quantity in the Calabrian Wonderful spray-dried peels and in the lowest amount in the other cultivar peels. Total phenols were higher than total flavonoids in all the samples analyzed (Figure 1B). These data evidenced that the Wonderful cultivar had the majority of total phenols and flavonoids. These data highlighted that the geographic conditions, in which a determined type of cultivar grows, can drive the synthesis of bio-compounds, shifting the metabolism from primary to secondary. Data from Ramakrishna and Ravishankar [29] showed how drought conditions increased, in different plants and in different part of the plants, the amount of total flavonoids and phenolic acids that were used as antioxidants to overcome stress conditions. Vaneková et al. [30] showed how environmental factors such as altitude, habitat type, and sunlight exposure influenced the synthesis of total phenols and flavonoids in seven different cultivars of berries. The distribution of drylands is quite accentuated in southern and Mediterranean countries, and Calabria in particular is dominated by climate conditions mainly characterized by dry summers and mild wet winters that, as demonstrated by Fialho et al. [31], in pomegranates increased secondary metabolites with nutraceutical properties. The data of this research are in line with literature findings and highlight that the major quantity of total phenols and flavonoids contained in Wonderful peel cultivated in Calabria could be the result of the microclimatic conditions, which in turn affect metabolism, increasing antioxidants as well as antioxidant activities, and the soluble protein amount, which can have a double function of working as osmolytes or as antioxidative enzymes, as demonstrated by Kosová et al. [32]. In the spray-dried peel, the greatest amount of these compounds was found, evidencing that the innovative methodology used to dry the peels did not denature the bio-compounds but rather concentrated them. Vitamins were contained in greater amounts in the Wonderful cultivar than in the other cultivars, and were more concentrated in PSD and CFD. Vitamin E was the most abundant in all the pomegranate peel samples, except for the peel of the Indian cultivars. Vitamins have a great role as antioxidants and have important health benefits when consumed with the diet. The PSD contained a huge amount of vitamins (Figure 2). It was demonstrated that vitamins are enzymatic cofactors and act as antioxidants. Vitamin C increases under stress conditions to protect plants from oxidative stress by acting to detoxify reactive oxygen species (ROS) by direct scavenging or by acting as cofactors in the enzymatic reactions that involved ascorbate peroxidase and glutathione reductase enzymes [33]. Vitamin E, which is the most abundant vitamin, is a major single oxygen scavenger that provides protection against lipid peroxidation [34]. In support of the above findings, the activities of the antioxidant enzymes were greater in PSD than in the other samples. All the activities (DPPH, ABTS, and TAC) were expressed more in PSD and CFD than in the peels of the other samples (Figure 3). These data evidence that the innovative dry process did not affect the biological compounds and the enzymes.

Single phenolic acids were higher in Wonderful peels than in the peels of the other cultivars. In PSD, the greatest amount of single phenolic acids was detected (Table 1). Ellagic acid was the most abundant, followed in ranking by 2-5 dihydroxy-benzoic, gallic, protocatechuic, *p*-coumaric, chlorogenic, and ferulic acids. It has been widely demonstrated that ellagic acid (EA) is a potent antioxidant with antimicrobial, anti-inflammatory, neuroprotective, antihepatotoxic, anticholestatic, antifibrogenic, anticarcinogenic, cytotoxic, and antiviral effects [35]. Recently, Reis Jordão et al. [36] evidenced that ellagic acid can be a promising alternative treatment for hypertension and cardiovascular disease, and Pei et al. [37] demonstrated that EA can be used to prevent diabetic cardiac dysfunction. The doses of EA generally tested in the prevention health treatments were 30 mg/kg. PSD peel contained a great amount of EA (240 mg/g), suggesting its possible use as a nutraceutical supplement for the prevention of numerous diseases. Additionally, gallic,

2-5 dihydroxy-benzoic, protocatechuic, and ferulic acids have been found in a number of phytomedicines with diverse biological and pharmacological activities, including radical scavenging, apoptosis of cancer cells, antihyperglycemic, antioxidant effects, and cardioprotective activity [38–41]. Among the single flavonoids (Table 2), procyanidin 2, punicalagin, procyanidin 1, and pelargonidin were, in this order, the most abundant compounds in PSD. Conversely, in the freeze-dried Calabrian and South African Wonderful peel, a lesser amount of single flavonoids than PSD was found, but in a greater quantity than the Indian cultivars. CFD had a greater amount of punicalagin (65 mg/g), procyanidin 1 (1.6 mg/g), procyanidin 2 (1.6 mg/g), and delphinidin than SAFD. IC contained a great amount of procyanidin 2 only in respect to CFD and SAFD.

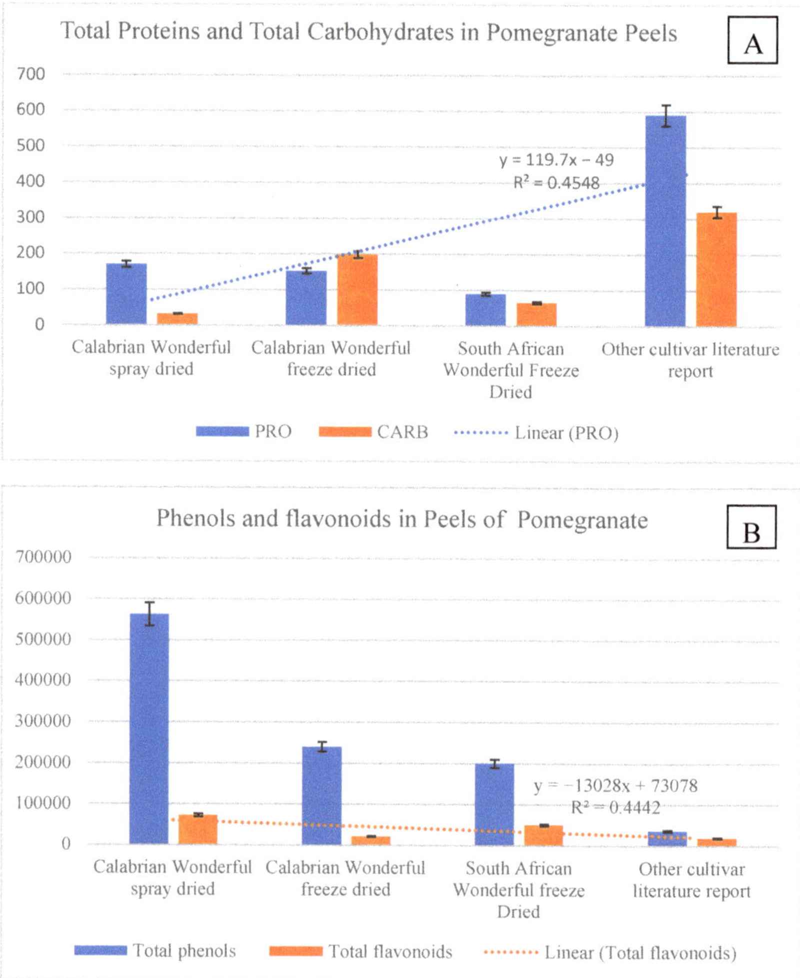

Figure 1. Soluble proteins and total carbohydrates (**A**) and total phenols and total flavonoids (**B**) in peels of different pomegranate cultivars dried differently. PSD (spray-dried Wonderful peel, experimental data); CFD (Calabrian Wonderful peel freeze-dried, experimental data); SAFD (South African Wonderful peel freeze-dried, literature data); IC (Indian cultivar peel freeze-dried, literature data). The experimental data are the mean of six replicates. Soluble protein (mg BSE g^{-1} DW), carbohydrates (mg g^{-1} DW), total phenols (μg TAE g^{-1} DW), total flavonoids (μg quercetin g^{-1} DW).

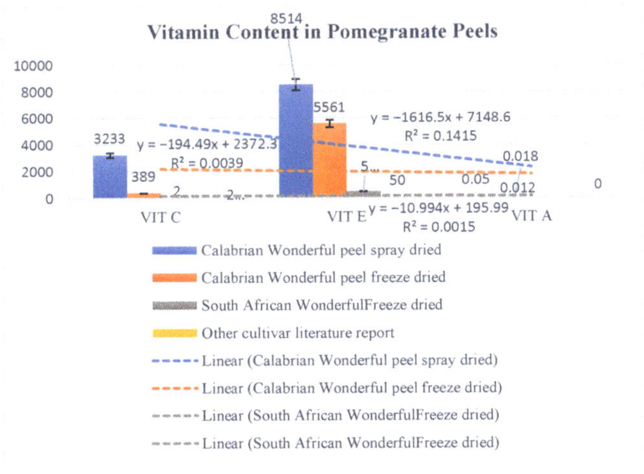

Figure 2. Vitamin A (μg retinol g^{-1} DW), C (μg ascorbate g^{-1} DW), and E (μg α-tocopherol g^{-1} DW) in peels of different pomegranate cultivars dried differently. PSD (spray-dried Wonderful peel, experimental data); CFD (Calabrian Wonderful peel freeze-dried, experimental data); SAFD (South African Wonderful peel freeze-dried, literature data); IC (Indian cultivar peel freeze-dried, literature data). The experimental data are the mean of six replicates.

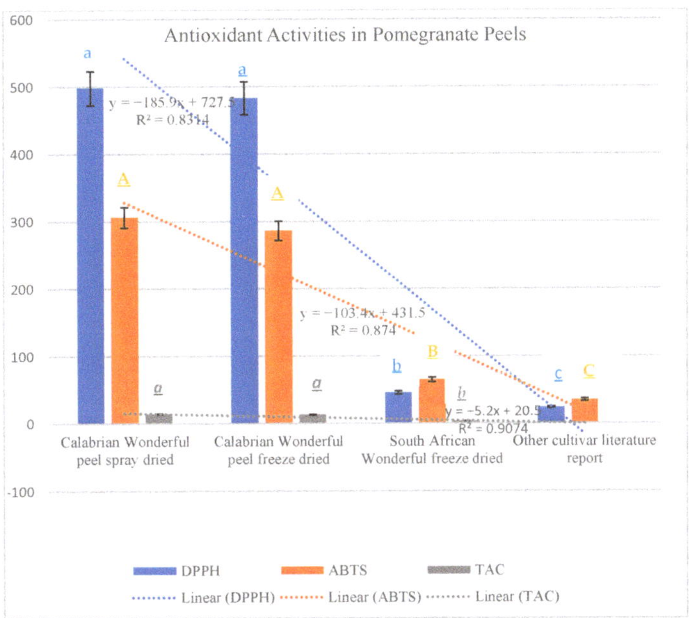

Figure 3. Antioxidant activities expressed as 2,2-diphenyl-1-picryl-hydrazyl-hydrate (DPPH), total antioxidant capacity (TAC), and 2,2′-azino-bis-3-ethylbenzothiazoline-6-sulfonic acid (ABTS) in peels of different pomegranate cultivars dried differently. PSD (spray-dried Wonderful peel, experimental data); CFD (Calabrian Wonderful peel freeze-dried, experimental data); SAFD (South African Wonderful peel freeze-dried, literature data); IC (Indian cultivar peel freeze-dried, literature data). The experimental data are the mean of six replicates. Different letters indicate significant differences $p \leq 0.05$. Lowercase (DPPH), capital (ABTS), and italic (TAC).

Table 1. Single phenolic acids contained in the peels of different pomegranate cultivars dried differently. PSD (spray-dried Wonderful peel, experimental data); CFD (Calabrian Wonderful peel freeze-dried, experimental data); SAFD (South African Wonderful peel freeze-dried, literature data); IC (Indian cultivar peel freeze-dried, literature data). The experimental data are the mean of six replicates.

	PSD	CFD	SAFD	IC
	mg/g SS	mg/g SS	mg/g SS	mg/g SS
Phenolic acids				
Gallic	11.1 a	0.2 c	1.2 b	0.2 c
Protocatechuic	7.9 a	1.4 b	nd	nd
Syringic	0.6	nd	nd	nd
p-coumaric	5.8 a	4 b	0.09 c	0.07 d
m-coumaric	4 a	0.6 b	nd	nd
o-coumaric	0.8 ab	nd	1.2 a	0.5 b
Trans-cinnamic	1.2 b	0.4 c	2.0 a	0.3 c
3-hydroxycinnamic	0.6	nd	nd	nd
Trans-4-hydroxycinnamic acid	0.6 b	1 a	0.3 b	0.05 c
Sinapic acid	0.7 a	0.2 b	0.05 c	0.01 d
2,5 dihydroxy-benzoic acid	16.4 a	0.8 b	0.4 b	0.04 c
Vanillic acid	0.8 b	2.6 a	0.8 b	0.2 c
Chlorogenic acid	5.6 a	6 a	1.4 b	0.5 c
Ferulic acid	4 a	0.4 b	0.3 b	0.02 c
Ellagic acid	240 a	8 b	2 c	0.5 d

Different letters in the same row indicate significant differences $p \leq 0.05$.

Table 2. Single flavonoids in peels of different pomegranate cultivars dried differently. PSD (spray-dried Wonderful peel, experimental data); CFD (Calabrian Wonderful peel freeze-dried, experimental data); SAFD (South African Wonderful peel freeze-dried, literature data); IC (Indian cultivar peel freeze-dried, literature data). The experimental data are the mean of six replicates.

	PSD	CFD	SAFD	OCLR
	mg/g SS	mg/g SS	mg/g SS	mg/g SS
Flavonoids				
Procyanidin B2	178 a	1.6 c	1.2 d	8.9 b
Pelargonidin	5.8 a	nd	nd	nd-
Cyanidin 3 *O*-glucoside	12.2 a	4 b	4 b	0.15 c
Catechin	12 a	3 a	0.03 c	1.4 b
Epicatechin	1.37 a	nd	0.017 c	0.07 b
Delphinidin	0.8 b	173 a	0.39 c	0.42 c
Myricetin	1.2 a	1.4 a	nd	nd
Luteolin	nd	1	nd	0.0025
Naringin	0.9	nd	nd	nd
Apigenin-7-neohesperosside	0.2 b	1.2 a	0.34 b	nd
Spiraeoside	1.0 a	0.6 b	0.5 b	nd
Quercetin	3 a	2 a	0.3 b	0.02 c
Kaempferol	1.2 a	0.2 b	0.05 b	0.1 b
Procyanidin B1	13 a	1.6 b	1.2 b	nd
Vicenin 2	nd	2	nd	nd
Rutin	0.3 b	3 a	0.56 b	0.021 c
Quercetin-3 beta-D glucoside	1.3 a	nd	0.18 b	0.05 c
Apigenin	2 a	2 a	0.7 b	0.037 c
Others				
Erythrocin	0.9	nd	nd	nd
Punicalagin	86 a	65 b	40	28
Tocopherol	2.4	nd	nd	nd

Different letters in the same row indicate significant differences $p \leq 0.05$.

Pearson's correlation was used to determine the degree of correlation between selected reference data and variables (Table 3). As expected, TP was positively correlated with all the variables except for CARB. Vitamins and proteins were also positively correlated with all the variables except for carbohydrates. Total flavonoids did not correlate with the antioxidant activities and CARB. In addition, strong positive correlations between the antioxidant assays were reported at r = 0.998, $p = 0.05$, between DPPH and ABTS; at r = 0.996, $p = 0.05$, between DPPH and TAC; and at r = 0.988, $p = 0.05$, between ABTS and TAC (Table 3). This agrees with the results in Figure 1B suggesting that the high phenolic content in peel extracts determines the strong antioxidant activity.

Table 3. Pearson's correlations (r) between total phenols (TP), total flavonoids (TF), vitamin A (VIT A), vitamin C (VIT C), vitamin E (VIT E), 2,2-diphenyl-1-picryl-hydrazyl-hydrate (DPPH), total antioxidant capacity (TAC), 2,2′-azino-bis-3-ethylbenzothiazoline-6-sulfonic acid (ABTS), total protein (PRO), and total carbohydrates (CARB). Values in bold are different from 0 with a significance level alpha = 0.05.

Variables	TP	TF	VIT A	VIT C	VIT E	DPPH	ABTS	TAC	PRO	CARB
TP	1	0.848	**0.954**	0.940	0.883	0.766	0.803	0.727	0.867	−0.649
TF	0.848	**1**	0.787	0.796	0.502	0.314	0.374	0.253	0.488	−0.232
VIT A	**0.954**	0.787	**1**	**0.998**	0.878	0.722	0.750	0.705	0.792	−0.783
VIT C	0.940	0.796	**0.998**	**1**	0.850	0.682	0.710	0.666	0.754	−0.770
VIT E	0.883	0.502	0.878	0.850	**1**	**0.964**	**0.972**	**0.958**	**0.972**	−0.880
DPPH	0.766	0.314	0.722	0.682	**0.964**	**1**	**0.998**	**0.996**	**0.979**	−0.825
ABTS	0.803	0.374	0.750	0.710	**0.972**	**0.998**	**1**	**0.988**	**0.990**	−0.809
TAC	0.727	0.253	0.705	0.666	**0.958**	**0.996**	**0.988**	**1**	**0.958**	−0.859
PRO	0.867	0.488	0.792	0.754	**0.972**	**0.979**	**0.990**	**0.958**	**1**	−0.760
CARB	−0.649	−0.232	−0.783	−0.770	−0.880	−0.825	−0.809	−0.859	−0.760	**1**

PCA analysis confirmed this assertion, and evidenced that TP, TF, VIT C, and VIT A were mainly correlated with PSD (Figure 4). No correlation between SAFD and IC was evidenced. Single phenolic acids correlated only with PSD (Figure 5), whereas single flavonoids were mainly correlated with PSD and in part with CFD. Rutin, luteolin, delphinidin, and apigenin were the single flavonoids in the highest quantities and correlated with CFD (Figure 6).

Procyanidins B1 and B2 were discovered to inhibit human colorectal adenocarcinoma and to improve the survival of chronic disease patients by reducing the complications of cardiovascular disease and metabolic syndrome, improving the overall quality of life [42]. Additionally, punicalagin is a flavonoid with proven antioxidant, hepatoprotective, anti-atherosclerotic, and antitumoral activity [43]. An ethical study was performed in 50 subjects (25 treated with supplements and 25 with placebo) to identify clinical features induced by 25 mg dried pomegranate (*Punica granatum*) fruit extract (which in turn contained 3.75 mg procyanidins) and 8.75 mg punicalagin–ellagic acid. Results evidenced after 60 days of treatment that the values for systemic oxidative stress, plasmatic antioxidant capacity, and skin antioxidant power increased significantly [44]. Other authors evidenced that the daily intake of pomegranate juice, rich in flavonoids and phenols, decreased the susceptibility of low-density lipoproteins (LDLs) to aggregate, and in cultured human coronary artery endothelial cells exposed to high shear stress, it down-regulated the expression of redox-sensitive genes and increased the functioning of blood endothelial cells [7]. Considering that peels contain many more phenols and flavonoids than juice, as already demonstrated by the previous study of Derakhshana et al. [45] and Russo et al. [46] carried out on different cultivars, it is possible to conclude that pomegranate peels represent a resource comparable to the fruit, if not better, that can be used as a source of bio-compounds with high added value in the nutraceutical field to formulate new supplements with beneficial effects on human health.

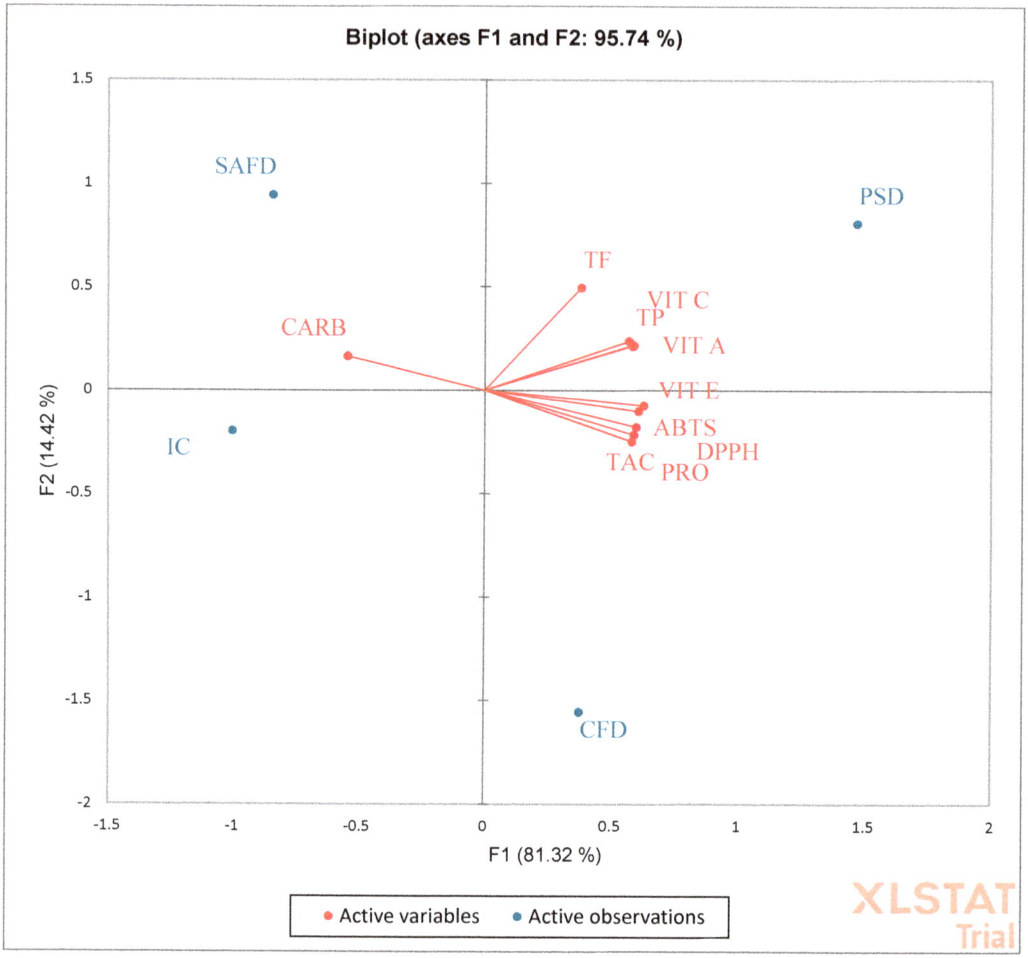

Figure 4. Total phenols (TP), total flavonoids (TF), proteins (PRO), vitamin C (VIT C), vitamin A (VIT A), vitamin E (VIT E), total carbohydrates (CARB), soluble proteins, 2,2-diphenyl-1-picryl-hydrazyl-hydrate (DPPH), total antioxidant capacity (TAC), and 2,2′-azino-bis-3-ethylbenzothiazoline-6-sulfonic acid (ABTS) contained in peels of different pomegranate cultivars dried differently. PSD (spray-dried Wonderful peel, experimental data); CFD (Calabrian Wonderful peel freeze-dried, experimental data); SAFD (South African Wonderful peel freeze-dried, literature data); IC (Indian cultivar peel freeze-dried, literature data).

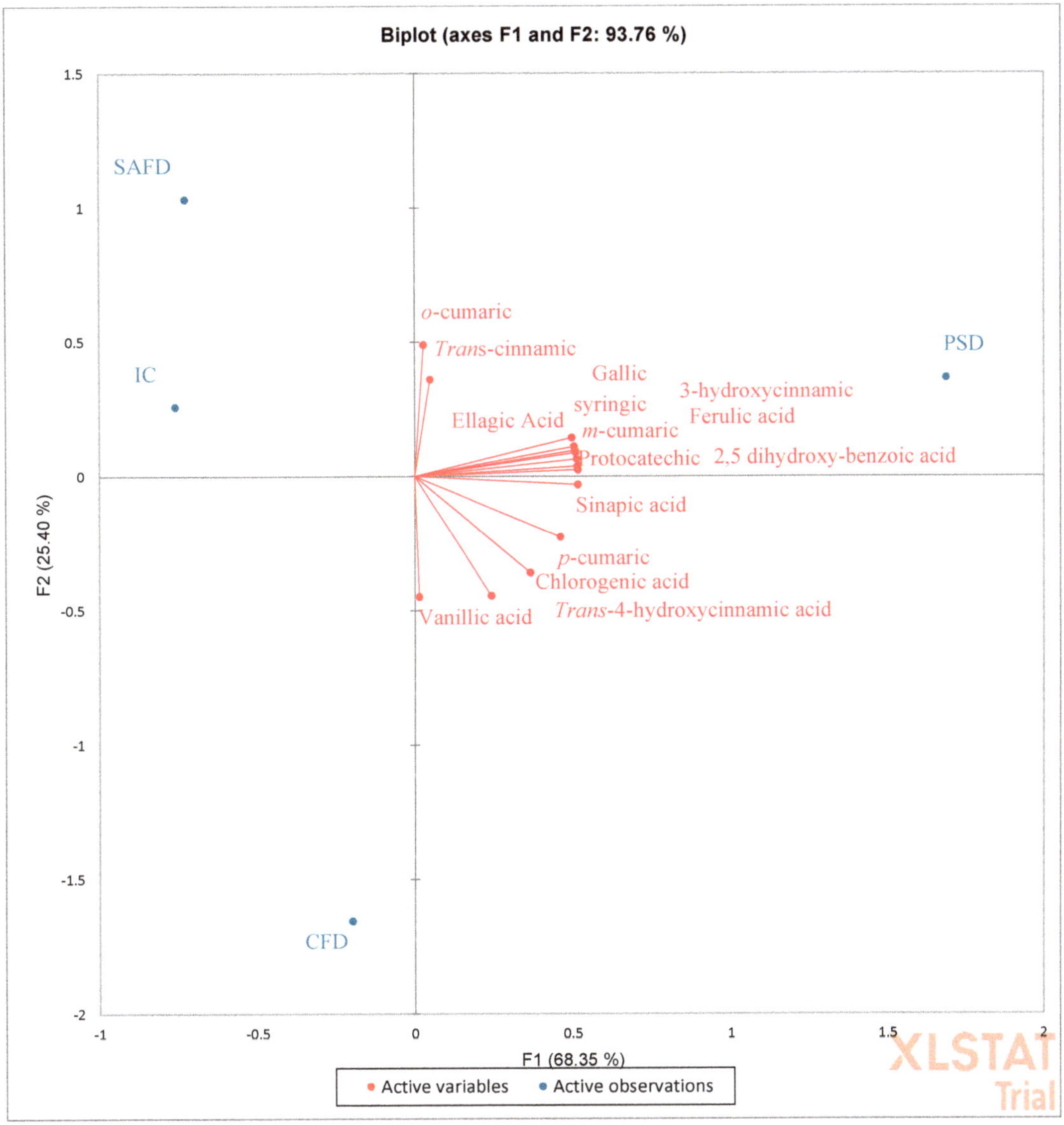

Figure 5. PCA (principal component analysis) diagram of single phenolic acids contained in the peels of different pomegranate cultivars dried differently. PSD (spray-dried Wonderful peel, experimental data); CFD (Calabrian Wonderful peel freeze-dried, experimental data); SAFD (South African Wonderful peel freeze-dried, literature data); IC (Indian cultivar peel freeze-dried, literature data).

Figure 6. PCA (principal component analysis) diagram of single flavonoids contained in the peels of different pomegranate cultivars dried differently. PSD (spray-dried Wonderful peel, experimental data); CFD (Calabrian Wonderful peel freeze-dried, experimental data); SAFD (South African Wonderful peel freeze-dried, literature data); IC (Indian cultivar peel freeze-dried, literature data).

4. Conclusions

In short, the data evidenced that pomegranate peel is a valuable raw material rich in bioactive compounds. The amount and the type of bioactive compounds depends on the cultivars, but more so on the area where the plant grows.

Same cultivars grown in different conditions can have different amounts and varieties of phenolic compounds and diverse antioxidant activities, evidencing that the pedoclimatic variables drive the metabolism of the cultivars, increasing or decreasing specific secondary metabolites implicated in physiological adjustment of plants to adapt to stress conditions or changing conditions induced by climate. In addition, the spray-drying system developed by Gioia Succhi is innovative because it is able to perfectly concentrate and conserve the bioactive compounds with beneficial effects on human health in respect to the freeze-drying system, which has been already demonstrated by numerous studies to be much more relevant than the oven-drying procedure (40–60 °C). According to the achieved results, the high antioxidant capacity of pomegranate peel, exalted by the innovative method of drying, highlights its use as a supplement to preserve human health from various diseases. The innovative spray-drying method appears to be advantageous from an economic point of view, as it is able to condense the bioactive compounds considerably, providing a concentrated peel powder ready to be used in the nutraceutical field.

Author Contributions: F.M.: formal analysis; B.P.: methodology, formal analysis; F.C.: formal analysis; A.M. (Angela Maffia): statistical analysis; C.M.: formal analysis; A.M. (Adele Muscolo): project administration, funding acquisition, writing—original draft, writing—review and editing. All authors have read and agreed to the published version of the manuscript.

Funding: This research received no external funding.

Institutional Review Board Statement: Not applicable.

Informed Consent Statement: Not applicable.

Conflicts of Interest: The authors declare no conflict of interest. The funders had no role in the design of the study; in the collection, analyses, or interpretation of data; in the writing of the manuscript; or in the decision to publish the results.

Sample Availability: Samples of the compounds are available from the authors.

References

1. Sarkhosh, A.; Zamani, Z.; Fatahi, R.; Ebadi, A. RAPD markers reveal polymorphism among some Iranian pomegranate (*Punica granatum* L.) genotypes. *Sci. Hortic.* **2006**, *111*, 24–29. [CrossRef]
2. Derakhshana, Z.; Ferrante, M.; Tadi, M.; Ansari, F.; Heydarif, A.; Zam, W.; Khaddour, A. Anti-virulence effects of aqueous pomegranate peel extract on *E. coli* urinary tract infection. *Progr. Nutr.* **2019**, *19*, 98–104. [CrossRef]
3. Mastrogiovanni, F.; Mukhopadhya, A.; Lacetera, N.; Ryan, M.T.; Romani, A.; Bernini, R.; Sweeney, T. Anti-Inflammatory Effects of Pomegranate Peel Extracts on In Vitro Human Intestinal Caco-2 Cells and Ex Vivo Porcine Colonic Tissue Explants. *Nutrients* **2019**, *11*, 548. [CrossRef] [PubMed]
4. Alexandre, E.M.C.; Silva, S.; Santos, S.A.O.; Silvestre, A.J.D.; Duarte, M.F.; Saraiva, J.A.; Pintado, M. Antimicrobial activity of pomegranate peel extracts performed by high pressure and enzymatic assisted extraction. *Food Res. Int.* **2019**, *115*, 167–176. [CrossRef]
5. Usha, T.; Middha, S.K.; Sidhalinghamurthy, K.R. Pomegranate Peel and Its Anticancer Activity: A Mechanism-Based Review. In *Plant-Derived Bioactives*; Swamy, M., Ed.; Springer: Singapore, 2020; pp. 223–250. [CrossRef]
6. Aviram, M.; Rosenblat, M. Pomegranate Protection against Cardiovascular Diseases. Evidence-Based Complement. *Altern. Med.* **2012**, *2012*, 382763. [CrossRef]
7. Wang, D.; Özen, C.; Abu-Reidah, I.M.; Chigurupati, S.; Patra, J.K.; Horbańczuk, J.O.; Jozwik, A.; Tzvetkov, N.T.; Uhrin, P.; Atanasov, A.G. Vasculoprotective Effects of Pomegranate (*Punica granatum* L.). *Front. Pharmacol.* **2018**, *9*, 544. [CrossRef]
8. Zuraini, N.Z.A.; Sekar, M.; Wu, Y.S.; Gan, S.H.; Bonam, S.R.; Rani, N.N.I.M.; Begum, M.Y.; Lum, P.T.; Subramaniyan, V.; Fuloria, N.K.; et al. Promising Nutritional Fruits Against Cardiovascular Diseases: An Overview of Experimental Evidence and Understanding Their Mechanisms of Action. *Vasc. Health Risk Manag.* **2021**, *17*, 739–769. [CrossRef] [PubMed]
9. Tito, A.; Colantuono, A.; Pirone, L.; Pedone, E.; Intartaglia, D.; Giamundo, G.; Conte, I.; Vitaglione, P.; Apone, F. Pomegranate Peel Extract as an Inhibitor of SARS-CoV-2 Spike Binding to Human ACE2 Receptor (in vitro): A Promising Source of Novel Antiviral Drugs. *Front. Chem.* **2021**, *9*, 638187. [CrossRef]
10. Czieczor, L.; Bentkamp, C.; Damerow, L.; Blanke, M. Non-invasive determination of the quality of pomegranate fruit. *Postharvest Biol. Technol.* **2018**, *136*, 74–79. [CrossRef]
11. Kumar, N.; Neeraj, D. Study on physico-chemical and antioxidant properties of pomegranate peel. *J. Pharmacogn. Phytochem.* **2018**, *7*, 2141–2147.

12. Chang, S.K.; Alasalvar, C.; Shahidi, F. Superfruits: Phytochemicals, antioxidant efficacies, and health effects—A comprehensive review. *Crit. Rev. Food Sci. Nutr.* **2019**, *59*, 1580–1604. [CrossRef] [PubMed]
13. Karimi, M.; Sadeghi, R.; Kokini, J. Pomegranate as a promising opportunity in medicine and nanotechnology. *Trends Food Sci. Technol.* **2017**, *69*, 59–73. [CrossRef]
14. Hmid, I.; Elothmani, D.; Hanine, H.; Oukabli, A.; Mehinagic, E. Comparative study of phenolic compounds and their antioxidant attributes of eighteen pomegranate (*Punica granatum* L.) cultivars grown in Morocco. *Arab. J. Chem.* **2017**, *10*, S2675–S2684. [CrossRef]
15. Bassiri-Jahromi, S.; Doostkam, A. Comparative evaluation of bioactive compounds of various cultivars of pomegranate (*Punica granatum*) in different world regions. *AIMS Agric. Food* **2019**, *4*, 41–55. [CrossRef]
16. Lee, C.J.; Chen, L.G.; Liang, W.L.; Wang, C.C. Anti-inflammatory effects of *Punica granatum* Linne in vitro and in vivo. *Food Chem.* **2010**, *118*, 315–322. [CrossRef]
17. Panichayupakaranant, P.; Tewtrakul, S.; Yuenyongsawad, S. Antibacterial, anti-inflammatory and anti-allergic activities of standardised pomegranate rind extract. *Food Chem.* **2010**, *123*, 400–403. [CrossRef]
18. Fischer, U.A.; Carle, R.; Kammerer, D.R. Identification and quantification of phenolic compounds from pomegranate (*Punica granatum* L.) peel, mesocarp, aril and differently produced juices by HPLC-DAD–ESI/MSn. *Food Chem.* **2011**, *127*, 807–821. [CrossRef] [PubMed]
19. Li, Y.; Guo, C.; Yang, J.; Wei, J.; Xu, J.; Cheng, S. Evaluation of antioxidant properties of pomegranate peel extract in comparison with pomegranate pulp extract. *Food Chem.* **2006**, *96*, 254–260. [CrossRef]
20. Hossin, F.L.A. Effect of Pomegranate (*Punica granatum*) Peels and It's Extract on Obese Hypercholesterolemic Rats. *Pak. J. Nutr.* **2009**, *8*, 1251–1257. [CrossRef]
21. Lin, L.-T.; Chen, T.-Y.; Lin, S.-C.; Chung, C.-Y.; Lin, T.-C.; Wang, G.-H.; Anderson, R.; Lin, C.-C.; Richardson, C.D. Broad-spectrum antiviral activity of chebulagic acid and punicalagin against viruses that use glycosaminoglycans for entry. *BMC Microbiol.* **2013**, *13*, 187. [CrossRef]
22. Mphahlele, R.R.; Fawole, O.A.; Makunga, N.P.; Opara, U.L. Effect of drying on the bioactive compounds, antioxidant, antibacterial and antityrosinase activities of pomegranate peel. *BMC Complement. Altern. Med.* **2016**, *16*, 143. [CrossRef] [PubMed]
23. Bhat, M.; Thakur, N.S.; Jindal, N. Studies on the effect of drying methods and packaging on quality and shelf life of dried wild pomegranate arils. *Asian J. Dairy Food Res.* **2014**, *33*, 18–24. [CrossRef]
24. Kumar, N.; Pratibha; Neeraj; Sami, R.; Khojah, E.; Aljahani, A.H.; Al-Mushhin, A.A.M. Effects of drying methods and solvent extraction on quantification of major bioactive compounds in pomegranate peel waste using HPLC. *Sci. Rep.* **2022**, *12*, 8000. [CrossRef] [PubMed]
25. Muscolo, A.; Papalia, T.; Settineri, G.; Mallamaci, C.; Panuccio, M.R. Sulfur bentonite-organic-based fertilizers as tool for improving bio-compounds with antioxidant activities in red onion. *J. Sci. Food Agric.* **2019**, *100*, 785–793. [CrossRef] [PubMed]
26. Aremu, S.O.; Nweze, C.O. Determination of vitamin A content from selected Nigerian fruits using spectrophotometric method. *Bangladesh J. Sci. Ind. Res.* **2017**, *52*, 153–158. [CrossRef]
27. Davies, S.H.; Masten, S.J. Spectrophotometric method for ascorbic acid using dichlorophenolindophenol: Elimination of the interference due to iron. *Anal. Chim. Acta* **1991**, *248*, 225–227. [CrossRef]
28. Prieto, P.; Pineda, M.; Aguilar, M. Spectrophotometric Quantitation of Antioxidant Capacity through the Formation of a Phosphomolybdenum Complex: Specific Application to the Determination of Vitamin E. *Anal. Biochem.* **1999**, *269*, 337–341. [CrossRef] [PubMed]
29. Akula, R.; Ravishankar, G.A. Influence of abiotic stress signals on secondary metabolites in plants. *Plant Signal. Behav.* **2011**, *6*, 1720–1731. [CrossRef]
30. Vaneková, Z.; Vanek, M.; Škvarenina, J.; Nagy, M. The Influence of Local Habitat and Microclimate on the Levels of Secondary Metabolites in Slovak Bilberry (*Vaccinium myrtillus* L.) Fruits. *Plants* **2020**, *9*, 436. [CrossRef]
31. Fialho, L.; Ramôa, S.; Parenzan, S.; Guerreiro, I.; Catronga, H.; Soldado, D.; Guerreiro, O.; García, V.G.; e Silva, P.O.; Jerónimo, E. Effect of regulated deficit irrigation on pomegranate fruit quality at harvest and during cold storage. *Agric. Water Manag.* **2021**, *251*, 106869. [CrossRef]
32. Kosová, K.; Vítámvás, P.; Urban, M.O.; Prášil, I.T.; Renaut, J. Plant Abiotic Stress Proteomics: The Major Factors Determining Alterations in Cellular Proteome. *Front. Plant Sci.* **2018**, *9*, 122. [CrossRef]
33. Laspina, N.V.; Groppa, M.D.; Tomaro, M.D.; Benavides, M.P. Nitric oxide protects sunflower leaves against Cd-induced oxidative stress. *Plant Sci.* **2005**, *169*, 323–330. [CrossRef]
34. Muñoz, P.; Munné-Bosch, S. Vitamin E in Plants: Biosynthesis, Transport, and Function. *Trends Plant Sci.* **2019**, *24*, 1040–1051. [CrossRef] [PubMed]
35. Vattem, D.; Shetty, K. Biological functionality of ellagic acid: A review. *J. Food Biochem.* **2005**, *29*, 234–266. [CrossRef]
36. Jordão, J.B.R.; Porto, H.K.P.; Lopes, F.M.; Batista, A.C.; Rocha, M.L. Protective Effects of Ellagic Acid on Cardiovascular Injuries Caused by Hypertension in Rats. *Planta Med.* **2017**, *83*, 830–836. [CrossRef]
37. Pei, S.; Zhao, H.; Chen, L.; He, X.; Hua, Q.; Meng, X.; Shi, R.; Zhang, J.; Zhang, H.; Liu, R.; et al. Preventive Effect of Ellagic Acid on Cardiac Dysfunction in Diabetic Mice through Regulating DNA Hydroxymethylation. *J. Agric. Food Chem.* **2022**, *70*, 1902–1910. [CrossRef]

38. Badhani, B.; Sharma, N.; Kakkar, R. Gallic acid: A versatile antioxidant with promising therapeutic and industrial applications. *RSC Adv.* **2015**, *5*, 27540–27557. [CrossRef]
39. Kakkar, S.; Bais, S. A Review on Protocatechuic Acid and Its Pharmacological Potential. *Int. Sch. Res. Not.* **2014**, *2014*, 952943. [CrossRef]
40. Sarker, U.; Oba, S. Polyphenol and flavonoid profiles and radical scavenging activity in leafy vegetable Amaranthus gangeticus. *BMC Plant Biol.* **2020**, *20*, 499. [CrossRef]
41. Vázquez-Ruiz, Z.; Toledo, E.; Vitelli-Storelli, F.; Goni, L.; de la O, V.; Bes-Rastrollo, M.; Martínez-González, M. Effect of Dietary Phenolic Compounds on Incidence of Cardiovascular Disease in the SUN Project; 10 Years of Follow-Up. *Antioxidants* **2022**, *11*, 783. [CrossRef]
42. Dasiman, R.; Md Nor, N.; Eshak, Z.; Mutalip, S.S.M.; Suwandi, N.R.; Bidin, H. A Review of Procyanidin: Updates on Current Bioactivities and Potential Health Benefits. *Biointerface Res. Appl. Chem.* **2022**, *12*, 5918–5940. [CrossRef]
43. Venusova, E.; Kolesarova, A.; Horky, P.; Slama, P. Physiological and Immune Functions of Punicalagin. *Nutrients* **2021**, *13*, 2150. [CrossRef] [PubMed]
44. Buonocore, D.; Nobile, V.; Cestone, E.; Santin, G.; Bottone, M.G.; Marzatico, F.; Lazzeretti, A.; Tocabens, P. Resveratrol-procyanidin blend: Nutraceutical and antiaging efficacy evaluated in a placebo-controlled, double-blind study. *Clin. Cosmet. Investig. Dermatol.* **2012**, *5*, 159–165. [CrossRef] [PubMed]
45. Derakhshan, Z.; Ferrante, M.; Tadi, M.; Ansari, F.; Heydari, A.; Hosseini, M.S.; Conti, G.O.; Sadrabad, E.K. Antioxidant activity and total phenolic content of ethanolic extract of pomegranate peels, juice and seeds. *Food Chem. Toxicol.* **2018**, *114*, 108–111. [CrossRef]
46. Russo, M.; Fanali, C.; Tripodo, G.; Dugo, P.; Muleo, R.; Dugo, L.; De Gara, L.; Mondello, L. Analysis of phenolic compounds in different parts of pomegranate (*Punica granatum*) fruit by HPLC-PDA-ESI/MS and evaluation of their antioxidant activity: Application to different Italian varieties. *Anal. Bioanal. Chem.* **2018**, *410*, 3507–3520. [CrossRef]

molecules

MDPI

Article

Wistin Exerts an Anti-Inflammatory Effect via Nuclear Factor-κB and p38 Signaling Pathways in Lipopolysaccharide-Stimulated RAW264.7 Cells

Jangeun An [1], Gyoungah Ryu [1], Seong-Ah Shin [1], Huiji Kim [1], Mi-Jeong Ahn [1], Jun Hyuck Lee [2,3] and Chang Sup Lee [1,*]

1 College of Pharmacy and Research Institute of Pharmaceutical Sciences, Gyeongsang National University, Jinju 52828, Korea
2 Research Unit of Cryogenic Novel Material, Korea Polar Research Institute, Incheon 21990, Korea
3 Department of Polar Sciences, University of Science and Technology, Incheon 21990, Korea
* Correspondence: changsup@gnu.ac.kr; Tel.: +82-55-772-2432

Abstract: Inflammation is an immune response to cellular damage caused by various stimuli (internal or external) and is essential to human health. However, excessive inflammatory responses may be detrimental to the host. Considering that the existing drugs for the treatment of inflammatory diseases have various side effects, such as allergic reactions, stomach ulcers, and cardiovascular problems, there is a need for research on new anti-inflammatory agents with low toxicity and fewer side effects. As 4′,6-dimethoxyisoflavone-7-*O*-β-d-glucopyranoside (wistin) is a phytochemical that belongs to an isoflavonoid family, we investigated whether wistin could potentially serve as a novel anti-inflammatory agent. In this study, we found that wistin significantly reduced the production of nitric oxide and intracellular reactive oxygen species in lipopolysaccharide-stimulated RAW 264.7 cells. Moreover, wistin reduced the mRNA levels of pro-inflammatory enzymes (inducible nitric oxide synthase (iNOS) and cyclooxygenase (COX-2)) and cytokines (interleukin (IL)-1β and IL-6) and significantly reduced the protein expression of pro-inflammatory enzymes (iNOS and COX-2). Furthermore, wistin reduced the activation of the nuclear factor-κB and p38 signaling pathways. Together, these results suggest that wistin is a prospective candidate for the development of anti-inflammatory drugs.

Keywords: inflammation; phytochemical; wistin

Citation: An, J.; Ryu, G.; Shin, S.-A.; Kim, H.; Ahn, M.-J.; Lee, J.H.; Lee, C.S. Wistin Exerts an Anti-Inflammatory Effect via Nuclear Factor-κB and p38 Signaling Pathways in Lipopolysaccharide-Stimulated RAW264.7 Cells. *Molecules* **2022**, *27*, 5719. https://doi.org/10.3390/molecules27175719

Academic Editor: Nour Eddine Es-Safi

Received: 10 August 2022
Accepted: 2 September 2022
Published: 5 September 2022

1. Introduction

Inflammation is the body's immune response and defense mechanism against tissue damage caused by exposure to harmful or toxic external agents, infections, and physical injuries [1]. Inflammation resulting from infection or injury is associated with inflammatory responses such as immune cell recruitment and accumulation, release of inflammatory mediators, and changes in blood vessel permeability [2]. Generally, there are two types of inflammation: acute and chronic. Acute inflammation is a rapid process that repairs quickly to minimize damage and restore tissue homeostasis [3]. However, in chronic conditions, the inflammatory response continues, resulting in severe organ damage [4]. Chronic inflammation promotes the progression of several diseases, including cardiovascular disease, inflammatory bowel disease, rheumatoid arthritis and diabetes [5]. Crohn's disease (CD) is the most prevalent IBD syndrome treated using 6-mercaptopurine (6-MP) and its prodrug azathioprine (AZA) [6]. As immunosuppressive drugs, 6-MP and AZA inhibit inflammation by blocking the body's immune response; however, they may cause disease recurrence and side effects, such as hepatotoxicity [6]. Sulfasalazine, a medication for patients with rheumatoid arthritis, has immunomodulatory properties and suppresses pro-inflammatory cytokines; however, it can cause gastrointestinal side effects, such as headache, dizziness, rash, and bone marrow suppression [7]. Non-steroidal anti-inflammatory drugs (NSAIDs) are anti-inflammatory drugs used worldwide to treat inflammatory conditions [8]. Among NSAIDs, ibuprofen (IBU) is a non-selective inhibitor of both cyclooxygenase (COX)-1 and -2 isozymes [9]; however, it is associated with the risk of cardiovascular and gastrointestinal complications [10,11]. Naproxen is another NSAID used to treat osteoarthritis, migraine, and rheumatoid arthritis [10]. However, NSAIDs have several side effects, including gastrointestinal toxicity, cardiovascular risk, kidney damage, and hepatotoxicity [8]. Therefore, it is necessary to develop safer anti-inflammatory treatment strategies.

When the human body is infected by pathogens in damaged tissue, immune cells recognize pathogen-associated molecular patterns (PAMPs) and damage-associated molecular patterns (DAMPs) using pattern recognition receptors (PRRs) and promote inflammatory signaling pathways [12]. The Toll-like receptor (TLR) family is mainly expressed in immune cells as major PRRs that play key roles in the induction of innate immune responses and first-line defense [13,14]. Lipopolysaccharide (LPS) is a well-known inflammatory PAMP molecule that exists in the cell walls of gram-negative bacteria [13]. The TLR4–LPS interaction activates the nuclear factor-κB (NF-κB) and mitogen-activated protein kinase (MAPK) pathways [15,16]. TLR4-mediated modulation of myeloid differentiation factor 88 (MyD88) results in activation of NF-κB. Furthermore, NF-κB is regulated by the IκB and IκB kinase (IKK) complex [17], which causes IκB phosphorylation. The phosphorylation of the IκB protein results in the phosphorylation of p65, which translocates to the nucleus [18]. In addition, NF-κB is regulated by phosphatidylinositol 3-kinase (PI3K)-Akt [19]. The MAPK pathway includes extracellular signal-regulated kinase (ERK1/2), c-Jun N-terminal kinase (JNK), and p38, which result in further activation of the transcription factor activator protein 1 (AP-1) [16]. Activation of the NF-κB and MAPK pathways has been reported to increase the expression of inflammatory enzymes (inducible nitric oxide synthase (iNOS) and cyclooxygenase (COX-2)) and cytokines (interleukin (IL)-1β and IL-6) in immune cells [16,20]. Therefore, understanding the mechanism by which signaling pathways are controlled at the molecular level might help modulate the inflammatory response and develop new modulators of inflammation.

Phytochemicals extracted from plants have emerged as new agents for the treatment of chronic inflammatory diseases [21]. Current anti-inflammatory disease treatments (non-steroidal anti-inflammatory drugs and glucocorticoids) have many side effects, including tissue damage, cardiovascular problems, and liver complications. As plant-derived drugs have been reported to have fewer toxic effects, it has been suggested that they could be potential modulators of inflammation [22,23]. Human papillomavirus (HPV) infection can cause chronic inflammation due to the release of pro-inflammatory cytokines [24,25]. Sinecatechin extracted from green tea has been used as a medication to treat HPV infections [24,25]. Silymarin extracted from the seeds of milk thistle (*Silybum marianum*) has anti-inflammatory activity; it inhibits NF-κB activity and improves liver function in patients with hepatitis B (HBV) [26–28]. Eupatilin, a plant-derived drug extracted from *Artemisia asiatica Nakai*, is used to treat gastritis by mediating anti-inflammatory effects and promoting the regeneration of offended mucosa [29]. Among various isoflavones, genistein is known for its antioxidant and anti-inflammatory properties. Genistein prevents endothelial inflammatory damage by inhibiting the NF-κB pathway, which mediates the transcription of proinflammatory cytokines [30]. Although investigations are required to confirm its safety and efficacy, low- (5–15 mg/kg/day) and high-dose (160 mg/kg/day) genistein treatments for patients with mucopolysaccharidosis (MPS) III reportedly have no major side effects [31]. Another isoflavone, daidzein, inhibits the JNK, PARP, and NF-κB signaling pathways to express pro-inflammatory cytokines [32]. In addition, plant polyphenols, particularly flavonoids, have exhibited anti-inflammatory activity both in vitro and in vivo [32]. Compound 4′,6-dimethoxyisooflavone-7-*O*-β-d-glucopyranoside (wistin), belonging to the isoflavone family, is an agonist of PPARγ and PPARα in adipocytes and hepatocytes, respectively [33,34]. However, its anti-inflammatory role has not been investigated. Therefore, in this study, we aimed to identify its anti-inflammatory effects and elucidate the molecular mechanisms underlying its anti-inflammatory effects in LPS-stimulated RAW264.7 cells.

2. Results

2.1. Effects of Wistin on Cell Viability in LPS-Induced RAW 264.7 Cells

Because isoflavones have been reported to be cytotoxic at high doses, we attempted to confirm whether wistin is cytotoxic at high concentrations [35]. The cytotoxicity of wistin (Figure 1a) on the viability of RAW 264.7 cells was evaluated using the 3-(4, 5-dimethylthiazolyl-2)-2, 5-diphenyltetrazolium bromide) (MTT) assay. Cotreatment of wistin (50 μM, 100 μM, and 150 μM) and LPS (0.1 μg/mL) did not show cytotoxicity in RAW 264.7 at 24 h (Figure 1b).

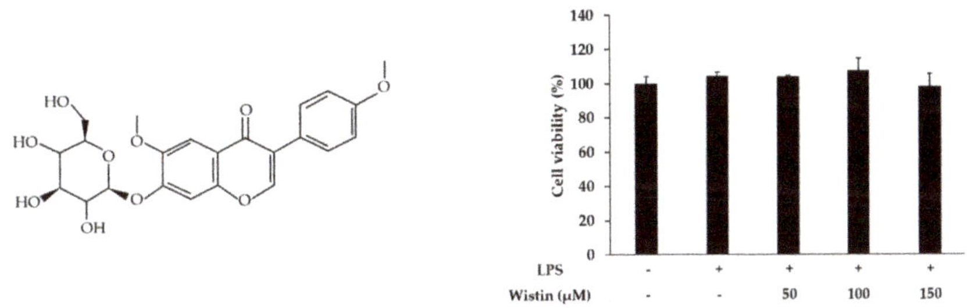

(a) (b)

Figure 1. The effects of wistin on cell viability in LPS-induced RAW 264.7. (**a**) The chemical structure of wistin. (**b**) The cells were treated with the indicated concentrations of wistin (0, 50, 100, 150 μM) for 30 min prior to treatment with LPS (0.1 μg/mL) for 24 h, and then cell viability was examined using MTT assay. The data are presented as the means ± SD; $n = 3$.

2.2. Effects of Wistin on the Production of Pro-Inflammatory Mediators in LPS-Induced RAW 264.7 Cells

Prolonged inflammatory processes increase the production of nitric oxide (NO) and reactive oxygen species (ROS), leading to tissue dysfunction [36]. ROS can regulate pro-inflammatory gene expression, and NO is an important pro-inflammatory mediator in inflammatory signaling [37]. To examine the anti-inflammatory effects of wistin, we investigated the production of NO and ROS. Wistin showed a significant decrease in LPS-induced NO production in a dose-dependent manner compared to that in the control group (Figure 2a). In addition, dose-dependent inhibition of intracellular ROS generation by wistin was identified using a microplate reader, fluorescence-activated cell sorting (FACS), and fluorescence microscopy (Figure 2b–e). These data suggest that wistin can reduce the production of pro-inflammatory mediators (NO and ROS) induced by LPS.

2.3. Effects of Wistin on Pro-Inflammatory Enzymes and Cytokine Gene Expression in LPS-Induced RAW 264.7 Cells

Next, we investigated the involvement of wistin in the modulation of mRNA levels of inflammatory enzymes (iNOS and COX-2) and pro-inflammatory cytokines (IL-1β and IL-6) at indicated time points after LPS treatment [38]. Wistin significantly decreased the mRNA expression levels of inflammatory enzymes (iNOS and COX-2) (Figure 3a,b). The expression levels of pro-inflammatory cytokines (IL-1β and IL-6) were not detected at 0 h after LPS treatment and were the highest at 12 h after LPS treatment. In addition, wistin-treated groups showed a significant decrease in the mRNA expression levels of pro-inflammatory cytokines at each time point compared to LPS treatment groups. Therefore, these results suggest that wistin could modulate pro-inflammatory enzymes and cytokine gene expression induced by LPS.

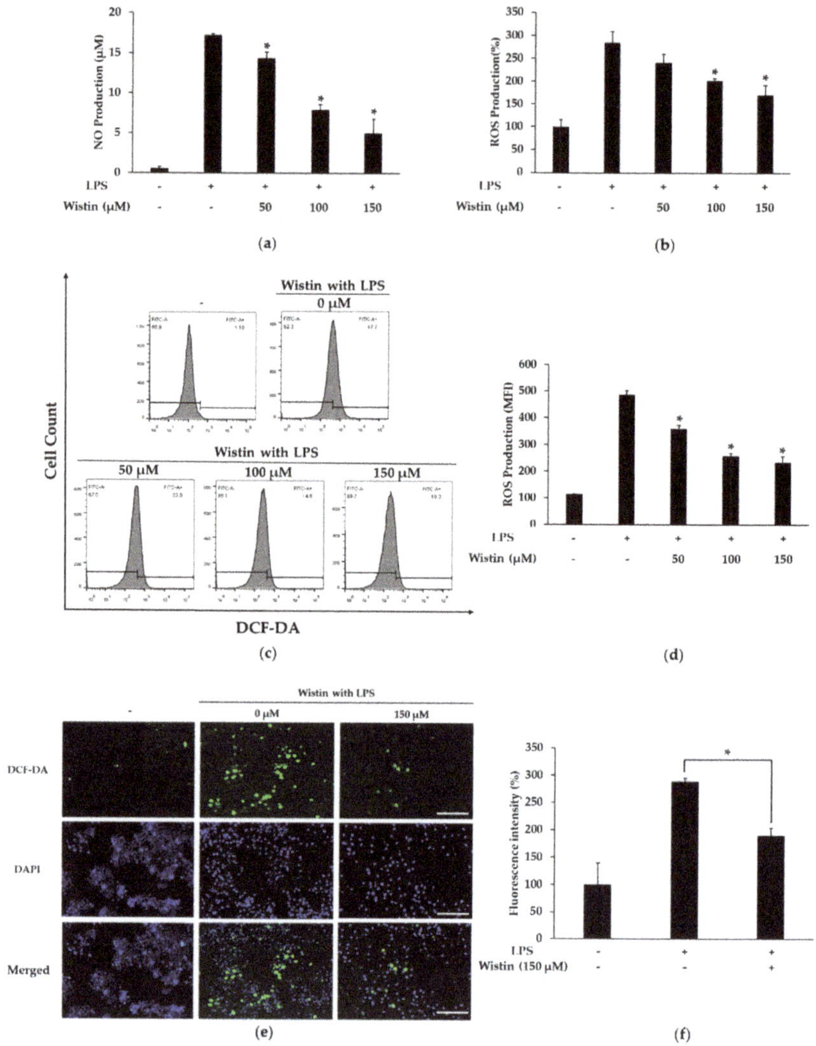

Figure 2. The effect of wistin on NO and ROS production in LPS-stimulated RAW264.7. The cells were treated with the indicated concentrations of wistin for 30 min prior to treatment with LPS (0.1 μg/mL) for 24 h. (**a**) The NO production was measured in treated wistin in the culture supernatant. ROS production was measured on a microplate reader (**b**), FACS (**c**,**d**), and fluorescence microscope (**e**). (**f**) Quantitative analysis of ROS production using ImageJ software. The scale bar represents 100 μm. * $p < 0.05$ compared with the LPS-treated group. MFI: mean fluorescence intensity. The data are presented as the means ± SD; $n = 3$.

2.4. Effects of Wistin on the Protein Expression Level of Pro-Inflammatory Enzymes in LPS-Induced RAW 264.7 Cells

The expression of iNOS and COX-2 regulates key inflammatory mediators [39]. Therefore, we investigated whether wistin exhibits anti-inflammatory effects by inhibiting iNOS and COX-2 protein expression. As shown in Figure 4a,b, wistin significantly reduced the protein expression levels of iNOS and COX-2 compared to those in the LPS group. Therefore, these results suggest that wistin could also inhibit the expression of pro-inflammatory enzymes at the protein level.

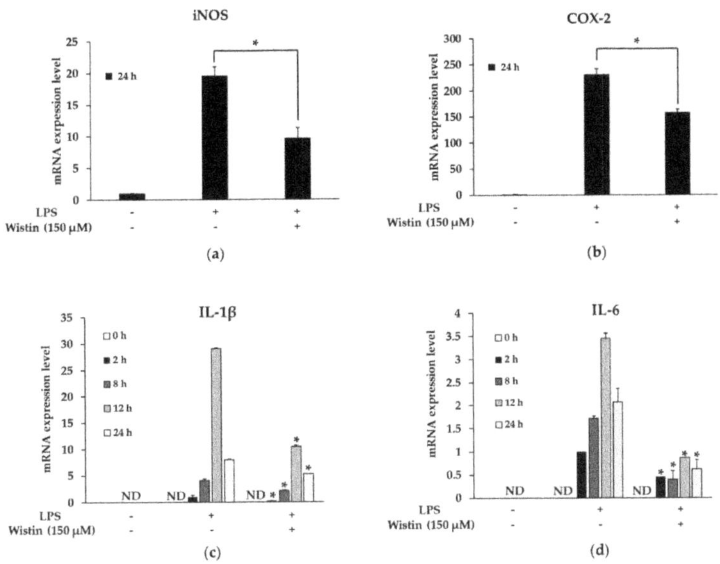

Figure 3. The effects of wistin on mRNA expression level in LPS-induced RAW 264.7 cells. The cells were treated with the indicated concentrations of wistin for 30 min prior to treatment with LPS (0.1 μg/mL). Then, LPS-induced (**a**) iNOS, (**b**) COX-2, (**c**) IL-1β, and (**d**) IL-6 mRNA level were measured using a quantitative reverse transcription polymerase chain reaction (qRT-PCR). * $p < 0.05$ compared with the LPS-treated group. ND: not detected. The data are presented as the means ± SD; $n = 3$.

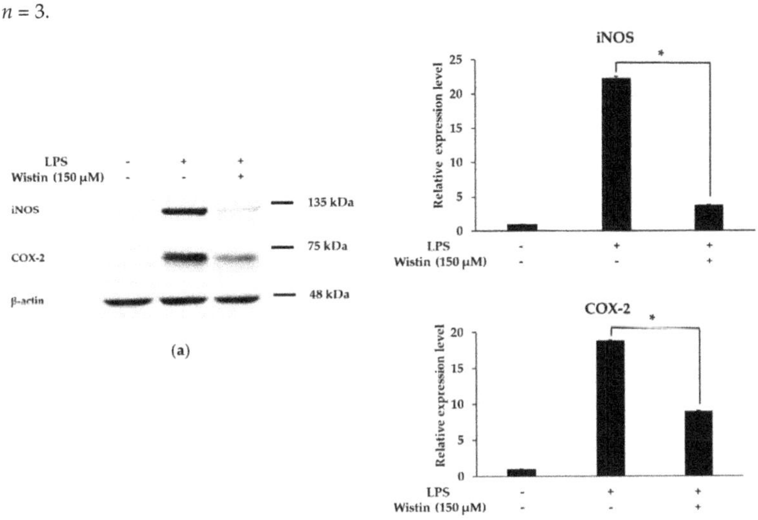

Figure 4. The effects of wistin on protein expression of iNOS and COX-2 in LPS-induced RAW 264.7 cells. The cells were treated with the indicated concentrations of wistin for 30 min prior to treatment with LPS (0.1 μg/mL) for 24 h. (**a**) The protein levels of pro-inflammatory enzymes iNOS and COX-2 were determined by Western blot. (**b**) Quantitative analysis of the iNOS/β-actin and COX-2/β-actin using image J. * $p < 0.05$ compared with the LPS-treated group. The data are presented as the means ± SD; $n = 2$.

2.5. Effects of Wistin on the Activation of AKT/NF-κB Pathway in LPS-Induced RAW 264.7 Cells

The NF-κB pathway regulates the expression and production of pro-inflammatory enzymes and cytokines [40]. Therefore, we examined the phosphorylation level of LPS-induced AKT and NF-κB (p65 subunit) following wistin (150 μM) treatment for 2 h. Wistin significantly reduced the phosphorylation level of AKT and p65 compared to that in the LPS group (Figure 5a–d). Furthermore, in the absence of LPS stimulation, p65 (red in the merged image) was present in the cytoplasm. Upon LPS stimulation, p65 (pink in merged images) was translocated to the nucleus, whereas wistin treatment reduced the nuclear translocation of p65 (red and pink in merged images) (Figure 5e). These results suggest that wistin suppresses the AKT/NF-κB signaling pathway.

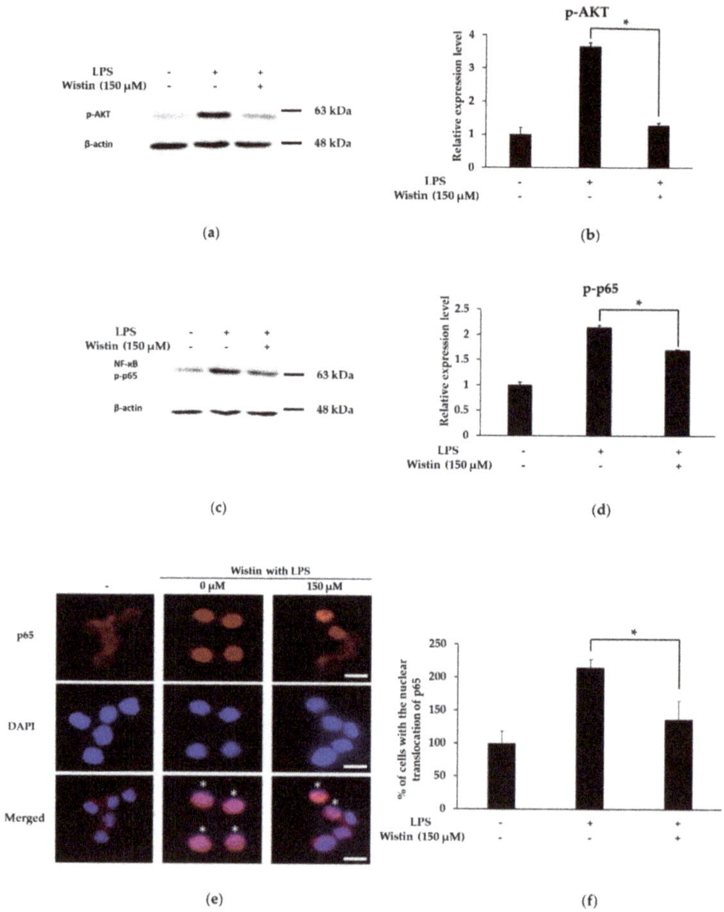

Figure 5. The effects of wistin on p-AKT, NF-κB (p-p65 subunit) in LPS-induced RAW 264.7 cells. The cells were treated with the indicated concentrations of wistin for 30 min prior to treatment with LPS (0.1 μg/mL) for 2 h. Then, the phosphorylation of AKT (**a**) and p65 (**c**) were measured using a Western blot. Quantitative analysis of the p-AKT/β-actin (**b**) and p-p65/β-actin (**d**) using image J. (**e**) The effects of wistin on the nuclear translocation of p65 (red) using fluorescence microscopy. The merged images were acquired by overlaying two channels (p65 (red) and DAPI (blue)). * Indicates translocation of p65 from the cytoplasm to the nucleus. The scale bar represents 10 μm. (**f**) Quantitative analysis of the nuclear translocation of p65 using ImageJ software. * $p < 0.05$ compared with the LPS-treated group. The data are presented as the means ± SD; $n = 3$.

2.6. Effects of Wistin on the Activation of MAPK Pathway in LPS-Induced RAW 264.7 Cells

The MAPK pathway is known for its role in the modulation of inflammatory responses [41]. We investigated the role of wistin in the phosphorylation of three MAPKs (p38, c-Jun N-terminal kinase (JNK), and extracellular signal-regulated kinase (ERK)). Phosphorylation of p38 was decreased by wistin, but that of ERK and JNK was not affected compared to that in the LPS group (Figure 6). Therefore, these results suggest that wistin could work by negatively regulating the p38 MAPK pathway.

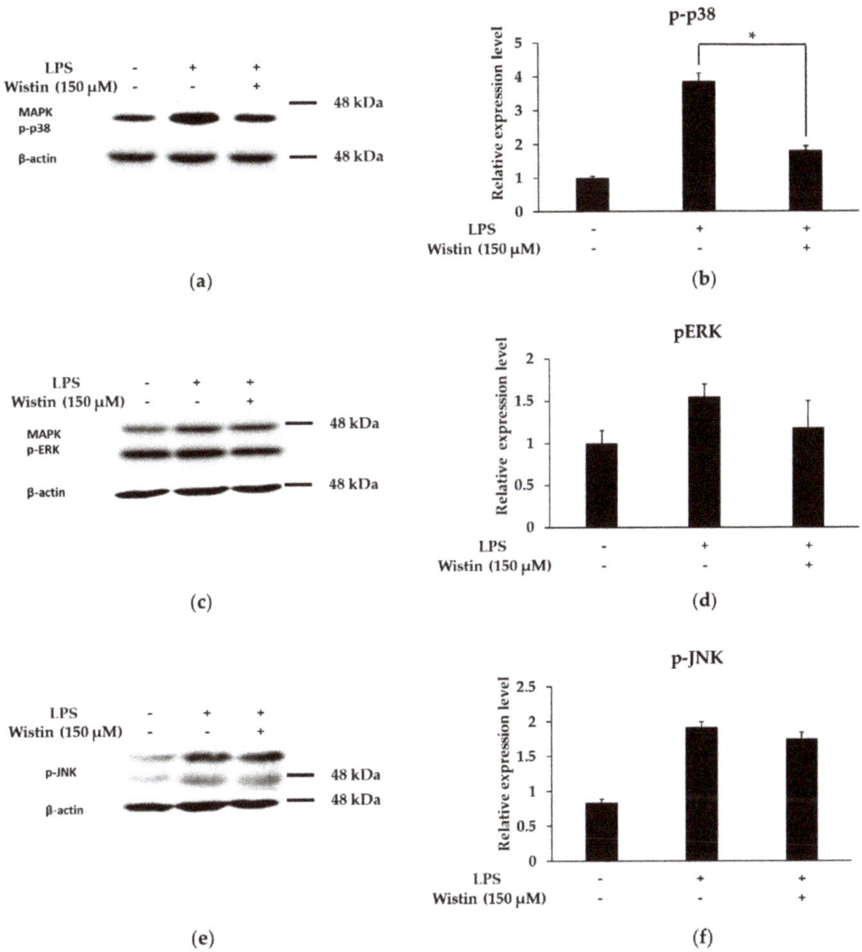

Figure 6. The effects of wistin on MAPK (p-p38, p-ERK, and p-JNK) in LPS-induced RAW 264.7cells. The cells were treated with the indicated concentrations of wistin for 30 min prior to treatment with LPS (0.1 μg/mL) for 2 h. Phosphorylation of p38 (**a**), ERK, (**c**) and JNK (**e**) was measured using a Western blot. Quantitative analysis of the p-p38/β-actin (**b**), p-ERK/β-actin (**d**), and p-JNK/β-actin (**f**) using image J. * $p < 0.05$ compared with the LPS-treated group. The data are presented as the means ± SD; $n = 2$.

3. Discussion

Inflammation is a protective response to harmful stimuli (PAMP and DAMP) by the immune system [42]. However, chronic inflammation can cause various diseases, such as cardiovascular disease, inflammatory bowel disease, rheumatoid arthritis, and diabetes [5]. Ibuprofen, naproxen, and other NSAIDs are conventional anti-inflammatory drugs used to treat rheumatism, osteoarthritis, arteriosclerosis, neuroinflammatory diseases, and other inflammatory diseases [8,10]. However, these drugs have several side effects, such as an increased risk of gastric mucosal injury, renal injury, and other medical complications [8,10]. Therefore, it is important to identify novel anti-inflammatory agents that can overcome the shortcomings of conventional anti-inflammatory drugs. Several plants have been used as folk medicines for the treatment and prevention of diseases [43]. Currently, sinecatechin extracted from green tea and silymarin extracted from milk thistle seeds (*Silybum marianum*) are used as plant-derived anti-inflammatory drugs for treating HPV and HBV infections, respectively [24–27]. Furthermore, as plant-derived anti-inflammatory drugs, eupatilin and JOINS tablets are used to treat gastritis and knee osteoarthritis [44,45]. Isoflavones have antioxidant, anticancer, antibacterial, and anti-inflammatory properties [32]. Our study highlights the potential of wistin as an anti-inflammatory agent with fewer side effects via modulation of the inflammatory signaling pathway. However, further in vivo and clinical studies of wistin in this regard are warranted.

During inflammation, NO plays an important role in the regulation of immune and inflammatory responses [46]. Additionally, the high production of ROS during inflammation can lead to cell damage through the oxidation of DNA, RNA, and proteins [47]. iNOS and COX-2 are pro-inflammatory mediators that are regulated by pro-inflammatory transcription factors [48]. When we examined the effects of wistin on NO and ROS generation, we found that wistin reduced the levels of both NO and ROS. In addition, wistin significantly decreased the mRNA expression of inflammatory enzymes (iNOS and COX-2) and inflammatory cytokines (IL-1β and IL-6). Furthermore, wistin decreased the protein expression of iNOS and COX-2. Therefore, these results suggest that wistin exerts anti-inflammatory effects by modulating inflammatory enzymes, inflammatory mediators, and cytokines.

NF-κB signaling is a well-known inflammatory pathway that regulates the expression of pro-inflammatory cytokines (IL-1β and IL-6) and pro-inflammatory enzymes (iNOS and COX-2) [49]. p65 (RelA) is a component of NF-κB, and phosphorylation of p65 induces the expression of a variety of genes [50]. In addition, AKT regulates NF-κB by phosphorylating the IκB kinase (IKK) complex, which phosphorylates the p65 subunit [19]. Wistin reduced the phosphorylation of AKT and p65 in LPS-stimulated RAW264.7. In addition, wistin decreased the translocation of p65 from the cytosol to the nucleus. These results suggest that wistin exerts anti-inflammatory effects by inhibiting the AKT/NF-κBsignaling pathway.

The MAPK pathway (ERK, JNK, and p38) regulates pro-inflammatory mediators [41]. In particular, p38 can induce NF-κB activation to induce the expression of pro-inflammatory cytokines [51]. p38 MAPK activates mitogen- and stress-activated protein kinases (MSK), which can activate NF-κB (p65 subunit) signaling [52]. Wistin decreased LPS-induced p38 phosphorylation but has no effect on LPS-induced EKR and JNK phosphorylation. These results suggest that wistin could have anti-inflammatory effects by inhibiting p38 pathways in LPS-stimulated RAW264.7.

In conclusion, we demonstrated that wistin exerts anti-inflammatory effects by downregulating pro-inflammatory mediators in LPS-mediated signaling. Wistin could regulate pro-inflammatory mediators, including NO, ROS, pro-inflammatory cytokines, and enzymes, by inhibiting the NF-κB and p38 signaling pathways (Figure 7). Similarly, genistein and daidzein inhibit inflammatory mediators via the NF-κB and MAPK signaling pathways [30,32]. They belong to the isoflavonoid family and exhibit versatile pharmacological activities [30,32]. Therefore, wistin may be developed as a plant-derived anti-inflammatory agent with fewer side effects than other conventional anti-inflammatory drugs.

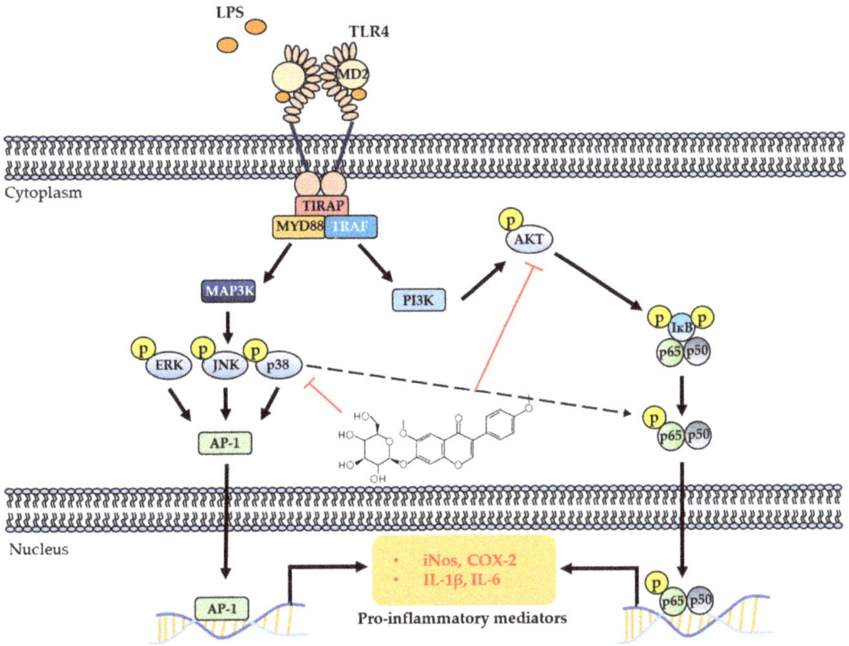

Figure 7. Wistin suppressed LPS-stimulated inflammation by inhibiting the phosphorylation of the p65 and p38 signaling pathways.

4. Materials and Methods

4.1. Reagents

Wistin was purchased from Indofine Chemical Company, Inc. (Hillsborough, NJ, USA) and was dissolved in dimethyl sulfoxide (DMSO). Dulbecco's modified Eagle's medium (DMEM, high glucose) was purchased from Hyclone Laboratories Inc. (Marlborough, MA, USA), and fetal bovine serum (FBS) was purchased from Corning (Corning, NY, USA). Penicillin-streptomycin-glutamine was purchased from Gibco (Waltham, MA, USA). MTT, LPS from *Escherichia coli* O127:B8, N-(1-naphtyhyl) ethylenediamine dihydrochloride, phosphoric acid, sulfanilic acid, and nitrite ion standard solutions were purchased from Sigma-Aldrich Co. (St. Louis, MO, USA). Dulbecco's phosphate-buffered saline (DPBS) and Hank's balanced salt solution (HBSS) were purchased from WELGENE, Inc. (Gyeongsan, Korea). 2′7-Dicholrodihydrofluorecsein diacetate (DCF-DA) was purchased from Cayman Chemical Company (Ann Arbor, MI, USA). The quantitative reverse transcription polymerase chain reaction (qRT-PCR) primers were purchased from Macrogen (Seoul, Korea). Primary antibodies ((iNOS (#13120), COX-2(#12282), phospho-p65 (#13346), phospho-JNK (#9255), phospho-ERK (#4370), and β-actin (#8457)) were purchased from Cell Signaling Technology, Inc. (Danvers, MA, USA). Goat anti-rabbit IgG (5220-0036) and goat anti-mouse IgG (5220-0341) antibodies were purchased from SeraCare Life Sciences, Inc. (Gaithersburg, MD, USA).

4.2. Cell Culture

The RAW264.7 cell line (mouse origin) was obtained from Dr. Sung Ho Ryu's lab (POSTECH, Korea). The cells were cultured in DMEM supplemented with 10% FBS and 1% penicillin–streptomycin and incubated at 37 °C with 5% CO_2.

4.3. Cell Viability

RAW 264.7 cells were plated in a 48-well plate at a density of 3×10^4 cells/well. After 24 h, RAW 264.7 cells were pre-treated with 50, 100, and 150 μM of wistin for 30 min, followed by LPS (0.1 μg/mL) treatment and incubation for 24 h. After that, the DMEM medium was removed, and 1mg/mL of MTT solution was added to each well and was further incubated for 2 h at 37 °C. The solution was removed from the wells and the formed formazan crystals were dissolved in 200 μL of DMSO. Finally, absorbance was measured at 570 nm using a Varioskan LUX Multimode Microplate Reader (Thermo Fisher Scientific Co., Waltham, MA, USA).

4.4. Measurement of NO Production

The Griess test was used to determine NO production. RAW 264.7 cells were plated into a 24-well plate at a density of 6×10^4 cells/well. After 24 h, RAW 264.7 cells were pre-treated with 50, 100, and 150 μM of wistin for 30 min, followed by treatment with LPS (0.1 μg/mL) and further incubation for 24 h. The supernatant from each well was transferred to a 96-well plate, followed by the addition of Griess reagent in a 1:1 ratio. After 15 min, absorbance was measured at 540 nm using a microplate reader (Varioskan LUX Multimode Microplate Reader, Thermo Fisher Scientific Co.).

4.5. Measurement of ROS Production

ROS production was determined using the 2′,7′-dichlorofluorescin-diacetate (DCF-DA) assay. RAW 264.7 cells were plated into a 24-well plate at a density of 6×10^4 cells/well. After 24 h, RAW264.7 cells were pre-treated with 50, 100, and 150 μM wistin for 30 min, followed by LPS (0.1 μg/mL) treatment and further incubation for 24 h. The solution was completely removed from all the wells before incubation with a DCF-DA probe for 30 min at 37 °C. The cells were washed three times with HBSS before measuring fluorescence. The fluorescence intensity was measured at excitation and emission wavelengths of 485 nm and 530 nm, respectively, using a microplate reader (Varioskan LUX Multimode Microplate Reader, Thermo Fisher Scientific Co.). ROS in intact cells were detected with fluorescence microscopy using a Zeiss microscope (Zeiss, Jena, Germany). ROS production was further analyzed using flow cytometry (BD FACSVerse™, BD Bioscience).

4.6. QRT-PCR

RAW264.7 cells plated into a 6-well plate at a density of 3×10^5 cells/well were pre-treated with 150 μM of wistin for 30 min, followed by LPS (0.1 μg/mL) treatment. After 0, 2, 8, 12, and 24 h, cells were washed with DPBS and harvested. Total RNA was isolated using the TRIzol reagent (Invitrogen, Waltham, MA, USA). cDNA was synthesized from total RNA using a SimpliAmp Thermal Cycler (Applied Biosystems Co., Waltham, MA, USA), and then qRT-PCR was performed on a Step One Plus Real-time PCR System Cycler (Applied Biosystems Co.) using a Power SYBR Green PCR mix Cycler (Applied Biosystems Co.). The PCR primer sequences (forward and reverse) used are listed in Table 1, and β-actin was used as the reference gene.

Table 1. Primers used in the quantitative reverse transcription polymerase chain reaction.

Name of the Primer	Primer Sequence
iNos	Forward 5′-GAACGGAGAACGTTGGATTTG-3′ Reverse 5′-TCAGGTCACTTTGGTAGGATTT-3′
Cox-2	Forward 5′-GAAGATTCCCTCCGGTGTTT-3′ Reverse 5′-CCCTTCTCACTGGCTTATGTAG-3′
Tnf-α	Forward 5′-ACGTGGAACTGGCAGAAGAG-3′ Reverse 5′-GGTCTGGGCCATAGAACTGA-3′

Table 1. *Cont.*

Name of the Primer	Primer Sequence
Il-6	Forward 5′-TCTGAAGGACTCTGGCTTTG-3′ Reverse 5′-GATGGATGCTACCAAACTGGA-3′
Il-1β	Forward 5′-AGGTCAAAGGTTTGGAAGCA-3′ Reverse 5′-TGAAGCAGCTATGGCAACTG-3′
β-actin	Forward 5′-ATGGAGGGGAATACAGCCC-3′ Reverse 5′-TTCTTTGCAGCTCCTTCGTT-3′

4.7. Western Blot Analysis

Protein isolation was performed using RAW264.7 cells plated in a 60 mm culture dish at a density of 7×10^5 cells/well. After 24 h, RAW264.7 cells were pre-treated with 150 μM wistin for 30 min, followed by LPS (0.1 μg/mL) treatment. The protein was isolated at two time points, 2 and 24 h. Cells were harvested and lysed with lysis buffer (10 mM Tris pH 7.4, 150 mM NaCl, 1 mM EDTA pH 8, 1% Triton X-100, 1% sodium deoxycholate, 30 mM NaF, 1.5 mM NaVO4, 1 mM PMSF, and 1 mg/mL each of aprotinin, leupeptin, and pepstatin A) and sonicated for 10 s. The cell lysates were centrifuged at 13,000 rpm for 10 min at 4 °C (Centrifuge 5424 R (rotor type: FA-45-30-11), Eppendorf, Hamburg, Germany) and quantified with the help of the Bradford (Abcam, Cambridge, UK) assay. The proteins (15 μg) were subjected to sodium dodecyl sulfate–polyacrylamide gel electrophoresis and transferred to a nitrocellulose membrane. The membranes were blocked with 5% skim milk in Tris-buffered saline supplemented with 0.1% Tween-20 (TBS-T) at room temperature (RT) for 30 min. Membranes were then incubated with suitable primary antibodies overnight at RT with gentle shaking. The following day, the membranes were washed with TBS-T for 30 min. Membranes were then incubated with secondary horseradish peroxidase (HRP)-conjugated anti-rabbit/mouse IgG antibodies for 1 h at RT. Finally, the membranes were detected using an ECL reagent (Cytiva, Malborough, MS, USA), and the bands were visualized using Chemidoc (iBright™ CL1500, Invitrogen). Band intensities were analyzed using ImageJ software. β-actin was used as the loading control.

4.8. Immunofluorescence

RAW264.7 cells plated into a 4-well plate with a density of 6×10^4 cells/well were pre-treated with 150 μM of wistin for 30 min, followed by LPS (0.1 μg/mL) treatment. After 2 h, cells were washed with PBS and fixed with 4% paraformaldehyde (Sigma-Aldrich; Merck KGaA) in PBS for 20 min. The cells were then incubated with 0.2% Triton X-100 and 0.1% citrate (Sigma-Aldrich; Merck KGaA) in PBS for 5 min, washed with PBS, and blocked with 2% bovine serum albumin (Sigma-Aldrich; Merck KGaA) for 30 min, followed by probing with rabbit anti-p65 NF-κB antibody (1:500; cat. no. 8242; Cell Signaling Technology, Inc.) for 2 h. After washing with 2% bovine serum albumin, the cells were incubated with fluorescein goat anti-rabbit IgG (H+L) cross-adsorbed secondary antibody Alexa Fluor 555 (4 μg/mL; cat. no. A-21428; Invitrogen Inc., Middlesex County, MA, USA) for 1 h. After washing with 2% bovine serum albumin, the nuclei were counterstained with DAPI solution (1 mg/mL) for 5 min in the dark, and fluorescence was visualized using a fluorescence microscope (Zeiss microscope, Carl Zeiss AC).

4.9. Statistical Analysis

Results are presented as mean ± standard deviation (SD). Two statistical analysis methods were used. ANOVA and Dunnett's post hoc tests were conducted to compare three or more groups. The Student's *t*-test was conducted to compare two groups. Statistical significance was set at $p < 0.05$(*).

Author Contributions: J.A., G.R., S.-A.S. and H.K. performed the experiments. J.A. and C.S.L. wrote the manuscript with guidance from C.S.L., M.-J.A. and J.H.L. provided intellectual contribution to this study.All authors have read and agreed to the published version of the manuscript.

Funding: This research was supported by the National Research Foundation of Korea (NRF) grant funded by the Korea government (MSIT) (Grant Number: 2020R1F1A1070844) and the Korea Polar Research Institute (KOPRI) grant funded by the Ministry of Oceans and Fisheries (KOPRI project No. * PE21900).

Institutional Review Board Statement: Not applicable.

Informed Consent Statement: Not applicable.

Data Availability Statement: Not applicable.

Conflicts of Interest: The authors declare no conflict of interest.

References

1. Singh, N.; Baby, D.; Rajguru, J.P.; Patil, P.B.; Thakkannavar, S.S.; Pujari, V.B. Inflammation and cancer. *Ann. Afr. Med.* **2019**, *18*, 121–126. [CrossRef] [PubMed]
2. Chen, L.; Deng, H.; Cui, H.; Fang, J.; Zuo, Z.; Deng, J.; Li, Y.; Wang, X.; Zhao, L. Inflammatory responses and inflammation-associated diseases in organs. *Oncotarget* **2018**, *9*, 7204–7218. [CrossRef] [PubMed]
3. Yao, C.; Narumiya, S. Prostaglandin-cytokine crosstalk in chronic inflammation. *Br. J. Pharmacol.* **2019**, *176*, 337–354. [CrossRef] [PubMed]
4. Zhong, J.; Shi, G. Editorial: Regulation of Inflammation in Chronic Disease. *Front. Immunol.* **2019**, *10*, 737. [CrossRef]
5. Panigrahy, D.; Gilligan, M.M.; Serhan, C.N.; Kashfi, K. Resolution of inflammation: An organizing principle in biology and medicine. *Pharmacol. Ther.* **2021**, *227*, 107879. [CrossRef]
6. Kozuch, P.L.; Hanauer, S.B. Treatment of inflammatory bowel disease: A review of medical therapy. *World J. Gastroenterol.* **2008**, *14*, 354–377. [CrossRef]
7. PLoSker, G.L.; Croom, K.F. Sulfasalazine: A review of its use in the management of rheumatoid arthritis. *Drugs* **2005**, *65*, 1825–1849. [CrossRef]
8. Bindu, S.; Mazumder, S.; Bandyopadhyay, U. Non-steroidal anti-inflammatory drugs (NSAIDs) and organ damage: A current perspective. *Biochem. Pharmacol.* **2020**, *180*, 114147. [CrossRef]
9. Gonzalez-Rey, M.; Bebianno, M.J. Non-steroidal anti-inflammatory drug (NSAID) ibuprofen distresses antioxidant defense system in mussel Mytilus galloprovincialis gills. *Aquat. Toxicol.* **2011**, *105*, 264–269. [CrossRef]
10. Parolini, M. Toxicity of the Non-Steroidal Anti-Inflammatory Drugs (NSAIDs) acetylsalicylic acid, paracetamol, diclofenac, ibuprofen and naproxen towards freshwater invertebrates: A review. *Sci. Total Environ.* **2020**, *740*, 140043. [CrossRef]
11. Rainsford, K.D. Ibuprofen: Pharmacology, efficacy and safety. *Inflammopharmacology* **2009**, *17*, 275–342. [CrossRef] [PubMed]
12. Tang, D.; Kang, R.; Coyne, C.B.; Zeh, H.J.; Lotze, M.T. PAMPs and DAMPs: Signal 0s that spur autophagy and immunity. *Immunol. Rev.* **2012**, *249*, 158–175. [CrossRef]
13. Lu, Y.C.; Yeh, W.C.; Ohashi, P.S. LPS/TLR4 signal transduction pathway. *Cytokine* **2008**, *42*, 145–151. [CrossRef] [PubMed]
14. Kawai, T.; Akira, S. TLR signaling. *Semin. Immunol.* **2007**, *19*, 24–32. [CrossRef] [PubMed]
15. Miao, F.; Shan, C.; Ning, D. Walnut oil alleviates LPS-induced intestinal epithelial cells injury by inhibiting TLR4/MyD88/NF-kappaB pathway activation. *J. Food Biochem.* **2021**, *45*, e13955. [CrossRef]
16. Kaminska, B. MAPK signalling pathways as molecular targets for anti-inflammatory therapy–from molecular mechanisms to therapeutic benefits. *Biochim. Biophys. Acta* **2005**, *1754*, 253–262. [CrossRef]
17. Sharif, O.; Bolshakov, V.N.; Raines, S.; Newham, P.; Perkins, N.D. Transcriptional profiling of the LPS induced NF-kappaB response in macrophages. *BMC Immunol.* **2007**, *8*, 1. [CrossRef]
18. Giridharan, S.; Srinivasan, M. Mechanisms of NF-kappaB p65 and strategies for therapeutic manipulation. *J. Inflamm. Res.* **2018**, *11*, 407–419. [CrossRef]
19. Bai, D.; Ueno, L.; Vogt, P.K. Akt-mediated regulation of NFkappaB and the essentialness of NFkappaB for the oncogenicity of PI3K and Akt. *Int. J. Cancer* **2009**, *125*, 2863–2870. [CrossRef]
20. Liu, T.; Zhang, L.; Joo, D.; Sun, S.C. NF-kappaB signaling in inflammation. *Signal Transduct. Target. Ther.* **2017**, 2. [CrossRef]
21. Kang, K.S. Phytochemical Constituents of Medicinal Plants for the Treatment of Chronic Inflammation. *Biomolecules* **2021**, *11*, 672. [CrossRef] [PubMed]
22. Mueller, A.L.; Brockmueller, A.; Kunnumakkara, A.B.; Shakibaei, M. Modulation of Inflammation by Plant-Derived Nutraceuticals in Tendinitis. *Nutrients* **2022**, *14*, 30. [CrossRef] [PubMed]
23. Vonkeman, H.E.; van de Laar, M.A. Nonsteroidal anti-inflammatory drugs: Adverse effects and their prevention. *Semin. Arthritis. Rheum.* **2010**, *39*, 294–312. [CrossRef]
24. Nguyen, H.P.; Doan, H.Q.; Brunell, D.J.; Rady, P.; Tyring, S.K. Apoptotic gene expression in sinecatechins-treated external genital and perianal warts. *Viral Immunol.* **2014**, *27*, 556–558. [CrossRef] [PubMed]

25. Fernandes, J.V.; Fernandes, T.A.A.d.M.; de Azevedo, J.C.V.; Cobucci, R.N.O.; Cobucci, R.N.; de Carvalho, M.G.F.; Andrade, V.S.; de Araújo, J.M.G. Link between chronic inflammation and human papillomavirus-induced carcinogenesis (Review). *Oncol. Lett.* **2015**, *9*, 1015–1026. [CrossRef]
26. Lovelace, E.S.; Wagoner, J.; MacDonald, J.; Bammler, T.; Bruckner, J.; Brownell, J.; Beyer, R.P.; Zink, E.M.; Kim, Y.M.; Kyle, J.E.; et al. Silymarin Suppresses Cellular Inflammation By Inducing Reparative Stress Signaling. *J. Nat. Prod.* **2015**, *78*, 1990–2000. [CrossRef]
27. Fried, M.W.; Navarro, V.J.; Afdhal, N.; Belle, S.H.; Wahed, A.S.; Hawke, R.L.; Doo, E.; Meyers, C.M.; Reddy, K.R.; Silymarinin, N.; et al. Effect of silymarin (milk thistle) on liver disease in patients with chronic hepatitis C unsuccessfully treated with interferon therapy: A randomized controlled trial. *JAMA* **2012**, *308*, 274–282. [CrossRef]
28. Mayer, K.E.; Myers, R.P.; Lee, S.S. Silymarin treatment of viral hepatitis: A systematic review. *J. Viral Hepat.* **2005**, *12*, 559–567. [CrossRef]
29. Ryoo, S.B.; Oh, H.K.; Yu, S.A.; Moon, S.H.; Choe, E.K.; Oh, T.Y.; Park, K.J. The effects of eupatilin (stillen(R)) on motility of human lower gastrointestinal tracts. *Korean J. Physiol. Pharmacol.* **2014**, *18*, 383–390. [CrossRef]
30. Saleh, H.A.; Yousef, M.H.; Abdelnaser, A. The Anti-Inflammatory Properties of Phytochemicals and Their Effects on Epigenetic Mechanisms Involved in TLR4/NF-kappaB-Mediated Inflammation. *Front. Immunol.* **2021**, *12*, 606069. [CrossRef]
31. Goh, Y.X.; Jalil, J.; Lam, K.W.; Husain, K.; Premakumar, C.M. Genistein: A Review on its Anti-Inflammatory Properties. *Front. Pharmacol.* **2022**, *13*, 820969. [CrossRef]
32. Yu, J.; Bi, X.; Yu, B.; Chen, D. Isoflavones: Anti-Inflammatory Benefit and Possible Caveats. *Nutrients* **2016**, *8*, 361. [CrossRef] [PubMed]
33. Sanada, M.; Hayashi, R.; Imai, Y.; Nakamura, F.; Inoue, T.; Ohta, S.; Kawachi, H. 4′,6-dimethoxyisoflavone-7-O-beta-D-glucopyranoside (wistin) is a peroxisome proliferator-activated receptor gamma (PPARgamma) agonist that stimulates adipocyte differentiation. *Anim. Sci. J.* **2016**, *87*, 1347–1351. [CrossRef] [PubMed]
34. Suzuki, M.; Nakamura, F.; Taguchi, E.; Nakata, M.; Wada, F.; Takihi, M.; Inoue, T.; Ohta, S.; Kawachi, H. 4′,6-Dimethoxyisoflavone-7-O-beta-D-glucopyranoside (wistin) is a peroxisome proliferator-activated receptor alpha (PPARalpha) agonist in mouse hepatocytes. *Mol. Cell Biochem.* **2018**, *446*, 35–41. [CrossRef] [PubMed]
35. Bernatoniene, J.; Kazlauskaite, J.A.; Kopustinskiene, D.M. Pleiotropic Effects of Isoflavones in Inflammation and Chronic Degenerative Diseases. *Int. J. Mol. Sci.* **2021**, *22*, 5656. [CrossRef] [PubMed]
36. Agita, A.; Alsagaff, M.T. Inflammation, Immunity, and Hypertension. *Acta Med. Indones.* **2017**, *49*, 158–165. [PubMed]
37. Ma, Y.; Tang, T.; Sheng, L.; Wang, Z.; Tao, H.; Zhang, Q.; Zhang, Y.; Qi, Z. Aloin suppresses lipopolysaccharideinduced inflammation by inhibiting JAK1STAT1/3 activation and ROS production in RAW264.7 cells. *Int. J. Mol. Med.* **2018**, *42*, 1925–1934. [CrossRef]
38. Park, J.Y.; Chung, T.W.; Jeong, Y.J.; Kwak, C.H.; Ha, S.H.; Kwon, K.M.; Abekura, F.; Cho, S.H.; Lee, Y.C.; Ha, K.T.; et al. Ascofuranone inhibits lipopolysaccharide-induced inflammatory response via NF-kappaB and AP-1, p-ERK, TNF-alpha, IL-6 and IL-1beta in RAW 264.7 macrophages. *PLoS ONE* **2017**, *12*, e0171322. [CrossRef]
39. Murakami, A.; Ohigashi, H. Targeting NOX, INOS and COX-2 in inflammatory cells: Chemoprevention using food phytochemicals. *Int. J. Cancer* **2007**, *121*, 2357–2363. [CrossRef]
40. Yu, H.; Lin, L.; Zhang, Z.; Zhang, H.; Hu, H. Targeting NF-kappaB pathway for the therapy of diseases: Mechanism and clinical study. *Signal. Transduct. Target. Ther.* **2020**, *5*, 209. [CrossRef]
41. Thalhamer, T.; McGrath, M.A.; Harnett, M.M. MAPKs and their relevance to arthritis and inflammation. *Rheumatology (Oxford)* **2008**, *47*, 409–414. [CrossRef] [PubMed]
42. Chen, C.C.; Lin, M.W.; Liang, C.J.; Wang, S.H. The Anti-Inflammatory Effects and Mechanisms of Eupafolin in Lipopolysaccharide-Induced Inflammatory Responses in RAW264.7 Macrophages. *PLoS ONE* **2016**, *11*, e0158662. [CrossRef] [PubMed]
43. Adedapo, A.; Adewuyi, T.; Sofidiya, M. Phytochemistry, anti-inflammatory and analgesic activities of the aqueous leaf extract of Lagenaria breviflora (Cucurbitaceae) in laboratory animals. *Rev. Biol. Trop.* **2013**, *61*, 281–290. [CrossRef] [PubMed]
44. Song, E.H.; Chung, K.S.; Kang, Y.M.; Lee, J.H.; Lee, M.; An, H.J. Eupatilin suppresses the allergic inflammatory response in vitro and in vivo. *Phytomedicine* **2018**, *42*, 1–8. [CrossRef] [PubMed]
45. Hartog, A.; Hougee, S.; Faber, J.; Sanders, A.; Zuurman, C.; Smit, H.F.; van der Kraan, P.M.; Hoijer, M.A.; Garssen, J. The multicomponent phytopharmaceutical SKI306X inhibits in vitro cartilage degradation and the production of inflammatory mediators. *Phytomedicine* **2008**, *15*, 313–320. [CrossRef]
46. Guzik, T.J.; Korbut, R.; Adamek-Guzik, T. Nitric oxide and superoxide in inflammation and immune regulation. *J. Physiol. Pharmacol.* **2003**, *54*, 469–487. [PubMed]
47. Checa, J.; Aran, J.M. Reactive Oxygen Species: Drivers of Physiological and Pathological Processes. *J. Inflamm. Res.* **2020**, *13*, 1057–1073. [CrossRef]
48. Moita, E.; Gil-Izquierdo, A.; Sousa, C.; Ferreres, F.; Silva, L.R.; Valentao, P.; Dominguez-Perles, R.; Baenas, N.; Andrade, P.B. Integrated analysis of COX-2 and iNOS derived inflammatory mediators in LPS-stimulated RAW macrophages pre-exposed to Echium plantagineum L. bee pollen extract. *PLoS ONE* **2013**, *8*, e59131. [CrossRef]
49. Xiao, K.; Liu, C.; Tu, Z.; Xu, Q.; Chen, S.; Zhang, Y.; Wang, X.; Zhang, J.; Hu, C.A.; Liu, Y. Activation of the NF-kappaB and MAPK Signaling Pathways Contributes to the Inflammatory Responses, but Not Cell Injury, in IPEC-1 Cells Challenged with Hydrogen Peroxide. *Oxid. Med. Cell Longev.* **2020**, *2020*, 5803639. [CrossRef]

50. Hoesel, B.; Schmid, J.A. The complexity of NF-kappaB signaling in inflammation and cancer. *Mol. Cancer* **2013**, *12*, 86. [CrossRef]
51. Tang, Y.; Sun, M.; Liu, Z. Phytochemicals with protective effects against acute pancreatitis: A review of recent literature. *Pharm. Biol.* **2022**, *60*, 479–490. [CrossRef] [PubMed]
52. Salminen, A.; Kauppinen, A.; Kaarniranta, K. Emerging role of NF-kappaB signaling in the induction of senescence-associated secretory phenotype (SASP). *Cell Signal.* **2012**, *24*, 835–845. [CrossRef] [PubMed]

molecules

MDPI

Article

Selectivity Tuning by Natural Deep Eutectic Solvents (NADESs) for Extraction of Bioactive Compounds from *Cytinus hypocistis*—Studies of Antioxidative, Enzyme-Inhibitive Properties and LC-MS Profiles

Gokhan Zengin [1], María de la Luz Cádiz-Gurrea [2,*], Álvaro Fernández-Ochoa [2,*], Francisco Javier Leyva-Jiménez [3,4], Antonio Segura Carretero [2], Malwina Momotko [5], Evren Yildiztugay [6], Refik Karatas [1], Sharmeen Jugreet [7], Mohamad Fawzi Mahomoodally [7,8,9] and Grzegorz Boczkaj [10,11]

1 Department of Biology, Science Faculty, Selcuk University, Konya 42130, Turkey
2 Department of Analytical Chemistry, Faculty of Sciences, University of Granada, Fuentenueva s/n, 18071 Granada, Spain
3 Department of Analytical Chemistry and Food Science and Technology, University of Castilla-La Mancha, Ronda de Calatrava 7, 13071 Ciudad Real, Spain
4 Regional Institute for Applied Scientific Research (IRICA), Area of Food Science, University of Castilla-La Mancha, Avenida Camilo Jose Cela, 10, 13071 Ciudad Real, Spain
5 Department of Process Engineering and Chemical Technology, Faculty of Chemistry, Gdansk University of Technology, G. Narutowicza St. 11/12, 80-0233 Gdansk, Poland
6 Department of Biotechnology, Science Faculty, Selcuk University, Konya 42130, Turkey
7 Department of Health Sciences, Faculty of Medicine and Health Sciences, University of Mauritius, Réduit 80837, Mauritius
8 Center for Transdisciplinary Research, Department of Pharmacology, Saveetha Institute of Medical and Technical Science, Saveetha Dental College, Chennai 600077, India
9 Centre of Excellence for Pharmaceutical Sciences (Pharmacen), North West University, Potchefstroom 2520, South Africa
10 Department of Sanitary Engineering, Faculty of Civil and Environmental Engineering, Gdansk University of Technology, G. Narutowicza St. 11/12, 80-0233 Gdansk, Poland
11 Advanced Materials Center, Gdansk University of Technology, G. Narutowicza St. 11/12, 80-233 Gdansk, Poland
* Correspondence: mluzcadiz@ugr.es (M.d.l.L.C.-G.); alvaroferochoa@ugr.es (Á.-F.O.)

Citation: Zengin, G.; Cádiz-Gurrea, M.d.l.L.; Fernández-Ochoa, Á.; Leyva-Jiménez, F.J.; Carretero, A.S.; Momotko, M.; Yildiztugay, E.; Karatas, R.; Jugreet, S.; Mahomoodally, M.F.; et al. Selectivity Tuning by Natural Deep Eutectic Solvents (NADESs) for Extraction of Bioactive Compounds from *Cytinus hypocistis*—Studies of Antioxidative, Enzyme-Inhibitive Properties and LC-MS Profiles. *Molecules* **2022**, *27*, 5788. https://doi.org/10.3390/molecules27185788

Academic Editor: Nour Eddine Es-Safi

Received: 23 August 2022
Accepted: 4 September 2022
Published: 7 September 2022

Abstract: In the present study, the extracts of *Cytinus hypocistis* (L.) L using both traditional solvents (hexane, ethyl acetate, dichloromethane, ethanol, ethanol/water, and water) and natural deep eutectic solvents (NADESs) were investigated in terms of their total polyphenolic contents and antioxidant and enzyme-inhibitive properties. The extracts were found to possess total phenolic and total flavonoid contents in the ranges of 26.47–186.13 mg GAE/g and 0.68–12.55 mg RE/g, respectively. Higher total phenolic contents were obtained for NADES extracts. Compositional differences were reported in relation to antioxidant potential studied by several assays (DPPH: 70.19–939.35 mg TE/g, ABTS: 172.56–4026.50 mg TE/g; CUPRAC: 97.41–1730.38 mg TE/g, FRAP: 84.11–1534.85 mg TE/g). Application of NADESs (choline chloride—urea 1:2, a so-called Reline) allowed one to obtain the highest number of extracts having antioxidant potential in the radical scavenging and reducing assays. NADES-B (protonated by HCl L-proline-xylitol 5:1) was the only extractant from the studied solvents that isolated a specific fraction without chelating activity. Reline extract exhibited the highest acetylcholinesterase inhibition compared to NADES-B and NADES-C (protonated by H_2SO_4 L-proline-xylitol 5:1) extracts, which showed no inhibition. The NADES extracts were observed to have higher tyrosinase inhibitory properties compared to extracts obtained by traditional organic solvents. Furthermore, the NADES extracts were relatively better inhibitors of the diabetic enzymes. These findings provided an interesting comparison in terms of total polyphenolic content yields, antioxidant and enzyme inhibitory properties (cholinesterase, amylase, glucosidase, and tyrosinase) between traditional solvent extracts and NADES extracts, used as an alternative. While the organic solvents showed better antioxidant activity, the NADES extracts were found to have some other improved properties, such as higher total phenolic content and enzyme-inhibiting properties, suggesting functional prospects for their use in phytonutrient extraction and fractionation. The obtained results could also be used to give a broad overview of the different biological potentials of *C. hypocistis*.

Keywords: NADES; total polyphenolic content; antioxidants; enzyme inhibition; functional food; natural medicine; Alzheimer cholinesterase inhibitors

1. Introduction

The genus *Cytinus*, composed of endophytic parasitic plants (family: *Cytinaceae*), bears eight recognised species distributed around two centres of diversity: one in southern Africa and Madagascar, and one in the Mediterranean region [1].

Indeed, folkloric medicine has dedicated substantial consideration to this genus. These plants have been used traditionally for treating dysentery, including their ability to soothe inflammations of the eyes and throat. Some ethnobotanical reviews have also noted the use of *Cytinus* juice as an astringent, a haemostatic, and a tonic substance. They are also used as a scar-healing agent, whereby the scalp pulp is applied daily on corns and calluses, skin and swollen mucous membranes as an astringent and anti-inflammatory therapy [1,2].

Some studies have also pointed out these plants' beneficial potential and suggested their antimicrobial effects over a range of bacterial strains and antioxidant activities [2–4]. Furthermore, they have been highlighted as good sources of biologically active ingredients of cosmeceutical interest [2,3,5]. In fact, their biological activities have been correlated with their high tannin content. For instance, hydrolysable tannins were found to be the active cytotoxic compounds identified in three *Cytinus* taxa and were assessed against a wide variety of cancer cell lines [6]. In another study, the tested extract of *C. hypocistis* was found to exhibit anti-inflammatory activity and effective cytotoxicity againcst tumour cells, while it showed the lowest cytotoxicity on a non-tumour cell line, and interestingly, hydrolysable tannins and flavonoids were also identified as the main groups in the extract [7].

Plants possess a diverse range of such bioactive constituents. However, their availability strongly depends on the extraction techniques used, among other factors. Even though to date numerous methods have been developed and upgraded, there is still a need to achieve a standardised solution with high consideration for the extraction of bioactive compounds from plants [8].

In addition, there is growing pressure to investigate alternative solvents that retain the technological advantages of organic solvents while posing less risk to human health and the environment. Deep eutectic solvents (DESs) and their specialised form obtained from compounds of natural origin—natural DESs (NADESs)—have shown the most promise in the field of green chemistry because they are abundant, inexpensive, recyclable, and appealing for a wide range of applications (food, cosmetic, and pharmaceutical). DESs have already proved to have several advantages in separation science, especially in terms of unusual selectivity useful in chromatography [9,10] and extraction [11,12], as well as membrane processes [13,14]. Many studies have effectively employed NADES extraction to gain high-quality extracts from numerous plants, including medicinal plants [15–17]. NADESs' potential biological activity, bioavailability, and the availability of a variety of solvent combinations for its preparation are also intriguing characteristics. Therefore, NADES extraction is a cutting-edge technique that has piqued the interest of researchers and already exhibited great promise in the extraction and isolation of bioactive compounds from plants [17].

Therefore, the aim of the present study is to compare characteristics of extracts obtained by means of NADESs with classic organic solvents. For this purpose, the *Cytinus hypocistis* (L.) L. extracts were analysed in respect to their LC-MS profiles, the total phenolic (TPC) and flavonoid (TFC) contents, and antioxidant as well as enzyme-inhibitive properties.

2. Results and Discussion

2.1. Phytochemical Profiles

Natural products are important sources for drug development. Thus, it is of crucial importance to develop effective methods to extract and isolate these bioactive products. Indeed, the lab-intensive and laborious extraction and isolation processes have been a major challenge in the application of natural products in drug development. There is an urgent need to develop efficient and selective methods for this [18]. In this respect, different types of solvents have been widely used for the extraction of phytochemicals, whereby dried plant powders are used to extract bioactive phytochemicals and remove the Interference of water concomitantly. The solvents used for the extraction of biomolecules from plants are chosen based on the polarity of the desired solute. For example, a solvent with the same polarity as the solute will effectively dissolve the solute. Several solvents can be used sequentially to limit the number of analogous compounds in the desired yield [19].

Solubility, bioavailability, and stability are all factors in the pharmacological efficacy of plant extracts and their bioactive principles. Natural deep eutectic solvents (NADESs) are considered as green solvents to enhance the extraction performance of plant metabolites [16]. As functional liquid media, NADESs can dissolve both natural and synthetic substances with low water solubility. Hence, they are alternative candidates for applications with some organic solvents, as well as ionic liquids [20], indicating the enormous potential for NADESs to be utilised in the development of pharmaceutical formulations, such as nutraceuticals derived from plant-based metabolites [16]. Thus, in this study, both traditional solvents and NADESs were used to prepare *C. hypocistis* extracts and to compare their overall performance in terms of their bioactive content yields and biological activities.

In the phytochemical studies, the investigation of the polyphenol content present in plant extracts is an important part of assessing their biological properties. In this study, the extracts were found to possess TPC and TFC in the range of 26.47–186.13 mg GAE/g and 0.68–12.55 mg RE/g, respectively. Interestingly, the extracts obtained by NADESs yielded higher TPC (167.57–186.13 mg GAE/g), followed by ethyl acetate, water, ethanol, and ethanol/water extracts (123.51–127.83 mg GAE/g). On the other hand, the dichloromethane and hexane extracts yielded low TPC. In the TFC assay, the highest yield was obtained by ethyl acetate and ethanol/water extracts, followed by water and ethanol extracts, while the least TFC was yielded by hexane, NADES-C, and dichloromethane extracts (Figure 1, Table S1). The variation in total polyphenol content clearly varied with the polarity of the solvents used. However, not only the polarity of the solvents, but also other parameters such as pH, extraction time, methods, and temperature can affect the extraction yield and total phenolic content [21–23].

2.2. Characterization of Polar Bioactive Compounds from C. hypocistis Extracts by UPLC-ESI-QTOF-MS

Following the described LC-MS method, all extracts were analysed, resulting in a total of 148 detected compounds. Figure 2 shows the base peak chromatograms performed for each extraction condition and Table S2 summarises all information about detected compounds such as retention time, m/z ratio, error in ppm, molecular formula, and name of each proposed compound. In addition, peak numbers were assigned according to their elution order. The peak areas of the detected compounds are given in Table S3.

It is worth to note that to our knowledge, there is little reference to a comprehensive characterisation of *C. hypocistis* extracts [1]. For this reason, our work is especially relevant. Considering the accurate mass spectra information and data previously reported by literature, 136 compounds were tentatively annotated in this study. Only one common molecular feature was detected among all the different extractions; however, this molecular feature could not be annotated and remained as an unknown compound (Table 1). This could be explained mainly by the differences in the extraction efficiencies of the different solvents used, also considering the different physicochemical properties of the compounds present in the matrix.

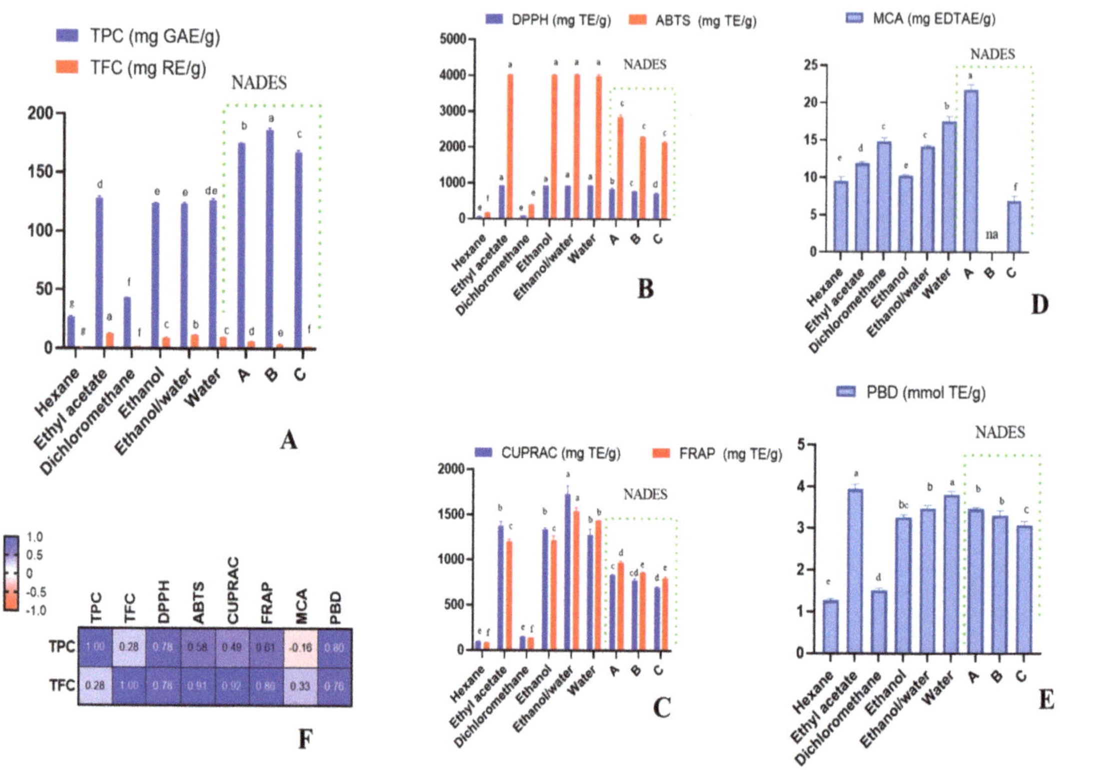

Figure 1. Total phenolic and flavonoid content (**A**), radical scavenging ability (**B**), reducing power (**C**), metal-chelating ability (MCA) (**D**), total antioxidant ability (by phosphomolybdenum assay (PBD)) (**E**), and Pearson's correlations between total bioactive compounds and antioxidant assays ($p < 0.05$) (**F**). na: not active; different letters in column for same assays indicate significant differences in the extracts ($p < 0.05$).

Overall, the tentative characterisation allowed one to classify the compounds in five major groups: gallotannins, ellagitannins, flavonoids, fatty acids, and other compounds, with it being important to note that gallotannins were the most important one, with 61 compounds included. These were mainly annotated as mono-, di-, tri-, tetra-, penta-, hexa-, and hepta-galloyl hexoside. In addition, different isomers were also annotated for each of these types of chemical structures. Silva et al., reported in 2020 from mono- to penta-galloyl hexosides in different parts (petals, stalks, and nectar) of *C. hypocistis* [24]. However, all these compounds were previously reported in other sources such as *Magnifera indica* L. kernels and peels [25,26], *Rhodiola crenulate* roots [27], *Rhodiola rosea* roots, leaves, stems and flowers [27], Paeonia plants [28], and *Pistacia vera* leaves [29], among others [30]. Table 1 shows that a major number of these gallotannins were found in ethyl acetate, ethanol, water/ethanol, and water extracts. In recent years, several authors have reported a wide range of biological properties of these galloyl hexoside derivatives [27,29,31,32]. Among other gallic acid derivatives presented in the gallotannins group, three isomers from neochebulagic acid corresponding to peaks 60, 61, and 70 have been also annotated. These compounds were reported in *Terminalia chebula* Retz. Playing a role in the intestinal glucose transport [33]. Moreover, several compounds previously found in *Trapa quadrispinosa* pericarps were detected in our extracts, such as peaks 49, 67, 75, and 87 (digalloyl-lactonised valoneoyl-d-glucose isomers); peaks 71, 92, and 98 (trigalloyl-lactonised valoneoyl glucose isomers); and peaks 78, 85, and 102 (galloyl-penta-hydroxy-benzoic-brevifolincarboxyl-glucose isomers) [34].

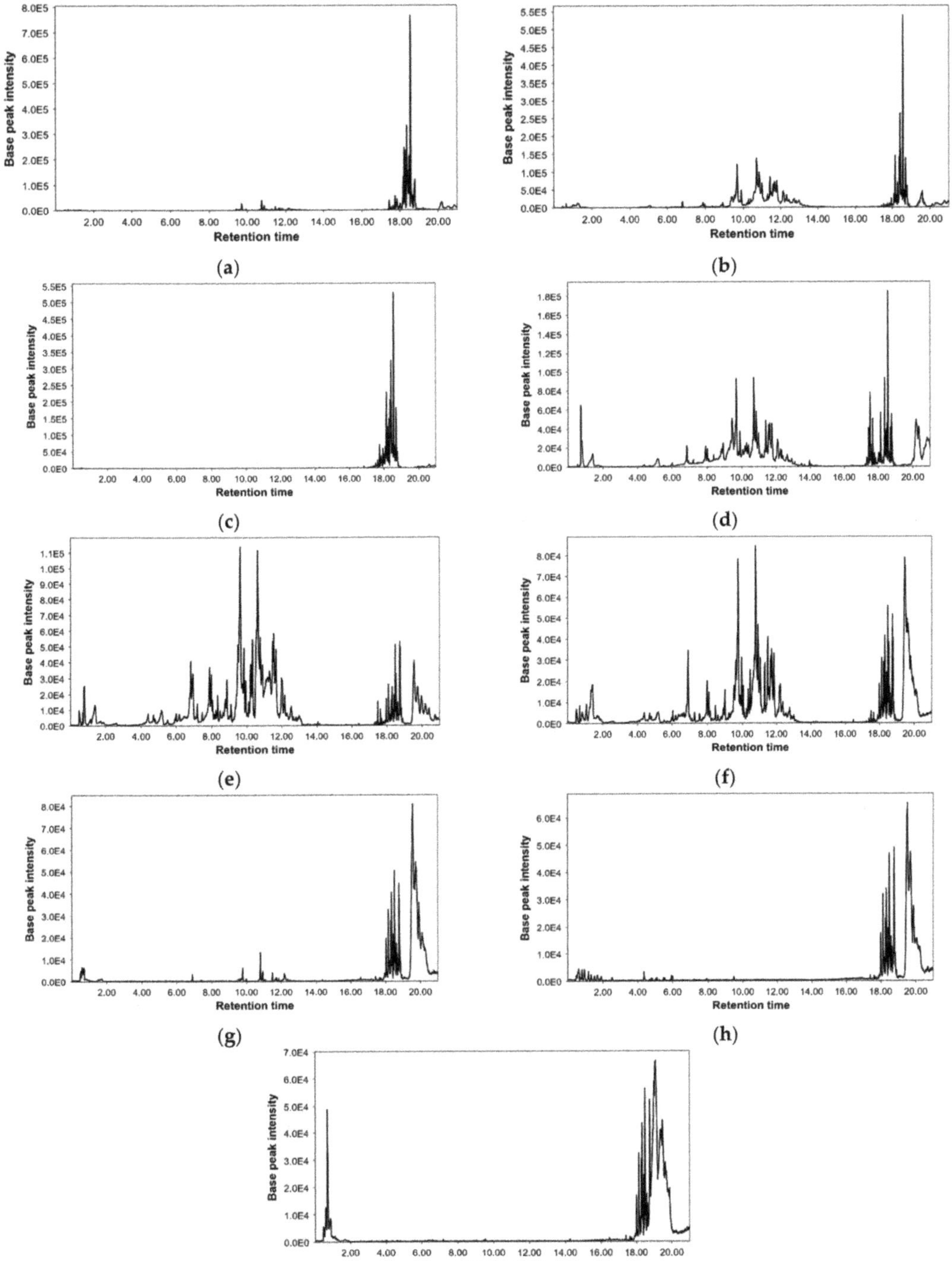

Figure 2. Base peak chromatograms from Cytinus (**a**) hexane, (**b**) ethyl acetate, (**c**) dichloromethane, (**d**) ethanol, (**e**) ethanol/water, (**f**) water, (**g**) NADES-A, (**h**) NADES-B, and (**i**) NADES-C extracts by UPLC-MS.

Table 1. Chemical characterisation of the tested extracts.

Peak	Compound	Hexane	Ethyl Acetate	Dichloromethane	Ethanol	Ethanol/Water	Water	NADES-A	NADES-B	NADES-C
1	Hydroxy-pseudouric acid	+	+	+	ND	+	+	ND	+	ND
2	Galloyl-galactarolactone	+	+	ND	ND	+	+	ND	ND	ND
3	Unknown 1	+	+	+	+	+	+	+	+	+
4	Unknown 2	ND	ND	ND	ND	ND	ND	ND	+	ND
5	Proline	ND	ND	ND	ND	ND	ND	+	ND	+
6	Disaccharide	+	+	ND	+	+	+	ND	ND	ND
7	Unknown 3	ND	ND	ND	ND	ND	ND	ND	ND	+
8	Glucose	ND	ND	ND	+	+	ND	ND	ND	ND
9	Unknown 4	ND	ND	ND	ND	ND	ND	ND	ND	+
10	Quinic acid	ND	+	ND	+	+	+	ND	ND	ND
11	Galloyl-diglucose	ND	ND	ND	+	+	ND	ND	ND	ND
12	Galloylglucose isomer 1	+	+	ND	+	+	+	ND	+	ND
13	Galloylglucose isomer 2	+	+	ND	+	+	+	ND	+	ND
14	Galloylglucose isomer 3	+	+	ND	+	+	+	ND	+	ND
15	Pyrogallol	ND	+	ND	+	+	+	+	+	+
16	Gallic acid	ND	+	ND	+	+	+	+	+	+
17	Galloylglucose isomer 4	ND	+	ND	+	+	+	ND	+	ND
18	Fukiic acid	ND	+	ND	ND	ND	+	ND	+	ND
19	Digalloylglucose isomer 1	ND	+	ND	+	+	+	ND	+	ND
20	Digalloylglucose isomer 2	ND	ND	ND	+	+	+	ND	+	ND
21	Digalloylglucose isomer 3	ND	+	ND	+	+	+	ND	+	ND
22	Digalloylglucose isomer 4	ND	+	ND	+	+	+	ND	+	ND
23	Brevifolin carboxylic acid	ND	ND	ND	ND	+	+	ND	ND	ND
24	Digallate isomer 1	ND	ND	ND	+	+	+	ND	ND	ND
25	Unknown 5	+	+	ND	ND	ND	ND	ND	ND	ND

Table 1. *Cont.*

Peak	Compound	Hexane	Ethyl Acetate	Dichloromethane	Ethanol	Ethanol/Water	Water	NADES-A	NADES-B	NADES-C
26	Trigalloyl-glucoside isomer 1	ND	+	ND	+	+	+	+	ND	ND
27	HHDP-galloylglucose isomer 1	ND	+	ND	+	+	+	ND	ND	ND
28	Galloylnorbergenin isomer 1	+	+	ND	+	+	+	+	ND	ND
29	Brevifolin	ND	ND	ND	+	+	+	+	ND	ND
30	Galloylnorbergenin isomer 2	ND	+	ND	+	+	+	ND	ND	ND
31	Digalloyl-HHDP-glucose isomer 1	ND	ND	ND	+	+	+	ND	ND	ND
32	Galloylnorbergenin isomer 3	+	+	ND	+	+	+	ND	ND	ND
33	Trigalloyl-glucoside isomer 2	ND	+	ND	+	+	+	ND	+	ND
34	HHDP-galloylglucose isomer 2	ND	+	ND	+	+	+	ND	ND	ND
35	Trigalloyl-glucoside isomer 3	ND	+	ND	+	+	+	ND	+	ND
36	Digalloyl-HHDP-glucose isomer 2	ND	+	ND	+	+	+	ND	ND	ND
37	HHDP-galloylglucose isomer 3	ND	+	ND	+	+	+	ND	ND	ND
38	Trigalloyl-glucoside isomer 4	ND	+	ND	+	+	+	ND	ND	ND
39	Terflavin B isomer 1	ND	+	ND	+	ND	ND	ND	ND	ND
40	Galloyl-HHDP-glucose isomer 1	ND	+	ND	+	+	+	ND	ND	ND

Table 1. *Cont.*

Peak	Compound	Hexane	Ethyl Acetate	Dichloromethane	Ethanol	Ethanol/Water	Water	NADES-A	NADES-B	NADES-C
41	Tetragalloyl-glucoside isomer 1	ND	+	ND	+	+	+	ND	ND	ND
42	Balanophotannin E isomer 1	ND	+	ND	+	+	+	ND	ND	ND
43	Trigalloyl-HHDP-glucose isomer 1	ND	ND	ND	ND	+	ND	ND	ND	ND
44	Terflavin B isomer 2	ND	+	ND	+	ND	ND	ND	ND	ND
45	Galloflavin	+	+	ND	ND	ND	ND	ND	ND	ND
46	Ellagic acid	+	+	ND	+	+	+	+	+	ND
47	Terflavin B isomer 3	+	+	ND	+	ND	ND	ND	ND	ND
48	Trigalloyl-HHDP-glucose isomer 2	+	+	ND	+	+	+	ND	ND	ND
49	Digalloyl-lactonised valoneoyl-d-glucose isomer 1	+	+	ND	+	ND	ND	ND	ND	ND
50	Trigalloyl-brevifolincarboxyl-glucose isomer 1	ND	+	ND	+	+	+	ND	ND	ND
51	Tetragalloyl-glucoside isomer 2	+	+	ND	+	+	+	+	ND	+
52	Catechin	ND	+	ND	+	+	+	ND	ND	ND
53	Tetragalloyl-glucoside isomer 3	+	+	ND	+	+	+	+	ND	+
54	Digallate isomer 2	ND	+	ND	+	+	+	ND	ND	ND
55	Tetragalloyl-glucoside isomer 4	+	+	ND	+	+	+	+	ND	ND

Table 1. *Cont.*

Peak	Compound	Hexane	Ethyl Acetate	Dichloromethane	Ethanol	Ethanol/Water	Water	NADES-A	NADES-B	NADES-C
56	Digallate isomer 3	ND	+	ND	+	+	+	ND	ND	ND
57	Terflavin B isomer 4	+	+	ND	+	+	ND	ND	ND	ND
58	Galloyl-HHDP-glucose isomer 2	+	+	ND	+	+	+	+	ND	ND
59	Tetragalloyl-glucoside isomer 5	+	+	ND	+	+	+	+	ND	ND
60	Neochebulagic acid isomer 1	ND	+	ND	+	+	+	ND	ND	ND
61	Neochebulagic acid isomer 2	ND	+	ND	+	+	+	ND	ND	ND
62	Isorhamnetin glucoside isomer 1	+	+	ND	+	+	+	ND	ND	ND
63	Epicatechin	ND	+	ND	+	+	+	ND	ND	ND
64	Quercetin	ND	+	ND	ND	+	ND	ND	ND	ND
65	Balanophotannin E isomer 2	ND	+	ND	+	+	+	ND	ND	ND
66	Unknown 6	+	+	ND	+	+	+	ND	ND	ND
67	Digalloyl-lactonised valoneoyl-d-glucose isomer 2	ND	+	ND	+	ND	ND	ND	ND	ND
68	Trigalloyl-brevifolincarboxyl-glucose isomer 2	+	+	ND	+	+	+	ND	ND	ND
69	Pentagalloyl-glucose isomer 1	+	+	ND	+	+	+	ND	ND	ND
70	Neochebulagic acid isomer 3	ND	ND	ND	ND	+	ND	ND	ND	ND

Table 1. *Cont.*

Peak	Compound	Hexane	Ethyl Acetate	Dichloromethane	Ethanol	Ethanol/Water	Water	NADES-A	NADES-B	NADES-C
71	Trigalloyl-lactonised valoneoyl glucose isomer 1	ND	+	ND	+	+	ND	ND	ND	ND
72	(Galloyl)galloyl-tetragalloylglucose isomer 1	ND	+	ND	+	+	+	ND	ND	ND
73	Trigalloyl-brevifolincarboxyl-glucose isomer 3	+	+	ND	+	+	+	ND	ND	ND
74	Trigalloyl-DHHDP-glucose isomer 1	+	+	ND	ND	ND	ND	ND	ND	ND
75	Digalloyl-lactonised valoneoyl-d-glucose isomer 3	+	+	ND	+	+	+	+	ND	ND
76	Castalagin	+	+	ND	ND	ND	ND	ND	ND	ND
77	Trigalloyl-HHDP-glucose isomer 3	+	+	ND	+	+	+	+	ND	ND
78	Galloyl-penta-hydroxy-benzoic-brevifolincarboxyl-glucose isomer 1	+	+	ND	+	+	+	+	ND	ND
79	Pentagalloyl-glucose isomer 2	+	+	ND	+	+	+	+	ND	ND
80	Trisgalloyl HHDP glucose isome isomer 1	+	+	ND	ND	ND	ND	ND	ND	ND
81	Amurensisin	+	+	ND	+	+	+	ND	ND	ND

Table 1. *Cont.*

Peak	Compound	Hexane	Ethyl Acetate	Dichloromethane	Ethanol	Ethanol/Water	Water	NADES-A	NADES-B	NADES-C
82	Trigalloyl-DHHDP-glucose isomer 2	ND	+	ND	+	ND	+	ND	ND	ND
83	Digalloyl-lactonised valoneoyl-d-glucose isomer 4	+	+	ND	+	ND	ND	ND	ND	ND
84	Pentagalloyl-glucose isomer 3	+	+	ND	+	+	+	+	ND	ND
85	Galloyl-penta-hydroxy-benzoic-brevifolincarboxyl-glucose isomer 2	+	+	ND	+	+	+	ND	ND	ND
86	Trigalloyl-brevifolincarboxyl-glucose isomer 4	+	+	ND	+	+	ND	ND	ND	ND
87	Phyllanthusiin C isomer 1	+	+	ND	+	+	+	ND	ND	ND
88	Trigalloyl-DHHDP-glucose isomer 3	+	+	ND	ND	+	+	ND	ND	ND
89	Ethyl gallate	ND	ND	ND	+	+	+	ND	ND	ND
90	Trigalloyl-brevifolincarboxyl-glucose isomer 5	+	+	ND	+	+	+	ND	ND	ND
91	Trisgalloyl HHDP glucose isome isomer 2	ND	+	ND	ND	ND	+	ND	ND	ND

Table 1. *Cont.*

Peak	Compound	Hexane	Ethyl Acetate	Dichloromethane	Ethanol	Ethanol/Water	Water	NADES-A	NADES-B	NADES-C
92	Trigalloyl-lactonised valoneoyl glucose isomer 2	+	+	ND	+	ND	ND	ND	ND	ND
93	(Galloyl)galloyl-tetragalloylglucose isomer 2	+	+	ND	+	+	+	+	ND	ND
94	Digalloyl-HHDP-iso DHDG-glucose isomer 1	+	+	ND	+	ND	ND	ND	ND	ND
95	Balanophotannin E isomer 3	ND	+	ND	+	+	ND	ND	ND	ND
96	Isorhamnetin glucoside isomer 2	ND	+	ND	+	+	ND	ND	ND	ND
97	Hexagalloyl-glucose isomer 1	+	+	ND	+	+	+	+	ND	ND
98	Trigalloyl-lactonised valoneoyl glucose isomer 3	+	+	ND	+	+	ND	ND	ND	ND
99	Ellagic acid derivative	+	+	ND	ND	ND	+	ND	ND	ND
100	(Galloyl)galloyl-tetragalloylglucose isomer 3	+	+	ND	+	+	+	ND	ND	ND
101	(Galloyl)galloyl-tetragalloylglucose isomer 4	+	+	ND	+	+	+	ND	ND	ND
102	Galloyl-penta-hydroxy-benzoic-brevifolincarboxyl-glucose isomer 3	ND	+	ND	+	+	+	ND	ND	ND

Table 1. *Cont.*

Peak	Compound	Hexane	Ethyl Acetate	Dichloromethane	Ethanol	Ethanol/Water	Water	NADES-A	NADES-B	NADES-C
103	Hexagalloyl-glucose isomer 2	+	+	ND	+	+	+	+	ND	ND
104	Galloyl-HHDP-glucose isomer 3	+	+	ND	+	+	+	+	ND	ND
105	Hexagalloyl-glucose isomer 3	+	+	ND	+	+	+	+	ND	ND
106	Digalloyl-HHDP-iso DHDG-glucose isomer 2	ND	+	ND	ND	+	+	ND	ND	ND
107	Tetragalloyl-hydroxybenzoyl-glucopyranoside isomer 1	ND	+	ND	+	+	ND	ND	ND	ND
108	Heptagalloyl hexose isomer 1	+	+	ND	+	+	+	ND	ND	ND
109	Galloylmyricetin	+	+	ND	+	+	+	+	ND	ND
110	Heptagalloyl hexose isomer 2	+	+	ND	+	+	+	+	ND	ND
111	Phyllanthusiin C isomer 2	ND	+	ND	ND	+	+	ND	ND	ND
112	Digalloyl-HHDP-iso DHDG-glucose isomer 3	ND	+	ND	ND	+	+	ND	ND	ND
113	Trigalloyl-brevifolincarboxyl-glucose isomer 6	ND	+	ND	+	+	ND	ND	ND	ND
114	Heptagalloyl hexose isomer 3	+	+	ND	+	+	+	ND	ND	ND
115	Tetragalloyl-hydroxybenzoyl-glucopyranoside isomer 2	+	+	ND	+	+	+	ND	ND	ND

Table 1. *Cont.*

Peak	Compound	Hexane	Ethyl Acetate	Dichloromethane	Ethanol	Ethanol/Water	Water	NADES-A	NADES-B	NADES-C
116	Tetragalloyl-hydroxybenzoyl-glucopyranoside isomer 3	+	+	ND	+	ND	+	ND	ND	ND
117	Unknown 7	ND	ND	ND	+	ND	ND	ND	ND	ND
118	Trihydroxy-octadecenoic acid	ND	+	+	ND	ND	ND	ND	ND	ND
119	Hydroxyretinoic acid	+	ND	+	ND	ND	ND	ND	ND	ND
120	Unknown 8	+	+	+	ND	ND	ND	ND	ND	ND
121	Hexadecanedioic acid	+	+	+	ND	ND	ND	ND	ND	ND
122	Hydroxyeicosatrienoic acid	+	+	+	+	ND	ND	ND	ND	ND
123	Valerenic acid	+	+	+	+	ND	ND	ND	ND	ND
124	Hydroxylinoleic acid	+	+	+	ND	ND	ND	ND	ND	ND
125	Linoleic acid	ND	+	+	+	ND	ND	ND	ND	ND
126	Hydroxylinolenic acid	+	+	+	+	+	ND	ND	ND	ND
127	Dodecenyl-succinic anhydride	+	+	+	+	ND	ND	ND	ND	ND
128	Retinoic acid	+	+	+	ND	ND	ND	ND	ND	ND
129	Oleic acid	+	+	+	+	+	ND	ND	+	ND
130	Pentadecenoic acid	+	+	+	+	+	ND	ND	ND	ND
131	Oleiyl glucoside	ND	+	+	ND	ND	ND	ND	ND	ND
132	Linolenic acid	+	+	+	+	+	+	ND	ND	ND
133	Stearic acid	+	+	ND	+	ND	ND	ND	ND	ND
134	Unknown 9	+	+	+	+	+	ND	ND	ND	ND
135	Unknown 10	+	+	+	+	ND	ND	ND	ND	ND

Table 1. *Cont.*

Peak	Compound	Hexane	Ethyl Acetate	Dichloromethane	Ethanol	Ethanol/Water	Water	NADES-A	NADES-B	NADES-C
136	Eicosapentaenoic acid	+	+	+	+	ND	+	ND	ND	ND
137	Palmitoleic acid	+	+	+	ND	+	+	+	ND	+
138	Linoleic acid	+	+	+	+	+	+	ND	ND	+
139	Oxodecanedioic acid	+	ND	ND	ND	ND	ND	ND	ND	ND
140	Methyl arachidonate	+	+	+	+	ND	ND	ND	ND	ND
141	Arjungenin	ND	ND	+	ND	ND	ND	ND	ND	ND
142	Heptadecenoic acid	+	ND	ND	ND	+	+	ND	ND	ND
143	Unknown 11	ND	+	+	+	ND	ND	ND	ND	ND
144	Unknown 12	+	+	+	+	+	+	ND	ND	ND
145	Glycerylmonooleate	+	ND	ND	ND	ND	ND	ND	ND	ND
146	Palmitic acid	+	ND	ND	+	+	+	+	+	+
147	Hydroxydocosanoic acid	+	+	+	+	+	ND	ND	ND	ND
148	Dodecenylsuccinic acid	ND	+	ND	ND	+	ND	ND	ND	ND

+: present; ND: nondetected.

As the second most important group, ellagitannins contains 30 compounds, which were annotated as different mono-, di-, and tri-galloyl-DHHDP-glucose isomers and digalloyl-HHDP-iso DHDG-glucose isomers. Besides these compounds, other ellagitannins such as terflavin B, phyllanthussin C, geraniin, and balanophlorotannin E isomers have been previously reported in various *Terminalia* species and in *Trapa* species [34,35]. In case of our species, only peaks corresponding to m/z 937 and 783 were annotated [24].

Regarding flavonoids, catechin and epicatechin as flavan-3-ols were found; quercetin and two isorhamnetin glucoside isomers as flavonols were also detected, mostly with water, ethanol, their mixture, and ethylacetate.

On the other hand, hexane and dichloromethane extracts presented the major number of fatty acids (Peaks 118, 119, 121–132, 135–141, 144–148). Regarding NADESs, the three extracts showed the lower number of features, and many of the features obtained could not be annotated such as peaks 7 and 9 (unknowns 3 and 4) were only found in NADES-C extract. The same happened for peak 2 (unknown 2); this compound was only detected in NADES-B. Among them, NADES-A presented high number of features corresponding mainly to gallotannins. This highlights the potential of NADESs to obtain extracts containing additional bioactive compounds and to "tailor" the properties of extracts. Secondly, extraction by two NADESs sequentially should allow for the fractionation of bioactive compounds.

Other compounds have been also annotated, although they have not been classified in specific groups due to the low number and the high range of structures. For instance, organic acids (quinic and fukiic acids); simple phenols (gallic and ellagic acids); and two alkyl-phenylketones, namely brevifolin and brevifolin carboxylic acid [36], were included in this group. In addition, three isomers from galloylnorbergenin, antioxidant isocoumarins that were previously found in leaves of *Diospyros gilletii* De Wild [37] were also included.

2.3. Antioxidant Effects

Antioxidants are important chemical substances that occur naturally in food and can reduce or prevent oxidative stress of the physiological system as the body continuously produces free radicals. Oxidative stress plays a key role in the development of chronic and degenerative diseases such as cancer, autoimmune diseases, and neurodegenerative and cardiovascular diseases. The human body has a variety of mechanisms to counter oxidative stress by producing antioxidants that are either naturally produced in situ or supplied externally through foods such as plants, as a rich source of naturally produced antioxidants. Hence, antioxidants acting as free radical scavengers can eventually help to avert and repair cellular damage generated by these radicals [38,39].

Herein, moderate-to-very-potent antioxidant activity was noted for *C. hypocistis* extracts. For instance, in the radical scavenging assays, the ethanol/water, ethanol, ethyl acetate, water, and NADES-A extracts demonstrated very high antioxidant potential (DPPH: 829.11–939.35 mg TE/g; ABTS: 2830.66–4026.50 mg TE/g), followed by NADES-C and NADES-B (DPPH: 701.49 and 767.55 mg TE/g; ABTS: 2134.94 and 2285.15 mg TE/g, respectively). On the other hand, a much lower scavenging ability was displayed by the hexane and dichloromethane extracts (DPPH: 70.19 and 93.25 mg TE/g; 172.56 and 398.03 mg TE/g, respectively), compared to the other extracts. The same trend was observed with the reducing assays, whereby the ethanol/water, ethanol, ethyl acetate, water, and NADES-A extracts showed significant reducing capacity in the range of 1730.38–1377.38 mg TE/g and 968.98–1534.85 mg TE/g in CUPRAC and FRAP assays, respectively, whereas a lower reducing activity was noted for NADES-B and NADES-C and a much lower content for hexane and dichloromethane extracts (CUPRAC: 97.41–774.94 mg TE/g; FRAP: 84.11–860.90 mg TE/g) (Figure 1, Table S4). It is clear that most valuable bioactive components are extracted by relatively polar extracts, while low-polarity solvents such as n-hexane and dichloromethane are not effective in this case. In Figure 1, Pearson's correlation analysis indicates a linear correlation ($R > 0.7$) between total flavanoid content and radical scavenging and reducing power assays.

While only NADES-B did not possess metal-chelating activity, all the other extracts showed metal-chelating potential in the range of 6.87–21.76 mg EDTAE/g. The extracts also demonstrated total antioxidant capacity in phosphomolybdenum assay (1.27–3.94 mmol TE/g). The lowest total antioxidant capacity was revealed for hexane and dichloromethane extracts (Figure 1). This part of the research reveals a possibility to isolate metal-chelating components by step extraction using NADES-B at the first stage, followed by a second extractant that is effective to extract this specific group of compounds present in the studied plant.

2.4. Enzyme Inhibitory Effects

Low levels of the neurotransmitter acetylcholine, oxidative stress, and inflammation in the central nervous system (CNS) are hallmarks of Alzheimer's disease (AD), a progressive neurodegenerative disease. To date, patients diagnosed with AD are only offered enzyme inhibitors (acetylcholinesterase/butyrylcholinesterase, or AChE/BChE) for treatment [40]. Hence, as mechanism of pharmacological action, these cholinesterase inhibitors are able to modify cholinergic signalling by disrupting the degradation of acetylcholine [41].

All extracts, except NADES-B and NADES-C extracts, possessed anti-AChE activity (7.32–15.16 mg GALAE/g). Interestingly, the highest anti-AChE activity was revealed for NADES-A. It is particularly important observation, as components of NADES-A—choline chloride and urea—are nontoxic; thus, such obtained extracts could be further used without removal of NADES. On the other hand, only the extracts prepared in the traditional way, with solvents hexane, ethyl acetate, dichloromethane, ethanol, and ethanol/water, displayed anti-BChE potential (1.39–2.13 mg GALAE/g). The water extract and the NADES extracts showed no anti-BChE activity. It follows from the ionic nature of the used NADESs, which in this case was not favourable for extraction. The ethanol extract demonstrated relatively higher BChE inhibitory effect compared to the other extracts (Figure 3).

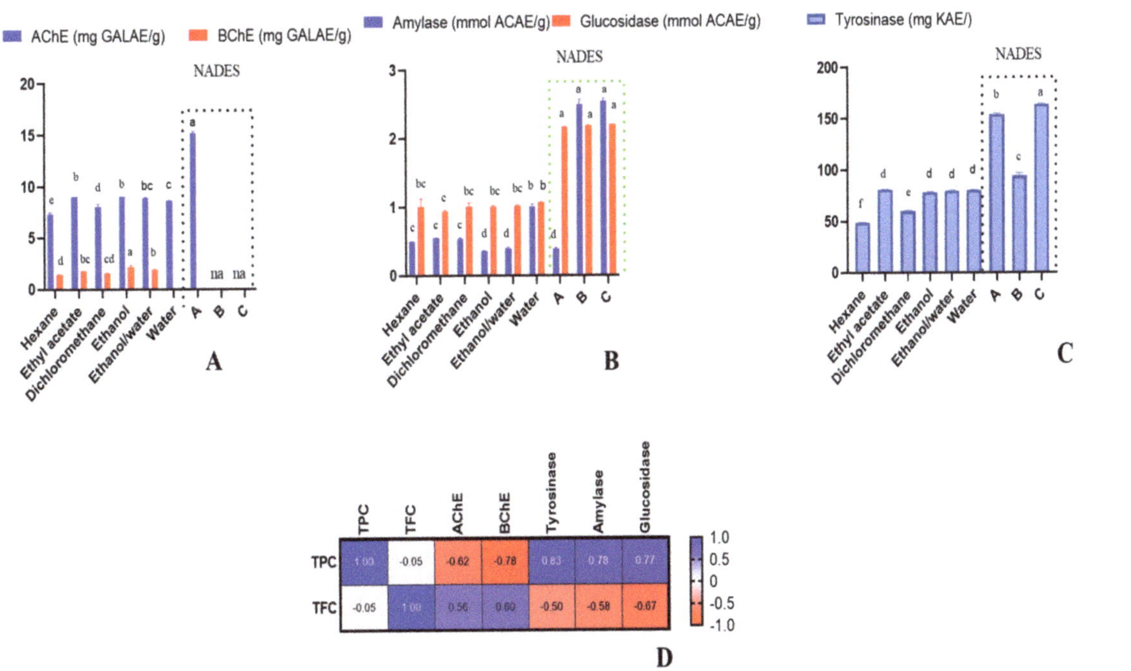

Figure 3. Cholinesterase inhibitory effects (**A**), amylase and glucosidase inhibitory effects (**B**), tyrosinase inhibitory effects (**C**), Pearson's correlations between total bioactive compounds and enzyme inhibitory assays ($p < 0.05$) (**D**). na: not active. Different letters in column for same assays indicate significant differences in the extracts ($p < 0.05$).

Tyrosinase is the rate-limiting enzyme in melanin synthesis. Melanin is synthesised in human melanocytes when tyrosine is hydroxylated to l-DOPA, which is then oxidised to dopaquinone and polymerises to form melanin. Melasma, melanoma, and freckles are just some of the dermatological conditions that can develop when melanin production increases too rapidly. However, tyrosinase in plant-based foods oxidises phenolic compounds into quinones. The former reacts with amino acids and proteins to produce brown/black pigments, a process known as enzymatic browning, which is one of the most pressing problems in the food industry and the source of 50 percent of the industry's economic losses. In addition, browning reduces the food's nutritional value and safety because it leads to the loss of vitamin C, antioxidants, and other nutrients, and can even lead to the production of antinutritional and toxic substances. Consequently, tyrosinase inhibition is seen as an efficient method for preventing hyperpigmentation in the pharmaceutical industry and delaying enzymatic browning, which is helpful in the food industry [42].

In the current investigation, all the studied extracts were found to possess antityrosinase activity (49.14–153.97 mg KAE/g). However, NADES-A exhibited the highest inhibitory activity against tyrosinase, while the hexane extract displayed the lowest (Figure 3). On this basis, it is clear that polar NADES-A, as well as other polar solvents, should be preferred for the extraction of bioactive compounds responsible for tyrosinase activity.

Interestingly, in the study by Zucca et al. [2], the ethanolic extract of *C. hypocistis* showed the highest tyrosinase inhibition activity, compared to cyclohexane and water extracts, probably due to being predominantly rich in polyphenols, in most part hydrolysable tannins. *C. hypocistis* aerial part extract was also reported to show tyrosinase inhibition of 80%, when tested at 50 µg/mL [43]. Furthermore, a linear correlation was obtained between enzymatic activities and increasing TPC and TFC. The reason could be a specific class of polyphenols acting against tyrosinase through a competitive inhibition mechanism, thus interfering with the biological function of tyrosinase, which is a polyphenoloxidase [43].

The inhibition of the carbohydrate-digesting enzymes alpha-glucosidase and alpha-amylase is an important strategy for controlling blood glucose levels in patients with Type 2 diabetes and borderline diabetes, because it significantly reduces the postprandial rise in blood glucose [44]. Even though drugs such as voglibose, acarbose, and miglitol are commercially available as those enzymes' inhibitors and are also used in practice, they produce undesirous effects such as abdominal discomfort, bloating, and diarrhoea.

In addition, many chronic diseases such as diabetes are associated with oxidative stress, during which reactive oxygen species (O_2^-, H_2O_2 and OH^-) are generated. The role of free radicals in the onset and development of diabetes has also been established. Therefore, compounds that possess both antidiabetic and antioxidant properties without causing serious side effects would be of great value [45].

In the present investigation, all extracts were found to inhibit both carbohydrate-hydrolysing enzymes (Amylase: 0.35–2.54 mmol ACAE/g; glucosidase: 0.93–2.20 mmol ACAE/g). Remarkably, the NADES extracts were found to be better inhibitors of amylase and glucosidase compared the other extracts (Figure 3). This could be due to the higher TPC in the NADES extracts, and this fact was also confirmed by Pearson's correlation analysis (Figure 3). In fact, it has been previously suggested that phenolics are involved in the modulation of the activity of starch digestive enzymes [46].

2.5. Data Mining

To gain more insight into the tested extracts and biological activity assays, we performed PCA analysis. The results are given in Figure 4. Firstly, we examined the relationship between the tested extracts based on the biological activity results. We obtained a good distribution, and the tested extracts were very well-separated based on the biological activity results. Two components (PC1: 47.3% and PC2: 33.8) accounted for 811% of the total components. Two extracts (hexane and dichloromethane) exhibited the lowest biological abilities and were distributed in the same axis. In addition, polar extracts (ethanol, ethanol/water, water) and NADES-A had similar biological abilities and were placed in the same group. The PCA plot also confirmed a strong correlation between total flavonoid and antioxidant properties, which were very close to each other on the PCA plot. In addition to biological activity results, we investigated the similarities/differences of the tested extracts based on their chemical profiles. Two components were used in the analysis to determine the distribution of the tested solvents (PC1: 60.1% and PC2: 18.8%). In Figure 4b, the used nonpolar and polar solvents and NADESs were clearly separated, and these results were very similar to the distribution from the biological activity results. Taken together, we concluded that there is a good connection between chemical compounds and biological activities of *Cytinus* extracts.

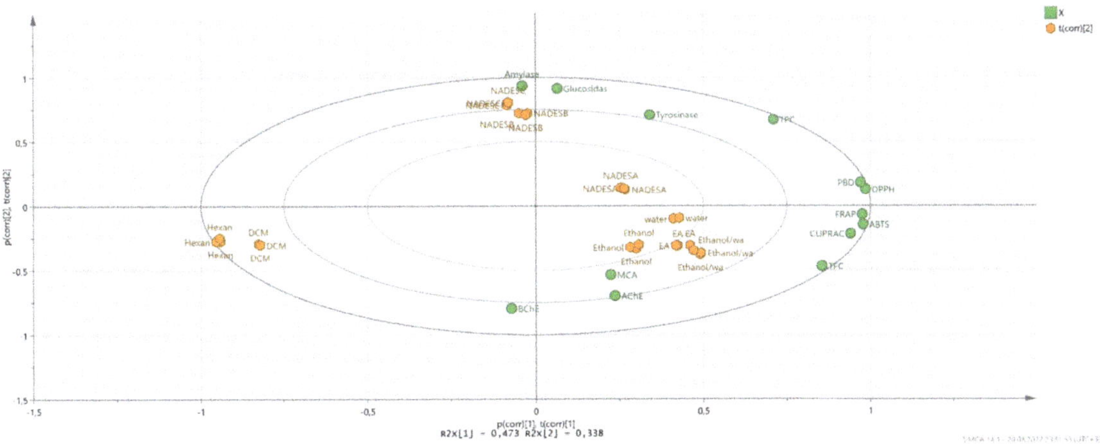

(a)

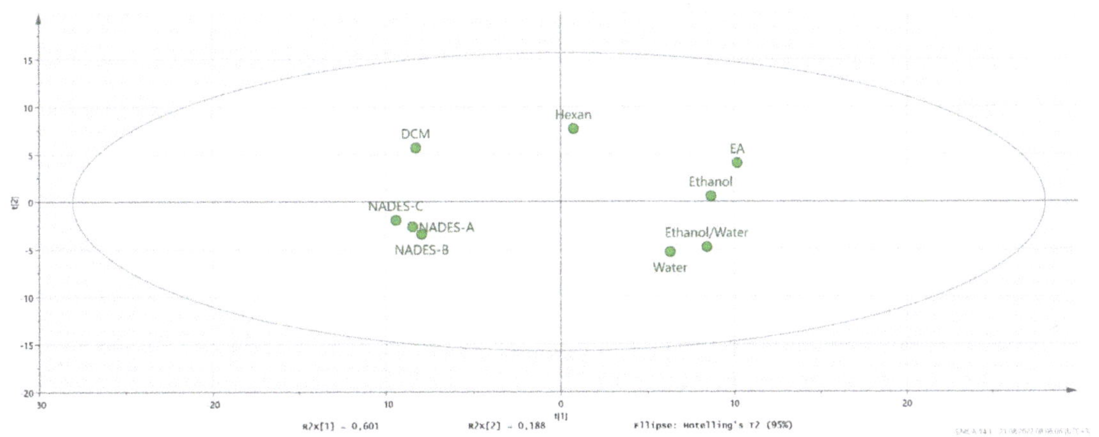

(b)

Figure 4. Principal component analysis between tested extracts and biological activities (**a**). Distribution of the tested extracts in principal component analysis by using chemical compound peak areas (**b**).

3. Materials and Methods

3.1. Materials

The investigated DESs were prepared from reagents of >99% purity (Sigma Aldrich, Burlington, VT, USA), while hydrochloric and sulphuric acid for DES protonation were of analytical reagent grade (POCH, Gliwice, Poland).

3.2. Apparatus

DES: chemicals were precisely prepared on a weight basis using a AS.310.R2 analytical balance (Radwag, Radom, Poland). A 06-MSH-PRO-T magnetic stirrer (Chemland, Stargard, Poland) was used to prepare DESs.

3.3. Preparation of NADESs

Three NADESs having different properties were used. NADES-A—choline chloride—urea 1:2; NADES-B protonated by HCl L-proline-xylitol 5:1; NADES-C, protonated by H_2SO_4 L-proline-xylitol 5:1.

Two synthesis routes were used.

NADES-A was prepared by simple mixing of compounds at 60 °C. The synthesis of the deep eutectic solvents based on protonated L-proline (NADES-A and -B) involved dissolving L-proline in an acid solution (the amount of acid with respect to L-proline was equimolar). Next, a predetermined amount of xylitol was added to the solution. Thus, prepared solution was placed in a rotary evaporator model Rotavapor R-300 (Buchi, Flawil, Switzerland) and water was distilled off under reduced pressure. Studies on the synthesis of this type of DESs and their physicochemical characteristics were the subject of a separate paper [47].

3.4. Plant Material and Preparation of Extracts

Cytinus hypocistis samples were collected at the flowering season in June 2020 (Anamur, Mersin, Turkey). The plants were identified by one of the authors (Dr. Evren Yildiztugay) and voucher specimens were deposited at the herbarium of Selcuk University, Konya, Turkey. The samples' aerial parts were dried in the shade at room temperature for about 7 days, and then ground into a powder using a mill. All of the samples were kept in a dark place.

In this study, the extracts were prepared using traditional solvents (n-hexane, ethyl acetate, dichloromethane, ethanol, ethanol/water (70%), and water) and NADEs. Maceration was employed as the extraction method to obtain n-hexane, ethyl acetate, dichloromethane, EtOH, and EtOH/water extracts. The plant materials (10 g) were macerated overnight at room temperature with 200 mL of these solvents. Finally, the solvents were evaporated from the mixtures. To obtain water extracts, the plant materials (10 g) were kept with 200 mL of boiled water, and then the extracts were filtered and lyophilised. In the preparation of NADE extracts, the plant materials (10 g) were mixed with the NADESs for 20 min at 25 °C in an ultrasonic bath. The extracts were filtered and all extracts were stored at 4 °C until further analysis was required.

3.5. Chemical Reagents

All chemicals were of HPLC-MS grade and used as received. Acetic acid and acetonitrile for UPLC were purchased from Fluka (Sigma-Aldrich, Steinheim, Germany) and Lab-Scan (Gliwice, Sowinskiego, Poland), respectively. For solutions, ultrapure water was obtained with a Milli-Q system Millipore (Bedford, MA, USA), and absolute ethanol was purchased from VWR chemicals (Radnor, PA, USA).

3.6. UPLC-ESI-QTOF-MS Conditions

Cytinus extracts were redissolved at 5 mg/mL in the same extraction solvent and filtered by 0.22 μm. The compounds were separated using an ACQUITY UPLC H-Class System (Waters Corp., Milford, MA, USA) with a reversed-phase column (ACQUITY UPLC BEH

Shield RP18, 130Å, 1.7 μm, 2.1 mm × 150 mm) at a flow rate of 0.7 mL/min and using a injection volume of 10 μL. The mobile phases were acidified water (0.5% acetic acid, *v*/*v*) and acetonitrile as solvents A and B, respectively. The following multi-step linear gradient was used in order to achieve an efficient separation: 0.00 min [A:B 99/1], 2.33 min [A:B 99/1], 4.37 min [A:B 93/7], 8.11 min [A:B 86/14], 12.19 min [A:B 76/24], 15.99 min [A:B 60/40], 18.31 min [A:B 2/98], 21.03 min [A:B 2/98], 22.39 min [A:B 99/1] and 25.00 [A:B 99/1].

The UPLC was coupled to an electrospray-quadrupole-time of flight mass spectrometer (ESI-QTOF-MS) Synapt G2 (Waters Corp., Milford, MA, USA) working in negative-ionisation mode in a *m*/*z* range from 50 to 1200 *m*/*z*. The MS acquisition was based on two parallel scan functions switching between them continuously. The first function was operated at low collision energy in the gas cell (4 eV) and the other at an elevated collision energy (MS^E energy linear ramp: from 20 to 60 eV). Leu-enkephalin was injected for mass calibration continuously. Other MS parameters were, as follows, source temperature 100 °C; scan duration 0.1 s; resolution 20,000 FWHM; desolvation temperature 500 °C; desolvation gas flow 700 L/h; capillary voltage 2.2 kV; cone voltage 30 V; cone gas flow 50 L/h. Finally, the acquired data were processed using MZmine 2.53 open-source software and Sirius 4.4.29.

3.7. UPLC-ESI-QTOF-MS Data Processing

Firstly, the raw data files were transformed to. mzML format using MSConvert software. The converted data were processed using the open-source software MZmine 2.53 (Pluskal et al., 2020). A noise level of 1.0×10^3 was selected. ADAP chromatogram builder method was used under the following parameters: MS level: 1; min number of scans: 9; group intensity threshold: 1.0×10^3; min highest intensity: 1.0×10^4; *m*/*z* tolerance: 10 ppm. After that, the chromatogram was deconvoluted using the Wavelets (ADAP) algorithm and the following parameters: S/N threshold: 50; min feature height: 5E4; coefficient/area threshold: 110; peak duration range: 0.05–0.3 min; RT wavelet range: 0–0.30. An isotopic peak grouper algorithm was also applied (*m*/*z* tolerance: 10 ppm, RT tolerance: 0.02 min, maximum charge: 2). The obtained features were aligned between samples using the "Join Aligner" algorithm using a *m*/*z* tolerance of 10 ppm and a RT tolerance of 0.1 min. The molecular features, which were also detected in blank samples, were filtered from the final dataset. Finally, the molecular formulas of the final features were predicted using Sirius 4.4.29 (Dührkop et al., 2019) and the biological identities were annotated by comparing the MS/MS spectra of different databases (e.g., MoNA, Massbank, HMDB, FoodDB, etc.), with the fragments detected in the MSE scans.

3.8. Determination of Total Polyphenol and Flavonoids Contents

Total phenolic and flavonoid contents were calculated with the Folin-Ciocalteu and $AlCl_3$ assays, respectively [48]. Gallic acid equivalents (mg GAEs/g dry extract) and rutin equivalents (mg REs/g dry extract) were used to describe the outcomes of the two tests.

3.9. Antioxidant and Enzyme Inhibitory Assays

The antioxidant and enzyme inhibitory activity of comfrey root extracts was assessed according to methods presented previously [49,50]. Data were expressed as: mg Trolox equivalents (TE)/g extract in ferric ion reducing antioxidant power (FRAP), cupric ion reducing antioxidant capacity (CUPRAC), ABTS and DPPH radical scavenging activity; mg EDTA equivalents (EDTAE)/g extract in the metal chelating ability (MCA), mmol TE/g extract in the phosphomolybdenum assay (PBD); mg galanthamine equivalents (GALAE)/g extract in AChE and BChE assays; mg kojic acid equivalents (KAE)/g extract in tyrosinase inhibitory assay; and mmol acarbose equivalents (ACAE)/g extract in amylase and glucosidase assays.

3.10. Data Analysis

All analyses were performed in triplicate and results were reported as means ± SD. Pearson's correlation coefficients were calculated between total bioactive components and biological activity parameters. Pearson's correlation was performed by GraphPad version 9.0. The relationship between species, chemical compounds and bioactivities was also assessed using principal component analysis (PCA). PCA analysis was performed by SIMCA version 14.0.

4. Conclusions

In the present study, nine extracts of *C. hypocistis* obtained using traditional solvents and NADESs were investigated for their total polyphenolic contents, antioxidant and enzyme inhibitory properties. The extracts were found to be richer in TPC than TFC. In particular, the NADES extracts were found to yield higher TPC compared to the other extracts. On the other hand, generally, ethanol/water, ethanol, and water showed very potent and better antioxidant potential compared to the other extracts, whereas hexane and dichloromethane exhibited weaker antioxidant potential in almost all antioxidant assays. While NADES-A extract displayed the highest anti-AChE activity, none of the NADES extracts displayed inhibition against the butylcholinesterase. With the exception of the water extract, all the other traditional solvent extracts showed dual cholinesterase inhibitory properties. Remarkably, the NADES extracts were found to have an enhanced antityrosinase effect compared to the traditional solvent extracts. Similarly, the NADES extracts were found to be better glucosidase inhibitors, and the NADES-B and -C extracts showed higher antiamylase activity in comparison with the other studied extracts. It is worth highlighting that this study demonstrated NADESs to have brought some improved abilities in bioactive content yields and bioactivity compared to the traditional extracts, especially in terms of TPC and some of the enzyme inhibitory activities. However, the traditional solvents were much better in extracting the antioxidant compounds. Hence, the current investigation enabled a comparison between traditional solvents and NADESs and suggested the potential of NADESs as an alternative to traditional organic solvents for higher extraction of phytonutrients and some better biological performance, although the traditional solvents were found to be more effective in yielding higher antioxidant activity. The paper proved the importance of natural medicine where sources of pro-health components are taken from plants. NADESs as mixtures obtained from compounds of natural origin, in respect to the studied examples, are nontoxic; thus, their extracts could be used as components of functional foods as well as food additives without the need of NADES removal.

Supplementary Materials: The following supporting information can be downloaded at: https://www.mdpi.com/article/10.3390/molecules27185788/s1, Table S1: Total phenolic (TPC) and flavonoid content (TFC) of the tested extracts; Table S2: Proposed annotated compounds by UPLC-ESI-QTOF-MS in all Cytinus extracts; Table S3: Compound areas extracted for each Cytinus extract; Table S4: Antioxidant properties of the tested extracts; Table S5: Enzyme inhibitory of the tested extracts.

Author Contributions: Conceptualisation, G.Z., M.d.l.L.C.-G. and G.B.; methodology, G.Z., M.d.l.L.C.-G., Á.F.-O., F.J.L.-J., M.M., E.Y., R.K. and G.B.; software, G.Z., M.d.l.L.C.-G. and G.B.; validation, R.K., S.J. and M.F.M.; formal analysis, G.Z.; investigation, G.Z., S.J. and M.F.M.; resources, G.Z., E.Y. and R.K.; data curation, G.Z., M.d.l.L.C.-G. and M.F.M.; writing—original draft preparation, G.Z., M.d.l.L.C.-G., S.J. and M.F.M.; writing—review and editing, M.M. and G.B.; visualisation, G.Z.; supervision, G.B., A.S.C.; project administration, G.Z. and G.B.; funding acquisition, G.Z. and A.S.C. All authors have read and agreed to the published version of the manuscript.

Funding: M. Momotko and G. Boczkaj gratefully acknowledge the financial support from the National Science Centre, Warsaw, Poland—decision no. UMO-2018/30/E/ST8/00642.

Institutional Review Board Statement: Not applicable.

Informed Consent Statement: Not applicable.

Data Availability Statement: Not applicable.

Acknowledgments: The authors M.d.l.L.C.-G. and Á.F.-O. would like to thank the Regional Ministry of Economy, Knowledge, Enterprise and Universities of Andalusia for the contract for Young Researchers (PAIDI) at the University of Granada. F.J.L.-J. thanks the Spanish Ministry of Science and Innovation for the postdoctoral contract Juan de la Cierva-Formación (FJC2020-044298-I).

Conflicts of Interest: The authors declare no conflict of interest.

References

1. Sanjust, E.; Rinaldi, A.C. Cytinus under the Microscope: Disclosing the Secrets of a Parasitic Plant. *Plants* **2021**, *10*, 146. [CrossRef]
2. Zucca, P.; Pintus, M.; Manzo, G.; Nieddu, M.; Steri, D.; Rinaldi, A.C. Antimicrobial, antioxidant and anti-tyrosinase properties of extracts of the Mediterranean parasitic plant *Cytinus hypocistis*. *BMC Res. Notes* **2015**, *8*, 562. [CrossRef]
3. Maisetta, G.; Batoni, G.; Caboni, P.; Esin, S.; Rinaldi, A.C.; Zucca, P. Tannin profile, antioxidant properties, and antimicrobial activity of extracts from two Mediterranean species of parasitic plant Cytinus. *BMC Complementary Altern. Med.* **2019**, *19*, 82. [CrossRef] [PubMed]
4. Mandrone, M.; Bonvicini, F.; Lianza, M.; Sanna, C.; Maxia, A.; Gentilomi, G.; Poli, F. Sardinian plants with antimicrobial potential. Biological screening with multivariate data treatment of thirty-six extracts. *Ind. Crops Prod.* **2019**, *137*, 557–565. [CrossRef]
5. Silva, A.R.; Pinela, J.; García, P.A.; Ferreira, I.C.; Barros, L. *Cytinus hypocistis* (L.) L.: Optimised heat/ultrasound-assisted extraction of tannins by response surface methodology. *Sep. Purif. Technol.* **2021**, *276*, 119358. [CrossRef]
6. Magiatis, P.; Pratsinis, H.; Kalpoutzakis, E.; Konstantinidou, A.; Davaris, P.; Skaltsounis, A.-L. Hydrolyzable tannins, the active constituents of three Greek Cytinus taxa against several tumor cell lines. *Biol. Pharm. Bull.* **2001**, *24*, 707–709. [CrossRef]
7. Silva, A.R.; Ayuso, M.; Pereira, C.; Dias, M.I.; Kostić, M.; Calhelha, R.C.; Soković, M.; García, P.A.; Ferreira, I.C.; Barros, L. Evaluation of parasite and host phenolic composition and bioactivities—The Practical Case of *Cytinus hypocistis* (L.) L. and *Halimium lasianthum* (Lam.) Greuter. *Ind. Crops Prod.* **2022**, *176*, 114343. [CrossRef]
8. Koçak, E.; Pazır, F. Effect of Extraction Methods on Bioactive Compounds of Plant Origin. *Turk. J. Agric. Food Sci. Technol.* **2018**, *6*, 663–675.
9. Momotko, M.; Łuczak, J.; Przyjazny, A.; Boczkaj, G. A natural deep eutectic solvent-protonated L-proline-xylitol-based stationary phase for gas chromatography. *J. Chromatogr. A* **2022**, *1676*, 463238. [CrossRef]
10. Momotko, M.; Łuczak, J.; Przyjazny, A.; Boczkaj, G. First deep eutectic solvent-based (DES) stationary phase for gas chromatography and future perspectives for DES application in separation techniques. *J. Chromatogr. A* **2021**, *1635*, 461701. [CrossRef]
11. Faraz, N.; Haq, H.U.; Balal Arain, M.; Castro-Muñoz, R.; Boczkaj, G.; Khan, A. Deep eutectic solvent based method for analysis of Niclosamide in pharmaceutical and wastewater samples—A green analytical chemistry approach. *J. Mol. Liq.* **2021**, *335*, 116142. [CrossRef]
12. Serna-Vázquez, J.; Ahmad, M.Z.; Boczkaj, G.; Castro-Muñoz, R. Latest Insights on Novel Deep Eutectic Solvents (DES) for Sustainable Extraction of Phenolic Compounds from Natural Sources. *Molecules* **2021**, *26*, 5037. [CrossRef] [PubMed]
13. Khajavian, M.; Vatanpour, V.; Castro-Muñoz, R.; Boczkaj, G. Chitin and derivative chitosan-based structures—Preparation strategies aided by deep eutectic solvents: A review. *Carbohydr. Polym.* **2022**, *275*, 118702. [CrossRef]
14. Castro-Muñoz, R.; Msahel, A.; Galiano, F.; Serocki, M.; Ryl, J.; Hamouda, S.B.; Hafiane, A.; Boczkaj, G.; Figoli, A. Towards azeotropic MeOH-MTBE separation using pervaporation chitosan-based deep eutectic solvent membranes. *Sep. Purif. Technol.* **2022**, *281*, 119979. [CrossRef]
15. Grozdanova, T.; Trusheva, B.; Alipieva, K.; Popova, M.; Dimitrova, L.; Najdenski, H.; Zaharieva, M.M.; Ilieva, Y.; Vasileva, B.; Miloshev, G. Extracts of medicinal plants with natural deep eutectic solvents: Enhanced antimicrobial activity and low genotoxicity. *BMC Chem.* **2020**, *14*, 73. [CrossRef] [PubMed]
16. Hikmawanti, N.P.E.; Ramadon, D.; Jantan, I.; Mun'im, A. Natural deep eutectic solvents (Nades): Phytochemical extraction performance enhancer for pharmaceutical and nutraceutical product development. *Plants* **2021**, *10*, 2091. [CrossRef]
17. Pavlić, B.; Mrkonjić, Ž.; Teslić, N.; Kljakić, A.C.; Pojić, M.; Mandić, A.; Stupar, A.; Santos, F.; Duarte, A.R.C.; Mišan, A. Natural Deep Eutectic Solvent (NADES) Extraction Improves Polyphenol Yield and Antioxidant Activity of Wild Thyme (*Thymus serpyllum* L.) Extracts. *Molecules* **2022**, *27*, 1508. [CrossRef]
18. Zhang, Q.-W.; Lin, L.-G.; Ye, W.-C. Techniques for extraction and isolation of natural products: A comprehensive review. *Chin. Med.* **2018**, *13*, 20. [CrossRef] [PubMed]
19. Altemimi, A.; Lakhssassi, N.; Baharlouei, A.; Watson, D.G.; Lightfoot, D.A. Phytochemicals: Extraction, isolation, and identification of bioactive compounds from plant extracts. *Plants* **2017**, *6*, 42. [CrossRef]
20. Liu, Y.; Friesen, J.B.; McAlpine, J.B.; Lankin, D.C.; Chen, S.-N.; Pauli, G.F. Natural deep eutectic solvents: Properties, applications, and perspectives. *J. Nat. Prod.* **2018**, *81*, 679–690. [CrossRef]
21. Do, Q.D.; Angkawijaya, A.E.; Tran-Nguyen, P.L.; Huynh, L.H.; Soetaredjo, F.E.; Ismadji, S.; Ju, Y.-H. Effect of extraction solvent on total phenol content, total flavonoid content, and antioxidant activity of Limnophila aromatica. *J. Food Drug Anal.* **2014**, *22*, 296–302. [CrossRef] [PubMed]
22. Dewi, S.R.; Stevens, L.A.; Pearson, A.E.; Ferrari, R.; Irvine, D.J.; Binner, E.R. Investigating the role of solvent type and microwave selective heating on the extraction of phenolic compounds from cacao (*Theobroma cacao* L.) pod husk. *Food Bioprod. Process.* **2022**, *134*, 210–222. [CrossRef]

23. Belwal, T.; Ezzat, S.M.; Rastrelli, L.; Bhatt, I.D.; Daglia, M.; Baldi, A.; Devkota, H.P.; Orhan, I.E.; Patra, J.K.; Das, G. A critical analysis of extraction techniques used for botanicals: Trends, priorities, industrial uses and optimization strategies. *TrAC Trends Anal. Chem.* **2018**, *100*, 82–102. [CrossRef]
24. Silva, A.R.; Pinela, J.; Dias, M.I.; Calhelha, R.C.; Alves, M.J.; Mocan, A.; García, P.A.; Barros, L.; Ferreira, I.C. Exploring the phytochemical profile of *Cytinus hypocistis* (L.) L. as a source of health-promoting biomolecules behind its in vitro bioactive and enzyme inhibitory properties. *Food Chem. Toxicol.* **2020**, *136*, 111071. [CrossRef]
25. Luo, F.; Fu, Y.; Xiang, Y.; Yan, S.; Hu, G.; Huang, X.; Huang, G.; Sun, C.; Li, X.; Chen, K. Identification and quantification of gallotannins in mango (*Mangifera indica* L.) kernel and peel and their antiproliferative activities. *J. Funct. Foods* **2014**, *8*, 282–291. [CrossRef]
26. Namngam, C.; Pinsirodom, P.; Boonyuen, S. Fractionation, antioxidant and inhibitory activity of Thai mango seed kernel extracts. *Czech J. Food Sci.* **2018**, *36*, 8–15. [CrossRef]
27. Olennikov, D.N.; Chirikova, N.K.; Vasilieva, A.G.; Fedorov, I.A. LC-MS profile, gastrointestinal and gut microbiota stability and antioxidant activity of Rhodiola rosea herb metabolites: A comparative study with subterranean organs. *Antioxidants* **2020**, *9*, 526. [CrossRef]
28. Tong, N.-N.; Zhou, X.-Y.; Peng, L.-P.; Liu, Z.-A.; Shu, Q.-Y. A comprehensive study of three species of Paeonia stem and leaf phytochemicals, and their antioxidant activities. *J. Ethnopharmacol.* **2021**, *273*, 113985. [CrossRef] [PubMed]
29. Gok, H.N.; Pekacar, S.; Orhan, D.D. Investigation of Enzyme Inhibitory Activities, Antioxidant Activities, and Chemical Properties of Pistacia vera Leaves Using LC-QTOF-MS and RP-HPLC. *Iran. J. Pharm. Res.* **2022**, *21*, e127033. [CrossRef]
30. Gan, R.-Y.; Kong, K.-W.; Li, H.-B.; Wu, K.; Ge, Y.-Y.; Chan, C.-L.; Shi, X.-M.; Corke, H. Separation, identification, and bioactivities of the main gallotannins of red sword bean (*Canavalia gladiata*) coats. *Front. Chem.* **2018**, *6*, 39. [CrossRef] [PubMed]
31. Ho, K.-V.; Roy, A.; Foote, S.; Vo, P.H.; Lall, N.; Lin, C.-H. Profiling anticancer and antioxidant activities of phenolic compounds present in black walnuts (*Juglans nigra*) using a high-throughput screening approach. *Molecules* **2020**, *25*, 4516. [CrossRef] [PubMed]
32. Quintana, S.E.; Salas, S.; García-Zapateiro, L.A. Bioactive compounds of mango (*Mangifera indica*): A review of extraction technologies and chemical constituents. *J. Sci. Food Agric.* **2021**, *101*, 6186–6192. [CrossRef]
33. Wang, H.; Fowler, M.I.; Messenger, D.J.; Ordaz-Ortiz, J.J.; Gu, X.; Shi, S.; Terry, L.A.; Berry, M.J.; Lian, G.; Wang, S. Inhibition of the intestinal postprandial glucose transport by gallic acid and gallic acid derivatives. *Food Funct.* **2021**, *12*, 5399–5406. [CrossRef] [PubMed]
34. Liang, X.; Jiang, Y.; Guo, Z.; Fang, S. Separation, UPLC-QTOF-MS/MS analysis, and antioxidant activity of hydrolyzable tannins from water caltrop (*Trapa quadrispinosa*) pericarps. *LWT* **2020**, *133*, 110010. [CrossRef]
35. Chang, Z.; Zhang, Q.; Liang, W.; Zhou, K.; Jian, P.; She, G.; Zhang, L. A Comprehensive Review of the Structure Elucidation of Tannins from *Terminalia* Linn. *Evid.-Based Complement. Altern. Med.* **2019**, *2019*, 8623909. [CrossRef]
36. Fidelis, M.; de Moura, C.; Kabbas Junior, T.; Pap, N.; Mattila, P.; Mäkinen, S.; Putnik, P.; Bursać Kovačević, D.; Tian, Y.; Yang, B. Fruit seeds as sources of bioactive compounds: Sustainable production of high value-added ingredients from by-products within circular economy. *Molecules* **2019**, *24*, 3854. [CrossRef]
37. Tameye, N.S.J.; Akak, C.M.; Happi, G.M.; Frese, M.; Stammler, H.-G.; Neumann, B.; Lenta, B.N.; Sewald, N.; Nkengfack, A.E. Antioxidant norbergenin derivatives from the leaves of *Diospyros gilletii* De Wild (Ebenaceae). *Phytochem. Lett.* **2020**, *36*, 63–67. [CrossRef]
38. Lobo, V.; Patil, A.; Phatak, A.; Chandra, N. Free radicals, antioxidants and functional foods: Impact on human health. *Pharmacogn. Rev.* **2010**, *4*, 118. [CrossRef]
39. Pham-Huy, L.A.; He, H.; Pham-Huy, C. Free radicals, antioxidants in disease and health. *Int. J. Biomed. Sci.* **2008**, *4*, 89. [PubMed]
40. Szwajgier, D.; Baranowska-Wójcik, E.; Winiarska-Mieczan, A.; Gajowniczek-Ałasa, D. Honeys as Possible Sources of Cholinesterase Inhibitors. *Nutrients* **2022**, *14*, 2969. [CrossRef] [PubMed]
41. Pope, C.; Karanth, S.; Liu, J. Pharmacology and toxicology of cholinesterase inhibitors: Uses and misuses of a common mechanism of action. *Environ. Toxicol. Pharmacol.* **2005**, *19*, 433–446.
42. Yu, Z.-Y.; Xu, K.; Wang, X.; Wen, Y.-T.; Wang, L.-J.; Huang, D.-Q.; Chen, X.-X.; Chai, W.-M. Punicalagin as a novel tyrosinase and melanin inhibitor: Inhibitory activity and mechanism. *LWT* **2022**, *161*, 113318. [CrossRef]
43. Chiocchio, I.; Mandrone, M.; Sanna, C.; Maxia, A.; Tacchini, M.; Poli, F. Screening of a hundred plant extracts as tyrosinase and elastase inhibitors, two enzymatic targets of cosmetic interest. *Ind. Crops Prod.* **2018**, *122*, 498–505. [CrossRef]
44. Tundis, R.; Loizzo, M.; Menichini, F. Natural products as α-amylase and α-glucosidase inhibitors and their hypoglycaemic potential in the treatment of diabetes: An update. *Mini Rev. Med. Chem.* **2010**, *10*, 315–331. [CrossRef] [PubMed]
45. Alqahtani, A.S.; Hidayathulla, S.; Rehman, M.T.; ElGamal, A.A.; Al-Massarani, S.; Razmovski-Naumovski, V.; Alqahtani, M.S.; El Dib, R.A.; AlAjmi, M.F. Alpha-amylase and alpha-glucosidase enzyme inhibition and antioxidant potential of 3-oxolupenal and katononic acid isolated from Nuxia oppositifolia. *Biomolecules* **2019**, *10*, 61. [CrossRef]
46. Aleixandre, A.; Gil, J.V.; Sineiro, J.; Rosell, C.M. Understanding phenolic acids inhibition of α-amylase and α-glucosidase and influence of reaction conditions. *Food Chem.* **2022**, *372*, 131231. [CrossRef]
47. Janicka, P.; Przyjazny, A.; Boczkaj, G. Novel "acid tuned" deep eutectic solvents based on protonated L-proline. *J. Mol. Liq.* **2021**, *333*, 115965. [CrossRef]
48. Zengin, G.; Aktumsek, A. Investigation of antioxidant potentials of solvent extracts from different anatomical parts of *Asphodeline anatolica* E. Tuzlaci: An endemic plant to Turkey. *Afr. J. Tradit. Complement. Altern. Med.* **2014**, *11*, 481–488. [CrossRef] [PubMed]

49. Uysal, S.; Zengin, G.; Locatelli, M.; Bahadori, M.B.; Mocan, A.; Bellagamba, G.; De Luca, E.; Mollica, A.; Aktumsek, A. Cytotoxic and enzyme inhibitory potential of two Potentilla species (*P. speciosa* L. and *P. reptans* Willd.) and their chemical composition. *Front. Pharmacol.* **2017**, *8*, 290.
50. Grochowski, D.M.; Uysal, S.; Aktumsek, A.; Granica, S.; Zengin, G.; Ceylan, R.; Locatelli, M.; Tomczyk, M. In vitro enzyme inhibitory properties, antioxidant activities, and phytochemical profile of *Potentilla thuringiaca*. *Phytochem. Lett.* **2017**, *20*, 365–372. [CrossRef]

molecules

MDPI

Review

A Comprehensive Literature Review on Cardioprotective Effects of Bioactive Compounds Present in Fruits of *Aristotelia chilensis* Stuntz (Maqui)

Lyanne Rodríguez [1,†], Andrés Trostchansky [2,†], Hermine Vogel [3], Irene Wood [2], Iván Palomo [1], Sergio Wehinger [1,*] and Eduardo Fuentes [1,*]

1 Thrombosis Research Center, Medical Technology School, Department of Clinical Biochemistry and Immunohaematology, Faculty of Health Sciences, Universidad de Talca, Talca 3480094, Chile
2 Departamento de Bioquímica and Centro de Investigaciones Biomédicas (CEINBIO), Facultad de Medicina, Universidad de la República, Montevideo 11200, Uruguay
3 Departamento de Horticultura, CENATIV, Facultad de Ciencias Agrarias, Universidad de Talca, Talca 3480094, Chile
* Correspondence: snunez@utalca.cl (S.W.); edfuentes@utalca.cl (E.F.)
† These authors contributed equally to this work.

Abstract: Some fruits and vegetables, rich in bioactive compounds such as polyphenols, flavonoids, and anthocyanins, may inhibit platelet activation pathways and therefore reduce the risk of suffering from CVD when consumed regularly. *Aristotelia chilensis* Stuntz (Maqui) is a shrub or tree native to Chile with outstanding antioxidant activity, associated with its high content in anthocyanins, polyphenols, and flavonoids. Previous studies reveal different pharmacological properties for this berry, but its cardioprotective potential has been little studied. Despite having an abundant composition, and being rich in bioactive products with an antiplatelet role, there are few studies linking this berry with antiplatelet activity. This review summarizes and discusses relevant information on the cardioprotective potential of Maqui, based on its composition of bioactive compounds, mainly as a nutraceutical antiplatelet agent. Articles published between 2000 and 2022 in the following bibliographic databases were selected: PubMed, ScienceDirect, and Google Scholar. Our search revealed that Maqui is a promising cardiovascular target since extracts from this berry have direct effects on the reduction in cardiovascular risk factors (glucose index, obesity, diabetes, among others). Although studies on antiplatelet activity in this fruit are recent, its rich chemical composition clearly shows that the presence of chemical compounds (anthocyanins, flavonoids, phenolic acids, among others) with high antiplatelet potential can provide this berry with antiplatelet properties. These bioactive compounds have antiplatelet effects with multiple targets in the platelet, particularly, they have been related to the inhibition of thromboxane, thrombin, ADP, and GPVI receptors, or through the pathways by which these receptors stimulate platelet aggregation. Detailed studies are needed to clarify this gap in the literature, as well as to specifically evaluate the mechanism of action of Maqui extracts, due to the presence of phenolic compounds.

Keywords: Maqui; platelets; phenolic compounds and cardiovascular

Citation: Rodríguez, L.; Trostchansky, A.; Vogel, H.; Wood, I.; Palomo, I.; Wehinger, S.; Fuentes, E. A Comprehensive Literature Review on Cardioprotective Effects of Bioactive Compounds Present in Fruits of *Aristotelia chilensis* Stuntz (Maqui). *Molecules* **2022**, 27, 6147. https://doi.org/10.3390/molecules27196147

Academic Editor: Nour Eddine Es-Safi

Received: 21 August 2022
Accepted: 13 September 2022
Published: 20 September 2022

1. Introduction

Cardiovascular diseases (CVDs: acute myocardial infarction, cerebrovascular disease, and peripheral arterial thrombosis) are responsible for approximately 30% of deaths worldwide [1,2]. Platelets play a relevant role in the atherosclerotic process in physiopathologic and thrombotic events. It has been observed that some bioactive compounds present in fruits and vegetables, when consumed regularly, can inhibit platelet aggregation and thus reduce the risk of CVD [3].

Epidemiological studies have shown that modifiable cardiovascular risk factors (CVRFs) increase the probability of suffering from CVD [4]. There are non-modifiable factors such as age and genetic predisposition in addition to modifiable CVRFs such as smoking, dyslipidemias, hypertension, diabetes, metabolic syndrome, and overweight/obesity [5–7]. It has been previously described how inflammation and thrombosis are involved in the onset and progression of non-communicable diseases [8].

Modifications in the population's lifestyle have effects on CVD. Consuming a healthy diet, rich in fruit and vegetables, with high concentrations of proven bioactive protective and antioxidant compounds, has shown to be a promising action in the prevention of CVD [9,10]. Diet and lifestyle are modifiable risk factors that can have a significant impact on an individual's likelihood of developing non-communicable diseases. It has become clear that a person's nutritional status is an important factor in preparing the immune system to deal with any disease [8].

The ethnomedicinal use of natural products and their natural bioactive compounds has increased for the treatment and prevention of CVD [11]. The traditional Mediterranean diet, as well as medicinal plants, have also been reported to exert cardioprotective and antiplatelet effects in the primary and secondary prevention of CVD [12]. According to estimates by the World Health Organization (WHO), approximately 80% of the world's population uses traditional medicinal herbs for their primary health care. Additionally, the consumption of berries, for example, chokeberries (*Aronia melanocarpa*), blueberries (*Vaccinium sect. Oxycoccus*), sea buckthorn berries (*Hippophae rhamnoides*), and grapes (*Vitis*), as well as their various derivative commercial drugs, has been linked to the prevention of CVD, such as atherosclerosis in elderly men [13].

Foods, fruits, vegetables, grains, and fermented beverages such as wine and beer, with antithrombotic and anti-inflammatory properties, contain a large number of phytochemicals such as phenolic compounds, carotenes, alkaloids, terpenes, peptides, and bioactive lipid molecules. Many of these compounds exhibit potent inhibition or modulation of several key proinflammatory and prothrombotic mediator signaling pathways, such as platelet-activating factor (PAF), thrombin, collagen, ADP, arachidonic acid, and related eicosanoids [8]. It has been described that the modulation of the intracellular oxidative state through the consumption of antioxidants in the diet could be a promising approach to reducing the risk of CVD [14].

Polyphenols are distinct in the Mediterranean diet because they can modulate platelet function through different mechanisms of action, e.g., modulating thromboxane formation. It is well known that phenolic compounds are the main components of many plants, and have gained increasing public and scientific interest due to their beneficial effects on health as antioxidants [14]. Therefore, plants with high polyphenol content and antioxidant activity are worth studying. In this group of plants of great interest, we have *Aristotelia chilensis* (Stuntz) also known as Maqui, which is a plant native to Chile, a member of the family Eleocarpaceae, which is distributed across the world in tropical and temperate areas of Asia, Oceania, and South America. Maqui has stood out for the presence of phenolic compounds and its anticancer, antimutagenic, anti-inflammatory, and antioxidant activities. Although some authors have discussed Maqui's cardioprotective activity, there exist few studies linking this berry with an antiplatelet effect.

Maqui: Relevance and Traditional Uses

Maqui is a sacred medicinal plant to the indigenous Mapuche people, native to Chile and the Argentinian border [15]. Its light to dark purple berries ripen between December and February [16–18]. Besides its medicinal use, Maqui is consumed and prepared as food (juice, pulp, jam) or liquor [19]. Maqui has a high phenol content and, depending on the solvent used for its extraction, the phenolic content of the fruit can reach up to 51 g GAE/kg [16,19]. Aqueous and ethanolic extracts of Maqui fruit have been used since ancient times for medicinal purposes to treat digestive disorders, inflammation, and migraines, while leaf extracts exert antiseptic, anti-inflammatory, and indigestion protective properties [20]. Most of the actions described for this berry are related to the high content of anthocyanins and polyphenols in the ripe fruit [21,22].

Many collectors and businessmen have opted for plantations of this native fruit, which is mainly exported to Europe and Asia. Around 170,000 hectares are planted throughout Chile (estimate of the area of wild Maqui indicated by collectors in the agricultural census), but the growing demand for Maqui is not satisfied by wild production. In recent years, the demand from the food and pharmaceutical industry for Maqui berries has increased, which prompted the domestication of the species to achieve greater availability and avoid the destruction of wild populations [23]. Studies of the genetic structure of natural populations were carried out, looking for genotypes to guide the selection of clones in addition to establishing the best agronomic parameters for their selection and cultivation [24,25]. For this, studies of the genetic diversity of different Maqui populations using molecular marker techniques such as chloroplast microsatellites and amplified fragment length polymorphisms (AFLPs) guided the domestication of this species [23,24]. More than 60 clones of Maqui trees were planted at the Experimental Station of the University of Talca (Panguilemo) and genetic variability was evaluated using molecular markers [24]. The most outstanding clones were planted in other experimental stations (Los Niches, Chillan, and Río Negro), which allowed a more detailed evaluation of them [23,26]. From these works, the clones Luna Nueva, Morena, and Perla Negra present relevant agronomic characteristics (harvest yield, fruit size, and early maturation) [24,27], making them suitable for further nutraceutical evaluations.

In wild plantations, about 50% of the fruit is immature at harvest, while in the domesticated tree this is only 20% [27]. Currently, only the ripe fruit is used in the food and pharmaceutical industry [19,28], while the immature fruit has no application and is discarded. Likewise, Maqui leaves constitute agroindustrial waste and contain large amounts of total phenols [29].

The genetic differences in Maqui clones significantly influence their biological potential [24]. Although the biological properties of Maqui have been extensively studied, few reports analyze its antiplatelet activity. However, when evaluating Maqui's composition, some compounds present in this berry have been shown to exert inhibition of platelet activation, secretion, and aggregation, thus reducing the risk of CVD [30,31].

In the current review, we will discuss and summarize the studies that relate Maqui with cardiovascular protection, specifically in the prevention of CVRFs. On the other hand, we will discuss how Maqui's bioactive compounds mainly present in its fruit modulate platelet function as a mechanism of CVD protection.

2. Platelets and Atherothrombosis

Platelets are anucleated cells (1.5–3.0 μm in diameter) that originate from the fragmentation of the cytoplasm of the megakaryocyte through endomitosis. This process leads to the formation of platelets that enter the bloodstream [32,33].

Platelets circulate as discoid-shaped elements. In response to vascular damage, they emit pseudopodia, secreting the content of their granules and remodeling their membrane [32]. They are formed by a plasma membrane expressing important glycoproteic receptors: collagen receptors (glycoprotein (GP) IIb/IIa, GPIb/IX, GPVI, GPIa/IIa, GPIV), and non-glycoproteic receptors: adenosine diphosphate (ADP) receptors (P2Y1, P2Y12), thrombin receptors (PAR1, PAR4), thromboxane A2 receptor (TxA2), serotonin receptor, and prostacyclin I_2 receptor. The receptor–agonist interaction participates in the outside/inside or inside/outside platelet signaling leading to platelet activation and/or inhibition [32,34].

Platelets, in their functional aspects, involve the processes of adhesion, activation, secretion, and platelet aggregation (Figure 1). Adhesion occurs in response to vascular damage; platelet membrane receptors interact with their respective ligands and bind to the injured wall resulting in platelets' morphological changes and secretion of their alpha and dense granule content [35]. Then, platelets begin the aggregation process which is characterized by platelet–platelet binding to form a platelet aggregate. Their functionality can be modified by CVRFs, e.g., developing a platelet hyperactivation state [36]. This condition has been described as an increase in the capacity of activation, secretion, and aggregation against low concentrations of agonists, due to alterations in different signaling pathways [37,38].

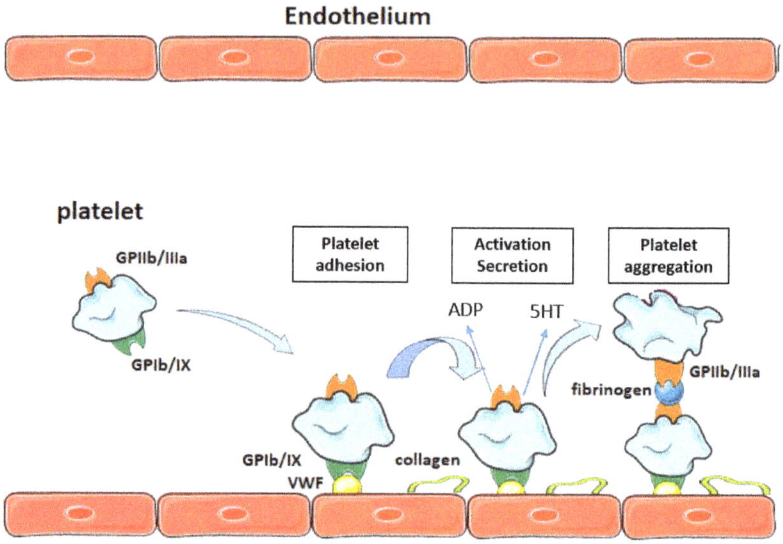

Figure 1. Platelet function in the process of primary hemostasis. ADP, adenosine diphosphate; VWF, von Willebrand factor; GP, glycoprotein; 5HT: serotonin.

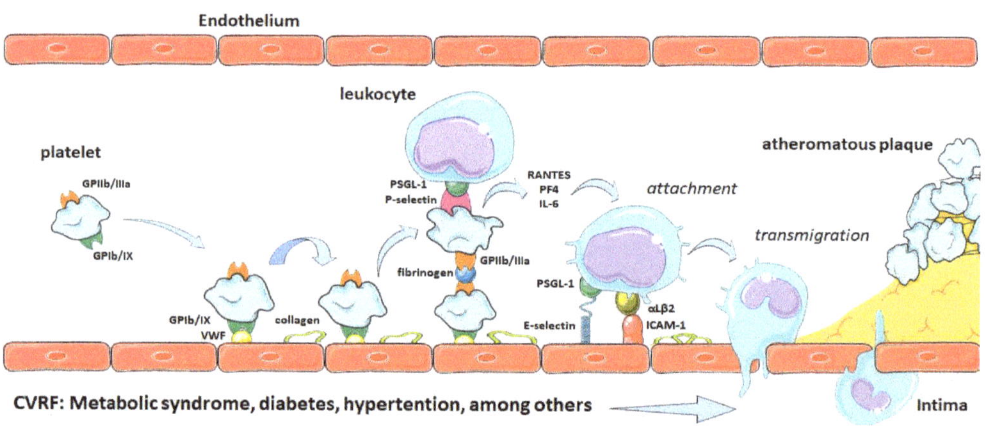

Figure 2. Participation of platelets in the atherothrombotic process. VWF, von Willebrand factor; GP, glycoprotein; IL, interleukin, PF4, platelet factor 4; PSGL-1, glycoprotein ligand-1 P-selectin; RANTES, beta-regulatory chemokine; VCAM-1, vascular cell-1 adhesion molecule.

Participation of Platelets in Atherothrombosis

Platelets represent the bridge between an inflammatory process and thrombosis, a fundamental process for the development of atherothrombosis and complication of atherosclerotic plaque [39]. Plaque rupture leads to the formation of a thrombus rich in platelets, a process characteristic of arterial thrombosis (Figure 2). During the atherothrombotic process, platelets participate in both the initial and final stages [40]. In the initial phase, platelets adhere to the CVRF-damaged endothelium, secreting and exposing molecules that amplify the inflammatory process [35,40]. In the final stage, after the plaque ruptures, the platelets adhere at endothelium formatting to aggregates and thus contribute significantly to thrombus formation [41].

It has been observed that some bioactive compounds in fruits and vegetables, polyphenols and flavonoids, when consumed regularly may inhibit platelet activation and therefore reduce the risk of CVD. The importance of natural antioxidants to provide cardiovascular protection has been highlighted before [3,42,43], as well as the reports about the antithrombotic activity of fruits and vegetables [44,45]. In the case of Maqui, these studies are scarce [20,30].

3. Chemical Characterization of Maqui

The development of chromatographic and spectrophotometric techniques such as high-performance liquid chromatography (HPLC), coupled with diode detectors (DADs) and a mass spectrometer (MS) have favored advances in the identification and quantification of anthocyanins and other polyphenolic compounds [46,47].

The main bioactive compounds already reported in Maqui include phenolic acids (caffeic and gallic), flavonols (quercetin, rutin, and myricetin), flavonoids (catechin and epicatechin), and anthocyanins (delphinidin and its derivatives, malvidin, petunidin, cyanidin, and peonidin) [18] (Figure 3).

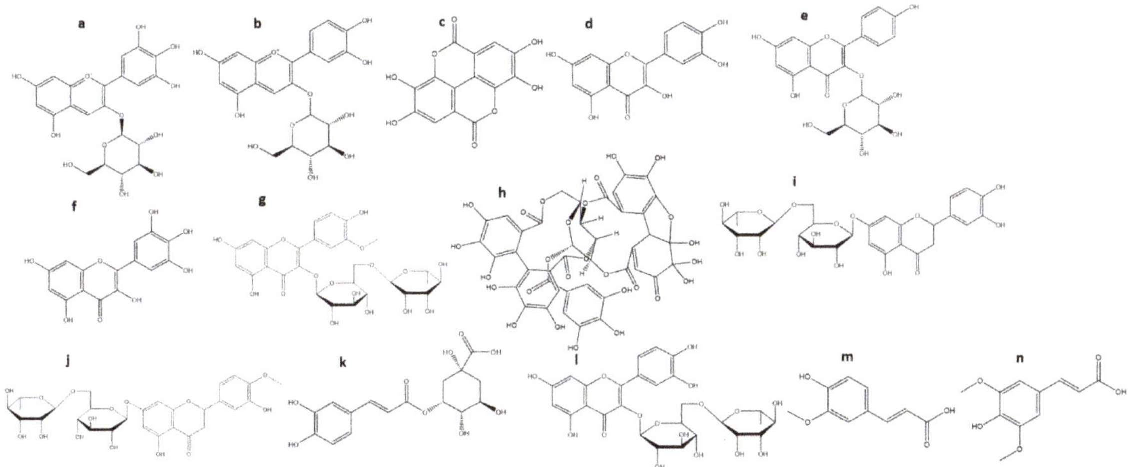

Figure 3. The chemical structure of the main chemical compounds present in the fruits of Maqui. Chemical structures corresponding to **a**, Delphinidin-3-glucoside; **b**, Cyanidin-3-glucoside; **c**, Ellagic acid; **d**, Quercetin; **e**, Kaempferol-3-glucoside; **f**, Myricetin; **g**, Isorhamnetin-3 -rutinoside; **h**, Granatin B; **i**, Eriodic-tyol-7-rutinoside; **j**, Hesperetin-7-rutinoside; **k**, 5-*O*-caffeoylquinic acid; **l**, Rutin; **m**, Ferulic acid and **n**, Sinapic acid.

Aqueous extracts of the fruits of Maqui contain a high content of phenolic acids (ferulic acid, gallic acid, caffeic acid), flavonoids (quercetin, myricetin, kaempferol, delphinidin, and cyanidin) and tannins (ellagitannins) [48]. On the other hand, the hydroalcoholic extract of the Maqui fruit presents high values of phenols and the presence of hydrophilic compounds, coumaric, gentisic, ferulic, and gallic acids, delphinidin-3,5-*O*-diglucoside, cyanidin-3-*O*-glucoside, and proanthocyanidin B, as well as hydrophobic compounds, catechins, quercetin, rutin, and anthocyanidins [49–51]. Cespedes et al. evaluated an ethanolic extract by extracting first with EtOH/H_2O (6:4) followed by a subsequent extraction with H_2O. This extract presented a composition based on gentisic acid, ferulic acid, gallic acid, p-coumaric acid, sinapic acid, 4-hydroxybenzoic acid, delphinidin, cyanidin, vanillic acid, delphinidin gallate, gallocatechin gallate, quercetin, rutin, myricetin, catechin and epicatechin, and anthocyanin glycosides [52]. Another study indicates the phytochemical characterization of several Maqui berries by HPLC-DAD-ESI/MS^n. Eight glycosylated anthocyanins, derived from cyanine and delphinine, have been reported: delphinidin-3-*O*-sambubioside-5-*O*-glucoside, delphinidin-3,5-*O*-diglucoside, cyanidin-3,5-*O*-diglucoside, cyanidin-3-*O*-sambubioside-5-*O*-glucoside [53]. Sonication has been reported as an ideal alternative to obtain extracts from Maqui berries with high bioactivity, increasing the content of anthocyanins in Maqui berries, 3-glucosides, 3,5-*O*-diglucosides, 3-*O*-sambubiosides, and 3-*O*-sambubioside-5-*O*-glucosides of delphinidin and cyanidin as determined by HPLC with photodiode array and MS detection [46]. Maqui fruit extracts' main components are summarized in Table 1.

Only a few studies refer to the phytochemical analysis of leaf extracts of Maqui. The presence of indole and quinoline alkaloids has been reported in the leaves [19], being identified as aristoteline, serratoline, aristone, horbatinol, and horbatina [54,55]. Studies performed on ethanolic extracts of Maqui leaves reported the presence of some polyphenols such as gallic acid, coumaric acid, quercetin, myricetin, rutin, pelargonidin, and catechin. Interestingly, these compounds have been associated with the prevention of some CVD [56].

Table 1. Anthocyanins and phenolic compounds identified and quantified in Maqui fruit extracts.

Anthocyanins (TA)	**% Average (Range)**	**References**
Delphinidin-3-*O*-sambubioside-5-*O*-glucoside	32.4 (15–49)	[21,46,52,53,61]
Delphinidin-3-*O*-glucoside (**a**)	18.6 (11–28)	[21,46,52,53,61]
Delphinidin-3,5-*O*-diglucoside	18.3 (14–24)	[46,52,53,61]
Delphinidin-3-*O*-sambubioside	10 (6–16)	[21,46,52,53,61]
Cyanidin-3-*O*-glucoside (**b**)	11 (6–16)	[46,52,61]
Cyanidin3,5-diglucoside	10.8 (7–14)	[21,46,52,53]
Cyanidin-3-*O*-sambubioside-5-*O*-glucoside	9 (7–11)	[52,61]
Cyanidin-3-*O*-sambubioside	6.5 (6–7)	[21,46,52,53]
Cyanidin-*O*-glucoside-5-*O*-rhamnoside	1.5 (1–2)	[21,53]
Phenolic Compounds (TP)		
Ellagic acid (**c**)	30	[22]
Ellagic acid rhamnoside	8	[62]
Ellagic acid hexoside	2.8 (2–3.5)	[22,53,62]
Quercetin-*O*-galloyl-*O*-hexoside	24	[22]
Quercetin-3-*O*-rutinoside	10 (7–13)	[53,62]
Quercetin-3-*O*-arabinoside	6 (5–7)	[22,53,62]
Quercetin-3-*O*-galactoside	6 (5–7)	[22,53,62]
Quercetin-3-*O*-rhamnoside	5.5 (4.1–6)	[53,62]
Quercetin-3-*O*-xyloside	3.5 (2–5)	[22,53,62]
Quercetin-3-*O*-glucoside	3 (2–4)	[22,53,62]
Quercetin (**d**)	2	[22]
Kaempferol-3-*O*-glucoside (**e**)	18	[53]
Kaempferol-3-*O*-galactoside	12	[53]
Kaempferol-3-*O*-rutinoside	2	[53]
Myricetin-3-*O*-glucoside	13 (6–20)	[22,53]
Myricetin (**f**)	8	[22]
Myricetin-3-*O*-galactoside	6.3 (4–10)	[22,53,62]
Myricetin-3-*O*-galoyl-*O*-glucoside	4 (2–6)	[22,53,62]
Myricetin-3-*O*-galoyl-*O*-glucoside	4 (2–6)	[22,53,62]
Isorhamnetin-3-*O*-rutinoside (**g**)	2	[53]
Granatin B (**h**)	20	[62]
Eriodictyol-7-*O*-rutinoside (**i**)	11	[62]
Hesperetin-7-*O*-rutinoside (**j**)	11	[62]
5-*O*-caffeoylquinic acid (**k**)	8.5 (5–12)	[22,53,62]
Rutin (**l**)	6	[22]
Ferulic acid (**m**)	4	[62]
Sinapic acid (**n**)	3	[62]

Average % (range): percentage in which the compound is present in the Maqui fruit.

A study carried out in different regions of Chile showed that the total anthocyanin concentrations (TA) vary between 660 and 1500 mg cyanidin-3-*O*-glucoside/100 g of dried fruit, while the TP ranges between 1070 and 2050 mg GAE/100 g of dried fruit [17]. The total content of anthocyanins and polyphenols in Maqui is highly variable, depending on growing conditions, harvest times, plant genotype [17], and the different extraction procedures [57]. Maqui fruits have higher total polyphenol (TP) levels and antioxidant activity than other species recognized for their high phenolic content such as blueberries, pomegranates, blackberries, and red raspberries [58,59]. If we compare the Maqui with other berries, for example, Euterpe oleracea, better known as açai, which has also stood out for being one of the most nutritious fruits in South America, the phenolic content does not exceed the values referenced for Maqui [60].

4. Cardioprotective Role of Maqui

Food and/or pharmaceutical supplements have become attractive alternatives to reducing CVRFs [63]. Some non-communicable diseases, including CVD, could be prevented by improving the lifestyle of the population, including the consumption of a healthy diet [64]. Studies have highlighted the importance of natural antioxidants present in vegetables, to provide cardiovascular protection [42]. An investigation showed the cardioprotective effect of proanthocyanidins present in a grape seed extract evaluated in mice [65]. Some works refer to the antioxidant activity of the methanolic extract of the fruit of Maqui and the cardioprotective effects on acute ischemia/reperfusion induced in the hearts of rats [20]. It has been described that a diet rich in fruits and vegetables favors adequate platelet function, with positive effects on cardiovascular health [30]. Undoubtedly, the Maqui compared to other fruits presents a wide biological potential, being a promising target for study mainly in the cardiovascular area (Figure 4).

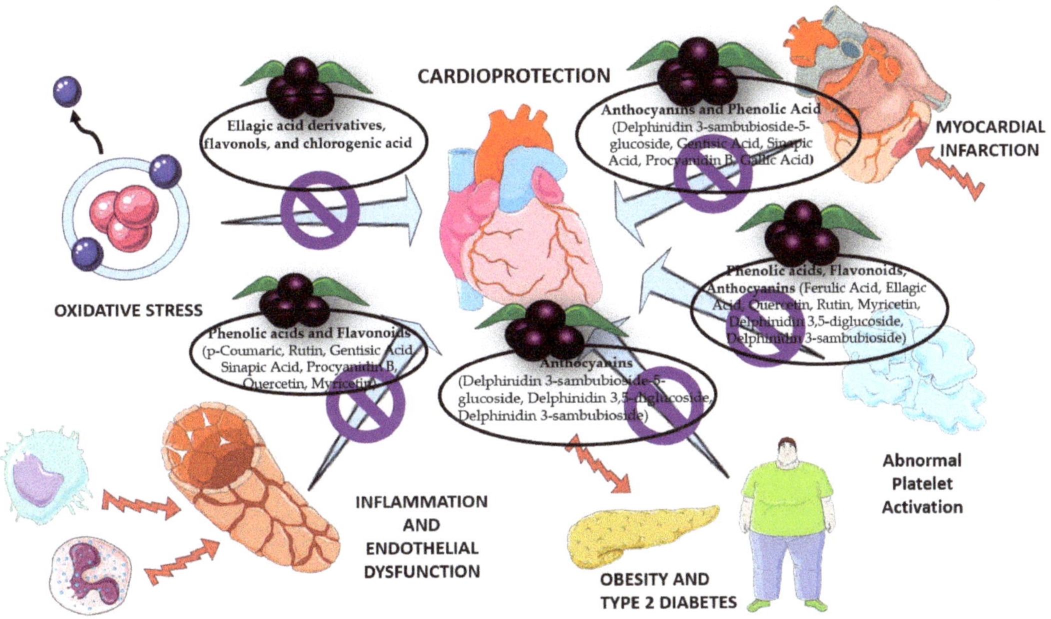

Figure 4. Cardioprotective effect of chemical compounds identified in Maqui.

4.1. Antioxidant Effect

Investigations highlight the antioxidant potential of Maqui, which is more potent compared to other berries; the antioxidant capacity expressed as Trolox equivalents is higher in Maqui berries compared to *Vaccinium floribundum* [61]. Other authors also point out that this berry has a better potential for radical scavenging and antioxidant reactivity using in vitro antioxidant capacity tests [66]. Some works refer to the antioxidant activity of the methanolic extract of the Maqui fruit, and the cardioprotective effects on acute ischemia/reperfusion induced in an in vivo rat model. In this study, the methanolic extract protected the animals from cardiac damage due to the incidence of reperfusion dysrhythmias and non-recovery of sinus rhythm. The antioxidant effect of this extract was also evaluated with ORAC, FRAP, and DPPH assays presenting an IC_{50} of 1.62 ppm against DPPH [20].

It has been shown that the antioxidant activity of the Maqui berry is directly related to cognitive protection and a decrease in oxidative stress markers. Methanol/water extracts and their partitions (acetone and ethyl acetate) of three Maqui varieties were studied for antitumor effects in HT-29, Caco-2, and NF-κ colon cancer cells. The inhibition of cell growth and nitric oxide (NO) production by the most active extracts was dose-dependent and significant effects were observed at concentrations of 25.0 and 10.0 ppm, respectively [67].

Maqui berries have shown better results against different in vitro tests of antioxidant activity [68]. A new drink based on lemon juice and Maqui has been compared against other mixtures with açai and blackthorn. The Maqui mixture was the most interesting in terms of antioxidant capacity. An ABTS assay showed that Maqui-containing mixtures presented better results (8.35 $\pm$ 0.55 mM Trolox), supporting the strong antioxidant activity of this berry. Açai blends showed lower capacity compared to Maqui blends (3.88 $\pm$ 0.58 mM Trolox). This study also showed the most potent reducing capacity of Maqui compared to açai against the DPPH radical (3.07 mM $\pm$ 0.09 mM Trolox and 0.12 $\pm$ 0.19 mM Trolox, respectively) [69].

4.2. *Effect on Inflammation and Endothelial Dysfunction*

Maqui has received attention due to its broad biological potential, including reducing inflammation. In vitro studies have mainly shown that Maqui limits adipogenesis and inflammatory pathways [61] in addition to affecting the production of NO [70]. A hydroalcoholic extract of this species, as well as rutin, improved endothelium-dependent relaxation and reduced plasma levels of cholesterol, LDL, and triglycerides. Additionally, Maqui and rutin improve the bioavailability of NO [71]. Another study evidenced the protective effect of a hydroethanolic extract of Maqui and its flavonoids (rutin and quercetin) on endothelial dysfunction induced by high glucose and pyrogallol, through a greater generation and bioavailability of NO [72]. An aqueous extract of Maqui berries modulated NO production in a hyperglycemic and/or hyperinsulinemic microenvironment, in addition to improving endothelial function, through the possible vasorelaxant properties of NO [70]. Maqui methanolic extracts under inflammatory conditions suppress NO production, through the regulation of inducible nitric oxide synthases (iNOSs). These results correlate with the inhibition of the expressions of NF-κB and COX-2 proteins [67]. On the other hand, the methanolic extract of the Maqui fruit decreases lipid oxidation and reduces the concentration of thiobarbituric acid reactive species (lipid peroxidation index) [20]. In human endothelial cell cultures, Maqui juice dose-dependently protects intracellular oxidative stress induced by hydrogen peroxide, suggesting that it might have antiatherogenic properties [66].

4.3. *Effect on Diabetes and Obesity*

Maqui dietary supplementation has been shown to have positive effects on fasting glucose and insulin levels in human and mouse models of type 2 diabetes and obesity. Maqui standardized extract showed a significant reduction in glycemia in 10 volunteers with moderate glucose intolerance [73]. It was shown that obese mice induced by a high-fat diet and supplemented with lyophilized Maqui had better insulin response capacity, less weight gain, and greater thermogenic activity [74].

The effect of Maqui anthocyanins on a murine model of type 2 diabetes was studied. Oral administration of these anthocyanins improved fasting blood glucose levels and glucose tolerance in hyperglycemic obese C57BL/6J mice, fed a high-fat diet. On the other hand, it also improved the negative regulation of the gluconeogenic enzyme glucose-6-phosphatase stimulated by insulin and decreased glucose production. Additionally, delphinidin-3-*O*-sambubioside-5-*O*-glucoside was shown to dose-dependently lower fasting blood glucose levels in obese C57BL/6J mice [75].

Maqui berries are commercially extracted to produce a standardized polyphenolic extract, which contains 25% delphinidin, the most abundant anthocyanin in this fruit.

The extract has other constituents, phenolic acids, flavonols (quercetin, rutin, myricetin, and flavanols (catechins and epicatechins). Maqui berry extract is marketed worldwide as Delphinol® and is distributed by the company Maqui New Life (MNL, trademark owner), based in Santiago, Chile, and Cham, Switzerland [76]. Delphinol® is a nutritional supplement with the ability to naturally control postprandial glycemia through the inhibition of the sodium and glucose cotransporter in the small intestine. An investigation carried out on ten volunteers with moderate glucose intolerance, using a double-blind, placebo-controlled, crossover study model, showed the effect of Delphinol® on glucose. This supplement significantly inhibited non-desired changes in postprandial blood glucose levels, 60 and 90 min after ingestion of boiled rice [73]. Alvarado et al. showed that the effects of Delphinol® on basal glycemia and insulinemia could be related to the inhibition of intestinal glucose transporters, as well as an incretin-mediated effect or by improving sensitivity to insulin [77].

Additionally, the role of Maqui extracts has been highlighted not only in preventing diabetes but also in preventing and treating obesity. A high-fat diet supplemented with a regular dose of Maqui berries shows better insulin response and less weight gain. In addition, a differential expression of genes involved in de novo lipogenesis, fatty acid oxidation, the formation of multilocular lipid droplets, and thermogenesis in subcutaneous white adipose tissue (scWAT) is evidenced. These findings highlighted the role of this berry in preventing or treating type 2 diabetes and obesity-related diseases, as well as their metabolic complications [74].

Alpha-glucosidase inhibition is considered one of the measures to regulate type 2 diabetes [69]. Gironés-Vilaplana et al. (2014) reported that EC_{50} values of glucosidase inhibition for açai and Maqui berries were in the range of 0.33–2.14 mg/mL, while for strawberry pomace the values were not so influential, which shows once again the potential of Maqui over other berries [69,78]. Maqui has shown potent postprandial glycemic lowering effects at a single dose of approximately 1000 μmol GAE of polyphenols derived from Maqui berry extract and lemon juice, following carbohydrate (glucose and rice) ingestion [79]. A similar dose was also effective in lowering postprandial blood glucose (induced by ingestion of 25 g glucose) by ingestion of coffee containing 2.5 mmol/L chlorogenic acid (a total amount of 1000 μmol chlorogenic acid) [79,80]. Table 2 summarizes the clinical studies that link Maqui with cardiovascular protection. These results show the effect of different extracts of this berry on cardiovascular risk factors (glucose index, obesity, diabetes, among others). Some studies showed the biological activity of the Maqui extract, while others relate the biological potential with the presence of some compounds of a mainly phenolic nature.

Table 2. Cardioprotective role of Maqui: clinical studies.

Plant Part(s)	Type of Extract	Dose or Concentration	Mechanism of Action or Effect of Extract and/or Pure Compound Characterization of the Extract	Characterization	Identified Compound	Reference
			Cardioprotection			
Fruit	Methanolic extract of ripe fruits of Maqui.	100, 10, and 1 ppm/kg of rat body weight	Methanolic extract of ripe fruits of Maqui has antioxidant activity and cardioprotective effect on acute ischemia/reperfusion performed in rat hearts. The extract protected from heart damage due to the incidence of reperfusion dysrhythmias and non-recovery of sinus rhythm. It also prevents harmful events in the heart of the animal by reducing lipid oxidation and reducing the concentration of substances reactive to thiobarbituric acid, and lipid peroxidation index.	No		[20]

Table 2. *Cont.*

Plant Part(s)	Type of Extract	Dose or Concentration	Mechanism of Action or Effect of Extract and/or Pure Compound Characterization of the Extract	Characterization	Identified Compound	Reference
Fruit	Maqui berry extract	150 mg standardized	**Oxidative stress biomarkers** Delphinol reduces levels of Ox-LDL and urinary F2-isoprostanes (8-iso-prostaglandin F2α).	No		[81]
Fruit	Maqui berry powder (ground whole fruit rich in anthocyanins)	50 and 100 mg/kg	**Oxidative stress markers** The administration of the aqueous extract of Maqui berry prevents the cognitive deficit caused by chronic exposure to ozone. Decreases levels of oxidative stress markers and superoxide enzymatic activity in animals exposed to ozone through four oxidative stress markers: 4HNE, MDA, Nt3, and AGEs in brain areas involved in learning and memory processes.	Yes	Ellagic acid derivatives, flavonols, and chlorogenic acid.	[82]
Fruit	Hydroethanolic extract of a Chilean berry Maqui. Pure compounds: rutin and quercetin	500 µg/mL (extract) 50 µM (quercetin) and 10 µM (rutin)	**Endothelial dysfunction and oxidative stress** Maqui berry extracts, quercetin, and rutin protect against endothelial dysfunction induced by high glucose and pyrogallol through increased generation and bioavailability of NO.	Yes	p-Coumaric acid, rutin, gentisic acid, sinapic acid, procyanidin B; gallic acid, quercetin, myricetin, delphinidin-3-*O*-glucoside, cyanidin-3-*O*-glucoside, delphinidin-3,5-*O*-diglucoside, delphinidin-3-*O*-sambubioside, cyanidin-3-*O*-sambubioside, proanthocyanidin B, proanthocyanidin blend, catechin epicatechin blend, p-coumaric acid and p-hydroxybenzoic acid blend, cyanidin catechin blend and free sugar blend.	[72]
Fruit	Hydroalcoholic extract of Maqui. Pure compound: rutin	50 mg (extract) 30 mg/kg (rutin)	**Vascular reactivity, hyperglycemia, and dyslipidemia** Maqui reduces plasma levels of cholesterol, LDL and triglycerides. Rutin lowers blood sugar and enhances endothelium-dependent relaxation. Maqui and rutin improved the bioavailability of nitric oxide.	Yes	Gentisic acid, ferulic acid, gallic acid, p-coumaric acid, sinapic acid, 4-hydroxybenzoic acid, delphinidin, cyanidin, vanillic acid, quercetin, myricetin, mixed catechin and epicatechin, delphinidin, delphinidin-3-*O*-sambubioside-5-*O*-glucoside, delphinidin-3,5-*O*-diglucoside, cyanidin-3-*O*-sambubioside-5-*O*-glucoside, cyanidin-3,5-*O*-diglucoside, delphinidin-3-*O*-sambubioside, delphinidin-3-*O*-glucoside, cyanidin-3-*O*-sambubioside, and proanthocyanidin B.	[71]
Fruit	Maqui extract enriched with anthocyanins. Pure compound: delphinidin-3-*O*-sambubioside-5-*O*-glucoside	125–500 mg/kg (extract) 2–10 µg/mL and 5–100 µg/mL (delphinidin-3-*O*-sambubioside-5-*O*-glucoside)	**Type 2 diabetes** Oral administration of anthocyanins reduces fasting blood glucose levels and glucose tolerance in hyperglycemic obese C57BL/6J mice fed a high-fat diet. Oral administration of delphinidin-3-*O*-sambubioside-5-*O*-glucoside dose-dependently lowered fasting blood glucose levels in obese C57BL/6J mice (2–10 µg/mL). It also decreases glucose production in rat liver cells (50–100 µg/mL).	Yes	Delphinidin-3-*O*-sambubioside-5-*O*-glucoside, delphinidin-3,5-*O*-diglucoside, delphinidin3-*O*-sambubioside, delphinidin-3-*O*-glucoside, cyanidin-3-*O*-sambubioside, cyanidin-3-*O*-glucoside, cyanidin-3-*O*-sambubioside-5-*O*-glucoside + cyanidin-3,5-*O*-diglucoside.	[75]

Table 2. *Cont.*

Plant Part(s)	Type of Extract	Dose or Concentration	Mechanism of Action or Effect of Extract and/or Pure Compound Characterization of the Extract	Characterization	Identified Compound	Reference
Fruit	Standardized extract of berries of Maqui. Pure compound: delphinidin	20 mg/kg (extract) 50 μM (delphinidin)	**Postprandial blood glucose** Delphinol® lowers blood glucose and postprandial insulin. Daily oral application of Delphinol® for four months reduces fasting blood glucose levels and lowers postprandial glycemia due to sodium–glucose cotransporter inhibition in the small intestine.	No		[73]
Fruit	Maqui	250 mL containing approx. 1000 μmol GAE of polyphenols)	**Postprandial blood glucose** Reduction in the glycemic peak mediated by the glucose + Maqui + lemon mixture. Mixing glucose + Maqui + lemon reduces the glycemic peak of glucose (20.5 ± 8.4%) compared to glucose. This amount represents a reduction of 36.7 ± 15.0 mg/dL in postprandial blood glucose.	No		[75]
Fruit	Maqui extract	20 mg of freeze-dried Maqui/mL of filtered tap water	**Obesity** Maqui extract prevents diet-induced obesity and its associated comorbidities. Reduced fasting glucose. Improves insulin response and reduces weight gain, and also a differential expression of genes involved in de novo lipogenesis.	Yes	Delphinidin-3-*O*-sambubioside-5-*O*-glucoside, delphinidin-3-*O*-sambubioside, cyanidin-3-*O*-sambubioside-5-*O*-glucoside, cyanidin-3-*O*-glucoside, cyanidin-3-*O*-sambubioside.	[74]

5. Antiplatelet Activity of the Compounds in Maqui

Several fruits such as red grapes, strawberries, kiwis, and pineapples have been shown to exert antiplatelet effects [63]. The most commonly investigated berries with antiplatelet potential have been grapes, aronia berries, and sea buckthorn berries, which contain phenolic compounds such as hippuric acid, pyrogallol, catechol, and resorcinol which inhibit collagen-stimulated platelets at a concentration of 100 μM [83,84]. A combination of extracts from different berries (blueberries, strawberry puree, cranberries, blackcurrant puree, and raspberry juice) showed an antiplatelet effect when using ADP and collagen as agonists, in platelets from healthy adults who consumed moderate amounts of berries for 8 weeks [84,85]. Another in vivo study, in whole blood, showed that the combination of grape seed and skin extracts reduced collagen-induced platelet aggregation to a greater extent than either extract alone, showing the importance of synergistic inhibitory actions at physiological levels [84,86]. Cranberry juice also inhibited ADP- and collagen-induced platelet aggregation after four days of consumption four times a day [87]. In addition, strawberry extract (0.1–1 mg/mL) in an in vivo model inhibits platelet aggregation induced by arachidonic acid (AA) and ADP in a dose-dependent manner [44]. Results of in vitro and ex vivo studies have reported that polyphenols present in red grape and purple grape juices inhibit platelet aggregation induced by ADP, thrombin, collagen, epinephrine, and AA [85,88–91]. Chlorogenic acid, a polyphenol present in cherries, apples, kiwis, eggplants, plums, and coffee, also exhibits antiplatelet activity [92].

The few studies about antiplatelet activity in Maqui are recent [93]. The aggregation and platelet secretion induced by ADP and collagen were significantly inhibited by leaf and immature fruit extracts of the varieties "Luna Nueva" and "Morena". These extracts also reduced oxidative stress, an effect that might be related to the high content of antioxidant compounds [20]. The chemical characterization allowed us to identify several compounds with known antiplatelet potential, such as caffeic acid, quercetin, isorhamnetin, kaempferol, and rutin, among others. The varieties differ in the composition and concentration of phenolic compounds, supporting the fact that extracts from different genotypes and parts of the tree, e.g., immature compared to mature fruit, vary in their antiplatelet activities. The participation of each bioactive compound with antiplatelet activity was also investigated using Pearson's statistical analysis. The results showed that the levels of phenolic compounds are responsible for the antiplatelet effects of Maqui [93].

Although this berry has been attributed with a wide biological potential, a gap can be seen in the literature regarding its antiplatelet properties. However, its rich chemical composition clearly shows that the presence of chemical compounds with high antiplatelet potential can give Maqui a prominent potential in the development of new antiplatelet agents. Table 3 summarizes some in vitro and in vivo studies of the antiplatelet activity of bioactive compounds identified in Maqui, i.e., anthocyanins, flavonoids, and phenolic acids that have important antiaggregant activity against several agonists, e.g., TRAP-6, ADP, collagen, and arachidonic acid (AA).

5.1. Anthocyanins

The anthocyanins delphinidin-3-*O*-rutinoside, cyanidin-3-*O*-glucoside, cyanidin-3-*O*-rutinoside, and malvidin-3-*O*-glucoside inhibit platelet aggregation [30,31,96,98,111,118] using TRAP-6, epinephrine, collagen, and ADP as agonists [30,96,98,101,109–111,116,118]. The effects were even greater in mixtures of these compounds, possibly due to the synergism of polyphenols and anthocyanins.

The use of pure chemical compounds in studies on platelet function is vital to elucidate the mechanism of antiplatelet action. Yan et al. determined the effects of delphinidin-3-*O*-glucoside on platelet activation in both in vitro and in vivo models of thrombosis, concluding a dose-dependent antiplatelet effect in human platelet-rich plasma (PRP) and washed platelets, activated with collagen, ADP, and TRAP-6 [31]. Delphinidin-3-*O*-glucoside also inhibited the growth of the thrombus in mice and the exposure of P-selectin in the platelet membrane. Furthermore, the authors reported that this compound inhibited the phosphorylation of the adenosine-activated protein kinase (AMPK) [31]. The concentrations of anthocyanins used in the cited experiments were within the physiological levels and the pharmacological dose (0.5–50 μM) [30].

In the presence of anthocyanins, thrombus formation in mice and human in vivo perfusion chambers was reduced at low and high shear rates [31]. In a model of intravital microscopy thrombosis, anthocyanins prolonged the time for thrombus formation at a dose of 0.5 μM and delayed vessel occlusion significantly at 5 μM and 50 μM. The release of α and dense granules, the expression of P-selectin, the cluster of differentiation (CD) 63 (CD63), CD40L, and the secretion of the cytosolic proteins were observed. In addition, delphinidin-3-*O*-glucoside decreased the expression of the $\alpha_2b\beta_3$ integrin. Overall, the cited study suggests that the daily consumption of anthocyanins may play a fundamental role in the protection against CVD, which could be related to the inhibition of platelet activity [31].

5.2. Flavonols

Quercetin is one of the main flavonoids identified in several natural sources, including Maqui [119]. Both antiplatelet potential and other biological properties have been attributed to this flavonol. Quercetin inhibits platelet aggregation induced by collagen, with an IC_{50} = 6 μg/mL at a concentration of 5 μg/mL of collagen [120]. The antiplatelet mechanism of this flavonoid has been related to the inhibition of collagen-stimulated tyrosine phosphorylation, a key component of the collagen signaling pathway through glycoprotein VI, Syk [121]. This compound has not only shown potent activity against collagen; previous studies report that it inhibits platelet aggregation induced by AA (substrate of cyclooxygenase-1 (COX-1)) and by U46619 (synthetic mimetic of TxA_2) [122].

Table 3. Antiplatelet activity of anthocyanins and phenolic compounds in Maqui.

Compounds	In Vitro	In Vivo	Reference
	Anthocyanins		
Delphinidin-3-O-glucoside (5–50 μg/mL)	Inhibition of platelet aggregation with collagen (10 μg/mL) and TRAP-6 (100 μM) at 0.5 μM and 50 μM in washed platelets. Inhibition of platelet aggregation with ADP (5 μM), collagen (10 μg/mL) and TRAP-6 (100 μM) at 0.5 μM and 50 μM in platelet-rich plasma Dose-dependent reduction in activated GPIIb/IIIa expression. Inhibition of platelet adhesion and aggregation in perfusion chamber assays at low and high shear rates. Decreased platelet deposition, thrombus formation, and vessel occlusion.		[94]
	Inhibition of platelet aggregation with ADP (5 μM), collagen (2 μg/mL), and TRAP (100 μM). Inhibition of the activation and secretion of P-selectin, CD63, CD40L, αllbβ3, and fibrinogen with ADP (200 μM), collagen (10 μg mL), thrombin (1 U/mL), and TRAP (250 μM). Mechanism: inhibition of the phosphorylation of MAPK induced by collagen (25 μg/mL).	Inhibition of collagen-induced thrombus formation (100 μg/mL), using controlled flow. Inhibition of thrombus formation induced by $FeCl_3$ at 50 μg/mL, using intravital microscopy.	[31,95]
Cyanidin-3-O-glucoside (5–50 μg/mL)	Inhibition of the activation and secretion of P-selectin, CD63, CD40L, αllbβ3, fibrinogen with collagen (10 μg/mL), thrombin (2 U/mL), and TRAP (250 μM). Inhibition of platelet aggregation with collagen (2.5 μg/mL), thrombin (0.1 U/mL), and TRAP (100 μM). Mechanism: via receiver GPVI collagen (2.5 μg/mL). Inhibition of the phosphorylation of tyrosine protein induced by collagen (2.5 μg/mL) at 5–50 μM.	Inhibition of the formation of the thrombus induced by collagen (0.5–50 μM) and $FeCl_3$ at 5–50 μM.	[95,96]
	Inhibition of platelet aggregation with collagen (10 μg/mL) and TRAP-6 (100 μM) at 0.5 μM-and 50 μM in washed platelets. Inhibition of platelet aggregation with ADP (5 μM), collagen (10 μg/mL) and TRAP-6 (100 μM) at 0.5 μM and 50 μM in platelet-rich plasma Dose-dependent reduction in activated GPIIb/IIIa expression. Inhibition of platelet adhesion and aggregation in perfusion chamber assays at low and high shear rates. Decreased platelet deposition, thrombus formation, and vessel occlusion.		[94]
	Inhibition of platelet granules (P-selectin, CD40L, 5-HT, RANTES, and TGF-β1) with thrombin (0.5 U/mL).	Attenuated serum levels of PF4 and β-TG in mice fed high-fat diets at a dose of 1000 mg/kg.	[97]
	Flavonols		
Quercetin- 4″-O-β-D-glucoside		Inhibition of platelet aggregation with collagen (50 μL) at 150 mg. Mechanism: inhibition of the phosphorylation of protein tyrosine kinase Syk and PLCγ2 induced by collagen (25 μg/mL) at 150 mg.	[98]
Quercetin	Inhibition of platelet aggregation with AA (100 μM), ADP (20 μM), collagen (10 μg/mL) at 13 μM. It inhibits ATP release with ADP (7 μM) and epinephrine (7 μM) at 2.5 μM. Mechanism: inhibits the formation of TxA_2 and PG induced by AA (100 μM) at 5 μM.		[99]
	Inhibition of platelet aggregation with 100 μg/mL of AA (100 μM) and collagen (10 μg/mL) at 100 μg/mL.	Relaxation in the thoracic aorta of the rat is induced by norepinephrine (3 μM) at 100 μM.	[100]
	Inhibition of platelet aggregation with collagen (0.5–5 μL/mL) at IC_{50}: 2.37–8.69. Inhibition of the mobilization of Ca^{2+} induced by collagen (5 μL/mL) at 15 μM. Mechanism: inhibits the GPVI signaling pathways, phosphorylation of tyrosine protein, and PI3 kinase induced by collagen (25 μL/mL) at 25 μM.		[101]
	Inhibition of platelet aggregation with collagen (0.5–5 μL/mL) at IC_{50}: 2.37–8.69. Inhibition of the mobilization of Ca^{2+} induced by collagen (5 μL/mL) at 15 μM. Mechanism: inhibits the GPVI signaling pathways, phosphorylation of tyrosine protein, and PI3 kinase induced by collagen (25 μL/mL) at 25 μM.		[102]
	Inhibition of platelet aggregation with AA (150 μM) IC_{50}: 18 μM. The increase in cAMP stimulated by PGI_2 (0.5 nM) decreased at 50 μM. Mechanism: inhibition of the activity of COX-1 and lipoxygenase at 10 μM and 50 μM.		[103]

Table 3. *Cont.*

Compounds	In Vitro	In Vivo	Reference
	Anthocyanins		
Kaempferol	Inhibition of thrombin (40 mU) and FXa (20 mU) (68 ± 1.6% and 52 ± 2.4%, respectively). Attenuated fibrin polymer formation in turbidity and phosphorylation of ERK 1/2, p38, JNK 1/2, and phosphoinositide PI3K/PKB (AKT) in cells stimulated with thrombin (0.5 U/mL). Inhibition of platelet aggregation stimulated by collagen/epinephrine (34.6%). Mechanism: inhibition of phosphorylation of ERK 1/2, p38, JNK 1/2, and PI3K/PKB.	Decreased thrombus formation in 3 animal models (collagen/epinephrine and thrombin-induced acute thromboembolism, $FeCl_3$-induced model, and carotid arterial thrombus model).	[104]
	Decreased collagen adhesion in resting platelets and activated platelets with thrombin at a dose of 5 μg/mL. Inhibition of platelets activated by thrombin and fibrinogen (40%). Inhibition of platelet aggregation with collagen (5 μg/mL) and AA (0.5 μmol/L) at 50 μg/kg. Thrombin-stimulated reduction of enzymatic lipid peroxidation in platelets.		[105]
Myricetin	Inhibition of platelet aggregation with collagen (5 μg/mL) and AA (0.5 μmol/L) at 50 μg/kg. Thrombin-stimulated reduction in enzymatic lipid peroxidation in platelets.		[106]
	Inhibition of platelet aggregation and secretion of alpha granules. with TRAP-6 (10 μM) and collagen (1 μg/mL) at 15 and 30 μM. Decreased fibrinogen binding induced by CRP (1 μg/mL) and TRAP-6 (10 μM) at 15 μM. Reduction in adhesion on collagen and thrombus formation without affecting hemostasis in vivo. Mechanism: inhibition of ERp5 and PDI.		[107]
	Dose-dependent (20–30 μM) inhibition of platelet aggregation, granule secretion and activation (activation of αIIbβ3 integrin and P-selectin exposure), generation of ROS, and induced intracellular Ca^{2+} mobilization by CRP (0.1 μg/mL) and collagen (1 μg/mL). Mechanism: inhibition of GPVI during cell activation.	Reduction in ischemia/reperfusion-induced acute infarction in a mouse model of stroke. Blocked $FeCl_3$-induced arterial thrombus formation in vivo and thrombus formation on collagen-coated surfaces under low shear rate.	[108]
Rutin	Inhibition of platelet aggregation with collagen at 250 μM (1 μg/mL). The mobilization of Ca^{2+} induced by collagen (1 μg/mL) decreases to 250 μM. Mechanism: inhibition of the PLC phosphorylation and formation of TxA2, inhibits collagen-induced phosphorylation of P47 at 250 μM.		[109]
	Flavanones		
Eriodictyol	Inhibition of platelet aggregation with collagen (2 μg/mL) and AA (0.5 mmol/L) at 50 μM.		[110]
Hesperetin	Concentration-dependent inhibition of platelet aggregation induced by collagen (5 μg/mL) and AA (0.5 μmol/L) (IC_{50}: 20.5 and at IC_{50}: 69.2, respectively). Inhibition mobilization of cytosolic Ca^{2+} induced by collagen (10 μg/mL) at 20–50 μM. Inhibition of the secretion of serotonin with collagen (5 μg/mL) and AA (0.5 μmol/L) at IC_{50}: 10.5 and at IC_{50}: 25.2, respectively. Mechanism: inhibitionPLC-γ2 phosphorylation. Inhibition of COX-1 activity.		[111]
		Atherosclerosis inhibition	[112]
	Phenolic acids		
Ferulic acid	Inhibition of platelet aggregation induced by ADP, thrombin (0.5 U/mL), AA (2 mM), collagen (2 μg/mL), and U46619 (2 μM) at 50–200 μM. Inhibition of mobilization of cytosolic Ca^{2+} and TXB_2 production. Increased the levels of cAMP and cGMP and phosphorylated VASP. Decreased phospho-MAPK and PDE. Mechanism: activation of cAMP and cGMP signaling.	Decreased pulmonary thrombosis and prolonged tail bleeding and coagulation time in mice without altering coagulation parameters.	[113]
	Inhibition of platelet activation (serotonin secretion) stimulated by thrombin, collagen/epinephrine, and decreased clot retraction activity at 10 μg. Mechanism: decreased granule secretion, prolongation of the intrinsic coagulation cascade, and upregulation of αIIbβ3/FIB/AKT signaling expressions.	Decreased thrombosis in acute thromboembolism model and decreased αIIbβ3/ FIB expression and AKT phosphorylation in thrombin-stimulated platelet activation.	[114]

Table 3. *Cont.*

Compounds	In Vitro	In Vivo	Reference
	Anthocyanins		
Caffeic acid	Inhibition of platelet aggregation with ADP (8 μmol/L) and collagen (1.5 μg/mL) at 0.5 mmol/L.		[115]
	Inhibition of the activation and secretion of P-selectin with TRAP (25 μmol/L) at 100 μmol/L.		[83]
	Inhibition of platelet aggregation with collagen (2 μg/mL) at 15–25 μM. Mechanism: inhibition of the phosphorylation of cGMP/VAS Ser /VASP Ser157 at 15–25 μM. Decreases PKC and phosphorylation of P47 at 15–25 μM.		[116]
Ellagic acid	Inhibition of platelet aggregation with collagen (1 μg/mL) at IC_{50}: 50 μM. The mobilization of Ca^{2+} induced by collagen (1 μg/mL) decreases at 50 μM. Mechanism: inhibition of the PLCγ2-PKC cascade, OH* formation, MAPKs, and Akt induced by collagen (1 μg/mL) at 50 μM.		[117]

Previous works show the cardioprotective potential of quercetin and kaempferol, as well as their derivatives. Quercetin 3-*O*-[(6-*O*-E-feruloyl)-β-D-glucopyranosyl-(1→2)]-β-D-galactopyranoside-7-*O*-β-D-glucuropyranoside) and kaempferol 3-*O*-[(6-*O*-E-caffeoyl)-β-D-glucopyranosyl-(1→2)]-β-D-galactopyranoside-7-*O*-(2-*O*-E-caffeoyl′)-β-D–glucuropyranoside isolated from *Lens culinaris* Medik. showed potent antiplatelet action. Results revealed decreased collagen adhesion of resting platelets and thrombin-activated platelets after incubation with quercetin and kaempferol derivatives [105]. On the other hand, kaempferol has shown its potential to reduce and prevent thrombosis. The background shows that this flavonoid inhibits fibrin polymer formation, attenuates phosphorylation of extracellular signal-regulated kinase (ERK)1/2, p38, c-Jun N-terminal kinase (JNK)1/2, and phosphoinositide 3-kinase (PI3K)/PKB (AKT) in thrombin-stimulated cells and decreases collagen/epinephrine-stimulated platelet aggregation by 34.6%. Additionally, kaempferol protected mice from thrombosis in models of acute thromboembolism induced by collagen/epinephrine and thrombin, as well as carotid artery thrombus induced by $FeCl_3$ [104]. Another flavonol that has been identified in Maqui and that stands out for its antiplatelet action is myricetin. This compound reduces the ability of platelets to spread over collagen and form thrombi in vitro. This effect has been attributed mainly to the inhibition of PDI and ERp5 by binding to myricetin, forming non-covalent bonds [107]. Additionally, myricetin has been shown at physiologically relevant concentrations to inhibit platelet aggregation induced by TRAP-6 and AA. It also inhibits fibrinogen binding and collagen-related peptide-induced alpha granule secretion [123]. The inhibitory effect against several platelet agonists suggests that this flavonoid can act on molecules common to different pathways [107,123].

Isorhamnetin was one of the compounds that presented a positive correlation between its concentration and the antiplatelet activity of Maqui extracts [93]. The effect of isorhamnetin on mitochondrial function, platelet adhesion, and thrombus formation was evaluated under conditions of controlled blood flow in the Badimon perfusion chamber. This flavonol showed antiplatelet activity induced by collagen, and thrombin receptor activator peptide-6 (TRAP-6), with IC_{50} values of 8.1 ± 2.6 μM and 16.1 ± 11.1 μM, respectively [124]. It also decreased the mitochondrial membrane potential and reduced platelet deposition on the thrombus, confirming its antithrombotic effect [124].

Isorhamnetin and tamarixetine, quercetin methylated metabolites, stand out for their antiplatelet potential, even showing better results than aspirin in an aggregometry test. These compounds inhibit human platelet aggregation and suppress activating processes, including granule secretion, αIIbβ3 integrin function, calcium mobilization, and spleen tyrosine kinase (Syk). They also attenuated thrombus formation in an in vitro microfluidic model, while isorhamnetin inhibited thrombosis in a mouse model of laser injury [125].

5.3. Flavones

Eriodictyol, a flavone present in Maqui, inhibits platelet aggregation stimulated by collagen and AA, in platelet-rich plasma (IC50 = 912.9 ± 37 μM and IC50 = 1027.3 ± 551 μM, respectively) [110]. On the other hand, hesperetin can selectively inhibit collagen- and AA-mediated signal transduction, with IC50 of 20.5 ± 3.5 and 69.2 ± 5.1 μM, respectively. The proposed mechanism suggests the inhibition of PLC-γ2 phosphorylation and the activity of COX-1 [111].

5.4. Phenolic Acids

Caffeic acid, identified in Maqui, has also been noted for its antiplatelet properties. Studies have shown that this compound possesses antithrombotic activity on mouse brain arterioles in vivo, and inhibits platelet aggregation in vitro stimulated by various agonists (ADP and thrombin) [126,127]. It has been described that this polyphenol is a potent compound that increases the level of cAMP-dependent protein phosphorylation in collagen–platelet interactions [128]. Studies report its antithrombotic action at doses of 1.25–5 mg/kg, an effect related to its capacity to suppress phosphorylation of ERK, p38, and JNK, which leads to cAMP elevation, and it negatively regulates P-selectin expression and activation of αIIbβ3 [126–128]. Meanwhile, Nam et al. showed that this polyphenol decreases the production and release of thrombogenic molecules in human platelets. This effect was not only mediated by TxA_2 but also by the decrease in serotonin released by collagen by inhibiting the phosphorylation of JNK1 [129]. This phenolic compound at 25–100 μM further inhibited ADP-induced platelet aggregation, P-selectin expression, ATP release, Ca^{2+} mobilization, and αIIbβ3 integrin activation [127].

The potential of ferulic acid has been related to the activation of cAMP and cGMP signaling [113,114]. This compound inhibits dose-dependent (50–200 μM) platelet aggregation induced by platelet agonists (ADP, thrombin, collagen, AA, and U46619). Additionally, it attenuates intracellular Ca^{2+} mobilization and TxA2 production. It also increases cAMP, cGMP, and vasodilator-stimulated phosphoprotein (VASP) levels while decreasing phospho-MAPK and phosphodiesterase (PDE) in washed rat platelets [113]. Studies report that ferulic acid has an antithrombotic effect in the in vivo model of acute thromboembolism and decreases the expression of αIIbβ3/FIB and AKT phosphorylation in thrombin-stimulated platelet activation [114].

Both ellagic acid and ferulic acid inhibit platelet activation in vitro, induced by ADP and collagen [130]. Phenolic acid derivatives have also been shown to be novel antiplatelet targets, even more potent than their phenolic precursors. Dihydrocaffeic acid and dihydroferulic acid at doses of 0.01–100 μg/mL 1 μM decreased ADP-stimulated platelet activation, measured as P-selectin expression and fibrinogen binding [131].

AA, arachidonic acid; ADP, adenosine diphosphate; ATP, adenosine triphosphate; Akt, protein kinase B; CD63, a membrane protein associated with lysosome 3; CD40L, a ligand of the membrane protein; COX-1, cyclooxygenase; CRP, collagen-related peptide; ERK: extracellular signal-regulated kinase; FIB, plasma fibrinogen; FXa, factor Xa; cGMP, cyclic guanosine monophosphate; GP, glycoprotein; 5-HT, serotonin; αllbβ3, GPVI, glycoprotein VI; GPIIb/IIIa, glycoprotein IIb–IIIa; JNK: c-Jun N-terminal kinase, MAPK, protein kinase activated by mitogens; MW, molecular weight; OH*, hydroxyl radical; PAF, platelet activating factor; PDE: phosphodiesterase; PG, glycoprotein; PF4, platelet factor 4; PI3, phosphoinositol 3 kinase; PKB, protein kinase B; PKC, protein kinase C; PLC γ2, phospholipase C; Rantes, regulatory chemokine beta; ROS, reactive oxygen species; Ser, serotonin; Syk, tyrosine protein kinase; β-TG, beta thromboglobulin; TGF-β1, transforming growth factor-beta 1; TRAP-6, thrombin receptor activator for peptide 6; TxA_2, thromboxane; VASP, phosphoprotein stimulated by vasodilator.

The secretion of α and dense granules, the release of adenosine triphosphate (ATP), and the mobilization of Ca^{2+} are the main markers of activation reported to date by mechanisms involving the AA-derived pathway and GPVI receptor, among others. Some receptors for collagen (GPVI), thrombin (protein activated receptors, PARs), and TxA_2 activate phospholipase C (PLC). The aforementioned receptors generate diacylglycerol (DAG) and inositol triphosphate (IP3), activating the protein kinase C (PKC) and intracellular Ca^{2+} release, leading to granule secretion [98,101,102]. On the other hand, adenylyl cyclase (AC) favors the conversion of ATP into cyclic adenosine monophosphate (cAMP). The activation of phosphoinositide 3-kinase/protein kinase B (PI3K/Akt) could mediate the phosphorylation of endothelial nitric oxide synthase (eNOS), increasing the production of platelet nitric oxide (NO). This leads to the production of cyclic guanosine monophosphate (cGMP), and stimulates the activation of mitogen-activated protein kinases (MAPK), thus promoting granule secretion and activation of platelets [31,118].

Considering the studies about the antiplatelet activity of previous tested chemical compounds that have also been identified in Maqui, and the reviewed inhibition mechanisms of platelet activation, we propose a scheme of how the bioactive compounds present in Maqui extracts exert their effects (Figure 5).

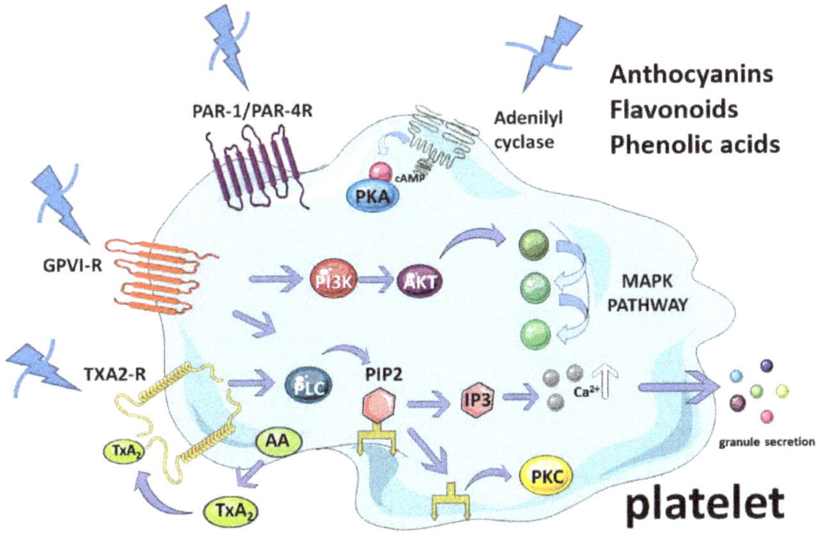

Figure 5. Antiplatelet mechanistic proposal for Maqui. AC, adenylate cyclase; AKT, protein kinase B; cAMP, cyclic adenosine monophosphate; DAG, diacylglycerol; PI_3, phosphoinositol 3 kinase; PIP2, phosphatidylinositol bisphosphate; GP, glycoprotein; MAP, mitogen-activated protein kinase; P2Y12/P2P1, ADP receptor; TxA_2R, thromboxane receptor.

6. Limitations and Future Perspective

Inhibition of platelet function has long been used to prevent and treat CVD [63]. Although antiplatelet drugs currently exist, this therapy has been accompanied by side effects, such as bleeding. Recent efforts focus on the search for and development of new therapeutic agents, together with the healthy habits that we must promote to contribute to adequate cardiovascular health.

In general, healthy eating is promoted, which includes minimally processed foods and foods rich in bioactive products such as fruits, nuts, seeds, beans, vegetables, whole grains, vegetable oils, yogurt, and fish [132]. Epidemiological studies have provided evidence for the protective role of healthy diets in the prevention of CVD. For example, eating two or three kiwis a day for 28 days reduces platelet aggregation induced by collagen and ADP. On the other hand, raw garlic and some of its preparations are recognized as antiplatelet agents [63]. As the above, Maqui plays a fundamental role among healthy fruits due to its wide biological potential, as highlighted in this work.

Maqui is known to be rich in phenolic compounds, and phenolic compounds have a well-recognized role in reducing the risk of chronic diseases, increasing healthy life years, and promoting healthy aging [133]. The protective mechanisms of foods rich in polyphenols may not only depend on their content of nutrients and bioactive compounds but also include their food matrix properties that affect glycemic load and energy density, among others [132].

The extraction procedure affects the chemical composition of the Maqui extracts; variables such as temperature, sonication, extraction time, and solvent are some of the parameters that would influence the chemical profile identified for this species [119]. Concerning the solvent used for extractions in Maqui matrices, it has been shown that hydroalcoholic mixtures are the most used solvents for the extraction of bioactive components from berries [134]. The biological potential of aqueous and ethanolic extracts of Maqui, which have a high content of polyphenols, has been reported [28,46]. Hydroalcoholic extract presented higher polyphenolic content, with antioxidant and antimicrobial properties meaning it is the most promising extract for pharmaceutical purposes [51]. As previously described, Maqui has the highest phenol content compared to other berries, and this depends substantially on the solvent used for extraction [16,19].

Sonication is an ideal alternative to obtain extracts from Maqui berries with high bioactivity [49] while the extraction time can also modify the antioxidant capacity of Chilean Maqui berries [48]. Three operating conditions of the extraction process were compared by modifying the type of solvent (methanol, ethanol, and acetone), solvent concentration (20, 60, and 100%), and extraction time (15, 127.5, and 240 min). Results show that after a certain extraction time there is a final equilibrium between the solutes of the matrix and the extraction solvent [28].

Metabolites such as polyphenols, flavonoids, and anthocyanins are the main ones that have been identified in Maqui extracts. Polyphenols may possess antiplatelet properties, but their coadministration may not be safe [135]. It is important to make the correct and precise determination of both the composition and the amounts of the phenolic compounds that we consume to avoid toxicity or unwanted side effects. Of the total number of trials with polyphenols in the last 20 years, 20% analyzed vascular and endothelial responses, and trials on platelet function and thrombosis are lacking [135,136]. In vivo and trial studies evaluating potential polyphenol–drug interactions are needed to address this limitation [123].

On the other hand, studies refer to the protective effects of flavonoids against drug-induced toxicity. It was concluded that flavonoids, both dietary and derived from plant medicines, can exert protective effects against drug-induced toxicity [135,137,138]. These compounds are generally recognized as safe, due to the long history of use and consumption of foods rich in flavonoids. The total daily intake of flavonoids in Europe is estimated to be around 428 $\pm$ 49 mg, with 136 $\pm$ 14 mg being monomeric [135,139], while flavonoid-rich beverages and vegetables can reach 1000 mg/day of flavonoids [135,140].

Although the benefits of phenols have been demonstrated in different in vitro and in vivo models, there are few reports evaluating the possible effects of Maqui extracts on platelet aggregation [93]. Although there are clinical studies that show that Maqui extract or capsules are safe for human consumption [74,75], studies that evaluate their administration over longer periods are required.

Although in vitro studies help to understand the possible health contributions of Maqui berries, these may not be fully applicable to humans, as in the case of anthocyanins, such as delphinidin and anthocyanin, which have been shown to have limited bioavailability. In animal and human studies, anthocyanins are poorly absorbed and thus show low bioavailability [73]. Furthermore, several groups have suggested that dietary cyanidins and delphinidins may be subject to extensive biotransformation in humans, most likely involving the colonic microbiota [73,141].

Despite the above, it appears that anthocyanins are more bioavailable than is perceived, and their metabolites are present in the circulation for $\leq$48 h after ingestion. A study carried out on eight male volunteers showed that cyanidin-3-*O*-glucoside, a main component of the Maqui berry, has a non-negligible bioavailability of 12.38 $\pm$ 1.38%. Blood concentrations of cyanidin-3-*O*-glucoside and cyanidin-3-*O*-glucoside conjugates were observed to appear between 1 and 2 h after the intake of 500 mg of cyanidin-3-*O*-glucoside [142].

There are few in vivo and in vitro studies comparing Maqui with other fruits and berries to determine the benefits and weaknesses of Maqui compared to other foods with similar bioactive compound composition. There is a lack of studies to establish how much Maqui should be consumed daily to see this effect at physiologically relevant concentrations.

The natural bioactive ingredients of Maqui have antiplatelet effects with multiple targets on the platelet, while the synergistic effects of the phenolic compounds could enhance the antiplatelet activity. Due to the above, there exists an interest in the cardiovascular benefits of Maqui berries, so daily consumption of this berry can prevent the development of CVD, although detailed studies are needed to reinforce its clinical utility.

7. Conclusions

Due to the prevalence of CVD, new antiplatelet drugs are needed to prevent and treat arterial thrombosis as well as other CVDs. Some bioactive compounds, such as polyphenols and anthocyanins in fruits and vegetables, have been reported to inhibit platelet activation and so reduce the risk of CVD.

The broad chemical profile of Maqui (flavonoids, anthocyanins, and phenolic acids) is directly related to its high biological potential. The current knowledge about Maqui's antioxidant, anti-inflammatory, and hypoglycemic effects suggests that a diet including Maqui could aid in the prevention of CVD, with more studies being required to prove this hypothesis. The fundamental mechanisms which this species influences have been mainly related to the inhibition of lipid peroxidation, decrease in cholesterol and blood glucose levels, as well as a decrease in oxidative stress. Studies are needed to establish how much Maqui should be consumed daily to see this effect at physiologically relevant concentrations.

Additionally, this species can be studied in detail for antiplatelet purposes, since to date, there have been few authors who have highlighted this potential in the fruits of Maqui. Our findings highlighted that the main mechanism by which the compounds identified in this species act is related to the metabolic pathways of the AA and GPVI receptors. Although anthocyanins are the main phenolic compounds in this berry, antiplatelet activity may be directly related to the presence of a specific compound or enhanced by the synergy of several phenolic compounds.

This review allowed us to investigate the antiplatelet and cardioprotective activity of the bioactive compounds present in Maqui and highlight the areas in which much remains to be investigated. Without a doubt, we have shown that Maqui is an interesting target in the search for new antiplatelet therapies.

Author Contributions: L.R. and A.T.: conceptualization, conducting the research, writing the original draft. H.V. and I.W.: conceptualization. I.P.: conceptualization and writing the original draft. E.F. and S.W.: conceptualization and conducting the research. All authors have read and agreed to the published version of the manuscript.

Funding: This work has been funded by ANID/REDES190112 "INTERNATIONAL NETWORK ON THE STUDY OF ENDOPLASMIC RETICULUM STRESS IN PLATELET FOR PREVENT CARDIOVASCULAR DISEASE IN GLUCOLIPOTOXIC MILIEU".

Acknowledgments: We would like to thank ANID-FONDECYT N°1211136, ANID-FONDECYT N°1220339, ANID-FONDEQUIP EQM200049, ANID/FONDECYT N°3220099 and Interuniversity Center for Healthy Aging, Code RED211993. Andres Trostchansky thanks Comisión Sectorial de investigación Científica (CSIC), Uruguay Grupos N°536 and Proyecto Donaciones Especiales (Medical Plus), Fundacion Manuel Perez, Uruguay.

Conflicts of Interest: The authors declare no conflict of interest.

References

1. World Medical Association. World Medical Association Declaration of Helsinki: Ethical principles for medical research involving human subjects. *JAMA* **2013**, *310*, 2191–2194. [CrossRef] [PubMed]
2. Benjamin, E.J.; Blaha, M.J.; Chiuve, S.E.; Cushman, M.; Das, S.R.; Deo, R.; de Ferranti, S.D.; Floyd, J.; Fornage, M.; Gillespie, C. Heart disease and stroke statistics—2017 update: A report from the American Heart Association. *Circulation* **2017**, *135*, e146–e603. [CrossRef] [PubMed]
3. Zuraini, N.Z.A.; Sekar, M.; Wu, Y.S.; Gan, S.H.; Bonam, S.R.; Rani, N.N.I.M.; Begum, M.Y.; Lum, P.T.; Subramaniyan, V.; Fuloria, N.K.; et al. Promising nutritional fruits against cardiovascular diseases: An overview of experimental evidence and understanding their mechanisms of action. *Vasc. Health Risk Manag.* **2021**, *17*, 739. [CrossRef] [PubMed]
4. Kalantar-Zadeh, K.; Block, G.; Humphreys, M.H.; Kopple, J.D. Reverse epidemiology of cardiovascular risk factors in maintenance dialysis patients. *Kidney Int.* **2003**, *63*, 793–808. [CrossRef] [PubMed]
5. Albala, C.; PGIGSR. Funcionalidad en los Adultos Mayores: Fragilidad y Dependencia. In *Envejecimiento: Demografía, Salud e Impacto Social*; Talca, U.D., Ed.; Universidad de Talca: Talca, Chile, 2016.
6. Goff, D.C., Jr.; Lloyd-Jones, D.M.; Bennett, G.; Coady, S.; D'Agostino, R.B., Sr.; Gibbons, R.; Greenland, P.; Lackland, D.T.; Levy, D.; O'Donnell, C.J. 2013 ACC/AHA guideline on the assessment of cardiovascular risk: A report of the American College of Cardiology/American Heart Association Task Force on Practice Guidelines. *J. Am. Coll. Cardiol.* **2014**, *63*, 2935–2959. [CrossRef] [PubMed]
7. Gensini, G.F.; Comeglio, M.; Colella, A. Classical risk factors and emerging elements in the risk profile for coronary artery disease. *Eur. Heart J.* **1998**, *19*, A53–A61.
8. Tsoupras, A.; Lordan, R.; Zabetakis, I. Thrombosis and COVID-19: The Potential role of nutrition. *Front. Nutr.* **2020**, *7*, 177. [CrossRef]
9. Palomo, I.F.; Torres, G.I.; Alarcón, M.A.; Maragaño, P.J.; Leiva, E.; Mujica, V. High prevalence of classic cardiovascular risk factors in a population of university students from south central Chile. *Rev. Española Cardiol.* **2006**, *59*, 1099–1105. [CrossRef]
10. Badimon, L.; Vilahur, G. Coronary Atherothrombotic Disease: Progress in Antiplatelet Therapy. *Rev. Española De Cardiol.* **2008**, *61*, 501–513. [CrossRef]
11. Eduardo, F.; Andrés, T.; Lívia Mateus, R.; Mario Roberto, M., Jr.; Iván, P. Antiplatelet Effects of Bioactive Compounds Present in Tomato Pomace. *Curr. Drug Targets* **2021**, *22*, 1716–1724.
12. Irfan, M.; Kwon, T.-H.; Lee, D.-H.; Hong, S.-B.; Oh, J.-W.; Kim, S.-D.; Rhee, M.H. Antiplatelet and Antithrombotic Effects of Epimedium koreanum Nakai. *Evid. Based Complementary Altern. Med.* **2021**, *2021*, 7071987. [CrossRef]
13. Olas, B. Dietary supplements with antiplatelet activity: A solution for everyone? *Adv. Nutr.* **2018**, *9*, 51–57. [CrossRef] [PubMed]
14. Meshkini, A.; Tahmasbi, M. Antiplatelet aggregation activity of walnut hull extract via suppression of reactive oxygen species generation and caspase activation. *J. Acupunct. Meridian Stud.* **2017**, *10*, 193–203. [CrossRef] [PubMed]
15. Rodríguez, R. Elaeocarpaceae Juss. Ex DC. *Flora Chile* **2005**, *2*, 15–17.
16. Misle, E.; Garrido, E.; Contardo, H.; González, W. Maqui (*Aristotelia chilensis* (Mol.) Stuntz) the amazing chilean tree: A review. *J. Ofagricultural Sci. Technol. B* **2011**, *1*, 473–482.
17. Fredes, C.; Yousef, G.G.; Robert, P.; Grace, M.H.; Lila, M.A.; Gómez, M.; Gebauer, M.; Montenegro, G. Anthocyanin profiling of wild maqui berries (*Aristotelia chilensis* [Mol.] Stuntz) from different geographical regions in Chile. *J. Sci. Food Agric.* **2014**, *94*, 2639–2648. [CrossRef]
18. Zúñiga, G.E.; Tapia, A.; Arenas, A.; Contreras, R.A.; Zúñiga-Libano, G. Phytochemistry and biological properties of *Aristotelia chilensis* a Chilean blackberry: A review. *Phytochem. Rev.* **2017**, *16*, 1081–1094. [CrossRef]
19. Brauch, J.E.; Buchweitz, M.; Schweiggert, R.M.; Carle, R. Detailed analyses of fresh and dried maqui (*Aristotelia chilensis* (Mol.) Stuntz) berries and juice. *Food Chem.* **2016**, *190*, 308–316. [CrossRef]
20. Céspedes, C.L.; El-Hafidi, M.; Pavon, N.; Alarcon, J. Antioxidant and cardioprotective activities of phenolic extracts from fruits of Chilean blackberry *Aristotelia chilensis* (Elaeocarpaceae), Maqui. *Food Chem.* **2008**, *107*, 820–829. [CrossRef]
21. Gironés-Vilaplana, A.; Mena, P.; García-Viguera, C.; Moreno, D.A. A novel beverage rich in antioxidant phenolics: Maqui berry (*Aristotelia chilensis*) and lemon juice. *LWT-Food Sci. Technol.* **2012**, *47*, 279–286. [CrossRef]
22. Genskowsky, E.; Puente, L.A.; Pérez-Álvarez, J.A.; Fernández-López, J.; Muñoz, L.A.; Viuda-Martos, M. Determination of polyphenolic profile, antioxidant activity and antibacterial properties of maqui [*Aristotelia chilensis* (Molina) Stuntz] a Chilean blackberry. *J. Sci. Food Agric.* **2016**, *96*, 4235–4242. [CrossRef] [PubMed]
23. Vogel, H.; González, B.; Catenacci, G.; Doll, U. Domestication and Sustainable Production of Wild Crafted Plants with Special Reference to the Chilean Maqui Berry (Aristotelia Chilensis). *Julius-Kühn-Archiv* **2016**, *453*, 50–52.
24. Salgado, P.; Prinz, K.; Finkeldey, R.; Ramírez, C.C.; Vogel, H. Genetic Variability of Aristotelia Chilensis ("Maqui") Based on Aflp and Chloroplast Microsatellite Markers. *Genetic Resources and Crop. Evolution* **2017**, *64*, 2083–2091. [CrossRef]
25. Vogel, H.; Peñailillo, P.; Doll, U.; Contreras, G.; Catenacci, G.; González, B. Maqui (*Aristotelia chilensis*): Morpho-phenological characterization to design high-yielding cultivation techniques. *J. Appl. Res. Med. Aromat. Plants* **2014**, *1*, 123–133. [CrossRef]
26. Brauch, J.E.; Reuter, L.; Conrad, J.; Vogel, H.; Schweiggert, R.M.; Carle, R. Characterization of anthocyanins in novel Chilean maqui berry clones by HPLC–DAD–ESI/MS n and NMR-spectroscopy. *J. Food Compos. Anal.* **2017**, *58*, 16–22. [CrossRef]
27. González, B.; Vogel, H.; Razmilic, I.; Wolfram, E. Polyphenol, anthocyanin and antioxidant content in different parts of maqui fruits (*Aristotelia chilensis*) during ripening and conservation treatments after harvest. *Ind. Crops Prod.* **2015**, *76*, 158–165. [CrossRef]

28. Makinistian, F.G.; Sette, P.; Gallo, L.; Bucalá, V.; Salvatori, D. Optimized aqueous extracts of maqui (*Aristotelia chilensis*) suitable for powder production. *J. Food Sci. Technol.* **2019**, *56*, 3553–3560. [CrossRef]
29. Rivera-Tovar, P.R.; Mariotti-Celis, M.S.; Pérez-Correa, J.R. Maqui (*Aristotelia chilensis* (Mol.) Stuntz) and murta (*Ugni molinae* Turcz): Native Chilean sources of polyphenol compounds. *Mini-Rev. Org. Chem.* **2019**, *16*, 261–276. [CrossRef]
30. Rechner, A.R.; Kroner, C. Anthocyanins and colonic metabolites of dietary polyphenols inhibit platelet function. *Thromb. Res.* **2005**, *116*, 327–334. [CrossRef]
31. Yang, Y.; Shi, Z.; Reheman, A.; Jin, J.W.; Li, C.; Wang, Y.; Andrews, M.C.; Chen, P.; Zhu, G.; Ling, W. Plant food delphinidin-3-glucoside significantly inhibits platelet activation and thrombosis: Novel protective roles against cardiovascular diseases. *PLoS ONE* **2012**, *7*, e37323. [CrossRef]
32. Palomo, I.; Pereira, J.; Palma, J. *Hematología: Fisiopatología y Diagnóstico*; Universidad de Talca: Talca, Chile, 2005.
33. Evstatiev, R.; Bukaty, A.; Jimenez, K.; Kulnigg-Dabsch, S.; Surman, L.; Schmid, W.; Eferl, R.; Lippert, K.; Scheiber-Mojdehkar, B.; Michael Kvasnicka, H. Iron deficiency alters megakaryopoiesis and platelet phenotype independent of thrombopoietin. *Am. J. Hematol.* **2014**, *89*, 524–529. [CrossRef]
34. Jennings, L.K. Mechanisms of platelet activation: Need for new strategies to protect against platelet-mediated atherothrombosis. *Thromb. Haemost.* **2009**, *101*, 248–257. [CrossRef] [PubMed]
35. Spencer, F.A.; Becker, R.C. Platelets: Structure, function, and their fundamental contribution to hemostasis and pathologic thrombosis. In *Textbook of Coronary Thrombosis and Thrombolysis*; Springer: Berlin/Heidelberg, Germany, 1997; pp. 31–49.
36. Robless, P.A.; Okonko, D.; Lintott, P.; Mansfield, A.O.; Mikhailidis, D.P.; Stansby, G.P. Increased platelet aggregation and activation in peripheral arterial disease. *Eur. J. Vasc. Endovasc. Surg.* **2003**, *25*, 16–22. [CrossRef] [PubMed]
37. Ghoshal, K.; Bhattacharyya, M. Overview of platelet physiology: Its hemostatic and nonhemostatic role in disease pathogenesis. *Sci. World J.* **2014**, *2014*, 781857. [CrossRef]
38. El Haouari, M.; Rosado, J.A. Platelet signalling abnormalities in patients with type 2 diabetes mellitus: A review. *Blood Cells Mol. Dis.* **2008**, *41*, 119–123. [CrossRef] [PubMed]
39. Geisler, T.; Anders, N.; Paterok, M.; Langer, H.; Stellos, K.; Lindemann, S.; Herdeg, C.; May, A.E.; Gawaz, M. Platelet response to clopidogrel is attenuated in diabetic patients undergoing coronary stent implantation. *Diabetes Care* **2007**, *30*, 372–374. [CrossRef]
40. Badimon, L.; Vilahur, G. Thrombosis formation on atherosclerotic lesions and plaque rupture. *J. Intern. Med.* **2014**, *276*, 618–632. [CrossRef]
41. Fuentes, Q.E.; Fuentes, Q.F.; Andrés, V.; Pello, O.M.; de Mora, J.F.; Palomo, G.I. Role of platelets as mediators that link inflammation and thrombosis in atherosclerosis. *Platelets* **2013**, *24*, 255–262. [CrossRef]
42. Vilahur, G.; Badimon, L. Antiplatelet properties of natural products. *Vasc. Pharmacol.* **2013**, *59*, 67–75. [CrossRef]
43. Meerson, F.Z.; Kagan, V.E.; Kozlov, Y.P.; Belkina, L.M.; Arkhipenko, Y.V. *The Role of Lipid Peroxidation in Pathogenesis of Ischemic Damage and the Antioxidant Protection of the Heart*; Springer: Berlin/Heidelberg, Germany, 1982.
44. Alarcón, M.; Fuentes, E.; Olate, N.; Navarrete, S.; Carrasco, G.; Palomo, I. Strawberry extract presents antiplatelet activity by inhibition of inflammatory mediator of atherosclerosis (sP-selectin, sCD40L, RANTES, and IL-1β) and thrombus formation. *Platelets* **2015**, *26*, 224–229. [CrossRef]
45. Fuentes, E.J.; Astudillo, L.A.; Gutiérrez, M.I.; Contreras, S.O.; Bustamante, L.O.; Rubio, P.I.; Moore-Carrasco, R.; Alarcón, M.A.; Fuentes, J.A.; González, D.E. Fractions of aqueous and methanolic extracts from tomato (*Solanum lycopersicum* L.) present platelet antiaggregant activity. *Blood Coagul. Fibrinolysis* **2012**, *23*, 109–117. [CrossRef] [PubMed]
46. Escribano-Bailón, M.T.; Alcalde-Eon, C.; Muñoz, O.; Rivas-Gonzalo, J.C.; Santos-Buelga, C. Anthocyanins in berries of maqui [*Aristotelia chilensis* (Mol.) Stuntz]. *Phytochem. Anal.* **2006**, *17*, 8–14. [CrossRef] [PubMed]
47. Giusti, M.M.; Rodríguez-Saona, L.E.; Griffin, D.; Wrolstad, R.E. Electrospray and tandem mass spectroscopy as tools for anthocyanin characterization. *J. Agric. Food Chem.* **1999**, *47*, 4657–4664. [CrossRef] [PubMed]
48. Quispe-Fuentes, I.; Vega-Gálvez, A.; Campos-Requena, V.H.J.A. Antioxidant compound extraction from maqui (*Aristotelia chilensis* [Mol] Stuntz) berries: Optimization by response surface methodology. *Antioxidants* **2017**, *6*, 10. [CrossRef]
49. Ruiz, A.; Hermosin-Gutierrez, I.; Mardones, C.; Vergara, C.; Herlitz, E.; Vega, M.; Dorau, C.; Winterhalter, P.; von Baer, D. Polyphenols and antioxidant activity of calafate (*Berberis microphylla*) fruits and other native berries from Southern Chile. *J. Agric. Food Chem.* **2010**, *58*, 6081–6089. [CrossRef] [PubMed]
50. Martins, N.; Barros, L.; Santos-Buelga, C.; Henriques, M.; Silva, S.; Ferreira, I.C.F.R. Evaluation of bioactive properties and phenolic compounds in different extracts prepared from *Salvia officinalis* L. *Food Chem.* **2015**, *170*, 378–385. [CrossRef]
51. López de Dicastillo, C.; López-Carballo, G.; Gavara, R.; Muriel Galet, V.; Guarda, A.; Galotto, M.J. Improving polyphenolic thermal stability of Aristotelia chilensis fruit extract by encapsulation within electrospun cyclodextrin capsules. *J. Food Processing Preserv.* **2019**, *43*, e14044. [CrossRef]
52. Céspedes, C.L.; Valdez-Morales, M.; Avila, J.G.; El-Hafidi, M.; Alarcón, J.; Paredes-López, O. Phytochemical profile and the antioxidant activity of Chilean wild black-berry fruits, *Aristotelia chilensis* (Mol) Stuntz (Elaeocarpaceae). *Food Chem.* **2010**, *119*, 886–895. [CrossRef]
53. Gironés-Vilaplana, A.; Baenas, N.; Villaño, D.; Speisky, H.; García-Viguera, C.; Moreno, D.A. Evaluation of Latin-American fruits rich in phytochemicals with biological effects. *J. Funct. Foods* **2014**, *7*, 599–608. [CrossRef]
54. Paz Robles, C.; Badilla Vidal, N.; Suarez, S.; Baggio, R. Hobartine: A tetracyclic indole alkaloid extracted from *Aristotelia chilensis* (maqui). Acta Crystallogr. *Sect. C Struct. Chem.* **2014**, *70*, 1075–1078. [CrossRef]

55. Muñoz, O.; Christen, P.; Cretton, S.; Backhouse, N.; Torres, V.; Correa, O.; Costa, E.; Miranda, H.; Delporte, C. Chemical study and anti-inflammatory, analgesic and antioxidant activities of the leaves of *Aristotelia chilensis* (Mol.) Stuntz, Elaeocarpaceae. *J. Pharm. Pharmacol.* **2011**, *63*, 849–859. [CrossRef] [PubMed]
56. Vidal, J.; Avello, L.; Loyola, C.; Campos, P.; Aqueveque, M.; Dungan, S.R.; Galotto, L.; Guarda, M. Microencapsulation of maqui (*Aristotelia chilensis* Molina Stuntz) leaf extracts to preserve and control antioxidant properties. *Chil. J. Agric. Res.* **2013**, *73*, 17–23. [CrossRef]
57. Di Lorenzo, A.; Sobolev, A.P.; Nabavi, S.F.; Sureda, A.; Moghaddam, A.H.; Khanjani, S.; Di Giovanni, C.; Xiao, J.; Shirooie, S.; Sokeng, A.J.T. Antidepressive effects of a chemically characterized maqui berry extract (*Aristotelia chilensis* (molina) stuntz) in a mouse model of Post-stroke depression. *Food Chem. Toxicol.* **2019**, *129*, 434–443. [CrossRef] [PubMed]
58. Kulkarni, A.P.; Aradhya, S.M. Chemical changes and antioxidant activity in pomegranate arils during fruit development. *Food Chem.* **2005**, *93*, 319–324. [CrossRef]
59. Wang, S.Y.; Chen, C.-T.; Wang, C.Y. The influence of light and maturity on fruit quality and flavonoid content of red raspberries. *Food Chem.* **2009**, *112*, 676–684. [CrossRef]
60. Matta, F.V.; Xiong, J.; Lila, M.A.; Ward, N.I.; Felipe-Sotelo, M.; Esposito, D. Chemical Composition and Bioactive Properties of Commercial and Non-Commercial Purple and White Açaí Berries. *Foods* **2020**, *9*, 1481. [CrossRef]
61. Schreckinger, M.E.; Wang, J.; Yousef, G.; Lila, M.A.; de Mejia, E.G. Antioxidant capacity and in vitro inhibition of adipogenesis and inflammation by phenolic extracts of *Vaccinium floribundum* and *Aristotelia chilensis*. *J. Agric. Food Chem.* **2010**, *58*, 8966–8976. [CrossRef]
62. Gironeés-Vilaplana, A.; Valentaão Pc Moreno, D.A.; Ferreres, F.; García-Viguera, C.; Andrade, P.B. New beverages of lemon juice enriched with the exotic berries maqui, açaí, and blackthorn: Bioactive components and in vitro biological properties. *J. Agric. Food Chem.* **2012**, *60*, 6571–6580. [CrossRef]
63. Fuentes, E.; Palomo, I. Antiplatelet effects of natural bioactive compounds by multiple targets: Food and drug interactions. *J. Funct. Foods* **2014**, *6*, 73–81. [CrossRef]
64. Fuentes, F.; Alarcón, M.; Badimon, L.; Fuentes, M.; Klotz, K.-N.; Vilahur, G.; Kachler, S.; Padró, T.; Palomo, I.; Fuentes, E. Guanosine exerts antiplatelet and antithrombotic properties through an adenosine-related cAMP-PKA signaling. *Int. J. Cardiol.* **2017**, *248*, 294–300. [CrossRef]
65. Prior, R.L.; Gu, L. Occurrence and biological significance of proanthocyanidins in the American diet. *Phytochemistry* **2005**, *66*, 2264–2280. [CrossRef] [PubMed]
66. Miranda-Rottmann, S.; Aspillaga, A.A.; Pérez, D.D.; Vasquez, L.; Martinez, A.L.; Leighton, F.J. Juice and phenolic fractions of the berry Aristotelia chilensis inhibit LDL oxidation in vitro and protect human endothelial cells against oxidative stress. *J. Agric. Food Chem.* **2002**, *50*, 7542–7547. [CrossRef] [PubMed]
67. Céspedes-Acuña, C.L.; Xiao, J.; Wei, Z.-J.; Chen, L.; Bastias, J.M.; Avila, J.G.; Alarcon-Enos, J.; Werner-Navarrete, E.; Kubo, I. Antioxidant and anti-inflammatory effects of extracts from Maqui berry Aristotelia chilensis in human colon cancer cells. *J. Berry Res.* **2018**, *8*, 275–296. [CrossRef]
68. Ortiz, T.; Argüelles-Arias, F.; Begines, B.; García-Montes, J.M.; Pereira, A.; Victoriano, M.; Vázquez-Román, V.; Pérez, J.L.; Callejon, R.M.; de Miguel, M.; et al. Native Chilean Berries Preservation and In Vitro Studies of a Polyphenol Highly Antioxidant Extract from Maqui as a Potential Agent against Inflammatory Diseases. *Antioxidants* **2021**, *10*, 843. [CrossRef]
69. Gironés-Vilaplana, A.; Villaño, D.; Moreno, D.A.; García-Viguera, C. New isotonic drinks with antioxidant and biological capacities from berries (maqui, açaí and blackthorn) and lemon juice. *Int. J. Food Sci. Nutr.* **2013**, *64*, 897–906. [CrossRef] [PubMed]
70. Martin, K. Effect of Hot Water Extracts of Maqui Berry on Human Aortic Endothelial Cells Exposed to a Hyperglycemic Environment. *Curr. Dev. Nutr.* **2020**, *4*, 435. [CrossRef]
71. Fuentes, O.; Fuentes, M.; Badilla, S.; Troncoso, F. Maqui (*Aristotelia chilensis*) and rutin (quercetin-3-*O*-rutinoside) protects against the functional impairment of the endothelium-dependent vasorelaxation caused by a reduction of nitric oxide availability in diabetes. *Boletín Latinoam. Caribe Plantas Med. Aromáticas* **2013**, *12*, 220–229.
72. Fuentes, O.; Cespedes, C.; Sepulveda, R. Aristotelia chilensis, rutin and quercetin amielorates acute vascular endothelial dysfunction in rat thoracic aorta exposed to oxidative stress. *Boletín Latinoam. Caribe Plantas Med. Aromáticas* **2015**, *14*, 11–20.
73. Hidalgo, J.; Flores, C.; Hidalgo, M.; Perez, M.; Yañez, A.; Quiñones, L.; Caceres, D.; Burgos, R.J.P.M. Delphinol® standardized maqui berry extract reduces postprandial blood glucose increase in individuals with impaired glucose regulation by novel mechanism of sodium glucose cotransporter inhibition. *Panminerva Med.* **2014**, *56*, 1–7.
74. Sandoval, V.; Femenias, A.; Martínez-Garza, Ú.; Sanz-Lamora, H.; Castagnini, J.M.; Quifer-Rada, P.; Lamuela-Raventós, R.M.; Marrero, P.F.; Haro, D.; Relat, J. Lyophilized maqui (*Aristotelia chilensis*) berry induces browning in the subcutaneous white adipose tissue and ameliorates the insulin resistance in high fat diet-induced obese mice. *Antioxidants* **2019**, *8*, 360. [CrossRef]
75. Rojo, L.E.; Ribnicky, D.; Logendra, S.; Poulev, A.; Rojas-Silva, P.; Kuhn, P.; Dorn, R.; Grace, M.H.; Lila, M.A.; Raskin, I. In vitro and in vivo anti-diabetic effects of anthocyanins from Maqui Berry (*Aristotelia chilensis*). *Food Chem.* **2012**, *131*, 387–396. [CrossRef] [PubMed]
76. Watson, R.R.; Schönlau, F. Nutraceutical and antioxidant effects of a delphinidin-rich maqui berry extract Delphinol®: A review. *Minerva Cardioangiol* **2015**, *63*, 1–12. [PubMed]

77. Alvarado, J.L.; Leschot, A.; Olivera-Nappa, Á.; Salgado, A.-M.; Rioseco, H.; Lyon, C.; Vigil, P. Delphinidin-Rich Maqui Berry Extract (Delphinol®) Lowers Fasting and Postprandial Glycemia and Insulinemia in Prediabetic Individuals during Oral Glucose Tolerance Tests. *BioMed Res. Int.* **2016**, *2016*, 9070537. [CrossRef]
78. Šaponjac, V.T.; Gironés-Vilaplana, A.; Djilas, S.; Mena, P.; Ćetković, G.; Moreno, D.A.; Čanadanović-Brunet, J.; Vulić, J.; Stajčić, S.; Vinčić, M. Chemical composition and potential bioactivity of strawberry pomace. *RSC Adv.* **2015**, *5*, 5397–5405. [CrossRef]
79. Ávila, F.; Jiménez-Aspee, F.; Cruz, N.; Gómez, C.; González, M.A.; Ravello, N. Additive effect of maqui (Aristotelia chilensis) and lemon (Citrus × limon) juice in the postprandial glycemic responses after the intake of high glycemic index meals in healthy men. *NFS J.* **2019**, *17*, 8–16. [CrossRef]
80. Johnston, K.L.; Clifford, M.N.; Morgan, L. Coffee acutely modifies gastrointestinal hormone secretion and glucose tolerance in humans: Glycemic effects of chlorogenic acid and caffeine. *Am. J. Clin. Nutr.* **2003**, *78*, 728–733. [CrossRef]
81. Davinelli, S.; Bertoglio, J.C.; Zarrelli, A.; Pina, R.; Scapagnini, G. A randomized clinical trial evaluating the efficacy of an anthocyanin–maqui berry extract (Delphinol®) on oxidative stress biomarkers. *J. Am. Coll. Nutr.* **2015**, *34*, 28–33. [CrossRef]
82. Bribiesca-Cruz, I.; Moreno, D.A.; García-Viguera, C.; Gallardo, J.M.; Segura-Uribe, J.J.; Pinto-Almazán, R.; Guerra-Araiza, C. Maqui Berry (*Aristotelia Chilensis*) Extract Improves Memory and Decreases Oxidative Stress in Male Rat Brain Exposed to Ozone. *Nutr. Neurosci.* **2021**, *24*, 477–489. [CrossRef]
83. Ostertag, L.M.; O'Kennedy, N.; Horgan, G.W.; Kroon, P.A.; Duthie, G.G.; de Roos, B. In vitro anti-platelet effects of simple plant-derived phenolic compounds are only found at high, non-physiological concentrations. *Mol. Nutr. Food Res.* **2011**, *55*, 1624–1636. [CrossRef]
84. Olas, B. The multifunctionality of berries toward blood platelets and the role of berry phenolics in cardiovascular disorders. *Platelets* **2017**, *28*, 540–549. [CrossRef]
85. Erlund, I.; Koli, R.; Alfthan, G.; Marniemi, J.; Puukka, P.; Mustonen, P.; Mattila, P.; Jula, A. Favorable effects of berry consumption on platelet function, blood pressure, and HDL cholesterol. *Am. J. Clin. Nutr.* **2008**, *87*, 323–331. [CrossRef] [PubMed]
86. Shanmuganayagam, D.; Beahm, M.R.; Osman, H.E.; Krueger, C.G.; Reed, J.D.; Folts, J.D. Grape seed and grape skin extracts elicit a greater antiplatelet effect when used in combination than when used individually in dogs and humans. *J. Nutr.* **2002**, *132*, 3592–3598. [CrossRef] [PubMed]
87. Wilson, T.; Bauer, B.A. Advising consumers about dietary supplements: Lessons from cranberry products. *J. Diet. Suppl.* **2009**, *6*, 377–384. [CrossRef] [PubMed]
88. Polagruto, J.A.; Gross, H.B.; Kamangar, F.; Kosuna, K.-i.; Sun, B.; Fujii, H.; Keen, C.L.; Hackman, R.M. Platelet reactivity in male smokers following the acute consumption of a flavanol-rich grapeseed extract. *J. Med. Food* **2007**, *10*, 725–730. [CrossRef] [PubMed]
89. Keevil, J.G.; Osman, H.E.; Reed, J.D.; Folts, J.D. Grape juice, but not orange juice or grapefruit juice, inhibits human platelet aggregation. *J. Nutr.* **2000**, *130*, 53–56. [CrossRef]
90. Freedman, J.E.; Parker Iii, C.; Li, L.; Perlman, J.A.; Frei, B.; Ivanov, V.; Deak, L.R.; Iafrati, M.D.; Folts, J.D. Select flavonoids and whole juice from purple grapes inhibit platelet function and enhance nitric oxide release. *Circulation* **2001**, *103*, 2792–2798. [CrossRef]
91. Eccleston, C.; Baoru, Y.; Tahvonen, R.; Kallio, H.; Rimbach, G.H.; Minihane, A.M. Effects of an antioxidant-rich juice (sea buckthorn) on risk factors for coronary heart disease in humans. *J. Nutr. Biochem.* **2002**, *13*, 346–354. [CrossRef]
92. Fuentes, E.; Caballero, J.; Alarcón, M.; Rojas, A.; Palomo, I. Chlorogenic acid inhibits human platelet activation and thrombus formation. *PLoS ONE* **2014**, *9*, e90699. [CrossRef]
93. Rodríguez, L.; Trostchansky, A.; Wood, I.; Mastrogiovanni, M.; Vogel, H.; González, B.; Maróstica Junior, M.; Fuentes, E.; Palomo, I. Antiplatelet activity and chemical analysis of leaf and fruit extracts from Aristotelia chilensis. *PLoS ONE* **2021**, *16*, e0250852. [CrossRef]
94. Yang, Y.; Shi, Z.; Reheman, A.; Jin, W.; Li, C.; Zhu, G.; Wang, Y.; Freedman, J.J.; Ling, W.; Ni, H. Anthocyanins Inhibit Platelet Activation and Attenuate Thrombus Growth in Both Human and Murine Thrombosis Models. *Blood* **2010**, *116*, 3197. [CrossRef]
95. Rodríguez, L.; Mendez, D.; Montecino, H.; Carrasco, B.; Arevalo, B.; Palomo, I.; Fuentes, E. Role of *Phaseolus vulgaris* L. in the Prevention of Cardiovascular Diseases—Cardioprotective Potential of Bioactive Compounds. *Plants* **2022**, *11*, 186. [CrossRef] [PubMed]
96. Yao, Y.; Chen, Y.; Adili, R.; McKeown, T.; Chen, P.; Zhu, G.; Li, D.; Ling, W.; Ni, H.; Yang, Y. Plant-Based Food Cyanidin-3-Glucoside Modulates Human Platelet Glycoprotein Vi Signaling and Inhibits Platelet Activation and Thrombus Formation. *J. Nutr.* **2017**, *147*, 1917–1925. [CrossRef] [PubMed]
97. Zhou, F.-H.; Deng, X.-J.; Chen, Y.-Q.; Ya, F.-L.; Zhang, X.-D.; Song, F.; Li, D.; Yang, Y. Anthocyanin cyanidin-3-glucoside attenuates platelet granule release in mice fed high-fat diets. *J. Nutr. Sci. Vitaminol.* **2017**, *63*, 237–243. [CrossRef] [PubMed]
98. Hubbard, G.P.; Wolffram, S.; Lovegrove, J.A.; Gibbins, J.M. Ingestion of Quercetin Inhibits Platelet Aggregation and Essential Components of the Collagen-Stimulated Platelet Activation Pathway in Humans. *J. Thromb. Haemost.* **2004**, *2*, 2138–2145. [CrossRef] [PubMed]
99. Tzeng, S.-H.; Ko, W.-C.; Ko, F.-N.; Teng, C.-M. Inhibition of platelet aggregation by some flavonoids. *Thromb. Res.* **1991**, *64*, 91–100. [CrossRef]
100. Chung, M.-I.; Gan, K.-H.; Lin, C.-N.; Ko, F.-N.; Teng, C.-M. Antiplatelet effects and vasorelaxing action of some constituents of Formosan plants. *J. Nat. Prod.* **1993**, *56*, 929–934. [CrossRef]

101. Hubbard, G.P.; Stevens, J.M.; Cicmil, M.; Sage, T.; Jordan, P.A.; Williams, C.M.; Lovegrove, J.A.; Gibbins, J.M. Quercetin Inhibits Collagen-Stimulated Platelet Activation through Inhibition of Multiple Components of the Glycoprotein Vi Signaling Pathway. *J. Thromb. Haemost.* **2003**, *1*, 1079–1088. [CrossRef]
102. Beretz, A.; Cazenave, J.-P.; Anton, R. Inhibition of aggregation and secretion of human platelets by quercetin and other flavonoids: Structure-activity relationships. *Agents Actions* **1982**, *12*, 382–387. [CrossRef]
103. Landolfi, R.; Mower, R.L.; Steiner, M. Modification of platelet function and arachidonic acid metabolism by bioflavonoids: Structure-activity relations. *Biochem. Pharmacol.* **1984**, *33*, 1525–1530. [CrossRef]
104. Choi, J.-H.; Park, S.-E.; Kim, S.-J.; Kim, S. Kaempferol inhibits thrombosis and platelet activation. *Biochimie* **2015**, *115*, 177–186. [CrossRef]
105. Rolnik, A.; Żuchowski, J.; Stochmal, A.; Olas, B. Quercetin and kaempferol derivatives isolated from aerial parts of Lens culinaris Medik as modulators of blood platelet functions. *Ind. Crops Prod.* **2020**, *152*, 112536.
106. Tong, Y.; Zhou, X.-M.; Wang, S.-J.; Yang, Y.; Cao, Y.-L. Analgesic activity of myricetin isolated from Myrica rubra Sieb. et Zucc. leaves. *Arch. Pharmacal. Res.* **2009**, *32*, 527–533. [CrossRef] [PubMed]
107. Gaspar, R.S.; da Silva, S.A.; Stapleton, J.; Fontelles, J.L.d.L.; Sousa, H.R.; Chagas, V.T.; Alsufyani, S.; Trostchansky, A.; Gibbins, J.M.; Paes, A.M.D.A. Myricetin, the main flavonoid in Syzygium cumini leaf, is a novel inhibitor of platelet thiol isomerases PDI and ERp5. *Front. Pharmacol.* **2020**, *10*, 1678. [CrossRef] [PubMed]
108. Oh, T.W.; Do, H.J.; Jeon, J.-H.; Kim, K. Quercitrin inhibits platelet activation in arterial thrombosis. *Phytomedicine* **2021**, *80*, 153363. [CrossRef] [PubMed]
109. Sheu, J.-R.; Hsiao, G.; Chou, P.-H.; Shen, M.-Y.; Chou, D.-S. Mechanisms involved in the antiplatelet activity of rutin, a glycoside of the flavonol quercetin, in human platelets. *J. Agric. Food Chem.* **2004**, *52*, 4414–4418. [CrossRef]
110. Koleckar, V.; Brojerova, E.; Rehakova, Z.; Kubikova, K.; Cervenka, F.; Kuca, K.; Jun, D.; Hronek, M.; Opletalova, V.; Opletal, L. In vitro antiplatelet activity of flavonoids from Leuzea carthamoides. *Drug Chem. Toxicol.* **2008**, *31*, 27–35. [CrossRef]
111. Jin, Y.R.; Han, X.H.; Zhang, Y.H.; Lee, J.J.; Lim, Y.; Chung, J.-H.; Yun, Y.P. Antiplatelet Activity of Hesperetin, a Bioflavonoid, Is Mainly Mediated by Inhibition of Plc-Γ2 Phosphorylation and Cyclooxygenase-1 Activity. *Atherosclerosis* **2007**, *194*, 144–152. [CrossRef]
112. Sugasawa, N.; Katagi, A.; Kurobe, H.; Nakayama, T.; Nishio, C.; Takumi, H.; Higashiguchi, F.; Aihara, K.I.; Shimabukuro, M.; Sata, M.; et al. Inhibition of Atherosclerotic Plaque Development by Oral Administration of α-Glucosyl Hesperidin and Water-Dispersible Hesperetin in Apolipoprotein E Knockout Mice. *J. Am. Coll. Nutr.* **2019**, *38*, 15–22. [CrossRef]
113. Hong, Q.; Ma, Z.-C.; Huang, H.; Wang, Y.-G.; Tan, H.-L.; Xiao, C.-R.; Liang, Q.D.; Zhang, H.T.; Gao, Y. Antithrombotic activities of ferulic acid via intracellular cyclic nucleotide signaling. *Eur. J. Pharmacol.* **2016**, *777*, 1–8. [CrossRef]
114. Choi, J.H.; Park, J.K.; Kim, K.M.; Lee, H.J.; Kim, S. In vitro and in vivo antithrombotic and cytotoxicity effects of ferulic acid. Journal of biochemical and molecular toxicology. *J. Biochem. Mol. Toxicol.* **2018**, *32*, e22004. [CrossRef]
115. Fuentes, E.; Forero-Doria, O.; Carrasco, G.; Maricán, A.; Santos, L.S.; Alarcón, M.; Palomo, I. Effect of tomato industrial processing on phenolic profile and antiplatelet activity. *Molecules* **2013**, *18*, 11526–11536. [CrossRef] [PubMed]
116. Chen, T.; Lee, J.; Lin, K.; Shen, C.; Chou, D.; Sheu, J. Antiplatelet activity of caffeic acid phenethyl ester is mediated through a cyclic GMP-dependent pathway in human platelets. *Chin. J. Physiol.* **2007**, *50*, 121. [PubMed]
117. Chang, Y.; Chen, W.-F.; Lin, K.-H.; Hsieh, C.-Y.; Chou, D.-S.; Lin, L.-J.; Sheu, J.-R.; Chang, C.-C. Novel bioactivity of ellagic acid in inhibiting human platelet activation. *Evid. Based Complementary Altern. Med.* **2013**, *2013*, 595128. [CrossRef]
118. Song, F.; Zhu, Y.; Shi, Z.; Tian, J.; Deng, X.; Ren, J.; Andrews, M.C.; Ni, H.; Ling, W.; Yang, Y. Plant Food Anthocyanins Inhibit Platelet Granule Secretion in Hypercholesterolaemia: Involving the Signalling Pathway of Pi3k–Akt. *Thromb. Haemost.* **2014**, *111*, 981–991. [CrossRef] [PubMed]
119. Rodríguez-Pérez, L. *Antiplatelet Effect of Aristotelia chilensis (Maqui) Extracts through In Vitro Studies*; University of Talca: Talca, Chile, 2021.
120. Ro, J.-Y.; Ryu, J.-H.; Park, H.-J.; Cho, H.-J. Onion (*Allium cepa* L.) peel extract has anti-platelet effects in rat platelets. *SpringerPlus* **2015**, *4*, 17. [CrossRef] [PubMed]
121. Hubbard, G.P.; Wolffram, S.; de Vos, R.; Bovy, A.; Gibbins, J.M.; Lovegrove, J.A. Ingestion of Onion Soup High in Quercetin Inhibits Platelet Aggregation and Essential Components of the Collagen-Stimulated Platelet Activation Pathway in Man: A Pilot Study. *Br. J. Nutr.* **2006**, *96*, 482–488.
122. Vallance, T.M.; Ravishankar, D.; Albadawi, D.A.I.; Osborn, H.M.I.; Vaiyapuri, S. Synthetic Flavonoids as Novel Modulators of Platelet Function and Thrombosis. *Int. J. Mol. Sci.* **2019**, *20*, 3106. [CrossRef]
123. Fuentes, E.; Wehinger, S.; Trostchansky, A. Regulation of Key Antiplatelet Pathways by Bioactive Compounds with Minimal Bleeding Risk. *Int. J. Mol. Sci.* **2021**, *22*, 12380. [CrossRef]
124. Rodríguez, L.; Badimon, L.; Méndez, D.; Padró, T.; Vilahur, G.; Peña, E.; Carrasco, B.; Vogel, H.; Palomo, I.; Fuentes, E. Antiplatelet Activity of Isorhamnetin via Mitochondrial Regulation. *Antioxidants* **2021**, *10*, 666. [CrossRef]
125. Stainer, A.R.; Sasikumar, P.; Bye, A.P.; Unsworth, A.J.; Holbrook, L.M.; Tindall, M.; Lovegrove, J.A.; Gibbins, J.M. The metabolites of the dietary flavonoid quercetin possess potent antithrombotic activity, and interact with aspirin to enhance antiplatelet effects. *TH Open Companion J. Thromb. Haemost.* **2019**, *3*, e244. [CrossRef]
126. Nam, G.S.; Park, H.-J.; Nam, K.-S. The antithrombotic effect of caffeic acid is associated with a cAMP-dependent pathway and clot retraction in human platelets. *Thromb. Res.* **2020**, *195*, 87–94. [CrossRef] [PubMed]

127. Lu, Y.; Li, Q.; Liu, Y.Y.; Sun, K.; Fan, J.Y.; Wang, C.S.; Han, J.Y. Inhibitory effect of caffeic acid on ADP-induced thrombus formation and platelet activation involves mitogen-activated protein kinases. *Sci. Rep.* **2015**, *5*, 13824. [CrossRef] [PubMed]
128. Lee, D.-H.; Kim, H.-H.; Cho, H.-J.; Bae, J.-S.; Yu, Y.-B.; Park, H.-J. Antiplatelet effects of caffeic acid due to Ca2+ mobilization-inhibition via cAMP-dependent inositol-1, 4, 5-trisphosphate receptor phosphorylation. *J. Atheroscler. Thromb.* **2013**, *21*, 18994. [CrossRef] [PubMed]
129. Nam, G.S.; Nam, K.-S.; Park, H.-J. Caffeic Acid Diminishes the Production and Release of Thrombogenic Molecules in Human Platelets. *Biotechnol. Bioprocess Eng.* **2018**, *23*, 641–648. [CrossRef]
130. Kyriakidis, K.D.; Vartholomatos, E.G.; Markopoulos, G.S.; Sciences, H. Evaluation of Antiplatelet Activity of Phenolic Compounds by Flow Cytometry. *Eur. J. Med. Health Sci.* **2021**, *3*, 165–170. [CrossRef]
131. Baeza, G.; Bachmair, E.-M.; Wood, S.; Mateos, R.; Bravo, L.; De Roos, B.J.F. The colonic metabolites dihydrocaffeic acid and dihydroferulic acid are more effective inhibitors of in vitro platelet activation than their phenolic precursors. *Food Funct.* **2017**, *8*, 1333–1342. [CrossRef]
132. Lutz, M.; Fuentes, E.; Ávila, F.; Alarcón, M.; Palomo, I. Roles of Phenolic Compounds in the Reduction of Risk Factors of Cardiovascular Diseases. *Molecules* **2019**, *24*, 366. [CrossRef]
133. Rahman, M.M.; Rahaman, M.S.; Islam, M.R.; Rahman, F.; Mithi, F.M.; Alqahtani, T.; Almikhlafi, M.A.; Alghamdi, S.Q.; Alruwaili, A.S.; Hossain, M.S.J.M. Role of phenolic compounds in human disease: Current knowledge and future prospects. *Molecules* **2021**, *27*, 233. [CrossRef]
134. Vázquez-Espinosa, M.; V González de Peredo, A.; Ferreiro-González, M.; Carrera, C.; Palma, M.; Barbero, G.F.; Espada-Bellido, E. Assessment of ultrasound assisted extraction as an alternative method for the extraction of anthocyanins and total phenolic compounds from Maqui berries (*Aristotelia chilensis* (Mol.) Stuntz). *Agronomy* **2019**, *9*, 148. [CrossRef]
135. Khan, H.; Jawad, M.; Kamal, M.A.; Baldi, A.; Xiao, J.; Nabavi, S.M.; Daglia, M. Evidence and prospective of plant derived flavonoids as antiplatelet agents: Strong candidates to be drugs of future. *Food Chem. Toxicol.* **2018**, *119*, 355–367. [CrossRef]
136. Marino, M.; Del Bo, C.; Martini, D.; Porrini, M.; Riso, P.J.F. A review of registered clinical trials on dietary (poly) phenols: Past efforts and possible future directions. *Foods* **2020**, *9*, 1606. [CrossRef] [PubMed]
137. Jiménez-Aguilar, D.M.; Grusak, M.A. Analysis, Minerals, vitamin C, phenolics, flavonoids and antioxidant activity of Amaranthus leafy vegetables. *J. Food Compos. Anal.* **2017**, *58*, 33–39. [CrossRef]
138. Carrasco-Pozo, C.; Mizgier, M.L.; Speisky, H.; Gotteland, M.J. Differential protective effects of quercetin, resveratrol, rutin and epigallocatechin gallate against mitochondrial dysfunction induced by indomethacin in Caco-2 cells. *Chem. Interact.* **2012**, *195*, 199–205. [CrossRef] [PubMed]
139. Vogiatzoglou, A.; Mulligan, A.A.; Lentjes, M.A.; Luben, R.N.; Spencer, J.P.; Schroeter, H.; Khaw, K.-T.; Kuhnle, G. Flavonoid intake in European adults (18 to 64 years). *PLoS ONE* **2015**, *10*, e0128132. [CrossRef]
140. Wu, X.; Beecher, G.R.; Holden, J.M.; Haytowitz, D.B.; Gebhardt, S.E.; Prior, R.L.J. Concentrations of anthocyanins in common foods in the United States and estimation of normal consumption. *J. Agric. Food Chem.* **2006**, *54*, 4069–4075. [CrossRef]
141. Frank, T.; Netzel, G.; Kammerer, D.R.; Carle, R.; Kler, A.; Kriesl, E.; Bitsch, I.; Bitsch, R.; Netzel, M.J. Consumption of Hibiscus sabdariffa L. aqueous extract and its impact on systemic antioxidant potential in healthy subjects. *J. Sci. Food Agric.* **2012**, *92*, 2207–2218. [CrossRef]
142. Czank, C.; Cassidy, A.; Zhang, Q.; Morrison, D.J.; Preston, T.; Kroon, P.A.; Botting, N.P.; Kay, C.D. Human metabolism and elimination of the anthocyanin, cyanidin-3-glucoside: A 13C-tracer study. *Am. Clin. Nutr.* **2013**, *97*, 995–1003. [CrossRef]

molecules

MDPI

Article

Comparison of the Antihypertensive Activity of Phenolic Acids

Myeongnam Yu [1], Hyun Joo Kim [2], Huijin Heo [1], Minjun Kim [1], Yesol Jeon [1], Hana Lee [1,*] and Junsoo Lee [1,*]

1 Department of Food Science and Biotechnology, Chungbuk National University, Cheongju 28644, Korea
2 Department of Central Area Crop Science, National Institute of Crop Science, Rural Development Administration, Suwon 16613, Korea
* Correspondence: dlgksk0514@naver.com (H.L.); junsoo@chungbuk.ac.kr (J.L.)

Abstract: Phenolic acids, found in cereals, legumes, vegetables, and fruits, have various biological functions. We aimed to compare the antihypertensive potential of different phenolic acids by evaluating their ACE inhibitory activity and cytoprotective capacity in EA.hy 926 endothelial cells. In addition, we explored the mechanism underlying the antihypertensive activity of sinapic acid. Of all the phenolic acids studied, sinapic acid, caffeic acid, coumaric acid, and ferulic acid significantly inhibited ACE activity. Moreover, gallic acid, sinapic acid, and ferulic acid significantly enhanced intracellular NO production. Based on the results of GSH depletion, ROS production, and MDA level analyses, sinapic acid was selected to study the mechanism underlying the antihypertensive effect. Sinapic acid decreases endothelial dysfunction by enhancing the expression of antioxidant-related proteins. Sinapic acid increased phosphorylation of eNOS and Akt in a dose-dependent manner. These findings indicate the potential of sinapic acid as a treatment for hypertension.

Keywords: phenolic acids; sinapic acid; ACE inhibition; hypertension; endothelial dysfunction

1. Introduction

Cardiovascular diseases (CVDs), including coronary artery disease, atherosclerosis, and hypertension, are a group of diseases that acutely threaten human health [1]. Endothelial dysfunction, a hallmark of hypertension, can be caused by oxidative stress. Nitric oxide (NO) is a crucial mediator of endothelium-dependent relaxation in blood pressure regulation [2]. Increased radicals rapidly react with NO, resulting in altered NO bioavailability and impaired endothelial relaxation [3]. Reactive oxygen species (ROS) affect the structure and function of vascular media. Vascular remodeling by ROS leads to enhanced medial thickness [4]. Therefore, preventing oxidation and ROS generation may help improve hypertension-related diseases [5]. Nuclear factor-E2-related factor 2 (Nrf2), an antioxidant transcription factor important in CVD resistance [6], is highly sensitive to oxidative damage. Nrf2 promotes the transcription of antioxidant genes, including heme oxygenase-1 (HO-1), NADPH quinone oxidoreductase (NQO-1), and glutamate-cysteine ligase catalytic subunit (GCLC). A previous study demonstrated that accumulation of Nrf2 in the nucleus and activation of protein kinase B (Akt) accompanied HO-1 and NQO-1 expression [7]. In the endothelium, many growth factors and hormones act as agonists to induce the activation of Akt and phosphorylation of endothelial nitric oxide synthase (eNOS), which increases NO production [8]. As various endothelial signaling pathways converge on Akt, it may be an ideal target protein for eNOS responses [9]. Therefore, the Akt/eNOS and Nrf2 signaling pathways are crucial checkpoints for the induction of phase II enzymes and treatment of endothelial dysfunction.

Phenolic compounds are the most abundant phytochemicals in plant-based foods. Phenolic acids are a major class of phenolic compounds that can suppress ROS, thus reducing oxidative stress to biomolecules within cells [10]. Phenolic acids exert various biological activities, including antioxidant, anticancer, antidiabetic, anti-inflammatory, and antihypertension [11]. The ameliorative effect of phenolic acids on chronic diseases may

Citation: Yu, M.; Kim, H.J.; Heo, H.; Kim, M.; Jeon, Y.; Lee, H.; Lee, J. Comparison of the Antihypertensive Activity of Phenolic Acids. *Molecules* **2022**, 27, 6185. https://doi.org/10.3390/molecules27196185

Academic Editor: Nour Eddine Es-Safi

Received: 31 August 2022
Accepted: 19 September 2022
Published: 21 September 2022

be due to their high antioxidative potential [12]. Gallic acid suppressed hypertension in L-NAME-treated mice and spontaneously hypertensive rats [13,14]. Ferulic acid reduced oxidative injury by increasing the bioavailability of NO in arterial vasculature [15]. Moreover, chlorogenic acid and caffeic acid lowered blood pressure and decreased the properties of enzymes associated with the pathogenesis of hypertension [16]. However, information on the comparative efficacy of phenolic acids in modulating endothelial dysfunction and hypertension is limited. This study aimed to compare the inhibitory effect of phenolic acids on endothelial dysfunction against the oxidative damages in EA.hy 926 endothelial cells.

2. Materials and Methods

2.1. Chemicals

Griess reagent, benzoic acid, hydroxybenzoic acid, caffeic acid, cinnamic acid, coumaric acid, ferulic acid, gallic acid, protocatechuic acid, sinapic acid, syringic acid, vanillic acid, quercetin, angiotensin-converting enzyme (ACE), captopril, hydrogen peroxide (H_2O_2), and diacetyldichlorofluorescein were purchased from Sigma-Aldrich (St. Louis, MO, USA). Antibodies against p-eNOS, eNOS, p-Akt, Akt, Nrf-2, NQO-1, PCNA, HO-1, GCLC, and β-actin were obtained from Cell Signaling Technology (Beverly, MA, USA). Dulbecco's modified Eagle's medium (DMEM), fetal bovine serum (FBS), and penicillinstreptomycin were purchased from Hyclone (General Electric Healthcare Life Sciences, Mississauga, Canada). The structure of phenolic acids is indicated in Table 1.

Table 1. The structure of phenolic acids.

Name	Structure	Name	Structure
Benzoic acid	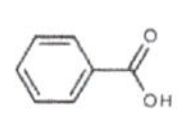	Gallic acid	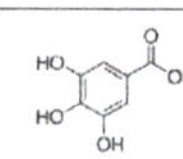
Hydroxybenzoic acid	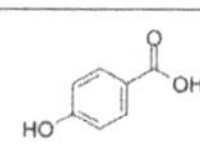	Protocatechuic acid	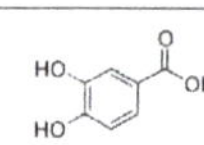
Caffeic acid	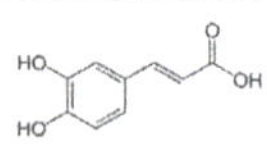	Sinapic acid	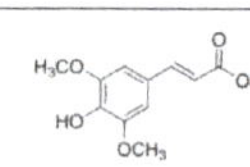
Cinnamic acid	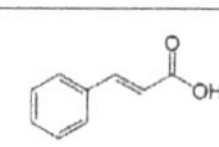	Syringic acid	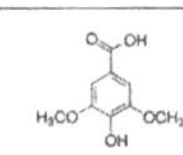
Coumaric acid	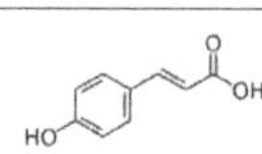	Vanillic acid	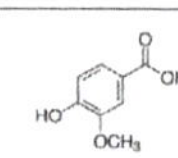
Ferulic acid	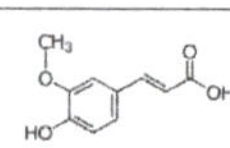		

2.2. ACE Inhibitory Activity Assay

ACE inhibitory activity of phenolic acids was determined according to the method reported by Cushman and Cheung (1971), using ACE (0.1 U/mL) and hippuryl-His-Leu (5 mM) [17]. The absorbance was measured at 228 nm using a spectrophotometer (BioTek, Inc., Winooski, VT, USA). Captopril was used as a positive control. The inhibition rate was calculated using the following formula.

$$\text{ACE inhibition rate (\%)} = (OD_{control} - OD_{sample})/OD_{control} \times 100$$

2.3. Cell Culture and Sample Treatment

EA.hy 926 cells were incubated in DMEM supplemented with 10% FBS at 37 °C in humidified air with 5% CO_2. Endothelial cells were seeded at a density of 6 × 10^5 cells/mL in a 96-well plate. The cells were pre-treated with a serum-free medium containing 50 μM phenolic acids for 1 h and then exposed to 600 μM H_2O_2 with phenolic acids for 24 h. Cell cytotoxicity was determined using a thiazolyl blue tetrazolium bromide reagent.

2.4. Measurement of Intracellular ROS, GSH, Malondialdehyde, and NO Levels

Endothelial cells were treated with 50 μM phenolic acid and 600 μM H_2O_2. Next, the cells were washed with PBS and stained with 25 μM diacetyldichlorofluorescein. The fluorescence intensity was analyzed. Glutathione and malondialdehyde levels were measured using the DTNB-GSSG reductase recycling and TBARS assays, respectively. Nitric oxide levels were measured using Griess reagent.

2.5. Western Blot Analysis

The EA.hy 926 endothelial cells were cultured in a 6-well plate with or without sinapic acid at a density of 6 × 10^5 cells/mL. Total proteins and nuclear proteins were extracted using the Pro-Prep™ protein extraction solution (iNtRON Biotechnology, Seongnam, Korea) and NE-PER® nuclear and cytoplasmic extraction reagents (Thermo Fisher Scientific, Inc., Worcester, MA, USA), respectively. Membranes were incubated with primary and secondary antibodies (1:2000 dilution for β-actin; 1:1000 dilution for p-eNOS, eNOS, p-Akt, Akt, NQO-1, PCNA, HO-1, GCLC, anti-mouse, and anti-rabbit; 1:500 dilution for Nrf-2). The bands were visualized using X-ray film.

2.6. Statistical Analysis

Data were analyzed using Duncan's multiple comparison test and Tukey's post hoc test using SAS (version 8.1; SAS Institute, Cary, NC, USA) and GraphPad Prism software (version 5; GraphPad Software Inc., La Jolla, CA, USA).

3. Results and Discussion

3.1. Effect of Phenolic Acids on ACE Inhibition and NO Production

ACE plays a crucial role in regulating blood pressure [18]. Many synthetic ACE inhibitors are currently being used for the treatment of hypertension. However, these drugs may cause adverse effects. Most natural compounds are safe and do not cause adverse effects. A previous study reported that plant phenolics have the potential to inhibit ACE in vitro [19]. Zhang et al. (2018) demonstrated the ACE inhibition effect of phenolic extracts and fractions derived from lentils, black soybean, and black turtle bean [20]. To confirm the antihypertensive effect of phenolic acids, we measured the ACE inhibitory activity. As shown in Figure 1, among the selected phenolic acids, sinapic acid showed the highest ACE inhibition rate (89%), followed by caffeic acid (78%). In this study, we used the EA.hy 926 endothelial cell line to evaluate the effect of phenolic acids on NO production. Treatment with phenolic acids (50 μM) did not affect the cytotoxicity of endothelial cells (Figure 2A). Reduced NO levels contribute to hypertension and endothelial dysfunction. NO plays an essential role in the vasorelaxation of large arteries [21]. We found that treatment with gallic acid, sinapic acid, and ferulic acid significantly increased NO production by 85.1, 50.5, and 31.9%, respectively, compared with that in the control group cells (Figure 2B). These results indicate that phenolic acids may improve endothelial dysfunction, consequently regulating blood pressure.

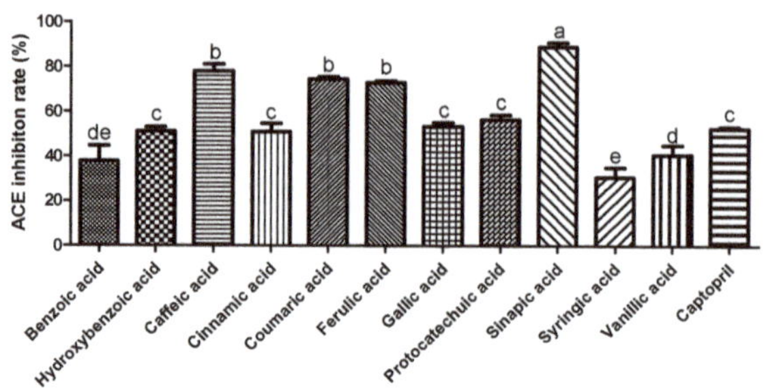

Figure 1. Inhibitory effect of selected phenolic acids (10 mM) on angiotensin I converting enzyme. Captopril (1.15 μM) was used as positive control. Each value was expressed as the mean ± standard error ($n = 3$). Different letters above the bars indicate significant differences based on the Duncan's test ($p < 0.05$).

(A)

(B)

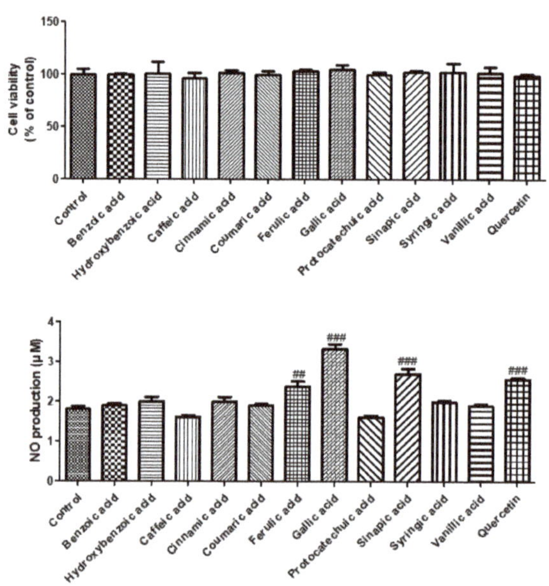

Figure 2. Effect of selected phenolic acids (50 μM) on cell cytotoxicity (**A**) and NO production (**B**) in EA.hy 926 cells. Quercetin (25 μM) was used as positive control. Each value was expressed as the mean ± standard error ($n = 3$). Statistical significance was analyzed using the Tukey test. $^{##}$ $p < 0.01$ and $^{###}$ $p < 0.001$ versus nontreated cells.

3.2. *Cytoprotective Effect of Phenolic Acids against Hydrogen Peroxide*

Excessive ROS levels lead to endothelial dysfunction and elevated blood pressure [22]. MDA, a marker of oxidative damage, can cause an abnormal physiological state in the body [23]. GSH, an active peptide with good antioxidant activity, can modulate oxidative balance and suppress oxidative damage [24]. In this study, we investigated the cytoprotective effects of phenolic acids on H_2O_2-induced oxidative stress in EA.hy 926 endothelial cells. Treatment with H_2O_2 (600 μM) decreased cell viability by 24.8%. However, treatments by caffeic, ferulic, gallic, and sinapic acid markedly increased the cell viability by 43.4, 43.6, 35.5, and 39.1%, respectively, compared to H_2O_2-induced cells (Figure 3A). To examine whether phenolic acids protect endothelial cells against oxidative damage, we measured ROS, GSH, and MDA levels (Figure 3B–D). Sinapic acid markedly reduced ROS generation by 44.1% compared to that in H_2O_2-treated cells. Caffeic acid, cinnamic acid, coumaric acid, ferulic acid, gallic acid, sinapic acid, and syringic acid significantly enhanced the GSH levels. Our findings show that H_2O_2 treatment increased ROS levels and decreased intracellular GSH levels, whereas treatment with phenolic acids significantly reduced oxidative damage-induced ROS production and GSH depletion. In addition, we investigated the effect of phenolic acids on oxidative stress-induced lipid peroxidation in EA.hy 926 cells. Among the phenolic acids, sinapic acid showed the strongest inhibitory effect on lipid peroxidation. Lee and Lee (2021) reported that protocatechuic acid and gallic acid significantly decreased ROS levels, thereby regulating insulin resistance [25]. Caffeic acid and chlorogenic acid decreased blood pressure in hypertensive rats by increasing GSH and reducing MDA levels [16]. Taken together, these results suggest that sinapic acid plays a crucial role in the protection of endothelial cells by regulating ROS, MDA, and GSH levels.

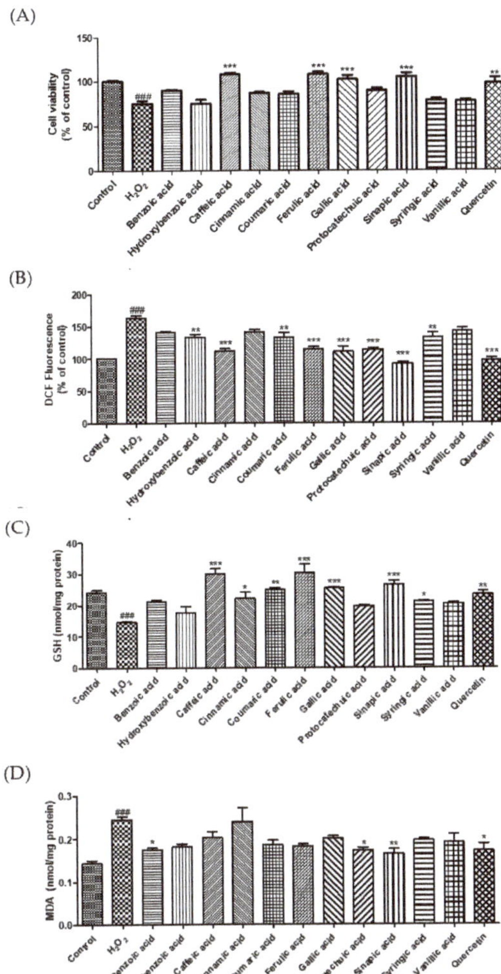

Figure 3. Effect of selected phenolic acids (50 μM) on cell viability (**A**), the generation of reactive oxygen species (**B**), glutathione (**C**), and malondialdehyde (**D**) against hydrogen peroxide (600 μM) in EA.hy 926 cells. Quercetin (25 μM) was used as positive control. Each value was expressed as the mean ± standard error ($n = 3$). Statistical significance was analyzed using the Tukey test. ### $p < 0.001$ versus nontreated cells. * $p < 0.05$, ** $p < 0.01$, and *** $p < 0.001$ versus hydrogen-peroxide-treated cells.

3.3. *Effects of Sinapic Acid on the Expression of Phase II Enzymes and the Activation of Nrf2*

Based on our results, sinapic acid was selected for exploring the mechanism underlying the antihypertensive effect of phenolic acid. We measured the protein expression levels of HO-1, NQO-1, GCLC, and Nrf2. As shown in Figure 4, treatment with sinapic acid enhanced HO-1, NQO-1, and GCLC expression levels in a dose-dependent manner. In addition, sinapic acid significantly increased the nuclear translocation of Nrf2. The Nrf2 pathway is important for protection against various stressors [26]. Cytotoxicity caused by t-BHP-induced oxidative damage was recovered by caffeic acid via an increase in the expression of detoxifying enzymes, including HO-1 and GCLC [27]. Luo et al. (2018) reported that HO-1 ameliorates oxidative stress-induced endothelial aging by modulating eNOS activation [28]. Ginsenoside Rg3 upregulates the Nrf2 signaling pathway via Akt activation and improves endothelial dysfunction [29]. Moreover, sinapic acid reduces renal apoptosis, inflammation, and oxidative damage [30]. These results suggest that sinapic acid-mediated endothelial cell protection against oxidative damage may be associated with the antioxidative properties of sinapic acid.

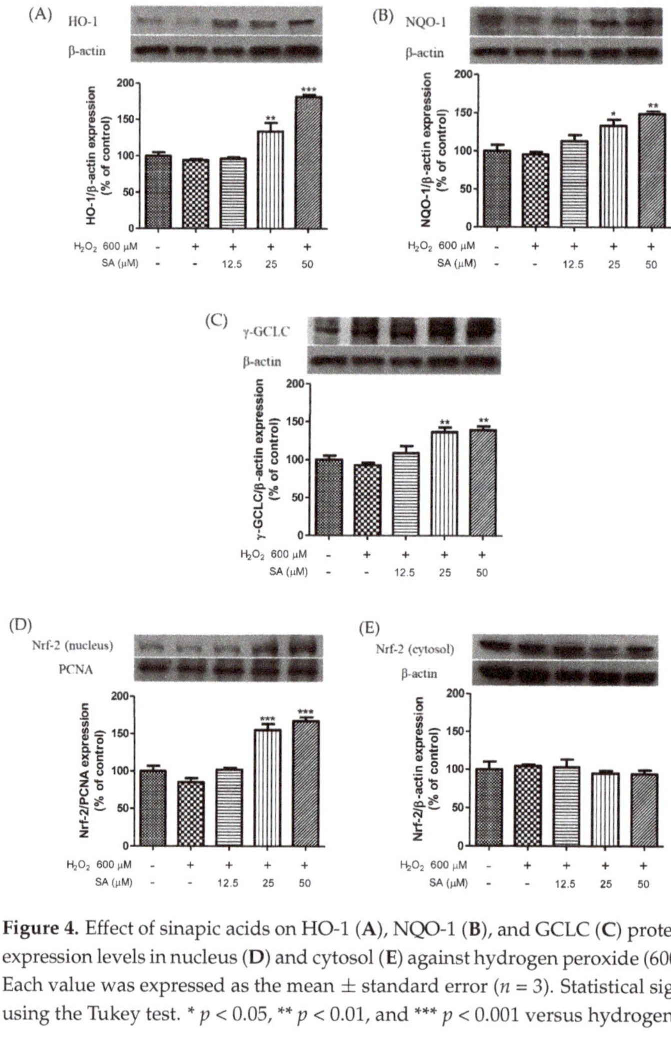

Figure 4. Effect of sinapic acids on HO-1 (**A**), NQO-1 (**B**), and GCLC (**C**) protein expression and Nrf-2 expression levels in nucleus (**D**) and cytosol (**E**) against hydrogen peroxide (600 μM) in EA.hy 926 cells. Each value was expressed as the mean ± standard error (n = 3). Statistical significance was analyzed using the Tukey test. * $p < 0.05$, ** $p < 0.01$, and *** $p < 0.001$ versus hydrogen-peroxide-treated cells.

3.4. Effects of Phenolic Acids on Endothelial Dysfunction

NO is essential for maintaining vascular function in the endothelium. Phosphorylation of eNOS can regulate NO production [31] and is essential for the improvement of CVD [32]. Akt mediates NO production via phosphorylation of eNOS, which promotes endothelial cell migration and angiogenesis [33]. A previous study reported that eNOS phosphorylation facilitates vasorelaxation via the PI3K/Akt signaling pathway in HUVECs [34]. Therefore, phosphorylation of eNOS and Akt is important for the treatment of endothelial dysfunction. As shown in Figure 5, treatment with H_2O_2 (600 μM) significantly reduced the phosphorylation of eNOS and Akt. However, sinapic acid treatment at concentrations of 12.5, 25, and 50 μM enhanced the phosphorylation of eNOS by 14.1, 26.3, and 48%, respectively, compared to that in the H_2O_2-treated group. Sinapic acid increased Akt phosphorylation in a dose-dependent manner. Chen et al. (2020) reported that phenolic acids extracted from ginseng protect against vascular endothelial cell injury via the activation of the PI3K/Akt/eNOS pathway [35]. Yan et al. (2020) reported that gallic acid attenuated vascular dysfunction and hypertension in angiotensin II-induced C57BL/6J mice by suppressing eNOS degradation [36]. Taken together, our results showed that sinapic acid may be effective in the treatment of endothelial dysfunction via phosphorylation of eNOS and Akt.

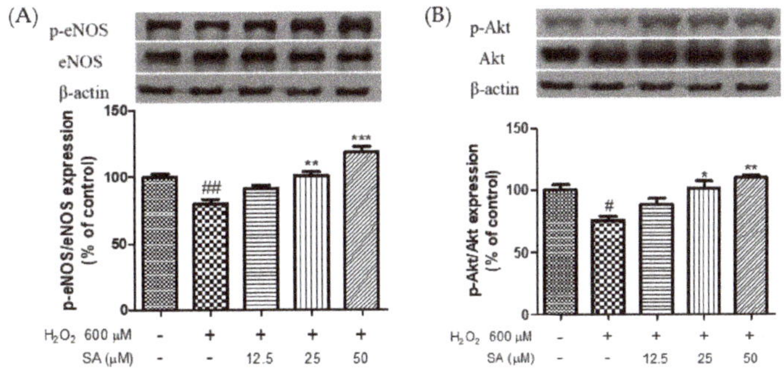

Figure 5. Effect of sinapic acids on p-eNOS (**A**) and p-Akt (**B**) protein expression against hydrogen peroxide (600 μM) in EA.hy 926 cells. Each value was expressed as the mean ± standard error (n = 3). Statistical significance was analyzed using the Tukey test. # $p < 0.05$ and ## $p < 0.01$ versus nontreated cells. * $p < 0.05$, ** $p < 0.01$, and *** $p < 0.001$ versus hydrogen-peroxide-treated cells.

4. Conclusions

This study showed that phenolic acids significantly protected endothelial cells against H_2O_2-induced oxidative damage by modulating NO, GSH, MDA, and ROS levels. Sinapic acid alleviated endothelial dysfunction by enhancing HO-1, NQO-1, GCLC, p-Akt, and p-eNOS expression levels, as well as activating Nrf2 nuclear translocation. Overall, these results illustrate that sinapic acid, which exists abundantly in cereals, spices, vegetables, oil seed crops, citrus, and berry fruits, has the potential as a treatment option for hypertension. However, further in vivo studies and clinical trials are needed to determine the underlying mechanism of action.

Author Contributions: Conceptualization and writing—original draft preparation: M.Y.; project administration: H.J.K.; methodology: H.H.; data curation: M.K. and Y.J.; writing—review and editing: H.L.; supervision: J.L. All authors have read and agreed to the published version of the manuscript.

Funding: This research was funded by the Rural Development Administration, grant number PJ01415004.

Institutional Review Board Statement: Not applicable.

Informed Consent Statement: Not applicable.

Data Availability Statement: Not applicable.

Conflicts of Interest: The authors declare no conflict of interest.

Sample Availability: Not available.

References

1. Wu, L.; Ashraf, M.H.N.; Facci, M.; Wang, R.; Paterson, P.G.; Ferrie, A.; Juurlink, B.H.J. Dietary Approach to Attenuate Oxidative Stress, Hypertension, and Inflammation in the Cardiovascular System. *Proc. Natl. Acad. Sci. USA* **2004**, *101*, 7094–7099. [CrossRef]
2. Loperena, R.; Harrison, D.G. Oxidative Stress and Hypertensive Diseases. *Med. Clin. N. Am.* **2017**, *101*, 169–193. [CrossRef]
3. Zhang, C.; Hein, T.W.; Wang, W.; Miller, M.W.; Fossum, T.W.; McDonald, M.M.; Humphrey, J.D.; Kuo, L. Upregulation of Vascular Arginase in Hypertension Decreases Nitric Oxide-Mediated Dilation of Coronary Arterioles. *Hypertension* **2004**, *44*, 935–943. [CrossRef]
4. Folkow, B.; Grimby, G.; Thulesius, O. Adaptive Structural Changes of the Vascular Walls in Hypertension and Their Relation to the Control of the Peripheral Resistance. *Acta Physiol. Scand.* **1958**, *44*, 255–272. [CrossRef]
5. Lu, X.; Zhang, L.; Sun, Q.; Song, G.; Huang, J. Extraction, Identification and Structure-Activity Relationship of Antioxidant Peptides from Sesame (*Sesamum indicum* L.) Protein Hydrolysate. *Food Res. Int.* **2019**, *116*, 707–716. [CrossRef]
6. Howden, R. Nrf2 and Cardiovascular Defense. *Oxidative Med. Cell. Longev.* **2013**, *2013*, e104308. [CrossRef]
7. Zhang, Y.; Guan, L.; Wang, X.; Wen, T.; Xing, J.; Zhao, J. Protection of Chlorophyllin against Oxidative Damage by Inducing HO-1 and NQO1 Expression Mediated by PI3K/Akt and Nrf2. *Free Radic. Res.* **2008**, *42*, 362–371. [CrossRef] [PubMed]
8. Shi, F.; Wang, Y.-C.; Zhao, T.-Z.; Zhang, S.; Du, T.-Y.; Yang, C.-B.; Li, Y.-H.; Sun, X.-Q. Effects of Simulated Microgravity on Human Umbilical Vein Endothelial Cell Angiogenesis and Role of the PI3K-Akt-ENOS Signal Pathway. *PLoS ONE* **2012**, *7*, e40365. [CrossRef]
9. Iaccarino, G.; Ciccarelli, M.; Sorriento, D.; Cipolletta, E.; Cerullo, V.; Iovino, G.L.; Paudice, A.; Elia, A.; Santulli, G.; Campanile, A.; et al. AKT Participates in Endothelial Dysfunction in Hypertension. *Circulation* **2004**, *109*, 2587–2593. [CrossRef]
10. Sun, J.; Chu, Y.-F.; Wu, X.; Liu, R.H. Antioxidant and Antiproliferative Activities of Common Fruits. *J. Agric. Food Chem.* **2002**, *50*, 7449–7454. [CrossRef]
11. Saibabu, V.; Fatima, Z.; Khan, L.A.; Hameed, S. Therapeutic Potential of Dietary Phenolic Acids. *Adv. Pharmacol. Sci.* **2015**, *2015*, e823539. [CrossRef] [PubMed]
12. Srinivasan, M.; Sudheer, A.R.; Menon, V.P. Ferulic Acid: Therapeutic Potential Through Its Antioxidant Property. *J. Clin. Biochem. Nutr.* **2007**, *40*, 92–100. [CrossRef] [PubMed]
13. Kang, N.; Lee, J.-H.; Lee, W.; Ko, J.-Y.; Kim, E.-A.; Kim, J.-S.; Heu, M.-S.; Kim, G.H.; Jeon, Y.-J. Gallic Acid Isolated from Spirogyra Sp. Improves Cardiovascular Disease through a Vasorelaxant and Antihypertensive Effect. *Environ. Toxicol. Pharmacol.* **2015**, *39*, 764–772. [CrossRef]
14. Jin, L.; Lin, M.Q.; Piao, Z.H.; Cho, J.Y.; Kim, G.R.; Choi, S.Y.; Ryu, Y.; Sun, S.; Kee, H.J.; Jeong, M.H. Gallic Acid Attenuates Hypertension, Cardiac Remodeling, and Fibrosis in Mice with NG-Nitro-L-Arginine Methyl Ester-Induced Hypertension via Regulation of Histone Deacetylase 1 or Histone Deacetylase 2. *J. Hypertens.* **2017**, *35*, 1502–1512. [CrossRef]
15. Alam, M.A. Anti-Hypertensive Effect of Cereal Antioxidant Ferulic Acid and Its Mechanism of Action. *Front. Nutr.* **2019**, *6*, 121. [CrossRef]
16. Agunloye, O.M.; Oboh, G.; Ademiluyi, A.O.; Ademosun, A.O.; Akindahunsi, A.A.; Oyagbemi, A.A.; Omobowale, T.O.; Ajibade, T.O.; Adedapo, A.A. Cardio-Protective and Antioxidant Properties of Caffeic Acid and Chlorogenic Acid: Mechanistic Role of Angiotensin Converting Enzyme, Cholinesterase and Arginase Activities in Cyclosporine Induced Hypertensive Rats. *Biomed. Pharmacother.* **2019**, *109*, 450–458. [CrossRef]
17. Cushman, D.W.; Cheung, H.S. Spectrophotometric Assay and Properties of the Angiotensin-Converting Enzyme of Rabbit Lung. *Biochem. Pharmacol.* **1971**, *20*, 1637–1648. [CrossRef]
18. Nasution, S.A. The Use of ACE Inhibitor in Cardiovascular Disease. *Acta Med. Indones.* **2006**, *38*, 60–64.
19. Ojeda, D.; Jiménez-Ferrer, E.; Zamilpa, A.; Herrera-Arellano, A.; Tortoriello, J.; Alvarez, L. Inhibition of Angiotensin Convertin Enzyme (ACE) Activity by the Anthocyanins Delphinidin- and Cyanidin-3-O-Sambubiosides from Hibiscus Sabdariffa. *J. Ethnopharmacol.* **2010**, *127*, 7–10. [CrossRef]
20. Zhang, Y.; Pechan, T.; Chang, S.K.C. Antioxidant and Angiotensin-I Converting Enzyme Inhibitory Activities of Phenolic Extracts and Fractions Derived from Three Phenolic-Rich Legume Varieties. *J. Funct. Foods* **2018**, *42*, 289–297. [CrossRef]
21. Joannides, R.; Haefeli, W.E.; Linder, L.; Richard, V.; Bakkali, E.H.; Thuillez, C.; Lüscher, T.F. Nitric Oxide Is Responsible for Flow-Dependent Dilatation of Human Peripheral Conduit Arteries in Vivo. *Circulation* **1995**, *91*, 1314–1319. [CrossRef] [PubMed]
22. Li, Y.; Li, Y.; Fang, Z.; Huang, D.; Yang, Y.; Zhao, D.; Hang, M.; Wang, J. The Effect of Malus Doumeri Leaf Flavonoids on Oxidative Stress Injury Induced by Hydrogen Peroxide (H_2O_2) in Human Embryonic Kidney 293 T Cells. *BMC Complement. Med. Ther.* **2020**, *20*, 276. [CrossRef] [PubMed]
23. Liu, B.; Zhang, C.; Zhang, J.; Zhao, X. Wu Shan Shen Cha (Malus Asiatica Nakai. Leaves)-Derived Flavonoids Alleviate Alcohol-Induced Gastric Injury in Mice via an Anti-Oxidative Mechanism. *Biomolecules* **2019**, *9*, 169. [CrossRef]
24. Xiang, Y.; Ye, W.; Huang, C.; Lou, B.; Zhang, J.; Yu, D.; Huang, X.; Chen, B.; Zhou, M. Brusatol Inhibits Growth and Induces Apoptosis in Pancreatic Cancer Cells via JNK/P38 MAPK/NF-Kb/Stat3/Bcl-2 Signaling Pathway. *Biochem. Biophys. Res. Commun.* **2017**, *487*, 820–826. [CrossRef]

25. Lee, H.; Lee, J. Anti-Diabetic Effect of Hydroxybenzoic Acid Derivatives in Free Fatty Acid-Induced HepG2 Cells via MiR-1271/IRS1/PI3K/AKT/FOXO1 Pathway. *J. Food Biochem.* **2021**, *45*, e13993. [CrossRef]
26. Jadeja, R.N.; Upadhyay, K.K.; Devkar, R.V.; Khurana, S. Naturally Occurring Nrf2 Activators: Potential in Treatment of Liver Injury. *Oxid. Med. Cell Longev.* **2016**, *2016*, 3453926. [CrossRef]
27. Yang, S.-Y.; Pyo, M.C.; Nam, M.-H.; Lee, K.-W. ERK/Nrf2 Pathway Activation by Caffeic Acid in HepG2 Cells Alleviates Its Hepatocellular Damage Caused by t-Butylhydroperoxide-Induced Oxidative Stress. *BMC Complement. Altern. Med.* **2019**, *19*, 139. [CrossRef]
28. Luo, W.; Wang, Y.; Yang, H.; Dai, C.; Hong, H.; Li, J.; Liu, Z.; Guo, Z.; Chen, X.; He, P.; et al. Heme Oxygenase-1 Ameliorates Oxidative Stress-Induced Endothelial Senescence via Regulating Endothelial Nitric Oxide Synthase Activation and Coupling. *Aging* **2018**, *10*, 1722–1744. [CrossRef]
29. Wang, X.; Chen, L.; Wang, T.; Jiang, X.; Zhang, H.; Li, P.; Lv, B.; Gao, X. Ginsenoside Rg3 Antagonizes Adriamycin-Induced Cardiotoxicity by Improving Endothelial Dysfunction from Oxidative Stress via Upregulating the Nrf2-ARE Pathway through the Activation of Akt. *Phytomedicine* **2015**, *22*, 875–884. [CrossRef]
30. Unsal, V.; Kolukcu, E.; Firat, F.; Gevrek, F. The Protective Effects of Sinapic Acid on Acute Renal Ischemia/Reperfusion Injury. *Turk. J. Biochem.* **2021**, *46*, 563–571. [CrossRef]
31. Lee, J.H.; Parveen, A.; Do, M.H.; Lim, Y.; Shim, S.H.; Kim, S.Y. Lespedeza Cuneata Protects the Endothelial Dysfunction via ENOS Phosphorylation of PI3K/Akt Signaling Pathway in HUVECs. *Phytomedicine* **2018**, *48*, 1–9. [CrossRef]
32. Förstermann, U.; Münzel, T. Endothelial Nitric Oxide Synthase in Vascular Disease: From Marvel to Menace. *Circulation* **2006**, *113*, 1708–1714. [CrossRef] [PubMed]
33. Kawasaki, K.; Smith, R.S., Jr.; Hsieh, C.M.; Sun, J.; Chao, J.; Liao, J.K. Activation of the phosphatidylinositol 3-kinase/protein kinase Akt pathway mediates nitric oxide-induced endothelial cell migration and angiogenesis. *Mol. Cell. Biol.* **2003**, *23*, 5726–5737. [CrossRef]
34. Anwar, M.A.; Samaha, A.A.; Ballan, S.; Saleh, A.I.; Iratni, R.; Eid, A.H. *Salvia fruticosa* induces Vasorelaxation in Rat Isolated Thoracic Aorta: Role of the PI3K/Akt/eNOS/NO/cGMP Signaling Pathway. *Sci. Rep.* **2017**, *7*, 686. [CrossRef]
35. Chen, X.; Yao, F.; Song, J.; Fu, B.; Sun, G.; Song, X.; Fu, C.; Jiang, R.; Sun, L. Protective Effects of Phenolic Acid Extract from Ginseng on Vascular Endothelial Cell Injury Induced by Palmitate via Activation of PI3K/Akt/ENOS Pathway. *J. Food Sci.* **2020**, *85*, 576–581. [CrossRef]
36. Yan, X.; Zhang, Q.-Y.; Zhang, Y.-L.; Han, X.; Guo, S.-B.; Li, H.-H. Gallic Acid Attenuates Angiotensin II-Induced Hypertension and Vascular Dysfunction by Inhibiting the Degradation of Endothelial Nitric Oxide Synthase. *Front. Pharmacol.* **2020**, *11*, 1121. [CrossRef]

molecules

MDPI

Article

Protective Effects of PollenAid Plus Soft Gel Capsules' Hydroalcoholic Extract in Isolated Prostates and Ovaries Exposed to Lipopolysaccharide

Annalisa Chiavaroli [1], Simonetta Cristina Di Simone [1], Alessandra Acquaviva [1], Maria Loreta Libero [1], Claudia Campana [1], Lucia Recinella [1], Sheila Leone [1], Luigi Brunetti [1], Giustino Orlando [1], Nilofar [1], Irene Vitale [1], Stefania Cesa [2], Gokhan Zengin [3,*], Luigi Menghini [1,*] and Claudio Ferrante [1]

[1] Department of Pharmacy, Botanic Garden "Giardino dei Semplici", "G. d'Annunzio" University, via dei Vestini 31, 66100 Chieti, Italy
[2] Department of Drug Chemistry and Technology, Sapienza University of Rome, 00185 Rome, Italy
[3] Department of Biology, Science Faculty, Selcuk University, Konya 42130, Turkey
* Correspondence: gokhanzengin@selcuk.edu.tr (G.Z.); luigi.menghini@unich.it (L.M.)

Citation: Chiavaroli, A.; Di Simone, S.C.; Acquaviva, A.; Libero, M.L.; Campana, C.; Recinella, L.; Leone, S.; Brunetti, L.; Orlando, G.; Nilofar; et al. Protective Effects of PollenAid Plus Soft Gel Capsules' Hydroalcoholic Extract in Isolated Prostates and Ovaries Exposed to Lipopolysaccharide. *Molecules* **2022**, *27*, 6279. https://doi.org/10.3390/molecules27196279

Academic Editor: Nour Eddine Es-Safi

Received: 31 August 2022
Accepted: 17 September 2022
Published: 23 September 2022

Abstract: Pollen extract represents an innovative approach for the management of the clinical symptoms related to prostatitis and pelvic inflammatory disease (PID). In this context, the aims of the present work were to analyze the phenolic composition of a hydroalcoholic extract of PollenAid Plus soft gel capsules, and to evaluate the extract's cytotoxic effects, in human prostate cancer PC3 cells and human ovary cancer OVCAR-3 cells. Additionally, protective effects were investigated in isolated prostate and ovary specimens exposed to lipopolysaccharide (LPS). The phytochemical investigation identified catechin, chlorogenic acid, gentisic acid, and 3-hydroxytyrosol as the prominent phenolics. The extract did not exert a relevant cytotoxic effect on PC3 and OVCAR-3 cells. However, the extract showed a dose-dependent inhibition of pro-inflammatory IL-6 and TNF-α gene expression in prostate and ovary specimens, and the extract was effective in preventing the LPS-induced upregulation of CAT and SOD gene expression, which are deeply involved in tissue antioxidant defense systems. Finally, a docking approach suggested the capability of catechin and chlorogenic acid to interact with the TRPV1 receptor, playing a master role in prostate inflammation. Overall, the present findings demonstrated anti-inflammatory and antioxidant effects of this formulation; thus, suggesting its capability in the management of the clinical symptoms related to prostatitis and PID.

Keywords: Graminex pollen; hydroalcoholic extract; phenolic compounds; inflammation; oxidative stress; gene expression; TRPV1

1. Introduction

Prostatitis and pelvic inflammatory disease (PID) are common chronic conditions in the population, caused by pathogenic infection [1,2]. Herbal extracts endowed with antioxidant/anti-inflammatory effects have been long considered as a reliable strategy to blunt the burden of oxidative stress and inflammation in prostate and ovary tissue [3–5]. Pollen extract represents an innovative approach for the management of the clinical symptoms related to prostatitis [6], being also able to relieve inflammation and hyperplasia of the prostate [7], with anticancer potential most likely associated with antioxidant and antimutagenic effects [8]. In this case, pollen appears to relieve pain in patients with benign prostatic hyperplasia, at least in the early stages. Its administration together with chemotherapeutic agents has been seen to increase the number of people who have experienced a significant therapeutic effect [9]. Due to its content in phytoestrogens, pollen has also been shown to improve the symptoms of polycystic ovary syndrome in rats [10], although there is still a lack of scientific literature about the effects of pollen in PID.

Pollen represents the set of microgametophytes produced by spermatophytes in the male cones, in the case of gymnosperms, and in the anthers, the fertile part of the stamens, in the case of angiosperms. Pollen has different shapes and colors and dimensions between 2.5 and 250 μm. Its composition varies mainly according to the geographical origins and the botanical species visited by the insect, together with less relevant but still important factors such as climatic conditions and the type of soil [8]. From the literature, it appears that pollen grains deriving from various plant species contain about 200 active substances, including proteins and amino acids, carbohydrates, lipids and fatty acids, enzymes and coenzymes, nucleic acids, phenolic compounds, vitamins, and minerals [9].

Pollen, as well as other bee products, has been used since ancient times as a food for its nutritional value and for a wide spectrum of therapeutic activities, of which the best known are antifungal, antibacterial, antiviral, antioxidant, and anti-inflammatory. Evidence has been reported on the activity of phenolic compounds in pollen extracts against Gram-positive and Gram-negative bacteria, fungi, and yeasts [10–12]. The content in phenolic compounds has been also related to the anti-inflammatory and antioxidant properties of pollen [13].

In this context, the aims of the present work were to analyze the phenolic composition of an innovative formulation containing Graminex G60^{TM} Flower Pollen Extract, a mixture of standardized and dry pollen of rye grass (*Secale cereale* L.), corn (*Zea mays* L.), and timothy (*Phleum pratense* L.), and NAX^{TM} 7% paste, both suspended in extra virgin olive oil (EVO) as amber soft gel capsules (PollenAid Plus), and to evaluate cytotoxic activity of the hydroalcoholic extract from this formulation on immortalized human prostate cancer PC3 cells and human ovary cancer OVCAR-3 cells. The effect of the extract on cell viability was also investigated in a myoblast C2C12 cell line, which was chosen as a non-tumor comparison cell model. The protective effects of this extract were also investigated in isolated prostate and ovary tissues exposed to *Escherichia coli* lipopolysaccharide (LPS), a reliable experimental model of tissue inflammation [14]. In this context, we measured the gene expression of pro-inflammatory factors, including inteleukin-6 (IL-6) and tumor necrosis factor α (TNF-α). The gene expression of superoxide dismutase (SOD) and catalase, which are deeply involved in antioxidant response, was measured in both tissues, as well. Finally, an in silico study was conducted for unraveling, albeit partially, the mechanisms of action underlying the observed effects and putative interactions against transient receptor potential vanilloid 1 (TRPV1), an ion channel present on sensory neurons and localized in particular on small neurons and type C amielin fibers responsible for nociceptive transmission. TRPV1 is activated by a number of harmful stimuli and its activity is regulated by numerous inflammatory mediators including prostaglandins, bradykinins, and serotonin. In addition to the sensitization of TRPV1 by inflammatory mediators, the activation of TRPV1 stimulates the release of inflammatory molecules associated with the transmission of pain such as substance P and bradykinin, which in turn contribute to the peripheral sensitization of TRPV1 as well as the activation of mast cells and the perpetuation of the state of neurogenic inflammation. The administration of substances capable of acting on some of the actors underlying the pathophysiology of pain, such as TRPV1, and at the same time counteracting neurogenic inflammation is a rational approach in the treatment of chronic pelvic pain pathologies.

2. Results

In the present study, 17 compounds were identified in the hydroalcoholic extract and quantified through HPLC-DAD-MS. The quantification was carried out by comparison with pure standards (Figure 1). Among assayed compounds, 3-hydroxytyrosol, catechin, gentisic acid, and chlorogenic acid were the main phytochemicals (Table S1). The quantification of such compounds in the extract is consistent with their previous identification in the plants of origin of the pollen [15,16], despite the presence of the vehicle (extra-virgin olive oil: EVO); thus, indicating the EVO as a reliable vehicle which displays multiple advantages: biocompatibility, health-promoting effects, and sustainability. The determination of phenolic compounds is consistent with our previous study of Graminex pollen using different analytical conditions [17]. The presence of phenolic compounds in the extract makes rational the evaluation of protective effects in prostate and ovary cells and tissues, as described below.

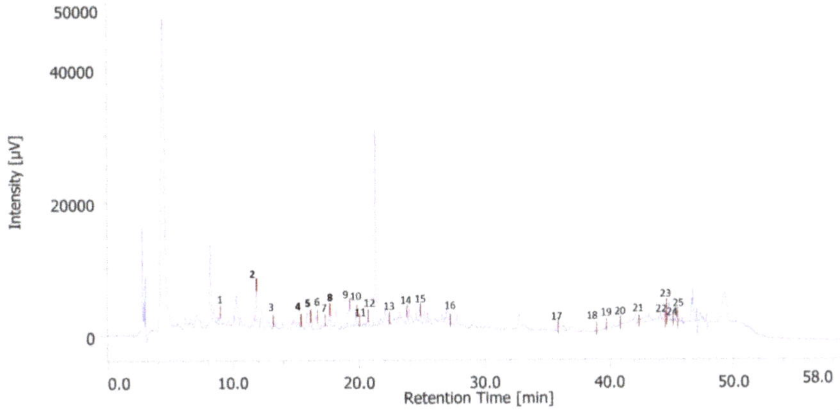

(A)

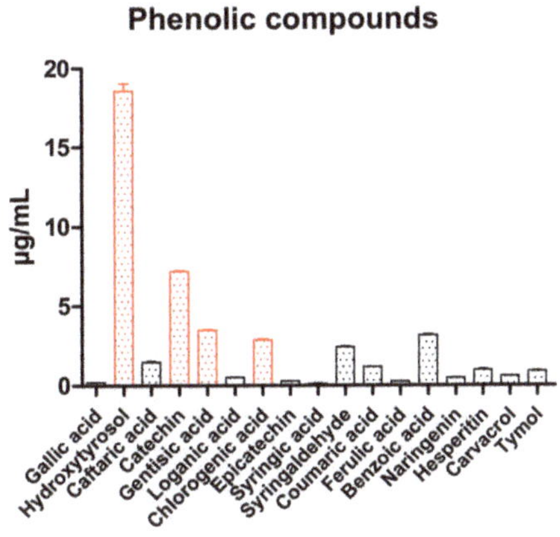

(B)

Figure 1. (**A**) Chromatogram related to the analysis of the hydroalcoholic extract from Graminex pollen. (**B**) Phenolic compounds identified and quantified in the extract. 3-Hydroxytyrosol (peak #2), catechin (peak #4), gentisic acid (peak #5), and chlorogenic acid (peak #8) were the most abundant phenolic compounds present in the extract.

Regarding the pharmacological study, the extract (10–2000 µg/mL) was tested on different cell lines, namely human prostate cancer PC3 cells and human ovary cancer OVCAR-3 cells, to investigate cytotoxic properties against tumor cells. Additionally, the extract was also added to the medium of myoblast C2C12 cells, to determine the susceptibility of a non-tumor cell line to scalar concentrations. Intriguingly, all three cell lines displayed a similar response after exposure to the extract. Indeed, the cell viability was slightly reduced at the highest tested concentration (2000 µg/mL). However, the cell viability was >70% compared to the control (ctrl) group, in all three cell models (Figure 2); thus, ruling out any significant cytotoxic effect towards both tumor and non-tumor cells.

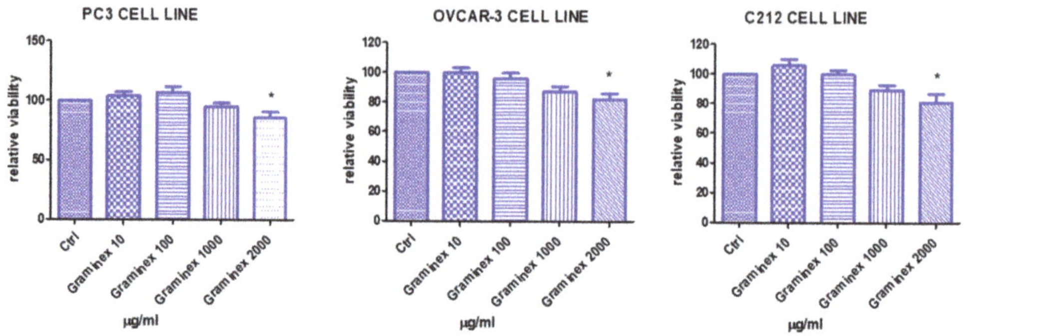

Figure 2. Effects of the extract on human prostate cancer PC3 cells, human ovary cancer OVCAR-3 cells, and the non-tumoral C2C12 myoblast cell line. At the highest tested concentration, the extract induced a mild reduction of cell viability in all tested cell lines (ANOVA, $p < 0.05$; * $p < 0.05$ vs. ctrl group). However, the cell viability was always over 70% compared to the respective ctrl group; thus, suggesting biocompatibility in the concentration range 10–2000 μg/mL. This range was considered as biocompatible for the subsequent ex vivo determination in the prostate and ovary tissues.

The extract (10–2000 μg/mL) was also tested in isolated prostate and ovary specimens challenged with *E. coli* LPS, chosen as pro-inflammatory stimulus [14,17]. In this context, it is notable that *E. coli* infection has been related to both prostatitis and PID [18,19]. The LPS stimulus induced the upregulation of TNF-α and IL-6 both in prostate and ovary tissues (Figures 3 and 4). The extract treatment was effective in reverting the increased gene expression of both cytokines; thus, demonstrating anti-inflammatory effects in both tissues. In the case of prostate tissues, this study is also consistent with previous clinical observations about the capability of Graminex pollen to contrast the inflammatory component of prostatitis [20,21]. The anti-inflammatory effects also agree with previous studies highlighting the inhibition of IL-8 production [22] and cyclooxygenase (COX)-2 and inducible nitric oxide synthase (iNOS) activities, measured as prostaglandin E_2 and nitrites levels, respectively [23–25].

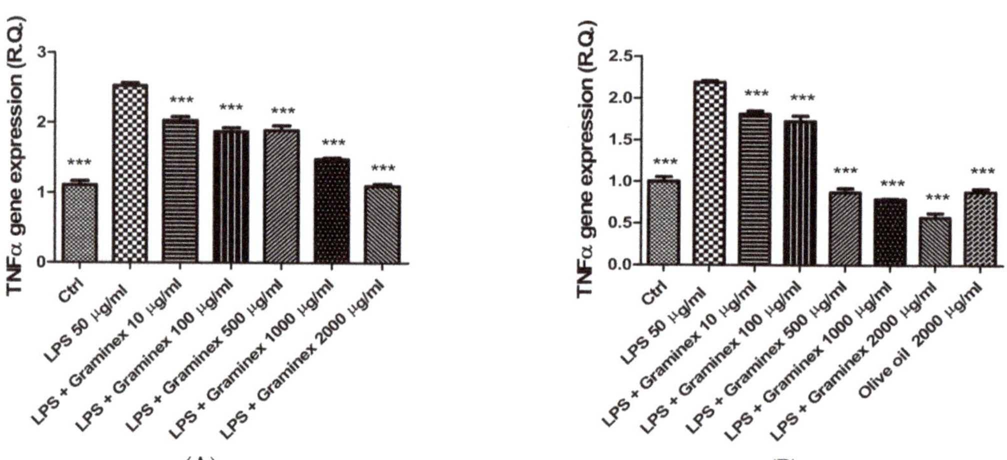

Figure 3. Inhibitory effects of the hydroalcoholic extract (10–2000 μg/mL) on TNF-α gene expression in isolated prostate (**A**) and ovary (**B**) specimens. ANOVA, $p < 0.0001$; *** $p < 0.001$ vs. ctrl group.

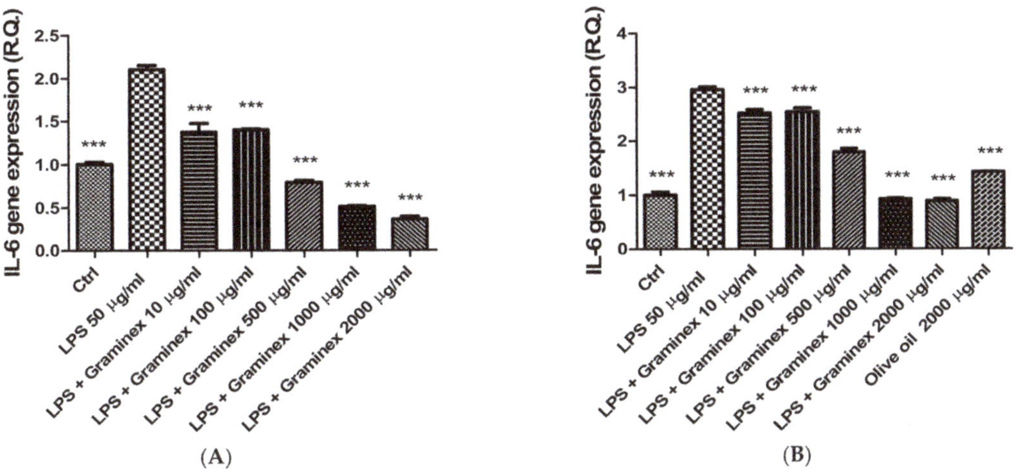

Figure 4. Inhibitory effects of the hydroalcoholic extract (10–2000 µg/mL) on IL-6 gene expression in isolated prostate (**A**) and ovary (**B**) specimens. ANOVA, $p < 0.0001$; *** $p < 0.001$ vs. ctrl group.

In prostate and ovary specimens, LPS stimulus (50 µg/mL) was also effective in increasing the gene expression of both CAT and SOD (Figures 5 and 6), which are deeply involved in the antioxidant response. Indeed, the extract was able to prevent the LPS-induced upregulation of CAT and SOD gene expression. Additionally, after extract administration, the gene expression of both enzymes was even lower than the one displayed by the control (ctrl) group. Previously, LPS stimulus has been found to alter CAT and SOD levels, with both inhibitory and stimulatory effects. We cannot exclude that these discrepancies could depend, albeit partially, on the employed experimental models [26,27]. Therefore, also considering the intrinsic scavenging/reducing properties and ability of Graminex to blunt LPS-induced lipoperoxidation in isolated prostate [17], we hypothesize that the extract effects on SOD and CAT gene expression could be related to antioxidant effects, which can be mediated, albeit partially, by polyphenolic compounds.

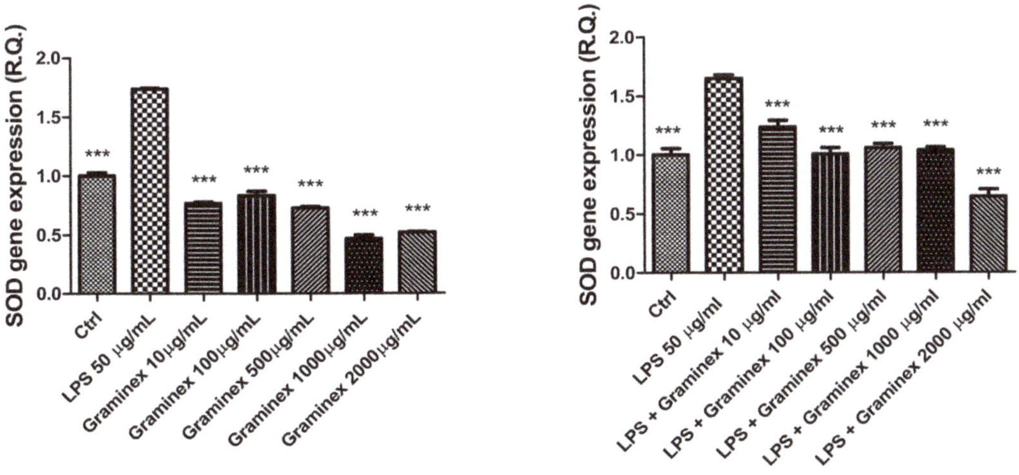

Figure 5. Inhibitory effects of the hydroalcoholic extract (10–2000 µg/mL) on SOD gene expression in isolated prostate (**A**) and ovary (**B**) specimens. ANOVA, $p < 0.0001$; *** $p < 0.001$ vs. ctrl group.

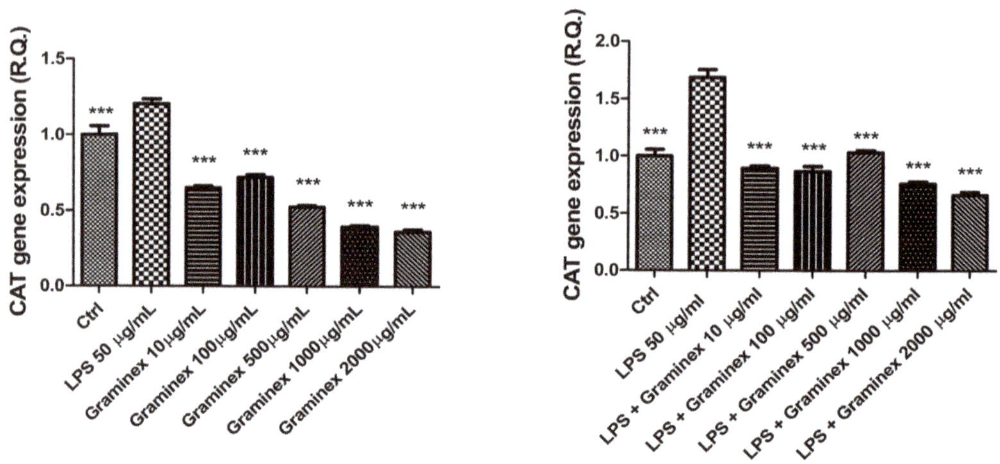

(A) (B)

Figure 6. Inhibitory effects of the hydroalcoholic extract (10–2000 μg/mL) on CAT gene expression in isolated prostate (**A**) and ovary (**B**) specimens. ANOVA, $p < 0.0001$; *** $p < 0.001$ vs. ctrl group.

In order to explore the mechanisms of action underlying the observed effects, an in silico study was conducted on the platform STITCH, considering the main phytochemicals present in the extract; namely catechin, 3-hydroxytyrosol, chlorogenic acid, and gentisic acid (2,5-dihydroxybenzoic acid). Catechin was predicted to interact with IL-6, cyclooxygenase-2 (COX-2, PGTS2), and with iNOS (Figure 7). This is partly consistent with our findings of anti-inflammatory effects by the extract, in both prostate and ovary tissue, and with the literature data [5]. Intriguingly, 3-hydroxytyrosol and chlorogenic acid were predicted to interact with BCL-2 and caspase-3, respectively. Previous studies showed the capability of 3-hydroxytyrosol and chlorogenic acid to reduce BCL-2 and caspase-3 gene and protein expression, respectively [28,29], while gentisic acid could interact with fibroblast growth factor 1 (FGF1), whose levels are increased in prostate cancer [30]. This could explain, albeit partially, the mild reduction (<30%) of cell viability in all considered cell lines at the highest tested concentration.

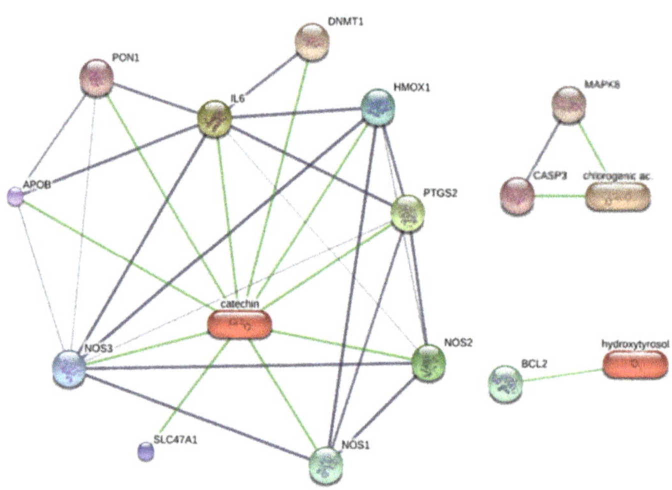

Figure 7. *Cont.*

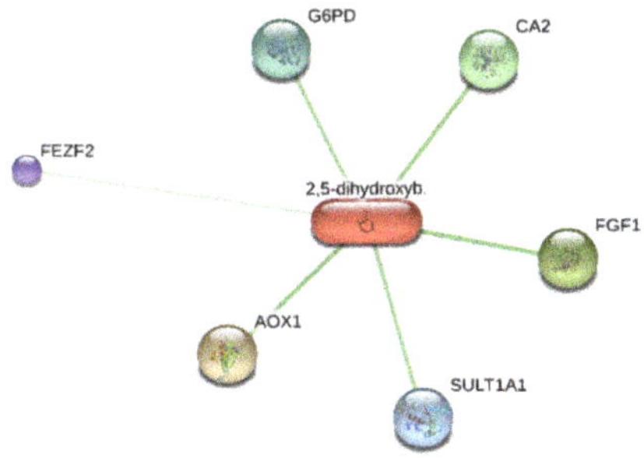

Figure 7. Target component analysis conducted on the STITCH bioinformatics platform for unraveling putative interactions between prominent extracts' phytochemicals and putative proteins involved in inflammatory and cytotoxicity effects.

Finally, a docking approach was conducted to explore the putative interactions between the extract's phytochemicals and the TRPV1 receptor, whose expression is increased in prostate inflammation [31]. Among phytochemicals detected in the extract, chlorogenic acid and catechin showed micromolar affinity (10–12 μM) towards the TRPV1 receptor (Figure 8); thus, suggesting direct interactions that could be crucial in mediating the observed anti-inflammatory properties. According to these predictions, further in vitro studies are needed to unravel the effects of catechin and chlorogenic acid on TRPV1 expression and activity.

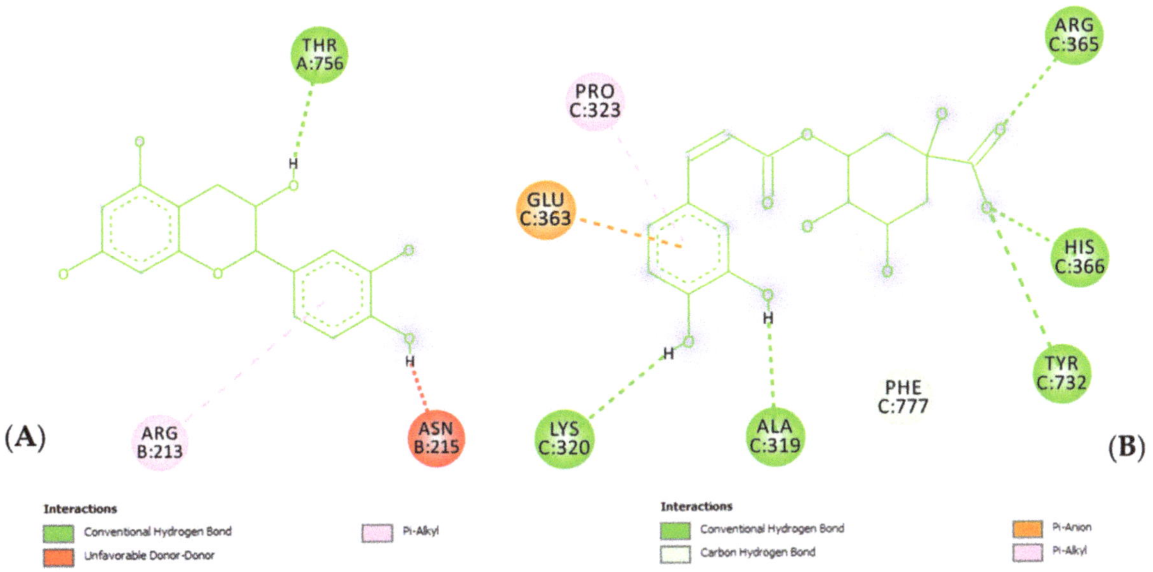

Figure 8. (**A**) Putative interactions between catechin and TRPV1 receptor (PDB ID: 7LR0). Free energy binding (ΔG) and putative affinity (K_i) are −6.8 kcal/mol and 10.5 μM, respectively. (**B**) Putative interactions between chlorogenic acid and TRPV1 receptor (PDB ID: 7LR0). Free energy binding (ΔG) and putative affinity (K_i) are −6.7 kcal/mol and 12.5 μM, respectively.

Color Analysis

The data obtained by the color analysis of the olive oil used as a vehicle and of the formulated product used to fill the soft capsules are reported in Table 1. The high luminance (70.46) of the pale yellow extra-virgin olive oil was changed to a very low value (14.96), which accounts for the very dark brown color of the formulated product. This drastic change, not shown by the only a* parameter, a weak red parameter, was accompanied by a drastic decrease of b* (positive, yellow parameter) and correlated saturation (C*ab), as well as the nuance turning from pale yellow to dark orange.

Table 1. CIEL*a*b* parameters of the carrier olive oil and of the filling formulated product.

	Olive Oil	PollenAid Plus Fill	
L*	70.46	14.96	
a*	5.55	5.42	
b*	111.60	13.87	
C^*_{ab}	111.74	14.89	
h_{ab}	87.15	68.67	
ΔL*		Respect to olive oil:	−55.50 Darker
		Respect to Graminex pollen powder:	−71.09 Darker
Δa*		Respect to olive oil:	−0.14 More green
		Respect to Graminex pollen powder	+3.71 More red
Δb*		Respect to olive oil:	−97.73 More blue
		Respect to Graminex pollen powder:	+1.91 More yellow
ΔC^*_{ab}		Respect to olive oil:	−96.84 More opaque
		Respect to Graminex pollen powder:	+2.80 Less opaque
Δh_{ab}		Respect to olive oil:	−8.48 More red
		Respect to Graminex pollen powder:	−13.21 More red
ΔE		Respect to olive oil:	+112.39
		Respect to Graminex pollen powder:	+71.21

The calculated color differences of the Graminex G60^{TM} Flower Pollen Extract used for the formulation showed a much lower luminance (ΔL*, −71), a little redder color and a more yellow sample, much darker and browner, but less opaque in respect to the pollen powder used in the formulation, whose CIEL*a*b* parameters, reported in our previous work [15], were L* 86.05; a* 1.71; b* 11.96; C^*_{ab} 12.09; h_{ab} 81.88. On the other hand, if slightly less important differences were shown between the formulated product and the extra virgin olive oil in terms of L* (ΔL* −55), and not-relevant changes of a* were registered, more significant differences were, in contrast, shown by the b* parameter (Δb*, −98) so that, on the whole, the sample appeared opaque, dark, and completely without color.

It seems of particular concern to compare matrices so different in superficial characteristics (solid powder, oily, sticky paste) and coming from different compositions and mixtures. Moreover, to our knowledge only one study is available, in which microscopic analysis, NIR spectroscopy, e-nose and e-tongue methods, as well as color analysis, were applied to perform a discrimination of bee pollens. As the authors reported, chemical composition largely depended on the botanical origin and can change due to the oxidation process. Authors also reported that dominance of positive a* and b* parameters could account for carotenoid and flavonoid compounds [32].

This statement agrees with the analyses we performed on the carrier oil and on the formulation after 9 months of storage (Figure 9), which showed slight modification of oil, towards a greener color, and of the formulation to a less intense brown, which could both account for a slight discoloration of carotenoids, both coming from the oil and from the pollen in the case of the formulated mixture. We reported in our previous works the carotenoid bleaching in powder infant formulas and in powder allium samples evaluated by color analyses [33,34].

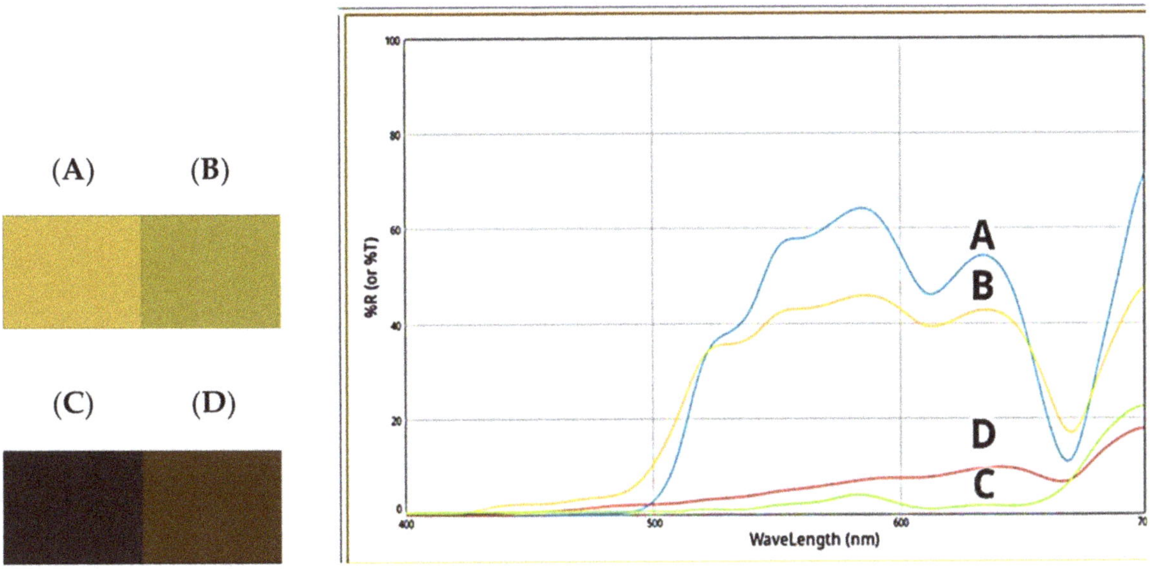

Figure 9. Color palette of olive oil ((**A**): t°; (**B**) t 9 months) and of the formulated product ((**C**): t°; (**D**): t 9 months) and relative reflectance curves.

3. Material and Methods

3.1. Samples

PollenAid Plus soft gel capsules were kindly provided by IdiPharma (Catania, Italy). The formulation contains Graminex G60™ Flower Pollen Extract (45.86%), NAX™ 7% paste (2.29%), soy lecithin (3.06%), yellow beeswax (2.93%), and extra virgin olive oil (45.86) as the vehicle. All the ingredients are contained in oval dark amber capsules. Graminex G60™ Flower Pollen Extract is a water-soluble extract of rye pollen grown in the USA. It is standardized to 6% amino acids and exhibits both antioxidant and anti-inflammatory activities. It is also non-allergenic, solvent-free, vegan, and non-GMO. Graminex G60™ Flower Pollen Extract was studied in our previous paper [17]. NAX™ 7% paste is a lipid-soluble paste of rye pollen grown in the USA. It is standardized to 7% phytosterols including β-sitosterol plus essential fatty acids such as ω-3 and ω-6. Applications include women's health, skin health, and heart health, and it can be directly formulated into soft gels. It is non-allergenic, solvent-free, vegan, and non-GMO. The extraction of this formulation was carried out by diluting 100 µL in 1 mL of a hydroalcoholic solution, constituted by 500 µL of ultrapure water and 500 µL of methanol. Subsequently, ultrasound-assisted extraction (UAE) was carried out. The operative conditions were 60 °C for 20 min at full power.

3.2. HPLC-DAD-MS

The identification and quantification of phenolic compounds were conducted through HPLC-DAD-MS analysis. The HPLC apparatus consisted of two PU-2080 PLUS chromatographic pumps, a DG-2080-54 line degasser, a mix-2080-32 mixer, UV, diode array (DAD) detector, a mass spectrometer (MS) detector (expression compact mass spectrometer (CMS), Advion, Ithaca, NY, USA), an AS-2057 PLUS autosampler, and a CO-2060 PLUS column thermostat (all from Jasco, Tokyo, Japan). Integration was performed by ChromNAV2 chromatography software. The separation was conducted within 60 min of the chromatographic run, starting from the following separation conditions: 97% water with 0.1% formic acid, 3% methanol with 0.1% formic acid. Details of the gradient are reported in Table 2. The separation was performed on an Infinity lab Poroshell 120-SB reverse phase column (C18, 150 × 4.6 mm i.d., 2.7 µm) (Agilent, Santa Clara, CA, USA). Column temperature was set at

30 °C. The injection volume was 5 µL. Quantitative determination of phenolic compounds was performed via DAD detector at 254 nm, through 7-point calibration curves, with linearity coefficients (R2) > 0.999, in the concentration range 2–140 µg/mL. The area under the curve from the HPLC chromatogram was used to quantify the analyte concentrations in the extract. Details of the phytochemical identification are included in Table 3. The extract was also qualitatively analyzed with an MS detector in positive and negative ion mode. MS signal identification was realized through comparison with a standard solution and MS spectra present in the MassBank Europe database. The statistical analysis was performed using GraphPad Prism version 5.01 software (San Diego, CA, USA).

Table 2. Gradient elution conditions.

TIME (min)	Composition A% (Water + Formic Acid 0.1%)	Composition B% (Methanol + Formic Acid 0.1%)	Flow (mL/min)
1	97	3	0.6
5	77	23	0.6
12	73	27	0.6
18	57	43	0.6
25	52	48	0.6
32	50	50	0.6
34	50	50	0.6
37	35	65	0.6
40	5	95	0.6
47	10	90	0.6
48	10	90	0.6

Table 3. Mass to charge (m/z) ratios, retention times, and quantities of the investigated phenolic compounds. DAD detector was set at 254 nm.

	Standard	*m/z*	Retention Time (min)
1	Gallic acid	170.15	8.967
2	3-Hydroxytyrosol	154.16	11.85
3	Caftaric acid	312.23	13.19
4	Catechin	290.27	15.383
5	Gentisic acid	154.12	16.147
6	4-Hydroxybenzoic acid	138.12	16.633
7	Loganic acid	376.36	17.257
8	Chlorogenic acid	354.31	17.66
9	Vanillic acid	168.15	19.22
10	Caffeic acid	180.16	19.73
11	Epicatechin	290.27	19.973
12	Syringic acid	198.17	20.673
13	Syringaldehyde	182.17	22.32
14	*p*-Coumaric acid	164.16	23.73
15	*t*-Ferulic acid	194.18	24.793
16	Benzoic acid	122.12	27.153

Table 3. *Cont.*

	Standard	m/z	Retention Time (min)
17	*t*-Cinnamic acid	148.15	35.817
18	Naringenin	272.25	38.87
19	2.3-Dimethylbenzoic acid	150.17	39.7
20	Hesperetin	302.28	40.773
21	Kaempferol	286.24	42.333
22	Carvacrol	150.22	44.393
23	Thymol	150.22	44.5
24	Flavone	222.24	45.077
25	3-Hydroxyflavone	238.24	45.36

3.3. Cell Cultures

The effects of the extract (100–500 µg/mL) on myocyte C2C12, human prostate PC3 cancer, and human ovary OVCAR-3 cancer cell viability were determined through the 3-(4,5-dimethylthiazol-2-yl)-2,5-diphenyltetrazolium bromide (MTT) test. The experimental conditions are fully described in our previous papers [35,36].

3.4. Ex Vivo Studies

Adult C57/BL6 and female mice (3-month-old, weight 20–25 g) were housed in Plexiglass cages (2–4 animals per cage; 55 cm × 33 cm × 19 cm) and maintained under standard laboratory conditions (21 ± 2 °C; 55 ± 5% humidity) on a 14/10 h light/dark cycle, with ad libitum access to water and food. Housing conditions and experimentation procedures were strictly in agreement with the European Community ethical regulations (EU Directive no. 26/2014) on the care of animals for scientific research. In agreement with the recognized principles of "replacement, refinement and reduction in animals in research", colon specimens were obtained as residual material from vehicle-treated mice randomized in our previous experiments, approved by the local ethical committee ('G. d'Annunzio' University, Chieti, Italy) and Italian Health Ministry (Project no. 885/2018-PR).

Isolated prostate and ovary specimens were maintained in a humidified incubator with 5% CO_2 at 37 °C for 4 h (incubation period), in RPMI buffer with added bacterial LPS (10 µg/mL), as previously described [17]. During the incubation period, the tissues were challenged with scalar concentrations of the extract (10–2000 µg/mL).

3.5. Gene Expression Analysis

Total RNA was extracted from both prostate and ovary specimens using TRI reagent (Sigma-Aldrich, St. Louis, MO, USA), according to the manufacturer's protocol, and reverse transcribed using a High Capacity cDNA Reverse Transcription Kit (Thermo Fischer Scientific, Waltman, MA, USA). Gene expression of TNF-α, IL-6, CAT, and SOD was determined by quantitative real-time PCR using TaqMan probe-based chemistry, as previously described [14]. PCR primers and TaqMan probes were purchased from Thermo Fisher Scientific Inc. The elaboration of data was conducted with the Sequence Detection System (SDS) software version 2.3 (Thermo Fischer Scientific). Relative quantification of gene expression was performed by the comparative $2^{-\Delta\Delta Ct}$ method [37].

3.6. In Silico Studies

Human proteins targeted by extract components were predicted using the bioinformatics platform STITCH. Docking calculations were conducted using AutoDock Vina PyRx 0.8 software. Crystal structure of the target protein was derived from the Protein Data Bank (PDB) with PDB ID as follows: 7LR0 (TRPV1). In order to prepare the protein for the docking simulation, all the water molecules and the co-crystalized heteromolecules were removed, followed by addition of hydrogen atoms and neutralization using Kollman united-atom charges. The dimensions of the grid box were 60 × 60 × 60 with 0.375 Å distance between points. Autodock4 and Lamarckian genetic algorithms were used to dock 250 conformations for each test compound (Molinspiration database). The Discovery Studio 2020 visualizer was employed to investigate the protein–ligand non-bonding interactions.

3.7. Statistical Analysis

Statistical analyses were performed using GraphPad Prism version 5.01 software (San Diego, CA, USA). Means ± S.E.M. were determined for each experimental group and analyzed by one-way analysis of variance (ANOVA), followed by Newman–Keuls comparison multiple test. Statistical significance was set at $p < 0.05$. The number of animals randomized for each experimental group was calculated on the basis of the "resource equation" $N = (E + T)/T$ $(10 \leq E \leq 20)$ [38].

3.8. Color Analysis

The samples under examination (olive oil as vehicle present at about 46% in the formulation, and PollenAid Plus Fill as formulated product used to fill the capsules) were subjected to colorimetric analysis using a spectrometer X-Rite, equipped with full-spectrum D65 illuminant and an observer angle and an observer angle of 10°. Cylindrical coordinates C^*_{ab} and h_{ab} were calculated from a* and b* as is customary in the literature [15].

4. Conclusions

In the present study, phytochemical and pharmacological investigations were conducted on an innovative formulation of PollenAid Plus Soft Gel capsules, which contain phenolic compounds with well-established anti-inflammatory and antioxidant activities. The relative instability shown by the colorimetric analyses performed at nine months of storage on the mixture used to fill the soft capsules, probably due to a carotenoid bleaching, indicates the importance of protecting the obtained mixture. For this reason, the formulation is contained in amber soft gel capsules. This formulation also covers the unappealing dark brown color of the mixture.

For the other analyses and to perform the biological assays, a hydroalcoholic extract was prepared from this commercial formulation and tested in the experimental paradigm. In this regard, the extract was rich in phenolic compounds, with 3-hydroxytyrosol, catechin, gentisic acid, and chlorogenic acid being the prominent phytochemicals. In in vitro models constituted by prostate and ovary cancer cells, the extract altered cell viability only at the highest concentration, with a slight reduction that cannot be considered toxic because it was <30% compared to the control treated group. However, in isolated prostate and ovary specimens exposed to LPS, the extract displayed significant reduction of IL-6 and TNF-α gene expression, which demonstrate anti-inflammatory effects. In the same ex vivo models, the extract was effective in restoring the mRNA levels of both SOD and CAT, which are deeply involved in the endogenous antioxidant mechanism. Lastly, an in silico study predicted the putative targets of the main phytochemicals present in the extracts. Particularly, catechin was predicted to be the main phenolic compound influencing the anti-inflammatory effects of the extract, whereas 3-hydroxytyrosol, chlorogenic acid, and gentisic acid could be the main phytochemicals responsible for the slight reduction of cell viability induced by the extract, at the highest tested concentration. Overall, the present findings demonstrated anti-inflammatory and antioxidant effects of this formulation in both prostate and tissue; thus, suggesting its capability in the management of the inflammatory component of both bacterial prostatitis and PID. Intriguingly, docking runs suggest the TRPV1 as a putative target, and the predicted micromolar interactions between catechin and chlorogenic acid towards this receptor support future studies for a better comprehension of the molecular mechanisms.

Supplementary Materials: The following supporting information can be downloaded at: https://www.mdpi.com/article/10.3390/molecules27196279/s1, Table S1: Quantitative analysis of the extract.

Author Contributions: C.F., L.M., G.O., S.C. and G.Z.: methodology, formal analysis, investigation, writing—original draft, revision, funding. A.A., A.C., L.R., S.C., S.C.D.S., S.L., I.V., C.C. and M.L.L.: methodology, formal analysis, investigation. L.B.: visualization, supervision, revision. N.: investigation. All authors have read and agreed to the published version of the manuscript.

Funding: This research received no external funding.

Institutional Review Board Statement: Not applicable.

Informed Consent Statement: Not applicable.

Data Availability Statement: The datasets generated and analyzed in the current study are available from the corresponding author on reasonable request.

Acknowledgments: The present study is also part of the third mission activities of the Botanic Garden "Giardino dei Semplici" of "G. d'Annunzio" University. The authors gratefully acknowledge Idipharma, Catania (Italy) for the donation of the formulation used in the present study.

Conflicts of Interest: The authors declare no conflict of interest.

Sample Availability: Samples of the compounds are available from the authors.

References

1. Steenkamp, V.; Gouws, M.C.; Gulumian, M.; Elgorashi, E.E.; van Staden, J. Studies on antibacterial, anti-inflammatory and antioxidant activity of herbal remedies used in the treatment of benign prostatic hyperplasia and prostatitis. *J. Ethnopharmacol.* **2006**, *103*, 71–75. [CrossRef] [PubMed]
2. Wang, D.; Jiang, Y.; Feng, J.; Gao, J.; Yu, J.; Zhao, J.; Liu, P.; Han, Y. Evidence for the Use of Complementary and Alternative Medicine for Pelvic Inflammatory Disease: A Literature Review. *Evid.-Based Complement. Altern. Med.* **2022**, *2022*, 1364297. [CrossRef] [PubMed]
3. Zhang, L.-J.; Zhu, J.-Y.; Sun, M.-Y.; Song, Y.-N.; Rahman, K.; Peng, C.; Zhang, M.; Ye, Y.-M.; Zhang, H. Anti-inflammatory effect of Man-Pen-Fang, a Chinese herbal compound, on chronic pelvic inflammation in rats. *J. Ethnopharmacol.* **2017**, *208*, 57–65. [CrossRef] [PubMed]
4. Orlando, G.; Chiavaroli, A.; Adorisio, S.; Delfino, D.; Brunetti, L.; Recinella, L.; Leone, S.; Zengin, G.; Acquaviva, A.; Angelini, P.; et al. Unravelling the Phytochemical Composition and the Pharmacological Properties of an Optimized Extract from the Fruit from *Prunus mahaleb* L.: From Traditional Liqueur Market to the Pharmacy Shelf. *Molecules* **2021**, *26*, 4422. [CrossRef] [PubMed]
5. Ferrante, C.; Chiavaroli, A.; Angelini, P.; Venanzoni, R.; Angeles Flores, G.; Brunetti, L.; Petrucci, M.; Politi, M.; Menghini, L.; Leone, S.; et al. Phenolic Content and Antimicrobial and Anti-Inflammatory Effects of *Solidago virga-aurea, Phyllanthus niruri, Epilobium angustifolium, Peumus boldus,* and *Ononis spinosa* Extracts. *Antibiotics* **2020**, *9*, 783. [CrossRef]
6. Cai, T.; Gallelli, L.; Cione, E.; Verze, P.; Palmieri, A.; Mirone, V.; Bonkat, G.; Wagenlehner, F.M.; Johansen, T.E.B. The efficacy and tollerability of pollen extract in combination with hyaluronic acid and vitamins in the management of patients affected by chronic prostatitis/chronic pelvic pain syndrome: A 26 weeks, randomized, controlled, single-blinded, phase III study. *Minerva Urol. Nephrol.* **2021**. [CrossRef]
7. Csikós, E.; Horváth, A.; Ács, K.; Papp, N.; Balázs, V.L.; Dolenc, M.S.; Kenda, M.; Glavač, N.K.; Nagy, M.; Protti, M.; et al. Treatment of Benign Prostatic Hyperplasia by Natural Drugs. *Molecules* **2021**, *26*, 7141. [CrossRef]
8. Denisow, B.; Denisow-Pietrzyk, M. Biological and therapeutic properties of bee pollen: A review. *J. Sci. Food Agric.* **2016**, *96*, 4303–4309. [CrossRef]
9. Komosinska-Vassev, K.; Olczyk, P.; Kaźmierczak, J.; Mencner, L.; Olczyk, K. Bee Pollen: Chemical Composition and Therapeutic Application. *Evid.-Based Complement. Altern. Med.* **2015**, *2015*, 297425. [CrossRef]
10. Naseri, L.; Khazaei, M.R.; Khazaei, M. Synergic effect of bee pollen and metformin on proliferation and apoptosis of granulosa cells: Rat model of polycystic ovary syndrome. *J. Food Biochem.* **2022**, *46*, e13635. [CrossRef]
11. Kaškonienė, V.; Venskutonis, R.; Čeksteryte, V. Antibacterial Activity of Honey and Beebread of Different Origin Against S. aureus and S. epidermidis. *Food Technol. Biotechnol.* **2007**, *45*, 201–208.
12. Kačániová, M.; Vuković, N.; Chlebo, R.; Haščík, P.; Rovná, K.; Cubon, J.; Dżugan, M.; Pasternakiewicz, A. The antimicrobial activity of honey, bee pollen loads and beeswax from Slovakia. *Arch. Biol. Sci.* **2012**, *64*, 927–934. [CrossRef]
13. Choi, E.-M. Antinociceptive and antiinflammatory activities of pine (*Pinus densiflora*) pollen extract. *Phytother. Res.* **2007**, *21*, 471–475. [CrossRef]

14. Menghini, L.; Ferrante, C.; Leporini, L.; Recinella, L.; Chiavaroli, A.; Leone, S.; Pintore, G.; Vacca, M.; Orlando, G.; Brunetti, L. An Hydroalcoholic Chamomile Extract Modulates Inflammatory and Immune Response in HT29 Cells and Isolated Rat Colon. *Phytother. Res.* **2016**, *30*, 1513–1518. [CrossRef] [PubMed]
15. Kaur, P.; Sandhu, K.S.; Bangar, S.P.; Purewal, S.S.; Kaur, M.; Ilyas, R.A.; Asyraf, M.R.M.; Razman, M.R. Unraveling the Bioactive Profile, Antioxidant and DNA Damage Protection Potential of Rye (*Secale cereale*) Flour. *Antioxidants* **2021**, *10*, 1214. [CrossRef] [PubMed]
16. Cairone, F.; Carradori, S.; Locatelli, M.; Casadei, M.A.; Cesa, S. Reflectance colorimetry: A mirror for food quality—A mini review. *Eur. Food Res. Technol.* **2020**, *246*, 259–272. [CrossRef]
17. Locatelli, M.; Macchione, N.; Ferrante, C.; Chiavaroli, A.; Recinella, L.; Carradori, S.; Zengin, G.; Cesa, S.; Leporini, L.; Leone, S.; et al. Graminex Pollen: Phenolic Pattern, Colorimetric Analysis and Protective Effects in Immortalized Prostate Cells (PC3) and Rat Prostate Challenged with LPS. *Molecules* **2018**, *23*, 1145. [CrossRef]
18. Mitchell, C.M.; Anyalechi, G.E.; Cohen, C.R.; Haggerty, C.L.; Manhart, L.E.; Hillier, S.L. Etiology and Diagnosis of Pelvic Inflammatory Disease: Looking Beyond Gonorrhea and Chlamydia. *J. Infect. Dis.* **2021**, *224*, S29–S35. [CrossRef]
19. Zhanel, G.G.; Zhanel, M.A.; Karlowsky, J.A. Oral Fosfomycin for the Treatment of Acute and Chronic Bacterial Prostatitis Caused by Multidrug-Resistant *Escherichia coli*. *Can. J. Infect. Dis. Med Microbiol.* **2018**, *2018*, 1404813. [CrossRef]
20. Togo, Y.; Ichioka, D.; Miyazaki, J.; Maeda, Y.; Kameyama, K.; Yasuda, M.; Hiyama, Y.; Takahashi, S.; Nagae, H.; Hirota, S.; et al. Oral administration of cernitin pollen extract (Cernilton®) for 30 days might be useful to avoid unnecessary biopsy in prostate biopsy candidates: A preliminary study. *Int. J. Urol.* **2018**, *25*, 479–485. [CrossRef]
21. Wagenlehner, F.M.; Schneider, H.; Ludwig, M.; Schnitker, J.; Brähler, E.; Weidner, W. A Pollen Extract (Cernilton) in Patients with Inflammatory Chronic Prostatitis–Chronic Pelvic Pain Syndrome: A Multicentre, Randomised, Prospective, Double-Blind, Placebo-Controlled Phase 3 Study. *Eur. Urol.* **2009**, *56*, 544–551. [CrossRef] [PubMed]
22. Cai, T.; Verze, P.; La Rocca, R.; Palmieri, A.; Tiscione, D.; Luciani, L.G.; Mazzoli, S.; Mirone, V.; Malossini, G. The Clinical Efficacy of Pollen Extract and Vitamins on Chronic Prostatitis/Chronic Pelvic Pain Syndrome Is Linked to a Decrease in the Pro-Inflammatory Cytokine Interleukin-8. *World J. Men's Health* **2017**, *35*, 120–128. [CrossRef] [PubMed]
23. Shirahama, T.; Sakakura, C. Overexpression of cyclooxygenase-2 in squamous cell carcinoma of the urinary bladder. *Clin. Cancer Res.* **2001**, *7*, 558–561. [PubMed]
24. Moita, E.; Gil-Izquierdo, A.; Sousa, C.; Ferreres, F.; Silva, L.R.; Valentão, P.; Domínguez-Perles, R.; Baenas, N.; Andrade, P.B. Integrated Analysis of COX-2 and iNOS Derived Inflammatory Mediators in LPS-Stimulated RAW Macrophages Pre-Exposed to *Echium plantagineum* L. Bee Pollen Extract. *PLoS ONE* **2013**, *8*, e59131. [CrossRef]
25. Tsikas, D. Assessment of lipid peroxidation by measuring malondialdehyde (MDA) and relatives in biological samples: Analytical and biological challenges. *Anal. Biochem.* **2017**, *524*, 13–30. [CrossRef]
26. Kwatra, M.; Ahmed, S.; Gangipangi, V.K.; Panda, S.R.; Gupta, N.; Shantanu, P.; Gawali, B.; Naidu, V. Lipopolysaccharide exacerbates chronic restraint stress-induced neurobehavioral deficits: Mechanisms by redox imbalance, ASK1-related apoptosis, autophagic dysregulation. *J. Psychiatr. Res.* **2021**, *144*, 462–482. [CrossRef]
27. Peker, E.G.G.; Kaltalioglu, K. Cinnamaldehyde and eugenol protect against LPS-stimulated oxidative stress and inflammation in Raw 264.7 cells. *J. Food Biochem.* **2021**, *45*, e13980. [CrossRef]
28. Imran, M.; Nadeem, M.; Gilani, S.A.; Khan, S.; Sajid, M.W.; Amir, R.M. Antitumor Perspectives of Oleuropein and Its Metabolite Hydroxytyrosol: Recent Updates. *J. Food Sci.* **2018**, *83*, 1781–1791. [CrossRef]
29. Shah, M.-A.; Kang, J.-B.; Park, D.-J.; Kim, M.-O.; Koh, P.-O. Chlorogenic acid alleviates neurobehavioral disorders and brain damage in focal ischemia animal models. *Neurosci. Lett.* **2021**, *760*, 136085. [CrossRef]
30. Kwabi-Addo, B.; Ozen, M.; Ittmann, M. The role of fibroblast growth factors and their receptors in prostate cancer. *Endocr.-Relat. Cancer* **2004**, *11*, 709–724. [CrossRef]
31. Funahashi, Y.; Takahashi, R.; Mizoguchi, S.; Suzuki, T.; Takaoka, E.; Ni, J.; Wang, Z.; DeFranco, D.B.; De Groat, W.C.; Tyagi, P.; et al. Bladder overactivity and afferent hyperexcitability induced by prostate-to-bladder cross-sensitization in rats with prostatic inflammation. *J. Physiol.* **2019**, *597*, 2063–2078. [CrossRef] [PubMed]
32. Sipos, L.; Végh, R.; Bodor, Z.; Zaukuu, J.-L.Z.; Hitka, G.; Bázár, G.; Kovacs, Z. Classification of Bee Pollen and Prediction of Sensory and Colorimetric Attributes—A Sensometric Fusion Approach by e-Nose, e-Tongue and NIR. *Sensors* **2020**, *20*, 6768. [CrossRef] [PubMed]
33. Cesa, S.; Casadei, M.A.; Cerreto, F.; Paolicelli, P. Infant Milk Formulas: Effect of Storage Conditions on the Stability of Powdered Products towards Autoxidation. *Foods* **2015**, *4*, 487–500. [CrossRef] [PubMed]
34. Recinella, L.; Chiavaroli, A.; Masciulli, F.; Fraschetti, C.; Filippi, A.; Cesa, S.; Cairone, F.; Gorica, E.; De Leo, M.; Braca, A.; et al. Protective Effects Induced by a Hydroalcoholic *Allium sativum* Extract in Isolated Mouse Heart. *Nutrients* **2021**, *13*, 2332. [CrossRef] [PubMed]
35. Menghini, L.; Leporini, L.; Vecchiotti, G.; Locatelli, M.; Carradori, S.; Ferrante, C.; Zengin, G.; Recinella, L.; Chiavaroli, A.; Leone, S.; et al. *Crocus sativus* L. stigmas and byproducts: Qualitative fingerprint, antioxidant potentials and enzyme inhibitory activities. *Food Res. Int.* **2018**, *109*, 91–98. [CrossRef]
36. Orlando, G.; Leone, S.; Ferrante, C.; Chiavaroli, A.; Mollica, A.; Stefanucci, A.; Macedonio, G.; Dimmito, M.P.; Leporini, L.; Menghini, L.; et al. Effects of Kisspeptin-10 on Hypothalamic Neuropeptides and Neurotransmitters Involved in Appetite Control. *Molecules* **2018**, *23*, 3071. [CrossRef]

37. Livak, K.J.; Schmittgen, T.D. Analysis of relative gene expression data using real-time quantitative PCR and the $2^{-\Delta\Delta CT}$ Method. *Methods* **2001**, *25*, 402–408. [CrossRef]
38. Charan, J.; Kantharia, N.D. How to calculate sample size in animal studies? *J. Pharmacol. Pharmacother.* **2013**, *4*, 303–306. [CrossRef]

molecules

MDPI

Review

New Insights into Dietary Pterostilbene: Sources, Metabolism, and Health Promotion Effects

Sanjushree Nagarajan [1], Sundhar Mohandas [1], Kumar Ganesan [2], Baojun Xu [3,*] and Kunka Mohanram Ramkumar [1,*]

[1] Department of Biotechnology, School of Bioengineering, SRM Institute of Science & Technology, Kattankulathur 603 203, India
[2] School of Chinese Medicine, LKS Faculty of Medicine, University of Hong Kong, 10 Sassoon Road, Hong Kong 999077, China
[3] Food Science and Technology Programme, Department of Life Sciences, BNU-HKBU United International College, Zhuhai 519087, China
* Correspondence: baojunxu@uic.edu.cn (B.X.); ramkumak@srmist.edu.in (K.M.R.); Tel.: +86-756-3620636 (B.X.); Fax: +86-756-3620882 (B.X.)

Abstract: Pterostilbene (PTS), a compound most abundantly found in blueberries, is a natural analog of resveratrol. Several plant species, such as peanuts and grapes, produce PTS. While resveratrol has been extensively studied for its antioxidant properties, recent evidence also points out the diverse therapeutic potential of PTS. Several studies have identified the robust pharmacodynamic features of PTS, including better intestinal absorption and elevated hepatic stability than resveratrol. Indeed, due to its higher bioavailability paired with reduced toxicity compared to other stilbenes, PTS has become an attractive drug candidate for the treatment of several disease conditions, including diabetes, cancer, cardiovascular disease, neurodegenerative disorders, and aging. This review article provides an extensive summary of the nutraceutical potential of PTS in various disease conditions while discussing the crucial mechanistic pathways implicated. In particular, we share insights from our studies about the Nrf2-mediated effect of PTS in diabetes and associated complications. Moreover, we elucidate the important sources of PTS and discuss in detail its pharmacokinetics and the range of formulations and routes of administration used across experimental studies and human clinical trials. Furthermore, this review also summarizes the strategies successfully used to improve dietary availability and the bio-accessibility of PTS.

Keywords: pterostilbene; resveratrol; antioxidant; bioavailability; cancer; diabetes; Nrf2

Citation: Nagarajan, S.; Mohandas, S.; Ganesan, K.; Xu, B.; Ramkumar, K.M. New Insights into Dietary Pterostilbene: Sources, Metabolism, and Health Promotion Effects. *Molecules* **2022**, *27*, 6316. https://doi.org/10.3390/molecules27196316

Academic Editor: Nour Eddine Es-Safi

Received: 6 September 2022
Accepted: 21 September 2022
Published: 25 September 2022

1. Introduction

Pterostilbene (PTS) (trans-3,5-dimethoxy-4′-hydroxystilbene) is a natural polyphenol and a dimethyl ether analog of resveratrol [1]. PTS is produced by plants as a secondary metabolite that serves to respond to environmental challenges, including UV radiation, drought, fluctuating temperature extremes, grazing pressures, and fungal infections, and PTS serves as an important mediator of disease resistance [2,3]. Similar to resveratrol, PTS also behaves as a phytoalexin, conferring crucial anti-pathogenic defense to plants [4,5]. The daily consumption of PTS is determined by its dietary intake. Based on the type of blueberry ingested, the content of PTS is estimated to range from 99 ng to 520 ng/gram of fruit [6]. Though berries are the most evident source of PTS, it has been reported to be present in various other food sources, including peanuts.

As a natural dietary component, PTS has been documented to exhibit an increased bioavailability compared to other stilbene compounds, which further highlights the need to study the clinical potential of this compound in medical conditions [7,8]. Various evidence has demonstrated the effect of PTS in countering oxidative damage and inflammation, imparting preventive and therapeutic benefits in experimental disease models [4,7,9].

Indeed, through its antioxidant and anti-inflammatory activity, PTS has been reported to regulate pathogenic pathways associated with carcinogenesis, hematologic diseases, neurological disorders, vascular dysfunction, aging disorders, and diabetes [7].

Among the identified polyphenols, resveratrol has been determined to exhibit relatively poor oral bioavailability and undergoes rapid first-pass metabolism. On the contrary, methylated polyphenols such as PTS have been documented to possess better intestinal absorption and elevated hepatic stability. Considering the potential drawbacks that are exhibited due to the unfavorable pharmacodynamics of resveratrol, much focus has shifted towards understanding and characterizing the pharmacokinetics and therapeutic potential of PTS. Overall, this review provides evidence that PTS is a promising, novel, potent, and safe drug candidate for treating various diseases and disorders.

2. Potential Dietary Sources of PTS

Various human diet crops have been documented to produce PTS at varying levels (Table 1). Red sandalwood, also referred to as Heartwood (*Pterocarpus santalinus*), was the first identified source of PTS [3]. Of note, contrary to flavonoids, which are produced by many plants, stilbenes are synthesized by only a few plant species (Table 1). PTS has been reported to be abundantly found in Indian Kino (*Pterocarpus marsupium*), *Guibourtia tessmannii*, and *Vaccinium* spp. berries and at relatively lower levels in the leaves of grape (*Vitis vinifera*) and blueberry fruits [9,10]. The concentration ranges from 9.9 to 15.1 mg/kg of blueberries, 0.2 to 4.7 mg/g of the weight of the skin of fungus-infected grapes, 99 to 151 ng/g dried sample of rabbit-eye blueberry (*Vaccinium ashei*), and 520 ng/g dried sample of deerberries (*Vaccinium stamineum*) [11]. Additionally, peanut (*Arachis hypogaea*) has also been identified as a source of PTS [12] (Table 1). However, it should be noted that the amount of PTS in many of these food sources may be insufficient to provide documented health benefits. Dietary supplements of the formulated pure compound offer a solution to provide sufficient nutritional levels. Moreover, considering the growing interest in PTS as a promising nutraceutical compound, further research that aims to identify the compound in fresh and processed food products by employing standardized extraction and analytical methods is warranted.

Table 1. Potential Sources of Pterostilbene.

Source	Concentration Range	Reference
Blueberries	9.9–15.1 mg/kg of fresh weight	[13]
Blueberries	15 µg/100 g of weight	[11]
Vaccinium berries	99–520 ng/g of dry sample in *Vaccinium ashei* and *V. stamineum*	[11]
Fungal infected grapes	0.2–4.7 mg/g of fresh weight	[14]
Rabbit-eye blueberry	99–151 ng/g of dry sample	[11]
Deerberries	520 ng/g of dry sample	[11]
Peanut	NA	[12]

Biosynthesis and Nutraceutical Availability

The biosynthesis of PTS is facilitated through the conversion of the amino acids phenylalanine or tyrosine, which are products of the shikimate pathway. These amino acids are converted to coumarate and then to p-coumaroyl-CoA, resulting in the production of precursor stilbenes. While stilbene synthase converts precursor stilbenes to resveratrol, an O-methyl transferase further methylates two of the resveratrol hydroxyl groups of resveratrol to form PTS [15].

The availability of PTS in plants was found to vary within species based on various genetic and environmental factors [16]. Interestingly, fungal infection induced the elevated production of PTS in certain food crops, such as grapes [17]. Further, exposure to ultraviolet light amplified the production of resveratrol, while it reduced the production of PTS in grapevines [18]. The concentration of PTS was observed to be relatively high in the fruit skins or epidermal tissues of plants. This could be an evolutionarily conserved effect to protect the plant from harmful microbes that could cause infection through the penetration of the plant epidermis [19]. The variability in the nutraceutical availability within a single plant indicates that different conditions are required to produce resveratrol and PTS. The transgenic alteration of phenolic metabolism has been proposed as a key strategy to improve the yield of PTS from dietary sources. Notably, PTS was produced even by species that do not produce the compound through transgenic alteration. Tobacco (*Nicotiana tabacum* L.) and *Arabidopsis thaliana* (L.) Heynh. were transformed to produce PTS by employing a stilbene synthase transgene from peanut along with an O-methyltransferase transgene from *Sorghum* bicolor (L.) Moench [10]. Of note, the production of PTS in tobacco was accompanied by reduced flavonoid levels, indicating that stilbenes and flavonoids compete for p-coumaroyl-CoA in their biosynthesis pathways.

Additionally, PTS production has been reported to be amplified in food crops that already produce the compound through metabolic engineering. In grapevine cell cultures that were transformed to constitutively express *V. vinifera* O-methyltransferase, PTS production was observed to be elevated [20]. An effort was also made to employ stilbene-synthesizing gut bacteria to provide a constant PTS supply to animals without the dietary ingestion of the compound [21].

3. Analytical Aspects

Stilbenoids are a class of non-flavonoid polyphenolic compounds with a molecular weight of approximately ~200–300 g/mol [22]. Some of the members of the stilbene family include resveratrol, PTS, and 3′-hydroxy PTS [22]. Stilbenoids are characterized by a C6-C2-C6 skeleton and the presence of phenyl groups that are linked by ethene double bonds [22]. With a molecular weight of 256.29 g/mol, PTS is a 3,5-dimethoxy analog of resveratrol [22]. PTS is a methoxybenzene and a diether due to the presence of trans-stilbene with methoxy groups at the 3′ and 5′ positions and a hydroxy group at the 4′ position [22]. Even though PTS exists in both cis and trans structures, it is most abundant in its monomeric, lipid-soluble trans form [9]. Upon the equimolar administration of resveratrol and PTS in rats, systemic exposure and the plasma concentration (higher C_{max} and $AUC_{0\text{–}inf}$) were greater in PTS when compared to resveratrol, whereas the total body clearance of resveratrol was greater than that of PTS [4]. The two methoxy groups in PTS have been identified as responsible for increasing its oral absorption and bioavailability compared to resveratrol [22].

4. Pharmacokinetics

PTS is often consumed in berries, grapes, nuts, and wine. In the form of a dietary polyphenol, PTS has exhibited safety at a high dose of 3 g/kg body weight for 28 days, not leading to any toxicity in mice [23]. Indeed, PTS was found to exhibit dose-dependent pharmacokinetics. An increased intravenous dose (25 mg/kg) reduced the elimination of PTS and was associated with almost twice the rate of reduced clearance due to the saturation of PTS metabolism [24]. When administrating PTS through the oral route, escalating the dosage from 15 mg/kg to 30 or 60 mg/kg doubled the bioavailability (F) and prolonged the mean residence time. This trend is attributed to the absorption and limited elimination of the compound [25]. At a dose of 2.5 mg/kg, rapid absorption and moderate bioavailability were observed with the sublingual administration of PTS [25].

4.1. Absorption

PTS has high membrane permeability due to its specific features, including lipophilicity, low polar surface area, rotatable bonds, and hydrogen-bond acceptors and donors [3]. PTS has poor solubility in water (around 21 μg/mL), which can be overcome by solubilizing it with piperazine at a 2:1 stoichiometric molar ratio. Piperazine–PTS cocrystals had six times greater solubility than PTS alone [26]. Moreover, administering PTS after a meal was observed to increase its oral absorption, as food consumption leads to the acceleration of bile secretion, which enhances the aqueous solubility of drugs co-administered with food [24]. The bioavailability of PTS has also been reported to be enhanced by using it in combination with a 2-hydroxypropyl-β-cyclodextrin (15 mg/kg) solution [25]. Notably, following oral administration, PTS exhibited higher bioavailability than resveratrol, with greater total plasma levels of its metabolites and parent compound [4].

4.2. Distribution

The apparent volume of distribution (measured by the V_{ss} value) of PTS (V_{ss} = 5.3 L/kg) following intravenous dosing in rats was greater than that of the total body water (V_{ss} = 0.7 L/kg), indicating substantial tissue distribution [4]. The distribution of PTS has been found in the liver, kidney, heart, lungs, and brain [27]. The primary metabolites of PTS are the sulfate and glucuronide conjugates [4]. After intravenous administration in male rats, sulfate conjugates seem to be more extensive than glucuronide conjugates [4]. After 1–2 h of oral administration, the entero-hepatic recycling of the metabolite with an increased concentration of glucuronidated PTS was reported [2,28].

4.3. Metabolism

The cytochrome P450 superfamily comprises phase I enzymes, which are in charge of the biotransformation of compounds to reduce their toxicity and increase their polarity, which in turn facilitates the elimination of the drug from the system. Meanwhile, phase-II-mediated enzymes are involved in the biotransformation of xenobiotic metabolites that are the products of phase I metabolism. Notably, phase II detoxification enzymes are crucially involved in promoting drug conjugation and antioxidant reactions [3]. PTS is predominantly cleared through the phase-II-drug-metabolizing pathway by glucuronidation and sulfation [3]. The involvement of this metabolizing pathway is supported by increased concentrations of glucuronide and sulfate PTS conjugates in the systemic circulation when compared to its parent compound form [4].

4.4. Excretion

Around 99% of PTS is excreted through non-renal pathways, while 0.219% goes through the hepatic pathway [8]. A very small fraction of it was found to be excreted in urine [8]. When comparing intravenous doses of resveratrol and PTS in male rats, the clearance of PTS was much less than that of resveratrol, indicating a longer therapeutic availability [3]. Interestingly, the analysis of the urine samples in cannulated rats identified the parent PTS and its glucuronidated metabolite, which had been previously identified only in the systemic circulation [8]. Increasing the dose of PTS from 2.5 mg/kg to 25 mg/kg decreased the clearance rate by almost half, which is attributed to the saturation or partial saturation of PTS metabolism [24].

4.5. Toxicity

The administration of PTS, even in high doses, was observed to be nontoxic in mouse models. Four groups of mice were fed PTS doses ranging from 0 to 3000 mg/kg body weight/day for 4 weeks, and there were no significant alterations in the consumption of food or water or in weight gain [23]. Additionally, the in vivo administration of PTS attenuated tumorigenesis and metastasis with negligible toxicity [29]. Pharmacologically, PTS, when intravenously administered, has been noted to be safe, as the compound did not exhibit toxicity specific to any organ [30]. In humans, PTS has been observed to exhibit safety at doses up to 250 mg/day [31].

5. Major Pathways Associated with PTS

5.1. Antioxidative Pathway: Activation of Nrf2 Signaling

The antioxidant activity of PTS has been extensively studied and implicated in anti-carcinogenesis, the modulation of neurological disorders, the attenuation of vascular diseases, and diabetes management [7]. Extensive evidence has indicated that PTS reduces oxidative stress by attenuating the production of the superoxide anion and hydrogen peroxide, which are implicated in the initiation and progression of various pathogenic processes [7]. Nrf2, a nuclear transcription factor, is one of the major players in regulating cytoprotective and antioxidant genes, which also includes phase II metabolic and antioxidant enzymes [32]. The stimulation of Nrf2 signaling has been identified to produce anti-cancer, anti-diabetic, cardioprotective, and neuroprotective effects [33]. Kelch-like ECH-associated protein 1 (Keap-1) is a negative regulator of Nrf2 and targets the transcription factor for ubiquitylation and degradation. Our lab investigated the protective properties of PTS in pancreatic β-cell apoptosis through an Nrf2-mediated mechanism [32] (Figure 1). We found that PTS activates the Nrf2 pathway, thereby triggering the expression of Nrf2 downstream target genes to facilitate cellular protection in INS-1E cells. In particular, we demonstrated that PTS binds to the arginine residues of Keap-1 and facilitates its disassociation from Nrf2 [34]. Interestingly, PTS has also been reported to mediate the recruitment and interaction of the P62 autophagic cargo with Keap-1, thereby inducing the autophagic degradation of Keap-1 [35]. The major downstream targets activated following PTS administration were antioxidative enzymes, including HO 1, SOD, catalase, and GPX. Moreover, we observed the upregulation of anti-apoptotic gene expression, with the associated downregulation of the expression of the pro-apoptotic mediators Bax (Bcl-2 associated X protein) and caspase-3. Our evidence was strongly suggestive of the protective effect of PTS administration against hyperglycemia-induced oxidative damage in pancreatic β-cells [32] (Figure 1).

Our lab inferred that PTS protects β-cells in Streptozotocin (STZ)-induced diabetic mice, an effect accompanied by the induction of Nrf2 and the consequent upregulation of its target genes. We also investigated the anti-peroxidative role of PTS in the STZ-induced diabetic model. In addition, PTS normalized the circulatory concentration of VLDL and LDL while reducing lipid peroxidation in STZ-induced diabetic mice. Notably, the livers of diabetic mice indicated collapsed hepatic microvesicles on H&E staining due to altered lipid metabolism [36]. PTS administration reduced structural and functional alterations in the hepatic tissue, indicating its protective function in diabetic dyslipidemia mediated via Nrf2 activation [37] (Figure 1).

By countering oxidative damage, PTS treatment inhibited human retinal endothelial cell proliferation and delayed the progression of diabetic retinopathy [38]. In the livers of IUGR piglets, PTS attenuated liver injury caused by Nrf2 activation and the consequent induction of the antioxidant response [39]. Notably, adrenocorticotropic hormone (ACTH) was observed to interfere with Nrf2 signaling in metastatic cells. PTS reduced ACTH activity and was effective against various melanoma cell lines, including MelJuso, A2058, and MeWo [30]. PTS blocked cellular inflammation and oxidative stress in azoxymethane-induced colon carcinogenesis, thereby reducing tumorigenesis. Through the activation of Nrf2, PTS countered the induction of NF-κB (nuclear factor-kappaB) and diminished the levels of oxidative stress mediators, including inducible nitric oxide synthase (iNOS), Cyclo-oxygenase-2 (COX-2), and aldolase reductase in an AOM-induced colon cancer rodent model [40]. Furthermore, by maintaining glutathione, catalase, SOD, and GSH peroxidase activity through the Nrf2–antioxidant response, PTS showed anti-cancer activity in a UVB-stimulated skin cancer model [41]. In the innate immune system, neutrophils produce reactive oxygen species (ROS) to destroy pathogens with the help of NADPH oxidase, which produces a superoxide anion. The overproduction of ROS can cause tissue damage that is observed in diseases such as rheumatoid arthritis and ischemic injury. PTS lowered the neutrophil count in arthritic animals and facilitated a mild decrease in ROS production, with a limited effect on neutrophil activity [42].

In a dose-dependent manner, PTS has been identified to exhibit a potent antioxidant effect against several free radicals, including 2,2-Diphenyl-1-picryl-hydrazyl (DPPH), 2,2′-Azino-bis 3-ethylbenzothiazoline-6-sulfonic acid (ABTS), hydroxyl, superoxide, and hydrogen peroxide. Furthermore, PTS treatment is associated with increased antioxidant enzymes, such as SOD and GPX, via Nrf2 activation in neuronal cells [43]. In Alzheimer's disease models, PTS increased PPAR-α, a modulator of neural antioxidant activities [43].

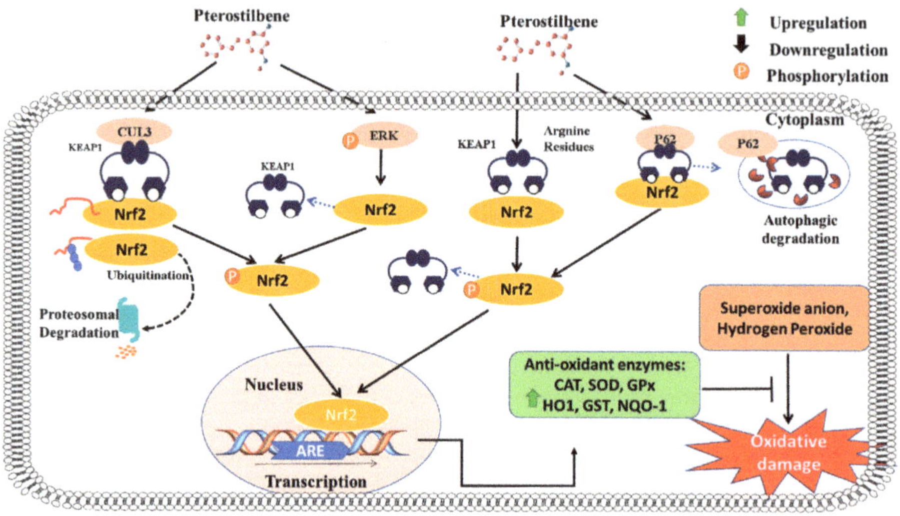

Figure 1. Nrf-2-mediated antioxidant pathway of pterostilbene: Activation and phosphorylation of Nrf-2 signaling is the major mechanism through which the antioxidative response is induced by PTS. Ubiquitination mediated by Cullin-3 (CUL-3) leads to the proteasomal degradation of Nrf-2. PTS inhibits the ubiquitin–proteasome system, thereby increasing the accumulation of Nrf-2. PTS also enables the phosphorylation of Nrf-2, which is critical in the nuclear translocation of the transcription factor. Moreover, PTS phosphorylates and activates the ERK signaling pathway, which mediates the dissociation of Keap-1, resulting in Nrf-2 activation. Furthermore, PTS stimulates the binding of Keap-1 and p62, which enhances the activation of Nrf2. Following its activation and nuclear translocation, Nrf-2 binds to ARE and induces the expression of antioxidant enzymes, which in turn critically attenuate oxidative damage in host cells.

5.2. *Pro- and Anti-Apoptotic Pathways*

PTS inhibited cell proliferation and acted as an active inducer of apoptosis in certain cancerous cell lines [8]. Moreover, PTS treatment induced caspase release and O_2^- production, which depolarizes the mitochondrial membrane, triggering the intrinsic mitochondrial-derived apoptosis of cancerous cells [44,45]. Chakraborty et al. identified that PTS modified markers associated with mitochondrial apoptosis and improved the expression of the antioxidant enzymes GPx, GR, and GSH in an in vitro prostate cancer model [46]. Moreover, PTS induced apoptosis in gastric adenocarcinoma cells through the increased upregulation of cytochrome C, Bad, Bax, and caspases [47]. Genomic analysis revealed that PTS treatment in pancreatic cancer upregulated pro-apoptotic genes and anti-proliferative markers [48].

Interestingly, PTS was reported to inhibit the effects of apoptosis in vascular endothelial cells [49]. Apoptosis induces plaque instability in atherosclerosis, where oxLDL (oxidized low-density lipoprotein) triggers the apoptosis of VEC by activating lectin-like oxLDL receptor-1. PTS inhibits the apoptosis induced by oxLDL and stimulates cytoprotective autophagic cell death in VECs, thereby dampening the atherosclerotic effect of oxLDL [50]. Notably, this effect was achieved by increasing the accumulation of intracellular calcium, followed by the subsequent activation of the AMPK α1 subunit (AMPKα1) [50]. PTS also suppressed the oxidative damage induced by oxLDL by reducing the mitochondrial membrane potential and lowering the levels of pro-apoptotic proteins such as Bax and p53 [49]. PTS administration to cochlear cells obtained from STZ-induced diabetic rats demonstrated the protection of the cochlea from ototoxicity through the inhibition of apoptosis [51]. Interesting work from our lab deduced that cytoprotection by PTS against cytokine-induced cellular damage in MIN6 mouse pancreatic cells involves the activation of Nrf2 signaling, associated with the inhibition of pro-apoptotic signaling through the attenuation of the BAX/Bcl-2 ratio and the reduced activity of caspase-3 [36]. Our evidence indicates that the PTS-mediated anti-apoptotic effect is also a consequence of Nrf-2 activation (Figure 2).

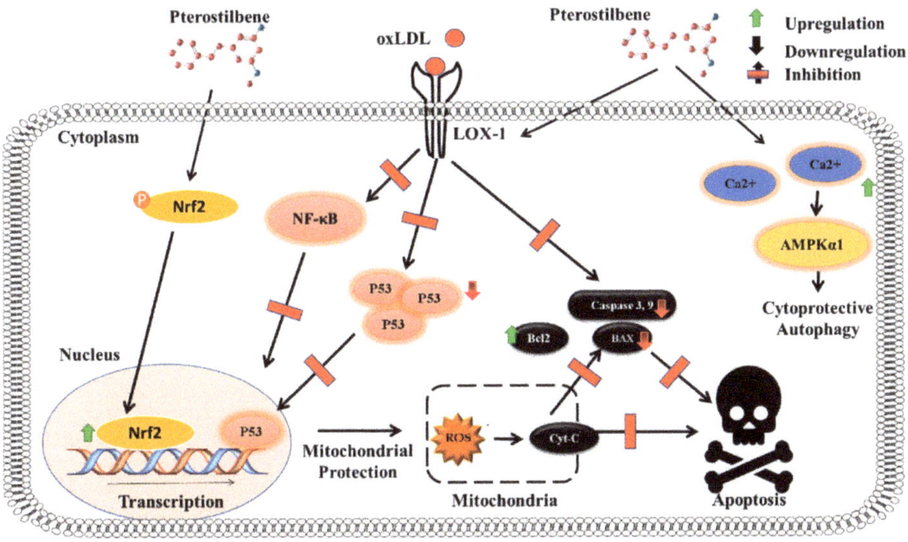

Figure 2. Anti-apoptotic pathway of pterostilbene: PTS mediates its cytoprotective effect mainly through the activation of Nrf-2, which in turn protects the mitochondrial functioning, thereby causing a reduction in the induction of pro-apoptotic factors such as cytochrome C, BAX, and caspase-3. In vascular endothelial cells, PTS protects against the initiation of apoptotic signaling by countering the effect of oxLDL in activating its receptor lectin-like oxidized low-density lipoprotein (LOX-1), thereby preventing the accumulation of P53 as well as the activation of NFκB. Moreover, PTS increases the intracellular calcium levels and promotes the cytoprotective autophagy of the cell, consequently preventing the deleterious effect of apoptosis.

5.3. *Anti-Inflammatory Pathway*

PTS possesses strong anti-inflammatory properties [27], with iNOS, COXs, leukotrienes, NF-κB, Tumor Necrosis Factor Alpha (TNF-α), and Interleukin-1 beta (IL-1β) reported as its primary targets [52]. Endoplasmic reticulum (ER) stress plays a vital role in inducing endothelial cell inflammation. In human umbilical vein endothelial cells (HUVECs) stimulated by TNF-α, PTS attenuated inflammatory cytokine production and inhibited monocyte adhesion. Importantly, PTS treatment also countered the ER-stress-related molecules stimulated by TNF-α [53]. In HT-29 colon cancer cell lines, PTS mediated the anti-inflammatory pathway through the inhibition of the protein kinase cascade activated by p38 mitogen and led to the suppression of pro-inflammatory cytokine production [54]. In canine chondrocytes, treatment with PTS decreased Matrix Metalloproteinase (MMP)-3, sGAG, and TNF-α, thereby exhibiting anti-inflammatory properties [8]. When combined with cyclodextrin, PTS treatment inhibited biofilm formation by *F. nucleatum* in periodontitis [55]. Importantly, PTS was identified to inhibit the NF-κB-induced inflammatory response by preventing its nuclear translocation. Indeed, PTS treatment in the TPA-induced mouse epidermis led to decreased IkappaB kinase (IKK) activity and improved the retention of IkappaBa (IKB-α), which ultimately blocked the nuclear translocation of NF-κB [56]. Another significant pathway through which PTS inhibited inflammation was through the marked attenuation of the transcription factor activator protein-1 (AP-1) by affecting the binding of the c-JUN subunit to AP-1 response elements [56] (Figure 3).

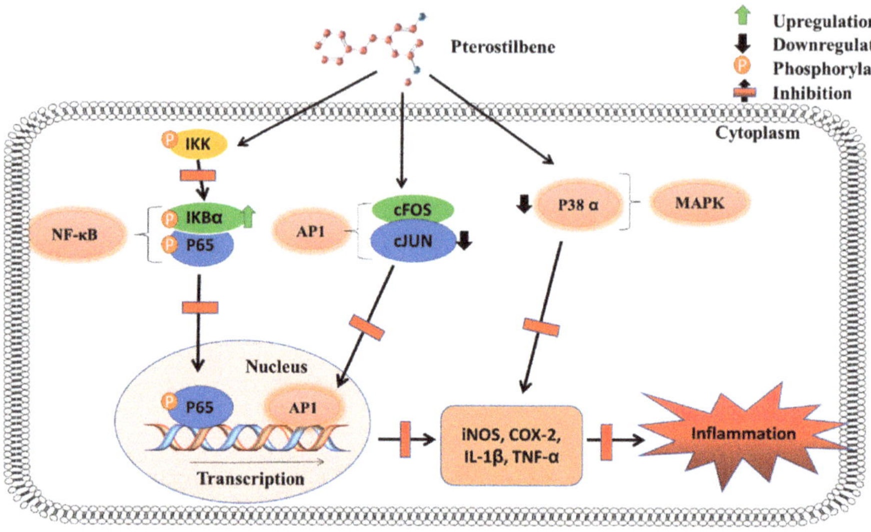

Figure 3. Anti-inflammatory pathway of pterostilbene: PTS mediates its anti-inflammatory effect mainly through the inhibition of the transcription factors NFκB and AP-1, which leads to the attenuation of downstream pro-inflammatory mediators, including TNF-α. Further, PTS also inhibits P38 mitogen-activated protein kinase (MAPK) signaling and prevents the induction of iNOS and COX-2.

While studying the anti-inflammatory effects of PTS on ischemia/reperfusion injury in a middle cerebral artery occlusion (MCAO) rodent model, it was found that treatment with PTS suppressed the swelling and disintegration of cells, the infiltration of macrophages and monocytes, and the degranulation of polymorphonuclear leukocytes, thereby exhibiting a neuroprotective effect through an anti-inflammatory mechanism [57]. Moreover, PTS dampened the astrocyte-mediated inflammatory and oxidative damage caused by ischemia/reperfusion injury through the inhibition of NF-κB [58]. The effect of PTS on lipopolysaccharide-induced pulmonary fibrosis was identified to involve the activation of Keap-1/Nrf2, the inhibition of caspase-dependent A20/NF-κB and NLRP3 signaling pathways, and the suppression of inflammation [59]. Furthermore, PTS exhibited a protective role in arthritis induced by Freund's adjuvant (CFA) in rats by suppressing inflammatory mediators and cytokines [60].

6. Therapeutic Properties of PTS

Various therapeutic properties of PTS have been documented since its discovery. The phytonutrient is reported to have potent anti-cancer, anti-inflammatory, immunomodulatory, anti-diabetic, antioxidant, analgesic, anti-obesity, neuroprotective, and anti-aging properties [9] (Figure 4).

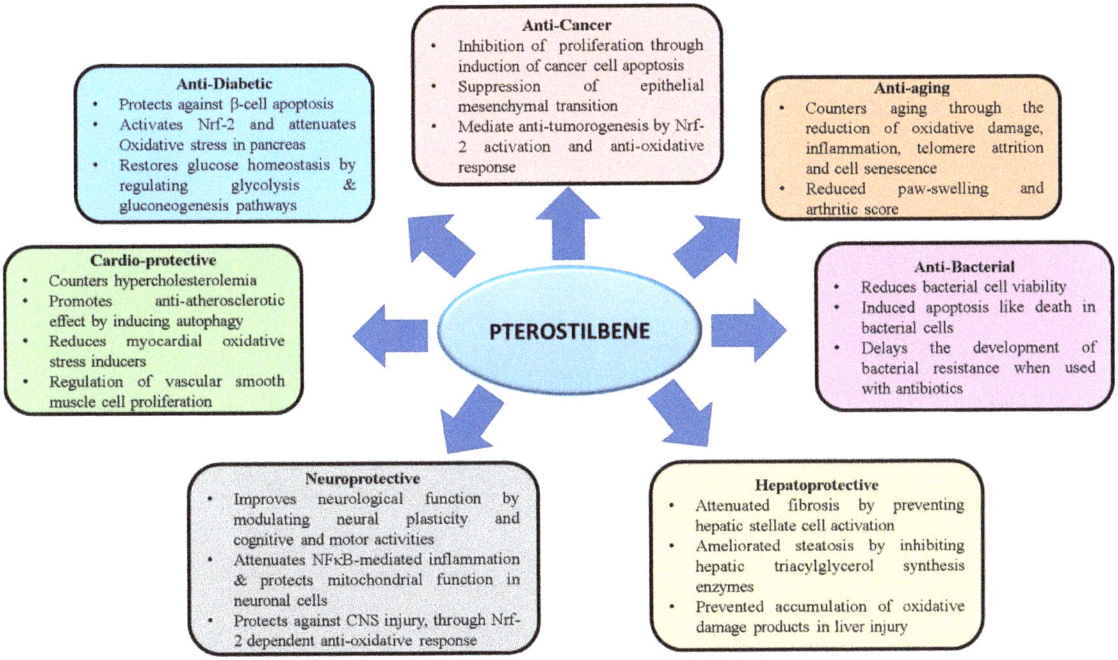

Figure 4. Therapeutic properties of PTS in various disease conditions.

6.1. Anti-Cancer Activity of PTS

Several experimental studies have demonstrated the inhibitory effects of PTS against various cancer cells, including stomach, skin, lung, liver, breast, colon, pancreas, oral, lymph, cervical, endometrial, hematological melanoma, prostate, leukemia, and myeloma tumor cells [29,30] (Table 2). PTS has been identified to be useful in preventing and treating cancer by regulating pro-apoptotic or non-apoptotic anti-cancer activities [33]. The development and progression of cancer involve various factors: a group of drug-metabolizing enzymes, cytochrome P450. These enzymes mediate the metabolic activation of several pro-carcinogens and play a crucial role in the inactivation and activation of anti-cancer drugs. CYP1A1 and CYP1B1 are members of the CYP1 superfamily and have important roles in cancer progression. PTS acts as an efficient inhibitor of CYP1A1, CYP1A2, and CYP1B2 in a competitive manner. The anti-proliferative mechanisms of PTS are seen in different concentrations for different cell types [3].

Table 2. Effect of Pterostilbene in Experimental Models of Various Disease Conditions.

Disease Condition	Experimental Model	Effect of Pterostilbene	Reference
Cancer	Endometrial cancer xenograft model	Reduced weight and volume of tumor	[61]
	DMBA-induced mammary alveolar precancerous lesions in mice	Reduced lesions	[62]
	MDA-MB-231 (breast cancer) xenograft model	Suppressed tumor growth	[63]
	UVB-induced skin cancer in mice	Nrf2-dependent antioxidant response	[41]
	Hematological cancer xenograft model	(a) Increased caspase activation (b) Reduced cell proliferation	[64,65]

Table 2. *Cont.*

Disease Condition	Experimental Model	Effect of Pterostilbene	Reference
	Azoxymethane-induced colon cancer model	Reduced iNOS levels and attenuated crypt formation	[66]
	MIA PaCa-2 xenograft model	Inhibited tumor growth and prominent central necrosis	[48]
	HPV-E6-positive cervical cancer mouse model	(a) Increased apoptosis (b) Downregulated E6 and VEGF expression	[67]
	Breast cancer xenograft mouse model	When coupled with Vitamin E: (a) Inhibited Akt (b) Downregulated cell-cycle proteins	[68]
	Melanoma xenograft mouse model	ACTH downregulation led to decreased Nrf2-mediated defenses	[69]
	Large B-cell lymphoma xenograft mouse model	(a) Reduced mitochondrial membrane potential (b) Increased apoptosis	[64]
	Biliary cancer xenograft mouse model	(a) Inhibited proliferation (b) Induced autophagy	[70]
	Multiple myeloma mouse xenograft	(a) Inhibited cell progression (b) Increased ROS production for apoptosis (c) Improved extracellular-signal-regulated kinases 1/2 and c-Jun N-terminal kinase signaling	[65]
	Xenograft of glioma stem cells in mice	Attenuated GRP78, suppressing tumorogenesis	[71]
Diabetes	STZ-induced diabetic rats	Protected rats from ototoxicity through the inhibition of apoptosis	[51]
	STZ- and Nicotinamide-induced diabetic rats	(a) Increased hepatic glycolytic enzyme hexokinase (b) Reduced the levels of glycogenic enzymes and enhanced peripheral utilization of glucose	[72]
	STZ-induced diabetic rats	(a) Regulated NF-κB signal pathway and inhibited oxidative stress and inflammation (b) Improved renal damage	[73]
	STZ-induced diabetic mouse	(a) Normalized plasma VLDL, LDL, and HDL (b) Reduced lipid peroxidation	[36]
	Diabetic rats	(a) Enhanced the peripheral utilization of glucose (b) Elevated the levels of hepatic hexokinase and hepatic phosphofructokinase	[74]

Table 2. *Cont.*

Disease Condition	Experimental Model	Effect of Pterostilbene	Reference
Liver injury	Dimethyl nitrosamine-induced rats with liver fibrosis	(a) Hepatoprotective activities (b) Inhibited TGF-b1/Smad signaling	[75]
	Zucker rats with liver steatosis	(a) Reduced insulin and hepatic triacylglycerol levels (b) Improved fatty acid profile	[76]
	Hamsters with a High-fat diet supplemented with 8% blueberry by-product	(a) Low plasma LDL (b) Increased phosphorylation of adenosine monophosphate-activated protein kinase	[77]
	Hypercholesterolemic hamsters	Cytoprotective macroautophagy	[7]
	CCl_4-stimulated hepatic fibrosis rat models	(a) Reduced levels of α-smooth muscle cell actin, desmins, MMP2, and MMP9 (b) Downregulated pro-fibrogenesis through the suppression of TGF-β1 in liver tissue	[78]
	Acetaminophen-exposed rats	Suppressed Acetaminophen-induced oxidative stress	[79]
Diseases of the Central Nervous System	Amyloid precursor protein (APP)/Presenilin 1 (PS1) SERCA mouse model with Alzheimer's disease	Reduced amyloid-beta content Improved pathological changes	[80]
	Common carotid artery occlusion mouse model to study cerebral ischemia/reperfusion injury	Upregulated antioxidant activity through HO-1	[81]
	SAMP8-Alzheimer's disease mouse model	Increased peroxisome proliferator-activated receptor-α expression	[82]
	Middle cerebral artery occlusion rodent model	(a) Suppressed the swelling and disintegration of cells and attenuated the infiltration of macrophages and monocytes (b) Attenuated the degranulation of polymorphonuclear leukocytes in neural tissue	[57]
Cardiovascular disease	Monocrotaline-treated rats with reduced cardiac function	(a) Prevented reduction in stroke volume and cardiac output (b) Reduced lipid peroxidation and total phospholamban (c) Increased SERCA expression in the right ventricle	[83]
	Apo-lipoprotein-E-deficient mice	Reduced atherosclerosis by inhibiting lipid peroxidation and enhanced antioxidants	[84]
	Fischer-344 rat model of coronary artery ligation	Reduced myocardial infarction size by 22%	[85]
Arthritis	Bovine type II collagen-stimulated rat arthritis model	Reduced pathological process of arthritis when coupled with physical exercise	[86]
	Freund's adjuvant (CFA)-induced arthritis rat model	(a) Alleviated the swelling of paw and reduced arthritic score (b) Improved body weight	[60]
	Injection of heat-killed *Mycobacterium butyricum* in Lewis rats	Lowered number of neutrophils, thereby downregulating neutrophil oxidative burst	[42]

Rimando et al. studied the cancer chemopreventive activity of PTS using a mouse mammary gland culture model and showed that PTS (ED50 = 4.8 μM) markedly reduced DMBA-induced mammary alveolar precancerous lesions through its peroxy-radical scavenging antioxidant activity [62]. In mutant p53-breast cancer cell lines MDA-MB-231 and T-47D, PTS facilitated the reduction in oncogenic β-catenin, mTOR, and mutant p53, as well as increased the expression of the pro-apoptotic Bax protein [87]. Moreover, in MDA-MB-231 xenograft mouse models, PTS suppressed the epithelial-to-mesenchymal transition (EMT) through the upregulation of miR-205 and the consequent reduction in pro-EMT src signaling [63]. Notably, PTS exhibited additive anti-cancer effects in combination with other natural compounds. A combination of α-tocopherol succinate (42 and 99 IU/kg) and PTS (40 μg/kg) attenuated the invasive capability of MDA-MB-231 cells [68]. The treatment of ER-positive breast cancer with a combination of tamoxifen (5 μM) and PTS (10 and 20 μM) also exhibited additive effects. In the studied cell lines (MCF7 and ZR-751), the suppression of cancer cell proliferation along with an elevation in apoptotic activity was observed when this combination was employed [45].

In the human colorectal adenocarcinoma cell line HT-29, PTS (≥10 μM) inhibited cell proliferation while inducing G1 cell arrest. Moreover, PTS treatment stimulated apoptosis through the attenuation of the STAT3 and AKT kinase signaling pathways [88]. In a rodent model of azoxymethane (AOM)-induced colon cancer, the intake of PTS (40 ppm) through the diet for 45 weeks led to a reduction in tumorigenesis and diminished the levels of proliferating cell nuclear antigen (PCNA), cyclin D1, and β-catenin [89]. PTS was also observed to mediate the anti-cancer effect through the stimulation of Nrf2 signaling and its target genes (HO-1 and GR), which counter the effect of NF-κB-mediated pro-inflammatory signaling. Studies have shown that the overexpression of iNOS and COX-2 is markedly correlated with the progression of colon cancer. Moreover, in an in vitro study using the HT-29 colon cancer model, PTS inhibited the transcriptional expression of augmented iNOS levels and moderated the inhibition of COX-2 in a concentration-dependent manner [40,89].

PTS in combination with quercetin at 20 mg/kg/day inhibited the metastatic activity in B16-F10 melanoma by reducing the adhesion of B16-F10 cells to the endothelium and also downregulated the levels of Bcl-2 in cancerous cells [90]. PTS (10 to 50 μM) suppressed the cancer cell proliferation and initiated apoptotic signaling through the induction of lysosomal membrane permeabilization in A375 melanoma cells [91]. Moreover, through the attenuation of iNOS and COX-2 expression, PTS prevented DMBA- and TPA-induced skin tumor formation [92]. Similarly, in rodent models of UVB-induced skin cancer, the anti-cancer activity of PTS was observed to include the prominent induction of Nrf2-mediated antioxidant signaling, resulting in glutathione level maintenance and the improved activities of catalase, SOD, and GPX [41]. Intravenously administered PTS suppresses human melanoma and pancreatic cancer growth in small animals. Evidence indicates an indirect mechanism of cancer growth inhibition, where PTS inhibits pituitary adrenocorticotropic hormone production, mediates the downregulation of glucocorticoid receptors, and stimulates the Nrf2-dependent cancer antioxidant defense system and the stress-related neuroendocrine signaling mechanism [30,69].

Multiple myeloma (MM) models, xenograft mouse models for hematological cancers, and several diffuse large B-cell lymphoma (DLBCL) models have been utilized to study the anti-cancer effect of PTS [64,65]. In the DLBCL cell line, the viability of the cancer cells was largely dependent on the concentration of PTS and was associated with reduced mitochondrial membrane potential, elevated free-radical generation, and caspase-mediated apoptosis when PTS was intravenously administered [64]. In MM cell lines, a similar concentration-dependent suppression of the proliferation of cancer cells was observed through increased caspase activation, further highlighting the anti-cancer properties of PTS [65]. Additionally, PTS treatment was reported to show benefits against Cholangiocarcinoma (CCA), also known as biliary tract cancer, as evidenced by its cytotoxic effects, mediated through autophagy and the inhibition of CCA tumor growth, in two different CCA cell lines [70].

In endometrial cancer cells, the combination of PTS and megestrol acetate produced a synergistic effect through the inhibition of cell-cycle regulators, including cyclin D1, cyclin B1, and CDK4 [93]. An open-label randomized Phase II clinical trial is underway to study the effect of PTS with megestrol acetate in endometrial cancer patients or patients with endometrial complex atypical hyperplasia who are scheduled for hysterectomy (ClinicalTrials.gov Identifier: NCT03671811) (Table 3). Furthermore, PTS suppresses cell-cycle progression and apoptosis in ovarian cancer cells (OVCAR-8 and Caov-3 cells) through the inhibition of the STAT3 pathway. PTS decreased the expression of cell-cycle and anti-apoptotic proteins involved in the STAT3 pathway, including Mcl-1, Bcl-2, and cyclin D1 [94].

Table 3. Clinical Trials Involving Pterostilbene.

Clinical Trial	Subjects	The Drug Used/Diet	Status	Findings	Reference
Studying the Effects of ElevATP on Body Composition and Athletic Performance of Healthy Individuals	Healthy 18–35-year-old males	ElevATP with and without caffeine and Vitamins	Completed	Improved strength and power output in the lower body	ClinicalTrials.gov Identifier: NCT02819219
Study of Pharmacokinetics and Safety of Basis™ in Subjects with Acute Kidney Injury	Patients with Acute Kidney Injury	Basis	Completed	Nicotinamide riboside with pterostilbene increased whole-blood NAD + levels	ClinicalTrials.gov Identifier: NCT03176628
Effect of PTS on Cholesterol and Hypertension	Patients with hyperlipidemia and increased Blood Pressure	1. PTS (low dosage of 50 mg and high dosage of 125 mg) 2. Low dose combination of PTS and grape extract (50 mg + 100 mg)	Completed	PTS increased LDL and reduced blood pressure in adults [95]	ClinicalTrials.gov Identifier: NCT01267227
Evaluating Safety and Benefits of Basis™ Among Elderly Subjects	60–80-year-old healthy subjects	Basis™	Completed	Yet to be published	ClinicalTrials.gov Identifier: NCT02678611
A Trial of Nicotinamide/PTS Supplement in Amyotrophic Lateral Sclerosis	Patients with Amyotrophic Lateral Sclerosis	EH301 (nicotinamide riboside/PTS)	Recruiting	Yet to be published	ClinicalTrials.gov Identifier: NCT05095571
Effect of Blueberries in Postmenopausal Women with Elevated Blood Pressure for Improving Vascular Endothelial Function	Postmenopausal women with elevated blood pressure, hypertension, and endothelial dysfunction	Blueberry powder	Active Phase 2 Phase 3	Yet to be published	ClinicalTrials.gov Identifier: NCT03370991
Supplementing Wild Blueberries to Study Cardiovascular Health in Middle-aged/Older Men	45- to 70-year-old men with hypertension and endothelial dysfunction	Blueberry powder	Recruiting Phase 1 Phase 2	Yet to be published	ClinicalTrials.gov Identifier: NCT04530916

Table 3. *Cont.*

Clinical Trial	Subjects	The Drug Used/Diet	Status	Findings	Reference
Effect of High-intensity Training and Daily Consumption of Basis™ on Muscle Metabolism and Exercise	18- to 25-year-old healthy men	NRPT (125 mg nicotinamide riboside and 25 mg PTS)	Phase 1	Yet to be published	ClinicalTrials.gov Identifier: NCT04050358
Study of Treating Megestrol Acetate with or without PTS in Patients with Endometrial Cancer Undergoing Hysterectomy	Women with Atypical Endometrial Hyperplasia and Endometrial Carcinoma	Megestrol Acetate with or without PTS	Recruiting Phase 2	Yet to be published	ClinicalTrials.gov Identifier: NCT03671811
Protection of Basis™ in Acute Kidney Injury	Patients with Acute Kidney Injury	Nicotinamide riboside + PTS	Recruiting Phase 2	Yet to be published	ClinicalTrials.gov Identifier: NCT04342975
Studying the Benefit of Supplementation with Short-term Curcumin and Multi-polyphenol	Non-smokers in good health not taking medications or dietary supplements	Polyresveratrol supplementation (100 mg of trans-PTS, 100 mg of curcumin phytosome, 100 mg of quercetin phytosome, 100 mg of green tea phytosome, and 100 mg of trans-resveratrol)	Unknown	Not Published	ClinicalTrials.gov Identifier: NCT02998918
Studying effects of Nicotinamide Riboside and PTS on Muscle Regeneration in Elderly Humans	55- to 80-year- old individuals	Nicotinamide riboside/PTS-NRPT (500 mg/100 mg twice daily)	Completed	Not published	ClinicalTrials.gov Identifier: NCT03754842
Evaluating Sirtuin Supplements to Benefit Elderly Trauma Patients to study recovery of function	Individuals 65 years and older presenting to trauma bay	Nicotinamide riboside and PTS	Unknown	Not published	ClinicalTrials.gov Identifier: NCT03635411
Effects of a Seven-day BASIS™ Supplementation on Menopausal Syndrome, Estradiol levels and Measurements of the Urinary Vitamin B3	Women 35 years or older	BASIS™	Completed	Not published	ClinicalTrials.gov Identifier: NCT04841499

There are scientific reports on the therapeutic potential of PTS in attenuating hepatocellular carcinoma (HCC), which is the second-most prominent cause of cancer-related mortality. In a recent study, PTS treatment was reported to inhibit tumor growth and cell proliferation in a dose-dependent manner in an animal model of HCC [96]. A combination of diethylnitrosamine and carbon tetrachloride was used to induce HCC in mouse livers. PTS treatment was demonstrated to upregulate caspase-3 activity and thereby induce apoptosis in HCC tumor tissue. Interestingly, PTS was identified to reduce HCC proliferation through a reduction in SOD2 and the induction of ROS-mediated mitochondrial apoptotic pathways [96]. Further, it was observed that PTS conferred protection against HCC proliferation and inhibited Hepatitis B virus proliferation in several HCC cell lines. Of note, PTS exhibited antiviral and anti-cancer activity in HCC cells that were resistant to Sorafenib (anti-cancer drug) and Lamivudine, an antiretroviral drug [97]. Importantly, the researchers identified that PTS exhibited anti-cancer and anti-retroviral effects through the potent inhibition of ribonucleotide reductase (RR), which plays a critical role in cellular DNA synthesis. Additionally, PTS treatment was demonstrated to markedly inhibit the growth of an HCC xenograft in nude mice with minimal toxicity [97]. PTS was reported to suppress the invasion and growth of HCC by down-regulating the expression of Metastasis-Associated Protein 1 (MTA1) and histone deacetylase 1 (HDAC1) while upregulating the acetylation of the tumor suppressor protein PTEN [98]. The epigenetic-level regulation by PTS could open new avenues in understanding its anti-cancer activity.

6.2. *Anti-Diabetic Activity of PTS*

Diabetes is a disease characterized by uncontrolled sugar levels due to insufficient secretion and the improper action of insulin, usually disturbing the metabolism of fats, carbohydrates, or proteins [99]. Various rodent models have demonstrated the anti-diabetic effect of PTS (Table 2). The compound has been reported to strongly influence glucose homeostasis by decreasing systemic glucose levels while increasing insulin concentrations [3,100]. Indeed, the findings from our lab indicated that PTS treatment markedly regulated blood glucose by improving insulin secretion in STZ-induced diabetic mice [37]. In particular, we observed that PTS-mediated glucose regulation is achieved by regulating glucose metabolism enzymes in the liver of STZ-induced diabetic mice [37]. The oral administration of PTS to diabetic rats elevated the levels of the hepatic glycolytic enzyme hexokinase, reduced the levels of glycogenic enzymes glucose-6-phosphatase and fructose-1,6-bisphosphatase, and thereby improved the peripheral utilization of glucose [72]. PTS also improved the antioxidant capacity in diabetic rats by upregulating GST, SOD, GPX, and catalase levels and counteracting ROS accumulation [100]. These mechanisms protect renal and hepatic cells from the deleterious effects of hyperglycemia-induced oxidative stress. Hence, PTS exhibits anti-diabetic activity by reducing hyperglycemia, but it also protects liver and kidney cells from hyperglycemia-associated damage [100].

While investigating PTS-mediated protection against β-cell apoptosis in STZ-induced diabetic rodents, our lab demonstrated that PTS treatment improved glucose homeostasis while attenuating the pro-inflammatory cytokine response. The cytoprotection of β-cells by PTS treatment was conferred by an Nrf2-mediated mechanism, as evidenced by the attenuation of caspase-3 activity and the BAX/Bcl-2 ratio (Figure 2). We also found the inhibition of iNOS and reduced nitric oxide (NO) synthesis in the diabetic pancreas. PTS significantly alleviated the function of pancreatic β-cell cells and improved their survival in the background of cytokine stress, thereby preventing the pathogenic features of STZ-induced diabetes [36]. Furthermore, our proteomic study demonstrated the molecular mechanisms involved following PTS administration in diabetic rodents by employing electrospray ionization tandem mass spectrometry (LC-MS/MS). Our findings indicated that the administration of PTS normalized the levels of 315 proteins that were modulated in diabetic mice. Outstandingly, a major proportion of these proteins were involved in the regulation of redox imbalance, the antioxidative stress response, the unfolded protein response, and ER degradation pathways, indicating that PTS treatment plays a crucial role in the rehabilitation of defective metabolic processes and stress sensors in diabetes [101].

Lipid peroxidation is a characteristic of diabetes, and lipid peroxidation products can damage DNA and contribute to extra-pancreatic tissue damage in diabetes. PTS significantly reduced lipid peroxidation levels and was reported to scavenge DPPH free radicals and peroxyl radicals (ROO*). In a tertiary-butyl hydroperoxide (TBHP)-induced oxidative damage rodent model, free radicals, including hydroxyl, superoxide, and hydrogen peroxide, were attenuated by PTS in a concentration-dependent manner [102]. Notably, in diabetic-nephropathy-induced rats, PTS ameliorated renal damage by dampening the NF-κB inflammatory signaling pathway and inhibiting oxidative stress [73]. Diabetic retinopathy is associated with pathogenic alterations in the structure of the retina, mediated through high glucose levels. PTS was observed to reduce ROS generation while increasing the SOD levels to scavenge free radicals and thereby suppress diabetic retinopathy [38].

6.3. Therapeutic Effect of PTS in Liver Diseases

Liver fibrosis results from the overt and chronic accumulation of extracellular matrix proteins, resulting in the scarring of hepatic tissue and the marked disruption of hepatic vasculature, which can ultimately lead to cirrhosis [103]. Along with an increased risk of mortality, cirrhosis is also a risk factor for developing hepatocellular carcinoma [103]. Studies employing both acute and chronic liver injury models have identified the alleviation of liver injury post-PTS administration (Table 2). Of note, PTS treatment administered to a dimethylnitrosamine (DMN)-induced liver fibrosis model in Sprague-Dawley rats reduced DMN-induced changes and attenuated pro-fibrogenic hepatic stellate cell activation. PTS also exhibited hepatoprotective activity through the inhibition of TGF-b1/Smad signaling [75].

NAFLD is a chronic progressive liver disorder in metabolic syndromes such as obesity and insulin resistance caused by excessive fat accumulation (hepatic steatosis). PTS administration in Zucker rats showed reduced insulin resistance and attenuated hepatic triacylglycerol levels, thus reducing liver steatosis. Importantly, treatment with PTS reduced hepatic steatosis from grade 2 to grade 1. Of note, PTS was observed to reduce the triacylglycerol synthesis capacity of the liver through a reduction in fatty acid disposal and through the inhibition of triacylglycerol synthesis enzymes such as DGAT2. The PTS-treated rats had improved fatty acid profiles, attributed to its delipidating effect [76]. Furthermore, PTS and its derivative 3′-Hydroxy-pterostilbene reduced NAFLD pathogenesis induced by free fatty acids and a fat-rich diet through the upregulation of SIRT1/AMPK and insulin signaling pathways and the downregulation of the protein expression of SREBP-1, which results in the activation of the β-oxidation of fatty acids and the consequent reduction in fatty acid synthesis. Moreover, PTS was also observed to promote the growth of vital beneficial microbiota, such as Oscillospira, while down-regulating the population of potentially pathogenic bacteria, such as Allobaculum, Phascolarctobacterium, and Staphylococcus [104].

Zhang et al. employed an IUGR-induced liver injury model and demonstrated increased circulating alanine transaminase activity, an elevated hepatocyte apoptosis rate, and ROS generation and accumulation. PTS administration reduced these pathogenic processes by preventing the accumulation of hepatic superoxide anions, 8-hydroxy-2 deoxyguanosine, and 4-hydroxynonenal-modified protein by stimulating the translocation of Nrf2 to the nucleus and inducing the antioxidant enzyme SOD2 [39]. Further, in a study investigating the efficacy of PTS against obesity, it was found that PTS formed three hydrogen bonds with the amino acids of PPAR-α, thereby inducing its expression in the livers of BBPX hamsters and lowering the plasma LDL concentration. Furthermore, PTS was also observed to regulate fatty acid oxidation by dose-dependently elevating the phosphorylation of 5′-AMPK [77].

6.4. Effects of PTS on Diseases of the Central Nervous System

The antioxidant and anti-inflammatory properties of PTS have been reported to be therapeutic for the aging brain. Evidence from experimental studies indicates that PTS confers protective benefits against Alzheimer's disease (AD) and vascular dementia [105] (Table 2). PTS treatment potently modulated cognitive impairment and cellular stress. This effect was closely linked to the presence of methoxy groups, which increases lipophilicity. It also positively modulates cellular stress markers by upregulating PPAR-α expression [82].

Joseph et al. administered PTS to rats at various doses and measured its concentration in blood plasma and brain tissue (hippocampus). The amount of PTS in the hippocampus was directly related to the intake of PTS and alleviated cognitive function through the modulation of neural plasticity and motor activity. When administered at high doses, the compound was detected in the serum and brain tissue; however, low doses were only found in the serum and not detected in brain tissue [105]. Dose studies are warranted to further understand the threshold dosage for PTS to cross the BBB.

The induction of the NF-kB signaling pathway is a vital pathogenic component in neurodegenerative diseases. Through the downregulation of NF-kB, PTS limited the inflammatory response in the CNS [82]. Cerebral ischemia/reperfusion injury is a period of impaired blood supply to the brain during an ischemic stroke. Five days of PTS treatment (10 mg/kg) in a common carotid artery occlusion model markedly elevated the membrane potential of mitochondria and induced cytochrome c expression, as well as complex I and IV activity. PTS attenuated the ROS generated by mitochondria and reduced the cytochrome c levels in the cytosol. Considering that HO-1 signaling exhibits protection in Parkinson's, Alzheimer's, and other neurodegenerative diseases, the upregulation of HO-1 expression by PTS exhibited cerebral protective effects [81].

PTS confers neuroprotection to neuronal Sh-SY5Y cells by reviving estrogen-receptor-α-induced signaling [106]. A reduction in high-glucose-induced CNS injury and mitochondrial-dysfunction-derived oxidative stress was observed upon PTS administration due to the activation of Nrf2 in hippocampal neuronal cells [107]. However, high doses of PTS or resveratrol inhibited the physiological immune response to pathogens [108–110]. Further dose-dependent studies are required to identify the appropriate PTS dose to achieve the therapeutic effect.

6.5. Effects of PTS on Cardiovascular Diseases

Hypercholesterolemia is associated with an increased risk of cardiovascular diseases. PTS treatment was demonstrated to reduce atherosclerosis and myocardial infarction in animal models of cardiovascular diseases (Table 2). PTS treatment lowered plasma lipoproteins and cholesterol, protecting vascular endothelial cells from oxidation and promoting cytoprotective macroautophagy [7]. pTeroPure, a highly purified trans-PTS patented by Chromadex, Irvine, CA, has been proven to significantly reduce blood pressure in adults [2]. The combination of PTS and hydroxypropyl-β-cyclodextrin improved cardiac function in an experimental monocrotaline (MCT)-induced/arterial-hypertension-provoked right-heart-failure model through the induction of the antioxidative response. In particular, PTS enabled the rehabilitation of glutathione metabolism and restored redox homeostasis in the right ventricle of MCT-treated rats. At higher doses, PTS attenuated lipoperoxidation and total phospholamban while increasing the levels of sarcoplasmic reticulum calcium ATPase (SERCA) in the right ventricles of diseased rodents [83].

An elevation of mechanical stress in the endothelium puts the heart at risk of injury to its vasculature and thrombogenesis, which is worsened by oxidative stress. The endogenous antioxidative response of the vascular system is responsible for exerting a protective effect by attenuating oxidative damage; however, the antioxidant capacity may become exhausted due to increased and chronic exposure to ROS, creating an imbalance between oxidant and antioxidant activities. In an ischemia/reperfusion-induced myocardial damage experimental model, PTS showed a cardioprotective effect by reducing myocardial peroxynitrite, superoxide production, malondialdehyde content, and NADPH oxidase enzyme expression and by increasing the antioxidant SOD activity to protect against oxidative stress [111].

The unchecked proliferation of vascular smooth muscle cells leads to atherosclerosis and the consequent development of vascular stenosis [112]. In atherosclerosis, PTS has been reported to exhibit protective effects through the modulation of vascular smooth muscle cells (VSMCs) and VECs through the blocking of an Akt (a serine/threonine kinase)-dependent pathway. In a platelet-derived growth factor (PDGF)-BB-induced VSMC proliferation model, PTS treatment downregulated the promoters of DNA synthesis and VSMC proliferation, including cyclin-dependent kinase (CDK)-2, CDK-4, cyclin E, cyclin D1, retinoblastoma (Rb), and proliferative cell nuclear antigen (PCNA) [113].

6.6. Effects of PTS on Aging

Polyphenols have been extensively documented to protect against aging and age-related diseases such as atherosclerosis, arthritis, cataracts, osteoporosis, diabetes, and neurodegenerative and cardiovascular disorders. Studies indicate that PTS acts as an anti-aging agent by regulating hallmark features, including oxidative damage, inflammation, telomere attrition, and cell senescence [43] (Table 2). Owing to its ability to cross the BBB, PTS can localize within the brain and provide potential therapeutic benefits against age-related neurodegenerative disorders [108]. Indeed, PTS countered lipopolysaccharide-induced microglial activation in rodents and ameliorated learning and memory impairments [114]. PTS was also demonstrated to effectively reverse aging-associated behavioral deficits in rats. Indeed, the concentration of PTS in the rat hippocampus was directly correlated with dopamine release and working memory [105].

Employing the SAMP8 mouse, which is increasingly being recognized as an effective model of accelerated aging in the background of sporadic and age-related AD, Chang et al. demonstrated that dietary doses of PTS exhibited more potency when compared to resveratrol in modulating cognitive behavior and cellular stress [82]. Notably, the study attributed the anti-aging effect to the activation of PPAR-α. Moreover, PTS was also reported to extend the lifespan of SAMP8 mice, an effect attributed to c-Jun N-terminal protein kinase inhibition [115].

Ocular surface inflammation is a multifactorial disease that is particularly prevalent among the elderly. PTS has been reported to restore the imbalance between oxygenases and antioxidative enzymes through the attenuation of COX-2 and the upregulation of SOD1 and peroxiredoxin-4 (PRDX4) activities in the background of hyperosmotic stress [116]. PTS treatment is associated with a reduction in oxidative damage mediators, including malondialdehyde (MDA), 4-hydroxynonenal (4-HNE), aconitase-2, and 8-hydroxydeoxyguanosine (8-OHdG) levels, in a human corneal epithelial cell model induced by hyperosmotic medium stress [116]. Blueberry consumption has been reported to prolong the lifespan and improve thermo-tolerance in C. elegans [117]. In D. melanogaster, blueberry extracts upregulated the expression of the antioxidant enzymes SOD and catalase, which were mainly attributed to lifespan extension [118]. Dietary supplementation with blueberries, which contain polyphenols such as PTS, alleviated the damaging effect of aging on motor behavior and neuronal signaling and lowered the amyloid-beta content in a transgenic AD rodent model [119]. In an open-label, single-arm, monocentric study investigating the efficacy of Pon skin brightening and PTS anti-aging, a cream formulation containing 0.4% PTS was highly effective in reducing aging markers and brightening the skin tone of study participants. Furthermore, employing an in vitro experiment, PTS was reported to exhibit anti-tyrosinase activity and inhibit melanogenesis, which could have contributed to the reduction in the markers of skin aging [120].

Arthritis is characterized by the painful swelling of joints, which worsens with age. In a Freund's adjuvant (CFA)-induced arthritis model in rats, PTS significantly reduced paw swelling, the arthritic score, and body weight. Interestingly, it also helped restore the healthy gut microbiota ecosystem by reducing the relative abundance of Helicobacter, Desulfovibrio, Lachnospiraceae, and Mucispirillum. Considering the evidence of PTS in suppressing inflammation through intestinal bacteria alterations, studies investigating its therapeutic potential against inflammatory bowel disorders could be of clinical value [60].

6.7. Antibacterial Effect of PTS

Stilbene compounds, including PTS, are natural antibacterial agents due to their low hydrophilicity, enabling them to penetrate hydrophobic biological membranes [121]. PTS with cyclodextrin exerts antimicrobial effects by inducing bacterial cell content leaks, resulting in a reduction in bacterial cell viability. It also inhibits *F. nucleatum* biofilm formation, making it a potential candidate for treating periodontitis [55]. Bacillus cereus, a foodborne pathogen contaminating uncooked food, was tested with PTS. Following treatment, apoptosis-like cell death (ALD) was induced and increased intracellular ROS in bacterial cells. Additionally, an improvement in the beneficial gut microbiota Bacteroidetes was also documented [122].

PTS, along with gentamicin, was tested against six Gram-positive and Gram-negative bacteria, and the combination was found to be synergistic against three susceptible strains, *Staphylococcus aureus* ATCC 25923, *Escherichia coli* O157, and *Pseudomonas aeruginosa* 15442. However, no significant difference was observed from gentamicin treatment alone. Bacterial growth was fully diminished after 2–8 h treatment with PTS and gentamicin, exhibiting the potential to delay the development of bacterial resistance by utilizing lower concentrations of antibacterial agents [123]. Methicillin-resistant S. aureus (MRSA) is a multi-drug-resistant S. aureus strain, whose biofilm thickness was reduced from 18 to 10 μm when treated with PTS. Topical administration ameliorated the abscess formation induced by MRSA, thereby lowering the bacterial burden and improving the architecture of the skin [124]. PTS has also been used to treat infections with Staphylococcus spp. or Enterococcus faecalis in biofilms due to a reduction in the growth capacity of Gram-positive cocci [121].

6.8. Therapeutic Effects of PTS against COVID-19 Infection

The COVID-19 pandemic, caused by severe acute respiratory syndrome coronavirus 2 (SARS-CoV2), triggered a major setback to global human health and the economy in the 21st century [125]. Resveratrol was tested and proved to have effective therapeutic value against MERS-CoV infection by decreasing cell death [126]. Screening studies have indicated that stilbenes inhibit complex formation between the spike protein and ACE-2 receptor, thereby blocking viral entry into the host cell [127]. Based on these studies, stilbene derivatives could be considered important drug candidates for COVID-19 [127]. PTS has been demonstrated to actively inhibit SARS-CoV-2 virus replication in infected African green monkey kidney cells. Antiviral activity was seen for up to five rounds of replication, which indicates a long-lasting therapeutic effect.

Moreover, in human primary bronchial epithelial cells isolated from healthy volunteers, PTS showed an antiviral effect for up to 48 h after infection. These data promote the use of PTS as a potent drug against COVID-19 and warrant further clinical trials to prove its antiviral efficacy early in COVID-19 [125]. Furthermore, PTS, co-administered with zinc, has also been identified as a potential COVID-19 adjuvant therapy for managing moderate–severe disease [128]. However, further clinical trials are required to back the pharmacotherapeutic potential of PTS. COVID-19 patients with comorbidities related to metabolic syndromes, such as diabetes, obesity, hypertension, and cardiovascular disease, have low levels of the HO antioxidant enzyme. Higher HO-1 expression has been associated with reduced susceptibility to COVID-19 infection [129]. Considering that COVID-19 patients are susceptible to the induction of overt inflammatory processes and cytokine storms, PTS treatment could exert anti-inflammatory and cytoprotective effects by increasing HO-1 expression [130].

7. Enhancement of PTS Bioavailability

The phenolic group at the 4′ position of PTS is an attractive target for phase 2 conjugative enzymes, which negatively impacts its bioavailability. PTS is also insoluble in water, limiting its concentration in the aqueous environment [77]. Safe, bioavailable formulations of PTS can help increase its bioavailability in the body and reduce known and unknown side effects [30,131]. Some of the ongoing trials for PTS include using dietary blueberry supplements, 250 mg of Nicotinamide Riboside, and 50 mg of PTS, commercially known as Basis™, developed by Elysium, or ElevATP, a combination of polyphenols developed by FutureCeuticals (Table 3). Further clinical trials are warranted to improve the therapeutic bioavailability and optimal administration methods of PTS.

Crystal engineering has been a promising field for bioavailability enhancement due to the advantageous, unique solubility and dissolution rates of drugs [132]. Formulations of cocrystals can improve the dissolution profile and bioavailability of PTS with improved performance. Notably, picolinic acid, an endogenous L-tryptophan metabolite, exhibits neuroprotective, immunological, and anti-proliferative properties. The oral administration of PTS–picolinic acid cocrystals to rats revealed a 10-fold amplification of the bioavailability of PTS when compared to solid oral forms [131]. PTS–caffeine cocrystal solubility was also observed to have over 27 times higher solubility [132].

In the form of a prodrug, a bis (hydroxymethyl) propionate-based analog of PTS exhibited superior tumor inhibitory activity in cisplatin-resistant oral squamous (CAR) cancer cell lines. An increase in absorption, a reduction in metabolism, and the maintenance of high concentrations of PTS were some of the observed benefits of prodrug administration. The concentration of PTS in the blood was significantly higher when the amino acids isoleucine or β-alanine were utilized. Oral gavage administration of PTS with megestrol acetate at 10 mg/kg/day reduced the tumor weight and volume in endometrial cancer rats [61].

Notably, the chronic administration of PTS facilitated burn-wound healing in diabetes, facilitated by a significant reduction in diabetes-induced oxidative stress and the suppression of hypoxia-induced factor1α (HIF1α) activity [130]. Liposome-engulfed PTS is also efficient for the topical administration of the drug [41]. Poly (3-acrylamidophenyl boric acid-b-PTS) [p(AAPBA-b-PTE)] is a round nanoparticle of PTS, with a size ranging from 150 to 250 nm. Notably, they possess an appropriate pH and sensitivity to glucose. In vitro and in vivo studies showed that PTS nanoparticles were nontoxic and safe. Upon administration, these nanoparticles were observed to reduce glucose levels, reverse micro-inflammation, and improve the antioxidant profile in mice [133]. Further, antibody-4arm-polyethylene glycol-PT conjugate, as an antibody–drug conjugate, was used for the targeting and co-delivery of drugs to tumors [134]. A composition with Zein/fucoidan composite nanoparticles as carriers of PTS was prepared using an anti-solvent precipitation method. The findings indicated that this method provided a better-controlled release and could be utilized as a potential carrier for the protective encapsulation of PTS [135]. Poly(2-oxazoline)–PTS block-copolymer nanoparticles are also being used for dual anti-cancer drug deliveries in chemotherapy [136].

Solubilizing PTS in 2-hydroxypropyl-b-cyclodextrin (HP-b-CD) improved its bioavailability by 3.7 times [25]. Furthermore, a lipid-based encapsulation system has been used to enhance the stability of PTS in the aqueous phase. In particular, encapsulation in nano-emulsions made of flaxseed oil and olive oil was studied by Sun et al. The bio-accessibility was high in the flaxseed-oil nano-emulsion. However, greater amounts of intact PTS were transported across intestinal enterocytes in olive-oil nano-emulsions [137]. The oral administration of zinc pectinate beads, prepared through the ionic gelation method, successfully distributed PTS to the colonic tissue [138]. Interestingly, physical exercise promoted the biosynthesis of the anti-inflammatory mediator MaR1, leading to a preventive effect in rheumatoid arthritis. The oral administration of PTS, when accompanied by moderate physical activity, attenuated the pathological process of rheumatoid arthritis in a bovine type II collagen (BIIC)-stimulated rat model [86].

8. Conclusions and Future Prospects

Numerous studies that have evaluated PTS for its therapeutic potential have demonstrated its role as a promising candidate drug for health benefits in a broad spectrum of disease conditions. Various experimental studies have confirmed that PTS has anti-cancer, anti-diabetic, anti-hypertensive, antimicrobial, anti-aging, anti-atherosclerotic, and neuroprotective properties. PTS has been documented to exert its beneficial effects mainly by modulating antioxidant, anti-apoptotic, and anti-inflammatory pathways. Of particular interest, the activation of the Nrf2 signaling pathway by PTS has been an essential focus of our lab. Our findings indicate its potential as a promising drug candidate for diabetes and associated complications. Even at higher doses, PTS did not exhibit toxicity in animal studies, providing further encouragement to explore the use of the compound in more human clinical trials. However, considering that some of the studies have employed the co-administration of PTS in combination with other compounds to improve its therapeutic efficiency, the potential effect of drug interactions should be considered.

Although PTS has been identified to exert marked therapeutic benefits, most findings have been proven only in experimental models. Human clinical trials have been largely limited due to the less-than-desired bioavailability of the compound. To overcome this limitation, various strategies have been implemented, which involve modifying the administration routes and formulations of PTS, including cocrystals, prodrugs, nanoparticles, lipid-based encapsulation, and beads. The potential effects of many drug interactions with PTS remain unclear. Of note, the mode of administration of PTS seems to play an important role in its bioavailability, as the administration of intravenous doses shows a higher distribution when compared to oral intake. Further research that carefully considers the dose, drug interactions, administration route, disease-specific formulations, and short- and long-term biomedical implications is warranted before the clinical adoption of this promising natural compound.

Author Contributions: Conceptualization, K.M.R., B.X., K.G., S.N. and S.M.; methodology and analysis, S.N. and S.M.; writing—original draft preparation, S.N. and S.M.; writing—review and editing, K.M.R., S.M., K.G. and B.X.; supervision, K.M.R. and B.X. All authors have read and agreed to the published version of the manuscript.

Funding: This research was funded by the Science and Engineering Research Board (SERB) (Grant No. EMR/2016/006196), Government of India.

Institutional Review Board Statement: Not applicable.

Informed Consent Statement: Not applicable.

Data availability Statement: Not applicable.

Acknowledgments: The authors gratefully acknowledge the Science and Engineering Research Board (SERB) (Grant No. EMR/2016/006196), Government of India, for financial assistance.

Conflicts of Interest: The authors declare no conflict of interest.

References

1. Wang, P.; Sang, S. Metabolism and Pharmacokinetics of Resveratrol and Pterostilbene. *Biofactors* **2018**, *44*, 16–25. [CrossRef] [PubMed]
2. Estrela, J.M.; Ortega, A.; Mena, S.; Rodriguez, M.L.; Asensi, M. Pterostilbene: Biomedical Applications. *Crit. Rev. Clin. Lab. Sci.* **2013**, *50*, 65–78. [CrossRef] [PubMed]
3. Kosuru, R.; Rai, U.; Prakash, S.; Singh, A.; Singh, S. Promising Therapeutic Potential of Pterostilbene and Its Mechanistic Insight Based on Preclinical Evidence. *Eur. J. Pharmacol.* **2016**, *789*, 229–243. [CrossRef] [PubMed]
4. Kapetanovic, I.M.; Muzzio, M.; Huang, Z.; Thompson, T.N.; McCormick, D.L. Pharmacokinetics, Oral Bioavailability, and Metabolic Profile of Resveratrol and Its Dimethylether Analog, Pterostilbene, in Rats. *Cancer Chemother. Pharmacol.* **2011**, *68*, 593–601. [CrossRef]
5. Langcake, P.; Cornford, C.A.; Pryce, R.J. Identification of Pterostilbene as a Phytoalexin from Vitis Vinifera Leaves. *Phytochemistry* **1979**, *18*, 1025–1027. [CrossRef]
6. Rodríguez-Bonilla, P.; López-Nicolás, J.M.; Méndez-Cazorla, L.; García-Carmona, F. Development of a Reversed Phase High Performance Liquid Chromatography Method Based on the Use of Cyclodextrins as Mobile Phase Additives to Determine Pterostilbene in Blueberries. *J. Chromatogr. B Anal. Technol. Biomed. Life Sci.* **2011**, *879*, 1091–1097. [CrossRef]
7. McCormack, D.; McFadden, D. A Review of Pterostilbene Antioxidant Activity and Disease Modification. *Oxid. Med. Cell Longev.* **2013**, *2013*, 575482. [CrossRef]
8. Remsberg, C.M.; Yáñez, J.A.; Ohgami, Y.; Vega-Villa, K.R.; Rimando, A.M.; Davies, N.M. Pharmacometrics of Pterostilbene: Preclinical Pharmacokinetics and Metabolism, Anticancer, Antiinflammatory, Antioxidant and Analgesic Activity. *Phytother. Res.* **2008**, *22*, 169–179. [CrossRef]
9. Chan, E.W.C.; Wong, C.W.; Tan, Y.H.; Foo, J.P.Y.; Wong, S.K.; Chan, H.T. Resveratrol and Pterostilbene: A Comparative Overview of Their Chemistry, Biosynthesis, Plant Sources and Pharmacological Properties. *J. App. Pharm. Sci.* **2019**, *9*, 124–129. [CrossRef]
10. Rimando, A.M.; Pan, Z.; Polashock, J.J.; Dayan, F.E.; Mizuno, C.S.; Snook, M.E.; Liu, C.-J.; Baerson, S.R. In Planta Production of the Highly Potent Resveratrol Analogue Pterostilbene via Stilbene Synthase and O-Methyltransferase Co-Expression. *Plant Biotechnol. J.* **2012**, *10*, 269–283. [CrossRef]

11. Rimando, A.M.; Kalt, W.; Magee, J.B.; Dewey, J.; Ballington, J.R. Resveratrol, Pterostilbene, and Piceatannol in Vaccinium Berries. *J. Agric. Food Chem.* **2004**, *52*, 4713–4719. [CrossRef] [PubMed]
12. Sobolev, V.S.; Khan, S.I.; Tabanca, N.; Wedge, D.E.; Manly, S.P.; Cutler, S.J.; Coy, M.R.; Becnel, J.J.; Neff, S.A.; Gloer, J.B. Biological Activity of Peanut (Arachis Hypogaea) Phytoalexins and Selected Natural and Synthetic Stilbenoids. *J. Agric. Food Chem.* **2011**, *59*, 1673–1682. [CrossRef] [PubMed]
13. Aiyer, H.S.; Warri, A.M.; Woode, D.R.; Hilakivi-Clarke, L.; Clarke, R. Influence of Berry Polyphenols on Receptor Signaling and Cell-Death Pathways: Implications for Breast Cancer Prevention. *J. Agric. Food Chem.* **2012**, *60*, 5693–5708. [CrossRef] [PubMed]
14. Adrian, M.; Jeandet, P.; Douillet-Breuil, A.C.; Tesson, L.; Bessis, R. Stilbene Content of Mature Vitis Vinifera Berries in Response to UV-C Elicitation. *J. Agric. Food Chem.* **2000**, *48*, 6103–6105. [CrossRef] [PubMed]
15. Duke, S.O. Benefits of Resveratrol and Pterostilbene to Crops and Their Potential Nutraceutical Value to Mammals. *Agriculture* **2022**, *12*, 368. [CrossRef]
16. Valletta, A.; Iozia, L.M.; Leonelli, F. Impact of Environmental Factors on Stilbene Biosynthesis. *Plants* **2021**, *10*, 90. [CrossRef]
17. Sarig, P.; Zutkhi, Y.; Monjauze, A.; Lisker, N.; Ben-Arie, R. Phytoalexin Elicitation in Grape Berries and Their Susceptibility ToRhizopus Stolonifer. *Physiol. Mol. Plant Pathol.* **1997**, *50*, 337–347. [CrossRef]
18. Douillet-Breuil, A.C.; Jeandet, P.; Adrian, M.; Bessis, R. Changes in the Phytoalexin Content of Various Vitis Spp. in Response to Ultraviolet C Elicitation. *J. Agric. Food Chem.* **1999**, *47*, 4456–4461. [CrossRef]
19. Shrikanta, A.; Kumar, A.; Govindaswamy, V. Resveratrol Content and Antioxidant Properties of Underutilized Fruits. *J. Food Sci. Technol.* **2015**, *52*, 383–390. [CrossRef]
20. Martínez-Márquez, A.; Morante-Carriel, J.A.; Ramírez-Estrada, K.; Cusidó, R.M.; Palazon, J.; Bru-Martínez, R. Production of Highly Bioactive Resveratrol Analogues Pterostilbene and Piceatannol in Metabolically Engineered Grapevine Cell Cultures. *Plant Biotechnol. J.* **2016**, *14*, 1813–1825. [CrossRef]
21. Jeong, Y.J.; An, C.H.; Woo, S.G.; Jeong, H.J.; Kim, Y.-M.; Park, S.-J.; Yoon, B.D.; Kim, C.Y. Production of Pinostilbene Compounds by the Expression of Resveratrol O-Methyltransferase Genes in *Escherichia coli*. *Enzym. Microb. Technol.* **2014**, *54*, 8–14. [CrossRef] [PubMed]
22. Tsai, H.-Y.; Ho, C.-T.; Chen, Y.-K. Biological Actions and Molecular Effects of Resveratrol, Pterostilbene, and 3′-Hydroxypterostilbene. *J. Food Drug Anal.* **2017**, *25*, 134–147. [CrossRef] [PubMed]
23. Ruiz, M.J.; Fernández, M.; Picó, Y.; Mañes, J.; Asensi, M.; Carda, C.; Asensio, G.; Estrela, J.M. Dietary Administration of High Doses of Pterostilbene and Quercetin to Mice Is Not Toxic. *J. Agric. Food Chem.* **2009**, *57*, 3180–3186. [CrossRef] [PubMed]
24. Liu, Y.; You, Y.; Lu, J.; Chen, X.; Yang, Z. Recent Advances in Synthesis, Bioactivity, and Pharmacokinetics of Pterostilbene, an Important Analog of Resveratrol. *Molecules* **2020**, *25*, 5166. [CrossRef]
25. Yeo, S.C.M.; Ho, P.C.; Lin, H.-S. Pharmacokinetics of Pterostilbene in Sprague-Dawley Rats: The Impacts of Aqueous Solubility, Fasting, Dose Escalation, and Dosing Route on Bioavailability. *Mol. Nutr. Food Res.* **2013**, *57*, 1015–1025. [CrossRef]
26. Bethune, S.J.; Schultheiss, N.; Henck, J.-O. Improving the Poor Aqueous Solubility of Nutraceutical Compound Pterostilbene through Cocrystal Formation. *Cryst. Growth Des.* **2011**, *11*, 2817–2823. [CrossRef]
27. Choo, Q.-Y.; Yeo, S.C.M.; Ho, P.C.; Tanaka, Y.; Lin, H.-S. Pterostilbene Surpassed Resveratrol for Anti-Inflammatory Application: Potency Consideration and Pharmacokinetics Perspective. *J. Funct. Foods* **2014**, *11*, 352–362. [CrossRef]
28. Manach, C.; Scalbert, A.; Morand, C.; Rémésy, C.; Jiménez, L. Polyphenols: Food Sources and Bioavailability. *Am. J. Clin. Nutr.* **2004**, *79*, 727–747. [CrossRef]
29. McCormack, D.; McFadden, D. Pterostilbene and Cancer: Current Review. *J. Surg. Res.* **2012**, *173*, e53–e61. [CrossRef]
30. Obrador, E.; Salvador-Palmer, R.; Jihad-Jebbar, A.; López-Blanch, R.; Dellinger, T.H.; Dellinger, R.W.; Estrela, J.M. Pterostilbene in Cancer Therapy. *Antioxidants* **2021**, *10*, 492. [CrossRef]
31. Riche, D.M.; McEwen, C.L.; Riche, K.D.; Sherman, J.J.; Wofford, M.R.; Deschamp, D.; Griswold, M. Analysis of Safety from a Human Clinical Trial with Pterostilbene. *J. Toxicol.* **2013**, *2013*, 463595. [CrossRef] [PubMed]
32. Elango, B.; Devibalan, S.; Veerapazham, T.; Palanisamy, R.; Paulmurugan, R.; Mohanram, R. Therapeutic Potential of Pterostilbene against Pancreatic β-Cell Apoptosis through Nrf2 Mechanism. *Br. J. Pharmacol.* **2014**, *171*, 1747–1757. [CrossRef]
33. Chen, R.-J.; Kuo, H.-C.; Cheng, L.-H.; Lee, Y.-H.; Chang, W.-T.; Wang, B.-J.; Wang, Y.-J.; Cheng, H.-C. Apoptotic and Nonapoptotic Activities of Pterostilbene against Cancer. *Int. J. Mol. Sci.* **2018**, *19*, 287. [CrossRef] [PubMed]
34. Bhakkiyalakshmi, E.; Dineshkumar, K.; Karthik, S.; Sireesh, D.; Hopper, W.; Paulmurugan, R.; Ramkumar, K.M. Pterostilbene-Mediated Nrf2 Activation: Mechanistic Insights on Keap1:Nrf2 Interface. *Bioorg. Med. Chem.* **2016**, *24*, 3378–3386. [CrossRef] [PubMed]
35. Xu, J.; Liu, J.; Li, Q.; Mi, Y.; Zhou, D.; Meng, Q.; Chen, G.; Li, N.; Hou, Y. Pterostilbene Alleviates Aβ1-42 -Induced Cognitive Dysfunction via Inhibition of Oxidative Stress by Activating Nrf2 Signaling Pathway. *Mol. Nutr. Food Res.* **2021**, *65*, e2000711. [CrossRef]
36. Sireesh, D.; Ganesh, M.-R.; Dhamodharan, U.; Sakthivadivel, M.; Sivasubramanian, S.; Gunasekaran, P.; Ramkumar, K.M. Role of Pterostilbene in Attenuating Immune Mediated Devastation of Pancreatic Beta Cells via Nrf2 Signaling Cascade. *J. Nutr. Biochem.* **2017**, *44*, 11–21. [CrossRef] [PubMed]
37. Bhakkiyalakshmi, E.; Sireesh, D.; Sakthivadivel, M.; Sivasubramanian, S.; Gunasekaran, P.; Ramkumar, K.M. Anti-Hyperlipidemic and Anti-Peroxidative Role of Pterostilbene via Nrf2 Signaling in Experimental Diabetes. *Eur. J. Pharmacol.* **2016**, *777*, 9–16. [CrossRef]

38. Shen, H.; Rong, H. Pterostilbene Impact on Retinal Endothelial Cells under High Glucose Environment. *Int. J. Clin. Exp. Pathol.* **2015**, *8*, 12589–12594.
39. Zhang, H.; Chen, Y.; Chen, Y.; Ji, S.; Jia, P.; Xu, J.; Li, Y.; Wang, T. Pterostilbene Attenuates Liver Injury and Oxidative Stress in Intrauterine Growth–Retarded Weanling Piglets. *Nutrition* **2021**, *81*, 110940. [CrossRef]
40. Chiou, Y.-S.; Tsai, M.-L.; Nagabhushanam, K.; Wang, Y.-J.; Wu, C.-H.; Ho, C.-T.; Pan, M.-H. Pterostilbene Is More Potent than Resveratrol in Preventing Azoxymethane (AOM)-Induced Colon Tumorigenesis via Activation of the NF-E2-Related Factor 2 (Nrf2)-Mediated Antioxidant Signaling Pathway. *J. Agric. Food Chem.* **2011**, *59*, 2725–2733. [CrossRef]
41. Sirerol, J.A.; Feddi, F.; Mena, S.; Rodriguez, M.L.; Sirera, P.; Aupí, M.; Pérez, S.; Asensi, M.; Ortega, A.; Estrela, J.M. Topical Treatment with Pterostilbene, a Natural Phytoalexin, Effectively Protects Hairless Mice against UVB Radiation-Induced Skin Damage and Carcinogenesis. *Free Radic. Biol. Med.* **2015**, *85*, 1–11. [CrossRef] [PubMed]
42. Perecko, T.; Drabikova, K.; Lojek, A.; Ciz, M.; Ponist, S.; Bauerova, K.; Nosal, R.; Harmatha, J.; Jancinova, V. The Effects of Pterostilbene on Neutrophil Activity in Experimental Model of Arthritis. *Biomed. Res. Int.* **2013**, *2013*, 106041. [CrossRef] [PubMed]
43. Li, Y.-R.; Li, S.; Lin, C.-C. Effect of Resveratrol and Pterostilbene on Aging and Longevity. *Biofactors* **2018**, *44*, 69–82. [CrossRef] [PubMed]
44. Alosi, J.A.; McDonald, D.E.; Schneider, J.S.; Privette, A.R.; McFadden, D.W. Pterostilbene Inhibits Breast Cancer in Vitro through Mitochondrial Depolarization and Induction of Caspase-Dependent Apoptosis. *J. Surg. Res.* **2010**, *161*, 195–201. [CrossRef] [PubMed]
45. Mannal, P.; McDonald, D.; McFadden, D. Pterostilbene and Tamoxifen Show an Additive Effect against Breast Cancer in Vitro. *Am. J. Surg.* **2010**, *200*, 577–580. [CrossRef]
46. Chakraborty, A.; Gupta, N.; Ghosh, K.; Roy, P. In Vitro Evaluation of the Cytotoxic, Anti-Proliferative and Anti-Oxidant Properties of Pterostilbene Isolated from Pterocarpus Marsupium. *Toxicol. Vitr.* **2010**, *24*, 1215–1228. [CrossRef]
47. Mh, P.; Yh, C.; V, B.; K, N.; Ct, H. Pterostilbene Induces Apoptosis and Cell Cycle Arrest in Human Gastric Carcinoma Cells. *J. Agric. Food Chem.* **2007**, *55*, 7777–7785. [CrossRef]
48. McCormack, D.E.; Mannal, P.; McDonald, D.; Tighe, S.; Hanson, J.; McFadden, D. Genomic Analysis of Pterostilbene Predicts Its Antiproliferative Effects against Pancreatic Cancer In Vitro and In Vivo. *J. Gastrointest. Surg.* **2012**, *16*, 1136–1143. [CrossRef]
49. Zhang, L.; Zhou, G.; Song, W.; Tan, X.; Guo, Y.; Zhou, B.; Jing, H.; Zhao, S.; Chen, L. Pterostilbene Protects Vascular Endothelial Cells against Oxidized Low-Density Lipoprotein-Induced Apoptosis in Vitro and in Vivo. *Apoptosis* **2012**, *17*, 25–36. [CrossRef]
50. Zhang, L.; Cui, L.; Zhou, G.; Jing, H.; Guo, Y.; Sun, W. Pterostilbene, a Natural Small-Molecular Compound, Promotes Cytoprotective Macroautophagy in Vascular Endothelial Cells. *J. Nutr. Biochem.* **2013**, *24*, 903–911. [CrossRef]
51. Özdaş, S.; Taştekin, B.; Gürgen, S.G.; Özdaş, T.; Pelit, A.; Erkan, S.O.; Tuhanioğlu, B.; Gülnar, B.; Görgülü, O. Pterostilbene Protects Cochlea from Ototoxicity in Streptozotocin-Induced Diabetic Rats by Inhibiting Apoptosis. *PLoS ONE* **2020**, *15*, e0228429. [CrossRef] [PubMed]
52. Dvorakova, M.; Landa, P. Anti-inflammatory activity of natural stilbenoids: A review. *Pharmacol. Res.* **2017**, *124*, 126–145. [CrossRef] [PubMed]
53. Liu, J.; Fan, C.; Yu, L.; Yang, Y.; Jiang, S.; Ma, Z.; Hu, W.; Li, T.; Yang, Z.; Tian, T.; et al. Pterostilbene Exerts an Anti-Inflammatory Effect via Regulating Endoplasmic Reticulum Stress in Endothelial Cells. *Cytokine* **2016**, *77*, 88–97. [CrossRef] [PubMed]
54. Paul, S.; Rimando, A.M.; Lee, H.J.; Ji, Y.; Reddy, B.S.; Suh, N. Anti-Inflammatory Action of Pterostilbene Is Mediated through the P38 Mitogen-Activated Protein Kinase Pathway in Colon Cancer Cells. *Cancer Prev. Res.* **2009**, *2*, 650–657. [CrossRef] [PubMed]
55. Lim, Y.R.I.; Preshaw, P.M.; Lim, L.P.; Ong, M.M.A.; Lin, H.-S.; Tan, K.S. Pterostilbene Complexed with Cyclodextrin Exerts Antimicrobial and Anti-Inflammatory Effects. *Sci. Rep.* **2020**, *10*, 9072. [CrossRef] [PubMed]
56. Cichocki, M.; Paluszczak, J.; Szaefer, H.; Piechowiak, A.; Rimando, A.M.; Baer-Dubowska, W. Pterostilbene Is Equally Potent as Resveratrol in Inhibiting 12-O-Tetradecanoylphorbol-13-Acetate Activated NFkappaB, AP-1, COX-2, and INOS in Mouse Epidermis. *Mol. Nutr. Food Res.* **2008**, *52* Suppl. S1, S62–S70. [CrossRef]
57. Yan, W.; Ren, D.; Feng, X.; Huang, J.; Wang, D.; Li, T.; Zhang, D. Neuroprotective and Anti-Inflammatory Effect of Pterostilbene against Cerebral Ischemia/Reperfusion Injury via Suppression of COX-2. *Front. Pharmacol.* **2021**, *12*, 770329. [CrossRef]
58. Liu, H.; Wu, X.; Luo, J.; Wang, X.; Guo, H.; Feng, D.; Zhao, L.; Bai, H.; Song, M.; Liu, X.; et al.; et al. Pterostilbene Attenuates Astrocytic Inflammation and Neuronal Oxidative Injury after Ischemia-Reperfusion by Inhibiting NF-KB Phosphorylation. *Front. Immunol.* **2019**, *10*, 2408. [CrossRef]
59. Yang, H.; Hua, C.; Yang, X.; Fan, X.; Song, H.; Peng, L.; Ci, X. Pterostilbene Prevents LPS-Induced Early Pulmonary Fibrosis by Suppressing Oxidative Stress, Inflammation and Apoptosis in Vivo. *Food Funct.* **2020**, *11*, 4471–4484. [CrossRef]
60. Rui, Z.; Zhang, L.; Li, X.; Han, J.; Yuan, Y.; Ding, H.; Liu, Y.; Ding, X. Pterostilbene Exert an Anti-Arthritic Effect by Attenuating Inflammation, Oxidative Stress, and Alteration of Gut Microbiota. *J. Food. Biochem.* **2022**, *46*, e14011. [CrossRef]
61. Azzolini, M.; Mattarei, A.; La Spina, M.; Fanin, M.; Chiodarelli, G.; Romio, M.; Zoratti, M.; Paradisi, C.; Biasutto, L. New Natural Amino Acid-Bearing Prodrugs Boost Pterostilbene's Oral Pharmacokinetic and Distribution Profile. *Eur. J. Pharm. Biopharm.* **2017**, *115*, 149–158. [CrossRef] [PubMed]
62. Rimando, A.; Cuendet, M.; Desmarchelier, C.; Mehta, R.; Pezzuto, J.; Duke, S. Cancer Chemopreventive and Antioxidant Activities of Pterostilbene, a Naturally Occurring Analogue of Resveratrol. *J. Agric. Food Chem.* **2002**, *50*, 3453–3457. [CrossRef] [PubMed]

63. Su, C.-M.; Lee, W.-H.; Wu, A.T.H.; Lin, Y.-K.; Wang, L.-S.; Wu, C.-H.; Yeh, C.-T. Pterostilbene Inhibits Triple-Negative Breast Cancer Metastasis via Inducing MicroRNA-205 Expression and Negatively Modulates Epithelial-to-Mesenchymal Transition. *J. Nutr. Biochem.* **2015**, *26*, 675–685. [CrossRef] [PubMed]
64. Kong, Y.; Chen, G.; Xu, Z.; Yang, G.; Li, B.; Wu, X.; Xiao, W.; Xie, B.; Hu, L.; Sun, X.; et al. Pterostilbene Induces Apoptosis and Cell Cycle Arrest in Diffuse Large B-Cell Lymphoma Cells. *Sci. Rep.* **2016**, *6*, 37417. [CrossRef] [PubMed]
65. Xie, B.; Xu, Z.; Hu, L.; Chen, G.; Wei, R.; Yang, G.; Li, B.; Chang, G.; Sun, X.; Wu, H.; et al. Pterostilbene Inhibits Human Multiple Myeloma Cells via ERK1/2 and JNK Pathway In Vitro and In Vivo. *Int. J. Mol. Sci.* **2016**, *17*, 1927. [CrossRef] [PubMed]
66. Suh, N.; Paul, S.; Hao, X.; Simi, B.; Xiao, H.; Rimando, A.M.; Reddy, B.S. Pterostilbene, an Active Constituent of Blueberries, Suppresses Aberrant Crypt Foci Formation in the Azoxymethane-Induced Colon Carcinogenesis Model in Rats. *Clin. Cancer Res.* **2007**, *13*, 350–355. [CrossRef]
67. Chatterjee, K.; Mukherjee, S.; Vanmanen, J.; Banerjee, P.; Fata, J.E. Dietary Polyphenols, Resveratrol and Pterostilbene Exhibit Antitumor Activity on an HPV E6-Positive Cervical Cancer Model: An in Vitro and in Vivo Analysis. *Front. Oncol.* **2019**, *9*, 352. [CrossRef]
68. Tam, K.-W.; Ho, C.-T.; Tu, S.-H.; Lee, W.-J.; Huang, C.-S.; Chen, C.-S.; Wu, C.-H.; Lee, C.-H.; Ho, Y.-S. α-Tocopherol Succinate Enhances Pterostilbene Anti-Tumor Activity in Human Breast Cancer Cells in Vivo and in Vitro. *Oncotarget* **2017**, *9*, 4593–4606. [CrossRef]
69. Benlloch, M.; Obrador, E.; Valles, S.L.; Rodriguez, M.L.; Sirerol, J.A.; Alcácer, J.; Pellicer, J.A.; Salvador, R.; Cerdá, C.; Sáez, G.T.; et al. Pterostilbene Decreases the Antioxidant Defenses of Aggressive Cancer Cells In Vivo: A Physiological Glucocorticoids- and Nrf2-Dependent Mechanism. *Antioxid. Redox. Signal.* **2016**, *24*, 974–990. [CrossRef]
70. Wang, D.; Guo, H.; Yang, H.; Wang, D.; Gao, P.; Wei, W. Pterostilbene, An Active Constituent of Blueberries, Suppresses Proliferation Potential of Human Cholangiocarcinoma via Enhancing the Autophagic Flux. *Front. Pharmacol.* **2019**, *10*, 1238. [CrossRef]
71. Huynh, T.-T.; Lin, C.-M.; Lee, W.-H.; Wu, A.T.H.; Lin, Y.-K.; Lin, Y.-F.; Yeh, C.-T.; Wang, L.-S. Pterostilbene Suppressed Irradiation-Resistant Glioma Stem Cells by Modulating GRP78/MiR-205 Axis. *J. Nutr. Biochem.* **2015**, *26*, 466–475. [CrossRef] [PubMed]
72. Pari, L.; Satheesh, M.A. Effect of Pterostilbene on Hepatic Key Enzymes of Glucose Metabolism in Streptozotocin- and Nicotinamide-Induced Diabetic Rats. *Life Sci.* **2006**, *79*, 641–645. [CrossRef] [PubMed]
73. Zhang, Y.; Ren, S.; Ji, Y.; Liang, Y. Pterostilbene Ameliorates Nephropathy Injury in Streptozotocin-Induced Diabetic Rats. *Pharmacology* **2019**, *104*, 71–80. [CrossRef]
74. Grover, J.K.; Vats, V.; Yadav, S. Effect of Feeding Aqueous Extract of Pterocarpus Marsupium on Glycogen Content of Tissues and the Key Enzymes of Carbohydrate Metabolism. *Mol. Cell Biochem.* **2002**, *241*, 53–59. [CrossRef] [PubMed]
75. Lee, M.-F.; Liu, M.-L.; Cheng, A.-C.; Tsai, M.-L.; Ho, C.-T.; Liou, W.-S.; Pan, M.-H. Pterostilbene Inhibits Dimethylnitrosamine-Induced Liver Fibrosis in Rats. *Food Chem.* **2013**, *138*, 802–807. [CrossRef] [PubMed]
76. Aguirre, L.; Palacios-Ortega, S.; Fernández-Quintela, A.; Hijona, E.; Bujanda, L.; Portillo, M.P. Pterostilbene Reduces Liver Steatosis and Modifies Hepatic Fatty Acid Profile in Obese Rats. *Nutrients* **2019**, *11*, 961. [CrossRef]
77. Kim, H.; Seo, K.-H.; Yokoyama, W. Chemistry of Pterostilbene and Its Metabolic Effects. *J. Agric. Food Chem.* **2020**, *68*, 12836–12841. [CrossRef]
78. Zhan, J.; Hu, T.; Shen, J.; Yang, G.; Ho, C.-T.; Li, S. Pterostilbene Is More Efficacious than Hydroxystilbenes in Protecting Liver Fibrogenesis in a Carbon Tetracholride-Induced Rat Model. *J. Funct. Foods* **2021**, *84*, 104604. [CrossRef]
79. Fan, X.; Wang, L.; Huang, J.; Lv, H.; Deng, X.; Ci, X. Pterostilbene Reduces Acetaminophen-Induced Liver Injury by Activating the Nrf2 Antioxidative Defense System via the AMPK/Akt/GSK3β Pathway. *CPB* **2018**, *49*, 1943–1958. [CrossRef]
80. Blueberry Supplementation Enhances Signaling and Prevents Behavioral Deficits in an Alzheimer Disease Model: Nutritional Neuroscience. Volume 6. No. 3. Available online: https://www.tandfonline.com/doi/abs/10.1080/1028415031000111282 (accessed on 5 May 2022).
81. Yang, Y.; Wang, J.; Li, Y.; Fan, C.; Jiang, S.; Zhao, L.; Di, S.; Xin, Z.; Wang, B.; Wu, G.; et al. HO-1 Signaling Activation by Pterostilbene Treatment Attenuates Mitochondrial Oxidative Damage Induced by Cerebral Ischemia Reperfusion Injury. *Mol. Neurobiol.* **2015**, *53*, 2339–2353. [CrossRef]
82. Chang, J.; Rimando, A.; Pallas, M.; Camins, A.; Porquet, D.; Reeves, J.; Shukitt-Hale, B.; Smith, M.A.; Joseph, J.A.; Casadesus, G. Low-Dose Pterostilbene, but Not Resveratrol, Is a Potent Neuromodulator in Aging and Alzheimer's Disease. *Neurobiol. Aging* **2012**, *33*, 2062–2071. [CrossRef] [PubMed]
83. Lacerda, D.; Türck, P.; Campos-Carraro, C.; Hickmann, A.; Ortiz, V.; Bianchi, S.; Belló-Klein, A.; de Castro, A.L.; Bassani, V.L.; da Rosa Araujo, A.S. Pterostilbene Improves Cardiac Function in a Rat Model of Right Heart Failure through Modulation of Calcium Handling Proteins and Oxidative Stress. *Appl. Physiol. Nutr. Metab.* **2020**, *45*, 987–995. [CrossRef] [PubMed]
84. Wu, X.; Kang, J.; Xie, C.; Burris, R.; Ferguson, M.E.; Badger, T.M.; Nagarajan, S. Dietary Blueberries Attenuate Atherosclerosis in Apolipoprotein E-Deficient Mice by Upregulating Antioxidant Enzyme Expression. *J. Nutr.* **2010**, *140*, 1628–1632. [CrossRef] [PubMed]
85. Ahmet, I.; Spangler, E.; Shukitt-Hale, B.; Juhaszova, M.; Sollott, S.J.; Joseph, J.A.; Ingram, D.K.; Talan, M. Blueberry-Enriched Diet Protects Rat Heart from Ischemic Damage. *PLoS ONE* **2009**, *4*, e5954. [CrossRef]

86. Yang, G.; Sun, J.; Lu, K.; Shan, S.; Li, S.; Sun, C. Pterostilbene Coupled with Physical Exercise Effectively Mitigates Collagen-Induced Articular Synovial by Correcting the PI3K/Akt/NF-KB Signal Pathway. *J. Agric. Food Chem.* **2021**, *69*, 13821–13830. [CrossRef]
87. Elsherbini, A.M.; Sheweita, S.A.; Sultan, A.S. Pterostilbene as a Phytochemical Compound Induces Signaling Pathways Involved in the Apoptosis and Death of Mutant P53-Breast Cancer Cell Lines. *Nutr. Cancer* **2021**, *73*, 1976–1984. [CrossRef]
88. Priego, S.; Feddi, F.; Ferrer, P.; Mena, S.; Benlloch, M.; Ortega, A.; Carretero, J.; Obrador, E.; Asensi, M.; Estrela, J.M. Natural Polyphenols Facilitate Elimination of HT-29 Colorectal Cancer Xenografts by Chemoradiotherapy: A Bcl-2- and Superoxide Dismutase 2-Dependent Mechanism. *Mol. Cancer Ther.* **2008**, *7*, 3330–3342. [CrossRef]
89. Paul, S.; DeCastro, A.J.; Lee, H.J.; Smolarek, A.K.; So, J.Y.; Simi, B.; Wang, C.X.; Zhou, R.; Rimando, A.M.; Suh, N. Dietary Intake of Pterostilbene, a Constituent of Blueberries, Inhibits the β-Catenin/P65 Downstream Signaling Pathway and Colon Carcinogenesis in Rats. *Carcinogenesis* **2010**, *31*, 1272–1278. [CrossRef]
90. Ferrer, P.; Asensi, M.; Segarra, R.; Ortega, A.; Benlloch, M.; Obrador, E.; Varea, M.T.; Asensio, G.; Jordá, L.; Estrela, J.M. Association between Pterostilbene and Quercetin Inhibits Metastatic Activity of B16 Melanoma. *Neoplasia* **2005**, *7*, 37–47. [CrossRef]
91. Mena, S.; Rodríguez, M.L.; Ponsoda, X.; Estrela, J.M.; Jäättela, M.; Ortega, A.L. Pterostilbene-Induced Tumor Cytotoxicity: A Lysosomal Membrane Permeabilization-Dependent Mechanism. *PLoS ONE* **2012**, *7*, e44524. [CrossRef]
92. Tsai, M.-L.; Lai, C.-S.; Chang, Y.-H.; Chen, W.-J.; Ho, C.-T.; Pan, M.-H. Pterostilbene, a Natural Analogue of Resveratrol, Potently Inhibits 7,12-Dimethylbenz[a]Anthracene (DMBA)/12-O-Tetradecanoylphorbol-13-Acetate (TPA)-Induced Mouse Skin Carcinogenesis. *Food Funct.* **2012**, *3*, 1185–1194. [CrossRef] [PubMed]
93. Wen, W.; Lowe, G.; Roberts, C.M.; Finlay, J.; Han, E.S.; Glackin, C.A.; Dellinger, T.H. Pterostilbene, a Natural Phenolic Compound, Synergizes the Antineoplastic Effects of Megestrol Acetate in Endometrial Cancer. *Sci. Rep.* **2017**, *7*, 12754. [CrossRef] [PubMed]
94. Wen, W.; Lowe, G.; Roberts, C.M.; Finlay, J.; Han, E.S.; Glackin, C.A.; Dellinger, T.H. Pterostilbene Suppresses Ovarian Cancer Growth via Induction of Apoptosis and Blockade of Cell Cycle Progression Involving Inhibition of the STAT3 Pathway. *Int. J. Mol. Sci.* **2018**, *19*, 1983. [CrossRef]
95. Riche, D.M.; Riche, K.D.; Blackshear, C.T.; McEwen, C.L.; Sherman, J.J.; Wofford, M.R.; Griswold, M.E. Pterostilbene on Metabolic Parameters: A Randomized, Double-Blind, and Placebo-Controlled Trial. *Evid. Based Complement. Altern. Med.* **2014**, *2014*, 459165. [CrossRef]
96. Guo, L.; Tan, K.; Wang, H.; Zhang, X. Pterostilbene Inhibits Hepatocellular Carcinoma through P53/SOD2/ROS-Mediated Mitochondrial Apoptosis. *Oncol. Rep.* **2016**, *36*, 3233–3240. [CrossRef] [PubMed]
97. Wang, R.; Xu, Z.; Tian, J.; Liu, Q.; Dong, J.; Guo, L.; Hai, B.; Liu, X.; Yao, H.; Chen, Z.; et al. Pterostilbene Inhibits Hepatocellular Carcinoma Proliferation and HBV Replication by Targeting Ribonucleotide Reductase M2 Protein. *Am. J. Cancer Res.* **2021**, *11*, 2975–2989. [PubMed]
98. Qian, Y.-Y.; Liu, Z.-S.; Yan, H.-J.; Yuan, Y.-F.; Levenson, A.S.; Li, K. Pterostilbene Inhibits MTA1/HDAC1 Complex Leading to PTEN Acetylation in Hepatocellular Carcinoma. *Biomed. Pharmacother.* **2018**, *101*, 852–859. [CrossRef]
99. Baquer, N.Z.; Gupta, D.; Raju, J. Regulation of Metabolic Pathways in Liver and Kidney during Experimental Diabetes: Effects of Antidiabetic Compounds. *Indian J. Clin. Biochem.* **1998**, *13*, 63–80. [CrossRef]
100. Satheesh, M.A.; Pari, L. The Antioxidant Role of Pterostilbene in Streptozotocin-Nicotinamide-Induced Type 2 Diabetes Mellitus in Wistar Rats. *J. Pharm. Pharmacol.* **2010**, *58*, 1483–1490. [CrossRef]
101. Dornadula, S.; Thiruppathi, S.; Palanisamy, R.; Umapathy, D.; Suzuki, T.; Mohanram, R.K. Differential Proteomic Profiling Identifies Novel Molecular Targets of Pterostilbene against Experimental Diabetes. *J. Cell. Physiol.* **2019**, *234*, 1996–2012. [CrossRef]
102. Acharya, J.D.; Ghaskadbi, S.S. Protective Effect of Pterostilbene against Free Radical Mediated Oxidative Damage. *BMC Complement. Altern. Med.* **2013**, *13*, 238. [CrossRef] [PubMed]
103. Schuppan, D.; Afdhal, N.H. Liver Cirrhosis. *Lancet* **2008**, *371*, 838–851. [CrossRef]
104. Tsai, H.-Y.; Shih, Y.-Y.; Yeh, Y.-T.; Huang, C.-H.; Liao, C.-A.; Hu, C.-Y.; Nagabhushanam, K.; Ho, C.-T.; Chen, Y.-K. Pterostilbene and Its Derivative 3′-Hydroxypterostilbene Ameliorated Nonalcoholic Fatty Liver Disease Through Synergistic Modulation of the Gut Microbiota and SIRT1/AMPK Signaling Pathway. *J. Agric. Food Chem.* **2022**, *70*, 4966–4980. [CrossRef] [PubMed]
105. Joseph, J.A.; Fisher, D.R.; Cheng, V.; Rimando, A.M.; Shukitt-Hale, B. Cellular and Behavioral Effects of Stilbene Resveratrol Analogues: Implications for Reducing the Deleterious Effects of Aging. *J. Agric. Food Chem.* **2008**, *56*, 10544–10551. [CrossRef]
106. Song, Z.; Han, S.; Pan, X.; Gong, Y.; Wang, M. Pterostilbene Mediates Neuroprotection against Oxidative Toxicity via Oestrogen Receptor α Signalling Pathways. *J. Pharm. Pharmacol.* **2015**, *67*, 720–730. [CrossRef] [PubMed]
107. Yang, Y.; Fan, C.; Wang, B.; Ma, Z.; Wang, D.; Gong, B.; Di, S.; Jiang, S.; Li, Y.; Li, T.; et al. Pterostilbene Attenuates High Glucose-Induced Oxidative Injury in Hippocampal Neuronal Cells by Activating Nuclear Factor Erythroid 2-Related Factor 2. *Biochim. Biophys. Acta Mol. Basis Dis.* **2017**, *1863*, 827–837. [CrossRef] [PubMed]
108. Poulose, S.M.; Thangthaeng, N.; Miller, M.G.; Shukitt-Hale, B. Effects of Pterostilbene and Resveratrol on Brain and Behavior. *Neurochem. Int.* **2015**, *89*, 227–233. [CrossRef]
109. Sinha, K.; Chaudhary, G.; Kumar Gupta, Y. Protective Effect of Resveratrol against Oxidative Stress in Middle Cerebral Artery Occlusion Model of Stroke in Rats. *Life Sci.* **2002**, *71*, 655–665. [CrossRef]
110. Zhang, F.; Shi, J.-S.; Zhou, H.; Wilson, B.; Hong, J.-S.; Gao, H.-M. Resveratrol Protects Dopamine Neurons against Lipopolysaccharide-Induced Neurotoxicity through Its Anti-Inflammatory Actions. *Mol. Pharmacol.* **2010**, *78*, 466–477. [CrossRef]

111. Kosuru, R.; Kandula, V.; Rai, U.; Prakash, S.; Xia, Z.; Singh, S. Pterostilbene Decreases Cardiac Oxidative Stress and Inflammation via Activation of AMPK/Nrf2/HO-1 Pathway in Fructose-Fed Diabetic Rats. *Cardiovasc. Drugs Ther.* **2018**, *32*, 147–163. [CrossRef]
112. Mechanisms of atherosclerosis–a review. *Adv. Nephrol. Necker Hosp.* **1990**, *19*, 79–86.
113. Park, E.-S.; Lim, Y.; Hong, J.-T.; Yoo, H.-S.; Lee, C.-K.; Pyo, M.-Y.; Yun, Y.-P. Pterostilbene, a Natural Dimethylated Analog of Resveratrol, Inhibits Rat Aortic Vascular Smooth Muscle Cell Proliferation by Blocking Akt-Dependent Pathway. *Vasc. Pharmacol.* **2010**, *53*, 61–67. [CrossRef] [PubMed]
114. Hou, Y.; Xie, G.; Miao, F.; Ding, L.; Mou, Y.; Wang, L.; Su, G.; Chen, G.; Yang, J.; Wu, C. Pterostilbene Attenuates Lipopolysaccharide-Induced Learning and Memory Impairment Possibly via Inhibiting Microglia Activation and Protecting Neuronal Injury in Mice. *Prog. Neuro-Psychopharmacol. Biol. Psychiatry* **2014**, *54*, 92–102. [CrossRef] [PubMed]
115. Porquet, D.; Casadesús, G.; Bayod, S.; Vicente, A.; Canudas, A.M.; Vilaplana, J.; Pelegrí, C.; Sanfeliu, C.; Camins, A.; Pallàs, M.; et al. Dietary Resveratrol Prevents Alzheimer's Markers and Increases Life Span in SAMP8. *AGE* **2013**, *35*, 1851–1865. [CrossRef] [PubMed]
116. Li, J.; Deng, R.; Hua, X.; Zhang, L.; Lu, F.; Coursey, T.G.; Pflugfelder, S.C.; Li, D.-Q. Blueberry Component Pterostilbene Protects Corneal Epithelial Cells from Inflammation via Anti-Oxidative Pathway. *Sci. Rep.* **2016**, *6*, 19408. [CrossRef]
117. Wilson, M.A.; Shukitt-Hale, B.; Kalt, W.; Ingram, D.K.; Joseph, J.A.; Wolkow, C.A. Blueberry Polyphenols Increase Lifespan and Thermotolerance in Caenorhabditis Elegans. *Aging Cell* **2006**, *5*, 59–68. [CrossRef]
118. Peng, C.; Zuo, Y.; Kwan, K.M.; Liang, Y.; Ma, K.Y.; Chan, H.Y.E.; Huang, Y.; Yu, H.; Chen, Z.-Y. Blueberry Extract Prolongs Lifespan of Drosophila Melanogaster. *Exp. Gerontol.* **2012**, *47*, 170–178. [CrossRef]
119. Joseph, J.A.; Denisova, N.A.; Arendash, G.; Gordon, M.; Diamond, D.; Shukitt-Hale, B.; Morgan, D. Blueberry Supplementation Enhances Signaling and Prevents Behavioral Deficits in an Alzheimer Disease Model. *Nutr. Neurosci.* **2003**, *6*, 153–162. [CrossRef]
120. Majeed, M.; Majeed, S.; Jain, R.; Mundkur, L.; Rajalakshmi, H.; Lad, P.S.; Neupane, P. An Open-Label Single-Arm, Monocentric Study Assessing the Efficacy and Safety of Natural Pterostilbene (Pterocarpus Marsupium) for Skin Brightening and Antiaging Effects. *Clin. Cosmet. Investig. Dermatol.* **2020**, *13*, 105–116. [CrossRef]
121. Vaňková, E.; Paldrychová, M.; Kašparová, P.; Lokočová, K.; Kodeš, Z.; Maťátková, O.; Kolouchová, I.; Masák, J. Natural Antioxidant Pterostilbene as an Effective Antibiofilm Agent, Particularly for Gram-Positive Cocci. *World J. Microbiol. Biotechnol.* **2020**, *36*, 101. [CrossRef]
122. Shih, Y.-H.; Tsai, P.-J.; Chen, Y.-L.; Pranata, R.; Chen, R.-J. Assessment of the Antibacterial Mechanism of Pterostilbene against Bacillus Cereus through Apoptosis-like Cell Death and Evaluation of Its Beneficial Effects on the Gut Microbiota. *J. Agric. Food Chem.* **2021**, *69*, 12219–12229. [CrossRef] [PubMed]
123. Lee, W.X.; Basri, D.F.; Ghazali, A.R. Bactericidal Effect of Pterostilbene Alone and in Combination with Gentamicin against Human Pathogenic Bacteria. *Molecules* **2017**, *22*, 463. [CrossRef] [PubMed]
124. Yang, S.-C.; Tseng, C.-H.; Wang, P.-W.; Lu, P.-L.; Weng, Y.-H.; Yen, F.-L.; Fang, J.-Y. Pterostilbene, a Methoxylated Resveratrol Derivative, Efficiently Eradicates Planktonic, Biofilm, and Intracellular MRSA by Topical Application. *Front. Microbiol.* **2017**, *8*, 1103. [CrossRef] [PubMed]
125. ter Ellen, B.M.; Dinesh Kumar, N.; Bouma, E.M.; Troost, B.; van de Pol, D.P.I.; van der Ende-Metselaar, H.H.; Apperloo, L.; van Gosliga, D.; van den Berge, M.; Nawijn, M.C.; et al. Resveratrol and Pterostilbene Inhibit SARS-CoV-2 Replication in Air–Liquid Interface Cultured Human Primary Bronchial Epithelial Cells. *Viruses* **2021**, *13*, 1335. [CrossRef] [PubMed]
126. Lin, S.-C.; Ho, C.-T.; Chuo, W.-H.; Li, S.; Wang, T.T.; Lin, C.-C. Effective Inhibition of MERS-CoV Infection by Resveratrol. *BMC Infect. Dis.* **2017**, *17*, 144. [CrossRef]
127. Das, A.; Pandita, D.; Jain, G.K.; Agarwal, P.; Grewal, A.S.; Khar, R.K.; Lather, V. Role of Phytoconstituents in the Management of COVID-19. *Chem. Biol. Interact.* **2021**, *341*, 109449. [CrossRef] [PubMed]
128. Kelleni, M.T. Resveratrol-Zinc Nanoparticles or Pterostilbene-Zinc: Potential COVID-19 Mono and Adjuvant Therapy. *Biomed Pharmacother.* **2021**, *139*, 111626. [CrossRef]
129. Hsieh, Y.-H.; Chen, C.W.S.; Schmitz, S.-F.H.; King, C.-C.; Chen, W.-J.; Wu, Y.-C.; Ho, M.-S. Candidate Genes Associated with Susceptibility for SARS-Coronavirus. *Bull. Math. Biol.* **2010**, *72*, 122–132. [CrossRef]
130. Hu, W.; Yu, H.; Zhou, X.; Li, M.; Xiao, L.; Ruan, Q.; Huang, X.; Li, L.; Xie, W.; Guo, X.; et al. Topical Administration of Pterostilbene Accelerates Burn Wound Healing in Diabetes through Activation of the HIF1α Signaling Pathway. *Burns* **2021**, *48*, 1452–1461. [CrossRef]
131. Bofill, L.; Barbas, R.; de Sande, D.; Font-Bardia, M.; Ràfols, C.; Albertí, J.; Prohens, R. A Novel, Extremely Bioavailable Cocrystal of Pterostilbene. *Cryst. Growth Des.* **2021**, *21*, 2315–2323. [CrossRef]
132. Schultheiss, N.; Bethune, S.; Henck, J.-O. Nutraceutical Cocrystals: Utilizing Pterostilbene as a Cocrystal Former. *CrystEngComm* **2010**, *12*, 2436–2442. [CrossRef]
133. Zhao, X.; Shi, A.; Ma, Q.; Yan, X.; Bian, L.; Zhang, P.; Wu, J. Nanoparticles Prepared from Pterostilbene Reduce Blood Glucose and Improve Diabetes Complications. *J. Nanobiotechnol.* **2021**, *19*, 191. [CrossRef] [PubMed]
134. Liu, K.-F.; Liu, Y.-X.; Dai, L.; Li, C.-X.; Wang, L.; Liu, J.; Lei, J.-D. A Novel Self-Assembled PH-Sensitive Targeted Nanoparticle Platform Based on Antibody–4arm-Polyethylene Glycol–Pterostilbene Conjugates for Co-Delivery of Anticancer Drugs. *J. Mater. Chem. B* **2018**, *6*, 656–665. [CrossRef] [PubMed]

135. Liu, Q.; Chen, J.; Qin, Y.; Jiang, B.; Zhang, T. Zein/Fucoidan-Based Composite Nanoparticles for the Encapsulation of Pterostilbene: Preparation, Characterization, Physicochemical Stability, and Formation Mechanism. *Int. J. Biol. Macromol.* **2020**, *158*, 461–470. [CrossRef]
136. Romio, M.; Morgese, G.; Trachsel, L.; Babity, S.; Paradisi, C.; Brambilla, D.; Benetti, E.M. Poly(2-Oxazoline)–Pterostilbene Block Copolymer Nanoparticles for Dual-Anticancer Drug Delivery. *Biomacromolecules* **2018**, *19*, 103–111. [CrossRef]
137. Sun, Y.; Xia, Z.; Zheng, J.; Qiu, P.; Zhang, L.; McClements, D.J.; Xiao, H. Nanoemulsion-Based Delivery Systems for Nutraceuticals: Influence of Carrier Oil Type on Bioavailability of Pterostilbene. *J. Funct. Foods* **2015**, *13*, 61–70. [CrossRef]
138. Ansari, M.; Sadarani, B.; Majumdar, A. Colon Targeted Beads Loaded with Pterostilbene: Formulation, Optimization, Characterization and in Vivo Evaluation. *Saudi Pharm. J.* **2019**, *27*, 71–81. [CrossRef]

molecules MDPI

Article

Cytotoxic Effect of *Rosmarinus officinalis* Extract on Glioblastoma and Rhabdomyosarcoma Cell Lines

Eleni Kakouri [1], Olti Nikola [2], Charalabos Kanakis [1], Kyriaki Hatziagapiou [2,3], George I. Lambrou [2], Panayiotis Trigas [4], Christina Kanaka-Gantenbein [2] and Petros A. Tarantilis [1,*]

1 Laboratory of Chemistry, Department of Food Science & Human Nutrition, School of Food and Nutritional Sciences, Agricultural University of Athens, Iera Odos 75, 11855 Athens, Greece
2 Choremeio Research Laboratory, First Department of Pediatrics, National and Kapodistrian University of Athens, Thivon & Levadias 8, 11527 Athens, Greece
3 Physiotherapy Department, Faculty of Health and Care Sciences, State University of West Attica, Agiou Spiridonos 28, 12243 Athens, Greece
4 Laboratory of Systematic Botany, Department of Crop Science, School of Plant Sciences, Agricultural University of Athens, Iera Odos 75, 11855 Athens, Greece
* Correspondence: ptara@aua.gr; Tel.: +30-210-529-4262

Abstract: *Rosmarinus officinalis* is a well-studied plant, known for its therapeutic properties. However, its biological activity against several diseases is not known in detail. The aim of this study is to present new data regarding the cytotoxic activity of a hydroethanolic extract of *Rosmarinus officinalis* on glioblastoma (A172) and rhabdomyosarcoma (TE671) cancer cell lines. The chemical composition of the extract is evaluated using liquid chromatography combined with time-of-flight mass spectrometry, alongside its total phenolic content and antioxidant activity. The extract showed a promising time- and dose-dependent cytotoxic activity against both cell lines. The lowest IC_{50} values for both cell lines were calculated at 72 h after treatment and correspond to 0.249 ± 1.09 mg/mL for TE671 cell line and 0.577 ± 0.98 mg/mL for A172 cell line. The extract presented high phenolic content, equal to 35.65 ± 0.03 mg GAE/g of dry material as well as a strong antioxidant activity. The IC_{50} values for the antioxidant assays were estimated at 12.8 ± 2.7 μg/mL (DPPH assay) and 6.98 ± 1.9 μg/mL (ABTS assay). The compound detected in abundance was carnosol, a phenolic diterpene, followed by the polyphenol rosmarinic acid, while the presence of phenolic compounds such as rhamnetin glucoside, hesperidin, cirsimaritin was notable. These preliminary results suggest that *R. officinalis* is a potential, alternative source of bioactive compounds to further examine for abilities against glioblastoma and rhabdomyosarcoma.

Keywords: *Rosmarinus officinalis*; phenolic compounds; chemical analysis; glioblastoma; rhabdomyosarcoma; cancer

Citation: Kakouri, E.; Nikola, O.; Kanakis, C.; Hatziagapiou, K.; Lambrou, G.I.; Trigas, P.; Kanaka-Gantenbein, C.; Tarantilis, P.A. Cytotoxic Effect of *Rosmarinus officinalis* Extract on Glioblastoma and Rhabdomyosarcoma Cell Lines. *Molecules* **2022**, *27*, 6348. https://doi.org/10.3390/molecules27196348

Academic Editor: Nour Eddine Es-Safi

Received: 30 August 2022
Accepted: 21 September 2022
Published: 26 September 2022

1. Introduction

Rosmarinus officinalis L. (Lamiaceae), commonly known as rosemary, is a much-branched, evergreen small shrub, usually 50–100 cm tall. It is native in the Mediterranean region and widely cultivated elsewhere for its essential oil, as well as ornamental purposes. Most Greek populations are probably naturalized and originated from cultivated plants, but at least some populations are considered native in the country [1].

Rosemary is considered a typical spice of the Mediterranean diet and it has been characterized as a functional ingredient [2,3]. Traditionally, rosemary leaves have been used against muscle, joint and rheumatism pain [4], as a stimulant and diaphoretic and for its flatulence-relieving properties [5,6]. Headaches, epilepsy, dysmennorhea, inflammation and spasmolytic conditions were also treated with rosemary [7,8]. Nowadays rosemary is among the most studied medicinal plants and its essential oil and extracts' therapeutic activity has been evaluated against various diseases [9,10]. In particular, *R. officinalis* extracts have been studied for their antioxidant, anticancer, anti-inflammatory, and antimicrobial activity. Ameliorating the status of metabolic and central nervous system (CNS) disorders has also been evaluated [11–14].

Glioblastoma is an aggressive, malignant cancer of the CNS that originates from the glial cells, characterized by poor survival rate. One reason may be an intrinsic or acquired resistance to radiation and chemotherapy, as many brain tumors could intrinsically manifest a multidrug resistance (MDR) phenotype, thus resulting in relapses or disease progression [15,16]. Rhabdomyosarcoma forms at the soft tissues and more frequently affects the skeletal muscle tissue. It is generally considered a disease of childhood, as most cases are observed between the ages of 0–18 years old. Localized disease is associated with a good prognosis and an overall 5-year survival rate of over 80% with combined surgery, radiation therapy, and chemotherapy. However, in metastatic disease, prognosis is poor with a 5-year event-free survival rate of less than 30% [17,18]. It is the third most common extracranial tumor of the pediatric population, accounting for 4.5% of all cases of childhood cancers.

Both are considered rare types of cancers. Rare-type cancers comprise 22% of the reported cases of cancer [19]. Generally, among the difficulties that accompany a rare disease is the reluctance of pharmaceutical industries to invest time and, of course, a considerable amount of money for the development of a specific therapeutic treatment, since it will be addressed only to a small population. Therefore, one of the ongoing challenges is the continuous gaining of experimental data that will significantly contribute and facilitate the design of specific pharmaceutical formulations. However, independent of the cost required for the research of new pharmacologically active compounds, it should be taken under consideration that many cancer cells are resistant to current therapy due to mutations. Although current therapeutic approaches aim to alleviate symptoms, increase life expectancy and maintain the progression of the disease in remission, they are not few the cases of synthetic formulations leading to severe side effects that are not associated to the disease itself. Consequently, because of limitations regarding the many side effects that impair quality of daily life, cancer drug resistance, rapid increase in the percentage of cancer mortality and numerous new cases diagnosed, scientists are driven towards the development of new therapeutic agents, with fewer or no side effects, to be used as monotherapy or together with current available treatment. To this end, natural products and, in particular, those found in abundance in nature or are easy to cultivate, consist a new area of research, since most of the times the cost of the raw material is affordable and the side effects are usually minimized [20,21].

Given the acquired knowledge from Traditional medicine and the continuous interest in *R. officinalis* as a potential therapeutic agent, the present study aims to evaluate the cytotoxic effect of a hydroethanolic extract of *R. officinalis* against A172 glioblastoma and TE671 rhabdomyosarcoma cell lines, since its effect against these two cancer types has not been previously reported. The biologic activity of a plant is attributed to its chemical profile. However, the chemical profile is strongly dependent on many parameters [22,23]. Therefore, although the chemical characterization of *R. officinalis* extracts has already been given in previous studies [24–26], here, we present again the chemistry of the extract used, alongside its total phenolic content and its antioxidant activity.

2. Results

2.1. Total Phenolic Content and Antioxidant Activity

The extract contains a considerable number of phenolic compounds that corresponds to 35.65 ± 0.03 GAE/g. The extract also exhibited a notable antioxidant activity. The IC_{50} value calculated for the DPPH assay was 12.8 ± 2.7 μg/mL while for the ABTS assay the IC_{50} value was estimated at 6.98 ± 1.9 μg/mL.

2.2. Identification of Secondary Metabolites by LC/Q-TOF/HRMS Analysis

Although the chemistry of *Rosmarinus officinalis* is known, we report again its chemical profile, since not only does it depends upon the area, season, and extraction method but it also is essential for explaining its biological activity on A172 and TE671 cell lines.

Characterization of the compounds presented in *R. officinalis* extract was performed with the LC/Q-TOF/HRMS analysis. Most of the compounds identified were flavonoids and phenolic terpenes. Data obtained from the ESI (+) and the ESI (−) ionization mode are summarized in Table 1. Information regarding the generated ms/ms fragmentation process is given as Supplementary Materials. Identification of the compounds detected was based on data obtained from the MassHunter Workstation Software and literature data [26–30].

Table 1. Tentatively identified compounds of *Rosmarinus officinalis* leaves at the positive and negative ionization mode.

			ESI (+)				ESI (−)			
Peak Number	Identification	Molecular Formula	Observed Mass	Mass Error (Δ*m*)	$[M+H]^+$ (*m/z*)	t_R	Observed Mass	Mass Error (Δ*m*)	$[M-H]^-$ (*m/z*)	t_R
1	caffeic acid hexoside	$C_{15}H_{18}O_9$	343.1023	0.00	163.0387; 145.0273; 135.0428	2.32	341.0875	−0.91	179,0340; 161,0237; 135.0442	1.97
2	caffeic acid	$C_9H_8O_4$	181.0496	0.39	163.0385; 135.0444; 117.0328	2.68		n.d		
3	chlorogenic acid	$C_{16}H_{18}O_9$	355.1023	0.00	163.0385; 145.0264; 135.0424	2.97	353.0873	−1.44	191.0547; 179.0336; 173.0451; 135.0446	2.90
4	tuberonic acid	$C_{12}H_{18}O_4$	227.1278	0.08	209.1138; 191.1068; 163.1114;	3.56	739.1672	0.54	449.0852; 339.0510; 177.0177	4.78
5	rhamnetin hexoside	$C_{22}H_{22}O_{12}$	479.1181	−0.63	317.0648; 302.0425; 163.0381	7.05		n.d.		
6	hesperidin	$C_{28}H_{34}O_{15}$	611.1968	−0.41	303.0857; 285.0757; 195.0284; 153.0180	7.81	609.1453	−1.51	300.0268; 271.0241; 255.0292; 151.0032	5.97
7	apigenin glucoside	$C_{21}H_{20}O_{10}$	433.1129	−0.05	271.0602; 119.0468	7.88	463.0871	−1.51	300.0267; 271.0240; 151.0029	6.45
8	hispidulin rutinoside	$C_{28}H_{32}O_{15}$	609.1821	1.15	463.1221; 301.0702; 269.0288	8.19	593.1509	−0.50	327.0473; 285.0388; 255.0288; 227.0343; 151.0054	7.06
9	rosmarinic acid hexoside	$C_{24}H_{26}O_{13}$		n.d			521.1292	−1.67	359.0800; 179.0334; 133.0305	7.32
10	rosmarinic acid	$C_{18}H_{16}O_8$	361.0918	0.28	181.0473; 163.0386; 135.0341	8.25	359.0764	2.34	197.0445; 179.0337; 161.0236	8.30
11	umbelliferone	$C_9H_6O_3$	163.0391	0.80	145.0279; 117.0331	8.52		n.d		

Table 1. *Cont.*

Peak Number	Identification	Molecular Formula	ESI (+) Observed Mass	ESI (+) Mass Error (Δ*m*)	ESI (+) [M+H]$^+$ (*m/z*)	ESI (+) t_R	ESI (−) Observed Mass	ESI (−) Mass Error (Δ*m*)	ESI (−) [M-H]$^-$ (*m/z*)	ESI (−) t_R
12	luteolin-acetyl-glucuronide	$C_{23}H_{20}O_{13}$	n.d				503.0828	−0.64	399.0726; 285.0390; 199.0381; 151.0016; 133.0285	9.56
13	methyl rosmarinic acid	$C_{19}H_{18}O_8$	n.d				393.09220	−1.85	359.0758 373.0922; 179.0341; 135.0442	9.59
14	cirsimaritin hexoside	$C_{23}H_{24}O_{11}$	477.1395	0.84	300.0861; 282.0507	9.59	n.d			
15	cirsimaritin	$C_{17}H_{14}O_6$	315.0866	0.92	300.0615; 282.0512	13.09	313.0712	−1.79	298.0467; 283.0241	12.99
16	rosmanol	$C_{20}H_{26}O_5$	347.1857	0.29	301.1785; 283.1676	13.83	345.1703	−1.30	301.1791; 283.1691	13.43
17	methyl umbelliferone	$C_{10}H_8O_3$	177.0546	−0.11	149.0230; 93.0310	14.45	n.d.			
18	salvigenin	$C_{18}H_{16}O_6$	329.1020	0.12	296.0680; 268.0727	16.85	285.0392	−4.56	267.0258; 213.0525; 151.9210 133.0281	9.14
19	rosmadial	$C_{20}H_{24}O_5$	n.d				343.1544	−2.04	300.0996	17.67
20	epirosmanol methyl ether	$C_{21}H_{28}O_5$	n.d				359.1856	−2.22	329.1742; 283.1695; 285.1781	17.96
21	carnosol	$C_{20}H_{26}O_4$	331.1900	−1.15	285.1844; 243.1364	18.51	329.1748	−3.13	285.1852	18.51
22	carnosol isomer	$C_{20}H_{26}O_4$	331.1902	−0.54	285.1848; 243.1385	18.62	n.d			
23	rosmaridiphenol	$C_{20}H_{28}O_3$	317.2112	0.00	299.1998; 285.1872; 281.1906	19.97	n.d.			

n.d: not detected.

2.3. Evaluation of Cytotoxicity

Both cell lines were exposed to increased concentrations of the extract ranging from 6.25–0.04 mg/mL. The extract exhibited its cytotoxic effect in a dose- and time-dependent manner. Significant differences were observed between the control group and the treated cells, quite at the same range of concentrations. For TE671 cells, the range of the concentrations that reduces cell growth and proliferation ranged from 6.25 mg/mL to 0.39 mg/mL. Note that at the concentration of 0.19 mg/mL, no significant differences were observed at 24 and 48 h of treatment, where proliferation seems to begin. On the contrary, this effect was not observed at 72 h (Figure 1A).

For the A172 cell line, this effect was evident at the concentrations from 6.25 mg/mL to 0.78 mg/mL while at the concentration of 0.39 mg/mL, no statistically significant differences were observed in comparison to the control group, while proliferation of cancer cells had begun.

Interestingly, regarding dose-dependent results and the concentration of 0.78 mg/mL, the effect of the extract was maximal at 72 h after treatment (Figure 1B).

In addition, common for both cell lines is the fact that, for TE671 cells and for the concentrations ranging from 6.25–0.39 mg/mL, the degree of the cytotoxic effect of the extract was the same. For A172 cells, the same was observed for the concentrations ranging from 6.25–1.56 mg/mL. Furthermore, when cells were treated with 0.78 mg/mL and at 24 and 48 h, although the extracts' cytotoxic activity was still evident, at the same time proliferation of cells had begun slightly. On the contrary, at 72 h of treatment, cancer-cells' viability had not considerably increased with respect to that of 24 and 48 h.

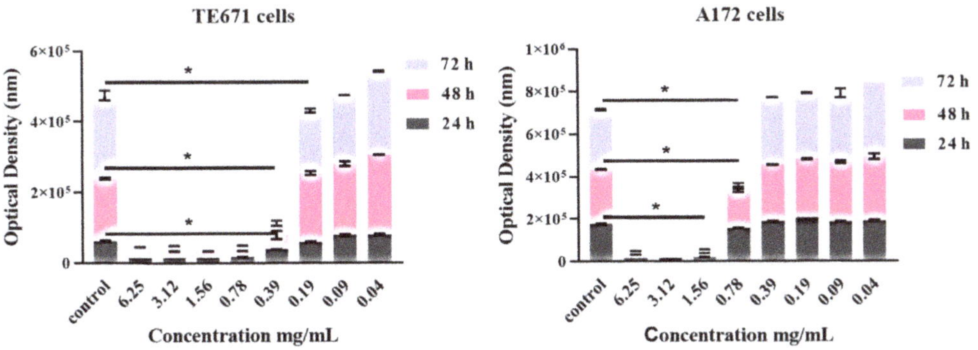

Figure 1. Dose-dependent and time-dependent effect of *R. officinalis* extract on TE671 (**A**) and A172 (**B**) cells. Data are presented as the mean ± standard error of the mean (SEM) ($n = 8$). The asterisk (*) indicates significant differences between untreated and treated cells. The grey color corresponds to 24 h of treatment, the pink to 48 h and the light blue to 72 h.

IC_{50} value, thus, half of the maximal concentration of the tested extract required to inhibit growth and proliferation of cancer cells, was estimated. Dose–response curves regarding all the time points were constructed using a four-parameter logistic model. Normalized results are presented as log10 concentration in Figure 2A for the TE671 cancer cell line and in Figure 2B for the A172 cancer cell line. In the case of TE671 cells, the lowest IC_{50} value was estimated at 0.249 ± 1.09 mg/mL at 72 h after treatment with the extract. The IC_{50} values at 24 and 48 h were calculated at 0.287 ± 1.22 mg/mL and 0.274 ± 1.4 mg/mL, respectively. Regarding A172 cells, the lowest IC_{50} value was observed at 0.577 ± 0.98 mg/mL at 72 h after treatment. For the first 24 h, the IC_{50} value was calculated at 0.952 ± 1.11 mg/mL and after 48 h of treatment the corresponding value was found to be 0.871 ± 1.36 mg/mL. IC_{50} values decreased with increasing exposure time. The calculated values demonstrated that TE671 cells are more sensitive to the extract, since the IC_{50} value is lower than that of A172 cells. In addition, as is presented in Figure 2, the behavior of TE671 cells at all time points is almost the same, given the fact that IC_{50}s do not differ considerably. On the contrary, for A172 cells, those values are rather close for the first 24 and 48 h; however, at 72 h, IC_{50} significantly decreases. This might be attributed to the population doubling time, which reached 80 h for TE671 cells and 40 h for A172 cells.

Microscopical investigation of TE671 is presented in Figure 3A–C. More precisely, in Figure 3A, cells are confluent since they have undergone any treatment, while at the concentration of 0.39 mg/mL cells are significantly reduced (Figure 3B). In Figure 3C, which corresponds to the concentration of 0.19 mg/mL, cells proliferation has begun. In the case of A172 cells, the same behavior was observed (Figure 4A–C). Figure 4A represents those cells that have received no treatment. At the concentration of 0.78 mg/mL, cells are less confluent (Figure 4B), while at the concentration of 0.39 mg/mL, proliferation of cells is evident (Figure 4C).

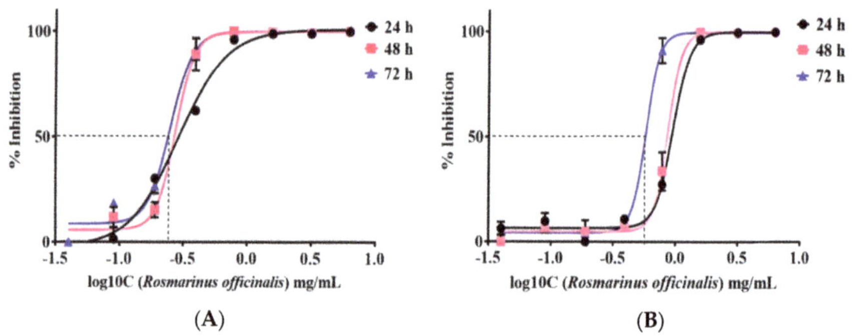

Figure 2. IC_{50} of *R. officinalis* extract on TE671 cells (**A**) and A172 cells at 24, 48 and 72 h (**B**). The lowest IC_{50} value for TE671 cell line was 0.249 ± 1.09 mg/mL, calculated at 72 h and 0.577 ± 0.98 mg/mL for A172 cell line, calculated at 72 h. Cancer -cell viability increases as concentration of the drug decreases.

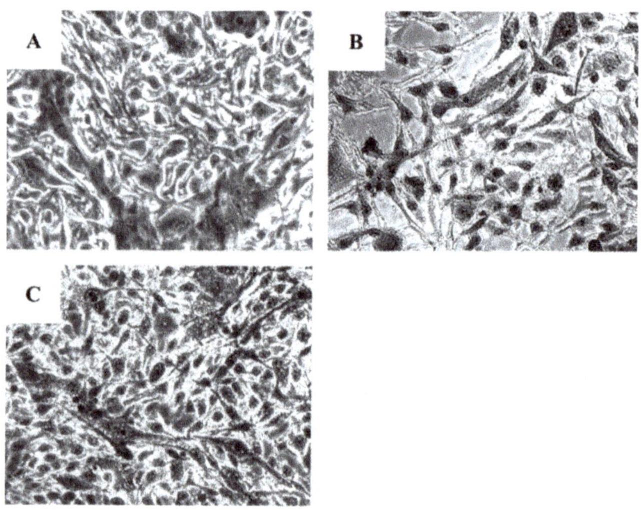

Figure 3. Microscopic images of the TE671 rhabdomyosarcoma cells, grown for 72 h in DMEM with no other treatment (**A**), cells treated with 0.39 mg/mL of the extract (**B**) and cells treated with 0.19 mg/mL of the extract (**C**). Images were captured at ×200 magnification.

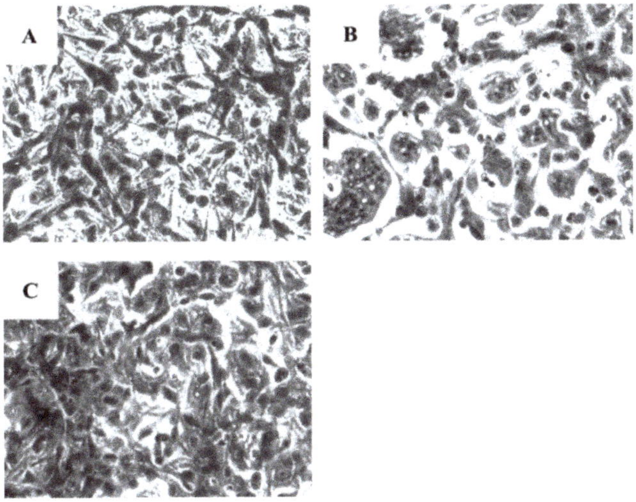

Figure 4. Microscopic images of the A172 glioblastoma cells, grown for 72 h in DMEM with no other treatment (**A**), cells treated with 0.78 mg/mL of the extract (**B**) and cells treated with 0.39 mg/mL of the extract (**C**). Images were captured at ×200 magnification.

3. Discussion

R. officinalis is a plant known for its potent antioxidant activity as it has been evaluated in many studies and with different antioxidant assays. Such activity is mainly attributed to the presence of rosmarinic acid, carnosic acid, carnosol and rosmanol [25,31,32]. It has been proposed that the catechol group of these compounds is responsible for their antioxidant activity [25].

R. officinalis belongs to the Lamiaceae family, a well-known family which includes a variety of plant species that contain a plethora of bioactive compounds [33]. *R. officinalis* leaves' extracts have been studied for their chemical composition and the presence of multiple compounds that belong to flavonoids (apigenin, genkwanin, scutellarein), phenolic diterpenes (carnosol, rosmanol, epirosmanol, carnosic acid), triterpenes (ursolic acid, betulinic acid) and caffeic acid esters (rosmarinic acid) has been reported [24,25].

Results reported in previous studies are in accordance with data presented here. In particular, in our study, chemical analysis of the hydroethanolic extract of *R. officinalis* showed the presence of rosmarinic acid, hydroxycinnamic acids, flavonoids and phenolic terpenes. According to the relative abundance as generated by the MassHunter software, carnosol was the compound presented in abundance followed by rosmarinic acid. Rosmanol, epirosmanol and rosmaridiphenol are metabolites derived from carnosic acid. Carnosol is an oxidized derivate of carnosic acid, produced via a non-enzymatic reaction [34]. References do report both the presence of carnosic acid and carnosol in *R.officinalis* plants [29,30,35]. Furthermore, many studies indicate that plants of the genus *Rosmarinus* grown in the Mediterranean, area are a very rich source of carnosic acid [25,36,37]. However, under extreme environmental conditions and in order for the plant to protect itself from various exogenous invasions, oxidative stress is unavoidable. That means that abiotic-induced stress was possibly the main reason for the oxidation of carnosic acid to carnosol, as well as for the presence of other oxidation metabolites, as previously reported [34,36].

Three are the most studied compounds isolated from *R. officinalis* extracts, carnosic acid, carnosol and rosmarinic acid. Carnosic acid is a compound commonly found in Lamiaceae species and was first isolated from *Salvia officinalis* [38]. Later, it was also found in abundance in rosemary which is yet considered as the richest source of all the Lamiaceae family plants. Chemically, carnosic acid is a phenolic diterpene and has been studied for its health-promoting properties, namely, antioxidant, antitumor, chemo-preventive, anti-inflammatory and hypoglycemic [39]. Carnosol belongs to phenolic diterpenes. It is a strong antioxidant, anticancer, chemo-preventive and anti-inflammatory agent [12,34,40]. The third well-studied compound of the plant is rosmarinic acid. Rosmarinic acid has been documented as a strong antioxidant and antimicrobial compound and it has also been tested against different cancer cell lines and against skin-irritating conditions such as atopic dermatitis [41–44]. Although the above three compounds are usually found in abundance in *R. officinalis* extracts, the therapeutic activity of the plant specifically regarding cancer, is not attributed only to these [45]. It has been observed that extracts from *R. officinalis* exert better antitumor activity with respect to its isolated compounds, precisely carnosol, ursolic and carnosic acid [46,47]. Interestingly, in the study of González-Vallinas et al., (2014) [47], a combination of carnosic acid and carnosol presented a better antiproliferative activity probably due to the synergistic effect of the two compounds. Given the cost advantages for a pharmaceutical company regarding the use of an extract rather than purified compounds and taking into account the above-mentioned findings, in this study a hydroethanolic extract derived from the leaves of the plant was used, to evaluate its cytotoxic activity against A172 and TE671 cancer cell lines. According to our knowledge, this is the first time that the cytotoxicity of *R. officinalis* has been evaluated against these two specific cell lines.

In our study, we observed that treatment with *R. officinalis* manifested a threshold-like mechanism, as it appeared that up to certain concentrations the extract manifested similar toxicity as the control sample and, on the other hand, after a certain "step" (0.39 mg/mL) the extract becomes effective. This phenomenon was not only dose-dependent but also time-dependent, as it manifested the same behavior at 24 h, 48 h and 72 h. This type of action is reported for the first time. Studies concerning the effects of *R.officinalis* on prostate cancer cells [48,49], melanoma [50] and in hematopoietic, epithelial, and mesenchymal tumor cell types [51] manifested a gradual dose-dependent type of action. Thus, the most interesting conclusion from these observations is that *R. officinalis* acts on tumor cell survival differently, depending on the cell type. All studies agree that the extract is effective against tumor cells, yet the fact that it acts in a cell-dependent manner urges towards a more in-depth investigation into its mechanics.

R. officinalis anticancer effects against glioblastoma cell lines have been previously described. U87MG has been used and it was shown that an aqueous extract of the plant (1/75 v/v dilution) managed to inhibit cancer cell proliferation by 42%. On the contrary, rosemary extract boosted the viability of mouse embryonic fibroblasts cells (MEF) by 9.5%. Authors compared the efficacy of the extract with that of etoposide, a highly toxic agent that causes myelosuppression. Etoposide reduced cell viability to a higher degree with respect to *R. officinalis*. However, authors also showed that co-treatment with the extract and etoposide does not influence the chemotherapeutic agent toxicity but increases cells rate inhibition. Nevertheless, rosemary extract does not seem to inhibit growth in MEF cells to the same degree as etoposide [52].

Carnosol was examined for its potent cytotoxic activity on the U87MG glioblastoma cell line. Using a range of concentrations between 100 nM–60 μM, carnosol not only significantly inhibited in a dose-dependent manner cancer cell viability at 48 and 72 h of treatment, but also its anti-proliferative effect continued even after washing the substance. Furthermore, the compound did not promote the metastasis of cancer cells. The same effect was also observed when cells were treated with a mixture of carnosol and temozolomide, an alkylating agent, used to treat brain tumors. In addition to this, carnosol potentiated the cytotoxic effect of temozolomide. Of note, also, is that carnosol did not affect the proliferation of healthy cells. In addition, since U87MG cell lines express the p53 gene, the possible involvement of carnosol in the p53-activation pathway was investigated. A re-activation of p53 and the concomitant activation of BAX protein and deactivation of Bcl-2 were observed [53]. The results of our study are in accordance with the above-mentioned studies. The extract of *R. officinalis* inhibited the growth and proliferation of A172 glioblastoma cells. In contrast to the study of Giacomelli et al. (2016) [53], in our study, the extract exhibited its antiproliferative effect after 24 h of treatment and the peak of its effect was observed at 72 h. Nonetheless, its cytotoxic potency did not outweigh that of carnosol expressed as IC_{50} values.

Chemotherapeutic treatment of rhabdomyosarcoma includes the use of agents such as doxorubicin, vinblastine, and etoposide. Combination of these drugs with rosemary extracts allowed to diminish the concentration of the chemical agent, thus reducing its toxic effects [45,54]. Regarding the cytotoxic activity of *R. officinalis* on rhabdomyosarcoma cell lines, there is lack of literature data. According to our results, the extract used is capable of inhibiting cancer cell proliferation by exerting its best activity at 72 h after treatment.

In general, many studies report the use of secondary metabolites against cancer [55–60].

Regarding the family of phenolic compounds, those belong to the most studied biomolecules. The anticancer activity of phenolic compounds has been demonstrated in a variety of malignant cell lines such as HT-1080 fibrosarcoma cell line, HT-44 melanoma cells, HT-20, HT-29 and DLD-1 colon cancer cells, MCF-7, MDA-MB 468 and 231, T47D breast cancer cells, PC-3 and LNC prostate cancer cells, HS-22 lung cancer cells, SGC-7901 gastic cancer cells, cervical cancer cells (HeLa), human leukemia (HL-60) and NB-4 promyelocytic leukemia cells, adenocarcinomic human alveolar basal epithelial cells (A549) and OAW adenocarcinoma cancer cells [58,61].

In our study, visual observation of microscopic images of the cells demonstrated that the extract exerts its cytotoxic activity by reduction in cell population. In addition, reduction in cell size was observable, as well as a nuclei fragmentation, which confirmed the observed cytotoxicity through the photometric method. This observation gave us a hint for the type of cell death caused by the extract, yet with more investigations in need to confirm.

A lot of mechanisms have been proposed to explain the cytotoxic effect of phenolic compounds. For example, phenolics chemoprotective/ anticancer activity is mainly due to their antioxidant and anti-inflammatory properties and many studies relate a phenolic rich diet with minor incidence of cancer development [62–65]. Many researchers have pointed out the potential of these molecules to interfere with crucial signaling pathways of the proliferation, migration, differentiation, apoptosis and angiogenesis of cancer cells [61,66–68]. For example, cinnamic and benzoic acid induce their antiproliferative effect on melanoma and breast cancer cells by interrupting the S and G2/M phase, respectively. Furthermore, caffeic acid, 5-caffeoylquinic acid, di-caffeoylquinic acid, ferulin and p-coumaric acid exert a potent antiproliferative effect against various cancer cell types [61]. In addition, cell death is another point that has been evaluated using phenolic compounds. Arrest of the cell cycle at Go/G1 phage, morphological changes in cancer cells; activation of apoptosis regulators such as caspaces and Bax protein and p53 and p21 genes; downregulation of transcription factors, such as transcription factor kappa B (NF-kB) and Bcl-2 (B-cell lymphoma 2) gene; and inhibition of enzymes vital for DNA transcription are some examples that confirm the potential of phenolic compounds to accelerate cancer cell death [63,69].

4. Materials and Methods

4.1. Plant Material

Plant material of *R. officinalis* was collected from the Botanical Garden of Philodassiki Enossi Athinon, at the foothills of Mt Hymettus (Attica, Greece). The living collection established in the Botanical Garden originated from a native population located in Ritsona area (eastern Sterea Ellas, Greece). Voucher specimen was deposited at the Herbarium of the Agricultural University of Athens (ACA), with the following label: Greece, Sterea Ellas, prefecture of Attiki, Botanical Garden of Philodassiki Enossi Athinon, alt. 360 m, 37°57′ N, 23°47′ E, 20.09.2016, Trigas 6327, ACA.

4.2. Sampling Extraction

Four grams of dried *R. officinalis* leaves were extracted as previously described by Kakouri et al., (2019) [70], in an ultrasonic water bath using a hydroethanolic solution (70% v/v). Extraction took place in triplicate.

4.3. Total Phenolic Content and Antioxidant Activity

Total phenolic content was performed using Folin–Ciocalteu reagent (0.2N) and gallic acid to construct the calibration curve. The experiment took place as previously described by Kakouri et al., (2019) [70]. Results were expressed as mg of gallic acid equivalents (GAE) per gram of dry material, derived from threefold measurements and according to the following equation:

$$y = 0.0012x + 0.012\ (r = 0.998) \tag{1}$$

The antioxidant activity was estimated using the 2,2-Diphenyl-1-picrylhydrazyl (DPPH•) and the 2,2′-azinobis [3-ethylbenzthiazoline-6-acid] (ABTS•+) radical scavenging assays. The experimental procedure followed that of Kakouri et al., 2019 [70]. For both the assays trolox was used as standard antioxidant. Results were expressed as IC_{50} values and according to the following equation:

$$\%\ \text{Inhibition} = (\text{Acontrol} - \text{Asample})/\text{Acontrol} \times 100 \tag{2}$$

4.4. LC/Q-TOF/HRMS Conditions

To identify the chemical profile of *R. officinalis* extract an HPLC system (high performance liquid chromatography) consisting of a degasser, autosampler, quaternary pump, diode array detector, and column oven (Agilent Series 1260, Agilent Technologies, Santa Clara, CA, USA), coupled to a 6530 Q-TOF mass spectrometer (Agilent Technologies, Santa Clara, CA, USA) was used. Experimental conditions were adjusted as in the study of Kakouri et al., (2019) [70]. The extract was analyzed under the positive and negative ionization mode. The parameters set for the Q-TOF mass analysis follow those described at our previous analysis [70]. CID-ms/ms spectra were recorded on the auto MS/MS mode. Mass range was set to 50–1000 and collision energy was set at 40 V. Results were analyzed using the Agilent MassHunter Workstation software LC-MS Data Acquisition for 6530 series Q-TOF (version B07.00, Agilent Technologies, Santa Clara, CA, USA).

4.5. Evaluation of Cytotoxicity

4.5.1. Cells Treatment before and after Exposure to the Extract

The TE671 rhabdomyosarcoma cancer cell line was obtained from the European Collection of cell cultures (ECACC, London, UK). The A172 glioblastoma cell line was obtained from a male patient of 53 years old (ECACC, London, UK, Cat. Nr 88062428). Cells were grown in a cell-culture flask (75 cm^2 surface area) in Dulbecco's modified Eagle's Medium (DMEM) (ThermoFisher Scientific Inc. (Gibco), Waltham, MA, USA Cat. Nr. 10566016) enriched with glucose (4500 mg/mL), and 15% fetal bovine serum (FBS) (ThermoFisher Scientific Inc. (Gibco), Waltham, MA, USA Cat. Nr. 26140-079), L-glutamine (2 mM) (ThermoFisher Scientific Inc. (Gibco), Waltham, MA, USA Cat. Nr. A12860-01). A dual antibiotic solution of penicillin G (100IU) and streptomycin (100 μg/mL) was added (ThermoFisher Scientific Inc. (Gibco), Waltham, MA, USA Cat. Nr. 15140-122), in addition to an amphotericin B solution (ThermoFisher Scientific Inc. (Gibco), Waltham, MA, USA Cat. Nr. 14140-122). A Coulter counter apparatus was used to measure the number of cells inoculated in the experimental setup. Cells were seeded in 96-well plates (1.5×10^3 cells/mL) and were allowed to grow for 24 h until reaching ~80% confluence. After 24 h, cells were exposed to successive diluted concentrations of the extract (t = 0 h), ranging from 0.04 to 6.25 mg/mL, derived from the dried extract diluted de novo in DMSO (10% *v*/*v*). The final concentration of DMSO when the extract was added to cell culture was 1% *v*/*v*. A control well with the same concentration of DMSO was used to confirm that no toxic effect was observed. Cells were incubated for 24, 48 and 72 h.

Experiments were performed in 96-well plates (CellStar® Sigma-Aldrich Chemie GmbH, Taufkirchen, DE Cat. Nr. M3687-60EA, Saint Louis, MO, USA). Plate set up was as follows: a column contained only cell culture medium, a column of cell culture and the staining chemical, a column with cultured cells and a column with cultured cells plus the staining chemical. The remaining wells were used for the testing of the extract in various concentrations. As blank were used those wells containing cell culture medium only, cells and no staining agent or drug, whereas as positive control were used those wells with cultured cells without staining agent. All experiments were performed in triplicate.

4.5.2. Alamar Blue Assay

Assessment of cell viability after incubation with testing agents was performed with resazurin reduction experiments, using Alamar Blue viability assay. Cell viability at each time point (24, 48, 72 h) was quantified by adding Alamar Blue (Gibco, Invitrogen Inc. Carlsbad, CA, USA) to each well. Treated cells were supplemented with 10% alamar blue reagent and incubated for 6 h at 37 °C. Wells that contained only alamar blue were considered as blank while positive control were considered those wells that contained the untreated cells. Percent viability was calculated according to the following formula:

$$V\ (\%) = (OD1 - OD2 - Blank)/(OD3 - OD2) \times 100 \quad (3)$$

where OD1 stands for the optical density in nm for treated with the chemical cells, OD2 stands for the optical density in nm for those well containing nutrient medium and the chemical and OD3 stands for the cells that were not exposed to the chemical. Optical density was read at 570 nm. Results were expressed as IC_{50} values; thus, the concentration of the chemical that causes 50% inhibition with respect to the untreated cells.

4.5.3. Giemsa Staining

Cells were colored with the Giemsa stain. Briefly, 100 μL of pure ethanol were added to a 96-well plate containing the treated cells after removing the nutrient medium. Cells were left in ethanol for five minutes and then 100 μL of the Giemsa stain were added. Plates remained for 15 min at room temperature, followed by stain removal and cell washing with 100 μL of NaCl 0.9% (*v*/*v*). Cells were microscopically observed at 24, 48 and 72 h of incubation.

4.5.4. Data Analysis

The GraphPad Prism (version 8.4.2, GraphPad Software for Windows, San Diego, CA, USA) was used to calculate the IC_{50} value according to a four-parameter logistic model. Dose- and time-dependent effect of the tested extract with respect to the control group were also calculated with GraphPad Prism (version 8.4.2). All data were presented as mean ± standard error of the mean (SEM). Statistical differences between untreated and treated cells were evaluated with the Student *t*-test. p values < 0.05 were considered statistically significant and confidence intervals were at ±95% (±95% CI). Normalized results are presented as log10 concentration.

5. Conclusions

In conclusion, the extract examined in this study highlights, for the first time, the cytotoxic effect of *R. officinalis* against TE671 and A172 cancer cell lines. The extract inhibited cancer cell proliferation in a dose- and time-dependent manner, with TE671 cells being more susceptible to the treatment with the extract. A further approach of this study would be to determine the mechanism by which *R. officinalis* exhibited its cytotoxic activity. However, our results are a first approach to the use of the plant as a candidate therapeutic agent. As expected, the extract presented notable antioxidant activity, high total phenolic content and a rich chemical profile. Given the importance of phenolic compounds as potent antioxidant molecules and the connection of antioxidant activity with cancer treatment, this Mediterranean plant is certainly worth further investigation.

Supplementary Materials: The following supporting information can be downloaded at: https://www.mdpi.com/article/10.3390/molecules27196348/s1, File S1: Detailed description of the ms/ms fragmentation process.

Author Contributions: Conceptualization, E.K., G.I.L. and P.A.T.; methodology, E.K., O.N., C.K. and K.H.; investigation, E.K., C.K., G.I.L., P.T., C.K.-G. and P.A.T.; data curation, E.K. and C.K.; writing—original draft preparation, E.K.; writing—review and editing, E.K.; O.N., C.K., K.H., G.I.L., P.T., C.K.-G. and P.A.T.; supervision, P.A.T. All authors have read and agreed to the published version of the manuscript.

Funding: This research received no external funding.

Institutional Review Board Statement: Not applicable.

Informed Consent Statement: Not applicable.

Data Availability Statement: Not applicable.

Conflicts of Interest: The authors declare no conflict of interest.

Sample Availability: Samples of *R. officinalis* extract are available from the authors.

References

1. Strid, A. Atlas of the Aegean Flora. Part 1: Text & Plates. Part 2: Maps. Berlin: Botanic Garden and Botanical Museum, Freie Universität Berlin. Englera; 33 (1, 2). *Edinb. J. Botany* **2016**, *73*, 371–373.
2. Naito, Y.; Oka, S.; Yoshikawa, T. Inflammatory Response in the Pathogenesis of Atherosclerosis and Its Prevention by Rosmarinic Acid, a Functional Ingredient of Rosemary. In *Food Factors in Health Promotion and Disease Prevention*; ACS Publications: Washington, WA, USA, 2003; Volume 851, pp. 208–212. [CrossRef]
3. Kaur, R.; Gupta, B.T.; Bronlund, J.; Kaur, L. The potential of rosemary as a functional ingredient for meat products—A review. *Food Rev. Int.* **2021**. [CrossRef]
4. Andrade, J.M.; Faustino, C.; Garcia, C.; Ladeiras, D.; Reis, C.P.; Rijo, P. *Rosmarinus officinalis* L.: An update review of its phytochemistry and biological activity. *Future Sci. OA* **2018**, *4*, FSO283. [CrossRef] [PubMed]
5. Peng, C.H.; Su, J.D.; Chyau, C.C.; Sung, T.Y.; Ho, S.S.; Peng, C.C.; Peng, R.Y. Supercritical fluid extracts of rosemary leaves exhibit potent anti-inflammation and anti-tumor effects. *Biosci. Biotechnol. Biochem.* **2007**, *71*, 2223–2232. [CrossRef] [PubMed]
6. Peter, K.V.; Shylaja, M.R. Introduction to herbs and spices: Definitions, trade and applications. In *Handbook of Herbs and Spices*, 2nd ed.; Peter, K.V., Ed.; Woodhead Publishing Series in Food Science, Technology and Nutrition; Woodhead Publishing: Cambridge, UK, 2012; Volume 1, pp. 1–24.
7. Faheem, M.; Ameer, S.; Khan, A.W.; Haseeb, M.; Raza, Q.; Shah, F.A.; Khusro, A.; Aari, C.; Khayani-Sahibzada, M.U.; Batiha, G.; et al. A comprehensive review on antiepileptic properties of medicinal plants. *Arab. J. Chem.* **2022**, *15*, 103478. [CrossRef]
8. Ribeiro-Santos, R.; Carvalho-Costa, D.; Cavaleiro, C.; Costa, H.S.; Albuquerque, T.G.; Castilho, M.C.; Ramos, F.; Melo, N.R.; Sanches-Silva, A. A novel insight on an ancient aromatic plant: The rosemary (*Rosmarinus officinalis* L.). *Trends Food Sci. Technol.* **2015**, *45*, 355–368. [CrossRef]
9. Alvi, S.S.; Ahmad, P.; Ishrat, M.; Iqbal, D.; Khan, M.S. Secondary Metabolites from Rosemary (*Rosmarinus officinalis* L.): Structure, Biochemistry and Therapeutic Implications Against Neurodegenerative Diseases. In *Natural Bio-active Compounds*; Swamy, M., Akhtar, M., Eds.; Springer: Singapore, 2019; pp. 1–24. [CrossRef]
10. Borges, R.S.; Ortiz, B.; Pereira, A.; Keita, H.; Carvalho, J. *Rosmarinus officinalis* essential oil: A review of its phytochemistry, anti-inflammatory activity, and mechanisms of action involved. *J. Ethnopharmacol.* **2019**, *229*, 29–45. [CrossRef] [PubMed]
11. González-Trujano, M.E.; Peña, E.I.; Martínez, A.L.; Moreno, J.; Guevara-Fefer, P.; Déciga-Campos, M.; López-Muñoz, F.J. Evaluation of the antinociceptive effect of *Rosmarinus officinalis* L. using three different experimental models in rodents. *J. Ethnopharmacol.* **2007**, *111*, 476–482. [CrossRef]
12. Johnson, J.J. Carnosol: A promising anti-cancer and anti-inflammatory agent. *Cancer Lett.* **2011**, *305*, 1–7. [CrossRef]
13. González-Vallinas, M.; Molina, S.; Vicente, G.; de la Cueva, A.; Vargas, T.; Santoyo, S.; García-Risco, M.R.; Fornari, T.; Reglero, G.; Ramírez de Molina, A. Antitumor effect of 5-fluorouracil is enhanced by rosemary extract in both drug sensitive and resistant colon cancer cells. *Pharmacol. Res.* **2013**, *72*, 61–68. [CrossRef]
14. de Oliveira, J.R.; Camargo, S.E.A.; de Oliveira, L.D. *Rosmarinus officinalis* L. (rosemary) as therapeutic and prophylactic agent. *J. Biomed. Sci.* **2019**, *26*, 5. [CrossRef] [PubMed]
15. Hatziagapiou, K.; Braoudaki, M.; Karpusas, M.; Tzortzatou-Stathopoulou, F. Evaluation of antitumor activity of gefitinib in pediatric glioblastoma and neuroblastoma cells. *Clin. Lab.* **2011**, *57*, 781–784. [PubMed]
16. Lambrou, G.I.; Hatziagapiou, K.; Vlahopoulos, S. Inflammation and tissue homeostasis: The NF-kappaB system in physiology and malignant progression. *Mol. Biol. Rep.* **2020**, *47*, 4047–4063. [CrossRef] [PubMed]
17. Lambrou, G.I.; Zaravinos, A.; Adamaki, M.; Spandidos, D.A.; Tzortzatou-Stathopoulou, F.; Vlachopoulos, S. Pathway simulations in common oncogenic drivers of leukemic and rhabdomyosarcoma cells: A systems biology approach. *Int. J. Oncol.* **2012**, *40*, 1365–1390. [CrossRef] [PubMed]
18. Zaravinos, C.T.; Tsartsalis, A.; Tagka, A.N.; Kotoulas, A.; Geronikolou, S.A.; Braoudaki, M.; Lambrou, G.I. Systems Approaches in the Common Metabolomics in Acute Lymphoblastic Leukemia and Rhabdomyosarcoma Cells: A Computational Approach. *Adv. Exp. Med. Biol.* **2021**, *1338*, 55–66.

19. Pillai, R.K.; Jayasree, K. Rare cancers: Challenges & issues. *Indian J. Med. Res.* **2017**, *145*, 17–27. [CrossRef] [PubMed]
20. Medina-Franco, J.L. New Approaches for the Discovery of Pharmacologically-Active Natural Compounds. *Biomolecules* **2019**, *9*, 115. [CrossRef] [PubMed]
21. Rodrigues, T.; Reker, D.; Schneider, P.; Schneider, G. Counting on natural products for drug design. *Nat. Chem.* **2016**, *8*, 531–541. [CrossRef] [PubMed]
22. Heleno, S.A.; Martins, A.; Queiroz, M.J.; Ferreira, I.C. Bioactivity of phenolic acids: Metabolites versus parent compounds: A review. *Food Chem.* **2015**, *173*, 501–513. [CrossRef] [PubMed]
23. Shahidi, F.; Ambigaipalan, P. Phenolics and polyphenolics in foods, beverages and spices: Antioxidant activity and health effects—A review. *J. Funct. Foods* **2015**, *18*, 820–897. [CrossRef]
24. Zeng, H.H.; Tu, P.F.; Zhou, K.; Wang, H.; Wang, B.H.; Lu, J.F. Antioxidant properties of phenolic diterpenes from *Rosmarinus officinalis*. *Acta Pharmacol. Sin.* **2001**, *22*, 1094–1098. [PubMed]
25. del Baño, M.J.; Lorente, J.; Castillo, J.; Benavente-García, O.; del Río, J.A.; Ortuño, A.; Quirin, K.W.; Gerard, D. Phenolic diterpenes, flavones, and rosmarinic acid distribution during the development of leaves, flowers, stems, and roots of *Rosmarinus officinalis*. Antioxidant activity. *J. Agric. Food Chem.* **2003**, *51*, 4247–4253. [CrossRef] [PubMed]
26. Almela, L.; Sánchez-Muñoz, B.; Fernández-López, J.A.; Roca, M.J.; Rabe, V. Liquid chromatograpic-mass spectrometric analysis of phenolics and free radical scavenging activity of rosemary extract from different raw material. *J. Chromatogr. A* **2006**, *1120*, 221–229. [CrossRef] [PubMed]
27. Okamura, N.; Haraguchi, H.; Hashimoto, K.; Yagi, A. Flavonoids in *Rosmarinus officinalis* leaves. *Phytochemistry* **1994**, *37*, 1463–1466. [CrossRef]
28. Boudiar, T.; Lozano-Sánchez, J.; Harfi, B.; Del Mar Contreras, M.; Segura-Carretero, A. Phytochemical characterization of bioactive compounds composition of Rosmarinus eriocalyx by RP-HPLC-ESI-QTOF-MS. *Nat. Prod. Res.* **2019**, *33*, 2208–2214. [CrossRef]
29. Borrás-Linares, I.; Stojanović, Z.; Quirantes-Piné, R.; Arráez-Román, D.; Švarc-Gajić, J.; Fernández-Gutiérrez, A.; Segura-Carretero, A. *Rosmarinus officinalis* leaves as a natural source of bioactive compounds. *Int. J. Mol. Sci.* **2014**, *15*, 20585–20606. [CrossRef]
30. Kontogianni, V.G.; Tomic, G.; Nikolic, I.; Nerantzaki, A.A.; Sayyad, N.; Stosic-Grujicic, S.; Stojanovic, I.; Gerothanassis, I.P.; Tzakos, A.G. Phytochemical profile of *Rosmarinus officinalis* and *Salvia officinalis* extracts and correlation to their antioxidant and anti-proliferative activity. *Food Chem.* **2013**, *136*, 120–129. [CrossRef]
31. Pérez-Fons, L.; Garzón, M.T.; Micol, V. Relationship between the antioxidant capacity and effect of rosemary (*Rosmarinus officinalis* L.) polyphenols on membrane phospholipid order. *J. Agric. Food Chem.* **2010**, *58*, 161–171. [CrossRef]
32. Nieto, G.; Ros, G.; Castillo, J. Antioxidant and Antimicrobial Properties of Rosemary (*Rosmarinus officinalis*, L.): A Review. *Medicines* **2018**, *5*, 98. [CrossRef]
33. Hossain, M.B.; Rai, D.K.; Brunton, N.P.; Martin-Diana, A.B.; Barry-Ryan, C. Characterization of phenolic composition in Lamiaceae spices by LC-ESI-MS/MS. *J Agric. Food Chem.* **2010**, *58*, 10576–10581. [CrossRef]
34. Loussouarn, M.; Krieger-Liszkay, A.; Svilar, L.; Bily, A.; Birtić, S.; Havaux, M. Carnosic Acid and Carnosol, Two Major Antioxidants of Rosemary, Act through Different Mechanisms. *Plant Physiol.* **2017**, *175*, 1381–1394. [CrossRef] [PubMed]
35. Bellumori, M.; Innocenti, M.; Congiu, F.; Cencetti, G.; Raio, A.; Menicucci, F.; Mulinacci, N.; Michelozzi, M. Within-Plant Variation in *Rosmarinus officinalis* L. Terpenes and Phenols and Their Antimicrobial Activity against the Rosemary Phytopathogens Alternaria alternata and *Pseudomonas viridiflava*. *Molecules* **2021**, *26*, 3425. [CrossRef] [PubMed]
36. Luis, J.; Johnson, C. Seasonal variations of rosmarinic and carnosic acids in rosemary extracts. Analysis of their in vitro antiradical activity. *Span. J. Agric. Res.* **2005**, *3*, 106–112. [CrossRef]
37. Munné-Bosch, S.; Alegre, L. Drought-induced changes in the redox state of alpha-tocopherol, ascorbate, and the diterpene carnosic acid in chloroplasts of Labiatae species differing in carnosic acid contents. *Plant Physiol.* **2003**, *131*, 1816–1825. [CrossRef] [PubMed]
38. Linde, H. Ein neues Diterpen aus *Salvia officinalis* L. und eine Notiz zur Konstitution von Pikrosalvin. *HCA* **1964**, *47*, 1234–1239. [CrossRef]
39. Birtić, S.; Dussort, P.; Pierre, F.X.; Bily, A.C.; Roller, M. Carnosic acid. *Phytochemistry* **2015**, *115*, 9–19. [CrossRef]
40. Akihisa, T.; Yasukawa, K.; Tokuda, H. Potentially Cancer Chemopreventive and Anti-Inflammatory Terpenoids from Natural Sources. *Stud. Nat. Prod. Chem.* **2003**, *29*, 73–126. [CrossRef]
41. Lee, J.; Jung, E.; Koh, J.; Kim, Y.S.; Park, D. Effect of rosmarinic acid on atopic dermatitis. *J. Dermatol.* **2008**, *35*, 768–771. [CrossRef] [PubMed]
42. Jordán, M.J.; Lax, V.; Rota, M.C.; Lorán, S.; Sotomayor, J.A. Relevance of carnosic acid, carnosol, and rosmarinic acid concentrations in the in vitro antioxidant and antimicrobial activities of *Rosmarinus officinalis* (L.) methanolic extracts. *J. Agric. Food Chem.* **2012**, *60*, 9603–9608. [CrossRef]
43. Benedec, D.; Hanganu, D.; Oniga, I.; Tiperciuc, B.; Olah, N.K.; Raita, O.; Bischin, C.; Silaghi-Dumitrescu, R.; Vlase, L. Assessment of rosmarinic acid content in six Lamiaceae species extracts and their antioxidant and antimicrobial potential. *Pak. J. Pharm. Sci.* **2015**, *28*, 2297–2303.
44. Ivanov, M.; Kostić, M.; Stojković, D.; Soković, M. Rosmarinic acid–Modes of antimicrobial and antibiofilm activities of common plant polyphenol. *S. Afr. J. Bot.* **2022**, *146*, 521–527. [CrossRef]
45. Plouzek, C.A.; Ciolino, H.P.; Clarke, R.; Yeh, G.C. Inhibition of P-glycoprotein activity and reversal of multidrug resistance in vitro by rosemary extract. *Eur. J. Cancer* **1999**, *35*, 1541–1545. [CrossRef] [PubMed]

46. Huang, M.T.; Ho, C.T.; Wang, Z.Y.; Ferraro, T.; Lou, Y.R.; Stauber, K.; Ma, W.; Georgiadis, C.; Laskin, J.D.; Conney, A.H. Inhibition of skin tumorigenesis by rosemary and its constituents carnosol and ursolic acid. *Cancer Res.* **1994**, *54*, 701–708.
47. González-Vallinas, M.; Molina, S.; Vicente, G.; Zarza, V.; Martín-Hernández, R.; García-Risco, M.R.; Fornari, T.; Reglero, G.; Ramírez de Molina, A. Expression of MicroRNA-15b and the glycosyltransferase GCNT3 correlates with antitumor efficacy of rosemary diterpenes in colon and pancreatic cancer. *PLoS ONE* **2014**, *9*, e98556. [CrossRef]
48. Jaglanian, A.; Termini, D.; Tsiani, E. Rosemary (*Rosmarinus officinalis* L.) extract inhibits prostate cancer cell proliferation and survival by targeting Akt and mTOR. *Biomed. Pharmacother.* **2020**, *131*, 110717. [CrossRef] [PubMed]
49. Jang, Y.-G.; Hwang, K.-A.; Choi, K.-C. Rosmarinic Acid, a Component of Rosemary Tea, Induced the Cell Cycle Arrest and Apoptosis through Modulation of HDAC2 Expression in Prostate Cancer Cell Lines. *Nutrients* **2018**, *10*, 1784. [CrossRef] [PubMed]
50. Lin, K.I.; Lin, C.C.; Kuo, S.M.; Lai, J.C.; Wang, Y.Q.; You, H.L.; Hsu, M.L.; Chen, C.H.; Shiu, L.Y. Carnosic acid impedes cell growth and enhances anticancer effects of carmustine and lomustine in melanoma. *Biosci. Rep.* **2018**, *38*, BSR20180005. [CrossRef]
51. Levine, C.B.; Bayle, J.; Biourge, V.; Wakshlag, J.J. Cellular effects of a turmeric root and rosemary leaf extract on canine neoplastic cell lines. *BMC Vet. Res.* **2017**, *13*, 388. [CrossRef]
52. Ozdemir, M.D.; Gokturk, D. The Effect of *Rosmarinus officinalis* and Chemotherapeutic Etoposide on Glioblastoma (U87 MG) Cell Culture. *Turk. Neurosurg.* **2018**, *28*, 853–857. [CrossRef]
53. Giacomelli, C.; Natali, L.; Trincavelli, M.L.; Daniele, S.; Bertoli, A.; Flamini, G.; Braca, A.; Martini, C. New insights into the anticancer activity of carnosol: p53 reactivation in the U87MG human glioblastoma cell line. *Int. J. Biochem. Cell Biol.* **2016**, *74*, 95–108. [CrossRef]
54. Almakhatreh, M.; Hafez, E.; Tousson, E.; Masoud, A. Biochemical and Molecular Studies on the Role of Rosemary (*Rosmarinus officinalis*) Extract in Reducing Liver and Kidney Toxicity Due to Etoposide in Male Rats. *Asian J. Pharm. Sci.* **2019**, *7*, 1–11. [CrossRef]
55. Mann, J. Natural products in cancer chemotherapy: Past, present and future. *Nat. Rev. Cancer* **2002**, *2*, 143–148. [CrossRef] [PubMed]
56. Cragg, G.M.; Newman, D.J. Plants as a source of anti-cancer agents. *J. Ethnopharmacol.* **2005**, *100*, 72–79. [CrossRef] [PubMed]
57. Ververidis, F.; Trantas, E.; Douglas, C.; Vollmer, G.; Kretzschmar, G.; Panopoulos, N. Biotechnology of flavonoids and other phenylpropanoid-derived natural products. Part I: Chemical diversity, impacts on plant biology and human health. *Biotechnol. J.* **2007**, *2*, 1214–1234. [CrossRef]
58. Abotaleb, M.; Samuel, S.M.; Varghese, E.; Varghese, S.; Kubatka, P.; Liskova, A.; Büsselberg, D. Flavonoids in Cancer and Apoptosis. *Cancers* **2018**, *11*, 28. [CrossRef]
59. Talebi, M.; Talebi, M.; Farkhondeh, T.; Simal-Gandara, J.; Kopustinskiene, D.M.; Bernatoniene, J.; Samarghandian, S. Emerging cellular and molecular mechanisms underlying anticancer indications of chrysin. *Cancer Cell Int.* **2021**, *21*, 214. [CrossRef] [PubMed]
60. Talebi, M.; Kakouri, E.; Talebi, M.; Tarantilis, P.; Farkhondeh, T.; İlgün, S.; Pourbagher-Shahri, A.M.; Samarghandian, S. Nutraceuticals-based therapeutic approach: Recent advances to combat pathogenesis of Alzheimer's disease. *Expert Rev. Neurother.* **2021**, *21*, 625–642. [CrossRef]
61. Anantharaju, P.G.; Gowda, P.C.; Vimalambike, M.G.; Madhunapantula, S.V. An overview on the role of dietary phenolics for the treatment of cancers. *Nutr. J.* **2016**, *15*, 99. [CrossRef] [PubMed]
62. García-Lafuente, A.; Guillamón, E.; Villares, A.; Rostagno, M.A.; Martínez, J.A. Flavonoids as anti-inflammatory agents: Implications in cancer and cardiovascular disease. *Inflamm. Res.* **2009**, *58*, 537–552. [CrossRef] [PubMed]
63. Chahar, M.K.; Sharma, N.; Dobhal, M.P.; Joshi, Y.C. Flavonoids: A versatile source of anticancer drugs. *Pharmacogn. Rev.* **2011**, *5*, 1–12. [CrossRef] [PubMed]
64. Zhou, Y.; Zheng, J.; Li, Y.; Xu, D.P.; Li, S.; Chen, Y.M.; Li, H.B. Natural Polyphenols for Prevention and Treatment of Cancer. *Nutrients* **2016**, *8*, 515. [CrossRef] [PubMed]
65. Maleki, S.J.; Crespo, J.F.; Cabanillas, B. Anti-inflammatory effects of flavonoids. *Food Chem.* **2019**, *299*, 125124. [CrossRef]
66. Rayan, A.; Raiyn, J.; Falah, M. Nature is the best source of anticancer drugs: Indexing natural products for their anticancer bioactivity. *PLoS ONE* **2017**, *12*, e0187925. [CrossRef]
67. Kikuchi, H.; Yuan, B.; Hu, X.; Okazaki, M. Chemopreventive and anticancer activity of flavonoids and its possibility for clinical use by combining with conventional chemotherapeutic agents. *Am. J. Cancer Res.* **2019**, *9*, 1517. [PubMed]
68. Allegra, A.; Tonacci, A.; Pioggia, G.; Musolino, C.; Gangemi, S. Anticancer Activity of *Rosmarinus officinalis* L.: Mechanisms of Action and Therapeutic Potentials. *Nutrients* **2020**, *12*, 1739. [CrossRef] [PubMed]
69. Batra, P.; Sharma, A.K. Anti-cancer potential of flavonoids: Recent trends and future perspectives. *3 Biotech* **2013**, *3*, 439–459. [CrossRef] [PubMed]
70. Kakouri, E.; Kanakis, C.; Trigas, P.; Tarantilis, P.A. Characterization of the chemical composition of Drimia numidica plant parts using high-resolution mass spectrometry: Study of their total phenolic content and antioxidant activity. *Anal. Bioanal. Chem.* **2019**, *411*, 3135–3150. [CrossRef] [PubMed]

molecules

MDPI

Article

Phytochemicals from Red Onion, Grown with Eco-Sustainable Fertilizers, Protect Mammalian Cells from Oxidative Stress, Increasing Their Viability

Maria Laura Matrella [1,†], Alessio Valletti [1,†], Federica Marra [2,†], Carmelo Mallamaci [2,†], Tiziana Cocco [1] and Adele Muscolo [2,*]

1 Department of Basic Medical Sciences, Neurosciences and Sense Organs, Biochemistry Section, University of Bari "Aldo Moro", 70124 Bari, Italy
2 Department of AGRARIA, Mediterranea University, Feo di Vito, 89122 Reggio Calabria, Italy
* Correspondence: amuscolo@unirc.it; Tel.: +39-09651694364
† The authors contributed equally to the work.

Abstract: Red onion, a species of great economic importance rich in phytochemicals (bioactive compounds) known for its medicinal properties, was fertilized with sulphur-bentonite enriched with orange residue or olive pomace, with the aim of producing onion enriched in health beneficial compounds. There is a worldwide great demand of minimally processed food or food ingredients with functional properties because of a new awareness of how important healthy functional nutrition is in life. Phytochemicals have the capacity to regulate most of the metabolic processes resulting in health benefits. Red onion bioactive compound quantity and quality can vary according to cultivation practices. The main aims of the current research were to determine the chemical characteristics of the crude extracts from red onion bulbs differently fertilized and to evaluate their biological activity in normal and oxidative stress conditions. The lyophilized onion bulbs have been tested in vitro on two cellular models, i.e., the H9c2 rat cardiomyoblast cell line and primary human dermal fibroblasts, in terms of viability and oxygen radical homeostasis. The results evidenced different phytochemical compositions and antioxidant activities of the extracts obtained from red onions differently fertilized. Sulphur-bentonite fertilizers containing orange waste and olive pomace positively affected the red onion quality with respect to the red onion control, evidencing that sulphur-bentonite-organic fertilization was able to stimulate plant a secondary metabolism inducing the production of phytochemicals with healthy functions. A positive effect of the extracts from red onions treated with fertilizers—in particular, with those containing orange waste, such as the reduction of oxidative stress and induction of cell viability of H9c2 and human fibroblasts—was observed, showing a concentration- and time-dependent profile. The results evidenced that the positive effects were related to the phenols and, in particular, to chlorogenic and p-coumaric acids and to the flavonol kaempferol, which were more present in red onion treated with low orange residue than in the other treated ones.

Keywords: red onion; phytochemicals; polyphenols; oxidative stress; H9c2 rat cardiomyoblast; primary human fibroblasts

Citation: Matrella, M.L.; Valletti, A.; Marra, F.; Mallamaci, C.; Cocco, T.; Muscolo, A. Phytochemicals from Red Onion, Grown with Eco-Sustainable Fertilizers, Protect Mammalian Cells from Oxidative Stress, Increasing Their Viability. *Molecules* **2022**, *27*, 6365. https://doi.org/10.3390/molecules27196365

Academic Editor: Nour Eddine Es-Safi

Received: 11 August 2022
Accepted: 21 September 2022
Published: 27 September 2022

1. Introduction

Nowadays, there is an increasing attention on the food we eat. There is a worldwide great demand of minimally processed food or food ingredients with functional properties because of a new awareness of how a healthy and sustainable living is important. Bioactive food compounds, also known as phytochemicals, have the capacity to regulate most of the metabolic processes resulting in health benefits. So far, about 10,000 phytochemicals have been identified, but a large percentage remains still unknown. The identified phytochemicals include tannins, flavones, triterpenoids, steroids, saponins, and alkaloids. Numerous studies have associated the protective and beneficial roles of phytochemicals with their antioxidant activity, since the overproduction of oxidants (reactive oxygen species and reactive nitrogen species) in the human body is the cause of cellular aging [1], and of many chronic diseases [2]. Antioxidant phytochemicals exist widely in fruits, vegetables, cereal grains, edible macrofungi, microalgae, and medicinal plants. Among the vegetables rich in bioactive compounds, *Allium cepa* L. (the common onion) is one of the oldest plants cultivated around the world and consumed as a vegetable and spice. It is greatly appreciated as a medicinal plant in traditional medicine for its high content of phytochemicals, including polyphenols, flavonoids, and sulphur-based compounds. These secondary metabolites, widely contained in onions, have a different mode of action and biosynthetic pathways but are all able to promote beneficial health effects. A regular onion bulb intake is reported to have profound radical scavenging activity and several beneficial effects on health [3], such as preventing cardiovascular diseases [4], diabetes [5], cancers [6], and neurodegeneration [7]. Therefore, foods in our diet that can aid in the prevention of these diseases are of major interest to both the scientific and public communities.

Epidemiological data evidenced that a high intake of onions was positively correlated with a low risk of carcinoma [8,9]. Hertog and Katan [10] showed that a high consumption of quercetin-rich onion was associated with a 50% cancer risk reduction of the digestive and respiratory tracts. Organosulphur compounds such as diallyl disulfide (DDS), S-allylcysteine (SAC), and S-methylcysteine (SMC) have been demonstrated to inhibit colon and renal carcinogenesis [11,12]. Phytochemicals act through two different mechanisms: cancer cell apoptosis induction [13] and gene transcription inhibition [14].

The quantity and quality of bioactive compounds contained in the onion bulb can vary according to the variety and cultivation practices. Among the varieties, it was well-demonstrated that *A. cepa L. var. tropeana* (red onion) contains more phytochemicals than white onion [15]. With respect to the cultivation conditions, numerous researchers have evidenced that onion is a sulphur-loving crop and that sulphur increased the bulb yield quality and flavors. Other works indicated an increase in onion quality when organic fertilizers were used [16]. Muscolo et al. [17] showed that the use of sulphur-organic-based fertilizers increased, in red onion, the production of bioactive organosulphur compounds and antioxidants with respect to the type and concentration of sulphur-organic-based fertilizer used.

Based on the above findings, the main aims of the current research were to: (1) determine the chemical characteristics, phytochemical amount, and profile of red onion bulbs differently fertilized and (2) evaluate their biological activity in normal and oxidative stress conditions. The lyophilized onion bulbs were tested in two cellular models, i.e., the H9c2 rat cardiomyoblast cell line and primary human control, and *parkin*-mutant fibroblasts in terms of viability and oxygen radical homeostasis. H9c2 cells are a valid alternative for primary cardiomyocytes where oxidative stress is an important pathophysiological pathway, affecting multiple aspects of cardiac functionality, including signal transduction, cell cycle arrest, apoptosis, and necrosis [18,19]. In Parkinson's disease (PD), oxidative stress plays a significant role in the cascade, consequently leading to the degeneration of dopaminergic neurons. Moreover, other aspects of the degenerative process, such as mitochondrial malfunction, excitotoxicity, nitric oxide toxicity, and inflammation, are all linked to oxidative stress [20]. Our goal was to link the protective benefits of red onion to the phytochemical content and specific class of compounds in order to emphasize the medicinal worth of these onions, which may be used in a health prevention program.

2. Results and Discussion

2.1. Red Onion Chemical Properties

The treatments with fertilizer pads, SB, SBOR and SBOP, influenced positively, but to different extents, the properties of red onions compared to the control (CTR). Pads containing orange positively affected the red onion quality, followed by SBOP and SB. These were due to the presence of organic components in the pads, as reported in previous publications [21,22] that evidenced a great level of flavonoids in organically grown Welsh onions and red onion. Ren et al. [21] also found high amounts of phenolics, total flavonoids, and anthocyanins, as well antioxidant activities, in two different onion varieties grown under organic production. Muscolo et al. [17] evidenced a positive effect of sulphur bentonite-organic-based fertilizers on secondary metabolite (SMs) production in red onions, suggesting that sulphur bentonite organic fertilization was able to stimulate the plant's secondary metabolism, inducing the production of phytochemicals that can be useful in preserving human health. Human natural antioxidant systems, if perfectly working, are able to mitigate damage to important biomolecules, such as DNA, proteins, lipids, and carbohydrates, avoiding the insurgence of diseases [23]. The additional intake of antioxidants with the diet represents a very important way to prevent the diseases caused by oxidative stresses. There is, nowadays, a growing interest to enrich the human diet with functional foods naturally rich in antioxidant compounds. Polyphenols represent the most important natural antioxidant compounds with beneficial effects on human health [24].

Our results evidenced in red onion bulb the greatest increase in polyphenols (Table 1) in the presence of SBOR at both concentrations (low and high); SBOP also increased the quantity of polyphenols with respect to the control but less than SBOR. In contrast, an inverse trend was observed for the total flavonoids (Table 1) that increased more in the presence of SBOP than SBOR LP and HP. Anthocyanins were the highest in all fertilized red onion bulbs. Phenolic acids (Figure 1) found in the CTR and fertilized onions were caffeic and chlorogenic. Gallic acid was present only in the fertilized onions (Figure 1), while *p*-coumaric acid in the CTR and, in the greatest amount, in SBORLP. Caffeic and chlorogenic acids did not show significant differences with respect to the CTR, except for the onion fertilized with SBORHP. *p*-coumaric and gallic acids are antioxidants with diverse physiological functions that are beneficial for human health with ascertained anticancer, anti-inflammatory, and antimicrobial properties [25–27]. The mechanisms of action of polyphenols are various and complex and depend on their chemical structures. The antioxidant property of *p*-coumaric acid is ascribed to its phenyl hydroxyl group (-OH) that enables it to donate hydrogen or electrons. In vivo studies on the *p*-coumaric mechanism of action evidenced, on a rat model, that it was able to reduce basal oxidative DNA damage, inducing glutathione (GSH) and glutathione S-transferase Mu 2 (GST-M2) in colonic mucosa. Additionally, it was demonstrated that *p*-coumaric acid was capable of decreasing the expression of the inflammatory mediators, such as TNF-α and IL-6, regulating the production of cytokines [28]. Nasr Bouzaiene et al. [29] showed how the proliferation of human lung (A549) and colon (HT29-D4) cancer cells was significantly inhibited by ferulic, caffeic, and *p*-coumaric acids. These inhibitory effects were likely to be mediated by the suppression of DNA synthesis induced by the phenolic acids in MCF-7. Caffeic acid, among the phenolic acids, was found more able to block the many modulators involved in tumor progression, including NF-kB, COX-2, TNF-a, IL-6, Nrf2, iNOS, NFAT and HIF-1α, repressing cancer angiogenesis and therefore recognized as an inducer of tumor cell death and performer of cancer growth blockage [30].

Table 1. Total phenols (mg·GAE·g^{-1} DW), flavonoids (mg·rutin·g^{-1} DW), and anthocyanins (mg·cyanidin-3-glucosideg^{-1} DW) found in red onion bulbs differently fertilized: control (CTR), sulphur bentonite (SB), sulphur bentonite-low percentage orange residue (SBOR LP), sulphur bentonite-high percentage orange residue (SBOR HP), and sulphur bentonite-olive pomace (SBOP). Data are the mean of three replicates ± the standard error.

	Total Phenols	Flavonoids	Anthocyanins
CTR	28 ± 1 ^{c}	2.1 ± 0.1 ^{c}	23 ± 1 ^{b}
SB	37 ± 2 ^{b}	3.7 ± 0.2 ^{b}	37 ± 2 ^{a}
SBORLP	48 ± 2 ^{a}	3.9 ± 0.3 ^{b}	37 ± 2 ^{a}
SBORHP	43 ± 3 ^{a}	3.8 ± 0.2 ^{b}	37 ± 2 ^{a}
SBOP	37 ± 2 ^{b}	5.0 ± 0.5 ^{a}	37 ± 2 ^{a}

Means followed by different letters in the same column are significantly different (Tukey's test at $p < 0.05$).

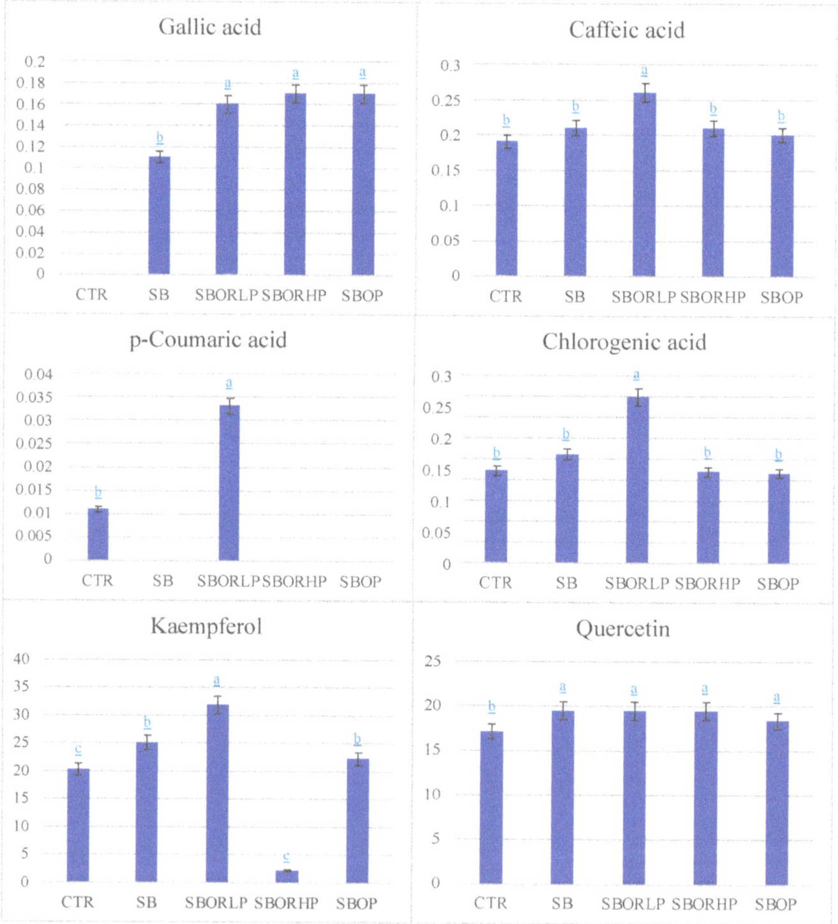

Figure 1. Phenolic acids and flavonols (mg·100·g^{-1} FW) found in red onion bulbs differently fertilized: control (CTR), sulphur bentonite (SB), sulphur bentonite-low percentage orange residue (SBOR LP), sulphur bentonite-high percentage orange residue (SBOR HP), and sulphur bentonite—olive pomace (SBOP). Data are the mean of three replicates ± the standard error. Different letters indicate significant differences at $p < 0.05$.

Anthocyanidins (Figure 2), increased in treated onions compared to the CTR. Equally, S methyl-cysteine sulfoxide and the majority of organosulphides (Table 2) increased with respect to the CTR, mostly in red onions treated with SBOR pads and particularly with SBORLP. Anthocyanidins have health-promoting effects linked with antioxidant, anti-inflammatory, and anticarcinogenic properties. Their antioxidant nature was observed in all neurological diseases through MMP2, MMP3 and MMP9 metalloproteinase inhibition; reactive oxygen species generation inhibition; endogenous antioxidants modulation as superoxide dismutase and glutathione; the formation and aggregation of beta-amyloid (β-A) protein inhibition; and brain protective action through the modulation of brain-derived neurotrophic factor (BDNF), important for neural plasticity [31]. Additionally, organosulphur compounds have a well-recognized antiproliferative activity in several tumor cell lines that is mediated by the induction of apoptosis and alterations of the cell cycle. Organosulphur compounds generally act by modulating the activity of several metabolizing enzymes that activate (cytochrome P450s) or detoxify (glutathione S-transferases) carcinogens and inhibit the formation of DNA adducts in several target tissues [32]. Their low amounts found in SBORHP and SBOP treated onions can be related to the contemporary increase in other SMs with antioxidant properties. This suggests that the fertilizers used were able to influence the biosynthesis and accumulation of other SMs, evidencing that these fertilizers are capable of redirecting the metabolism to consequently regulate the production of specific bioactive constituents, as already reported by [33].

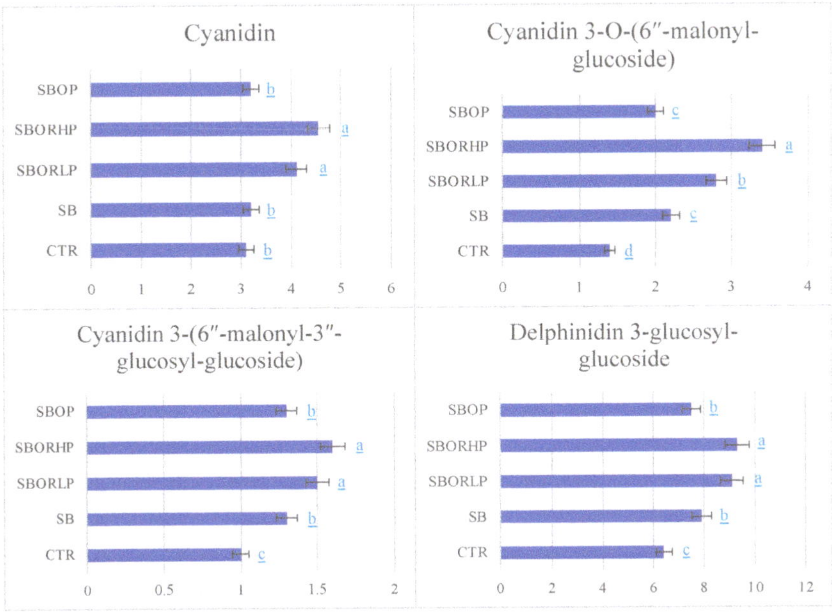

Figure 2. Anthocyanidins (mg·100·g^{-1} FW) found in red onion bulbs differently fertilized: control (CTR), sulphur bentonite (SB), sulphur bentonite-low percentage orange residue (SBOR LP), sulphur bentonite-high percentage orange residue (SBOR HP), and sulphur bentonite-olive pomace (SBOP). Data are the mean of three replicates ± the standard error. Different letters indicate significant differences at $p < 0.05$.

Table 2. S-Methyl-L-cysteine sulfoxide (µg·g^{-1} FW) and the relative concentration µg·g^{-1} FW of volatile organic compounds in the onion bulbs differently fertilized: control (CTR), sulphur bentonite (SB), sulphur bentonite-low percentage orange residue (SBOR LP), sulphur bentonite-high percentage orange residue (SBOR HP), and sulphur bentonite-olive pomace (SBOP). Data are the mean of three replicates ± the standard error.

ID	S-methyl Cysteine Sulfoxide	Trisulfide Dipropyl	Diallyl Disulfide	Disulfide Di-Isopropyl	Disulfide Methyl-L-Propenyl	Isopropyl Mercaptan	Heptane-6-Methyl 4-5 Dithia-1-Heptene
CTR	110 ± 9 ^{e}	3.32 ± 0.5 ^{b}	nd	36.7 ± 2 ^{b}	0.38 ± 0.04 ^{c}	53 ± 3 ^{b}	1.0 ± 0.2 ^{c}
SB	590 ± 12 ^{a}	1.95 ± 0.2 ^{c}	nd	34.3 ± 3 ^{b}	nd	61 ± 4 ^{a}	1.0 ± 0.1 ^{c}
SBORLP	390 ±13 ^{b}	10.3 ± 1 ^{a}	0.32 ± 9 ^{a}	53.6 ± 2 ^{a}	1.44 ± 0.05 ^{a}	53 ± 2 ^{b}	8.4 ± 1 ^{a}
SBORHP	140 ± 12 ^{c}	3.32 ± 0.6 ^{b}	nd	35.3 ± 1 ^{b}	0.73 ± 0.02 ^{b}	51 ± 3 ^{b}	1.5 ± 0.2 ^{b}
SBOP	180 ± 12 ^{d}	1.63 ± 0.5 ^{e}	nd	36.7 ± 3 ^{b}	0.07 ± 0.001 ^{d}	54 ± 2 ^{b}	0.6 ± 0.02 ^{d}

Means followed by different letters in the same column are significantly different (Tukey's test at $p < 0.05$).

The in vitro antioxidant capacity, determined with DPPH, ABTS and ORAC (Figure 3), increased in red onion grown mainly with SBOR and SBOP than the CTR (Figure 3). Specifically, ORAC was the highest in bulbs of red onion grown with SBOR LP, while DPPH and ABTS were the highest in bulbs of red onion grown with SBOR, both LP and HP. Cavalheiro et al. [34] demonstrated an increase in the antioxidant activities in bulbs treated with organic fertilizers. The antioxidant activities are generally related to the chemical composition of the plants in terms of the typology of antioxidant compounds. Each single compound has its own biological activity with different effects on human health [35]. Flavonoids can scavenge free radicals and can form complexes with catalytic metal ions rendering them inactive. There is also evidence of an additional mechanism by which total phenols protect against oxidative stress by producing hydrogen peroxide (H_2O_2), which can then help to regulate immune response actions, such as cellular growth [36].

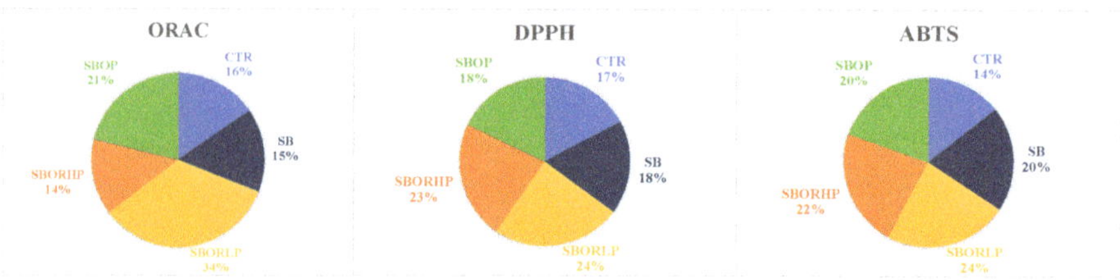

Figure 3. Antioxidant activities (ORAC, DPPH and ABTS) detected in red onion bulbs differently fertilized: control (CTR), sulphur bentonite (SB), sulphur bentonite-low percentage orange residue (SBOR LP), sulphur bentonite-high percentage orange residue (SBOR HP), and sulphur bentonite-olive pomace (SBOP). Data are the mean of three replicates.

Pearson's correlation (Figure 4) evidenced that the total phenols were positively and significantly correlated with ABTS (r = 0.96), DPPH (r = 0.57), and ORAC (r = 0.62); the flavonoids correlated only with ABTS (r = 0.63), while the anthocyanins correlated with DPPH (r = 0.57) and ORAC (r = 0.87). Among the single phenolic acids, caffeic acid correlated with all the antioxidant activities, gallic acid correlated with DPPH and ABTS and chlorogenic acid with DPPH and ORAC, while *p*-coumaric acid correlated only with ORAC (r = 0.85). Among the flavonoids, the flavanol quercetin correlated with DPPH and ABTS. S-methyl cysteine sulfoxide was not involved in the antioxidative system; conversely the organic volatile compounds correlated with the antioxidant activities, mostly with DPPH and ORAC and less with ABTS.

	DPPH	ORAC	ABTS
Polyphenols	0.940299802	0.616902765	0.965711962
Flavonoids	0.33383836	0.360681611	0.635338087
Anthocyanins	0.572527528	0.297317658	0.872871561
Gallic acid	0.689725403	0.399517529	0.904049752
Caffeic acid	0.810770907	0.824307181	0.704333803
p-Coumaric acid	0.542708385	0.84607779	0.233561336
Chlorogenic acid	0.69700628	0.861174703	0.532264246
Kaempferol	−0.119279294	0.650539182	−0.144061809
Quercetin	0.695829741	0.222803166	0.906573945
S-methyl cysteine sulfoxide	0.043203487	0.148185378	0.2677205
Trisulfide dipropyl	0.731906984	0.832632659	0.452896898
Isopropyl mercaptan	0.667371374	−0.252738104	−0.171229373
Heptane-6-methyl 4-5 dithia-1-heptene	0.745511186	0.874011871	0.526441637
Diallyl disulfide	0.699755867	0.910535329	0.490990253
Disulfide di-isopropyl	0.667371374	0.937620603	0.430230466
Disulfide Methyl-1-propenyl	0.83902444	0.637777528	0.609854677
Cyanidin	0.217627384	0.210133242	0.299344802
Cyanidin 3-*O*-(6″-malonyl-glucoside)	0.062819821	0.149343851	0.232512342
Cyanidin 3-(6″-malonyl-3″-glucosyl-glucoside)	0.385481909	0.605032014	0.481273519
Delphinidin 3-glucosyl-glucoside	0.169861148	0.416487049	0.476419311

Figure 4. Pearson's correlations (r) between phytochemicals and antioxidant activities. The boxed dots show the significant correlations between values; the color shows the level of correlation (yellow boxed dots $p < 0.05$ and green boxed dots $p < 0.01$). The red dots indicate a negative correlation.

2.2. Effects Red Onion Phytochemicals in Terms of Cell Proliferation or Cytotoxicity

To evaluate the possible effects of the phytochemical contents of onion samples in terms of cell proliferation or cytotoxicity, in this work, the H9c2 cells, found to be closer to normal primary cardiomyocytes for their energy metabolism features, were successfully used as an in vitro cellular model [18]. The H9c2 cells were incubated with red onion samples, fertilized, and not with the different pads in a range of concentrations between 0.5 and 10 mg/mL, and the cell viability was determined 24, 48 and 72 h after treatment following the chemical reduction of 3-(4,5-dimethylthiazol-2-yl)-2,5-diphenyltet-razolium bromide (MTT) by mitochondrial reductases in live cells [37]. As reported in Figure 5, the fertilization of red onion with recycled sulphur bentonite pads modified the red onion samples' ability to affect the proliferation rate and/or the oxidative metabolism of H9c2 cells with respect to the 'CTR' one. No significant toxic effects on cell viability were detected in the different conditions for all the onion samples up to a concentration of 10 mg/mL, except for the 'CTR' and 'SBORLP' at the highest concentration. In the latter, the toxic effect was very strong, and it was already observed at 24 h of treatment. Furthermore, at 72 h of incubation time, a toxic effect was also observed with 'SBOP' at a low concentration.

Noteworthy is the significant increase in cell viability of H9c2 cells treated with 'SB', 'SBOR LP', and 'SBOR HP' samples as compared with the 'CTR' one, even at low concentrations and already after 24 h of treatment, which could be caused by an increase in the cells' number and/or by an improvement in the oxidative metabolism. The effect was noticeable as early as after 24 h of treatment at very low concentrations (0.5 and 1 mg/mL) and up to 72 h for the 'SB' sample. 'SBOP' pads reduced these effects, as indicated by the overall similar results obtained with the 'CTR' and 'SBOP' treatments. Red onion samples' capabilities to alter the proliferation rate and/or the oxidative metabolism were more evident after 24 h of treatment with 'SBOR HP' and after longer exposure times, 48 and 72 h, in the presence of 'SBOR LP'. Overall, these data show a positive effect on the cell viability of H9c2, possibly related to an increase in energy metabolism, in the presence of 'SB' alone or with the addition of orange residue both at low and high percentages as compared to the 'CRT', likely due to the greatest level in bioactive compounds. To evaluate if the different red onion samples were able to protect H9c2 in oxidative stress conditions, the cells were pretreated with them for 24, 48 and 72 h before the exposure for 45 min to tert-butyl hydroperoxide (TBHP), an exogenous oxidative stress inducer. According to the treatment conditions of the previous screening, the cells were treated with two concentrations of onion samples, 0.5 and 5 mg/mL, except for 'SBOR LP', for which the concentrations of 1 and 5 mg/mL were used.

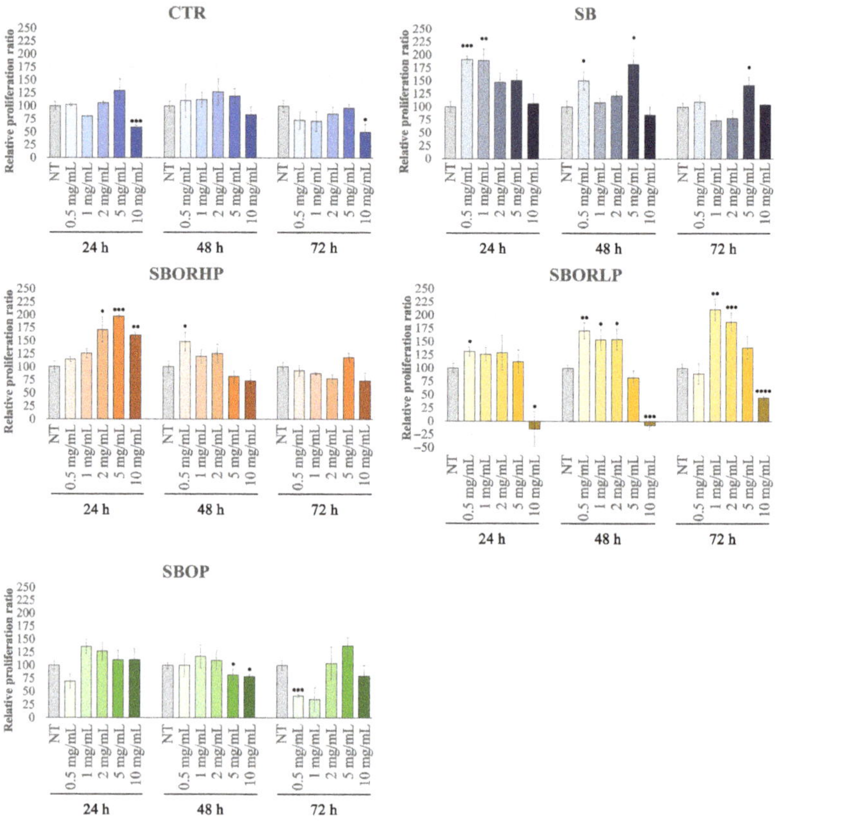

Figure 5. MTT assay performed on the H9c2 cell line. Cells have been treated for 24 h, 48 h and 72 h with different samples of red onions at different concentrations ranging from 0.5 mg/mL to 10 mg/mL. Data were the means ± SEM from at least 3 independent experiments under each condition and were expressed as the percentage of vehicle-treated cells. Statistical analyses were performed using Brown-Forsythe and Welch one-way analysis of variance, and mean comparisons were made using the unpaired *t*-test with Welch's correction. * $p < 0.05$, ** $p < 0.01$, *** $p < 0.001$, and **** $p < 0.0001$.

Firstly, the basal ROS levels after treatment with the onion samples for 24, 48 and 72 h were measured (Figure 6). No changes were observed in the ROS levels for all onion samples, except for an increase observed in the presence of 5 mg/mL 'SBOP' at 24 h of incubation, which returned to the basal level already at 48 h of incubation. A significant decrease of the basal ROS levels was, however, observed at the longest incubation time, with all the onion samples, albeit at different concentrations and, in particular, in the presence of 'SBOR HP' and 'SBOP' samples, at 0.5 mg/mL. Intriguingly, these same samples showed no effects at the highest concentration. On the contrary, the 'SB' and 'SBOR LP' samples showed the same effects elicited by the 'CTR' red onions, inducing a decrease of the basal ROS level at a higher concentration (5 mg/mL).

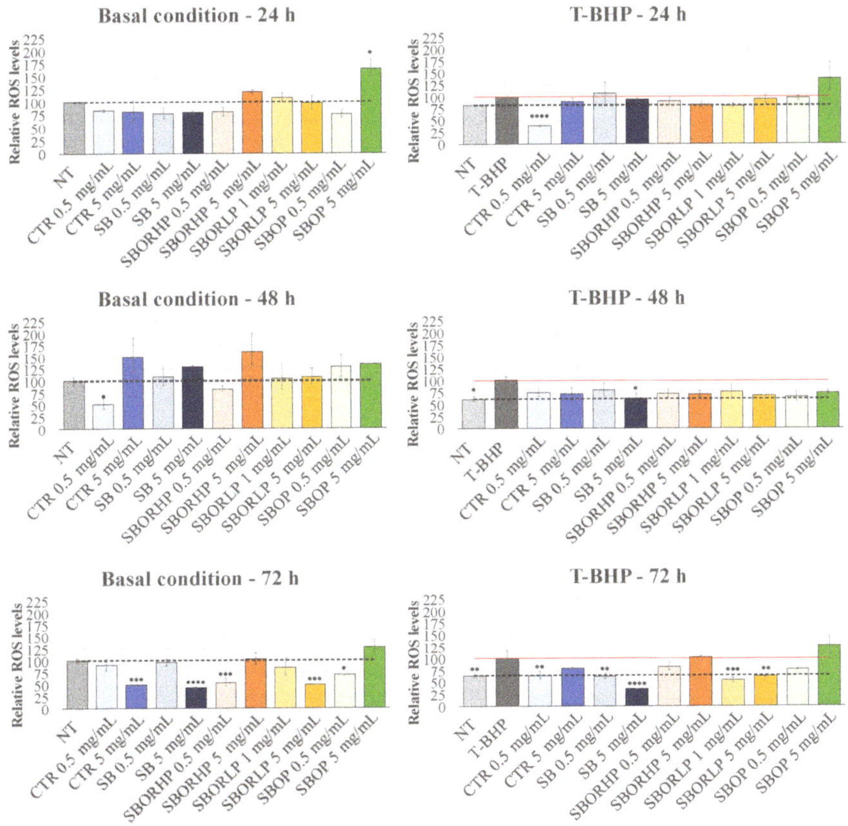

Figure 6. ROS detection performed on the H9c2 cell line. Cells were treated for 24 h, 48 h and 72 h with different samples of red onions at two different concentrations, and the ROS levels were measured by the DCF assay under the basal conditions and after exposure to the exogenous inducer of oxidative stress, T-BHP. Data were the, means ± SEM from at least 3 independent experiments under each condition and were expressed as the percentage of vehicle-treated cells (dotted line) or tert-Butyl hydroperoxide-treated-cells (T-BHP) (red line). Statistical analyses were performed using Brown–Forsythe and Welch one-way analysis of variance, and mean comparisons were made using the unpaired *t*-test with Welch's correction. * $p < 0.05$, ** $p < 0.01$, *** $p < 0.001$, **** $p < 0.0001$.

Next, the effects of red onion samples were measured on the ROS levels under TBHP-induced oxidative stress conditions (Figure 6). Even in this stress condition, only the longest pretreatment with the red onion samples showed an evident effect restoring the ROS basal levels or, in the case of the highest concentration of the 'SB' sample, further reducing them. Notably, 'SBORLP' showed the same capability of reducing the ROS levels to the basal ones at both concentrations used in the pretreatment. An early significant effect, at 24 h, was observed after incubation with the lower concentration of the 'CTR' extract and in the presence of 5 mg/mL of 'SB' at 48 h. The results that the different onion samples were able to decrease TBHP-induced oxidative stress could be ascribed either to a direct scavenger activity or to an enhancement of the activity of the antioxidant defenses that neutralize the ROS levels [38].

Finally, the 'SB', 'SBOR' and 'SBOP' treatments influenced, even if to different extents, the H9c2 viability and oxygen radical homeostasis with respect to the 'CTR' sample. In particular, the 'SBOR' treatments showed an interesting influence on the viability of H9c2 cells, dependent on the concentration of orange residue and time of exposure, requiring longer exposure times, in some cases, when the percentage was lower. Notably, 'SBORLP'-fertilized onions reduced the ROS levels in the basal and in oxidative stress conditions, confirming the results related to the in vitro antioxidant capacity determined by the ORAC, DPPH and ABTS assays. In particular, the results related to the oxygen radical homeostasis for the 'SBOR' treatments were not due only to the greatest content of the phenolic component present in these samples but also to the higher level of the organosulphides that have been shown to scavenge ROS and prevent damage caused by oxidative stress [39].

To assess the potential benefits of the differentially fertilized red onion samples in a pathological scenario, highly characterized primary human skin fibroblasts isolated from a healthy subject (control fibroblasts) and from a patient affected by early-onset Parkinson's disease (*parkin*-mutant) fibroblasts [40–48] were treated as previously described for H9c2 treatment. Indeed, *parkin*-mutant fibroblasts are representative of oxidative stress-correlated chronic diseases, as they display mitochondrial defects associated with deregulated reactive oxygen species (ROS) production, along with impaired energy metabolism and lipid oxidation [42]. As described in Figure 7, the incubation of control fibroblasts at low concentrations of the 'CTR' sample (0.5 and 1 mg/mL) showed an increase in the cell viability after 24, 48 and 72 h and a gradual decrease at the highest concentrations (5 and 10 mg/mL), resulting in a significant inhibition of cell proliferation. Furthermore, an increase in cell proliferation was also observed after 24 h of treatment in the presence of low concentrations of 'SBOR HP' and after 48 h and 72 h in the presence of 'SBOP'. Noteworthy, treatments with all the different onion samples at high concentrations and at long incubation times induced an inhibition of cell vitality of control fibroblasts, except for 'SBOP', which showed a protective action. In *parkin*-mutant fibroblasts, whereas the treatment with the 'CTR' sample induced a reduction in cellular viability, even at low concentrations and short incubation times, in the presence of all the other onion treatments, except for 'SBOR LP' at the highest concentrations, no change in the cellular vitality was observed. The lack of increase in the cellular vitality in the *parkin*-mutant fibroblasts, which was instead observed in control fibroblasts, could be due to the specific impairment of these cells. *Parkin*-mutant fibroblasts adapted to live in an environment characterized by a condition of oxidative stress showed a deficit in the mitochondrial biogenesis process, which could not lead to an increase in the cellular proliferation induced by the red onion samples. Finally, in the control fibroblasts, 'CTR', 'SBOR HP' and 'SBOP' onion extracts were able to increase the cellular vitality not observed in *parkin*-mutant fibroblasts. In these latter fibroblasts, the treatments with fertilized red onion samples could avoid the decrease in cellular vitality that was instead observed in the presence of the 'CTR' sample.

As described previously for the H9c2 cell line, it was evaluated if the different red onion samples were able to protect human fibroblasts in TBHP-induced oxidative stress conditions. According to the findings of the viability screening, cells were treated with 0.5 and 5 mg/mL for 24, 48 and 72 h (Figure 8A). In control fibroblasts, a decrease in the basal ROS levels was observed already at 24 h of incubation in the presence of high concentrations of 'CTR' but low concentrations of 'SB' samples. Furthermore, a decrease in the ROS basal levels was induced by 'SB' at 48 h of incubation, as well as by 'SBOR HP', 'SBOR LP', and at a higher extent, by 'SBOP'. It is possible to assume that the increase in the ROS basal level observed at 24 h of incubation in the presence of 0.5 mg/mL and 5 mg/mL of 'SBOR HP' and 'SBOP', respectively, might have induced an antioxidant enzymatic response, which, in turn, resulted in a ROS scavenger effect at 48 h of incubation. In *parkin*-mutant fibroblasts at 24 h of incubation, the 'CTR' induced a decrease at the basal ROS levels at low and high concentrations, and this effect persisted also at the highest concentrations and longest incubation times, similar to what was observed in the control cells. In addition, the decrease in the basal ROS levels was observed at 24 and 48 h of incubation in the presence of 5 mg/mL of 'SB' and at low and high concentrations of 'SBORHP' after 48 h of treatment. The scavenger effect of the 'CTR' sample, observed in *parkin*-mutant fibroblasts, especially in oxidative stress conditions, demonstrated the effectiveness of red onion already rich in bioactive compounds.

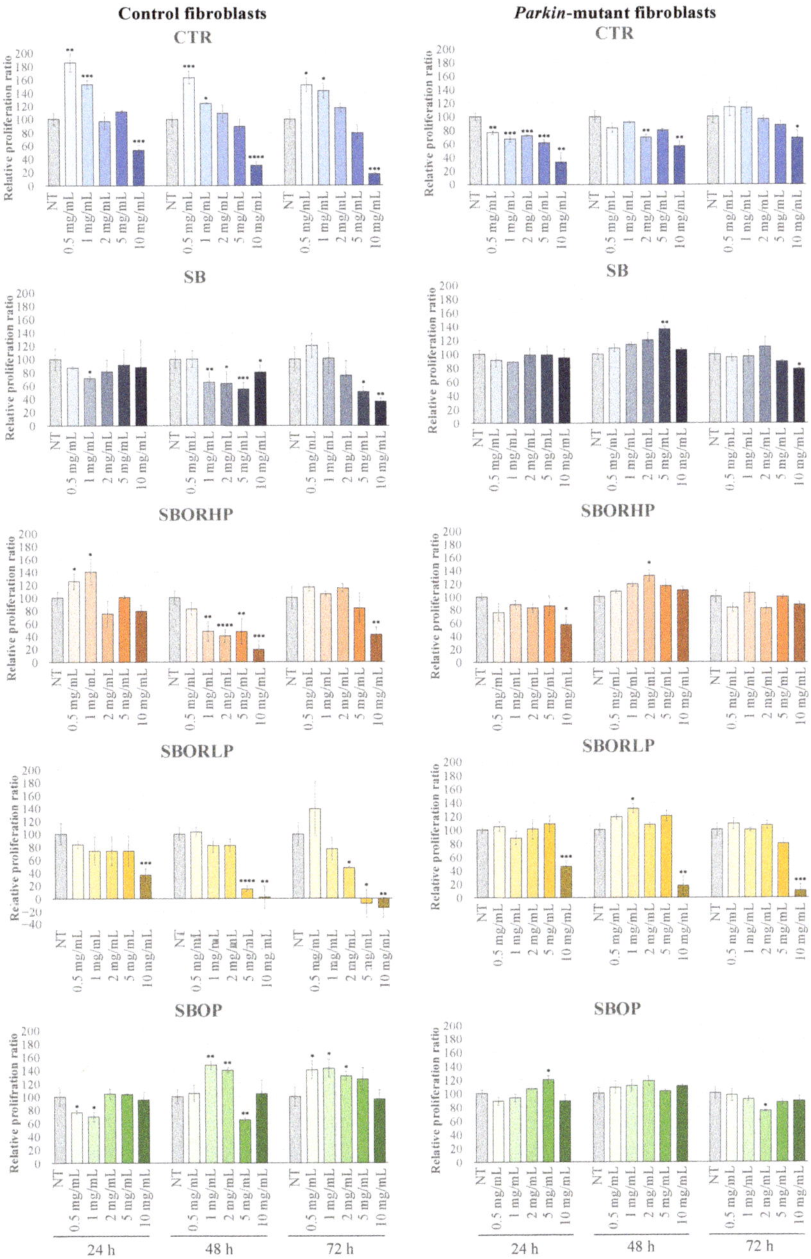

Figure 7. The MTT assay performed on the control and *parkin*-mutant fibroblasts. Cells were treated for 24 h, 48 h and 72 h with different samples of red onions at different concentrations ranging from 0.5 mg/mL to 10 mg/mL. Data were the means ± SEM from at least 3 independent experiments under each condition and were expressed as the percentage of vehicle-treated cells. Statistical analyses were performed using Brown-Forsythe and Welch one-way analysis of variance, and mean comparisons were made using the unpaired *t*-test with Welch's correction. * $p < 0.05$, ** $p < 0.01$, *** $p < 0.001$, and **** $p < 0.0001$.

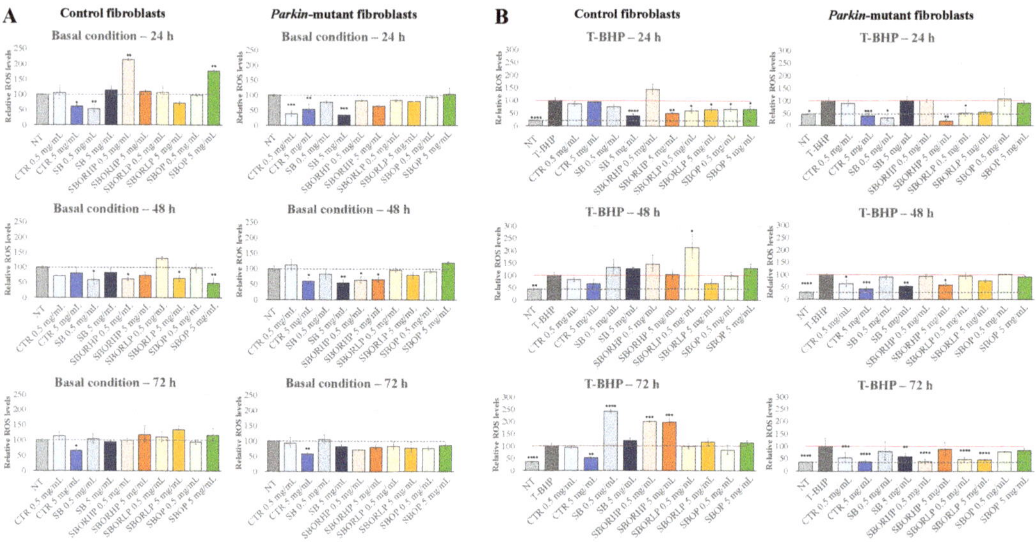

Figure 8. ROS detection performed on the control and *parkin*-mutant fibroblasts. Cells were treated for 24 h, 48 h, and 72 h with different samples of red onions at two different concentrations, and the ROS levels were measured by the DCF assay in the basal condition (**A**) and after exposure to the exogenous inducer of oxidative stress, T-BHP (**B**). Data were the means ± SEM from at least 3 independent experiments under each condition and were expressed as a percentage of vehicle-treated cells (dotted line) (**A**) or tert-Butyl hydroperoxide-treated cells (T-BHP) (red line) (**B**). Statistical analyses were performed using one-way analysis of variance, and mean comparisons were made using Fisher's LSD test. * $p < 0.05$, ** $p < 0.01$, *** $p < 0.001$, and **** $p < 0.0001$.

Next, the effects of red onion samples on the ROS levels under TBHP-induced oxidative stress conditions were assessed (Figure 8B). In this condition, the behavior observed in the control and *parkin*-mutant fibroblasts was substantially different, mainly the long incubation time. A significant decrease was observed in the control cells at 24 h of incubation with all the treated onion samples, albeit at different concentrations, as compared with the 'CTR'. In *parkin*-mutant fibroblasts in the same conditions, a reduction of TBHP-induced oxidative stress was observed at higher concentrations of the 'CTR' and in the presence of 'SB', 'SBOR HP' and 'SBOR LP' samples. At 48 and 72 h of incubation, no effect of red onion treatment on TBHP-induced oxidative stress was observed in the control cells, except for the reduction of the ROS level in the presence of a high concentration of 'CTR' over long time of incubation, whereas a reduction at 48 h of incubation was observed in *parkin*-mutant fibroblasts in the presence of a different concentration of 'CTR', 'SB' and 'SBOR HP' samples, as well as at 72 h with low and high concentrations of 'SBOR LP'. Furthermore, in the control cells, an increase in the ROS level with respect to TBHP-induced oxidative stress was observed with 0.5 mg/mL of 'SB' and with low and high concentrations of 'SBOR HP', highlighting a possible toxic effect, as already shown by the decrease in cellular viability in this condition. The scavenger effect observed in *parkin*-mutant fibroblasts and in basal and TBHP-induced oxidative stress conditions pointed out an effective protective role of red onion samples fertilized with sulphur bentonite containing orange, particularly at low concentrations with respect to the 'CTR' sample.

3. Materials and Methods

3.1. Chemicals

Metaphosphoric acid, 2,2-diphenyl-1-picrylhydrazyl (DPPH),NaOH, nitroblue tetrazolium, dichlorophenol-indophenol (DCPID),2,2′-azino-bis (3-ethylbenzothiazoline-6-sulfonic acid) diammoniumsalt (ABTS•+), 6-hydroxy-2,5,7,8-tetramethylchromane-2-carboxyl acid (Trolox), phenazine methosulphate, ethanol, gallic acid, ethylenediaminetetraaceticacid (EDTA), ferrozine, 2,4,6-tris(2-piridil)-s-triazina(TPTZ) and iron sulphate heptahydrate were purchased from Sigma Chemical Co. (St. Louis, MO, USA). Acetonitrile and acetic acid were HPLC-grade and were purchased from Merck (Darmstadt, Germany). All the phenolic standards were obtained from Extra Syntheses (Genay, France). Solvents and reagents for carotenoid detection were purchasedfrom Panreac (Barcelona, Spain). Other chemicals were of analyticalgrade purchased from Carlo Erba Reagents s.r.l. (Cornaredo, MI, Italy).

3.2. Red Onion Experimental Conditions

The experiment was conducted in triplicate for 3 months in the field (until the bulb is fully ripe) in alkaline sandy-loam soil with a pH of 8.5. The soils contained 3.09% organic matter, 0.17% nitrogen, 110 g·kg^{-1} $CaCO_3$ and 0.334 g·kg^{-1} SO_4. In each parcel, 30 uniform seedlings of red onion/m squared were transplanted. The soil was divided into parcels of 10 m squared. In each parcel, 30 uniform seedlings of red onion/m^2 were transplanted. Pads of sulphur bentonite (SB, 90%/10%) or SB with two percentages of orange residue (SBOR) conventionally called high (with a greater amount of orange residue, SBORHP) and low (with a lower amount of orange residue, SBORLP) or SBOP with olive pomace (OP) were used at a concentration of 16 gm^{-2} (corresponding to 476 kg S ha^{-1}), the dose generally used to lower the pH and to replenish the S. Sulphur was the major component of the fertilizers. The industrial process and the formulation of the pads are covered by an industrial secret drawn up in the agreement signed with Steel Belt System in 2015. Nonamended soils were used as the control CTR (Table 3). OP was used only at a low concentration, because previous experiments (data not shown) demonstrated a toxicity of OP on crops increasing its concentration (preliminary experiments carried out in a greenhouse, data not shown). SB and SB with OP or OR produced onions with greater bulb sizes with respect to the control (data not shown).

Table 3. Experimental design.

ID Sample	Description
CTR	Control unfertilized soil
SB	Onion grown with sulphur bentonite
SBOR HP	Onion grown with sulphur bentonite-high percentage of orange residue
SBOR LP	Onion grown with sulphur bentonite-low percentage of orange residue
SBOP	Onion grown with sulphur bentonite and olive pomace

During the experiment, all onion plants were regularly irrigated to maintain 70% of their field capacity. At harvest time (3 months), onions were collected and stored at −20 °C for chemical and biological determinations.

3.3. Extraction and Determination of Total Anthocyanins

The assessment of the total anthocyanin content was carried out by the pH differential method according to the Official Methods of Analysis of AOAC International, as previously described by Muscolo et al. [17]. Absorbance was measured using a 1800 UV-Vis Spectrophotometer (Shimadzu, Kyoto, Japan) at 510 and 700 nm in buffers at pH 1.0 and 4.5. Values were expressed as mg cyanidin-3-glucoside equivalent g^{-1} dry weight (DW) using 26,900 as the molar extinction coefficient.

3.4. Ethanolic Extracts

Five hundred milligrams of frozen onion samples were weighed and extracted at room temperature under continuous stirring for 1.5 h with ethanol (15 mL), as described in Muscolo et al. [17]. The samples were centrifuged at 2365× g for 15 min, and the supernatants were filtered dried and resuspended in 3 mL of ethanol.

3.5. Determination of Total Phenolic Compounds and Total Flavonoids

The Folin–Ciocalteu assay was used for evaluating the total phenol content as reported in Muscolo et al. [17]. The absorbance of the samples was recorded at 760 nm. A calibration curve was constructed with gallic acid, and the results were expressed as the gallic acid equivalent (GAE) in mg·g^{-1} DW.

The total flavonoid was detected according to the spectrophotometric method, as reported in Muscolo et al. [17]. One milliliter of extract was mixed with 1 mL of 20 g L^{-1} $AlCl_3$ methanolic solution. After incubation at room temperature for 15 min, the absorbance was measured at 430 nm. The flavonoid content was calculated from a calibration curve of rutin and expressed as mg rutin g^{-1} DW.

3.6. Determination of Antioxidant Activities

The antioxidant activity against the DPPH (2,2-diphenyl-1-picryl-hydrazyl-hydrate) radical was determined according to [49]. The DPPH concentration in the cuvette was chosen to give absorbance values of ≈1.0. The reaction mixtures were composed of 10 µL of each extract, 700 µL of DPPH, and ethanol up to a final volume of 1 mL. A blank without ethanol extract was prepared for each sample. The change in absorbance of the violet solution was recorded at 517 nm after 30 min of incubation at 37 °C. The inhibition I (%) of radical scavenging activity was calculated as I (%) = [(A0 − AS)/A0] × 100, where A0 is the absorbance of the control, and AS is the absorbance of the sample after 30 min of incubation. The results were expressed as Trolox equivalents (TE).

The ABTS (2,2′-Azino-bis(3-ethylbenzothiazoline-6-sulfonic acid) assay was performed according to [50]. Solutions of 7 mmol L^{-1} ABTS$^+$ (final concentration) and 2.45 mmol L^{-1} ammonium persulfate (final concentration) in phosphate-buffered saline (PBS) were mixed and kept in the dark at room temperature for 12–16 h. Before use, the absorbance of the ABTS$^+$ solution was fixed at 0.70 ± 0.02 at 734 nm. Aliquots of ethanol extract (25, 50 and 100 µL) were added to 0.5 mL of ABTS$^+$ solution and brought to a final volume of 600 µL with PBS. After 6 min of incubation in the dark at room temperature, the absorbance of the samples was recorded at 734 nm using a UV-Visible spectrophotometer. The inhibition I (%) of radical scavenging activity was calculated as I (%) = [(A0 − AS)/A0] × 100, where A0 is the absorbance of the control, and AS is the absorbance of the sample after 4 min of incubation. The results were expressed as µmol·L^{-1} TE using a Trolox (1–50 µmol·L^{-1}) calibration curve.

The oxygen radical absorbance capacity (ORAC) assay was performed according to [50]. A 20-µL aliquot of extract was added to 120 µL of fresh fluorescein solution (117 nmol L^{-1}). After a preincubation time of 15 min at 37 °C, 60 µL of freshly prepared AAPH solution (40 mmol L^{-1}) was added. Fluorescence was recorded every 30 s for 90 min (λex 485 nm, λem 520 nm). A blank using 20 µL of methanol instead of the sample was also analyzed. ORAC values were expressed as µmol TE mg^{-1} FW using a Trolox (10–100 µmol L^{-1}) calibration curve.

3.7. HPLC and Gas Chromatography/Mass Spectrometry (GC/MS) Analysis of Volatile Organic Compounds

Frozen onion samples (1 g) were incubated overnight in absolute methanol at 4 °C. Then, the methanol was separated from the pieces of onion and collected in a balloon. The onion pieces were homogenized with absolute methanol (10 mL) in a mortar and stirred 30 min at room temperature (25 °C). Samples were then centrifuged, and each supernatant was mixed with the other methanol. The precipitates were resuspended in methanol (10 mL), and the above operations were repeated twice. The methanolic phases were combined, reduced to a volume of 10 mL in a rotary evaporator, and stored at −18 °C until use. Methanolic extracts (1 mL) were diluted with dimethylformamide (1 mL) and filtered through an Iso-Disk P-34, 3 mm in diameter poly(tetrafluoroethylene) (PTFE) membrane, and 0.45 µm pore size supplied by Supelco. Diode array detection

(DAD)-HPLC (Shimadzu, Kyoto, Japan) separation of onion flavonoids was performed according to the method described by [49].

Reverse phase-diode array detector-high-performance liquid chromatography (RP-DAD-HPLC) analyses of the samples were carried out with a Shimadzu system (Kyoto, Japan) consisting of a LC-10AD pump system, a vacuum degasser, a quaternary solvent mixing, a SPD-M10 diode array detector, and a Rheodyne 7725i injector (Merck KGaA, Darmstadt, Germany). Separation of each compound was done on a 250 × 4.6 mm i.d., 5-μm Discovery C18 column, supplied by Supelco Park (Bellefonte, PA, USA) equipped with a 4.0 × 20-mm guard column. The column was placed in a column oven set at 25 °C. The injection loop was 20 μL, and the flow rate was 1.0 mL/min. The mobile phase consisted of a linear gradient of solvent A (acetonitrile) in 2% acidified water (acetic acid:H_2O, 2:98) as follows: 0–80% (0–55 min), 90% (55–70 min), 95% (70–80 min), 100% (80–90 min), and 0% (90–110 min). UV-Vis spectra were measured between 200 and 600 nm and simultaneous detection using a diode array at 278 and 325 nm. Compounds were identified using their retention time and UV spectra through comparisons with purified standards (Sigma Chemical Co., St. Louis, MO, USA). Anthocyanins were extracted from frozen onion tissues (0.5 g) homogenized in a mortar with 10 mL of methanol containing 1 mL L^{-1} HCl at room temperature for 2 h. The extracts were filtered through an Iso-Disk P-34, 3 mm in diameter PTFE membrane of 0.45-μm pore size (Supelco), and utilized for HPLC analysis. HPLC separation was carried out using a Spherisorb S5 ODS2, Merck KGaA, Darmstadt, Germany (250 mm × 4.6 mm i.d., 5 μm), as described by [51]. To detect S-methyl-L-cysteine sulfoxide (SMCSO), small pieces of frozen red onion (250 mg) were homogenized with 5 mL of distilled water and filtered through filter paper. SMCSO was quantified by HPLC after derivatization with o-phthalaldehyde, as reported in [52].

A red onion volatile organic compound (VOC) analysis was performed using a Thermo Fisher gas chromatograph (TRACE 1310, Thermo Fisher Scientific, Waltham, MA, USA) equipped with a single-quadrupole mass spectrometer (ISQ LT, Thermo Fisher Scientific, Waltham, MA, USA), as reported in [17].

3.8. Red Onion Sample Preparation for Cell Culture Treatments

The lyophilized onion samples were dissolved in high-glucose Dulbecco's modified Eagle's medium (DMEM) supplemented with 10% (*v*/*v*) fetal bovine serum (FBS), 1% (*v*/*v*) L-glutamine, and 1% (*v*/*v*) penicillin/streptomycin and incubated for 1 h at 37 °C. After this time, samples were centrifuged at 2600× *g* for 15 min, and the supernatant was filtered and sterilized through a 0.22-μm membrane filter. The final concentration of the stock solution was 10 mg/mL in DMEM. The onion extract (OE) concentrations tested for the cell culture treatments were: 0.5, 1, 2, 5 and 10 mg/mL.

3.9. Cells and Culture Conditions

The H9c2 cell line, derived from embryonic rat hearts (ATCC; Manassas, VA, USA), and primary fibroblasts from a patient affected by early-onset Parkinson's disease (PD) and from one healthy subject, obtained by explants from a skin punch biopsy after informed consent [40,41], were grown in high-glucose Dulbecco's modified Eagle's medium (DMEM) supplemented with 10% (*v*/*v*) fetal bovine serum (FBS), 1% (*v*/*v*) L-glutamine, and 1% (*v*/*v*) penicillin/streptomycin at 37 °C in a humidified atmosphere of 5% CO_2. For the treatment conditions, cells were seeded in 96-well plates and grown for 24 h. After that, the media was removed, and the cells were cultured at 37 °C in culture plates for 24, 48, and 72 h in fresh media containing different concentrations of red onion samples.

3.10. Cell Viability

Cell viability was assessed by the 3-(4,5-dimethylthiazol-2-yl)2,5-diphenyltetrazolium (MTT) assay after 24, 48 and 72 h of exposure of the cells and seeded in 96-well plates to the different concentrations of red onion samples. After the incubation, 150 μL DMEM and 15 μL of MTT (5 mg/mL) were added to each well. The plates were incubated for

3 h at 37 °C. The media was removed, and formazan crystals were dissolved in 150 μL of isopropanol with gentle shaking. The absorbance was measured at 570 nm by the Victor 2030 multilabel reader (PerkinElmer, Waltham, MA, USA).

3.11. Determination of Reactive Oxygen Species (ROS)

The H_2O_2 levels were determined by the cell permeant probe 2′-7′-dichlorodihydrofluorescin diacetate (H_2DCFDA). Briefly, after 24, 48 and 72 h of exposure to specific red onion sample concentrations (0, 0.5, 1 and 5 mg/mL), based on the results of the cell viability assay, the media were changed and oxidative stress was induced by 50 μM tert-butyl hydroperoxide (T-BHP) treatment (Sigma-Aldrich, B2633, St. Louis, MO, USA), as described in [53,54]. After 45 min of treatment, the cells were incubated in the dark, at 37 °C for 20 min with 10 μM H_2DCFDA. After that, the cells were washed and resuspended in 150 μL of PBS, and the H_2O_2-dependent oxidation of the fluorescent probe was measured by the Victor 2030 multilabel reader (PerkinElmer, Waltham, MA, USA) (at 507-nm excitation and 530-nm emission wavelength).

3.12. Statistical Analysis

Analysis of variance was carried out for all the data sets. One-way ANOVA with Tukey's honestly significant difference test was carried out to analyze the effects of fertilizers on each of the various parameters measured. The ANOVA and *t*-test were carried out using SPSS software (IBM Corp. 2012). The effects were significant at $p \leq 0.05$.

Raw data from MTT assays and ROS determinations were first analyzed in Microsoft Excel Spreadsheet software to calculate the relative proliferation and ROS levels ratios, respectively. Then, log-transformed data were imported to GraphPad Prism to apply statistical tests. The Welch and Brown–Forsythe versions of one-way ANOVA were used to compare samples treated with different red onions at different concentrations, while each comparison was evaluated by applying an unpaired *t*-test with Welch's correction. Statistical significance was set at $p < 0.05$.

4. Conclusions

These intriguing results may be explained by the different bioactive compounds identified in each fertilized red onion. SBOR onion, which showed the best positive effects on the cell culture treatment, contained the highest amount of total phenols, single phenolic acids, kaempferol, anthocyanidins, S-methyl cysteine sulfoxide, and volatile compounds correlated to a better in vitro antioxidant capacity as determined by the DPPH, ABTS and ORAC assays. The correlation data evidenced a diversity of action, at the metabolic level, of the different classes of secondary metabolites and mostly of the single compounds belonging to the different classes, evidencing that the chemical structure of a biocompound can determines its reactivity versus free radicals and other ROS, influencing the antioxidant activity.

Considering the data of the relative cellular vitality, the positive effects could be due to the total phenols and, in particular, to the great presence of specific phenolic acids such as chlorogenic and *p*-coumaric and, also, to the flavanol kaempferol, which were more correlated with ORAC and more present in red onions treated with SBOR(LP) than in the other fertilized red onions. In addition to their great ability as scavengers, free phenolic acids, unlike flavonoids, have a high bioavailability and good water solubility [55] and can be absorbed in the stomach, contrary to flavonoids that cannot be absorbed, and only their small quantity can be transported passively through the intestinal wall into the blood [56].

Our study describes, for the first time, the antioxidant effect of the bioactive phenolic fraction from red onion bulbs fertilized with sulphur-bentonite enriched with orange residue or olive pomace on rat cardiomyocytes and primary human fibroblasts. This work is a pilot study that highlighted significant and useful data of how sustainable fertilization can lead to the improvement of the quality of red onions that can be used to develop functional foods or nutraceuticals for the prevention and management of numerous diseases.

From this manuscript emerges how the use of sulphur-bentonite-based fertilizers can represent a tool to increase, also in other species, phytochemicals with beneficial effects on human health and how the antioxidant activity/capacity of functional foods is important in preventing and treating numerous diseases. The phytochemicals enhance the medical and economic values of crops with important consequences on the bio and green economy, creating new opportunities for business.

Further investigations are in progress to test the effects on the H9c2 rat cardiomyoblast cell line and primary human dermal fibroblasts, in terms of viability and oxygen radical homeostasis mammalian cells, of each single compound identified in the extracts, with the aim to verify if the positive effects are due to a single specific compound or to a synergic or additive effect of more compounds.

Author Contributions: Investigation, formal analysis, data curation, M.L.M.; conceptualization and methodology, A.V.; Investigation, formal analysis and data analysis, F.M.; Investigation, formal analysis and data analysis, C.M.; data curation, writing—review & editing, T.C.; project administration, funding acquisition, writing original draft, writing—review & editing, A.M. All authors have read and agreed to the published version of the manuscript.

Funding: This research was funded by Steel Belt Systems s.r.l., Venegono inferior, Varese, Italy (grant number 08092014).

Institutional Review Board Statement: This study involving human fibroblasts was approved by the local ethics committee at the University of Bari Medical School (Deliberation n. 847, 30 June 2011).

Informed Consent Statement: Informed consent was obtained from all subjects involved in the study as reported in Materials and Methods (3.8).

Data Availability Statement: Not applicable.

Acknowledgments: Not applicable.

Conflicts of Interest: The authors declare no conflict of interest. The funders had no role in the design of the study; in the collection, analyses, or interpretation of data; in the writing of the manuscript; or in the decision to publish the results.

Sample Availability: Samples of the compounds are available from all the authors.

References

1. Finkel, T.; Holbrook, N.J. Oxidants, Oxidative Stress and the Biology of Ageing. *Nature* **2000**, *408*, 239–247. [CrossRef] [PubMed]
2. Hajam, Y.A.; Rani, R.; Ganie, S.Y.; Sheikh, T.A.; Javaid, D.; Qadri, S.S.; Pramodh, S.; Alsulimani, A.; Alkhanani, M.F.; Harakeh, S.; et al. Oxidative Stress in Human Pathology and Aging: Molecular Mechanisms and Perspectives. *Cells* **2022**, *11*, 552. [CrossRef] [PubMed]
3. Borek, C. Antioxidant Health Effects of Aged Garlic Extract. *J. Nutr.* **2001**, *131*, 1010S–1015S. [CrossRef] [PubMed]
4. Colina-Coca, C.; González-Peña, D.; De Ancos, B.; Sánchez-Moreno, C. Dietary Onion Ameliorates Antioxidant Defence, Inflammatory Response, and Cardiovascular Risk Biomarkers in Hypercholesterolemic Wistar Rats. *J. Funct. Foods* **2017**, *36*, 300–309. [CrossRef]
5. Jini, D.; Sharmila, S. Green Synthesis of Silver Nanoparticles from *Allium cepa* and Its in Vitro Antidiabetic Activity. *Mater. Today Proc.* **2020**. [CrossRef]
6. Tsuboki, J.; Fujiwara, Y.; Horlad, H.; Shiraishi, D.; Nohara, T.; Tayama, S.; Motohara, T.; Saito, Y.; Ikeda, T.; Takaishi, K.; et al. Onionin A Inhibits Ovarian Cancer Progression by Suppressing Cancer Cell Proliferation and the Protumour Function of Macrophages. *Sci. Rep.* **2016**, *6*, 29588. [CrossRef]
7. Yang, E.-J.; Kim, G.-S.; Kim, J.A.; Song, K.-S. Protective Effects of Onion-Derived Quercetin on Glutamate-Mediated Hippocampal Neuronal Cell Death. *Pharmacogn. Mag.* **2013**, *9*, 302–308. [CrossRef]
8. Zhou, Y.; Zhuang, W.; Hu, W.; Liu, G.-J.; Wu, T.-X.; Wu, X.-T. Consumption of Large Amounts of Allium Vegetables Reduces Risk for Gastric Cancer in a Meta-Analysis. *Gastroenterology* **2011**, *141*, 80–89. [CrossRef]
9. Nicastro, H.L.; Ross, S.A.; Milner, J.A. Garlic and Onions: Their Cancer Prevention Properties. *Cancer Prev. Res. Phila. Pa* **2015**, *8*, 181–189. [CrossRef]
10. Hertog, M.; Katan, M. Quercetin in Foods, Cardiovascular Disease, and Cancer. *Flavonoids Health Dis.* **1998**, *20*, 447–467.
11. Hatono, S.; Jimenez, A.; Wargovich, M.J. Chemopreventive Effect of S-Allylcysteine and Its Relationship to the Detoxification Enzyme Glutathione S-Transferase. *Carcinogenesis* **1996**, *17*, 1041–1044. [CrossRef]

12. Fukushima, S.; Takada, N.; Wanibuchi, H.; Hori, T.; Min, W.; Ogawa, M. Suppression of Chemical Carcinogenesis by Water-Soluble Organosulphur Compounds. *J. Nutr.* **2001**, *131*, 1049S–1053S. [CrossRef]
13. Richter, M.; Ebermann, R.; Marian, B. Quercetin-Induced Apoptosis in Colorectal Tumor Cells: Possible Role of EGF Receptor Signaling. *Nutr. Cancer* **1999**, *34*, 88–99. [CrossRef]
14. Miodini, P.; Fioravanti, L.; Fronzo, G.D.; Cappelletti, V. The Two Phyto-Oestrogens Genistein and Quercetin Exert Different Effects on Oestrogen Receptor Function. *Br. J. Cancer* **1999**, *80*, 1150–1155. [CrossRef]
15. Marefati, N.; Ghorani, V.; Shakeri, F.; Boskabady, M.; Kianian, F.; Rezaee, R.; Boskabady, M.H. A Review of Anti-Inflammatory, Antioxidant, and Immunomodulatory Effects of *Allium cepa* and Its Main Constituents. *Pharm. Biol.* **2021**, *59*, 287–302. [CrossRef]
16. Kazimierczak, R.; Średnicka-Tober, D.; Barański, M.; Hallmann, E.; Góralska-Walczak, R.; Kopczyńska, K.; Rembiałkowska, E.; Górski, J.; Leifert, C.; Rempelos, L.; et al. The Effect of Different Fertilization Regimes on Yield, Selected Nutrients, and Bioactive Compounds Profiles of Onion. *Agronomy* **2021**, *11*, 883. [CrossRef]
17. Muscolo, A.; Papalia, T.; Settineri, G.; Mallamaci, C.; Panuccio, M. Sulphur Bentonite-organic-based Fertilizers as Tool for Improving Bio-compounds with Antioxidant Activities in Red Onion. *J. Sci. Food Agric.* **2019**, 100. [CrossRef]
18. Kuznetsov, A.V.; Javadov, S.; Sickinger, S.; Frotschnig, S.; Grimm, M. H9c2 and HL-1 Cells Demonstrate Distinct Features of Energy Metabolism, Mitochondrial Function and Sensitivity to Hypoxia-Reoxygenation. *Biochim. Biophys. Acta* **2015**, *1853*, 276–284. [CrossRef]
19. Rababa'h, A.M.; Guillory, A.N.; Mustafa, R.; Hijjawi, T. Oxidative Stress and Cardiac Remodeling: An Updated Edge. *Curr. Cardiol. Rev.* **2018**, *14*, 53–59. [CrossRef]
20. Wei, Z.; Li, X.; Li, X.; Liu, Q.; Cheng, Y. Oxidative Stress in Parkinson's Disease: A Systematic Review and Meta-Analysis. *Front. Mol. Neurosci.* **2018**, *11*, 236. [CrossRef]
21. Ren, F.; Reilly, K.; Kerry, J.P.; Gaffney, M.; Hossain, M.; Rai, D.K. Higher Antioxidant Activity, Total Flavonols, and Specific Quercetin Glucosides in Two Different Onion (*Allium cepa* L.) Varieties Grown under Organic Production: Results from a 6-Year Field Study. *J. Agric. Food Chem.* **2017**, *65*, 5122–5132. [CrossRef]
22. Hallmann, E.; Rembialkowska, E. Antioxidant compounds content in selected onion bulbs from organic and conventional cultivation. *J. Res. Appl. Agric. Eng. Pol.* **2006**, *51*, 42–46.
23. Fredotović, Ž.; Šprung, M.; Soldo, B.; Ljubenkov, I.; Budić-Leto, I.; Bilušić, T.; Čikeš-Čulić, V.; Puizina, J. Chemical Composition and Biological Activity of *Allium Cepa* L. and *Allium* × *Cornutum* (Clementi Ex Visiani 1842) Methanolic Extracts. *Molecules* **2017**, *22*, 448. [CrossRef]
24. Aguiar, J.; Gonçalves, J.L.; Alves, V.L.; Câmara, J.S. Chemical Fingerprint of Free Polyphenols and Antioxidant Activity in Dietary Fruits and Vegetables Using a Non-Targeted Approach Based on QuEChERS Ultrasound-Assisted Extraction Combined with UHPLC-PDA. *Antioxidants* **2020**, *9*, 305. [CrossRef]
25. Žilić, S.; Šukalović, V.H.-T.; Dodig, D.; Maksimović, V.; Maksimović, M.; Basić, Z. Antioxidant Activity of Small Grain Cereals Caused by Phenolics and Lipid Soluble Antioxidants. *J. Cereal Sci.* **2011**, *3*, 417–424. [CrossRef]
26. Kiokias, S.; Proestos, C.; Oreopoulou, V. Phenolic Acids of Plant Origin—A Review on Their Antioxidant Activity In Vitro (O/W Emulsion Systems) Along with Their in Vivo Health Biochemical Properties. *Foods* **2020**, *9*, 534. [CrossRef]
27. Roychoudhury, S.; Sinha, B.; Choudhury, B.P.; Jha, N.K.; Palit, P.; Kundu, S.; Mandal, S.C.; Kolesarova, A.; Yousef, M.I.; Ruokolainen, J.; et al. Scavenging Properties of Plant-Derived Natural Biomolecule Para-Coumaric Acid in the Prevention of Oxidative Stress-Induced Diseases. *Antioxidants* **2021**, *10*, 1205. [CrossRef]
28. Zhu, H.; Liang, Q.H.; Xiong, X.G.; Wang, Y.; Zhang, Z.H.; Sun, M.J.; Lu, X.; Wu, D. Anti-Inflammatory Effects of p-Coumaric Acid, a Natural Compound of *Oldenlandia diffusa*, on Arthritis Model Rats. *Evid. Based Complement. Altern. Med.* **2018**, *2018*, 5198594. [CrossRef]
29. Nasr Bouzaiene, N.; Kilani Jaziri, S.; Kovacic, H.; Chekir-Ghedira, L.; Ghedira, K.; Luis, J. The effects of caffeic, coumaric and ferulic acids on proliferation, superoxide production, adhesion and migration of human tumor cells in vitro. *Eur. J. Pharmacol.* **2015**, *766*, 99–105. [CrossRef]
30. Alam, M.; Ahmed, S.; Elasbali, A.M.; Adnan, M.; Alam, S.; Hassan, M.I.; Pasupuleti, V.R. Therapeutic Implications of Caffeic Acid in Cancer and Neurological Diseases. *Front. Oncol.* **2022**, *12*, 860508. [CrossRef]
31. Silva dos Santos, J.; Gonçalves Cirino, J.P.; de Oliveira Carvalho, P.; Ortega, M.M. The Pharmacological Action of Kaempferol in Central Nervous System Diseases: A Review. *Front. Pharmacol.* **2021**, *11*, 565700. [CrossRef]
32. Omar, S.H.; Al-Wabel, N.A. Organosulfur compounds and possible mechanism of garlic in cancer. *Saudi Pharm. J.* **2010**, *18*, 51–58. [CrossRef] [PubMed]
33. Yang, L.; Wen, K.-S.; Ruan, X.; Zhao, Y.-X.; Wei, F.; Wang, Q. Response of Plant Secondary Metabolites to Environmental Factors. *Molecules* **2018**, *23*, 762. [CrossRef] [PubMed]
34. Cavalheiro, T.R.T.; de Alcoforado, R.O.; de Silva, V.S.A.; Coimbra, P.P.S.; de Mendes, N.S.; Cavalcanti, E.D.C.; de Jurelevicius, D.A.; de Gonçalves, É.C.B.A. The Impact of Organic Fertilizer Produced with Vegetable Residues in Lettuce (*Lactuca sativa* L.) Cultivation and Antioxidant Activity. *Sustainability* **2021**, *13*, 128. [CrossRef]
35. Yashin, A.; Yashin, Y.; Xia, X.; Nemzer, B. Antioxidant Activity of Spices and Their Impact on Human Health: A Review. *Antioxidants* **2017**, *6*, 70. [CrossRef] [PubMed]

36. Li, Q.; Qiu, Z.; Wang, Y.; Guo, C.; Cai, X.; Zhang, Y.; Liu, L.; Xue, H.; Tang, J. Tea Polyphenols Alleviate Hydrogen Peroxide-Induced Oxidative Stress Damage through the Mst/Nrf2 Axis and the Keap1/Nrf2/HO-1 Pathway in Murine RAW264.7. *Cells. Exp. Ther. Med.* **2021**, *22*, 1473. [CrossRef] [PubMed]
37. van de Loosdrecht, A.A.; Beelen, R.H.; Ossenkoppele, G.J.; Broekhoven, M.G.; Langenhuijsen, M.M. A Tetrazolium-Based Colorimetric MTT Assay to Quantitate Human Monocyte Mediated Cytotoxicity against Leukemic Cells from Cell Lines and Patients with Acute Myeloid Leukemia. *J. Immunol. Methods* **1994**, *174*, 311–320. [CrossRef]
38. Mete, R.; Oran, M.; Topcu, B.; Oznur, M.; Seber, E.S.; Gedikbasi, A.; Yetisyigit, T. Protective Effects of Onion (*Allium cepa*) Extract against Doxorubicin-Induced Hepatotoxicity in Rats. *Toxicol. Ind. Health* **2016**, *32*, 551–557. [CrossRef]
39. Imai, J.; Ide, N.; Nagae, S.; Moriguchi, T.; Matsuura, H.; Itakura, Y. Antioxidant and Radical Scavenging Effects of Aged Garlic Extract and Its Constituents. *Planta Med.* **1994**, *60*, 417–420. [CrossRef]
40. Ferretta, A.; Gaballo, A.; Tanzarella, P.; Piccoli, C.; Capitanio, N.; Nico, B.; Annese, T.; Di Paola, M.; Dell'Aquila, C.; De Mari, M.; et al. Effect of Resveratrol on Mitochondrial Function: Implications in Parkin-Associated Familiar Parkinson's Disease. *Biochim. Biophys. Acta BBA Mol. Basis Dis.* **2014**, *1842*, 902–915. [CrossRef]
41. Tanzarella, P.; Ferretta, A.; Barile, S.N.; Ancona, M.; De Rasmo, D.; Signorile, A.; Papa, S.; Capitanio, N.; Pacelli, C.; Cocco, T. Increased Levels of CAMP by the Calcium-Dependent Activation of Soluble Adenylyl Cyclase in Parkin-Mutant Fibroblasts. *Cells* **2019**, *8*, 250. [CrossRef]
42. Pacelli, C.; De Rasmo, D.; Signorile, A.; Grattagliano, I.; di Tullio, G.; D'Orazio, A.; Nico, B.; Comi, G.P.; Ronchi, D.; Ferranini, E.; et al. Mitochondrial Defect and PGC-1α Dysfunction in Parkin-Associated Familial Parkinson's Disease. *Biochim. Biophys. Acta* **2011**, *1812*, 1041–1053. [CrossRef]
43. Vergara, D.; Gaballo, A.; Signorile, A.; Ferretta, A.; Tanzarella, P.; Pacelli, C.; Di Paola, M.; Cocco, T.; Maffia, M. Resveratrol Modulation of Protein Expression in Parkin-Mutant Human Skin Fibroblasts: A Proteomic Approach. Oxid. *Med. Cell. Longev.* **2017**, *2017*, 98243. [CrossRef]
44. Lippolis, R.; Siciliano, R.A.; Pacelli, C.; Ferretta, A.; Mazzeo, M.F.; Scacco, S.; Papa, F.; Gaballo, A.; Dell'Aquila, C.; De Mari, M.; et al. Altered Protein Expression Pattern in Skin Fibroblasts from Parkin-Mutant Early-Onset Parkinson's Disease Patients. *Biochim. Biophys. Acta BBA Mol. Basis Dis.* **2015**, *1852*, 1960–1970. [CrossRef]
45. Lobasso, S.; Tanzarella, P.; Vergara, D.; Maffia, M.; Cocco, T.; Corcelli, A. Lipid Profiling of Parkin-Mutant Human Skin Fibroblasts. *J. Cell. Physiol.* **2017**, *232*, 3540–3551. [CrossRef]
46. Guerra, F.; Girolimetti, G.; Beli, R.; Mitruccio, M.; Pacelli, C.; Ferretta, A.; Gasparre, G.; Cocco, T.; Bucci, C. Synergistic Effect of Mitochondrial and Lysosomal Dysfunction in Parkinson's Disease. *Cells* **2019**, *8*, 452. [CrossRef]
47. Pacelli, C.; Rotundo, G.; Lecce, L.; Menga, M.; Bidollari, E.; Scrima, R.; Cela, O.; Piccoli, C.; Cocco, T.; Vescovi, A.L.; et al. Parkin Mutation Affects Clock Gene-Dependent Energy Metabolism. *Int. J. Mol. Sci.* **2019**, *20*, 2772. [CrossRef]
48. Signorile, A.; Ferretta, A.; Pacelli, C.; Capitanio, N.; Tanzarella, P.; Matrella, M.L.; Valletti, A.; De Rasmo, D.; Cocco, T. Resveratrol Treatment in Human Parkin-Mutant Fibroblasts Modulates CAMP and Calcium Homeostasis Regulating the Expression of Mitochondria-Associated Membranes Resident Proteins. *Biomolecules* **2021**, *11*, 1511. [CrossRef]
49. Papalia, T.; Barreca, D.; Panuccio, M.R. Assessment of Antioxidant and Cytoprotective Potential of Jatropha (*Jatropha curcas*) Grown in Southern Italy. *Int. J. Mol. Sci.* **2017**, *18*, 660. [CrossRef]
50. Muscolo, A.; Romeo, F.; Marra, F.; Mallamaci, C. Recycling Agricultural, Municipal and Industrial Pollutant Wastes into Fertilizers for a Sustainable Healthy Food Production. *J. Environ. Manag.* **2021**, *300*, 113771. [CrossRef]
51. Fossen, T.; Cabrita, L.; Andersen, O.M. Colour and Stability of Pure Anthocyanins Influenced by PH Including the Alkaline Region. *Food Chem.* **1998**, *63*, 435–440. [CrossRef]
52. Muscolo, A.; Sidari, M.; Settineri, G.; Papalia, T.; Mallamaci, C.; Attinà, E. Influence of Soil Properties on Bioactive Compounds and Antioxidant Capacity of Brassica Rupestris Raf. *J. Soil Sci. Plant Nutr.* **2019**, 19. [CrossRef]
53. De Rasmo, D.; Signorile, A.; Larizza, M.; Pacelli, C.; Cocco, T.; Papa, S. Activation of the CAMP Cascade in Human Fibroblast Cultures Rescues the Activity of Oxidatively Damaged Complex I. *Free Radic. Biol. Med.* **2012**, *52*, 757–764. [CrossRef] [PubMed]
54. Signorile, A.; Santeramo, A.; Tamma, G.; Pellegrino, T.; D'Oria, S.; Lattanzio, P.; De Rasmo, D. Mitochondrial CAMP Prevents Apoptosis Modulating Sirt3 Protein Level and OPA1 Processing in Cardiac Myoblast Cells. *Biochim. Biophys. Acta Mol. Cell Res.* **2017**, *1864*, 355–366. [CrossRef] [PubMed]
55. Middleton, E.; Kandaswami, C.; Theoharides, T.C. The Effects of Plant Flavonoids on Mammalian Cells: Implications for Inflammation, Heart Disease, and Cancer. *Pharmacol. Rev.* **2000**, *52*, 673–751. [PubMed]
56. Wang, W.; Sun, C.; Mao, L.; Ma, P.; Liu, F.; Yang, J.; Gao, Y. The Biological Activities, Chemical Stability, Metabolism and Delivery Systems of Quercetin: A Review. *Trends Food Sci. Technol.* **2016**, *56*, 21–38. [CrossRef]

molecules

MDPI

Review

Polyphenol-Rich Ginger (*Zingiber officinale*) for Iron Deficiency Anaemia and Other Clinical Entities Associated with Altered Iron Metabolism

Soo Liang Ooi [1], Sok Cheon Pak [1,*], Ron Campbell [2] and Arumugam Manoharan [3]

[1] School of Dentistry and Medical Sciences, Charles Sturt University, Bathurst, NSW 2795, Australia
[2] The Oaks Medical Practice, The Oaks, NSW 2570, Australia
[3] Graduate School of Medicine, University of Wollongong, Wollongong, NSW 2522, Australia
* Correspondence: spak@csu.edu.au; Tel.: +61-2-6338-4952; Fax: +61-2-6338-4993

Abstract: Ginger (*Zingiber officinale*) is rich in natural polyphenols and may potentially complement oral iron therapy in treating and preventing iron deficiency anaemia (IDA). This narrative review explores the benefits of ginger for IDA and other clinical entities associated with altered iron metabolism. Through in vivo, in vitro, and limited human studies, ginger supplementation was shown to enhance iron absorption and thus increase oral iron therapy's efficacy. It also reduces oxidative stress and inflammation and thus protects against excess free iron. Ginger's bioactive polyphenols are prebiotics to the gut microbiota, promoting gut health and reducing the unwanted side effects of iron tablets. Moreover, ginger polyphenols can enhance the effectiveness of erythropoiesis. In the case of iron overload due to comorbidities from chronic inflammatory disorders, ginger can potentially reverse the adverse impacts and restore iron balance. Ginger can also be used to synthesise nanoparticles sustainably to develop newer and more effective oral iron products and functional ingredients for IDA treatment and prevention. Further research is still needed to explore the applications of ginger polyphenols in iron balance and anaemic conditions. Specifically, long-term, well-designed, controlled trials are required to validate the effectiveness of ginger as an adjuvant treatment for IDA.

Keywords: blood disorder; haemoglobin; natural product; nutraceutical; nutritional disease; phenolic compounds

Citation: Ooi, S.L.; Pak, S.C.; Campbell, R.; Manoharan, A. Polyphenol-Rich Ginger (*Zingiber officinale*) for Iron Deficiency Anaemia and Other Clinical Entities Associated with Altered Iron Metabolism. *Molecules* **2022**, *27*, 6417. https://doi.org/10.3390/molecules27196417

Academic Editor: Nour Eddine Es-Safi

Received: 31 July 2022
Accepted: 15 September 2022
Published: 28 September 2022

1. Introduction

Anaemia develops when the body's circulating erythrocytes or red blood cells (RBCs) fall below normal. According to the guidelines of the World Health Organization (WHO), anaemia is diagnosed with haemoglobin (Hb) concentration lower than the current cut-off level, defined as Hb <130 g/L for adult males, <120 g/L for non-pregnant women, and <110 g/L for children (6–59 months) [1]. The reduction in RBCs may lead to insufficient oxygen-carrying capacity of the blood to meet physiological needs, resulting in symptoms such as fatigue, weakness, shortness of breath, chest pain, reduced physical tolerance and restless leg syndrome [2]. In adults, anaemia may lead to increased morbidity and decreased work productivity and poor birth outcomes during pregnancy. In children, anaemia can cause impaired cognitive and behavioural development, and even increase mortality [1].

In 2019, the worldwide prevalence of anaemia for all ages was 22.8% (95% confidence interval (CI): 22.6–23.1) [3]. The global burden of disease measured in years living in disability for anaemia is 672.4 (95% CI: 447.2–981.5) per 100,000 population [4]. Women and children in low-income countries are the most vulnerable groups. This condition affected 29.9% (95% CI: 27.0–32.8) of women of reproductive age, of which 36.5% (95% CI: 34–39.1) of pregnant women suffered from anaemia compared to 29.6% (95% CI: 26.6–32.5) of non-pregnant women [5]. About 269 million children under five also had anaemia, with a global prevalence of 39.8% (95% CI: 36.0–43.8). The highest was in the African region, affecting 60.2% (95% CI: 56.6–63.7) of children [5].

Even in a wealthy society like Australia, anaemia remains a health risk. According to the Australian Health Survey last conducted in 2011–2012, 4.5% of the population aged 18 years and over had anaemia, with women having a relative risk of 2.56 times more than men [6]. The burden of anaemia is even higher among the Australian indigenous population, with a female-specific prevalence of 15.3% [7], signifying health inequality due to socioeconomic consequences.

Anaemia is not a disease per se but a manifestation of other underlying causes. Iron deficiency, which results from increased iron demands, deminished iron supply, blood loss or malabsorption of iron, is the most common cause of anaemia worldwide, accounting for nearly two-thirds of global anaemia cases [4,8]. Hence, iron deficiency anaemia (IDA) is a global health concern affecting millions worldwide, especially women and children in less developed regions. IDA is routinely treated with oral iron supplements such as ferrous sulphate but compliance issues due to gastrointestinal side effects often hamper its effectiveness [9,10].

In the broader context, ginger plants refer to all perennial flowering plants in the Zingiberaceae family, which include many aromatic herbs and spices such as turmeric, cardamom, and galangal. Currently, there are 1888 unique species in the Zingiberaceae family classified into 62 genera, of which 204 species belong to the *Zingiber* genera [11]. The ginger root commonly consumed worldwide is the rhizome of *Zingiber officinale* species. Accordingly, in this review, ginger only refers to the *Z. officinale* species.

Z. officinale has a long history of culinary and medicinal use, possibly even before formally recorded history. Wu [12] suggested that ginger cultivation originated around the Yangtze River and Yellow River basins in ancient China. However, this claim has yet to be widely accepted. The spice trade spread ginger to major civilisations from East Asia and India to the Greek, the Roman Empire, and beyond. Today, ginger is primarily used as a food, spice, herb, and flavouring agent, with a global trade volume of USD 1.06 billion in 2019, with China being the top exporter supplying over 57.8% of the world demand [13]. Three varieties of ginger are consumed as food and herbs: white (var. Roscoe), small white (var. Amarum) and red (var. Rubra), with *Z. officinale* Roscoe being the most common variety [14].

Ginger possesses several health-promoting properties and has been traditionally used in East Asia to ease fatigue and weaknesses. Contemporarily, ginger is considered a functional food that can confer health benefits beyond its nutritional values for preventing, managing, or treating disease [15–18]. As a rich source of natural polyphenols, ginger may potentially complement oral iron therapy in treating IDA and be a supportive dietary strategy for preventing IDA. Hence, there has been heightened commercial interest in using ginger, especially in China, as an ingredient for functional foods or ethnomedicine for IDA, as evidenced by the growing number of patents filled with the World Intellectual Property Organization in recent years (see Table A1 in Appendix A). However, there is a lack of research literature critically reviewing such potentials.

The objective of this narrative review is to inform translational research on the benefits of ginger and its bioactive polyphenols in the context of IDA and other clinical entities associated with altered iron metabolism based on available research. The ensuing sections first examine ginger's application as a functional food, followed by the pathophysiology of IDA and its treatment. These overviews provide context to support the subsequent review of the various beneficial properties of ginger and its polyphenols applicable to IDA based on pre-clinical and clinical evidence. To the authors' knowledge, this is the first attempt to comprehensively investigate the scope and depth of current literature on this underexplored topic.

2. Ginger as a Functional Food

2.1. Nutritional Composition and Traditional Use

Nutritional analysis has shown that ginger consists mainly of moisture, carbohydrate, protein, fibre, fat, and ash. It is rich in polyphenols and contains micronutrients including ascorbic acid, β-carotene, calcium, iron, and copper. However, it is worth noting that the nutritional composition of ginger can vary greatly depending on the varieties, origin, time of harvest, drying method, and storage condition. Table 1 shows the approximate nutritional composition of dried ginger powder reported in the literature [19,20]. Ginger is, however, valued beyond its nutritional benefits. It is believed that the Indian and Chinese populations have used ginger as a tonic for over 5000 years [21]. In Shen Nong Ben Cao Jing, the oldest surviving Chinese materia medica circa 100BC, ginger was classified as a middle category of herb with little or no toxicity and was mainly used in combination prescription to treat deficiency to prevent illness or resist worsening disease [22]. Incidentally, ginger is also used in traditional Ayurvedic medicine to treat many diseases such as diabetes, flatulence, intestinal colic, indigestion, infertility, inflammation, insomnia, nausea, rheumatism, stomach ache, and urinary tract infections [23].

Table 1. Nutritional composition of dried ginger powder as reported by Ajayi et al. [19] and Sangwan et al. [20].

Nutrient	Amount	Unit
Carbohydrate	39.70–58.21	%
Protein	11.65–12.05	%
Crude fibre	8.30–21.90	%
Fat	9.89–17.11	%
Moisture	3.95–4.63	%
Ash	4.95–7.45	%
β-carotene	0.68–0.81	mg/100g
Ascorbic acid	2.2–3.8	mg/100g
Polyphenols	11.8–12.5	mg/100g
Calcium	64.4–69.2	mg/100g
Iron	1.5–1.8	mg/100g
Copper	0.46–0.75	mg/100g

2.2. Phytochemistry and Health Benefits

With increasing interest in the therapeutic applications of natural products, ginger has received much attention in recent years, with numerous studies and reviews exploring its phytochemistry, pharmacological, and health-benefiting properties [14,21–26]. Overall, ginger is rich in phytochemicals, with over 300 identified constituents divided into three main categories: gingerols, volatile oils, and diarylheptanoids, as reported by Liu et al. [27]. The most notable are the phenolic compounds in gingerols, shogaols, and paradols, including 6-gingerol, 6-shogaol, 8-gingerol, 8-paradol, 10-gingerol, and many more [28]. These compounds are responsible for the unique pungent smell and taste of ginger. Other gingerol-related compounds include zingerone, gingerenone A, and 1-dehydro-10-gingerdione [29]. In addition, the volatile oils of ginger also contain terpene compounds such as β-bisabolene, α-curcumene, zingiberene, α-farnesene, and β-sesquiphellandrene [30]. The ginger's diarylheptanoid contents can be divided into linear diphenyl heptane and cyclic diphenyl heptane compounds with antioxidant activity [27].

Ginger is also known to contain many secondary metabolites of flavonoids and other phenolic components such as quercetin, rutin, catechin, epicatechin, kaempferol, naringenin, fisetin, morin, hesperidin, salicylic acid, and chlorogenic acid [31–33]. The concentrations of these secondary metabolites in ginger may vary significantly across samples as their contents are influenced by the environmental conditions (including light intensity, temperature, insects, etc.) where ginger is grown and methods of drying and storage. Higher phenolic and flavonoid content in ginger is known to increase its antioxidant activities [34,35]. Figure 1 provides an overview of ginger and its constituents.

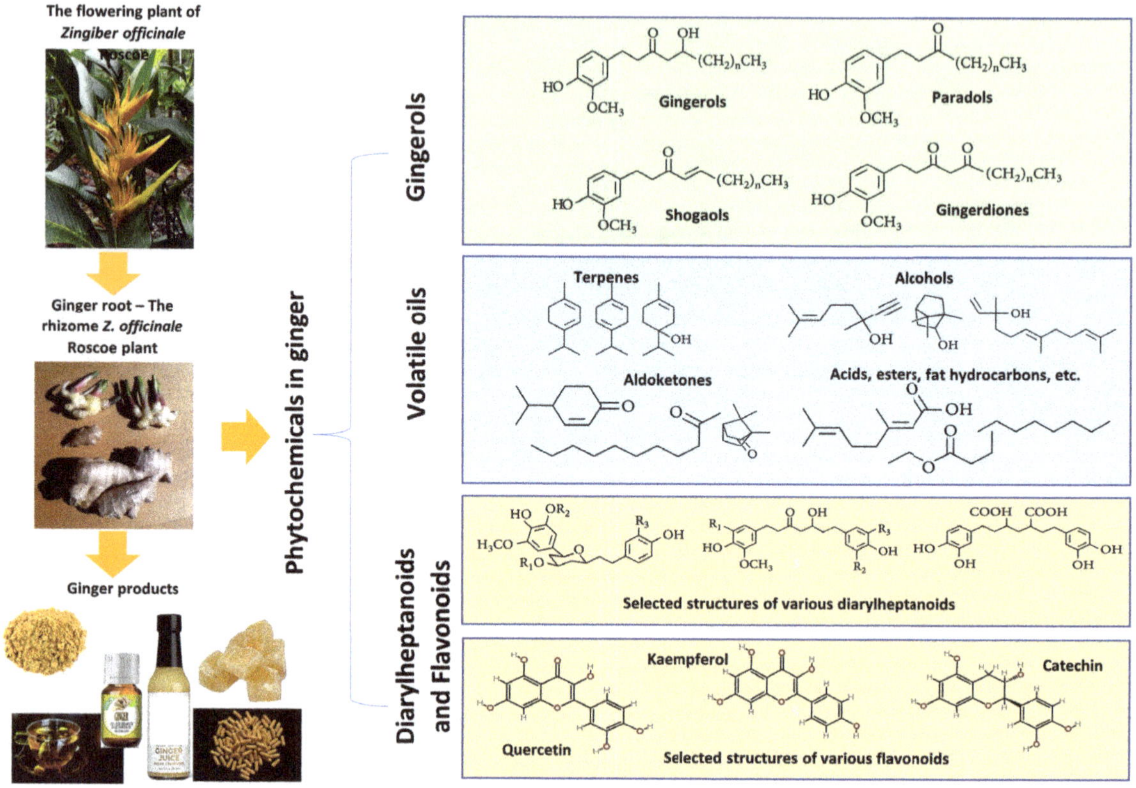

Figure 1. An overview of ginger and its main groups of active constituents of gingerols, volatile oils, diarylheptanoids, and flavanoids with some sample structural formulas. In the formulas of gingerols, $_n$ denotes the number of repeating units of CH_2. R_1, R_2, and R_3 in the selected diarylheptanoid structures are sites for functional groups or substituents (e.g., H, CH_3, OCH_3, or $COCH_3$).

Among these constituents, 6-gingerol is the most pharmacologically active compound [26], whereas 6-shogaol has a higher potency than 6-gingerol in terms of bioactivities but exists in a lesser quantity naturally [36]. By studying ginger's typical metabolic pathways in a mouse model, the in vivo effects were found to derive mainly from 6-gingerol and 6-shogaol, with hydrogenation, demethylation, glucuronidation, sulfation, and thiolation being their major metabolic reactions [16]. Shogaols are metabolised through more complicated pathways than gingerols, and these two compounds have different molecular targets even though both are reported to have potent antioxidant and anti-inflammatory actions [37].

The pharmacological effects of ginger and its active compounds have been recently reviewed by many authors, including Unuofin et al. [23], Mao et al. [28], and Choi et al. [38]. Notable effects include anti-diabetic [39–41], anti-emetic [42–44], anti-nauseant [45,46], anti-obesity [47–51], anti-inflammatory [52–55], antioxidant [56–60], nephroprotective [61–63], neuroprotective [64–66], gastroprotective [67,68], and anti-melanogenesis [69,70] effects. Moreover, ginger may also be protective against male infertility [71]. The anti-inflammatory and antioxidant activities of ginger extract, and 6-gingerol in particular, have also been shown to have antiproliferative and anticancer properties in another review by de Lima et al. [26]. For clinical application, a systematic review of 109 randomised controlled trials (RCTs) by Anh et al. [25] found evidence to support the use of ginger for improvement in nausea and vomiting in pregnancy, inflammation, metabolic syndromes, digestive function, and colorectal cancer markers.

2.3. The Growing Popularity of Ginger

Ginger is widely used for food processing in many forms, including as fresh ginger and dried ginger, as an oleoresin, as an essential oil, as an extract, or as a powder. The addition of ginger to food gives a spicy taste and serves as a natural antioxidant for shelf-life extension [72,73]. Furthermore, with many health-promoting properties supported by research evidence, it is not surprising that ginger is also a choice ingredient in various kinds of functional foods, such as candy, biscuits, herbal tea, and beverages, etc. Most notably, ginger candy is used to reduce the vomiting frequency among pregnant women in the first trimester [74]. Ginger tea, a spicy caffeine-free alternative to black tea or coffee, has also emerged as one of the most popular beverages in the world [75]. Ginger supplements also ranked 6th in the American top-selling herbal chart of the mainstream retail channels in 2020, with over USD 64 million in sales, a 39.3% growth compared to sales volume in 2019 [76]. Undeniably, the popularity of ginger as a functional food will continue to spread with the growing consumer awareness of its health benefits.

2.4. Safety

Safety data from an animal study showed that consuming up to 1 g/kg body weight per day of a standardised ethanol extract of ginger had no significant effects on blood glucose, blood coagulation, blood pressure, and heart rate in rats compared to controls [77]. Long-term (35 days) force-feeding of rats with ginger powder up to 2 g/kg body weight was also not associated with any mortality or abnormalities in general conditions, behaviour, growth, and food and water consumption, as shown by Rong et al. [78]. Acute and subacute toxicity studies in rats with an enriched ginger extract (8% gingerols) reported no mortality or clinical signs of toxicity at a dose level of 2000 mg/kg (LD_{50} > 2000 mg/kg). The repeated administration of ginger extract for 28 days in rats at 1000 mg/kg also did not induce any observable toxic effects, with the no observed adverse effect level (NOAEL) calculated as 1000 mg/kg daily [79].

An oral toxicity study of ginger essential oil (31% zingiberene) in Wistar rats found no adverse effect after 13 weeks of subchronic oral administration. The NOAEL for ginger essential oil was determined to be over 500 mg/kg per day [80]. In another study, Idang et al. [81] performed toxicological assessments of both ginger essential oil and ginger fixed oil in Wistar rats for 60 days and found signs of increased oxidative stress and some forms of pathologies in the livers and spleens of the experimental animals fed up to 0.2 mL/kg of ginger fixed oil. Hence, the authors cautioned against the long-term use of fixed oils derived from ginger. Additional study to validate and confirm the toxicity of ginger fixed oil is required.

In pregnant rats, Weidner and Sigwart [82] also showed that feeding with 1 g/kg body weight of a standardised ginger extract did not cause any maternal or developmental toxicity. In contrast, Wilkinson [83] reported that pregnant Sprague-Dawley rats' exposure to ginger tea increased early embryo loss but enhanced growth in surviving foetuses. Notwithstanding, a systematic review of 14 RCTs and 3 prospective clinical studies over more than 25 years found that ginger use during pregnancy does not pose a risk for the mother and the foetus [84]. Therefore, ginger consumption has no safety concern during pregnancy.

2.5. Adverse Events

In humans, a daily intake of ginger up to four grams is generally considered safe [85]. However, consumption at doses higher than six grams may increase the risk of gastrointestinal disturbances such as gastrointestinal reflux, heartburn, and diarrhoea [86]. In a systematic review of RCTs by Anh [25], 17 of 43 high-quality included trials provided adverse event information. Heartburn was the most common adverse event reported in 16 studies. Other gastrointestinal disturbances reported with ginger treatment were abdominal pain, bloating, gas, and epigastric distress. None of the events were considered severe [25].

Rare cases of allergic rhinoconjunctivitis were described in the literature concerning occupational exposure to ginger. For instance, Schmidt et al. [87] reported that a 60-year-old woman developed allergic rhinoconjunctivitis with red eyes, sneezing, nasal congestion and a runny nose due to exposure to dust particles containing ginger in her workplace. Similar incidences were reported in other occupational settings as well, with positive skin-prick tests against ginger confirming the reactions to be an immunoglobulin E antibody-mediated allergy [88,89]. The cysteine proteinase GP-I in ginger is thought to be the relevant allergen in these cases [90]. Moreover, occupational allergic contact dermatitis caused by ginger and other spices has also been reported [91]. Hence, ginger could contribute to allergic reactions, especially in patients with known hypersensitivity to spices.

Other potential side effects of ginger, when taken in large doses, may include prolonged pre-existing bleeding, central nervous system depression, and arrhythmia [86]. Therefore, properly dosing concentrated forms of ginger extract and derivatives are essential.

2.6. Drug Interactions

Ginger may also potentially interfere with the bioavailability of many oral pharmaceutical substances, either by increasing their absorption from the gastrointestinal tract or preventing their metabolism in the liver after absorption [92,93].

The most notable is the potential interaction between ginger and anticoagulant therapy as ginger is known to possess anti-platelet aggregation properties. Some case reports have suggested that ginger may interact with warfarin and increase the risk of prolonged bleeding [94–97]. Additionally, a prospective longitudinal study reported that ginger was associated with an increased risk of self-reported bleeding among patients taking warfarin, even though no significant risk of increasing blood clotting time was found [98]. However, a study investigating the effect of ginger on the pharmacokinetics and pharmacodynamics of warfarin in healthy volunteers co-administered with 0.4 g of ginger extract did not find evidence of interactions [99]. Furthermore, the effect of ginger on platelet aggregation and coagulation remains equivocal, according to a systematic review by Marx et al. [100] that included eight clinical trials and two observational studies. Hence, the European medicines agency did not find the case reports on the potential interactions between ginger and warfarin to be convincing [101].

Studies have shown that ginger may potentially interact with drugs such as crizotinib (an anti-cancer drug) [102], cyclosporine (an immunosuppressant) [103], metronidazole (an antibiotic) [104], and ketoconazole (an antifungal) [105]. With the increased use of ginger and its derivatives as nutraceuticals, more research is needed to identify and confirm potential ginger-drug interactions to reduce and avoid side effects induced by unfavourable interactions.

3. Pathophysiology of IDA and Its Treatment

3.1. Iron Homeostasis

The pathophysiology of IDA is closely related to the iron balance in the body. Iron is a vital trace mineral that plays a crucial role in many body functions, including oxygen transport, immune regulation, and enzyme catalyse reactions [106]. An overview of the iron cycle is depicted in Figure 2.

The total iron content in the body for a 70 kg adult male is about 3500 mg to 4000 mg, corresponding to a per kg bodyweight content of approximately 50 mg to 60 mg [106]. Around 65% of iron is contained within the Hb of RBCs to carry oxygen around the body [107]. RBCs have a typical life span of about 120 days. Old and worn RBCs are phagocytosed by the reticuloendothelial system, particularly the macrophages, in the spleen, liver, and bone marrow to release iron. In ferrous ion (Fe^{2+}) form, about 3 mg of free iron is carried in the bloodstream by transferrin to the bone marrow to be recycled into making new RBCs. Approximately 300 mg of iron is kept in bone marrow for erythropoiesis [108,109]. Excess iron of up to 1000 mg is stored in the liver hepatocytes in complex hemosiderin. As ferritin, about 300 mg of iron is also contained in cells and tissues of organs and muscles. Together, transferrin, ferritin, and hemosiderin store approximately 20 to 30% of the iron in the body as spare [108,109].

The body loses about 1 to 2 mg of iron each day through blood loss or peeling of epithelial cells. Such loss is unregulated as humans and mammals have no iron excretion mechanism and must be replenished via controlled uptake from diet to maintain iron balance [109]. Diet is the only source of iron supply for the body after birth, excluding exogenous therapeutic means. There are two types of dietary iron, animal-derived haem iron and non-haem iron, the dominant form of iron in plants and abundant in both animal-based and plant-based foods [110]. The organic haem iron, which accounts for about 10% of the dietary iron, is more readily absorbable by the body [108], but its uptake mechanisms remain unclear, with receptor-mediated endocytosis and direct haem transporters of intestinal enterocytes being the two most prevailing hypotheses [108,110–112].

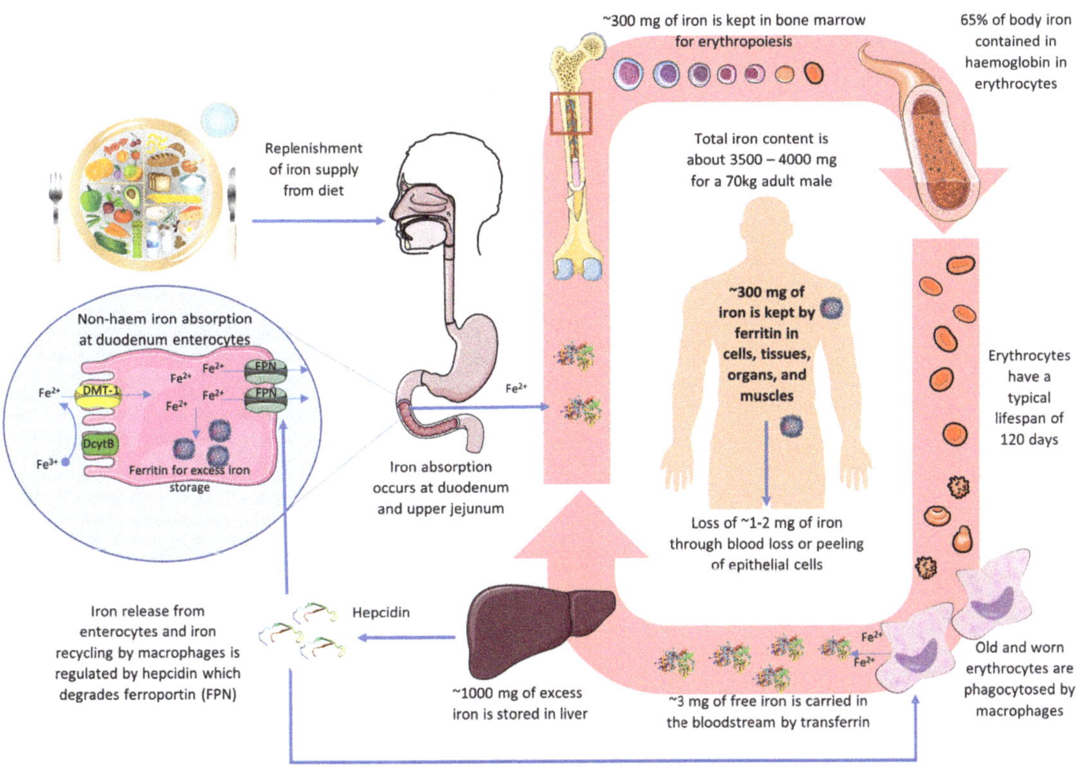

Figure 2. An overview of the iron cycle in humans depicting iron absorption, transportation, functioning, recycling, storage, and regulation.

The inorganic non-haem iron, which makes up 90% of the total iron in food, is absorbed in the duodenum and upper jejunum section of the small intestines through a complex process [108]. Non-haem iron can be in Fe^{2+}or ferric (Fe^{3+}) forms. First, Fe^{3+} must be reduced to Fe^{2+} ion by duodenal cytochrome B (DcytB), a trans-plasma membrane ferric reductase enzyme. Fe^{2+} can then traverse into the cytoplasm of the duodenal enterocytes by the divalent metal transporter 1 (DMT-1), the hydrogen-driven metal transporter residing at the apical membrane. An acidic micro-environment and the presence of ascorbate can facilitate the reduction of Fe^{3+} to Fe^{2+} and iron uptake across the apical membrane [106]. Iron in the enterocytes can be sequestered within the iron-buffering protein ferritin for later use. When iron is required by the body, it can be exported across the enterocyte basolateral membrane via ferroportin (FPN), an iron-regulated transporter protein. Iron stored as ferritin will be lost if not released by FPN as the enterocyte is sloughed after several days [110].

The body maintains its iron content in a tight balance; even though iron is physiologically essential, too much iron is toxic and detrimental to health. Systemic iron homeostasis is controlled through hepcidin, a liver peptide serving as the principal regulator [112]. Hepcidin induces FPN degradation to prevent export from iron-absorptive enterocytes and iron-recycling macrophages. High iron loading and inflammation will transactivate the hepcidin antimicrobial peptide (HAMP) gene in hepatocytes to encode hepcidin to reduce iron entry into circulation. In contrast, low iron status and increased erythropoietic demand will inhibit hepcidin transcription and promote iron export via FPN [109,112].

3.2. *IDA and Its Aetiology*

Iron deficiency anaemia can result from insufficient intake from the diet, decreased absorption, or blood loss [113]. Age, sex, lifestyle, and socioeconomic status may influence the adequacy of dietary iron intake. For example, growth spurts in children and adolescents or pregnancy may increase iron demand and cause iron deficiency without increased consumption of iron-rich foods. Poverty and poor diet can lead to malnutrition with low iron intake [114]. Decreased absorption can be due to dietary factors (e.g., high phytate diet or improper vegetarian or vegan diet), surgery, or gastrointestinal conditions (e.g., coeliac disease, inflammatory bowel disease, or gastritis) [113]. Blood loss due to heavy menstrual bleeding commonly leads to IDA in premenopausal women. Other sources of blood loss, such as injury, surgery, or occult gastrointestinal tract bleeding, also deplete the available RBCs and cause IDA [115].

There is also a reciprocal relationship between iron deficiency and inflammation, as shown in a 3-year prospective longitudinal study of 2141 relatively healthy older adults aged 70+ [116]. Posthoc analysis of the high sensitivity C-reactive proteins (CRP) and interleukin (IL)-6 levels, measured at 12, 24, and 36 months of follow-up, found baseline iron deficiency was associated with a more significant increase in IL-6 levels (mean difference in change: 0.52 ng/L, 95% CI: 0.03–1.00, $p = 0.04$) over 3 years. Additionally, iron deficiency at any yearly time point was associated with higher increases in CRP (mean difference in change: 1.62 mg/L, 95% CI: 0.98–2.26, $p < 0.001$) and IL-6 levels (mean difference in change: 1.33 ng/L, 95% CI: 0.87–1.79, $p < 0.001$) over 3 years. The results were independent of anaemia status as there was no interaction between iron deficiency and anaemia. As such, the findings suggest that iron deficiency may be involved in low-grade inflammation even in relatively healthy older adults [116].

Consequently, chronic inflammatory conditions such as cancer, chronic infections, immune-mediated diseases, and obesity can also reduce RBCs. An estimated 40% of all anaemia cases worldwide are due to chronic disease or inflammation as a contributing cause [117]. Anaemia of inflammation (AI) or anaemia of chronic disease is the second most common type of anaemia after iron deficiency. Unlike absolute IDA, whereby the body's iron store is depleted, the iron store of a patient with AI can remain normal. Hence, AI is also sometimes referred to as functional iron deficiency, which is characterised by the body's inability to mobilise the available iron for erythropoiesis [118]. However, IDA and AI may co-exist in some patients, with anaemia due to inflammatory bowel disease being one example. As such, the diagnosis and management of IDA require a systematic evaluation of the case history and haematological profile, plus an investigation of the potential underlying causes of blood loss [115,119].

3.3. Treatment of IDA

Repleting iron stores is the primary strategy in IDA treatment, with prompt treatment needed to alleviate fatigue, improve quality of life, and reduce cognitive impairment [114]. Research evidence has shown that iron supplements are more effective than dietary iron for restoring iron status and Hb recovery [120,121]. A meta-regression analysis with data from 41 RCTs by Casgrain et al. [120] showed that iron supplementation affected serum ferritin (SF) concentration linearly related to duration (+0.51 µg/L per week, 95% CI: 0.02–1.00, $p = 0.04$) and dose (+0.10 µg/L per g Fe, 95% CI: 0.01–0.20, $p = 0.036$). Hb concentration is also expected to increase significantly by 0.8 g/L for every 10 µg/L increase in baseline SF level with iron supplementation ($p = 0.02$). It should be noted that these effects were observed in the healthy adult population as per the inclusion criteria of Casgrain et al. [120]. Another systematic review and meta-analysis by Houston et al. [122] further demonstrated that iron supplementation is more effective than placebo in reducing self-reported fatigue (standardised mean difference: −0.38, 95% CI: −0.52 to −0.23) among iron-deficient adults with no anaemia. Hence, for patients with iron deficiency or straightforward and uncomplicated IDA, oral iron supplement (ferrous sulphate, ferrous gluconate, or ferrous fumarate) is considered the standard care [123]. However, for IDA with complications such as intolerance, intestinal malabsorption, and ongoing blood losses that exceed iron absorption capacity, intravenous iron therapy may be needed to rapidly restore iron supply [124].

3.4. Adverse Effects of Oral Iron Therapy

While oral iron therapy as the first-line treatment for IDA is indisputable, considerable side effects are reported. A systematic review and meta-analysis of 43 RCTs by Tolkien et al. [9] confirmed that ferrous sulphate was 2.32 times (95% CI: 1.74–3.08, $p < 0.0001$) and 3.05 times (95% CI: 2.07–4.48, $p < 0.0001$) more likely to cause gastrointestinal side effects than placebo and intravenous iron therapy, respectively. Notably, meta-regression from this study did not find a significant relationship between the odds ratios of gastrointestinal side effects and dose. The common gastrointestinal side effects reported in the RCTs are constipation, nausea, diarrhoea, abdominal pain, vomiting, heartburn, dark stools, and flatulence [9].

There are also other negative impacts of oral iron therapy on the gastrointestinal tract besides symptoms of discomfort. Free iron, Fe^2, is a divalent metal cation that can react with hydrogen peroxide in cells to produce hydroxyl radicals, which induce oxidative stress responses in cells. These radicals are cytotoxic to endothelial and smooth muscle cells and have deleterious impacts on health [110,125,126]. Hence, excess iron from oral iron supplements can lead to lipid peroxidation, which causes ferroptosis, mitochondrial damage, and endoplasmic reticulum dysfunction, leading to the destruction of the intestinal epithelial cells, affecting the intestinal mechanical barrier's integrity [126]. Such are the possible underlying mechanisms of frequent gastrointestinal side effects of oral iron therapy. Moreover, anaemic patients taking oral iron supplements also demonstrated increased systemic oxidative stress as measured by lipid peroxide, protein carbonyl, conjugated dienes, lipid hydroperoxide and oxidised glutathione levels accompanied by a reduced total antioxidant level [127]. Hence, it is advisable to increase dietary antioxidant intake while taking oral iron supplements.

Excessive unabsorbed iron may cause gut dysbiosis, as it can modify gut microbiota composition, promoting the growth of pathogenic bacteria at the expense of the healthy ones [128–130]. Specifically, increased luminal iron concentration appeared to favour the growth of *Escherichia coli*, *Salmonella*, and *Bacteroides* species of pathogenic bacteria while lowering the abundance of probiotics such as *Lactobacillus* and *Bifidobacterium* species [128]. The disruption of gut microbiota equilibrium can also lead to inflammation and chronic conditions such as inflammatory bowel disease and metabolic dysfunction [128,131], which, in turn, can also hamper iron absorption and reduce the effectiveness of iron supplementation [132,133].

Overuse of iron supplements also carries considerable risks. Intake of iron above 60 mg/kg body weight can result in severe toxicity leading to gastrointestinal tract injury and impaired cellular metabolism in the heart, liver, and central nervous system, which can be fatal without immediate medical attention [134,135]. Moreover, prolonged ingestion of iron supplements for years has been reported to cause iron overload, where the body stores too much iron, a condition that can cause severe organ damage [136].

4. Ginger and IDA

The following sections will explore available evidence from in vivo, ex vivo, in vitro, and human clinical studies to elucidate the potential benefits of ginger for IDA and its associated clinical manifestations of altered iron metabolism. A summary of the research findings is presented in Table 2.

Table 2. A summary of the research findings on ginger's beneficial properties in its applications for iron deficiency anaemia and associated clinical manifestations of altered iron metabolism.

Beneficial Property	Study Type	Research Findings	Reference
Iron absorption enhancement	Ex vivo	Ginger was the most potent spice for enhancing iron absorption by increasing uptake by 28.5 ± 2.09% in the jejunum of rats compared to control.	[137]
	In vitro	Adding ginger to food enhanced the bioaccessibility of dietary iron by 2- to 3-fold depending on the formulations.	[138]
	Human study	Ginger plus oral iron therapy improved haematological and iron parameters of anaemic patients better than oral iron therapy alone.	[139,140]
Antioxidant activity	In vivo	Adding ginger to the diet significantly increased the activities of antioxidant enzymes ($p < 0.05$) at the intestinal and gastric mucosa of rats, demonstrating enhanced protective effects against oxidative stress.	[141]
	In vitro	The polyphenols and diarylheptanoid derivatives of ginger contributed to both radical scavenging and inhibitory effects of autoxidation.	[142]
	In vitro	Both red and white ginger variants possessed antioxidant capacities against free iron radicals in rat brains, but red ginger was superior at inhibiting Fe^2-induced lipid peroxidation and chelating Fe^2.	[143]
	In vitro	Water-based extract of ginger showed relatively low antioxidant activities compared to other spices due to reduced phenolic contents produced from hydro-distillation extraction.	[144]
Anti-inflammatory action	Review	The bioactive compounds in ginger possessed broad anti-inflammatory properties that can block the activation of NF-$\kappa\beta$ by suppressing pro-inflammatory cytokines of IL-1, TNF-α and IL-6, thus preventing hepcidin production.	[145]
	Human study	Ginger plus oral iron therapy significantly reduced the inflammatory marker TNF-α ($p < 0.05$) in anaemic patients better than oral iron therapy alone.	[139,140]
Gut microbiota modulation	In vitro	Undigested ginger polyphenols significantly increased the abundances of *Bifidobacterium* ($p < 0.05$) and *Enterococcus* ($p < 0.01$) after faecal inoculated fermentation, accompanied by elevated levels of SCFA and decreased pH value.	[146]
	In vivo	Ginger supplementation could mitigate the detrimental impact of a high-fat diet in mice by promoting the abundance of *Bifidobacterium* genus and SCFA-producing bacteria (*Alloprevotella* and *Allobaculum*).	[147]

Table 2. *Cont.*

Beneficial Property	Study Type	Research Findings	Reference
	In vivo	Ginger treatment significantly reduced antibiotic-associated diarrhoea symptoms ($p < 0.05$) in rats with an associated increase in microbiota diversity and improved intestinal barrier integrity.	[148]
	Human study	Ginger juice consumption in healthy adults decreased the Prevotella-to-Bacteroides ratio and pro-inflammatory *Ruminococcus_1 and Ruminococcus_2* genus while increasing the Firmicutes-to-Bacteroidetes ratio, Proteobacteria and anti-inflammatory *Faecalibacterium*.	[149]
Erythropoiesis stimulation	In vivo	Ginger, with its bioactive compounds of 8-gingerol, 10-gingerol, 8-shogaol, and 10-shogaol, promoted the expression of Gata1 in erythroid cells of zebrafish embryos through the Bmp signalling pathway.	[150]
	In vivo	Ginger induced scl/runx1 expression through Bmp and Notch signalling pathways which up-regulated nitric oxide production for regeneration of haematopoietic stem/progenitor cells.	[151]
Iron overload prevention	In vivo	The bioactive lipids in ginger repressed some iron-related parameters, including reductions in 20% of ^{59}Fe absorption, 65% of pancreatic non-haem iron, and 40% to 50% of serum ferritin levels, compared to controls.	[152]
	In vivo	Ginger extract demonstrated strong protective effects against iron toxicity through its free radical scavenging activities in iron-overloaded rats.	[153]
	Case series	Ginger extract rich in 6-shogaol prevented iron overload in three patients with myelodysplastic syndrome. These patients had elevated serum ferritin (>300 g/µL) at baseline but achieved >40% reductions after three months through upregulation of hepcidin.	[154]
Ginger-synthesised iron nanoparticles	In vitro	Ginger was used to bio-reduce the metallic ions to nanoparticles (Fe^{3+} ions to FeNPs). Transmission electron microscopy showed that the FeNPs in ginger were in the range of 14.08–21.57 nm with almost spherical forms and demonstrated considerable radical scavenging properties and antimicrobial activities against Gram-positive and Gram-negative bacteria and fungi.	[155]
	In vitro	Ginger can be a suitable green material for synthesising iron nanoparticles with high antioxidant and antibacterial properties.	[156,157]

Abbreviations: bone morphogenetic protein (Bmp); GATA-binding factor 1 (Gata1); interleukin (IL); short-chain fatty acids (SCFA); tumour necrosis factor alpha (TNF-α).

4.1. Iron Absorption Enhancement

Prakash and Srinivasan [137] demonstrated in an animal study that various spices (ginger, capsaicin, and piperine) can improve the bioavailability of dietary iron. Groups of Wistar rats were fed diets containing different spices for eight weeks before being sacrificed. Everted segments of duodenum, jejunum, and ileum of small intestines isolated from these rats were examined for ex vivo iron uptake through incubation with a medium containing finger millet powder fortified with iron. Compared to the control group that was not fed with any spices, all spice-fed groups showed significantly higher iron uptake percentages ($p < 0.05$) in all sections of the small intestine. Ginger was the most potent among the three spices to increase iron uptake across all sections. The highest was 28.5 ± 2.09% uptake at the jejunum for ginger compared to 26.3 ± 1.30% for piperine, 22.6 ± 1.40% for capsaicin, and only 18.3 ± 0.39% for the control group. The suggested mechanism could be that the pungent spices altered mucosal permeation characteristics by increasing the absorptive surface [137].

Ginger can also be used as a food additive to enhance the bioavailability of non-haem iron. A study by Jaiswal et al. [138] compared the effects of iron bioaccessibility by adding various spices (ajwain, cumin, cinnamon, fennel, black pepper, and ginger) at 1 or 2% weight to wheat flour and different Indian bread formulations. The bioaccessibility of iron was measured through an in vitro dialysis method. Adding spices at 2% significantly enhanced iron bioaccessibility ($p < 0.0001$). Ginger, in particular, was shown to increase iron bioaccessibility in all food formulations by 2- to 3-folds. The authors attributed effects to the ascorbic acid and amino acids within ginger that favour iron absorption [138]. However, with its low vitamin C content (Table 1), iron bioaccessibility is likely to be assisted by other bioactive compounds in ginger.

The effects of ginger in enhancing iron availability have been confirmed in a human intervention study. Kulkarni et al. [139,140] conducted a human clinical study to demonstrate the potential use of ginger supplements in the treatment of IDA along with oral iron therapy. The study recruited 62 patients with anaemia following the WHO Hb cut-off levels, consisting of 12 males and 50 females, from a hospital in India. Their conditions were likely due to nutritional deficiency since those with chronic conditions, pregnancy, and blood donors were excluded. Participants were divided into two groups. The intervention group (n = 30) took 1.5 g of ginger powder with oral iron therapy, whereas the control group (n = 32) received only oral iron therapy as routine care. Fasting blood samples were collected at baseline and after 30 days. Pre- and post-treatment comparisons of haematological and iron parameters found significant increases ($p < 0.05$) in all parameters in both groups. However, the intervention group achieved a more remarkable improvement in percentage difference than the control group in Hb (+8.23% vs. +2.3%), iron status (+19.63% vs. +5.54%), total iron-binding capacity (−7.23% vs. −4.47%), and SF (+45.11% vs. +34.11%). However, the authors did not report whether there were significant differences in the mean differences between groups. Notwithstanding, based on the published sample size, means, and standard deviations, the p-values could be easily estimated. The mean difference between groups for serum iron levels was significantly different post-treatment ($p < 0.007$). Hence, the study demonstrated that ginger could assist in iron absorption and improve the efficacy of oral iron therapy for IDA.

The study by Kulkarni et al. [139,140] had several drawbacks. The first is its short duration of trial with only 30 days. Although patients with uncomplicated anaemia are expected to show improvement after 4 weeks of oral iron treatment, replenishment of iron store will take longer to achieve [123,124]. Secondly, the study did not track the adverse events experienced by the participants. Thus, there was no data on ginger's effects on any side effects of oral iron therapy. Thirdly, there could be selection bias in the study design as it is unclear how the group assignment was carried out. Hence, the results from this study need further validation with more well-designed clinical trials.

4.2. Antioxidant Activity

Another study by Prakash and Srinivasan [141] showed that ginger can significantly enhance the activities of antioxidant enzymes ($p < 0.05$), including superoxide dismutase (SOD), catalase (CAT), glutathione reductase (GR), and glutathione-S-transferase (GST), in both gastric and intestinal mucosa in vivo. For eight weeks, eight male Wistar rats were fed ad libitum with a basal diet enriched with 0.05% ginger powder. Compared to the control group provided with the basal diet only, the ginger group showed 48%, 11%, 67%, and 50% stimulation in the activities of CAT, SOD, GST, and GR, respectively, in the intestinal mucosa. In rats subjected to ethanol-induced oxidative stress, ginger treatment demonstrated higher SOD, GST, and GR activities in the gastric mucosa by 35%, 39%, and 30%, respectively, compared to controls. Moreover, the ginger-fed group also had 56% higher mucin content of gastric mucosa than the ethanol-treated controls. In short, this study illustrated the gastrointestinal protective effects of dietary ginger against oxidative stress.

The gingerol-related polyphenols and diarylheptanoids derived from the rhizomes of ginger possess remarkable free radical scavenging activities [142]. However, the antioxidant potency may vary across ginger varieties. In an in vitro experiment, Oboh et al. [143] studied the antioxidant effects of two types of gingers (red and white) against free iron radicals in rat brains. Although both variants possessed antioxidant capacities against Fe^2, the study found red ginger (*Z. officinale* var. Rubra) superior to white ginger (*Z. officinale* Roscoe) at inhibiting Fe^2-induced lipid peroxidation and chelating Fe^2, likely due to its higher ascorbic acid, phenol, and flavonoid contents.

Hinneburg et al. [144] demonstrated in another study that the contents of total phenols in various spices had significant positive correlations with their antioxidant properties in terms of iron reduction ($r^2 = 0.8871$, $p < 0.001$) and inhibition of lipid peroxidation ($r^2 = 0.7327$, $p < 0.01$). Specifically, the hydro-distilled ginger extract showed relatively low Fe^3 to Fe^2 reducing activity and Fe^2 chelating capacity compared to basil, parsley, juniper, cumin, and fennel extracts. The reduced antioxidant activities were attributed to the low total phenols/extractable compounds ratio of only 7.8% for ginger versus 59.7% for basil. The water-based extraction method used in Hinneburg et al. [144] could not preserve the antioxidant activity of essential oils from ginger. Hence, the variety, extraction and processing methods can greatly affect the antioxidant properties of ginger.

4.3. Anti-Inflammatory Action

The anti-inflammatory action of ginger and its potential application in AI were aptly reviewed by Kumar et al. [145]. Inflammation stimulates hepcidin production in response to pro-inflammatory cytokines such as IL-1, tumour necrotic factor (TNF)-α and, in particular, IL-6. As an acute phase protein, hepcidin's role is to inhibit iron absorption and thus minimise the free iron supply to invading pathogens. Hence, sustained elevation of hepcidin will cause hypoferremia leading to anaemia. Additionally, an increase in IL-6 activates the nuclear factor kappa B (NF-$\kappa\beta$) pathway, resulting in the synthesis of CRP from the hepatocytes. The rising CRP levels indicate systemic inflammation and can blunt the erythropoiesis stimulation response in AI, especially in chronic kidney disease [145]. The bioactive compounds in ginger, such as 6-gingerol, 6-shogaol and 6-paradol, are known to possess broad anti-inflammatory properties that can block the activation of NF-$\kappa\beta$ by suppressing pro-inflammatory cytokines [145,158].

The human intervention study on ginger and iron absorption reported by Kulkarni et al. [139,140] mentioned earlier also measured malondialdehyde (MDA) as the serum biomarker for oxidative stress and TNF-α as the inflammatory marker of the 62 participants receiving either ginger and iron treatment or oral iron therapy only. Both groups had an insignificant difference in mean MDA and TNF-α levels at baseline. After 30 days, both oxidative stress (MDA: −18.62%, $p < 0.001$) and inflammatory markers (TNF-α: −20.11%, $p < 0.05$) were significantly reduced in the oral iron plus ginger group. Conversely, in the control group taking only oral iron therapy, there was a significant decrease in post-trial MDA levels (−9.67%, $p < 0.05$) and a non-significant increase in TNF-α (+3.86%, $p > 0.05$) [140]. The estimated mean difference between groups was not significant in changes in MDA but was significant in TNF-α ($p < 0.05$). It can be inferred from these results that, compared to oral iron therapy alone, combining ginger with oral iron therapy can better alleviate oxidative stress and reduce inflammation in patients with IDA while correcting their anaemic condition.

4.4. Gut Microbiota Modulation

Both iron deficiency and excess iron can lead to dysbiosis, characterised by an imbalance in the gut microbial community, and are associated with diseases [128]. Recent research has found ginger to exact prebiotic effects that improve gut microbiota composition. In a study that stimulated digestion and fermentation in vitro, 85% of the polyphenols in a dry ginger powder were still detectable in the digestive fluids after simulated digestion [146]. These polyphenol constituents include 6-, 8-, 10-gingerols and 6-shogaol. The undigested ginger extract significantly modulated faecal microbiota structure following mixed-culture fermentation with faecal inoculation compared with the control group. After 12 h of fermentation, the abundances of the beneficial bacterial groups of *Bifidobacterium*

($p < 0.05$) and *Enterococcus* ($p < 0.01$) were significantly higher in the ginger group than in the control group. The study also found elevated levels of short-chain fatty acids (SCFA) accompanied by decreased pH value after fermentation with ginger extract compared to control. The results demonstrated that ginger and its polyphenol compounds could improve human health through gut microbiota modulation [146].

Ginger's effects on gut microbiota modulation were also confirmed in a mice model in an in vivo study [147]. Five-week-old C57BL/6J male mice were fed a high-fat diet with or without ginger supplementation for 16 weeks. With ginger treatment, mice on a high-fat diet showed lower body weight and amelioration of liver steatosis, low-grade inflammation, and insulin resistance compared to controls. Analysis of the gut microbiome showed an increase in *Bifidobacterium* genus and SCFA-producing bacteria (*Alloprevotella* and *Allobaculum*) and increases in faecal SCFA concentrations. As a high-fat diet promotes oxidative stress and chronic low-grade inflammation associated with metabolic dysfunction, this study demonstrated that ginger supplementation could mitigate the detrimental impact of a high-fat diet on gut microbiota composition to promote health.

Another in vivo study also found ginger to have therapeutic effects in relieving diarrhoea after antibiotic use through gut microbiota recovery [148]. The study used 5-week-old Sprague-Dawley rats treated with antibiotics by gavage for seven days before administering ginger extract for another seven days. The study found ginger treatment significantly reduced diarrhoea symptoms ($p < 0.05$) compared to the control group that did not receive ginger treatment. Furthermore, the ginger treatment also considerably increased microbiota diversity in the gut, showing accelerated recovery. Specifically, the abundance of Proteobacteria phyla was increased by antibiotics treatment but restored significantly after ginger administration ($p < 0.0001$). In contrast, antibiotic treatment depressed the abundance of Bacteroidetes phyla and saw significant improvement with ginger treatment ($p < 0.001$). The study found *Escherichia Shigella* decreased at the genus levels, whereas Bacteroides increased the most in relative abundance. Histopathological observation of the colon also revealed evidence of intestinal barrier integrity improvement with ginger treatment associated with restoring tight junction protein Zonula occludens-1.

Changes in gut microbiota compositions in humans were also observed after consumption of ginger in a recent RCT [149]. The study recruited 138 healthy adults. All participants were advised to consume their usual diet but avoid ginger-rich products, probiotics, and prebiotics for one week before starting the study. During the intervention period, the participants were randomly assigned to take either fresh ginger juice (ginger group, n = 68) or sterile 0.9% sodium chloride (control group, n = 66) daily for seven days. Blood serum and faecal samples were collected at baseline and after seven days. A total of 4 participants in the control group and 7 in the ginger group were lost to follow-up, with only 123 participants completing the study. The study found increased counts of intestinal bacterial species when comparing the taxonomic composition between the ginger and control groups. The ginger juice intervention decreased the relative abundance of the Prevotella-to-Bacteroides ratio and the pro-inflammatory *Ruminococcus_1* and *Ruminococcus_2* genus and increased the Firmicutes-to-Bacteroidetes ratio, *Proteobacteria* and anti-inflammatory *Faecalibacterium*. Hence, the study concluded that consuming ginger juice for a short period had substantial effects on the composition and function of gut microbiota in healthy people.

There is currently no known pre-clinical and clinical study on the effect of ginger on the gut microbiota in the case of anaemia or iron deficiency. Nevertheless, prebiotics, such as inulin, are known to affect gut microbiota to improve iron absorption in IDA [130]. Thus, it is logical to assume a similar positive impact of ginger on IDA through inferences. This is an area of further research.

4.5. Erythropoiesis Stimulation

Erythropoiesis, the process of producing RBCs, is impeded in IDA due to insufficient dietary iron intake, impairment in iron absorption as affected by inflammation, or an imbalance between surging iron requirements and iron available resulting from rapid growth or heavy blood loss [159]. In fact, the use of ginger for haematopoiesis has an ethnomedicinal origin. Dry ginger is used in traditional Chinese medicine as a warm herb to promote blood flow, remove blood stasis, and alleviate weakness and fatigue. As such, ginger is used in many blood tonic herbal formulas for treating blood circulation, anaemia, and haemorheological conditions [160,161]. Ginger is also a tonic food recommended for postpartum women during the one month confinement period immediately after delivery for recovery from blood loss. Specifically, in Southern China, traditional postpartum dietary practices include the 'ginger vinegar soup' made from sweet vinegar, ginger, egg, and pig's trotters [162]. This is an example of iron-rich food leveraging ginger as a functional ingredient to improve iron bioavailability and promote erythropoiesis. Thus, the ability of ginger to stimulate erythropoiesis can be an added benefit to the effectiveness of oral iron therapy for IDA treatment.

In a study that utilised zebrafish embryos to investigate the effect of ginger extract on haematopoiesis in vivo, Ferri-Lagneau et al. [150] found ginger, with its bioactive compounds of 8-gingerol, 10-gingerol, 8-shogaol, and 10-shogaol, promoted the expression of GATA-binding factor 1 (Gata1) in erythroid cells. Gata1 is an early marker and key regulator of erythropoiesis. Moreover, increases in the expression of haematopoietic progenitor markers cmyb and scl were also observed. The study also identified 10-gingerol as the most potent stimulator in promoting the primitive wave of erythropoiesis in early developing zebrafish embryos. The study further confirmed that the haematopoiesis effect of ginger was mediated through the bone morphogenetic protein (Bmp) signalling pathway.

In a subsequent study, the same group of researchers further demonstrated that ginger/10-gingerol can rescue the expression of haematopoietic stem/progenitor cells (HSPC) in zebrafish embryos with genetic defects [151]. Ginger was found to induce scl/runx1 expression through Bmp and Notch signalling pathways that led to arteriogenesis and HSPC formation. Bmp and Notch are known to regulate nitric oxide (NO) production, which plays an active role in the modulation of haematopoietic cell growth and differentiation. The study also showed that ginger produced a robust up-regulation of NO in the rescued mutant zebrafish embryos. Therefore, the combined effect of ginger on Bmp, Notch and NO production can be beneficial for the regulation of erythropoiesis for regeneration/recovery.

4.6. Iron Overload Prevention

Ginger's ability in modulating iron absorption in the case of overload was an unexpected finding in a study that used ginger nanoparticle-derived lipid vectors (GDLV) to deliver DMT-1 short-interference RNAs (siRNA) that suppressed DMT-1 mRNA expression to reduce iron absorption in an iron-loading mice model [152]. The study found that GDLV containing negative control appeared to repress some iron-related parameters similar to the DMT-1 siRNA treatment. The observed effects were reductions of 20% in ^{59}Fe absorption, approximately 65% of pancreatic non-haem iron, and 40 to 50% lower SF compared to controls. Hence, the authors suggested that the bioactive lipids in ginger could influence iron absorption and homeostasis.

In another animal model, Gholampour et al. [153] showcased the protective properties of ginger against the deleterious effects of iron overloading. To induce iron overload, male Wistar rats were given ferrous sulphate at 30 mg/kg/day, dissolved in 1 mL distilled water, intraperitoneally for 14 days. These rats showed significantly higher serum hepatic markers and bilirubin levels, elevated MDA levels, lower serum albumin levels, total protein, triglyceride, cholesterol, and glucose, decreased creatinine clearance and higher fractional excretion of sodium compared to controls ($p < 0.001$). The histopathological examination further confirmed their liver and kidney damage. A separate group of iron overloaded rats was fed with a hydroalcoholic ginger extract at 400 mg/kg/day dissolved in 1 mL distilled water and given by gavage for 11 days from the fourth day of ferrous sulphate injection. The feeding of ginger markedly reversed the adverse impacts of iron overload, as evidenced in the significantly higher levels of hepatic serum markers, renal functional markers and lipid peroxidation markers in this group compared to the iron only group ($p < 0.01$). Moreover, depleted serum total protein, albumin, glucose, triglycerides, and cholesterol were restored with bilirubin concentration decreased in the blood. Hence, ginger extract demonstrated strong protective effects against iron toxicity potentially through its free radical scavenging activities. The preservation of the liver and kidney was also corroborated through histological examinations.

The potential of ginger, especially its 6-shogaol derivative, in preventing iron overload is further demonstrated in a case series reported by Golombick et al. [154]. In this study, 6 early-stage, transfusion-independent patients with myelodysplastic syndrome (MDS) were given a daily supplement of 20 mg of a ginger extract standardised for 20% 6-shogaol. Blood and urine samples were collected monthly. At three months, the study found that 6-shogaol was able to reduce the SF levels (>40% reductions) of three of the patients who had elevated SF (>300 g/μL) at baseline. Two of the patients who had SF reduction repeated the study for another 3 months after a washout period. Again, a greater than 40% reduction in SF was observed in the repeat tests for both patients. The two patients were tested for their serum hepcidin levels at the repeat tests. Both patients demonstrated elevation of serum hepcidin that accompanied the SF reduction. Furthermore, one patient who had high liver function enzymes due to alcohol consumption also saw normalisation of liver function with a greater than 40% reduction in these enzymes at the end of the study. The restoration of liver function was achieved without changing alcohol consumption habits. The research concluded that ginger extract rich in 6-shogaol prevented iron overload in MDS patients through upregulation of hepcidin, potentially with liver function restoration.

4.7. *Ginger-Synthesised Iron Nanoparticles*

Nanotechnology, the understanding and control of matter generally in the 1–100 nm dimension, is gaining much medical research as it holds the potential for breakthroughs in preventing, diagnosing, and treating various diseases due to the unique physicochemical properties of nanomaterials [163,164]. Unsurprisingly, iron nanoparticle (FeNP) preparations have also been developed to overcome the inherent limitations of conventional ferrous and ferric iron formulations in the treatment of IDA. Pre-clinical studies showed that iron nanoparticles have high bioavailability, are non-toxic, and induce lesser side effects than conventional iron preparations for IDA, even though the delivery and safety issues in humans for therapeutic use required further research [165]. The high bioavailability of iron nanoparticles is also ideal for food fortifications as the nanoparticles do not cause unacceptable taste or colour in food vehicles. Hence, it is suggested that nanosized iron salts can have potential applications in food fortification to reduce IDA worldwide [166].

Ginger has been used in the green approach for metallic nanoparticles, including iron. The green synthesis approach is preferred to avoid the production of unwanted or harmful chemical by-products and achieve a cost-effective and sustainable supply of nanoparticles [167]. El-Refai et al. [155] used ginger and garlic extracts to synthesise silver, copper, iron, and zinc nanoparticles, and their antioxidant and antimicrobial activities were evaluated. The high flavonoid and phenolic contents in garlic and ginger water extracts revealed in the phytochemical analysis strongly support the potential of garlic and ginger to bio-reduce the metallic ions to their respective nanoparticles (e.g., Fe^{3+} ions to FeNPs). Transmission electron microscopy showed that the FeNPs in ginger were in the range of 14.08–21.57 nm with almost spherical forms. In comparison, the particle size of FeNPs in garlic ranged from 60.30 to 82.63 nm with tetragonal structures. All nanoparticles extracted in this study, including FeNPs from ginger, demonstrated considerable radical scavenging properties and antimicrobial activities against Gram-positive and Gram-negative bacteria and fungi [155]. Other researchers have similarly shown ginger to be a suitable green material for synthesising FeNPs with high antioxidant and antibacterial properties [156,157].

5. Conclusions

In summary, ginger with its rich polyphenols can support IDA treatment and prevention in many ways. It can improve iron bioavailability by enhancing iron absorption and thus increasing the efficacy of oral iron therapy. Its antioxidant and anti-inflammatory properties help in reducing oxidative stress and pro-inflammatory cytokine cascade and thus protect the gastrointestinal tract from the delirious effects of excess free iron. Ginger and its bioactive polyphenols can also serve as prebiotics to the gut microbiota to promote gut health and potentially reduce the unwanted side effects of oral iron therapy. Ginger can also stimulate erythropoiesis to generate the much-needed healthy RBCs for proper functioning. In the case of iron overload due to comorbidities from inflammatory disorders or chronic conditions, ginger can potentially reverse the adverse impacts and restore iron balance. Ginger can also be used to synthesise FeNPs sustainably to develop newer and more effective oral iron products and functional ingredients for IDA treatment and prevention.

There are, however, still many unknowns regarding the physiological effects of ginger and its active compounds in IDA. Much research is still needed to understand how the phenolic compounds of ginger can influence the mechanistic pathways of iron absorption and metabolism. More pre-clinical studies are required to further explore how ginger's antioxidant, anti-inflammatory, gut microbiota modulation, and erythropoiesis stimulation properties can affect IDA in areas such as side effects induced by oral iron therapy, gastrointestinal micro-environment and microbiota changes, inflammatory cytokine signalling and erythropoiesis effectiveness. Most importantly, there is a lack of clinical study on the effect of co-administration of ginger and oral iron therapy for IDA treatment other than one short human trial. There is a need for a longer-term, randomised, double-blind placebo-controlled trial to validate the effectiveness of ginger as an adjuvant treatment for IDA.

To conclude, polyphenol-rich ginger can play a much bigger role in addressing the global public health problem of IDA, but more research and development are needed to realise its full potential.

Author Contributions: Conceptualization, S.L.O. and S.C.P.; writing—original draft preparation, S.L.O.; writing—review and editing, S.C.P., R.C. and A.M.; visualization, S.L.O.; supervision, S.C.P.; project administration, S.C.P.; funding acquisition, S.L.O., S.C.P. and R.C. All authors have read and agreed to the published version of the manuscript.

Funding: This research was funded by the Australian Traditional-Medicine Society (ATMS) through the ATMS Research Grant 2021.

Institutional Review Board Statement: Not applicable.

Informed Consent Statement: Not applicable.

Data Availability Statement: Not applicable.

Conflicts of Interest: The authors declare no conflict of interest.

Abbreviations

The following abbreviations are used in this manuscript:

AI	Anaemia of inflammation
CAT	Catalase
CI	Confidence interval
CRP	C-reactive protein
DcytB	Duodenal cytochrome B
DMT-1	Divalent metal transporter 1
FeNP	Iron nanoparticle
FPN	Ferroportin
Gata1	GATA-binding factor 1
GR	Glutathione reductase

GST	Glutathione-S-transferase
HAMP	Hepcidin antimicrobial peptide
Hb	Haemoglobin
HSPC	Haematopoietic stem/progenitor cells
IDA	Iron deficiency anaemia
MDA	Malondialdehyde
NFκβ	Nuclear factor kappa B
NO	Nitric oxide
NOAEL	No observed adverse effect level
RBC	Red blood cell
RCT	Randomised controlled trial
SF	Serum ferritin
siRNA	Short interference RNA
SOD	Superoxide dismutase
TNF	Tumour necrotic factor
WHO	World Health Organization

Appendix A. International Patents

The following table summarises all patent publications (total = 16) found using the World Intellectual Property Organization online PATENTSCOPE database with the search term (Ginger AND "Iron deficiency").

Table A1. A summary list of patent publications related to functional foods or ethnomedicine indicated for iron deficiency with ginger as an ingredient.

Publication No	Date	Classification Code	Title	Country
101243891	20 August 2008	A23L 1/337	Sea tangle vegetarian stuffing boiled dumplings and its processing method	China
103947928	11 March 2014	A23L 1/10	Fleece-flower root nutrition eight-treasure porridge and its preparation method	China
104026495	10 September 2014	A23L 1/212	Haw flake containing pig blood and coarse cereals, and preparation method thereof	China
105495158	25 September 2014	A23L 1/315	Black-bone chicken sausage and preparation method thereof	China
104095016	15 October 2014	A61K 36/9068	Infantile iron-deficiency anemia treating cookie and preparing method thereof	China
104323303	4 November 2015	A23L 1/314	Method for making tomato beef stewed product	China
104643216	27 May 2015	A23L 2/02	Blood-replenishing and beautifying calcium blended lotus root juice and preparation method thereof	China
105664116	15 June 2016	A61K 36/9068	Traditional Chinese medicine for treating infant iron deficiency anemia as well as preparation method and application thereof	China
106362108	1 February 2017	A61K 36/9068	Traditional Chinese medicinal pill used for hematogenesis	China
106616937	10 May 2017	A23L 31/00	Stropharia rugosoannulata and black chicken can and preparation method thereof	China
106889422	27 June 2017	A61K 36/9068	Edible flour for tonifying blood and warming the uterus and production method thereof	China
107772293	9 March 2018	A23L 13/50	Body-nourishing black bone chicken	China
107772373	9 March 2018	A61K 36/9068	Decoction preventing and curing osteoporosis	China
108887607	27 November 2018	A23L 13/70	Spicy shredded pork with garlic sauce and making method thereof	China
108433084	24 August 2018	A23L 27/50	Soy sauce	China
113278488	20 August 2021	A61K 36/9068	Spartina alterniflora spleen-tonifying stomach-nourishing pericarpium citri reticulatae wine decocting pot and preparation process thereof	China

References

1. Chaparro, C.M.; Suchdev, P.S. Anemia epidemiology, pathophysiology, and etiology in low- and middle-income countries. *Ann. N. Y. Acad. Sci.* **2019**, *1450*, 15–31. [CrossRef] [PubMed]
2. Turner, J.; Parsi, M.; Badireddy, M. Anemia. In *StatPearls*, StatPearls Publishing: Tampa, FL, USA, 2022. Available online: https://www.ncbi.nlm.nih.gov/books/NBK499994/ (accessed on 13 March 2022).
3. Gardner, W.; Kassebaum, N. Global, regional, and national prevalence of anemia and its causes in 204 countries and territories, 1990–2019. *Curr. Dev. Nutr.* **2020**, *4*, 830. [CrossRef]
4. Safiri, S.; Kolahi, A.A.; Noori, M.; Nejadghaderi, S.A.; Karamzad, N.; Bragazzi, N.L.; Sullman, M.J.M.; Abdollahi, M.; Collins, G.S.; Kaufman, J.S.; et al. Burden of anemia and its underlying causes in 204 countries and territories, 1990–2019: Results from the Global Burden of Disease Study 2019. *J. Hematol. Oncol.* **2021**, *14*, 185. [CrossRef] [PubMed]
5. World Health Organization. *World Health Statistics 2021: Monitoring Health for the SDGs, Sustainable Development Goals*; World Health Organization: Geneva, Switzerland, 2021.
6. Australian Bureau of Statistics. Australian Health Survey 2011–2012: Biomedical Results for Chronic Diseases. 2013. Available online: https://www.abs.gov.au/statistics/health/health-conditions-and-risks/australian-health-survey-biomedical-results-chronic-diseases/latest-release (accessed on 15 March 2022).
7. Azzopardi, P.S.; Sawyer, S.M.; Carlin, J.B.; Degenhardt, L.; Brown, N.; Brown, A.D.; Patton, G.C. Health and wellbeing of indigenous adolescents in Australia: A systematic synthesis of population data. *Lancet* **2018**, *391*, 766–782. [CrossRef]
8. Mantadakis, E.; Chatzimichael, E.; Zikidou, P. Iron deficiency anemia in children residing in high and low-income countries: Risk factors, prevention, diagnosis and therapy. *Mediterr. J. Hematol. Infect. Dis.* **2020**, *12*, e2020041. [CrossRef]
9. Tolkien, Z.; Stecher, L.; Mander, A.P.; Pereira, D.I.; Powell, J.J. Ferrous sulfate supplementation causes significant gastrointestinal side-effects in adults: A systematic review and meta-analysis. *PLoS ONE* **2015**, *10*, 1–20. [CrossRef]
10. Gereklioglu, C.; Asma, S.; Korur, A.; Erdogan, F.; Kut, A. Medication adherence to oral iron therapy in patients with iron deficiency anemia. *Pak. J. Med. Sci.* **2016**, *32*, 604–607. [CrossRef]
11. WFO: World Flora Online. World Flora Online Taxonomic Backbone v.2022.04. 2022. Available online: http://www.worldfloraonline.org/downloadData (accessed on 2 September 2022).
12. Wu, D. A preliminary study on the origin of ginger. *Agric. Archaeol.* **1985**, *5*, 247–250.
13. The Observatory of Economic Complexity. Ginger (HS: 091010) Product Trade, Exporters and Importers. Available online: https://oec.world/en/profile/hs92/ginger (accessed on 10 March 2022).
14. Syafitri, D.M.; Levita, J.; Mutakin, M.; Diantini, A. A review: Is ginger (*Zingiber officinale* var. Roscoe) potential for future phytomedicine? *Indones. J. Appl. Sci.* **2018**, *8*, 8–13. [CrossRef]
15. Martirosyan, D.; Singharaj, B., Health claims and functional food: The future of functional foods under FDA and EFSA regulation. In *Functional Foods for Chronic Diseases*, 1st ed.; Martirosyan, D.M., Ed.; Food Science Publisher: Chicago, IL, USA, 2016; pp. 410–424.
16. He, L.; Qin, Z.; Li, M.; Chen, Z.; Zeng, C.; Yao, Z.; Yu, Y.; Dai, Y.; Yao, X. Metabolic profiles of ginger, a functional food, and its representative pungent compounds in rats by ultraperformance liquid chromatography coupled with quadrupole time-of-flight tandem mass spectrometry. *J. Agric. Food Chem.* **2018**, *66*, 9010–9033. [CrossRef]
17. Aromatic ginger (*Kaempferia galanga* L.) extracts with ameliorative and protective potential as a functional food, beyond its flavor and nutritional benefits. *Toxicol. Rep.* **2019**, *6*, 521–528. [CrossRef]
18. Ozkur, M.; Benlier, N.; Takan, I.; Vasileiou, C.; Georgakilas, A.G.; Pavlopoulou, A.; Cetin, Z.; Saygili, E.I. Ginger for healthy ageing: A systematic review on current evidence of its antioxidant, anti-inflammatory, and anticancer properties. *Oxidative Med. Cell. Longev.* **2022**, *2022*, 1–16. [CrossRef]
19. Ajayi, O.B.; Akomolafe, S.F.; Akinyemi, F.T. Food value of two varieties of ginger (*Zingiber officinale*) commonly consumed in Nigeria. *ISRN Nutr.* **2013**, *2013*, 359727. [CrossRef]
20. Sangwan, A.; Kawatra, A.; Sehgal, S. Nutritional composition of ginger powder prepared using various drying methods. *J. Food Sci. Technol.* **2014**, *51*, 2260–2262. [CrossRef]
21. Bode, A.M.; Dong, Z., The amazing and mighty ginger. In *Herbal Medicine: Biomolecular and Clinical Aspects*, 2nd ed.; Benzie, I.F.F., Wachtel-Galor, S., Eds.; CRC Press/Taylor & Francis: Boca Raton, FL, USA, 2011. Available online: https://www.ncbi.nlm.nih.gov/books/NBK92775/ (accessed on 2 September 2022).
22. Li, X.; Ao, M.; Zhang, C.; Fan, S.; Chen, Z.; Yu, L. Zingiberis rhizoma recens: A review of its traditional uses, phytochemistry, pharmacology, and toxicology. *Evid.-Based Complement. Altern. Med.* **2021**, *2021*, 6668990. [CrossRef]
23. Unuofin, J.O.; Masuku, N.P.; Paimo, O.K.; Lebelo, S.L. Ginger from farmyard to town: Nutritional and pharmacological applications. *Front. Pharmacol.* **2021**, *12*, 779352. [CrossRef]
24. Li, H.; Liu, Y.; Luo, D.; Ma, Y.; Zhang, J.; Li, M.; Yao, L.; Shi, X.; Liu, X.; Yang, K. Ginger for health care: An overview of systematic reviews. *Complement. Ther. Med.* **2019**, *45*, 114–123. [CrossRef]
25. Anh, N.H.; Kim, S.J.; Long, N.P.; Min, J.E.; Yoon, Y.C.; Lee, E.G.; Kim, M.; Kim, T.J.; Yang, Y.Y.; Son, E.Y.; et al. Ginger on human health: A comprehensive systematic review of 109 randomized controlled trials. *Nutrients* **2020**, *12*, 157. [CrossRef]
26. de Lima, R.M.T.; Reis, A.C.D.; de Menezes, A.A.P.M.; de Oliveira Santos, J.V.; de Oliveira Filho, J.W.G.; de Oliveira Ferreira, J.R.; de Alencar, M.V.O.B.; da Mata, A.M.O.F.; Khan, I.N.; Islam, A.; et al. Protective and therapeutic potential of ginger (*Zingiber officinale*) extract and [6]-gingerol in cancer: A comprehensive review. *Phytother. Res.* **2018**, *32*, 1885–1907. [CrossRef]

27. Liu, Y.; Liu, J.; Zhang, Y. Research progress on chemical constituents of Zingiber officinale Roscoe. *BioMed Res. Int.* **2019**, *2019*, 5370823. [CrossRef]
28. Mao, Q.Q.; Xu, X.Y.; Cao, S.Y.; Gan, R.Y.; Corke, H.; Beta, T.; Li, H.B. Bioactive compounds and bioactivities of ginger (*Zingiber officinale* Roscoe). *Foods* **2019**, *8*, 185. [CrossRef]
29. Rahmani, A.H.; Shabrmi, F.M.A.; Aly, S.M. Active ingredients of ginger as potential candidates in the prevention and treatment of diseases via modulation of biological activities. *Int. J. Physiol. Pathophysiol. Pharmacol.* **2014**, *6*, 125–136. Available online: https://www.ncbi.nlm.nih.gov/pmc/articles/PMC4106649/ (accessed on 2 September 2022).
30. Hasan, H.A.; Raauf, A.M.R.; Razik, B.M.A.; Hassan, B.A.R. Chemical composition and antimicrobial activity of the crude extracts isolated from *Zingiber officinale* by different solvents. *Pharm. Anal. Acta* **2012**, *3*. [CrossRef]
31. Ghasemzadeh, A.; Jaafar, H.Z.; Rahmat, A. Identification and concentration of some flavonoid components in malaysian young ginger (*Zingiber officinale* Roscoe) varieties by a high performance liquid chromatography method. *Molecules* **2010**, *15*, 6231–6243. [CrossRef]
32. Rahman, S.; Salehin, F.; Iqbal, A. In vitro antioxidant and anticancer activity of young *Zingiber officinale* against human breast carcinoma cell lines. *BMC Complement. Altern. Med.* **2011**, *11*, 76. [CrossRef]
33. Fahmi, A.; Hassanen, N.; Abdur-Rahman, M.; Shams-Eldin, E. Phytochemicals, antioxidant activity and hepatoprotective effect of ginger (*Zingiber officinale*) on diethylnitrosamine toxicity in rats. *Biomarkers* **2019**, *24*, 436–447. [CrossRef]
34. Ghasemzadeh, A.; Jaafar, H.Z.; Rahmat, A. Synthesis of phenolics and flavonoids in ginger (Zingiber officinale Roscoe) and their effects on photosynthesis rate. *Int. J. Mol. Sci.* **2010**, *11*, 4539–55. [CrossRef]
35. Ghasemzadeh, A.; Jaafar, H.Z.E.; Rahmat, A. Variation of the Phytochemical Constituents and Antioxidant Activities of *Zingiber officinale* var. rubrum Theilade Associated with Different Drying Methods and Polyphenol Oxidase Activity. *Molecules* **2016**, *21*, 780. [CrossRef] [PubMed]
36. Kou, X.; Wang, X.; Ji, R.; Liu, L.; Qiao, Y.; Lou, Z.; Ma, C.; Li, S.; Wang, H.; Ho, C.T. Occurrence, biological activity and metabolism of 6-shogaol. *Food Funct.* **2018**, *9*, 1310–1327. [CrossRef] [PubMed]
37. Sang, S.; Snook, H.D.; Tareq, F.S.; Fasina, Y. Precision research on ginger: The type of ginger matters. *J. Agric. Food Chem.* **2020**, *68*, 8517–8523. [CrossRef] [PubMed]
38. Choi, J.G.; Kim, S.Y.; Jeong, M.; Oh, M.S. Pharmacotherapeutic potential of ginger and its compounds in age-related neurological disorders. *Pharmacol. Ther.* **2018**, *182*, 56–69. [CrossRef]
39. Sampath, C.; Sang, S.; Ahmedna, M. In vitro and in vivo inhibition of aldose reductase and advanced glycation end products by phloretin, epigallocatechin 3-gallate and [6]-gingerol. *Biomed. Pharmacother.* **2016**, *84*, 502–513. [CrossRef]
40. Sampath, C.; Rashid, M.R.; Sang, S.; Ahmedna, M. Specific bioactive compounds in ginger and apple alleviate hyperglycemia in mice with high fat diet-induced obesity via Nrf2 mediated pathway. *Food Chem.* **2017**, *226*, 79–88. [CrossRef]
41. Ahmad, B.; Rehman, M.U.; Amin, I.; ur Rahman Mir, M.; Ahmad, S.B.; Farooq, A.; Muzamil, S.; Hussain, I.; Masoodi, M.; Fatima, B. Zingerone (4-(4-hydroxy-3-methylphenyl) butan-2-one) protects against alloxan-induced diabetes via alleviation of oxidative stress and inflammation: Probable role of NF-kB activation. *Saudi Pharm. J.* **2018**, *26*, 1137–1145. [CrossRef]
42. Cheng, Q.; Feng, X.; Meng, Q.; Li, Y.; Chen, S.; Wang, G.; Nie, K. [6]-gingerol ameliorates cisplatin-induced PICA by regulating the TPH/MAO-A/SERT/5-HT/5-HT3 receptor system in rats. *Drug Des. Dev. Ther.* **2020**, *14*, 4085–4099. [CrossRef]
43. Qian, Q.H.; Yue, W.; Chen, W.H.; Yang, Z.H.; Liu, Z.T.; Wang, Y.X. Effect of gingerol on substance P and NK_1 receptor expression in a vomiting model of mink. *Chin. Med. J.* **2010**, *123*, 478–484.
44. Qian, Q.; Yue, W.; Wang, Y.; Yang, Z.; Liu, Z.; Chen, W. Gingerol inhibits cisplatin-induced vomiting by down regulating 5-hydroxytryptamine, dopamine and substance P expression in minks. *Arch. Pharmacal Res.* **2009**, *32*, 565–573. [CrossRef]
45. Qian, W.; Cai, X.; Wang, Y.; Zhang, X.; Zhao, H.; Qian, Q.; Yang, Z.; Liu, Z.; Hasegawa, J. Effect of gingerol on cisplatin-induced pica analogous to emesis via modulating expressions of dopamine 2 receptor, dopamine transporter and tyrosine hydroxylase in the vomiting model of rats. *Yonago Acta Med.* **2016**, *59*, 100–110.
46. Tian, L.; Qian, W.; Qian, Q.; Zhang, W.; Cai, X. Gingerol inhibits cisplatin-induced acute and delayed emesis in rats and minks by regulating the central and peripheral 5-HT, SP, and DA systems. *J. Nat. Med.* **2020**, *74*, 353–370. [CrossRef]
47. Tzeng, T.F.; Chang, C.J.; Liu, I.M. 6-gingerol inhibits rosiglitazone-induced adipogenesis in 3T3-L1 adipocytes. *Phytother. Res.* **2014**, *28*, 187–192. [CrossRef]
48. Saravanan, G.; Ponmurugan, P.; Deepa, M.A.; Senthilkumar, B. Anti-obesity action of gingerol: Effect on lipid profile, insulin, leptin, amylase and lipase in male obese rats induced by a high-fat diet. *J. Sci. Food Agric.* **2014**, *94*, 2972–2977. [CrossRef]
49. Suk, S.; Seo, S.G.; Yu, J.G.; Yang, H.; Jeong, E.; Jang, Y.J.; Yaghmoor, S.S.; Ahmed, Y.; Yousef, J.M.; Abualnaja, K.O.; et al. A bioactive constituent of ginger, 6-shogaol, prevents adipogenesis and stimulates lipolysis in 3T3-L1 adipocytes. *J. Food Biochem.* **2016**, *40*, 84–90. [CrossRef]
50. Choi, J.; Kim, K.J.; Kim, B.H.; Koh, E.J.; Seo, M.J.; Lee, B.Y. 6-gingerol suppresses adipocyte-derived mediators of inflammation in vitro and in high-fat diet-induced obese zebra fish. *Planta Med.* **2017**, *83*, 245–253. [CrossRef]
51. Sampath, S.J.P.; Birineni, S.; Perugu, S.; Kotikalapudi, N.; Venkatesan, V. Therapeutic efficacy of 6-gingerol and 6-shogaol in promoting browning of white adipocytes vis-à-vis enhanced thermogenesis portrayed in high fat milieu. *Food Biosci.* **2021**, *42*, 101211. [CrossRef]
52. Zhang, F.L.; Zhou, B.W.; Yan, Z.Z.; Zhao, J.; Zhao, B.C.; Liu, W.F.; Li, C.; Liu, K.X. 6-gingerol attenuates macrophages pyroptosis via the inhibition of MAPK signaling pathways and predicts a good prognosis in sepsis. *Cytokine* **2020**, *125*, 154854. [CrossRef]

53. Alsahli, M.A.; Almatroodi, S.A.; Almatroudi, A.; Khan, A.A.; Anwar, S.; Almutary, A.G.; Alrumaihi, F.; Rahmani, A.H. 6-gingerol, a major ingredient of ginger attenuates diethylnitrosamine-induced liver injury in rats through the modulation of oxidative stress and anti-inflammatory activity. *Mediat. Inflamm.* **2021**, *2021*, 6661937. [CrossRef]
54. Zhu, Y.; Wang, C.; Luo, J.; Hua, S.; Li, D.; Peng, L.; Liu, H.; Song, L. The protective role of Zingerone in a murine asthma model via activation of the AMPK/Nrf2/HO-1 pathway. *Food Funct.* **2021**, *12*, 3120–3131. [CrossRef]
55. Lim, Y.J.; Min, H.Y.; Jang, W.G. Zingerone attenuates pi-induced vascular calcification via AMPK-mediated TIMP4 expression. *J. Lipid Atheroscler.* **2021**, *10*, 62–73. [CrossRef]
56. Danwilai, K.; Konmun, J.; Sripanidkulchai, B.O.; Subongkot, S. Antioxidant activity of ginger extract as a daily supplement in cancer patients receiving adjuvant chemotherapy: A pilot study. *Cancer Manag. Res.* **2017**, *9*, 11–18. [CrossRef]
57. Chen, H.; Fu, J.; Chen, H.; Hu, Y.; Soroka, D.N.; Prigge, J.R.; Schmidt, E.E.; Yan, F.; Major, M.B.; Chen, X.; et al. Ginger compound [6]-shogaol and its cysteine-conjugated metabolite (M2) activate Nrf2 in colon epithelial cells in vitro and in vivo. *Chem. Res. Toxicol.* **2014**, *27*, 1575–1585. [CrossRef]
58. Ji, K.; Fang, L.; Zhao, H.; Li, Q.; Shi, Y.; Xu, C.; Wang, Y.; Du, L.; Wang, J.; Liu, Q. Ginger oleoresin alleviated γ-ray irradiation-induced reactive oxygen species via the nrf2 protective response in human mesenchymal stem cells. *Oxidative Med. Cell. Longev.* **2017**, *2017*, 1480294. [CrossRef] [PubMed]
59. Hosseinzadeh, A.; Bahrampour Juybari, K.; Fatemi, M.J.; Kamarul, T.; Bagheri, A.; Tekiyehmaroof, N.; Sharifi, A.M. Protective effect of ginger (*Zingiber officinale* Roscoe) extract against oxidative stress and mitochondrial apoptosis induced by interleukin-1β in cultured chondrocytes. *Cells Tissues Organs* **2017**, *204*, 241–250. [CrossRef] [PubMed]
60. Romero, A.; Forero, M.; Sequeda-Castañeda, L.G.; Grismaldo, A.; Iglesias, J.; Celis-Zambrano, C.A.; Schuler, I.; Morales, L. Effect of ginger extract on membrane potential changes and AKT activation on a peroxide-induced oxidative stress cell model. *J. King Saud Univ.-Sci.* **2018**, *30*, 263–269. [CrossRef]
61. Elshopakey, G.E.; Almeer, R.; Alfaraj, S.; Albasher, G.; Abdelgawad, M.E.; Moneim, A.E.A.; Essawy, E.A. Zingerone mitigates inflammation, apoptosis and oxidative injuries associated with renal impairment in adriamycin-intoxicated mice. *Toxin Rev.* **2021**, 1–12. [CrossRef]
62. Song, S.; Dang, M.; Kumar, M. Anti-inflammatory and renal protective effect of gingerol in high-fat diet/streptozotocin-induced diabetic rats via inflammatory mechanism. *Inflammopharmacology* **2019**, *27*, 1243–1254. [CrossRef]
63. Rodrigues, F.A.; Prata, M.M.; Oliveira, I.C.; Alves, N.T.; Freitas, R.E.; Monteiro, H.S.; Silva, J.A.; Vieira, P.C.; Viana, D.A.; Libório, A.B.; et al. Gingerol fraction from Zingiber officinale protects against gentamicin-induced nephrotoxicity. *Antimicrob. Agents Chemother.* **2014**, *58*, 1872–1878. [CrossRef]
64. El-Akabawy, G.; El-Kholy, W. Neuroprotective effect of ginger in the brain of streptozotocin-induced diabetic rats. *Ann. Anat.-Anat. Anz.* **2014**, *196*, 119–128. [CrossRef]
65. Zhao, D.; Gu, M.Y.; Xu, J.L.; Zhang, L.J.; Ryu, S.Y.; Yang, H.O. Anti-neuroinflammatory effects of 12-dehydrogingerdione in LPS-activated microglia through inhibiting AKT/IKK/NF-κb pathway and activating Nrf-2/HO-1 pathway. *Biomol. Ther.* **2019**, *27*, 92–100. [CrossRef]
66. Sapkota, A.; Park, S.J.; Choi, J.W. Neuroprotective effects of 6-shogaol and its metabolite, 6-paradol, in a mouse model of multiple sclerosis. *Biomol. Ther.* **2019**, *27*, 152–159. [CrossRef]
67. Sistani Karampour, N.; Arzi, A.; Rezaie, A.; Pashmforoosh, M.; Kordi, F. Gastroprotective effect of Zingerone on ethanol-induced gastric ulcers in rats. *Medicina* **2019**, *55*, 64. [CrossRef]
68. Saiah, W.; Halzoune, H.; Djaziri, R.; Tabani, K.; Koceir, E.A.; Omari, N. Antioxidant and gastroprotective actions of butanol fraction of *Zingiber officinale* against diclofenac sodium-induced gastric damage in rats. *J. Food Biochem.* **2018**, *42*, e12456. [CrossRef]
69. Huang, H.C.; Chou, Y.C.; Wu, C.Y.; Chang, T.M. [8]-Gingerol inhibits melanogenesis in murine melanoma cells through down-regulation of the MAPK and PKA signal pathways. *Biochem. Biophys. Res. Commun.* **2013**, *438*, 375–381. [CrossRef]
70. Wang, L.X.; Qian, J.; Zhao, L.N.; Zhao, S.H. Effects of volatile oil from ginger on the murine B16 melanoma cells and its mechanism. *Food Funct.* **2018**, *9*, 1058–1069. [CrossRef]
71. Donkor, Y.O.; Abaidoo, C.S.; Tetteh, J.; Darko, N.D.; Atuahene, O.O.D.; Appiah, A.K.; Diby, T.; Maalman, R.S.E. The effect of *Zingiber officinale* (ginger) root ethanolic extract on the semen characteristics of adult male wistar rats. *Int. J. Anat. Res.* **2018**, *6*, 5481–5487. [CrossRef]
72. Marak, N.R.; Malemnganbi, C.C.; Marak, C.R.; Mishra, L.K. Functional and antioxidant properties of cookies incorporated with foxtail millet and ginger powder. *J. Food Sci. Technol.* **2019**, *56*, 5087–5096. [CrossRef]
73. Indiarto, R.; Subroto, E.; Angeline.; Selly. Ginger rhizomes (*Zingiber officinale*) functionality in food and health perspective: A review. *Food Res.* **2021**, *5*, 497–505. [CrossRef]
74. Anita, N.; Sartini.; Alam, G. Ginger candy (*Zingiber officinale*) reduces the frequency of vomiting of first-trimester pregnant women with emesis gravidarum. *Enferm. Clín.* **2020**, *30*, 536–538. [CrossRef]
75. Fortune Business Insights. Ginger Tea Market Size, Share & Industry Analysis, Forecast 2022–2029. 2021. Available online: https://www.fortunebusinessinsights.com/ginger-tea-market-106069 (accessed on 15 March 2022).
76. Smith, T.; Majid, F.; Eckl, V.; Reynolds, C.M. Herbal supplement sales in US increase by record-breaking 17.3% in 2020. *HerbalGram* **2021**, *131*, 52–65. Available online: http://herbalgram.org/resources/herbalgram/issues/131/table-of-contents/hg131-mkrpt/ (accessed on 2 September 2022).

77. Weidner, M.S.; Sigwart, K. The safety of a ginger extract in the rat. *J. Ethnopharmacol.* **2000**, *73*, 513–520. [CrossRef]
78. Rong, X.; Peng, G.; Suzuki, T.; Yang, Q.; Yamahara, J.; Li, Y. A 35-day gavage safety assessment of ginger in rats. *Regul. Toxicol. Pharmacol.* **2009**, *54*, 118–123. [CrossRef]
79. Benny, M.; Shylaja, M.R.; Antony, B.; Gupta, N.K.; Mary, R.; Anto, A.; Jacob, S. Acute and sub acute toxicity studies with ginger extract in rats. *Int. J. Pharm. Sci. Res.* **2021**, *12*, 2799. [CrossRef]
80. Jeena, K.; Liju, V.B.; Kuttan, R. A Preliminary 13-Week Oral Toxicity Study of Ginger Oil in Male and Female Wistar Rats. *Int. J. Toxicol.* **2011**, *30*, 662–670. [CrossRef]
81. Idang, E.O.; Yemitan, O.; Ogbuagu, E.O.; Yemitan, O.K.; Mbagwu, H.O.C.; Udom, G.J.; Udobang, J.A. Toxicological assessment of *Zingiber officinale* Roscoe (ginger) root oil extracts in albino rats. *Toxicol. Dig.* **2019**, *4*, 108–119.
82. Weidner, M.S.; Sigwart, K. Investigation of the teratogenic potential of a zingiber officinale extract in the rat. *Reprod. Toxicol. (Elmsford, N.Y.)* **2001**, *15*, 75–80. [CrossRef]
83. Wilkinson, J.M. Effect of ginger tea on the fetal development of Sprague-Dawley rats. *Reprod. Toxicol.* **2000**, *14*, 507–512. [CrossRef]
84. Stanisiere, J.; Mousset, P.Y.; Lafay, S. How safe is ginger rhizome for decreasing nausea and vomiting in women during early pregnancy? *Foods* **2018**, *7*, 50. [CrossRef]
85. Ryan, J.L.; Morrow, G.R. Ginger. *Oncol. Nurse Ed.* **2010**, *24*, 46–49. Available online: https://www.ncbi.nlm.nih.gov/pmc/articles/PMC5008850/ (accessed on 2 September 2022).
86. Modi, M.; Modi, K. Ginger Root. In *StatPearls*; StatPearls Publishing: Tampa, FL, USA, 2022. Available online: https://www.ncbi.nlm.nih.gov/books/NBK565886/ (accessed on 15 March 2022).
87. Schmidt, J.; Dahl, S.; Sherson, D.L. Allergic rhinoconjunctivitis caused by occupational exposure to ginger. *Ugeskr Laeger* **2015**, *177*, V12140723. [PubMed]
88. Cueva, B.; Izquierdo, G.; Crespo, J.F.; Rodriguez, J. Unexpected spice allergy in the meat industry. *J. Allergy Clin. Immunol.* **2001**, *108*, 144. [CrossRef]
89. van Toorenenbergen, A.W.; Dieges, P.H. Immunoglobulin E antibodies against coriander and other spices. *J. Allergy Clin. Immunol.* **1985**, *76*, 477–481. [CrossRef]
90. Gehlhaar, P.; de Olano, D.G.; Madrigal-Burgaleta, R.; Bartolomé, B.; Pastor-Vargas, C. Allergy to ginger with cysteine proteinase GP-I as the relevant allergen. *Ann. Allergy, Asthma Immunol.* **2018**, *121*, 624–625. [CrossRef] [PubMed]
91. Kanerva, L.; Estlander, T.; Jolanki, R. Occupational allergic contact dermatitis from spices. *Contact Dermat.* **1996**, *35*, 157–162. [CrossRef] [PubMed]
92. Kim, I.S.; Kim, S.Y.; Yoo, H.H. Effects of an aqueous-ethanolic extract of ginger on cytochrome P450 enzyme-mediated drug metabolism. *Die Pharm.* **2012**, *67*, 1007–1009.
93. Qiu, J.X.; Zhou, Z.W.; He, Z.X.; Zhang, X.; Zhou, S.F.; Zhu, S. Estimation of the binding modes with important human cytochrome P450 enzymes, drug interaction potential, pharmacokinetics, and hepatotoxicity of ginger components using molecular docking, computational, and pharmacokinetic modeling studies. *Drug Des. Dev. Ther.* **2015**, *9*, 841–866. [CrossRef]
94. Gressenberger, P.; Rief, P.; Jud, P.; Gütl, K.; Muster, V.; Ghanim, L.; Brodmann, M.; Gary, T. Increased bleeding risk in a patient with oral anticoagulant therapy and concomitant herbal intake—A case report. *eJIFCC* **2019**, *30*, 95–98.
95. Rubin, D.; Patel, V.; Dietrich, E. Effects of oral ginger supplementation on the INR. *Case Rep. Med.* **2019**, *2019*, 8784029. [CrossRef]
96. Lesho, E.P.; Saullo, L.; Udvari-Nagy, S. A 76-year-old woman with erratic anticoagulation. *Clevel. Clin. J. Med.* **2004**, *71*, 651–656. [CrossRef]
97. Krüth, P.; Brosi, E.; Fux, R.; Mörike, K.; Gleiter, C.H. Ginger-associated overanticoagulation by phenprocoumon. *Ann. Pharmacother.* **2004**, *38*, 257–260. [CrossRef]
98. Shalansky, S.; Lynd, L.; Richardson, K.; Ingaszewski, A.; Kerr, C. Risk of Warfarin-related bleeding events and supratherapeutic International Normalized Ratios associated with complementary and alternative medicine: A longitudinal analysis. *Pharmacother. J. Hum. Pharmacol. Drug Ther.* **2007**, *27*, 1237–1247. [CrossRef]
99. Jiang, X.; Williams, K.M.; Liauw, W.S.; Ammit, A.J.; Roufogalis, B.D.; Duke, C.C.; Day, R.O.; McLachlan, A.J. Effect of ginkgo and ginger on the pharmacokinetics and pharmacodynamics of warfarin in healthy subjects. *Br. J. Clin. Pharmacol.* **2005**, *59*, 425–432. [CrossRef]
100. Marx, W.; McKavanagh, D.; McCarthy, A.L.; Bird, R.; Ried, K.; Chan, A.; Isenring, L. The effect of ginger (*Zingiber officinale*) on platelet aggregation: A systematic literature review. *PLoS ONE* **2015**, *10*, e0141119. [CrossRef]
101. Committee on Herbal Medicinal Products. *Assessment Report on Zingiber officinale Roscoe, Rhizoma*; Technical Report EMA/HMPC/577856/2010; European Medicines Agency: London, UK, 2012. Available online: https://www.ema.europa.eu/en/documents/herbal-report/final-assessment-report-zingiber-officinale-roscoe-rhizoma_en.pdf (accessed on 2 September 2022).
102. Revol, B.; Gautier-Veyret, E.; Arrivé, C.; Fouilhé Sam-Laï, N.; McLeer-Florin, A.; Pluchart, H.; Pinsolle, J.; Toffart, A. Pharmacokinetic herb-drug interaction between ginger and crizotinib. *Br. J. Clin. Pharmacol.* **2020**, *86*, 1892–1893. [CrossRef]
103. Chiang, H.M.; Chao, P.D.L.; Hsiu, S.L.; Wen, K.C.; Tsai, S.Y.; Hou, Y.C. Ginger significantly decreased the oral bioavailability of cyclosporine in rats. *Am. J. Chin. Med.* **2006**, *34*, 845–855. [CrossRef]
104. Okonta, J.; Uboh, M.; Obonga, W. Herb-drug interaction: A case study of effect of ginger on the pharmacokinetic of metronidazole in rabbit. *Indian J. Pharm. Sci.* **2008**, *70*, 230–232. [CrossRef]
105. Mukkavilli, R.; Yang, C.; Singh Tanwar, R.; Ghareeb, A.; Luthra, L.; Aneja, R. Absorption, metabolic stability, and pharmacokinetics of ginger phytochemicals. *Molecules* **2017**, *22*, 553. [CrossRef]

106. Yiannikourides, A.; Latunde-Dada, G. A short review of iron metabolism and pathophysiology of iron disorders. *Medicines* **2019**, *6*, 85. [CrossRef]
107. Wallace, D.F. The regulation of iron absorption and homeostasis. *Clin. Biochem. Rev.* **2016**, *37*, 51–62.
108. Tandara, L.; Salamunic, I. Iron metabolism: Current facts and future directions. *Biochem. Med.* **2012**, *22*, 311–328. [CrossRef]
109. Dev, S.; Babitt, J.L. Overview of iron metabolism in health and disease. *Hemodial. Int.* **2017**, *21*, S6–S20. [CrossRef]
110. Anderson, G.J.; Frazer, D.M. Current understanding of iron homeostasis. *Am. J. Clin. Nutr.* **2017**, *106*, 1559S–1566S. [CrossRef]
111. West, A.R.; Oates, P.S. Mechanisms of heme iron absorption: Current questions and controversies. *World J. Gastroenterol.* **2008**, *14*, 4101–4110. [CrossRef]
112. Camaschella, C.; Nai, A.; Silvestri, L. Iron metabolism and iron disorders revisited in the hepcidin era. *Haematologica* **2020**, *105*, 260–272. [CrossRef]
113. Cappellini, M.D.; Musallam, K.M.; Taher, A.T. Iron deficiency anaemia revisited. *J. Intern. Med.* **2020**, *287*, 153–170. [CrossRef]
114. Jimenez, K.; Kulnigg-Dabsch, S.; Gasche, C. Management of iron deficiency anemia. *Gastroenterol. Hepatol.* **2015**, *11*, 241–250. Available online: https://www.ncbi.nlm.nih.gov/pmc/articles/PMC4836595/ (accessed on 2 September 2022).
115. Elstrott, B.; Khan, L.; Olson, S.; Raghunathan, V.; DeLoughery, T.; Shatzel, J.J. The role of iron repletion in adult iron deficiency anemia and other diseases. *Eur. J. Haematol.* **2020**, *104*, 153–161. [CrossRef]
116. Wieczorek, M.; Schwarz, F.; Sadlon, A.; Abderhalden, L.A.; de Godoi Rezende Costa Molino, C.; Spahn, D.R.; Schaer, D.J.; Orav, E.J.; Egli, A.; Bischoff-Ferrari, H.A. Iron deficiency and biomarkers of inflammation: A 3-year prospective analysis of the DO-HEALTH trial. *Aging Clin. Exp. Res.* **2022**, *34*, 515–525. [CrossRef]
117. Weiss, G.; Ganz, T.; Goodnough, L.T. Anemia of inflammation. *Blood* **2019**, *133*, 40–50. [CrossRef]
118. Pasricha, S.R.S.; Tye-Din, J.; Muckenthaler, M.U.; Swinkels, D.W. Iron deficiency. *Lancet* **2021**, *397*, 233–248. [CrossRef]
119. Short, M.W.; Domagalski, J.E. Iron deficiency anemia: Evaluation and management. *Am. Fam. Physician* **2013**, *87*, 98–104. Available online: https://www.aafp.org/pubs/afp/issues/2013/0115/p98.html (accessed on 2 September 2022).
120. Casgrain, A.; Collings, R.; Harvey, L.J.; Hooper, L.; Fairweather-Tait, S.J. Effect of iron intake on iron status: A systematic review and meta-analysis of randomized controlled trials. *Am. J. Cinical Nutr.* **2012**, *96*, 768–780. [CrossRef]
121. Neto, L.G.R.S.; dos Santos Neto, J.E.; Bueno, N.B.; de Oliveira, S.L.; da Rocha Ataide, T. Effects of iron supplementation versus dietary iron on the nutritional iron status: Systematic review with meta-analysis of randomized controlled trials. *Crit. Rev. Food Sci. Nutr.* **2019**, *59*, 2553–2561. [CrossRef] [PubMed]
122. Houston, B.L.; Hurrie, D.; Graham, J.; Perija, B.; Rimmer, E.; Rabbani, R.; Bernstein, C.N.; Turgeon, A.F.; Fergusson, D.A.; Houston, D.S.; et al. Efficacy of iron supplementation on fatigue and physical capacity in non-anaemic iron-deficient adults: A systematic review of randomised controlled trials. *BMJ Open* **2018**, *8*. [CrossRef] [PubMed]
123. National Blood Authority Australia. *Iron Product Choice and Dose Calculation for Adults: Guidance for Australian Health Providers (March 2016)*; National Blood Authority: Canberra, ACT, Australia, 2016.
124. Pasricha, S.R.S.; Flecknoe-Brown, S.C.; Allen, K.J.; Gibson, P.R.; McMahon, L.P.; Olynyk, J.K.; Roger, S.D.; Savoia, H.F.; Tampi, R.; Thomson, A.R.; et al. Diagnosis and management of iron deficiency anaemia: A clinical update. *Med J. Aust.* **2010**, *193*, 525–532. [CrossRef] [PubMed]
125. Scarcello, E.; Herpain, A.; Tomatis, M.; Turci, F.; Jacques, P.J.; Lison, D. Hydroxyl radicals and oxidative stress: The dark side of Fe corrosion. *Colloids Surfaces B Biointerfaces* **2020**, *185*, 110542. [CrossRef]
126. Qi, X.; Zhang, Y.; Guo, H.; Hai, Y.; Luo, Y.; Yue, T. Mechanism and intervention measures of iron side effects on the intestine. *Crit. Rev. Food Sci. Nutr.* **2020**, *60*, 2113–2125. [CrossRef]
127. Tiwari, A.K.M.; Mahdi, A.A.; Chandyan, S.; Zahra, F.; Godbole, M.M.; Jaiswar, S.P.; Srivastava, V.K.; Negi, M.P.S. Oral iron supplementation leads to oxidative imbalance in anemic women: A prospective study. *Clin. Nutr.* **2011**, *30*, 188–193. [CrossRef]
128. Yilmaz, B.; Li, H. Gut microbiota and iron: The crucial actors in health and disease. *Pharmaceuticals* **2018**, *11*, 1–20. [CrossRef]
129. Paganini, D.; Zimmermann, M.B. The effects of iron fortification and supplementation on the gut microbiome and diarrhea in infants and children: A review. *Am. J. Clin. Nutr.* **2017**, *106*, 1688S–1693S. [CrossRef]
130. Rusu, I.G.; Suharoschi, R.; Vodnar, D.C.; Pop, C.R.; Socaci, S.A.; Vulturar, R.; Istrati, M.; Moroșan, I.; Fărcaș, A.C.; Kerezsi, A.D.; et al. Iron supplementation influence on the gut microbiota and probiotic intake effect in iron deficiency—A literature-based review. *Nutrients* **2020**, *12*, 1993. [CrossRef]
131. Botta, A.; Barra, N.G.; Lam, N.H.; Chow, S.; Pantopoulos, K.; Schertzer, J.D.; Sweeney, G. Iron reshapes the gut microbiome and host metabolism. *J. Lipid Atheroscler.* **2021**, *10*, 160–183. [CrossRef]
132. Baumgartner, J.; Smuts, C.M.; Aeberli, I.; Malan, L.; Tjalsma, H.; Zimmermann, M.B. Overweight impairs efficacy of iron supplementation in iron-deficient South African children: A randomized controlled intervention. *Int. J. Obes. (2005)* **2013**, *37*, 24–30. [CrossRef]
133. Htet, M.K.; Fahmida, U.; Dillon, D.; Akib, A.; Utomo, B.; Thurnham, D.I. Is iron supplementation influenced by sub-clinical inflammation?: A randomized controlled trial among adolescent schoolgirls in Myanmar. *Nutrients* **2019**, *11*, 918. [CrossRef]
134. Yuen, H.W.; Becker, W. Iron toxicity. In *StatPearls*; StatPearls Publishing: Tampa, FL, USA, 2021. Available online: https://www.ncbi.nlm.nih.gov/books/NBK459224/ (accessed on 15 March 2022).
135. Abhilash, K.P.P.; Arul, J.J.; Bala, D. Fatal overdose of iron tablets in adults. *Indian J. Crit. Care Med.* **2013**, *17*, 311–313. [CrossRef]
136. Barton, J.C.; Lee, P.L.; West, C.; Bottomley, S.S. Iron overload and prolonged ingestion of iron supplements: Clinical features and mutation analysis of hemochromatosis-associated genes in four cases. *Am. J. Hematol.* **2006**, *81*, 760–767. [CrossRef]

137. Prakash, U.N.S.; Srinivasan, K. Enhanced intestinal uptake of iron, zinc and calcium in rats fed pungent spice principles—Piperine, capsaicin and ginger (*Zingiber officinale*). *J. Trace Elem. Med. Biol.* **2013**, *27*, 184–190. [CrossRef]
138. Jaiswal, A.; Pathania, V.; A, J.L. An exploratory trial of food formulations with enhanced bioaccessibility of iron and zinc aided by spices. *LWT* **2021**, *143*, 111122. [CrossRef]
139. Kulkarni, R.A.; Deshpande, A.; Saxena, K.; Varma, M.; Sinha, A.R. Ginger supplementary therapy for iron absorption in iron deficiency anemia. *Indian J. Tradit. Knowl.* **2012**, *11*, 78–80.
140. Kulkarni, R.A. A Study of Anti-Inflammatory and Antioxidant Effects of Zingiber Officinale in Tuberculosis Patients with Anemia. Ph.D. Thesis, Shri Aurobindo Institute of Medical Sciences, Indore, MP, India, 2010.
141. Prakash, U.N.S.; Srinivasan, K. Gastrointestinal protective effect of dietary spices during ethanol-induced oxidant stress in experimental rats. *Appl. Physiol. Nutr. Metab.* **2010**, *35*, 134–141. [CrossRef]
142. Masuda, Y.; Kikuzaki, H.; Hisamoto, M.; Nakatani, N. Antioxidant properties of gingerol related compounds from ginger. *BioFactors (Oxford)* **2004**, *21*, 293–296. [CrossRef]
143. Oboh, G.; Akinyemi, A.J.; Ademiluyi, A.O. Antioxidant and inhibitory effect of red ginger (Zingiber officinale var. Rubra) and white ginger (*Zingiber officinale* Roscoe) on Fe^{2+} induced lipid peroxidation in rat brain in vitro. *Exp. Toxicol. Pathol.* **2012**, *64*, 31–36. [CrossRef]
144. Hinneburg, I.; Dorman, H.J.D.; Hiltunen, R. Antioxidant activities of extracts from selected culinary herbs and spices. *Food Chem.* **2006**, *97*, 122–129. [CrossRef]
145. Kumar, S.; Saxena, K.; Uday, I.; Singh, N.; Saxena, R.; Singh, U.N. Anti-inflammatory action of ginger: A critical review in anemia of inflammation and its future aspects. *Int. J. Herb. Med.* **2013**, *1*, 16–20. Available online: https://www.florajournal.com/archives/2013/vol1issue4/PartA/2.1.pdf (accessed on 2 September 2022).
146. Wang, J.; Chen, Y.; Hu, X.; Feng, F.; Cai, L.; Chen, F. Assessing the effects of ginger extract on polyphenol profiles and the subsequent impact on the fecal microbiota by simulating digestion and fermentation in vitro. *Nutrients* **2020**, *12*, 3194. [CrossRef]
147. Wang, J.; Wang, P.; Li, D.; Hu, X.; Chen, F. Beneficial effects of ginger on prevention of obesity through modulation of gut microbiota in mice. *Eur. J. Nutr.* **2020**, *59*, 699–718. [CrossRef]
148. Ma, Z.J.; Wang, H.J.; Ma, X.J.; Li, Y.; Yang, H.J.; Li, H.; Su, J.R.; Zhang, C.E.; Huang, L.Q. Modulation of gut microbiota and intestinal barrier function during alleviation of antibiotic-associated diarrhea with Rhizoma: *Zingiber officinale* (Ginger) extract. *Food Funct.* **2020**, *11*, 10839–10851. [CrossRef]
149. Wang, X.; Zhang, D.; Jiang, H.; Zhang, S.; Pang, X.; Gao, S.; Zhang, H.; Zhang, S.; Xiao, Q.; Chen, L.; et al. Gut microbiota variation with short-term intake of ginger juice on human health. *Front. Microbiol.* **2021**, *11*. [CrossRef]
150. Ferri-Lagneau, K.F.; Moshal, K.S.; Grimes, M.; Zahora, B.; Lv, L.; Sang, S.; Leung, T. Ginger stimulates hematopoiesis via Bmp pathway in zebrafish. *PLoS ONE* **2012**, *7*, e39327. [CrossRef]
151. Ferri-Lagneau, K.F.; Haider, J.; Sang, S.; Leung, T. Rescue of hematopoietic stem/progenitor cells formation in plcg1 zebrafish mutant. *Sci. Rep.* **2019**, *9*, 244. [CrossRef]
152. Wang, X.; Zhang, M.; Woloshun, R.R.; Yu, Y.; Lee, J.K.; Flores, S.R.; Merlin, D.; Collins, J.F. Oral administration of ginger-derived lipid nanoparticles and dmt1 sirna potentiates the effect of dietary iron restriction and mitigates pre-existing iron overload in hamp ko mice. *Nutrients* **2021**, *13*, 1686. [CrossRef]
153. Gholampour, F.; Ghiasabadi, F.B.; Owji, S.M.; Vatanparast, J. The protective effect of hydroalcoholic extract of ginger (*Zingiber officinale* Rosc.) against iron-induced functional and histological damages in rat liver and kidney. *Avicenna J. Phytomedicine* **2017**, *7*, 542–553. Available online: https://www.ncbi.nlm.nih.gov/pmc/articles/PMC5745538/ (accessed on 2 September 2022).
154. Golombick, T.; Diamond, T.H.; Manoharan, A.; Ramakrishna, R.; Badmaev, V. Effect of the ginger derivative, 6-shogaol, on ferritin levels in patients with low to intermediate-1-risk myelodysplastic syndrome-A small, investigative study. *Clin. Med. Insights: Blood Disord.* **2017**, *10*, 1–4. [CrossRef] [PubMed]
155. El-Refai, A.A.; Ghoniem, G.A.; El-Khateeb, A.Y.; Hassaan, M.M. Eco-friendly synthesis of metal nanoparticles using ginger and garlic extracts as biocompatible novel antioxidant and antimicrobial agents. *J. Nanostructure Chem.* **2018**, *8*, 71–81. [CrossRef]
156. Kirdat, P.N.; Dandge, P.B.; Hagwane, R.M.; Nikam, A.S.; Mahadik, S.P.; Jirange, S.T. Synthesis and characterization of ginger (*Z. officinale*) extract mediated iron oxide nanoparticles and its antibacterial activity. *Mater. Today Proc.* **2020**, *43*, 2826–2831. [CrossRef]
157. Noor, R.; Yasmin, H.; Ilyas, N.; Nosheen, A.; Hassan, M.N.; Mumtaz, S.; Khan, N.; Ahmad, A.; Ahmad, P. Comparative analysis of iron oxide nanoparticles synthesized from ginger (*Zingiber officinale*) and cumin seeds (*Cuminum cyminum*) to induce resistance in wheat against drought stress. *Chemosphere* **2022**, *292*, 133201. [CrossRef]
158. Ooi, S.L.; Campbell, R.; Pak, S.C.; Golombick, T.; Manoharan, A.; Ramakrishna, R.; Badmaev, V.; Schloss, J. Is 6-shogaol an effective phytochemical for patients with lower-risk myelodysplastic syndrome? A narrative review. *Integr. Cancer Ther.* **2021**, *20*, 1–12. [CrossRef]
159. Goodnough, L.T.; Nemeth, E.; Ganz, T. Detection, evaluation, and management of iron-restricted erythropoiesis. *Blood* **2010**, *116*, 4754–4761. [CrossRef]
160. Huang, Q.; Feng, L.; Li, H.; Zheng, L.; Qi, X.; Wang, Y.; Feng, Q.; Liu, Z.; Liu, X.; Lu, L. Jian-Pi-Bu-Xue-Formula alleviates cyclophosphamide-induced myelosuppression via up-regulating NRF2/HO1/NQO1 signaling. *Front. Pharmacol.* **2020**, *11*, 1302. [CrossRef]

161. Lam, C.T.W.; Chan, P.H.; Lee, P.S.C.; Lau, K.M.; Kong, A.Y.Y.; Gong, A.G.W.; Xu, M.L.; Lam, K.Y.C.; Dong, T.T.X.; Lin, H.; et al. Chemical and biological assessment of Jujube (*Ziziphus jujuba*)-containing herbal decoctions: Induction of erythropoietin expression in cultures. *J. Chromatography. B Anal. Technol. Biomed. Life Sci.* **2016**, *1026*, 254–262. [CrossRef]
162. Chan, S.M.; Nelson, E.A.; Leung, S.S.; Cheung, P.C.; Li, C.Y. Special postpartum dietary practices of Hong Kong Chinese women. *Eur. J. Clin. Nutr.* **2000**, *54*, 797–802. [CrossRef]
163. Zhang, L.; Gu, F.X.; Chan, J.M.; Wang, A.Z.; Langer, R.S.; Farokhzad, O.C. Nanoparticles in medicine: Therapeutic applications and developments. *Clin. Pharmacol. Ther.* **2008**, *83*, 761–769. [CrossRef]
164. Zdrojewicz, Z.; Waracki, M.; Bugaj, B.; Pypno, D.; Cabała, K. Medical applications of nanotechnology. *Adv. Hyg. Exp. Med.* **2015**, *69*, 1196–1204. [CrossRef]
165. Saha, P.K.; Saha, L. Iron nanoparticles and its potential application: A literature review. *Indian J. Pharmacol.* **2021**, *53*, 339–340. [CrossRef]
166. Kumari, A.; Chauhan, A.K. Iron nanoparticles as a promising compound for food fortification in iron deficiency anemia: A review. *J. Food Sci. Technol.* **2021**, 1–17. [CrossRef]
167. Singh, J.; Dutta, T.; Kim, K.H.; Rawat, M.; Samddar, P.; Kumar, P. 'Green' synthesis of metals and their oxide nanoparticles: applications for environmental remediation. *J. Nanobiotechnol.* **2018**, *16*, 84. [CrossRef]

molecules

MDPI

Article

Lichen Extracts from Cetrarioid Clade Provide Neuroprotection against Hydrogen Peroxide-Induced Oxidative Stress

Isabel Ureña-Vacas, Elena González-Burgos *, Pradeep Kumar Divakar and María Pilar Gómez-Serranillos

Department of Pharmacology, Pharmacognosy and Botany, Faculty of Pharmacy, Universidad Complutense de Madrid, Plaza Ramón y Cajal s/n, Ciudad Universitaria, 28040 Madrid, Spain
* Correspondence: elenagon@ucm.es

Abstract: Oxidative stress is involved in the pathophysiology of many neurodegenerative diseases. Lichens have antioxidant properties attributed to their own secondary metabolites with phenol groups. Very few studies delve into the protective capacity of lichens based on their antioxidant properties and their action mechanism. The present study evaluates the neuroprotective role of *Dactylina arctica*, *Nephromopsis stracheyi*, *Tuckermannopsis americana* and *Vulpicida pinastri* methanol extracts in a hydrogen peroxide (H_2O_2) oxidative stress model in neuroblastoma cell line "SH-SY5Y cells". Cells were pretreated with different concentrations of lichen extracts (24 h) before H_2O_2 (250 μM, 1 h). Our results showed that *D. arctica* (10 μg/mL), *N. stracheyi* (25 μg/mL), *T. americana* (50 μg/mL) and *V. pinastri* (5 μg/mL) prevented cell death and morphological changes. Moreover, these lichens significantly inhibited reactive oxygen species (ROS) production and lipid peroxidation and increased superoxide dismutase (SOD) and catalase (CAT) activities and glutathione (GSH) levels. Furthermore, they attenuated mitochondrial membrane potential decline and calcium homeostasis disruption. Finally, high-performance liquid chromatography (HPLC) analysis revealed that the secondary metabolites were gyrophoric acid and lecanoric acid in *D. artica*, usnic acid, pinastric acid and vulpinic acid in *V. pinastri*, and alectoronic acid in *T. americana*. In conclusion, *D. arctica* and *V. pinastri* are the most promising lichens to prevent and to treat oxidative stress-related neurodegenerative diseases.

Keywords: lichens; neuroprotection; cetrarioid clade; oxidative stress

Citation: Ureña-Vacas, I.; González-Burgos, E.; Divakar, P.K.; Gómez-Serranillos, M.P. Lichen Extracts from Cetrarioid Clade Provide Neuroprotection against Hydrogen Peroxide-Induced Oxidative Stress. *Molecules* **2022**, *27*, 6520. https://doi.org/10.3390/molecules27196520

Academic Editor: Nour Eddine Es-Safi

Received: 25 August 2022
Accepted: 30 September 2022
Published: 2 October 2022

1. Introduction

Oxidative stress is an imbalance in cellular redox homeostasis caused by ROS overproduction and/or antioxidant system dysfunction. The brain is particularly susceptible to oxidative stress and this process is increased with aging. The brain consumes almost 20% of the total basal oxygen, it is rich in polyunsaturated *n*-3 fatty acids and redox-active transition metals, it has a low endogenous antioxidant defense (i.e., catalase), and neurotransmitters such as dopamine can auto-oxidize leading to free radicals [1,2]. Moreover, the mitochondrial electron transport chain consumes around 98% of oxygen and the residual oxygen is converted into radical superoxide ($O_2^{\bullet-}$) and the non-radical oxidant H_2O_2; excessive mitochondrial-derived ROS accumulation can lead to mitochondrial dysfunction [3]. Furthermore, hydrogen peroxide, the major redox metabolite, diffuses across membranes by water channels, oxidizes and damages biomolecules [4,5]. The main mechanism underlying the neurotoxic effects of hydrogen peroxide occurs through Fenton's reaction. In this reaction, ferrous iron and hydrogen peroxide react to yield hydroxyl radical. Hydroxyl radical is the most deleterious free radical, reacting with macromolecules by hydroxyl addition and hydrogen abstraction [6]. The altered cellular redox homeostasis causes oxidative injury to lipids, proteins, and DNA and modifications in cellular function that finally contribute to cell death mainly by apoptosis [3]. Oxidative stress is involved as a major pathophysiologic mechanism of age-related neurodegenerative diseases such as Parkinson's disease, Alzheimer's disease, and amyotrophic lateral sclerosis [7].

Antioxidants modulate the redox state of cells through single electron transfer (SET), hydrogen atom transfer (HAT), transition metals chelation and the up-regulation of enzymatic and non-enzymatic antioxidants. Antioxidants are potentially beneficial in the prevention and treatment of central nervous system pathologies associated with oxidative stress and constitute one of the most promising therapeutical strategies [8].

Lichens have aroused great pharmacological interest in recent years because they produce compounds unique to these species. These bioactive compounds are primarily phenol derivatives such as dibenzofurans, depsidones and depsides. Lichen extracts and their secondary metabolites have shown an interesting antioxidant activity [8,9]. However, studies focusing on therapeutic and protective strategy based on the antioxidant ability of lichens are very limited [9].

Recently, in previous studies of this group, the antioxidant activity of lichen extracts from the cetrarioid clade was evaluated using different in vitro methods (1,1-Diphenyl-2-picrylhydrazyl (DPPH), oxygen radical absorbance capacity (ORAC) and ferric-reducing antioxidant power (FRAP) assays) and multivariate statistical techniques. This study revealed that the lichen species *Dactylina arctica* (Hook) Nyl., *Nephromopsis stracheyi* (C. Bab.) Müll. Arg., *Tuckermannopsis americana* (Sprengel) Hale, and *Vulpicida pinastri* (Scop.) J.-E. Mattsson & M. J. Lai. were the ones with the highest antioxidant capacities [10].

The aim of the present work is to evaluate for the first time the neuroprotective activity, based on antioxidant properties, of the methanol lichen extracts *Dactylina arctica*, *Nephromopsis stracheyi*, *Tuckermannopsis americana* and *Vulpicida pinastri* in a hydrogen peroxide-induced oxidative stress model in a neuroblastoma cell line.

2. Results

2.1. Lichen Extracts from Cetrarioid Clade Promoted Neuronal Survival after H_2O_2-Induced Oxidative Stress

Initially, we evaluated the effect of the methanol extracts of the lichens *Dactylina arctica*, *Nephromopsis stracheyi*, *Tuckermannopsis americana* and *Vulpicida pinastri* on the human neuroblastoma SH-SY5Y cell viability using 3-[4,5-dimethylthiazol-2-yl]-2,5 diphenyl tetrazolium bromide (MTT) assay. As shown in Figure 1, *T. americana* did not cause cytotoxicity at any assayed concentrations. On the other hand, *N. stracheyi* significantly reduced cell viability at 50 μg/mL (38% of cell viability) whereas *D. arctica* and *V. pinastri* affected cell viability at 25 μg/mL (62.7% and 65.9% of cell viability, respectively) and at 50 μg/mL (43.9% and 52.8% of cell viability, respectively).

Next, we investigated the potential protective effect of non-toxic concentrations of lichen extracts against hydrogen peroxide-induced oxidative stress. Figure 2A demonstrates that cell viability of the 250 μM for 1 h H_2O_2-treated SH-SY5Y cells significantly decreased by 57.5% compared to control cells (100%). However, pretreatments with methanol lichen extracts of cetrarioid clade promoted neuronal survival compared to hydrogen peroxide-treated cells. With 24 h pretreatment, the percentage of cell viability was increased over 68.9% and 76.8% for *D. arctica* at 5 and 10 μg/mL, respectively, over 65.4% and 63.2% for *N. stracheyi* at 10 and 25 μg/mL, respectively, 58.2% for *T. americana* at 50 μg/mL, and 78.9% for *V. pinastri* at 5 μg/mL. Therefore, we chose the most protective concentrations of each lichen extract to delve into the protective mechanism of these extracts and identify which of them is the most active. Hence, the maximum cell viability protection was 5 μg/mL for *V. pinastri*, 10 μg/mL for *D. arctica* and *N. stracheyi* and 50 μg/mL for *T. americana*.

Figure 2B showed the effect of the most protective lichen extracts on cell morphology. Hydrogen peroxide (250 μM for 1 h) caused morphological changes toward a cellular SH-SY5Y rounding. By contrast, lichen extracts improved morphological changes of neuroblastoma cells as shown in the presence of cellular projections.

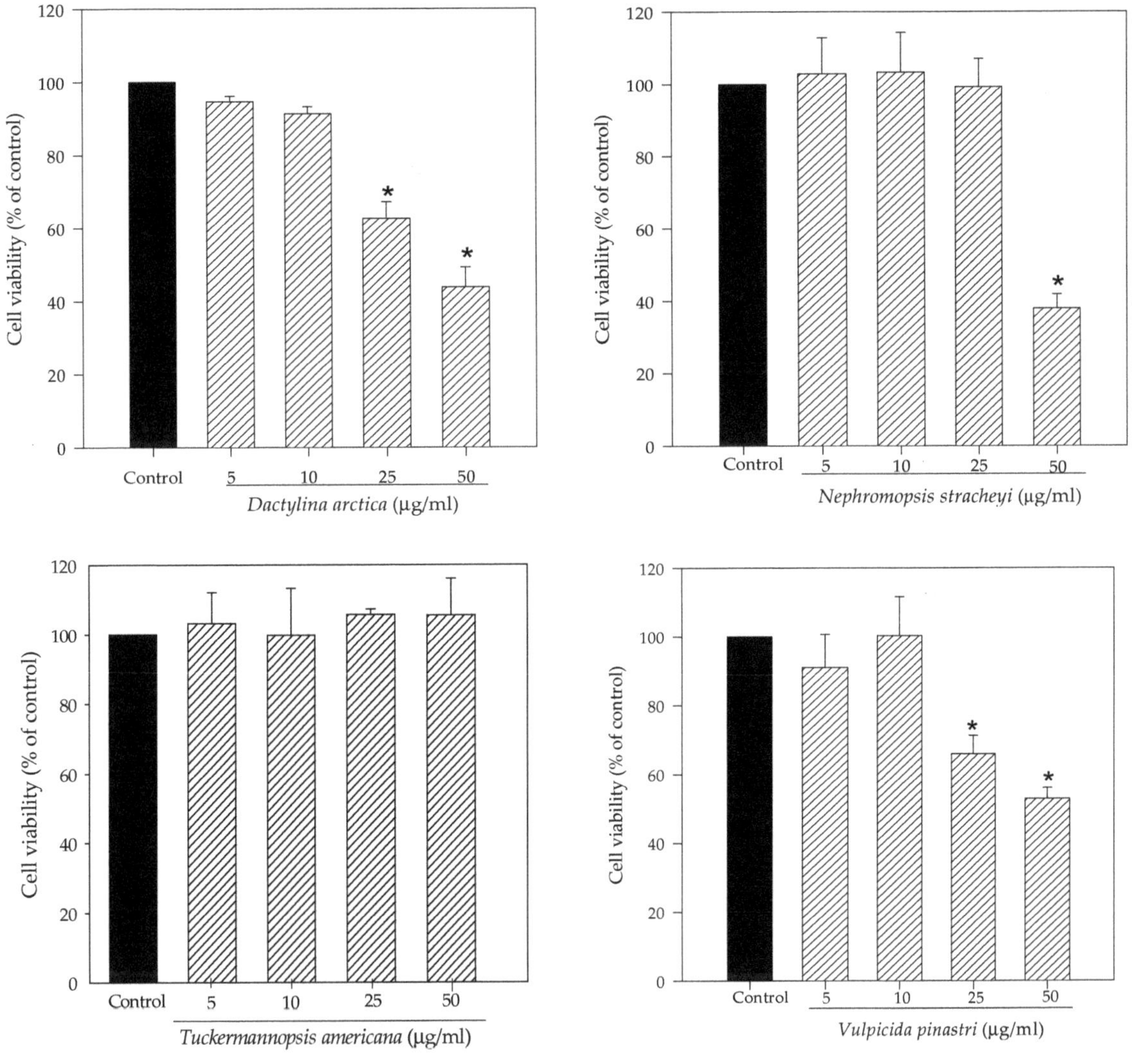

Figure 1. Effect of methanol lichen extracts of cetrarioid clade on cell viability. SH-SY5Y cells were treated with different concentrations of extracts from 5 to 50 µg/mL for 24 h. Cell viability was determined using MTT assay. Results are expressed as mean ± standard deviation (SD) (triplicate experiments). * $p < 0.05$ versus control.

2.2. Lichen Extracts from Cetrarioid Clade Reduced ROS Production after H_2O_2-Induced Oxidative Stress

Figure 3 revealed a significant increase in intracellular ROS production when SH-SY5Y cells were treated with hydrogen peroxide. 1 h treatment with H_2O_2 at 250 µM enhanced ROS generation by 155.5% compared to control cells (100%). On the other hand, 24 h pretreatments with methanol lichen extracts significantly reduced ROS production. In particular, the highest reduction was shown by *D. arctica* (42.4% of reduction compared to hydrogen peroxide treatment), followed by *T. americana* (41.1% of reduction versus H_2O_2), *N. stracheyi* (38.7% of reduction versus H_2O_2) and *V. pinastri* (35.3% of reduction versus H_2O_2).

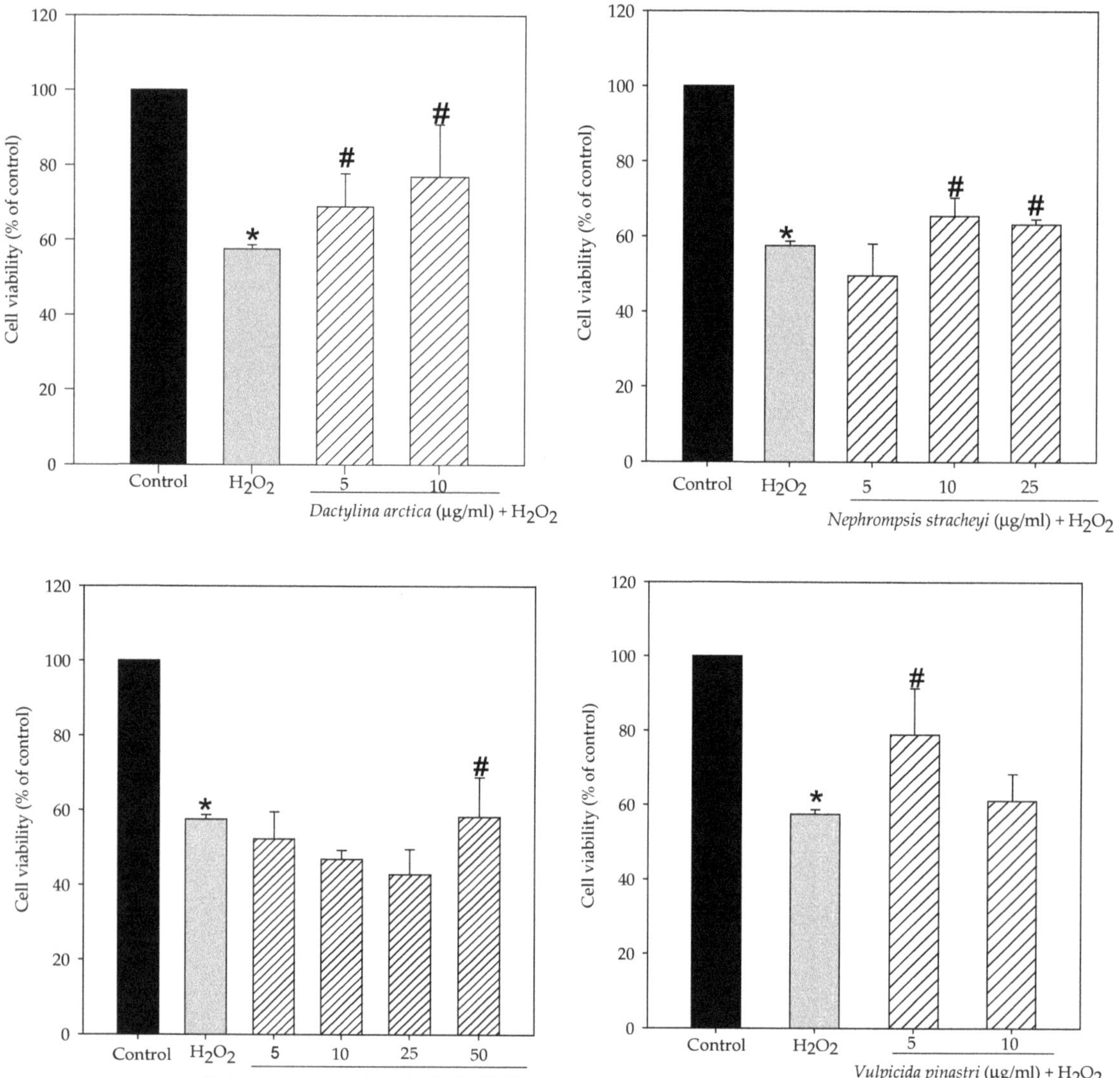

(A)

Figure 2. *Cont.*

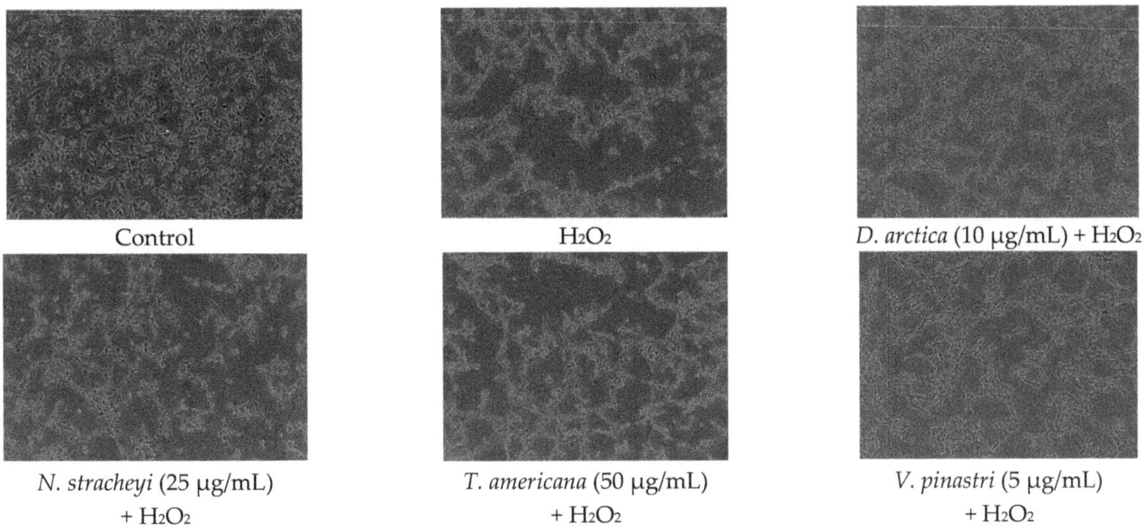

(**B**)

Figure 2. (**A**) Effect of methanol lichen extracts of cetrarioid clade on cytoprotection in stress oxidative models. SH-SY5Y cells were pretreated with non-cytotoxic concentrations of lichens for 24 h before H_2O_2 (250 µM, 1 h). Cell viability was determined using MTT assay. Results were expressed as mean ± SD (triplicate experiments). * $p < 0.01$ versus control; # $p < 0.01$ versus H_2O_2. (**B**) SH-SY5Y cells morphology after treatments.

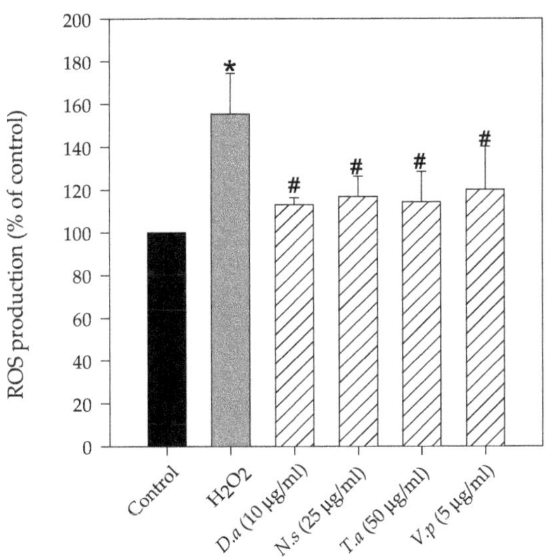

Figure 3. Effect of methanol lichen extracts of cetrarioid clade on intracellular ROS production. SH-SY5Y cells were pretreated with *D. arctica* (*D.a*), *N. stracheyi* (*N.s.*), *T. americana* (*T.a*) and *V. pinastri* (*V.p*) for 24 h before H_2O_2 (250 µM, 1 h). The levels of intracellular ROS production were measured using dichlorodihydrofluorescein diacetate (DCFH-DA) method. Results are expressed as mean ± SD (triplicate experiments). * $p < 0.01$ versus control; # $p < 0.01$ versus H_2O_2.

2.3. Lichen Extracts from Cetrarioid Clade Improved Oxidative Stress Markers and Antioxidant Enzyme Activity

Exposure to hydrogen peroxide significantly increased thiobarbituric acid reactive substances (TBARS) levels (174%), impaired GSH content (51%) and reduced SOD (68%) and CAT activity (58.5%) in SH-SY5Y cells compared to control cells (100%) (Figure 4). However, pretreatments with methanol lichen extracts of cetrarioid clade exhibited noteworthy protection against hydrogen peroxide-induced oxidative injury by improving the antioxidant status. Hence, *D. arctica* and *V. pinastri* significantly reduced lipid peroxidation levels (109% and 122%, respectively), restored GSH content (93% and 83.5%, respectively) and increased SOD activity (96% and 87%, respectively). Moreover, *T. americana* also augmented SOD enzyme activity by 92%. All the extracts significantly increased CAT activity, reverting H_2O_2 effects. *D. arctica, V. pinastri* and *T. americana* showed the highest values (100.9%, 97% and 95%, respectively), followed by *N. stracheyi*, with moderate CAT activity (81.6%).

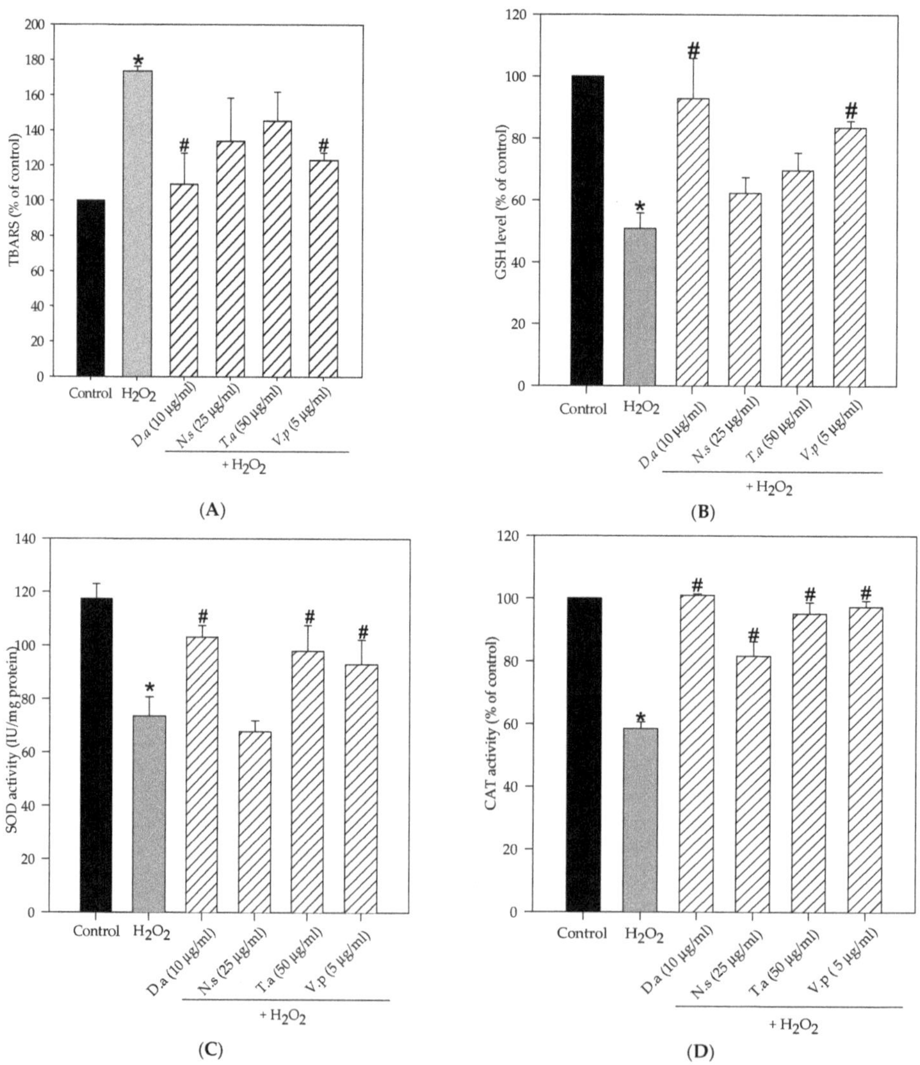

Figure 4. Lichen extracts from cetrarioid clade improved oxidative stress markers and antioxidant enzyme activity. SH-SY5Y cells were pretreated with *Dactylina arctica* (*D.a*), *Nephromopsis stracheyi* (*N.s*),

Tuckermannopsis americana (*T.a*) and *Vulpicida pinastri* (*V.p*) for 24 h before H_2O_2 (250 μM, 1 h). (**A**) Lipid peroxidation, (**B**) GSH levels and (**C**) SOD activity, (**D**) CAT activity. Results are expressed as mean ± SD (triplicate experiments). * $p < 0.05$ versus control; # $p < 0.05$ versus H_2O_2.

2.4. Lichen Extracts from Cetrarioid Clade Protected against H_2O_2-Induced Mitochondrial Dysfunction

Figure 5 shows the effect of lichen extracts on different mitochondrial parameters (mitochondrial membrane potential and calcium levels). Treatment with hydrogen peroxide (250 μM, 1 h) caused a significant decrease in the mitochondrial membrane potential (35% compared to 100% control cells) and a significant increase in mitochondrial calcium levels (1.14 relative to control) and cytosolic calcium levels (1007 nM compared to 516 nM control cells). However, pretreatments with lichen extracts prevented H_2O_2-induced mitochondrial changes. In particular, extracts of *D. arctica* (10 μg/mL) and *V. pinastri* (5 μg/mL) significantly increased mitochondrial membrane potential by 67% and 69%, respectively, and significantly reduced cytosolic calcium levels by 557 nM and 721 nM, respectively. Mitochondrial calcium levels were reduced in pretreated cells with selected concentrations of *D. arctica*, *V. pinastri* and *T. americana* extracts (1.08, 1.07, 1.05 relative to control, respectively).

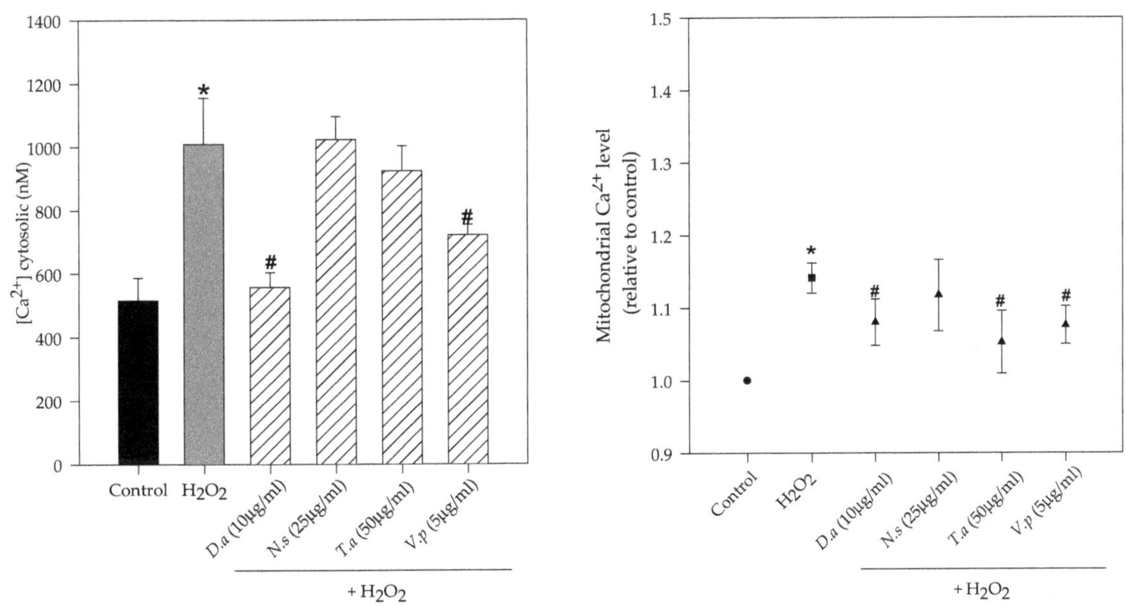

Figure 5. *Cont.*

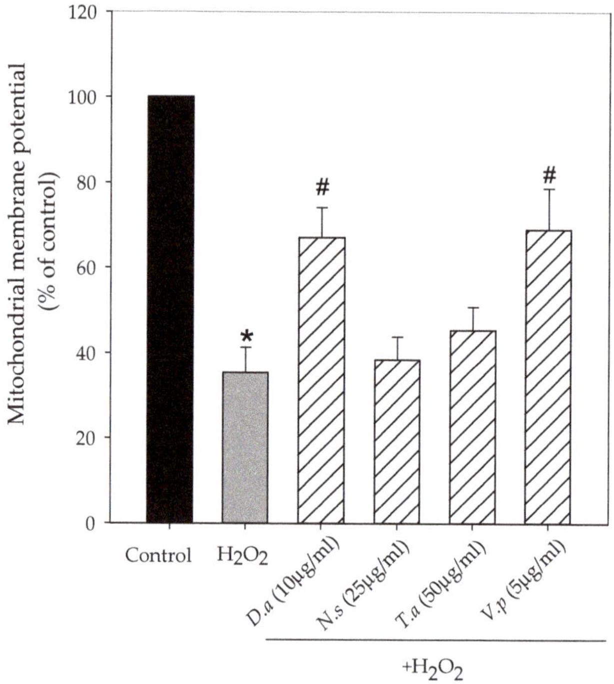

(C)

Figure 5. Effect of lichen extracts against H_2O_2 -induced mitochondrial dysfunction in SH-SY5Y. (**A**) on cytosolic calcium levels. (**B**) on mitochondrial calcium levels. (**C**) on mitochondrial membrane potential. Data are expressed as means ± SD (% of control) * $p < 0.001$ vs. control; # $p < 0.001$ vs. H_2O_2).

2.5. HPLC Profile of Lichen Extracts from Cetrarioid Clade

The most promising lichen extracts were analyzed using the HPLC-UV method, whose representative chromatograms are shown in Figure 6. Secondary metabolites were identified based on their retention times and ultraviolet spectra as compared to standards and previously reported lichen extracts. Table 1 reported retention times and the absorbance maxima (nm) UV spectrum. Results showed that the main compounds in *D. artica* were gyrophoric acid (GYR) and lecanoric acid (LEC). The lichen *T. americana* contained alectoronic acid (ALE). The compounds usnic acid (USN), pinastric acid (PIN) and vulpinic acid (VUL) were the majority in *V. pinastri*.

Table 1. Retention times and UV absorbance maxima (nm) of main secondary metabolites of studied lichens.

Compounds	Retention Time (t_R, min)	UV Detection (λ max, nm)
Alectoronic acid (ALE)	29.1 ± 0.001	214, 254, 316
Gyrophoric acid (GYR)	25.9 ± 0.003	212, 270, 304
Lecanoric acid (LEC)	19.4 ± 0.0312	212, 270, 304
Methyl orsellinate (Me-ORS)	13.9 ± 0.002	210, 262, 298
Orsellinic acid (ORS)	9.8 ± 0.011	210, 262, 298
Pinastric acid (PIN)	26.9 ± 0.047	<210, 246, 392
Usnic acid (USN)	32.3 ± 0.046	226/234, 282
Vulpinic acid (VUL)	25.1 ± 0.034	<210, 234, 282, 354

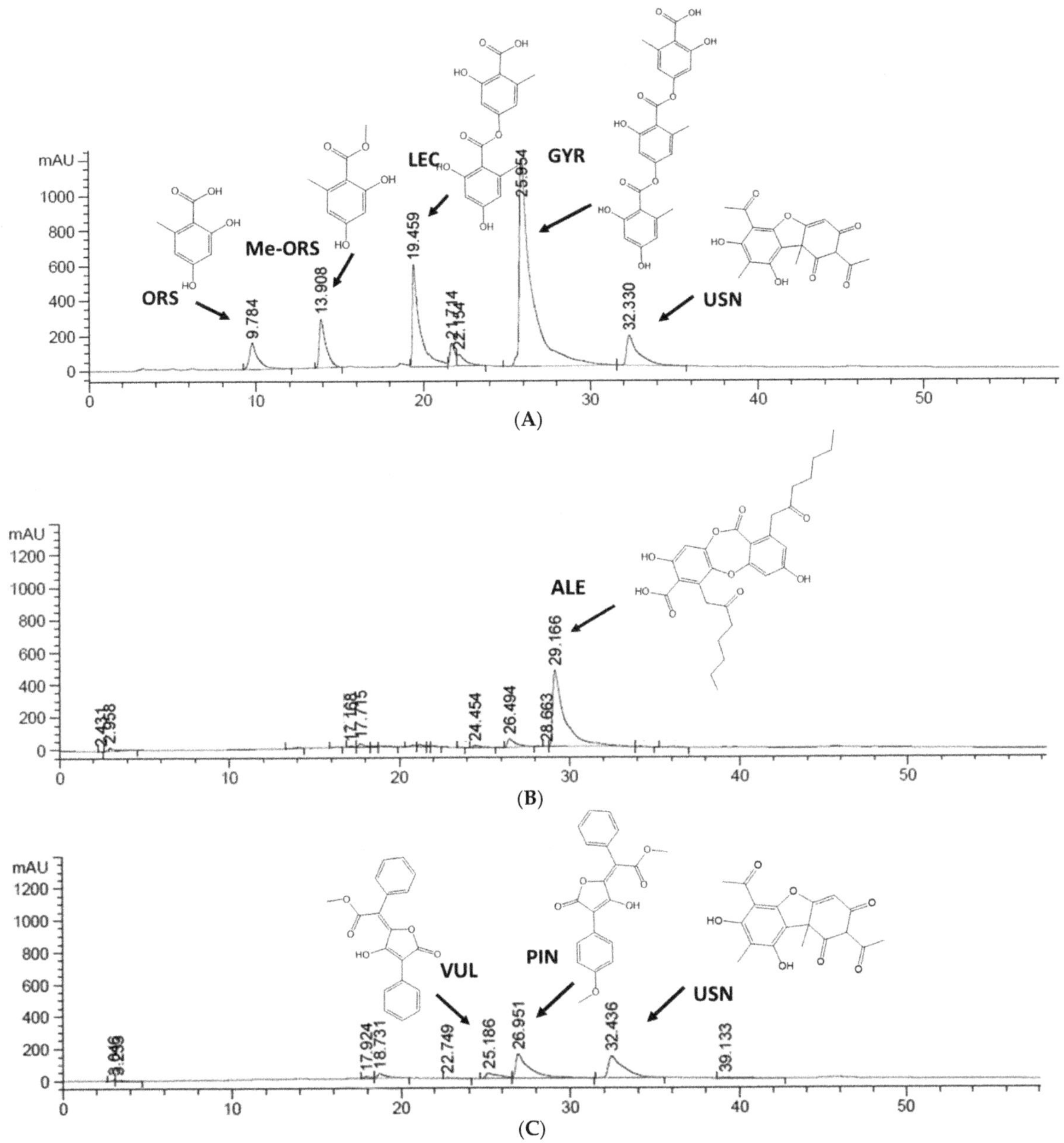

Figure 6. Representative HPLC chromatograms (λ = 254 nm) (**A**) *Dactylina artica* (**B**) *Tuckermannopsis americana* (**C**) *Vulpicida pinastri*.

3. Discussion

The present work demonstrated that pretreatments with methanol lichen extracts from cetrarioid clade provide neuroprotection against hydrogen peroxide in SH-SY5Y cells as evidenced in ROS reduction, improvement in oxidative stress biomarkers and antioxidant enzyme activity and mitochondrial protection.

The brain is a metabolically active organ that yields high ROS levels compared to other organs. Moreover, neurons are the most sensitive cell types to free radicals [11]. Overproduction of ROS can lead to oxidative macromolecules injury and consequently, to cell death. Activation of necrotic and apoptotic pathways by ROS induces cell death. Among the mechanisms responsible for ROS-causing apoptosis are receptor activation, caspase activation and mitochondrial dysfunction [12]. Oxidative stress (ROS/antioxidant imbalance) has been implicated in the initiation and progression of age-related neurodegenerative diseases [13]. Therefore, the prevention of oxidative stress is one of the most promising strategies for all those diseases that involve an alteration of redox homeostasis. H_2O_2 acts as an inducer of oxidative stress damage, increasing ROS levels and leading to cell death. The current study found that methanol lichen extracts from cetrarioid clade significantly attenuated hydrogen peroxide-induced ROS production in the human neuroblastoma SH-SY5Y cell line and consequently prevented cell death. Hydrogen peroxide can cross cell membranes and cause oxygen-derived free radicals. Hence, hydrogen peroxide can be converted into hydroxyl radicals in the presence of ferrous ions (Fenton reaction) [14]. Lichens contain phenolic compounds in their composition which can act as antioxidants through mechanisms such as radical scavenging activity and metal chelating activity [15–17].

Furthermore, these methanol lichen extracts mitigated changes in biomarkers of oxidative stress (lipid peroxidation reduction and GSH increase). The brain presents the highest rate of lipid metabolism in the body. In Alzheimer's disease and in Parkinson's diseases, there is an overproduction of ROS that induces the oxidation of lipid membrane constituents, leading to lipid hydroperoxides. This process, named lipid peroxidation, constitutes a hallmark for most neurodegenerative disorders. In fact, polyunsaturated fatty acids (PUFAs) are the main target of ROS attack, due to the high number of double bonds in their structure [18,19]. This early event in the brain causes cytotoxic and genotoxic effects. TBARS is a common biomarker of polyunsaturated fatty acids peroxidation. High amounts of lipid peroxidation products have been identified in post-mortem brains of people affected with neurodegenerative diseases. These lipid peroxidation products cause tissue injury and failures of antioxidant systems [20]. The lichen extracts of *V. pinastri* and *D. arctica* markedly reduced lipid peroxidation in neuroblastoma cells. Glutathione (GSH) is the major endogenous antioxidant defense. This primary antioxidant scavenges free radicals through its thiol group of its cysteine residue, and it functions as a co-substrate of the antioxidant enzymes selenium-glutathione peroxidase (GPx) and glutathione S-transferase (GST) [21]. GPx reduces lipid peroxides to alcohols and aldehydes. It has been reported that a reduction of GSH to its oxidized form provokes a decrease in intracellular GSH [22]. Restoring the levels of GSH is strongly related to the ROS and lipid peroxidation reduction [18]. Previous studies reported that fumarprotocetraric acid (depsidone), evernic acid (depside) and usnic acid (dibenzofuran-like) inhibited ROS generation, lipid peroxidation and glutathione depletion in neurons and the astrocytes cell model using hydrogen peroxide as an oxidative stress inductor [23,24]. In addition, the Parmeliaceae lichens *Cetraria islandica* and *Vulpicida canadensis* also showed protective effects against H_2O_2-induced injury in the human astrocytoma cell line U373-MG, as evidenced by reduced ROS production, increased GSH levels and the inhibition of lipid peroxidation [9].

Moreover, lichen extracts increase SOD and CAT activity. The enzyme SOD catalyzes the dismutation of superoxide anion to hydrogen peroxide which is then converted into oxygen and water by the action of the antioxidant enzymes catalase and glutathione peroxidase. The enzymatic activity of SOD and CAT showed to be significantly reduced in postmortem brain tissue at an advanced age [22]. Therefore, the results of this study showed that of the four methanol lichen extracts tested, *N. stracheyi* exerted its neuroprotective activity via ROS inhibition and increased CAT activity while *V. pinastri*, *D. arctica*, and *T. americana* prevented ROS overproduction and maintained enzymes activity. Upregulation of enzymes activity could be associated with an increase in its expression. Previously, it has been demonstrated that the depside fumarprotocetraric acid, isolated from *Cetraria islandica*, upregulated the antioxidant enzymes catalase, superoxide dismutase-1, and hemoxigenase-

1 expression which was related to Nrf2 signaling pathway activation [23]. Other lichens such as *Parmotrema perlatum* and *Hypotrachyna formosana* have also evidenced to reduce intracellular ROS generation, inhibit the peroxidation of lipids, and increase GSH levels and SOD activity [25].

Mitochondria are a major cellular organelle that play a key role in aging and degenerative diseases and are a target for oxidative damage. Mitochondria are the main source of ROS, particularly of superoxide radicals, through the complexes I and III of the respiratory chain [26,27]. An overproduction of mitochondria ROS may alter membrane permeability and calcium homeostasis as well as induce DNA mutations and injure the mitochondrial respiratory chain [28]. In our study, methanol lichen extracts of *V. pinastri* and *D. arctica* prevented mitochondrial changes by regulating calcium homeostasis and increasing mitochondrial membrane potential, suggesting a protective activity against H_2O_2. These extracts which target mitochondria are of great interest because they can pass across the mitochondrial phospholipid bilayer and reduce ROS damage at the heart of the source [29]. Other studies demonstrated that the depsidone fumarprotocetraric isolated from *Cetraria islandica* prevented mitochondrial membrane potential dissipation and mitochondrial calcium increase [9].

The analytical study by HPLC-UV revealed that the major compounds presented in *V. pinastri* were usnic acid, pinastric acid and vulpinic acid, and in *D. arctica* were gyrophoric acid, lecanoric acid and usnic acid, while in *T. americana* it was alectoronic acid. Lichen compounds are biosynthesized through three pathways: via the acetylpolymanolate pathway which produces depsides, depsidones and dibenzofurans, the shikimic pathway which produces pulvinic acids and the mevalonic acid pathway which is involved in terpenes formation. Therefore, gyrophoric acid and lecanoric acid are depsides, usnic acid is a dibenzofuran, alectoronic acid is a depsidone and vulpinic acid and pinastric acid are pulvinic acids [30,31]. All these lichen compounds have shown a great diversity of activities including anti-cancer (i.e., gyrophoric acid, vulpinic acid), antimicrobial (i.e., gyrophoric acid, usnic acid, vulpinic acid) photoprotective (i.e., gyrophoric acid) and neuroprotective (i.e., usnic acid) activities [32–37].

Based on the chemical structure of lichen compounds and its potential antioxidant activity, the depsides gyrophoric acid and lecanoric acid have carboxyl and hydroxyl groups that interact with several enzymatic active sites. Moreover, the aromatic rings of gyrophoric acid and lecanoric acid are responsible for their free radical scavenging properties [38,39]. Among the lichens investigated in this study, *D. artica* was the most active specie. Previous works have shown that *D. artica* has potent antioxidant properties (ORAC value 8.2 µmol TE/mg dry extract, DPPH value IC_{50} 346.3 µg/mL and FRAP value 29.6 µmol of Fe^{2+} eq/g sample) which are attributed to the anti-free radical properties of gyrophoric acid and lecanoric acid [10].

On the other hand, the antioxidant properties of *V. pinastri* are mainly due to the presence of vulpinic acid and pinastric acid. These pulvinic acids have a butanolide ring with an -OH group at the 4-position and a carboxylic acid function at the double bond. This double bond is involved in radical stabilization, and is a good descriptor of antioxidant properties [17].The antioxidant activity of pulvinic acids has been demonstrated using quantitative structure–activity relationship (QSAR) techniques combined with a multivariate analysis [40,41].

Regarding lichen compounds with a depsidone structure, previous studies revealed that they are potent hydroxyl and superoxide anion radical scavengers in polar environments but not good peroxyl radical scavengers [15]. Moreover, a better hydrogen-donating potency in those depsidones with no butyrolactone ring has been reported; this is the case of alectoronic acid, which has been identified in *T. americana* [42]. Finally, the compound usnic acid, presented in *D. arctica* and *V. pinastri*, has shown reducing potential in DPPH, ABTS and DMPD radical cation assays, and superoxide radical and peroxyl scavenging abilities [16,37]. In addition to this, the presence of a phenolic ring with functional groups of −CO, −COH and −COOH showed metal chelating ability, including Fe^{2+} ion [16].

4. Materials and Methods

4.1. Reagents

All reagents were acquired from Sigma-Aldrich (St. Louis, MO, USA) except for HPLC grade methanol and dimethyl sulfoxide (DMSO), that were purchased from Panreac (Barcelona, Spain). Molecular probes were obtained from Invitrogen-Thermo Fisher Scientific (Carlsbad, CA, USA).

4.2. Lichen Collection and Preparation of Methanol Extracts

The lichens *Dactylina arctica* (Central Siberia, Russia, July 1995; MAF-Lich 96262), *Nephromopsis stracheyi* (North Sikkim, India, August 2004; MAF-Lich 22748), *Tuckermannopsis americana* (Maine, USA, June 2010; MAF-Lich 19828) and *Vulpicida pinastri* (Alto del Peñón, Zamora, Spain, September 2017; MAF-Lich 22753) were identified and authenticated by Dr. P.K. Divakar and Professor A. Crespo and preserved in the Herbarium of the Faculty of Pharmacy (MAF), University Complutense of Madrid (Spain).

For the preparation of extracts, 2 mL of pure methanol was mixed with dry thalli samples (50 mg) and after being shaken for 20 s, every 15 min for 2 h, was left overnight. Methanol extracts were filtered (0.45 μm pore) and evaporated at room temperature. Dry residues were stored until their use.

4.3. Human Neuroblastoma Cell Line (SH-SY5Y Cells)

SH-SY5Y cells were grown in DMEM supplemented with 10% fetal bovine serum and 0.5% gentamicin at 37 °C and 5% CO_2/95% air. Confluence was between 80 and 90%.

4.4. Cell Treatments

Lichen extracts were dissolved in DMSO and PBS (1 mg/mL) as stock. Serial dilutions were then made with PBS. SH-SY5Y cells were pretreated with different concentrations of methanol lichen extracts for 24 h, before H_2O_2 (250 μM, 1 h). Final DMSO concentration was lower than 0.1% at the highest concentration.

4.5. Metabolic Activity Measurement

Survival rate and cytoprotection were determined using an MTT assay according to the method described by Mosmann [43] with some modifications. After treatments, a solution of MTT (2 mg/mL, 100 μL) was added to wells, and plates were incubated for 1 h. Then, the medium was removed, and formazan crystals were dissolved with DMSO (100 μL). Absorbance was measured at 550 nm with a Spectrostar BMG microplate reader.

4.6. Intracellular ROS Production

Intracellular ROS production was determined using a DCFH-DA assay as described by LeBel et al. (1992) [44]. Briefly, DCFH-DA dissolved in DMEM medium (1%) without phenol red was added to 96-well plates for 30 min. This solution was then removed, and cells were treated with non-cytotoxic lichen concentrations for 24 h before hydrogen peroxide (250 μM). Fluorescence was measured with a microplate reader (FLUOstar OPTIMA, BMG Labtech, Ortenberg, Germany) at excitation/emission wavelength 485/528 nm.

4.7. BCA Assay

The protein concentration was calculated using a bicinchoninic acid (BCA) assay. The colorimetric reaction was measured at 550 nm in a Spectrostar microplate reader (BMG Labtech, Ortenberg, Germany). Samples of total cellular extracts were mixed with a reaction solution [bicinchoninic acid and copper (II) sulfate]. The purple color proportionally increased with the amount of protein. A bovine serum albumin (BSA) curve was used to normalize protein [45].

4.8. Glutathione Levels

GSH content was determined following the Hissin and Hilf (1976) method with some modifications [46]. In 96-well plates, phosphate-EDTA buffer (pH 8.0, 150 µL) was mixed with total extract samples (50 µL). Then, O-phthalaldehyde (OPT) was added (20 µL) and samples were incubated for 15 min in the dark. Fluorescence was measured at an excitation/emission wavelength 360/460 nm. A standard curve of reduced GSH was used.

4.9. Antioxidant Enzymatic Activity

4.9.1. SOD Enzymatic Activity

Superoxide dismutase catalyzes the conversion of superoxide anions into oxygen and hydrogen peroxide. In 96-well plates, it was added to the reaction mixture consisting of total cellular extracts, EDTA, buffer phosphate (pH 7.8, with 0.2% Triton X-100), hydroxylamine chlorohydrate, and nitroblue tetrazolium (NBT). NBT reduction was measured at 530 nm each minute, during 15 min, using a SPECTROstar Omega microplate reader (BMG Labtech, Ortenberg, Germany) [47].

4.9.2. CAT Activity

In 96-well plates, total cell extracts were mixed with hydrogen peroxide (14 mM). Absorbance was measured at 240 nm wavelength for 1 min using a SPECTROstar Omega microplate reader (BMG Labtech, Ortenberg, Germany) following Aebi et al.'s method with some modifications [48].

4.10. TBARS Assay

Lipid peroxidation was determined by performing a TBARS assay [49]. After treatments, cell pellets were stored at 80 °C. On the day of the experiment, pellets were defrosted at room temperature and were mixed with TBA-TCA-HCl. This mixture was boiled at 100 °C for 10 min. Samples were placed on ice to stop the reaction. Then, samples were centrifuged at 4 °C (3000 rpm, 10 min) and supernatants were added into 96-well plates to measure absorbance at 530 nm using a SPECTROstar Omega microplate reader (BMG Labtech, Ortenberg, Germany). Results were expressed as a percentage of TBARS (100% of control).

4.11. Calcium Cytosolic Quantification

Calcium cytosolic was quantified using Indo-1/AM as a cell-permeant dye [50]. After treatments, a Krebs medium containing Indo-1/AM dye (3 mM) and calcium (1mM $CaCl_2$) was added to cells for 45 min at 37 °C. Then, the medium was removed, and cells were incubated with a dye-free Krebs medium for 15 min at 37 °C in the dark. Fluorescence was recorded in a microplate reader (FLUOstar OPTIMA, BMG Labtech, Ortenberg, Germany) at 350 nm excitation wavelength and at 410 nm emission wavelength. The formula for cytosolic calcium concentration was $[Ca^{2+}]_i = Kd \times [F - Fmin]/[Fmax - F]$, where Kd is the dissociation constant for Indo-1; F is the fluorescence signal for samples; Fmax is the maximum fluorescence signal after ionomycin addition and *Fmin* is calculated using this formula: $Fmin = AF + 1/12 \times (Fmax\text{-}AF)$, AF being the minimum fluorescence after adding $MnCl_2$.

4.12. Mitochondrial Calcium Quantification

Mitochondrial calcium was quantified using Rhod-2/AM as a fluorescent cationic probe [51]. After treatments, cells were incubated in a Krebs medium containing 0.1% BSA, 1 mM calcium and 10 mM Rhod-2/AM during 40 min at 37 °C. Then, cells were maintained in dye-free Krebs medium cells in this medium for 30 min at 37 °C in the dark. Basal fluorescence intensity was recorded using a microplate reader (FLUOstar OPTIMA, BMG Labtech, Ortenberg, Germany) at λ 552 nm excitation wavelength and λ 581 nm emission wavelength for 5 min at 37 °C. Then, maximum fluorescence was measured for 15 min after adding calcium ionophore A23187 (5 µM) in the same conditions as described above.

Mitochondrial calcium levels were the ratio between fluorescence measures before and after ionophore addition. Results were expressed relative to control.

4.13. Mitochondrial Membrane Potential (MMP)

Mitochondrial membrane potential was quantified using the fluorescent cationic dye tetramethylrhodaminemethylester (TMRM) following Correia et al.'s (2012) protocol with some modifications [52]. After cell treatments, the Krebs medium with calcium (1 mM $CaCl_2$) and TMRM (250 nM) was added. A FLUOstar OPTIMA (BMG Labtech, Ortenberg, Germany) microplate reader was used to measure basal fluorescence activity at λ 549 nm excitation and λ 573 nm emission at 37 °C for 45 min. Then, the maximum fluorescence value was estimated after adding FCCP (6 mM) and oligomycin (0.25 mg/mL). Mitochondrial membrane potential ($\Delta\psi m$) is the result by subtracting fluorescence basal values from maximum fluorescence values. Results were expressed as % of the control.

4.14. Secondary Metabolites Detection Using High Performance Liquid Chromatography

HPLC analysis was used for secondary metabolites detection using the method described by de Paz et al. (2010) [53]. The HPLC instrumentation was an Agilent 1260 instrument (Agilent Technologies, CA, USA) equipped with a photodiode array detector (190–800 nm) and a reversed-phase Mediterranean Sea 18 column (150 mm × 4.6 mm, 3 μm particle size; Teknokroma, Barcelona, Spain). Running conditions included: a mobile gradient phase [1% orthophosphoric acid in milli-Q water (A)/methanol (B)]; a flow rate of 0.6 mL/min; a column temperature of 40 °C and a UV spectrum between 190 and 400 nm. Secondary metabolites of lichens were identified by comparing the retention time and UV absorption spectra with standard compounds (commercialized and isolated previously by our research team) and other lichen species [9,54].

4.15. Statistical Analysis

All assays were measured in triplicate and data were expressed as mean ± SD. Statistical analysis was performed by SigmaPlot 11.0 using analysis of variance (ANOVA) and Tukey's post hoc test (5% significance level).

5. Conclusions

In conclusion, our findings provide evidence of the protective activity of methanol extracts obtained from cetrarioid clade against the neurotoxic effects of hydrogen peroxide in neuroblastoma cells. *D. arctica* and *V. pinastri* afford the highest protective effect. Future research should be aimed at studying the protective activity of isolated compounds from *D. arctica* and *V. pinastri*, delving into their mechanism of action.

Author Contributions: Conceptualization, I.U.-V., E.G.-B. and M.P.G.-S.; Methodology: I.U.-V.; Investigation, I.U.-V. and E.G.-B.; Formal analysis, I.U.-V. and E.G.-B.; Data Curation, I.U.-V. and E.G.-B.; Writing—Original Draft Preparation, I.U.-V. and E.G.-B.; Validation, E.G.-B.; Writing—Review & Editing, P.K.D. and M.P.G.-S., Supervision, E.G.-B. and M.P.G.-S., Project administration. P.K.D. and M.P.G.-S. All authors have read and agreed to the published version of the manuscript.

Funding: This study was supported by the Spanish Ministry of Science, Innovation and Universities (PID2019-105312GB-100) and the Santander-University Complutense of Madrid (PR87/19-22637). I. Ureña Vacas was supported by a grant CT42/18-CT43/18 from Complutense University of Madrid for predoctoral research.

Institutional Review Board Statement: Not applicable.

Informed Consent Statement: Not applicable.

Data Availability Statement: Not applicable.

Conflicts of Interest: The authors declare that they have no conflict of interest.

Sample Availability: Samples are not available from the authors.

References

1. Salim, S. Oxidative Stress and the Central Nervous System. *J. Pharmacol. Exp. Ther.* **2017**, *360*, 201–205. [CrossRef] [PubMed]
2. Cobley, J.N.; Fiorello, M.L.; Bailey, D.M. 13 reasons why the brain is susceptible to oxidative stress. *Redox Biol.* **2018**, *15*, 490–503. [CrossRef] [PubMed]
3. Huang, W.J.; Zhang, X.; Chen, W.W. Role of oxidative stress in Alzheimer's disease. *Biomed. Rep.* **2016**, *4*, 519–522. [CrossRef]
4. Gottfredsen, R.H.; Larsen, U.G.; Enghild, J.J.; Petersen, S.V. Hydrogen peroxide induce modifications of human extracellular superoxide dismutase that results in enzyme inhibition. *Redox Biol.* **2013**, *1*, 24–31. [CrossRef] [PubMed]
5. Sies, H. Hydrogen peroxide as a central redox signaling molecule in physiological oxidative stress: Oxidative eustress. *Redox Biol.* **2017**, *11*, 613–619. [CrossRef]
6. Zhao, Z. Iron and oxidizing species in oxidative stress and Alzheimer's disease. *Aging Med.* **2019**, *2*, 82–87. [CrossRef] [PubMed]
7. Huang, Z. Bcl-2 family proteins as targets for anticancer drug design. *Oncogene* **2000**, *19*, 6627–6631. [CrossRef] [PubMed]
8. Lü, J.M.; Lin, P.H.; Yao, Q.; Chen, C. Chemical and molecular mechanisms of antioxidants: Experimental approaches and model systems. *J. Cell Mol. Med.* **2010**, *14*, 840–860. [CrossRef]
9. Fernández-Moriano, C.; Divakar, P.K.; Crespo, A.; Gómez-Serranillos, M.P. Neuroprotective activity and cytotoxic potential of two Parmeliaceae lichens: Identification of active compounds. *Phytomedicine* **2015**, *22*, 847–855. [CrossRef]
10. Ureña-Vacas, I.; González-Burgos, E.; De Vita, S.; Divakar, P.K.; Bifulco, G.; Gómez-Serranillos, M.P. Phytochemical Characterization and Pharmacological Properties of Lichen Extracts from Cetrarioid Clade by Multivariate Analysis and Molecular Docking. *Evid.-Based Complement. Altern. Med.* **2022**, *2022*, 5218248. [CrossRef]
11. Gilgun-Sherki, Y.; Melamed, E.; Offen, D. Oxidative stress induced-neurodegenerative diseases: The need for antioxidants that penetrate the blood brain barrier. *Neuropharmacology* **2001**, *40*, 959–975. [CrossRef]
12. Ryter, S.W.; Kim, H.P.; Hoetzel, A.; Park, J.W.; Nakahira, K.; Wang, X.; Choi, A.M.K. Mechanisms of Cell Death in Oxidative Stress. *Antioxid. Redox Signal.* **2006**, *9*, 49–89. [CrossRef] [PubMed]
13. Kim, G.H.; Kim, J.E.; Rhie, S.J.; Yoon, S. The Role of Oxidative Stress in Neurodegenerative Diseases. *Exp. Neurobiol.* **2015**, *24*, 325–340. [CrossRef] [PubMed]
14. Ofoedu, C.E.; You, L.; Osuji, C.M.; Iwouno, J.O.; Kabuo, N.O.; Ojukwu, M.; Agunwah, I.M.; Chacha, J.S.; Muobike, O.P.; Agunbiade, A.O.; et al. Hydrogen Peroxide Effects on Natural-Sourced Polysacchrides: Free Radical Formation/Production, Degradation Process, and Reaction Mechanism-A Critical Synopsis. *Foods* **2021**, *10*, 699. [CrossRef]
15. Bay, M.V.; Nam, P.C.; Quang, D.T.; Mechler, A.; Hien, N.K.; Hoa, N.T.; Vo, Q.V. Theoretical Study on the Antioxidant Activity of Natural Depsidones. *ACS Omega* **2020**, *5*, 7895–7902. [CrossRef]
16. Hoa, N.T.; Van Bay, M.; Mechler, A.; Vo, Q.V. Is Usnic Acid a Promising Radical Scavenger? *ACS Omega* **2020**, *5*, 17715–17720. [CrossRef]
17. Le Roux, A.; Kuzmanovski, I.; Habrant, D.; Meunier, S.; Bischoff, P.; Nadal, B.; Thetiot-Laurent, S.A.L.; Le Gall, T.; Wagner, A.; Novič, M. Design and Synthesis of New Antioxidants Predicted by the Model Developed on a Set of Pulvinic Acid Derivatives. *J. Chem. Inf. Model.* **2011**, *51*, 3050–3059. [CrossRef]
18. Angelova, P.R.; Esteras, N.; Abramov, A.Y. Mitochondria and lipid peroxidation in the mechanism of neurodegeneration: Finding ways for prevention. *Med. Res. Rev.* **2021**, *41*, 770–784. [CrossRef]
19. Lin, M.C.; Liu, C.C.; Lin, Y.C.; Liao, C.S. Resveratrol Protects against Cerebral Ischemic Injury via Restraining Lipid Peroxidation, Transition Elements, and Toxic Metal Levels, but Enhancing Anti-Oxidant Activity. *Antioxidants* **2021**, *10*, 1515. [CrossRef]
20. Yoshida, Y.; Umeno, A.; Shichiri, M. Lipid peroxidation biomarkers for evaluating oxidative stress and assessing antioxidant capacity in vivo. *J. Clin. Biochem. Nutr.* **2013**, *52*, 9–16. [CrossRef]
21. Leong, P.K.; Ko, P.M. Induction of the Glutathione Antioxidant Response/Glutathione Redox Cycling by Nutraceuticals: Mechanism of Protection against Oxidant-induced Cell Death. *J. Nutraceuticals Food Sci.* **2016**, *1*, 1.
22. Salminen, L.E.; Paul, R.H. Oxidative stress and genetic markers of suboptimal antioxidant defense in the aging brain: A theoretical review. *Rev. Neurosci.* **2014**, *25*, 805–819. [CrossRef] [PubMed]
23. Fernández-Moriano, C.; Divakar, P.K.; Crespo, A.; Gómez-Serranillos, M.P. In vitro neuroprotective potential of lichen metabolite fumarprotocetraric acid via intracellular redox modulation. *Toxicol. Appl. Pharmacol.* **2017**, *316*, 83–94. [CrossRef] [PubMed]
24. Fernández-Moriano, C.; Divakar, P.K.; Crespo, A.; Gómez-Serranillos, M.P. Protective effects of lichen metabolites evernic and usnic acids against redox impairment-mediated cytotoxicity in central nervous system-like cells. *Food Chem. Toxicol.* **2017**, *105*, 262–277. [CrossRef] [PubMed]
25. Sieteiglesias, V.; González-Burgos, E.; Bermejo-Bescós, P.; Divakar, P.K.; Gómez-Serranillos, M.P. Lichens of Parmelioid Clade as Promising Multitarget Neuroprotective Agents. *Chem. Res. Toxicol.* **2019**, *32*, 1165–1177. [CrossRef]
26. Lenaz, G. The mitochondrial production of reactive oxygen species: Mechanisms and implications in human pathology. *IUBMB Life* **2001**, *52*, 159–164. [CrossRef]
27. Murphy, M.P. How mitochondria produce reactive oxygen species. *Biochem. J.* **2009**, *417*, 1–13. [CrossRef]
28. Guo, C.; Sun, L.; Chen, X.; Zhang, D. Oxidative stress, mitochondrial damage and neurodegenerative diseases. *Neural Regen. Res.* **2013**, *8*, 2003–2014. [CrossRef]
29. Oyewole, A.O.; Birch-Machin, M.A. Mitochondria-targeted antioxidants. *FASEB J.* **2015**, *29*, 4766–4771. [CrossRef]

30. Goga, M.; Elečko, J.; Marcinčinová, M.; Ručová, D.; Bačkorová, M.; Backor, M. Lichen Metabolites: An Overview of Some Secondary Metabolites and Their Biological Potential. In *Co-Evolution of Secondary Metabolites*; Springer: Berlin/Heidelberg, Germany, 2018; pp. 1–36.
31. Calcott, M.J.; Ackerley, D.F.; Knight, A.; Keyzers, R.A.; Owen, J.G. Secondary metabolism in the lichen symbiosis. *Chem. Soc. Rev.* **2018**, *47*, 1730–1760. [CrossRef]
32. Lauterwein, M.; Oethinger, M.; Belsner, K.; Peters, T.; Marre, R. In vitro activities of the lichen secondary metabolites vulpinic acid, (+)-usnic acid, and (−)-usnic acid against aerobic and anaerobic microorganisms. *Antimicrob. Agents Chemother.* **1995**, *39*, 2541–2543. [CrossRef] [PubMed]
33. Candan, M.; Yilmaz, M.; Tay, T.; Erdem, M.; Türk, A.O. Antimicrobial activity of extracts of the lichen Parmelia sulcata and its salazinic acid constituent. *Z. Nat. C J. Biosci.* **2007**, *62*, 619–621. [CrossRef] [PubMed]
34. Bačkorová, M.; Bačkor, M.; Mikeš, J.; Jendželovský, R.; Fedoročko, P. Variable responses of different human cancer cells to the lichen compounds parietin, atranorin, usnic acid and gyrophoric acid. *Toxicol. Vitr.* **2011**, *25*, 37–44. [CrossRef] [PubMed]
35. Koparal, A.T. Anti-angiogenic and antiproliferative properties of the lichen substances (−)-usnic acid and vulpinic acid. *Z. Nat. C J. Biosci.* **2015**, *70*, 159–164. [CrossRef]
36. Varol, M.; Türk, A.; Candan, M.; Tay, T.; Koparal, A.T. Photoprotective Activity of Vulpinic and Gyrophoric Acids toward Ultraviolet B-Induced Damage in Human Keratinocytes. *Phytother. Res.* **2016**, *30*, 9–15. [CrossRef]
37. Cetin Cakmak, K.; Gulçin, I. Anticholinergic and antioxidant activities of usnic acid-an activity-structure insight. *Toxicol. Rep.* **2019**, *6*, 1273–1280. [CrossRef]
38. Manojlovic, N.T.; Vasiljevic, P.J.; Maskovic, P.Z.; Juskovic, M.; Bogdanovic-Dusanovic, G. Chemical Composition, Antioxidant, and Antimicrobial Activities of Lichen *Umbilicaria cylindrica* (L.) Delise (Umbilicariaceae). *Evid.-Based Complement. Alternat. Med.* **2012**, *2012*, 452431. [CrossRef]
39. Mohammadi, M.; Bagheri, L.; Badreldin, A.; Fatehi, P.; Pakzad, L.; Suntres, Z.; van Wijnen, A.J. Biological Effects of Gyrophoric Acid and Other Lichen Derived Metabolites, on Cell Proliferation, Apoptosis and Cell Signaling pathways. *Chem. Biol. Interact.* **2022**, *351*, 109768. [CrossRef]
40. Martinčič, R.; Kuzmanovski, I.; Wagner, A.; Novič, M. Development of models for prediction of the antioxidant activity of derivatives of natural compounds. *Anal. Chim. Acta* **2015**, *868*, 23–35. [CrossRef]
41. Ahmadi, S.; Ghanbari, H.; Lotfi, S.; Azimi, N. Predictive QSAR modeling for the antioxidant activity of natural compounds derivatives based on Monte Carlo method. *Mol. Divers.* **2021**, *25*, 87–97. [CrossRef]
42. Lohézic-Le Dévéhat, F.; Tomasi, S.; Elix, J.A.; Bernard, A.; Rouaud, I.; Uriac, P.; Boustie, J. Stictic acid derivatives from the lichen Usnea articulata and their antioxidant activities. *J. Nat. Prod.* **2007**, *70*, 1218–1220. [CrossRef] [PubMed]
43. Mosmann, T. Rapid colorimetric assay for cellular growth and survival: Application to proliferation and cytotoxicity assays. *J. Immunol. Methods* **1983**, *65*, 55–63. [CrossRef]
44. LeBel, C.P.; Ischiropoulos, H.; Bondy, S.C. Evaluation of the probe 2′,7′-dichlorofluorescin as an indicator of reactive oxygen species formation and oxidative stress. *Chem. Res. Toxicol.* **1992**, *5*, 227–231. [CrossRef] [PubMed]
45. Smith, P.K.; Krohn, R.I.; Hermanson, G.T.; Mallia, A.K.; Gartner, F.H.; Provenzano, M.D.; Fujimoto, E.K.; Goeke, N.M.; Olson, B.J.; Klenk, D.C. Measurement of protein using bicinchoninic acid. *Anal. Biochem.* **1985**, *150*, 76–85. [CrossRef]
46. Hissin, P.J.; Hilf, R. A fluorometric method for determination of oxidized and reduced glutathione in tissues. *Anal. Biochem.* **1976**, *74*, 214–226. [CrossRef]
47. Beauchamp, C.; Fridovich, I. Superoxide dismutase: Improved assays and an assay applicable to acrylamide gels. *Anal. Biochem.* **1971**, *44*, 276–287. [CrossRef]
48. Aebi, H. Catalase in vitro. *Methods Enzym.* **1984**, *105*, 121–126. [CrossRef]
49. Mihara, M.; Uchiyama, M. Determination of malonaldehyde precursor in tissues by thiobarbituric acid test. *Anal. Biochem.* **1978**, *86*, 271–278. [CrossRef]
50. Resende, R.; Ferreiro, E.; Pereira, C.; Resende de Oliveira, C. Neurotoxic effect of oligomeric and fibrillar species of amyloid-beta peptide 1–42: Involvement of endoplasmic reticulum calcium release in oligomer-induced cell death. *Neuroscience* **2008**, *155*, 725–737. [CrossRef]
51. Takahashi, A.; Camacho, P.; Lechleiter, J.D.; Herman, B. Measurement of intracellular calcium. *Physiol. Rev.* **1999**, *79*, 1089–1125. [CrossRef]
52. Correia, S.C.; Santos, R.X.; Cardoso, S.M.; Santos, M.S.; Oliveira, C.R.; Moreira, P.I. Cyanide preconditioning protects brain endothelial and NT2 neuron-like cells against glucotoxicity: Role of mitochondrial reactive oxygen species and HIF-1α. *Neurobiol. Dis.* **2012**, *45*, 206–218. [CrossRef] [PubMed]
53. de Paz, G.A.; Raggio, J.; Gómez-Serranillos, M.P.; Palomino, O.M.; González-Burgos, E.; Carretero, M.E.; Crespo, A. HPLC isolation of antioxidant constituents from *Xanthoparmelia* spp. *J. Pharm. Biomed. Anal.* **2010**, *53*, 165–171. [CrossRef] [PubMed]
54. Yoshimura, I.; Kinoshita, Y.; Yamamoto, Y.; Huneck, S.; Yamada, Y. Analysis of secondary metabolites from lichen by high performance liquid chromatography with a photodiode array detector. *Phytochem. Anal.* **1994**, *5*, 197–205. [CrossRef]

molecules

MDPI

Review

Grape Polyphenols in the Treatment of Human Skeletal Muscle Damage Due to Inflammation and Oxidative Stress during Obesity and Aging: Early Outcomes and Promises

Adriana Capozzi [1,2], Cédric Saucier [2], Catherine Bisbal [1,*] and Karen Lambert [1,*]

1 PhyMedExp, INSERM U1046, CNRS UMR 9214, University of Montpellier, CEDEX 5, 34295 Montpellier, France
2 SPO, INRAE, Institute Agro, University of Montpellier, 34000 Montpellier, France
* Correspondence: catherine.bisbal@inserm.fr (C.B.); karen.lambert-cordillac@umontpellier.fr (K.L.); Tel.: +33-(0)4-1175-9891 (C.B. & K.L.)

Abstract: Today, inactivity and high-calorie diets contribute to the development of obesity and premature aging. In addition, the population of elderly people is growing due to improvements in healthcare management. Obesity and aging are together key risk factors for non-communicable diseases associated with several co-morbidities and increased mortality, with a major impact on skeletal muscle defect and/or poor muscle mass quality. Skeletal muscles contribute to multiple body functions and play a vital role throughout the day, in all our activities. In our society, limiting skeletal muscle deterioration, frailty and dependence is not only a major public health challenge but also a major socio-economic issue. Specific diet supplementation with natural chemical compounds such as grape polyphenols had shown to play a relevant and direct role in regulating metabolic and molecular pathways involved in the prevention and treatment of obesity and aging and their related muscle comorbidities in cell culture and animal studies. However, clinical studies aiming to restore skeletal muscle mass and function with nutritional grape polyphenols supplementation are still very scarce. There is an urgent need for clinical studies to validate the very encouraging results observed in animal models.

Keywords: grape polyphenols; resveratrol; sarcopenia; skeletal muscle; clinical trial

Citation: Capozzi, A.; Saucier, C.; Bisbal, C.; Lambert, K. Grape Polyphenols in the Treatment of Human Skeletal Muscle Damage Due to Inflammation and Oxidative Stress during Obesity and Aging: Early Outcomes and Promises. *Molecules* **2022**, 27, 6594. https://doi.org/10.3390/molecules27196594

Academic Editor: Nour Eddine Es-Safi

Received: 3 August 2022
Accepted: 28 September 2022
Published: 5 October 2022

1. Obesity and Aging: Two Major Healthcare Challenges to Solve

The population of elderly people is expanding worldwide with the older adults aged between 65–80 years being the fastest-growing portion, thanks to improvements in healthcare management which allows for increasing life expectancy [1]. Associated with this increase in lifespan, a global obesity epidemic is spreading due to life changes such as inactivity and high-calorie diets, favoring the growth of non-communicable diseases. Obesity and aging are together key risk factors for the development and progression of several chronic/non-communicable diseases (metabolic syndrome, Insulin Resistance (IR), Type 2 Diabetes (T2D) [2–7], sarcopenia [8,9], and frailty [10]. The World Health Organization (WHO) defines metabolic syndrome as a pathologic condition characterized by obesity (Body Mass Index (BMI) $\geq$30 kg/m^2), IR, hypertension, and hyperlipidemia [11]). T2D is IR associated with decreased insulin secretion by the pancreas. [12]. Frailty is a clinical syndrome in elderly people comprising an increased risk for poor health outcomes, falls, incident disability, hospitalization, and mortality [13].

1.1. Skeletal Muscle Alterations Are Central

Although non-communicable diseases associated with aging and obesity have different etiologies, development, and progression, they are all associated with skeletal muscle defect and/or poor muscle mass quality. Sarcopenia, defined as a loss of skeletal muscle mass and function [8], is frequently associated with aging and with a loss of independence, disability, frailty, and compromised quality of life, and, therefore, represents a high risk for morbidity and mortality [9]. Obese patients could also develop sarcopenia [7] and with the progression of obesity with aging [14], a growing number of obese sarcopenic patients is expected. Then, the management of skeletal muscle alteration during obesity and aging is mandatory.

Skeletal muscles are among the major tissues of the body, accounting for 40% of our total body weight and containing 50–75% of all body proteins. Moreover, skeletal muscles are responsible for more than 80% of glucose uptake after insulin stimulation, highlighting their central role in metabolism regulation. They ensure three main functions: posture and locomotion, thermoregulation, storage, and utilization of nutrients. Thus, they play a vital role in all our activities [15]. Growing evidence points to the central role of skeletal muscle in the systemic regulation of age-related diseases [16]. Indeed, functional and metabolic muscle alterations, as well as skeletal muscle mass decreasing, are all associated with the human mortality rate [17,18]. Then, impairment of skeletal muscle mass and/or function could lead to major pathologies, such as IR, T2D, and to weakness and disability which considerably decrease quality of life and are associated with a higher risk for morbidity and mortality [5,7,19,20].

1.2. Muscle Alterations

1.2.1. Muscle Alterations in Obesity

Obesity exerts multiple effects on skeletal muscle metabolism. In obese grade I insulin-resistant women, only skeletal muscle insulin-signaling alteration was found with no variation in subcutaneous adipose tissue [21]. Moreover, the correlation of muscle alteration with glucose infusion rate during the hyperinsulinemic-euglycemic clamp underlines the major role of skeletal muscle in IR development [21]. Obesity is accompanied by increased ectopically lipid deposition in non-adipose tissues including skeletal muscle [22,23]. This lipid overload affects several cell signaling pathways and is associated with metabolic effects on mitochondrial function, insulin response, and energetic metabolism [24,25] (Figure 1). After their cell translocation by fatty acid translocase (FAT/CD36), a receptor and transporter for free fatty acid (FFA), lipids are stored as intramuscular triacylglycerols (TAG) in lipid droplets (IMTG). Myotubes from obese people present increased FFA uptake and esterification into complex lipids with overexpression of FAT/CD36 [23,26,27]. TAG turnover is also highly altered due to decreased lipolysis as lower hormone-sensitive lipase (HSL) and higher adipose triglyceride lipase (ATGL) protein levels are detected in the skeletal muscle of obese subjects [28]. Increased IMTG is also associated with higher levels of lipotoxic intermediates such as diacylglycerols (DAG) and ceramides. DAG might inhibit insulin signaling via the activation of the protein kinase C (PKC), which, in turn, decreases the activities of the phosphoinositide 3-kinase (PI3K) and of the insulin receptor substrate-1 (IRS1) in the insulin signaling pathway [29]. Ceramides have several functions in the alteration of skeletal muscle metabolism. They have been described as inhibitors of Akt phosphorylation via the activation of protein phosphatase 2A (PP2A) or, on the opposite, as inductors of Akt phosphorylation on an inhibitory residue via the activation of the protein kinase Cζ (PKCζ). Ceramides act also at the mitochondrial level by decreasing mitochondrial respiration, inhibiting oxidative phosphorylation, and promoting mitochondrial fragmentation. These activities lead to an induction of reactive oxidative species (ROSP) [30]. Decreased mitochondrial content in muscle of obese patients has also been attributed to impairment of mitochondrial biogenesis, decreasing the ability to oxidize lipids [31]. More recently, it was shown that mitochondrial lipid oxidation was impaired due to a decreased activity of the mitochondrial protein carnitine palmitoyltransferase 1 (CPT1). CPT1 is involved in the transport of long-chain fatty acids into the mitochondria. Then decreased CPT1 activity results in decreased β oxidation [32]. Circulating FFA could also induce a chronic low-grade inflammation through activation of toll-like receptor 4 (TLR4) and nuclear factor-kappa B (NFκB), resulting in the release of several pro-inflammatory cytokines as interleukines (IL)

6, 8, and 15 and tumor necrosis factor (TNF) α. These cytokines, as secreted by muscle, are also known as myokines. They could exert an autocrine or paracrine effect. However, it must be kept in mind that there is not a consensus on their secretion and their role in skeletal muscle inflammation and skeletal muscle metabolism dysregulation during obesity [21,33] Importantly, lipids overload also affects muscle maintenance and regeneration [34]. FFA accumulation decreases AMPKα activity. AMPKα promoted myogenesis by regulating the expression of miR-206 and miR-206's target cyclin D1, which allows the regulation of the cell cycle and cell proliferation of muscular stem cells (satellite cells) during the skeletal muscle regeneration process [35]. By affecting general skeletal muscle function, mass, and quality, obesity reduces mobility [36] modifies lipid and carbohydrate metabolism and increases the risk of several comorbidities [37]. In fact, obesity accelerates the aging process and decreases life expectancy. Forty-year-old obese non-smoker females lost 7.1 years and forty-year-old obese non-smoker males lost 5.8 years of life expectancy [38].

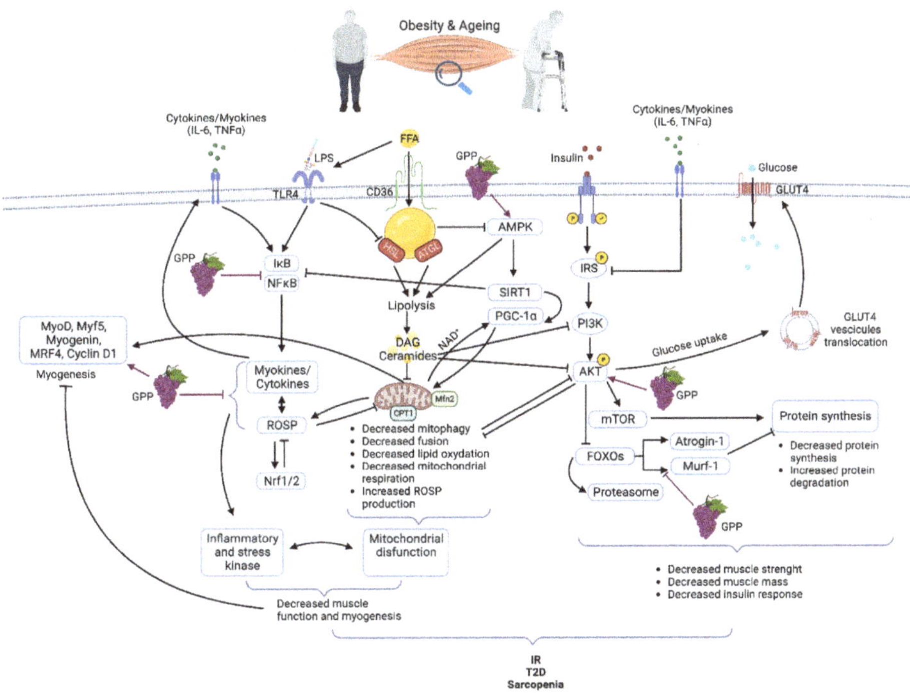

Figure 1. Main pathways involved in muscle dysfunction during obesity and aging and their modifications by grape polyphenols (GPP) (Created with BioRender.com).

1.2.2. Muscle Alterations in Aging

The gradual decline of all physiological functions of all organs including skeletal muscle characterizes normal aging. During aging, due to an alteration of muscle quality, the loss of muscle strength precedes the loss of muscle mass [39,40]. However, muscle mass is lost as early as the fifth decade with acceleration from the age of 70 and sarcopenia is a hallmark of the aging process. Like in obesity, the reduction of muscle quality and mass with aging is the consequence of the interplay of a multitude of mechanisms [41] including damage caused by ROSP [42] inflammation [43], lipid infiltration [44], proteostasis imbalance [45] and mitochondrial impairment [46]. In aging like in obesity, alteration in mitochondria content, fusion, fission, and function are associated with muscle loss and muscle lipid accumulation [46] (Figure 1). Progressive mitochondrial dysfunction and ROSP accumulation are central to the aging process [47]. Aged skeletal muscle accumulates dysfunctional mitochondria due to a defect in mitophagy, mitochondrial biogenesis, and dynamics. Reduced expression levels of genes such as nuclear respiratory factor 1/2 (Nrf1/2), AMP-activated protein kinase (AMPK), peroxisome proliferative activated receptor gamma, coactivator 1 alpha (PGC-1α), mitofusin 1 and 2 (Mfn1/2) causes a reduction of mitochondrial number, mitochondrial content, mtDNA copy number, and impairment in mitochondria morphology in the skeletal muscle. Throughout life, the accumulation of dysfunctional mitochondria producing high ROSP levels contributes to the establishment of oxidative stress and increased risk of IR and T2D. Moreover, these defective mitochondria are unable to sustain enough energy in the cells, which results in progressive functional decline and cell death. [48–50]. Enhancement of ectopic fat deposit during aging is also accompanied by a resulting heightened production of pro-inflammatory cytokines (IL-6, TNFα) associated with lipotoxicity and leading to an increased risk of IR and T2D [51]. Protein quality control pathways, autophagy, and proteasome activity decrease participate also in the skeletal muscle dysfunctions and the decreased muscle mass observed during aging [45,52]. Maintaining muscle mass is a balance between protein synthesis and protein degradation systems. Aged skeletal muscle shows a marked defect in the contraction-induced activation of the protein synthesis pathway phosphatidylinositol 3-kinase/protein kinase B/mammalian target of rapamycin (PI3K/Akt/mTOR). Concerning the proteasome pathway and the muscle-specific ubiquitin ligase muscle RING-finger protein-1 (MuRF1) and atrogin-1 mRNA, levels in aged muscle are increased or unchanged [53,54].

The aging process presents several common mechanisms for obesity. On the other side, obesity in elderly people accelerates the aging process. Obesity and aging both deregulate cell metabolism and create a vicious circle that precipitates the aging process and the development of associated comorbidities [41,55]. Several mechanisms operating at different levels of muscle physiology are implicated in the muscle defects observed in obesity and aging. However, oxidative stress and inflammation are central components at the onset of muscle defect regardless of its etiology [6,19,56–58].

2. Oxidative Stress and Inflammation: Two Essential Harms

Oxidative stress occurs when there is an imbalance between the production of ROSP in the cells and tissues and the antioxidant systems, which are responsible for their neutralization and removal [59]. Reactive oxidative species include derivatives of oxygen (Reactive Oxygen Species, ROS), nitrogen (Reactive Nitrogen Species, RNS), and sulfur (Reactive Sulfur Species, RSS), capable of oxidizing different substrates. The antioxidant defense system involves non-enzymatic scavengers provided by food as vitamins (trans retinol 2, vitamin A; ascorbic acid, vitamin C; α-tocopherol, vitamin E), carotenoids, polyphenols, and endogenous antioxidant enzymes superoxide dismutase (SOD), catalase, glutathione peroxidase (GPx), glutathione transferase (GST) (see [60,61] for complete reviews).

If an excess of ROSP can be extremely deleterious, their regular generation is necessary for the physiological maintenance of all the tissues of our body and among them, the skeletal muscle [62]. In fact, via oxidation of redox-sensitive protein-cysteine, ROSP act as second messengers. They are important signaling molecules regulating metabolism, cell growth and differentiation, cell repair, immunity, and so on [59,63]. ROSP production is thus beneficial and essential to cell and tissue function. As oxidative stress, inflammation is also essential for normal organ function. Inflammation is a protective biological response aimed at identifying and eliminating a threat. Inflammation could be triggered by infection or not and participates in the activation of the immune system. It is the first line of defense against pathogens, but it also allows for repairing cell damage and tissue injury [64,65]. ROSP production and inflammation have dual roles. They are activated during several physiological responses, and they play an essential role in cellular signaling and regulatory pathways. However, they must be tightly regulated to avoid the development of oxidative stress, tissue injury, and chronic inflammation, which are detrimental to normal cells and tissues. In fact, when ROSP are produced in excess or inadequately they can cause irreversible damage to cells by oxidizing plasma membranes, DNA, proteins, and lipids [66].

In the same way, uncontrolled inflammation could lead to chronic inflammation and inflammatory diseases that could affect absolutely all the organs as skeletal muscle [64,67]. Interestingly, ROSP and inflammation could regulate each other in a two-way reciprocal direction [51]. Both ROSP production and inflammation participate together in skeletal muscle tissue repair [52]. However, during obesity and aging, ROSP production and inflammation are increased while antioxidant systems and anti-inflammatory pathways decreased in skeletal muscle [21,52,53]. In obesity and aging, high plasma levels of free fatty acids (FFA) and increased concentrations of lipopolysaccharide (LPS) from gut microbiota (due to increased permeability of gut), bind to Toll-like Receptor 4 (TLR4) [68,69]. TLR4 is an innate immunity receptor present in the skeletal muscle that can activate NFκB and inflammation via the MyD88 pathway. Moreover, the E3 ubiquitin ligase RNF41 has been found to participate in the TLR4 inflammation pathway in the muscle of insulin-resistant grade I obese women [56]. An excess of FFA is also responsible for deleterious effects on mitochondria such as uncoupling of oxidative phosphorylation, energy failure, decreased clearance, decreased fission, and release of ROSP [70]. Mitochondrial dysfunction and inflammation/oxidative stress together are responsible for a decrease in myogenesis and muscle function [71]. In skeletal muscle, elevated ROSP levels concurrently inhibit anabolic pathways as PI3K/Akt/mTOR [72], contributing to muscle mass loss and atrophy [73,74], and activating several mechanisms of the catabolic pathways (Figure 1). Under physiological conditions, Akt phosphorylates and inhibits the Forkhead box O transcription factors (FOXOs), thus inhibiting the muscle-specific ubiquitin ligase MuRF1 and atrogin-1, of the ubiquitin-proteasome system. With the inhibition of Akt, atrogin-1, MuRF-1 and proteasome are activated resulting in protein degradation [75]. Activation of the cysteine proteases, calpain, and caspase 3, which play a key role in the initial breakdown of sarcomeres during atrophic conditions, also participates in protein breakdown and muscle atrophy. The expression of inflammatory myokines such as tumor necrosis factor-alpha (TNFα) and interleukin 6 (IL6) is induced due to an increase in the activity of the transcription factor nuclear factor kappa B (NFκB) (Figure 1). NFκB activity can be increased by ROSP but also by several other stimuli such as free fatty acids (FFA), advanced glycation products, and inflammatory cytokines induced by oxidative stress [19]. On the other way, high levels of ROSP activate the nuclear factor, erythroid 2-like 2 (Nrf2) pathway leading to increased transcription of genes coding for antioxidant proteins, and consequently inducing the antioxidant ROSP-fighting effects [76].

3. Grape Polyphenols: An Effective Tool

Adapted diet and physical activity are known for several years as central countermeasures to avoid the deleterious physiological effects of obesity and aging on tissue homeostasis and to promote a healthy life. Treatment of obesity is difficult, and initially based on lifestyle change, diet recommendations, and increased physical activity [77] but, whereas it is actually effective, it is associated with very low compliance. Aging is inevitable but it is desirable to age in a healthy way, therefore the purpose of our society is now to increase the rate of healthy aging to avoid harmful consequences on skeletal muscle mass and function and to limit frailty and dependence. As stated above, oxidative stress and chronic inflammation are core mechanisms associated with obesity and aging. Among the numerous natural chemical compounds tested for their antioxidant and anti-inflammatory properties, grape polyphenols present great interest.

Attention to the importance of dietary intake of polyphenols was ignited by the phenomenon called the 'French Paradox', first described by Serge Renaud from the University of Bordeaux in 1992. According to his observations, the French population when compared to other Western populations whose diet is rich in saturated fatty acids (e.g., the American population), shows a much lower incidence of coronary heart disease and associated mortality [78]. In fact, adherence to the Mediterranean diet which includes mainly plant-derived foods and red wine consumption has been associated with a lower risk of chronic diseases and mortality [79] and a lower frailty index in older adults [80]. The first explanation put forward to explain these associations was the moderate consumption of red wine. Undeniably, among fruits, grapes (but also red wine, grape seeds, and grape pomace) contain high amounts of polyphenols [81] (Table 1), although not being the richest source. This does not affect its great biological interest due to the exceptional variety of polyphenol families and molecules of known beneficial activity on human health that can all be found in it (Table 2).

Table 1. Polyphenols characterization, total polyphenol content (TPC) of major well-known vegetal sources of dietary polyphenols. TPC is expressed in mg/100 g of fresh weight (FW). Data in the table were extracted from the PhenolExplorer database [82–84].

Source	Family of Polyphenol	Amount	Mean TPC (Folin Assay)
Apple	Anthocyanins	0.93 mg/100 g FW	200.96 mg/100 g FW
	Dihydrochalcones	5.38 mg/100 g FW	
	Flavanols	24.12 mg/100 g FW	
	Flavonols	6.86 mg/100 g FW	
	Phenolic acids	19 mg/100 g FW	
Artichoke, heads, raw	Flavones	57.8 mg/100 g FW	1142.40 mg/100 g FW
	Phenolic acids	202.23 mg/100 g FW	
Blueberries	Flavonols	12.23 mg/100 g FW	151.33 mg/100 g FW
	Phenolic acids	162.47 mg/100 g FW	
	Phenolic acids	37.06 mg/100 g FW	
	Other polyphenols	0.45527 mg/100 g FW	
Cocoa, powder	Flavanols	511.62 mg/100 g FW	5624.23 mg/100 g FW
	Phenolic acids	37.06 mg/100 g FW	
	Other polyphenols	0.45527 mg/100 g FW	
Grape	Anthocyanins	72.1 mg/100 g FW	184.97 mg/100 g FW
	Flavanols	17.11 mg/100 g FW	
	Flavonols	3.08 mg/100 g FW	
	Phenolic acids	1.69 mg/100 g FW	
	Stilbenes	0.3362 mg/100 g FW	
Green tea	Flavanols	71.18 mg/100 g FW	61.86 mg/100 ml
	Flavonols	5.29 mg/100 g FW	
	Phenolic acids	12.53 mg/100 g FW	
Olives, green	Flavones	0.56 mg/100 g FW	161.24 mg/100 g FW
	Phenolic acids	134.94 mg/100 g FW	
	Other polyphenols	211.05 mg/100 g FW	
Persil, fresh	Other polyphenols	13.95 mg/100 g FW	89.27 mg/100 g FW
Strawberries	Anthocyanins	73.01 mg/100 g FW	289.20 mg/100 g FW
	Flavanols	9.1375 mg/100 g FW	
	Flavonols	2.32 mg/100 g FW	
	Phenolic acids	10.74 mg/100 g FW	
	Stilbenes	0.35 mg/100 g FW	

Table 2. Main effects of polyphenols present in grapes on mechanisms involved in obesity and aging.

Family and Subfamily	Compound	Effect and Mechanism	References
Flavonoids/Flavan-3-ols	*EGCG*	• **Antioxidant** Radical scavenging Metal ion chelation ↑ CAT, ↑ SOD1 e SOD2, ↑ GPx • **Anti-inflammatory** ↓ NFkB via ↓ Ikβ ↓ COX-2 ↓ IRF3 via ↓ TBK1 • **Anti-diabetic** ↓ insulin resistance ↑ lipid oxidation in muscle ↑ NFkB, ↑ AMPK, ↑ MAPK • **Anti-aging**/pro-apoptotic ↑ Beclin-1 and ↑ caspases	• Fraga et al. [85] Bernatoniene et al. [86] Meng et al. [57] • Youn et al. [87] • Casanova et al. [88] Li et al. [89] • Pallauf and Rimbach [90]
	Grape seed proanthocyanidins	• **Anti-diabetic** ↑ Nrf1, ↑SIRT1, and ↑PGC-1α, ↑ slow myosin heavy chain, ↑ succinic dehydrogenase and malate dehydrogenase activities, ↑ resistance to fatigue	• Xu et al. [91]
Flavonoids/Flavanols	*Quercetin*	• **Anti-inflammatory** ↑ Nrf2/ARE pathways ↑ Antioxidant enzymes ↓ TNF-α, ↓ IL-6, ↓ IL-1β, ↓ COX-2, ↓ iNOS, ↓ NFkB in adipocytes and macrophages	• Costa et al. [92] • Sato et al. [93]
	Myricetin	• **Anti-diabetic** ↑ glucose uptake, ↓ insulin resistance, ↑Akt and ↑AMPK signaling pathways	• Pandey et al. [94]
	Kaempferol	• **Anti-inflammatory** ↓ IL-6, IL-1β, 18 and TNF-α ↑ Nrf2 and synthesis targets Inhibition TLR4	• Alam et al. [95]
Flavonoids/Anthocyanes	*Anthocyanins*	• **Anti-inflammatory** ↓ COX-1 and COX-2 ↓ C-reactive protein	• Mozos et al. [96] • Sivamaruthi et al. [97]
	Cyanidin-3-O-glucoside	• **Antidyslipidemic** ↑ PPARs • **Anti-diabetic** ↑ Insulin sensitivity → ↑ PPARs ↑ Insulin secretion → ↓ IL-1β and IL-6 ↑ TLR4/IkBα pathway	• Jia et al. [98] • Geng et al. [99]
Flavonoids/Flavones	*Luteolin/Apigenin*	• **Anti-inflammatory** ↓ NO and ↓ PGE2	• Tian et al. [100]
	Apigenin	• **Anti-obesity** Radical scavenger ↑ Increase muscle fibers size ↑ number and volume mitochondria ↑ SOD and GPx	• Wang et al. [101]
Flavonoids/Isoflavones	*Daidzein*	• **Anti-diabetic** Inhibition α-amylase and α-glycosidase ↑ AMPK, ↑ GK, ↓ G6Pase, ↓ PEPCK, ↑ GLUT4, ↑ IRS1, ↑ IRS2, ↑ PPARγ • **Anti-inflammatory** ↑ PPARγ, ↓ TNFα, ↓ NFkB, ↓ IL-6, ↓ Ccl2, ↓ Cxcl2	• Park et al. [102] • Das et al. [103]
	Genistein	• **Anti-diabetic** ↑ AMPK in skeletal muscle ↑ insulin sensitivity ↑ lipid oxidation	• Guevara-Cruz et al. [104]

Table 2. *Cont.*

Family and Subfamily	Compound	Effect and Mechanism	References
Flavonoids/Flavanones	*Naringenin*	• **Anti-diabetic** ↑ Insulin secretion • **Anti-inflammatory** ↓ TNF-α and IL-6 ↑ SOD	• Rehman et al. [105]
	Hesperidin	• **Anti-diabetic** ↑ IRS, ↑ Akt, and ↑ GLUT4 in muscle cells	• Dhanya et al. [106]
Stilbenes	*Resveratrol*	• **Anti-diabetic** ↑ SIRT1 and ↑ PGC-1α ↑ mitochondrial activity (exercise mimetic effect) ↑ Akt and AMPK pathways → ↑ insulin sensitivity • **Anti-obesity** ↓ fat accumulation ↑ lipolysis • **Anti-aging** ↓ caspase 3	• Lagouge et al. [107] • Lagouge et al. [107] Kang et al. [108] • Huang et al. [109] • Bai et al. [110]
Phenolic acids	*Phenolic acids*	• **Anti-diabetic** ↑ GLUT2 in pancreatic β-cells ↑ PI3K/Akt and ↑ GLUT4 in adipose and muscle tissues ↓ α-glucosidase activity	• Kumar et al. [111] • Duboit et al. [96]
	Gallic acid/p-coumaric acid	• **Anti-diabetic and anti-obesity** ↓ TNF-α and ↓ PPAR γ in adipose tissue	• Abdel-Moneim et al. [112]
	Caffeic acid phenetyl ester	• **Anti-inflammatory** ↓ COX and ↓ LOX Inhibition detachment *arachidonic acid.*	• Silva et al. [113]
	Vanillic acid	• **Anti-obesity** ↓ PPAR and C/EBPα ↑ Lipid oxidation through ↑ AMPKα	• Jung et al. [114]
	Syringic acid	• **Anti-diabetic** ↑ PGC-1α and Nrf2 ↑ increased mitochondrial biogenesis. ↓ TNF-α, IL-1β, and IL-6	• Rashedinina et al. [115]

4. Structure and Function of Grape Polyphenol

Recently, the polyphenols intake has been estimated to be 1607 mg/d in a French well-balanced diet [81] which is above than previously found in epidemiological studies [116]. Grape is one of the major fruit crops produced worldwide and wine is the most widespread alcoholic beverage consumed. Moreover, the by-products of wine production are also a rich resource of biologically active molecules, which need to be emphasized since they possess the same phytochemicals of wine without the deleterious effect of alcohol [117].

Accumulating evidence shows that grape polyphenols regulate several mechanisms to prevent oxidative stress and inflammatory-mediated diseases [118] (Figure 1). Thanks to the phenolic groups, polyphenols can neutralize ROSP due to a transfer of electrons and/or hydrogen atoms to form phenoxyl radicals which are relatively more stable, thanks to resonance stabilization [42]. Polyphenols can also activate the endogenous antioxidant system via the ancestor Nrf2- Kelch-like ECH-associated protein 1 (KEAP1) signaling pathway [119,120]. Then, in aged muscle, grape polyphenols can increase the antioxidant enzymes such as catalase, SOD, glutathione reductase, and GPx leading to a reduction of muscle lipid damage associated with improvement in muscle function [121]. Resveratrol, and grape polyphenols, can increase the endurance capacity of muscle, which is related to increase reliance on mitochondrial lipid oxidation [107,122]. Polyphenols can increase mitochondria biogenesis and function due to their ability to activate sirtuin 1 (SIRT1) a class III histone deacetylase, NAD+-dependent (Figure 1). SIRT1 regulates several important processes such as gene expression, metabolism, and oxidative stress response via deacetylation of lysine groups [123]. More specifically, SIRT1 activates PGC-1α, mitochondrial transcription factor A (TFAM), nuclear receptor peroxisome proliferator-activated receptor (PPAR), and nuclear respiratory factor (NRF), all involved in mitochondria biogenesis and function. Moreover, SIRT1 not only modulates mitochondrial function but also regulates FOXO protein acetylation which in turn modulates manganese SOD and catalase expression increasing the defense against oxidative stress [124]. The energy sensor, AMPK, could also activate SIRT1 since it regulates the cellular levels of NAD+. Grape polyphenols can phosphorylate and activate AMPK [122]. Then, polyphenols can increase lipid oxidation since it is well known that AMPK controls malonylCoA level a major inhibitor of the CPT and lipid entry into mitochondria [125] but also AMPK controls mitochondrial biogenesis via the deacetylation and activation of PGC-1α by SIRT1 [107]. Indeed, grape polyphenols supplementation increases muscle PGC-1α alpha mRNA and maintains CPT1 mRNA after fructose ingestion in first-degree relatives of T2D patients, suggesting a conserved lipid oxidation capacity which could explain the body weight gain decrease in this group compared to placebo [126]. The inflammation pathway is also negatively regulated by SIRT1 through NFkB acetylation explaining in part the anti-inflammatory activity of polyphenols. On the other part, resveratrol has been found to alleviate obesity-induced skeletal muscle inflammation due to a shift in macrophage toward an anti-inflammatory profile and a decrease in TLR4 expression [127]. In the context of muscle atrophy, resveratrol has demonstrated an ability to decrease atrogin-1 and MuRF-1 expression depending on SIRT1 activity [128]. Then, polyphenols but mostly resveratrol have been described as exercise mimetic compounds according to their positive regulation of muscle mass [129]. Polyphenols are secondary metabolites of vegetables and fruits (largely in berries) with important physiological and defense functions as pigments and antioxidant molecules. They are accumulated in the berries, especially in the solid parts (mainly skin and seeds), in response to various internal and external stimuli (growth, free radicals excess, ultraviolet (UV) radiations, fungi, insects, and bacteria attack) [130]. Then, polyphenols content in grapes is highly variable depending on the cultivar, maturity, and fermentation processes during wine production [131]. They play an important role in maintaining the organoleptic characteristics of wines (color, taste, and astringency), and thanks to their antimicrobial activity they are used as preservatives in the food and cosmetic industries [132,133]. By highlighting their antioxidant properties [134,135], the French paradox stimulated the exploration of the effects that polyphenols from different sources could have on various diseases [136]. As we stated at the outset, the study of grape polyphenols on muscle is still in its infancy, and therefore studies specifically demonstrating the effects of polyphenols extracts obtained from grapes are still scarce. That is why in this section, we will report studies on main polyphenols (individual molecules or groups of molecules) found in grape, that can be also isolated from other sources [137]. In fact, even if isolated from other sources, these polyphenols show the same chemical structure, and we reasonably assume that they could exert the same mechanism of action. For each subfamily, we will cite studies focusing on their effects on obesity, aging, T2D, inflammation, and oxidative stress, with a particular look at pathways affecting muscle. The chemical structure of polyphenols is characterized by the presence of one or more hydroxyl groups and one or more aromatic rings with six carbon atoms.

Polyphenols are a wide class of compounds with more than 8000 molecules isolated and described [138]. Thus, for easier understanding, polyphenols are divided into two big families according to the number of aromatic rings, the way they are linked to each other, and the position and oxidation state of the hydroxyl groups: flavonoids and non-flavonoids, each of which contains other classes (Figure 2).

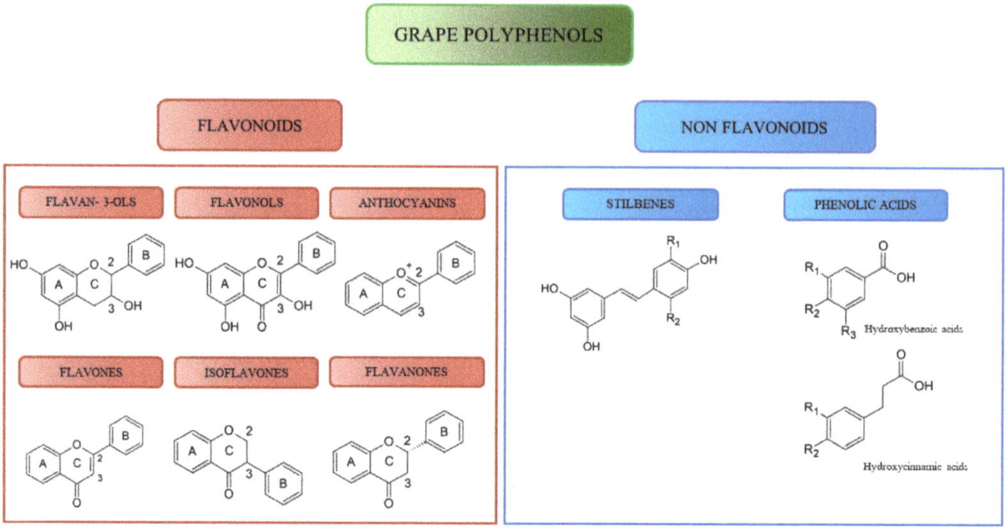

Figure 2. Polyphenols classification.

4.1. The Flavonoids

Flavonoids are known as plant pigments but also for their antioxidant, antimicrobial, and light-screening functions [139]. Moreover, thanks to their properties they are successfully used in the pharmaceutical and food sectors as preservatives and pigments [140]. Depending on the chemical structure, oxidation degree, or unsaturation of the heterocyclic ring C, flavonoids can be further classified into six mean groups, notably: flavanols, flavonols, anthocyanes, flavones, isoflavones, flavanones [141]. These molecules are generally water-soluble, and they can be found in glycosylated or aglycone form. Their basic structure is the flavone ring.

4.1.1. Flavan-3-ols

Flavanols are relevant molecules in grapes and wine for their contribution to color stabilization and their astringent and bitter properties [142]. They are generally referred to as catechins and in their structure, the double bond on the C ring is absent and, accordingly, two chiral carbons are found (C2 and C3, see Figure 2) [143]. The monomeric catechins can exist in the form of four stereoisomers, depending on the hydroxylation on the C4 of the C ring. The *trans* isomer is called catechin, the *cis* one is called epicatechin. When catechins and epicatechins polymerize, they form condensed tannins also known as proanthocyanidins [144]. The organoleptic and pharmacological properties of tannins are strictly related to their structure and polymerization degree. Molecules with a higher polymerization degree have stronger radical scavenger activity and moreover, they are more bioavailable thanks to increased resistance to acid hydrolysis in the stomach [145]. Indeed, grape seed proanthocyanidins have been found to promote health-benefit via inhibition of protein damage linked to their antioxidant capacities and enzyme inhibition [146]. Grape seed proanthocyanidins extract is able to activate AMPK in C2C12 myotubes, consequently switching on Nrf1, SIRT1, and PGC-1α. This mechanism is behind the increase in slow myosin heavy chain, decrease in fast myosin chains, increased activity of succinic dehydrogenase and malate dehydrogenase, and decreased activity of lactate dehydrogenase all of these leading to increased resistance to fatigue [91]. (-)-Epigallocatechin-3-gallate (EGCG) wields an antioxidant activity through a combination of mechanisms. On one side it acts as a radical scavenger or by chelating metal ions that favor the formation of radicals [85], on the other, it enhances the activity of endogenous antioxidant enzymes as well as inhibits the activity of pro-oxidant enzymes and pro- inflammatory TLR4 pathways [57,86,87]. EGCG has a role in mitigating the oxidative stress-based inflammation in T2D and obesity by regulating the NFkB, Adenosine Monophosphate Activated Protein Kinase (AMPK), and Mitogen-Activated Protein Kinases (MAPK) signaling pathways, thus decreasing IR and increasing muscle lipid oxidation [88,89]. EGCG has a beneficial effect on the aging process by promoting autophagy through activation of Beclin 1 and apoptosis by induction of caspase proteins [90], see Table 2.

4.1.2. Flavonols

The flavonols exhibit a double bond between the carbons C2 and C3 and the C3 is hydroxylated. In this position, the aglycones can be linked to different sugars (often glucose and rhamnose) [147]. They are found in big amounts in the grape skin because their biosynthesis is stimulated by the sun light, from which they protect the plant. Quercetin, one of the major flavonoids in grapes, reduces inflammatory signaling pathway by inhibiting inflammatory receptors in mice with obesity-induced skeletal muscle atrophy [148] and attenuates adipogenesis and fibrosis in a human muscle-derived mesenchymal progenitors cells model [149]. In human myotubes from a healthy donor, at a physiological dose, quercetin modestly increases the insulin signaling pathway and glycogen storage which could participate in improving insulin sensitivity [150]. Their signaling pathway expresses an anti-inflammatory activity through the activation of Nrf2/Antioxidant Responsive Element (ARE) pathways with a subsequent upregulated synthesis of antioxidant endogenous enzymes [92]. Furthermore, it was described that quercetin reduces inflammation in the adipocytes and macrophages by reducing the expressions of genes encoding for TNF-α, IL-6, IL-1β, Cyclooxygenase-2 (COX-2), Inducible Nitric Oxide Synthase (iNOS) and also by keeping away from the activation of NFkB [93]. Myricetin, another of the flavonoids most present in red grapes, has been shown to have anti-diabetic effects in numerous studies [94]. On C2C12 myotubes, it was demonstrated that myricetin increases glucose uptake thanks to the activation of AMPK and Akt signaling pathways, thus decreasing insulin resistance [151]. Kaempferol, as well, is known to be an anti-inflammatory compound. The different mechanisms were summarized by Alam et al. in 2020 and they include: decreased release of IL-6, IL-1β, 18, and TNF-α, activation of the Nrf2 pathway and synthesis of target enzymes, and inhibition of TLR4 [95].

4.1.3. Anthocyanes

The anthocyanes (or anthocyanins) are, with the chlorophyll and the carotenoids, the most important vegetal pigments [152]. In grapes, anthocyanins are located in the skin and exhibit a strong antioxidant power, to protect the plant from the damage caused by UV radiation [153]. The base structure is the 3,5,7,4′-tetrahydroxyflavylium cation or flavylium cation. The OH group in position 3 is always glycosylated and the one in position 5 is very frequently. There are many studies that show the beneficial antioxidant and anti-inflammatory effects of anthocyanin supplementation on obesity state meticulously reported by Sivamaruthi et al. [97]. To cite one, high fat diet obese mice supplemented with 250 mg/kg/d grape pomace extract, rich in anthocyanin, decreased the levels of plasma C-reactive protein after 12 weeks, thus exerting an anti-inflammatory activity [154]. Many studies have already described their disparate biological activities. Among them, anthocyanins have been shown to exhibit an inhibitory effect on the enzymes COX-1 and COX-2, thus reducing systematic and cardiovascular inflammation [96]. Moreover, glucose and lipid metabolism are also improved following anthocyanin supplementation. For example, cyanidin-3-*O*-glucoside supplementation increases the expression of Peroxisome Proliferator-Activated Receptors (PPARs), thus reducing dyslipidemia and increasing sensitivity to insulin in mice after eight weeks of supplementation by increasing lipid oxidation [98] and plays a role to retrieve IR in diabetes by re-establishing insulin secretion and decreasing IL-1β and IL-6 concentrations [99].

4.1.4. Flavones

These molecules, vegetal yellow pigments, have the fundamental skeleton of the flavone (also known as 2-phenylchromone), with the presence of a double bond between C2 and C3 and no hydroxyl group in C3. It is therefore an oxidized form of flavanones. Like other flavonoids, they mostly appear in the form of water-soluble glycosides. Luteolin and apigenin showed in lipopolysaccharide (LPS)-activated macrophages a strong anti-inflammatory effect thanks to the decreased production of NO and prostaglandin E2 [100]. It was also described as a beneficial effect of apigenin in mitigating obesity-induced atrophy in mice. After the treatment, the size of muscle fibers was enhanced, and mitochondria were increased in number and volume. The effect has been scribed to the counteracting of oxidative species and improving the activity of antioxidant enzymes such as SOD and GPx [101].

4.1.5. Isoflavones

Isoflavones, commonly known as phytoestrogens, are isomers of flavones in which the phenyl group (ring B) is linked to the C3 and not to the C2 of ring C. They are synthetized following a microbial attack or in stress conditions, thus acting as phytoalexins [155]. It described their anti-inflammatory activity and their effects on the mitigation of T2D. Daidzein acts as an inhibitor of the enzymes α-glucosidase and α-amylase, thus decreasing post-prandial glycemia [102]. Additionally, daidzein acts on different targets involved in the insulin response (AMPK, Glycerol Kinase (GK), Glucose 6-Phosphatase (G6Pase), Phosphoenolpyruvate Carboxykinase (PEPCK), PPARγ, Glucose Transporter 4 (GLUT4), Insulin Receptor Substrate 1 (IRS1), IRS2, etc.) and in the anti-inflammatory response (PPARγ, TNFα, NFkβ, IL-6, Chemokine ligand 2 (Ccl2), Chemokine (C-X-C motif) ligand 2 (Cxcl2), etc.) [103]. In obese patients, genistein (50 mg/day for 2 months) ameliorates IR associated with an increase in skeletal muscle AMPK activation, thus increasing fatty acid oxidation and insulin sensitivity [104].

4.1.6. Flavanones

All the molecules of this class have a structure based on the progenitor flavone, with a double bond between C2 and C3. We can find flavanones in grapes and some representative molecules such as naringenin and hesperetin which exhibit important anti-inflammatory, antioxidant, and antidiabetic action. In fact, naringenin is able to ameliorate hyperglycemia and to improve the secretion of insulin. Moreover, it has an action on the inflammatory status by decreasing cytokines like TNFα and IL-6 and increasing the activity of SOD [105]. Naringenin and hesperetin are aglycones, but they are often found in their glycosylated form called naringin (naringenin + neohexperidose) and hesperidin (hesperetin + rutinose). It was demonstrated that hesperidin has an antidiabetic action, exerted by upregulating IRS, Akt, and GLUT4 in muscle cells, which is higher than the aglycone hesperetin [106].

4.2. Non-Flavonoids

4.2.1. Stilbenes

Stilbenes are known as phytoalexins, protective compounds secreted by the plant following contact with a pathogen or an abiotic stress [156]. Stilbenes are diaryl ethers, ethenes substituted with a phenyl group on both carbon atoms of the double bond. Thus, there are two possible geometric isomers of stilbene, *trans* and *cis* [157]. Grape berries are an excellent source of trans-resveratrol, the most notable compound among stilbenes, with a concentration of 10–100 higher than in other berries [158]. Trans-resveratrol is the molecule belonging to this class that has been most investigated for its biological properties [159] due to the activation of Sirtuin 1 (SIRT1). Lagouge et al., in a groundbreaking paper, showed that resveratrol activates the deacetylase SIRT1 and the coactivator Peroxisome Proliferator-Activated Receptor-Gamma Coactivator-1α (PGC-1α), thus increasing mitochondrial activity, its decrease is a cause of aging and metabolic diseases, and miming caloric restriction and exercise [107]. Then, resveratrol can be used successfully against different pathologies [160]. Another beneficial effect on metabolism is given by the fact that resveratrol improves insulin sensitivity by activating Akt, and AMPK pathways and inhibiting NFkB favoring insulin signaling, lipid oxidation, and decreasing inflammation in rodents [107,108]. Besides its beneficial metabolic effect, resveratrol has been efficient in animal models or cell cultures for preventing muscle atrophy due to dexamethasone [128], mechanical unloading [129], cancer cachexia [161], and sarcopenia in obese rodents [109,110]. Interestingly, resveratrol has been shown to decrease oxidative stress and inflammation associated with aging without a reversal of muscle atrophy [162,163].

4.2.2. Phenolic Acids

Phenolic acids are one of the classes of phenolic compounds found in higher concentrations in the plant world [164]. This class of compounds is divided into hydroxybenzoic acids and hydroxycinnamic acids, in the function of the position of the carboxylic group on the aromatic ring (C1-C6 or C3-C6 structure) [143]. They are often found in conjugated form with tartaric acid or glucose, forming soluble compounds [165]. Phenolic acids are known to have a beneficial impact on human health, acting as oxidative species scavengers but also regulating some key signaling pathways. As an example, phenolic acids exert an adjuvant effect on diabetes by activating the Phosphatidylinositol 3-kinase (PI3K)/Akt pathway, increasing the translocation of GLUT4 in adipose and muscle tissues and thus insulin sensitivity [111]. Gallic acid and *p*-coumaric acid showed to have a potent hypoglycemic and lipid-lowering effect on diabetic rats, exerted by decreasing TNF-α and modulating PPAR-γ in adipose tissue [112] and by modulating muscle AMPK [166]. Caffeic acid phenetyl ester, a derivative of caffeic acid found in grapes, inhibits the enzymes COX and lipoxygenase (LOX), the main enzymes involved in inflammation [113]. To bring some examples of biological activities among the hydroxybenzoic acids, vanillic acid can be successfully employed in obesity because it decreases the adipogenic PPAR and CCAAT-enhancer-binding proteins α (C/EBPα) and increases lipid oxidation through AMPKα [114]. Syringic acid is an antidiabetic agent, which combines an antihyperglycemic effect with the mitigation of diabetic neuropathy. The effect is given by the activation of PGC-1α and Nrf1, with consequent increases in mitochondrial biogenesis, and decreased secretion of pro-inflammatory cytokines (IL-6, IL-1β, and TNF-α) [115].

It is evident nowadays that introducing dietary grape polyphenols through alimentation is essential for maintaining a good state of health. Several in vitro and in vivo studies and clinical trials demonstrated that their action on the body is exerted by regulating metabolism, weight, and muscle function, mitigating oxidative stress, inflammation, and chronic diseases [122,126,167].

Although their antioxidant action is considered their main mechanism of action, this action alone is not able to explain all the biological effects of grape polyphenols. In fact, it has been proven that they also act through the modulation of receptors [168], transcription factors during myogenesis [169,170], enzymes activities [171,172], and also through epigenetic modulation [173], and non-coding (nc) RNA regulation [174].

5. Metabolism of Polyphenols

Structures and activities of polyphenols could be altered by their interaction with other molecules contained within the food matrix and of course by hepatic and intestinal metabolism. Consequently, human plasma concentrations of polyphenols are not comparable to those concentrations described as necessary to achieve a great biological activity as demonstrated by in vitro studies. Their metabolism and bioavailability vary considerably from molecule to molecule and there is a strong hypothesis that polyphenols' metabolites produced in vivo are also responsible for the biological action of polyphenols [175].

The series of processes involved in the metabolization, and absorption of polyphenols begins in the oral cavity with saliva and then continues in the gastrointestinal tract involving the gut microbiome. After these processes, a part of the polyphenols will be absorbed in their original form, a part will be excreted in the feces and a part will be transformed into new molecules with biological effects. The destiny of polyphenols in the digestive system depends on their original chemical structure: different polyphenols will undergo different transformations.

Saliva is composed mainly of water, salts, enzymes, and proteins (albumin, α-amylase, sulphomucins, sialomucins, glycoproteins, sulfated cystatins, agglutinins, histatins, lysozymes, mucins, immunoglobulins, proline-rich proteins) among which amylase is the most abundant [176]. In the mouth, through mastication polyphenols are mixed with saliva and solubilized. Molecules such as tannins can form a complex and precipitate with the tannin-binding salivary proteins (TBSPs). These complexes remain stable during the transit in the stomach, while they are solubilized in the intestine in presence of bile salts [177]. Lipophilic polyphenols such as resveratrol, curcumin, and quercetin are poorly bioavailable because of their lack of solubility, thus limiting their antioxidant action in the body. Saliva has been described as able to solubilize lipophilic polyphenols, thus increasing their bioavailability and their antioxidant activity [178].

After the mechanical and chemical transformations in the mouth, polyphenols are transported to the gastrointestinal tract. Absorption can occur by passive transport or, much more frequently because of lipophilia, by carrier-mediated active transport. In rats, phenolic acids [179] and not glycosylated flavonoids [180] can be absorbed at the stomach level. Chen et al. described that among the total polyphenolic compounds only 5–10% can be directly absorbed in the small intestine while the rest must undergo transformations by enzymes in subsequent sections of the gastrointestinal tract before they can be absorbed [179]. For instance, glycosylated flavonoids, such as quercetin, are poorly absorbed due to their hydrophilic character. They must be deglycosylated by β-glucosidases of the small intestine and then absorbed as aglycones [181]. The human digestive tube is populated by a copious microbial population, counting more than 100 trillion different microorganisms, whose name is gut microbiota. Gut microbiota has a big impact on the polyphenols' absorption and bioavailability, because of their own enzymatic capacities. First, the *O*-glycosides are hydrolyzed to aglycones, which will further undergo reactions of glucuronidation or sulfonation [182]. For example, trans-piceid, which is chemically an *O*-glycoside of resveratrol, is hydrolyzed to free resveratrol by the gut microbiota [183]. Subsequently, both molecules are largely sulfonated or glucuronidated but also hydroxylated, to produce different derivatives such as dihydroresveratrol, dihydropiceid, and many others [184]. Moreover, the gut microbiota is able to perform catabolic reactions, like degradation of aromatic rings via carbon-carbon cleavage, decarboxylation, hydrogenation, ihydroxylation, demethylation, thus forming derivatives with simpler structures [182]. Cyanidin, taken as a model of anthocyanidin, is metabolized by the gut microbiota starting by the opening of the pyranic ring followed by a second carbon-carbon cleavage, giving protocatechuic acid and 2-(2,4,6-trihydroxyphenyl) acetic acid as final products [185]. Hydrolysable tannins are complex phenolic compounds, metabolized by the gut microbiota. The first enzymes involved are hydrolases (tannin acyl hydrolase) which release gallic acid (gallotannins) or ellagic acid (ellagitannins). Gallic acid is further transformed by decarboxylation and hydroxylation, while ellagic acid only transforms by dihydroxylation [186].

Ferulic acid is mostly found in its esterified form. Its methyl ester, for example, is readily demethylated to ferulic acid then its double bond is saturated by hydrogenation, the methoxy group on the carbon 3 is demethylated and the carbon 4 is dehydroxylated to obtain 3-phenylpropionic acid [187]. The polyphenols' metabolites thus formed will be partly absorbed into the systemic circulation and partly excreted as waste to terminate their biological activity. Apart from the metabolization carried out by the gut microbiota, they can enter the enterohepatic cycle and undergo phase I and phase II metabolism, eventually going back from the liver to the intestine through the bile [188]. Phase I metabolism is carried out in the liver by the cytochrome P450 (CYP450) superfamily of enzymes, and it involves reactions of oxidation, hydrolysis, and reduction. Phase II metabolism has the aim to conjugate polyphenols to augment their hydrophilicity and help their rapid elimination from the body. Phase II enzymes include UDP-glucuronosyltransferase, responsible for glucuronidation, N-acetyltransferase, which catalyzes the transfer of acetyl groups from acetyl-CoA to polyphenols, glutathione-S-transferase which leads the polyphenol to conjugation with a reduced glutathione molecule [189]. It was reported for instance, that naringin and naringenin are susceptible to phase I and phase II metabolism. In fact, they are firstly oxidized or demethylated by CYP450 and subsequently glucuronidated, sulfated and methylated. From the metabolism of the only naringin 32 metabolites are derived, some keeping the flavonoid structure and some only a phenolic one [190]. As another example, it is possible to find in the human intestine quercetin-3′-*O*-glucuronide and quercetin-4′-*O*-glucuronide as a result of phase II metabolism on quercetin [191]. Another well-studied molecule is resveratrol, which when it reaches the human gastrointestinal tract goes through reactions of sulfation and glucuronidation, phase II reactions, generating a variety of reported metabolites, such as trans-resveratrol-3-*O*-sulfate, trans-resveratrol-4′-*O*-sulfate, *trans*-resveratrol-3,4′-disulfate, *trans*-resveratrol-3-*O*-glucuronide and *trans*-resveratrol-4′-*O*-glucuronide [192].

6. Clinical Studies

The number of clinical studies evaluating skeletal muscle mass and function for patients with chronic/non-communicable diseases and following grape polyphenols administration is strongly limited compared to in vitro and animal studies. This relative absence of clinical trials is particularly difficult to understand considering the encouraging results obtained in vitro, in animal models, and during the scarce clinical studies yet realized some years ago [126,193–195]. In 2011, Brasnyo et al. studied the effect of supplementation of resveratrol (10 mg/d) in a group of 10 T2D subjects for four weeks [193]. This low dose of resveratrol improved insulin sensitivity, decreased blood glucose levels, decreased blood oxidative stress, and increased Akt phosphorylation in platelets without any data on muscles. The same year, Timmers et al. demonstrated the beneficial effect of nutritional supplementation of 150 mg/d of resveratrol for 30 d in 11 healthy obese men [194]. This supplementation clearly induced metabolic changes in obese humans, mimicking the effects of calorie restriction or endurance exercise. In fact, Timmers observed activation of AMPK in muscle biopsies of the resveratrol-treated subjects, increased PGC-1α and citrate synthase activity, attesting to an increased muscle mitochondrial activity. Interestingly, resveratrol decreased hepatic lipid content but increased intra-myocellular lipid content. Hokayem et al. showed that eight weeks of supplementation with a natural mixture of grape polyphenols at nutritional doses (2 g/d), efficiently prevents fructose-induced oxidative stress and IR in healthy overweight/obese first-degree relatives of T2D patients [126]. At the skeletal muscle level, grape polyphenols supplementation protected mitochondrial function, prevented oxidative stress, and tended to increase insulin sensitivity after fructose challenge Goh et al. showed that supplementation with 3 g resveratrol daily for 12 weeks regulates energy expenditure through increased skeletal muscle SIRT1 and AMPK protein expression in patients with T2D [195]. These interesting results indicate that resveratrol may have beneficial exercise-mimetic effects in patients with T2D [195]. However, no significant modification in systemic insulin sensitivity was observed during this study whereas a decrease in glycated hemoglobin was observed, suggesting an improvement in glucose tolerance. The number of volunteers included in this trial, 10 subjects, could be not enough to have a clear demonstration of grape polyphenols' effects on systemic insulin sensitivity. In seventeen well-controlled T2D subjects, supplementation of 150 mg/d of resveratrol for 30 d was also not able to change insulin sensitivity whereas it increased lipid-derivate mitochondrial respiration in muscle [196]. An 8-week supplementation of obese subjects with red wine polyphenols (600 mg/day) does not also improve obesity-associated IR [197]. The main limitation of this study was that total polyphenols intake was similar between the treated and placebo groups limiting the conclusions on the supplementation. Overall, the improvement of insulin sensitivity with grape polyphenols (resveratrol or polyphenols mixture) was mainly found in metabolically stressed patients at a systemic level but not in a normal state of health [198], whereas the improvement in mitochondrial function and oxidative stress were evident whatever the tissues and the metabolic status of patients. These data underline that in obese/T2D humans, grape polyphenols seem to act mainly as antioxidants.

Clinical trials for investigating polyphenols supplementation on aged muscle are even more limited. In an attempt to decrease functional limitations, most studies on elderly subjects have combined exercise training and polyphenols supplementation in order to amplify the benefits of exercise alone. One study found the beneficial effect of 500 mg/d of resveratrol supplementation associated with exercise on muscle force with higher mitochondrial density and muscle fatigue resistance in elderly subjects allowing to potentially reverse sarcopenia [199]. However, most of the studies report no effect of the combination [200] or even detrimental effect on muscle [201,202]. Recently, in a pilot study on elderly community-dwelling adults, Harper et al. found that 1000 mg/d of resveratrol permits a 33.1 m improvement in the 6-min walk distance associated with higher mitochondrial function and decreased inflammation [203]. This data could be related to the epidemiologic studies where an association between the Mediterranean diet with gait speed is found [204]. However, the beneficial effect of grape polyphenols supplementation on sarcopenia needs to be investigated. To date, no prospective study with data on sarcopenic or frailty subjects is available but some clinical trials are ongoing [124].

7. Conclusions

Restoration of skeletal muscle mass and function is vital to cure the comorbidities associated with obesity and aging. Grape polyphenols have therapeutic potential for such diseases mediated by oxidative stress and inflammation. Nevertheless, large-scale clinical trials are still necessary to better investigate the activity of these natural compounds at the skeletal muscle level. Moreover, the majority of clinical trials have studied the effect of one purified polyphenol, resveratrol, and not the activity of a crude extract or of a mixture of several molecules of polyphenols. Yet in vitro studies or animal models, even not concerning skeletal muscle, showed an interest in using a mixture of polyphenols. In fact, such a mixture of polyphenols is more active than each individual molecule alone as a synergy effect occurs between the molecules [205,206].

Another important point in using a mixture of polyphenols is that an extract in which many polyphenolic molecules are blended together in low concentrations has the advantage of reducing the toxicity that could be derived from the use of a single molecule in higher concentrations [207].

Additionally, the advantage of grapes over other sources of polyphenols is that grapes, after strawberries and lychees, are among the fruits with the highest polyphenols content (anthocyanins/anthocyanidins, flavonols) [208]. As shown in Table 1, moreover grape boasts a wide variety of phenolic molecules. Although stilbenes, such as resveratrol, have a lower relative abundance than the previously mentioned molecules in grapes, it is at least 10–100 times higher than in other berries [209], making grapes one of its primary sources. Grapes are consumed worldwide. They are the fourth most abundant fruit crop cultivated all over the world [210], so their supply as a source of extraction is easy and affordable for many countries. Moreover, there is an environmental and eco-friendly aspect to using grape polyphenols. Indeed, grape skins and grape seeds are waste by-products of wine or grape juice production. They are rich sources of polyphenols and could be used in the form of food supplements.

Translational data confirming the benefit of grape polyphenols in human health at the muscle level would allow the development of grape polyphenols supplementation in therapy and in the management of obese and aging people on a routine basis. This would improve the quality of life and decrease the economic cost of medical care for obese and elderly patients.

Author Contributions: A.C., C.B., C.S. and K.L. wrote the manuscript. All authors have read and agreed to the published version of the manuscript.

Funding: This work was supported by Region Occitanie: Green2Grape project and PhD grant (for Adriana Capozzi) from University of Montpellier (ED GAIA).

Conflicts of Interest: The authors declare no conflict of interest.

References

1. Zarulli, V.; Sopina, E.; Toffolutti, V.; Lenart, A. Health Care System Efficiency and Life Expectancy: A 140-Country Study. *PLoS ONE* **2021**, *16*, e0253450. [CrossRef]
2. Budreviciute, A.; Damiati, S.; Sabir, D.K.; Onder, K.; Schuller-Goetzburg, P.; Plakys, G.; Katileviciute, A.; Khoja, S.; Kodzius, R. Management and Prevention Strategies for Non-Communicable Diseases (NCDs) and Their Risk Factors. *Front. Public Health* **2020**, *8*, 574111. [CrossRef]
3. Divo, M.J.; Martinez, C.H.; Mannino, D.M. Ageing and the Epidemiology of Multimorbidity. *Eur. Respir. J.* **2014**, *44*, 1055–1068. [CrossRef]
4. Puspitasari, Y.M.; Ministrini, S.; Schwarz, L.; Karch, C.; Liberale, L.; Camici, G.G. Modern Concepts in Cardiovascular Disease: Inflamm-Aging. *Front. Cell Dev. Biol.* **2022**, *10*, 882211. [CrossRef]
5. The GBD 2015 Obesity Collaborators. Health Effects of Overweight and Obesity in 195 Countries over 25 Years. *N. Engl. J. Med.* **2017**, *377*, 13–27. [CrossRef] [PubMed]
6. Rohm, T.V.; Meier, D.T.; Olefsky, J.M.; Donath, M.Y. Inflammation in Obesity, Diabetes, and Related Disorders. *Immunity* **2022**, *55*, 31–55. [CrossRef] [PubMed]
7. Nunan, E.; Wright, C.L.; Semola, O.A.; Subramanian, M.; Balasubramanian, P.; Lovern, P.C.; Fancher, I.S.; Butcher, J.T. Obesity as a Premature Aging Phenotype—Implications for Sarcopenic Obesity. *GeroScience* **2022**, *44*, 1393–1405. [CrossRef] [PubMed]
8. Wilkinson, D.J.; Piasecki, M.; Atherton, P.J. The Age-Related Loss of Skeletal Muscle Mass and Function: Measurement and Physiology of Muscle Fibre Atrophy and Muscle Fibre Loss in Humans. *Ageing Res. Rev.* **2018**, *47*, 123–132. [CrossRef]
9. Tian, S.; Xu, Y. Association of Sarcopenic Obesity with the Risk of All-Cause Mortality: A Meta-Analysis of Prospective Cohort Studies: Sarcopenic Obesity and Mortality. *Geriatr. Gerontol. Int.* **2016**, *16*, 155–166. [CrossRef]
10. Afonso, C.; Sousa-Santos, A.R.; Santos, A.; Borges, N.; Padrão, P.; Moreira, P.; Amaral, T.F. Frailty Status Is Related to General and Abdominal Obesity in Older Adults. *Nutr. Res.* **2021**, *85*, 21–30. [CrossRef]
11. Huang, P.L. A Comprehensive Definition for Metabolic Syndrome. *Dis. Models Mech.* **2009**, *2*, 231–237. [CrossRef] [PubMed]
12. Petersmann, A.; Müller-Wieland, D.; Müller, U.A.; Landgraf, R.; Nauck, M.; Freckmann, G.; Heinemann, L.; Schleicher, E. Definition, Classification and Diagnosis of Diabetes Mellitus. *Exp. Clin. Endocrinol. Diabetes* **2019**, *127*, S1–S7. [CrossRef] [PubMed]
13. Xue, Q.-L. The Frailty Syndrome: Definition and Natural History. *Clin. Geriatr. Med.* **2011**, *27*, 1–15. [CrossRef] [PubMed]
14. Santos, A.L.; Sinha, S. Obesity and Aging: Molecular Mechanisms and Therapeutic Approaches. *Ageing Res. Rev.* **2021**, *67*, 101268. [CrossRef]
15. Frontera, W.R.; Ochala, J. Skeletal Muscle: A Brief Review of Structure and Function. *Calcif. Tissue Int.* **2015**, *96*, 183–195. [CrossRef]
16. Demontis, F.; Piccirillo, R.; Goldberg, A.L.; Perrimon, N. The Influence of Skeletal Muscle on Systemic Aging and Lifespan. *Aging Cell* **2013**, *12*, 943–949. [CrossRef]
17. Anker, S.D.; Ponikowski, P.; Varney, S.; Chua, T.P.; Clark, A.L.; Webb-Peploe, K.M.; Harrington, D.; Kox, W.J.; Poole-Wilson, P.A.; Coats, A.J. Wasting as Independent Risk Factor for Mortality in Chronic Heart Failure. *Lancet* **1997**, *349*, 1050–1053. [CrossRef]
18. Metter, E.J.; Talbot, L.A.; Schrager, M.; Conwit, R. Skeletal Muscle Strength as a Predictor of All-Cause Mortality in Healthy Men. *J. Gerontol. Ser. A Biol. Sci. Med. Sci.* **2002**, *57*, B359–B365. [CrossRef]
19. Gonzalez, A.; Simon, F.; Achiardi, O.; Vilos, C.; Cabrera, D.; Cabello-Verrugio, C. The Critical Role of Oxidative Stress in Sarcopenic Obesity. *Oxidative Med. Cell. Longev.* **2021**, *2021*, 4493817. [CrossRef] [PubMed]
20. Bielecka-Dabrowa, A.; Ebner, N.; Santos, M.R.; Ishida, J.; Hasenfuss, G.; Haehling, S. Cachexia, Muscle Wasting, and Frailty in Cardiovascular Disease. *Eur. J. Heart Fail.* **2020**, *22*, 2314–2326. [CrossRef]
21. Amouzou, C.; Breuker, C.; Fabre, O.; Bourret, A.; Lambert, K.; Birot, O.; Fédou, C.; Dupuy, A.-M.; Cristol, J.-P.; Sutra, T.; et al. Skeletal Muscle Insulin Resistance and Absence of Inflammation Characterize Insulin-Resistant Grade I Obese Women. *PLoS ONE* **2016**, *11*, e0154119. [CrossRef] [PubMed]
22. Aguer, C.; Mercier, J.; Kitzmann, M. Lipid Content and Response to Insulin Are Not Invariably Linked in Human Muscle Cells. *Mol. Cell. Endocrinol.* **2010**, *315*, 225–232. [CrossRef] [PubMed]
23. Aguer, C.; Foretz, M.; Lantier, L.; Hebrard, S.; Viollet, B.; Mercier, J.; Kitzmann, M. Increased FAT/CD36 Cycling and Lipid Accumulation in Myotubes Derived from Obese Type 2 Diabetic Patients. *PLoS ONE* **2011**, *6*, e28981. [CrossRef]
24. Mengeste, A.M.; Rustan, A.C.; Lund, J. Skeletal Muscle Energy Metabolism in Obesity. *Obesity* **2021**, *29*, 1582–1595. [CrossRef]
25. Borén, J.; Taskinen, M.-R.; Olofsson, S.-O.; Levin, M. Ectopic Lipid Storage and Insulin Resistance: A Harmful Relationship. *J. Intern. Med.* **2013**, *274*, 25–40. [CrossRef] [PubMed]
26. Bell, J.A.; Reed, M.A.; Consitt, L.A.; Martin, O.J.; Haynie, K.R.; Hulver, M.W.; Muoio, D.M.; Dohm, G.L. Lipid Partitioning, Incomplete Fatty Acid Oxidation, and Insulin Signal Transduction in Primary Human Muscle Cells: Effects of Severe Obesity, Fatty Acid Incubation, and Fatty Acid Translocase/CD36 Overexpression. *J. Clin. Endocrinol. Metab.* **2010**, *95*, 3400–3410. [CrossRef] [PubMed]

27. Løvsletten, N.G.; Rustan, A.C.; Laurens, C.; Thoresen, G.H.; Moro, C.; Nikolić, N. Primary Defects in Lipid Handling and Resistance to Exercise in Myotubes from Obese Donors with and without Type 2 Diabetes. *Appl. Physiol. Nutr. Metab.* **2020**, *45*, 169–179. [CrossRef] [PubMed]
28. Laurens, C.; Moro, C. Intramyocellular Fat Storage in Metabolic Diseases. *Horm. Mol. Biol. Clin. Investig.* **2016**, *26*, 43–52. [CrossRef] [PubMed]
29. Kitessa, S.; Abeywardena, M. Lipid-Induced Insulin Resistance in Skeletal Muscle: The Chase for the Culprit Goes from Total Intramuscular Fat to Lipid Intermediates, and Finally to Species of Lipid Intermediates. *Nutrients* **2016**, *8*, 466. [CrossRef] [PubMed]
30. Chaurasia, B.; Summers, S.A. Ceramides in Metabolism: Key Lipotoxic Players. *Annu. Rev. Physiol.* **2021**, *83*, 303–330. [CrossRef] [PubMed]
31. Holloway, G.P.; Thrush, A.B.; Heigenhauser, G.J.F.; Tandon, N.N.; Dyck, D.J.; Bonen, A.; Spriet, L.L. Skeletal Muscle Mitochondrial FAT/CD36 Content and Palmitate Oxidation Are Not Decreased in Obese Women. *Am. J. Physiol. Endocrinol. Metab.* **2007**, *292*, E1782–E1789. [CrossRef] [PubMed]
32. Schlaepfer, I.R.; Joshi, M. CPT1A-Mediated Fat Oxidation, Mechanisms, and Therapeutic Potential. *Endocrinology* **2020**, *161*, bqz046. [CrossRef]
33. Wu, H.; Ballantyne, C.M. Skeletal Muscle Inflammation and Insulin Resistance in Obesity. *J. Clin. Investig.* **2017**, *127*, 43–54. [CrossRef] [PubMed]
34. Akhmedov, D.; Berdeaux, R. The Effects of Obesity on Skeletal Muscle Regeneration. *Front. Physiol.* **2013**, *4*, 371. [CrossRef]
35. Jiang, A.; Guo, H.; Zhang, L.; Jiang, X.; Zhang, X.; Wu, W.; Liu, H. Free Fatty Acid Impairs Myogenic Differentiation through the AMPKα-MicroRNA 206 Pathway. *Mol. Cell. Biol.* **2022**, *42*, e00327-21. [CrossRef]
36. Tallis, J.; James, R.S.; Seebacher, F. The Effects of Obesity on Skeletal Muscle Contractile Function. *J. Exp. Biol.* **2018**, *221*, jeb163840. [CrossRef]
37. Abdelaal, M.; le Roux, C.W.; Docherty, N.G. Morbidity and Mortality Associated with Obesity. *Ann. Transl. Med.* **2017**, *5*, 161. [CrossRef]
38. Peeters, A.; Barendregt, J.J.; Willekens, F.; Mackenbach, J.P.; Mamun, A.A.; Bonneux, L.; NEDCOM, The Netherlands Epidemiology and Demography Compression of Morbidity Research Group. Obesity in Adulthood and Its Consequences for Life Expectancy: A Life-Table Analysis. *Ann. Intern. Med.* **2003**, *138*, 24. [CrossRef]
39. Goodpaster, B.H.; Park, S.W.; Harris, T.B.; Kritchevsky, S.B.; Nevitt, M.; Schwartz, A.V.; Simonsick, E.M.; Tylavsky, F.A.; Visser, M.; Newman, A.B.; et al. The Loss of Skeletal Muscle Strength, Mass, and Quality in Older Adults: The Health, Aging and Body Composition Study. *J. Gerontol. Ser. A Biol. Sci. Med. Sci.* **2006**, *61*, 1059–1064. [CrossRef] [PubMed]
40. Romanello, V.; Sandri, M. Mitochondrial Quality Control and Muscle Mass Maintenance. *Front. Physiol.* **2016**, *6*, 422. [CrossRef]
41. Tam, B.T.; Morais, J.A.; Santosa, S. Obesity and Ageing: Two Sides of the Same Coin. *Obes. Rev.* **2020**, *21*, e12991. [CrossRef] [PubMed]
42. Wu, M.; Luo, Q.; Nie, R.; Yang, X.; Tang, Z.; Chen, H. Potential Implications of Polyphenols on Aging Considering Oxidative Stress, Inflammation, Autophagy, and Gut Microbiota. *Crit. Rev. Food Sci. Nutr.* **2021**, *61*, 2175–2193. [CrossRef] [PubMed]
43. Peake, J.; Gatta, P.D.; Cameron-Smith, D. Aging and Its Effects on Inflammation in Skeletal Muscle at Rest and Following Exercise-Induced Muscle Injury. *Am. J. Physiol. Regul. Integr. Comp. Physiol.* **2010**, *298*, R1485–R1495. [CrossRef]
44. Visser, M.; Goodpaster, B.H.; Kritchevsky, S.B.; Newman, A.B.; Nevitt, M.; Rubin, S.M.; Simonsick, E.M.; Harris, T.B.; Health ABC Study. Muscle Mass, Muscle Strength, and Muscle Fat Infiltration as Predictors of Incident Mobility Limitations in Well-Functioning Older Persons. *J. Gerontol. Ser. A Biol. Sci. Med. Sci.* **2005**, *60*, 324–333. [CrossRef]
45. Milan, G.; Romanello, V.; Pescatore, F.; Armani, A.; Paik, J.-H.; Frasson, L.; Seydel, A.; Zhao, J.; Abraham, R.; Goldberg, A.L.; et al. Regulation of Autophagy and the Ubiquitin–Proteasome System by the FoxO Transcriptional Network during Muscle Atrophy. *Nat. Commun.* **2015**, *6*, 6670. [CrossRef] [PubMed]
46. Romanello, V. The Interplay between Mitochondrial Morphology and Myomitokines in Aging Sarcopenia. *Int. J. Mol. Sci.* **2020**, *22*, 91. [CrossRef] [PubMed]
47. Boengler, K.; Kosiol, M.; Mayr, M.; Schulz, R.; Rohrbach, S. Mitochondria and Ageing: Role in Heart, Skeletal Muscle and Adipose Tissue: Mitochondria and Ageing. *J. Cachexia Sarcopenia Muscle* **2017**, *8*, 349–369. [CrossRef]
48. Romanello, V.; Guadagnin, E.; Gomes, L.; Roder, I.; Sandri, C.; Petersen, Y.; Milan, G.; Masiero, E.; Del Piccolo, P.; Foretz, M.; et al. Mitochondrial Fission and Remodelling Contributes to Muscle Atrophy. *EMBO J.* **2010**, *29*, 1774–1785. [CrossRef] [PubMed]
49. Jang, J.Y.; Blum, A.; Liu, J.; Finkel, T. The Role of Mitochondria in Aging. *J. Clin. Investig.* **2018**, *128*, 3662–3670. [CrossRef]
50. Huang, D.-D.; Fan, S.-D.; Chen, X.-Y.; Yan, X.-L.; Zhang, X.-Z.; Ma, B.-W.; Yu, D.-Y.; Xiao, W.-Y.; Zhuang, C.-L.; Yu, Z. Nrf2 Deficiency Exacerbates Frailty and Sarcopenia by Impairing Skeletal Muscle Mitochondrial Biogenesis and Dynamics in an Age-Dependent Manner. *Exp. Gerontol.* **2019**, *119*, 61–73. [CrossRef]
51. Conte, M.; Martucci, M.; Sandri, M.; Franceschi, C.; Salvioli, S. The Dual Role of the Pervasive "Fattish" Tissue Remodeling With Age. *Front. Endocrinol.* **2019**, *10*, 114. [CrossRef]
52. Kaushik, S.; Cuervo, A.M. Proteostasis and Aging. *Nat. Med.* **2015**, *21*, 1406–1415. [CrossRef] [PubMed]
53. Sakuma, K.; Aoi, W.; Yamaguchi, A. Current Understanding of Sarcopenia: Possible Candidates Modulating Muscle Mass. *Pflugers Arch. Eur. J. Physiol.* **2015**, *467*, 213–229. [CrossRef] [PubMed]
54. Gumucio, J.P.; Mendias, C.L. Atrogin-1, MuRF-1, and Sarcopenia. *Endocrine* **2013**, *43*, 12–21. [CrossRef]

55. Jura, M.; Kozak, L.P. Obesity and Related Consequences to Ageing. *Age* **2016**, *38*, 23. [CrossRef] [PubMed]
56. Zuo, L.; Prather, E.R.; Stetskiv, M.; Garrison, D.E.; Meade, J.R.; Peace, T.I.; Zhou, T. Inflammaging and Oxidative Stress in Human Diseases: From Molecular Mechanisms to Novel Treatments. *Int. J. Mol. Sci.* **2019**, *20*, 4472. [CrossRef]
57. Meng, S.-J.; Yu, L.-J. Oxidative Stress, Molecular Inflammation and Sarcopenia. *Int. J. Mol. Sci.* **2010**, *11*, 1509–1526. [CrossRef]
58. Rotariu, D.; Babes, E.E.; Tit, D.M.; Moisi, M.; Bustea, C.; Stoicescu, M.; Radu, A.-F.; Vesa, C.M.; Behl, T.; Bungau, A.F.; et al. Oxidative Stress—Complex Pathological Issues Concerning the Hallmark of Cardiovascular and Metabolic Disorders. *Biomed. Pharmacother.* **2022**, *152*, 113238. [CrossRef] [PubMed]
59. Pizzino, G.; Irrera, N.; Cucinotta, M.; Pallio, G.; Mannino, F.; Arcoraci, V.; Squadrito, F.; Altavilla, D.; Bitto, A. Oxidative Stress: Harms and Benefits for Human Health. *Oxidative Med. Cell. Longev.* **2017**, *2017*, 8416763. [CrossRef]
60. Birben, E.; Sahiner, U.M.; Sackesen, C.; Erzurum, S.; Kalayci, O. Oxidative Stress and Antioxidant Defense. *World Allergy Organ. J.* **2012**, *5*, 9–19. [CrossRef] [PubMed]
61. Ali, S.S.; Ahsan, H.; Zia, M.K.; Siddiqui, T.; Khan, F.H. Understanding Oxidants and Antioxidants: Classical Team with New Players. *J. Food Biochem.* **2020**, *44*, e13145. [CrossRef] [PubMed]
62. Bouviere, J.; Fortunato, R.S.; Dupuy, C.; Werneck-de-Castro, J.P.; Carvalho, D.P.; Louzada, R.A. Exercise-Stimulated ROS Sensitive Signaling Pathways in Skeletal Muscle. *Antioxidants* **2021**, *10*, 537. [CrossRef] [PubMed]
63. Forrester, S.J.; Kikuchi, D.S.; Hernandes, M.S.; Xu, Q.; Griendling, K.K. Reactive Oxygen Species in Metabolic and Inflammatory Signaling. *Circ. Res.* **2018**, *122*, 877–902. [CrossRef] [PubMed]
64. Chen, L.; Deng, H.; Cui, H.; Fang, J.; Zuo, Z.; Deng, J.; Li, Y.; Wang, X.; Zhao, L. Inflammatory Responses and Inflammation-Associated Diseases in Organs. *Oncotarget* **2018**, *9*, 7204–7218. [CrossRef] [PubMed]
65. Cooke, J.P. Inflammation and Its Role in Regeneration and Repair: A Caution for Novel Anti-Inflammatory Therapies. *Circ. Res.* **2019**, *124*, 1166–1168. [CrossRef] [PubMed]
66. Juan, C.A.; Pérez de la Lastra, J.M.; Plou, F.J.; Pérez-Lebeña, E. The Chemistry of Reactive Oxygen Species (ROS) Revisited: Outlining Their Role in Biological Macromolecules (DNA, Lipids and Proteins) and Induced Pathologies. *Int. J. Mol. Sci.* **2021**, *22*, 4642. [CrossRef]
67. Nelke, C.; Dziewas, R.; Minnerup, J.; Meuth, S.G.; Ruck, T. Skeletal Muscle as Potential Central Link between Sarcopenia and Immune Senescence. *EBioMedicine* **2019**, *49*, 381–388. [CrossRef] [PubMed]
68. Guijarro-Muñoz, I.; Compte, M.; Álvarez-Cienfuegos, A.; Álvarez-Vallina, L.; Sanz, L. Lipopolysaccharide Activates Toll-like Receptor 4 (TLR4)-Mediated NF-KB Signaling Pathway and Proinflammatory Response in Human Pericytes. *J. Biol. Chem.* **2014**, *289*, 2457–2468. [CrossRef]
69. Lee, S.; Norheim, F.; Gulseth, H.L.; Langleite, T.M.; Kolnes, K.J.; Tangen, D.S.; Stadheim, H.K.; Gilfillan, G.D.; Holen, T.; Birkeland, K.I.; et al. Interaction between Plasma Fetuin-A and Free Fatty Acids Predicts Changes in Insulin Sensitivity in Response to Long-Term Exercise. *Physiol. Rep.* **2017**, *5*, e13183. [CrossRef]
70. Di Paola, M.; Lorusso, M. Interaction of Free Fatty Acids with Mitochondria: Coupling, Uncoupling and Permeability Transition. *Biochim. Biophys. Acta BBA Bioenerg.* **2006**, *1757*, 1330–1337. [CrossRef]
71. Lian, D.; Chen, M.-M.; Wu, H.; Deng, S.; Hu, X. The Role of Oxidative Stress in Skeletal Muscle Myogenesis and Muscle Disease. *Antioxidants* **2022**, *11*, 755. [CrossRef] [PubMed]
72. Kma, L.; Baruah, T.J. The Interplay of ROS and the PI3K/Akt Pathway in Autophagy Regulation. *Biotech. Appl. Biochem.* **2022**, *69*, 248–264. [CrossRef] [PubMed]
73. McCarthy, J.J.; Murach, K.A. Anabolic and Catabolic Signaling Pathways That Regulate Skeletal Muscle Mass. In *Nutrition and Enhanced Sports Performance*; Elsevier: Amsterdam, The Netherlands, 2019; pp. 275–290. ISBN 978-0-12-813922-6.
74. Jang, Y.C.; Rodriguez, K.; Lustgarten, M.S.; Muller, F.L.; Bhattacharya, A.; Pierce, A.; Choi, J.J.; Lee, N.H.; Chaudhuri, A.; Richardson, A.G.; et al. Superoxide-Mediated Oxidative Stress Accelerates Skeletal Muscle Atrophy by Synchronous Activation of Proteolytic Systems. *GeroScience* **2020**, *42*, 1579–1591. [CrossRef] [PubMed]
75. Nader, G.A. Molecular Determinants of Skeletal Muscle Mass: Getting the "AKT" Together. *Int. J. Biochem. Cell Biol.* **2005**, *37*, 1985–1996. [CrossRef]
76. Kitaoka, Y. The Role of Nrf2 in Skeletal Muscle on Exercise Capacity. *Antioxidants* **2021**, *10*, 1712. [CrossRef]
77. Huang, C.-J.; McAllister, M.J.; Slusher, A.L.; Webb, H.E.; Mock, J.T.; Acevedo, E.O. Obesity-Related Oxidative Stress: The Impact of Physical Activity and Diet Manipulation. *Sports Med. Open* **2015**, *1*, 32. [CrossRef]
78. Renaud, S.; de Lorgeril, M. Wine, Alcohol, Platelets, and the French Paradox for Coronary Heart Disease. *Lancet* **1992**, *339*, 1523–1526. [CrossRef]
79. Aune, D. Plant Foods, Antioxidant Biomarkers, and the Risk of Cardiovascular Disease, Cancer, and Mortality: A Review of the Evidence. *Adv. Nutr.* **2019**, *10*, S404–S421. [CrossRef]
80. Tanaka, T.; Talegawkar, S.A.; Jin, Y.; Bandinelli, S.; Ferrucci, L. Association of Adherence to the Mediterranean-Style Diet with Lower Frailty Index in Older Adults. *Nutrients* **2021**, *13*, 1129. [CrossRef]
81. Amiot, M.-J.; Latgé, C.; Plumey, L.; Raynal, S. Intake Estimation of Phytochemicals in a French Well-Balanced Diet. *Nutrients* **2021**, *13*, 3628. [CrossRef]
82. Neveu, V.; Perez-Jimenez, J.; Vos, F.; Crespy, V.; du Chaffaut, L.; Mennen, L.; Knox, C.; Eisner, R.; Cruz, J.; Wishart, D.; et al. Phenol-Explorer: An Online Comprehensive Database on Polyphenol Contents in Foods. *Database* **2010**, *2010*, bap024. [CrossRef] [PubMed]

83. Rothwell, J.A.; Perez-Jimenez, J.; Neveu, V.; Medina-Remon, A.; M'Hiri, N.; Garcia-Lobato, P.; Manach, C.; Knox, C.; Eisner, R.; Wishart, D.S.; et al. Phenol-Explorer 3.0: A Major Update of the Phenol-Explorer Database to Incorporate Data on the Effects of Food Processing on Polyphenol Content. *Database* **2013**, *2013*, bat070. [CrossRef] [PubMed]
84. Rothwell, J.A.; Urpi-Sarda, M.; Boto-Ordonez, M.; Knox, C.; Llorach, R.; Eisner, R.; Cruz, J.; Neveu, V.; Wishart, D.; Manach, C.; et al. Phenol-Explorer 2.0: A Major Update of the Phenol-Explorer Database Integrating Data on Polyphenol Metabolism and Pharmacokinetics in Humans and Experimental Animals. *Database* **2012**, *2012*, bas031. [CrossRef]
85. Fraga, C.G.; Galleano, M.; Verstraeten, S.V.; Oteiza, P.I. Basic Biochemical Mechanisms behind the Health Benefits of Polyphenols. *Mol. Asp. Med.* **2010**, *31*, 435–445. [CrossRef] [PubMed]
86. Bernatoniene, J.; Kopustinskiene, D. The Role of Catechins in Cellular Responses to Oxidative Stress. *Molecules* **2018**, *23*, 965. [CrossRef]
87. Youn, H.S.; Lee, J.Y.; Saitoh, S.I.; Miyake, K.; Kang, K.W.; Choi, Y.J.; Hwang, D.H. Suppression of MyD88- and TRIF-Dependent Signaling Pathways of Toll-like Receptor by (−)-Epigallocatechin-3-Gallate, a Polyphenol Component of Green Tea. *Biochem. Pharmacol.* **2006**, *72*, 850–859. [CrossRef] [PubMed]
88. Casanova, E.; Salvadó, J.; Crescenti, A.; Gibert-Ramos, A. Epigallocatechin Gallate Modulates Muscle Homeostasis in Type 2 Diabetes and Obesity by Targeting Energetic and Redox Pathways: A Narrative Review. *Int. J. Mol. Sci.* **2019**, *20*, 532. [CrossRef]
89. Li, Y.; Zhao, S.; Zhang, W.; Zhao, P.; He, B.; Wu, N.; Han, P. Epigallocatechin-3-O-Gallate (EGCG) Attenuates FFAs-Induced Peripheral Insulin Resistance through AMPK Pathway and Insulin Signaling Pathway in Vivo. *Diabetes Res. Clin. Pract.* **2011**, *93*, 205–214. [CrossRef]
90. Pallauf, K.; Rimbach, G. Autophagy, Polyphenols and Healthy Ageing. *Ageing Res. Rev.* **2013**, *12*, 237–252. [CrossRef]
91. Xu, M.; Chen, X.; Huang, Z.; Chen, D.; Yu, B.; Chen, H.; Luo, Y.; Zheng, P.; Yu, J.; He, J. Grape Seed Proanthocyanidin Extract Promotes Skeletal Muscle Fiber Type Transformation via AMPK Signaling Pathway. *J. Nutr. Biochem.* **2020**, *84*, 108462. [CrossRef]
92. Costa, L.G.; Garrick, J.M.; Roquè, P.J.; Pellacani, C. Mechanisms of Neuroprotection by Quercetin: Counteracting Oxidative Stress and More. *Oxidative Med. Cell. Longev.* **2016**, *2016*, 2986796. [CrossRef] [PubMed]
93. Sato, S.; Mukai, Y. Modulation of Chronic Inflammation by Quercetin: The Beneficial Effects on Obesity. *JIR* **2020**, *13*, 421–431. [CrossRef]
94. Pandey, K.B.; Rizvi, S.I. Role of Red Grape Polyphenols as Antidiabetic Agents. *Integr. Med. Res.* **2014**, *3*, 119–125. [CrossRef] [PubMed]
95. Alam, W.; Khan, H.; Shah, M.A.; Cauli, O.; Saso, L. Kaempferol as a Dietary Anti-Inflammatory Agent: Current Therapeutic Standing. *Molecules* **2020**, *25*, 4073. [CrossRef] [PubMed]
96. Mozos, I.; Flangea, C.; Vlad, D.C.; Gug, C.; Mozos, C.; Stoian, D.; Luca, C.T.; Horbańczuk, J.O.; Horbańczuk, O.K.; Atanasov, A.G. Effects of Anthocyanins on Vascular Health. *Biomolecules* **2021**, *11*, 811. [CrossRef]
97. Sivamaruthi, B.S.; Kesika, P.; Chaiyasut, C. The Influence of Supplementation of Anthocyanins on Obesity-Associated Comorbidities: A Concise Review. *Foods* **2020**, *9*, 687. [CrossRef]
98. Jia, Y.; Wu, C.; Kim, Y.-S.; Yang, S.O.; Kim, Y.; Kim, J.-S.; Jeong, M.-Y.; Lee, J.H.; Kim, B.; Lee, S.; et al. A Dietary Anthocyanin Cyanidin-3-O-Glucoside Binds to PPARs to Regulate Glucose Metabolism and Insulin Sensitivity in Mice. *Commun. Biol.* **2020**, *3*, 514. [CrossRef]
99. Geng, X.; Ji, J.; Liu, Y.; Li, X.; Chen, Y.; Su, L.; Zhao, L. Cyanidin-3-O-Glucoside Supplementation Ameliorates Metabolic Insulin Resistance via Restoration of Nitric Oxide-Mediated Endothelial Insulin Transport. *Mol. Nutr. Food Res.* **2022**, *66*, 2100742. [CrossRef]
100. Tian, C.; Liu, X.; Chang, Y.; Wang, R.; Lv, T.; Cui, C.; Liu, M. Investigation of the Anti-Inflammatory and Antioxidant Activities of Luteolin, Kaempferol, Apigenin and Quercetin. *S. Afr. J. Bot.* **2021**, *137*, 257–264. [CrossRef]
101. Wang, D.; Yang, Y.; Zou, X.; Zhang, J.; Zheng, Z.; Wang, Z. Antioxidant Apigenin Relieves Age-Related Muscle Atrophy by Inhibiting Oxidative Stress and Hyperactive Mitophagy and Apoptosis in Skeletal Muscle of Mice. *J. Gerontol. Ser. A* **2020**, *75*, 2081–2088. [CrossRef]
102. Park, M.-H.; Ju, J.-W.; Park, M.; Han, J. Daidzein Inhibits Carbohydrate Digestive Enzymes in Vitro and Alleviates Postprandial Hyperglycemia in Diabetic Mice. *Eur. J. Pharmacol.* **2013**, *712*, 48–52. [CrossRef] [PubMed]
103. Das, D.; Sarkar, S.; Bordoloi, J.; Wann, S.B.; Kalita, J.; Manna, P. Daidzein, Its Effects on Impaired Glucose and Lipid Metabolism and Vascular Inflammation Associated with Type 2 Diabetes: Prophylactic Role of Daidzein in Type 2 Diabetes. *BioFactors* **2018**, *44*, 407–417. [CrossRef] [PubMed]
104. Guevara-Cruz, M.; Godinez-Salas, E.T.; Sanchez-Tapia, M.; Torres-Villalobos, G.; Pichardo-Ontiveros, E.; Guizar-Heredia, R.; Arteaga-Sanchez, L.; Gamba, G.; Mojica-Espinosa, R.; Schcolnik-Cabrera, A.; et al. Genistein Stimulates Insulin Sensitivity through Gut Microbiota Reshaping and Skeletal Muscle AMPK Activation in Obese Subjects. *BMJ Open Diabetes Res. Care* **2020**, *8*, e000948. [CrossRef]
105. Rehman, K.; Khan, I.I.; Akash, M.S.H.; Jabeen, K.; Haider, K. Naringenin Downregulates Inflammation-mediated Nitric Oxide Overproduction and Potentiates Endogenous Antioxidant Status during Hyperglycemia. *J. Food Biochem.* **2020**, *44*, e13422. [CrossRef] [PubMed]
106. Dhanya, R.; Jayamurthy, P. In Vitro Evaluation of Antidiabetic Potential of Hesperidin and Its Aglycone Hesperetin under Oxidative Stress in Skeletal Muscle Cell Line. *Cell Biochem. Funct.* **2020**, *38*, 419–427. [CrossRef]

107. Lagouge, M.; Argmann, C.; Gerhart-Hines, Z.; Meziane, H.; Lerin, C.; Daussin, F.; Messadeq, N.; Milne, J.; Lambert, P.; Elliott, P.; et al. Resveratrol Improves Mitochondrial Function and Protects against Metabolic Disease by Activating SIRT1 and PGC-1α. *Cell* **2006**, *127*, 1109–1122. [CrossRef]
108. Kang, W.; Hong, H.J.; Guan, J.; Kim, D.G.; Yang, E.-J.; Koh, G.; Park, D.; Han, C.H.; Lee, Y.-J.; Lee, D.-H. Resveratrol Improves Insulin Signaling in a Tissue-Specific Manner under Insulin-Resistant Conditions Only: In Vitro and in Vivo Experiments in Rodents. *Metabolism* **2012**, *61*, 424–433. [CrossRef]
109. Huang, Y.; Zhu, X.; Chen, K.; Lang, H.; Zhang, Y.; Hou, P.; Ran, L.; Zhou, M.; Zheng, J.; Yi, L.; et al. Resveratrol Prevents Sarcopenic Obesity by Reversing Mitochondrial Dysfunction and Oxidative Stress via the PKA/LKB1/AMPK Pathway. *Aging* **2019**, *11*, 2217–2240. [CrossRef]
110. Bai, C.-H.; Alizargar, J.; Peng, C.-Y.; Wu, J.-P. Combination of Exercise Training and Resveratrol Attenuates Obese Sarcopenia in Skeletal Muscle Atrophy. *Chin. J. Physiol.* **2020**, *63*, 101. [CrossRef]
111. Kumar, N.; Goel, N. Phenolic Acids: Natural Versatile Molecules with Promising Therapeutic Applications. *Biotechnol. Rep.* **2019**, *24*, e00370. [CrossRef]
112. Abdel-Moneim, A.; El-Twab, S.M.A.; Yousef, A.I.; Reheim, E.S.A.; Ashour, M.B. Modulation of Hyperglycemia and Dyslipidemia in Experimental Type 2 Diabetes by Gallic Acid and P-Coumaric Acid: The Role of Adipocytokines and PPARγ. *Biomed. Pharmacother.* **2018**, *105*, 1091–1097. [CrossRef] [PubMed]
113. Silva, H.; Lopes, N.M.F. Cardiovascular Effects of Caffeic Acid and Its Derivatives: A Comprehensive Review. *Front. Physiol.* **2020**, *11*, 595516. [CrossRef] [PubMed]
114. Jung, Y.; Park, J.; Kim, H.; Sim, J.; Youn, D.; Kang, J.; Lim, S.; Jeong, M.; Yang, W.M.; Lee, S.; et al. Vanillic Acid Attenuates Obesity via Activation of the AMPK Pathway and Thermogenic Factors in Vivo and in Vitro. *FASEB J.* **2018**, *32*, 1388–1402. [CrossRef]
115. Rashedinia, M.; Alimohammadi, M.; Shalfroushan, N.; Khoshnoud, M.J.; Mansourian, M.; Azarpira, N.; Sabahi, Z. Neuroprotective Effect of Syringic Acid by Modulation of Oxidative Stress and Mitochondrial Mass in Diabetic Rats. *BioMed Res. Int.* **2020**, *2020*, 8297984. [CrossRef] [PubMed]
116. Pérez-Jiménez, J.; Fezeu, L.; Touvier, M.; Arnault, N.; Manach, C.; Hercberg, S.; Galan, P.; Scalbert, A. Dietary Intake of 337 Polyphenols in French Adults. *Am. J. Clin. Nutr.* **2011**, *93*, 1220–1228. [CrossRef]
117. Faustino, M.; Veiga, M.; Sousa, P.; Costa, E.; Silva, S.; Pintado, M. Agro-Food Byproducts as a New Source of Natural Food Additives. *Molecules* **2019**, *24*, 1056. [CrossRef]
118. Magrone, T.; Magrone, M.; Russo, M.A.; Jirillo, E. Recent Advances on the Anti-Inflammatory and Antioxidant Properties of Red Grape Polyphenols: In Vitro and In Vivo Studies. *Antioxidants* **2019**, *9*, 35. [CrossRef]
119. Bellezza, I.; Giambanco, I.; Minelli, A.; Donato, R. Nrf2-Keap1 Signaling in Oxidative and Reductive Stress. *Biochim. Biophys. Acta BBA Mol. Cell Res.* **2018**, *1865*, 721–733. [CrossRef]
120. Kicinska, A.; Jarmuszkiewicz, W. Flavonoids and Mitochondria: Activation of Cytoprotective Pathways? *Molecules* **2020**, *25*, 3060. [CrossRef]
121. Annunziata, G.; Jimenez-García, M.; Tejada, S.; Moranta, D.; Arnone, A.; Ciampaglia, R.; Tenore, G.C.; Sureda, A.; Novellino, E.; Capó, X. Grape Polyphenols Ameliorate Muscle Decline Reducing Oxidative Stress and Oxidative Damage in Aged Rats. *Nutrients* **2020**, *12*, 1280. [CrossRef] [PubMed]
122. Lambert, K.; Hokayem, M.; Thomas, C.; Fabre, O.; Cassan, C.; Bourret, A.; Bernex, F.; Feuillet-Coudray, C.; Notarnicola, C.; Mercier, J.; et al. Combination of Nutritional Polyphenols Supplementation with Exercise Training Counteracts Insulin Resistance and Improves Endurance in High-Fat Diet-Induced Obese Rats. *Sci. Rep.* **2018**, *8*, 2885. [CrossRef] [PubMed]
123. Rodgers, J.T.; Lerin, C.; Gerhart-Hines, Z.; Puigserver, P. Metabolic Adaptations through the PGC-1α and SIRT1 Pathways. *FEBS Lett.* **2008**, *582*, 46–53. [CrossRef]
124. Iside, C.; Scafuro, M.; Nebbioso, A.; Altucci, L. SIRT1 Activation by Natural Phytochemicals: An Overview. *Front. Pharmacol.* **2020**, *11*, 1225. [CrossRef]
125. Collier, C.A.; Bruce, C.R.; Smith, A.C.; Lopaschuk, G.; Dyck, D.J. Metformin Counters the Insulin-Induced Suppression of Fatty Acid Oxidation and Stimulation of Triacylglycerol Storage in Rodent Skeletal Muscle. *Am. J. Physiol. Endocrinol. Metab.* **2006**, *291*, E182–E189. [CrossRef]
126. Hokayem, M.; Blond, E.; Vidal, H.; Lambert, K.; Meugnier, E.; Feillet-Coudray, C.; Coudray, C.; Pesenti, S.; Luyton, C.; Lambert-Porcheron, S.; et al. Grape Polyphenols Prevent Fructose-Induced Oxidative Stress and Insulin Resistance in First-Degree Relatives of Type 2 Diabetic Patients. *Diabetes Care* **2013**, *36*, 1454–1461. [CrossRef]
127. Shabani, M.; Sadeghi, A.; Hosseini, H.; Teimouri, M.; Babaei Khorzoughi, R.; Pasalar, P.; Meshkani, R. Resveratrol Alleviates Obesity-Induced Skeletal Muscle Inflammation via Decreasing M1 Macrophage Polarization and Increasing the Regulatory T Cell Population. *Sci. Rep.* **2020**, *10*, 3791. [CrossRef]
128. Alamdari, N.; Aversa, Z.; Castillero, E.; Gurav, A.; Petkova, V.; Tizio, S.; Hasselgren, P.-O. Resveratrol Prevents Dexamethasone-Induced Expression of the Muscle Atrophy-Related Ubiquitin Ligases Atrogin-1 and MuRF1 in Cultured Myotubes through a SIRT1-Dependent Mechanism. *Biochem. Biophys. Res. Commun.* **2012**, *417*, 528–533. [CrossRef] [PubMed]
129. Momken, I.; Stevens, L.; Bergouignan, A.; Desplanches, D.; Rudwill, F.; Chery, I.; Zahariev, A.; Zahn, S.; Stein, T.P.; Sebedio, J.L.; et al. Resveratrol Prevents the Wasting Disorders of Mechanical Unloading by Acting as a Physical Exercise Mimetic in the Rat. *FASEB J.* **2011**, *25*, 3646–3660. [CrossRef]

130. Tuladhar, P.; Sasidharan, S.; Saudagar, P. Role of Phenols and Polyphenols in Plant Defense Response to Biotic and Abiotic Stresses. In *Biocontrol Agents and Secondary Metabolites*; Elsevier: Amsterdam, The Netherlands, 2021; pp. 419–441. ISBN 978-0-12-822919-4.
131. Benbouguerra, N.; Valls-Fonayet, J.; Krisa, S.; Garcia, F.; Saucier, C.; Richard, T.; Hornedo-Ortega, R. Polyphenolic Characterization of Merlot, Tannat and Syrah Skin Extracts at Different Degrees of Maturity and Anti-Inflammatory Potential in RAW 264.7 Cells. *Foods* **2021**, *10*, 541. [CrossRef] [PubMed]
132. Savoia, D. Plant-Derived Antimicrobial Compounds: Alternatives to Antibiotics. *Future Microbiol.* **2012**, *7*, 979–990. [CrossRef] [PubMed]
133. Cherubim, D.J.; Martins, C.V.; Fariña, L.; Lucca, R.A. Polyphenols as Natural Antioxidants in Cosmetics Applications. *J. Cosmet. Dermatol.* **2020**, *19*, 33–37. [CrossRef]
134. Frankel, E.N.; German, J.B.; Kinsella, J.E.; Parks, E.; Kanner, J. Inhibition of Oxidation of Human Low-Density Lipoprotein by Phenolic Substances in Red Wine. *Lancet* **1993**, *341*, 454–457. [CrossRef]
135. Frankel, E.N.; Waterhouse, A.L.; Kinsella, J.E. Inhibition of Human LDL Oxidation by Resveratrol. *Lancet* **1993**, *341*, 1103–1104. [CrossRef]
136. Sun, A.Y.; Simonyi, A.; Sun, G.Y. The "French Paradox" and beyond: Neuroprotective Effects of Polyphenols. *Free Radic. Biol. Med.* **2002**, *32*, 314–318. [CrossRef]
137. Gutiérrez-Escobar, R.; Aliaño-González, M.J.; Cantos-Villar, E. Wine Polyphenol Content and Its Influence on Wine Quality and Properties: A Review. *Molecules* **2021**, *26*, 718. [CrossRef]
138. Watson, R.R. *Polyphenols: Mechanisms of Action in Human Health and Disease*, 1st ed.; Elsevier: San Diego, CA, USA, 2018; ISBN 978-0-12-813006-3.
139. Baskar, V.; Venkatesh, R.; Ramalingam, S. Flavonoids (Antioxidants Systems) in Higher Plants and Their Response to Stresses. In *Antioxidants and Antioxidant Enzymes in Higher Plants*; Gupta, D.K., Palma, J.M., Corpas, F.J., Eds.; Springer International Publishing: Cham, Switzerand, 2018; pp. 253–268. ISBN 978-3-319-75087-3.
140. Ruiz-Cruz, S.; Chaparro-Hernández, S.; Ruiz, K.L.H.; Cira-Chávez, L.A.; Estrada-Alvarado, M.I.; Ortega, L.E.G.; de Ornelas-Paz, J.J.; Mata, M.A.L. Flavonoids: Important Biocompounds in Food. In *Flavonoids—From Biosynthesis to Human Health*; Justino, G.C., Ed.; InTech: Rijeka, Croatia, 2017; ISBN 978-953-51-3423-7.
141. Shen, N.; Wang, T.; Gan, Q.; Liu, S.; Wang, L.; Jin, B. Plant Flavonoids: Classification, Distribution, Biosynthesis, and Antioxidant Activity. *Food Chem.* **2022**, *383*, 132531. [CrossRef]
142. Aron, P.M.; Kennedy, J.A. Flavan-3-Ols: Nature, Occurrence and Biological Activity. *Mol. Nutr. Food Res.* **2008**, *52*, 79–104. [CrossRef]
143. Tsao, R. Chemistry and Biochemistry of Dietary Polyphenols. *Nutrients* **2010**, *2*, 1231–1246. [CrossRef] [PubMed]
144. Padilla-González, G.F.; Grosskopf, E.; Sadgrove, N.J.; Simmonds, M.S.J. Chemical Diversity of Flavan-3-Ols in Grape Seeds: Modulating Factors and Quality Requirements. *Plants* **2022**, *11*, 809. [CrossRef] [PubMed]
145. Latos-Brozio, M.; Masek, A. Structure-Activity Relationships Analysis of Monomeric and Polymeric Polyphenols (Quercetin, Rutin and Catechin) Obtained by Various Polymerization Methods. *Chem. Biodivers.* **2019**, *16*, e1900426. [CrossRef]
146. Habib, H.M.; El-Fakharany, E.M.; Kheadr, E.; Ibrahim, W.H. Grape Seed Proanthocyanidin Extract Inhibits DNA and Protein Damage and Labile Iron, Enzyme, and Cancer Cell Activities. *Sci. Rep.* **2022**, *12*, 12393. [CrossRef] [PubMed]
147. Jackson, R.S. Grapevine Structure and Function. In *Wine Science*; Elsevier: Amsterdam, The Netherlands, 2008; pp. 50–107. ISBN 978-0-12-373646-8.
148. Le, N.H.; Kim, C.-S.; Park, T.; Park, J.H.Y.; Sung, M.-K.; Lee, D.G.; Hong, S.-M.; Choe, S.-Y.; Goto, T.; Kawada, T.; et al. Quercetin Protects against Obesity-Induced Skeletal Muscle Inflammation and Atrophy. *Mediat. Inflamm.* **2014**, *2014*, 834294. [CrossRef] [PubMed]
149. Ohmae, S.; Akazawa, S.; Takahashi, T.; Izumo, T.; Rogi, T.; Nakai, M. Quercetin Attenuates Adipogenesis and Fibrosis in Human Skeletal Muscle. *Biochem. Biophys. Res. Commun.* **2022**, *615*, 24–30. [CrossRef]
150. Eseberri, I.; Laurens, C.; Miranda, J.; Louche, K.; Lasa, A.; Moro, C.; Portillo, M.P. Effects of Physiological Doses of Resveratrol and Quercetin on Glucose Metabolism in Primary Myotubes. *Int. J. Mol. Sci.* **2021**, *22*, 1384. [CrossRef]
151. Ding, Y.; Dai, X.; Zhang, Z.; Li, Y. Myricetin Attenuates Hyperinsulinemia-Induced Insulin Resistance in Skeletal Muscle Cells. *Eur. Food Res. Technol.* **2012**, *234*, 873–881. [CrossRef]
152. Sousa, C. Anthocyanins, Carotenoids and Chlorophylls in Edible Plant Leaves Unveiled by Tandem Mass Spectrometry. *Foods* **2022**, *11*, 1924. [CrossRef]
153. Chalker-Scott, L. Environmental Significance of Anthocyanins in Plant Stress Responses. *Photochem. Photobiol.* **1999**, *70*, 1–9. [CrossRef]
154. Hogan, S.; Canning, C.; Sun, S.; Sun, X.; Zhou, K. Effects of Grape Pomace Antioxidant Extract on Oxidative Stress and Inflammation in Diet Induced Obese Mice. *J. Agric. Food Chem.* **2010**, *58*, 11250–11256. [CrossRef]
155. Křížová, L.; Dadáková, K.; Kašparovská, J.; Kašparovský, T. Isoflavones. *Molecules* **2019**, *24*, 1076. [CrossRef]
156. Valletta, A.; Iozia, L.M.; Leonelli, F. Impact of Environmental Factors on Stilbene Biosynthesis. *Plants* **2021**, *10*, 90. [CrossRef]
157. Pecyna, P.; Wargula, J.; Murias, M.; Kucinska, M. More Than Resveratrol: New Insights into Stilbene-Based Compounds. *Biomolecules* **2020**, *10*, 1111. [CrossRef] [PubMed]

158. Babazadeh, A.; Taghvimi, A.; Hamishehkar, H.; Tabibiazar, M. Development of New Ultrasonic–Solvent Assisted Method for Determination of Trans-Resveratrol from Red Grapes: Optimization, Characterization, and Antioxidant Activity (ORAC Assay). *Food Biosci.* **2017**, *20*, 36–42. [CrossRef]
159. Baur, J.A.; Sinclair, D.A. Therapeutic Potential of Resveratrol: The in Vivo Evidence. *Nat. Rev. Drug Discov.* **2006**, *5*, 493–506. [CrossRef] [PubMed]
160. Koushki, M.; Amiri-Dashatan, N.; Ahmadi, N.; Abbaszadeh, H.-A.; Rezaei-Tavirani, M. Resveratrol: A Miraculous Natural Compound for Diseases Treatment. *Food Sci. Nutr.* **2018**, *6*, 2473–2490. [CrossRef] [PubMed]
161. Shadfar, S.; Couch, M.E.; McKinney, K.A.; Weinstein, L.J.; Yin, X.; Rodríguez, J.E.; Guttridge, D.C.; Willis, M. Oral Resveratrol Therapy Inhibits Cancer-Induced Skeletal Muscle and Cardiac Atrophy In Vivo. *Nutr. Cancer* **2011**, *63*, 749–762. [CrossRef]
162. Jackson, J.R.; Ryan, M.J.; Alway, S.E. Long-Term Supplementation With Resveratrol Alleviates Oxidative Stress but Does Not Attenuate Sarcopenia in Aged Mice. *J. Gerontol. Ser. A* **2011**, *66A*, 751–764. [CrossRef]
163. Sirago, G.; Toniolo, L.; Crea, E.; Giacomello, E. A Short-Term Treatment with Resveratrol Improves the Inflammatory Conditions of Middle-Aged Mice Skeletal Muscles. *Int. J. Food Sci. Nutr.* **2022**, *73*, 630–637. [CrossRef]
164. Ali, G.; Neda, G. Flavonoids and Phenolic Acids: Role and Biochemical Activity in Plants and Human. *J. Med. Plants Res.* **2011**, *5*, 6697–6703. [CrossRef]
165. Rashmi, H.B.; Negi, P.S. Phenolic Acids from Vegetables: A Review on Processing Stability and Health Benefits. *Food Res. Int.* **2020**, *136*, 109298. [CrossRef]
166. Doan, K.V.; Ko, C.M.; Kinyua, A.W.; Yang, D.J.; Choi, Y.-H.; Oh, I.Y.; Nguyen, N.M.; Ko, A.; Choi, J.W.; Jeong, Y.; et al. Gallic Acid Regulates Body Weight and Glucose Homeostasis Through AMPK Activation. *Endocrinology* **2015**, *156*, 157–168. [CrossRef]
167. Cory, H.; Passarelli, S.; Szeto, J.; Tamez, M.; Mattei, J. The Role of Polyphenols in Human Health and Food Systems: A Mini-Review. *Front. Nutr.* **2018**, *5*, 87. [CrossRef] [PubMed]
168. Cipolletti, M.; Solar Fernandez, V.; Montalesi, E.; Marino, M.; Fiocchetti, M. Beyond the Antioxidant Activity of Dietary Polyphenols in Cancer: The Modulation of Estrogen Receptors (ERs) Signaling. *Int. J. Mol. Sci.* **2018**, *19*, 2624. [CrossRef] [PubMed]
169. Hong, K.-B.; Lee, H.-S.; Hong, J.S.; Kim, D.H.; Moon, J.M.; Park, Y. Effects of Tannase-Converted Green Tea Extract on Skeletal Muscle Development. *BMC Complement. Med. Ther.* **2020**, *20*, 47. [CrossRef] [PubMed]
170. Nikawa, T.; Ulla, A.; Sakakibara, I. Polyphenols and Their Effects on Muscle Atrophy and Muscle Health. *Molecules* **2021**, *26*, 4887. [CrossRef] [PubMed]
171. Dudoit, A.; Benbouguerra, N.; Richard, T.; Hornedo-Ortega, R.; Valls-Fonayet, J.; Coussot, G.; Saucier, C. α-Glucosidase Inhibitory Activity of Tannat Grape Phenolic Extracts in Relation to Their Ripening Stages. *Biomolecules* **2020**, *10*, 1088. [CrossRef]
172. Aleixandre, A.; Gil, J.V.; Sineiro, J.; Rosell, C.M. Understanding Phenolic Acids Inhibition of α-Amylase and α-Glucosidase and Influence of Reaction Conditions. *Food Chem.* **2022**, *372*, 131231. [CrossRef]
173. Arora, I.; Sharma, M.; Sun, L.Y.; Tollefsbol, T.O. The Epigenetic Link between Polyphenols, Aging and Age-Related Diseases. *Genes* **2020**, *11*, 1094. [CrossRef]
174. Shirazi-Tehrani, E.; Chamasemani, A.; Firouzabadi, N.; Mousaei, M. NcRNAs and Polyphenols: New Therapeutic Strategies for Hypertension. *RNA Biol.* **2022**, *19*, 575–587. [CrossRef]
175. Scalbert, A.; Morand, C.; Manach, C.; Rémésy, C. Absorption and Metabolism of Polyphenols in the Gut and Impact on Health. *Biomed. Pharmacother.* **2002**, *56*, 276–282. [CrossRef]
176. Roblegg, E.; Coughran, A.; Sirjani, D. Saliva: An All-Rounder of Our Body. *Eur. J. Pharm. Biopharm.* **2019**, *142*, 133–141. [CrossRef]
177. Morzel, M.; Canon, F.; Guyot, S. Interactions between Salivary Proteins and Dietary Polyphenols: Potential Consequences on Gastrointestinal Digestive Events. *J. Agric. Food Chem.* **2022**, *70*, 6317–6327. [CrossRef] [PubMed]
178. Ginsburg, I.; Kohen, R.; Koren, E. Saliva: A 'Solubilizer' of Lipophilic Antioxidant Polyphenols. *Oral Dis.* **2013**, *19*, 321–322. [CrossRef] [PubMed]
179. Chen, L.; Cao, H.; Xiao, J. Polyphenols. In *Polyphenols: Properties, Recovery, and Applications*; Elsevier: Amsterdam, The Netherlands, 2018; pp. 45–67. ISBN 978-0-12-813572-3.
180. Chen, L.; Teng, H.; Xie, Z.; Cao, H.; Cheang, W.S.; Skalicka-Woniak, K.; Georgiev, M.I.; Xiao, J. Modifications of Dietary Flavonoids towards Improved Bioactivity: An Update on Structure–Activity Relationship. *Crit. Rev. Food Sci. Nutr.* **2018**, *58*, 513–527. [CrossRef] [PubMed]
181. Walle, T. Absorption and Metabolism of Flavonoids. *Free Radic. Biol. Med.* **2004**, *36*, 829–837. [CrossRef]
182. Catalkaya, G.; Venema, K.; Lucini, L.; Rocchetti, G.; Delmas, D.; Daglia, M.; De Filippis, A.; Xiao, H.; Quiles, J.L.; Xiao, J.; et al. Interaction of Dietary Polyphenols and Gut Microbiota: Microbial Metabolism of Polyphenols, Influence on the Gut Microbiota, and Implications on Host Health. *Food Front.* **2020**, *1*, 109–133. [CrossRef]
183. Bode, L.M.; Bunzel, D.; Huch, M.; Cho, G.-S.; Ruhland, D.; Bunzel, M.; Bub, A.; Franz, C.M.; Kulling, S.E. In Vivo and in Vitro Metabolism of Trans-Resveratrol by Human Gut Microbiota. *Am. J. Clin. Nutr.* **2013**, *97*, 295–309. [CrossRef]
184. Cortés-Martín, A.; Selma, M.V.; Tomás-Barberán, F.A.; González-Sarrías, A.; Espín, J.C. Where to Look into the Puzzle of Polyphenols and Health? The Postbiotics and Gut Microbiota Associated with Human Metabotypes. *Mol. Nutr. Food Res.* **2020**, *64*, 1900952. [CrossRef]
185. Stevens, J.F.; Maier, C.S. The Chemistry of Gut Microbial Metabolism of Polyphenols. *Phytochem. Rev.* **2016**, *15*, 425–444. [CrossRef]

186. Selma, M.V.; Tomás-Barberán, F.A.; Romo-Vaquero, M.; Cortés-Martín, A.; Espín, J.C. Understanding Polyphenols' Health Effects Through the Gut Microbiota. In *Dietary Polyphenols*; Tomás-Barberán, F.A., González-Sarrías, A., García-Villalba, R., Eds.; Wiley: Hoboken, NJ, USA, 2020; pp. 497–531. ISBN 978-1-119-56375-4.
187. Russell, W.R.; Scobbie, L.; Chesson, A.; Richardson, A.J.; Stewart, C.S.; Duncan, S.H.; Drew, J.E.; Duthie, G.G. Anti-Inflammatory Implications of the Microbial Transformation of Dietary Phenolic Compounds. *Nutr. Cancer* **2008**, *60*, 636–642. [CrossRef]
188. Serreli, G.; Deiana, M. In Vivo Formed Metabolites of Polyphenols and Their Biological Efficacy. *Food Funct.* **2019**, *10*, 6999–7021. [CrossRef]
189. Hoda, M.; Hemaiswarya, S.; Doble, M. Pharmacokinetics and Pharmacodynamics of Polyphenols. In *Role of Phenolic Phytochemicals in Diabetes Management*; Springer: Singapore, 2019; pp. 159–173. ISBN 9789811389962.
190. Yang, Y.; Trevethan, M.; Wang, S.; Zhao, L. Beneficial Effects of Citrus Flavanones Naringin and Naringenin and Their Food Sources on Lipid Metabolism: An Update on Bioavailability, Pharmacokinetics, and Mechanisms. *J. Nutr. Biochem.* **2022**, *104*, 108967. [CrossRef] [PubMed]
191. Chalet, C.; Rubbens, J.; Tack, J.; Duchateau, G.S.; Augustijns, P. Intestinal Disposition of Quercetin and Its Phase-II Metabolites after Oral Administration in Healthy Volunteers. *J. Pharm. Pharmacol.* **2018**, *70*, 1002–1008. [CrossRef]
192. Springer, M.; Moco, S. Resveratrol and Its Human Metabolites—Effects on Metabolic Health and Obesity. *Nutrients* **2019**, *11*, 143. [CrossRef]
193. Brasnyó, P.; Molnár, G.A.; Mohás, M.; Markó, L.; Laczy, B.; Cseh, J.; Mikolás, E.; Szijártó, I.A.; Mérei, Á.; Halmai, R.; et al. Resveratrol Improves Insulin Sensitivity, Reduces Oxidative Stress and Activates the Akt Pathway in Type 2 Diabetic Patients. *Br. J. Nutr.* **2011**, *106*, 383–389. [CrossRef]
194. Timmers, S.; Konings, E.; Bilet, L.; Houtkooper, R.H.; van de Weijer, T.; Goossens, G.H.; Hoeks, J.; van der Krieken, S.; Ryu, D.; Kersten, S.; et al. Calorie Restriction-like Effects of 30 Days of Resveratrol Supplementation on Energy Metabolism and Metabolic Profile in Obese Humans. *Cell Metab.* **2011**, *14*, 612–622. [CrossRef] [PubMed]
195. Goh, K.P.; Lee, H.Y.; Lau, D.P.; Supaat, W.; Chan, Y.H.; Koh, A.F.Y. Effects of Resveratrol in Patients with Type 2 Diabetes Mellitus on Skeletal Muscle SIRT1 Expression and Energy Expenditure. *Int. J. Sport Nutr. Exerc. Metab.* **2014**, *24*, 2–13. [CrossRef]
196. Timmers, S.; de Ligt, M.; Phielix, E.; van de Weijer, T.; Hansen, J.; Moonen-Kornips, E.; Schaart, G.; Kunz, I.; Hesselink, M.K.C.; Schrauwen-Hinderling, V.B.; et al. Resveratrol as Add-on Therapy in Subjects With Well-Controlled Type 2 Diabetes: A Randomized Controlled Trial. *Diabetes Care* **2016**, *39*, 2211–2217. [CrossRef] [PubMed]
197. Woerdeman, J.; Del Rio, D.; Calani, L.; Eringa, E.C.; Smulders, Y.M.; Serné, E.H. Red Wine Polyphenols Do Not Improve Obesity-associated Insulin Resistance: A Randomized Controlled Trial. *Diabetes Obes. Metab.* **2018**, *20*, 206–210. [CrossRef]
198. Segrestin, B.; Delage, P.; Nemeth, A.; Seyssel, K.; Disse, E.; Nazare, J.-A.; Lambert-Porcheron, S.; Meiller, L.; Sauvinet, V.; Chanon, S.; et al. Polyphenol Supplementation Did Not Affect Insulin Sensitivity and Fat Deposition During One-Month Overfeeding in Randomized Placebo-Controlled Trials in Men and in Women. *Front. Nutr.* **2022**, *9*, 854255. [CrossRef]
199. Alway, S.E.; McCrory, J.L.; Kearcher, K.; Vickers, A.; Frear, B.; Gilleland, D.L.; Bonner, D.E.; Thomas, J.M.; Donley, D.A.; Lively, M.W.; et al. Resveratrol Enhances Exercise-Induced Cellular and Functional Adaptations of Skeletal Muscle in Older Men and Women. *J. Gerontol. Ser. A* **2017**, *72*, 1595–1606. [CrossRef] [PubMed]
200. McDermott, M.M.; Leeuwenburgh, C.; Guralnik, J.M.; Tian, L.; Sufit, R.; Zhao, L.; Criqui, M.H.; Kibbe, M.R.; Stein, J.H.; Lloyd-Jones, D.; et al. Effect of Resveratrol on Walking Performance in Older People With Peripheral Artery Disease: The RESTORE Randomized Clinical Trial. *JAMA Cardiol.* **2017**, *2*, 902. [CrossRef]
201. Gliemann, L.; Schmidt, J.F.; Olesen, J.; Biensø, R.S.; Peronard, S.L.; Grandjean, S.U.; Mortensen, S.P.; Nyberg, M.; Bangsbo, J.; Pilegaard, H.; et al. Resveratrol Blunts the Positive Effects of Exercise Training on Cardiovascular Health in Aged Men: Adverse Effects of Resveratrol on Cardiovascular Health. *J. Physiol.* **2013**, *591*, 5047–5059. [CrossRef]
202. Olesen, J.; Gliemann, L.; Biensø, R.; Schmidt, J.; Hellsten, Y.; Pilegaard, H. Exercise Training, but Not Resveratrol, Improves Metabolic and Inflammatory Status in Skeletal Muscle of Aged Men: Resveratrol and Exercise Training in Aged Human Subjects. *J. Physiol.* **2014**, *592*, 1873–1886. [CrossRef]
203. Harper, S.A.; Bassler, J.R.; Peramsetty, S.; Yang, Y.; Roberts, L.M.; Drummer, D.; Mankowski, R.T.; Leeuwenburgh, C.; Ricart, K.; Patel, R.P.; et al. Resveratrol and Exercise Combined to Treat Functional Limitations in Late Life: A Pilot Randomized Controlled Trial. *Exp. Gerontol.* **2021**, *143*, 111111. [CrossRef]
204. Buchanan, A.; Villani, A. Association of Adherence to a Mediterranean Diet with Excess Body Mass, Muscle Strength and Physical Performance in Overweight or Obese Adults with or without Type 2 Diabetes: Two Cross-Sectional Studies. *Healthcare* **2021**, *9*, 1255. [CrossRef] [PubMed]
205. De La Cruz Cortés, J.P.; Vallejo-Carmona, L.; Arrebola, M.M.; Martín-Aurioles, E.; Rodriguez-Pérez, M.D.; Ortega-Hombrados, L.; Verdugo, C.; Fernández-Prior, M.Á.; Bermúdez-Oria, A.; González-Correa, J.A. Synergistic Effect of 3′,4′-Dihidroxifenilglicol and Hydroxytyrosol on Oxidative and Nitrosative Stress and Some Cardiovascular Biomarkers in an Experimental Model of Type 1 Diabetes Mellitus. *Antioxidants* **2021**, *10*, 1983. [CrossRef]
206. Kurin, E.; Atanasov, A.; Donath, O.; Heiss, E.; Dirsch, V.; Nagy, M. Synergy Study of the Inhibitory Potential of Red Wine Polyphenols on Vascular Smooth Muscle Cell Proliferation. *Planta Med.* **2012**, *78*, 772–778. [CrossRef]
207. Brglez Mojzer, E.; Knez Hrnčič, M.; Škerget, M.; Knez, Ž.; Bren, U. Polyphenols: Extraction Methods, Antioxidative Action, Bioavailability and Anticarcinogenic Effects. *Molecules* **2016**, *21*, 901. [CrossRef]

208. Brat, P.; Mennen, L.; Georgé, S.; Scalbert, A.; Bellamy, A.; Amiot-Carlin, M.-J.; Du Chaffaut, L. Determination of the polyphenol content of fruits and vegetables. Establishment of a database and estimation of the polyphenol intake in the french diet. *Acta Hortic.* **2007**, *744*, 61–70. [CrossRef]
209. Chuang, C.-C.; McIntosh, M.K. Potential Mechanisms by Which Polyphenol-Rich Grapes Prevent Obesity-Mediated Inflammation and Metabolic Diseases. *Annu. Rev. Nutr.* **2011**, *31*, 155–176. [CrossRef]
210. Shahbandeh, M. Statista: Global Production of Fruit by Variety Selected 2020. Available online: https://www.statista.com/statistics/264001/worldwide-production-of-fruit-by-variety (accessed on 2 August 2022).

molecules

MDPI

Article

Comprehensive Lichenometabolomic Exploration of *Ramalina conduplicans* Vain Using UPLC-Q-ToF-MS/MS: An Identification of Free Radical Scavenging and Anti-Hyperglycemic Constituents

Tatapudi Kiran Kumar [1,2,†], Bandi Siva [1,†], Ajay Anand [1], Komati Anusha [1], Satish Mohabe [3,4], Araveeti Madhusudana Reddy [3], Françoise Le Devehat [5], Ashok Kumar Tiwari [1,2], Joël Boustie [5,*] and Katragadda Suresh Babu [1,2,*]

1 Centre for Natural Products & Traditional Knowledge, CSIR-Indian Institute of Chemical Technology, Uppal Road, Tarnaka, Hyderabad 500007, India
2 Academy of Scientific and Innovative Research (AcSIR), Ghaziabad 201002, India
3 Department of Botany, Yogi Vemana University, Vemanapuram, Kadapa 516003, India
4 Faculty of Sciences & IT, Madhyanchal Professional University, Ratibad, Bhopal 462044, India
5 Institut des Sciences Chimiques de Rennes, Université Rennes, CNRS, ISCR-UMR6226, 35000 Rennes, France
* Correspondence: joel.boustie@univ-rennes1.fr (J.B.); suresh@iict.res.in (K.S.B.)
† These authors contributed equally to this work.

Abstract: In this study, we propose ultra-performance liquid chromatography coupled with quadrupole/time-of-flight mass spectrometry (UPLC-QToF-MS/MS)-guided metabolite isolation as a choice analytical approach to the ongoing structure–activity investigations of chemical isolates from the edible lichen, *Ramalina conduplicans* Vain. This strategy led to the isolation and identification of a new depside (**5**) along with 13 known compounds (**1–4**, **6–14**), most of which being newly described in this lichen species. The structures of the isolates were established by detailed analysis of their spectral data (IR, NMR, and Mass). The acetone extract was further analyzed by UPLC-Q-ToF-MS/MS in a negative ionization mode, which facilitated the identification and confirmation of 18 compounds based on their fragmentation patterns. The antioxidant capacities of the lichen acetone extract (AE) and isolates were measured by tracking DPPH and ABTS free radical scavenging activities. Most isolates displayed marked radical scavenging activities against ABTS while moderate activities were observed against DPPH radical scavenging. Except for atranol (**14**), oxidative DNA damage was limited by all the tested compounds, with a marked protection for the novel isolated compound (**5**), as previously noted for the acetone extract ($p < 0.001$). Furthermore, compound (**4**) and acetone extract (AE) have inhibited intestinal α-glucosidase enzyme significantly ($p < 0.01$). Although some phytochemical studies were already performed on this lichen, this study provided new insights into the isolation and identification of bioactive compounds, illustrating interest in future novel analytical techniques.

Keywords: *R. conduplicans*; lichen; secondary metabolites; antioxidant; DNA damage; α-glucosidase inhibition

Citation: Kumar, T.K.; Siva, B.; Anand, A.; Anusha, K.; Mohabe, S.; Reddy, A.M.; Le Devehat, F.; Tiwari, A.K.; Boustie, J.; Babu, K.S. Comprehensive Lichenometabolomic Exploration of *Ramalina conduplicans* Vain Using UPLC-Q-ToF-MS/MS: An Identification of Free Radical Scavenging and Anti-Hyperglycemic Constituents. *Molecules* **2022**, *27*, 6720. https://doi.org/10.3390/molecules27196720

Academic Editor: Nour Eddine Es-Safi

Received: 24 August 2022
Accepted: 30 September 2022
Published: 9 October 2022

1. Introduction

Lichens are structurally complex and self-sustaining unique consortia comprised of a fungus host (mycobiont) living with algae or cyanobacteria (photobiont partner) in the framework of a unique symbiotic type of relationship. In recent years, much attention has been paid to the biological roles of lichen metabolites because of their potential applications in perfumery, cosmetics, creative crafts, the dye industry, and the pharmaceutical sector [1,2]. Moreover, many lichens and their extracts and metabolites have been utilized as ingredients in ethnic food preparations and specialties, along with ethnomedicinal applications [3]. For example, a mixture of lichens called Yangben in the Rai and Limbu communities of East Nepal is mainly composed of *Ramalina* species [4]. Among these fruticose epiphytic species, *Ramalina conduplicans* Vain is common and one of the most widely-used edible lichen of the Ramalinaceae family, this is distributed in Central and Southeastern Asian countries [5].

In Southwestern China, people used to prepare their traditional cold dishes with this lichen at their marriage banquets [6], and it also has a long history of consumption as a spice in many places in India and as a traditional food by selected communities in East Nepal [7,8]. In addition to its useful edible properties, crude extracts from this lichen are used as ethnomedicine to counteract inflammation, anthelminthic [9], and act as an anti-diabetic [10], along with antibiotic activities [3,11,12]. Many studies on this lichen have focused on their nutritional value along with the important trace elements [13] and antioxidant properties of *R. conduplicans* [9,14] concerning sekikaic acid and homosekikaic acid [15]. However, systematic investigations of its constituents for their bioactive potentials have not been carried out to date.

Therefore, the antioxidant and alpha-glucosidase inhibiting properties of metabolites from *Ramalina conduplicans* were investigated here as part of our ongoing exploration of natural flora for the isolation of bioactive secondary metabolites [16,17]. Accordingly, we have designed a strategy and workflow based on the Total Ion Current Chromatography (TIC) of the acetone extract (AE) to recognize and to isolate compounds from *R. conduplicans* by UPLC-Q-ToF-MS/MS. AE and all the isolates were assayed for their antioxidant free-radical scavenging properties, including DNA damage protection and anti-hyperglycemic potential, through α-glucosidase inhibition.

2. Results and Discussion

The *R. conduplicans* sample was identified by morphological characteristics and thallus reactions: K+ pale yellow, KC−, P−, and also negative reaction of the medulla to calcium hypochlorite solution (C−) (Figure S2, Supplementary Materials). These usual spot tests are based on the presence of lichen metabolites, but have to be supplemented by accurate analytical studies to reveal the metabolite content.

2.1. Chemical Profiling, Isolation, and Structure Elucidation

A HPTLC (Figure S3, Supplementary Materials) co-migration with standards and the UPLC-PDA profile (Figure S4, Supplementary Materials) of the acetone extract of *R. conduplicans* suggested the presence of a dozen of visible compounds, among which salazinic acid, usnic acid, sekikaic acid, homosekikaic acid, and divaricatic acid were identified against standards and appeared to represent the most abundant compounds. Initial LC-QToF-MS^E analyses of the acetone extract of *R. conduplicans* indicated the presence of depsides, depsidones, and monophenolic acids based on High-Resolution Mass Spectroscopy (HRMS). Molecular formulae for $C_{10–35}H_{10–50}O_{2–15}$ were generated from mass ranges m/z 150–750 coupled with the fragment ions and their MS spectral data (accurate mass and fragmentation pattern) and compared to online databases (DNP, Reaxys, SciFinder).

Mass spectrometry (MS) and, particularly, quadruple time-of-flight coupled to Liquid Chromatography (UPLC-Q-ToF-MS) has been widely utilized for profiling metabolites due to its superiority in high-resolution mass, precision, and sensitivity [18], and was helpful to clearly discriminate between the depsides, depsidones, simple phenol acids, dibenzofurans, and hydroxyl fatty acids based on the fragmentation of lichen molecules [16]. Therefore, the acquired TIC of the *R. conduplicans* extracts, obtained within 16 min, were analyzed from spectra obtained in negative mode and, thus, are effective for characterizing trace components (Figure 1). Metabolite assignments were made based on their polarity related to their retention time (Rt) and molecular formulae from accurate molecular weight measurement, along with adducts $[M - H]^-$/ fragment ions and Ring Double Bond Equivalence (RDBE). In the present study, a total of 18 compounds were clearly characterized from the crude extract of *R. conduplicans* by molecular formulae generated by ToF-MS/MS and MS/MS including their fragmentation profiles, as reported in the literature and presented in Table S1 (Supplementary Materials).

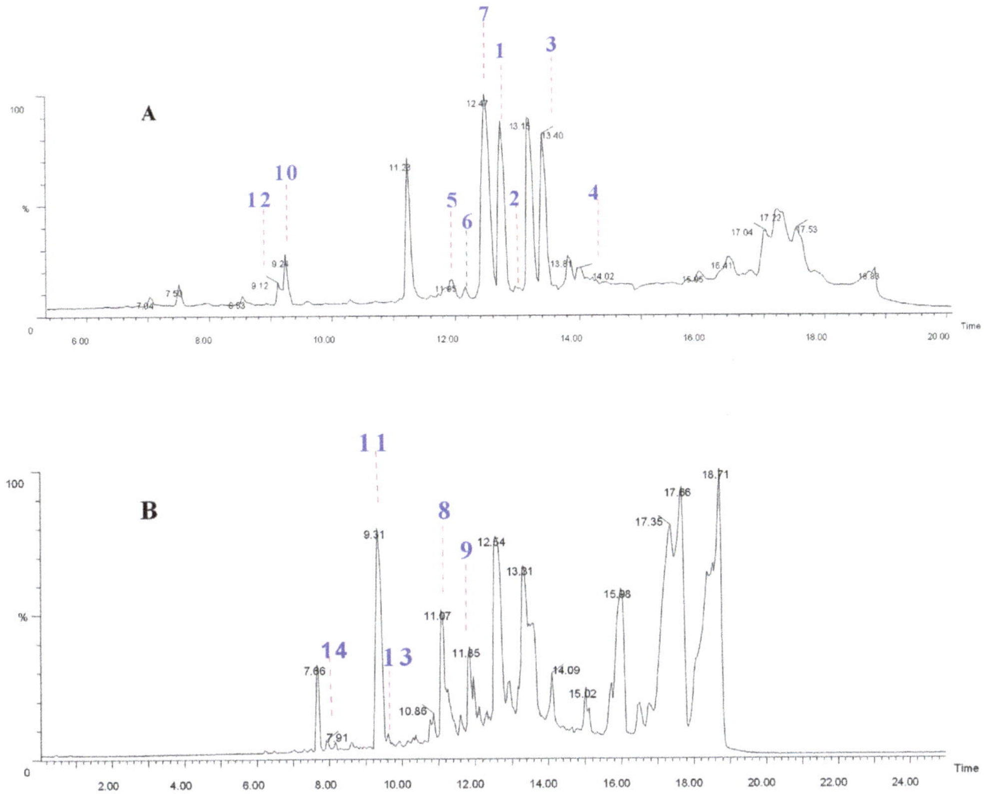

Figure 1. TIC of (**A**) *R. conduplicans* acetone extract and (**B**) enriched fraction-4.

Based on the fragmentation of isolates, we have identified compounds (**1–5**) and (**7–9**) along with atranorin belonging to depsides. The literature clearly indicates that sekikaic acid is an abundant molecule in *Ramalina* species [19]. Sekikaic acid (**1**) is a *m*-depside corresponding to the esterification of two divaricatinic acid units and is found with Rt at 11.98 min and m/z 417.1547 ($C_{22}H_{25}O_8$) with fragments m/z 209 and m/z 225 corresponding to the A ring and B ring, respectively [20]. Compounds **1**, **3**, **5**, **7**, **8**, and **9**, having a common fragment m/z 209 (Figures S47 and S48, Supplementary Materials), clearly indicate the difference in locating the other Bring. These depsides can be considered as ester derivatives of divaricatinic acid (**11**) while compound **2** is a divaric acid derivative (recognized at Rt 7.50 min, m/z 195.0657). The other identified monoaromatic compounds correspond to 2,4-di-*O*-methyldivaric acid (**6**), 4-*O*-methyldivaricatic acid (**10**), divaricatinic acid (**11**), olivetolic acid (**12**), divarinolmonomethylether (**13**), and atranol (**14**). In this run, three additional compounds were ionized and fragmented (Rt = 8.53 min, Rt = 11.88 min, and Rt = 13.17 min) and not determined. The fragmentation feature of Compound **5** (m/z 401.1954 $[M-H]^-$ (calcd. for $[C_{23}H_{28}O_6]^-$ 401,1964)) suggested the coupling of a divaricatinic acid moiety to an olivetol monomethylether moiety (Figure S17, Supplementary Materials). Based on these fragmentation studies, we assigned compounds as shown in the Supporting Information section and in Table S1, including the monoaromatic divaric acid, along with the common and already-described atranorin (depside), usnic acid (related to dibenzofurans), and salazinic acid (a depsidone). The structures were concluded through MS/MS fragmentation patterns and compared with in-house standards.

Subsequently, the acetone extract was subjected to column chromatography to give eight fractions (I to VIII). An LC–MS^E analysis of all fractions revealed the presence of depsides in III–VI fractions (Figures S4–S6, Supplementary Materials). Thus, the targeted isolation and purification of III–VI fractions yielded the isolation of one new depside (**5**), along with other known depsides (**1–4** and**7–9**) and monoaromatic compounds (**6** and **10–14**). The spectra and fragmentation patterns of these molecules were shown in the Supporting Information section (Figures S7–S46, Supplementary Materials).

The structures of the isolated compounds (Figure 2) were determined by a combination of spectroscopic data (HRESIMS, ^{1}H and ^{13}C NMR) and in comparison with the reported literature data. They were identified as sekikaic acid (**1**) [21], 4′-*O*-methylnorhomosekikaic acid (**2**) [22], homosekikaic acid (**3**) [22], hyperhomosekikaic acid (**4**) [23], 2,4-dimethyldivaric acid (**6**) [24], divaricatic acid (**7**) [25], decarboxydivaricatic acid (**8**) [26], decarboxystenosporic acid (**9**) [26], methyldivaricatinate (**10**) [24], divaricatinic acid (**11**) [21], olivetolic acid (**12**) [27], divarinolmonomethylether (**13**) [21], and atranol (**14**) [28].

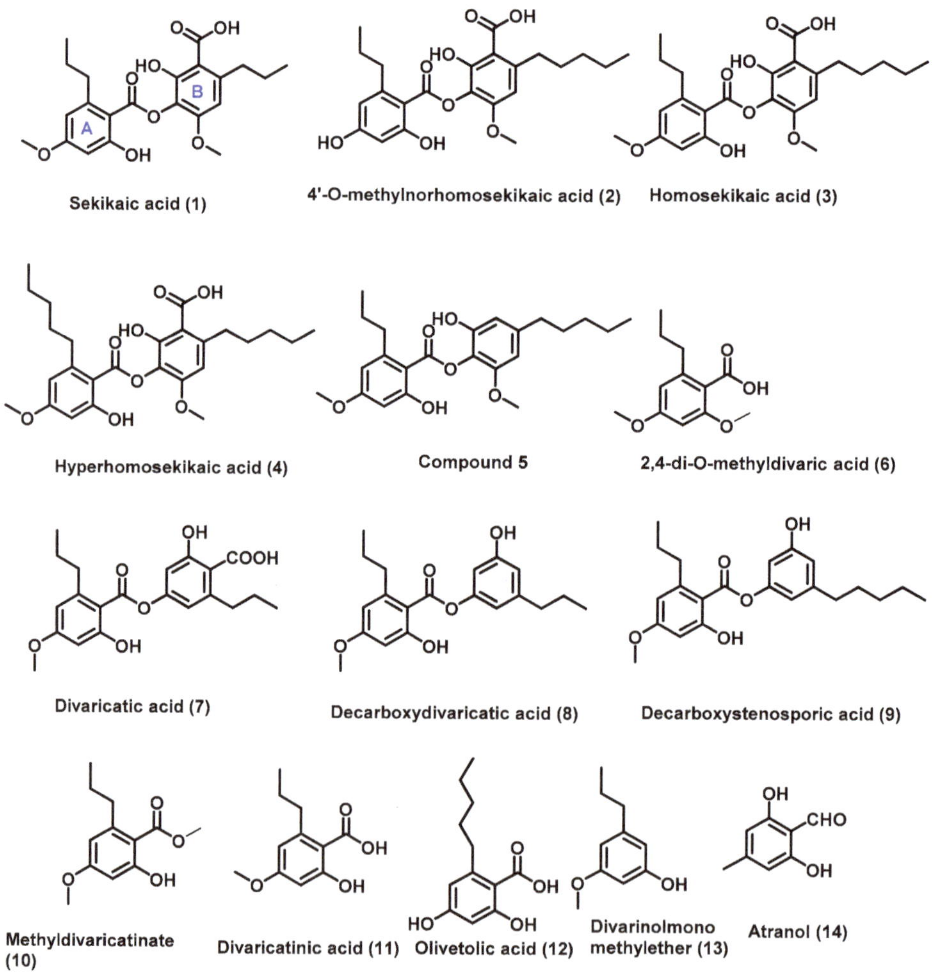

Figure 2. Isolated compounds (**1–14**) from *Ramalina conduplicans* Vain.

Compound **5** was isolated as white amorphous powder and identified as a new compound. Its molecular formula was established as $C_{23}H_{29}O_6$ based on a HRESIMS ion at m/z 401.1954 $[M-H]^-$ (calcd. for $[C_{23}H_{28}O_6]^-$, 401.1964). The ^{1}H and ^{13}C NMR data of **5** (Table 1) showed the presence of four aromatic protons, (δ_H 6.53 (d, J = 1.8 Hz, 1H), 6.51 (d, J = 1.8 Hz, 1H), 6.46 (d, J = 2.6 Hz), and 6.40 (d, J = 2.6 Hz); δ_C 120.0, 110.7, 105.4, and 100.8) and one ester carbonyl (δ_C 170.3). In addition, two methoxyl groups (δ_H 3.86 (3H, s), 3.81 (3H, s)) and two n-alkane side chains of two methylene groups that were adjacent to a benzene ring (δ_H 3.02–2.93 (m, 2H), 2.62–2.51 (m, 2H)) were also distinguished from the NMR spectra, respectively (Table S2, Supplementary Materials). These spectral features, together with the characteristic ester carbonyl group at C-7 (δ_C 170.3) in the ^{13}C NMR spectrum, strongly imply that **5** is a depside-type derivative [16,29].

Table 1. NMR data of compound **5** (400 & 100 MHz, acetone-d_6) *.

S no	^{1}H NMR of 5	^{13}C NMR of 5
1	–	107.35
2	–	166.10
3	6.46 (d, J = 2.6 Hz, 1H)	112.07
4	–	166.10
5	6.40 (d, J = 2.6 Hz, 1H)	100.65
6	–	149.91
7	–	170.47
1′	6.53 (d, J = 1.8 Hz, 1H)	110.80
2′	–	151.36
3′	–	154.12
4′	–	143.54
5′	6.51 (d, J = 1.8 Hz, 1H)	105.49
6′	–	150.38
1″	3.0–2.93 (m, 2H)	40.06
2″	1.82–1.68 (m, 2H)	26.58
3″	0.93 (t, J = 7.6, 3H)	15.60
1″	2.62–2.51 (m, 2H)	37.69
2‴	1.67–1.59 (m, 2H)	32.89
3‴	1.41–1.30 (m, 2H)	33.25
4‴	1.41–1.30 (m, 2H)	24.18
5‴	0.93 (t, J = 7.6 Hz, 3H)	15.30
OMe-7′	3.81 (s, 3H)	57.31
OMe-8	3.86 (s, 3H)	56.83

* = values are assigned with the comparison of sekikaic acid data and COSY/NOESY correlations.

A comparison of ^{1}H NMR and ^{13}C NMR data from**5** with those of 4′-*O*-methylnorhomosekikaic acid, which were isolated from the same species, indicated an overall similarity, except for the absence of a COOH group and the presence of two additional methylenes. This reasoning was further supported by its ^{13}C NMR spectrum, which showed the absence of a carbonyl COOH group, and its ^{1}H NMR spectrum indicated the presence of an additional aromatic proton at 6.53 (d, J = 1.8 Hz, 1H). A comprehensive analysis of 2D NMR (COSY, and HSQC) data, especially the ^{1}H–^{1}H COSY spectrum, revealed two discrete spin systems, including -CH-CH_2-CH_3- (from H-1″, H-2″ and H-3″) and -CH-CH_2-CH_2-CH_2-CH_3 (from H-1‴ to 5‴), as drawn with bold lines in Figure 3. The position of the *n*-pentyl group at C-6′ and *n*-propyl chain at C-6 was confirmed on the basis of the NOESY correlations (H-1‴/H-5′, H-1‴/H-1′ and H-1″/H-5) (Figure 3) and in comparison with the sekikaic acid data. In addition, the MS/MS spectrum of **5** showed (Figure 4) product ions *m*/*z* 209, thereby indicating the breakage of the C–O bond between two aromatic rings supported by the fragments at *m*/*z* 165 and 137. Based on these spectral characteristics, the structure of **5** was established and trivially named as decarboxyhomosekikaic acid.

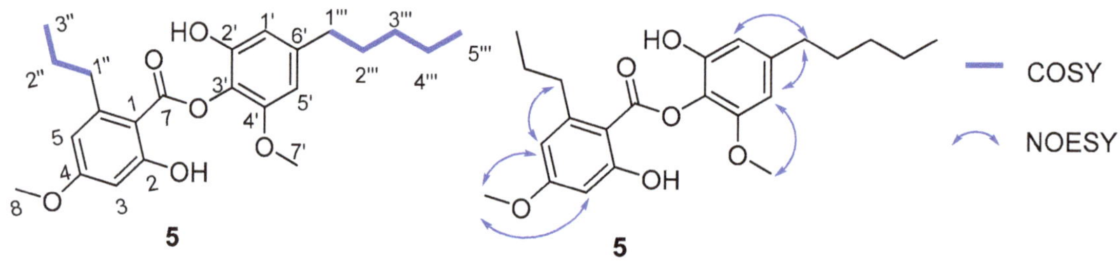

Figure 3. Key COSY and NOESY correlations of compound **5**.

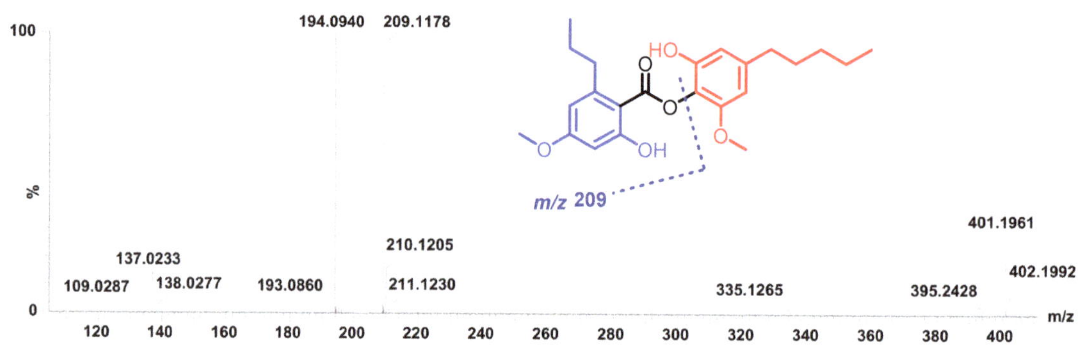

Figure 4. MS/MS spectrum and proposed fragmentation of compound **5**.

3. Biological Activity

3.1. Assessment of Compounds and Extract for Free Radicals Scavenging and Antioxidant Activity

The amphiphilic nature of the ABTS$^{•+}$ cation was used to identify both hydrophilic and hydrophobic antioxidants in dietary materials, whereas the DPPH- radical was used to measure an antioxidant's reducing power [30]. These fundamental chemical experiments reveal the radical scavenging and reduction characteristics of the potential antioxidant candidates.

Acetone extract (AE) and all isolated compounds (**1–14**) scavenged ABTS$^{•+}$ and DPPH^{-} radicals and the results are presented in Table 2. The results have demonstrated that AE and all the other compounds potently neutralized ABTS$^{•+}$ radicals (more than 70%) and have shown activity equal to the ascorbic acid standard, except for compound **13**, which only scavenged radicals by 50%.The pattern and potentials in decreasing order of ABTS$^{•+}$ scavenging potentials were observed as follows:olivetolic acid (**12**) > compound **5** > divarinolmonomethylether (**13**) > decarboxydivaricatic acid (**8**) > 4-*O*-methylnorhomo sekikaic acid (**2**) > atranol (**14**) > divaricatic acid (**7**) > decarboxystenosporic acid (**9**) > sekikaic acid (**1**) > 2,4-dimethyldivaric acid (**6**) > homo sekikaic acid (**3**) > methyldivaricatinate (**10**). In the case of the DPPH-radical scavenging assay, acetone extract (AE) and compound **14** scavenged DPPH-radicals potently by more than 50%, whereas **1**, **2**, **3**, **4**, **5**, **6**, **12**, and **13** countervailed DPPH-radicals (20–40%) moderately. It is important to mention that potent ABTS$^{•+}$ scavenging activities were observed in all compounds, but DPPH scavenging activity was detected to be moderate in all compounds except compound **14** (Table 2). As ABTS$^{•+}$ is a planar radical, it can be used to identify antioxidants even with low redox potentials. However, due to the steric barrier of the N$^{\bullet}$ radical, they may react slowly or not at all when tested on DPPH radicals [31]. This might be the reason why extracts and compounds scavenge ABTS$^{\bullet+}$ more potently than DPPH-radicals. To check whether this radical scavenging activity is related to antioxidant properties, we challenged genomic DNA with hydrogen peroxide (H_2O_2)-induced oxidative damage.

Table 2. Free radical scavenging activities of AE and compounds (**1–14**) of *Ramalina conduplicans*.

Compound Name (Code)	**DPPH Assay % Scavenging (SC_{50}, μg/mL)**	**ABTS Assay % Scavenging, (SC_{50}, μg/mL)**
Sekikaic acid (**1**)	37.75 ± 0.65	99.05 ± 0.00 (2.45)
4-*O*-methylnorhomosekikaic acid (**2**)	36.78 ± 1.57	98.57 ± 0.00 (1.40)
Homosekikaic acid (**3**)	38.28 ± 1.22	98.10 ± 0.00 (2.81)
Hyperhomosekikaic acid (**4**)	27.32 ± 0.34	79.52 ± 1.02 (17.44)
Compound **5**	46.29 ± 3.70	100.48 ± 0.68 (0.44)
2,4-dimethyldivaric acid (**6**)	29.67 ± 0.89	100.20 ± 0.34 (2.46)
Divaricatic acid (**7**)	8.57 ± 1.65	99.28 ± 1.02 (2.09)
Decarboxydivaricatic acid (**8**)	17.17 ± 1.74	100.28 ± 0.00 (0.75)
Decarboxystenosporic acid (**9**)	13.04 ± 0.73	96.7 ± 0.5 (2.41)
Methyl divaricatinate (**10**)	6.50 ± 0.76	100.0 ± 1.0 (2.90)
Divaricatinic acid (**11**)	ND	100.0 ± 0.5 (2.63)
Olivetolic acid (**12**)	41.94 ± 1.11	92.7 ± 0.0 (0.13)
Divarinolmonomethylether (**13**)	21.12 ± 1.21	51.7 ± 2.5 (0.57)
Atranol (**14**)	74.66 ± 2.59 (18.65)	88.9 ± 7.3 (2.05)
Acetone Extract (AE)	50.64	97.3
Ascorbic Acid	93.25 ± 1.23 (3.96)	99.02 ± 0.03 (0.47)

ND = Not determined. The activity is expressed as % scavenging with regard to ascorbic acid scavenging activity. The SC_{50} is indicated for the most active compounds.

3.2. *Protective Effect of R. conduplicans AE and Isolated Compounds on Oxidative DNA Damage*

The Fenton's reaction produces the hydroxyl radical, which is a ROS that is detrimental to the human body. Hydroxyl radicals react with different nucleobases, thereby inducing the formation of mutated bases that eventually lead to DNA damage [32]. Figure 5 demonstrated that FR damaged DNA significantly ($p < 0.001$) compared to the control (DMSO + DNA). Though all compounds showed significant protection against hydroxyl radical-induced DNA damage ($p < 0.001$, cpd **10**: $p < 0.05$), compound **14** could not prevent the oxidative damage to DNA (Figures 5 and S49). The genoprotective activity of these compounds and the AE may be attributed to the presence of free radical scavenging potential.

3.3. *Assessment of In Vitro Antihyperglycemic Activity of Compounds and Extract as Intestinal α-Glucosidase Enzyme Inhibition*

The α-glucosidase enzyme is a key enzyme that catalyses disaccharide digestion. The inhibition of α-glucosidase in the intestine slows digestion and the overall rate of glucose absorption into the blood. This has proven to be one of the most effective ways for lowering post-prandial blood glucose levels and, as a result, preventing the onset of late diabetes complications [33]. Sekikaic acid (**1**) was already recognized to inhibit α-glucosidase along with usnic acid and salazinic acid from other *Ramalina* species, but it is not the most effective compound [34]. As per Figure 6, it was stated that acetone extract (AE) and compound **4** have displayed better α-glucosidase inhibition ($p < 0.01$) than Acarbose.

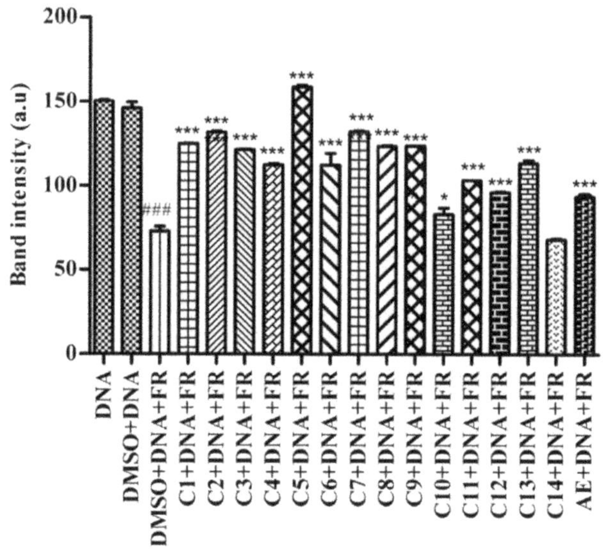

Figure 5. In vitro DNA damage assay. Compounds (**1–14**) and *R. conduplicans.* AE were incubated with DNA and Fenton's Reagent and DNA damage was recorded with Agarose Gel electrophoresis. Respective graphical representation. ### $p < 0.001$; vs. control (DMSO + DNA). *** $p < 0.001$, * $p < 0.05$; vs.DMSO + DNA + FR, One-way ANOVA followed by Tukey's multiple comparison test was used to calculate values. Values are represented as mean ± SD, $n = 3$. AE = Acetone Extract, FR = Fenton's Reagent.

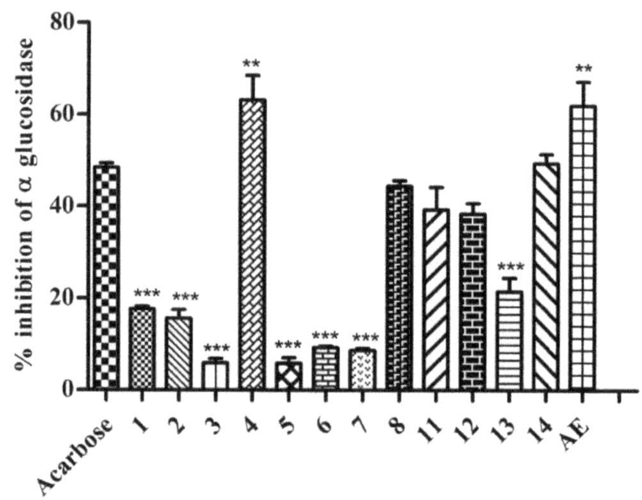

Figure 6. Intestinal α-glucosidase inhibitory assay. Compounds (**1–8**), (**11–14**), and acetone extract (AE) were incubated with α-glucosidase enzyme and the release of *p*-nitrophenol was determined. *** $p < 0.001$, ** $p < 0.01$; vs. Acarbose. One-way ANOVA followed by Tukey's multiple comparison test was applied to compare differences. Values are represented as mean ± SD, $n = 3$. AE = Acetone Extract. Activity was not detected for compounds **9** and **10**.

On the other hand, compounds **8**, **11**, **12**, and **14** demonstrated inhibitory effects comparable to those of the standard Acarbose (Figure 6). As contrasted activities can be observed between structurally-related compounds, structure–activity relationships can be considered. This is the case between depsides **3** and **4,** suggesting a positive influence of the C1-pentyl chain with regard to substitution by a C1-propyl chain. When this length modification of the alkyl chain occurs on the B ring of decarboxylated derivatives (active compound **8** versus inactive compound **9**) the opposite influence can be observed. The presence of a C6′-carboxylic group lowers the α-glucosidase inhibitory activity as compound **7** is less active than compound **8**. Methylation of the carboxylic function of the monoaromatic divaricatinic acid **11** resulted in a complete loss of activity. Nevertheless, most of the tested compounds were found with some activity, such as compounds **1**, **2**, **3**, **5**, **6**, **7**, and **13,** which displayed mild to moderate enzyme inhibition ($p < 0.001$). These results are to be pooled with the growing number of reports on the antidiabetic potential of lichen extracts or molecules [34–36]. The combination of activities with different mechanisms of action is of particular interest to develop potent antihyperglycemic effects. Lowering glucose absorption and limiting oxidative damages due to hyperglycemia, as expected from the lichen extract, could be promising. The challenge is to use standardized extracts that were previously checked to be safe for acute and chronic intake.

4. Materials and Methods

4.1. General

The NMR spectra were recorded on a Bruker FT-400 MHz NMR spectrometer and samples were dissolved in deuterated acetone-d_6. Mass data were acquired on aXevo™ G2 XS-ESI-QTof mass spectrometer (Waters Corp., Manchester, UK). For thin layer chromatography (TLC) analysis, precoated Merck plates (silica gel 60 F_{254}) were utilized. Silica gel (100–200 mesh) (Qing-dao Marine Chemical, Inc., Qingdao, China) was chosen for column chromatographic separation. Semi-preparative chromatography was performed on a Gilson HPLC (Middleton, WI, USA) instrument equipped with a 321 binary pump, GX-281 liquid handler, and UV-155 detector with X Select HSS T3 (250 mm × 100 mm, 5 μm) (Waters Corp., Drinagh, Ireland) as a stationary phase using a Trilution LC v2.1 platform. Formic acid (Optima™ Mass spec grade) (Thermo Fisher Scientific, Geel, Belgium), HPLC-grade acetonitrile, LiChrosolv (Merck, Darmstadt, Germany), and ultra-pure water (Millipore System, Randolph, MA, USA) were used.

4.2. Instrumental UPLC Conditions

The instrumental conditions were set-up as per our recent report (Reddy et al., 2019) with slight modifications. Chromatographic separation was performed on an Acquity H Class UPLC system (Waters, Milford, MA, USA) with a conditioned auto sampler using an ACQUITY UPLC CSH Phenyl-Hexyl column (100 mm × 2.1 mm id., 1.7 μm particle size) (Waters, Milford, MA, USA). Column temperature was maintained at 40 °C. High-resolution masses of secondary metabolites were measured after UPLC separation. A mobile phase consisting of water with 0.1% formic acid in water (solvent A) and acetonitrile with 0.1% formic acid (solvent B) was pumped at a flow rate of 0.4 mL/min. The gradient elution program was as follows: 0 min, 5% B; 3.00 min, 20% B; 5.00min, 35% B; 7.50 min, 50% B; 10.00 min, 70% B; 12.50 min, 95% B; 17.00 min 95% B; and 21.00 min 5% B. The equilibration time was 4.0 min and the injection volume was 2 μL. The LC-QTof-MS^E mode was applied to analyze the samples in both TIC as well as the MS/MS mode, where the collision energy was ramped at 15–45 eV. Eluted compounds were detected from m/z 50 to 1200 using a Xevo G2-XS Q-Tof mass spectrometer (Waters, Manchester, UK), which was connected to Electro-spray ionization (ESI) interface with a negative ion mode using the following instrument settings: capillary voltage, 2.0 KV; sample cone, 40 V; source temperature, 120 °C; desolvation temperature 350 °C; cone gas flow rate 50 L/h; desolvation gas (N_2) flow rate 850 L/h, argon as CID gas for MS/MS experiments. All analyses were performed using lock spray, which ensured accuracy and reproducibility. Leucine–Enkephalin (5 ng/mL) was used as a lock mass, generating a reference ion in the negative mode at m/z 554.2615, and was introduced by a lock spray at 10 μL/min for accurate mass acquisition. Data acquisition was achieved using MassLynx ver. 4.1.

Acquiring data in this manner provided information on intact precursor ions as well as fragment ions.

4.3. Lichen Sample Collection and Identification

The lichen, *Ramalina conduplicans,* was collected from tree bark in Bichpuri Range, Bijrani Zone of Corbett National Park, alt. N 29°26′40″ E79°04′06 (1283 m) in the month of May 2019. The morphological features of lichen thallus and ascomata were observed under Magnüs MS 24/13, and spot tests for color reaction were carried out by 10% aqueous solution of potassium hydroxide (K), Steiner's stable *p*-phenylene diamine solution (PD), and calcium hypochlorite solution (C). For the anatomical investigation of fruiting bodies, a light microscope from ZEISS Axiostar was used. The lichen substances were identified with thin layer chromatography in solvent system 'A' following White and James's methods (1985). The standard literature [37] was referred to for identification of lichen samples. The voucher specimens (Satish Mohabe & A. Madhusudhana Reddy 7658YVUH) of species were deposited at the Herbarium, Department of Botany, Yogi Vemana University, Kadapa, Andhra Pradesh. The corresponding data are shown in Figures S1–S3, Supporting Materials.

4.4. Extraction and Isolation

The sorted-out lichen *Ramalina conduplicans* (300 g) was shade-dried, powdered, and extracted with acetone (6L) at room temperature for 48h. The result was that acetone extract was evaporated to dryness under reduced pressure, thereby affording a syrupy residue (20 g). This crude extract was subjected to gradient column chromatography (SiO_2, 60–120 mesh) and eluted with a hexane/EtOAc mixture of increasing polarity with 10% intervals that yielded 8 fractions. These eight fractions were reconstituted in acetonitrile and subjected to UPLC Q-ToF MS^E. Based on the TIC profile, we selected fractions 3–6 for purification (mass profile shown in supporting information, discussion in results section). All these fractions were subjected to semi-preparative HPLC (X Select HSST3 OBD Prep Column,5 µm, 10 mm × 250 mm), 0.1 % formic acid with water (solvent A) and acetonitrile (solvent B) as mobile phase at flow rate 4 mL/min, detected at 254 nm. Semi-preparative HPLC were conducted by gradient elution programs to obtain compounds as follows: Fraction 3 (quantity 70 mg, loop volume 250 µL was eluted by 0 min, 5% B; 5 min, 5% B; 10.00 min, 35% B; 16.00 min, 60% B; 25 min, 95% B; 30 min, 95% B; 5% B; 30.50 min, 5% B; 35.00 min. to yield **3** (homosekikaic acid, 7 mg), **7** (divaricatic acid, 4 mg), **11** (divaricatinic acid, 3 mg), **12** (olivetolic acid, 5 mg), and **14** (atranol, 2 mg). Fraction 4 (quantity 50 mg, loop volume 250 µL was eluted by 0 min, 30% B; 5 min, 30% B; 10.00 min, 50% B; 23.00 min, 95% B; 27.00 min, 95% B; 27.50 min, 30% B; 30% B; 30.00 min at flow rate 4 mL/min, detected at 254 nm to yield **10** (methyldivaricatinate, 3 mg), **8** (decarboxydivaricatic acid, 5 mg), **9** (decarboxystenosporic acid, 2 mg), and **4** (hyperhomosekikaic acid, 1mg). Fraction 5 (quantity 25 mg, loop volume 250 µL was eluted by 0 min, 5% B; 8.50 min, 30% B; 15.00 min, 50% B; 22.00 min, 95% B; 28.00 min, 95% B; 29.0 min, 5% B; 5% B; 35.00 min at flow rate 4 mL/min, detected at 254 nm) to yield **13** (divarinolmonomethylether, 3 mg) and compound **5** (2 mg). Fraction 6 (quantity 40 mg, loop volume 250 µL was eluted by 0 min, 10% B; 8.50 min, 40% B; 18.00 min, 55% B; 25.00 min, 75% B; 32.00 min, 95% B; 36.0 min, 95% B; 10% B; 37.00 min, 10% B; 42.00 min at flow rate 4 mL/min, detected at 254 nm) to yield **1** (sekikaic acid, 5 mg), **2** (4-*O*-methylnorhomosekikaic acid, 7 mg) and 2,4-di-*O*-methyldivaric acid **6** (2 mg). Physicochemical data are shown in the Supporting Materials.

4.5. In Vitro Antihyperglycemic and Antioxidant Assay

4.5.1. DPPH Radical Scavenging Activity

A DPPH radical scavenging assay was carried out as previously reported [38]. Scavenging of 2,2-diphenyl-1-picryhydrazyl (DPPH) radicals by the acetone extract (AE) (50 µg of 2 mg/mL solution dissolved in DMSO) and compounds (**1**–**14**) (50 µg of 2 mg/mL solution dissolved in DMSO) was measured in 100 mM Tris-HCl buffer (pH 7.4) by recording the absorbance at 517 nm spectrophotometrically. Ascorbic acid (50 µg of 2 mg/mL solution dissolved in DMSO) served as the standard. The results were expressed as %-scavenging and calculated by using the following formula: $(A_c - A_t)/100 \times A_c$, where A_c was the absorbance of control and A_t was the absorbance of the test sample. Different concentrations of compounds were evaluated to obtain 50% scavenging activity (SC_{50}). The SC_{50} was calculated based on the equation obtained from regression analysis.

4.5.2. ABTS Radical Scavenging Activity

Scavenging of the 2,2′-azino-bis(3-ethylbenzothiazoline-6-sulphonic acid) radical cation ($ABTS^{\bullet+}$) was performed as per the earlier method [39]. Acetone extract (AE) (20 µg of 2 mg/mL solution dissolved in DMSO) and compounds (**1–14**) (20 µg of 2 mg/mL solution dissolved in DMSO) were incubated with $ABTS^{\bullet+}$ solution in 6.8 mM phosphate buffer (pH 8.0) as described earlier. The discoloration of the $ABTS^{\bullet+}$ solution was determined by measuring the absorbance at 734 nm spectrophotometrically. Ascorbic acid (20 µg of 2 mg/mL solution dissolved in DMSO) served as the standard. The activity was expressed as %-scavenging and calculated as follows: $(A_c - A_t)/(100 \times A_c)$, where A_c was the absorbance of control and A_t was the absorbance of the test sample. The SC_{50} of compounds was calculated as per the above formula.

4.5.3. Free Radical Induced DNA Damage

The protective effect of acetone extract (AE) and compounds (**1–14**) on oxidative DNA damage was evaluated as per the previous method [40]. A total of 2 µL calf-thymus DNA mixed with 5 µL of 39 mM Tris buffer (pH 7.4) and 5 µL (10 µg) acetone extract and compounds (**1–14**) (10 µg of 2 mg/mL solution dissolved in DMSO) mixture was incubated at room temperature for 20 min. The reaction was initiated by adding 5 µL $FeCl_3$ (500 µM) and 10 µL H_2O_2 (0.8 M) and incubated for 10 min at 37 °C. The reaction was stopped by adding 3 µL DNA loading dye. Finally, the mixture was subjected to 0.8% agarose gel electrophoresis in TAE (40 mM Tris, 20 mM acetic acid and 0.5 M EDTA) buffer (pH 7.2). A total of 3 µL of Ethidium bromide was added to agarose solution to stain DNA bands. The image was viewed under transilluminating UV light and photographed (Bio-Rad, ChemiDocTM XRS, Hercules, CA, USA with Image LabTM software (ver. 6.0.1, build34, standard edition, 2017). The band intensity of the DNA was measured by using ImageJ software (ver. 1.4.3.67, Broken Symmetry Software, Scottsdale, AZ, USA).

4.5.4. Intestinal α-Glucosidase Inhibition

An intestinal α-glucosidase enzyme inhibition assay was performed as per the previous method [36]. A total of 20 µL (40 µg) of acetone extract and compounds (**1–14**) (40 µg of 2 mg/mL solution dissolved in DMSO) were incubated with 50 µL of rat intestinal α-glucosidase enzyme (89.93 mM, prepared in 0.9% NaCl) in 100 mM phosphate buffer (pH 6.8) for 10 min. After the incubation period, 50 µL of substrate (4-nitroplenyl α-D-glucopyranoside) solution was added. The release of *p*-nitrophenol from substrate was measured by recording the absorbance at 405 nm spectrophotometrically. Acarbose (40 µg of 2 mg/mL solution dissolved in DMSO) was taken as the standard. The activity was expressed and calculated as follows: $(A_c - A_t)/100 \times A_c$, where A_c was the absorbance of control and A_t was the absorbance of the test sample.

4.5.5. Statistical Analysis

Comparisons within the groups were done by applying one-way ANOVA followed by a post-test Tukey's Multiple comparison test. Statistical significance was set at $p < 0.05$. Data analysis was performed by using GraphPad Prism ver. 5.01 (GraphPad Software Inc., San Diego, CA, USA).

5. Conclusions

A novel UPLC-QToF-MS/MS-guided strategy was proposed here for the isolation and characterization of one new depside, decarboxyhomosekikaic acid, along with 13 known metabolites from *Ramalinaconduplicans*—most of them being minor metabolites that were reported on for the first time from this species. In the whole experimental design, UPLC-QToF-MS/MS was selected for multiple purposes, including targeting, finding, profiling, and isolating active constituents. Three hitherto unreferenced compounds were detected in this lichen, with their molecular formulae being deduced from HR-QToF-MS. Although in minute amounts, one isolate could be identified as an additional homosekikaic derivative. The expected major compounds atranorin, usnic acid, salazinic acid, and sekikaic acid were also obtained. However, efforts for isolating, identifying, and testing mainly targeted alkyldepsides- and monoaromatic-related compounds.

These compounds were tested for their antioxidant and α-glucosidase inhibition potential. Most of them, and the crude acetone extract (AE), have displayed antioxidant potential by scavenging ABTS and DPPH radicals and protected DNA from oxidative damage. Five compounds, and particularly hyperhomosekicaic acid, exhibited a comparable or better α-glucosidase inhibition to that of the acarbose standard. On the basis of these results, it is suggested that these lichen substances have a great potential to be used as bioresources or as structural models for novel bioactive candidate compounds. Docking experiments are necessary to document the structure–activities observed in this study along with pharmacomodulation studies to evaluate the antidiabetic properties. Acetone extract unexpectedly showed a comparable effect to that of the Acarbose standard, though it was not sufficient to consider its hypoglycemic activity in the context of the traditional use made of this edible lichen [10].

It should be kept in mind that activities obtained from the crude extract or from any of the active metabolites cannot be claimed to support a preventive or a therapeutic activity as no clinical assay has been carried out to validate an effect with a standardized dosage. Unexpected side effects can occur when preparations differ from the real traditional use, and toxicity trials have to be carried out at once.

Supplementary Materials: The following supporting information can be downloaded at: https://www.mdpi.com/article/10.3390/molecules27196720/s1. Detailed Materials and Method. Table S1: LC-QToF-MSE data of compounds from acetone extract of *R. conduplicans*. Physicochemical data of isolated compounds. Figure S1: Authenticate *Ramalina conduplicans*. Figure S2: Shows the spot tests for identification of *Ramalina conduplicans*. Figure S3: Depicts the HPTLC profiles for *R. conduplicans* extract. Figure S4: UPLC-PDA chromatogram of acetone extract of *R. conduplicans*. Figure S5: Total Ion Chromatogram (TIC) of fractions and pure compounds. Figure S6–S46: Spectral data (FTIR spectra, HRESIMS, MS/MS spectra and fragmentation pattern, ^{1}H & ^{13}C NMR, DEPT, HSQC, DQF-COSY and NOESY spectra) of compounds **1–14**, salazinic acid, usnic acid and atronarin. Figure S47: TIC of *R.conduplicans*. Figure S48: Common fragmentation of depsides. Figure S49. Invitro DNA damage assay.

Author Contributions: K.S.B., the biologist A.K.T. and J.B. conceived and designed the experiments; T.K.K., B.S. and K.S.B. performed the experiments; K.S.B., F.L.D. and J.B. analyzed the data; A.M.R. and S.M. contributed to the material identification reagents/materials and tools; A.A. and K.A. performed in vitro biological experiments and analyzed data; B.S., A.K.T., K.S.B. and J.B. wrote/revised the paper. All authors have read and agreed to the published version of the manuscript.

Funding: This research received no external funding.

Institutional Review Board Statement: Not applicable.

Informed Consent Statement: Not applicable.

Data Availability Statement: The data presented in this study are available on request from the corresponding authors.

Acknowledgments: This research was performed as a part of the Indo-French IRP 'Natural Products and Synthesis for Affordable Health' and acknowledgments are due to CSIR (India) and CNRS (France). Authors thank to Director, IICT for his constant encouragement and support and to S. Labarre for English revision. B.S. thanks CSIR for fellowship program and financial support. IICT communication no. IICT/Pubs./2022/082.

Conflicts of Interest: The authors declare no conflict of interest.

Sample Availability: Samples of the compounds **1–14** are available from the authors, and the samples were deposited in the institutional depository.

References

1. Shukla, V.; Joshi, G.P.; Rawat, M.S.M. Lichens as a potential natural source of bioactive compounds: A review. *Phytochem. Rev.* **2010**, *9*, 303–314. [CrossRef]
2. Ingelfinger, R.; Henke, M.; Roser, L.; Ulshöfer, T.; Calchera, A.; Singh, G.; Parnham, M.J.; Geisslinger, G.; Fürst, R.; Schmitt, I.; et al. Unraveling the pharmacological potential of lichen extracts in the context of cancer and inflammation with a broad screening approach. *Front. Pharmacol.* **2020**, *11*, 1322. [CrossRef]
3. Zhao, Y.; Wang, M.; Xu, B.A. Comprehensive review on secondary metabolites and health-promoting effects of edible lichen. *J. Funct. Foods* **2021**, *80*, 104283. [CrossRef]
4. Bhattarai, T.; Subba, D.; Subba, R. Nutritional value of some edible lichens of east Nepal. *Angew. Bot.* **1999**, *73*, 11–14.
5. Awasthi, D.D. *A Compendium of the Macrolichens from India, Nepal and Sri Lanka*; Bishen Singh Mahendra Pal Singh: Dehradun, India, 2000.
6. Wang, L.S.; Narui, T.; Harada, H.; Culberson, C.F.; Culberson, W.L. Ethnic uses of lichens in Yunnan, China. *Bryologist* **2001**, *104*, 345–349. [CrossRef]
7. Hanus, L.O.; Temina, M.; Dembitsky, V. Biodiversity of the chemical constituents in the epiphytic lichenized ascomycete *Ramalinalacera* grown on different substrates *Crataegussinaicus*, *Pinus halepensis*, and *Quercus calliprinos*. *Biomed. Pap. Med. Fac. Univ. Palacky Olomouc. Czech Repub.* **2008**, *152*, 203–208. [CrossRef]
8. Upreti, D.K.; Divakar, P.K.; Nayaka, S. Commercial and ethnic use of lichens in India. *Econ. Bot.* **2005**, *59*, 269–273. [CrossRef]
9. Kumar, S.P.; Kekuda, T.P.; Vinayaka, K.S.; Sudharshan, S.J. Anthelmintic and antioxidant efficacy of two macrolichens of Ramalinaceae. *Pharmacogn. J.* **2009**, *1*, 238–242.
10. Vinitha, M.T.; Veranja, K. Potential of lichen compounds as antidiabetic agents with antioxidative properties: A review. *Oxid. Med. Cell. Longev.* **2017**, *2017*, 2079697.
11. Devi, A.B.; Mohabe, S.; Nayaka, S.; Reddy, M. *In-vitro* antimicrobial activity of lichen *Ramalina conduplicans* Vain collected from Eastern Ghats, India. *Sci. Res. Rep.* **2016**, *6*, 99–108.
12. Ankith, G.N.; Rajesh, M.R.; Karthik, K.N.; Avinash, H.C.; Kekuda, P.T.; Vinayaka, K.S. Antibacterial and antifungal activity of three Ramalina species. *J. Drug Deliv. Ther.* **2017**, *7*, 27–32. [CrossRef]
13. Liu, X.Y. Determination of trace element of 4 lichens in Yunnan. *Stud. Trans. Elem. Health* **2003**, *20*, 30–31.
14. Ramya, K.; Thirunalasundari, T. Lichens: A myriad hue of bioresources with medicinal properties. *Int. J. Life Sci.* **2017**, *5*, 387–393.
15. Luo, H.; Wei, X.; Yamamoto, Y.; Liu, Y.; Wang, L.; Jung, J.S.; Hur, J.S. Antioxidant activities of edible lichen *Ramalina conduplicans* and its free radical-scavenging constituents. *Mycoscience* **2010**, *51*, 391–395. [CrossRef]
16. Kumar, K.; Siva, B.; Sarma, V.; Mohabe, S.; Reddy, A.M.; Boustie, J.; Tiwari, A.K.; Rao, N.R.; Babu, K.S. UPLC-MS/MS quantitative analysis and structural fragmentation study of five *Parmotrema* lichens from the Eastern Ghats. *J. Pharm. Biomed. Anal.* **2018**, *156*, 45–57. [CrossRef]
17. Reddy, S.D.; Siva, B.; Kumar, K.; Babu, V.S.P.; Sravanthi, V.; Boustie, J.; Nayak, V.L.; Tiwari, A.K.; Rao, C.; Sridhar, B.; et al. Comprehensive analysis of secondary metabolites in *Usnealongissima* (Lichenized Ascomycetes, Parmeliaceae) using UPLC-ESI-QTOF-MS/MS and Pro-apoptotic activity of barbatic acid. *Molecules* **2019**, *24*, 2270. [CrossRef]
18. Olivier-Jimenez, D.; Chollet-Krugler, M.; Rondeau, D.; Beniddir, M.A.; Ferron, S.; Delhaye, T.; Allard, P.-M.; Wolfender, J.L.; Sipman, H.; Lücking, R.; et al. A database of high-resolution MS/MS spectra for lichen metabolites. *Sci. Data* **2019**, *6*, 294. [CrossRef]
19. Moreira, A.S.; Braz-Filho, R.; Mussi-Dias, V.; Vieira, I.J. Chemistry and biological activity of Ramalina lichenized fungi. *Molecules* **2015**, *20*, 8952–8987. [CrossRef]
20. Musharraf, S.G.; Kanwal, N.; Thadhani, V.M.; Choudhary, M.I. Rapid identification of lichen compounds based on the structure–fragmentation relationship using ESI-MS/MS analysis. *Anal. Methods* **2015**, *7*, 6066–6076. [CrossRef]
21. Linh, N.T.T.; Danova, A.; Truong, T.L.; Chavasiri, W.; Phung, N.K.P.; Chi, H.B.L. Chemical constituents of chloroform extract from the lichen *Ramalina peruviana* Arch (Ramalinaceae). *Vietnam J. Chem.* **2020**, *58*, 231–236. [CrossRef]
22. Culberson, C.F.; Culberson, W.L.; Johnson, A. Orcinol-type depsides and depsidones in the lichens of the *Cladoniachlorophaea* group (Ascomycotina, Cladoniaceae). *Bryologist* **1985**, *88*, 380. [CrossRef]
23. Gunasekaran, S.; Rajan, V.P.; Ramanathan, S.; Murugaiyah, V.; Samsudin, M.W.; Din, L.B. Antibacterial and antioxidant activity of lichens *Usnearubrotincta*, *Ramalinadumeticola*, *Cladoniaverticillata* and their chemical constituents. *Malays. J. Anal. Sci.* **2016**, *20*, 1–13. [CrossRef]

24. Sun, W.; Zhuang, C.; Li, X.; Zhang, B.; Lu, X.; Zheng, Z.; Dong, Y. Varic acid analogues from fungus as PTP1B inhibitors: Biological evaluation and structure–activity relationships. *Bioorg. Med. Chem. Lett.* **2017**, *27*, 3382–3385. [CrossRef] [PubMed]
25. Brandão, L.F.; Alcantara, G.B.; Matos, M.; Bogo, D.; Freitas, D.; Oyama, N.M.; Honda, N.K. Cytotoxic evaluation of phenolic compounds from lichens against melanoma cells. *Chem. Pharm. Bull.* **2013**, *61*, 176–183.
26. Elix, J.A.; Wardlaw, J.H. New depsides from the lichen *Neofusceliadepsidella*. *Aust. J. Chem.* **1997**, *50*, 1145–1150. [CrossRef]
27. Ismed, F.; Farhan, A.; Bakhtiar, A.; Zaini, E.; Nugraha, Y.P.; Putra, O.D.; Uekusa, H. Crystal structure of olivetolic acid: A natural product from *Cetreliasanguinea* (Schaer.). *Acta Crystallogr. E Crystallogr. Comm.* **2016**, *72*, 1587–1589. [CrossRef]
28. Bouges, H.; Monchot, A.; Antoniotti, S. Enzyme-catalysed conversion of atranol and derivatives into dimeric hydrosoluble materials: Application to the preparation of a low-atranol oakmoss absolute. *Cosmetics* **2018**, *5*, 69. [CrossRef]
29. Bauer, J.; Waltenberger, B.; Noha, S.M.; Schuster, D.; Rollinger, J.M.; Boustie, J.; Chollet, M.; Stuppner, H.; Werz, O. Discovery of depsides and depsidones from lichen as potent inhibitors of microsomal prostaglandin E2 synthase-1 using pharmacophore models. *ChemMedChem* **2012**, *7*, 2077–2081. [CrossRef]
30. Munteanu, I.G.; Apetrei, C. Analytical methods used in determining antioxidant activity: A review. *Int. J. Mol. Sci.* **2021**, *22*, 3380. [CrossRef]
31. Siddeeg, A.; Al-Kehayez, N.M.; Abu-Hiamed, H.A.; Al-Sanea, E.A.; Al-Farga, A.M. Mode of action and determination of antioxidant activity in the dietary sources: An overview. *Saudi J. Biol. Sci.* **2021**, *28*, 1633–1644. [CrossRef]
32. Lloyd, D.R.; Phillips, D.H. Oxidative DNA damage mediated by copper (II), iron (II) and nickel (II) Fenton reactions: Evidence for site-specific mechanisms in the formation of double-strand breaks, 8-hydroxydeoxyguanosine and putative intrastrand cross-links. *Mutat. Res. Fundam. Mol. Mech. Mutagen.* **1999**, *424*, 23–36. [CrossRef]
33. Komati, A.; Anand, A.; Nagendla, N.K.; Madhusudana, K.; Mudiam, M.K.; Babu, K.S.; Tiwari, A.K. *Bombax ceiba* calyx displays antihyperglycemic activity via improving insulin secretion and sensitivity: Identification of bioactive phytometabolomes by UPLC-QTof-MS/MS. *J. Food Sci.* **2022**, *87*, 1865–1881. [CrossRef] [PubMed]
34. Verma, N.; Behera, B.C.; Sharma, B.O. Glucosidase inhibitory and radical scavenging properties of lichen metabolites salazinic acid, sekikaic acid and usnic acid. *Hacet. J. Biol. Chem.* **2012**, *40*, 7–21.
35. Schinkovitz, A.; Le Pogam, P.; Derbre, S.; Roy-Vessieres, E.; Blanchard, P.; Thirumaran, S.L.; Breard, D.; Aumond, M.C.; Zehl, M.; Urban, E.; et al. Secondary metabolites from lichen as potent inhibitors of advanced glycation end products and vasodilative agents. *Fitoterapia* **2018**, *131*, 182–188. [CrossRef] [PubMed]
36. Duong, T.H.; Hang, T.X.H.; Le Pogam, P.; Tran, T.N.; Mac, D.D.; Dinh, M.H.; Sichaem, J. α-Glucosidase inhibitory depsidones from the lichen *Parmotrematsavoense*. *Planta Med.* **2020**, *86*, 776–781. [PubMed]
37. Orange, A.; James, P.W.; White, F.J. *Microchemical methods for the identification of lichens*; British Lichen Society: London, UK, 2001.
38. Tiwari, A.K.; Manasa, K.; Kumar, D.A.; Zehra, A. Raw horse gram seeds possess more in vitro antihyperglycaemic activities and antioxidant properties than their sprouts. *Nutrafoods* **2013**, *12*, 47–54. [CrossRef]
39. Deepthi, S.; Anusha, K.; Anand, A.; Manasa, A.; Babu, K.S.; Mudiam, M.K.R.; Tiwari, A.K. Micronutrients and phytochemicals content in various rice (*Oryzasativa Linn.*) samples control carbohydrate digestion variedly and present differential antioxidant activities: An in vitro appraisal. *Indian J. Tradit. Knowl.* **2020**, *19*, 821–831.
40. Chang, M.; Bellaoui, M.; Boone, C.; Brown, G.W. A genome-wide screen for methyl methanesulfonate-sensitive mutants reveals genes required for S phase progression in the presence of DNA damage. *Proc. Natl. Acad. Sci. USA* **2002**, *99*, 16934–16939. [CrossRef]

molecules MDPI

Article

Antiglycation Effects of Adlay Seed and Its Active Polyphenol Compounds: An In Vitro Study

Cheng-Pei Chung [1,2,†], Shih-Min Hsia [3,4,5,6,†], Wen-Szu Chang [7], Din-Wen Huang [7,8], Wen-Chang Chiang [7], Mohamed Ali [9], Ming-Yi Lee [1,10] and Chi-Hao Wu [11,*]

1 Department of Nutrition and Health Sciences, College of Human Ecology, Chang Gung University of Science and Technology, Taoyuan 333324, Taiwan
2 Research Center for Food and Cosmetic Safety, College of Human Ecology, Chang Gung University of Science and Technology, Taoyuan 333324, Taiwan
3 School of Nutrition and Health Sciences, College of Nutrition, Taipei Medical University, Taipei 11031, Taiwan
4 Graduate Institute of Metabolism and Obesity, College of Nutrition, Taipei Medical University, Taipei 110301, Taiwan
5 School of Food and Safety, Taipei Medical University, Taipei 110301, Taiwan
6 Nutrition Research Center, Taipei Medical University Hospital, Taipei 110301, Taiwan
7 College of Bioresources and Agriculture, National Taiwan University, Taipei 10617, Taiwan
8 School of Life Science, Huizhou University, No. 46 Yanda Road, Huizhou 516007, China
9 Clinical Pharmacy Department, Faculty of Pharmacy, Ain Shams University, Cairo 11566, Egypt
10 Research Center for Chinese Herbal Medicine, College of Human Ecology, Chang Gung University of Science and Technology, Taoyuan 333324, Taiwan
11 Graduate Programs of Nutrition Science, School of Life Science, National Taiwan Normal University, Taipei 106209, Taiwan
* Correspondence: chwu@ntnu.edu.tw; Tel.: +886-2-7749-1427
† These authors contributed equally to this work.

Citation: Chung, C.-P.; Hsia, S.-M.; Chang, W.-S.; Huang, D.-W.; Chiang, W.-C.; Ali, M.; Lee, M.-Y.; Wu, C.-H. Antiglycation Effects of Adlay Seed and Its Active Polyphenol Compounds: An In Vitro Study. *Molecules* **2022**, *27*, 6729. https://doi.org/10.3390/molecules27196729

Academic Editor: Nour Eddine Es-Safi

Received: 29 August 2022
Accepted: 5 October 2022
Published: 9 October 2022

Abstract: This study aimed to evaluate the antiglycation effects of adlay on protein glycation using in vitro glycation assays. Adlay seed was divided into the following four parts: the hull (AH), testa (AT), bran (AB), and polished adlay (PA). A solvent extraction technique and column chromatography were utilized to investigate the active fractions and components of adlay. Based on a BSA-glucose assay, the ethanolic extracts of AT (ATE) and AB (ABE) revealed a greater capacity to inhibit protein glycation. ATE was further consecutively partitioned into four solvent fractions with *n*-hexane, ethyl acetate (ATE-Ea), 1-butanol (ATE-BuOH), and water. ATE-BuOH and -Ea show marked inhibition of glucose-mediated glycation. Medium–high polarity subfractions eluted from ATE-BuOH below 50% methanol with Diaion HP-20, ATE-BuOH-c to -f, exhibited superior antiglycation activity, with a maximum inhibitory percentage of 88%. Two phenolic compounds, chlorogenic acid and ferulic acid, identified in ATE-BuOH with HPLC, exhibited potent inhibition of the individual stage of protein glycation and its subsequent crosslinking, as evaluated by the BSA-glucose assay, BS-methylglyoxal (MGO) assay, and G.K. peptide-ribose assay. In conclusion, this study demonstrated the antiglycation properties of ATE in vitro that suggest a beneficial effect in targeting hyperglycemia-mediated protein modification.

Keywords: adlay; adlay testa; antiglycation; phenolic acid

1. Introduction

Advanced glycation end products (AGEs) are a heterogeneous group of reactive, crosslinking compounds produced from nonenzymatic glycation, also known as the Maillard reaction, which occurs between reducing sugars and the amino groups of proteins, nucleic acids, and phospholipids [1]. Protein glycation is formed from Schiff bases, followed by Amadori rearrangements. Amadori products are the initial products of AGEs and can undergo additional glycoxidative modifications [2]. In addition to the above reaction, glucose and Schiff bases can undergo auto-oxidation to form reactive 1,2-dicarbonyl compounds, such as methylglyoxal (MGO) and 3-deoxyglucosone [1,3]. Amadori products also undergo glycoxidation to yield glyoxal and glucosone [2]. The formation of AGEs is relatively slow under physiological status, but significantly accelerated in hyperglycemia conditions. Thus, AGEs may play an important role in the pathogenesis of diabetic complications and aging-related diseases [3,4], and may even have an undefined relationship with COVID-19 morbidity and mortality [5].

AGEs adversely affect our body through several mechanisms [2,3]. The first is the modification of intracellular proteins, including the protein regulation of gene transcription. The second mechanism is through these AGE precursors that diffuse out of the cell and modify the extracellular matrix. These modifications dampen the signaling transduction between the matrix and cells, leading to cellular dysfunction. Finally, the third mechanism is these AGE precursors that modify circulating proteins in the blood, such as albumin. The glycated proteins then bind to the receptor for AGEs (RAGE) and activate AGE–RAGE signaling pathways, evoking oxidative stress and an inflammatory reaction [6].

There are several known mechanisms of AGE inhibition [6]. First, AGE formation can be reduced by tight glycemic control. Carbonyl-trapping agents can block AGE formation, reducing the harmful effects of reactive carbonyl species (RCS). Metal ion chelators can also reduce glycoxidative stress due to suppressing the redox reaction, and finally, AGE levels in vivo can be decreased by crosslink breakers. Aminoguanidine (AG), an AGE inhibitor, is often used as a positive control in antiglycation studies. AG is a nucleophilic agent that traps RCS, such as MGO, by forming non-toxic stable adducts [7]. On the other hand, phenolic antioxidants have demonstrated antiglycative properties primarily by scavenging free radicals and chelating metal ions [6,8]. Theoretically, if an inhibitor possesses more than one of the mechanisms mentioned above, it would make an ideal AGE inhibitor for mitigating diabetic complications and other chronic diseases [9].

Adlay (*Coix lachryma-jobi* L. var. *ma-yuen Stapf*), commonly known as Job's tears, is widely cultivated in Asia and is utilized as a Chinese folk medicine, as well as a nutritious food [10]. Structurally, adlay seeds consist of the following four parts from the outside to the inside: the hull (AH), testa (AT), bran (AB), and polished adlay (PA). Recent studies have indicated that adlay and its solvent extracts consist of more than 30 ingredients with 20 biofunctional effects, based on clinical and experimental studies [10–17]. In addition, adlay has been shown to regulate blood sugar [14], blood pressure [18], immunity [19], uterine contractions [20], anti-influenza viruses [21], and osteoporosis preventive activities [12]. However, there are still limited studies that explore the effect of adlay on protein glycation and AGE formation. Thus, this study aimed to evaluate the antiglycation potential of adlay using in vitro glycation assays and to investigate the active fractions and compounds of adlay.

2. Results

2.1. Effects of the Individual Parts of Adlay on Protein Glycation According to BSA-Glucose Assay

In the BSA-glucose assay, glucose was used as the glycating agent, and BSA served as the amine group donor, as the glycated target of glucose. This assay aimed to determine whether adlay could inhibit post-Amadori glycation based on the development of AGE-related fluorescence [8]. Firstly, the adlay seeds were divided into the following four parts: hull (AH), testa (AT), bran (AB), and polished adlay (PA). A solvent extraction technique with column chromatography was used to investigate the active fractions and components of adlay (Figure 1). Ethanol extracts of the hull, testa, bran, and polished adlay were referred to as AHE, ATE, ABE, and PAE, respectively. The results showed that among the individual parts of the adlay, ATE and ABE demonstrated greater inhibitory capacities against protein glycation (Figure 2).

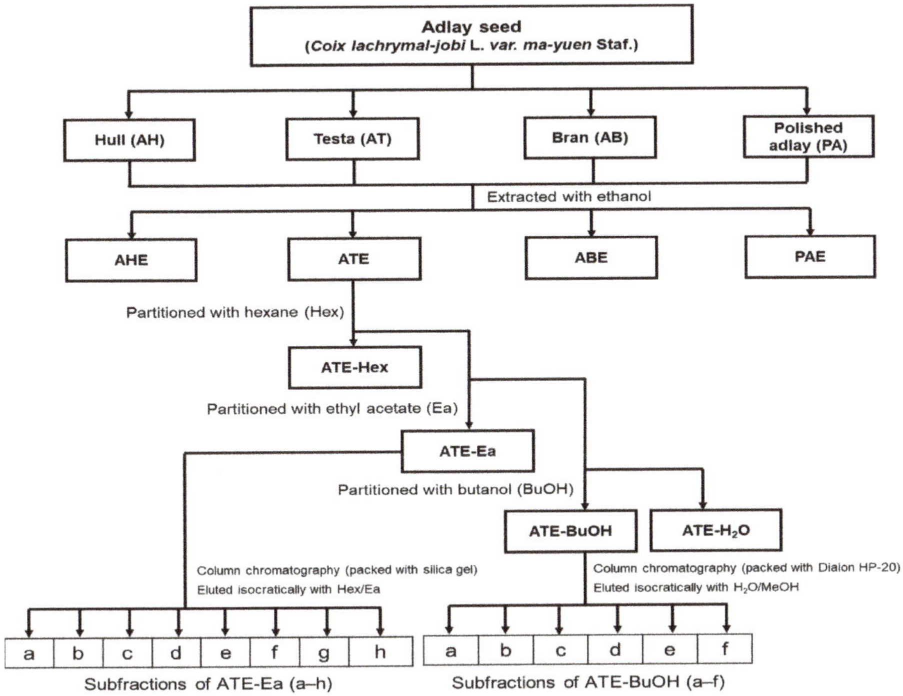

Figure 1. Partitioning scheme for preparing and isolating active fractions and components from adlay ethanolic extracts. ATE-Ea was dissolved in Ea and subjected to column chromatography on a silica gel with a Hex/Ea gradient from 0 to 100% EA (every 10%) to yield a–h. ATE-BuOH was dissolved in MeOH and subjected to column chromatography on a Diaion HP-20 resin with an H_2O/MeOH gradient from 0 to 100% MeOH (every 25%)to yield a–f.

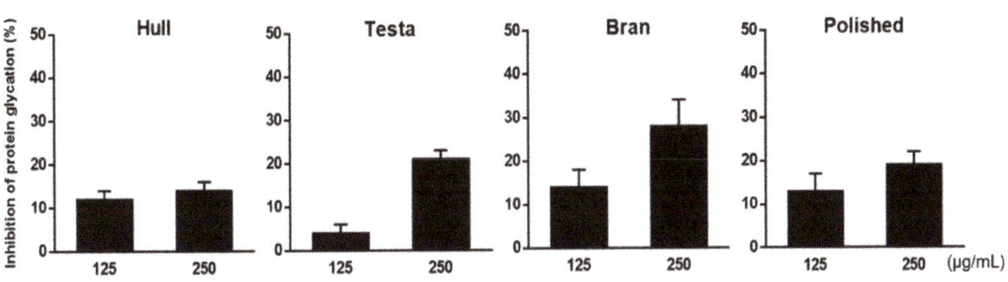

Figure 2. Effects of ethanolic extracts of adlay hull (AHE), adlay testa (ATE), adlay bran (ABE), and polished adlay (PAE) on glucose-mediated development of fluorescence of AGEs (BSA-glucose assay). Results are means ± SD for n = 3.

2.2. *Effects of ATE and ABE Subfractions on Protein Glycation*

ATE and ABE were further consecutively partitioned into H_2O, 1-butanol (BuOH), ethyl acetate (Ea), and n-hexane (Hex) fractions. As shown in Figure 3, ATE-BuOH and ATE-Ea exhibited greater antiglycation properties than the other fractions (Figure 3a), and the same findings were found for the ABE fractions (Figure 3b). The ATE-BuOH fractions were obtained and chromatographically isolated to subfractions (a–f), whereas the ATE-Ea fractions were isolated to subfractions (a–h). Inhibitory percentages of ATE-BuOH subfractions -a to -f against protein glycation were as follows: −7%, 0%, 88%, 85%, 56%, and 53%, respectively, at a concentration of 250 μg/mL (Figure 4a). Similarly, the inhibitory potency of the ATE-Ea subfractions (a to h) was 15% to 37% (Figure 4b). In addition, the effects of ATE-BuOH-C and ATE-BuOH-D on the middle stage of glycation were determined by the BSA-MGO assay. MGO belongs to the group of RCS, which is a critical precursor in the formation of AGEs. Figure 5 shows that ATE-BuOH-c and ATE-BuOH-d exhibited significant inhibition of 26% and 30%, respectively, at a concentration of 250 µg/mL. These data suggested the RCS-trapping capacities of ATE-BuOH-c and -d and the main antiglycation components that possibly existed in the ATE-BuOH subfractions.

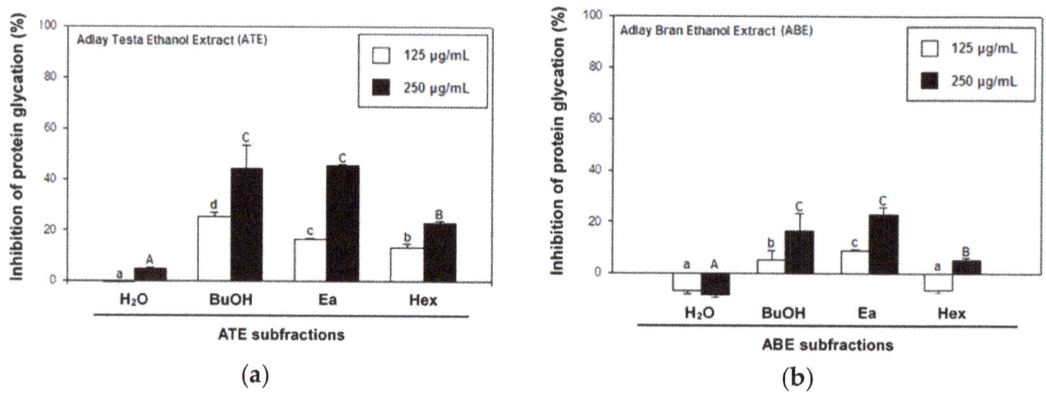

Figure 3. Effects of different solvent fractions of ATE (**a**) and ABE (**b**) on glucose-mediated development of fluorescence of AGEs (BSA-glucose assay). Groups with different letter superscripts are significantly different ($p < 0.05$). a–c, treated with 125 µg/mL; A–C, treated with 250 µg/mL.

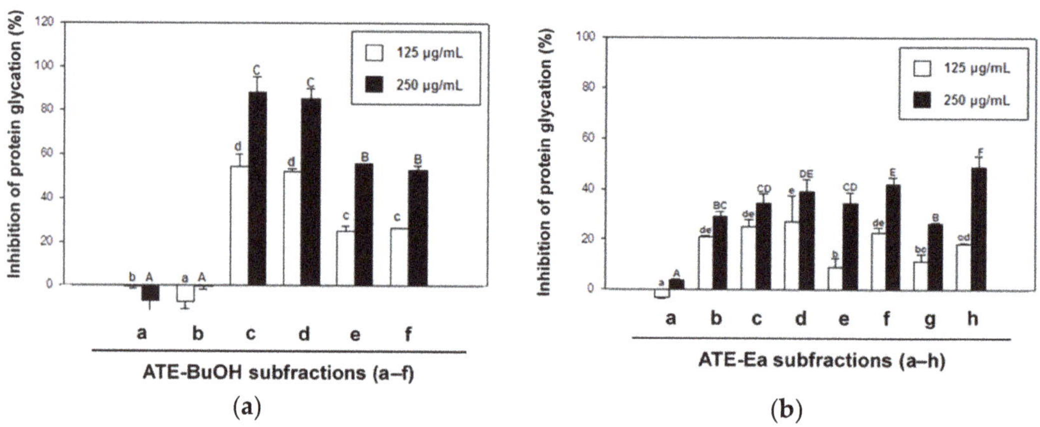

Figure 4. Effects of ATE-BuOH (**a**) and ATE-Ea (**b**) subfractions on glucose-mediated development of fluorescence of AGEs (BSA-glucose assay). Results are means ± SD for $n = 3$. Groups with different letter superscripts are significantly different ($p < 0.05$). a–h, treated with 125 µg/mL; A–F, treated with 250 µg/mL.

2.3. *Effect of ATE-BuOH-Containing Phenolics on the Individual Stage of Protein Glycation*

According to our previous study, phenolic acids were major components in ATE-BuOH [22]. In addition, medium–high polarity ethanol extracts of the hull, testa, branty subfractions, eluted from ATE-BuOH below 50% methanol with Diaion HP-20 resin, possessed more significant antiglycation (Figure 4a) as compared with ATE-EA, especially ATE-BuOH-c and –d (Figure 5).

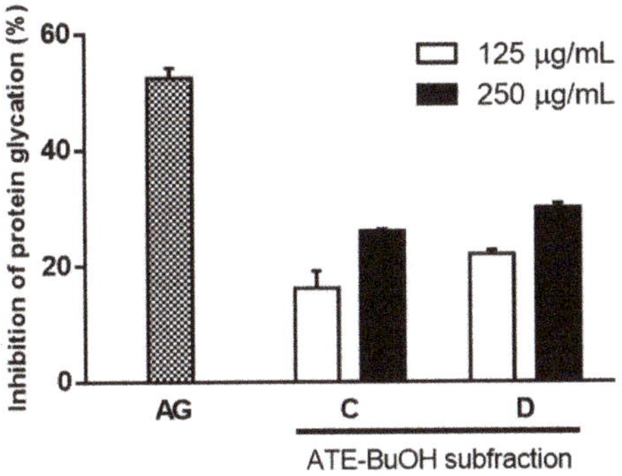

Figure 5. Effects of ATE-BuOH subfractions C and D on MGO-mediated development of fluorescence of AGEs (BSA-MGO assay). Results are means ± SD for $n = 3$. AG, aminoguanidine. C, the subfraction C of ATE-BuOH. D, the subfraction D of ATE-BuOH.

A high-performance liquid chromatography (HPLC) analysis was carried out to investigate the chemical composition of ATE-BuOH. As shown in Figure 6, chlorogenic acid, caffeic acid, *p*-coumaric acid, and ferulic acid were identified in ATE-BuOH with contents of 1.01 ± 0.03, 1.32 ± 0.04, 9.51 ± 0.94, and 2.54 ± 0.68 mg/g, respectively. However, gallic acid was not detectable in ATE-BuOH.

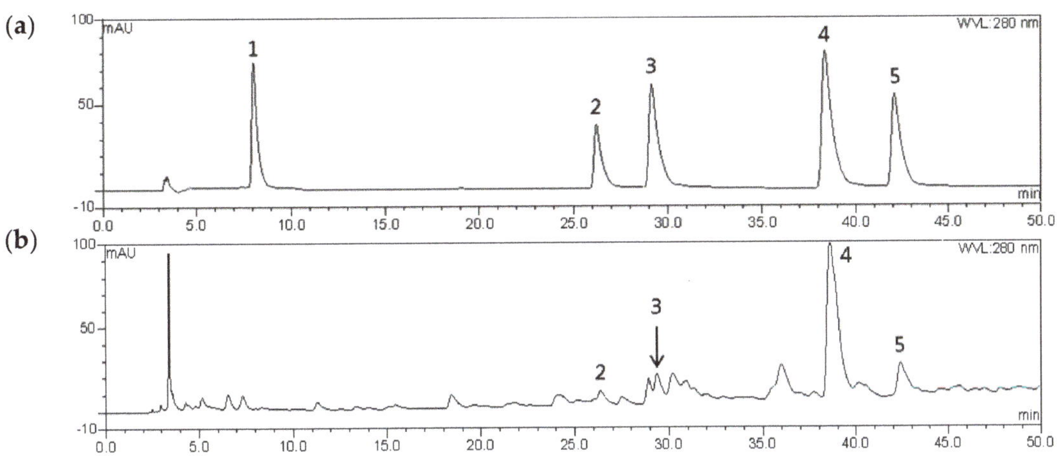

Figure 6. High−performance liquid chromatography (HPLC) chromatographic profiles recorded at 280 nm of (**a**) phenolic compound mixed standards (each 0.1 mg/mL) and (**b**) the ATE-BuOH fraction (10 m(g/mL) (1: gallic acid, 2: chlorogenic acid, 3: caffeic acid, 4: p-coumaric acid, and 5: ferulic acid).

In the BSA-glucose assay, chlorogenic acid and ferulic acid exhibited 20% and 28% inhibitory activity, respectively (Figure 7), indicating that these phenolics aid in reducing glucose-mediated protein modification. The BSA-MGO assay also determined the inhibition of MGO-mediated protein glycation by these phenolics. The results showed that chlorogenic acid and ferulic acid exhibited significant inhibition of 24% and 15%, respectively. Meanwhile, *p*-coumaric acid had a slight impact on antiglycation (Figure 7). In contrast, the other phenolic compounds identified in ATE-BuOH, such as caffeic acid and 6-methoxy-2-benzoxazolinone [22], showed no antiglycation effect at a concentration of 100 µM (data not shown).

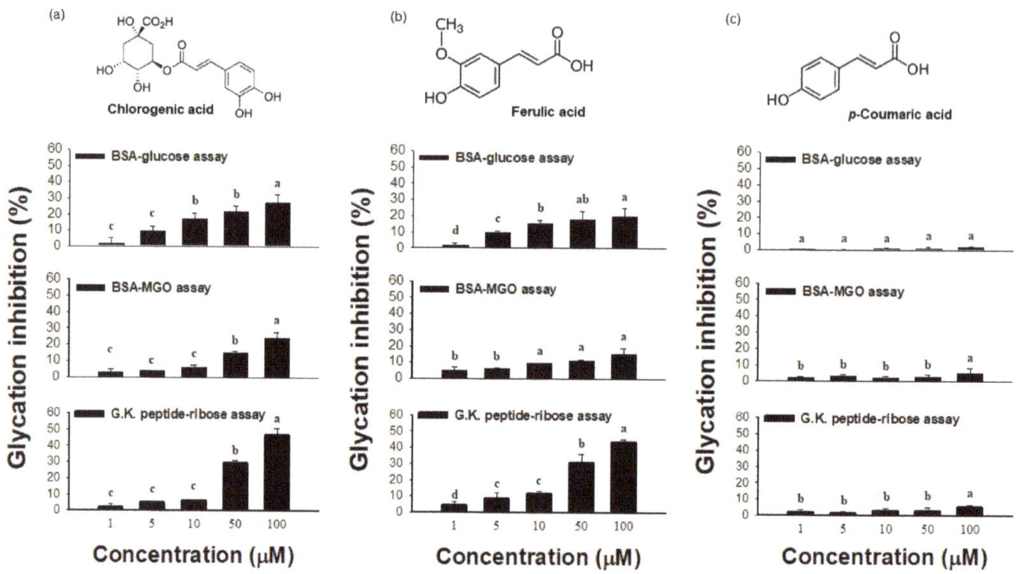

Figure 7. Chemical structures and the effects of chlorogenic acid (**a**), ferulic acid (**b**), and *p*-coumaric acid (**c**) identified in ATE-BuOH fractions on protein glycation and crosslinking. Dose responses of glycation inhibition (%) on individual stages of protein glycation were determined by model systems of the BSA-glucose assay (upper panel), the BSA-MGO assay (middle panel), and the G.K.-ribose assay (bottom panel). Data are the means ± SD for $n = 6$. Groups with different letters (a–d) are significantly different from each other in individual assays ($p < 0.05$).

G.K. peptide-ribose assay was used to generate peptides with advanced Maillard reaction products with dimerization through lysine-lysine crosslinking [23]. Rahbar et al. [23] and our previous study [8] demonstrated that incubation of G.K. peptides with ribose resulted in late glycation product formation. Therefore, the present study utilized this model system to evaluate the inhibitory effect of phenolics on protein crosslinking. As shown in Figure 7 (lower panel), chlorogenic acid and ferulic acid exhibited substantial anti-crosslinking activities (47% and 43%, respectively, at a concentration of 100 μM).

3. Discussion

Protein glycation and subsequent AGE formation in the body have been evidenced as a risk factor in the development of diabetic macrovascular and microvascular complications and age-related diseases [24,25]. Clinical observation has revealed that patients with complicated diabetes have 40 to 100% higher AGE levels than healthy subjects [26]. Therefore, investigating AGE inhibitors, especially natural anti-glycation agents with fewer adverse effects, may be a beneficial approach to preventing diabetic complications.

The main question addressed by this study was whether adlay could inhibit protein glycation. Adlay seeds have long been consumed as a food supplement and herbal medicine in traditional Chinese medicine [10,12]. Although the health-promoting effects and therapeutic potential of adlay have been reported [10,12], studies on the potential of adlay to act against protein glycation and AGE formation are limited. This study used a classic in vitro glycation assay to evaluate the effect of individual parts of adlay (hull, testa, bran, and polished adlay) on glucose-mediated BSA glycation. Ethanol was chosen as the initial extraction solvent based on safety considerations for human consumption [27]. Moreover, most active ingredients, including polyphenols, phytosterols, and coixol, found in adlay have been extracted by ethanol with a high yield in previous studies [22,28,29]. For the first time, this study demonstrated that ATE and ABE exhibited better glycation inhibitory effects (Figure 2), suggesting antiglycating agents are possibly present in the bran and testa of adlay.

Using a solvent extraction technique, two ATE-BuOH subfractions, ATE-BuOH-C and ATE-BuOH-D, exhibited superior antiglycation activities (Figures 4a and 5). Phenolic compounds in adlay seed were analyzed by HPLC in our previous studies [13,22,28,29]. Caffeic acid, chlorogenic acid, *p*-coumaric acid, ferulic acid, gallic acid, *p*-hydroxybenzoic acid, syringic acid, and vanillic acid were identified in ATE-BuOH and/or ATE-Ea fractions. In the present investigation, *p*-coumaric acid (9.51 $\pm$ 0.94 mg/g), along with chlorogenic acid (1.01 $\pm$ 0.03 mg/g), caffeic acid (1.32 $\pm$ 0.04 mg/g), and ferulic acid (2.54 $\pm$ 0.68 mg/g) were identified in ATE-BuOH through HPLC analysis (Figure 6). Notably, chlorogenic acid and ferulic acid from ATE-BuOH exhibited significant potent inhibition of the individual stage of protein glycation, especially in the reduction in protein crosslinking (Figure 7). After partially purified by column chromatography, the ATE-BuOH-c and –d (eluent from 50 to 25% methanol) showed superior inhibitory effects on antiglycation (Figures 3a, 4a and 5). A previous study has demonstrated that the major components in subfractions eluted from 50 to 25% methanol from ATE-BuOH were caffeic acid (9.02 mg/g subfractions), chlorogenic acid (30.30 mg/g subfractions), and ferulic acid (0.05 mg/g subfractions) [22]. These results suggest that the antiglycation properties of ATE were at least partly related to its phenolic acid content.

Research has demonstrated that antioxidative polyphenols show potent antiglycation activities [1,6,8]. Several studies have drawn attention to the positive correlation between the free radical scavenging activity and antiglycation capacity, which may be due to the interruption of ROS formation during glycation [6,8]. In addition, the antioxidant ability of phenolic acids, such as caffeic acid and chlorogenic acid, depends on the number of hydroxyl groups in the molecule that would be strengthened by steric hindrance [30]. The present study showed that chlorogenic acid and ferulic acid are potent inhibitors that act at the individual stage of glycation, including post-Amadori glycation, as evidenced by the decreased development of AGE-related fluorescence in BSA-glucose assay (Figure 7, upper panel). Notably, these phenolic acids exhibited MGO-trapping activity and anti-crosslinking action, as evidenced by the MGO-BSA assay (Figure 7, middle panel) and G.K. peptide-ribose assay (Figure 7, bottom panel). However, certain compounds, such as caffeic acid, had no inhibitory effect on glycation (data not shown), which was consistent with a former study [31].

Since MGO is a critical intermediate and precursor during AGE formation, the MGO-BSA assay was utilized to determine the middle stage of protein glycation. MGO served as the glycating agent, and BSA provided the amine source targeted by the glycating agents. Albumin was used as the protein because it is the most abundant protein in serum. A clinical study noted that the serum levels of glycated albumins are two to three times higher in diabetic patients than in healthy people [32]. It is known that MGO can readily react with proteins that contain lysine residues and crosslink with proteins, along with ROS production [33]. MGO-glycated BSA was detected by specific fluorescence formation. This study showed that ATE-BuOH subfractions and their active phenolics (chlorogenic acid and ferulic acid) achieved nearly 30% inhibition of MGO-mediated protein modification (Figures 5 and 7), indicating the possible MGO-trapping potential of adlay.

Nagaraj and his colleagues [34] indicated that protein crosslinking is the major end result of the Maillard reaction. In this study, a synthetic G.K. peptide that contained lysine residue was incubated with ribose. This chemical model was designed to generate peptides with AGEs that dimerize through lysine-lysine crosslinking and increase the late glycation product formation, as determined by intrinsic fluorescence formation [23]. Chlorogenic acid and ferulic acid were found to be anti-crosslinking agents (Figure 7, lower panel). Taken together, the data suggested that the presence of methoxy groups at the C3 position of the aromatic ring in ferulic acid and the quinic acid groups in chlorogenic acid may contribute to the antiglycation actions.

4. Materials and Methods

4.1. Preparation of Adlay Extracts

The adlay utilized in this experiment was purchased from a local farmer in Taichung, and the variety of adlay was Taichung Shuenyu no. 4 (TCS 4) of *Coix lachryma-jobi* L. var. *ma-yuen Stapf*. Sample preparation and extraction used previously reported methods from our laboratory [22]. Briefly, adlay seeds were air-dried at 40 °C and ground to obtain AH, and further ground to separate dehulled adlay and AT by an electric fan. After these procedures, the whole grain contained 7.2 ± 0.4% water, while that in AH and AT contained 5.2 ± 0.1% and 7.2 ± 0.1%, respectively. A total of 12 kg of AT was extracted with 120 L of ethanol 3 times at room temperature for 24 h, and filtered through #1 filter paper (Cytiva, Washington, D.C., USA). The filtrate was then concentrated to dryness to yield approximately 538 g of dried ATE (Figure 1). The dried extracts were stored at −20 °C until use. ATE was then suspended in a 10% methanol solution and partitioned with *n*-hexane to obtain the lipid-rich ATE-Hex fraction, yielding approximately 144 g. The defatted ATE was further partitioned with Ea to obtain about 160 g of ATE-Ea. The remaining ATE was partitioned with butanol to obtain about 50 g of ATE-BuOH. Finally, the residual extract was the ATE-H_2O fraction. ATE-Ea was dissolved in Ea and subjected to column chromatography on a silica gel to yield ATE-Ea subfractions a–h. The gradient used for elution was a hexane/Ea gradient from 0 to 100% EA (every 10%), and the sub-fractions with similar chemical compositions on thin-layer chromatography (TLC) were combined and concentrated to dryness under a vacuum at 60 °C. Subfractions were stored at −20 °C until use. On the other hand, ATE-BuOH was dissolved in methanol and subjected to column chromatography on a Diaion HP-20 resin to yield ATE-BuOH subfractions a–f. The gradient used for elution was an H_2O/MeOH gradient from 0 to 100% MeOH (every 25%) and was washed with 100% EA. Sub-fractions with similar chemical compositions on TLC were combined and concentrated to dryness in a vacuum at 60 °C and stored at −20 °C until use.

4.2. BSA-Glucose Assay

The BSA-glucose assay was devised by Matsuura et al. [35] to screen for effective AGE inhibitors from natural product extracts. A total of 50 mL of 50 mM phosphate buffer (pH 7.4) and 200 mM glucose were mixed as a stock solution. Then, 500 µL of reaction mixture containing 400 µg of BSA and 200 mM glucose, with or without 10 µL of plant extracts, in 50 mM phosphate buffer at a pH of 7.4 in an Eppendorf tube was prepared. The reaction mixture was heated at 60 °C on a heat block for 48 h. A blank, unreacted solution sample without inhibitors was kept at 4 °C until measurement. Samples were cooled to room temperature. After cooling, 100 µL aliquots were transferred to new 1.5 mL plastic tubes, and 10 µL of TCA was added to each tube. The tubes were centrifuged at 15,000 rpm at 4 °C for 4 min. The supernatant was removed and AGE-BSA precipitate was dissolved in 800 µL of PBS. The fluorescence of the samples was measured at an excitation wavelength of 370 nm and an emission wavelength of 440 nm, and inhibitory activity was calculated as a percentage using the following formula:

$$I\% = [1 - (\text{fluorescence of BSA + glucose + inhibitor} - \text{fluorescence of BSA + inhibitor})/(\text{fluorescence of BSA + glucose}) - (\text{fluorescence of BSA})] \times 100\%$$

4.3. BSA-MGO Assay

The middle stage of protein glycation was determined using the BSA-MGO assay described by Wu et al. [8], with a slight modification. BSA (50 mg/mL) was incubated with 100 mM MGO under sterile conditions in 0.1 M phosphate buffer (pH 7.4) at 37 °C for 9 days. Sample fractions and subfractions were added to the model system at final concentrations of 125 µg/mL and 250 µg/mL, respectively. The sample solutions were kept at 4 °C until measurement. After cooling, 100 µL of aliquots were transferred to 1.5 mL plastic tubes, and 10 µL of 100% TCA was added to each tube. Tubes were centrifuged at 15,000 rpm at 4 °C for 4 min. The supernatant was removed, and the precipitate of MGO-BSA was dissolved in 800 μL of PBS. The fluorescence of the samples was measured at the excitation and emission maxima of 330 and 410 nm, respectively, and compared with that of the unincubated blank that contained the protein, MGO, and inhibitors. The percent inhibition was calculated as follows:

$$\text{I\%} = [1 - (\text{fluorescence of BSA + MGO + inhibitor} - \text{fluorescence of BSA + inhibitor})/(\text{fluorescence of BSA + MGO}) - (\text{fluorescence of BSA})] \times 100\% \quad (1)$$

4.4. G.K. Peptide—Ribose Assay

This test was used to evaluate the ability of flavonoids to inhibit the crosslinking of G.K. peptides (last glycation products) in the presence of ribose, using the method described by Rahbar et al. [23]. The G.K. peptide (80 mg/mL) was incubated with 0.8 M ribose under sterile conditions in 0.5 M sodium phosphate buffer (pH 7.4) at 37 °C for 24 h. The adlay samples or phenolic acids were added at final concentrations of 125 μg/mL and 250 μg/mL, respectively. At the end of the incubation period, samples were analyzed for specific fluorescence (excitation, 340 nm; emission, 420 nm). The % inhibition by different concentrations of inhibitor was calculated as described above.

4.5. HPLC Analysis

The HPLC analysis method according to a previous report [22], with a slight modification, was used and carried out on an UltiMate 3000 pump equipped with an autosampler column compartment system and a variable wavelength detector (DIONEX, manufacturer, Sunnyvale, CA, USA). A 4.6 mmØ × 250 mm (5 μM) reverse-phase C18 column was used (HiQ sil C18HS). Gradient elution was performed with 2% acetic acid$_{(aq)}$ (v/v, solvent A) and 0.5% acetic acid$_{(aq)}$/acetonitrile (1:1, v/v, solvent B) at a constant flow rate of 1 mL/min. The initial condition was at 5% B, and the eluent was as follow: 0–10 min B increased from 5 to 10%, 10–40 min B increased from 10 to 40% and 40–55 min B increased from 40 to 55%. The waste was washed by increasing B from 55 to 100% within 10 min, and the eluent was adjusted to the initial condition (5% B) within 15 min. The absorbance at 280 nm was recorded. Phenolic standards, dissolved in DMSO, were analyzed, and peak areas of serial diluted standards were calculated for calibration curves. ATE-BuOH was analyzed using the HPLC system, and the contents of phenolic acids were compared with the retention times and calibration curves.

4.6. Statistics

All the experiments were performed in triplicate, and the values are expressed as the mean ± SD. Statistical analysis was performed via one-way analysis of variance (ANOVA), and multiple comparisons were performed utilizing Duncan's multiple range test. All statistical analyses were performed using the computer software SPSS 12.0. $p < 0.05$ was considered statistically significant.

5. Conclusions

Adlay testa exhibits antiglycative activities that affect protein glycation and the protein's subsequent crosslinking. Adlay inhibits glycation, perhaps mainly due to its antioxidant phenolic acid compounds of chlorogenic acid and ferulic acid. Therefore, adlay seeds may be a potential candidate for the future development of alternative therapeutics for AGE-related diseases.

Author Contributions: Conceptualization, S.-M.H., C.-P.C., W.-S.C., D.-W.H., W.-C.C. and C.-H.W.; experimentation, W.-S.C., C.-P.C., D.-W.H., M.-Y.L. and C.-H.W.; data analysis and figure preparation, S.-M.H., C.-P.C., W.-S.C., D.-W.H. and C.-H.W.; methodology and resources, C.-P.C., W.-S.C., D.-W.H., W.-C.C., M.A., M.-Y.L. and C.-H.W.; writing—original draft preparation, W.-S.C. and C.-H.W.; writing—review and editing, C.-H.W.; editing and approval of the final version of the manuscript, S.-M.H. and C.-H.W. All authors have read and agreed to the published version of the manuscript.

Funding: This study was supported by grants MOST106-2311-B-003-006 from the Ministry of Science and Technology, ZRRPF3M0081 from the Research Center for Food and Cosmetic Safety, and ZRRPF3L0091 and ZRRPF3M0091 from the Research Center for Chinese Herbal Medicine, Chang Gung University of Science and Technology.

Institutional Review Board Statement: Not applicable.

Informed Consent Statement: Not applicable.

Data Availability Statement: Data are contained within the article.

Conflicts of Interest: The authors declare no conflict of interest.

Sample Availability: Samples of the crude extracts prepared from adlay seed are available from the authors.

References

1. Yeh, W.J.; Hsia, S.M.; Lee, W.H.; Wu, C.H. Polyphenols with antiglycation activity and mechanisms of action: A review of recent findings. *J. Food Drug Anal.* **2017**, *25*, 84–92. [CrossRef] [PubMed]
2. Vistoli, G.; De Maddis, D.; Cipak, A.; Zarkovic, N.; Carini, M.; Aldini, G. Advanced glycoxidation and lipoxidation end products (AGEs and ALEs): An overview of their mechanisms of formation. *Free Radic. Res.* **2013**, *47*, 3–27. [CrossRef]
3. Takeuchi, M. Toxic AGEs (TAGE) theory: A new concept for preventing the development of diseases related to lifestyle. *Diabetol. Metab. Syndr.* **2020**, *12*, 105. [CrossRef] [PubMed]
4. Gill, V.; Kumar, V.; Singh, K.; Kumar, A.; Kim, J.J. Advanced glycation end products (AGEs) may be a striking link between modern diet and health. *Biomolecules* **2019**, *9*, 888. [CrossRef] [PubMed]
5. Sellegounder, D.; Zafari, P.; Rajabinejad, M.; Taghadosi, M.; Kapahi, P. Advanced glycation end products (AGEs) and its receptor, RAGE, modulate age-dependent COVID-19 morbidity and mortality. A review and hypothesis. *Int. Immunopharmacol.* **2021**, *98*, 107806. [CrossRef] [PubMed]
6. González, I.; Morales, M.A.; Rojas, A. Polyphenols and AGEs/RAGE axis. Trends and challenges. *Food Res. Int.* **2020**, *129*, 108843. [CrossRef]
7. Thornalley, P.J. Use of aminoguanidine (Pimagedine) to prevent the formation of advanced glycation endproducts. *Arch. Biochem. Biophys.* **2003**, *419*, 31–40. [CrossRef] [PubMed]
8. Wu, C.H.; Yen, G.C. Inhibitory effect of naturally occurring flavonoids on the formation of advanced glycation endproducts. *J. Agric. Food Chem.* **2005**, *53*, 3167–3173. [CrossRef] [PubMed]
9. Jahan, H.; Choudhary, M.I. Glycation, carbonyl stress and AGEs inhibitors: A patent review. *Expert Opin. Ther. Pat.* **2015**, *25*, 1267–1284. [PubMed]
10. Zeng, Y.; Yang, J.; Chen, J.; Pu, X.; Li, X.; Yang, X.; Yang, L.; Ding, Y.; Nong, M.; Zhang, S.; et al. Actional mechanisms of active ingredients in functional food adlay for human health. *Molecules* **2022**, *27*, 4808. [CrossRef] [PubMed]
11. Kuo, C.C.; Chen, H.H.; Chiang, W.J. Adlay (yi yi; "soft-shelled job's tears"; the seeds of *Coix lachryma-jobi* L. var. ma-yuen Stapf) is a potential cancer chemopreventive agent toward multistage carcinogenesis processes. *Tradit. Complement. Med.* **2012**, *2*, 267–275. [CrossRef]
12. Yang, R.S.; Lu, Y.H.; Chiang, W.; Liu, S.H. Osteoporosis prevention by adlay (yì yĭ: The seeds of *Coix lachryma-jobi* L. var. ma-yuen Stapf) in a mouse model. *J. Tradit. Complement. Med.* **2013**, *3*, 134–138. [CrossRef] [PubMed]
13. Huang, D.W.; Wu, C.H.; Shih, C.K.; Liu, C.Y.; Shih, P.H.; Shieh, T.M.; Lin, C.I.; Chiang, W.C.; Hsia, S.M. Application of the solvent extraction technique to investigation of the anti-inflammatory activity of adlay bran. *Food Chem.* **2014**, *145*, 445–453. [CrossRef] [PubMed]
14. Tseng, Y.H.; Chang, C.W.; Chiang, W.; Hsieh, S.C. Adlay Bran oil suppresses hepatic gluconeogenesis and attenuates hyperlipidemia in type 2 diabetes rats. *J. Med. Food* **2019**, *22*, 22–28. [CrossRef] [PubMed]
15. Zhang, C.; Zhang, W.; Shi, R.; Tang, B.; Xie, S. *Coix lachryma-jobi* extract ameliorates inflammation and oxidative stress in a complete Freund's adjuvant-induced rheumatoid arthritis model. *Pharm. Biol.* **2019**, *57*, 792–798. [CrossRef] [PubMed]
16. Chiang, H.; Lu, H.F.; Chen, J.C.; Chen, Y.H.; Sun, H.T.; Huang, H.C.; Tien, H.H.; Huang, C. Adlay seed (*Coix lacryma-jobi* L.) extracts exhibit a prophylactic effect on diet-induced metabolic dysfunction and nonalcoholic fatty liver disease in mice. *Evid. Based Complement. Altern. Med.* **2020**, *2020*, 9519625. [CrossRef] [PubMed]

17. Chung, C.P.; Lee, M.Y.; Hsia, S.M.; Chiang, W.; Kuo, Y.H.; Hsu, H.Y.; Lin, Y.L. Suppression on allergic airway inflammation of dehulled adlay (*Coix lachryma-jobi* L. var. ma-yuen Stapf) in mice and anti-degranulation phytosterols from adlay bran. *Food Funct.* **2021**, *12*, 12788–12799. [CrossRef]
18. Yeh, W.J.; Ko, J.; Cheng, W.Y.; Yang, H.Y. Dehulled adlay consumption modulates blood pressure in spontaneously hypertensive rats and overweight and obese young adults. *Nutrients* **2021**, *13*, 2305. [CrossRef] [PubMed]
19. Chen, H.J.; Shih, C.K.; Hsu, H.Y.; Chiang, W. Mast cell-dependent allergic responses are inhibited by ethanolic extract of adlay (*Coix lachryma-jobi* L. var. ma-yuen Stapf) testa. *J. Agric. Food Chem.* **2010**, *58*, 2596–2601. [CrossRef] [PubMed]
20. Hsia, S.M.; Kuo, Y.H.; Chiang, W.; Wang, P.S. Effects of adlay hull extracts on uterine contraction and Ca^{2+} mobilization in the rat. *Am. J. Physiol. Endocrinol. Metab.* **2008**, *295*, E719–E726. [CrossRef]
21. Nagai, E.; Iwai, M.; Koketsu, R.; Okuno, Y.; Suzuki, Y.; Morimoto, R.; Sumitani, H.; Ohshima, A.; Enomoto, T.; Isegawa, Y. Anti-influenza virus activity of adlay tea components. *Plant Foods Hum. Nutr.* **2019**, *74*, 538–543. [CrossRef] [PubMed]
22. Huang, D.W.; Chung, C.P.; Kuo, Y.H.; Lin, Y.L.; Chiang, W. Identification of compounds in adlay (*Coix lachryma-jobi* L. var. ma-yuen Stapf) seed hull extracts that inhibit lipopolysaccharide-induced inflammation in RAW 264.7 macrophages. *J. Agric. Food Chem.* **2009**, *57*, 10651–10657. [CrossRef]
23. Rahbar, S.; Yerneni, K.K.; Scott, S.; Gonzales, N.; Lalezari, I. Novel inhibitors of advanced glycation endproducts (part II). *Mol. Cell Biol. Res. Commun.* **2000**, *3*, 360–366. [CrossRef] [PubMed]
24. Rowan, S.; Bejarano, E.; Taylor, A. Mechanistic targeting of advanced glycation end-products in age-related diseases. *Biochim. Biophys. Acta. Mol. Basis Dis.* **2018**, *1864*, 3631–3643. [CrossRef] [PubMed]
25. Wu, X.Q.; Zhang, D.D.; Wang, Y.N.; Tan, Y.Q.; Yu, X.Y.; Zhao, Y.Y. AGE/RAGE in diabetic kidney disease and ageing kidney. *Free Radic. Biol. Med.* **2021**, *171*, 260–271. [CrossRef] [PubMed]
26. Huebschmann, A.G.; Regensteiner, J.G.; Vlassara, H.; Reusch, J.E.B. Diabetes and advanced glycation glycoxidation end products. *Diabetes Care* **2006**, *29*, 1420–1432. [CrossRef] [PubMed]
27. Sultana, B.; Anwar, F.; Ashraf, M. Effect of extraction solvent/technique on the antioxidant activity of selected medicinal plant extracts. *Molecules* **2009**, *14*, 2167–2180. [CrossRef] [PubMed]
28. Huang, Y.J.; Chen, Y.C.; Chen, H.Y.; Chiang, Y.F.; Ali, M.; Chiang, W.; Chung, C.P.; Hsia, S.M. Ethanolic extracts of adlay testa and hull and their active biomolecules exert relaxing effect on uterine muscle contraction through blocking extracellular calcium influx in ex vivo and in vivo studies. *Biomolecules* **2021**, *11*, 887. [CrossRef] [PubMed]
29. Huang, Y.J.; Chang, C.C.; Wang, Y.Y.; Chiang, W.C.; Shih, Y.H.; Shieh, T.M.; Wang, K.L.; Ali, M.; Hsia, S.M. Adlay testa (*Coix lachryma-jobi* L. var. Ma-yuen Stapf.) Ethanolic extract and its active components exert anti-proliferative effects on endometrial cancer cells via cell cycle arrest. *Molecules* **2021**, *26*, 1966. [CrossRef]
30. Kim, D.O.; Lee, C.Y. Comprehensive study on vitamin C equivalent antioxidant capacity (VCEAC) of various polyphenolics in scavenging a free radical and its structural relationship.Crit. Rev. *Food Sci. Nutr.* **2004**, *44*, 253–273. [CrossRef] [PubMed]
31. Wu, C.H.; Huang, H.W.; Lin, J.A.; Huang, S.M.; Yen, G.C. The proglycation effect of caffeic acid leads to the elevation of oxidative stress and inflammation in monocytes, macrophages and vascular endothelial cells. *J. Nutr. Biochem.* **2011**, *22*, 585–594. [CrossRef] [PubMed]
32. Woodside, J.V.; Yarnell, J.W.; McMaster, D.; Young, I.S.; Harmon, D.L.; McCrum, E.E.; Patterson, C.C.; Gey, K.F.; Whitehead, A.S.; Evans, A. Effect of B-group vitamins and antioxidant vitamins on hyperhomocysteinemia: a double-blind, randomized, factorial-design, controlled trial. *Am. J. Clin. Nutr.* **1998**, *67*, 858–866. [CrossRef] [PubMed]
33. de Bari, L.; Scirè, A.; Minnelli, C.; Cianfruglia, L.; Kalapos, M.P.; Armeni, T. Interplay among oxidative stress, methylglyoxal pathway and S-glutathionylation. *Antioxidants* **2020**, *10*, 19. [CrossRef] [PubMed]
34. Nagaraj, R.H.; Shipanova, I.N.; Faust, F.M. Protein cross-linking by the Maillard reaction. Isolation, characterization, and in vivo detection of a lysine-lysine cross-link derived from methylglyoxal. *J. Biol. Chem.* **1996**, *271*, 19338–19345. [CrossRef] [PubMed]
35. Matsuura, N.; Aradate, A.; Sasaki, C.; Kojima, H.; Ohara, M.; Hasegawa, J.; Ubukata, M. Screening system for the Maillard reaction inhibitor from natural product extracts. *J. Health Sci.* **2002**, *48*, 520–526. [CrossRef]

molecules MDPI

Review

Grape Pomace Polyphenols as a Source of Compounds for Management of Oxidative Stress and Inflammation—A Possible Alternative for Non-Steroidal Anti-Inflammatory Drugs?

Veronica Sanda Chedea [1,†], Ștefan Octavian Macovei [2,†], Ioana Corina Bocsan [3,*], Dan Claudiu Măgureanu [2], Antonia Mihaela Levai [4], Anca Dana Buzoianu [3] and Raluca Maria Pop [3]

1 Research Department, Research Station for Viticulture and Enology Blaj (SCDVV Blaj), 515400 Blaj, Romania
2 Faculty of Medicine, "Iuliu Hatieganu" University of Medicine and Pharmacy, 400012 Cluj-Napoca, Romania
3 Department of Pharmacology, Toxicology and Clinical Pharmacology, Faculty of Medicine, "Iuliu Hatieganu" University of Medicine and Pharmacy, No. 23, Marinescu Street, 400012 Cluj Napoca, Romania
4 Department Mother and Child, Faculty of Medicine, "Iuliu Hatieganu" University of Medicine and Pharmacy, No. 3–5, Clinicilor Street, 400012 Cluj Napoca, Romania
* Correspondence: bocsan.corina@umfcluj.ro
† These authors contributed equally to this work.

Abstract: Flavonoids, stilbenes, lignans, and phenolic acids, classes of polyphenols found in grape pomace (GP), were investigated as an important alternative source for active substances that could be used in the management of oxidative stress and inflammation. The benefic antioxidant and anti-inflammatory actions of GP are presented in the literature, but they are derived from a large variety of experimental *in vitro* and *in vivo* settings. In these *in vitro* works, the decrease in reactive oxygen species, malondialdehyde, and thiobarbituric acid reactive substances levels and the increase in glutathione levels show the antioxidant effects. The inhibition of nuclear factor kappa B and prostaglandin E2 inflammatory pathways and the decrease of some inflammatory markers such as interleukin-8 (IL-8) demonstrate the anti-inflammatory actions of GP polyphenols. The *in vivo* studies further confirmed the antioxidant (increase in catalase, superoxide dismutase and glutathione peroxidase levels and a stimulation of endothelial nitric oxide synthase -eNOS gene expression) and anti-inflammatory (inhibition of IL-1α, IL-1β, IL-6, interferon-γ, TNF-α and C-reactive protein release) activities. Grape pomace as a whole extract, but also different individual polyphenols that are contained in GP can modulate the endogenous pathway responsible in reducing oxidative stress and chronic inflammation. The present review analyzed the effects of GP in oxidative stress and inflammation, suggesting that it could become a valuable therapeutic candidate capable to reduce the aforementioned pathological processes. Grape pomace extract could become an adjuvant treatment in the attempt to reduce the side effects of the classical anti-inflammatory medication like non-steroidal anti-inflammatory drugs (NSAIDs).

Keywords: grape pomace; polyphenols; antioxidant; anti-inflammatory

Citation: Chedea, V.S.; Macovei, Ș.O.; Bocsan, I.C.; Măgureanu, D.C.; Levai, A.M.; Buzoianu, A.D.; Pop, R.M. Grape Pomace Polyphenols as a Source of Compounds for Management of Oxidative Stress and Inflammation—A Possible Alternative for Non-Steroidal Anti-Inflammatory Drugs?. *Molecules* **2022**, *27*, 6826. https://doi.org/10.3390/molecules27206826

Academic Editor: Nour Eddine Es-Safi

Received: 15 September 2022
Accepted: 9 October 2022
Published: 12 October 2022

1. Introduction

The study of plant-derived compounds received increased attention among researchers worldwide in the attempt to discover new bioactive molecules with marked therapeutic actions and minimal adverse effects. A healthy diet and regular exercise protect the human body from cardiovascular problems, diabetes, and cancer [1]. The definition of a healthy diet is permanently evolving and changing to comprise all the knowledge with respect to different foods, nutrients or food components in health and disease. Thus, a healthy food approach refers to diets rich in plant-based foods (fresh fruits, vegetables, legumes, seeds, whole grains, nuts) and low in animal-based products (e.g., fatty and processed meats) [2]. Therefore, WHO recommends consuming 400 g (5 portions) of fruit and vegetables to our diet [3] because they possess significant levels of phytochemicals like flavonoids, stilbenes, lignans, and phenolic acids found in different concentrations [1,4,5].

Grapes, with a global production of around 78 million tons in 2020 [6], and with around 75% of total production going into the wine industry, are such an example. One of the main reasons for grapes research is the quantity of waste resulting from the juice and winemaking process [7–9]. An important by-product of the winemaking industry is grape pomace (GP), and due to the important consumption of wine (27 million tons in 2019) [6], there are also important quantities of GP generation (8.49 million tons) [10]. Disposal of this by-product causes an ecological problem because of pollution and other hazardous issues that come with producing large quantities in a short time and little space for deposit [11]. To overcome these, different solutions were proposed for by-products valorization, including the recovery of containing bioactive compounds like phenolic compounds [12–14]. Polyphenols have beneficial effects on various diseases, especially because of their anti-inflammatory and antioxidant effects via different mechanisms [15–17]. Accordingly, it was observed that GP, because of its various and rich quantity of phenolics has potential antioxidant, anti-inflammatory, anti-microbial, anti-cancer effects and also beneficial cardiovascular and hepatic effects [18–21]. Knowing that oxidative stress and inflammation processes are common to many diseases, this review will focus especially on these pathophysiological processes since their management is a key step in the attempt to prevent or reduce disease progression. Thus, their modulation by polyphenols from GP as reported in both *in vitro* and *in vivo* studies will be addressed. Also, another important aim of this review is to compare the anti-inflammatory effects of GP polyphenols with the one of non-steroidal anti-inflammatory drugs (NSAIDs). Non-steroidal anti-inflammatory drugs are used for medical purposes since ancient times, being one of the most prescribed drug classes in the world. They possess anti-inflammatory, antipyretic, and analgesic effects and therefore are prescribed for many diseases, especially chronic diseases like rheumatoid arthritis, osteoarthritis, ankylosing spondylitis, and chronic pain [22–24]. Unfortunately, the therapeutic effects can be associated with pulmonary, renal, gastrointestinal, cardiovascular, and hepatic adverse reactions that might lead to severe complications [25–28]. Thus, there is an urgent need in identifying novel antioxidant and anti-inflammatory agents, therapeutically potent and with minimal side effects that could be used instead of classical drugs or as add-on therapy, and GP could be a source of such kind of compounds.

2. Oxidative Stress Process

An imbalance between the synthesis and accumulation of oxygen reactive species (ROS) in cells and tissues and the capacity of an organism to eliminate these reactive compounds results in oxidative stress [29]. Thus, organism homeostasis can be altered if higher accumulation of free radicals occurs. ROS are normally generated through different reactions like enzymatic and non-enzymatic, and they can have exogenous or endogenous sources. The enzymatic reactions responsible for ROS generation are characterized by phagocytosis, cytochrome P450 reactions, mitochondrial respiratory chain, and cyclooxygenase-synthesis of prostaglandin [30]. The non-enzymatic system involves the reaction of free oxygen with organic molecules or through tissue-radiation exposure [29]. The exogenous sources of free radicals production are represented by radiation [30], air pollution, and cigarette smoke while endogenous sources are represented by the enzymes systems: mitochondrial respiratory chain, nicotinamide adenine dinucleotide phosphate (NADPH) oxidase family, cyclooxygenase, and lipoxygenase [31–34].

ROS production direct measurement is a very complex and difficult process mostly because of ROS sources and species variety, low steady-state concentrations, high reactivity, and of the detection methods involved in the measurement analysis. Also, the oxidative stress assessment is performed by the indirect evaluation of ROS-induced damage on biological targets like DNA, proteins, membrane lipids, and others [35]. Additionally, ROS generation differs significantly in tissue localization, activation mechanisms, and functions in diseases. As a result, ROS levels must be kept within a range that enables organisms to function normally. These concentration ranges may vary between tissues. However, the balance between physiological redox states and oxidative stress is fragile and is based on relative rates of ROS production and destruction. All of these make it difficult to establish an absolute scale that can offer the values of ROS concentrations in physiological and pathological conditions [36]. Usually, ROS in low and moderate concentrations doesn't possess a significant threat to homeostasis, their beneficial role being known in different physiological processes (e.g., the synthesis of different cellular structures, the help of defense system to neutralize pathogen agents) [29]. If the free radicals are present in large quantities, they cross the physiological barrier and can lead to several health issues, such as cardiovascular, neurological, kidney, respiratory, and rheumatoid diseases [32,34,37]. To counter the side effects of ROS, the organism possess antioxidant mechanisms represented by enzymatic systems (superoxide dismutase—SOD, catalase—CAT, glutathione peroxidase—GPx) and non-enzymatic ones (glutathione, vitamins A, C, and E) [32,33,37]. Superoxide dismutase (SOD) is responsible for O_2^- reduction to H_2O_2. Also, another role of this enzyme is to prevent the formation of other free radicals (peroxynitrite—$ONOO^-$). Further, the H_2O_2 molecule will be converted by catalase (CAT) in water and oxygen [33,38]. Glutathione peroxidase (GPx) is another enzyme with a role in degrading H_2O_2 and hydroperoxide molecules using glutathione (GSH) as a proton donor [32].

Another category of endogenous substances resulting from internal metabolism with consequences on human health due to an imbalance between synthesis and elimination is reactive nitrogen species (RNS). RNS are already known to take part in the pathophysiological process of different diseases (diabetes, Parkinson's disease, pulmonary, cardiovascular, rheumatological, liver, and neurological diseases) [39]. RNS are synthesized through the interaction of nitric oxide molecules with other reactive oxygen species. Nitric oxide (NO) is a versatile natural molecule found in organisms resulting from the breakdown of arginine to citrulline [40]. Nitric oxide possesses antimicrobial activity, promotes vascular relaxation with reducing blood pressure, and cell signaling. NO also acts as a scavenger molecule interacting with superoxide anion (O_2^-) leading to peroxynitrite ($ONOO^-$) formation. $ONOO^-$ leads to endothelial damage, DNA oxidation, and lipid peroxidation [40,41]. Nitrogen dioxide (NO_2^-) is another RNS, that results from the interaction of NO with peroxyl radical, which produces lipid peroxidation, ascorbic acid oxidation, and alters tyrosine structure and function leading to the 3-nitrotyrosine formation (3NT) [40,42].

Even though the role of ROS and RNS in damaging the cells and signal transduction is well known, there are still several highly debated issues that need to be resolved. When present in low quantities, ROS and RNS operate as regulatory mediators in signaling processes; nevertheless, when present in high concentrations, they are toxic to living organisms by inactivating critical cellular components. This indicates that the concentrations of ROS and RNS control the change from their favorable to unfavorable effects, although the concentrations at which this change occurs are not well known [36].

The imbalance between the increased ROS and RNS production and decrease of antioxidant molecules will eventually lead to chronic inflammation [43]. During the oxidative stress process, the reactive oxygen/nitrogen species can initiate intracellular signaling and through which specific proinflammatory genes are expressed [43,44]. Generally, between oxidative stress and inflammation, there is a state of interdependency, with one potentiating the other via different mechanisms. For example, during oxidative stress at the brain level, lipids and proteins suffer alterations through oxidation which conduct to disruptions in neurons' communication with inflammation being stimulated [45]. During oxidative stress, DNA suffers damage resulting in different metabolites (8-oxo-7 8-dihydro-2′-deoxyguanosine, 8 oxo-guanine). It was demonstrated that 8-oxo-guanine presence stimulates the nuclear factor kappa B (Nf-κB) [46]. The activation of Nf-κB enhances the pro-inflammatory response through increased synthesis of inflammatory molecules (cytokines, chemokines) and also activation of immune cells [47–49]. Cytokines manage also to increase the ROS levels in non-immune cells. For example, it was shown that IL-6 stimulates the NADPH in lung cancer cell lines leading to an increase in ROS levels [47].

Oxidative stress leads to the release of arachidonic acid which under cyclooxygenase and lipoxygenase enzymes reactions, results in prostaglandins and leukotrienes synthesis [49]. Also, oxidative stress can be enhanced by inflammation. During inflammation, immune cells responsible for phagocytosis (neutrophils and macrophages) produce reactive oxygen and nitrogen species to dissolve the pathogen. In the case of an exaggerated inflammatory response, these reactive species exit from phagocytic cells resulting in tissue injury outside the inflammatory site [47].

Thus, inflammation and oxidative stress, are tightly interconnected [50] initiation and maintenance of many pathological conditions, their prevention and control could provide a safe alternative in chronic disease management.

3. Inflammation

Inflammation represents a pathophysiological mechanism of defense that acts in case of homeostasis perturbations provoked by infectious agents or trauma. Immune cells enact this mechanism which has a role in locating the pathological agent, digestion, and resolution of the inflammation with restoring homeostasis. Thus, inflammation could be considered a protective response, but an uncontrolled inflammation may be potentially harmful and may lead to many acute and chronic diseases [51].

Inflammation could be acute or chronic depending on the interval of time from the onset of homeostasis impairment until the development of the entire process and the appearance of clinical symptoms. Acute inflammatory diseases present a rapid and non-specific immune response which usually lasts up to 2 weeks with the resolution of the inflammation process [52]. If the process is not healed, this can lead to a prolonged immune response called chronic inflammation, which lasts from months to years [53]. Thus, chronic inflammations may require a long-lasting management and they become a burden not only for individuals but also for society due to higher costs and health assistance. For all these reasons, there is a perpetual race to discover new molecules with pharmacological properties that may limit the chronic evolution of inflammation.

The etiology of it is also variable, different agents like infectious agents or trauma could trigger cells belonging to both innate and adaptive immunity. The most important aspect of acute inflammation is the recognition of the pathogen or damaged tissue through circulating molecules, which signals innate immune cells and leads to a cascade of biological reactions [54,55]. These reactions are mediated by several proinflammatory molecules, such as cytokines and chemokines, which act interconnected in a signaling network, leading to the recruitment of more immune cells and mediators at the site of inflammation [56,57]. Vasodilation also occurs at the site of inflammation to bring the immune cells there. This process is facilitated by local mediators (nitric oxide, prostaglandins) produced by endothelial and inflammatory cells, leading to the translocation of vascular fluid into interstitial space and enhancing the migration of immune cells [58,59].

Neutrophils represent the first and the most important type of cells belonging to innate immunity that act at the site of inflammation. Neutrophils create a toxic environment by releasing cytotoxic compounds from their vesicles, such as proteases and reactive oxygen and nitrogen species, that destroy the pathogen and the surrounding tissue [55,60]. After the pathogen is destroyed, the resolution process begins, and monocytes are recruited for wound healing. Monocytes block other new inflammatory processes with possible downside effects for the host [55].

Clinical signs and symptoms that characterize inflammation reflect these pathophysiological processes. Locally at the inflammation site, heat, edema, redness, pain, and impaired function could be observed [57]. Cellular injury inducing an immune response activates the intrinsic blood coagulation pathway [61]. Coagulation is activated to create the fibrin clot, which has a role in the isolation of the inflammation process, but also it enhances the pro-inflammatory response [62]. Another mechanism that intervenes in innate immunity is the complement system, which consists in several proteins that could be activated. These proteins, through a series of chain reactions, create a so-called "membrane complex attack" that disrupts the microbe's cell membrane and induces death [58,59,62].

The acute inflammatory response may transit to a chronic response if the neutrophils fail to eliminate the pathogen agent in the first place. After this, the innate immune response is followed by an adaptive immune response characterized by the presence of macrophages and lymphocytes. Besides them, fibroblasts and plasma cells can be found at the site of chronic inflammation [63]. Monocytes are a group of cells that migrate from blood to different tissues and differentiate themselves into macrophages. Macrophages work in innate immune response alongside neutrophils and in the adaptive immune response through activating lymphocytes [59,63]. Lymphocytes T (Ly T) are activated by macrophages and will be divided into different populations with specific role in the adaptive immune response. However, their primary function is to enhance the immune system, stimulating all immune cells to give a specific defense response against the etiological agent [63]. Lymphocytes B (Ly B), activated by Ly T cells, differentiate in plasma cells which produce antibodies against different types of antigens [64]. The chronic process is associated with tissue destruction, and the repair process is maintained by fibroblasts that secrete collagen, representing the main component needed for wound healing [65]. Thus, acute inflammation is characterized by vasodilatation, and innate immune cells, while the chronic one is described by the involvement of adaptive immune cells, and fibroblast proliferation with significant changes in wound healing. Chronic inflammation is an important component of many diseases, such as atherosclerosis, diabetes, metabolic disorders, cancer, and autoimmune conditions, so the conversion of acute inflammation to a chronic one is a desirable outcome.

4. Grape Pomace Generation

Historically, the first evidence of wine production was found in south Caucasus and dated back as far as 6000 BC. From there, traces of wine production were discovered in Syria, Palestine, and Egypt. It spread around the Mediterranean basin during Roman times and further in Europe as long as the Roman Empire was further invading [20,66].

In European Union, as of 2020, 3.2 million hectares were used for grapevine growing and grape production, about 45% of total global grapevine growing surfaces [67]. Most of the grapes produced are sent to the wine industry, which leads to large quantities of waste products, such as GP. Grape pomace is generated through the vinification process which comprises all the technological phases from grape harvesting to wine bottling. Destemming is the first process after harvesting grapes, removing stems and grape stalks. After this, crushing takes place in which grapes are mashed into juice, skins, pulp, and seeds [66]. Maceration follows when juice (must) remains alongside grape skins and pulp seeds for different periods depending on what type of wine is produced. In the case of aromatic white wine, the maceration takes a shorter time, while for red wine it will be longer [20,66].Afterwards, in the pressing process, the must is extracted from grape waste. Therefore, there are two types of GP, red and white, according to the grapes used in the vinification process. The main difference between red and white wine is that white must is separated from pomace before fermentation. In contrast, in the case of red wine, pressing happens only after fermentation. Also, while pressing, the wine will gain its color and flavor [12,20,68]. After the pressing process, where must (red or white) is extracted, the remaining GP has different properties. For example, in terms of pressing and fermentation, white GP is sweet due to its higher sugar content because it didn't undergo fermentation. At the same time, red GP is a fermented matrix containing different concentrations of alcohol [20]. At the same time, the polyphenols content varies depending on the vinification technology, respectively red or white.

Grape Pomace Polyphenols

From the chemical structure point of view, polyphenols are compounds made from several hydroxyl groups attached to an aromatic ring. These molecules can vary from simple to complex structures [69]. Also, these compounds are synthesized in plants and exist as glycosides, which are further formed from the basic polyphenol structure and glycosylic radical (sugar fragment) [70]. GP polyphenols are divided into two classes: flavonoids and non-flavonoids (Figure 1). Grape pomace flavonoids are further subdivided into flavonols, anthocyanins, flavanols, proanthocyanidins, and anthocyanidins, while GP non-flavonoids in stilbenes and phenolic acids [71,72].

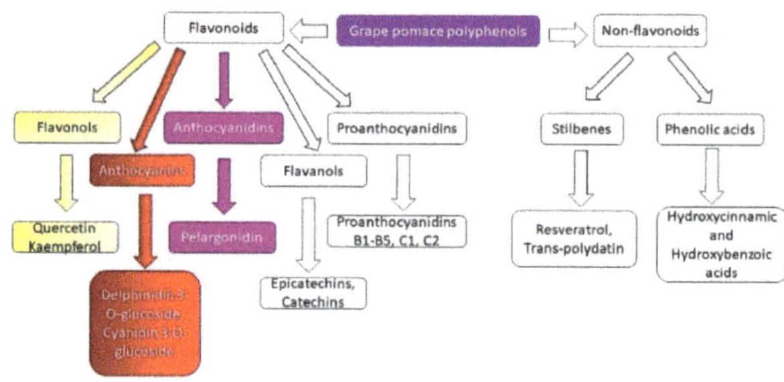

Figure 1. Grape pomace polyphenols classes.

The main structure of a flavonoid is made up of two phenyl radicals (rings A and B) linked to a heterocyclic ring, which contains an atom of oxygen (ring C). Based on the oxidation state and hydroxyl radicals' distribution pattern on the heterocyclic ring, flavonoids are further divided into the classes mentioned before [69,70] (Figure 2).

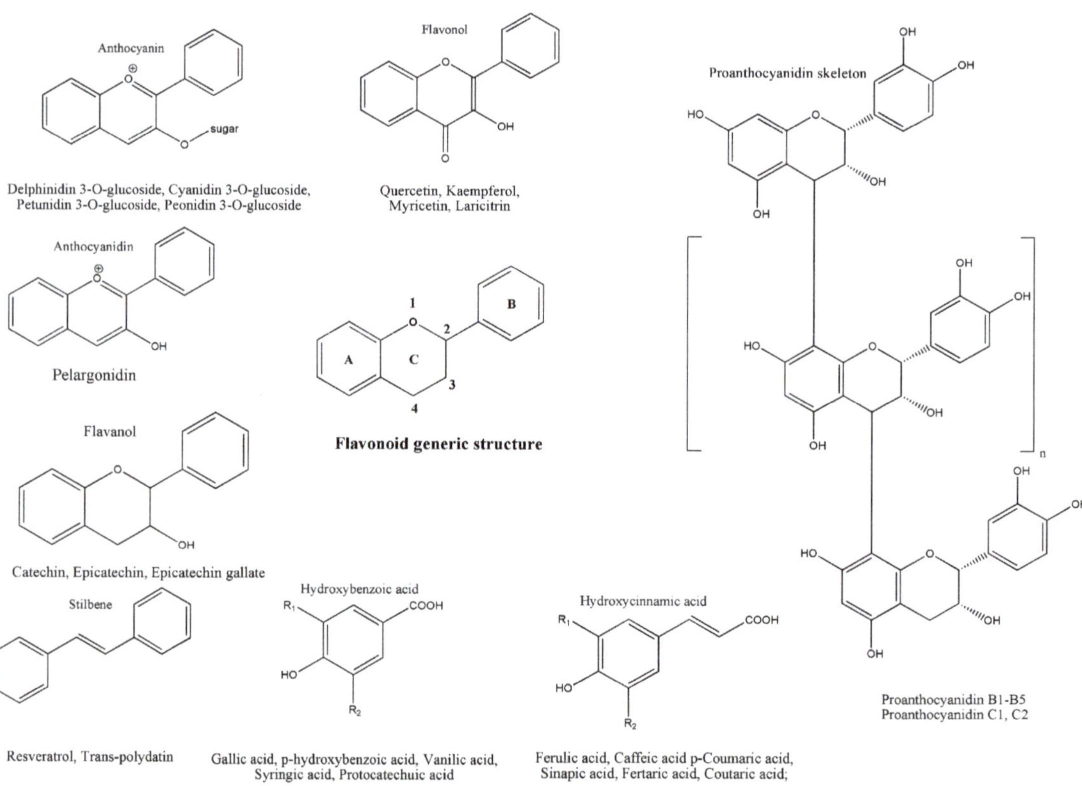

Figure 2. Flavonoid generic structure, structures of the main grape pomace polyphenols classes, as well as the most representative compounds.

Flavonols structure presents a C2−C3 double bond with a hydroxyl radical at the C3 position and ring B coupled to the C2 position. Flavanols structure presents with a hydroxyl radical situated at the C3 position, and their ring B is attached to the C2 position [70,73,74] (Figure 2). Proanthocyanidins, also called procyanidins or condensed tannins, are formed from subunits of flavanols that bond together in dimers (usually referred to as B-series—B1, B2, B3, B4, and B5) or trimers (known as C-series—C1, and C2). This series of procyanidins are being found in grape skins and seeds [75] (Figure 2). Anthocyanins present double bonds in the heterocyclic ring, their aromatic ring B being bonded to the C2 position. They represent the glycosylated form of anthocyanidins (aglycone) which result from the bond between the hydroxyl group at C3 and the sugar fragment and are the most abundant polyphenols in the peel of red grapes [70,74] (Figure 2). It is known that anthocyanins are only found in red GP because they act as a natural colorant, giving the red grape specific color. High contents of anthocyanins are found in red GP also because of the red grape skin thickness as compared to the white ones [76,77].

Compared to flavonoids, non-flavonoid polyphenols have one aromatic ring as a basic structure. Non-flavonoid molecules found in GP are phenolic acids, and stilbenes [69,70,78]. Phenolic acids are further divided into hydroxybenzoic and hydroxycinnamic acids. As representants of hydroxybenzoic acids, there are gallic, p-hydroxybenzoic, and syringic acids, and for the hydroxycinnamic acids, caffeic, p-coumaric, ferulic and synaptic acids are the most found in GP [69,70,78,79]. Stilbenes are formed from two aromatic rings bounded through the ethylene radical. The most known and studied stilbene is resveratrol. Besides GP, stilbenes are reported to be found also in grapes, and wine [69,70,78] (Figure 2).

The polyphenolic composition from different assortments of GP may differ based on the grape cultivar, type of soil, weather, geographical location, and winemaking process [76,80]. The content of polyphenols found in GP differ from study to study; some suggest that red GP possess the highest content of polyphenols, and other suggest otherwise, but the principal idea is that no matter which GP is analyzed, all of them possess high quantities of polyphenols. For instance, Kammerer et al. 2004 studied the polyphenol composition of 14 different red and white GP [79]. The study did not find significant differences in composition between red and white varieties except for the presence of anthocyanins found in red ones [79].

5. Grape Pomace Polyphenols Benefic Actions

Grape pomace polyphenols research studies have grown in the last decades, given their potential benefic effects on promoting human health. Some of their benefic actions are observed in oxidative stress and inflammation aiming at homeostasis restoration. Regarding the antioxidant effect, polyphenols can modulate the endogenous pathway responsible for combating oxidative stress. These effects can be achieved by polyphenols capacity to activate the nuclear factor E2 and to up-regulate superoxide dismutase, catalase, glutathione, glutathione peroxidase, and heme-oxidase 1 [81,82] or their capacity to scavenge and chelate reactive oxygen species involved in ROS production [83] (Figure 3). In inflammation, polyphenols are reported to inhibit the mitogen-activated kinase pathway, Nf-kB, anddown-regulate cytokines and chemokines [81,82]. Polyphenols also inhibit cyclooxygenase and lipoxygenase, which are involved in the arachidonic acid signaling pathway, being responsible for synthesizing prostaglandin, thromboxane A2, and leukotrienes which further increase inflammatory response [73,74,82] (Figure 3).

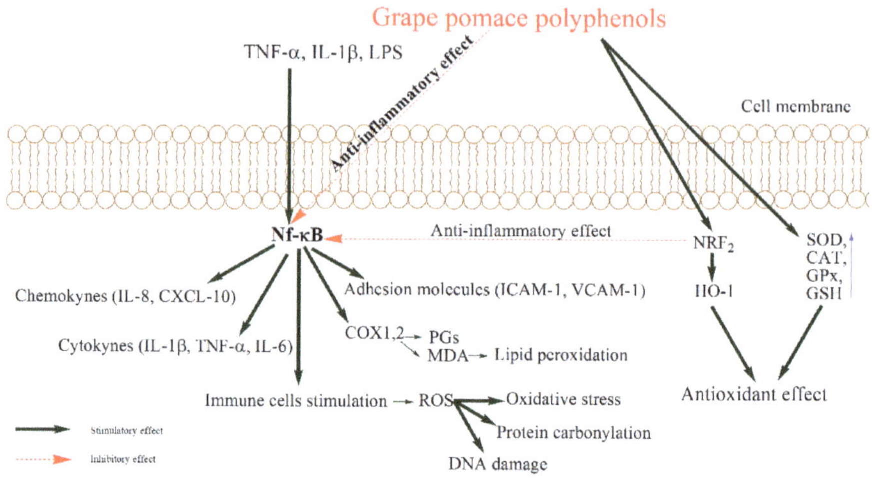

Abbreviations: CAT - catalase; COX - cyclooxygenase; DNA - deoxyribonucleic acid; GPx - glutathione peroxidase; GSH - glutathione; HO-1 - heme oxygenase 1; ICAM -1 - intercellular adhesion molecule 1; LPS - lipopolysaccharide; MDA - malondyaldehide; Nf-κB - nuclear factor kappa-light-chain-enhancer of activated B cells; NRF_2 - nuclear factor erythroid 2-related factor 2; PGs - prostaglandins; ROS -reactive oxygen species; SOD - superoxyde dismutase; TNF-α - tumor nevrosis factor α; VCAM - 1 - vascular cell adhesion protein 1;

Figure 3. Proposed antioxidant and anti-inflammatory actions of polyphenols from grape pomace.

Further, there are presented the *in vitro* and *in vivo* beneficial effects of different red and white GP in oxidative stress and inflammatory conditions.

5.1. *In Vitro Beneficial Actions of Grape Pomace in Oxidative Stress and Inflammation*

The *in vitro* studies, as presented in Table 1, can offer the possibility to investigate and identify the diversity of related diseases in which GP exerts the optimum antioxidant and anti-inflammatory effects.

The *in vitro* beneficial action of GP was studied by Goutzourelas et al. (2015) [84]. They investigated an extract of red GP on muscle and endothelial cells using non-cytotoxic doses to check the effect of GP polyphenols extract on cells' antioxidant enzymes [84]. The red GP extract was investigated as a mixture of compounds that contained phenolic acids (caftaric acid, gallic acid), anthocyanins, flavanols (epicatechin and catechin), flavonols (quercetin), and anthocyanidins. It was observed that GP treatment increased Glutathione S-transferase (GST) and GSH levels in both cell lines. CAT levels were decreased in endothelial cells, while in muscle cells it showed no significant differences. SOD and HO-1 presented no differences in any population. An explanation for these inconstant findings, in which some of the antioxidant enzymes are not modified, is the ability of GP to enhance other antioxidant systems (GSC, GSH) [84] (Table 1). Another *in vitro* study, of Pop et al. (2022), investigated the antioxidant effect of red GP (mixture of Pinot Noir, Cabernet Sauvignon, Fetească Neagră, and Mamaia cultivars) and white GP (mixture of Sauvignon Blanc and Muscat Ottonel cultivars) added to a mouthwash on both H_2O_2 exposed and non-exposed fibroblast cells [90]. They observed that both red grape pomace (RGP) and white grape pomace (WGP) decreased ROS levels in a dose-dependent matter (100 < 200 < 300 μg/mL). Similar to the non-exposed condition, in the presence of H_2O_2, red GP and white GP led to a significant decrease in ROS levels, the only difference being that while red GP effect was dose-dependent, and white GP produced a non-dependent action [90]. Moreover, they also studied the anti-inflammatory effects of these extracts on lipopolysaccharides (LPS) induced inflammation in cells [90]. It was observed that while in the case of white GP a dose of 100 μg/mL was sufficient to induce a significant reduction of interleukin (IL) -8 levels, for red GP was necessary a higher dose of 200 μg/mL. At the dose of 300 μg/mL, both extracts significantly reduced IL-8 levels, but not even the highest dose did significantly reduce the levels of IL-6. In the case of IL-1β, the lowest dose, 100 μg/mL, reduced its level to a similar one found in the non-exposed cells, while the doses of 200 and 300 μg/mL reduced, even more, the levels of IL-1β [90].

Table 1. *In vitro* beneficial actions of grape pomace in oxidative stress and inflammation.

Grape Pomace Variety	Models	Polyphenols Content	Antioxidant and Anti-Inflammatory Effects	References
		Red Grape Pomace Variety		
Batiki Tyrnavou variety (Greece)	Tert-butyl hydroperoxide induced-oxidative stress in C2C12 muscle cells and EA.hy926 endothelial cells	Flavanols, (catechin and epicatechin), anthocyanidins, anthocyanins, flavonols (quercetin) phenolic acids (gallic acid and caftaric acid)	-decreased ROS levels in muscle cells -decreased TBARS and carbonyls levels in both cells line -increased GSH levels in both cells line;	[84]
Tinta Cao and Cabernet Franc (USA)	CA77 cell line	-	-decreased CGRP levels	[85]
Tempranillo variety (University of Burgos, Spain) -WGPI-gastrointestinal digestion; WPF—colonic fermentation	Hyperglycemic treatment in EA.hy926 endothelial cells	Phenolic acids, flavanols, stillbenes, flavonols	-increased mRNA Nrf2 levels, pNrf2/Nrf2 ratio; -increased pAkt/Akt ratio; -decreased pIκBα/IκBα and pIKK/IKK ratio, mRNA COX2, NOX4, SOD2 levels; -increased mRNA SIRT1, HO-1, CAT, NQO1 levels, GSH/GSSG ratio; -increased mRNA GS, GR levels in WGPI; -increased phospho-p38-MAPK/p38-MAPK ratio in WPF; -decreased mRNA NF-κB levels and pNF-κB p65/NF-κB p65 ratio in WPF; -increased mRNA GCLC, GS, GR, SOD, GPx1 levels in WPF;	[86]
Carignan variety (Northern Tunisia)	-6-hydroxydopamine-induced oxidative stress in mesencephalic cells (dopaminergic cells) -6-hydroxydopamine-induced oxidative stress in dopaminergic cells derived from stem cells	-	-increased cell viability in mesencephalic primary cells; -decreased ROS production in stem cells; -decreased phospho-NF-κB p65 translocation	[87]
		White grape pomace variety		
Batiki Tyrnavou variety (Central Greece)	Bovine spermatozoa incubated with different GP concentrations	-	-decreased MDA levels;	[88]
		Red and White pomace variety		
White grape pomace (Moscato branco) and mixed grape pomace (red + white) Enzymatic hydrolysis treated fractions	IL-1β treated Caco-2 cells	Quercetin, catechin, resveratrol, gallic and caffeic acids, trans-resveratrol, rutin and procyanidin B2	-decreased ROS in all fractions (100–200 μg/mL); -decreased NF-κB, PGE2 levels in all fractions; -significantly greater decrease of IL-8 levels in mixed grape pomace with or without enzymatic hydrolysis;	[89]

Abbreviations: CAT—catalase; CGRP—calcitonin gene-related peptide; COX 2—cyclooxygenase 2; GR—glutathione reductase; GS—glutathione syntase; GCLC—glutamate-cysteine ligase catalytic subunit; GPx1—glutathione peroxidase 1; GSH—glutathione; GSSG—glutathione disulfide; HO-1—heme oxygenase 1; MDA—malondialdehyde; NF-κB—nuclear factor kappa-light-chain-enhancer of activated B cells; NF-κB p65—nuclear factor kappa-light-chain-enhancer of activated B cells transcription factor; NOX4—nicotinamide adenine dinucleotide phosphate oxidase 4; NQO1 —nicotinamide adenine dinucleotide plus hydrogen quinone oxidoreductase 1 mRNA Nrf2—messenger ribonucleic acid nuclear factor erythroid 2-related factor 2; p38-MAPK—p38mitogen-activated protein kinase PGE2—prostaglandin E2; pAkt—phosphorylated protein kinase B; pIκBα—phosphorylated inhibitor of kappa B; pIKK—phosphorylated IκB kinase; pNrf2—phosphorylated Nrf2 ROS—reactive oxygen species; SIRT1—sirtuin 1; SOD2—superoxide dismutase 2; TBARS—thiobarbituric acid reactive substances; TNF-α—tumor necrosis factor alpha.

Marzulli et al. (2018), treated mononuclear cells with phorbol 12-myristate 13-acetate (PMA) to activate inflammation, and with different GPs (red Negroamaro cultivar or white Koshu cultivar) extracts (water, ethanol), to observe their immunomodulatory effects [91]. In terms of cytokine release, all GP fractions and extracts increased anti-inflammatory (IL-10) and pro-inflammatory (IL-12, IL-1β, IL-6, tumor necrosis factor-alpha (TNF-α)) cytokines. The water extracts of both GPs managed to increase T regulatory cells and forkhead box P3 (FoxP3) protein, which is responsible for the genes activity control that are involved in the immune system regulation. Another benefic effect of GPs extracts is FoxP3 increase which is a marker with a role in stabilizing the T regulatory cells' function. All extracts lowered the release of granzyme (GrB) compared to PMA treated group [91]. GrB is an enzyme secreted by cytolytic T cells with role in cell necrosis leading to harmful effects on homeostasis [91]. Regarding intracellular cytokines, the water extract of red Negroamaro GP increased TNF-α and IL-10 content in monocytes, while the red Negroamaro GP ethanol extract increased IL-12 and IL-10 levels in lymphocytes. Further, the white Koshu GP water extract increased monocyte levels of IL-10 and IL-12, while the white Koshu GP ethanol extract increased lymphocyte levels of TNF-α and IL-10. IL-10 was increased by both water or ethanolic, red or white GP extracts and as underlined by authors [91], the release of IL-10 by T cells and monocytes is a key step in maintaining the immune homeostasis. In conclusion, GPs extracts could induce immune homeostasis through the anti-inflammatory IL-10 secretion which counterbalances the pro-inflammatory cytokines (IL-12 and TNF-α) [91]. Another study that reinforces the anti-inflammatory effects of RGP from *Vitis vinifera* L. cv. Montepulciano d'Abruzzo on LPS-stimulated macrophages is that of Mollica et al. (2021). They observed that the extract significantly inhibited the release of cytokines (IL-6, TNF-α, and IL-1β), the maximal inhibitory action being at the dose of 100 μg/mL [92].

The possible potential impact of GP extracts on *in vitro* calcitonin gene-related peptide (CGRP) secretion was investigated as a potential mechanism to influence migraine [85]. The treatment of CA 77 cells with different red GP extracts showed a significant decrease in CGRP levels. CGRP is a gene that represents a key mediator of migraine-induced inflammation [85]. The results suggest that GP extracts had anti-inflammatory effect preventing the release of CGRP in migraine [8885].

White GP extract and a mixture of red and white GP extract, in different concentrations (100, 200, 500 μg/mL dry extract *w*/*v*), were added to Caco-2 cells after treatment with an inflammation inducer (IL-1β) to observe the effects on IL-8 secretion and NF-κB expression [93]. Grape pomaces were hydrolyzed enzymatically to determine if anti-inflammatory effects would be augmented. Both white and red GP contained quercetin, catechin, resveratrol, gallic and caffeic acids, trans-resveratrol, rutin, and procyanidin B2 [93]. All GP fractions (100, 200 μg/mL dry extract *w*/*v*) with or without enzymatic transformation decreased ROS levels, while treatment with GP extracts in higher concentration (500 μg/mL dry extract *w*/*v*) showed a considerable increase in ROS levels. Furthermore, NF-κB expression and prostaglandin E2 (PGE2) levels were significantly reduced in all fractions. At the same time, IL-8 secretion revealed a more substantial drop in enzymatically treated fractions of mixed GP, presenting beneficial effects of enzyme hydrolysis. The mixed GP had a more potent anti-inflammatory effect due to the high content of anthocyanins found in red GP [89].

Concerning the benefic antioxidant and anti-inflammatory GP actions, the literature presents a large variety of experimental settings that can be considered for future *in vivo* research. Also, we can observe that there is still space for other hypotheses, for both red and white GPs, but especially for the white ones which were much less investigated.

Further, in the next step the GP effects *in vivo* studies were analyzed. The *in vivo* studies usually use rodents to induce different models of inflammation, but fish and lamb were also introduced.

5.2. *In Vivo Beneficial Actions of Grape Pomace in Oxidative Stress and Inflammation*

The effect of GP extracts on the pathophysiology of oxidative stress and inflammation in various types of diseases can be well documented using different *in vivo* experimental models. These types of studies are very important in deciding whether the GP can be further used in safe conditions in human clinical trials.

The antioxidant and anti-inflammatory effects of both fresh and fermented GP extracts (*Vitis vinifera* L. cultivars, Fetească neagră, and Pinot noir, from Romania) were investigated using and a rat model of induced inflammation by turpentine oil [94]. The administration of turpentine oil increased the total oxidative status, oxidative stress index and reduced total antioxidant reactivity [94]. Treatment with GP decreased total oxidative status and oxidative stress index in a dose-dependent manner, but total antioxidant reactivity was not modified. All GP's fractions significantly reduced malondialdehyde (MDA) levels.

Total thiols were considerably lessened by turpentine, but GP managed to increase them in a concentration-dependent way. The same results were observed in the case of NOx production. 3NT was also increased by turpentine, but GP varieties decreased the levels. Due to higher phenolic content, the fresh extract showed a higher antioxidant effect. MDA is a lipid peroxidation waste product with hazardous potential for normal homeostasis. Thiols, under oxidative stress, manage to form disulphide bonds between them to reduce oxidative stress. NO presents a dual effect based on its concentrations. Small doses possess an antioxidant effect, while high doses can cause an increase in oxidative stress through the synthesis of new and stronger radicals. 3NT is a waste product resulting from tyrosine nitration induced by reactive nitrogen species [94]. The authors concluded that GP extracts could be used considered a potential agent in nutraceuticals formulation.

An interesting study that evaluates the effects of red GP flour dietary inclusion on growth, anti-inflammatory, antioxidant, innate-adaptive immunity, and on immune genes expression was performed on *Labeo rohita* fish against *Flavobacterium columnaris* induced infection [95]. Treatment with 200 and 300 mg GP flour showed a significant increase in GSH, SOD, and GPx activities as compared to regular diet or 100 mg GP supplementation, in both infected and uninfected groups. Regarding GP action on innate-adaptive immune activity, higher doses of GP (200, 300 mg) increased phagocytosis, alternative-complement pathway activity, raised IgM levels, and serum lysozyme (Lyz) activity when compared to regular diet or 100 mg GP supplementation in infected or uninfected group. In terms of immune-related genes, Lyz, (β-2 microglobulin) β-2M, 3rd component complement (CC3), and immunoglobulin M (IgM) gene expression pointed out a significant growth in infected fish with 200, 300 mg GP supplementation compared to other groups. However, the uninfected group treated with the same doses of GP showed higher gene expression than the infected group. Antioxidant related-genes were measured, and SOD, GPx, nuclear factor erythroid 2-related factor 2 (Nrf2), and (natural killer-cell enhancing factor β) NKEF-β were remarkably higher in all groups treated with raised doses of GP compared to 100 mg GP diet or regular diet in infected or uninfected groups. Furthermore, uninfected groups treated with high doses of GP showed a more significant increase in SOD and GPx expression levels. As for pro-inflammatory-related genes, IL-1β and TNF-α were not modified in any group. Hepcidin and toll-like receptor-22 (TLR22) expression were increased in infected and uninfected groups treated with a high dose of GP [95].

Therefore, in Rajković et al. (2022), GP was given to piglets to assess their positive effects on the animal organism without antibiotics side effects [96]. During the experiment tissue samplings (liver, jejunum, ileum) were collected on days 27/28 and 55/56, while blood samples were taken on days 6, days 27/28, and 55/56. Regarding antioxidant enzymes, GPx (GPx1-liver, GPx-2 jejunum, and ileum) wasn't different between diets, but the enzyme activity was significantly increased on days 55/56 compared to 27/28 [96]. About, SOD and Manganese Superoxide Dismutase (Mn-SOD) enzymes, there weren't any differences between diets in jejunum, ileum, and liver, but there was an increase between sampling dates, in days 55/56 compared to 27/28 in the liver. The copper superoxide dismutase system (Cu-SOD or SOD1) presented no distinction between any sampling days in the liver or ileum. CAT activity wasn't affected by any of the diets in the jejunum and liver, but there were differences between sampling days in the liver and ileum. TBARS concentrations weren't affected by diets in any organs, only in the jejunum between sampling days (decreased levels on days 55/56 compared to 27/28). GPx2 and SOD1 gene expression were modified at the jejunum level (decreased in days 55/56 compared to 27/28), while CAT expression presented the same results at the ileum level. In the liver, the authors have observed differences between samples for SOD1, CAT, and GPx1 in the liver, with a higher expression on days 27/28 compared to 55/56. In terms of inflammation, pig major acute phase-protein serum levels presented a decrease on days 55/56 and 27/28 compared to day 6 [96]. MDA serum levels decreased through sampling days while for SOD different fluctuations were noticed, without showing any significant values on day 55/56 versus other time points. As a speculative explanation for the variation of antioxidant enzymes decreasing it can be stated that the systemic presence of antioxidant substances can lead to a decreasing need for endogenous antioxidant enzymes production [96].

Another important direction in GP research is to check whether it is suitable to be used as an adjuvant treatment in different pathologies to reduce conventional drugs side effects. Thus, in Mossa et al. (2015) study, cypermethrin was given to female rats to observe toxic effects on the liver and kidneys, and white GP was added to check whether it can counter these toxic effects [97]. The assessment of kidneys and liver biomarkers showed a dose-dependent fall in liver enzymes: aspartate transaminase (AST), alanine transaminase (ALT), gamma-glutamyl transferase (GGT), and alkaline phosphatase (ALP), and a decrease kidneys urea nitrogen and creatine. Also, total proteins and albumin revealed a significant increase in GP treated group. The histological analysis pointed out significant changes due to inflammatory infiltrate in cypermethrin groups, while the GP supplemented group had regressed values for all biomarkers. Similar results were also observed in histological studies of kidneys samples. This may be due to the antioxidant effects of the white GP [97]. This study offers important evidence regarding the use of GP extract with hepatorenal protective activity and encourages future studies to investigate whether it can be used to reduce other drugs adverse reactions.

So far, the existing studies on GP suggest that through its anti-inflammatory and antioxidant effects, GP can be considered a potent agent that can contribute to the restoration of homeostasis to control levels or that can reduce different drug side effects (Table 2).

Table 2. *In vivo* beneficial actions of grape pomace in oxidative stress and inflammation.

Grape Pomace	Models	Polyphenols Content	Antioxidant and Anti-Inflammatory Effects	References
		Red grape pomace variety		
Tempranillo variety (Burgos, Spain)	Spontaneously hypertensive rats	Proanthocyanidin, anthocyanins, quercetin	-TAC increased; -decreased lipid peroxidation and carbonyl groups; -increased NO levels -increased HO-1, SOD2, eNOS gene expression;	[98]
Alicante and Pinot varieties (France) polyphenol-enriched Alicante	Dextran Sulfate Sodium-Induced colitis in Wistar male	Anthocyanins	-improved histological score; -decreased MPO activity; -increased SOD activity in polyphenol-enriched Alicante; -decreased IL-1α, IL-6 IFN-γ levels; -decreased IL-1β levels in Alicante and Pinot; -decreased IL6, ICAM-1, MMP-9 gene expression; -decreased IL-1β, iNOS gene expression in Alicante and Pinot; -decreased TNFα, NFκB p65, COX2 gene expression in polyphenol-enriched Alicante;	[99]
Malbec variety (Gualtallary, Mendoza, Argentina)	High fructose diet-induced Metabolic syndrome in Wistar rats	Quercetin, epicatechin, catechin, trans-resveratrol, ferulic, gallic, caffeic, syringic, p-coumaric acids	-reduced CRP levels; -reduced NADPH oxidase activity; -increased adiponectin; -reduced resistin; -increased insulin sensitivity;	[100]
Dimrit grapes variety	96 laying Hens given different GP concentrations	Catechin, Epicatechin, Gallocatechin, Epigallocatechin, Phenolic acids, Gallic acid, Caffeic acid, p-cumaric acid	-decreased plasma MDA levels; -decreased egg yolk MDA levels;	[101]
Red wine grape pomace	18 crossbreed lambs given different GP diets (5%, 10%)	-	-increased TAOC and SOD, GPX activity in longissimus dorsi muscle; -decreased ROS and MDA levels in longissimus dorsi muscle;	[102]
Red wine grape pomace	24 crossbreed ram lambs under pen conditions given GP diets (5%, 10%)	-	-decreased MDA and ROS levels in lamb testes; -increased CAT, SOD, GPx4 activity in lamb testes; -increased TAOC (GP 10%); -increased SOD, GPx4 mRNA expression (GP 10%); -increased CAT, SOD, GPx4 protein abundance;	[103]
Tempranillo variety	Wistar rats given high-fat diet	-	-decreased IL-1β and TNF-α levels; -increased FRAP plasma and liver levels; -increased liver GSH/GSSG ratio; -decreased plasma and liver MDA and carbonyl groups levels; -decreased 8-hydroxydeoxyguanosine plasma levels;	[104]

Table 2. *Cont.*

Grape Pomace	Models	Polyphenols Content	Antioxidant and Anti-Inflammatory Effects	References
Muscat Bailey A variety (Gyeongsangbuk-do, Korea)	High-fat diet induced obesity in male C57BL/6J mice	Catechins, resveratrol, flavonoids,	-decreased TNF-α, PAI-1 levels; -decreased liver NF-κB, IL-6 and TNF-α levels;	[105]
Cabernet Franc (Chrysalis Vineyards, Virginia)	C57BL/6NCr mice given high-fat diet	-	-decreased TNF-α, IFN-γ, IL-12β, PAI-1 and resistin levels	[106]
Red wine pomace	78 crossbreed piglet given apple or grape pomace	Flavanols	-decreased NF-κB mRNA expression in the stomach; -increased NF-κB and TNF-α mRNA expression in the liver and muscle; -decreased TNF-α and IL-10 mRNA expression in ileum; -increased IL-10 mRNA expression in the jejunum, colon, and liver;	[107]
Red dried grape pomace (*Vitis vinifera* L. variety)	20 Fresian cows given GP diet	Flavonoids, gallic acid, epicatechin	-lower MDA levels in the cheese from cow's milk that received GP; -lower thrombogenic index in cow's milk that received GP;	[108]
Pinotage variety (Bellevue Wine Estate, Stellenbosch, South Africa)	40 lambs given GP diets at different c% (0, 5, 10, 15, 20)	Proanthocyanidins, tannins	-increased antioxidant activity (15, 20% diets) within first 3 days of meat storage; -decreased TBARS levels of all diets from day 5 to day 9 of meat storage; -decreased carbonyl content in stored meat (10, 20%);	[109]
Pinotage variety (Bellevue, Beyers Kloof, Western Cape Province, South Africa)	Angus steer given dried grape pomace or dried citrus pulp	Proanthocyanidins, tannins	-decreased TBARS and carbonyl levels; -increased antioxidant activity;	[110]
Cencibel variety (Grupo Matarromera San Bernardo-Valbuena de Duero, Valladolid, Spain) Enzymatic hydrolysis treated fractions—tannase and carbohydrase enzyme complex—separately or combined	300 Cobb chicks given different GP c% (5, 10) diets—hydrolyzed/un-hydrolyzed	Gallic acid, Catechin, Epicatechin, Procyanidin B1, Procyanidin B2 Epicatechin-O-gallate;	-decreased MDA levels;	[111]
Moschato variety Tyrnavos (Larissa prefecture, Greece)	30 female broilers given GP diet for 15 or 35 days;	-	-15 days GP diet: decreased TBARS plasma levels, increased GSH levels in kidney and spleen, decreased TBARS levels in pancreas and intestine, decreased CARB levels in the kidney; -35 days GP diet: increased GSH erythrocytes levels, decreased TBARS plasma levels, increased GSH levels in kidney, spleen, heart, lung, and liver, increased TAC levels in liver, spleen, and kidney, decreased H_2O_2 decomposition in the intestine, decreased TBARS levels in spleen, quadriceps muscle, and heart, decreased CARB levels in spleen and kidney;	[112]
Cencibel variety	180 broiler chicks given different GP diets doses (15, 30, 60 mg/kg) or Vitamin E;	Condensed tannins	-decreased MDA levels in refrigerated breast meat;	[113]
Cencibel variety (Vinícola de Castilla S.A.,Manzanares, Ciudad Real, Spain)	120 broiler chicks given different GP diets doses (5, 15, 30 mg/kg) or Vitamin E;	-	-decreased MDA levels in refrigerated breast and thigh meat (day 7); -decreased MDA levels in refrigerated breast meat (day 4);	[114]
Moschato variety (Tyrnavos Larissa, Greece)	24 piglets given GP diet (blood and tissue samples taken at 15 and 30 days post-diet)	-	-15 days GP diet: decreased TAC plasma activity, decreased CARB levels in spleen, brain and liver, decreased TBARS levels in brain, kidneys, stomach, heart, lungs, quadriceps muscle, and spleen, increased TAC levels in stomach and pancreas, decreased TAC levels in the brain, increased GSH levels, heart, liver, spleen, stomach, pancreas, lungs, brain and quadriceps muscle; increased H_2O_2 decomposition activity in kidneys and decreased in lungs and stomach; -30 days GP diet: decreased CAT erythrocytes activity, decreased CARB levels in spleen, brain, liver, lungs, quadriceps muscle, stomach, and pancreas, decreased TBARS levels in brain, liver, heart, lungs, quadriceps muscle, spleen, and pancreas; increased TAC levels in the quadriceps muscle, kidneys, lungs, stomach, and pancreas, decreased TAC levelsthe in brain, increased GSH levels heart, liver, pancreas, lungs, brain and quadriceps muscle, kidneys; decreased TAC levels in stomach and spleen; increased H_2O_2 decomposition activity in kidneys, quadriceps muscle, pancreas and decreased in lungs and brain;	[115]
Cencibel variety (La Mancha, España)	70 broiler chicks given GP diets doses (0, 30, 60 mg/kg)	Condensed tannins, hydrolysable tannins;	-reduced TBARS levels in raw chicken patties (storage day 13, 20); - reduced TBARS levels in cooked chicken patties (storage day 3, 6, 13, 20); -reduced TBARS levels in raw chicken patties (60 mg/kg–6 months storage); -reduced TBARS levels in cooked chicken patties (30, 60 mg/kg–6 months storage);	[116]

Table 2. *Cont.*

Grape Pomace	Models	Polyphenols Content	Antioxidant and Anti-Inflammatory Effects	References
Moschato variety (Tyrnavos Larissa, Greece)	28 lambs given GP diet (blood and tissue samples taken at 27 and 55 days post-diet)	-	-27 days GP diet: increased CAT erythrocytes activity; decreased protein carbonyls level in the liver; increased TBARS activity in the brain; -55 days GP diet: reduced TBARS activity in liver, spleen, and heart; increased TBARS activity in the brain; decreased TAC in brain and liver; GSH levels increased in quadriceps muscle and spleen; decreased GSH levels in the liver;	[117]
Carignan varietyl (Northern Tunisia	Adult mice given 6-hydroxydopamine stereotaxic injection in midbrain (Parkinson disease model)	-	-increased SOD1 brain levels; -decreased neurons depletion in substantia nigra; -ameliorated motor impairment;	[92]
White grape pomace variety				
Koshu variety (Japan and Italy) Fermented or un-fermented fractions	Female rats induced-allergic reactions (asthma and passive cutaneous anaphylaxis)	-	-decreased serum IgE levels; -decreased eosinophils levels in bronchial lavage; -decreased cutaneous reaction in time and dose-dependent manners; -decreased cutaneous reaction compared to Tannat or Negroamaro GP (red varieties)	[118]

Abbreviations: AOPP—advanced oxidation protein product; CARB—protein carbonyls; CAT—catalase; COX 2—cyclooxygenase 2; CRP—C-reactive protein; DPPH—2,2-diphenyl-1-picrylhydrazyl; FRAP—ferric ion antioxidant reducing power; GP—grape pomace; GPx—glutathione peroxidase; γ-GCS—γ-synthase glutamyl cysteine; GSH—glutathione; GSSG—glutathione disulfide; GST—glutathione-s-transferase; HO-1—heme oxygenase 1; ICAM-1—Intercellular Adhesion Molecule 1; IFN-γ—interferon gamma; MDA—malondialdehyde; MMP-9—matrix metalloproteinase 9; MPO—myeloperoxidase activity; NADPH—nicotinamide adenine dinucleotide phosphate; NF-κB p65—nuclear factor kappa-light-chain-enhancer of activated B cells transcription factor; eNOS-endothelial nitric oxide synthase; iNOS—inducible nitric oxide synthase; NO—nitric oxide; oxLDL- Oxidized low-density lipoprotein; PAI-1—Plasminogen activator inhibitor-1; ROS—reactive oxygen species; SOD—superoxide dismutase; TAC—total antioxidant capacity; TAOC—total antioxidant capacity; TAS—total antioxidant status; TBARS—thiobarbituric acid reactive substances; TNF-α—tumor necrosis factor alpha.

6. Non-Steroidal Anti-Inflammatory Drugs

Non-steroidal anti-inflammatory drugs (NSAIDs) represent a group of chemically distinct compounds that act through the same mechanism producing a reversible inhibition of cyclooxygenase (COX) 1 and 2 isoenzymes, enzymes that produce prostaglandins (PGs). The main representatives of the group of NSAIDs that act non-selectively on COX-1 and COX-2 are aspirin, ibuprofen, diclofenac, indomethacin, naproxen, ketorolac, piroxicam and meloxicam [119]. The main difference between these two isoenzymes is represented by the fact that while COX-1 is a constitutive enzyme which produce regularly PGs that have a protective role mainly on the stomach and kidney, COX-2 expression is induced by inflammatory stress and PGs resulted from this pathway lead to the swelling and pain associated with inflammation [120]. The ability of NSAIDs to reduce pain and swelling associated with several inflammatory diseases and the fact that NSAIDs are over-the-counter drugs, contributed to their large-scale use, being of the most sold drugs worldwide [121]. Besides these beneficial effects, NSAIDs have multiple adverse reactions, mainly related to COX-1 inhibition. Thus, the most common adverse reactions are the gastrointestinal ones. Long-term usage of NSAIDs could lead especially to gastric ulcer, but can also alter renal function and sodium exchange and could lead to hypertension and/or renal failure. On the cardiovascular system, COX inhibition could lead to heart failure exacerbation. On hepatic activity, COXs function alteration could lead to acute liver injury [121]. Hypersensitivity reactions are also described and manifest clinically by NSAIDs-exacerbated respiratory diseases (rhinosinusitis, bronchial asthma, pneumonitis), NSAIDs-exacerbated cutaneous diseases (urticaria, photo-contact dermatitis, angioedema), NSAIDs-induced diseases (nephritis, aseptic meningitis) and anaphylaxis [122–125]. Because of these multiple and heterogenous adverse effects associated with COX-1 inhibition, pharmacology researchers discovered other compounds from the class of NSAIDs, celecoxib, which selectively inhibits COX-2. However, although studies have shown a reduced adverse effect on the gastrointestinal system, namely a lower risk of gastric ulcer, this compound more frequently leads to acute coronary syndromes [121]. Besides the anti-inflammatory activity of

NSAIDs, another area of interest was represented by their antioxidant activity. In this way, Costa et al. (2006) focused on the antioxidant activity via the *in vitro* scavenging ability of ROS and reactive nitrogen species (RNS) of ibuprofen, flurbiprofen, fenbufren, fenoprofen, naproxen, ketoprofen and indoprofen [126]. They observed that the greatest scavenging activity of ROS for O^{2-} was equal for fenbufen, flurbiprofen, indoprofen, ketoprofen, for H_2O_2 was equal for ketoprofen, indoprofen, fenbufen and for HO was equal for fenoprofen and ibuprofen. In the case of RNS, for NO the greatest scavenging activity was that of indoprofen, and for $ONOO^-$ was indoprofen [126].

However, in the *in vivo* studies no antioxidant activity was observed, but, on the contrary, a prooxidant effect was described. One example is represented by the study of Nawaz et al. (2021) who observed the effect of NSAIDs treatment on antioxidant status and oxidative stress in patients with rheumatoid arthritis [127]. They highlighted that the group of patients with rheumatoid arthritis under the treatment with NSAIDs, compared with the other three groups (control group, patients without rheumatoid arthritis and under the treatment with NSAIDs, patients with rheumatoid arthritis who did not take NSAIDs), showed the highest oxidative stress and the lowest free radical scavenging ability [127].

Taking into consideration all of the above, namely the multitude of adverse reactions associated with NSAIDs treatment, it is necessary to find a potent substitute for these drugs, one that has strong anti-inflammatory effects with as few adverse reactions as possible.

7. Comparative Effect of Polyphenols Found in Grape Pomace Versus Non-Steroidal Anti-Inflammatory Drugs in Oxidative Stress and Inflammation

Several diseases use NSAIDs to decrease vasodilatation, cell adhesion to vascular endothelium, cells migration to inflammation site, cytokine synthesis, and further tissue damage. However, as we already mention before, NSAIDs also possess several adverse reactions (gastrointestinal bleeding, hepatic, renal toxicity), which can cause a limitation in use for more extended periods. This led to studying different natural compounds to compare their effects with different NSAIDs (Table 3).

Unfortunately, we were unable to find studies in which the anti-inflammatory and antioxidant effects of GP whole extract were compared with those induced by NSAIDs, but only of several other compounds extracted from GP, such as resveratrol, flavonol (quercetin) and hydroxybenzoic acid (gallic acid). Therefore, this section will examine the differences between *in vitro* and *in vivo* effects of NSAIDs and different phenolic compounds that are also found in GP.

Thus, Zhang et al. (2021), subjected rat glial cells to LPS induced-inflammation and treated them with different doses of resveratrol (0.1–20 μM) or ibuprofen to see the differences [136]. Resveratrol showed a higher decrease in TNF-α (0.1–20 μM) and ROS production (20 μM) compared to ibuprofen [136]. In another study, horse immune cells were cultured with resveratrol or different NSAIDs (flunixin meglumine or phenylbutazone) to observe the effects on pro-inflammatory cytokine production. Resveratrol managed to lower interferon (IFN)-γand TNF-α levels, similar to of NSAIDs [137].

As for *in vivo* studies, resveratrol patches were given to rats injected with carrageenan [138]. Diclofenac gel was used to compare the effects of the two substances. Resveratrol and diclofenac decreased paw swelling, but resveratrol caused a much longer inflammation reduction when compared to diclofenac. Also, the resveratrol group presented the lowest paw swelling compared to the diclofenac and control groups [138].

Another polyphenol found in GP and recognized for its anti-inflammatory and antioxidant effects is represented by quercetin, which is a flavonol. In Wei et al. study (2019), osteoarthritis was surgically induced in the left knee in rats, that were also treated with quercetin or celecoxib (CXB) as a reference drug [139]. In terms of antioxidant effects, quercetin and CXB groups managed to increase SOD serum and synovial fluid levels compared to the untreated group. Metalloproteinases (MMPs) are a family of enzymes responsible for cartilage degradation and osteoarthritis appearance [139]. Tissue inhibitor metalloproteinase (TIMP) is accountable for cartilage repair through the downregulation of MMPs [139]. In this study, quercetin and CXB groups decreased MMP-13 in serum, synovial fluid, and synovium, while TIMP-1 levels increased in these groups compared to the untreated osteoarthritis group [139]. In this work, quercetin presented comparable effects with celecoxib, both having protective effects [139]. Heydari Nasrabadi et al. (2022) also analyzed the effects of quercetin in monosodium iodoacetate-induced osteoarthritis in rats injection [140]. In this experiment, ibuprofen was used as a reference drug. In histopathological studies of the rats' knees, the inflammatory cells were significantly decreased in the treated groups (quercetin and ibuprofen) compared to the untreated group. In addition, MMP-3 and MMP-13 were measured, and the results showed that the quercetin group reduced, even more, the levels of MMPs compared to the ibuprofen group [140].

Table 3. Grape pomace polyphenols versus non-steroidal anti-inflammatory drugs in oxidative stress and inflammation.

Administrated Polyphenol	NSAID	Model	Anti-Inflammatory and Antioxidant Effects of Polyphenols	Differences of Results Between Polyphenols and AINS	References
RES 50 mg/kg	Diclofenac 3 mg/kg	Adjuvant-induced arthritis in rats	-reduced TNF-α, IL-1β, TBARS and NOx levels; -reduced NF-κB p65 expression;	-	[128]
RES 50 mg/kg	Diclofenac 3 mg/kg	Adjuvant-induced arthritis in rats	-reduced TNF-α, IL-1β, TBARS and NOx levels; -attenuated histological changes (cartilage damage, pannus formation, cellular infiltration, synovial proliferation) -reduced NF-κB p65 expression;	-diclofenac group decreased paw volume	[129]
RES 10/50 mg/kg	Celecoxib 5 mg/kg	Adjuvant-induced arthritis in rats	-reduced paw volume; -decreased lymphocyte proliferation; -decreased COX2 expression; -decreased PGE2 levels;	-resveratrol exhibited similar results to celecoxib	[130]
RES 100 mg/kg	Celecoxib	Lipopolysaccharide induced-sepsis in rats	-no significant reduction of prostaglandin plasma and kidneys levels; -no significant reduction of mRNA MCP-1 and IL-6 levels;	-celecoxib tackle the inflammation effects compared to resveratrol	[131]
RES 10 mg/kg	Ibuprofen 30 mg/kg	Experimental arthritis and periodontitis in rats	-longer reduction in paw swelling; -higher gingival IL-4 levels;	-	[132]
RES 5/10 mg/kg	Etoricoxib 10 mg/kg	Experimental osteoarthritis induced in rats	-increased time of paw withdraw mechanical, heat and cold hyperalgesia test in RES groups; -increased spontaneous rats' movement in RES groups; -decreased serum TNF-α, IL-10 levels; -decreased serum IL-1β levels (RES 10 mg/kg); -decreased synovial TNF-α, IL-10 and IL-1β levels; -decreased cartilage mRNA TNF-α, IL-10 expression; -decreased cartilage mRNA IL-1β expression (RES 10 mg/kg); -decreased cartilage protein TNF-α, IL-10, IL-1β, IL-6, MMP-13 expression; -decreased mRNA iNOS expression; -decreased mRNA COX2 expression (RES 10 mg/kg); -decreased protein iNOS, COX2 expression;	-etoricoxib manage to present same anti-inflammatory effects as RES 10 mg/kg treated group;	[133]
Quercetin (75 mg/kg)	Phenylbutazone (80 mg/kg) and Indomethacin (6 mg/kg)	Freund's complete adjuvant-induced arthritis; carrageenan-induced paw edema	-reduced paw edema (carrageenan experiment); -reduced paw volume in the acute phase; -reduced paw volume in the chronic phase (days 8, 9, 10, 14, 15, 16)	-higher anti-inflammatory effect compared to phenylbutazone in carrageenan experiment;	[134]
Quercetin (80 mg/kg)	Phenylbutazone (80 mg/kg)	Experimental induced-arthritis in rats	-reduced paw volume in the acute phase; -reduced paw volume in the chronic phase (day 9, 10, 14, 16, 19, 23, 26 and 30)	-	[135]

Abbreviations: COX2—cyclooxygenase 2; iNOS—inducible nitric oxide synthase; MCP-1—monocyte chemoattractant protein-1; MMP-13—mettaloproteinase 13; NF-κB p65—nuclear factor kappa-light-chain-enhancer of activated B cells transcription factor; NOx—serum total nitrate/nitrite; PGE2—prostaglandin E2; RES—resveratrol; TBARS—thiobarbituric acid reactive substances; TNF-α—tumor necrosis factor alpha.

When taking into consideration the oxidative stress induced by the treatment with NSAIDs, other researchers focused on finding a substance that could limit this pro-oxidative side effect. In this matter, a great phenolic compound found in high quantities in GP, gallic acid, has proven to be a potent adjuvant in this case. Thereby, Moradi et al. (2021) evaluated if gallic acids taken alongside diclofenac had an impact on the oxidative status of 30 Wistar rats divided into 5 groups [141]. They observed that while only under diclofenac treatment there was a significant increase in urea, AST, ALT, creatinine, uric acid serum levels, and in IL-1β expression, in the group which received gallic acid after diclofenac there were registered reduced levels of these biochemical parameters and gene expression [141]. Moreover, the treatment with gallic acid alongside diclofenac led to a decrease in nitrate content, serum and renal MDA levels, and protein carbonyl level. Furthermore, on antioxidant enzymes activity, treatment with diclofenac and gallic acid led to an increase in renal SOD and CAT activities and renal GSH levels. The histopathological examination of the kidney highlighted that the group which received both diclofenac and gallic acid presented a reduced lymphocytic cell infiltration, focal hemorrhage, and vacuolar degeneration of tubular epithelial cells in comparison with the group which received only diclofenac [141]. Another study that reinforces these observations is that of Esmaeilzadeh et al. (2020) who investigated the hepatoprotective activity of gallic acid in diclofenac-induced liver toxicity in 30 Wistar rats [142]. They observed that while diclofenac led to an increase in GOT, GPT, ALP, and total bilirubin levels, gallic acid reduced the levels of these parameters. Moreover, at a dose of 100 mg/kg gallic acid, there was observed a significantly decreased compared with a dose of 50 mg/kg gallic acid [142]. The same difference between these two doses was also observed in the increase of plasma antioxidant capacity, hepatic SOD, GPx, GSH, and CAT activities and in the decrease of nitrite content, protein carbonyl levels, serum, and liver MDA levels compared with the group which received only diclofenac. The histopathological examination of the liver, both doses of gallic acid led to a significantly reduced in lymphocytic cell infiltration and liver degeneration [142].

These *in vivo* studies also suggest that GP is a valuable candidate that could be used as a potential therapeutic agent capable of reducing oxidative stress and inflammation and also as an adjuvant treatment in the attempt to reduce the side effects.

8. Conclusions

GP extracts due to their rich content in flavonols, anthocyanins, anthocyanidins, flavanols, proanthocyanidins, stilbenes, and phenolic acids are offering new perspectives as possible therapeutic agents. In vivo and *in vitro* studies showed promising results for both GP whole extracts and different types of polyphenols that are contained in it, in reducing oxidative stress and inflammatory markers in different models of chronic inflammation.

The first evidence in the direction of this possible therapeutic use was represented by in vitro studies. In these works, the antioxidant effects were demonstrated especially by the decrease in ROS, MDA, and TBARS levels and by the increase in GSH levels. The anti-inflammatory effects were given by the inhibition of some inflammatory pathways such as NF-kB and PGE2, an inhibition that led to a decrease in levels of inflammatory markers such as IL-8. These antioxidant and anti-inflammatory effects are also validated in *in vivo* studies. In terms of antioxidant activity, in addition to the effects already observed in the *in vitro* studies, an increase in CAT, SOD, and GPx4 levels and stimulation of eNOS gene expression were also seen. Similarly, the *in vivo* studies brought additional data regarding the anti-inflammatory activity, observing an inhibition of the release of several inflammatory markers such as IL-1α, IL-1β, IL-6, IFN-γ, TNF-α, and CRP. Anyway, it is still necessary to implement this research in clinical trials to be able to conclude the possible use of GP as an antioxidant and anti-inflammatory therapeutical agent. Further, most of the studies have shown that GP extracts are more effective than a single polyphenol, possibly because of polyphenols synergistic action that interferes with more than one pathophysiological mechanism. Compared to the antioxidant and anti-inflammatory effects of NSAIDs, the literature currently offers only the effects of single polyphenols, such as resveratrol, quercetin, and gallic acid. However, single polyphenols possess similar anti-inflammatory and antioxidant effects as the classical NSAIDs. Thereby, further comparative studies are needed to investigate if the entire complex of polyphenols that exists in GP may induce a similar or a better effect. The question that rises from these studies is whether the best results can be obtained when single isolated phenolic compounds with a known concentration are used or whether a standardized mixture of these compounds is used? If this hypothesis is validated, in the future GP could become an add-on therapeutic measure that could be used for better control of chronic inflammation than monotherapy with NSAIDs.

Author Contributions: Conceptualization, V.S.C., Ş.O.M., I.C.B. and R.M.P.; methodology, Ş.O.M., D.C.M. and A.M.L.; software A.M.L.; validation, V.S.C., I.C.B. and A.D.B.; formal analysis, V.S.C. and I.C.B.; investigation, Ş.O.M., D.C.M. and A.M.L.; resources, A.D.B.; data curation, V.S.C.; writing—original draft preparation, V.S.C., Ş.O.M. and R.M.P.; writing—review and editing, R.M.P. and A.D.B.; visualization, I.C.B.; supervision, R.M.P.; project administration, V.S.C.; funding acquisition, I.C.B. All authors have read and agreed to the published version of the manuscript.

Funding: This research received no external funding.

Institutional Review Board Statement: Not applicable.

Informed Consent Statement: Not applicable.

Data Availability Statement: Not applicable.

Conflicts of Interest: The authors declare no conflict of interest.

References

1. Arfaoui, L. Dietary Plant Polyphenols: Effects of Food Processing on Their Content and Bioavailability. *Molecules* **2021**, *26*, 2959. [CrossRef]
2. Cena, H.; Calder, P.C. Defining a Healthy Diet: Evidence for The Role of Contemporary Dietary Patterns in Health and Disease. *Nutrients* **2020**, *12*, 334. [CrossRef]
3. Healthy Diet. World Health Organization. Available online: https://www.who.int/news-room/fact-sheets/detail/healthy-diet (accessed on 1 August 2022).
4. Di Lorenzo, C.; Colombo, F.; Biella, S.; Stockley, C.; Restani, P. Polyphenols and Human Health: The Role of Bioavailability. *Nutrients* **2021**, *13*, 273. [CrossRef] [PubMed]
5. Tsao, R. Chemistry and Biochemistry of Dietary Polyphenols. *Nutrients* **2010**, *2*, 1231–1246. [CrossRef]
6. FAOSTAT. Available online: https://www.fao.org/faostat/en/#data/QCL (accessed on 9 September 2022).
7. Fontana, A.R.; Antoniolli, A.; Bottini, R.R. Grape Pomace as a Sustainable Source of Bioactive Compounds: Extraction, Characterization, and Biotechnological Applications of Phenolics. *J. Agric. Food Chem.* **2013**, *61*, 8987–9003. [CrossRef] [PubMed]
8. Zacharof, M.P. Grape Winery Waste as Feedstock for Bioconversions: Applying the Biorefinery Concept. *Waste Biomass Valorization* **2017**, *8*, 1011–1025. [CrossRef]
9. Koutelidakis, A. Grape Pomace: A Challenging Renewable Resource of Bioactive Phenolic Compounds with Diversified Health Benefits. *MOJ Food Process. Technol.* **2016**, *3*, 262–265. [CrossRef]
10. Castellanos-Gallo, L.; Ballinas-Casarrubias, L.; Espinoza-Hicks, J.C.; Hernández-Ochoa, L.R.; Muñoz-Castellanos, L.N.; Zermeño-Ortega, M.R.; Borrego-Loya, A.; Salas, E. Grape Pomace Valorization by Extraction of Phenolic Polymeric Pigments: A Review. *Processes* **2022**, *10*, 469. [CrossRef]
11. Moccia, F.; Agustin-Salazar, S.; Verotta, L.; Caneva, E.; Giovando, S.; D'Errico, G.; Panzella, L.; D'Ischia, M.; Napolitano, A. Antioxidant Properties of Agri-Food Byproducts and Specific Boosting Effects of Hydrolytic Treatments. *Antioxidants* **2020**, *9*, 438. [CrossRef]
12. Beres, C.; Costa, G.N.S.; Cabezudo, I.; da Silva-James, N.K.; Teles, A.S.C.; Cruz, A.P.G.; Mellinger-Silva, C.; Tonon, R.V.; Cabral, L.M.C.; Freitas, S.P. Towards Integral Utilization of Grape Pomace from Winemaking Process: A Review. *Waste Manag.* **2017**, *68*, 581–594. [CrossRef]
13. Tamelová, B.; Malaťák, J.; Velebil, J.; Gendek, A.; Aniszewska, M. Energy Utilization of Torrefied Residue from Wine Production. *Materials* **2021**, *14*, 1610. [CrossRef]
14. Dwyer, K.; Hosseinian, F.; Rod, M. The Market Potential of Grape Waste Alternatives. *J. Food Res.* **2014**, *3*, 91. [CrossRef]
15. Fraga, C.G.; Croft, K.D.; Kennedy, D.O.; Tomás-Barberán, F.A. The Effects of Polyphenols and other Bioactives on Human Health. *Food Funct.* **2019**, *10*, 514–528. [CrossRef] [PubMed]
16. Tournour, H.H.; Segundo, M.A.; Magalhães, L.M.; Barreiros, L.; Queiroz, J.; Cunha, L.M. Valorization of Grape Pomace: Extraction of Bioactive Phenolics with Antioxidant Properties. *Ind. Crops Prod.* **2015**, *74*, 397–406. [CrossRef]

17. Ferri, M.; Bin, S.; Vallini, V.; Fava, F.; Michelini, E.; Roda, A.; Minnucci, G.; Bucchi, G.; Tassoni, A. Recovery of Polyphenols from Red Grape Pomace and Assessment of their Antioxidant and Anti-Cholesterol Activities. *New Biotechnol.* **2016**, *33*, 338–344. [CrossRef]
18. Rasines-Perea, Z.; Ky, I.; Cros, G.; Crozier, A.; Teissedre, P.-L.; Rasines-Perea, Z.; Ky, I.; Cros, G.; Crozier, A.; Teissedre, P.-L. Grape Pomace: Antioxidant Activity, Potential Effect Against Hypertension and Metabolites Characterization after Intake. *Diseases* **2018**, *6*, 60. [CrossRef]
19. Calabriso, N.; Massaro, M.; Scoditti, E.; Verri, T.; Barca, A.; Gerardi, C.; Giovinazzo, G.; Carluccio, M.A. Grape Pomace Extract Attenuates Inflammatory Response in Intestinal Epithelial and Endothelial Cells: Potential Health-Promoting Properties in Bowel Inflammation. *Nutrients* **2022**, *14*, 1175. [CrossRef]
20. Vosloban, C.M.; Chedea, V.S. *Grape Pomace in Health and Disease Prevention*; Chedea, V.S., Ed.; Nova Science Publishers, Inc.: New York, NY, USA, 2022; Chapter 1; pp. 6–20.
21. Muñoz-Bernal, Ó.A.; Coria-Oliveros, A.J.; de la Rosa, L.A.; Rodrigo-García, J.; del Rocío Martínez-Ruiz, N.; Sayago-Ayerdi, S.G.; Alvarez-Parrilla, E. Cardioprotective Effect of Red Wine and Grape Pomace. *Food Res. Int.* **2021**, *140*, 110069. [CrossRef]
22. Munjal, A.; Allam, A.E. Indomethacin. In *xPharm: The Comprehensive Pharmacology Reference*; StatPearls Publishing: Treasure Island, FL, USA, 2021; pp. 1–5. Available online: https://www.ncbi.nlm.nih.gov/books/NBK555936/ (accessed on 31 August 2022).
23. Hall, D.; Bolinske, T.; Sinatra, E. Naproxen. In *The Essence of Analgesia and Analgesics*; StatPearls Publishing: Treasure Island, FL, USA, 2022; pp. 221–225. Available online: https://www.ncbi.nlm.nih.gov/books/NBK525965/ (accessed on 31 August 2022).
24. Alfaro, R.A.; Davis, D.D. Diclofenac. In *xPharm: The Comprehensive Pharmacology Reference*; StatPearls Publishing: Treasure Island, FL, USA, 2022; pp. 1–7. Available online: https://www.ncbi.nlm.nih.gov/books/NBK557879/ (accessed on 31 August 2022).
25. Bindu, S.; Mazumder, S.; Bandyopadhyay, U. Non-Steroidal Anti-Inflammatory Drugs (NSAIDS) and Organ Damage: A Current Perspective. *Biochem. Pharmacol.* **2020**, *180*, 114147. [CrossRef] [PubMed]
26. García-Rayado, G.; Navarro, M.; Lanas, A. NSAID Induced Gastrointestinal Damage and Designing Gi-Sparing NSAIDs. *Expert Rev. Clin. Pharmacol.* **2018**, *11*, 1031–1043. [CrossRef]
27. Hnepa, Y.Y.; Chopey, I.V.; Chubirko, K.I.; Bratasyuk, A.M. Short- and Long-Term Effects of NSAIDS on the Gastrointestinal Mucosa: Complex Analysis of Benefits and Complications Prevention. *Wiad. Lek.* **2021**, *74*, 1011–1018. [CrossRef]
28. Schjerning, A.M.; McGettigan, P.; Gislason, G. Cardiovascular Effects And Safety of (Non-Aspirin) NSAIDs. *Nat. Rev. Cardiol.* **2020**, *17*, 574–584. [CrossRef] [PubMed]
29. Pizzino, G.; Irrera, N.; Cucinotta, M.; Pallio, G.; Mannino, F.; Arcoraci, V.; Squadrito, F.; Altavilla, D.; Bitto, A. Oxidative Stress: Harms and Benefits for Human Health. *Oxid. Med. Cell. Longev.* **2017**, *2017*, 8416763. [CrossRef] [PubMed]
30. Tee, J.K.; Ong, C.N.; Bay, B.H.; Ho, H.K.; Leong, D.T. Oxidative Stress by Inorganic Nanoparticles. *WIREs Nanomed. Nanobiotechnol.* **2016**, *8*, 414–438. [CrossRef] [PubMed]
31. Sies, H. Oxidative stress: Eustress and distress in redox homeostasis. In *Stress: Physiology, Biochemistry, and Pathology*; Handbook of Stress Series; Fink, G., Ed.; Academic Press: London, UK; Elsevier: London, UK, 2019; Volume 3, Chapter 13; pp. 153–163.
32. Sharifi-Rad, M.; Anil Kumar, N.V.; Zucca, P.; Varoni, E.M.; Dini, L.; Panzarini, E.; Rajkovic, J.; Tsouh Fokou, P.V.; Azzini, E.; Peluso, I.; et al. Lifestyle, Oxidative Stress, and Antioxidants: Back and Forth in the Pathophysiology of Chronic Diseases. *Front. Physiol.* **2020**, *11*, 694. [CrossRef]
33. Shankar, K.; Mehendale, H.M. Oxidative Stress. In *Encyclopedia of Toxicology*, 3rd ed.; Wexler, P., Ed.; Elsevier: London, UK; Academic Press: London, UK, 2014; Volume 3, pp. 735–737.
34. Klaunig, J.E.; Wang, Z.; Pu, X.; Zhou, S. Oxidative Stress and Oxidative Damage in Chemical Carcinogenesis. *Toxicol. Appl. Pharmacol.* **2011**, *254*, 86–99. [CrossRef] [PubMed]
35. Mrakic-Sposta, S.; Gussoni, M.; Montorsi, M.; Porcelli, S.; Vezzoli, A. A Quantitative Method to Monitor Reactive Oxygen Species Production by Electron Paramagnetic Resonance in Physiological and Pathological Conditions. *Oxid. Med. Cell. Longev.* **2014**, *2014*, 306179. [CrossRef] [PubMed]
36. Di Meo, S.; Reed, T.T.; Venditti, P.; Victor, V.M. Role of ROS and RNS Sources in Physiological and Pathological Conditions. *Oxid. Med. Cell. Longev.* **2016**, *2016*, 1245049. [CrossRef] [PubMed]
37. Phaniendra, A.; Jestadi, D.B.; Periyasamy, L. Free Radicals: Properties, Sources, Targets, and their Implication in Various Diseases. *Indian J. Clin. Biochem.* **2015**, *30*, 11–26. [CrossRef]
38. Gagné, F. Oxidative Stress. In *Biochemical Ecotoxicology: Principles and Methods*; Elsevier: London, UK, 2014; Chapter 6; pp. 103–115.
39. Adams, L.B.; Dinauer, M.C.; Morgenstern, D.E.; Krahenbuhl, J.L. Comparison of the Roles of Reactive Oxygen and Nitrogen Intermediates in the Host Response to Mycobacterium Tuberculosis Using Transgenic Mice. *Tuber. Lung Dis.* **1997**, *78*, 237–246. [CrossRef]
40. Ozcan, A.; Ogun, M. Biochemistry of Reactive Oxygen and Nitrogen Species. In *Basic Principles and Clinical Significance of Oxidative Stress*; Gowder, S.J.T., Ed.; IntechOpen: London, UK, 2015; Chapter 3; pp. 37–58. [CrossRef]
41. Salisbury, D.; Bronas, U. Reactive Oxygen and Nitrogen Species: Impact on Endothelial Dysfunction. *Nurs. Res.* **2015**, *64*, 53–66. [CrossRef] [PubMed]
42. Tharmalingam, S.; Alhasawi, A.; Appanna, V.P.; Lemire, J.; Appanna, V.D. Reactive Nitrogen Species (RNS)-Resistant Microbes: Adaptation and Medical Implications. *Biol. Chem.* **2017**, *398*, 1193–1208. [CrossRef]
43. Norouzi, F.; Abareshi, A.; Asgharzadeh, F.; Beheshti, F.; Hosseini, M.; Farzadnia, M.; Khazaei, M. The Effect of Nigella Sativa on Inflammation-Induced Myocardial Fibrosis in Male Rats. *Res. Pharm. Sci.* **2017**, *12*, 74–81. [CrossRef]

44. Schiotis, R.E.; Goşa, D. Nigella Sativa as an Adjunctive Therapy for Rheumatic Diseases. In *Future Perspectives on Nigella Sativa: Characterization and Pharmacological Properties*; Pop, R.M., Ed.; Nova Science Publishers, Inc.: New York, NY, USA, 2018; Chapter 5; pp. 149–180.
45. Hussain, T.; Tan, B.; Yin, Y.; Blachier, F.; Tossou, M.C.B.; Rahu, N. Oxidative Stress and Inflammation: What Polyphenols Can Do for Us? *Oxid. Med. Cell. Longev.* **2016**, *2016*, 7432797. [CrossRef] [PubMed]
46. McGarry, T.; Biniecka, M.; Veale, D.J.; Fearon, U. Hypoxia, Oxidative Stress and Inflammation. *Free Radic. Biol. Med.* **2018**, *125*, 15–24. [CrossRef]
47. Biswas, S.K. Does the Interdependence between Oxidative Stress and Inflammation Explain the Antioxidant Paradox? *Oxid. Med. Cell. Longev.* **2016**, *2016*, 5698931. [CrossRef] [PubMed]
48. Elmarakby, A.A.; Sullivan, J.C. Relationship between Oxidative Stress and Inflammatory Cytokines in Diabetic Nephropathy. *Cardiovasc. Ther.* **2012**, *30*, 49–59. [CrossRef]
49. Caliri, A.W.; Tommasi, S.; Besaratinia, A. Relationships among Smoking, Oxidative Stress, Inflammation, Macromolecular Damage, and Cancer. *Mutat. Res. Rev.* **2021**, *787*, 108365. [CrossRef] [PubMed]
50. Donia, T.; Khamis, A. Management of Oxidative Stress and Inflammation in Cardiovascular Diseases: Mechanisms and Challenges. *Environ. Sci. Pollut. Res.* **2021**, *28*, 34121–34153. [CrossRef] [PubMed]
51. Sarkar, D.; Fisher, P.B. Molecular Mechanisms of Aging-Associated Inflammation. *Cancer Lett.* **2006**, *236*, 13–23. [CrossRef] [PubMed]
52. Hannoodee, S.; Nasuruddin, D.N. Acute Inflammatory Response. *Nature* **2021**, *206*, 20. [CrossRef]
53. Pahwa, R.; Goyal, A.; Jialal, I. *Chronic Inflammation In StatPearl*; StatPearls Publishing: Treasure Island, FL, USA, 2022. Available online: https://pubmed.ncbi.nlm.nih.gov/29630225/ (accessed on 31 August 2022).
54. Coates, M.; Lee, M.J.; Norton, D.; MacLeod, A.S. The Skin and Intestinal Microbiota and Their Specific Innate Immune Systems. *Front. Immunol.* **2019**, *10*, 2950. [CrossRef]
55. Varela, M.L.; Mogildea, M.; Moreno, I.; Lopes, A. Acute Inflammation and Metabolism. *Inflammation* **2018**, *41*, 1115–1127. [CrossRef] [PubMed]
56. Vaillant, A.A.J.; Qurie, A. Interleukin. In *StatPearl*; StatPearls Publishing: Treasure Island, FL, USA, 2021. Available online: https://www.ncbi.nlm.nih.gov/books/NBK499840/ (accessed on 31 August 2022).
57. Ashley, N.T.; Weil, Z.M.; Nelson, R.J. Inflammation: Mechanisms, Costs, and Natural Variation. *Annu. Rev. Ecol. Evol. Syst.* **2012**, *43*, 385–406. [CrossRef]
58. Sherwood, E.R.; Toliver-Kinsky, T. Mechanisms of the Inflammatory Response. *Best Pract. Res. Clin. Anaesthesiol.* **2004**, *18*, 385–405. [CrossRef] [PubMed]
59. Arulselvan, P.; Fard, M.T.; Tan, W.S.; Gothai, S.; Fakurazi, S.; Norhaizan, M.E.; Kumar, S.S. Role of Antioxidants and Natural Products in Inflammation. *Oxid. Med. Cell. Longev.* **2016**, *2016*, 5276130. [CrossRef] [PubMed]
60. Castanheira, F.V.S.; Kubes, P. Review Series Human Neutrophils Neutrophils and NETs in Modulating Acute and Chronic Inflammation. *Blood* **2019**, *133*, 2178–2185. [CrossRef] [PubMed]
61. Müller, F.; Mutch, N.J.; Schenk, W.A.; Smith, S.A.; Esterl, L.; Spronk, H.M.; Schmidbauer, S.; Gahl, W.A.; Morrissey, J.H.; Renné, T. Platelet Polyphosphates are Proinflammatory and Procoagulant Mediators in Vivo. *Cell* **2009**, *139*, 1143–1156. [CrossRef] [PubMed]
62. Margetic, S. Inflammation and Haemostasis. *Biochem. Med.* **2012**, *22*, 49–62. [CrossRef]
63. Medzhitov, R. Recognition of Microorganisms and Activation of The Immune Response. *Nature* **2007**, *449*, 819–826. [CrossRef]
64. Cain, D.; Kondo, M.; Chen, H.; Kelsoe, G. Effects of Acute and Chronic Inflammation on B-Cell Development and Differentiation. *J. Investig. Dermatol.* **2009**, *129*, 266–277. [CrossRef] [PubMed]
65. Fleit, H.B. Chronic Inflammation. In *Pathobiology of Human Disease: A Dynamic Encyclopedia of Disease Mechanisms*; McManus, L.M., Mitchell, R.N., Eds.; Academic Press: London, UK; Elsevier Inc.: London, UK, 2014; pp. 300–314.
66. Soleas, G.J.; Diamandis, E.P.; Goldberg, D.M. Wine as a Biological Fluid: History, Production, and Role in Disease Prevention. *J. Clin. Lab. Anal.* **1997**, *11*, 287–313. [CrossRef]
67. Vineyards in the EU-Statistics-Statistics Explained. Available online: https://ec.europa.eu/eurostat/statistics-explained/index.php?title=Vineyards_in_the_EU_-_statistics&oldid=566726 (accessed on 1 August 2022).
68. Eisenman, L. The Home Winemakers Manual. 1999, pp. 7–100. Available online: http://michaeljbowe.com/TSG/SG/SF_0007.pdf (accessed on 1 August 2022).
69. Ozcan, T.; Akpinar-Bayizit, A.; Yilmaz-Ersan, L.; Delikanli, B. Phenolics in Human Health. *Int. J. Chem. Eng. Appl.* **2014**, *5*, 393–396. [CrossRef]
70. Singla, R.K.; Dubey, A.K.; Garg, A.; Sharma, R.K.; Fiorino, M.; Ameen, S.M.; Haddad, M.A.; Al-Hiary, M. Natural Polyphenols: Chemical Classification, Definition of Classes, Subcategories, and Structures. *J. AOAC Int.* **2019**, *102*, 1397–1400. [CrossRef]
71. Durazzo, A.; Lucarini, M.; Souto, E.B.; Cicala, C.; Caiazzo, E.; Izzo, A.A.; Novellino, E.; Santini, A. Polyphenols: A Concise Overview on the Chemistry, Occurrence, and Human Health. *Phytother. Res.* **2019**, *33*, 2221–2243. [CrossRef]
72. Chedea, V.S.; Pop, R.M.; Pop, E.A. Grape Anthocyanins in Inflammation. In *Anthocyanins: Antioxidant Properties, Sources and Health Benefits*; Lorenzo Rodriguez, J.M., Barba, F.J., Munekata, P., Eds.; Nova Science Publishers, Inc.: New York, NY, USA, 2020; Chapter 11; pp. 295–315.
73. Panche, A.N.; Diwan, A.D.; Chandra, S.R. Flavonoids: An Overview. *J. Nutr. Sci.* **2016**, *5*, e47. [CrossRef]

74. Santhakumar, A.B.; Battino, M.; Alvarez-Suarez, J.M. Dietary Polyphenols: Structures, Bioavailability and Protective Effects Against Atherosclerosis. *Food Chem. Toxicol.* **2018**, *113*, 49–65. [CrossRef] [PubMed]
75. Yu, J.; Ahmedna, M. Functional Components of Grape Pomace: Their Composition, Biological Properties and Potential Applications. *Int. J. Food Sci. Technol.* **2013**, *48*, 221–237. [CrossRef]
76. de la Cerda-Carrasco, A.; López-Solís, R.; Nuñez-Kalasic, H.; Peña-Neira, Á.; Obreque-Slier, E. Phenolic Composition and Antioxidant Capacity of Pomaces from Four Grape Varieties (*Vitis vinifera* L.). *J. Sci. Food Agric.* **2015**, *95*, 1521–1527. [CrossRef]
77. Jin, Q.; O'Hair, J.; Stewart, A.C.; O'Keefe, S.F.; Neilson, A.P.; Kim, Y.T.; McGuire, M.; Lee, A.; Wilder, G.; Huang, H. Compositional Characterization of Different Industrial White and Red Grape Pomaces in Virginia and the Potential Valorization of the Major Components. *Foods* **2019**, *8*, 667. [CrossRef]
78. da Silva Mendonça, J.; de Cássia Avellaneda Guimarães, R.; Zorgetto-Pinheiro, V.A.; Di Pietro Fernandes, C.; Marcelino, G.; Bogo, D.; de Cássia Freitas, K.; Hiane, P.A.; de Pádua Melo, E.S.; Vilela, M.L.B.; et al. Natural Antioxidant Evaluation: A Review of Detection Methods. *Molecules* **2022**, *27*, 3563. [CrossRef]
79. Kammerer, D.; Claus, A.; Carle, R.; Schieber, A. Polyphenol Screening of Pomace from Red and White Grape Varieties (*Vitis vinifera* L.) by HPLC-DAD-MS/MS. *J. Agric. Food Chem.* **2004**, *52*, 4360–4367. [CrossRef] [PubMed]
80. Muncaciu, M.L.; Zamora Marín, F.; Pop, N.; Babeş, A.C. Comparative Polyphenolic Content of Grape Pomace Flours from "Fetească Neagră" and "Italian Riesling" Cultivars. *Not. Bot. Horti Agrobot. Cluj-Napoca* **2017**, *45*, 532–539. [CrossRef]
81. Tresserra-Rimbau, A.; Lamuela-Raventos, R.M.; Moreno, J.J. Polyphenols, Food and Pharma. Current Knowledge and Directions for Future Research. *Biochem. Pharmacol.* **2018**, *156*, 186–195. [CrossRef] [PubMed]
82. Yahfoufi, N.; Alsadi, N.; Jambi, M.; Matar, C. The Immunomodulatory and Anti-Inflammatory Role of Polyphenols. *Nutrients* **2018**, *10*, 1618. [CrossRef] [PubMed]
83. Kumar, S.; Pandey, A.K. Chemistry and Biological Activities of Flavonoids: An Overview. *Sci. World J.* **2013**, *2013*, 162750. [CrossRef] [PubMed]
84. Goutzourelas, N.; Stagos, D.; Housmekeridou, A.; Karapouliou, C.; Kerasioti, E.; Aligiannis, N.; Skaltsounis, A.L.; Spandidos, D.A.; Tsatsakis, A.M.; Kouretas, D. Grape Pomace Extract Exerts Antioxidant Effects Through an Increase in GCS Levels and GST Activity in Muscle and Endothelial Cells. *Int. J. Mol. Med.* **2015**, *36*, 433–441. [CrossRef] [PubMed]
85. Slavin, M.; Bourguignon, J.; Jackson, K.; Orciga, M.A. Impact of Food Components on In Vitro Calcitonin Gene-Related Peptide Secretion—A Potential Mechanism for Dietary Influence on Migraine. *Nutrients* **2016**, *8*, 406. [CrossRef] [PubMed]
86. Gerardi, G.; Cavia-Saiz, M.; Rivero-Pérez, M.D.; González-SanJosé, M.L.; Muñiz, P. Modulation of Akt-P38-MAPK/Nrf2/SIRT1 and NF-Kb Pathways by Wine Pomace Product in Hyperglycemic Endothelial Cell Line. *J. Funct. Foods* **2019**, *58*, 255–265. [CrossRef]
87. Ben Youssef, S.; Brisson, G.; Doucet-Beaupré, H.; Castonguay, A.M.; Gora, C.; Amri, M.; Lévesque, M. Neuroprotective Benefits of Grape Seed and Skin Extract in a Mouse Model of Parkinson's Disease. *Nutr. Neurosci.* **2021**, *24*, 197–211. [CrossRef]
88. Sapanidou, V.G.; Margaritis, I.; Siahos, N.; Arsenopoulos, K.; Dragatidou, E.; Taitzoglou, I.A.; Zervos, I.A.; Theodoridis, A.; Tsantarliotou, M.P. Antioxidant Effect of A Polyphenol-Rich Grape Pomace Extract on Motility, Viability and Lipid Peroxidation of Thawed Bovine Spermatozoa. *J. Biol. Res.* **2014**, *21*, 19. [CrossRef] [PubMed]
89. Martins, I.M.; Macedo, G.A.; Macedo, J.A. Biotransformed Grape Pomace as a Potential Source of Anti-Inflammatory Polyphenolics: Effects in Caco-2 Cells. *Food Biosci.* **2020**, *35*, 100607. [CrossRef]
90. Pop, A.; Bogdan, C.; Fizesan, I.; Iurian, S.; Carpa, R.; Bacali, C.; Vlase, L.; Benedec, D.; Moldovan, M.L. In Vitro Evaluation of Biological Activities of Canes and Pomace Extracts from Several Varieties of Vitis vinifera L. for Inclusion in Freeze-Drying Mouthwashes. *Antioxidants* **2022**, *11*, 218. [CrossRef] [PubMed]
91. Marzulli, G.; Magrone, T.; Kawaguchi, K.; Kumazawa, Y.; Jirillo, E. Fermented Grape Marc (FGM): Immunomodulating Properties and its Potential Exploitation in the Treatment of Neurodegenerative Diseases. *Curr. Pharm. Des.* **2012**, *18*, 43–50. [CrossRef] [PubMed]
92. Mollica, A.; Scioli, G.; Della Valle, A.; Cichelli, A.; Novellino, E.; Bauer, M.; Kamysz, W.; Llorent-Martínez, E.J.; De Córdova, M.L.F.; Castillo-López, R.; et al. Phenolic Analysis and In Vitro Biological Activity of Red Wine, Pomace and Grape Seeds Oil Derived from Vitis vinifera L. cv. Montepulciano d'Abruzzo. *Antioxidants* **2021**, *10*, 1704. [CrossRef]
93. Martins, I.M.; Macedo, G.A.; Macedo, J.A.; Chen, Q.; Blumberg, J.B.; Oliver Chen, C.-Y.; Oliver Chen, C.; Mayer, J. Tannase Enhances the Anti-Inflammatory Effect of Grape Pomace in Caco-2 1 Cells Treated with IL-1β. *J. Funct. Foods* **2017**, *29*, 69–76. [CrossRef]
94. Balea, Ş.S.; Pârvu, A.E.; Pârvu, M.; Vlase, L.; Dehelean, C.A.; Pop, T.I. Antioxidant, Anti-Inflammatory and Antiproliferative Effects of the Vitis vinifera L. var. Fetească Neagră and Pinot Noir Pomace Extracts. *Front. Pharmacol.* **2020**, *11*, 990. [CrossRef] [PubMed]
95. Harikrishnan, R.; Devi, G.; Van Doan, H.; Balasundaram, C.; Esteban, M.Á.; Abdel-Tawwab, M. Impact of Grape Pomace Flour (GPF) on Immunity and Immune-Antioxidant-Anti-Inflammatory Genes Expression in Labeo Rohita against Flavobacterium Columnaris. *Fish Shellfish Immunol.* **2021**, *111*, 69–82. [CrossRef]
96. Rajković, E.; Schwarz, C.; Kapsamer, S.B.; Schedle, K.; Reisinger, N.; Emsenhuber, C.; Ocelova, V.; Roth, N.; Frieten, D.; Dusel, G.; et al. Evaluation of a Dietary Grape Extract on Oxidative Status, Intestinal Morphology, Plasma Acute-Phase Proteins and Inflammation Parameters of Weaning Piglets at Various Points of Time. *Antioxidants* **2022**, *11*, 1428. [CrossRef] [PubMed]

97. Mossa, A.T.H.; Ibrahim, F.M.; Mohafrash, S.M.M.; Abou Baker, D.H.; El Gengaihi, S. Protective Effect of Ethanolic Extract of Grape Pomace against the Adverse Effects of Cypermethrin on Weanling Female Rats. *Evid.-Based Complement. Altern. Med.* **2015**, *2015*, 381919. [CrossRef]
98. Del Pino-García, R.; Rivero-Pérez, M.D.; González-SanJosé, M.L.; Croft, K.D.; Muñiz, P. Antihypertensive and Antioxidant Effects of Supplementation with Red Wine Pomace in Spontaneously Hypertensive Rats. *Food Funct.* **2017**, *8*, 2444–2454. [CrossRef] [PubMed]
99. Boussenna, A.; Cholet, J.; Goncalves-Mendes, N.; Joubert-Zakeyh, J.; Fraisse, D.; Vasson, M.P.; Texier, O.; Felgines, C. Polyphenol-Rich Grape Pomace Extracts Protect against Dextran Sulfate Sodium-Induced Colitis In Rats. *J. Sci. Food Agric.* **2016**, *96*, 1260–1268. [CrossRef]
100. Rodriguez Lanzi, C.; Perdicaro, D.J.; Antoniolli, A.; Fontana, A.R.; Miatello, R.M.; Bottini, R.; Vazquez Prieto, M.A. Grape Pomace and Grape Pomace Extract Improve Insulin Signaling in High-Fat-Fructose Fed Rat-Induced Metabolic Syndrome. *Food Funct.* **2016**, *7*, 1544–1553. [CrossRef] [PubMed]
101. Kara, K.; Kocaoℓlu Güçlü, B.; Baytok, E.; Şentürk, M. Effects of Grape Pomace Supplementation to Laying Hen Diet on Performance, Egg Quality, Egg Lipid Peroxidation and some Biochemical Parameters. *J. Appl. Anim. Res.* **2016**, *44*, 303–310. [CrossRef]
102. Zhao, J.X.; Li, Q.; Zhang, R.X.; Liu, W.Z.; Ren, Y.S.; Zhang, C.X.; Zhang, J.X. Effect of Dietary Grape Pomace on Growth Performance, Meat Quality and Antioxidant Activity in Ram Lambs. *Anim. Feed Sci. Technol.* **2018**, *236*, 76–85. [CrossRef]
103. Zhao, J.; Jin, Y.; Du, M.; Liu, W.; Ren, Y.; Zhang, C.; Zhang, J. The Effect of Dietary Grape Pomace Supplementation on Epididymal Sperm Quality and Testicular Antioxidant Ability in Ram Lambs. *Theriogenology* **2017**, *97*, 50–56. [CrossRef] [PubMed]
104. Gerardi, G.; Cavia-Saiz, M.; Rivero-Pérez, M.D.; González-SanJosé, M.L.; Muñiz, P. Wine Pomace Product Modulates Oxidative Stress and Microbiota in Obesity High-Fat Diet-Fed Rats. *J. Funct. Foods* **2020**, *68*, 103903. [CrossRef]
105. Cho, S.J.; Jung, U.J.; Park, H.J.; Kim, H.J.; Park, Y.B.; Kim, S.R.; Choi, M.S. Combined Ethanol Extract of Grape Pomace and Omija Fruit Ameliorates Adipogenesis, Hepatic Steatosis, and Inflammation in Diet-Induced Obese Mice. *Evid.-Based Complement. Altern. Med.* **2013**, *2013*, 212139. [CrossRef]
106. Li, H.; Parry, J.; Weeda, S.; Ren, S.; Castonguay, T.W.; Guo, T.L. Grape Pomace Aqueous Extract (GPE) Prevents High Fat Diet-Induced Diabetes and Attenuates Systemic Inflammation. *Food Nutr. Sci.* **2016**, *7*, 647–660. [CrossRef]
107. Sehm, J.; Lindermayer, H.; Meyer, H.H.D.; Pfaffl, M.W. The Influence of Apple- and Red-Wine Pomace Rich Diet on mRNA Expression of Inflammatory and Apoptotic Markers in Different Piglet Organs. *Anim. Sci.* **2006**, *82*, 877–887. [CrossRef]
108. Ianni, A.; Di Maio, G.; Pittia, P.; Grotta, L.; Perpetuini, G.; Tofalo, R.; Cichelli, A.; Martino, G. Chemical–Nutritional Quality and Oxidative Stability of Milk and Dairy Products Obtained from Friesian Cows Fed With a Dietary Supplementation of Dried Grape Pomace. *J. Sci. Food Agric.* **2019**, *99*, 3635–3643. [CrossRef]
109. Chikwanha, O.C.; Moelich, E.; Gouws, P.; Muchenje, V.; Nolte, J.V.E.; Dugan, M.E.R.; Mapiye, C. Effects of Feeding Increasing Levels of Grape (Vitis Vinifera Cv. Pinotage) Pomace on Lamb Shelf-Life and Eating Quality. *Meat Sci.* **2019**, *157*, 107887. [CrossRef] [PubMed]
110. Tayengwa, T.; Chikwanha, O.C.; Gouws, P.; Dugan, M.E.R.; Mutsvangwa, T.; Mapiye, C. Dietary Citrus Pulp and Grape Pomace as Potential Natural Preservatives for Extending Beef Shelf Life. *Meat Sci.* **2020**, *162*, 108029. [CrossRef]
111. Chamorro, S.; Viveros, A.; Rebolé, A.; Rica, B.D.; Arija, I.; Brenes, A. Influence of Dietary Enzyme Addition on Polyphenol Utilization and Meat Lipid Oxidation of Chicks Fed Grape Pomace. *Food Res. Int.* **2015**, *73*, 197–203. [CrossRef]
112. Makri, S.; Kafantaris, I.; Stagos, D.; Chamokeridou, T.; Petrotos, K.; Gerasopoulos, K.; Mpesios, A.; Goutzourelas, N.; Kokkas, S.; Goulas, P.; et al. Novel Feed Including Bioactive Compounds from Winery Wastes Improved Broilers' Redox Status in Blood and Tissues of Vital Organs. *Food Chem. Toxicol.* **2017**, *102*, 24–31. [CrossRef]
113. Brenes, A.; Viveros, A.; Goñi, I.; Centeno, C.; Sáyago-Ayerdy, S.G.; Arija, I.; Saura-Calixto, F. Effect of Grape Pomace Concentrate and Vitamin E on Digestibility of Polyphenols and Antioxidant Activity in Chickens. *Poult. Sci.* **2008**, *87*, 307–316. [CrossRef]
114. Goñi, I.; Brenes, A.; Centeno, C.; Viveros, A.; Saura-Calixto, F.; Rebolé, A.; Arija, I.; Estevez, R. Effect of Dietary Grape Pomace and Vitamin E on Growth Performance, Nutrient Digestibility, and Susceptibility to Meat Lipid Oxidation in Chickens. *Poult. Sci.* **2007**, *86*, 508–516. [CrossRef] [PubMed]
115. Kafantaris, I.; Stagos, D.; Kotsampasi, B.; Hatzis, A.; Kypriotakis, A.; Gerasopoulos, K.; Makri, S.; Goutzourelas, N.; Mitsagga, C.; Giavasis, I.; et al. Grape Pomace Improves Performance, Antioxidant Status, Fecal Microbiota and Meat Quality of Piglets. *Animal* **2018**, *12*, 246–255. [CrossRef]
116. Sáyago-Ayerdi, S.G.; Brenes, A.; Viveros, A.; Goñi, I. Antioxidative Effect of Dietary Grape Pomace Concentrate on Lipid Oxidation of Chilled and Long-Term Frozen Stored Chicken Patties. *Meat Sci.* **2009**, *83*, 528–533. [CrossRef] [PubMed]
117. Kafantaris, I.; Kotsampasi, B.; Christodoulou, V.; Kokka, E.; Kouka, P.; Terzopoulou, Z.; Gerasopoulos, K.; Stagos, D.; Mitsagga, C.; Giavasis, I.; et al. Grape Pomace Improves Antioxidant Capacity and Faecal Microflora of Lambs. *J. Anim. Physiol. Anim. Nutr.* **2017**, *101*, e108–e121. [CrossRef]
118. Tominaga, T.; Kawaguchi, K.; Kanesaka, M.; Kawauchi, H.; Jirillo, E.; Kumazawa, Y. Suppression of Type-I Allergic Responses by Oral Administration of Grape Marc Fermented with *Lactobacillus plantarum*. *Immunopharmacol. Immunotoxicol.* **2010**, *32*, 593–599. [CrossRef] [PubMed]
119. Bacchi, S.; Palumbo, P.; Sponta, A.; Coppolino, M.F. Clinical Pharmacology of Non-Steroidal Anti-Inflammatory Drugs: A Review. *Antiinflamm. Antiallergy Agents Med. Chem.* **2012**, *11*, 52–64. [CrossRef]

120. Vane, J.R.; Botting, R.M. Anti-Inflammatory Drugs and their Mechanism of Action. *Inflamm. Res.* **1998**, *47* (Suppl. S2), S78–S87. [CrossRef] [PubMed]
121. Moore, N.; Duong, M.; Gulmez, S.E.; Blin, P.; Droz, C. Pharmacoepidemiology of Non-Steroidal Anti-Inflammatory Drugs. *Therapie* **2019**, *74*, 271–277. [CrossRef]
122. Plaza-Serón, M.D.C.; Garciá-Martín, E.; Agúndez, J.A.; Ayuso, P. Hypersensitivity Reactions to Nonsteroidal Anti-Inflammatory Drugs: An Update on Pharmacogenetics Studies. *Pharmacogenomics* **2018**, *19*, 1069–1086. [CrossRef]
123. Kowalski, M.L.; Asero, R.; Bavbek, S.; Blanca, M.; Blanca-Lopez, N.; Bochenek, G.; Brockow, K.; Campo, P.; Celik, G.; Cernadas, J.; et al. Classification and Practical Approach to the Diagnosis and Management of Hypersensitivity to Nonsteroidal Anti-Inflammatory Drugs. *Allergy* **2013**, *68*, 1219–1232. [CrossRef] [PubMed]
124. Kowalski, M.L.; Makowska, J.S.; Blanca, M.; Bavbek, S.; Bochenek, G.; Bousquet, J.; Bousquet, P.; Celik, G.; Demoly, P.; Gomes, E.R.; et al. Hypersensitivity to Nonsteroidal Anti-Inflammatory Drugs (Nsaids)-Classification, Diagnosis and Management: Review of the EAACI/ENDA(#) and GA2LEN/HANNA*. *Allergy* **2011**, *66*, 818–829. [CrossRef] [PubMed]
125. Dona, I.; Salas, M.; Perkins, J.; Barrionuevo, E.; Gaeta, F.; Cornejo-Garcia, J.; Campo, P.; Torres, M. Hypersensitivity Reactions to Non-Steroidal Anti-Inflammatory Drugs. *Curr. Pharm. Des.* **2016**, *22*, 6784–6802. [CrossRef]
126. Costa, D.; Moutinho, L.; Lima, J.L.F.C.; Fernandes, E. Antioxidant Activity and Inhibition of Human Neutrophil Oxidative Burst Mediated by Arylpropionic Acid Non-Steroidal Anti-Inflammatory Drugs. *Biol. Pharm. Bull.* **2006**, *29*, 1659–1670. [CrossRef] [PubMed]
127. Nawaz, H.; Ali, A.; Rehman, T.; Aslam, A. Chronological Effects of Non-Steroidal Anti-Inflammatory Drug Therapy on Oxidative Stress and Antioxidant Status in Patients with Rheumatoid Arthritis. *Clin. Rheumatol.* **2021**, *40*, 1767–1778. [CrossRef] [PubMed]
128. El-Ghazaly, M.A.; Fadel, N.A.; Abdel-Naby, D.H.; Abd El-Rehim, H.A.; Zaki, H.F.; Kenawy, S.A. Amelioration of Adjuvant-Induced Arthritis by Exposure to Low Dose Gamma Radiation and Resveratrol Administration in Rats. *Int. J. Radiat. Biol.* **2020**, *96*, 857–867. [CrossRef] [PubMed]
129. El-Ghazaly, M.A.; Fadel, N.A.; Abdel-Naby, D.H.; Abd El-Rehim, H.A.; Zaki, H.F.; Kenawy, S.A. Potential Anti-Inflammatory Action of Resveratrol and Piperine in Adjuvant-Induced Arthritis: Effect on Pro-Inflammatory Cytokines and Oxidative Stress Biomarkers. *Egypt. Rheumatol.* **2020**, *42*, 71–77. [CrossRef]
130. Chen, X.; Lu, J.; An, M.; Ma, Z.; Zong, H.; Yang, J. Anti-Inflammatory Effect of Resveratrol on Adjuvant Arthritis Rats with Abnormal Immunological Function via the Reduction of Cyclooxygenase-2 and Prostaglandin E2. *Mol. Med. Rep.* **2014**, *9*, 2592–2598. [CrossRef] [PubMed]
131. Willenberg, I.; Meschede, A.K.; Gueler, F.; Jang, M.S.; Shushakova, N.; Schebb, N.H. Food Polyphenols Fail to Cause a Biologically Relevant Reduction of COX-2 Activity. *PLoS ONE* **2015**, *10*, e0139147. [CrossRef]
132. Correâ, M.G.; Pires, P.R.; Ribeiro, F.V.; Pimentel, S.P.; Cirano, F.R.; Napimoga, M.H.; Casati, M.Z.; Casarin, R.C.V. Systemic Treatment with Resveratrol Reduces the Progression of Experimental Periodontitis and Arthritis in Rats. *PLoS ONE* **2018**, *13*, e0204414. [CrossRef]
133. Wang, Z.M.; Chen, Y.C.; Wang, D.P. Resveratrol, a Natural Antioxidant, Protects Monosodium Iodoacetate-Induced Osteoarthritic Pain in Rats. *Biomed. Pharmacother.* **2016**, *83*, 763–770. [CrossRef]
134. Rotelli, A.E.; Guardia, T.; Juárez, A.O.; De La Rocha, N.E.; Pelzer, L.E. Comparative Study of Flavonoids in Experimental Models of Inflammation. *Pharmacol. Res.* **2003**, *48*, 601–606. [CrossRef]
135. Guardia, T.; Ester Rotelli, A.; Osvaldo Juarez, A.; Eugenia Pelzer, L. Anti-Inflammatory Properties of Plant Flavonoids. Effects of Rutin, Quercetin and Hesperidin on Adjuvant Arthritis in Rat. *Il Farmaco* **2001**, *56*, 683–687. [CrossRef]
136. Zhang, Y.; Anoopkumar-Dukie, S.; Mallik, S.B.; Davey, A.K. SIRT1 and SIRT2 Modulators Reduce LPS-induced Inflammation in HAPI Microglial Ccells and Protect SH-SY5Y Neuronal Cells in vitro. *J. Neural Transm.* **2021**, *128*, 631–644. [CrossRef]
137. Siard, M.H.; McMurry, K.E.; Adams, A.A. Effects of Polyphenols Including Curcuminoids, Resveratrol, Quercetin, Pterostilbene, and Hydroxypterostilbene on Lymphocyte Pro-Inflammatory Cytokine Production of Senior Horses in vitro. *Vet. Immunol. Immunopathol.* **2016**, *173*, 50–59. [CrossRef]
138. Kalita, B.; Das, M.K.; Sarma, M.; Deka, A. Sustained Anti-inflammatory Effect of Resveratrol-Phospholipid Complex Embedded Polymeric Patch. *AAPS PharmSciTech* **2017**, *18*, 629–645. [CrossRef] [PubMed]
139. Wei, B.; Zhang, Y.; Tang, L.; Ji, Y.; Yan, C.; Zhang, X. Protective Effects of Quercetin Against Inflammation and Oxidative Stress in a Rabbit Model of Knee Osteoarthritis. *Drug Dev. Res.* **2019**, *80*, 360–367. [CrossRef] [PubMed]
140. Heydari Nasrabadi, M.; Parsivand, M.; Mohammadi, N.; Asghari Moghaddam, N. Comparison of *Elaeagnus angustifolia* L. Extract and Quercetin on Mouse Model of Knee Osteoarthritis. *J. Ayurveda Integr. Med.* **2022**, *13*, 100529. [CrossRef] [PubMed]
141. Moradi, A.; Abolfathi, M.; Javadian, M.; Heidarian, E.; Roshanmehr, H.; Khaledi, M.; Nouri, A. Gallic Acid Exerts Nephroprotective, Anti-Oxidative Stress, and Anti-Inflammatory Effects Against Diclofenac-Induced Renal Injury in Malerats. *Arch. Med. Res.* **2021**, *52*, 380–388. [CrossRef] [PubMed]
142. Esmaeilzadeh, M.; Heidarian, E.; Shaghaghi, M.; Roshanmehr, H.; Najafi, M.; Moradi, A.; Nouri, A. Gallic Acid Mitigates Diclofenac-Induced Liver Toxicity by Modulating Oxidative Stress and Suppressing IL-1β Gene Expression in Male Rats. *Pharm. Biol.* **2020**, *58*, 590–596. [CrossRef]

molecules

MDPI

Article

Synthesis and Characterization of Curcumin-Loaded Nanoparticles of Poly(Glycerol Sebacate): A Novel Highly Stable Anticancer System

Alessio Massironi [1,*], Stefania Marzorati [1], Alessandra Marinelli [2], Marta Toccaceli [2], Stefano Gazzotti [3], Marco Aldo Ortenzi [3], Daniela Maggioni [3], Katia Petroni [2,*] and Luisella Verotta [1]

1 Department of Environmental Science and Policy, Università degli Studi di Milano, Via Celoria 2, 20133 Milano, Italy
2 Department of Biosciences, Università degli Studi di Milano, Via Celoria 26, 20133 Milano, Italy
3 Department of Chemistry, Università degli Studi di Milano, Via Golgi 19, 20133 Milano, Italy
* Correspondence: alessio.massironi@unimi.it (A.M.); katia.petroni@unimi.it (K.P.)

Abstract: The research for alternative administration methods for anticancer drugs, towards enhanced effectiveness and selectivity, represents a major challenge for the scientific community. In the last decade, polymeric nanostructured delivery systems represented a promising alternative to conventional drug administration since they ensure secure transport to the selected target, providing active compounds protection against elimination, while minimizing drug toxicity to non-target cells. In the present research, poly(glycerol sebacate), a biocompatible polymer, was synthesized and then nanostructured to allow curcumin encapsulation, a naturally occurring polyphenolic phytochemical isolated from the powdered rhizome of *Curcuma longa* L. Curcumin was selected as an anticancer agent in virtue of its strong chemotherapeutic activity against different cancer types combined with good cytocompatibility within healthy cells. Despite its strong and fascinating biological activity, its possible exploitation as a novel chemotherapeutic has been hampered by its low water solubility, which results in poor absorption and low bioavailability upon oral administration. Hence, its encapsulation within nanoparticles may overcome such issues. Nanoparticles obtained through nanoprecipitation, an easy and scalable technique, were characterized in terms of size and stability over time using dynamic light scattering and transmission electron microscopy, confirming their nanosized dimensions and spherical shape. Finally, biological investigation demonstrated an enhanced cytotoxic effect of curcumin-loaded PGS-NPs on human cervical cancer cells compared to free curcumin.

Keywords: curcumin; poly(glycerol sebacate); nanoparticles; drug delivery system; human cervical cancer

Citation: Massironi, A.; Marzorati, S.; Marinelli, A.; Toccaceli, M.; Gazzotti, S.; Ortenzi, M.A.; Maggioni, D.; Petroni, K.; Verotta, L. Synthesis and Characterization of Curcumin-Loaded Nanoparticles of Poly(Glycerol Sebacate): A Novel Highly Stable Anticancer System. *Molecules* **2022**, *27*, 6997. https://doi.org/10.3390/molecules27206997

Academic Editor: Nour Eddine Es-Safi

Received: 12 September 2022
Accepted: 12 October 2022
Published: 18 October 2022

1. Introduction

Cancer is still the second leading cause of death in the world after cardiovascular diseases and, despite the continued efforts of scientists, its incidence and mortality rates have not yet been stopped [1]. Therefore, pursuing novel strategies and less toxic cancer treatments still remains a major challenge for the scientific community.

Researchers, to improve the selectivity and bioavailability of chemotherapeutic agents, have developed several drug delivery systems (DDSs) designed to allow secure transport to the selected target, thus providing molecules protection against elimination while minimizing drug toxicity to non-target cells [2–6]. During their administration, anticancer agents are generally distributed non-specifically throughout the body, affecting both tumor and healthy cells, resulting in inefficient treatment due to excessive side effects and low internalization of anticancer agents into tumor tissues [7]. On the contrary, among several advantages over traditional chemotherapeutic administration, DDSs can deliver a drug more selectively to a specific target site with less frequent and less concentrated dosing [5]. Moreover, in cancer cells, the passive diffusion is maximized due to the presence of abnormal vascular architecture necessary to serve fast-growing cancer cells. Such phenomenon is named enhanced permeation and retention effect (EPR effect) [6,8]. Accordingly, DDSs may exploit the EPR effect thanks to the nanometer size of the carrier which enhances their penetration and thus incorporation into the tumor target thanks to leaky vasculature [8].

Thanks to their unique physicochemical versatility, biodegradability and biocompatibility, biodegradable polymers are widely investigated for the preparation of micro- and nanoparticles as drug carriers for anticancer agents [3,9]. Among the polymers that can be used for this purpose, poly(glycerol sebacate) (PGS) has received notable attention due to its unique physical features [10]. Both monomers used for PGS synthesis via polycondensation, i.e., glycerol and sebacic acid, are bio-based, biocompatible and approved by the Food and Drug Administration. PGS synthesis is inexpensive, and the obtained product is generally soft and has flexible mechanical properties that make it suitable for working with soft tissue and organs in a mechanically dynamic environment [11]. Originally designed as a biodegradable polymer with improved elastic mechanical properties and biocompatibility, research on PGS-based medical applications has uncovered several unique properties that have bolstered its use as a biomaterial [12]. Nowadays, PGS is commonly exploited to develop 3D structures such as scaffolds for tissue engineering [13] but its possible application as a nanodelivery system has been explored less [14,15]. Moreover, several studies demonstrate that the hydrophobicity degree of a carrier is one of the major determinants to achieve correct and complete drug delivery in specific environments (e.g., lymphatic system) [4,14–16]. In this context, the aim of the present study was the development of highly stable PGS nanostructures designed to ensure secure loading of curcumin extract (standardized as 95% in curcuminoids composed by: curcumin 85%, demethoxycurcumin 14% and bis-demethoxycurcumin 1%) (Figure 1) as a novel anticancer system. Curcumin, [1,7-bis(4-hydroxy-3-methoxyphenyl)-1,6-heptadiene-3,5-dione], a naturally occurring polyphenolic phytochemical isolated from the powdered rhizome of *Curcuma longa*, was selected as an anticancer agent in virtue of its strong chemotherapeutic activity against different cancers types combined with good cytocompatibility against healthy cells [17–19].

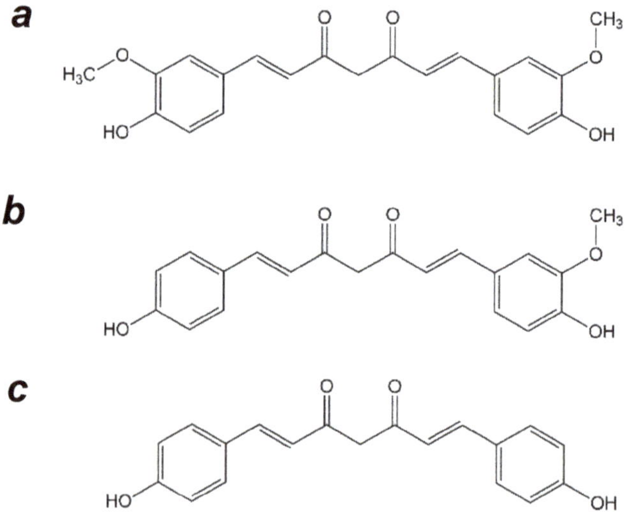

Figure 1. Curcuminoids structures (**a**) curcumin, (**b**) demethoxycurcumin and (**c**) bis-demethoxycurcumin.

Several studies indicate that curcumin can modulate all kinds of cancer hallmarks, including uncontrolled cell proliferation, cancer-associated inflammation, cancer cell death, signaling pathways, cancer angiogenesis, and metastasis [20]. However, its possible exploitation as a novel chemotherapeutic is hampered by its low water solubility [21–23], which results in poor absorption and low bioavailability upon oral administration; its possible encapsulation within PGS nanoparticles represents a novel and promising administration alternative combining its strong anticancer activity with the biocompatibility and high hydrophobicity of the polymer. Nanoprecipitation was selected as the formulation technique since it offers several advantages for producing smaller nanoparticles with narrow unimodal size distribution [24]. Furthermore, it is scalable and rapid to perform [25]. The obtained NPs were characterized through dynamic light scattering and transmission electron microscopy analysis to evaluate nanoparticle size and morphology and monitoring their stability over time.

Cervical cancer is the leading cause of death for cancer in women resulting in over 340,000 deaths worldwide [26]. Most cases of cervical cancer can be attributed to persistent human papilloma virus (HPV) infection, which is preventable thanks to safe and effective anti-HPV vaccination. Screening programs allow the identification of cervical pre-cancer lesions, thus allowing prompt treatment and cure of early stage cervical cancer with surgical interventions, chemotherapy and/or radiotherapy [27]. However, recurrent cervical cancer due to resistance to chemotherapy still represents a major challenge. In this context, polyphenols, such as nanoencapsulated curcumin, are currently viewed as a possible adjuvant therapy for overcoming chemoresistance in cancer cells, since they affect multiple targets, including cell death [28]. Therefore, the biological activity of the PGS nanosystem has been investigated in vitro against human cervical cancer HeLa cells to evaluate the pro-apoptotic anticancer activities of curcumin-loaded PGS-NPs compared to free curcumin. Our results show that curcumin-loaded PGS-NPs display higher cytotoxicity, anti-HPV activity and can activate apoptosis in HeLa cervical cancer cells.

2. Results

2.1. Synthesis and Characterization of Poly(Glycerol Sebacate)

PGS was synthesized via a polycondensation reaction between glycerol and sebacic acid to first form a linear aliphatic polyester (Scheme 1a) that can be further cured to yield a crosslinked thermoset elastomer (Scheme 1b).

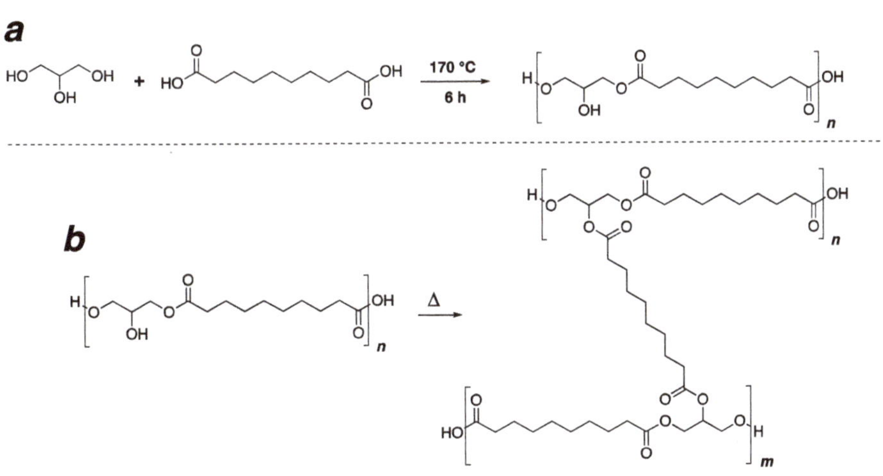

Scheme 1. Synthesis of PGS. (**a**) synthesis of linear PGS through polycondensation of glycerol and sebacic acid. (**b**) curing reaction to yield a thermoset.

The main hurdle in the synthesis of PGS derives from the nature of the monomers. While glycerol can be described as an A3 type monomer, sebacic acid is a difunctional species that can be described as a B2 type monomer. This difference in the number of reactive functionalities between the two molecules results in difficulties in proper control of the reactivity during the polycondensation reaction. In this regard, even if it is true that the primary alcohol groups in glycerin are more reactive than the secondary ones, a non-proper management of the reaction conditions can result in the formation of an undesired crosslinked material. Literature reports on the synthesis of PGS usually rely on an equimolar mixture of the two monomers, to target a linear polyester. The reaction is carried out for very long times at low temperatures, in order to limit the occurrence of crosslinking [29]. During this step, the molecular weight of the linear segments has limited control through different reaction times. After this first condensation step, the crosslinking reaction is carried out to yield a thermoset. The mechanical properties of the crosslinked product can be controlled through the curing conditions that can give access to a significant amount of different products with different properties [30].

This classical procedure for the synthesis of PGS requires long times, lasting several days. On the contrary, the procedure reported in this paper is based on a different approach. The two monomers are loaded in different molar amounts, with an excess of sebacic acid to partially make up for the excess of alcoholic functionalities. The reaction was carried out for 6 h at 170 °C resulting in a linear polymer in the form of a waxy solid that was further employed for the preparation of nanoparticles.

PGS was characterized through size exclusion chromatography (SEC) for the determination of the molecular weight and polydispersity. Molecular weight data of the polymer were detected as follows: (M_n) = 3100 Da; (M_w) = 12,000 Da; Đ = 4.0, expressed as polystyrene equivalents. The high polydispersity of the polymer was attributed to the occurrence of branching reactions. These kinds of reactions are competitive with a linear chain growth and affect the microstructure of the product resulting in a complex mixture of species. However, the product was filtered before the analysis in order to eliminate every possible crosslinked fraction.

DSC analysis pointed out the semicrystalline nature of the polymer, with $T_g = -11.3$ °C; $T_m = 8.1$ °C.

NMR and FT-IR were also carried out in order to depict the structural features of the polymer. Spectral data were in good agreement with the literature [31]. Spectra are shown in the supporting information.

2.2. PGS-NPs Fabrication

Loaded and unloaded PGS-NPs were prepared by nanoprecipitation according to a general procedure [24], where an acetone solution of PGS and curcumin was precipitated in deionized water allowing the formation of NPs. In order to evaluate the effect of polymer concentration over nanoparticles stability and size, PGS-NPs were formulated at different polymer concentrations in the starting organic solution (before nanoprecipitation); namely, $PGS_{0.1}C_{0.01}$, [PGS]: 0.1 mg/mL; $PGS_{0.5}C_{0.05}$, [PGS]: 0.5 mg/mL; $PGS_{1.0}C_{0.1}$, [PGS]: 1.0 mg/mL; $PGS_{5.0}C_{0.5}$, [PGS]: 5.0 mg/mL by keeping constant the curcumin-to-PGS ratio (curcumin/PGS ratio equal to 10% by weight). Concentration values refer to the final suspension water volume, after nanoprecipitation and after ethanol evaporation. For unloaded PGS-NPs (reported as Blank in Table 1) the same procedures, as well as the same PGS concentrations, were used but in the absence of curcumin. All the adopted experimental conditions led to the formation of monodisperse nanosized (loaded and unloaded) PGS particles, displaying a narrow monomodal size distribution in the nanometer range, as shown in Figure 2. Table 1 reports the maxima values of the distribution curves together with the polydispersion index (PdI) derived from the cumulant analysis.

Table 1. Effect of polymer concentration on PGS-NPs particle diameter at t = 0, after 7 days and after 14 days from sample preparation.

Run	Mean Diameter (nm)		
	0 Days	**7 Days**	**14 Days**
$PGS_{0.1}Blank$	89 ± 1 PdI: 0.21	92 ± 2 PdI: 0.21	96 ± 2 PdI: 0.04
$PGS_{0.1}C_{0.01}$	81 ± 1 PdI: 0.25	225 ± 14 PdI: 0.73 (presence of macroaggregates)	-
$PGS_{0.5}Blank$	123 ± 1 PdI: 0.01	120 ± 3 PdI: 0.01	119 ± 3 PdI: 0.08
$PGS_{0.5}C_{0.05}$	106 ± 2 PdI:0.21	136 ± 4 PdI:0.82 (presence of macroaggregates)	-
$PGS_{1.0}Blank$	164 ± 4 PdI: 0.01	165 ± 3 PdI: 0.01	163 ± 4 PdI: 0.03
$PGS_{1.0}C_{0.1}$	158 ± 4 PdI:0.03	159 ± 6 PdI:0.03	130 ± 4 PdI: 0.05
$PGS_{5.0}Blank$	234 ± 6 PdI: 0.02	229 ± 6 PdI: 0.03	230 ± 6 PdI: 0.01
$PGS_{5.0}C_{0.5}$	211 ± 2 PdI: 0.001	173 ± 4 PdI: 0.02	169 ± 2 PdI: 0.02

PdI: Polydispersity index.

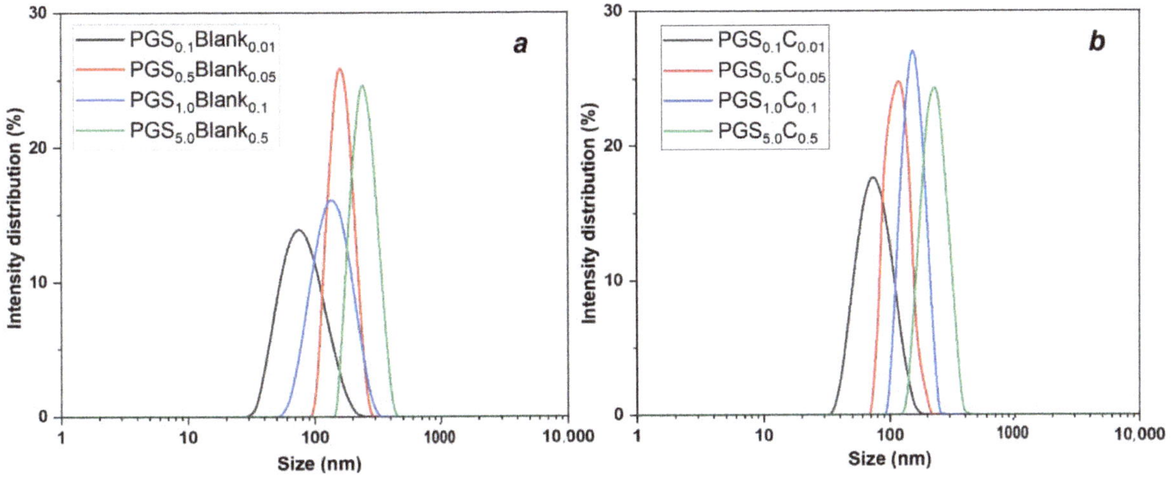

Figure 2. DLS size distribution (intensity) of unloaded (**a**) and curcumin-loaded PGS-NPs (**b**) formed at different polymer concentrations (0.1, 0.5, 1.0 and 5.0 mg/mL).

Results relevant to individual experiments are reported in Table 1. It was observed that the composition of the organic phase, determined by PGS and curcumin concentration, strongly affected the diameter distribution of the particles. Increasing the polymer concentration increased the PGS-NPs diameter.

On the contrary, the presence of curcuminoids seemed to not influence the formation of the NPs since only slight differences in terms of NP size were observed as a consequence of their loading. However, curcumin strongly affected particles stability, since all unloaded PGS-NPs (reported as Blank in Table 1) demonstrated good stability up to 14 days from the preparation of the samples, while an increase in size and polydispersity of the two curcumin-loaded samples (namely $PGS_{0.1}C_{0.01}$ and $PGS_{0.5}C_{0.05}$) was observed after 7 days from samples preparation (Table 1 and Figure 3). After 7 days, both formulations presented macroscopic yellow aggregates indicating the loss of their colloidal stability.

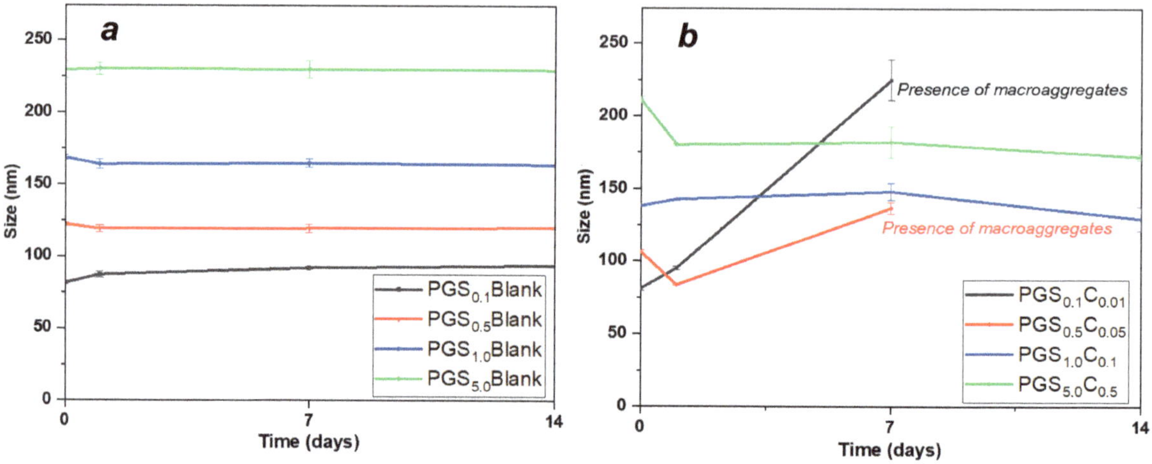

Figure 3. Hydrodynamic diameter of blank (**a**) and curcumin-loaded (**b**) PGS-NPs over time from sample preparation.

On the contrary, curcumin-loaded samples ($PGS_{5.0}C_{0.5}$ and $PGS_{1.0}C_{0.1}$) demonstrated better stability over 14 days, where only slight reductions in particles size distribution were observed, thus comparable to the blank formulations.

The encapsulation efficiency of the most promising formulations ($PGS_{1.0}C_{0.1}$ and $PGS_{5.0}C_{0.5}$ purified by centrifugation) were equal to 99.8% (as measured by ultra-high-performance liquid chromatography, UPLC) for both developed nanoparticles, confirming the complete encapsulation of the active compound within the polymeric matrix. These values were particularly high compared to those reported in the literature, where the encapsulation efficiency of curcumin-loaded NPs spanned from 5% to 95% [32–34]. Since almost the whole curcumin was entrapped into the NPs, no further purification of colloidal suspensions was deemed necessary. Most importantly, no additives such as Tween-80 or PEG, commonly used to maximize the loading, were employed [35,36].

Due to the high proven stability, two formulations (namely $PGS_{1.0}C_{0.1}$, $PGS_{5.0}C_{0.5}$) as well as the corresponding unloaded formulations ($PGS_{1.0}$Blank and $PGS_{5.0}$Blank) were selected as the most promising candidates for the in vitro release studies and biological assays.

2.3. PGS-NPs Stability Analysis in Cells Culture Medium

The colloidal stability of $PGS_{1.0}C_{0.1}$ and $PGS_{5.0}C_{0.5}$ was investigated in cells culture medium by diluting the samples in water at 37 °C and in Dulbecco's Modified Eagle Medium (DMEM) supplemented with fetal bovine serum (FBS) at the highest concentration tested during the planned biological investigation ([PGS]:0.2 mg/mL). A reduction in particle diameter was observed for both nanosystems (Table 2), possibly due to the more diluted conditions. Indeed, the higher dilution can cause NPs to be on average more distant from one another, giving rise to a less probable interaction among particles over time in comparison to higher concentrations where collisions and interactions are maximized. After a 24 h incubation in water at 37 °C, a reduction in particle size distribution was observed for $PGS_{1.0}C_{0.10}$, suggesting lower stability at higher temperatures compared to $PGS_{5.0}C_{0.5}$, where no differences were observed even at 37 °C after 7 days of incubation.

The results of particle stability tests in cell culture medium and deionized water at 37 °C are reported in Table 2 and the monitoring of size over time is shown in Figure 4.

Table 2. $PGS_{1.0}C_{0.1}$, $PGS_{5.0}C_{0.5}$ recorded mean diameters at different times from sample preparation in deionized water and DMEM.

Run in Deionized Water	Mean Diameter (nm)			
	Day 0	Day 1	Day 2	Day 3
$PGS_{1.0}C_{0.1}$ *[0.2 mg/mL]*	108 ± 3 PdI: 0.23	87 ± 1 PdI: 0.05	85 ± 2 PdI: 0.06	84 ± 2 PdI: 0.06
$PGS_{5.0}C_{0.5}$ *[0.2 mg/mL]*	136 ± 2 PdI: 0.07	136 ± 2 PdI: 0.07	136 ± 1 PdI: 0.02	135 ± 1 PdI: 0.02
Run in DMEM	**Mean Diameter (nm)**			
	Day 0	**Day 1**	**Day 2**	**Day 3**
$PGS_{1.0}C_{0.1}$ *[0.2 mg/mL]*	148 ± 2 PdI: 0.20	137 ± 2 PdI: 0.21	129 ± 1 PdI: 0.22	127 ± 1 PdI: 0.24
$PGS_{5.0}C_{0.5}$ *[0.2 mg/mL]*	137 ± 1 PdI: 0.17	129 ± 1 PdI: 0.19	129 ± 1 PdI: 0. 21	129 ± 1 PdI: 0.19

PdI: Polydispersity index.

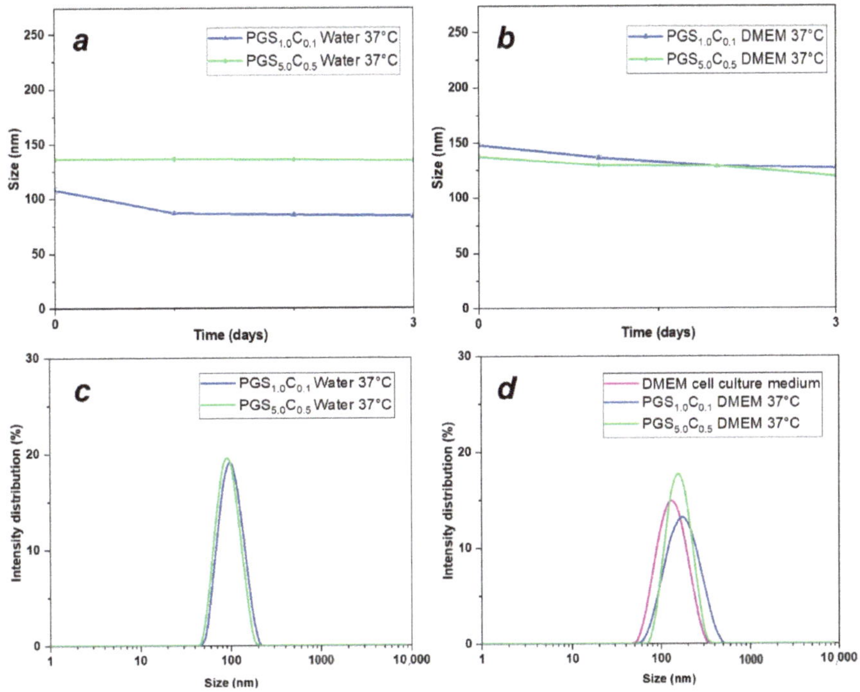

Figure 4. Hydrodynamic diameter of curcumin-loaded PGS-NPs over time from sample preparation in water (**a**) and in DMEM (**b**). DLS size distribution (intensity) of selected loaded PGS-NPs after their dilution in deionized water (**c**) and DMEM cell culture medium (**d**) at 37 °C.

The same experiment was conducted in cell culture medium. The background noise of cell culture medium caused an increase in the polydispersity index for all the analyzed samples. Indeed, a broad particle distribution of circa 120 nm was observed by DLS analysis of the sole DMEM. This may be due to the presence of several species in the medium such as amino acids, antibiotics and in particular of fetal bovine serum (FBS) which could strongly affect the particle size analysis. The DLS analysis of $PGS_{1.0}C_{0.1}$ NPs was heavily affected by the background noise of DMEM and did not allow for correct particle size distribution measurements. On the contrary, $PGS_{5.0}C_{0.5}$ revealed the presence of homogeneous particle size distribution with an unchanged maximum compared to the one observed in pure water, thus confirming the good colloidal stability even in the cell medium. In this context, $PGS_{5.0}C_{0.5}$ was selected as the most promising curcumin-loaded preparation to test anticancer activity. Our results evidence the importance of particle size since bigger PGS-NPs ensured better stability of the whole system, especially when encapsulating curcumin.

2.4. PGS-NPs Morphological Characterization

The selected $PGS_{5.0}C_{0.5}$ and $PGS_{5.0}$Blank samples were characterized by transmission electron microscopy, confirming their spherical shape (Figure 5). The dimensions of $PGS_{5.0}C_{0.5}$ and $PGS_{5.0}$Blank estimated by TEM analysis indicated an average size of 121 ± 11 nm and 124 ± 13 nm, respectively. Moreover, TEM analysis allowed confirmation of monomodal size distribution for both developed NPs. A few large aggregates were present in the micrographs as black spots of higher size compared to most of the obtained PGS-NPs, possibly formed during the particle deposition on formvar grids.

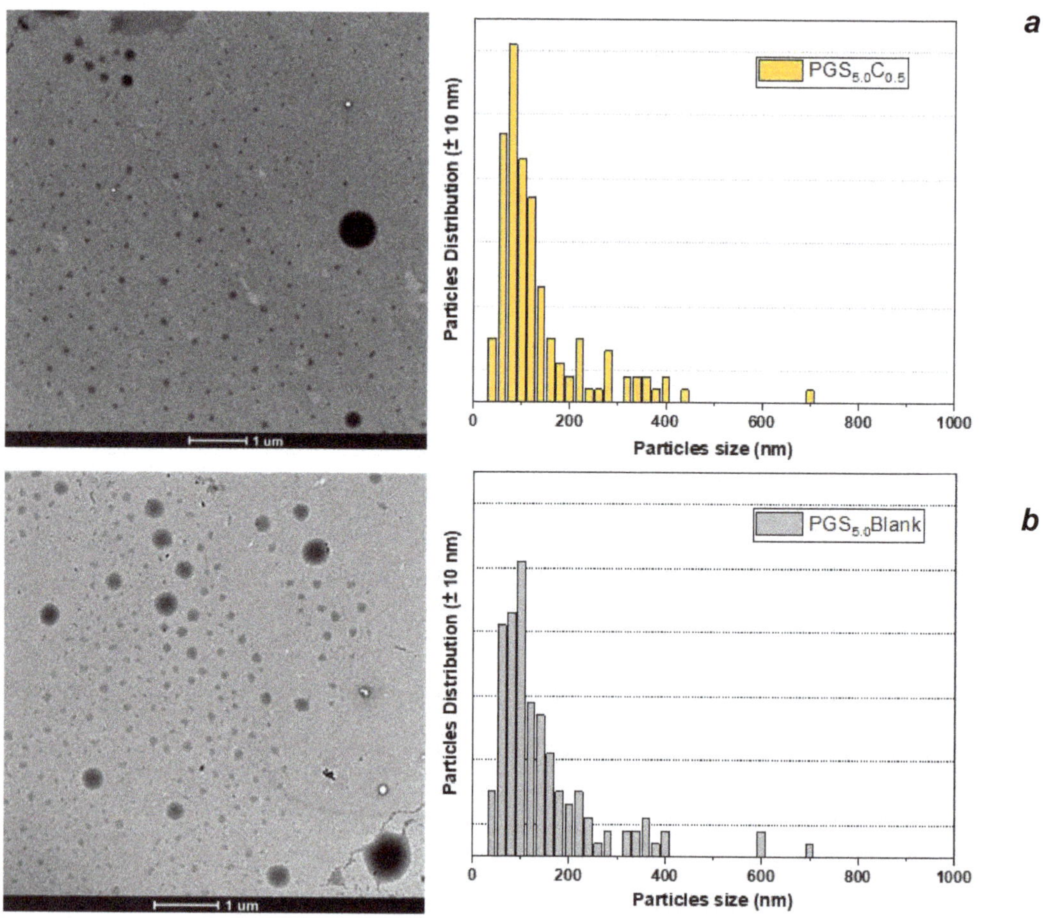

Figure 5. $PGS_{5.0}C_{0.5}$ (**a**) and $PGS_{5.0}$Blank (**b**) TEM images and relevant size distribution graphs calculated from TEM images.

2.5. Curcumin Release in Cell Culture Medium

The experimental solubility of curcumin in water at room temperature, at 37 °C and in DMEM at 37 °C were determined by UPLC as 2×10^{-4} mg/mL, 7×10^{-4} mg/mL and 1×10^{-3} mg/mL, respectively. Then, curcumin release from developed nanostructures was investigated. Figure 6 shows the release profile of curcumin from $PGS_{5.0}C_{0.5}$ nanoparticles through the dialysis membrane and curcumin solubility limit in medium. After 24 h, 4.4% of the total loaded curcumin was released from $PGS_{5.0}C_{0.5}$ in DMEM at 37 °C, reaching the experimental curcumin solubility limit in the medium (black dotted line in Figure 6). On the other hand, due to the lower solubility of curcumin in water, the plateaus corresponding to 3.2% and 0.8% of release were reached after 60 h in water at 37 °C and room temperature, respectively. The low release of curcumin from $PGS_{5.0}C_{0.5}$ was ascribable to the low solubility of curcumin in aqueous solvents.

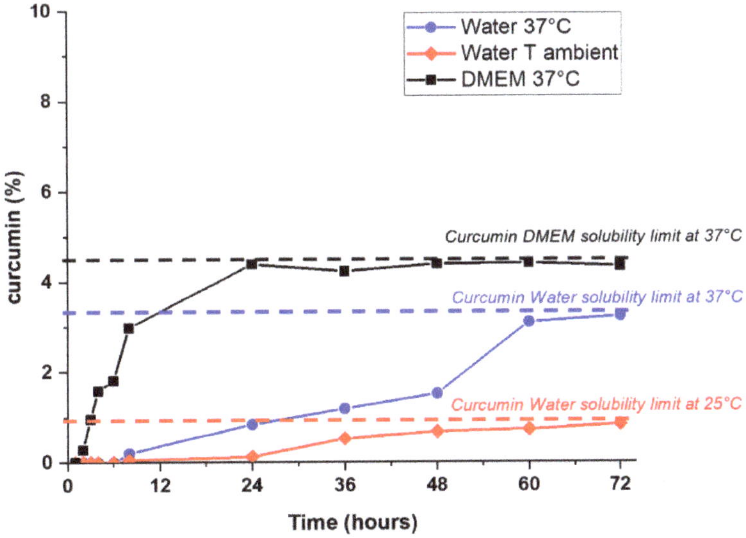

Figure 6. Release kinetics of curcumin from $PGS_{5.0}C_{0.5}$ NPs in DMEM at 37 °C (blank), water at 37 °C (blue), and water at room temperature (red).

2.6. Cytotoxic Effects of Unloaded and Curcumin-Loaded PGS-NPs in HeLa Cervical Cancer Cells

HeLa cells were treated with different concentrations of free curcumin dissolved in DMSO and of unloaded ($PGS_{5.0}$Blank) and curcumin-loaded PGS-NPs ($PGS_{5.0}C_{0.5}$) for 24, 48 and 72 h (Figure 7). No significant decrease in cell viability was observed when HeLa cells were treated with $PGS_{5.0}$Blank, indicating that PGS-NPs are not cytotoxic even at higher concentrations (Figure 7a). Instead, $PGS_{5.0}C_{0.5}$ showed dose-dependent cytotoxicity similar to that obtained with free curcumin dissolved in DMSO (Figure 7b), thus suggesting that the water solubility of curcumin-loaded NPs is similar to that of free curcumin. To better compare the cytotoxic efficacy of curcumin-loaded PG-NPs ($PGS_{5.0}C_{0.5}$) vs. free curcumin, the 50% inhibitory concentrations (IC_{50}) were calculated. Interestingly, the IC_{50} value of curcumin-loaded PGS-NPs at 72 h (15.95 μM) was significantly lower than that of free curcumin (21.27 μM, Figure 7c,d), suggesting a higher cytotoxic effect of curcumin-loaded PGS-NPs compared to free curcumin. No significant reduction in cell viability was observed when NIH-3T3 healthy fibroblast cells were treated with $PGS_{5.0}$Blank and different concentrations of $PGS_{5.0}C_{0.5}$, suggesting that unloaded and curcumin-loaded PGS-NPs have no cytotoxic effect on non-cancerous cells (Figure S5).

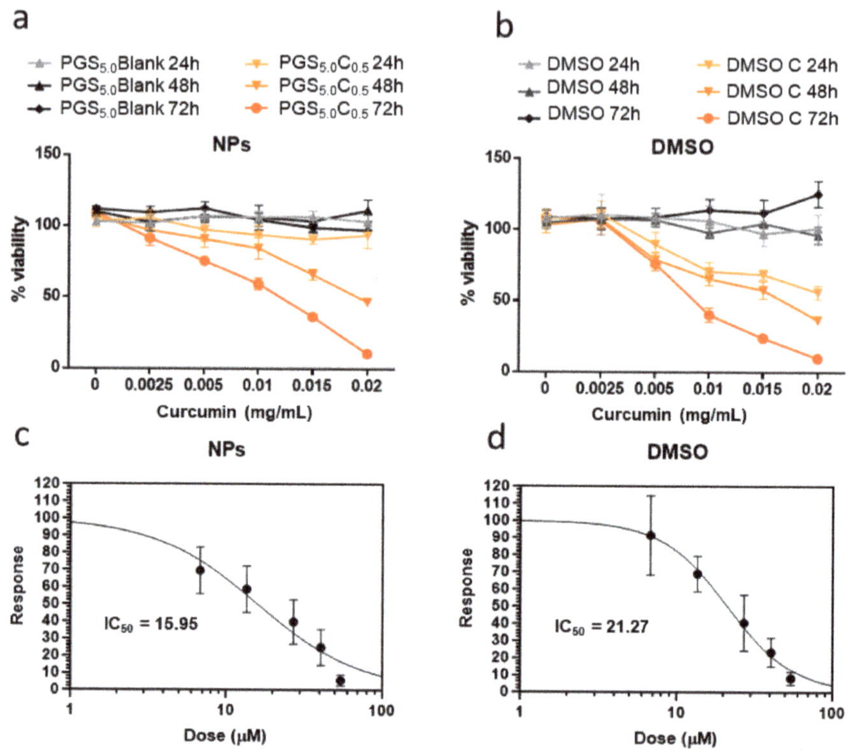

Figure 7. Cytotoxicity assay in HeLa cervical cancer cells with different concentrations of $PGS_{5.0}$Blank and $PGS_{5.0}C_{0.5}$ dissolved in sterile dH_2O (**a**) and of free curcumin dissolved in DMSO (DMSO-C, **b**). Values are means ± SEM (n = 12). IC_{50} values calculated are based on cell viability of HeLa cells at 72 h after treatment with $PGS_{5.0}C_{0.5}$ (**c**) and free curcumin dissolved in DMSO (**d**).

2.7. Effect of Curcumin-Loaded PGS-NPs in Inducing Apoptosis of HeLa Cells

The induction of apoptosis in HeLa cells treated with 0.005 and 0.01 mg/mL of $PGS_{5.0}C_{0.5}$ for 24, 48 and 72 h was then investigated. $PGS_{5.0}C_{0.5}$ provoked a dose-dependent up-regulation of genes involved in apoptosis (p53 and Bax, Figure 8a,b) and cell cycle arrest (p21, Figure 8c) compared to non-treated cells. No up-regulation was observed when HeLa cells were treated with $PGS_{5.0}$Blank (data not shown). Western blot analysis confirmed that $PGS_{5.0}C_{0.5}$ increased cleaved caspase-3 and PARP levels and induced apoptosis (Figure 8e). A significant reduction of the transcript levels of the viral HPV E6 oncogene was also observed, suggesting an inhibitory effect of $PGS_{5.0}C_{0.5}$ on its expression (Figure 8d). Overall, these findings suggest that $PGS_{5.0}C_{0.5}$ NPs are capable of inducing apoptosis by activating cell cycle arrest and apoptosis in HeLa cervical cancer cells.

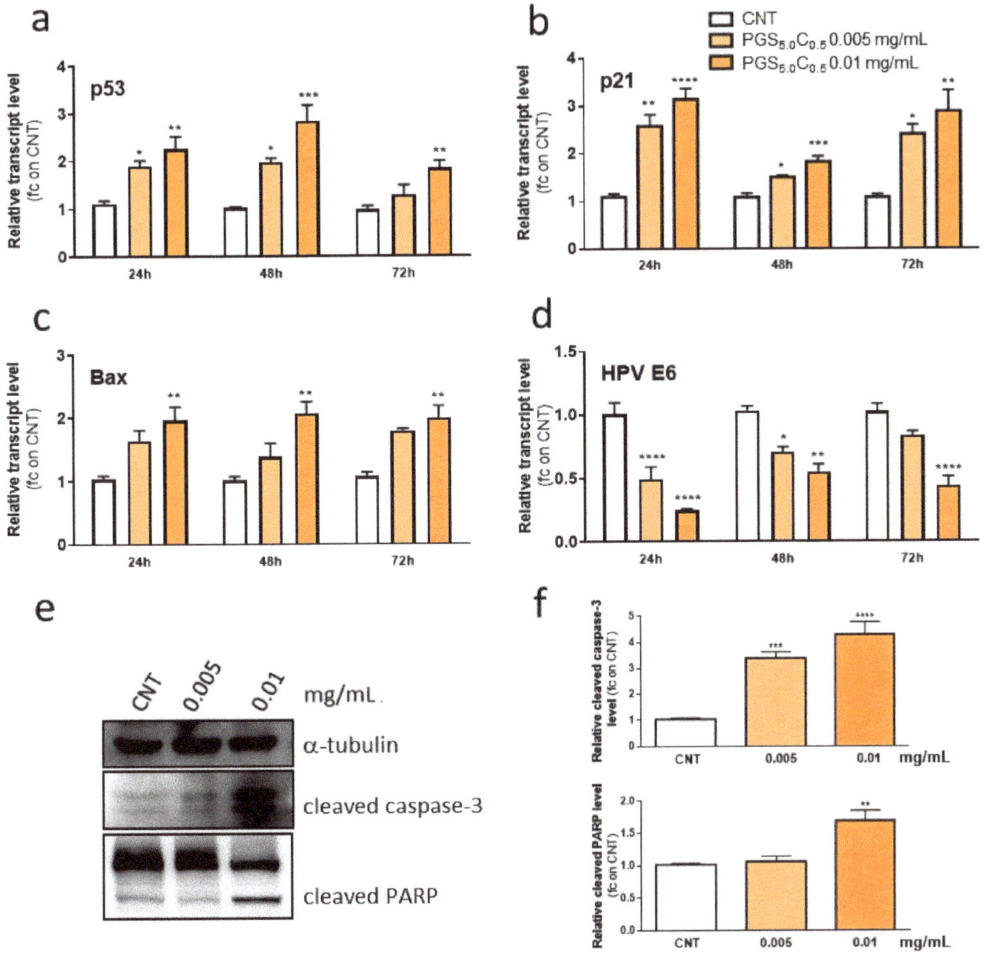

Figure 8. Effect of curcumin-loaded PGS-NPs on the induction of apoptosis. Real-time RT-PCR analysis of transcript levels of p53 (**a**), p21 (**b**), Bax (**c**) and HPV E6 genes (**d**) in HeLa cells treated with 0.005 or 0.01 mg/mL of $PGS_{5.0}C_{0.5}$ for 24, 48 and 72 h. Each transcript level is expressed as fold change (fc) with respect to control samples treated with vehicle (CNT). Values are means ± SEM (n = 6) (**e**) Western blot panel and (**f**) relative quantification of cleaved caspase-3 and PARP levels in HeLa cells treated with 0.005 or 0.01 mg/mL of $PGS_{5.0}C_{0.5}$ for 72 h. Data were analyzed with ImageJ and normalized to α-tubulin and expressed as fold change over CNT. Values are means ± SEM (n = 3). * $p < 0.05$, ** $p < 0.01$, *** $p < 0.001$, **** $p < 0.0001$ indicate significant differences (two-way ANOVA followed by Tukey's multiple comparisons test) versus CNT. CNT (blank): Control; $PGS_{5.0}C_{0.5}$ (light orange) [PGS] 0.005 mg/mL [Curcumin] 0.0005 mg/mL; $PGS_{5.0}C_{0.5}$ (dark orange) [PGS] 0.01 mg/mL [Curcumin] 0.001 mg/mL).

3. Materials and Methods

3.1. Materials

Glycerol and sebacic acid were purchased from Sigma-Aldrich Chemicals (Milan, Italy). Acetone was purchased from Carlo Erba (Milan, Italy). Curcumin extract (standardized as 95% in curcuminoids composed of: curcumin 85%, demethoxycurcumin 14% and bis-demethoxycurcumin 1%, (chromatograms relevant to curcuminoids UPLC analyses are reported in the supplementary information) was kindly donated by Indena S.p.A. DMEM cell culture medium supplemented with 10% (*v*/*v*) fetal bovine serum was purchased from Sigma-Aldrich, (St. Louis, MO, USA), penicillin and streptomycin from Gibco/BRL (Carlsbad, CA, USA) and L-glutamine from Life Technologies (Carlsbad, CA, USA). HPLC grade acetonitrile, water, and formic acid were purchased from Sigma-Aldrich (Milan, Italy).

3.2. Polymer Preparation

3.2.1. PGS Synthesis

Glycerol (0.116 mol, 10.69 g) and sebacic acid (0.145 mol, 29.36 g) were placed in a 250 mL three-necked round bottom flask and placed under nitrogen flow. The reaction was carried out in a closed oven at 170 °C for 6 h, providing mechanical stirring (40 rpm) and then the polymer was left cooling overnight under nitrogen flow.

3.2.2. PGS Characterization

The molecular weight of synthesized polymers was evaluated using a size exclusion chromatography (SEC) system having a Waters 1515 isocratic high-performance liquid chromatography (HPLC) pump and a four Waters Styragel column set (HR3-HR4-HR5-HR2) with a UV detector Waters 2487 Dual λ Absorbance Detector set at 230 nm, using a flow rate of 1 mL/min and 60 μL as the injection volume. The samples were prepared by dissolving 50 mg of polymer in 1 mL of anhydrous CH_2Cl_2 and filtering the solution through 0.45 μm filters. Given the relatively high loading, a check was performed using a lower concentration of polymer (5 mg/mL) to verify that no column overloading had occurred. Higher loadings were preferred as the UV signal of PGS was relatively weak. Molecular weight data are expressed in polystyrene (PS) equivalents. The calibration was built using 16 monodispersed PS standards, having a peak molecular weight ranging from 1,600,000 Da to 106 g/mol (i.e., ethylbenzene). For all analyses, 1,2-dichlorobenzene was used as an internal reference.

Molecular weight data of the synthesized PGS were detected as follows: $(Mn)^{-}$ = 3100 Da; $(Mw)^{-}$ = 12,000 Da; D = 3.98. All data are expressed as polystyrene equivalents.

Differential scanning calorimetry (DSC) analyses were conducted using a Mettler Toledo DSC1 on samples weighing from 5 to 10 mg each. Melting and crystallization temperatures were measured using the following temperature cycles: (1) heating from −50 °C to 150 °C at 10 °C/min; (2) cooling from 150 to −50 °C at 10 °C/min; (3) heating from −50 to 150 °C at 10 °C/min. The first two cycles were run to erase the thermal history of the samples. Glass transition temperature (Tg) and melting temperature (Tm) were determined during the second heating scan.

Thermal transition data of the synthesized PGS were detected as follows: Tg = −11.3 °C; Tm = 8.1 °C.

Fourier transform infrared spectroscopy (FT-IR) Using a FT-IR Spectrometer (Spectrum 100, PerkinElmer) with an attenuated total reflection (ATR) was used to register spectra for PGS samples. FT-IR spectrum of PGS is reported in the supporting information file.

^{1}H and ^{13}C NMR spectra for PGS sample were recorded using a Bruker Ultrashield 400 MHz. The chemical shifts are reported in ppm and referred to TMS as internal standards. All samples were prepared by dissolving 6–8 mg of polymer into 1 mL of DMSO d_6. Spectra are shown in the supporting information file.

3.3. PGS-NPs Preparation

3.3.1. Formulation Method

Curcumin-loaded PGS-NPs were prepared by nanoprecipitation according to a general procedure [24]. Briefly, in a glass vial equipped with a magnetic stirrer, selected amounts of PGS and curcumin powder (curcumin/PGS ratio equal to 10% by weight) were dissolved in 4 mL ethanol (organic phase, Table 1). The resulting mixture was stirred at room temperature until complete dissolution was reached. The organic phase was dropped by means of a 22G needle syringe, kept in vertical position and without piston, in 10 mL of deionized water under moderate stirring. The as-formed NPs were kept under nitrogen flow for 3 h to achieve complete ethanol evaporation, assessed by a 4 mL volume decrease. For unloaded PGS-NPs (reported as Blank in Table 3) the same procedure was repeated in the absence of curcumin.

Table 3. Tested formulation parameters for the preparation of PGS-based NPs by nanoprecipitation.

	Organic Phase Composition	
Sample	**PGS**	**Curcumin**
$PGS_{0.1}Blank$	1 mg [0.1 mg/mL]	\
$PGS_{0.1}C_{0.01}$	1 mg [0.1 mg/mL]	0.1 mg [0.01 mg/mL]
$PGS_{0.5}Blank$	5 mg [0.5 mg/mL]	\
$PGS_{0.5}C_{0.05}$	5 mg [0.5 mg/mL]	0.5 mg [0.05 mg/mL]
$PGS_{1.0}Blank$	10 mg [1 mg/mL]	\
$PGS_{1.0}C_{0.1}$	10 mg [1 mg/mL]	1 mg [0.1 mg/mL]
$PGS_{5.0}Blank$	50 mg [5 mg/mL]	\
$PGS_{5.0}C_{0.5}$	50 mg [5 mg/mL]	5 mg [0.5 mg/mL]

3.3.2. Procedure Optimization

In order to evaluate the effect of the different formulation parameters on the drug entrapment and release kinetics of curcumin, various formulation parameters were investigated (polymer concentration and purification method). Data relevant to individual experiments are summarized in Table 3.

3.4. Curcumin Encapsulation Efficiency

Encapsulation efficiency (the percentage of drug successfully entrapped into the nanoparticle) was estimated by using Equation (1):

$$Encapsulation\ efficiency = \frac{mg\ of\ encapsulated\ Curcumin}{mg\ of\ total\ Curcumin} \times 100 \quad (1)$$

To determine the encapsulation efficiency of PG-NPs, suspensions were firstly ultracentrifuged (VWR Microstar 17 centrifuge, Germany) at 16,000 rpm and 24 °C for 60 min from the aqueous medium containing a known amount of free curcumin. The supernatant was separated, filtered (0.2 µm nylon syringe filters) and injected in a UPLC (ultra-high-performance liquid chromatography) system for released curcumin quantification (chromatographic method described below).

3.5. PGS-NPs Morphological Characterization

Most promising unloaded and loaded PGS-NPs were characterized through transmission electron microscopy (TEM). Samples for TEM imaging were prepared by dropping PGS-NPs suspensions on single-side-polished copper grids coated with a formvar film and dried at room temperature. Samples were observed using a Talos L120C transmission electron microscope (Thermo Fisher Scientific, Milan, Italy) equipped with a 4K digital camera Ceta CMOS (Thermo Fisher Scientific). Obtained images were elaborated using Nanoscope software and the particle size distribution was calculated. A minimum of 150 nanoparticles were selected in each acquired TEM image.

Dynamic light scattering (DLS) analysis was carried out by using a Nanozetasizer (Malvern Instruments) equipped with a 4.0 mW He–Ne laser operating at 633 nm and an avalanche photodiode detector in order to determine the average size of the obtained PGS-NPs. All the measurements were repeated at least 3 times. Colloidal stability of PGS-NPs at different concentrations (0.2 and 0.1 mg/mL) was investigated by means of DLS at 37 °C. PGS-NP size was monitored at different times from sample preparation (time 0, 24, 48, 72 h). The same experiment was carried out using the Dulbecco's Modified Eagle

Medium (DMEM, Euroclone, Milano, Italy) supplemented as described below and stored at 37 °C in a humidified 5% CO_2 atmosphere.

3.6. Curcumin In Vitro Release Studies

Release kinetic studies were performed on $PGS_{5.0}C_{0.5}$ nanoparticles by using a dialysis tubing cellulose membrane (molecular weight cut-off = 14,000 Da). Curcumin-loaded NPs were diluted in an appropriate volume of water or DMEM cell culture medium and then placed into dialysis tubes. 5 mL of sample was dialyzed against 20 mL of water or DMEM at 37 °C. At specific time intervals (1, 2, 3, 6, 24, 36, 48, 60 and 72 h), 1 mL of the medium was analyzed by means of UPLC (chromatographic method described below) to determine the amount of released curcumin. Consequently, the same volume of fresh medium was replaced. All the measurements were performed in triplicate. The same experiment was performed at 4 °C and at room temperature to determine the influence of temperature during the curcumin release. The in vitro release profile was measured by using Equation (2):

$$Release_{curcumin}\ (\%) = \frac{C_{Released\ (t)}}{C_{Total}} \times 100 \tag{2}$$

where $C_{released}$ represents the concentration (mg/mL) of released curcumin at time point (t), and C_{Total} (mg/mL) corresponds to the total amount of curcumin loaded into PGS-NPs.

3.7. UPLC Analysis

Curcumin was quantified through a Waters ACQUITY UPLC system equipped with a quaternary solvent manager system, autosampler, thermostated column compartment and a PDA detector. The analytical separation was performed using an ACQUITY UPLC® (Waters corp., Milford, MA, USA) BEH C18 column (1.7 µm × 2.1 mm × 50 mm). Curcumin release was analyzed using a mobile phase composed of water (+0.1% of formic acid) (A) and acetonitrile (+0.1% of formic acid) (B). The flow rate was set at 0.25 mL min^{-1} and the linear gradient elution was: 0 min, 70% A; 10 min, 30% A; 11 min, 70% A; with a re-equilibration time of 3 min before the next injection. The column temperature was maintained at 34 °C and the wavelength set at 425 nm, corresponding to the maximum absorption of curcumin. Before samples injections, five dilutions of a curcumin methanolic solution (0.2 mg/mL) were prepared in the range 0.001–0.2 mg/mL. Standard solutions were filtered (0.2 µm nylon filters) and injected three times into the UPLC system. Curcumin was eluted as a sharp peak at retention time of 5.5 min. In the operative concentration range the trend was linear for each compound, with no saturation effects that could bend the linearity. The area under each peak was quantified by instrumental software and plotted versus the concentration. The best fit of experimental data in the plot "Peak area vs. [curcumin]" was then used for released curcumin quantification in each sample. Chromatograms relevant to curcuminoids UPLC analysis are reported in the supplementary information.

3.8. Cell Culture Assays—Cytotoxicity Study

3.8.1. Cell Culture and MTT Assay

Human cervical cancer $HPV18^+$ HeLa cell lines and the healthy fibroblast NIH-3T3 cell line were cultured at 37 °C in a humidified atmosphere containing 5% CO_2 in DMEM supplemented with 10% (v/v) fetal bovine serum (Sigma-Aldrich, St. Louis, MO, USA), 100 Units/mL penicillin, 100 µg/mL streptomycin (Gibco/BRL, Carlsbad, CA, USA) and 2 mM L-glutamine (Life Technologies, Carlsbad, CA, USA). The inhibitory effect of unloaded and curcumin-loaded PGS-NPs on cervical cancer HeLa cells and NIH-3T3 fibroblasts was analyzed by cell viability assay using 3-(4,5-dimethylthiazol-2-yl)-2,5-diphenyltetrazolium bromide (MTT) test as previously described [37]. Briefly, HeLa cells were seeded in a 96-well plate (about 25,000 cells per well for HeLa and 5000 cells per well for NIH-3T3), incubated for attachment for 24 h at 37 °C and then treated with free curcumin dissolved in DMSO and with unloaded and curcumin-loaded PGS-NPs in 12 replicates at different concentrations (0, 0.0025, 0.005, 0.01, 0.015, 0.02 mg/mL of curcumin). At different time points (24, 48 and 72 h) each well was supplemented with 10% (*v*/*v*) MTT solution and incubated for 4 h. Formazan crystals were dissolved in acidified isopropanol (isopropanol, 0.1 N HCl, 0.1% Tween-20). The optical density was measured at 570 nm by using a microplate reader (Tecan Infinite F200PRO). The 50% inhibitory concentration (IC_{50}) of curcumin-loaded PGS-NPs and free curcumin on HeLa cells was calculated at 72 h using Graph Pad Prism software.

3.8.2. RNA Extraction and Real-Time RT-PCR Analysis

Total RNA was isolated from cells (1×10^6 cells per well in 6-well plate) treated with unloaded and curcumin-loaded PGS-NPs (0.005 and 0.01 mg/mL) using Direct-Zol™ RNA MiniPrep kit (Zymo Research, Irvine, CA, USA). About 1 µg of RNA was reverse-transcribed with the iScript cDNA Synthesis Kit (Bio-Rad, Hercules, CA, USA) and real-time RT-PCR was performed with the QuantiNova SYBR Green kit (Qiagen, Hilden, Germany) for gene expression in a CFX96 Real-time PCR detection system (Bio-Rad, Hercules, CA, USA). Transcript levels were normalized against the *GAPDH* gene and values were expressed as fold change over control sample treated with vehicle (CNT). Primer sequences are reported in Table 4. Data are presented as mean ± SEM.

Table 4. Primer sequences used for real-time RT-PCR.

Primer	Primer Sequence 5′–3′
hs-HPV18-E6 fw	GTGCCAGAAACCGTTGAATCC
hs-HPV18-E6 rv	AGTCTTTCCTGTCGTGCTCG
hs-p53 fw	GAGGGATGTTTGGGAGATGTAA
hs-p53 rv	CCCTGGTTAGTACGGTGAAGTG
hs-p21 fw	GTCACTGTCTTGTACCCTTGTG
hs-p21 rv	AGAAATCTGTCATGCTGGTCTGC
hs-Bax fw	CATGGGCTGGACATTGGACTT
hs-Bax rv	AGGGACATCAGTCGCTTCAGT
hs-Gapdh fw	GCCTCAAGATCATCAGCAATGC
hs-Gapdh rv	CCACGATACCAAAGTTGTCATGG

fw, forward; rv, reverse.

3.8.3. Western Blotting

HeLa cells (1×10^6 cells per well in 6-well plate) treated with curcumin-loaded NPs (0.005 and 0.01 mg/mL of curcumin-loaded NPs) for 72 h were washed with PBS and then lysed in a buffer containing 50 mM Tris–HCl, pH 7.2, 0.1% sodium deoxycholate, 1% Triton X-100, 5 mM EDTA, 5 mM EGTA, 150 mM NaCl, 40 mM NaF, 2.175 mM $NaVO_4$, 0.1% SDS, 0.1% aprotinin and 1 mM PMSF. Forty micrograms of protein underwent SDS–PAGE following transfer on PVDF membranes (Amersham™ Hybond® P, Dasser Germany). Bands were detected using Immobilon ECL Western Blotting Substrate (ThermoFisher Scientific, Waltham, MA, USA) on a Chemidoc digital imaging machine (Bio-Rad). An anti-Cleaved Caspase-3 primary antibody (1:1000 in 1% BSA, Cell Signaling Technology, Inc. MA, USA) and an anti-PARP-1 primary antibody (1:1000 in 1% BSA, Santa Cruz Biotechnology, Dallas, TX, USA), were used both followed by a goat anti-rabbit horseradish peroxidase-conjugated secondary antibody (1:10,000; Abcam Cambridge CB2 0AX UK). Internal loading control was performed using the anti-α-Tubulin mouse primary antibody (1:10,000 in 1% BSA; Sigma-Aldrich) followed by a goat anti-mouse horseradish peroxidase-conjugated secondary antibody (1:10,000, Sigma-Aldrich). Densitometry quantification was performed with ImageJ Software (NIH) and expressed as ratio of cleaved caspase-3 and PARP to α-tubulin.

3.8.4. Statistics

The statistical significance of differences in cell assays, gene and protein expression analyses were determined by one-way ANOVA followed by Tukey's post hoc tests for multiple comparisons. The statistical analysis was carried out using the software GraphPad Prism 6 (GraphPad, San Diego, CA, USA). A value of $p < 0.05$ was considered significant.

4. Conclusions

Poly(glycerol sebacate) was, for the first time, exploited for the encapsulation of curcumin, with the aim to design a nanosized anticancer system characterized by high stability even under physiological conditions and strong biological activity through an easy, scalable and reproducible technique. Different experimental conditions have been tested, aimed to define the most promising formulations in terms of particle stability and size. Indeed, the curcumin-loaded sample $PGS_{5.0}C_{0.5}$ and the respective unloaded formulation $PGS_{5.0}$Blank, were selected in virtue of their stability even when diluted in simulated cells medium. Stability and morphological analysis were performed through DLS and TEM analysis which allowed confirmation of their nanosized dimension of about 150 nm and spherical shape. Finally, curcumin-loaded PGS-NPs displayed a higher cytotoxic effect compared to free curcumin, thanks to the higher solubility in physiological conditions. Consistent with this, the developed curcumin-loaded PGS-NPs were capable of inducing apoptosis by promoting the expression of *p53*, its target gene, *p21*, involved in cell cycle arrest and the pro-apoptotic *Bax* gene. Finally, a reduction of *HPV E6* expression was observed, suggesting that curcumin-loaded PGS-NPs can exert a significant anti-HPV activity, thus possibly preventing pre-cancer lesions. Overall, our results show that curcumin-loaded PGS-NPs may represent a possible adjuvant therapy for delaying/treating cervical cancer cells.

Supplementary Materials: The following supporting information can be downloaded at: https://www.mdpi.com/article/10.3390/molecules27206997/s1.

Author Contributions: Conceptualization, L.V. and K.P.; investigation, A.M. (Alessio Massironi), D.M., A.M. (Alessandra Marinelli), M.T., S.G.; writing—original draft preparation, A.M. (Alessio Massironi), A.M. (Alessandra Marinelli), M.T., S.G.; writing—review and editing, S.M., K.P., L.V., M.A.O.; supervision, L.V, S.M., K.P., M.A.O.; funding acquisition: K.P., D.M., L.V., S.M., M.A.O. All authors have read and agreed to the published version of the manuscript.

Funding: This research was partially funded by the Università degli Studi di Milano (Linea 3—Piano di Sostegno alla Ricerca 2020 Seal of Excellence project "To-PoST" grant RV_PSR_SOE_2020_GDICA to L.V, S.M, M.A.O.; the Linea 2—Piano di Sostegno alla Ricerca 2020 Seal of Excellence grant PSR2020_DIP_005_PI_ACOLO to D.M.), and by the Fondazione Umberto Veronesi (Grant 2014 to K.P.; Grant Fellowship to A.M2).

Institutional Review Board Statement: Not applicable.

Informed Consent Statement: Not applicable.

Data Availability Statement: Not applicable.

Acknowledgments: The authors thank Nadia Santo for TEM analyses.

Conflicts of Interest: The authors declare no conflict of interest.

Sample Availability: Samples of the compounds are not available from the authors.

References

1. Cancer: Data and Statistics. Available online: https://www.euro.who.int/en/health-topics/noncommunicable-diseases/cancer/data-and-statistics (accessed on 1 September 2022).
2. Cai, S.; Yang, Q.; Bagby, T.R.; Forrest, M.L. Lymphatic drug delivery using engineered liposomes and solid lipid nanoparticles. *Adv. Drug Deliv. Rev.* **2011**, *63*, 901–908. [CrossRef] [PubMed]
3. Zhao, K.; Li, D.; Shi, C.; Ma, X.; Rong, G.; Kang, H.; Wang, X.; Sun, B. Biodegradable Polymeric Nanoparticles as the Delivery Carrier for Drug. *Curr. Drug Deliv.* **2016**, *13*, 494–499. [CrossRef] [PubMed]
4. Masserini, M. Nanoparticles for Brain Drug Delivery. *ISRN Biochem.* **2013**, *2013*, 238428. [CrossRef] [PubMed]
5. Owens, D.E.; Peppas, N.A. Opsonization, biodistribution, and pharmacokinetics of polymeric nanoparticles. *Int. J. Pharm.* **2006**, *307*, 93–102. [CrossRef]

6. Stylianopoulos, T. EPR-effect: Utilizing size-dependent nanoparticle delivery to solid tumors. *Ther. Deliv.* **2013**, *4*, 421–423. [CrossRef]
7. Cho, K.; Wang, X.; Nie, S.; Chen, Z.; Shin, D.M. Therapeutic nanoparticles for drug delivery in cancer. *Clin. Cancer Res.* **2008**, *14*, 1310–1316. [CrossRef]
8. Maeda, H.; Nakamura, H.; Fang, J. The EPR effect for macromolecular drug delivery to solid tumors: Improvement of tumor uptake, lowering of systemic toxicity, and distinct tumor imaging in vivo. *Adv. Drug Deliv. Rev.* **2013**, *65*, 71–79. [CrossRef]
9. Chiellini, E.; Covolan, V.L.; Orsini, L.M.; Solaro, R. Polymeric nanoparticles based on polylactide and related copolymers. In *Macromolecular Symposia*; Wiley: Weinheim, Germany, 2003; Volume 197, pp. 345–354.
10. Vogt, L.; Ruther, F.; Salehi, S.; Boccaccini, A.R. Poly(Glycerol Sebacate) in Biomedical Applications—A Review of the Recent Literature. *Adv. Healthc. Mater.* **2021**, *10*, 2002026. [CrossRef]
11. Wu, Z.; Jin, K.; Wang, L.; Fan, Y. A Review: Optimization for Poly(glycerol sebacate) and Fabrication Techniques for Its Centered Scaffolds. *Macromol. Biosci.* **2021**, *21*, 2100022. [CrossRef]
12. Loh, X.J.; Abdul Karim, A.; Owh, C. Poly(glycerol sebacate) biomaterial: Synthesis and biomedical applications. *J. Mater. Chem. B* **2015**, *3*, 7641–7652. [CrossRef]
13. Sha, D.; Wu, Z.; Zhang, J.; Ma, Y.; Yang, Z.; Yuan, Y. Development of modified and multifunctional poly(glycerol sebacate) (PGS)-based biomaterials for biomedical applications. *Eur. Polym. J.* **2021**, *161*, 110830. [CrossRef]
14. Louage, B.; Tack, L.; Wang, Y.; De Geest, B.G. Poly(glycerol sebacate) nanoparticles for encapsulation of hydrophobic anti-cancer drugs. *Polym. Chem.* **2017**, *8*, 5033–5038. [CrossRef]
15. Sivanesan, D.; Verma, R.S.; Prasad, E. 5FU encapsulated polyglycerol sebacate nanoparticles as anti-cancer drug carriers. *RSC Adv.* **2021**, *11*, 18984–18993. [CrossRef] [PubMed]
16. Hawley, A.E.; Davis, S.S.; Illum, L. Targeting of colloids to lymph nodes: Influence of lymphatic physiology and colloidal characteristics. *Adv. Drug Deliv. Rev.* **1995**, *17*, 129–148. [CrossRef]
17. Tamvakopoulos, C.; Dimas, K.; Sofianos, Z.D.; Hatziantoniou, S.; Han, Z.; Liu, Z.L.; Wyche, J.H.; Pantazis, P. Metabolism and anticancer activity of the curcumin analogue, dimethoxycurcumin. *Clin. Cancer Res.* **2007**, *13*, 1269–1277. [CrossRef]
18. Subramani, P.A.; Panati, K.; Narala, V.R. Curcumin Nanotechnologies and Its Anticancer Activity. *Nutr. Cancer* **2017**, *69*, 381–393. [CrossRef]
19. Allegra, A.; Innao, V.; Russo, S.; Gerace, D.; Alonci, A.; Musolino, C. Anticancer Activity of Curcumin and Its Analogues: Preclinical and Clinical Studies. *Cancer Investig.* **2017**, *35*, 1–22. [CrossRef]
20. Aggarwal, B.B.; Kumar, A.; Bharti, A.C. Anticancer potential of curcumin: Preclinical and clinical studies. *Anticancer Res.* **2003**, *23*, 363–398.
21. Basniwal, R.K.; Khosla, R.; Jain, N. Improving the anticancer activity of curcumin using nanocurcumin dispersion in water. *Nutr. Cancer* **2014**, *66*, 1015–1022. [CrossRef]
22. Yallapu, M.M.; Khan, S.; Maher, D.M.; Ebeling, M.C.; Sundram, V.; Chauhan, N.; Ganju, A.; Balakrishna, S.; Gupta, B.K.; Zafar, N.; et al. Anti-cancer activity of curcumin loaded nanoparticles in prostate cancer. *Biomaterials* **2014**, *35*, 8635–8648. [CrossRef]
23. D'Angelo, N.A.; Noronha, M.A.; Kurnik, I.S.; Câmara, M.C.C.; Vieira, J.M.; Abrunhosa, L.; Martins, J.T.; Alves, T.F.R.; Tundisi, L.L.; Ataide, J.A.; et al. Curcumin encapsulation in nanostructures for cancer therapy: A 10-year overview. *Int. J. Pharm.* **2021**, *604*, 120534. [CrossRef] [PubMed]
24. Hornig, S.; Heinze, T.; Becer, C.R.; Schubert, U.S. Synthetic polymeric nanoparticles by nanoprecipitation. *J. Mater. Chem.* **2009**, *19*, 3838–3840. [CrossRef]
25. Errico, C.; Goñi-De-Cerio, F.; Alderighi, M.; Ferri, M.; Suarez-Merino, B.; Soroka, Y.; Frŭié-Zlotkin, M.; Chiellini, F. Retinyl palmitate-loaded poly(lactide-co-glycolide) nanoparticles for the topical treatment of skin diseases. *J. Bioact. Compat. Polym.* **2012**, *27*, 604–620. [CrossRef]
26. Sung, H.; Ferlay, J.; Siegel, R.L.; Laversanne, M.; Soerjomataram, I.; Jemal, A.; Bray, F. Global Cancer Statistics 2020: Globocan Estimates of Incidence and Mortality Worldwide for 36 Cancers in 185 Countries. *CA Cancer J. Clin.* **2021**, *71*, 209–249. [CrossRef] [PubMed]
27. Burmeister, C.A.; Khan, S.F.; Schäfer, G.; Mbatani, N.; Adams, T.; Moodley, J.; Prince, S. Cervical cancer therapies: Current challenges and future perspectives. *Tumour Virus Res.* **2022**, *13*, 200238. [CrossRef]
28. Maleki Dana, P.; Sadoughi, F.; Asemi, Z.; Yousefi, B. The role of polyphenols in overcoming cancer drug resistance: A comprehensive review. *Cell. Mol. Biol. Lett.* **2022**, *27*, 1–26. [CrossRef]
29. Wang, Y.; Ameer, G.A.; Sheppard, B.J.; Langer, R. A tough biodegradable elastomer. *Nat. Biotechnol.* **2002**, *20*, 602–606. [CrossRef]
30. Chen, Q.Z.; Bismarck, A.; Hansen, U.; Junaid, S.; Tran, M.Q.; Harding, S.E.; Ali, N.N.; Boccaccini, A.R. Characterisation of a soft elastomer poly(glycerol sebacate) designed to match the mechanical properties of myocardial tissue. *Biomaterials* **2008**, *29*, 47–57. [CrossRef]
31. Rai, R.; Tallawi, M.; Grigore, A.; Boccaccini, A.R. Synthesis, properties and biomedical applications of poly(glycerol sebacate) (PGS): A review. *Prog. Polym. Sci.* **2012**, *37*, 1051–1078. [CrossRef]
32. Rafiee, Z.; Nejatian, M.; Daeihamed, M.; Jafari, S.M. Application of different nanocarriers for encapsulation of curcumin. *Crit. Rev. Food Sci. Nutr.* **2019**, *59*, 3468–3497. [CrossRef]
33. Mukerjee, A.; Vishwanatha, J.K. Formulation, characterization and evaluation of curcumin-loaded PLGA nanospheres for cancer therapy. *Anticancer Res.* **2009**, *29*, 3867–3875. [PubMed]

34. Sari, T.P.; Mann, B.; Kumar, R.; Singh, R.R.B.; Sharma, R.; Bhardwaj, M.; Athira, S. Preparation and characterization of nanoemulsion encapsulating curcumin. *Food Hydrocoll.* **2015**, *43*, 540–546. [CrossRef]
35. De Moraes Carvalho, D.; Takeuchi, K.P.; Geraldine, R.M.; de Moura, C.J.; Torres, M.C.L. Production, solubility and antioxidant activity of curcumin nanosuspension. *Food Sci. Technol.* **2015**, *35*, 115–119. [CrossRef]
36. Deng, L.; Kang, X.; Liu, Y.; Feng, F.; Zhang, H. Effects of surfactants on the formation of gelatin nanofibres for controlled release of curcumin. *Food Chem.* **2017**, *231*, 70–77. [CrossRef] [PubMed]
37. Petroni, K.; Trinei, M.; Fornari, M.; Calvenzani, V.; Marinelli, A.; Micheli, L.A.; Pilu, R.; Matros, A.; Mock, H.P.; Tonelli, C.; et al. Dietary cyanidin 3-glucoside from purple corn ameliorates doxorubicin-induced cardiotoxicity in mice. *Nutr. Metab. Cardiovasc. Dis.* **2017**, *27*, 462–469. [CrossRef] [PubMed]

molecules

Review

Selected Plant-Derived Polyphenols as Potential Therapeutic Agents for Peripheral Artery Disease: Molecular Mechanisms, Efficacy and Safety

Guglielmina Froldi * and Eugenio Ragazzi

Department of Pharmaceutical and Pharmacological Sciences, University of Padova, 35131 Padova, Italy
* Correspondence: g.froldi@unipd.it

Abstract: Vascular diseases, such as peripheral artery disease (PAD), are associated with diabetes mellitus and a higher risk of cardiovascular disease and even death. Surgical revascularization and pharmacological treatments (mainly antiplatelet, lipid-lowering drugs, and antidiabetic agents) have some effectiveness, but the response and efficacy of therapy are overly dependent on the patient's conditions. Thus, the demand for new cures exists. In this regard, new studies on natural polyphenols that act on key points involved in the pathogenesis of vascular diseases and, thus, on PAD are of great urgency. The purpose of this review is to take into account the mechanisms that lead to endothelium dysfunction, such as the glycoxidation process and the production of advanced glycation end-products (AGEs) that result in protein misfolding, and to suggest plant-derived polyphenols that could be useful in PAD. Thus, five polyphenols are considered, baicalein, curcumin, mangiferin, quercetin and resveratrol, reviewing the literature in PubMed. The key molecular mechanisms and preclinical and clinical studies of each selected compound are examined. Furthermore, the safety profiles of the polyphenols are outlined, together with the unwanted effects reported in humans, also by searching the WHO database (VigiBase).

Keywords: baicalein; curcumin; mangiferin; quercetin; resveratrol; diabetes mellitus; endothelial dysfunction; peripheral artery disease; polyphenols; cardiovascular diseases

Citation: Froldi, G.; Ragazzi, E. Selected Plant-Derived Polyphenols as Potential Therapeutic Agents for Peripheral Artery Disease: Molecular Mechanisms, Efficacy and Safety. *Molecules* **2022**, *27*, 7110. https://doi.org/10.3390/molecules27207110

Academic Editor: Nour Eddine Es-Safi

Received: 30 September 2022
Accepted: 18 October 2022
Published: 21 October 2022

1. Introduction

Non-communicable diseases, such as cardiovascular diseases, are continuously monitored and investigated for prevalence, pathophysiology, outcome and prevention [1]; however, the impact of peripheral artery disease (PAD) is not adequately considered. Atherosclerotic occlusive disease of the lower extremities affects approximately 10% of people in their 60s and 70s, and approximately 20% of individuals over the age of 80 years [2,3]. Patients with diabetes mellitus (DM), those who smoke, the elderly and those affected by cardiovascular diseases are mainly at risk of encountering PAD [2,4]. A calculator to predict the lifetime risk of PAD is available online [5].

There is convincing evidence to support the fact that the primary pathophysiology of PAD is triggered by the obstruction of atherosclerotic plaques of the lower extremities causing a consequent decrease in blood flow (ankle brachial index, ABI ≤ 0.9) [6]. Leg pain during exercise due to poor blood flow (intermittent claudication) is a major symptom of PAD; in fact, increasing blood supply to the compromised limb after surgical revascularization can improve symptoms and hemodynamic parameters in the patients [7]. Primarily, antiplatelet therapy is recommended to reduce cardiovascular complications in PAD, whereas β-adrenergic inhibitors should be used with caution, if clinically indicated [8]. Other agents used in patients with PAD are standard lipid-lowering therapies, mainly the administration of statins [8]. In general, several therapeutic agents can improve endothelial vasodilator function and insulin resistance: pioglitazone, metformin, angiotensin-converting enzyme (ACE) inhibitors and angiotensin II-receptor antagonists

(AT1-receptor blockers) [8]. Additionally, various vasoactive drugs could be used to treat symptoms, such as improving walking distances [2]. Furthermore, propionyl-L-carnitine has been suggested to alleviate PAD symptoms through a metabolic pathway, thus improving exercise performance [9–11]. In addition, gene therapy using encoding genes of various types of growth factors is at the beginning stage [12].

Although these approaches have some success, the demand for other types of remedies remains unmet, and, for this reason, a large quantity of food supplements are commercially proposed, often without scientific evidence, to decrease PAD symptoms.

1.1. Mechanisms of PAD

Diabetes mellitus, especially if not properly treated, can cause damage to different organs and tissues of the human body. Microvascular complications include retinopathy, neuropathy and nephropathy, while macrovascular complications include cardiovascular diseases, stroke and peripheral vascular disease that can lead to bruising or lesions that do not heal, gangrene and, in severe cases, limb amputation. Peripheral artery disease (PAD), also called peripheral vascular disease, is caused by the narrowing of the lumen of blood vessels that carry blood to the arms, legs, stomach and kidneys. In people affected by DM, the risk of PAD increases with age, with the duration of diabetes and in the presence of neuropathy. Other factors associated with cardiovascular diseases, such as high C-reactive protein and homocysteine levels, are also associated with an increased risk of PAD [13,14]. Peripheral arteriopathy is characterized by two types of symptoms: intermittent claudication (or intermittent pain, which may occur during exercise or walking, but resolves with rest) and pain at rest (which is caused by chronic hypoxia/ischemia in the limb, with inadequate blood flow to the affected limb). This pathology is an important risk factor for lower-extremity amputation. Unfortunately, its diagnosis is often made too late and occurs when symptoms are marked, delaying a preventive treatment [15]. Insulin resistance, commonly manifested in type 2 diabetes mellitus (T2DM), is a consequence of several risk factors represented by obesity, sedentary behavior and aging; therefore, hyperglycemia, diabetes, hypertension and dyslipidemia are concomitant diseases often observed in patients affected by PAD [10]. In patients with DM, the diagnosis of symptomatic PAD is approximately twice as high as in patients not affected by diabetes [16]. Recently, Lilja et al. showed that men with DM developed symptomatic PAD more frequently compared to women (15.5% vs. 8.9%) [17], even if not all studies have observed significant differences in the prevalence of PAD between sexes.

Several molecular pathways are involved in the pathogenesis of arterial insufficiency, such as (i) endothelial dysfunction, e.g., the presence of increased plasma levels of asymmetric dimethylarginine (ADMA) is associated with endothelial vasodilator dysfunction, causing arterial stiffness and a reduction in $NO^{\bullet}$ production [18,19]; (ii) endoplasmic reticulum (ER) dysfunction, e.g., vascular endothelial cells of hyperglycemic subjects are characterized by an altered, rough endoplasmic reticulum and protein folding [20,21]; (iii) promotion of inflammation by secretion of cytokines, such as TNF–α, IL–1, IL–6 and IL–8, and chemotactic stimulus for monocytes and macrophages [22]; (iv) mitochondrial dysfunction and increased oxidative stress-induced damage, also linked to the activation of the transcription nuclear factor–κB (NF–κB) [23,24]; and (v) interaction between advanced glycation end-products (AGEs) and their receptors, causing inflammation and endothelial dysfunction [25,26].

The differences in endothelium function between women and men have been reported in various studies. The beneficial vascular effects of estrogens are related to the modification of the functional state of the endothelium [27]. The impaired endothelium-dependent vasodilation of the coronary or peripheral vasculature is positively correlated with an increased risk of cardiovascular events and is an independent predictor of vascular morbidity and mortality.

1.2. Role of AGEs

In conditions of persistent hyperglycemia, molecular rearrangements occur between tissue proteins and glucose or other reducing carbohydrates, leading to the irreversible formation of AGEs, which significantly contribute to the development of complications associated with DM, including PAD [28]. It has also been observed that the accumulation of AGEs worsens endothelial function [29,30].

The formation of AGEs begins with the non-enzymatic glycation of free amino groups by sugars and aldehydes, which leads to a succession of rearrangements of intermediate compounds and, finally, to irreversibly bound products known as AGEs [24]. Glycation and oxidative stress are intricately linked, and both phenomena are referred to as glycoxidation [28]. Persistent hyperglycemia and oxidative stress accelerate the formation of AGEs [31]. These have mainly been detected in long-lived proteins, with post-translational AGE modifications mainly occurring at the side-chain amino groups of the lysine and arginine residues. Glycation causes the irreversible modification of the protein structure and consequent loss of functionality, leading to detrimental effects in tissues, i.e., vasculature. Glycated proteins become less prone to proteolysis as a consequence, e.g., the accumulation of glycated collagen causes the thickening of blood vessels. Recently, authors showed that AGEs can cause the apoptosis of endothelial progenitor cells via nicotinamide adenine dinucleotide phosphate (NADPH) oxidase and promote atherosclerosis [32]. Nucleotides and lipids are also possible targets of glycation, causing the DNA mutation and modification of the integrity of the cell membrane, triggering cell death. Thus, it can be believed that compounds that counteract the effect of macromolecule damage (glycation-protective effects) can slow down the progression of vascular damage in the legs and, furthermore, also in other districts, such as the brain, protecting the health of humans [33]. Recently, several natural compounds, i.e., flavonoids and more generally polyphenols, have received attention for their endothelial protective effects [26,34].

1.3. Protective Activity of Polyphenols

Natural polyphenols are a heterogeneous group of secondary metabolites produced in plants that have several functions in the plant kingdom, such as the protection against pathogens, including microorganisms, insects and fungi, and protective activity against ultraviolet radiation and harmful environmental factors; moreover, they can also act as attractants of beneficial organisms [35,36]. These secondary compounds are biosynthesized through the shikimic acid and phenylpropanoid pathways and are believed to be participative in adapting plants in a stressed situation due to environmental changes [37]. These compounds, according to their chemical structures, are divided into various subclasses, such as phenolic acids, flavonoids, tannins, coumarins, lignans, quinones, stilbenes and curcuminoids [38].

New research on plant-derived polyphenols as a guide for exploring innovative synthetic compounds can provide new knowledge to develop new drugs in PAD treatment. In terms of this point, plant-derived natural products approved for therapeutic use in the last 30 years are used for the treatment of various diseases and modulate a wide range of molecular targets. Recent products approved as drugs by the FDA are Veregen® (sinecatechins; green tea (*Camellia sinensis* L.) leaf extract), Fulyzaq® (crofelemer; extract from the red latex of *Croton lechleri* Müll. Arg.) and Grastek® (Timothy grass (*Phleum pratense* L.) pollen allergen extract). Therefore, natural products still represent an important pool for the identification of new pharmacological tools at present.

The purpose of this review is to summarize existing evidence from preclinical and clinical studies on the contribution of natural polyphenols as vasodilators and, in general, as favorable agents for blood circulation, suggesting their potential use in PAD and against an associated causative disease: diabetes. Obviously, vasodilators can be useful not only in PAD, but also in several other cardiovascular diseases, such as hypertension, angina pectoris, heart failure and others. When searching PubMed for "natural polyphenols and vasodilators" or "natural polyphenols and vasodilation" using the criteria "English language" and "full texts", a total of 273 entries were found (23 September 2022), suggesting a wide research interest in the topic (Figure S1). However, when the including criterion "clinical trial" was applied, only six articles remained, of which five referred to the administration in healthy subjects of plant extracts, i.e., grape seed extract [39], apple extract [40], cocoa powder [41], a mixture of three plants (*Kaempferia parviflora*, *Punica granatum* and *Moringa oleifera*) [42] and a beverage containing numerous fruits and vegetables [43]. Only one trial focused on the administration of a single compound, resveratrol, administered in patients after myocardial infarction [44]. Furthermore, another search for "peripheral artery disease and natural compounds" produced only 35 items in the time interval from 1962 up to July 2022 (Figure S2). From these findings, several polyphenolic compounds have been selected and studied in detail in this review. We also wanted to highlight some observations on the importance of physical exercise, as a non-pharmacological approach, in preventing and slowing the onset of circulatory problems, partly derived from a sedentary lifestyle.

2. Prevention and Protective Factors: The Risk-Reduction Approach

2.1. Moderate Physical Exercise

Cross-sectional and longitudinal trials indicate that aerobic exercise can prevent and reverse age-associated arterial stiffness [45]. Moderate exercise tends to be a stimulator of $NO^{\bullet}$ release and favors endothelial function, which in turn reduces the cardiovascular risk profile in patients affected by DM [46]. Recently, a clinical trial was conducted on 33 healthy middle-aged and older subjects (67 ± 1 years), randomly divided into two groups, of which one had aerobic exercise training (AT), while the second was a sedentary control [19]. Circulating apelin and adropin levels gradually increased during the AT intervention and significantly increased from baseline at weeks 4, 6 and 8. Furthermore, plasma ADMA levels significantly decreased during the 8-week AT intervention. Therefore, this study suggested that exercise training induces favorable changes in the time course of $NO^{\bullet}$ production, participating in AT-induced improvements of central arterial stiffening with advancing age [19]. The data indicate that exercise training can increase endothelium-derived nitric oxide activity in patients with an impaired endothelial function [45], and positively modulates inflammation and the atherosclerotic process; consequently, these facts can attenuate the progression of lower-limb myopathy, with subsequent improvements in the patients' functional capacity and health-related quality of life [47,48].

A multicenter observational prospective cohort study on 500 PAD patients (53% of them also affected by DM) who underwent endovascular treatment showed that the implementation of home-based exercise, monitored by pedometers, significantly decreased the risk of major adverse events, including death and the amputation of the target limb [49]. Supervised exercise therapy (SET), implemented several times a week and gradually, is, to date, an approach that should be offered to patients affected by symptomatic PAD [50], and is considered a class IA (highest level) recommendation, according to the 2016 AHA/ACC Guideline on the Management of Patients with PAD [51]. Unfortunately, the long-term participation in these programs is low and, therefore, this approach is underused.

A recent study conducted on PAD patients showed that exercise induces an exaggerated increase in arterial mean pressure compared to controls, but popliteal artery blood flow remains impaired, and concomitantly inflammatory and oxidative stress markers increase [52]. These observations, although derived from a small group of subjects (8 PAD patients), suggest that the improvement of exercise tolerance should be considered as a therapeutic target for people affected by PAD and, furthermore, suggest that the use of anti-inflammatory and antioxidant agents may be a promising approach.

2.2. Plant-Derived Compounds

Among natural polyphenols, flavonoids are present in most plants, having antioxidant activity. Phenols and flavonoids can directly scavenge reactive oxygen species (ROS) as a result of their ability to donate hydrogen atoms or electrons. Every plant contains a unique combination of phytochemicals, which can cause different beneficial effects in individuals. Green tea, grapes, apples, ginkgo biloba, soybean, turmeric, berries and onions are rich in flavonoids and are used as natural food supplements. Many studies also confirmed the protective effects of polyphenol antioxidants against diabetic vascular complications [53–56], such as the effects of green tea and cocoa polyphenols on endothelial dysfunction in patients with DM, red-wine polyphenols on microvascular dysfunction and citrus-fruit consumption on vascular protection [57,58]. Virgin olive oil rich in polyphenols has a cardioprotective effect suggested with a consumption between 20 and 30 g/day [59]. The ingestion of 40 g of dark chocolate in 22 heart transplant subjects improved coronary circulation and decreased platelet adhesion 2 h after supplementation [60,61]. However, the heterogeneity of the results does not allow one to make a definitive evaluation on chocolate's usefulness in DM [62]. A randomized, double-blind, placebo-controlled cross-over trial is in development to investigate whether cocoa flavanols improve blood pressure and vascular reactivity in patients with T2DM [63]. Overall, the European Food Safety Authority (EFSA) stated that 200 mg daily of cocoa flavanols (10 g of high-flavanol dark chocolate) could be beneficial for endothelium-dependent vasodilation in healthy populations [64]. Therefore, there are several epidemiological studies and various prospective clinical trials suggesting that a diet rich in polyphenols reduces the risk of incurring in cardiovascular diseases, but a clinical evaluation of specific phenols is lacking.

The present proposal focused on the study of the activity of five selected plant-derived polyphenolic compounds in the treatment of vascular damage, especially linked to hyperglycemia. With this aim in mind, the selected polyphenols investigated were baicalein, curcumin, mangiferin, quercetin and resveratrol (Figure 1). Of course, to plan a factual intervention against vascular damage, particularly with respect to PAD, known risk factors, including smoking, hyperglycemia, hypertension, dyslipidemia, chronic kidney disease and depression, should be identified and limited as much as possible.

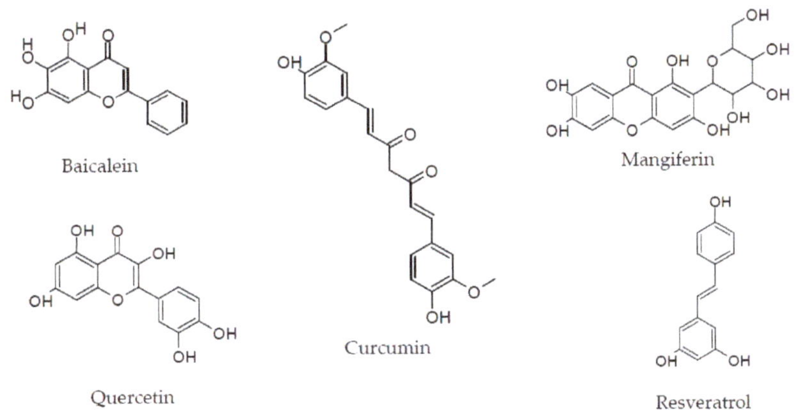

Figure 1. Chemical structure of the considered polyphenols.

3. Specific Polyphenolic Compounds of Interest

The existing knowledge from preclinical and clinical studies, as well as the recognized side effects, on the selected plant-derived polyphenols considered in this review are summarized in Tables 1 and 2.

3.1. Baicalein

3.1.1. Chemistry and Sources

Baicalein (5,6,7–trihydroxyflavone) is a naturally occurring polyphenol found in several medicinal plants, in particular in the roots of *Scutellaria baicalensis* G. and *Scutellaria lateriflora* L., and in the seeds, leaves, fruits and bark of *Oroxylum indicum* (L.) Kurz [65,66], all species that grow in several Asian countries. Baicalein is the aglycone of baicalin (baicalein 7-O-glucuronide), which is also found as such in plants and, moreover, is formed after the biotransformation of baicalein in vivo [65,67].

3.1.2. Activity: Preclinical Studies

Baicalein and baicalin, and extracts containing them, are proposed for various pharmacological effects, such as anti-inflammatory, antiviral, antibacterial, anticancer, antineurodegenerative and protective, against cardiovascular diseases [65,68,69]. The molecular mechanisms of action of baicalein, as well as those of other flavonoids, are mainly linked to the antioxidant activity, which occurs in different steps of the oxidative process; for example, it has demonstrated a role as scavengers of free radicals already formed in the medium (including lipid peroxyl radicals), as chelators of metal ions and by removing oxidatively altered biomolecules [70,71]. The role of the OH groups in polyphenols is dependent on the global geometry of the molecule and interactions among neighbor groups; a detailed conformational analysis and linked antioxidant mechanisms have been recognized for baicalein [70]. Recent studies on the structure–activity relationship suggest that the baicalein moiety is relevant for bioactivities [72]; however, the molecular mechanisms of baicalein are multiple and follow many molecular pathways that require further evaluation. The data confirm the beneficial role of *Scutellaria baicalensis*, which is widely used in traditional Chinese medicine to treat hypertension, respiratory infections, inflammation and diarrhea [73,74]. A study using three structurally related polyphenols, such as baicalin, baicalein and wogonin, showed the inhibition of endothelial cell barrier disruption, suggesting their protective activity against vascular inflammatory diseases [75]. Recently, studies have suggested that baicalein exhibits potential antidiabetic activities in metabolic syndrome [76]; the effect is related to the inhibition of α–glucosidase activity [77,78].

The in vitro and in vivo data suggest that baicalein is able to reduce vascular inflammation induced by high glucose levels. In human umbilical vein endothelial cells (HUVECs), baicalein (5–10 μM) was able to protect the cells from membrane disruption caused by a 25 mM glucose concentration; the polyphenol (10 μM) reduced the expression of the chemokines MPC–1 and IL–8, as well as ROS formation [75]. Furthermore, the alteration of vascular permeability induced by high glucose administration in mice was counteracted by baicalein (4.5–8.9 μg/mouse, i.v.) [75]. The vascular protective effect of baicalein was also demonstrated in vitro and in vivo as mediated by the inhibition of the high-mobility group box 1 (HMGB1) signaling pathway [79]. The role of baicalein on vascular function was assessed further by an in vitro study on the hybridoma endothelial cell line EA.hy926 [80]. The results indicate that baicalein is able to exploit protective effects against the oxidative stress of the endothelium, which is linked to the risk of diabetic angiopathy.

Baicalein has hypotensive effects on hypertensive rats, improving blood pressure and endothelium function [81]. Zhang et al., using streptozotocin and a high-fat-diet-induced diabetic rats, demonstrated that a 4–week treatment with baicalein (150 mg/kg/day) can reduce the level of blood glucose and improve insulin resistance, dyslipidemia and inflammation [82]. The effects have been attributed to the modulation of the gut microbiota, leading to increased levels of short-chain fatty acids (including acetate, propionate and butyrate), which are capable of improving gut barrier activity by stimulating epithelial growth and innate reactivity to invading bacteria [82].

A computational study revealed that baicalein is among the most promising candidates for the development of useful flavonoid derivatives in the treatment of DM [83]. Bioactivity is believed to derive from the structure that allows free-radical scavenging properties, reduces oxidized compounds, chelates metals and inhibits enzymes [83]. The experimental data suggest that baicalein can reduce oxidative stress, the expressions of iNOS and TGF–β1, as well as counteracting NF–κB activation [84]. Through a molecular modeling approach, together with microscopic and spectroscopic analyses, baicalein has also been proven to contrast the formation of AGEs and amyloid fibrils, phenomena that contribute to the loss of protein function and are connected to tissue damage [24].

In LPS-stimulated HUVECs, baicalein inhibits the expression of inflammatory cytokines IL–1β, IL–6 and TNF–α, as well as monocyte chemoattractant protein 1 (MCP–1) [85]. The authors also demonstrated that the inhibitory activity of baicalein occurs through the TLR4/NF–κB signaling pathway. The toll-like receptor 4 (TLR4)/NF–κB cascade pathway has been considered as a pivotal mechanism that leads to endothelial inflammation, activated by inflammatory signals, such as bacterial toxins; the fact that this pathway was modulated by baicalein provides an explanation for its potential as an anti-inflammatory agent.

Huan et al. indicated that baicalein exerts an anti-angiogenic effect on the inflammation microenvironment by inhibiting the transcriptional activity of activator protein-1 (AP–1) [86]. Tsai et al. observed that baicalein inhibits the expression of the lectin-like oxidized-LDL receptor 1 (LOX–1) protein in HUVECs, therefore protecting against oxidized-LDL atherogenic effects [87]. In this experimental model, baicalein also reduced ROS formation determined by the exposure to oxidized-LDL, as well as the consequent inflammation, by modulating AMPK/PKC/NADPH oxidase/NF–κB signaling [87].

However, Machha et al. demonstrated that acute exposure to baicalein alters the vascular tone in isolated rat aorta, due to the inhibition of endothelium-derived nitric oxide [88]. A dual effect of baicalein has been suggested, evaluating the endothelium-dependent contraction versus a direct relaxation produced by the substance in a rat mesenteric artery [89]. Conversely, the chronic oral administration of baicalein improves endothelium-dependent relaxation in spontaneously hypertensive rat aorta [90]. Considering vascular protection, baicalein (10 mg/kg/day for two weeks, orally) was shown to attenuate intimal hyperplasia in an in vivo model of vascular injury induced in the carotid artery of a rat [91]. As demonstrated in rat vascular smooth-muscle cells, the effect was due to the inhibition of proliferation via the MAPK, NF–κB, PI 3–kinase pathways, and the interaction with cell cycle machinery [91]. Using RAW264.7, HUVEC and MOVAS cells, Zhang et al. observed the anti-inflammatory activity of baicalein, attributed to the activation of the AMPK/Mfn–2 axis, together with the inhibition of downstream MAPKs/NF–κB signaling transduction [84].

Vascular calcification, often observed in patients with hypertension, atherosclerosis and DM, was also evaluated as a target of the action of baicalein [92]. Experiments on primary rat vascular smooth-muscle cells (VSMCs) indicated that baicalein decreased the mineralization rate, as well as calcium deposition and alkaline phosphatase activity [92]. Furthermore, Runt-related transcription factor 2 (Runx2, a transcription factor associated with osteoblast differentiation, also a regulator of the calcification of vascular smooth-muscle cells) and bone morphogenetic protein 2 (BMP–2, an osteogenic protein implicated in vascular calcification) expressions were negatively regulated in calcified VSMCs treated with baicalein. In vivo experiments, performed on a rat model of vascular calcification, demonstrated that baicalein was capable of inhibiting vascular calcification through multiple mechanisms, including the prevention of apoptosis, suppression of Runx2–BMP–2 signaling pathways and the preservation of the vascular contractility phenotype through the increased production of α–SM22 and α–SMA (vascular smooth-muscle markers) [92].

3.1.3. Activity: Clinical Studies

The pharmacokinetic characteristics of baicalein administered orally were evaluated in a cohort of 36 healthy Chinese subjects [93]. The study was a Phase-I single-center, randomized, double-blind, placebo-controlled, multiple ascending dose trial in which baicalein was administered at doses of 200, 400 and 600 mg once daily on days 1 and 10, 3 times daily on days 4 to 9. The absorption of the drug was rapid, with peak plasma levels evident within 2 h after administration. Treatment was found to be safe and well-tolerated; adverse events were mild and spontaneously resolved. Another study, consisting of multiple phases, considered a multiple design involving a total of 110 subjects, including a randomized, double-blind, placebo-controlled, single ascending dose study with doses of baicalein ranging from 100 to 800 mg [94]. In addition to the pharmacokinetic evaluation, the effect of food on the disposition of the drug was considered. The published clinical studies, although limited in sample dimension, indicate the favorable bioavailability and a good safety profile of baicalein, when used at suggested doses. However, to date, no studies on clinical effectiveness in the cardiovascular field are available.

3.1.4. Safety Profile

Baicalein is considered safe and has a good tolerability profile in humans, as per se it did not have adverse reports in VigiBase [95], the WHO global database of potential side effects of medicinal products.

Mild side effects were reported in a placebo-controlled study in which 68 healthy subjects were treated with a single dose of baicalein orally, 100 to 800 mg [94]. The most common side effects were increased levels of the C-reactive protein and triglycerides, and proteinuria; these undesirable effects did not depend on the dose of baicalein administered [94]. In a clinical trial, multiple doses of baicalein (200, 400 or 800 mg twice daily) in healthy volunteers produced various adverse effects, such as abdominal pain, constipation, and increased alanine transaminase and aspartate aminotransferase levels [96].

In fact, some concerns arose from reports on the possible liver toxicity of the use of plant extracts rich in flavonoids. Recently, it was shown in vitro that baicalein can inhibit human UDP-glucuronosyltransferases1A1, the enzyme that is primarily responsible for glucuronidation and the elimination of bilirubin [97], which can cause jaundice and severe liver disease. In general, further clinical trials involving a higher number of enrolled subjects are needed to produce a clearer picture of the safety profile of baicalein in humans.

3.2. Curcumin

3.2.1. Chemistry and Sources

Curcumin, also known as diferuloylmethane [(1*E*,6*E*)-1,7-bis(4-hydroxy-3-methoxyphenyl) hepta-1,6-diene-3,5-dione], is a diarylheptanoid that represents the main component of the turmeric rhizome (*Curcuma longa* L. and other *Curcuma* spp.), and has caused extensive investigations from biological and chemical points of view [98]. Turmeric is a widely used medicinal plant in Asian countries, appreciated for its antioxidant and anti-inflammatory activities, as well as for other multiple properties [99].

3.2.2. Activity: Preclinical Studies

Curcumin inhibits proinflammatory cytokines, TNF–α, ICAM–1, NOX2 and cyclo–oxygenase–2 expressions, and reduces the leucocyte-endothelium interaction in diabetes-induced rat vascular inflammatory models [100,101]. Curcumin has been widely evaluated with respect to the modulation of endothelial function [102]. Curcumin also presents a neuroprotective potential in the pathogenesis of Alzheimer's disease [103–105]. The many described biological activities of curcumin have been linked to its chemical instability that leads to the rapid autoxidation responsible for its antioxidant behavior. The oxidation of curcumin produces intermediate metabolites that conduct to a final bicyclopentadione [106,107]. Moreover, it has been demonstrated that electrophilic metabolites can covalently adduct to cellular protein targets, in particular those of the NF–κB pathway, which explains the anti-inflammatory activity of the natural compound. The methoxy groups of the molecule have also been shown to possess a role in its anti-inflammatory biological activity [108].

Ischemia and reperfusion injury in skeletal muscles, obtained by clamping the femoral artery and vein, is an experimental model that can provide information on the damage occurring in ischemic events of the limbs. Rats subjected to 4 h femoral occlusion followed by 2 h reperfusion, which received curcumin (100 mg/kg i.p., 1 h before reperfusion), presented a preservation of superoxide dismutase and catalase activities, as well as of the glutathione content in muscle tissues, compared to injured but not treated animals [109]. In the same context, the muscle and plasma levels of malondialdehyde and protein oxidation were reduced, as well as the plasma levels of inflammatory cytokines (TNF–α and IL–1β).

The data suggest that curcumin has protective activity against the ischemic damage of skeletal muscles [109].

Using a mouse model of hindlimb ischemia obtained by the ligation of the left femoral artery, Liu et al. observed that curcumin (100 mg/kg, i.p. administration 1 h before the ligation surgery) is capable of improving the running capacity on a treadmill of injured mice compared to the controls, evaluated 1 and 2 weeks following injury [110]. A histological examination of skeletal-muscle sections from the curcumin-treated group showed less muscle degeneration and fibrosis than in the control mice; also, curcumin treatment was able to significantly reduce macrophage infiltration and ischemia- induced inflammation. The regulation of the tissue inflammatory response has been associated with a decrease in the proinflammatory cytokines TNF–α, IL–1β and IL–6 as a result of the diminished expression of NF–κB, namely, of the subunit p65 [110]. These data suggest that curcumin could be a promising agent used against diseases of the ischemic limbs. Another in vivo experiment conducted on streptozotocin-induced diabetic mouse that received hindlimb ischemic surgery demonstrated that curcumin (1000 mg/kg in olive oil, applied by gavage for 14 days) can promote the significant recovery of blood flow detected by laser Doppler imaging, 7 and 14 days after ischemic surgery; histology evidenced neogenic capillary formation, as a consequence of improved angiogenesis and the proliferation of endothelial progenitor cells [111]. Zhang et al. studied the role of curcumin in an experimental PAD model of murine hindlimb ischemia obtained with surgical ligation and the excision of the femoral artery. Mice received 1000 mg/kg of curcumin (in 300 μL of olive oil, gavage) once a day for two weeks [112]. In animals treated with curcumin, the laser Doppler perfusion imaging system showed perfusion recovery 14, 21 and 28 days after ischemia induction, significantly better than in the control. Histology of the ischemic gastrocnemius muscle after 28 days revealed a significant increase in capillary density. Furthermore, 7 days after surgery, a 5-fold increase in the expression of microRNA (miR)–93 (a mediator of neovascularization in PAD) was detected in the muscle tissue, suggesting the role of curcumin in angiogenesis [112].

Another possible favorable mechanism of action of curcumin in the damage of large arteries could be the inhibition of hypoxia-inducible factor 1α (HIF–1α), implicated in the progression of atherosclerosis. Ouyang et al. demonstrated that in a culture of human macrophages, the protein level of HIF–1α, induced by hypoxia, is reduced in a concentration-dependent way by treatment with curcumin (20–40 μM), which acts in a proteasome-dependent manner without affecting the mRNA level of the factor [113]. Moreover, it has been shown that curcumin causes the inhibition of the expression of HIF–1α through ERK signaling pathway. Following HIF–1α inhibition, curcumin also affects macrophage apoptosis induced by hypoxia; also, it significantly reduces the hypoxia-induced release of proinflammatory cytokines IL–6 and TNF–α from macrophages [113].

Curcumin has been evaluated under the aspect of AGEs inhibition, a process that is strictly related to arterial senescence and degeneration. A review has recently taken into account this issue [114]. Curcumin has been shown to reduce in vitro the generation of AGEs, as well as to counteract in vivo AGE formation. The possible mechanisms involved include methylglyoxal trapping, but also the capacity to prevent AGE formation, due to the antioxidant activity and clearance of free radicals. Curcumin can also act on the expression of AGE receptors by suppressing RAGE (a receptor that mediates the detrimental effects of AGEs) and enhancing AGE–R1 (a distinct receptor that mediates the detoxification of AGEs) [115]. It should be mentioned that curcumin may act as favorable modulator of PAD also influencing several pathways linked to DM, the disease whose complications often evolve toward vascular manifestations. A recent overview has considered by bioinformatic methodology the many targets that can be affected by curcumin due to its antioxidant and anti-inflammatory properties, which can modulate the disease's progression [116].

3.2.3. Activity: Clinical Studies

From a strict, clinical point of view, the data on the effectiveness of curcumin in PAD are limited. A meta-analysis of 5 randomized clinical trials conducted on a total of 192 healthy subjects with doses of curcumin ranging from 25 to 200 mg, showed that curcumin supplementation is able to improve vascular function, determined as the flow-mediated dilation of the brachial artery [117]. Alidadi et al. reported 7 randomized clinical trials conducted on humans, ranging from healthy subjects to obese people and type 2 diabetic patients, with a sample size ranging from 21 to 88 subjects [118]. Curcumin doses ranged from 150 mg/day to 2 g/day, over periods of 8–16 weeks. In healthy subjects, curcumin appeared to improve endothelial function, determined by a non-invasive ultrasound-based technique that measured flow-mediated dilation. On the contrary, in pathological conditions, curcumin treatment did not show an appreciable improvement in endothelial function, also due to the limited number of patients evaluated [118].

Despite its interesting biological properties, curcumin has the limitation of being poorly absorbed by the gastrointestinal tract due to its high hydrophobicity [119]. Several formulations were investigated to enhance curcumin bioavailability [120,121], in particular micro- and nano-formulations. An optimized curcumin formulation was developed and investigated [122]. It consisted of poly (propylene sulfide) (PPS) microparticles used for curcumin encapsulation and the delivery of the active principle on demand at the site of oxidative stress. An oil-in-water emulsion was obtained, which permitted, in the presence of ROS, the transition of hydrophobic PPS into more hydrophilic sulfoxide and sulfone, thus releasing curcumin. Following an in vitro evaluation of the system, the authors tested the formulation in vivo (curcumin-loaded PPS microspheres: 5 mg/kg of curcumin with 10.3 mg/kg of PPS, administered i.m. in the gastrocnemius and adductor muscles) in a streptozotocin-induced mouse model of DM in which hindlimb ischemia was induced [122]. The results show that curcumin microspheres were able to selectively reduce ROS levels in the tissue of the ischemic limbs, significantly increasing blood perfusion and the length of the vasculature, providing evidence for a favorable action of curcumin in the treatment of diabetic PAD. Another more recent attempt to produce "on demand" treatment for PAD has been reported [123]. It consists of PVAX (vanillyl alcohol-incorporated copolyoxalate) nanoparticles loaded with curcumin that release the drug at a pathologically elevated level of H_2O_2, as occurs at an ischemic site. After optimizing in vitro the formulation of curcumin–PVAX nanoparticles and demonstrating their antioxidant, anti-inflammatory and angiogenic activities, the authors evaluated the activity of the product in a mouse model of unilateral hindlimb ischemia. The in vivo results, obtained with the i.m. injection of curcumin–PVAX nanoparticles into the ischemic area on days 0, 3 and 7, showed significant blood-flow recovery and neovascularization, with almost complete blood-flow restoration 12 days after ischemia. The histological examination demonstrated the suppression of tissue damage and inflammatory responses, with an extensive expression of angiogenic factor CD31 [123]. Several other formulations of curcumin, including nano-formulations or encapsulation, have appeared on the market to ensure the optimal intake of the active compound [124,125].

3.2.4. Safety Profile

Regarding curcumin toxicity, the compound is considered safe and has a good tolerability profile in humans [126–128]. The reported side effects mainly include gastrointestinal disturbances. However, in Italy, several cases of acute hepatitis have been reported to be associated with formulations that provide high bioavailability and high doses of curcumin [129]. In fact, the availability of many different formulations of curcumin with an enhanced bioavailability profile has focused on the question of safety. For instance, Fuloria et al. considered the curcumin toxicity profile that emerged during the most recent preclinical and clinical studies; excessive amounts of curcumin may lead to alterations in testosterone levels in men, influence blood clotting and contrast iron absorption [125].

In VigiBase curcumin has 71 potential side effects [95], such as gastrointestinal disorders (i.e., diarrhea and nausea), nervous system disorders (e.g., asthenia and others) and hepatobiliary disorders (e.g., acute liver failure, increase in alanine aminotransferase and aspartate aminotransferase, and others). More trials are needed to understand the safety of curcumin for pharmacological use, also reducing the number of additives, and considering customized microencapsulation [125].

3.3. *Mangiferin*

3.3.1. Chemistry and Sources

Mangiferin is a natural C-glucoside xanthone (2-β-D-glucopyranosyl-1,3,6,7-tetrahydroxy-9H-xanthen-9-one) contained in many plant species, but especially in the fruit, kernels and the leaves of the mango tree (*Mangifera indica* L.), a native plant of India [130,131]. The mango tree also contains similar phenolic components, isomangiferin and homomangiferin, which contribute to the beneficial effects of the plant extracts [132,133]. Mangiferin has four hydroxyl groups in the xanthone nucleus and, thus, can be considered among the occurring phenolic compounds present in higher plants providing antioxidant activity.

3.3.2. Activity: Preclinical Studies

Mangiferin has been reported to be an antidiabetic, anti-inflammatory, antimicrobial, immunomodulator, anticancer and hypocholesterolemic agent [133]. It has high antioxidant activities due to its hydroxyl groups and redox-active aromatic system of the catechol moiety [131,134]. In addition to the described scavenger activity on ROS, mangiferin can modulate the expression of several genes involved in inflammation and apoptosis, including the induction of the antioxidant Nrf2 pathway [135,136]. Furthermore, it can protect mitochondrial membranes against lipid peroxidation and prevent hydroxyl radical formation by inhibiting Fenton-type reactions [137]. Mangiferin exhibits a strong inhibition of oxidative stress associated with the endoplasmic reticulum by reducing ROS production and attenuating inositol-requiring enzyme 1 (IRE1) phosphorylation [138]. Mangiferin has also been shown to be an inhibitor of the NF–κB signaling pathway [131]. Furthermore, it has been evaluated as a possible pharmacophore structure for the development of new compounds with pharmacological activity in multiple pathological conditions [134], possibly related to inflammation and DM.

In HUVECs, mangiferin (20 μM) counteracts paraquat-induced endothelium damage, preserving the p120–catenin protein level [139]. Furthermore, it protects endothelial cells from oxidative injury induced by H_2O_2 or glycated protein–iron chelate, suggesting a protective role against pathologies linked to oxidative stress [140]. In an experimental model of high glucose/hypoxia-induced angiogenesis, mangiferin was effective in inhibiting angiogenesis by reducing hypoxia-inducible factor–1α, vascular endothelial growth factor and matrix metallopeptidase (MMP)–2 and MMP–9 [141]. Instead, Daud et al. observed that mango extract and mangiferin stimulated the migration of bovine endothelial aortic cells in a modified Boyden chamber assay, suggesting a role for the polyphenol in the promotion of the formation of new blood vessels [142].

The rationale of mangiferin use in DM is also related to its inhibition of AGE formation, hence counteracting the vascular damage typical of the disease. In a rat model of streptozotocin-induced diabetes, Hou et al. observed a sustained suppression of AGE production and a decrease in the protein expression of RAGE receptors; another relevant effect of mangiferin was the inhibition of NF–κB with a reduction in inflammatory cytokines [143]. Using an animal model of a mouse fed with a high-fat diet, Jiang et al. observed that mangiferin (5–20 mg/kg) could reduce plasma lipids and aorta wall thickening [144]. In oxidized-LDL-induced HUVEC injury, the compound was able to alleviate cellular dysfunction, reducing ROS levels, increasing the release of $NO^{\bullet}$ and activating the PTEN/Akt/eNOS signaling pathway [144]. Recently, a meta-analysis considering 19 studies on diabetic animals, mainly rats and mice, showed that mangiferin intake up to 422 mg/kg reduced blood glucose levels in a dose-dependent manner [145].

An extract of *Mangifera indica* administered to LDL-receptor-deficient mice for 2 weeks produced a reduction in plasma and liver cholesterol levels, ROS production in spleen mononuclear cells and increased plasma total antioxidant capacity [146]. Mangiferin activity in PAD was also indirectly investigated using ethanolic extracts of mango seed (EEMI) in an acute hindlimb ischemia-reperfusion model [147]. Streptozotocin-treated diabetic rats underwent a femoral artery ligation and, then, received EEMI (0.2–0.4 g/kg) for 14 days. Blood flow was observed to be significantly higher in treated animals than in the controls. The plasma levels of malondialdehyde, IL–6, TNF–α and IL–1β were reduced, while glutathione and IL–10 levels were increased in EEMI-treated animals, suggesting anti-inflammatory modulation [147]. In general, given the above reported experimental data, mangiferin appears to offer promise in the prevention and treatment of vascular disease also linked to diabetes and dyslipidemia.

3.3.3. Activity: Clinical Studies

A 12-week, double-blind, placebo-controlled clinical study was conducted on 104 overweight patients with hyperlipidemia subdivided into 2 groups, administering a dose of 150 mg/day of mangiferin or placebo for 12 weeks. Treatment with mangiferin produced a significant decrease in the serum levels of triglycerides and free fatty acids, and the insulin-resistance index [148]. However, clinical trials conducted on the use of mangiferin in cardiovascular diseases and, moreover, in patients with PAD are lacking.

3.3.4. Safety Profile

Mangiferin in high concentrations may cause damage to the mitochondrial respiratory chain, since free radicals are also needed for normal cellular activity [131]. The oral administration of 0.9 g of mangiferin has been reported to be harmless to adults [131,149]. Mangiferin per se did not have any adverse reports in VigiBase [95]; otherwise, *Mangifera indica* has 62 reports of unwanted effects, entirely from the Americas [95]. The side effects were mainly gastrointestinal (nausea and diarrhea), nervous system (dizziness) and skin (pruritus and erythema) disorders [95]. Overall, at moderate doses, the compound has been reported to be safe for humans.

3.4. *Quercetin*

3.4.1. Chemistry and Sources

Quercetin is pentahydroxyflavone (3,3′,4′,5,7-pentahydroxyflavone) having five hydroxyl groups belonging to the group of flavonols found in many fruits (e.g., apples, grapes, berries and citrus fruits) and vegetables (e.g., onions, broccoli and Italian chicory) [150], and widely used as a food supplement in various commercial products (generally, 500 mg, twice daily) for circulation, immune system function and respiratory function [151]. Natural derivatives of quercetin, such as isoquercetin and rutin, exist in natural sources, for instance, in onions and citrus foods [152]. However, the low solubility in water and limited bioavailability of quercetin have limited the medical use of the compound. Several enzymatically modified derivatives of quercetin have been obtained and investigated, mainly for their favorable bioactivity and bioavailability. Quercetin derivatives have also been obtained by using engineered *Escherichia coli* and other microorganisms [152]. The structure–activity relationships of quercetin and its derivatives have been recently investigated in detail [153], including the antioxidant, anti-inflammatory and antidiabetic properties.

3.4.2. Activity: Preclinical Studies

Quercetin shows prominent antioxidant potential and is considered an effective free-radical scavenger mainly based on several in vitro studies [154–156]. The antioxidant activity of quercetin is believed to be linked to the regulation of GSH levels [150]; moreover, the hydroxyl groups on the side phenyl ring of the molecule can bind to amino acid residues of key enzymes, such as acetylcholinesterase and butyrylcholinesterase, both linked to oxidative properties. Quercetin also increases the levels of endogenous antioxidant enzymes, e.g., catalase, GSH peroxidase and superoxide dismutase. More specifically, considering the vascular effects, quercetin induces the vasodilation of isolated rat arteries [157–159] and, in vivo, antihypertensive effects on rats fed with a high-fat, high-sucrose diet [158] and in spontaneously hypertensive animals, without effect in normotensive animals [160]. Generally, this flavonoid showed antiangiogenic activity in several experimental models [161–163].

The protective effect of quercetin against the activation of ER stress was attributed to the upregulation of markers, such as the 78 kDa glucose-regulated protein (GRP78), a molecular chaperone, and the C/EBP-homologous protein (CHOP) in unresolved diabetic and experimental ER-stress conditions [164]. Furthermore, quercetin pretreatment decreased the expression of tunicamycin-induced ER-stress markers in HUVECs [164]. In another study, mitochondrial-targeted quercetin activities were observed to be a mechanism of protection against neurodegenerative diseases [165]. Quercetin has been reported to protect against hydrogen peroxide-induced pheochromocytoma cell neurodegeneration [166].

In general, quercetin has shown anti-inflammatory action in several experimental models, mainly through mechanisms that inhibit NF–κB and cofactors in the chromatin of proinflammatory genes [167]. Quercetin also increases glucose uptake from the blood by inducing the glucose transporter GLUT4, and promotes glucose storage by the liver [156,168].

3.4.3. Activity: Clinical Studies

Few trials have examined the effect of quercetin supplementation on blood pressure, with discordant results; however, generally, a reduction in systolic blood pressure both in healthy subjects and hypertensive patients has been reported [169–172]. Recently, a meta-analysis showed significant s in both systolic and diastolic blood pressure with quercetin supplementation; the treatments ranged from 4 to 10 weeks, with doses ≥500 mg/day [173]. Another meta-analysis showed a significant reduction in blood pressure and, furthermore, for participants receiving quercetin for at least 8 weeks, a decrease in triglycerides in trials with a parallel design [174]. A double-blinded, placebo-controlled cross-over study conducted on 93 overweight or obese subjects aged 25 to 65 years with metabolic syndrome traits showed that quercetin administered at 150 mg/day (6-week-treatment period) reduced systolic blood pressure and plasma oxidized-LDL levels in overweight subjects with a high-CVD-risk phenotype [171].

In the scientific literature, no clinical studies have been reported on the use of quercetin in the treatment of PAD.

3.4.4. Safety Profile

Egert et al. reported that daily supplementation with 150 mg of quercetin/day, administered orally to volunteer subjects with a high-CVD-risk phenotype for 6 weeks, was safe [171]. In general, the clinical studies reported that the orally administered quercetin use was safe and well-tolerated [173,174].

Quercetin per se has few adverse effects reported in VigiBase [95]; a total of 45 unwanted effects are described, mainly in the Americas and Europe. Nervous system disorders (e.g., dizziness and headache), respiratory disorders (e.g., dyspnea), pruritus and drug interaction are the most recurrent effects [95]. Nevertheless, in general, moderate doses of quercetin are described as safe.

3.5. Resveratrol

3.5.1. Chemistry and Sources

Resveratrol is a stilbene derivative (3,5,4′–trihydroxystilbene) occurring both as *trans*- and *cis*-isomers that are present in a variable percentage in several natural sources, the *trans*-form being the most abundant and mainly responsible for the cardiovascular effects [175]. Resveratrol is a phytoalexin (a class of antimicrobials synthesized by plants under pathogen infection) found in many plant foods. Its presence is well-known in red grapes (skin) and red wine, but even in tea and berries and also in various medicinal plants (e.g.,

Polygonum spp. roots) used in popular medicine as treatments for allergic and inflammatory diseases [176].

3.5.2. Activity: Preclinical Studies

Resveratrol has antioxidant and free-radical scavenger activities that may be responsible for its several biological activities, such as anti-inflammatory, antiatherosclerosis and anticarcinogenic effects [177,178]. The structural determinants of the antioxidant activity of resveratrol have been linked to the presence of hydroxylic functions, in particular, but not only, the hydroxyl group at 4′ position, as demonstrated by an investigation on the derivatives [179].The dose-dependent biphasic hormetic effects of resveratrol have been reported: at low concentrations, it acts as an antioxidant that protects tissues from oxidative stress, while at high concentrations, it may be a pro-oxidant that increases oxidative stress [180]. Similarly, low and high concentrations can provide chemoprevention or cytotoxicity, respectively, against cancer cells [181]. However, resveratrol is especially known for its beneficial effects in cardiovascular diseases. Several authors have shown that it causes vasodilation in different types of isolated arteries obtained from various animal species (e.g., guinea pigs, pigs, rats and sheep) [182–185]. Studies showed that the anti-inflammatory activity of resveratrol is mainly mediated by antiadrenergic and antiprostaglandin activation [176]. Resveratrol reduced the sensitivity of myofilaments to free calcium in vascular smooth muscles and enhanced acetylcholine-stimulated calcium increase in the endothelium, promoting $NO^{\bullet}$ production and thus vasorelaxation [182]. At nanomolar concentrations, it induces the endothelial production of $NO^{\bullet}$ by activating the estrogen receptor-α (ERα)–Cav–1–c–SRC interaction, resulting in $NO^{\bullet}$ production through a Gα–protein-coupled mechanism [186]. It down-regulates VEGF/fetal liver kinase-1 (Flk-1) (VEGF receptor-2) expression and, therefore, modulates hyperpermeability and junction disruption in glomerular endothelial cells. In addition, resveratrol ameliorates high-glucose-induced hyperpermeability mediated by overexpressed caveolin-1 in aortic endothelial cells [187]. A recent study using a palmitate-induced insulin-resistance model revealed that resveratrol suppresses IKKβ/NF–κB phosphorylation, TNF–α and IL–6 production, and restores the IRS–1/Akt/eNOS signaling pathway in endothelial cells [188]. Resveratrol has been reported to block the TNF–α-induced activation of NF–κB in coronary arterial endothelial cells and inhibit inflammatory mediators [189], exerting the effect through the action on the IKK cascade, attributing to this mechanism its antioxidant properties. The report by Kim et al. demonstrated that resveratrol, as well as hesperidin and naringenin, reduces high-glucose-induced ICAM–1 expression via the p38 MAPK signaling pathway, contributing to the inhibition of monocyte adhesion to endothelial cells [190]. Recently, resveratrol showed an inhibitory effect against NF–κB p65 and proinflammatory mediators, including TNF–α, ICAM–1 and MCP–1 in endothelial cell lines [191]. Resveratrol confers a protective effect against high-glucose-induced oxidative stress in endothelial cells and vascular protection in high-fat-diet mice, through the Nrf2 pathway [192].

In several experimental models, in vitro and in vivo, resveratrol improved glucose homeostasis and insulin sensitivity [193,194]. An in vivo study conducted on diabetic rats demonstrated that the compound elicits antidiabetic potential by stimulating intracellular glucose uptake and the modulation of sirtuin-1 activity [195]. Resveratrol also relieves the status of diabetic nephropathy, kidney and oxidative stress in diabetic rats [196]. Resveratrol inhibits ATP-dependent K^+ channels and voltage-dependent delayed-rectifier K^+ channels in β-cells, suggesting its beneficial role in delaying the onset of insulin resistance and improving insulin secretion [197].

3.5.3. Activity: Clinical Studies

Resveratrol is mainly known since it offers a possible explanation for the so-called "French paradox", which is the low frequency of heart disease in the French population despite the relatively high-fat dietary use, believed to be linked to red wine consumption, as a source of resveratrol [198]. Several clinical trials considered resveratrol both in healthy volunteers and patients with various cardiovascular diseases, but they all considered a limited number of participants. Patients with coronary artery disease received 10 mg/day of resveratrol for 3 months, showing an increase in flow-mediated vasodilation and, in general, an improvement in cardiovascular parameters [44]. A clinical study using resveratrol revealed that oral, 100 mg consumption for 12 weeks may support the prevention of cardiovascular disease and atherosclerosis by stimulating endothelial function [199]. Furthermore, resveratrol also modulates $NO^{\bullet}$ metabolism and contributes to improved vascular function in hypertensive and dyslipidemic patients [200]. In addition, a crossover, double-blind, placebo-controlled study in which 22 healthy adults received 250 and 500 mg of resveratrol revealed dose-dependent increases in cerebral blood flow [201]. Another study also showed that the acute administration of 75 mg of resveratrol increased neurovascular coupling and cognitive performance in 36 subjects affected by T2DM, improving cerebral perfusion [202]. A meta-analysis concerning 3 clinical trials for a total of 50 DM subjects treated with resveratrol at doses of 10, 150 and 1000 mg daily, for a period of 4 to 5 weeks, did not show any favorable effects on the glycosylated hemoglobin A1c level or on insulin resistance [203]. Recently, a double-blind and randomized clinical trial, entitled "Resveratrol to Improve Outcomes in Older People With PAD" (RESTORE), was conducted on 66 patients with PAD treated with 125 mg/day or 500 mg/day of resveratrol, or placebo for 6 months [204]. However, the trial did not show reliable confirmation that resveratrol improves walking performance detected by the 6min walk test among patients with PAD [204]. Other clinical studies enrolling a higher number of patients affected by PAD are required to evaluate the efficacy of resveratrol on this disease.

3.5.4. Safety Profile

Resveratrol exhibited the systemic inhibition of P450 cytochromes when taken in high doses [205]. Furthermore, the ingestion of 25 mg/kg of resveratrol for 60 days in rats altered their thyroid function, causing a goitrogenic effect [206]. Other studies using oral doses of 200 mg/kg/day in rats and 600 mg/kg/day in dogs did not report adverse effects [207]. Few clinical trials have evaluated the safety and tolerability of resveratrol; Brown et al. reported gastrointestinal discomfort at doses of 2.5 and 5 g/day [208]. The administration of resveratrol with single doses of 0.5, 1, 2.5 or 5 g orally in 40 healthy volunteers caused minor adverse events that resolved spontaneously in a few days [209]. In general, 150 mg/day for adults is accepted to be safe [210].

Resveratrol per se has very few adverse reports in VigiBase [95]; a total of 20 unwanted effects have been described in the Americas, Europe and Oceania. Among these, the most common are general disorders (e.g., malaise and fatigue) and disorders of the gastrointestinal and nervous systems [95]. Therefore, low doses of resveratrol could be considered safe and potentially useful for vascular disorders.

4. Conclusions

Polyphenols have been extensively investigated for their beneficial effects on many clinical conditions, as well as cardiovascular-related diseases. Despite the profusion of experimental data suggesting potential mechanisms favorable for clinical use, trials on the side of peripheral vascular diseases are still hindered by the absence of large studies with sufficient statistical power in order to demonstrate the true efficacy of the compounds. The requirement of large resources to program a clinical study that does not necessarily allow for a unique patent outcome often discourages an industrial gamble, reducing the real possibility of developing new drugs in therapy. Resources for research on natural products from non-profit organizations are necessary.

Emerging polyphenolic compounds were selected based on their appearance in the literature and have been discussed here, with a focus on their role in improving endothelial and, in general, vascular functions; molecular mechanisms and preclinical and clinical evidence were highlighted. Overall, the data collected in this review suggest the potential role of selected polyphenols in the treatment of PAD, also concomitant with DM; however, clinical data on efficacy are still limited and need to be improved. Despite the extensive preclinical experimental results that confirm the potential role of baicalein, curcumin, quercetin, mangiferin and resveratrol against vascular diseases, their clinical effectiveness is still only preliminarily demonstrated, although there is evidence of safety for the selected doses. Natural-drug compounds that provide significant biological activities in the specific vascular district have been suggested, and the task of a chemical approach could be directed to optimize their bioavailability. However, caution should be devoted to the fact, as already observed that, for example, with curcumin, an absolute increase in bioavailability does not necessarily mean an improved benefit/risk ratio when the product is introduced for clinical use. Moreover, derivatives or analogs obtained on the structural basis of natural products do not necessarily produce to compounds that are safe for human use. The occurrence in the molecule of phenolic hydroxyl groups also raises the question of biological stability as drugs. Therefore, a new concept of the wise use of available natural resources should be applied, also in the prospective development of drugs, keeping in mind the limits and advantages of medicinal plants and relative natural compounds.

As a general consideration, a diet rich in polyphenols reduces the risk of cardiovascular adverse events, including PAD. Polyphenols, together with adequate moderate aerobic exercise, can help prevent and reverse age-associated arterial stiffness. In fact, exercise therapy is considered a class IA (highest level) recommendation for the treatment of patients with PAD. Unfortunately, the long-term participation in perspective clinical and population-based programs is scarce, and therefore this approach is still just outlined. The use of polyphenols, both as dietary intake and dietary supplements, could represent a favorable approach to maintaining the integrity of peripheral blood vessels and limiting the harmful effect of oxidants. The present evidence suggests the validity of further clinical trials to define the role of this class of compounds in the prevention and treatment of vascular artery disease.

Table 1. Plant-derived polyphenols with potential activity in preventing or improving cardiovascular diseases and, thus, also PAD: in vitro and in vivo studies using various cellular or animal models.

Chemistry	Natural Compounds	Food Sources/Medicinal Plants	Experimental Models	Findings	Cellular Mechanisms	References
Flavones	Baicalein	- *Scutellaria baicalensis* - *Scutellaria lateriflora* - *Oroxylum indicum*	- EA.hy926 cells - HUVECs - Mouse	- Improves vascular permeability and function - Inflammation inhibition	- Decreases AGEs and TNF–α - Decreases NF–κB activation - Increases AMPK activation - Decreases CAMs expression - Decreases MCP–1 and IL–8 - Decreases H_2O_2	[68,70,71,75,77–92]
Diarylheptanoids	Curcumin	- *Curcuma longa*	- Mesenteric arteries - Rat - Mouse(limb ischemia)	- Vasorelaxation - Blood-flow recovery - Vascular inflammation inhibition - Decreases hyperglycemia	- Decreases $O_2^{\bullet-}$ - PCK inhibition - Decreases ICAM–1 and NOX2 - Decreases ROS - Counteracts AGE formation - HIF–1α inhibition	[100,101,106–115,122,123]
Xanthones	Mangiferin	- *Mangifera indica* - *Anemarrhena asphodeloides* - Ferns	- HUVECs - RRCECs - Mice - Diabetic mice - Diabetic rats	- Improves endothelial function - Promotes new blood vessel formation - Improves blood flow - Antihyperglycemic effect - Inflammation inhibition	- Reduces ROS - Inhibits NF–κB activation - Inhibits HIF–1α - Counteracts AGEs - Cytokine inhibition - Inhibits endoplasmic reticulum stress	[135–147]
Flavonols	Quercetin	- Apples - Berries - Grapes - Onions - Tea - Red wine	- Collagen glycation - Glucose autoxidation - HAECs - Aortic rings (various animal species)	- Vasorelaxation - Glycation inhibition - Inhibition of endothelium dysfunction - Inflammation inhibition	- Increases eNOS - Decreases AGEs - Activation of AMPK - Inhibition of endothelin–1 - Inhibition of phosphodiesterase - Decreases COX2 expression and lipoxygenase	[150,154–168]

Table 1. *Cont.*

Chemistry	Natural Compounds	Food Sources/Medicinal Plants	Experimental Models	Findings	Cellular Mechanisms	References
Stilbenes	Resveratrol	- Berries - Blueberries - Cocoa - Nuts - Peanuts - Pomegranates - Red grapes - Red wine - Tea	- HUVECs - VSMC - Rat aortic rings - Mesenteric arteries - BSA assay - α–amylase - α–glucosidase - Diabetic rat retinas	- Vasorelaxation - Improves endothelial function - Reduces platelet aggregation - Inhibition of AGE formation - Inhibition of cell proliferation - Glycation inhibition - Antihyperglycemic effect	- cGMP increase (dSCA) - cAMP phosphodiesterase inhibition - Increases $NO^{\bullet}$ production - AMPK activation - SIRT1 activation - Scavenges ROS - Increases ERα–Cav–1–c–SRC interaction - Inhibits AGE-induced TGF–β1 mRNA increase - Trapping MGO - Inhibition of α–amylase and α–glucosidase activities - SIRT1 activator - Increases tight-junction protein occludin	[176,178–197]

AGEs, advanced glycation end-products; AMPK, adenosine monophosphate-activated protein kinase; CAMs, cell adhesion molecules; GO, glyoxal; HIF–1α, hypoxia-inducible factor 1α; HAECs, human aortic endothelial cells; HRPE cells, human retinal pigment epithelial cells; HUVECs, human umbilical vein endothelial cells; IL–6 or IL–8, interleukin–6 or –8; ICAM–1, intercellular adhesion molecule 1, MCP–1, monocyte chemoattractant protein–1; MGO, methylglyoxal; NOX2, phagocytic NADPH oxidase, NO, nitric oxide; eNOS, endothelial nitric oxide synthase; $O_2^{\bullet -}$, superoxide anion; VSMCs, vascular smooth-muscle cells; dSCA, denuded-sheep coronary arteries; PKC, protein kinase C; RRCECs, rat retinal capillary endothelial cells; SIRT1, silent mating- type information regulation 2 homolog 1.

Table 2. Plant-derived polyphenols with potential activity in preventing cardiovascular diseases and, thus, also PAD: clinical trials and adverse reactions.

Polyphenols	Clinical Trials	Systematic Review and Meta–Analysis	Findings	References
Baicalein	- Safety and PK evaluation in randomized, double-blind, placebo-controlled clinical trials (36–110 subjects)	– –	- No clinical data available on activity - Well-tolerated, no serious adverse effects	[95–97]
Curcumin	- Randomized clinical trials on healthy, obese or DM-affected subjects	- A meta-analysis on 192 healthy subjects	- Vascular-function improvement in healthy subjects - Modest effect on obese people - Good tolerability profile - Sporadic cases of hepatitis	[11,95,118,125–129]
Mangiferin	- Randomized controlled trial on hyperlipidemic subjects	– –	- Reduces triglycerides and free fatty acids, and insulin resistance index - Reported as safe in humans (few side effects with *Mangifera indica* extracts)	[95,148,149]
Quercetin	- A double-blinded, placebo-controlled cross-over study, on 93 overweight or obese subjects - Randomized controlled cross-over trial on 30 healthy subjects	- Meta-analysis on 841 normal and hypertensive subjects	- Reduces systolic blood pressure in healthy and hypertensive subjects - Reduces systolic blood pressure and plasma ox–LDL in hypertensive overweight subjects - No effect on heart disease risk factors - Reported as safe in moderate doses	[95,169–174]
Resveratrol	- Several clinical studies on healthy subjects and hypertensive and dyslipidemia or DM-affected patients - One pilot randomized clinical trial of 66 participants with PAD	- Meta-analysis on 50 DM subjects	- Improvement in cardiovascular parameters - Flow-mediated vasodilation - No clinical improvement in 6-minute walk test - No improvement of glycosylated - hemoglobin A1c levels and insulin resistance - Few side effects (gastrointestinal effects)	[44,95,199–210]

Supplementary Materials: The following supporting information can be downloaded at: https://www.mdpi.com/article/10.3390/molecules27207110/s1, Figure S1: Flowchart showing articles on polyphenols listed in PubMed from 1994 to 23 September 2022; Figure S2: Flowchart showing articles on peripheral artery disease and natural compounds listed in PubMed from 1963 to 23 September 2022.

Author Contributions: G.F. and E.R. wrote and revised the text. All authors have read and agreed to the published version of the manuscript.

Funding: This research received no external funding.

Institutional Review Board Statement: Not applicable.

Informed Consent Statement: Not applicable.

Data Availability Statement: Not applicable.

Conflicts of Interest: The authors declare no conflict of interest.

References

1. Riley, L.; Cowan, M. *Non-Communicable Diseases Progress Monitor 2022*; Licence: CC BY-NC-SA 3.0 IGO. Electronic Version; World Health Organization: Geneva, Switzerland, 2022; ISBN 978-92-4-004776-1.
2. Criqui, M.H.; Aboyans, V. Epidemiology of Peripheral Artery Disease. *Circ. Res.* **2015**, *116*, 1509–1526. [CrossRef] [PubMed]
3. Shu, J.; Santulli, G. Update on peripheral artery disease: Epidemiology and evidence-based facts. *Atherosclerosis* **2018**, *275*, 379–381. [CrossRef]
4. Antonelli-Incalzia, R.; Pedone, C.; McDermottb, M.M.; Bandinelli, S.; Miniati, B.; Lova, R.M.; Lauretanic, F.; Ferrucci, L. Association between nutrient intake and peripheral artery disease: Results from the InCHIANTI study. *Atherosclerosis* **2006**, *186*, 200–206. [CrossRef]
5. Johns Hopkins University Lifetime Risk and Prevalence of Lower Extremity Peripheral Artery Disease (PAD). Available online: http://ckdpcrisk.org/padrisk/ (accessed on 28 September 2022).
6. Aboyans, V.; Ricco, J.B.; Bartelink, M.L.E.L.; Björck, M.; Brodmann, M.; Cohnert, T.; Collet, J.P.; Czerny, M.; De Carlo, M.; Debus, S.; et al. 2017 ESC Guidelines on the Diagnosis and Treatment of Peripheral Arterial Diseases, in collaboration with the European Society for Vascular Surgery (ESVS): Document covering atherosclerotic disease of extracranial carotid and vertebral, mesenteric, renal, upper and lower extremity arteries. *Eur. Heart J.* **2018**, *39*, 763–816. [CrossRef] [PubMed]
7. Conte, M.S.; Bradbury, A.W.; Kolh, P.; White, J.V.; Dick, F.; Fitridge, R.; Mills, J.L.; Ricco, J.B.; Suresh, K.R.; Murad, M.H.; et al. Global Vascular Guidelines on the Management of Chronic Limb-Threatening Ischemia. *Eur. J. Vasc. Endovasc. Surg.* **2019**, *58*, S1–S109.e33. [CrossRef]
8. Halliday, A.; Bax, J.J. The 2017 ESC Guidelines on the Diagnosis and Treatment of Peripheral Arterial Diseases, in Collaboration with the European Society for Vascular Surgery (ESVS). *Eur. J. Vasc. Endovasc. Surg.* **2018**, *55*, 301–302. [CrossRef]
9. Brass, E.P.; Koster, D.; Hiatt, W.R.; Amato, A. A systematic review and meta-analysis of propionyl-l-carnitine effects on exercise performance in patients with claudication. *Vasc. Med.* **2013**, *18*, 3–12. [CrossRef]
10. Tama, B.; Fabara, S.P.; Zarrate, D.; Anas Sohail, A. Effectiveness of Propionyl-L-Carnitine Supplementation on Exercise Performance in Intermittent Claudication: A Systematic Review. *Cureus* **2021**, *13*, e17592. [CrossRef] [PubMed]
11. Kamoen, V.; Vander Stichele, R.; Campens, L.; De Bacquer, D.; Van Bortel, L.; De Backer, T. Propionyl-L-carnitine for intermittent claudication. *Cochrane Database Syst. Rev.* **2021**, *12*, CD010117. [CrossRef]
12. Forster, R.; Liew, A.; Bhattacharya, V.; Shaw, J.; Stansby, G. Gene therapy for peripheral arterial disease. *Cochrane Database Syst. Rev.* **2018**, *10*(10), CD012058. [CrossRef]
13. Pradhan, A.D.; Shrivastava, S.; Cook, N.R.; Rifai, N.; Creager, M.A.; Ridker, P.M. Symptomatic peripheral arterial disease in women: Nontraditional biomarkers of elevated risk. *Circulation* **2008**, *117*, 823–831. [CrossRef] [PubMed]
14. Krause, D.; Burghaus, I.; Thiem, U.; Trampisch, U.S.; Trampisch, M.; Klaassen-Mielke, R.; Trampisch, H.-J.; Diehm, C.; Rudolf, H. The risk of peripheral artery disease in older adults-seven-year results of the getABI study. *Vasa* **2016**, *45*, 403–410. [CrossRef] [PubMed]
15. Anderson, T.J.; Phillips, S.A. Assessment and prognosis of peripheral artery measures of vascular function. *Prog. Cardiovasc. Dis.* **2015**, *57*, 497–509. [CrossRef] [PubMed]
16. Norgren, L.; Hiatt, W.R.; Dormandy, J.A.; Nehler, M.R.; Harris, K.A.; Fowkes, F.G.R. Inter-Society Consensus for the Management of Peripheral Arterial Disease (TASC II). *J. Vasc. Surg.* **2007**, *45*, 5–67. [CrossRef] [PubMed]
17. Lilja, E.; Bergwall, S.; Sonestedt, E.; Gottsäter, A.; Acosta, S. The association between dietary intake, lifestyle and incident symptomatic peripheral arterial disease among individuals with diabetes mellitus: Insights from the Malmö Diet and Cancer study. *Ther. Adv. Endocrinol. Metab.* **2019**, *10*, 2042018819890532. [CrossRef] [PubMed]
18. Shimomura, M.; Fujie, S.; Sanada, K.; Kajimoto, H.; Hamaoka, T.; Iemitsu, M. Relationship between plasma asymmetric dimethylarginine and nitric oxide levels affects aerobic exercise training-induced reduction of arterial stiffness in middle-aged and older adults. *Phys. Act. Nutr.* **2021**, *25*, 16–22. [CrossRef]

19. Fujie, S.; Sanada, K.; Hamaoka, T.; Iemitsu, M. Time-dependent relationships between exercise training-induced changes in nitric oxide production and hormone regulation. *Exp. Gerontol.* **2022**, *166*, 111888. [CrossRef]
20. Toma, L.; Stancu, C.S.; Sima, A.V. Endothelial dysfunction in diabetes is aggravated by glycated lipoproteins; novel molecular therapies. *Biomedicines* **2021**, *9*, 18. [CrossRef]
21. Mustapha, S.; Mohammed, M.; Azemi, A.K.; Jatau, A.I.; Shehu, A.; Mustapha, L.; Aliyu, I.M.; Danraka, R.N.; Amin, A.; Bala, A.A.; et al. Current status of endoplasmic reticulum stress in type II diabetes. *Molecules* **2021**, *26*, 4362. [CrossRef]
22. Sprague, A.H.; Khalil, R.A. Inflammatory cytokines in vascular dysfunction and vascular disease. *Biochem. Pharmacol.* **2009**, *78*, 539–552. [CrossRef]
23. Pizzimenti, M.; Riou, M.; Charles, A.L.; Talha, S.; Meyer, A.; Andres, E.; Chakfé, N.; Lejay, A.; Geny, B. The rise of mitochondria in peripheral arterial disease physiopathology: Experimental and clinical data. *J. Clin. Med.* **2019**, *8*, 2125. [CrossRef] [PubMed]
24. Fournet, M.; Bonté, F.; Desmoulière, A. Glycation damage: A possible hub for major pathophysiological disorders and aging. *Aging Dis.* **2018**, *9*, 880–900. [CrossRef] [PubMed]
25. Farmer, D.G.S.; Kennedy, S. RAGE, vascular tone and vascular disease. *Pharmacol. Ther.* **2009**, *124*, 185–194. [CrossRef] [PubMed]
26. Chen, J.; Jing, J.; Yu, S.; Song, M.; Tan, H.; Cui, B.; Huang, L. Advanced glycation endproducts induce apoptosis of endothelial progenitor cells by activating receptor RAGE and NADPH oxidase/JNK signaling axis. *Am. J. Transl. Res.* **2016**, *8*, 2169–2178. [PubMed]
27. Islam, R.A.; Khalsa, S.S.S.; Vyas, A.K.; Rahimian, R. Sex-specific impacts of exercise on cardiovascular remodeling. *J. Clin. Med.* **2021**, *10*, 3833. [CrossRef]
28. Twarda-clapa, A.; Olczak, A.; Białkowska, A.M.; Koziołkiewicz, M. Advanced glycation end-products (AGEs): Formation, chemistry, classification, receptors, and diseases related to AGEs. *Cells* **2022**, *11*, 1312. [CrossRef]
29. Stirban, A.; Gawlowski, T.; Roden, M. Vascular effects of advanced glycation endproducts: Clinical effects and molecular mechanisms. *Mol. Metab.* **2014**, *3*, 94–108. [CrossRef]
30. Ahmad, S.; Siddiqui, Z.; Rehman, S.; Khan, M.Y.; Khan, H.; Khanum, S.; Alouffi, S.; Saeed, M. A Glycation Angle to Look into the Diabetic Vasculopathy: Cause and Cure. *Curr. Vasc. Pharmacol.* **2017**, *15*, 352–364. [CrossRef]
31. Yuan, T.; Yang, T.; Chen, H.; Fu, D.; Hu, Y.; Wang, J.; Yuan, Q.; Yu, H.; Xu, W.; Xie, X. New insights into oxidative stress and inflammation during diabetes mellitus-accelerated atherosclerosis. *Redox Biol.* **2019**, *20*, 247–260. [CrossRef]
32. Brown, O.I.; Bridge, K.I.; Kearney, M.T. Nicotinamide adenosine dinucleotide phosphate oxidases in glucose homeostasis and diabetes-related endothelial cell dysfunction. *Cells* **2021**, *10*, 2315. [CrossRef]
33. Tony Wyss, C. Ageing, neurodegeneration and brain rejuvenation. *Nature* **2016**, *10*, 180–186. [CrossRef] [PubMed]
34. Santhakumar, A.B.; Battino, M.; Alvarez-Suarez, J.M. Dietary polyphenols: Structures, bioavailability and protective effects against atherosclerosis. *Food Chem. Toxicol.* **2018**, *113*, 49–65. [CrossRef] [PubMed]
35. Yang, L.; Wen, K.S.; Ruan, X.; Zhao, Y.X.; Wei, F.; Wang, Q. Response of plant secondary metabolites to environmental factors. *Molecules* **2018**, *23*, 762. [CrossRef] [PubMed]
36. Pratyusha, S. Phenolic Compounds in the Plant Development and Defense: An Overview. In *Plant Stress Physiology*; Hasanuzzaman, M., Nahar, K., Eds.; IntechOpen: London, UK, 2022. [CrossRef]
37. Ge, Y.; Chen, Y.; Li, C.; Zhao, J.; Wei, M.; Li, X.; Yang, S.; Mi, Y. Effect of sodium nitroprusside treatment on shikimate and phenylpropanoid pathways of apple fruit. *Food Chem.* **2019**, *290*, 263–269. [CrossRef]
38. Ozcan, T.; Akpinar-Bayizit, A.; Yilmaz-Ersan, L.; Delikanli, B. Phenolics in Human Health. *Int. J. Chem. Eng. Appl.* **2014**, *5*, 393–396. [CrossRef]
39. Schön, C.; Allegrini, P.; Engelhart-jentzsch, K.; Riva, A.; Petrangolini, G. Grape Seed Extract PositivelyModulates Blood Pressure and Perceived Stress: ARandomized, Double-Blind, Placebo-Controlled Study in Healthy Volunteers. *Nutrients* **2021**, *13*, 654. [CrossRef]
40. Saarenhovi, M.; Salo, P.; Scheinin, M.; Lehto, J.; Lovró, Z.; Tiihonen, K.; Lehtinen, M.J.; Junnila, J.; Hasselwander, O.; Tarpila, A.; et al. The effect of an apple polyphenol extract rich in epicatechin and flavan-3-ol oligomers on brachial artery flow-mediated vasodilatory function in volunteers with elevated blood pressure. *Nutr. J.* **2017**, *16*, 73. [CrossRef]
41. Decroix, L.; Tonoli, C.; Soares, D.D.; Descat, A.; Drittij-Reijnders, M.J.; Weseler, A.R.; Bast, A.; Stahl, W.; Heyman, E.; Meeusen, R. Acute cocoa Flavanols intake has minimal effects on exercise-induced oxidative stress and nitric oxide production in healthy cyclists: A randomized controlled trial. *J. Int. Soc. Sport. Nutr.* **2017**, *14*, 28. [CrossRef]
42. Jacob, J.; Gopi, S.; Divya, C. A Randomized Single Dose Parallel Study on Enhancement of Nitric Oxide in Serum and Saliva with the Use of Natural Sports Supplement in Healthy Adults. *J. Diet. Suppl.* **2018**, *15*, 161–172. [CrossRef]
43. Nemzer, B.V.; Rodriguez, L.C.; Hammond, L.; Disilvestro, R.; Hunter, J.M.; Pietrzkowski, Z. Acute reduction of serum 8-iso-PGF2-alpha and advanced oxidation protein products in vivo by a polyphenol-rich beverage; A pilot clinical study with phytochemical and in vitro antioxidant characterization. *Nutr. J.* **2011**, *10*, 67. [CrossRef]
44. Magyar, K.; Halmosi, R.; Palfi, A.; Feher, G.; Czopf, L.; Fulop, A.; Battyany, I.; Sumegi, B.; Toth, K.; Szabados, E. Cardioprotection by resveratrol: A human clinical trial in patients with stable coronary artery disease. *Clin. Hemorheol. Microcirc.* **2012**, *50*, 179–187. [CrossRef] [PubMed]
45. Ashor, A.W.; Lara, J.; Siervo, M.; Celis-Morales, C.; Mathers, J.C. Effects of exercise modalities on arterial stiffness and wave reflection: A systematic review and meta-analysis of randomized controlled trials. *PLoS ONE* **2014**, *9*, e110034 . [CrossRef]

46. Suganya, N.; Bhakkiyalakshmi, E.; Sarada, D.V.L.L.; Ramkumar, K.M. Reversibility of endothelial dysfunction in diabetes: Role of polyphenols. *Br. J. Nutr.* **2016**, *116*, 223–246. [CrossRef]
47. Peñín-Grandes, S.; Martín-Hernández, J.; Valenzuela, P.L.; López-Ortiz, S.; Pinto-Fraga, J.; del Río Solá, L.; Emanuele, E.; Lista, S.; Lucia, A.; Santos-Lozano, A. Exercise and the hallmarks of peripheral arterial disease. *Atherosclerosis* **2022**, *350*, 41–50. [CrossRef] [PubMed]
48. Prior, B.M.; Lloyd, P.G.; Yang, H.T.; Terjung, R.L. Exercise-induced vascular remodeling. *Exerc. Sport Sci. Rev.* **2003**, *31*, 26–33. [CrossRef] [PubMed]
49. Kawamura, K.; Ejiri, K.; Toda, H. Association between home-based exercise using a pedometer and clinical prognosis after endovascular treatment in patients with peripheral artery disease. *J. Cardiol.* **2022**. S0914-5087(22)00228-3. [CrossRef]
50. Ehrman, J.K.; Gardner, A.W.; Salisbury, D.; Lui, K.; Treat-Jacobson, D. Supervised Exercise Therapy for Symptomatic Peripheral Artery Disease: A review of current experience and practice-based recommendations. *J. Cardiopulm. Rehabil. Prev.* 2022, *epub ahead of print*. [CrossRef] [PubMed]
51. Gerhard-Herman, M.D.; Gornik, H.L.; Barrett, C.; Barshes, N.R.; Corriere, M.A.; Drachman, D.E.; Fleisher, L.A.; Fowkes, F.G.R.; Hamburg, N.M.; Kinlay, S.; et al. 2016 AHA/ACC Guideline on the Management of Patients with Lower Extremity Peripheral Artery Disease: Executive Summary: A Report of the American College of Cardiology/American Heart Association Task Force on Clinical Practice Guidelines. *J. Am. Coll. Cardiol.* **2017**, *69*, 1465–1508. [CrossRef] [PubMed]
52. Craig, J.C.; Hart, C.R.; Layec, G.; Kwon, O.S.; Richardson, R.S.; Trinity, J.D. Impaired hemodynamic response to exercise in patients with peripheral artery disease: Evidence of a link to inflammation and oxidative stress. *Am. J. Physiol. -Regul. Integr. Comp. Physiol.* 2022; *epub ahead of print*. [CrossRef] [PubMed]
53. Habauzit, V.; Morand, C. Evidence for a protective effect of polyphenols-containing foods on cardiovascular health: An update for clinicians. *Ther. Adv. Chronic Dis.* **2012**, *3*, 87–106. [CrossRef] [PubMed]
54. Hoffman, R.P. Vascular endothelial dysfunction and nutritional compounds in early type 1 diabetes. *Curr. Diabetes Rev.* **2014**, *10*, 201–207. [CrossRef]
55. Stromsnes, K.; Mas-Bargues, C.; Gambini, J.; Gimeno-Mallench, L. Protective Effects of Polyphenols Present in Mediterranean Diet on Endothelial Dysfunction. *Oxidative Med. Cell. Longev.* **2020**, *2020*, 2097096. [CrossRef]
56. Oak, M.H.; Auger, C.; Belcastro, E.; Park, S.H.; Lee, H.H.; Schini-Kerth, V.B. Potential mechanisms underlying cardiovascular protection by polyphenols: Role of the endothelium. *Free. Radic. Biol. Med.* **2018**, *122*, 161–170. [CrossRef]
57. Munir, K.M.; Chandrasekaran, S.; Gao, F.; Quon, M.J. Mechanisms for food polyphenols to ameliorate insulin resistance and endothelial dysfunction: Therapeutic implications for diabetes and its cardiovascular complications. *Am. J. Physiol. Endocrinol. Metab.* **2013**, *305*, E679–E686. [CrossRef]
58. Woodward, K.A.; Draijer, R.; Thijssen, D.H.J.; Low, D.A. Polyphenols and Microvascular Function in Humans: A Systematic Review. *Curr. Pharm. Des.* **2018**, *24*, 203–226. [CrossRef] [PubMed]
59. Donat-Vargas, C.; Sandoval-Insausti, H.; Peñalvo, J.L.; Moreno Iribas, M.C.; Amiano, P.; Bes-Rastrollo, M.; Molina-Montes, E.; Moreno-Franco, B.; Agudo, A.; Mayo, C.L.; et al. Olive oil consumption is associated with a lower risk of cardiovascular disease and stroke. *Clin. Nutr.* **2022**, *41*, 122–130. [CrossRef]
60. Del Rio, D.; Rodriguez-Mateos, A.; Spencer, J.P.E.; Tognolini, M.; Borges, G.; Crozier, A. Dietary (poly)phenolics in human health: Structures, bioavailability, and evidence of protective effects against chronic diseases. *Antioxid. Redox Signal.* **2013**, *18*, 1818–1892. [CrossRef]
61. Flammer, A.J.; Hermann, F.; Sudano, I.; Spieker, L.; Hermann, M.; Cooper, K.A.; Serafini, M.; Lüscher, T.F.; Ruschitzka, F.; Noll, G.; et al. Dark chocolate improves coronary vasomotion and reduces platelet reactivity. *Circulation* **2007**, *116*, 2376–2382. [CrossRef]
62. Tanghe, A.; Heyman, E.; Vanden Wyngaert, K.; Van Ginckel, A.; Celie, B.; Rietzschel, E.; Calders, P.; Shadid, S. Evaluation of blood pressure lowering effects of cocoa flavanols in diabetes mellitus: A systematic review and meta-analysis. *J. Funct. Foods* **2021**, *79*, 104399. [CrossRef]
63. Tanghe, A.; Celie, B.; Shadid, S.; Rietzschel, E.; Op 't Roodt, J.; Reesink, K.D.; Heyman, E.; Calders, P. Acute Effects of Cocoa Flavanols on Blood Pressure and Peripheral Vascular Reactivity in Type 2 Diabetes Mellitus and Essential Hypertension: A Protocol for an Acute, Randomized, Double-Blinded, Placebo-Controlled Cross-Over Trial. *Front. Cardiovasc. Med.* **2021**, *8*, 602086. [CrossRef]
64. EFSA Panel on Dietetic Products, Nutrition and Allergies (NDA). Scientific Opinion on the substantiation of a health claim related to cocoa flavanols and maintenance of normal endothelium-dependent vasodilation pursuant to Article 13(5) of Regulation (EC) No 1924/2006. *EFSA J.* **2012**, *10*, 2809. [CrossRef]
65. Dinda, B.; Dinda, S.; DasSharma, S.; Banik, R.; Chakraborty, A.; Dinda, M. Therapeutic potentials of baicalin and its aglycone, baicalein against inflammatory disorders. *Eur. J. Med. Chem.* **2017**, *131*, 68–80. [CrossRef]
66. Dinda, B.; Silsarma, I.; Dinda, M.; Rudrapaul, P. *Oroxylum indicum* (L.) Kurz, an important Asian traditional medicine: From traditional uses to scientific data for its commercial exploitation. *J. Ethnopharmacol.* **2015**, *161*, 255–278. [CrossRef]
67. Lu, Q.-Y.; Zhang, L.; Moro, A.; Chen, M.C.; Harris, D.M.; Eibl, G.; Go, V.-L.W. Detection of Baicalin Metabolites Baicalein and Oroxylin-A in Mouse Pancreas and Pancreatic Xenografts. *Tissue Eng.* **2007**, *23*, 571–576. [CrossRef]
68. Ran, Y.; Qie, S.; Gao, F.; Ding, Z.; Yang, S.; Tian, G.; Liu, Z.; Xi, J. Baicalein ameliorates ischemic brain damage through suppressing proinflammatory microglia polarization via inhibiting the TLR4/NF-κB and STAT1 pathway. *Brain Res.* **2021**, *1770*, 147626. [CrossRef]

69. Song, J.W.; Long, J.Y.; Xie, L.; Zhang, L.L.; Xie, Q.X.; Chen, H.J.; Deng, M.; Li, X.F. Applications, phytochemistry, pharmacological effects, pharmacokinetics, toxicity of *Scutellaria baicalensis* Georgi. And its probably potential therapeutic effects on COVID-19: A review. *Chin. Med.* **2020**, *15*, 102. [CrossRef] [PubMed]
70. Marković, Z.S.; Dimitrić Marković, J.M.; Milenković, D.; Filipović, N. Mechanistic study of the structure-activity relationship for the free radical scavenging activity of baicalein. *J. Mol. Model.* **2011**, *17*, 2575–2584. [CrossRef] [PubMed]
71. Salleh, N.N.H.N.; Othman, F.A.; Kamarudin, N.A.; Tan, S.C. The Biological Activities and Therapeutic Potentials of Baicalein Extracted from *Oroxylum indicum*: A Systematic Review. *Molecules* **2020**, *25*, 5677. [CrossRef]
72. Zhao, Z.; Nian, M.; Qiao, H.; Yang, X.; Wu, S.; Zheng, X. Review of bioactivity and structure-activity relationship on baicalein (5,6,7-trihydroxyflavone) and wogonin (5,7-dihydroxy-8-methoxyflavone) derivatives: Structural modifications inspired from flavonoids in *Scutellaria baicalensis*. *Eur. J. Med. Chem.* **2022**, *243*, 114733. [CrossRef] [PubMed]
73. Zhao, T.; Tang, H.; Xie, L.; Zheng, Y.; Ma, Z.; Sun, Q.; Li, X. *Scutellaria baicalensis* Georgi. (Lamiaceae): A review of its traditional uses, botany, phytochemistry, pharmacology and toxicology. *J. Pharm. Pharmacol.* **2019**, *71*, 1353–1369. [CrossRef]
74. Costine, B.; Zhang, M.; Chhajed, S.; Pearson, B.; Chen, S.; Nadakuduti, S.S. Exploring native *Scutellaria* species provides insight into differential accumulation of flavones with medicinal properties. *Sci. Rep.* **2022**, *12*, 13201. [CrossRef] [PubMed]
75. Ku, S.K.; Bae, J.S. Baicalin, baicalein and wogonin inhibits high glucose-induced vascular inflammation in vitro and in vivo. *BMB Rep.* **2015**, *48*, 519–524. [CrossRef]
76. Baradaran Rahimi, V.; Askari, V.R.; Hosseinzadeh, H. Promising influences of *Scutellaria baicalensis* and its two active constituents, baicalin, and baicalein, against metabolic syndrome: A review. *Phytother. Res.* **2021**, *35*, 3558–3574. [CrossRef]
77. Zhang, B.W.; Li, X.; Sun, W.L.; Xing, Y.; Xiu, Z.L.; Zhuang, C.L.; Dong, Y.S. Dietary Flavonoids and Acarbose Synergistically Inhibit α-Glucosidase and Lower Postprandial Blood Glucose. *J. Agric. Food Chem.* **2017**, *65*, 8319–8330. [CrossRef]
78. Froldi, G.; Djeujo, F.M.; Bulf, N.; Caparelli, E.; Ragazzi, E. Comparative Evaluation of the Antiglycation and Anti-α-Glucosidase Activities of Baicalein, Baicalin (Baicalein 7-O-Glucuronide) and the Antidiabetic Drug Metformin. *Pharmaceutics* **2022**, *14*, 2141. [CrossRef]
79. Kwak, S.; Ku, S.K.; Han, M.S.; Bae, J.S. Vascular barrier protective effects of baicalin, baicalein and wogonin in vitro and in vivo. *Toxicol. Appl. Pharmacol.* **2014**, *281*, 30–38. [CrossRef]
80. Wang, C.; Sun, Y.; Liu, W.; Liu, Y.; Afzal, S.; Grover, J.; Chang, D.; Münch, G.; Li, C.G.; Lin, S.; et al. Protective effect of the curcumin-baicalein combination against macrovascular changes in diabetic angiopathy. *Frontiers in Endocrinology* **2022**, *13*, 953305. [CrossRef]
81. El-Bassossy, H.M.; Hassan, N.A.; Mahmoud, M.F.; Fahmy, A. Baicalein protects against hypertension associated with diabetes: Effect on vascular reactivity and stiffness. *Phytomedicine* **2014**, *21*, 1742–1745. [CrossRef]
82. Zhang, B.; Sun, W.; Yu, N.; Sun, J.; Yu, X.; Li, X.; Xing, Y.; Yan, D.; Ding, Q.; Xiu, Z.; et al. Anti-diabetic effect of baicalein is associated with the modulation of gut microbiota in streptozotocin and high-fat-diet induced diabetic rats. *J. Funct. Foods* **2018**, *46*, 256–267. [CrossRef]
83. Bitew, M.; Desalegn, T.; Demissie, T.B.; Belayneh, A.; Endale, M.; Eswaramoorthy, R. Pharmacokinetics and drug-likeness of antidiabetic flavonoids: Molecular docking and DFT study. *PLoS ONE* **2021**, *16*, e0260853. [CrossRef]
84. Zhang, X.; Qin, Y.; Ruan, W.; Wan, X.; lv, C.; He, L.; Lu, L.; Guo, X. Targeting inflammation-associated AMPK//Mfn-2/MAPKs signaling pathways by baicalein exerts anti-atherosclerotic action. *Phytother. Res.* **2021**, *35*, 4442–4455. [CrossRef]
85. Wan, C.X.; Xu, M.; Huang, S.H.; Wu, Q.Q.; Yuan, Y.; Deng, W.; Tang, Q.Z. Baicalein protects against endothelial cell injury by inhibiting the TLR4/NF-κB signaling pathway. *Mol. Med. Rep.* **2018**, *17*, 3085–3091. [CrossRef]
86. Huang, Y.; Miao, Z.; Hu, Y.; Yuan, Y.; Zhou, Y.; Wei, L.; Zhao, K.; Guo, Q.; Lu, N. Baicalein reduces angiogenesis in the inflammatory microenvironment via inhibiting the expression of AP-1. *Oncotarget* **2017**, *8*, 883–899. [CrossRef]
87. Tsai, K.L.; Hung, C.H.; Chan, S.H.; Shih, J.Y.; Cheng, Y.H.; Tsai, Y.J.; Lin, H.C.; Chu, P.M. Baicalein protects against oxLDL-caused oxidative stress and inflammation by modulation of AMPK-alpha. *Oncotarget* **2016**, *7*, 72458–72468. [CrossRef] [PubMed]
88. Machha, A.; Achike, F.I.; Mohd, M.A.; Mustafa, M.R. Baicalein impairs vascular tone in normal rat aortas: Role of superoxide anions. *Eur. J. Pharmacol.* **2007**, *565*, 144–150. [CrossRef]
89. Chen, Z.Y.; Su, Y.L.; Lau, C.W.; Law, W.I.; Huang, Y. Endothelium-dependent contraction and direct relaxation induced by baicalein in rat mesenteric artery. *Eur. J. Pharmacol.* **1999**, *374*, 41–47. [CrossRef]
90. Machha, A.; Mustafa, M.R. Chronic treatment with flavonoids prevents endothelial dysfunction in spontaneously hypertensive rat aorta. *J. Cardiovasc. Pharmacol.* **2005**, *46*, 36–40. [CrossRef]
91. Peng, C.Y.; Pan, S.L.; Huang, Y.W.; Guh, J.H.; Chang, Y.L.; Teng, C.M. Baicalein attenuates intimal hyperplasia after rat carotid balloon injury through arresting cell-cycle progression and inhibiting ERK, Akt, and NF-κB activity in vascular smooth-muscle cells. *Naunyn-Schmiedeberg's Arch. Pharmacol.* **2008**, *378*, 579–588. [CrossRef]
92. Sulistyowati, E.; Hsu, J.H.; Lee, S.J.; Huang, S.E.; Sihotang, W.Y.; Wu, B.N.; Dai, Z.K.; Lin, M.C.; Yeh, J.L. Potential Actions of Baicalein for Preventing Vascular Calcification of Smooth Muscle Cells In Vitro and In Vivo. *Int. J. Mol. Sci.* **2022**, *23*, 5673. [CrossRef] [PubMed]
93. Li, L.; Gao, H.; Lou, K.; Luo, H.; Hao, S.; Yuan, J.; Liu, Z.; Dong, R. Safety, tolerability, and pharmacokinetics of oral baicalein tablets in healthy Chinese subjects: A single-center, randomized, double-blind, placebo-controlled multiple-ascending-dose study. *Clin. Transl. Sci.* **2021**, *14*, 2017–2024. [CrossRef]

94. Dong, R.; Li, L.; Gao, H.; Lou, K.; Luo, H.; Hao, S.; Yuan, J.; Liu, Z. Safety, tolerability, pharmacokinetics, and food effect of baicalein tablets in healthy Chinese subjects: A single-center, randomized, double-blind, placebo-controlled, single-dose phase I study. *J. Ethnopharmacol.* **2021**, *274*, 114052. [CrossRef]
95. WHO Global Database VigiBase. Available online: https://www.vigiaccess.org/ (accessed on 28 September 2022).
96. Pang, H.; Xue, W.; Shi, A.; Li, M.; Li, Y.; Cao, G.; Yan, B.; Dong, F.; Xiao, W.; He, G.; et al. Multiple-Ascending-Dose Pharmacokinetics and Safety Evaluation of Baicalein Chewable Tablets in Healthy Chinese Volunteers. *Clin. Drug Investig.* **2016**, *36*, 713–724. [CrossRef]
97. Yang, X.; Zhu, G.; Zhang, Y.; Wu, X.; Liu, B.; Liu, Y.; Yang, Q.; Du, W.; Liang, J.; Hu, J.; et al. Inhibition of Human UDP-Glucuronosyltransferases1A1–Mediated Bilirubin Glucuronidation by the Popular Flavonoids Baicalein, Baicalin, and Hyperoside Is Responsible for Herb (Shuang-Huang-Lian)-Induced JaundiceS. *Drug Metab. Dispos.* **2022**, *50*, 552–565. [CrossRef]
98. Priyadarsini, K.I. The chemistry of curcumin: From extraction to therapeutic agent. *Molecules* **2014**, *19*, 20091–20112. [CrossRef]
99. Hewlings, S.J.; Kalman, D.S. Curcumin: A review of its effects on human health. *Foods* **2017**, *6*, 92. [CrossRef]
100. Wongeakin, N.; Bhattarakosol, P.; Patumraj, S. Molecular mechanisms of curcumin on diabetes-induced endothelial dysfunctions: Txnip, ICAM-1, and NOX2 expressions. *BioMed. Res. Int.* **2014**, *2014*, 12–14. [CrossRef]
101. Rungseesantivanon, S.; Thenchaisri, N.; Ruangvejvorachai, P.; Patumraj, S. Curcumin supplementation could improve diabetes-induced endothelial dysfunction associated with decreased vascular superoxide production and PKC inhibition. *BMC Complement. Altern. Med.* **2010**, *10*, 57. [CrossRef]
102. Karimian, M.S.; Pirro, M.; Johnston, T.P.; Majeed, M.; Sahebkar, A. Curcumin and Endothelial Function: Evidence and Mechanisms of Protective Effects. *Curr. Pharm. Des.* **2017**, *23*, 2462–2473. [CrossRef] [PubMed]
103. Nam, S.M.; Choi, J.H.; Yoo, D.Y.; Kim, W.; Jung, H.Y.; Kim, J.W.; Yoo, M.; Lee, S.; Kim, C.J.; Yoon, Y.S.; et al. Effects of Curcumin (*Curcuma longa*) on Learning and Spatial Memory as Well as Cell Proliferation and Neuroblast Differentiation in Adult and Aged Mice by Upregulating Brain-Derived Neurotrophic Factor and CREB Signaling. *J. Med. Food* **2014**, *17*, 641–649. [CrossRef]
104. De Lorenzi, E.; Contardi, C.; Serra, M.; Bisceglia, F.; Franceschini, D.; Pagetta, A.; Zusso, M.; Di Martino, R.M.C.; Seghetti, F.; Belluti, F. Modulation of Amyloid β-induced Microglia Activation and Neuronal Cell Death by Curcumin and Analogues. *Int. J. Mol. Sci.* **2022**, *23*, 4381. [CrossRef] [PubMed]
105. Wu, X.; Zheng, X.; Tang, H.; Zhao, L.; He, C.; Zou, Y.; Song, X.; Li, L.; Yin, Z.; Ye, G. A network pharmacology approach to identify the mechanisms and molecular targets of curcumin against Alzheimer disease. *Medicine* **2022**, *101*, e30194. [CrossRef] [PubMed]
106. Gordon, O.N.; Luis, P.B.; Sintim, H.O.; Schneider, C. Unraveling curcumin degradation: Autoxidation proceeds through spiroepoxide and vinylether intermediates en route to the main bicyclopentadione. *J. Biol. Chem.* **2015**, *290*, 4817–4828. [CrossRef] [PubMed]
107. Edwards, R.L.; Luis, P.B.; Varuzza, P.V.; Joseph, A.I.; Presley, S.H.; Chaturvedi, R.; Schneider, C. The anti-inflammatory activity of curcumin is mediated by its oxidative metabolites. *J. Biol. Chem.* **2017**, *292*, 21243–21252. [CrossRef]
108. Yang, H.; Du, Z.; Wang, W.; Song, M.; Sanidad, K.; Sukamtoh, E.; Zheng, J.; Tian, L.; Xiao, H.; Liu, Z.; et al. Structure-Activity Relationship of Curcumin: Role of the Methoxy Group in Anti-inflammatory and Anticolitis Effects of Curcumin. *J. Agric. Food Chem.* **2017**, *65*, 4509–4515. [CrossRef]
109. Avci, G.; Kadioglu, H.; Sehirli, A.O.; Bozkurt, S.; Guclu, O.; Arslan, E.; Muratli, S.K. Curcumin protects against ischemia/reperfusion injury in rat skeletal muscle. *J. Surg. Res.* **2012**, *172*, e39–e46. [CrossRef]
110. Liu, Y.; Chen, L.; Shen, Y.; Tan, T.; Xie, N.; Luo, M.; Li, Z.; Xie, X. Curcumin ameliorates ischemia-induced limb injury through immunomodulation. *Med. Sci. Monit.* **2016**, *22*, 2035–2042. [CrossRef]
111. You, J.; Sun, J.; Ma, T.; Yang, Z.; Wang, X.; Zhang, Z.; Li, J.; Wang, L.; Ii, M.; Yang, J.; et al. Curcumin induces therapeutic angiogenesis in a diabetic mouse hindlimb ischemia model via modulating the function of endothelial progenitor cells. *Stem Cell Res. Ther.* **2017**, *8*, 182. [CrossRef]
112. Zhang, J.; Wang, Q.; Rao, G.; Qiu, J.; He, R. Curcumin improves perfusion recovery in experimental peripheral arterial disease by upregulating microRNA-93 expression. *Exp. Ther. Med.* **2019**, *17*, 798–802. [CrossRef]
113. Ouyang, S.; Yao, Y.H.; Zhang, Z.M.; Liu, J.S.; Xiang, H. Curcumin inhibits HIF-1α induced functional changes of macrophages. *Eur. Rev. Med. Pharmacol. Sci.* **2019**, *23*, 1816–1825.
114. Alizadeh, M.; Kheirouri, S. Curcumin against advanced glycation end products (AGEs) and AGEs-induced detrimental agents. *Crit. Rev. Food Sci. Nutr.* **2019**, *59*, 1169–1177. [CrossRef]
115. Tang, Y.; Chen, A. Laboratory Investigation. *Lab. Investig.* **2014**, *94*, 503–516. [CrossRef]
116. Mahmoudi, A.; Atkin, S.L.; Nikiforov, N.G.; Sahebkar, A. Therapeutic Role of Curcumin in Diabetes: An Analysis Based on Bioinformatic Findings. *Nutrients* **2022**, *14*, 3244. [CrossRef]
117. Changal, K.H.; Khan, M.S.; Bashir, R.; Sheikh, M.A. Curcumin Preparations Can Improve Flow-Mediated Dilation and Endothelial Function: A Meta-Analysis. *Complement. Med. Res.* **2020**, *27*, 272–281. [CrossRef]
118. Alidadi, M.; Liberale, L.; Montecucco, F.; Majeed, M.; Al-Rasadi, K.; Banach, M.; Jamialahmadi, T.; Sahebkar, A. Protective Effects of Curcumin on Endothelium: An Updated Review. *Adv. Exp. Med. Biol.* **2021**, *1291*, 103–119. [CrossRef]
119. Anand, P.; Kunnumakkara, A.B.; Newman, R.A.; Aggarwal, B.B. Bioavailability of curcumin: Problems and promises. *Mol. Pharm.* **2007**, *4*, 807–818. [CrossRef]
120. Stohs, S.J.; Chen, O.; Ray, S.D.; Ji, J.; Bucci, L.R.; Preuss, H.G. Promising Avenues for Curcumin-Based Research and Application: A Review. *Molecules* **2020**, *25*, 1397. [CrossRef]

121. Witika, B.A.; Makoni, P.A.; Matafwali, S.K.; Mweetwa, L.L.; Shandele, G.C.; Walker, R.B. Enhancement of biological and pharmacological properties of an encapsulated polyphenol: Curcumin. *Molecules* **2021**, *26*, 4244. [CrossRef]
122. Poole, K.M.; Nelson, C.E.; Joshi, R.V.; Martin, J.R.; Gupta, M.K.; Haws, S.C.; Kavanaugh, T.E.; Skala, M.C.; Duvall, C.L. ROS-responsive microspheres for on demand antioxidant therapy in a model of diabetic peripheral arterial disease. *Biomaterials* **2015**, *41*, 166–175. [CrossRef]
123. Jung, E.; Noh, J.; Kang, C.; Yoo, D.; Song, C.; Lee, D. Ultrasound imaging and on-demand therapy of peripheral arterial diseases using H_2O_2-Activated bubble generating anti-inflammatory polymer particles. *Biomaterials* **2018**, *179*, 175–185. [CrossRef]
124. Verma, K.; Tarafdar, A.; Kumar, D.; Kumar, Y.; Rana, J.S.; Badgujar, P.C. Formulation and characterization of nano-curcumin fortified milk cream powder through microfluidization and spray drying. *Food Res. Int.* **2022**, *160*, 111705. [CrossRef]
125. Fuloria, S.; Mehta, J.; Chandel, A.; Sekar, M.; Rani, N.N.I.M.; Begum, M.Y.; Subramaniyan, V.; Chidambaram, K.; Thangavelu, L.; Nordin, R.; et al. A Comprehensive Review on the Therapeutic Potential of *Curcuma longa* Linn. in Relation to its Major Active Constituent Curcumin. *Front. Pharmacol.* **2022**, *13*, 820806. [CrossRef]
126. Goel, A.; Jhurani, S.; Aggarwal, B.B. Multi-targeted therapy by curcumin: How spicy is it? *Mol. Nutr. Food Res.* **2008**, *52*, 1010–1030. [CrossRef]
127. Soleimani, V.; Sahebkar, A.; Hosseinzadeh, H. Turmeric (*Curcuma longa*) and its major constituent (curcumin) as nontoxic and safe substances: Review. *Phytother. Res.* **2018**, *32*, 985–995. [CrossRef]
128. Surma, S.; Sahebkar, A.; Urbański, J.; Penson, P.E.; Banach, M. Curcumin-The Nutraceutical with Pleiotropic Effects? Which Cardiometabolic Subjects Might Benefit the Most? *Front. Nutr.* **2022**, *9*, 865497. [CrossRef]
129. Lombardi, N.; Crescioli, G.; Maggini, V.; Ippoliti, I.; Menniti-Ippolito, F.; Gallo, E.; Brilli, V.; Lanzi, C.; Mannaioni, G.; Firenzuoli, F.; et al. Acute liver injury following turmeric use in Tuscany: An analysis of the Italian Phytovigilance database and systematic review of case reports. *Br. J. Clin. Pharmacol.* **2021**, *87*, 741–753. [CrossRef]
130. Haynes, L.J.; Taylor, D.R. C-glycosyl compounds. Part V. Mangiferin; the nuclear magnetic resonance spectra of xanthones. *J. Chem. Soc. C Org.* **1966**, 1685–1687. [CrossRef]
131. Saha, S.; Sadhukhan, P.; Sil, P.C. Mangiferin: A xanthonoid with multipotent anti-inflammatory potential. *BioFactors* **2016**, *42*, 459–474. [CrossRef]
132. Barreto, J.C.; Trevisan, M.T.S.; Hull, W.E.; Erben, G.; De Brito, E.S.; Pfundstein, B.; Würtele, G.; Spiegelhalder, B.; Owen, R.W. Characterization and quantitation of polyphenolic compounds in bark, kernel, leaves, and peel of mango (*Mangifera indica* L.). *J. Agric. Food Chem.* **2008**, *56*, 5599–5610. [CrossRef]
133. Imran, M.; Arshad, M.S.; Butt, M.S.; Kwon, J.H.; Arshad, M.U.; Sultan, M.T. Mangiferin: A natural miracle bioactive compound against lifestyle related disorders. *Lipids Health Dis.* **2017**, *16*, 84. [CrossRef]
134. Benard, O.; Chi, Y. Medicinal Properties of Mangiferin, Structural Features, Derivative Synthesis, Pharmacokinetics and Biological Activities. *Mini-Rev. Med. Chem.* **2015**, *15*, 582–594. [CrossRef]
135. Das, J.; Ghosh, J.; Roy, A.; Sil, P.C. Mangiferin exerts hepatoprotective activity against D-galactosamine induced acute toxicity and oxidative/nitrosative stress via Nrf2-NFκB pathways. *Toxicol. Appl. Pharmacol.* **2012**, *260*, 35–47. [CrossRef]
136. Pal, P.B.; Sinha, K.; Sil, P.C. Mangiferin, a Natural Xanthone, Protects Murine Liver in Pb(II) Induced Hepatic Damage and Cell Death via MAP Kinase, NF-κB and Mitochondria Dependent Pathways. *PLoS ONE* **2013**, *8*, e56894. [CrossRef]
137. Andreu, G.P.; Delgado, R.; Velho, J.A.; Curti, C.; Vercesi, A.E. Iron complexing activity of mangiferin, a naturally occurring glucosylxanthone, inhibits mitochondrial lipid peroxidation induced by Fe 2+-citrate. *Eur. J. Pharmacol.* **2005**, *513*, 47–55. [CrossRef]
138. Song, J.; Li, J.; Hou, F.; Wang, X.; Liu, B. Mangiferin inhibits endoplasmic reticulum stress-associated thioredoxin-interacting protein/NLRP3 inflammasome activation with regulation of AMPK in endothelial cells. *Metab. Clin. Exp.* **2015**, *64*, 428–437. [CrossRef]
139. He, Q.; Ai, J.; Huang, Y. [Relationship between endothelial damage and p120-catenin in paraquat intoxication and the protective effect of mangiferin]. *Zhonghua Wei Zhong Bing Ji Jiu Yi Xue* **2014**, *26*, 369–373. [CrossRef]
140. Venugopal, R.; Sakthisekaran, D.; Rajkapoor, B.; Nishigaki, I. In vitro protective effect of mangiferin against glycated protein-iron chelate induced toxicity in human umbelical vein endothelial cells. *J. Biol. Sci.* **2007**, *7*, 1227–1232. [CrossRef]
141. Shi, J.; Lv, H.; Tang, C.; Li, Y.; Huang, J.; Zhang, H. Mangiferin inhibits cell migration and angiogenesis via PI3K/AKT/mTOR signaling in high glucose- and hypoxia-induced RRCECs. *Mol. Med. Rep.* **2021**, *23*, 473. [CrossRef]
142. Daud, N.H.; Aung, C.S.; Hewavitharana, A.K.; Wilkinson, A.S.; Pierson, J.T.; Roberts-Thomson, S.J.; Shaw, P.N.; Monteith, G.R.; Gidley, M.J.; Parat, M.O. Mango extracts and the mango component mangiferin promote endothelial cell migration. *J. Agric. Food Chem.* **2010**, *58*, 5181–5186. [CrossRef]
143. Hou, J.; Zheng, D.; Fung, G.; Deng, H.; Chen, L.; Liang, J.; Jiang, Y.; Hu, Y. Mangiferin suppressed advanced glycation end products (AGEs) through NF-κB deactivation and displayed anti-inflammatory effect in streptozotocin and high fat diet-diabetic cardiomyopathy rats. *Can. J. Physiol. Pharmacol.* **2016**, *94*, 332–340. [CrossRef]
144. Jiang, F.; Zhang, D.L.; Jia, M.; Hao, W.H.; Li, Y. jun Mangiferin inhibits high-fat diet induced vascular injury via regulation of PTEN/AKT/eNOS pathway. *J. Pharmacol. Sci.* **2018**, *137*, 265–273. [CrossRef]
145. Wu, Y.; Liu, W.; Yang, T.; Li, M.; Qin, L.; Wu, L.; Liu, T. Oral administration of mangiferin ameliorates diabetes in animal models: A meta-analysis and systematic review. *Nutr. Res.* **2021**, *87*, 57–69. [CrossRef] [PubMed]

146. Dorighello, G.G.; Inada, N.M.; Paim, B.A.; Pardo-Andreu, G.L.; Vercesi, A.E.; Oliveira, H.C.F. *Mangifera indica* L. extract (Vimang®) reduces plasma and liver cholesterol and leucocyte oxidative stress in hypercholesterolemic LDL receptor deficient mice. *Cell Biol. Int.* **2018**, *42*, 747–753. [CrossRef] [PubMed]
147. RamPravinKumar, M.; Dhananjayan, K. Peripheral arterial disease: Effects of ethanolic extracts of seed kernels of mango (*Mangifera indica* L.) on acute hind limb ischemia-reperfusion injury in diabetic rats. *J. Tradit. Complement. Med.* **2021**, *11*, 520–531. [CrossRef] [PubMed]
148. Na, L.; Zhang, Q.; Jiang, S.; Du, S.; Zhang, W.; Li, Y.; Sun, C.; Niu, Y. Mangiferin supplementation improves serum lipid profiles in overweight patients with hyperlipidemia: A double-blind randomized controlled trial. *Sci. Rep.* **2015**, *5*, 10344. [CrossRef]
149. Hou, S.; Wang, F.; Li, Y.; Li, Y.; Wang, M.; Sun, D.; Sun, C. Pharmacokinetic study of mangiferin in human plasma after oral administration. *Food Chem.* **2012**, *132*, 289–294. [CrossRef]
150. Xu, D.; Hu, M.J.; Wang, Y.Q.; Cui, Y.L. Antioxidant activities of quercetin and its complexes for medicinal application. *Molecules* **2019**, *24*, 1123. [CrossRef]
151. Li, Y.; Yao, J.; Han, C.; Yang, J.; Chaudhry, M.T.; Wang, S.; Liu, H.; Yin, Y. Quercetin, inflammation and immunity. *Nutrients* **2016**, *8*, 167. [CrossRef]
152. Magar, R.T.; Sohng, J.K. A Review on Structure, Modifications and Structure-Activity Relation of Quercetin and Its Derivatives. *J. Microbiol. Biotechnol.* **2020**, *30*, 11–20. [CrossRef]
153. Alizadeh, S.R.; Ebrahimzadeh, M.A. O-Glycoside quercetin derivatives: Biological activities, mechanisms of action, and structure–activity relationship for drug design, a review. *Phytother. Res.* **2022**, *36*, 778–807. [CrossRef]
154. Babenkova, I.V.; Osipov, A.N.; Teselkin, Y.O. The Effect of Dihydroquercetin on Catalytic Activity of Iron (II) Ions in the Fenton Reaction. *Bull. Exp. Biol. Med.* **2018**, *165*, 347–350. [CrossRef]
155. Anand David, A.V.; Arulmoli, R.; Parasuraman, S. Overviews of biological importance of quercetin: A bioactive flavonoid. *Pharmacogn. Rev.* **2016**, *10*, 84–89. [CrossRef]
156. Hosseini, A.; Razavi, B.M.; Banach, M.; Hosseinzadeh, H. Quercetin and metabolic syndrome: A review. *Phytother. Res.* **2021**, *35*, 5352–5364. [CrossRef] [PubMed]
157. Duarte, J.; Pérez-Vizcaíno, F.; Zarzuelo, A.; Jiménez, J.; Tamargo, J. Vasodilator effects of quercetin in isolated rat vascular smooth muscle. *Eur. J. Pharmacol.* **1993**, *239*, 1–7. [CrossRef]
158. Yamamoto, Y.; Oue, E. Antihypertensive effect of quercetin in rats fed with a high-fat high-sucrose diet. *Biosci. Biotechnol. Biochem.* **2006**, *70*, 933–939. [CrossRef]
159. Chen, C.K.; Pace-Asciak, C.R. Vasorelaxing activity of resveratrol and quercetin in isolated rat aorta. *Gen. Pharmacol. Vasc. Syst.* **1996**, *27*, 363–366. [CrossRef]
160. Duarte, J.; Pérez-Palencia, R.; Vargas, F.; Angeles Ocete, M.; Pérez-Vizcaino, F.; Zarzuelo, A.; Tamargo, J. Antihypertensive effects of the flavonoid quercetin in spontaneously hypertensive rats. *Br. J. Pharmacol.* **2001**, *133*, 117–124. [CrossRef] [PubMed]
161. Chen, Y.; Li, X.X.; Xing, N.Z.; Cao, X. guang Quercetin inhibits choroidal and retinal angiogenesis in vitro. *Graefe's Arch. Clin. Exp. Ophthalmol.* **2008**, *246*, 373–378. [CrossRef]
162. Igura, K.; Ohta, T.; Kuroda, Y.; Kaji, K. Resveratrol and quercetin inhibit angiogenesis in vitro. *Cancer Lett.* **2001**, *171*, 11–16. [CrossRef]
163. Tan, W.F.; Lin, L.P.; Li, M.H.; Zhang, Y.X.; Tong, Y.G.; Xiao, D.; Ding, J. Quercetin, a dietary-derived flavonoid, possesses antiangiogenic potential. *Eur. J. Pharmacol.* **2003**, *459*, 255–262. [CrossRef]
164. Suganya, N.; Bhakkiyalakshmi, E.; Suriyanarayanan, S.; Paulmurugan, R.; Ramkumar, K.M. Quercetin ameliorates tunicamycin-induced endoplasmic reticulum stress in endothelial cells. *Cell Prolif.* **2014**, *47*, 231–240. [CrossRef]
165. Costa, L.G.; Garrick, J.M.; Roquè, P.J.; Pellacani, C. Mechanisms of Neuroprotection by Quercetin: Counteracting Oxidative Stress and More. *Oxidative Med. Cell. Longev.* **2016**, *2016*, 2986796. [CrossRef] [PubMed]
166. Bao, D.; Wang, J.; Pang, X.; Liu, H. Protective Effect of Quercetin against Oxidative Stress-Induced Cytotoxicity in Rat Pheochromocytoma (PC-12) Cells. *Molecules* **2017**, *22*, 1122. [CrossRef] [PubMed]
167. Ruiz, P.A.; Braune, A.; Hölzlwimmer, G.; Quintanilla-Fend, L.; Haller, D. Quercetin inhibits TNF-induced NF-κB transcription factor recruitment to proinflammatory gene promoters in murine intestinal epithelial cells. *J. Nutr.* **2007**, *137*, 1208–1215. [CrossRef] [PubMed]
168. Fang, X.K.; Gao, J.; Zhu, D.N. Kaempferol and quercetin isolated from *Euonymus alatus* improve glucose uptake of 3T3-L1 cells without adipogenesis activity. *Life Sci.* **2008**, *82*, 615–622. [CrossRef]
169. Conquer, J.A.; Maiani, G.; Azzini, E.; Raguzzini, A.; Holub, B.J. Supplementation with quercetin markedly increases plasma quercetin concentration without effect on selected risk factors for heart disease in healthy subjects. *J. Nutr.* **1998**, *128*, 593–597. [CrossRef] [PubMed]
170. Edwards, R.L.; Lyon, T.; Litwin, S.E.; Rabovsky, A.; Symons, J.D.; Jalili, T. Quercetin reduces blood pressure in hypertensive subjects. *J. Nutr.* **2007**, *137*, 2405–2411. [CrossRef]
171. Egert, S.; Bosy-Westphal, A.; Seiberl, J.; Kürbitz, C.; Settler, U.; Plachta-Danielzik, S.; Wagner, A.E.; Frank, J.; Schrezenmeir, J.; Rimbach, G.; et al. Quercetin reduces systolic blood pressure and plasma oxidised low-density lipoprotein concentrations in overweight subjects with a high-cardiovascular disease risk phenotype: A double-blinded, placebo-controlled cross-over study. *Br. J. Nutr.* **2009**, *102*, 1065–1074. [CrossRef]

172. Popiolek-Kalisz, J.; Fornal, E. The effects of quercetin supplementation on blood pressure–meta-analysis. *Curr. Probl. Cardiol.* **2022**, *47*, 101350. [CrossRef]
173. Serban, M.C.; Sahebkar, A.; Zanchetti, A.; Mikhailidis, D.P.; Howard, G.; Antal, D.; Andrica, F.; Ahmed, A.; Aronow, W.S.; Muntner, P.; et al. Effects of Quercetin on Blood Pressure: A Systematic Review and Meta-Analysis of Randomized Controlled Trials. *J. Am. Heart Assoc.* **2016**, *5*, e002713. [CrossRef]
174. Huang, H.; Liao, D.; Dong, Y.; Pu, R. Effect of quercetin supplementation on plasma lipid profiles, blood pressure, and glucose levels: A systematic review and meta-analysis. *Nutr. Rev.* **2020**, *78*, 615–626. [CrossRef]
175. Bradamante, S.; Barenghi, L.; Villa, A. Cardiovascular protective effects of resveratrol. *Cardiovasc. Drug Rev.* **2004**, *22*, 169–188. [CrossRef] [PubMed]
176. Ragazzi, E.; Froldi, G.; Fassina, G. Resveratrol activity on guinea pig isolated trachea from normal and albumin-sensitized animals. *Pharmacol. Res. Commun.* **1988**, *20*, 79–82. [CrossRef]
177. Pervaiz, S. Resveratrol: From grapevines to mammalian biology. *FASEB J.* **2003**, *17*, 1975–1985. [CrossRef] [PubMed]
178. Coppa, T.; Lazzè, M.C.; Cazzalini, O.; Perucca, P.; Pizzala, R.; Bianchi, L.; Stivala, L.A.; Forti, L.; MacCario, C.; Vannini, V.; et al. Structure-Activity relationship of resveratrol and its analogue, 4,4′-dihydroxy-trans-stilbene, toward the endothelin axis in human endothelial cells. *J. Med. Food* **2011**, *14*, 1173–1180. [CrossRef]
179. Stivala, L.A.; Savio, M.; Carafoli, F.; Perucca, P.; Bianchi, L.; Maga, G.; Forti, L.; Pagnoni, U.M.; Albini, A.; Prosperi, E.; et al. Specific Structural Determinants Are Responsible for the Antioxidant Activity and the Cell Cycle Effects of Resveratrol. *J. Biol. Chem.* **2001**, *276*, 22586–22594. [CrossRef]
180. Plauth, A.; Geikowski, A.; Cichon, S.; Wowro, S.J.; Liedgens, L.; Rousseau, M.; Weidner, C.; Fuhr, L.; Kliem, M.; Jenkins, G.; et al. Hormetic shifting of redox environment by pro-oxidative resveratrol protects cells against stress. *Free. Radic. Biol. Med.* **2016**, *99*, 608–622. [CrossRef]
181. Shaito, A.; Posadino, A.M.; Younes, N.; Hasan, H.; Halabi, S.; Alhababi, D.; Al-Mohannadi, A.; Abdel-Rahman, W.M.; Eid, A.H.; Nasrallah, G.K.; et al. Potential adverse effects of resveratrol: A literature review. *Int. J. Mol. Sci.* **2020**, *21*, 2084. [CrossRef]
182. Buluc, M.; Demirel-Yilmaz, E. Resveratrol decreases calcium sensitivity of vascular smooth muscle and enhances cytosolic calcium increase in endothelium. *Vasc. Pharmacol.* **2006**, *44*, 231–237. [CrossRef]
183. Jäger, U.; Nguyen-Duong, H. Relaxant effect of trans-resveratrol on isolated porcine coronary arteries. *Arzneimittelforschung* **1999**, *49*, 207–211. [CrossRef]
184. El-Mowafy, A.M. Resveratrol activates membrane-bound guanylyl cyclase in coronary arterial smooth muscle: A novel signaling mechanism in support of coronary protection. *Biochem. Biophys. Res. Commun.* **2002**, *291*, 1218–1224. [CrossRef]
185. Naderali, E.K.; Doyle, P.J.; Williams, G. Resveratrol induces vasorelaxation of mesenteric and uterine arteries from female guinea-pigs. *Clin. Sci. (Lond. Engl. 1979)* **2000**, *98*, 537–543. [CrossRef]
186. Klinge, C.M.; Wickramasinghe, N.S.; Ivanova, M.M.; Dougherty, S.M. Resveratrol stimulates nitric oxide production by increasing estrogen receptor αa-Src-caveolin-1 interaction and phosphorylation in human umbilical vein endothelial cells. *FASEB J.* **2008**, *22*, 2185–2197. [CrossRef]
187. Tian, C.; Zhang, R.; Ye, X.; Zhang, C.; Jin, X.; Yamori, Y.; Hao, L.; Sun, X.; Ying, C. Resveratrol ameliorates high-glucose-induced hyperpermeability mediated by caveolae via VEGF/KDR pathway. *Genes Nutr.* **2013**, *8*, 231–239. [CrossRef]
188. Liu, Z.; Jiang, C.; Zhang, J.; Liu, B.; Du, Q. Resveratrol inhibits inflammation and ameliorates insulin resistant endothelial dysfunction via regulation of AMP-activated protein kinase and sirtuin 1 activities. *J. Diabetes* **2016**, *8*, 324–335. [CrossRef] [PubMed]
189. Csiszar, A.; Smith, K.; Labinskyy, N.; Orosz, Z.; Rivera, A.; Ungvari, Z. Resveratrol attenuates TNF-α-induced activation of coronary arterial endothelial cells: Role of NF-κB inhibition. *Am. J. Physiol. Heart Circ. Physiol.* **2006**, *291*, 1694–1699. [CrossRef] [PubMed]
190. Kim, S.W.; Kim, C.E.; Kim, M.H. Flavonoids inhibit high glucose-induced up-regulation of ICAM-1 via the p38 MAPK pathway in human vein endothelial cells. *Biochem. Biophys. Res. Commun.* **2011**, *415*, 602–607. [CrossRef] [PubMed]
191. Zheng, X.; Zhu, S.; Chang, S.; Cao, Y.; Dong, J.; Li, J.; Long, R.; Zhou, Y. Protective effects of chronic resveratrol treatment on vascular inflammatory injury in streptozotocin-induced type 2 diabetic rats: Role of NF-kappa B signaling. *Eur. J. Pharmacol.* **2013**, *720*, 147–157. [CrossRef]
192. Ungvari, Z.; Bagi, Z.; Feher, A.; Recchia, F.A.; Sonntag, W.E.; Pearson, K.; de Cabo, R.; Csiszar, A. Resveratrol confers endothelial protection via activation of the antioxidant transcription factor Nrf2. *Am. J. Physiology. Heart Circ. Physiol.* **2010**, *299*, H18-24. [CrossRef]
193. Szkudelski, T.; Szkudelska, K. Resveratrol and diabetes: From animal to human studies. *Biochim. Et Biophys. Acta Mol. Basis Dis.* **2015**, *1852*, 1145–1154. [CrossRef]
194. Milne, J.C.; Lambert, P.D.; Schenk, S.; Carney, D.P.; Smith, J.J.; Gagne, D.J.; Jin, L.; Boss, O.; Perni, R.B.; Vu, C.B.; et al. Small molecule activators of SIRT1 as therapeutics for the treatment of type 2 diabetes. *Nature* **2007**, *450*, 712–716. [CrossRef]
195. Mohammad, G.; Abdelaziz, G.M.; Siddiquei, M.M.; Ahmad, A.; De Hertogh, G.; Abu El-Asrar, A.M. Cross-Talk between Sirtuin 1 and the Proinflammatory Mediator High-Mobility Group Box-1 in the Regulation of Blood-Retinal Barrier Breakdown in Diabetic Retinopathy. *Curr. Eye Res.* **2019**, *44*, 1133–1143. [CrossRef] [PubMed]
196. Li, K.X.; Ji, M.J.; Sun, H.J. An updated pharmacological insight of resveratrol in the treatment of diabetic nephropathy. *Gene* **2021**, *780*, 145532. [CrossRef] [PubMed]

197. Chen, W.P.; Chi, T.C.; Chuang, L.M.; Su, M.J. Resveratrol enhances insulin secretion by blocking KATP and KV channels of beta cells. *Eur. J. Pharmacol.* **2007**, *568*, 269–277. [CrossRef] [PubMed]
198. Kopp, P. Resveratrol, a phytoestrogen found in red wine. A possible explanation for the conundrum of the 'French paradox'? *Eur. J. Endocrinol.* **1998**, *138*, 619–620. [CrossRef]
199. Imamura, H.; Yamaguchi, T.; Nagayama, D.; Saiki, A.; Shirai, K.; Tatsuno, I. Resveratrol ameliorates arterial stiffness assessed by cardio-ankle vascular index in patients with type 2 diabetes mellitus. *Int. Heart J.* **2017**, *58*, 577–583. [CrossRef]
200. Carrizzo, A.; Puca, A.; Damato, A.; Marino, M.; Franco, E.; Pompeo, F.; Traficante, A.; Civitillo, F.; Santini, L.; Trimarco, V.; et al. Resveratrol improves vascular function in patients with hypertension and dyslipidemia by modulating NO metabolism. *Hypertension* **2013**, *62*, 359–366. [CrossRef]
201. Kennedy, D.O.; Wightman, E.L.; Reay, J.L.; Lietz, G.; Okello, E.J.; Wilde, A.; Haskell, C.F. Effects of resveratrol on cerebral blood flow variables and cognitive performance in humans: A double-blind, placebo-controlled, crossover investigation. *Am. J. Clin. Nutr.* **2010**, *91*, 1590–1597. [CrossRef]
202. Wong, R.H.X.; Raederstorff, D.; Howe, P.R.C. Acute resveratrol consumption improves neurovascular coupling capacity in adults with type 2 diabetes mellitus. *Nutrients* **2016**, *8*, 425. [CrossRef]
203. Jeyaraman, M.M.; Al-Yousif, N.S.H.; Singh Mann, A.; Dolinsky, V.W.; Rabbani, R.; Zarychanski, R.; Abou-Setta, A.M. Resveratrol for adults with type 2 diabetes mellitus. *Cochrane Database Syst. Rev.* **2020**, *2020*, CD011919. [CrossRef]
204. McDermott, M.M.; Leeuwenburgh, C.; Guralnik, J.M.; Tian, L.; Sufit, R.; Zhao, L.; Criqui, M.H.; Kibbe, M.R.; Stein, J.H.; Lloyd-Jones, D.; et al. Effect of Resveratrol on Walking Performance in Older People with Peripheral Artery Disease: The RESTORE Randomized Clinical Trial. *JAMA Cardiol.* **2017**, *2*, 902–907. [CrossRef]
205. Hyrsova, L.; Vanduchova, A.; Dusek, J.; Smutny, T.; Carazo, A.; Maresova, V.; Trejtnar, F.; Barta, P.; Anzenbacher, P.; Dvorak, Z.; et al. Trans-resveratrol, but not other natural stilbenes occurring in food, carries the risk of drug-food interaction via inhibition of cytochrome P450 enzymes or interaction with xenosensor receptors. *Toxicol. Lett.* **2019**, *300*, 81–91. [CrossRef]
206. Giuliani, C.; Iezzi, M.; Ciolli, L.; Hysi, A.; Bucci, I.; Di Santo, S.; Rossi, C.; Zucchelli, M.; Napolitano, G. Resveratrol has anti-thyroid effects both in vitro and in vivo. *Food Chem. Toxicol.* **2017**, *107*, 237–247. [CrossRef] [PubMed]
207. Johnson, W.D.; Morrissey, R.L.; Usborne, A.L.; Kapetanovic, I.; Crowell, J.A.; Muzzio, M.; McCormick, D.L. Subchronic oral toxicity and cardiovascular safety pharmacology studies of resveratrol, a naturally occurring polyphenol with cancer preventive activity. *Food Chem. Toxicol.* **2011**, *49*, 3319–3327. [CrossRef] [PubMed]
208. Brown, V.A.; Patel, K.R.; Viskaduraki, M.; Crowell, J.A.; Perloff, M.; Booth, T.D.; Vasilinin, G.; Sen, A.; Schinas, A.M.; Piccirilli, G.; et al. Repeat dose study of the cancer chemopreventive agent resveratrol in healthy volunteers: Safety, pharmacokinetics, and effect on the insulin-like growth factor axis. *Cancer Res.* **2010**, *70*, 9003–9011. [CrossRef] [PubMed]
209. Boocock, D.J.; Faust, G.E.S.; Patel, K.R.; Schinas, A.M.; Brown, V.A.; Ducharme, M.P.; Booth, T.D.; Crowell, J.A.; Perloff, M.; Gescher, A.J.; et al. Phase I dose escalation pharmacokinetic study in healthy volunteers of resveratrol, a potential cancer chemopreventive agent. *Cancer Epidemiol. Biomark. Prev.* **2007**, *16*, 1246–1252. [CrossRef]
210. EFSA NDA Panel (EFSA Panel on Dietetic Products, Nutrition and Allergies). Safety of synthetic trans-resveratrol as a novel food pursuant to Regulation (EC) No 258/97. *EFSA J.* **2016**, *14*, 4368. [CrossRef]

MDPI
St. Alban-Anlage 66
4052 Basel
Switzerland
www.mdpi.com

Molecules Editorial Office
E-mail: molecules@mdpi.com
www.mdpi.com/journal/molecules

www.ingramcontent.com/pod-product-compliance
Lightning Source LLC
Chambersburg PA
CBHW061459220926
51850CB00004B/127
* 9 7 8 3 0 3 6 5 9 0 7 4 5 *